FOOD CHEMISTRY

FOOD SCIENCE AND TECHNOLOGY

A Series of Monographs, Textbooks, and Reference Books

1. *Flavor Research: Principles and Techniques,* R. Teranishi, I. Hornstein, P. Issenberg, and E. L. Wick
2. *Principles of Enzymology for the Food Sciences,* John R. Whitaker
3. *Low-Temperature Preservation of Foods and Living Matter,* Owen R. Fennema, William D. Powrie, and Elmer H. Marth
4. *Principles of Food Science*
   Part I: Food Chemistry, edited by Owen R. Fennema
   Part II: Physical Methods of Food Preservation, Marcus Karel, Owen R. Fennema, and Daryl B. Lund
5. *Food Emulsions,* edited by Stig E. Friberg
6. *Nutritional and Safety Aspects of Food Processing,* edited by Steven R. Tannenbaum
7. *Flavor Research: Recent Advances,* edited by R. Teranishi, Robert A. Flath, and Hiroshi Sugisawa
8. *Computer-Aided Techniques in Food Technology,* edited by Israel Saguy
9. *Handbook of Tropical Foods,* edited by Harvey T. Chan
10. *Antimicrobials in Foods,* edited by Alfred Larry Branen and P. Michael Davidson
11. *Food Constituents and Food Residues: Their Chromatographic Determination,* edited by James F. Lawrence
12. *Aspartame: Physiology and Biochemistry,* edited by Lewis D. Stegink and L. J. Filer, Jr.
13. *Handbook of Vitamins: Nutritional, Biochemical, and Clinical Aspects,* edited by Lawrence J. Machlin
14. *Starch Conversion Technology,* edited by G. M. A. van Beynum and J. A. Roels

15. *Food Chemistry: Second Edition, Revised and Expanded,* edited by Owen R. Fennema
16. *Sensory Evaluation of Food: Statistical Methods and Procedures,* Michael O'Mahony
17. *Alternative Sweetners,* edited by Lyn O'Brien Nabors and Robert C. Gelardi
18. *Citrus Fruits and Their Products: Analysis and Technology,* S. V. Ting and Russell L. Rouseff
19. *Engineering Properties of Foods,* edited by M. A. Rao and S. S. H. Rizvi
20. *Umami: A Basic Taste,* edited by Yojiro Kawamura and Morley R. Kare
21. *Food Biotechnology,* edited by Dietrich Knorr
22. *Food Texture: Instrumental and Sensory Measurement,* edited by Howard R. Moskowitz
23. *Seafoods and Fish Oils in Human Health and Disease,* John E. Kinsella
24. *Postharvest Physiology of Vegetables,* edited by J. Weichmann
25. *Handbook of Dietary Fiber: An Applied Approach,* Mark L. Dreher
26. *Food Toxicology, Parts A and B,* Jose M. Concon
27. *Modern Carbohydrate Chemistry,* Roger W. Binkley
28. *Trace Minerals in Foods,* edited by Kenneth T. Smith
29. *Protein Quality and the Effects of Processing,* edited by R. Dixon Phillips and John W. Finley
30. *Adulteration of Fruit Juice Beverages,* edited by Steven Nagy, John A. Attaway, and Martha E. Rhodes
31. *Foodborne Bacterial Pathogens,* edited by Michael P. Doyle
32. *Legumes: Chemistry, Technology, and Human Nutrition,* edited by Ruth H. Matthews
33. *Industrialization of Indigenous Fermented Foods,* edited by Keith H. Steinkraus
34. *International Food Regulation Handbook: Policy • Science • Law,* edited by Roger D. Middlekauff and Philippe Shubik
35. *Food Additives,* edited by A. Larry Branen, P. Michael Davidson, and Seppo Salminen
36. *Safety of Irradiated Foods,* J. F. Diehl
37. *Omega-3 Fatty Acids in Health and Disease,* edited by Robert S. Lees and Marcus Karel
38. *Food Emulsions: Second Edition, Revised and Expanded,* edited by Kåre Larsson and Stig E. Friberg
39. *Seafood: Effects of Technology on Nutrition,* George M. Pigott and Barbee W. Tucker
40. *Handbook of Vitamins, Second Edition, Revised and Expanded,* edited by Lawrence J. Machlin
41. *Handbook of Cereal Science and Technology,* Klaus J. Lorenz and Karel Kulp
42. *Food Processing Operations and Scale-Up,* Kenneth J. Valentas, Leon Levine, and J. Peter Clark
43. *Fish Quality Control by Computer Vision,* edited by L. F. Pau and R. Olafsson
44. *Volatile Compounds in Foods and Beverages,* edited by Henk Maarse
45. *Instrumental Methods for Quality Assurance in Foods,* edited by Daniel Y. C. Fung and Richard F. Matthews
46. Listeria, *Listeriosis, and Food Safety,* Elliot T. Ryser and Elmer H. Marth

47. *Acesulfame-K,* edited by D. G. Mayer and F. H. Kemper
48. *Alternative Sweeteners: Second Edition, Revised and Expanded,* edited by Lyn O'Brien Nabors and Robert C. Gelardi
49. *Food Extrusion Science and Technology*, edited by Jozef L. Kokini, Chi-Tang Ho, and Mukund V. Karwe
50. *Surimi Technology*, edited by Tyre C. Lanier and Chong M. Lee
51. *Handbook of Food Engineering*, edited by Dennis R. Heldman and Daryl B. Lund
52. *Food Analysis by HPLC*, edited by Leo M. L. Nollet
53. *Fatty Acids in Foods and Their Health Implications*, edited by Ching Kuang Chow
54. Clostridium botulinum: *Ecology and Control in Foods*, edited by Andreas H. W. Hauschild and Karen L. Dodds
55. *Cereals in Breadmaking: A Molecular Colloidal Approach*, Anne-Charlotte Eliasson and Kåre Larsson
56. *Low-Calorie Foods Handbook*, edited by Aaron M. Altschul
57. *Antimicrobials in Foods: Second Edition, Revised and Expanded*, edited by P. Michael Davidson and Alfred Larry Branen
58. *Lactic Acid Bacteria,* edited by Seppo Salminen and Atte von Wright
59. *Rice Science and Technology,* edited by Wayne E. Marshall and James I. Wadsworth
60. *Food Biosensor Analysis,* edited by Gabriele Wagner and George G. Guilbault
61. *Principles of Enzymology for the Food Sciences: Second Edition,* John R. Whitaker
62. Carbohydrate Polyesters as Fat Substitutes, *edited by Casimir C. Akoh and Barry G. Swanson*
63. Engineering Properties of Foods: Second Edition, Revised and Expanded, *edited by M. A. Rao and S. S. H. Rizvi*
64. Handbook of Brewing, *edited by William A. Hardwick*
65. Analyzing Food for Nutrition Labeling and Hazardous Contaminants, *edited by Ike J. Jeon and William G. Ikins*
66. Ingredient Interactions: Effects on Food Quality, *edited by Anilkumar G. Gaonkar*
67. Food Polysaccharides and Their Applications, *edited by Alistair M. Stephen*
68. Safety of Irradiated Foods: Second Edition, Revised and Expanded, *J. F. Diehl*
69. Nutrition Labeling Handbook, *edited by Ralph Shapiro*
70. Handbook of Fruit Science and Technology: Production, Composition, Storage, and Processing, *edited by D. K. Salunkhe and S. S. Kadam*
71. Food Antioxidants: Technological, Toxicological, and Health Perspectives, *edited by D. L. Madhavi, S. S. Deshpande, and D. K. Salunkhe*
72. Freezing Effects on Food Quality, *edited by Lester E. Jeremiah*
73. Handbook of Indigenous Fermented Foods: Second Edition, Revised and Expanded, *edited by Keith H. Steinkraus*
74. Carbohydrates in Food, *edited by Ann-Charlotte Eliasson*
75. Baked Goods Freshness: Technology, Evaluation, and Inhibition of Staling, *edited by Ronald E. Hebeda and Henry F. Zobel*
76. Food Chemistry: Third Edition, *edited by Owen R. Fennema*
77. Handbook of Food Analysis: Volumes 1 and 2, *edited by Leo M. L. Nollet*

Additional Volumes in Preparation

Computerized Control Systems in the Food Industry, *edited by Gauri S. Mittal*

Techniques for Analyzing Food Aroma, *edited by Ray Marsili*

FOOD CHEMISTRY

THIRD EDITION

edited by

Owen R. Fennema
University of Wisconsin–Madison
Madison, Wisconsin

MARCEL DEKKER, INC. NEW YORK • BASEL • HONG KONG

Library of Congress Cataloging-in-Publication Data

Food chemistry / edited by Owen R. Fennema. — 3rd ed.
p. cm. — (Food science and technology)
Includes index.
ISBN 0-8247-9346-3 (cloth : alk. paper). — ISBN 0-8247-9691-8 (paper : alk. paper)
1. Food—Analysis. 2. Food—Composition. I. Fennema, Owen R. II. Series: Food science and technology (Marcel Dekker, Inc.) ; v. 76.
TX541.F65 1996
664'.001'54 —dc20 96-19500
CIP

The publisher offers discounts on this book when ordered in bulk quantities. For more information, write to Special Sales/Professional Marketing at the address below.

This book is printed on acid-free paper.

Marcel Dekker, Inc.
270 Madison Avenue, New York, New York 10016

Current printing (last digit):
10 9 8 7 6 5 4 3

PRINTED IN THE UNITED STATES OF AMERICA

Preface to the Third Edition

More than a decade has passed since the publication of the second edition of *Food Chemistry,* so the appropriateness of an updated version should be apparent. The purposes of the book remain unchanged: it is primarily a textbook for upper division undergraduates and beginning graduate students who have sound backgrounds in organic chemistry and biochemistry, and is secondarily a reference book. Information on food analysis is intentionally absent, except where its presence fits logically with the topic under discussion. As a textbook for undergraduates, it is designed to serve as the basis of a two-semester course on food chemistry with the assumption that the instructor will make selective reading assignments as deemed appropriate. Individual chapters in the book should be useful as the basis of graduate-level courses on specialized topics in food chemistry.

The third edition differs in several important respects from the second. The chapters prepared by first-time contributors are totally new. These cover such topics as proteins, dispersions, enzymes, vitamins, minerals, animal tissues, toxicants, and pigments. Chapters by contributors to the second edition have been thoroughly revised. For example, in the chapter "Water and Ice," a major addition deals with molecular mobility and glass transition phenomena. The result is a book that is more than 60% new, has greatly improved graphics, and is better focused on material that is unique to food chemistry.

Chapters have been added on the topics of dispersions and minerals. In the second edition, treatment of dispersions was accomplished in the chapters "Lipids," "Proteins," and "Carbohydrates," and minerals were covered in the chapter "Vitamins and Minerals." Although this was organizationally sound, the result was superficial treatment of dispersions and minerals. The new chapters on these topics provide depth of coverage that is more consistent with the remainder of the book. Associated with these changes is a chapter, written by a new contributor, that is now devoted solely to vitamins. It is my belief that this chapter represents the first complete, in-depth treatise on vitamins with an emphasis on food chemistry.

I would be remiss not to thank the contributors for their hard work and tolerance of my sometimes severe editorial oversight. They have produced a book that is of first-rate quality. After twenty years and two previous editions, I am finally satisfied that all major topics are covered appropriately with regard to breadth and depth of coverage, and that a proper focus on reactions pertaining specifically to foods has been achieved. This focus successfully distinguishes food chemistry from biochemistry in the same sense that biochemistry is distinct from, yet still dependent on, organic chemistry.

Although I have planned and edited this edition with great care, minor errors are inevitable, especially in the first printing. If these are discovered, I would very much appreciate hearing from you so that corrections can be effected promptly.

Owen R. Fennema

Preface to the Second Edition

Considerable time has passed since publication of the favorably received first edition so a new edition seems appropriate. The purpose of the book remains unchanged—it is intended to serve as a textbook for upper division undergraduates or beginning graduate students who have sound backgrounds in organic chemistry and biochemistry, and to provide insight to researchers interested in food chemistry. Although the book is most suitable for a two-semester course on food chemistry, it can be adapted to a one-semester course by specifying selective reading assignments. It should also be noted that several chapters are of sufficient length and depth to be useful as primary source materials for graduate-level specialty courses.

This edition has the same organization as the first, but differs substantially in other ways. The chapters on carbohydrates, lipids, proteins, flavors, and milk and the concluding chapter have new authors and are, therefore, entirely new. The chapter on food dispersions has been deleted and the material distributed at appropriate locations in other chapters. The remaining chapters, without exception, have been substantially modified, and the index has been greatly expanded, including the addition of a chemical index. Furthermore, this edition, in contrast to the first, is more heavily weighted in the direction of subject matter that is unique to food chemistry, i.e., there is less overlap with materials covered in standard biochemistry courses. Thus the book has undergone major remodeling and refinement, and I am indebted to the various authors for their fine contributions and for their tolerance of my sometimes severe editorial guidance.

This book, in my opinion, provides comprehensive coverage of the subject of food chemistry with the same depth and thoroughness that is characteristic of the better quality introductory textbooks on organic chemistry and biochemistry. This, I believe, is a significant achievement that reflects a desirable maturation of the field of food chemistry.

Owen R. Fennema

Preface to the First Edition

For many years, an acute need has existed for a food chemistry textbook that is suitable for food science students with backgrounds in organic chemistry and biochemistry. This book is designed primarily to fill the aforementioned need, and secondarily, to serve as a reference source for persons involved in food research, food product development, quality assurance, food processing, and in other activities related to the food industry.

Careful thought was given to the number of contributors selected for this work, and a decision was made to use different authors for almost every chapter. Although involvement of many authors results in potential hazards with respect to uneven coverage, differing philosophies, unwarranted duplication, and inadvertent omission of important materials, this approach was deemed necessary to enable the many facets of food chemistry to be covered at a depth adequate for the primary audience. Since I am acutely aware of the above pitfalls, care has been taken to minimize them, and I believe the end product, considering it is a first edition, is really quite satisfying—except perhaps for the somewhat generous length. If the readers concur with my judgment, I will be pleased but unsurprised, since a book prepared by such outstanding personnel can hardly fail, unless of course the editor mismanages the talent.

Organization of the book is quite simple and I hope appropriate. Covered in sequence are major constituents of food, minor constituents of food, food dispersions, edible animal tissues, edible fluids of animal origin, edible plant tissues and interactions among food constituents—the intent being to progress from simple to more complex systems. Complete coverage of all aspects of food chemistry, of course, has not been attempted. It is hoped, however, that the topics of greatest importance have been treated adequately. In order to help achieve this objective, emphasis has been given to broadly based principles that apply to many foods.

Figures and tables have been used liberally in the belief that this approach facilitates understanding of the subject matter presented. The number of references cited should be adequate to permit easy access to additional information.

To all readers I extend an invitation to report errors that no doubt have escaped my attention, and to offer suggestions for improvements that can be incorporated in future (hopefuly) editions.

Since enjoyment is an unlikely reader response to this book, the best I can hope for is that readers will find it enlightening and well suited for its intended purpose.

Owen R. Fennema

Contents

Contributors

James N. BeMiller Department of Food Science, Purdue University, West Lafayette, Indiana

Grady W. Chism Department of Food Science and Technology, The Ohio State University, Columbus, Ohio

Srinivasan Damodaran Department of Food Science, University of Wisconsin—Madison, Madison, Wisconsin

Owen R. Fennema Department of Food Science, University of Wisconsin—Madison, Madison, Wisconsin

E. Allen Foegeding Department of Food Science, North Carolina State University, Raleigh, North Carolina

Jesse F. Gregory III Department of Food Science and Human Nutrition, University of Florida, Gainesville, Florida

Norman F. Haard Department of Food Science and Technology, Institute of Marine Resources, University of California, Davis, California

Herbert O. Hultin Department of Food Science, University of Massachusetts, Amherst, Massachusetts

Theodore P. Labuza Department of Food Science and Nutrition, University of Minnesota, St. Paul, Minnesota

Tyre C. Lanier Department of Food Science, North Carolina State University, Raleigh, North Carolina

Robert C. Lindsay Department of Food Science, University of Wisconsin—Madison, Madison, Wisconsin

Dennis D. Miller Department of Food Science, Cornell University, Ithaca, New York

Wassef W. Nawar Department of Food Science, University of Massachusetts, Amherst, Massachusetts

Michael W. Pariza Department of Food Microbiology and Toxicology, Food Research Institute, University of Wisconsin—Madison, Madison, Wisconsin

Steven J. Schwartz* Department of Food Science, North Carolina State University, Raleigh, North Carolina

Harold E. Swaisgood Department of Food Science, North Carolina State University, Raleigh, North Carolina

Steven R. Tannenbaum Department of Chemistry, Division of Toxicology, Massachusetts Institute of Technology, Cambridge, Massachusetts

Petros Taoukis Department of Chemical Engineering, National Technical University of Athens, Athens, Greece

J. H. von Elbe Department of Food Science, University of Wisconsin—Madison, Madison, Wisconsin

Pieter Walstra Department of Food Science, Wageningen Agricultural University, Wageningen, The Netherlands

Roy L. Whistler Department of Biochemistry, Purdue University, West Lafayette, Indiana

John R. Whitaker Department of Food Science and Technology, University of California, Davis, California

*Present affiliation: The Ohio State University, Columbus, Ohio.

1

Introduction to Food Chemistry

Owen R. Fennema
University of Wisconsin—Madison, Madison, Wisconsin

Steven R. Tannenbaum
Massachusetts Institute of Technology, Cambridge, Massachusetts

1.1 WHAT IS FOOD CHEMISTRY?

Concern about food exists throughout the world, but the aspects of concern differ with location. In underdeveloped regions of the world, the bulk of the population is involved in food production, yet attainment of adequate amounts and kinds of basic nutrients remains an ever-present problem. In developed regions of the world, food production is highly mechanized and only a small fraction of the population is involved in this activity. Food is available in abundance, much of it is processed, and the use of chemical additives is common. In these fortunate localities, concerns about food relate mainly to cost, quality, variety, convenience, and the effects of processing and added chemicals on wholesomeness and nutritive value. All of these concerns fall within the realm of food science—a science that deals with the physical, chemical, and biological properties of foods as they relate to stability, cost, quality, processing, safety, nutritive value, wholesomeness, and convenience.

Food science is an interdisciplinary subject involving primarily bacteriology, chemistry, biology, and engineering. Food chemistry, a major aspect of food science, deals with the composition and properties of food and the chemical changes it undergoes during handling, processing, and storage. Food chemistry is intimately related to chemistry, biochemistry, physiological chemistry, botany, zoology, and molecular biology. The food chemist relies heavily on knowledge of the aforementioned sciences to effectively study and control biological substances as sources of human food. Knowledge of the innate properties of biological substances and mastery of the means of manipulating them are common interests of both food chemists and biological scientists. The primary interests of biological scientists include reproduction, growth, and changes that biological substances undergo under environmental conditions that are compatible or marginally compatible with life. To the contrary, food chemists are concerned primarily with biological substances that are dead or dying (postharvest physiology of plants and postmortem physiology of muscle) and changes they undergo when exposed to a very wide range of environmental conditions. For example, conditions suitable for sustaining residual life processes are of concern to food chemists during the marketing of fresh fruits and vegetables, whereas conditions incompatible with life processes are of major interest when long-term preservation of food is attempted. In addition, food chemists are concerned with the chemical properties of disrupted food tissues (flour, fruit and vegetable juices, isolated and modified constituents, and manufactured foods), single-cell sources of food (eggs and microorganisms), and one major biological fluid, milk. In summary, food chemists have much in common with biological scientists, yet they also have interests that are distinctly different and are of the utmost importance to humankind.

1.2 HISTORY OF FOOD CHEMISTRY

The origins of food chemistry are obscure, and details of its history have not yet been rigorously studied and recorded. This is not surprising, since food chemistry did not acquire a clear identity until the twentieth century and its history is deeply entangled with that of agricultural chemistry for which historical documentation is not considered exhaustive [5,14]. Thus, the following brief excursion into the history of food chemistry is incomplete and selective. Nonetheless, available information is sufficient to indicate when, where, and why certain key events in food chemistry occurred, and to relate some of these events to major changes in the wholesomeness of the food supply since the early 1800s.

Although the origin of food chemistry, in a sense, extends to antiquity, the most significant discoveries, as we judge them today, began in the late 1700s. The best accounts of developments during this period are those of Filby [12] and Browne [5], and these sources have been relied upon for much of the information presented here.

During the period of 1780–1850 a number of famous chemists made important discoveries, many of which related directly or indirectly to the chemistry of food. The works of Scheele, Lavoisier, de Saussure, Gay-Lussac, Thenard, Davy, Berzelius, Thomson, Beaumont, and Liebig contain the origins of modern food chemistry. Some may question whether these scientists, whose most famous discoveries bear little relationship to food chemistry, deserve recognition as major figures in the origins of modern food chemistry. Although it is admittedly difficult to categorize early scientists as chemists, bacteriologists, or food chemists, it is relatively easy to determine whether a given scientist made substantial contributions to a given field of science. From the following brief examples it is clearly evident that many of these scientists studied foods intensively and made discoveries of such fundamental importance to food chemistry that exclusion of their contributions from any historical account of food chemistry would be inappropriate.

Carl Wilhelm Scheele (1742–1786), a Swedish pharmacist, was one of the greatest chemists of all time. In addition to his more famous discoveries of chlorine, glycerol, and oxygen (3 years before Priestly, but unpublished), he isolated and studied the properties of lactose (1780), prepared mucic acid by oxidation of lactic acid (1780), devised a means of preserving vinegar by means of heat (1782, well in advance of Appert's "discovery"), isolated citric acid from lemon juice (1784) and gooseberries (1785), isolated malic acid from apples (1785), and tested 20 common fruits for the presence of citric, malic, and tartaric acids (1785). His isolation of various new chemical compounds from plant and animal substances is considered the beginning of accurate analytical research in agricultural and food chemistry.

The French chemist Antoine Laurent Lavoisier (1743–1794) was instrumental in the final rejection of the phlogiston theory and in formulating the principles of modern chemistry. With respect to food chemistry, he established the fundamental principles of combustion organic analysis, he was the first to show that the process of fermentation could be expressed as a balanced equation, he made the first attempt to determine the elemental composition of alcohol (1784), and he presented one of the first papers (1786) on organic acids of various fruits.

(Nicolas) Théodore de Saussure (1767–1845), a French chemist, did much to formalize and clarify the principles of agricultural and food chemistry provided by Lavoisier. He also studied CO_2 and O_2 changes during plant respiration (1804), studied the mineral contents of plants by ashing, and made the first accurate elemental analysis of alcohol (1807).

Joseph Louis Gay-Lussac (1778–1850) and Louis-Jacques Thenard (1777–1857) devised in 1811 the first method to determine percentages of carbon, hydrogen, and nitrogen in dry vegetable substances.

The English chemist Sir Humphrey Davy (1778–1829) in the years 1807 and 1808 isolated the elements K, Na, Ba, Sr, Ca, and Mg. His contributions to agricultural and food chemistry came largely through his books on agricultural chemistry, of which the first (1813) was *Elements of Agriculture Chemistry, in a Course of Lectures for the Board of Agriculture* [8]. His books served to organize and clarify knowledge existing at that time. In the first edition he stated,

> All the different parts of plants are capable of being decomposed into a few elements. Their uses as food, or for the purpose of the arts, depend upon compound arrangements of these elements, which are capable of being produced either from their organized parts, or from the juices they contain; and the examination of the nature of these substances is an essential part of agricultural chemistry.

In the fifth edition he stated that plants are usually composed of only seven or eight elements, and that [9] "the most essential vegetable substances consist of hydrogen, carbon, and oxygen in different proportion, generally alone, but in some few cases combined with azote [nitrogen]" (p. 121).

The works of the Swedish chemist Jons Jacob Berzelius (1779–1848) and the Scottish chemist Thomas Thomson (1773–1852) resulted in the beginnings of organic formulas, "without which organic analysis would be a trackless desert and food analysis an endless task" [12]. Berzelius determined the elemental components of about 2000 compounds, thereby verifying the law of definite proportions. He also devised a means of accurately determining the water content of organic substances, a deficiency in the method of Gay-Lussac and Thenard. Moreover, Thomson showed that laws governing the composition of inorganic substances apply equally well to organic substances, a point of immense importance.

In a book entitled *Considérations générales sur l'analyse organique et sur ses applications* [6], Michel Eugene Chevreul (1786–1889), a French chemist, listed the elements known to exist at that time in organic substances (O, Cl, I, N, S, P, C, Si, H, Al, Mg, Ca, Na, K, Mn, Fe) and cited the processes then available for organic analysis: (a) extraction with a neutral solvent, such as water, alcohol, or aqueous ether, (b) slow distillation, or fractional distillation,

(c) steam distillation, (d) passing the substance through a tube heated to incandescence, and (e) analysis with oxygen. Chevreul was a pioneer in the analysis of organic substances, and his classic research on the composition of animal fat led to the discovery and naming of stearic and oleic acids.

Dr. William Beaumont (1785–1853), an American Army surgeon stationed at Fort Mackinac, Mich., performed classic experiments on gastric digestion that destroyed the concept existing from the time of Hippocrates that food contained a single nutritive component. His experiments were performed during the period 1825–1833 on a Canadian, Alexis St. Martin, whose musket wound afforded direct access to the stomach interior, thereby enabling food to be introduced and subsequently examined for digestive changes [4].

Among his many notable accomplishments, Justus von Liebig (1803–1873) showed in 1837 that acetaldehyde occurs as an intermediate between alcohol and acetic acid during fermentation of vinegar. In 1842 he classified foods as either nitrogenous (vegetable fibrin, albumin, casein, and animal flesh and blood) or nonnitrogenous (fats, carbohydrates, and alcoholic beverages). Although this classification is not correct in several respects, it served to distinguish important differences among various foods. He also perfected methods for the quantitative analysis of organic substances, especially by combustion, and he published in 1847 what is apparently the first book on food chemistry, *Researches on the Chemistry of Food* [18]. Included in this book are accounts of his research on the water-soluble constituents of muscle (creatine, creatinine, sarcosine, inosinic acid, lactic acid, etc.).

It is interesting that the developments just reviewed paralleled the beginning of serious and widespread adulteration of food, and it is no exaggeration to state that the need to detect impurities in food was a major stimulus for the development of analytical chemistry in general and analytical food chemistry in particular. Unfortunately, it is also true that advances in chemistry contributed somewhat to the adulteration of food, since unscrupulous purveyors of food were able to profit from the availability of chemical literature, including formulas for adulterated food, and could replace older, less effective empirical approaches to food adulteration with more efficient approaches based on scientific principles. Thus, the history of food chemistry and the history of food adulteration are closely interwoven by the threads of several causative relationships, and it is therefore appropriate to consider the matter of food adulteration from a historical perspective [12].

The history of food adulteration in the currently more developed countries of the world falls into three distinct phases. From ancient times to about 1820 food adulteration was not a serious problem and there was little need for methods of detection. The most obvious explanation for this situation was that food was procured from small businesses or individuals, and transactions involved a large measure of interpersonal accountability. The second phase began in the early 1800s, when intentional food adulteration increased greatly in both frequency and seriousness. This development can be attributed primarily to increased centralization of food processing and distribution, with a corresponding decline in interpersonal accountability, and partly to the rise of modern chemistry, as already mentioned. Intentional adulteration of food remained a serious problem until about 1920, which marks the end of phase two and the beginning of phase three. At this point regulatory pressures and effective methods of detection reduced the frequency and seriousness of intentional food adulteration to acceptable levels, and the situation has gradually improved up to the present time.

Some would argue that a fourth phase of food adulteration began about 1950, when foods containing legal chemical additives became increasingly prevalent, when the use of highly processed foods increased to a point where they represented a major part of the diet of persons in most of the industrialized countries, and when contamination of some foods with undesirable by-products of industrialization, such as mercury, lead, and pesticides, became of public and regulatory concern. The validity of this contention is hotly debated and disagreement persists to this day. Nevertheless, the course of action in the next few years seems clear. Public concern over the safety and nutritional adequacy of the food supply has already led to some recent changes, both voluntary and involuntary, in the manner in which foods are produced, handled, and processed, and more such actions are inevitable as we learn more about proper handling practices for food and as estimates of maximum tolerable intake of undesirable constituents become more accurate.

The early 1800s was a period of especially intense public concern over the quality and safety of the food supply. This concern, or more properly indignation, was aroused in England by Frederick Accum's publication *A Treatise on Adulterations of Food* [1] and by an anonymous publication entitled *Death in the Pot* [3]. Accum claimed that "Indeed, it would be difficult to mention a single article of food which is not to be met within an adulterated state; and there are some substances which are scarcely ever to be procured genuine" (p. 14). He further remarked, "It is not less lamentable that the extensive application of chemistry to the useful purposes of life, should have been perverted into an auxiliary to this nefarious traffic [adulteration]" (p. 20).

Although Filby [12] asserted that Accum's accusations were somewhat overstated, the seriousness of intentional adulteration of food that prevailed in the early 1800s is clearly exemplified by the following not uncommon adulterants cited by both Accum and Filby:

Annatto: Adulterants included turmeric, rye, barley, wheat flour, calcium sulfate and carbonate, salt, and Venetian red (ferric oxide, which in turn was sometimes adulterated with red lead and copper).

Pepper, black: This important product was commonly adulterated with gravel, leaves, twigs, stalks, pepper dust, linseed meal, and ground parts of plants other than pepper.

Pepper, cayenne. Substances such as vermillion (α-mercury sulfide), ocher (native earthy mixtures of metallic oxides and clay), and turmeric were commonly added to overcome bleaching that resulted from exposure to light.

Essential oils: Oil of turpentine, other oils, and alcohol.

Vinegar: Sulfuric acid

Lemon juice: Sulfuric and other acids

Coffee: Roasted grains, occasionally roasted carrots or scorched beans and peas; also, baked horse liver.

Tea: Spent, redried tea leaves, and leaves of many other plants.

Milk: Watering was the main form of adulteration; also, the addition of chalk, starch, turmeric (color), gums, and soda was common. Occasionally encountered were gelatin, dextrin, glucose, preservatives (borax, boric acid, salicylic acid, sodium salicylate, potassium nitrate, sodium fluoride, and benzoate), and such colors as annatto, saffron, caramel, and some sulfonated dyes.

Beer: "Black extract," obtained by boiling the poisonous berries of *Cocculus indicus* in water and concentrating the fluid, was apparently a common additive. This extract imparted flavor, narcotic properties, additional intoxicating qualities, and toxicity to the beverage.

Wine: Colorants: alum, husks of elderberries, Brazil wood, and burnt sugar, among others. Flavors: bitter almonds, tincture of raisin seeds, sweet-brier, oris root, and others. Aging agents: bitartrate of potash, "oenathis" ether (heptyl ether), and lead salts. Preservatives: salicylic acid, benzoic acid, fluoborates, and lead salts. Antacids: lime, chalk, gypsum, and lead salts.

Sugar: Sand, dust, lime, pulp, and coloring matters.

Butter: Excessive salt and water, potato flour, and curds.

Chocolate: Starch, ground sea biscuits, tallow, brick dust, ocher, Venetian red (ferric oxide), and potato flour.
Bread: Alum, and flour made from products other than wheat.
Confectionery products: Colorants containing lead and arsenic.

Once the seriousness of food adulteration in the early 1800s was made evident to the public, remedial forces gradually increased. These took the form of new legislation to make adulteration unlawful, and greatly expanded efforts by chemists to learn about the native properties of foods, the chemicals commonly used as adulterants, and the means of detecting them. Thus, during the period 1820–1850, chemistry and food chemistry began to assume importance in Europe. This was possible because of the work of the scientists already cited, and was stimulated largely by the establishment of chemical research laboratories for young students in various universities and by the founding of new journals for chemical research [5]. Since then, advances in food chemistry have continued at an accelerated pace, and some of these advances, along with causative factors, are mentioned below.

In 1860, the first publicly supported agriculture experiment station was established in Weede, Germany, and W. Hanneberg and F. Stohmann were appointed director and chemist, respectively. Based largely on the work of earlier chemists, they developed an important procedure for the routine determination of major constituents in food. By dividing a given sample into several portions they were able to determine moisture content, "crude fat," ash, and nitrogen. Then, by multiplying the nitrogen value by 6.25, they arrived at its protein content. Sequential digestion with dilute acid and dilute alkali yielded a residue termed "crude fiber." The portion remaining after removal of protein, fat, ash, and crude fiber was termed "nitrogen-free extract," and this was believed to represent utilizable carbohydrate. Unfortunately, for many years chemists and physiologists wrongfully assumed that like values obtained by this procedure represented like nutritive value, regardless of the kind of food [20].

In 1871, Jean Baptiste Duman (1800–1884) suggested that a diet consisting of only protein, carbohydrate, and fat was inadequate to support life.

In 1862, the Congress of the United States passed the Land-Grant College Act, authored by Justin Smith Morrill. This act helped establish colleges of agriculture in the United States and provided considerable impetus for the training of agricultural and food chemists. Also in 1862, the United States Department of Agriculture was established and Isaac Newton was appointed the first commissioner.

In 1863, Harvey Washington Wiley became chief chemist of the U.S. Department of Agriculture, from which office he led the campaign against misbranded and adulterated food, culminating in passage of the first Pure Food and Drug Act in the United States (1906).

In 1887, agriculture experiment stations were established in the United States following enactment of the Hatch Act. Representative William H. Hatch of Missouri, Chairman of the House Committee on Agriculture, was author of the act. As a result, the world's largest national system of agriculture experiment stations came into existence, and this had a great impact on food research in the United States.

During the first half of the twentieth century, most of the essential dietary substances were discovered and characterized, namely, vitamins, minerals, fatty acids, and some amino acids.

The development and extensive use of chemicals to aid in the growth, manufacture, and marketing of foods was an especially noteworthy and contentious event in the middle 1900s, and more will be said about this subject in a later section.

This historical review, although brief, makes the current food supply seem almost perfect in comparison to that which existed in the 1800s.

1.3. APPROACH TO THE STUDY OF FOOD CHEMISTRY

It is desirable to establish an analytical approach to the chemistry of food formulation, processing, and storage stability, so that facts derived from the study of one food or model system can enhance our understanding of other products. There are four components to this approach: (a) determining those properties that are important characteristics of safe, high-quality foods, (b) determining those chemical and biochemical reactions that have important influences on loss of quality and/or wholesomeness of foods, (c) integrating the first two points so that one understands how the key chemical and biochemical reactions influence quality and safety, and (d) applying this understanding to various situations encountered during formulation, processing, and storage of food.

1.3.1 Quality and Safety Attributes

It is essential to reiterate that safety is the first requisite of any food. In a broad sense, this means a food must be free of any harmful chemical or microbial contaminant at the time of its consumption. For operational purposes this definition takes on a more applied form. In the canning industry, "commercial" sterility as applied to low-acid foods means the absence of viable spores of *Clostridium botulinum*. This in turn can be translated into a specific set of heating conditions for a specific product in a specific package. Given these heating requirements, one can then select specific time–temperature conditions that will optimize retention of quality attributes. Similarly, in a product such as peanut butter, operational safety can be regarded primarily as the absence of aflatoxins—carcinogenic substances produced by certain species of molds. Steps taken to prevent growth of the mold in question may or may not interfere with retention of some other quality attribute; nevertheless, conditions producing a safe product must be employed.

A list of quality attributes of food and some alterations they can undergo during processing and storage is given in Table 1. The changes that can occur, with the exception of those involving nutritive value and safety, are readily evident to the consumer.

1.3.2 Chemical and Biochemical Reactions

Many reactions can alter food quality or safety. Some of the more important classes of these reactions are listed in Table 2. Each reaction class can involve different reactants or substrates depending on the specific food and the particular conditions for handling, processing, or storage. They are treated as reaction classes because the general nature of the substrates or reactants is similar for all foods. Thus, nonenzymic browning involves reaction of carbonyl compounds, which can arise from existing reducing sugars or from diverse reactions, such as oxidation of ascorbic acid, hydrolysis of starch, or oxidation of lipids. Oxidation may involve lipids, proteins, vitamins, or pigments, and more specifically, oxidation of lipids may involve triacylglycerols in one food or phospholipids in another. Discussion of these reactions in detail will occur in subsequent chapters of this book.

1.3.3 Effect of Reactions on the Quality and Safety of Food

The reactions listed in Table 3 cause the alterations listed in Table 1. Integration of the information contained in both tables can lead to an understanding of the causes of food deterioration. Deterioration of food usually consists of a series of primary events followed by

TABLE 1 Classification of Alterations That Can Occur in Food During Handling, Processing, or Storage

Attribute	Alteration
Texture	Loss of solubility
	Loss of water-holding capacity
	Toughening
	Softening
Flavor	Development of:
	Rancidity (hydrolytic or oxidative)
	Cooked or caramel flavors
	Other off-flavors
	Desirable flavors
Color	Darkening
	Bleaching
	Development of other off-colors
	Development of desirable colors (e.g., browning of baked goods)
Nutritive value	Loss, degradation or altered bioavailability of proteins, lipids, vitamins, minerals
Safety	Generation of toxic substances
	Development of substances that are protective to health
	Inactivation of toxic substances

secondary events, which, in turn, become evident as altered quality attributes (Table 1). Examples of sequences of this type are shown in Table 3. Note particularly that a given quality attribute can be altered as a result of several different primary events.

The sequences in Table 3 can be applied in two directions. Operating from left to right one can consider a particular primary event, the associated secondary events, and the effect on a

TABLE 2 Some Chemical and Biochemical Reactions That Can Lead to Alteration of Food Quality or Safety

Types of reaction	Examples
Nonenzymic browning	Baked goods
Enzymic browning	Cut fruits
Oxidation	Lipids (off-flavors), vitamin degradation, pigment decoloration, proteins (loss of nutritive value)
Hydrolysis	Lipids, proteins, vitamins, carbohydrates, pigments
Metal interactions	Complexation (anthocyanins), loss of Mg from chlorophyll, catalysis of oxidation
Lipid isomerization	Cis $\rightarrow$ trans, nonconjugated $\rightarrow$ conjugated
Lipid cyclization	Monocyclic fatty acids
Lipid polymerization	Foaming during deep fat frying
Protein denaturation	Egg white coagulation, enzyme inactivation
Protein cross-linking	Loss of nutritive value during alkali processing
Polysaccharide synthesis	In plants postharvest
Glycolytic changes	Animal tissue postmortem, plant tissue postharvest

TABLE 3 Cause-and-Effect Relationships Pertaining to Food Alterations During Handling, Storage, and Processing

Primary causative event	Secondary event	Attribute influenced (see Table 1)
Hydrolysis of lipids	Free fatty acids react with protein	Texture, flavor, nutritive value
Hydrolysis of polysaccharides	Sugars react with proteins	Texture, flavor, color, nutritive value
Oxidation of lipids	Oxidation products react with many other constituents	Texture, flavor, color, nutritive value; toxic substances can be generated
Bruising of fruit	Cells break, enzymes are released, oxygen accessible	Texture, flavor, color, nutritive value
Heating of green vegetables	Cell walls and membranes lose integrity, acids are released, enzymes become inactive	Texture, flavor, color, nutritive value
Heating of muscle tissue	Proteins denature and aggregate, enzymes become inactive	Texture, flavor, color, nutritive value
Cis → trans conversions in lipids	Enhanced rate of polymerization during deep fat frying	Excessive foaming during deep fat frying; diminished bioavailability of lipids

quality attribute. Alternatively, one can determine the probable cause(s) of an observed quality change (column 3, Table 3) by considering all primary events that could be involved and then isolating, by appropriate chemical tests, the key primary event. The utility of constructing such sequences is that they encourage one to approach problems of food alteration in an analytical manner.

Figure 1 is a simplistic summary of reactions and interactions of the major constituents

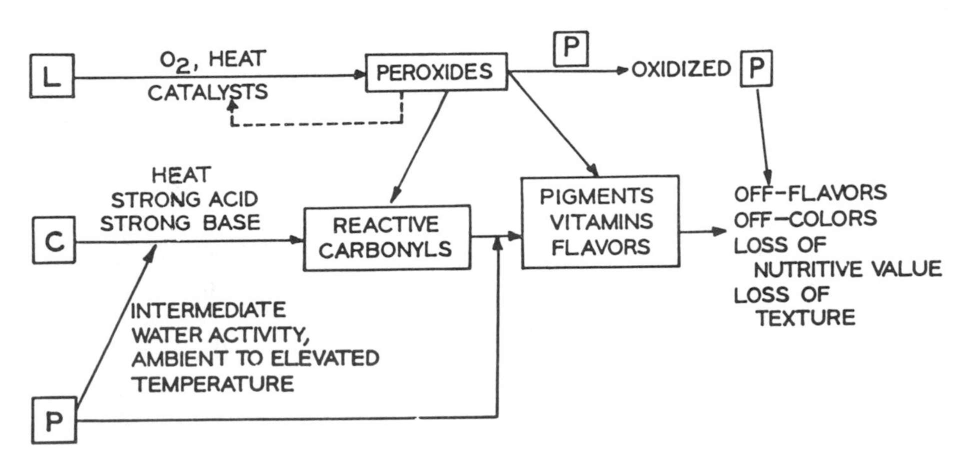

FIGURE 1 Summary of chemical interactions among major food constituents: L, lipid pool (triacylglycerols, fatty acids, and phospholipids); C, carbohydrate pool (polysaccharides, sugars, organic acids, and so on); P, protein pool (proteins, peptides, amino acids, and other N-containing substances).

TABLE 4 Important Factors Governing the Stability of Foods During Handling, Processing, and Storage

Product factors: chemical properties of individual constituents (including catalysts), oxygen content, pH, water activity, T_g and W_g
Environmental factors: temperature (T), time (t), composition of the atmosphere, chemical, physical or biological treatments imposed, exposure to light, contamination, physical abuse

Note. Water activity = p/p_o, where p is the partial pressure of water vapor above the food and p_o is the vapor pressure of pure water; T_g is the glass transition temperature; W_g is the product water content at T_g.

of food. The major cellular pools of carbohydrates, lipids, proteins, and their intermediary metabolites are shown on the left-hand side of the diagram. The exact nature of these pools is dependent on the physiological state of the tissue at the time of processing or storage, and the constituents present in or added to nontissue foods. Each class of compound can undergo its own characteristic type of deterioration. Noteworthy is the role that carbonyl compounds play in many deterioration processes. They arise mainly from lipid oxidation and carbohydrate degradation, and can lead to the destruction of nutritional value, to off-colors, and to off-flavors. Of course these same reactions lead to desirable flavors and colors during the cooking of many foods.

1.3.4 Analysis of Situations Encountered During the Storage and Processing of Food

Having before us a description of the attributes of high-quality, safe foods, the significant chemical reactions involved in the deterioration of food, and the relationship between the two, we can now begin to consider how to apply this information to situations encountered during the storage and processing of food.

The variables that are important during the storage and processing of food are listed in Table 4. Temperature is perhaps the most important of these variables because of its broad influence on all types of chemical reactions. The effect of temperature on an individual reaction can be estimated from the Arrhenius equation, $k = Ae^{-\Delta E/RT}$. Data conforming to the Arrhenius equation yield a straight line when log k is plotted versus $1/T$. Arrhenius plots in Figure 2 represent reactions important in food deterioration. It is evident that food reactions generally conform to the Arrhenius relationship over a limited intermediate temperature range but that deviations from this relationship can occur at high or low temperatures [21]. Thus, it is important to remember that the Arrhenius relationship for food systems is valid only over a range of temperature that has been experimentally verified. Deviations from the Arrhenius relationship can occur because of the following events, most of which are induced by either high or low temperatures: (a) enzyme activity may be lost, (b) the reaction pathway may change or may be influenced by a competing reaction(s), (c) the physical state of the system may change (e.g., by freezing), or (d) one or more of the reactants may become depleted.

Another important factor in Table 4 is time. During storage of a food product, one frequently wants to know how long the food can be expected to retain a specified level of quality. Therefore, one is interested in time with respect to the integral of chemical and/or microbiological changes that occur during a specified storage period, and in the way these changes combine to determine a specified storage life for the product. During processing, one is often

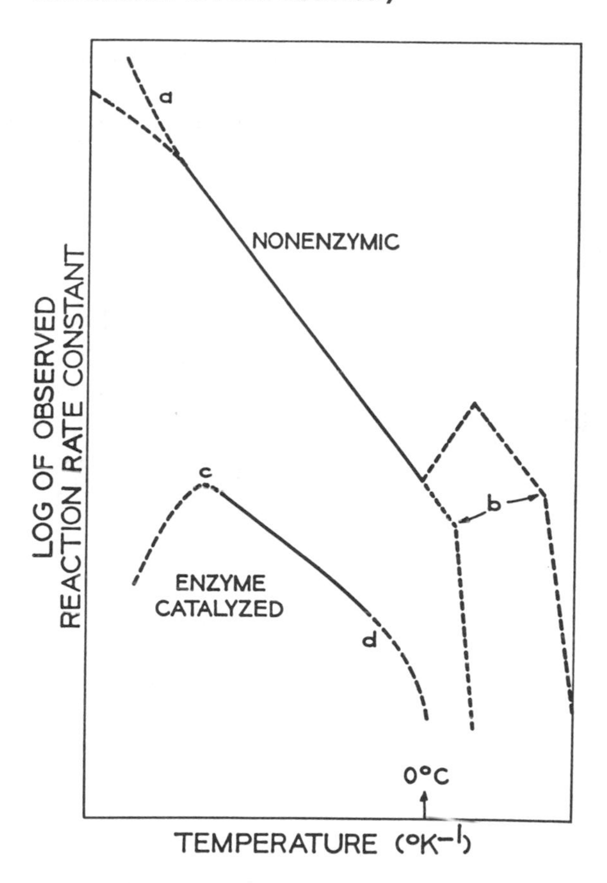

FIGURE 2 Conformity of important deteriorative reactions in food to the Arrhenius relationship. (a) Above a certain value of T there may be deviations from linearity due to a change in the path of the reaction. (b) As the temperature is lowered below the freezing point of the system, the ice phase (essentially pure) enlarges and the fluid phase, which contains all the solutes, diminishes. This concentration of solutes in the unfrozen phase can decrease reaction rates (supplement the effect of decreasing temperature) or increase reaction rates (oppose the effect of declining temperature), depending on the nature of the system (see Chap. 2). (c) For an enzymic reaction there is a temperature in the vicinity of the freezing point of water where subtle changes, such as the dissociation of an enzyme complex, can lead to a sharp decline in reaction rate.

interested in the time it takes to inactivate a particular population of microorganisms or in how long it takes for a reaction to proceed to a specified extent. For example, it may be of interest to know how long it takes to produce a desired brown color in potato chips during frying. To accomplish this, attention must be given to temperature change with time, that is, the rate of temperature change (dT/dt). This relationship is important because it determines the rate at which microorganisms are destroyed and the relative rates of competing chemical reactions. The latter is of interest in foods that deteriorate by more than one means, such as lipid oxidation and nonenzymic browning. If the products of the browning reaction are antioxidants, it is important to know whether the relative rates of these reactions are such that a significant interaction will occur between them.

Another variable, pH, influences the rates of many chemical and enzymic reactions. Extreme pH values are usually required for severe inhibition of microbial growth or enzymic processes, and these conditions can result in acceleration of acid- or base-catalyzed reactions. In contrast, even a relatively small pH change can cause profound changes in the quality of some foods, for example, muscle.

The composition of the product is important since this determines the reactants available for chemical transformation. Particularly important from a quality standpoint is the relationship that exists between composition of the raw material and composition of the finished product. For example, (a) the manner in which fruits and vegetables are handled postharvest can influence sugar content, and this, in turn, influences the degree of browning obtained during dehydration or deep-fat frying. (b) The manner in which animal tissues are handled postmortem influences the extents and rates of glycolysis and ATP degradation, and these in turn can influence storage life, water-holding capacity, toughness, flavor, and color. (c) The blending of raw materials may cause unexpected interactions; for example, the rate of oxidation can be accelerated or inhibited depending on the amount of salt present.

Another important compositional determinant of reaction rates in foods is water activity (a_w). Numerous investigators have shown a_w to strongly influence the rate of enzyme-catalyzed reactions [2], lipid oxidation [16,22], nonenzymic browning [10,16], sucrose hydrolysis [23], chlorophyll degradation [17], anthocyanin degradation [11], and others. As is discussed in Chapter 2, most reactions tend to decrease in rate below an a_w corresponding to the range of intermediate moisture foods (0.75–0.85). Oxidation of lipids and associated secondary effects, such as carotenoid decoloration, are exceptions to this rule; that is, these reactions accelerate at the lower end of the a_w scale.

More recently, it has become apparent that the glass transition temperature (T_g) of food and the corresponding water content (W_g) of the food at T_g are causatively related to rates of diffusion-limited events in food. Thus, T_g and W_g have relevance to the physical properties of frozen and dried foods, to conditions appropriate for freeze drying, to physical changes involving crystallization, recrystallization, gelatinization, and starch retrogradation, and to those chemical reactions that are diffusion-limited (see Chap. 2).

In fabricated foods, the composition can be controlled by adding approved chemicals, such as acidulants, chelating agents, flavors, or antioxidants, or by removing undesirable reactants, for example, removing glucose from dehydrated egg albumen.

Composition of the atmosphere is important mainly with respect to relative humidity and oxygen content, although ethylene and CO_2 are also important during storage of living plant foods. Unfortunately, in situations where exclusion of oxygen is desirable, this is almost impossible to achieve completely. The detrimental consequences of a small amount of residual oxygen sometimes become apparent during product storage. For example, early formation of a small amount of dehydroascorbic acid (from oxidation of ascorbic acid) can lead to Maillard browning during storage.

For some products, exposure to light can be detrimental, and it is then appropriate to package the products in light-impervious material or to control the intensity and wavelengths of light, if possible.

Food chemists must be able to integrate information about quality attributes of foods, deteriorative reactions to which foods are susceptible, and the factors governing kinds and rates of these deteriorative reactions, in order to solve problems related to food formulation, processing, and storage stability.

1.4 SOCIETAL ROLE OF FOOD CHEMISTS

1.4.1 Why Should Food Chemists Become Involved in Societal Issues?

Food chemists, for the following reasons, should feel obligated to become involved in societal issues that encompass pertinent technological aspects (technosocietal issues).

- Food chemists have had the privilege of receiving a high level of education and of acquiring special scientific skills, and these privileges and skills carry with them a corresponding high level of responsibility.
- Activities of food chemists influence adequacy of the food supply, healthfulness of the population, cost of foods, waste creation and disposal, water and energy use, and the nature of food regulations. Because these matters impinge on the general welfare of the public, it is reasonable that food chemists should feel a responsibility to have their activities directed to the benefit of society.
- If food chemists do not become involved in technosocietal issues, the opinions of others—scientists from other professions, professional lobbyists, persons in the news media, consumer activists, charlatans, antitechnology zealots—will prevail. Many of these individuals are less qualified than food chemists to speak on food-related issues, and some are obviously unqualified.

1.4.2 Types of Involvement

The societal obligations of food chemists include good job performance, good citizenship, and guarding the ethics of the scientific community, but fulfillment of these very necessary roles is not enough. An additional role of great importance, and one that often goes unfulfilled by food chemists, is that of helping determine how scientific knowledge is interpreted and used by society. Although food chemists and other food scientists should not have the only input to these decisions, they must, in the interest of wise decision making, have their views heard and considered. Acceptance of this position, which is surely indisputable, leads to the obvious question, "What exactly should food chemists do to properly discharge their responsibilities in this regard?" Several activities are appropriate.

1. Participate in pertinent professional societies.
2. Serve on governmental advisory committees, when invited.
3. Undertake personal initiatives of a public service nature.

The latter can involve letters to newspapers, journals, legislators, government regulators, company executives, university administrators, and others, and speeches before civic groups.

The major objectives of these efforts are to educate and enlighten the public with respect to food and dietary practices. This involves improving the public's ability to intelligently evaluate information on these topics. Accomplishing this will not be easy because a significant portion of the populace has ingrained false notions about food and proper dietary practices, and because food has, for many individuals, connotations that extend far beyond the chemist's narrow view. For these individuals, food may be an integral part of religious practice, cultural heritage, ritual, social symbolism, or a route to physiological well-being—attitudes that are, for the most part, not conducive to acquiring an ability to appraise foods and dietary practices in a sound, scientific manner.

One of the most contentious food issues, and one that has eluded appraisal by the public in a sound, scientific manner, is the use of chemicals to modify foods. "Chemophobia," the fear of chemicals, has afflicted a significant portion of the populace, causing food additives, in the minds of many, to represent hazards inconsistent with fact. One can find, with disturbing ease, articles in the popular literature whose authors claim the American food supply is sufficiently laden with poisons to render it unwholesome at best, and life-threatening at worst. Truly shocking, they say, is the manner in which greedy industrialists poison our foods for profit while an ineffectual Food and Drug Administration watches with placid unconcern. Should authors holding this viewpoint be believed? It is advisable to apply the following criteria when evaluating the validity of any journalistic account dealing with issues of this kind.

- Credibility of the author. Is the author, by virtue of formal education, experience, and acceptance by reputable scientists, qualified to write on the subject? The writer should, to be considered authoritative, have been a frequent publisher of articles in respected scientific journals, especially those requiring peer review. If the writer has contributed only to the popular literature, particularly in the form of articles with sensational titles and "catch" phrases, this is cause to exercise special care in assessing whether the presentation is scholarly and reliable.
- Appropriateness of literature citations. A lack of literature citations does not constitute proof of irresponsible or unreliable writing, but it should provoke a feeling of moderate skepticism in the reader. In trustworthy publications, literature citations will almost invariably be present and will direct the reader to highly regarded scientific publications. When "popular" articles constitute the bulk of the literature citations, the author's views should be regarded with suspicion.
- Credibility of the publisher. Is the publisher of the article, book, or magazine regarded by reputable scientists as a consistent publisher of high-quality scientific materials? If not, an extra measure of caution is appropriate when considering the data.

If information in poison-pen types of publications are evaluated by rational individuals on the basis of the preceding criteria, such information will be dismissed as unreliable. However, even if these criteria are followed, disagreement about the safety of foods still occurs. The great majority of knowledgeable individuals support the view that our food supply is acceptably safe and nutritious and that legally sanctioned food additives pose no unwarranted risks [7,13,15,19,24–26]. However, a relatively small group of knowledgeable individuals believes that our food supply is unnecessarily hazardous, particularly with regard to some of the legally sanctioned food additives, and this view is most vigorously represented by Michael Jacobson and his Center for Science in the Public Interest. This serious dichotomy of opinion cannot be resolved here, but information provided in Chapter 13 will help undecided individuals arrive at a soundly based personal perspective on food additives, contaminants in foods, and food safety.

In summary, scientists have greater obligations to society than do individuals without formal scientific education. Scientists are expected to generate knowledge in a productive, ethical manner, but this is not enough. They should also accept the responsibility of ensuring that scientific knowledge is used in a manner that will yield the greatest benefit to society. Fulfillment of this obligation requires that scientists not only strive for excellence and conformance to high ethical standards in their day-to-day professional activities, but that they also develop a deep-seated concern for the well-being and scientific enlightenment of the public.

REFERENCES

1. Accum, F. (1966). *A Treatise on Adulteration of Food, and Culinary Poisons, 1920,* Facsimile reprint by Mallinckrodt Chemical Works, St. Louis, MO.
2. Acker, L. W. (1969). Water activity and enzyme activity. *Food Technol. 23*(10):1257–1270.
3. Anonymous (1831). *Death in the Pot.* Cited by Filby, 1934 (Ref. 12).
4. Beaumont, W. (1833). *Experiments and Observations of the Gastric Juice and the Physiology of Digestion,* F. P. Allen, Plattsburgh, NY.
5. Browne, C. A. (1944). *A Source Book of Agricultural Chemistry.* Chronica Botanica Co., Waltham, MA.
6. Chevreul, M. E. (1824). Considérations générales sur l'analyse organique et sur ses applications. Cited by Filby, 1934 (Ref. 12).
7. Clydesdale, F. M., and F. J. Francis (1977). *Food, Nutrition and You,* Prentice-Hall, Englewood Cliffs, NJ.
8. Davy, H. (1813). *Elements of Agricultural Chemistry, in a Course of Lectures for the Board of Agriculture,* Longman, Hurst, Rees, Orme and Brown, London. Cited by Browne, 1944 (Ref. 5).
9. Davy, H. (1936). *Elements of Agricultural Chemistry,* 5th ed., Longman, Rees, Orme, Brown, Green and Longman, London.
10. Eichner, K., and M. Karel (1972). The influence of water content and water activity on the sugar-amino browning reaction in model systems under various conditions. *J. Agric. Food Chem. 20*(2):218–223.
11. Erlandson, J. A., and R. E. Wrolstad (1972). Degradation of anthocyanins at limited water concentration. *J. Food Sci. 37*(4):592–595.
12. Filby, F. A. (1934). *A History of Food Adulteration and Analysis,* George Allen and Unwin, London.
13. Hall, R. L. (1982). Food additives, in *Food and People* (D. Kirk and I. K. Eliason, eds.), Boyd and Fraser, San Francisco, pp. 148–156.
14. Ihde, A. J. (1964). *The Development of Modern Chemistry,* Harper and Row, New York.
15. Jukes, T. H. (1978). How safe is our food supply? *Arch. Intern. Med. 138*:772–774.
16. Labuza, T. P., S. R. Tannenbaum, and M. Karel (1970). Water content and stability of low-moisture and intermediate-moisture foods. *Food Techol. 24*(5):543–550.
17. LaJollo, F., S. R. Tannenbaum, and T. P. Labuza (1971). Reaction at limited water concentration. 2. Chlorophyll degradation. *J. Food Sci. 36*(6):850–853.
18. Liebig, J. von (1847). *Researches on the Chemistry of Food,* edited from the author's manuscript by William Gregory; Londson, Taylor and Walton, London. Cited by Browne, 1944 (Ref. 5).
19. Mayer, J. (1975). *A Diet for Living,* David McKay, Inc., New York.
20. McCollum, E. V. (1959). The history of nutrition. *World Rev. Nutr. Diet. 1*:1–27.
21. McWeeny, D. J. (1968). Reactions in food systems: Negative temperature coefficients and other abnormal temperature effects. *J. Food Technol. 3*:15–30.
22. Quast, D. G., and M. Karel (1972). Effects of environmental factors on the oxidation of potato chips. *J. Food Sci. 37*(4):584–588.
23. Schoebel, T., S. R. Tannenbaum, and T. P. Labuza (1969). Reaction at limited water concentration. 1. Sucrose hydrolysis. *J. Food Sci. 34*(4):324–329.
24. Stare, F. J., and E. M. Whelan (1978). *Eat OK—Feel OK,* Christopher Publishing House, North Quincy, MA.
25. Taylor, R. J. (1980). *Food Additives,* John Wiley & Sons, New York.
26. Whelan, E. M. (1993). *Toxic Terror,* Prometheus Books, Buffalo, NY.

2

Water and Ice

OWEN R. FENNEMA
University of Wisconsin—Madison, Madison, Wisconsin

PROLOGUE: WATER—THE DECEPTIVE MATTER OF LIFE AND DEATH

Unnoticed in the darkness of a subterranean cavern, a water droplet trickles slowly down a stalactite, following a path left by countless predecessors, imparting, as did they, a small but almost magical touch of mineral beauty. Pausing at the tip, the droplet grows slowly to full size, then plunges quickly to the cavern floor, as if anxious to perform other tasks or to assume different forms. For water, the possibilities are countless. Some droplets assume roles of quiet beauty—on a child's coat sleeve, where a snowflake of unique design and exquisite perfection lies unnoticed; on a spider's web, where dew drops burst into sudden brilliance at the first touch of the morning sun; in the countryside, where a summer shower brings refreshment; or in the city, where fog gently permeates the night air, subduing harsh sounds with a glaze of tranquility. Others lend themselves to the noise and vigor of a waterfall, to the overwhelming immensity of a glacier, to the ominous nature of an impending storm, or to the persuasiveness of a tear on a woman's cheek. For others the role is less obvious but far more critical. There is life—initiated and sustained by water in a myriad of subtle and poorly understood ways—or death inevitable, catalyzed under special circumstances by a few hostile crystals of ice; or decay at the forest's floor, where water works relentlessly to disassemble the past so life can begin anew. But the form of water most familiar to humans is none of these; rather, it is simple, ordinary, and uninspiring, unworthy of special notice as it flows forth in cool abundance from a household tap. "Humdrum," galunks a frog in concurrence, or so it seems as he views with stony indifference the watery milieu on which his very life depends. Surely, then, water's most remarkable feature is deception, for it is in reality a substance of infinite complexity, of great and unassessable importance, and one that is endowed with a strangeness and beauty sufficient to excite and challenge anyone making its acquaintance.

2.1 INTRODUCTION

On this planet, water is the only substance that occurs abundantly in all three physical states. It is the only common liquid and is the most widely distributed pure solid, being ever present

TABLE 1 Water Contents of Various Foods

Food	Water content (%)
Meat	
Pork, raw, composite of lean cuts	53–60
Beef, raw, retail cuts	50–70
Chicken, all classes, raw meat without skin	74
Fish, muscle proteins	65–81
Fruit	
Berries, cherries, pears	80–85
Apples, peaches, oranges, grapefruit	90–90
Rhubarb, strawberries, tomatos	90–95
Vegetables	
Avocado, bananas, peas (green)	74–80
Beets, broccoli, carrots, potatoes	85–90
Asparagus, beans (green), cabbage, cauliflower, lettuce	90–95

somewhere in the atmosphere as suspended ice particles, or on the earth's surface as various types of snow and ice. It is essential to life: as an important governor of body temperature, as a solvent, as a carrier of nutrients and waste products, as a reactant and reaction medium, as a lubricant and plasticizer, as a stabilizer of biopolymer conformation, as a likely facilitator of the dynamic behavior of macromolecules, including their catalytic (enzymatic) properties, and in other ways yet unknown. It is truly remarkable that organic life should depend so heavily on this small inorganic molecule, and, perhaps even more remarkable, that so few scientists are aware of this fact.

Water is the major component of many foods, each having its own characteristic allotment of this component (Table 1). Water in the proper amount, location, and orientation profoundly influences the structure, appearance, and taste of foods and their susceptibility to spoilage. Because most kinds of fresh foods contain large amounts of water, effective forms of preservation are needed if long-term storage is desired. Removal of water, either by conventional dehydration or by separation locally in the form of pure ice crystals (freezing), greatly alters the native properties of foods and biological matter. Furthermore, all attempts (rehydration, thawing) to return water to its original status are never more than partially successful. Ample justification exists, therefore, to study water and ice with considerable care.

2.2 PHYSICAL PROPERTIES OF WATER AND ICE

As a first step in becoming familiar with water, it is appropriate to consider its physical properties as shown in Table 2. By comparing water's properties with those of molecules of similar molecular weight and atomic composition (CH_4, NH_3, HF, H_2S, H_2Se, H_2Te) it is possible to determine if water behaves in a normal fashion. Based on this comparison, water is found to melt and boil at unusually high temperatures; to exhibit unusually large values for surface tension, permittivity (dielectric constant), heat capacity, and heats of phase transition (heats of fusion, vaporization, and sublimation); to have a moderately low value for density; to exhibit an unusual attribute of expanding upon solidification; and to possess a viscosity that in light of the foregoing oddities, is surprisingly normal.

In addition, the thermal conductivity of water is large compared to those of other liquids, and the thermal conductivity of ice is moderately large compared to those of other nonmetallic

TABLE 2 Physical Properties of Water and Ice

Property	Value
Molecular weight	18.0153
Phase transition properties	
Melting point at 101.3 kPa (1 atm)	0.000°C
Boiling point at 101.3 kPa (1 atm)	100.000°C
Critical temperature	373.99°C
Critical pressure	22.064 MPa (218.6 atm)
Triple point	0.01°C and 611.73 Pa (4.589 mm Hg)
Enthalpy of fusion at 0°C	6.012 kJ (1.436 kcal)/mol
Enthalpy of vaporization at 100°C	40.657 kJ (9.711 kcal)/mol
Enthalpy of sublimation at 0°C	50.91 kJ (12.16 kcal)/mol

	Temperature			
Other properties	20°C	0°C	0°C (ice)	−20°C (ice)
Density (g/cm^3)	0.99821	0.99984	0.9168	0.9193
Viscosity (pa · sec)	1.002×10^{-3}	1.793×10^{-3}	—	—
Surface tension against air (N/m)	72.75×10^{-3}	75.64×10^{-3}	—	—
Vapor pressure (kPa)	2.3388	0.6113	0.6113	0.103
Heat capacity (J/g · K)	4.1818	4.2176	2.1009	1.9544
Thermal conductivity (liquid) (W/m · K)	0.5984	0.5610	2.240	2.433
Thermal diffusitity (m^2/s)	1.4×10^{-7}	1.3×10^{-7}	11.7×10^{-7}	11.8×10^{-7}
Permittivity (dielectric constant)	80.20	87.90	~90	~98

Source: Mainly Ref. 69.

solids. Of greater interest is the fact that the thermal conductivity of ice at 0°C is approximately four times that of water at the same temperature, indicating that ice will conduct heat energy at a much greater rate than will immobilized water (e.g., in tissue). The thermal diffusivities of water and ice are of even greater interest since these values indicate the rate at which the solid and liquid forms of HOH will undergo changes in temperature. Ice has a thermal diffusivity approximately nine times greater than that of water, indicating that ice, in a given environment, will undergo a temperature change at a much greater rate than will water. These sizable differences in thermal conductivity and thermal diffusivity values of water and ice provide a sound basis for explaining why tissues freeze more rapidly than they thaw, when equal but reversed temperature differentials are employed.

2.3 THE WATER MOLECULE

Water's unusual properties suggest the existence of strong attractive forces among water molecules, and uncommon structures for water and ice. These features are best explained by considering the nature of first a single water molecule and then small groups of molecules. To form a molecule of water, two hydrogen atoms approach the two sp^3 bonding orbitals of oxygen and form two covalent sigma (σ) bonds (40% partial ionic character), each of which has a dissociation energy of 4.6×10^2 kJ/mol (110 kcal/mol). The localized molecular orbitals remain symmetrically oriented about the original orbital axes, thus retaining an approximate tetrahedral structure. A schematic orbital model of a water molecule is shown in Figure 1A and the appropriate van der Waals radii are shown in Figure 1B.

The bond angle of the isolated water molecule (vapor state) is 104.5° and this value is near the perfect tetrahedral angle of 109°28′. The O-H internuclear distance is 0.96 Å and the van der Waals radii for oxygen and hydrogen are, respectively, 1.40 and 1.2 Å.

At this point, it is important to emphasize that the picture so far presented is oversimplified. Pure water contains not only ordinary HOH molecules but also many other constituents in trace amounts. In addition to the common isotopes ^{16}O and ^{1}H, also present are ^{17}O, ^{18}O, ^{2}H

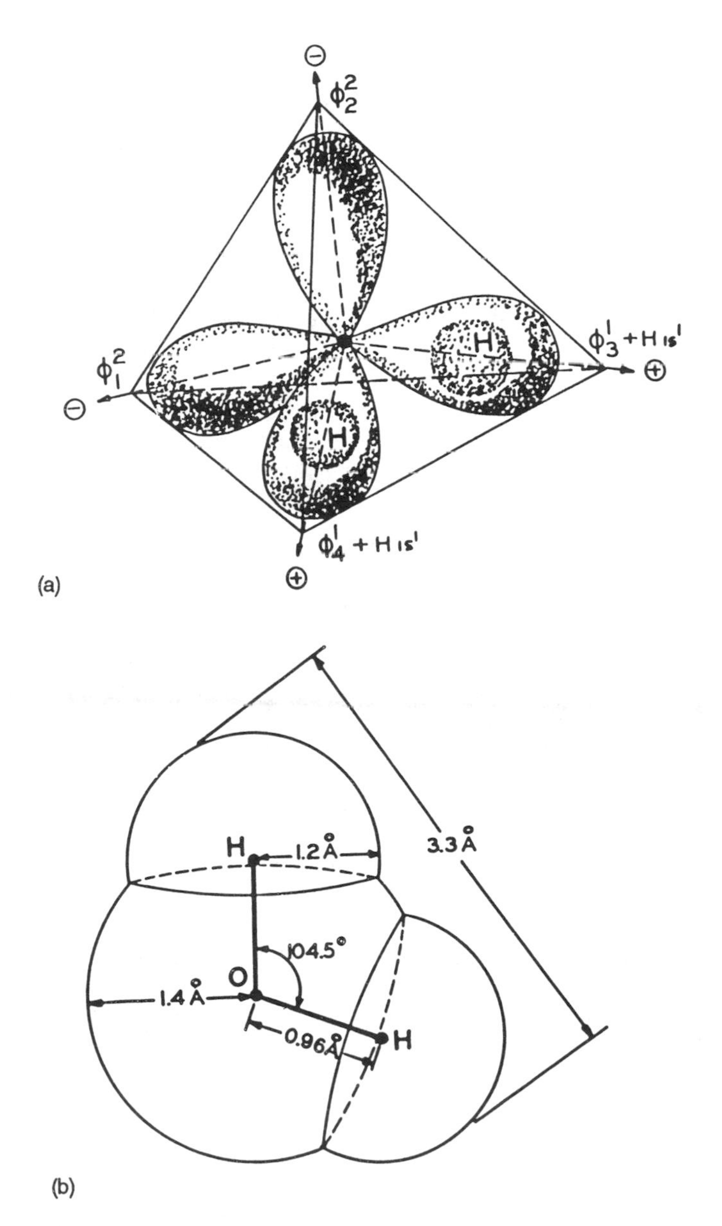

FIGURE 1 Schematic model of a single HOH molecule: (a) sp^3 configuration, and (b) van der Waals radii for a HOH molecule in the vapor state.

(deuterium) and ^{3}H (tritium), giving rise to 18 isotopic variants of molecular HOH. Water also contains ionic particles such as hydrogen ions (existing as H_3O^+), hydroxyl ions, and their isotopic variants. Water therefore consists of more than 33 chemical variants of HOH, but the variants occur in only minute amounts.

2.4 ASSOCIATION OF WATER MOLECULES

The V-like form of an HOH molecule and the polarized nature of the O-H bond result in an unsymmetrical charge distribution and a vapor-state dipole moment of 1.84D for pure water. Polarity of this magnitude produces intermolecular attractive forces, and water molecules therefore associate with considerable tenacity. Water's unusually large intermolecular attractive force cannot, however, be fully accounted for on the basis of its large dipole moment. This is not surprising, since dipole moments give no indication of the degree to which charges are exposed or of the geometry of the molecule, and these aspects, of course, have an important bearing on the intensity of molecular association.

Water's large intermolecular attractive forces can be explained quite adequately in terms of its ability to engage in multiple hydrogen bonding on a three-dimensional basis. Compared to covalent bonds (average bond energy of about 335 kJ/mol), hydrogen bonds are weak (typically 2–40 kJ/mol) and have greater and more variable lengths. The oxygen–hydrogen bond has a dissociation energy of about 13–25 kJ/mol.

Since electrostatic forces make a major contribution to the energy of the hydrogen bond (perhaps the largest contribution), and since an electrostatic model of water is simple and leads to an essentially correct geometric picture of HOH molecules as they are known to exist in ice, further discussion of the geometrical patterns formed by association of water molecules will emphasize electrostatic effects. This simplified approach, while entirely satisfactory for present purposes, must be modified if other behavioral characteristics of water are to be explained satisfactorily.

The highly electronegative oxygen of the water molecule can be visualized as partially drawing away the single electrons from the two covalently bonded hydrogen atoms, thereby leaving each hydrogen atom with a partial positive charge and a minimal electron shield; that is, each hydrogen atom assumes some characteristics of a bare proton. Since the hydrogen–oxygen bonding orbitals are located on two of the axes of an imaginary tetrahedron (Fig. 1a), these two axes can be thought of as representing lines of positive force (hydrogen-bond donor sites). Oxygen's two lone-pair orbitals can be pictured as residing along the remaining two axes of the imaginary tetrahedron, and these then represent lines of negative force (hydrogen-bond acceptor sites). By virtue of these four lines of force, each water molecule is able to hydrogen-bond with a maximum of four others. The resulting tetrahedral arrangement is depicted in Figure 2.

Because each water molecule has an equal number of hydrogen-bond donor and receptor sites, arranged to permit three-dimensional hydrogen bonding, it is found that the attractive forces among water molecules are unusually large, even when compared to those existing among other small molecules that also engage in hydrogen bonding (e.g., NH_3, HF). Ammonia, with its tetrahedral arrangement of three hydrogens and one receptor site, and hydrogen fluoride, with its tetrahedral arrangement of one hydrogen and three receptor sites, do not have equal numbers of donor and receptor sites and therefore can form only two-dimensional hydrogen-bonded networks involving less hydrogen bonds per molecule than water.

Conceptualizing the association of a few water molecules becomes considerably more complicated when one considers isotopic variants and hydronium and hydroxyl ions. The hydronium ion, because of its positive charge, would be expected to exhibit a greater hydrogen-bond donating potential than nonionized water (dashed lines are hydrogen bonds).

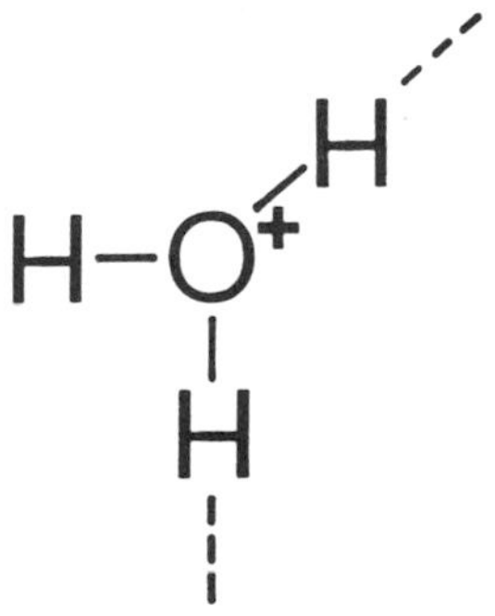

STRUCTURE 1 Structure and hydrogen-bond possibilities for a hydronium ion. Dashed lines are hydrogen bonds.

The hydroxyl ion, because of its negative charge, would be expected to exhibit a greater hydrogen-bond acceptor potential than un-ionized water (XH represents a solute or a water molecule).

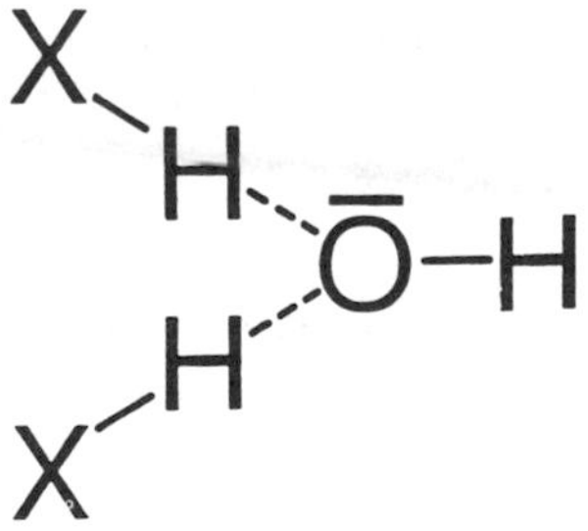

STRUCTURE 2 Structure and hydrogen-bond possibilities for a hydroxyl ion. Dashed lines are hydrogen bonds and XH represents a solute or a water molecule.

Water's ability to engage in three-dimensional hydrogen bonding provides a logical explanation for many of its unusual properties; its large values for heat capacity, melting point, boiling point, surface tension, and enthalpies of various phase transitions all are related to the extra energy needed to break intermolecular hydrogen bonds.

The permittivity (dielectric constant) of water is also influenced by hydrogen bonding. Although water is a dipole, this fact alone does not account for the magnitude of its permittivity. Hydrogen-bonded clusters of molecules apparently give rise to multimolecular di-

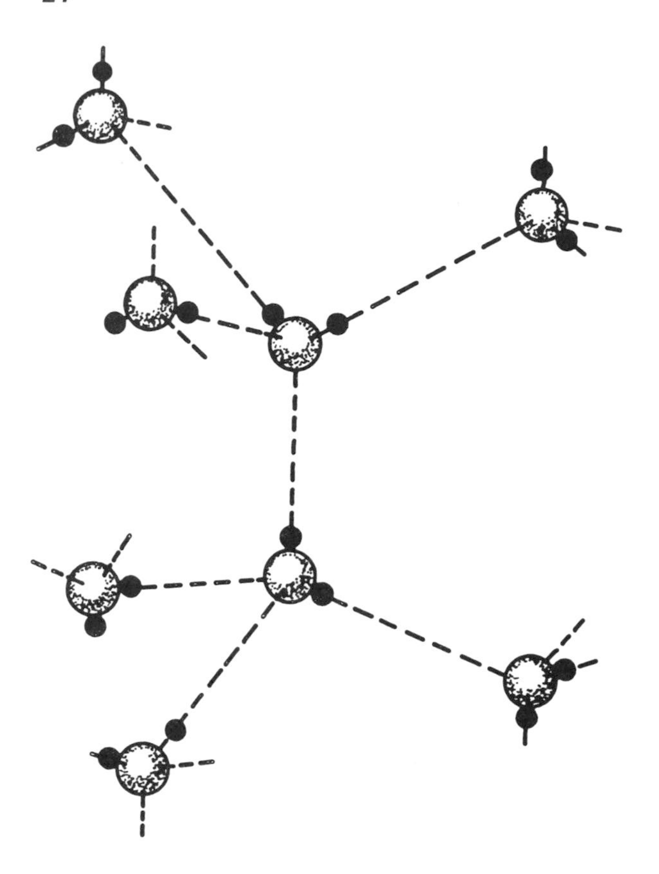

FIGURE 2 Hydrogen bonding of water molecules in a tetrahedral configuration. Open circles are oxygen atoms and closed circles are hydrogen atoms. Hydrogen bonds are represented by dashed lines.

poles, which effectively increase the permittivity of water. Water's viscosity is discussed in a later section.

2.5 STRUCTURE OF ICE

The structure of ice will be considered before the structure of water because the former is far better understood than the latter, and because ice's structure represents a logical extension of the information presented in the previous section.

2.5.1 Pure Ice

Water, with its tetrahedrally directed forces, crystallizes in an open (low density) structure that has been accurately elucidated. The O-O internuclear nearest-neighbor distance in ice is 2.76 Å and the O-O-O bond angle is about 109°, or very close to the perfect tetrahedral angle of 109°28′ (Fig. 3). The manner in which each HOH molecule can associate with four others (coordination number of four) is easily visualized in the unit cell of Figure 3 by considering molecule W and its four nearest neighbors 1, 2, 3, and W′.

When several unit cells are combined and viewed from the top (down the *c* axis) the hexagonal symmetry of ice becomes apparent. This is shown in Figure 4a. The tetrahedral

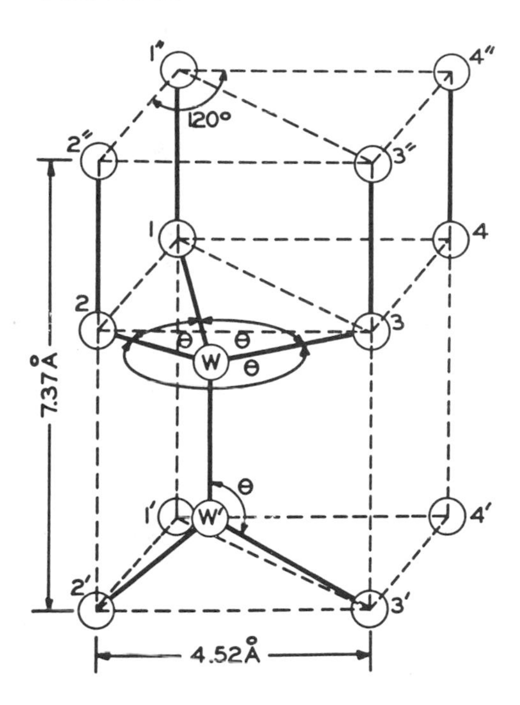

FIGURE 3 Unit cell of ordinary ice at 0°C. Circles represent oxygen atoms of water molecules. Nearest-neighbor internuclear O–O distance is 2.76 Å; θ is 109°.

substructure is evident from molecule W and its four nearest neighbors, with 1, 2, and 3 being visible, and the fourth lying below the plane of the paper directly under molecule W. When Figure 4a is viewed in three dimensions, as in Figure 4b, it is evident that two planes of molecules are involved (open and filled circles). These two planes are parallel, very close together, and they move as a unit during the "slip" or flow of ice under pressure, as in a glacier. Pairs of planes of this type comprise the "basal planes" of ice. By stacking several basal planes, an extended structure of ice is obtained. Three basal planes have been combined to form the structure shown in Figure 5. Viewed down the *c* axis, the appearance is exactly the same as that shown in Figure , indicating that the basal planes are perfectly aligned. Ice is monorefringent in this direction, whereas it is birefringent in all other directions. The *c* axis is therefore the optical axis of ice.

With regard to the location of hydrogen atoms in ice, it is generally agreed that:

1. Each line connecting two nearest neighbor oxygen atoms is occupied by one hydrogen atom located 1 ± 0.01 Å from the oxygen to which it is covalently bonded, and 1.76 ± 0.01 Å from the oxygen to which it is hydrogen bonded. This configuration is shown in Figure 6A.
2. If the locations of hydrogen atoms are viewed over a period of time, a somewhat different picture is obtained. A hydrogen atom on a line connecting two nearest neighbor oxygen atoms, X and Y, can situate itself in one of two possible positions—either 1 Å from X or 1 Å from Y. The two positions have an equal probability of being occupied. Expressed in another way, each position will, on the average, be occupied half of the time. This is possible because HOH molecules, except at extremely low temperatures, can cooperatively rotate, and hydrogen atoms can "jump" be-

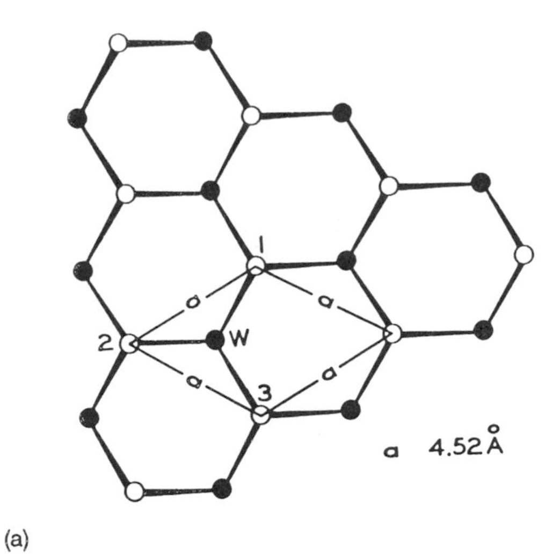

(a)

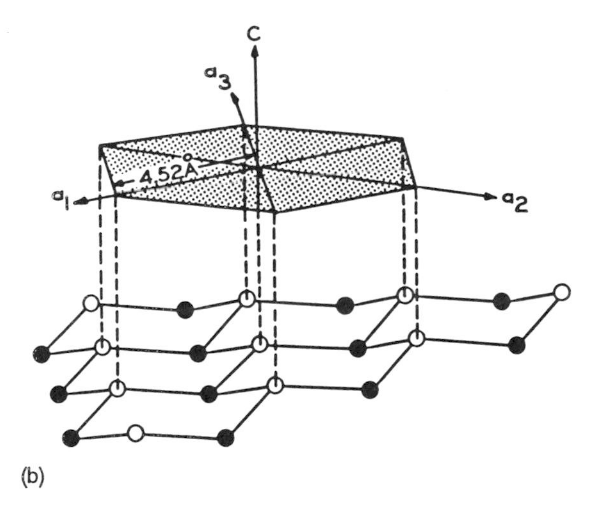

(b)

Figure 4 The "basal plane" of ice (combination of two layers of slightly different elevation). Each circle represents the oxygen atom of a water molecule. Open and shaded circles, respectively, represent oxygen atoms in the upper and lower layers of the basal planes. (a) Hexagonal structure viewed down the *c* axis. Numbered molecules relate to the unit cell in Figure 3. (b) Three-dimensional view of the basal plane. The front edge of view b corresponds to the bottom edge of view a. The crystallographic axes have been positioned in accordance with external (point) symmetry.

tween adjacent oxygen atoms. The resulting mean structure, known also as the half-hydrogen, Pauling, or statistical structure, is shown in Figure 6B.

With respect to crystal symmetry, ordinary ice belongs to the dihexagonal bipyramidal class of the hexagonal system. In addition, ice can exist in nine other crystalline polymorphic structures, and also in an amorphous or vitreous state of rather uncertain but largely noncrystal-

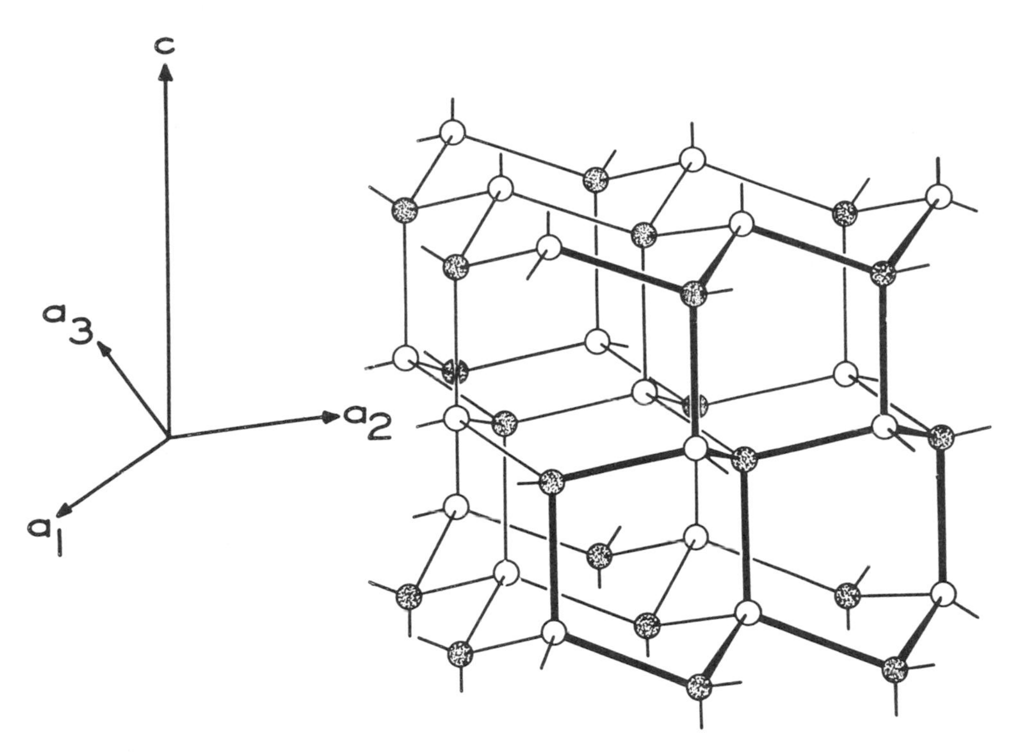

FIGURE 5 The extended structure of ordinary ice. Only oxygen atoms are shown. Open and shaded circles, respectively, represent oxygen atoms in upper and lower layers of a basal plane.

line structure. Of the eleven total structures, only ordinary hexagonal ice is stable under normal pressure at 0°C.

The structure of ice is not as simple as has been indicated. First of all, pure ice contains not only ordinary HOH molecules but also ionic and isotopic variants of HOH. Fortunately, the isotopic variants occur in such small amounts that they can, in most instances, be ignored, leaving for major consideration only HOH, H^+ (H_3O^+), and OH^-.

Second, ice crystals are never perfect, and the defects encountered are usually of the orientational (caused by proton dislocation accompanied by neutralizing orientations) or ionic types (caused by proton dislocation with formation of H_3O^+ and OH^-) (see Fig. 7). The presence of these defects provides a means for explaining the mobility of protons in ice and the small decrease in dc electrical conductivity that occurs when water is frozen.

In addition to the atomic mobilities involved in crystal defects, there are other types of activity in ice. Each HOH molecule in ice is believed to vibrate with a root mean amplitude (assuming each molecule vibrates as a unit) of about 0.4 Å at −10°C. Furthermore, HOH molecules that presumably exist in some of the interstitial spaces in ice can apparently diffuse slowly through the lattice.

Ice therefore is far from static or homogeneous, and its characteristics are dependent on temperature. Although the HOH molecules in ice are four-coordinated at all temperatures, it is necessary to lower the temperature to about −180°C or lower to "fix" the hydrogen atoms in one of the many possible configurations. Therefore, only at temperatures near −180°C or lower will

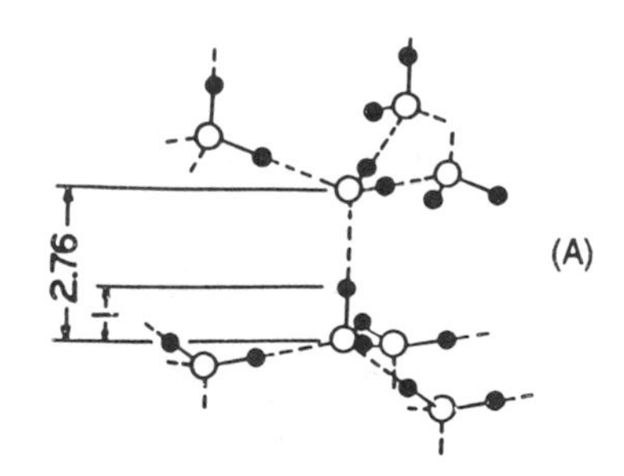

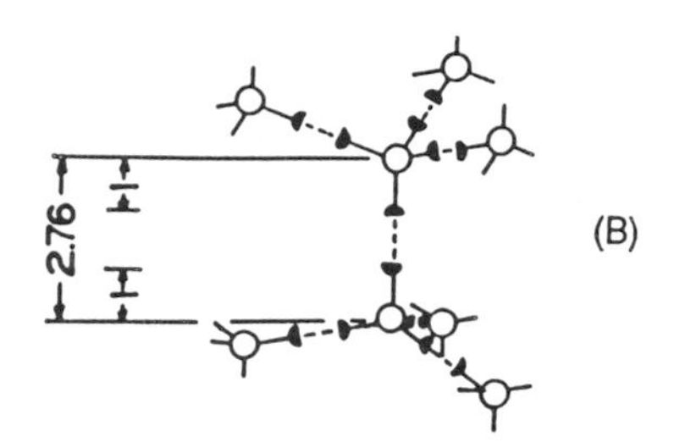

FIGURE 6 Location of hydrogen atoms (●) in the structure of ice. (A) Instantaneous structure. (B) Mean structure [known also as the half-hydrogen (◗), Pauling, or statistical structure]. Open circle is oxygen.

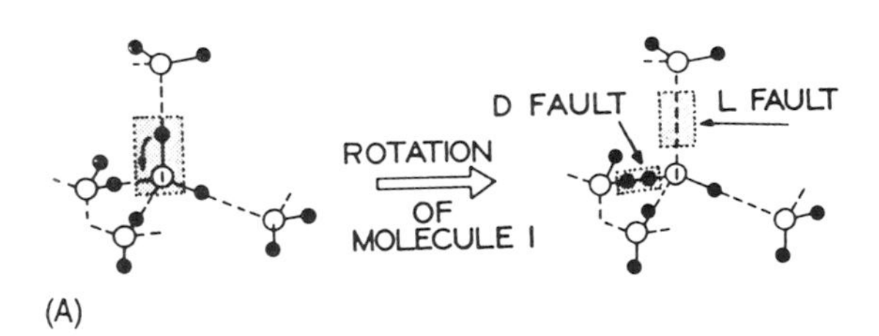

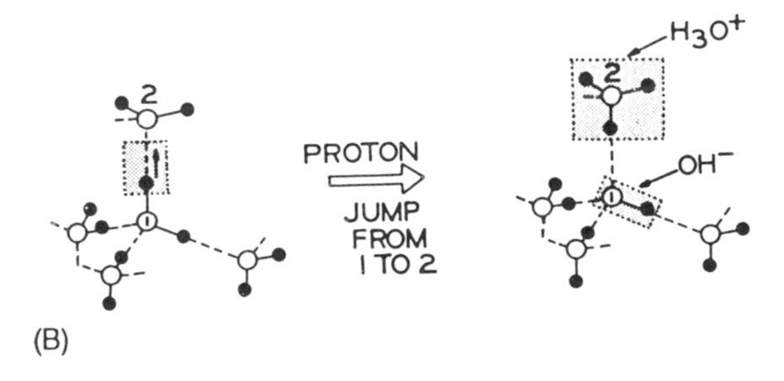

FIGURE 7 Schematic representation of proton defects in ice. (A) Formation of orientational defects. (B) Formation of ionic defects. Open and shaded circles represent oxygen and hydrogen atoms, respectively. Solid and dashed lines represent chemical bonds and hydrogen bonds, respectively.

all hydrogen bonds be intact, and as the temperature is raised, the mean number of intact (fixed) hydrogen bonds will decrease gradually.

2.5.2 Ice in the Presence of Solutes

The amount and kind of solutes present can influence the quantity, size, structure, location, and orientation of ice crystals. Consideration here will be given only to the effects of solutes on ice structure. Luyet and co-workers [75,77] studied the nature of ice crystals formed in the presence of various solutes including sucrose, glycerol, gelatin, albumin, and myosin. They devised a classification system based on morphology, elements of symmetry, and the cooling velocity required for development of various types of ice structure. Their four major classes are hexagonal forms, irregular dendrites, coarse spherulites, and evanescent spherulites.

The hexogonal form, which is most highly ordered, is found exclusively in foods, provided extremely rapid freezing is avoided and the solute is of a type and concentration that does not interfere unduly with the mobility of water molecules. Gelatin at high concentrations will, for example, result in more disordered forms of ice crystals.

2.6 STRUCTURE OF WATER

To some, it may seem strange to speak of structure in a liquid when fluidity is the essence of the liquid state. Yet it is an old and well-accepted idea [96] that liquid water has structure, obviously not sufficiently established to produce long-range rigidity, but certainly far more organized than that of molecules in the vapor state, and ample to cause the orientation and mobility of a given water molecule to be influenced by its neighbors.

Evidence for this view is compelling. For example, water is an "open" liquid, being only 60% as dense as would be expected on the basis of close packing that can prevail in nonstructured liquids. Partial retention of the open, hydrogen-bonded, tetrahedral arrangement of ice easily accounts for water's low density. Furthermore, the heat of fusion of ice, while unusually high, is sufficient to break only about 15% of the hydrogen bonds believed to exist in ice. Although this does not necessarily require that 85% of the hydrogen bonds existing in ice be retained in water (for example, more could be broken, but the change in energy could be masked by a simultaneous increase in van der Waals interactions), results of many studies support the notion that many water–water hydrogen bonds do exist.

Elucidation of the structure of pure water is an extremely complex problem. Many theories have been set forth, but all are incomplete, overly simple, and subject to weaknesses that are quickly cited by supporters of rival theories. That is, of course, a healthy situation, which will eventually result in an accurate structural picture (or pictures) of water. In the meantime, few statements can be made with any assurance that they will stand essentially unmodified in years to come. Thus, this subject will be dealt with only briefly.

Three general types of models have been proposed: mixture, interstitial, and continuum (also referred to as homogeneous or uniformist models) [5]. Mixture models embody the concept of intermolecular hydrogen bonds being momentarily concentrated in bulky clusters of water molecules that exist in dynamic equilibrium with other more dense species—momentarily meaning $\sim 10^{-11}$ sec [73].

Continuum models involve the idea that intermolecular hydrogen bonds are distributed uniformly throughout the sample, and that many of the bonds existing in ice simply become distorted rather than broken when ice is melted. It has been suggested that this permits a continuous network of water molecules to exist that is, of course, dynamic in nature [107,120].

The interstitial model involves the concept of water retaining an ice-like or clathrate-type structure with individual water molecules filling the interstitial spaces of the clathrates. In all three models, the dominant structural feature is the hydrogen-bonded association of liquid water in ephemeral, distorted tetrahedra. All models also permit individual water molecules to frequently alter their bonding arrangements by rapidly terminating one hydrogen bond in exchange for a new one, while maintaining, at constant temperature, a constant degree of hydrogen bonding and structure for the entire system.

The degree of intermolecular hydrogen bonding among water molecules is, of course, temperature dependent. Ice at 0°C has a coordination number (number of nearest neighbors) of 4.0, with nearest neighbors at a distance of 2.76 Å. With input of the latent heat of fusion, melting occurs; that is, some hydrogen bonds are broken (distance between nearest neighbors increases) and others are strained as water molecules assume a fluid state with associations that are, on average, more compact. As the temperature is raised, the coordination number increases from 4.0 in ice at 0°C, to 4.4 in water at 1.50°C, then to 4.9 at 83°C. Simultaneously, the distance between nearest neighbors increases from 2.76 Å in ice at 0°C, to 2.9 Å in water at 1.5°C, then to 3.05 Å at 83°C [7,80].

It is evident, therefore, that the ice-to-water transformation is accompanied by an increase in the distance between nearest neighbors (decreased density) and by an increase in the average number of nearest neighbors (increased density), with the latter factor predominating to yield the familiar net increase in density. Further warming above the melting point causes the density to pass through a maximum at 3.98°C, then gradually decline. It is apparent, then, that the effect of an increase in coordination number predominates at temperatures between 0 and 3.98°C, and that the effect of increasing distance between nearest neighbors (thermal expansion) predominates above 3.98°C.

The low viscosity of water is readily reconcilable with the type of structures that have been described, since the hydrogen-bonded arrangements of water molecules are highly dynamic, allowing individual molecules, within the time frame of nano- to picoseconds, to alter their hydrogen-bonding relationships with neighboring molecules, thereby facilitating mobility and fluidity.

2.7 WATER–SOLUTE INTERACTIONS

2.7.1 Macroscopic Level (Water Binding, Hydration, and Water Holding Capacity)

Before dealing with water–solute interactions at the molecular level, it is appropriate to discuss water-related phenomena referred to by terms such as water binding, hydration, and water holding capacity.

With respect to foods, the terms "water binding" and "hydration" are often used to convey a general tendency for water to associate with hydrophilic substances, including cellular materials. When used in this manner, the terms pertain to the macroscopic level. Although more specialized terms, such as "water binding potential," are defined in quantitative terms, they still apply only to the macroscopic level. The degree and tenacity of water binding or hydration depends on a number of factors including the nature of the nonaqueous constituent, salt composition, pH, and temperature.

"Water holding capacity" is a term that is frequently employed to describe the ability of a matrix of molecules, usually macromolecules present at low concentrations, to physically entrap large amounts of water in a manner that inhibits exudation. Familiar food matrices that entrap water in this way include gels of pectin and starch, and cells of tissues, both plant and animal.

Physically entrapped water does not flow from tissue foods even when they are cut or minced. On the other hand, this water behaves almost like pure water during food processing; that is, it is easily removed during drying, is easily converted to ice during freezing, and is available as a solvent. Thus, its bulk flow is severely restricted, but movement of individual molecules is essentially the same as that of water molecules in a dilute salt solution.

Nearly all of the water in tissues and gels can be categorized as physically entrapped, and impairment of the entrapment capability (water holding capacity) of foods has a profound effect on food quality. Examples of quality defects arising from impairment of water holding capacity are syneresis of gels, thaw exudate from previously frozen foods, and inferior performance of animal tissue in sausage resulting from a decline in muscle pH during normal physiological events postmortem.

Gel structures and water holding capacity are discussed more fully in other chapters.

2.7.2 Molecular Level: General Comments

Mixing of solutes and water results in altered properties of both constituents. Hydrophilic solutes cause changes in the structure and mobility of adjacent water, and water causes changes in the reactivity, and sometimes structure, of hydrophilic solutes. Hydrophobic groups of added solutes interact only weakly with adjacent water, preferring a nonaqueous environment.

The bonding forces existing between water and various kinds of solutes are of obvious interest, and these are summarized in Table 3.

2.7.3 Molecular Level: Bound Water

Bound water is not a homogeneous, easily identifiable entity, and because of this, descriptive terminology is difficult, numerous definitions have been suggested, and there is no consensus about which one is best. This term is controversial, frequently misused, and in general, poorly understood, causing increasing numbers of scientists to suggest that its use be terminated.

TABLE 3 Classifications of Types of Water–Solute Interactions

Type	Example	Strength of interaction compared to water–water hydrogen bond[a]
Dipole–ion	Water-free ion Water-charged group on organic molecule	Greater[b]
Dipole–dipole	Water–protein NH Water–protein CO Water–sidechain OH	Approx. equal
Hydrophobic hydration	Water + R[c] → R(hydrated)	Much less ($\Delta G > 0$)
Hydrophobic interaction	R(hydrated) + R(hydrated) → R_2 (hydrated) + H_2O	Not comparable[d] (> hydrophobic interaction; $\Delta G < 0$)

[a]About 12–25 kJ/mol.
[b]But much weaker than strength of single covalent bond.
[c]R is alkyl group.
[d]Hydrophobic interactions are entropy driven, whereas dipole–ion and dipole–dipole interactions are enthalpy driven.

Although this may be desirable, the term "bound water" is so common in the literature that it must be discussed.

The numerous definitions proposed for "bound water" should indicate why this term has created confusion [3,51]:

1. Bound water is the equilibrium water content of a sample at some appropriate temperature and low relative humidity.
2. Bound water is that which does not contribute significantly to permittivity at high frequencies and therefore has its rotational mobility restricted by the substance with which it is associated.
3. Bound water is that which does not freeze at some arbitrary low temperature (usually −40°C or lower).
4. Bound water is that which is unavailable as a solvent for additional solutes.
5. Bound water is that which produces line broadening in experiments involving proton nuclear magnetic resonance.
6. Bound water is that which moves with a macromolecule in experiments involving sedimentation rates, viscosity, or diffusion.
7. Bound water is that which exists in the vicinity of solutes and other nonaqueous substances and has properties differing significantly from those of "bulk" water in the same system.

All of these definitions are valid, but few will produce the same value when a given sample is analyzed.

From a conceptual standpoint it is useful to think of bound water as "water that exists in the vicinity of solutes and other nonaqueous constituents, and exhibits properties that are significantly altered from those of 'bulk water' in the same system." Bound water should be thought of as having "hindered mobility" as compared to bulk water, not as being "immobilized." In a typical food of high water content, this type of water comprises only a minute part of the total water present, approximately the first layer of water molecules adjacent to hydrophilic groups. The subject of bound water (hindered water) will be discussed further in the section dealing with molecular mobility (Mm) in frozen systems.

Interactions between water and specific classes of solutes will now be considered.

2.7.4 Interaction of Water with Ions and Ionic Groups

Ions and ionic groups of organic molecules hinder mobility of water molecules to a greater degree than do any other types of solutes. The strength of water–ion bonds is greater than that of water–water hydrogen bonds, but is much less than that of covalent bonds.

The normal structure of pure water (based on a hydrogen-bonded, tetrahedral arrangement) is disrupted by the addition of dissociable solutes. Water and simple inorganic ions undergo dipole–ion interactions. The example in Figure 8 involves hydration of the NaCl ion pair. Only first-layer water molecules in the plane of the paper are illustrated. In a dilute solution of ions in water, second-layer water is believed to exist in a structurally perturbed state because of conflicting structural influences of first-layer water and the more distant, tetrahedrally oriented "bulk-phase" water. In concentrated salt solutions, bulk-phase water would not exist and water structure would be dominated by the ions.

There is abundant evidence indicating that some ions in dilute aqueous solution have a net structure-breaking effect (solution is more fluid than pure water), whereas others have a net structure-forming effect (solution is less fluid than pure water). It should be understood that the term "net structure" refers to all kinds of structures, either normal or new types of water structure. From the standpoint of "normal" water structure, all ions are disruptive.

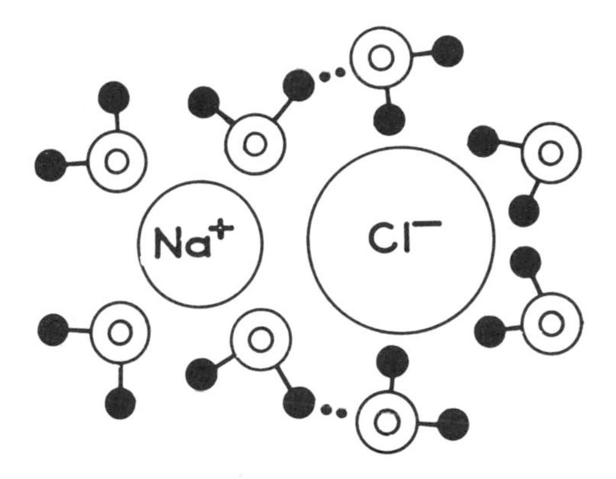

FIGURE 8 Likely arrangement of water molecules adjacent to sodium chloride. Only water molecules in plane of paper are shown.

The ability of a given ion to alter net structure is related closely to its polarizing power (charge divided by radius) or simply the strength of its electric field. Ions that are small and/or multivalent (mostly positive ions, such as Li^+, Na^+, H_3O^+, Ca^{2+}, Ba^{2+}, Mg^{2+}, Al^{3+}, F^-, and OH^-) have strong electric fields and are net structure formers. The structure imposed by these ions more than compensates for any loss in normal water structure. These ions strongly interact with the four to six first-layer water molecules, causing them to be less mobile and pack more densely than HOH molecules in pure water. Ions that are large and monovalent (most of the negatively charged ions and large positive ions, such as K^+, Rb^+, Cs^+, NH_4^+, Cl^-, Br^-, I^-, NO_3^-, BrO_3^-, IO_3^-, and ClO_4^-) have rather weak electric fields and are net structure breakers, although the effect is very slight with K^+. These ions disrupt the normal structure of water and fail to impose a compensating amount of new structure.

Ions, of course, have effects that extend well beyond their influence on water structure. Through their varying abilities to hydrate (compete for water), alter water structure, influence the permittivity of the aqueous medium, and govern the thickness of the electric double layer around colloids, ions profoundly influence the "degree of hospitality" extended to other nonaqueous solutes and to substances suspended in the medium. Thus, conformation of proteins and stability of colloids (salting-in, salting-out in accord with the Hofmeister or lyotropic series) are greatly influenced by the kinds and amounts of ions present [18,68].

2.7.5 Interaction of Water with Neutral Groups Possessing Hydrogen-Bonding Capabilities (Hydrophilic Solutes)

Interactions between water and nonionic, hydrophilic solutes are weaker than water–ion interactions and about the same strength as those of water–water hydrogen bonds. Depending on the strength of the water–solute hydrogen bonds, first-layer water may or may not exhibit reduced mobility and other altered properties as compared to bulk-phase water.

Solutes capable of hydrogen bonding might be expected to enhance or at least not disrupt the normal structure of pure water. However, in some instances it is found that the distribution and orientation of the solute's hydrogen-bonding sites are geometrically incompatible with those existing in normal water. Thus, these kinds of solutes frequently have a disruptive influence on the normal structure of water. Urea is a good example of a small hydrogen-bonding solute that for geometric reasons has a marked disruptive effect on the normal structure of water.

It should be noted that the total number of hydrogen bonds per mole of solution may not be significantly altered by addition of a hydrogen-bonding solute that disrupts the normal structure of water. This is possible since disrupted water–water hydrogen bonds may be replaced by water–solute hydrogen bonds. Solutes that behave in this manner have little influence on "net structure" as defined in the previous section.

Hydrogen bonding of water can occur with various potentially eligible groups (e.g., hydroxyl, amino, carbonyl, amide, imino, etc.). This sometimes results in "water bridges" where one water molecule interacts with two eligible hydrogen-bonding sites on one or more solutes. A schematic depiction of water hydrogen bonding (dashed lines) to two kinds of functional groups found in proteins is shown:

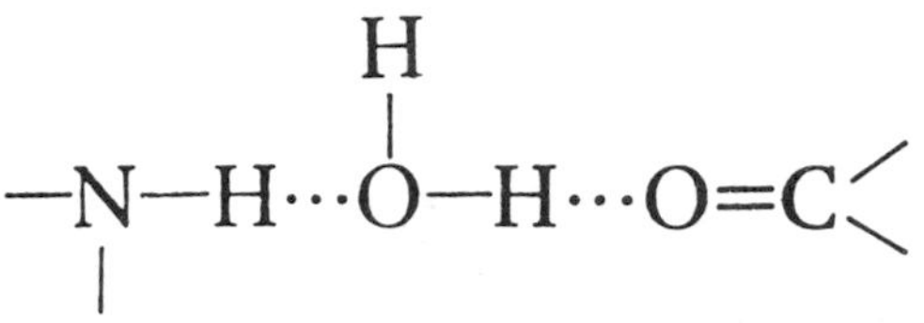

$$\text{—N—H}\cdots\overset{\text{H}}{\overset{|}{\text{O}}}\text{—H}\cdots\text{O=C}\langle$$

STRUCTURE 3 Hydrogen bonding (dotted lines) of water to two kinds of functional groups occurring in proteins.

A more elaborate example involving a three-HOH bridge between backbone peptide units in papain is shown in Figure 9.

It has been observed that hydrophilic groups in many crystalline macromolecules are separated by distances identical to the nearest-neighbor oxygen spacing in pure water. If this spacing prevails in hydrated macromolecules this would encourage cooperative hydrogen bonding in first- and second-layer water.

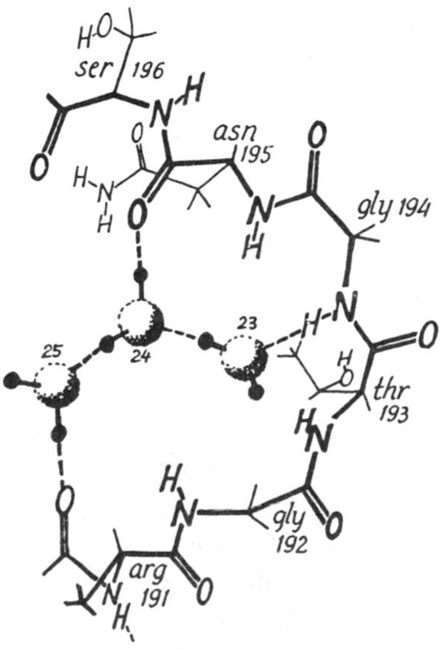

FIGURE 9 Examples of a three-molecule water bridge in papain; 23, 24, and 25 are water molecules. (From Ref. 4.)

2.7.6 Interaction of Water with Nonpolar Substances

The mixing of water and hydrophobic substances, such as hydrocarbons, rare gases, and the apolar groups of fatty acids, amino acids, and proteins is, not surprisingly, a thermodynamically unfavorable event ($\Delta G > 0$). The free energy is positive not because ΔH is positive, which is typically true for low-solubility solutes, but because $T\,\Delta S$ is negative [30]. This decrease in entropy occurs because of special structures that water forms in the vicinity of these incompatible apolar entities. This process has been referred to as hydrophobic hydration (Table 3 and Fig. 10a).

Because hydrophobic hydration is thermodynamically unfavorable, it is understandable that water would tend to minimize its association with apolar entities that are present. Thus, if two separated apolar groups are present, the incompatible aqueous environment will encourage them to associate, thereby lessening the water–apolar interfacial area—a process that is thermodynamically favorable ($\Delta G < 0$). This process, which is a partial reversal of hydrophobic hydration, is referred to as "hydrophobic interaction" and in its simplest form can be depicted as

$$\text{R (hydrated)} + \text{R (hydrated)} \rightarrow \text{R}_2\text{ (hydrated)} + \text{H}_2\text{O}$$

where R is an apolar group (Table 3 and Fig. 10b).

Because water and apolar groups exist in an antagonistic relationship, water structures itself to minimize contact with apolar groups. The type of water structure believed to exist in the layer next to apolar groups is depicted in Figure 11. Two aspects of the antagonistic relationship between water and hydrophobic groups are worthy of elaboration: formation of clathrate hydrates, and association of water with hydrophobic groups in proteins.

A clathrate hydrate is an ice-like inclusion compound wherein water, the "host" substance, forms a hydrogen-bonded, cage-like structure that physically entraps a small apolar molecule,

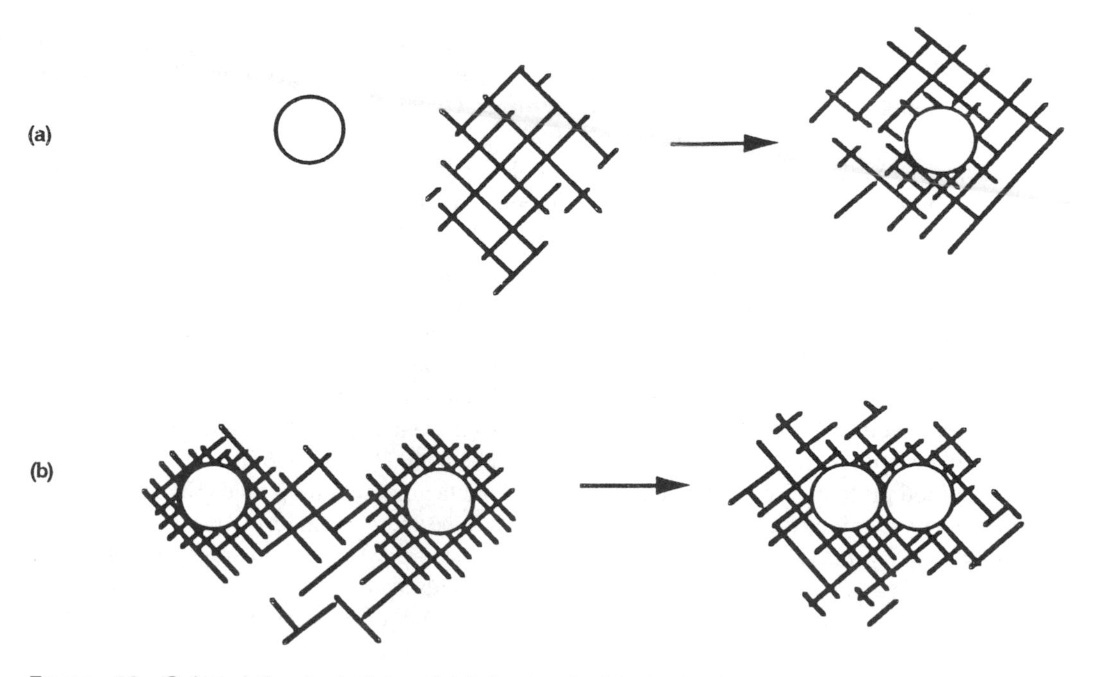

FIGURE 10 Schematic depiction of (a) hydrophobic hydration and (b) hydrophobic association. Open circles are hydrophobic groups. Hatched areas are water. (Adapted from Ref. 28.)

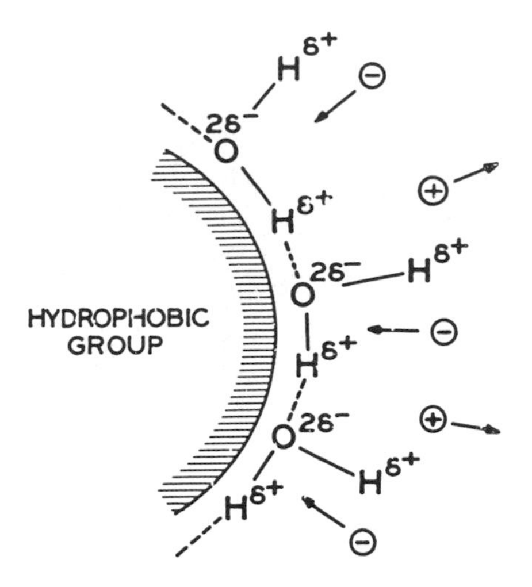

FIGURE 11 Proposed water orientation at a hydrophobic surface. (Adapted from Ref. 68.)

known as the "guest." These entities are of interest because they represent the most extreme structure-forming response of water to an apolar substance and because microstructures of a similar type may occur naturally in biological matter. Clathrate hydrates are, in fact, crystalline, they can easily be grown to visible size, and some are stable at temperatures above 0°C provided the pressure is sufficient.

The guest molecules of clathrate hydrates are low-molecular-weight compounds with sizes and shapes compatible with the dimensions of host water cages comprised of 20–74 water molecules. Typical guests include low-molecular-weight hydrocarbons and halogenated hydrocarbons; rare gases; short-chain primary, secondary, and tertiary amines; and alkyl ammonium, sulfonium, and phosphonium salts. Interaction between water and guest is slight, usually involving nothing more than weak van der Waals forces. Clathrate hydrates are the extraordinary result of water's attempt to avoid contact with hydrophobic groups.

There is evidence that structures similar to crystalline clathrate hydrates may exist naturally in biological matter, and if so, these structures would be of far greater importance than crystalline hydrates since they would likely influence the conformation, reactivity, and stability of molecules such as proteins. For example, it has been suggested that partial clathrate structures may exist around the exposed hydrophobic groups of proteins. It is also possible that clathrate-like structures of water have a role in the anesthetic action of inert gases such as xenon. For further information on clathrates, the reader is referred to Davidson [15].

Unavoidable association of water with hydrophobic groups of proteins has an important influence on protein functionality [5,124]. The extent of these unavoidable contacts is potentially fairly great because nonpolar side chains exist on about 40% of the amino acids in typical oligomeric food proteins. These nonpolar groups include the methyl group of alanine, the benzyl group of phenylalanine, the isopropyl group of valine, the mercaptomethyl group of cysteine, and the secondary butyl and isobutyl groups of the leucines. The nonpolar groups of other compounds such as alcohols, fatty acids, and free amino acids also can participate in hydrophobic interactions, but the consequences of these interactions are undoubtedly less important than those involving proteins.

Because exposure of protein nonpolar groups to water is thermodynamically unfavorable, association of hydrophobic groups or "hydrophobic interaction" is encouraged, and this occurrence is depicted schematically in Figure 12. Hydrophobic interaction provides a major driving

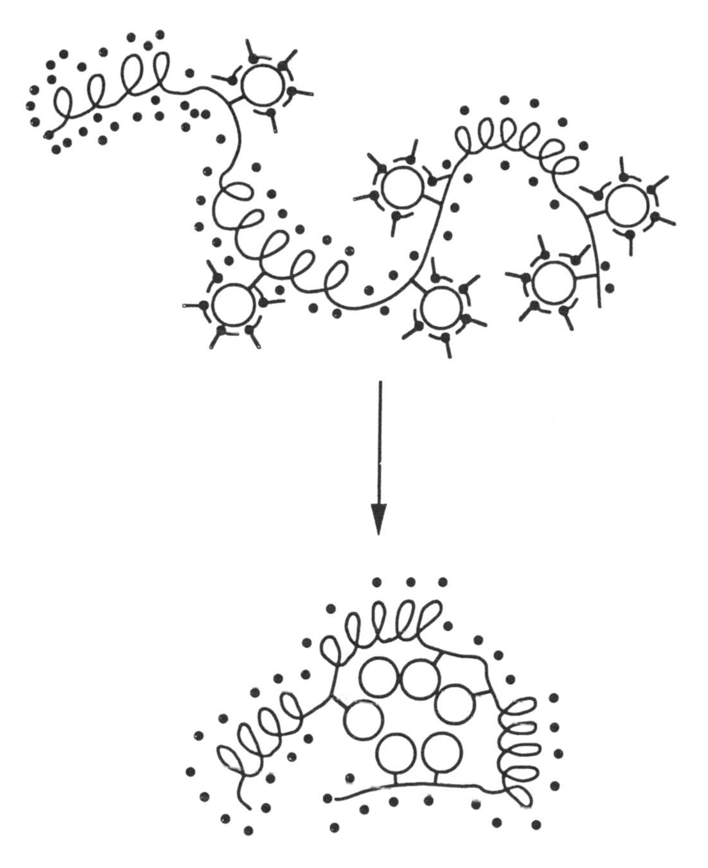

FIGURE 12 Schematic depiction of a globular protein undergoing hydrophobic interaction. Open circles are hydrophobic groups, "L-shaped" entities around circles are water molecules oriented in accordance with a hydrophobic surface, and dots represent water molecules associated with polar groups.

force for protein folding, causing many hydrophobic residues to assume positions in the protein interior. Despite hydrophobic interactions, it is estimated that nonpolar groups in globular proteins typically occupy about 40–50% of the surface area. Hydrophobic interactions also are regarded as being of primary importance in maintaining the tertiary structure of most proteins [19,85,123]. It is therefore of considerable importance that a reduction in temperature causes hydrophobic interactions to become weaker and hydrogen bounds to become stronger.

2.7.7 Details of Water Orientation Adjacent to Organic Molecules

Although determination of the arrangement of water molecules near organic molecules is experimentally difficult, this is an active field of research and useful data have been obtained. The hydrated pyranose sugar ring is shown in Figure 13 and a computer-simulated cross section of hydrated myosin is shown in Figure 14. Assuming a separation distance of 2.8 Å between hydration sites and full occupancy of these sites, about 360 HOH molecules would be in the primary hydration shell of myoglobin [71].

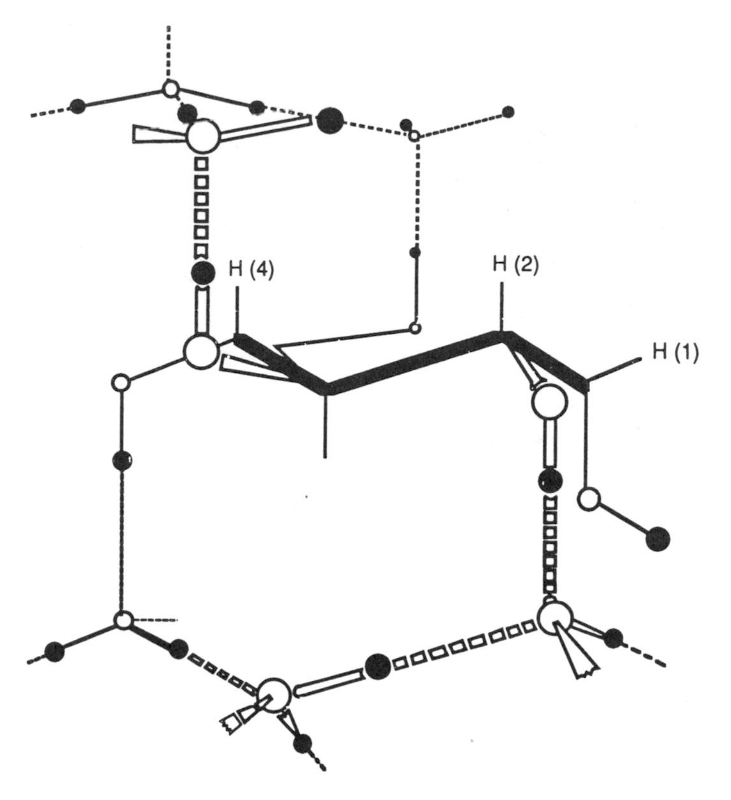

FIGURE 13 Association of α-D-glucose with tetrahedrally arranged water molecules. Heavy solid line represents the side view of the pyranose ring. Oxygens and hydrogens of water are represented by open and filled circles, respectively. Covalent and hydrogen bonds are represented by solid and broken lines, respectively. The hydroxymethyl protons [H(6)] are not shown. (From Ref. 122.)

2.7.8 Hydration Sequence of a Protein

It is instructive to consider water absorption by a dry food component and the location and properties of water at each stage of the process. A protein is chosen for this exercise because proteins are of major importance in foods, because they contain all of the major types of functional groups that are of interest during hydration, and because good data are available.

Shown in Table 4 are properties of globular proteins (based primarily on lysozyme) and the associated water at various stages of hydration. The corresponding sorption isotherm will be shown in Figure 19. The table is self-explanatory except for a few points. Hydration "zones" are referred to in both the table and the figure. These zones are useful aids to discussion but they are unlikely to actually exist (a continuum of water properties is much more likely).

A sample having a water content corresponding to the junction of Zones I and II is said to have a BET monolayer water content (the term "BET" comes from the names of the originators of the concept: Brunauer, Emmett, and Teller [8]). In this instance the BET monolayer water content is about 0.07 g HOH/g dry protein, and this corresponds to a p/p_0 value of about 0.2 (p is the partial pressure of water in the food and p_0 is the vapor pressure of pure water at the same temperature; p/p_0 is more commonly referred to as a_w). The BET monolayer value is of special importance because it often provides a good first estimate of the largest moisture content

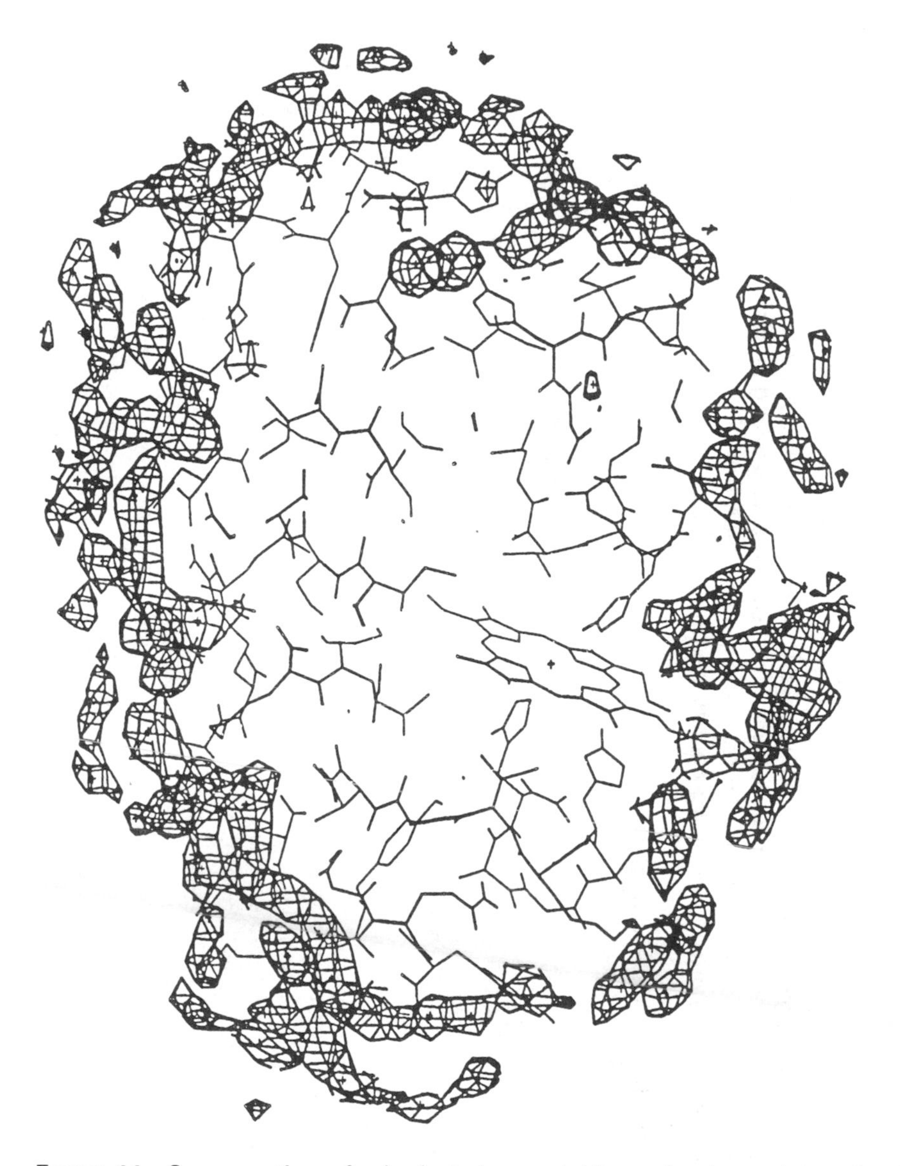

FIGURE 14 Cross-section of a hydrated myoglobin molecule as determined by molecular dynamics simulation. Mesh cages depict high probability sites of 1st-layer water molecules, stick figure represents the protein time-averaged structure. (From Ref. 72.)

a dry product can contain and still exhibit maximum stability. Although the BET value is commonly referred to as a monolayer, this is a faulty concept. For example, in starch the BET value corresponds to about one water molecule per anhydroglucose unit [126].

Also note that in Table 4 the term "true monolayer" is used. This term has a meaning quite different from BET monolayer. True monolayer refers to the water content at the junction of zones IIB and III (in this example, a water content of about 0.38 g HOH/g dry protein and a p/p_0 of about 0.85). This value corresponds to about 300 mol HOH per mol lysozyme and a moisture content of 27.5 wt%, with one HOH occupying, on average, 20 $Å^2$ of protein surface area [103]. This water content is significant because it represents the minimum water content

TABLE 4 Water/Protein Properties at Various Stages of Hydration[a]

Properties	Constitutional water[b]	Hydration shell (≤ 3 Å from surface)			Bulk-phase water: Free[c]	Bulk-phase water: Entrapped[d]
General description for lysozyme	Constitutional water is assumed to be present in the dry protein at the onset of the hydration process. Water is first absorbed at sites of ionized, carboxylic and amino side chains, with about 40 mol water/mol lysozyme associating in this manner. Further absorption of water results in gradual hydration of less attractive sites, mainly amide carbonyl groups of the protein backbone. Attainment of true monolayer hydration of the protein is achieved at 0.38 g H_2O/g dry protein, by water associating with sites that are still less attractive. At this point, there is, on average, 1 HOH/20 $Å^2$ of protein surface				Fully hydrated	Fully hydrated
Approximate water content:						
g H_2O/g dry protein (h)	< 0.01 h	0.01–0.07 h	0.07–0.25 h	0.25–0.58 h	> 0.38 h	> 0.38 h
mol H_2O/mol dry protein	< 8	8–56	56–200	200–304	> 304	> 304
wt% based on lysozyme	1%	1–6.5%	6.5–20%	20–27.5%	> 27.5%	> 27.5%
Location on isotherm[e]						
Relative vapor pressure (p/p_o)	< 0.02p/p_o	0.02–0.2p/p_o	0.2–0.75p/p_o	0.75–0.85p/p_o	> 0.85 p/p_o	> 0.85 p/p_o
Zone	Zone I, extreme left	Zone I	Zone IIA	Zone IIB	Zone III	Zone III
Water properties					Normal	Normal
Structure	Critical part of native protein structure	Water interacts principally with charged groups (~2 HOH/group) At 0.07 h: transition in surface water from disordered to ordered and/or from dispersed to clustered	Water interacts principally with polar protein surface groups (~1 HOH/ polar site) Water clusters centered on charged and polar sites Clusters fluctuate in size and/or arrangement	At 0.25 h: start of condensation of water onto weakly interacting unfilled patches of protein surface At 0.38 h: monolayer of water covers the surface of the protein and water		

		state; associated with completion of charged group hydration	At 0.15 h: long-range connectivity of the surface water is established	phase begins to form, and glass–rubber transition occurs		
Thermodynamic transfer properties[f]						
$\Delta\overline{G}$ (kJ/mol)	> |−6|	−6	−0.8	Close to bulk water	NA	NA
$\Delta\overline{H}$ (kJ/mol)	> |−17|	−70	−2.1	Close to bulk water	NA	NA
Approximate mobility (residence time)	10^{-2} to 10^{-8} sec	< 10^{-8} sec	< 10^{-9} sec	10^{-9} to 10^{-11} sec	10^{-11} to 10^{-12} sec	10^{-11} to 10^{-12} sec
Freezability	Unfreezable	Unfreezable	Unfreezable	Unfreezable	Normal	Normal
Solvent capability	None	None	None to slight	Slight to moderate	Normal	Normal
Protein properties						
Structure	Folded state stable	Water begins to plasticize amorphous regions	Further plasticization of amorphous regions			
Mobility	Enzymatic activity negligible	Enzymatic activity negligible	Internal protein motion (H exchange) increases from 1/1000 at 0.04 h to full solution rate at 0.15 h At 0.1–0.15 h: chymotrypsin and some other enzymes develop activity	At 0.38 h: lysozyme specific activity is 0.1 that in dilute solution	Maximum	Maximum

[a]Data from Rupley and Careri [103], Otting et al. [86], Lounnas and Pettit [71,72], Franks [30], and other sources. Based largely on lysozyme.
[b]Water molecules that occupy specific locations in the interior of the solute macromolecule.
[c]Macroscopic flow not physically constrained by a macromolecular matrix.
[d]Macroscopic flow physically constrained by a macromolecular matrix.
[e]See Figure 19.
[f]Partial molar values for transfer of water from bulk phase to hydration shell.

needed for "full hydration," that is, occupancy of all first-layer sites. Further added water will have properties that do not differ significantly from those of bulk water.

2.8 WATER ACTIVITY AND RELATIVE VAPOR PRESSURE

2.8.1 Introduction

It has long been recognized that a relationship, although imperfect, exists between the water content of food and its perishability. Concentration and dehydration processes are conducted primarily for the purpose of decreasing the water content of a food, simultaneously increasing the concentration of solutes and thereby decreasing perishability.

However, it has also been observed that various types of food with the same water content differ significantly in perishability. Thus, water content alone is not a reliable indicator of perishability. This situation is attributable, in part, to differences in the intensity with which water associates with nonaqueous constituents—water engaged in strong associations is less able to support degradative activities, such as growth of microorganisms and hydrolytic chemical reactions, than is weakly associated water. The term "water activity" (a_W) was developed to account for the intensity with which water associates with various nonaqueous constituents.

Food stability, safety, and other properties can be predicted far more reliably from a_W than from water content. Even so, a_W is not a totally reliable predictor. The reasons for this will be explained in the next section. Despite this lack of perfection, a_W correlates sufficiently well with rates of microbial growth and many degradative reactions to make it a useful indicator of product stability and microbial safety. The fact that a_W is specified in some U.S. federal regulations dealing with good manufacturing practices for food attests to its usefulness and credibility [44].

2.8.2 Definition and Measurement

The notion of substance "activity" was rigorously derived from the laws of equilibrium thermodynamics by G. N. Lewis, and its application to foods was pioneered by Scott [108,109]. It is sufficient here to state that

$$a_W = f/f_o \tag{1}$$

where f is the fugacity of the solvent (fugacity being the escaping tendency of a solvent from solution) and f_o is the fugacity of the pure solvent. At low pressures (e.g., ambient), the difference between f/f_o and p/p_o is less than 1%, so defining a_W in terms of p/p_o is clearly justifiable. Thus,

$$a_W = p/p_o \tag{2}$$

This equality is based on the assumptions of solution ideality and the existence of thermodynamic equilibrium. With foods, both assumptions are generally violated. Consequently, Equation 2 must be taken as an approximation and the proper expression is

$$a_W \approx p/p_o \tag{3}$$

Because p/p_o is the measured term and sometimes does not equal a_W, it is more accurate to use the term p/p_o rather than a_W. This practice will be followed here. "Relative vapor pressure" (RVP) is the name for p/p_o, and these terms will be used interchangeably. Despite the scientific soundness of using RVP rather a_W, the reader should be aware that the term a_W is in widespread use, appears in other chapters of this book and is not improper provided the user understands its true meaning.

Failure of the a_W–RVP approach to be a perfect estimator of food stability occurs for two basic reasons: violation of assumptions underlying Equation 2, and solute-specific effects. Violation of Equation 2 assumptions can, but usually does not, detract unduly from the usefulness of RVP as a technological tool. An exception occurs if dry products are prepared by absorption of water rather than desorption (hysteresis effects), and this will be discussed later. Violation of Equation 2 assumptions does, however, invalidate RVP as a tool for mechanistic interpretations in instances where the theoretical models used are based on these assumptions (often true of models for moisture sorption isotherms).

In a few instances that can be of great importance, solute-specific effects can cause RVP to be a poor indicator of food stability and safety. This can occur even when the assumptions underlying Equation 2 are fully met. In these situations, foods with the same RVP but different solute compositions will exhibit different stabilities and other properties. This is an important point that must not be overlooked by anyone relying on RVP as a tool for judging the safety or stability of food. Figure 15 is provided to reinforce this point. These data clearly indicate that the minimum p/p_0 for growth of *Staphylococcus aureus* is dependent on solute type.

RVP is related to percent equilibrium relative humidity (ERH) of the product environment as follows:

$$\text{RVP} = p/p_0 = \%\text{ERH}/100 \qquad (4)$$

Two aspects of this relationship should be noted. First, RVP is an intrinsic property of the sample whereas ERH is a property of the atmosphere in equilibrium with the sample. Second, the Equation 4 relationship is an equality only if equilibrium has been established between the product and its environment. Establishment of equilibrium is a time-consuming process even with very small samples (less than 1 g) and almost impossible in large samples, especially at temperatures below ~50°C.

The RVP of a small sample can be determined by placing it in a closed chamber for a time sufficient to achieve apparent equilibrium (constant weight) and then measuring either pressure or relative humidity in the chamber [33,52,90,119,125]. Various types of instruments are available for measuring pressure (manometers) and relative humidity (electric hygrometers, dew-point instruments). Knowledge of freezing point depression can also be used to determine RVP [26]. Based on collaborative studies, the precision of a_w determinations is about ±0.02.

If one desires to adjust a small sample to a specific RVP, this can be done by placing it in

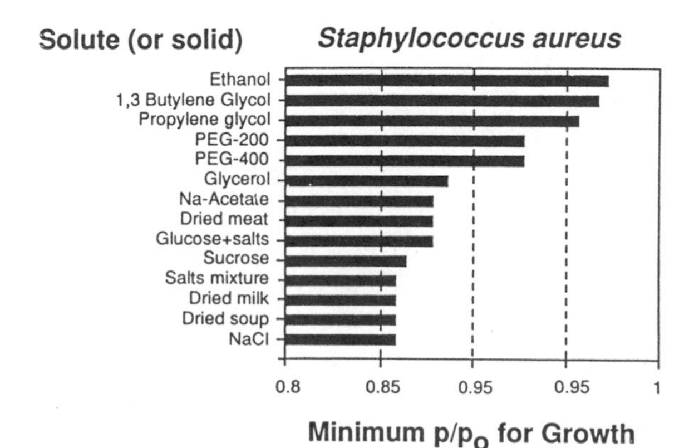

FIGURE 15 Minimum relative vapor pressure (RVP) for growth of *Staphylococcus aureus* as influenced by solute used to produce the RVP. Temperature is close to optimum for growth. PEG is polyethylene glycol. (From Ref. 11.)

a closed chamber at constant temperature, maintaining sample atmosphere at constant relative humidity by means of an appropriate saturated salt solution [69,119], and storing it until constant sample weight is achieved.

2.8.3 Temperature Dependence

Relative vapor pressure is temperature dependent, and the Clausius–Clapeyron equation in modified form provides a means for estimating this temperature dependence. This equation, although based on a_w, is applicable to RVP and has the following form [128]:

$$\frac{d \ln a_w}{d(1/T)} = \frac{-\Delta H}{R} \tag{5}$$

where T is absolute temperature, R is the gas constant, and ΔH is the isosteric net heat of sorption at the water content of the sample. By rearrangement, this equation can be made to conform to the generalized equation for a straight line. It then becomes evident that a plot of $\ln a_w$ versus $1/T$ (at constant water content) should be linear and the same should be true for $\ln p/p_0$ versus $1/T$.

Linear plots of $\ln p/p_0$ versus $1/T$ for native potato starch at various moisture contents are shown in Figure 16. It is apparent that the degree of temperature dependence is a function of moisture content. At a starting p/p_0 of 0.5, the temperature coefficient is 0.0034°C^{-1} over the temperature range 2–40°C. Based on the work of several investigators, temperature coefficients for p/p_0 (temperature range 5–50°C; starting p/p_0 0.5) range from 0.003 to 0.02°C^{-1} for high-carbohydrate or high-protein foods [128]. Thus, depending on the product, a 10°C change

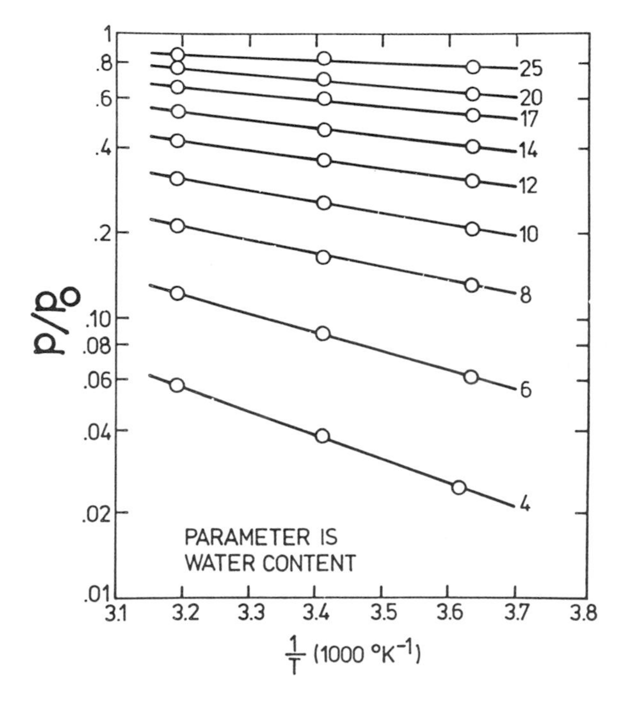

FIGURE 16 Relationship between relative water vapor pressure and temperature for native potato starch of different water contents. Water content values following each line are g HOH/g dry starch. (From Ref. 128.)

in temperature can cause a 0.03–0.2 change in p/p_0. This behavior can be important for a packaged food because it will undergo a change in RVP with a change in temperature, causing the temperature dependence of its stability to be greater than that of the same product unpackaged.

Plots of ln p/p_0 versus $1/T$ are not always linear over broad temperature ranges, and they generally exhibit sharp breaks with the onset of ice formation. Before showing data at subfreezing temperatures, it is appropriate to consider the definition of RVP as it applies to subfreezing temperatures. This is necessary because a question arises as to whether the denominator term (p_0) should be equated to the vapor pressure of supercooled water or to the vapor pressure of ice. The vapor pressure of supercooled water turns out to be the proper choice because (a) values of RVP at subfreezing temperatures can then, and only them, be accurately compared to RVP values at above-freezing temperatures and (b) choice of the vapor pressure of ice as p_0 would result, for samples containing ice, in a meaningless situation whereby RVP would be unity at all subfreezing temperatures. The second point results because the partial pressure of water in a frozen food is equal to the vapor pressure of ice at the same temperature [25,121].

Because the vapor pressure of supercooled water has been measured down to –15°C, and the vapor pressure of ice has been measured to much lower temperatures, it is possible to accurately calculate RVP values for frozen foods. This is clearly apparent when one considers the following relationship:

$$a_w = \frac{p_{ff}}{p_{o(SCW)}} = \frac{p_{ice}}{p_{o(SCW)}} \tag{6}$$

where p_{ff} is the partial pressure of water in partially frozen food, $p_{o(SCW)}$ is the vapor pressure of pure supercooled water, and p_{ice} is the vapor pressure of pure ice.

Presented in Table 5 are RVP values calculated from the vapor pressures of ice and supercooled water, and these values are identical to those of frozen foods at the same temperatures. Figure 17 is a plot of log p/p_0 versus $1/T$, illustrating that (a) the relationship is linear at

TABLE 5 Vapor Pressures and Vapor Pressure Ratios of Water and Ice

	Vapor pressure				
	Liquid water[a]		Ice[b] or food containing ice		
Temperature (°C)	(Pa)	(torr)	(Pa)	(torr)	(p_{ice}/p_{water})
0	611[b]	4.58	611	4.58	1.00
–5	421	3.16	402	3.02	0.95
–10	287	2.15	260	1.95	0.91
–15	191	1.43	165	1.24	0.86
–20	125	0.94	103	0.77	0.82
–25	80.7	0.61	63	0.47	0.78
–30	50.9	0.38	38	0.29	0.75
–40	18.9	0.14	13	0.098	0.69
–50	6.4	0.05	3.9	0.029	0.61

[a]Supercooled at all temperatures except 0°C. Observed data above –15°C, calculated below –15°C [79].
[b]Observed data from Ref. 69.

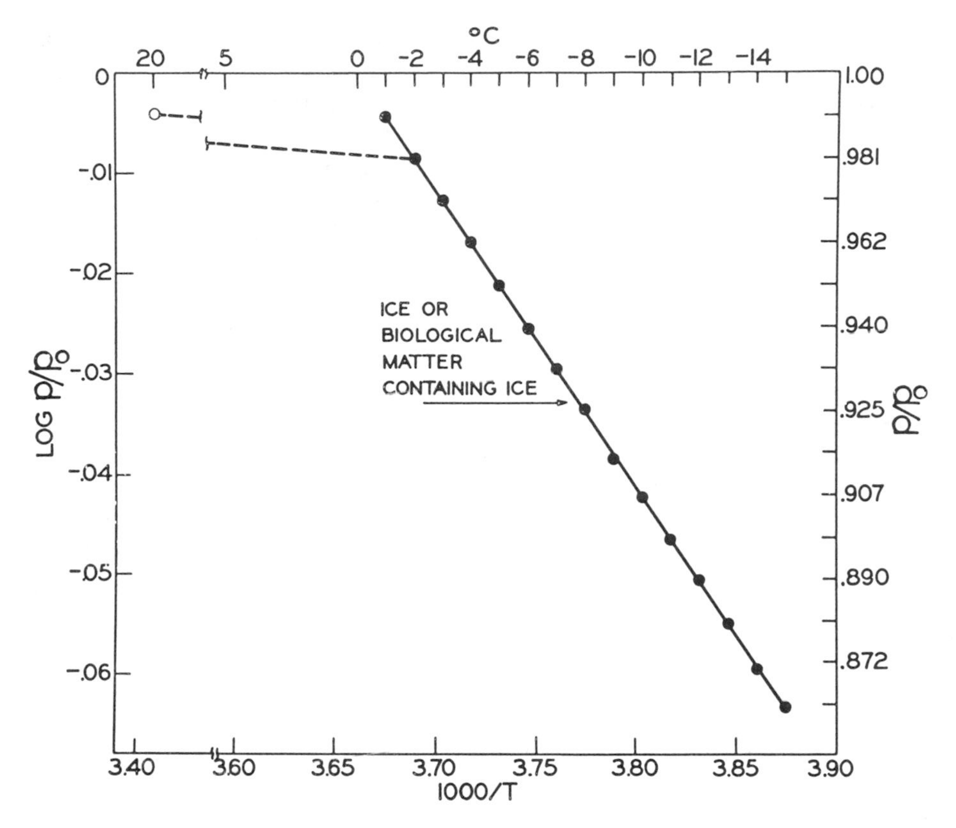

Figure 17 Relationship between relative vapor pressure and temperature for a complex food above and below freezing. (From Ref. 24.)

subfreezing temperatures, (b) the influence of temperature on RVP is typically far greater at subfreezing temperatures than at above-freezing temperatures, and (c) a sharp break occurs in the plot at the freezing point of the sample.

Two important distinctions should be noted when comparing RVP values at above- and below-freezing temperatures. First, at above-freezing temperatures, RVP is a function of sample composition and temperature, with the former factor predominating. At subfreezing temperatures, RVP becomes independent of sample composition and depends solely on temperature; that is, in the presence of an ice phase RVP values are not influenced by the kind or ratio of solutes present. As a consequence, any subfreezing event that is influenced by the kind of solute present (e.g., diffusion-controlled processes, catalyzed reactions, and reactions that are affected by the absence or presence of cryoprotective agents, by antimicrobial agents, and/or by chemicals that alter pH and oxidation–reduction potential) cannot be accurately forecast based on the RVP value [25]. Consequently, RVP values at subfreezing temperatures are far less valuable indicators of physical and chemical events than are RVP values at above-freezing temperatures. It follows that knowledge of RVP at a subfreezing temperature cannot be used to predict RVP at an above-freezing temperature.

Second, as the temperature is changed sufficiently to form or melt ice, the meaning of RVP, in terms of food stability, also changes. For example, in a product at –15°C ($p/p_0 = 0.86$), microorganisms will not grow and chemical reactions will occur slowly. However, at 20°C and

p/p_0 0.86, some chemical reactions will occur rapidly and some microorganisms will grow at moderate rates.

2.9 MOISTURE SORPTION ISOTHERMS

2.9.1 Definition and Zones

A plot of water content (expressed as mass of water per unit mass of dry material) of a food versus p/p_0 at constant temperature is known as a moisture sorption isotherm (MSI). Information derived from MSIs are useful (a) for concentration and dehydration processes, because the ease or difficulty of water removal is related to RVP, (b) for formulating food mixtures so as to avoid moisture transfer among the ingredients, (c) to determine the moisture barrier properties needed in a packaging material, (d) to determine what moisture content will curtail growth of microorganisms of interest, and (e) to predict the chemical and physical stability of food as a function of water content (see next section).

Shown in Figure 18 is a schematic MSI for a high-moisture food plotted to include the full range of water content from normal to dry. This kind of plot is not very useful because the data of greatest interest—those in the low-moisture region—are not shown in sufficient detail. Omission of the high-moisture region and expansion of the low-moisture region, as is usually done, yields an MSI that is much more useful (Fig. 19).

Several substances that have MSIs of markedly different shapes are shown in Figure 20. These are resorption (or adsorption) isotherms prepared by adding water to previously dried samples. Desorption isotherms are also common. Isotherms with a sigmoidal shape are characteristic of most foods. Foods such as fruits, confections, and coffee extract that contain large amounts of sugar and other small, soluble molecules and are not rich in polymeric materials exhibit a J-type isotherm shown as curve 1 in Figure 20. The shape and position of the isotherm are determined by several factors including sample composition, physical structure of the sample (e.g., crystalline or amorphous), sample pretreatments, temperature, and methodology.

Many attempts have been made to model MSIs, but success in achieving good conform-

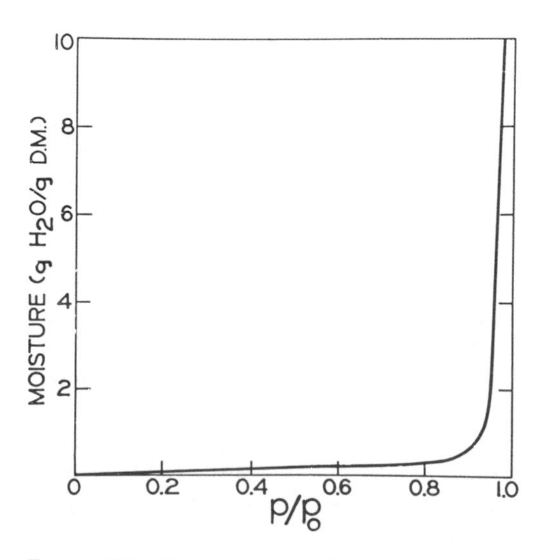

FIGURE 18 Schematic moisture sorption isotherm encompassing a broad range of moisture contents.

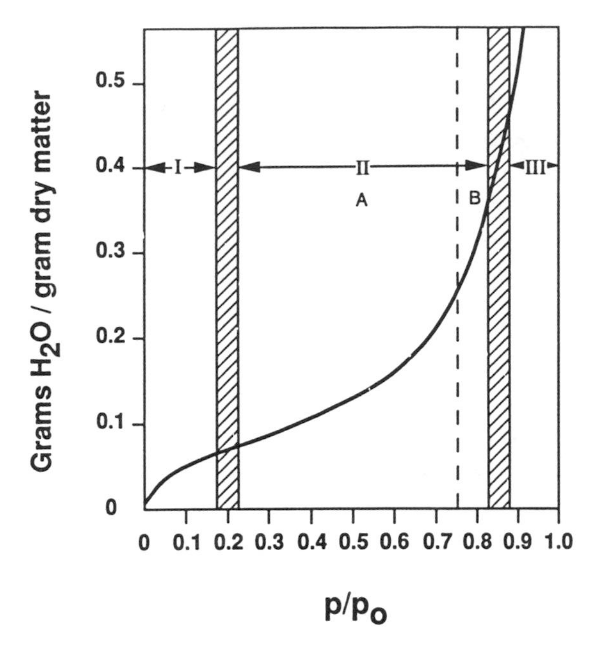

FIGURE 19 Generalized moisture sorption isotherm for the low-moisture segment of a food (20°C).

ance of a model to the full range of actual data for an MSI has been difficult. The oldest and best known model is that of Brunauer, Emmett, and Teller [8]. One of the best models is that developed by Guggenheim [36], Anderson [2], and DeBoer [16], and this is referred to as the GAB model.

As an aid to understanding the meaning and usefulness of sorption isotherms it is sometimes appropriate to divide them into zones as indicated in Figure 19. As water is added (resorption), sample composition moves from Zone I (dry) to Zone III (high moisture) and the properties of water associated with each zone differ significantly. These properties are described next and are summarized in Table 4.

Water present in Zone I of the isotherm is most strongly sorbed and least mobile. This water associates with accessible polar sites by water–ion or water–dipole interactions, is unfreezable at –40°C, has no ability to dissolve solutes, and is not present in sufficient amount to have a plasticizing effect on the solid. It behaves simply as part of the solid.

The high-moisture end of Zone I (boundary of Zones I and II) corresponds to the "BET monolayer" moisture content of the food. The BET monolayer value should be thought of as approximating the amount of water needed to form a monolayer over accessible, highly polar groups of the dry matter. In the case of starch, this amounts to one HOH per anhydroglucose unit [126]. Zone I water constitutes a tiny fraction of the total water in a high-moisture food material.

Water added in Zone II occupies first-layer sites that are still available. This water associates with neighboring water molecules and solute molecules primarily by hydrogen bonding, is slightly less mobile than bulk water, and most of it is unfreezable at –40°C. As water is added in the vicinity of the low-moisture end of Zone II, it exerts a significant plasticizing action on solutes, lowers their glass transition temperatures, and causes incipient swelling of the solid matrix. This action, coupled with the beginning of solution processes, leads to an

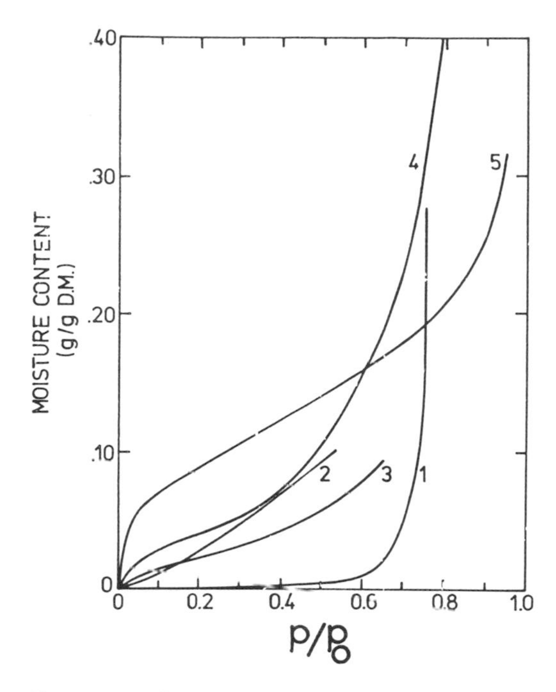

FIGURE 20 Resorption isotherms for various foods and biological substances. Temperature 20°C, except for number 1, which is 40°C: (1) confection (main component powdered sucrose), (2) spray-dried chicory extract, (3) roasted Columbian coffee, (4) pig pancreas extract powder, (5) native rico starch. (From Ref. 127.)

acceleration in the rate of most reactions. Water in Zones I and Zone II usually constitutes less than 5% of the water in a high-moisture food material.

In the vicinity of the junction of Zones II and III, water is sufficient to complete a true monolayer hydration shell for macromolecules such as globular proteins, and is sufficient to lower the glass transition temperature of macromolecules so that sample temperature and T_g are equal. Further addition of water (Zone III) causes a glass–rubber transition in samples containing glassy regions, a very large decrease in viscosity, a very large increase in molecular mobility, and commensurate increases in the rates of many reactions. This water is freezable, available as a solvent, and readily supports the growth of microorganism. Zone III water is referred to as bulk-phase water (Table 4). Additional water will have the properties of bulk-phase water and will not alter properties of existing solutes.

In gels or cellular systems, bulk-phase water is physically entrapped so that macroscopic flow is impeded. In all other respects this water has properties similar to that of water in a dilute salt solution. This is reasonable, since a typical water molecule added in Zone III is "insulated" from the effects of solutes molecules by several layers of Zone I and Zone II water molecules. The bulk-phase water of Zone III, either entrapped or free, usually constitutes more than 95% of the total water in a high-moisture food, a fact that is not evident from Figure 19.

As mentioned earlier, the zone boundaries indicated in Figure 19 are simply an aid to discussion rather than a reality. It is believed that water molecules can interchange rapidly within and between "zones" and that the concept of a continuum of water properties existing through Zones I–III is conceptually sounder than the notion of distinctly different properties existing in each zone. It is also of interest that addition of water to a dry material containing only a few water molecules will increase the mobility and lessen the residence time of these original water molecules [103]. However, addition of water to materials already having complete or near complete hydration shells is unlikely to have a significant effect on the properties of water originally present.

The important effects that these solute-induced differences in water properties have on stability of foods will be discussed in a later section. At this point, it will suffice to say that the most mobile fraction of water existing in any food sample governs stability.

2.9.2 Temperature Dependence

As mentioned earlier, RVP is temperature dependent; thus MSIs must also be temperature dependent. An example involving potato slices is shown in Figure 21. At any given moisture content, food p/p_0 increases with increasing temperature, in conformity with the Clausius–Clapeyron equation.

2.9.3 Hysteresis

An additional complication is that an MSI prepared by addition of water (resorption) to a dry sample will not necessarily be superimposable on an isotherm prepared by desorption. This lack

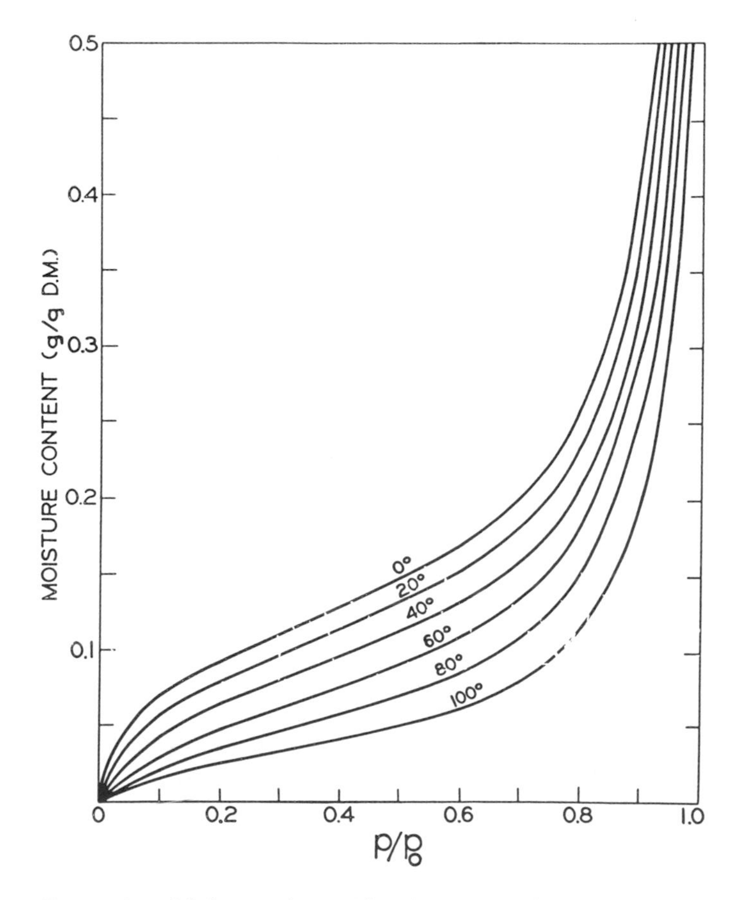

FIGURE 21 Moisture desorption isotherms for potatoes at various temperatures. (Redrawn from Ref. 35.)

of superimposability is referred to as "hysteresis," and a schematic example is shown in Figure 22. Typically, at any given p/p_0, the water content of the sample will be greater during desorption than during resorption. MSIs of polymers, glasses of low-molecular-weight compounds, and many foods exhibit hysteresis [46,47].

The magnitude of hysteresis, the shape of the curves, and the inception and termination points of the hysteresis loop can vary considerably depending on factors such as nature of the food, the physical changes it undergoes when water is removed or added, temperature, the rate of desorption, and the degree of water removal during desorption [45]. The effect of temperature is pronounced; hysteresis is often not detectable at high temperatures (~80°C) and generally becomes increasingly evident as the temperature is lowered [126].

Several largely qualitative theories have been advanced to explain sorption hysteresis [46,47]. These theories involve factors such as swelling phenomena, metastable local domains, chemisorption, phase transitions, capillary phenomena, and the fact that nonequilibrium states become increasingly persistent as the temperature is lowered. A definitive explanation (or explanations) of sorption hysteresis has yet to be formulated.

Sorption hysteresis is more than a laboratory curiosity. Labuza et al. [55] have conclusively established that lipid oxidation in strained meat from chicken and pork, at p/p_0 values in the range of 0.75–0.84, proceeds much more rapidly if the samples are adjusted to the desired p/p_0 value by desorption rather than resorption. The desorption samples, as already noted, would contain more water at a given p/p_0 than the resorption samples. This would cause the high-moisture sample to have a lower viscosity, which in turn would cause greater catalyst mobility, greater exposure of catalytic sites because of the swollen matrix, and somewhat greater oxygen diffusivity than in the lower moisture (resorption) sample.

In another study, Labuza et al. [56] found that the p/p_0 needed to stop the growth of several microorganisms is significantly lower if the product is prepared by desorption rather than resorption.

By now it should be abundantly clear that MSIs are highly product specific, that the MSI for a given product can be changed significantly by the manner in which the product is prepared, and that these points are of practical importance.

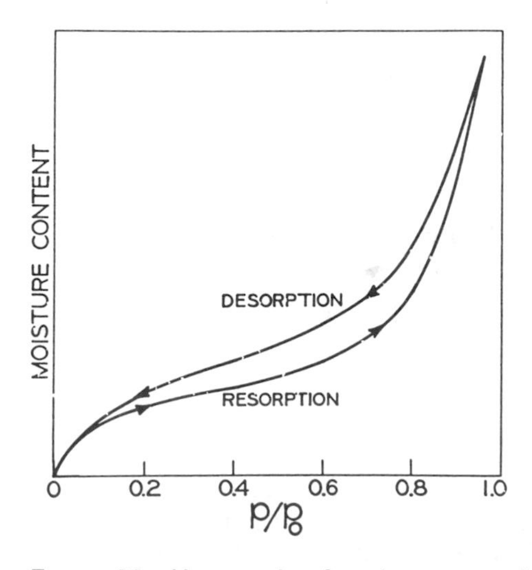

Figure 22 Hysteresis of moisture sorption isotherm.

2.10 RELATIVE VAPOR PRESSURE AND FOOD STABILITY

Food stability and p/p_0 are closely related in many situations. The data in Figure 23 and Table 6 provide examples of these relationships. Shown in Table 6 are various common microorganisms and the range of RVP permitting their growth. Also shown in this table are common foods categorized according to their RVP.

Data in Figure 23 are typical qualitative relationships between reaction rate and p/p_0 in the

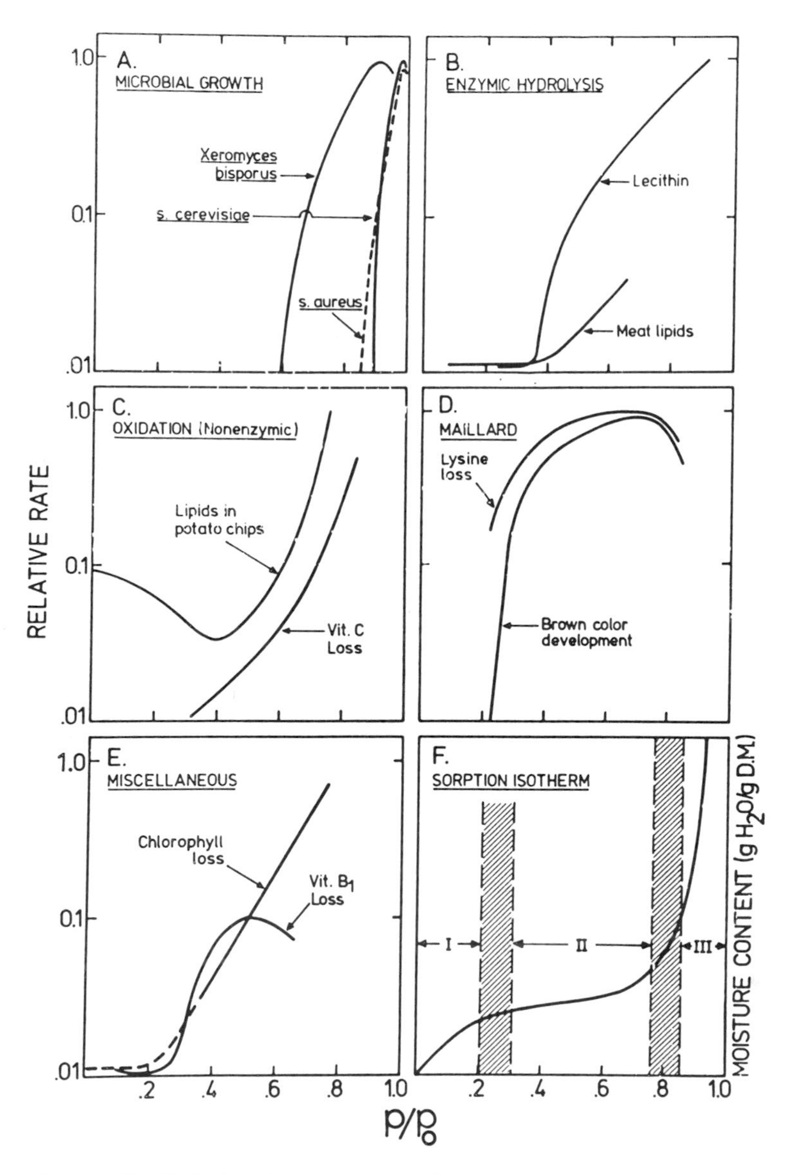

FIGURE 23 Relationships among relative water vapor pressure, food stability and sorption isotherms. (A) Microbial growth versus p/p_0. (B) Enzymic hydrolysis versus p/p_0. (C) Oxidation (nonenzymic) versus p/p_0. (D) Maillard browning versus p/p_0. (E) Miscellaneous reaction rates versus p/p_0. (F) Water content versus p/p_0. All ordinates are "relative rate" except for F. Data from various sources.

TABLE 6 Relative Vapor Pressure and Growth of Microorganisms in Food

Range of p/p_0	Microorganisms generally inhibited by lowest p/p_0 in this range	Foods generally within this range
1.00–0.95	*Pseudomonas, Escherichia, Proteus, Shigella, Klebsiella, Bacillus, Clostridium perfringens,* some yeasts	Highly perishable (fresh) foods and canned fruits, vegetables, meat, fish, and milk; cooked sausages and breads; foods containing up to approximately 40% (w/w) sucrose or 7% sodium chloride
0.95–0.91	*Salmonella, Vibrio parahaemolyticus, C. botulinum, Serratia, Lactobacillus, Pediococcus,* some molds, yeasts (*Rhodotorula, Pichia*)	Some cheeses (Cheddar, Swiss, Muenster, Provolone), cured meat (ham), some fruit juice concentrates; foods containing up to 55% (w/w) sucrose or 12% sodium chloride
0.91–0.87	Many yeasts (*Candida, Torulopsis, Hansenula*), *Micrococcus*	Fermented sausage (salami), sponge cakes, dry cheeses, margarine; foods containing up to 65% (w/w) sucrose (saturated) or 15% sodium chloride
0.87–0.80	Most molds (mycotoxigenic penicillia), *Staphylococcus aureus,* most *Saccharomyces (bailii)* spp., *Debaryomyces*	Most fruit juice concentrates, sweetened condensed milk, chocolate syrup, maple and fruit syrups; flour, rice, pulses containing 15–17% moisture; fruit cake; country-style ham, fondants, high-ratio cakes
0.80–0.75	Most halophilic bacteria, mycotoxigenic aspergilli	Jam, marmalade, marzipan, glacé fruits, some marshmallows
0.75–0.65	Xerophilic molds (*Aspergillus chevalieri, A. candidus, Wallemia sebi*), *Saccharomyces bisporus*	Rolled oats containing approximately 10% moisture; grained nougats, fudge, marshmallows, jelly, molasses, raw cane sugar, some dried fruits, nuts
0.65–0.60	Osmophilic yeasts (*Saccharomyces rouxii*), few molds (*Aspergillus echinulatus, Monascus bisporus*)	Dried fruits containing 15–20% moisture; some toffees and caramels; honey
0.50	No microbial proliferation	Pasta containing approximately 12% moisture; spices containing approximately 10% moisture
0.40	No microbial proliferation	Whole egg powder containing approximately 5% moisture
0.30	No microbial proliferation	Cookies, crackers, bread crusts, etc. containing 3–5% moisture
0.20	No microbial proliferation	Whole milk powder containing 2–3% moisture; dried vegetables containing approximately 5% moisture; corn flakes containing approximately 5% moisture; country style cookies, crackers

Source: Ref. 6

temperature range 25–45°C. For comparative purposes a typical isotherm, Figure 23F, is also shown. It is important to remember that the exact reaction rates and the positions and shapes of the curves (Fig. 23A–E) can be altered by sample composition, physical state and structure of the sample, composition of the atmosphere (especially oxygen), temperature, and by hysteresis effects.

For all chemical reactions in Figure 23, minimum reaction rates during desorption are typically first encountered at the boundary of Zones I and II of the isotherm (p/p_0 0.20–0.30), and all but oxidative reactions remain at this minimum as p/p_0 is further reduced. During desorption, the water content at the first-encountered rate minimum is the "BET monolayer" water content.

The unusual relationship between rate of lipid oxidation and p/p_0 at very low values of p/p_0 deserves comment (Fig. 23C). Starting at the extreme left of the isotherm, added water decreases the rate of oxidation until the BET monolayer value is attained. Clearly, overdrying of samples subject to oxidation will result in less than optimum stability. Karel and Yong [48] have offered the following interpretative suggestions regarding this behavior. The first water added to a very dry sample is believed to bind hydroperoxides, interfering with their decomposition and thereby hindering the progress of oxidation. In addition, this water hydrates metal ions that catalyze oxidation, apparently reducing their effectiveness.

Addition of water beyond the boundary of Zones I and II (Fig. 23C and Fig. 23F) results in increased rates of oxidation. Karel and Yong [48] suggested that water added in this region of the isotherm accelerates oxidation by increasing the solubility of oxygen and by allowing macromolecules to swell, thereby exposing more catalytic sites. At still greater p/p_0 values ($> \sim 0.80$) the added water may retard rates of oxidation, and the suggested explanation is that dilution of catalysts reduces their effectiveness.

It should be noted that curves for the Maillard reaction, vitamin B_1 degradation, and microbial growth all exhibit rate maxima at intermediate to high p/p_0 values (Fig. 23A, D, E). Two possibilities have been advanced to account for the decline in reaction rate that sometimes accompanies increases in RVP in foods having moderate to high moisture contents [20,54].

1. For those reactions in which water is a product, an increase in water content can result in product inhibition.
2. When the water content of the sample is such that solubility, accessibility (surfaces of macromolecules), and mobility of rate-enhancing constituents are no longer rate-limiting, then further addition of water will dilute rate-enhancing constituents and decrease the reaction rate.

Since the BET monolayer value of a food provides a good first estimate of the water content providing maximum stability of a dry product, knowledge of this value is of considerable practical importance. Determining the BET monolayer value for a specific food can be done with moderate ease if data for the low-moisture end of the MSI are available. One can then use the BET equation developed by Brunauer et al. [8] to compute the monolayer value

$$\frac{a_w}{m(1 - a_w)} = \frac{1}{m_1 c} + \frac{C - 1}{m_1 c} a_w \tag{8}$$

where a_w is water activity, m is water content (g H_2O/g dry matter), m_1 is the BET monolayer value, and C is a constant. In practice, p/p_0 values are used in Equation 8 rather than a_w values.

From this equation, it is apparent that a plot of $a_w/m(1 - a_w)$ versus a_w, known as a BET plot, should yield a straight line. An example for native potato starch, with a_w replaced by p/p_0, is shown in Figure 24. The linear relationship, as is generally acknowledged, begins to deteriorate at p/p_0 values greater than about 0.35.

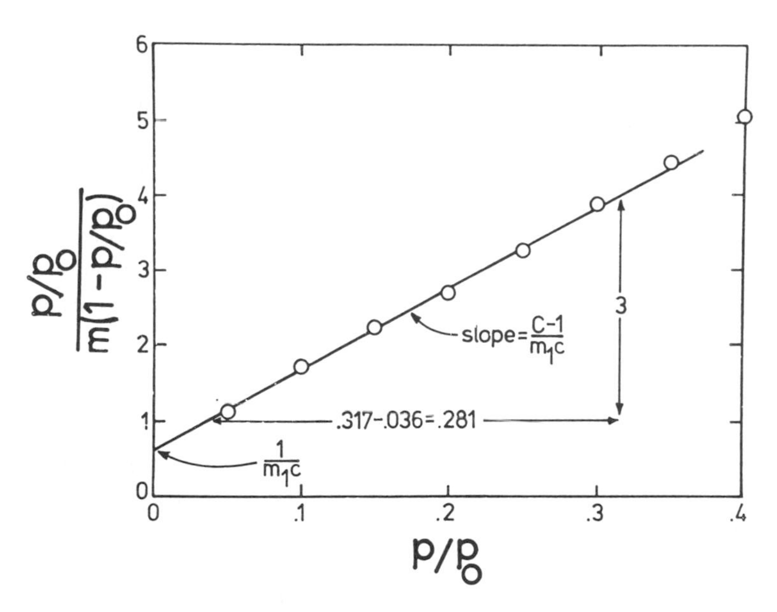

FIGURE 24 BET plot for native potato starch (resorption data, 20°C). (Data from Ref. 125.)

The BET monolayer value can be calculated as follows:

$$\text{Monolayer value} = m_1 = \frac{1}{(y \text{ intercept}) + (\text{slope})}$$

From Figure 24, the y intercept is 0.6. Calculation of the slope from Figure 24 yields a value of 10.7. Thus,

$$m_1 = \frac{1}{0.6 + 10.7} = 0.088 \text{ g } H_2O/\text{g dry matter}$$

In this particular instance the BET monolayer value corresponds to a p/p_0 of 0.2. The GAB equation yields a similar monolayer value [81].

In addition to chemical reactions and microbial growth, p/p_0 also influences the texture of dry and semidry foods. For example, suitably low RVPs are necessary if crispness of crackers, popped corn, and potato chips is to be retained; if caking of granulated sugar, dry milk, and instant coffee is to be avoided; and if stickiness of hard candy is to be prevented [53]. The maximum p/p_0 that can be tolerated in dry materials without incurring loss of desirable properties ranges from 0.35 to 0.5, depending on the product. Furthermore, suitably high water activities of soft-textured foods are needed to avoid undesirable hardness.

2.11 MOLECULAR MOBILITY (Mm) AND FOOD STABILITY

2.11.1 Introduction

Even though the RVP approach has served the food industry well, this should not preclude consideration of other approaches that can supplement or partially replace RVP as a tool for predicting and controlling food stability and processability. In recent years, evidence has become increasingly compelling that molecular mobility (Mm; translational or rotational motion) may be an attribute of foods that deserves attention because it is related causally to many important diffusion-limited properties of food.

Luyet and associates in the United States and Rey in France were apparently the first to draw attention to the relevance of Mm (glassy states, recrystallization, collapse temperatures during freeze drying) to properties of biological materials [74, 76, 78, 94]. John D. Ferry, a professor of chemistry at the University of Wisconsin, and his associates formulated many of the basic concepts pertaining to Mm in nonequilibrium systems consisting of synthetic, amorphous polymers [27,130]. In 1966, White and Cakebread [129] described the important role of glassy and supersaturated states in various sugar-containing foods and suggested that the existence of these states has an important influence on the stability and processability of many foods. Duckworth et al. [17] demonstrated the relevance of Mm to rates of nonenzymic browning and ascorbic acid oxidation, and thereby provided further evidence that the relationship between Mm and food stability is one of considerable importance.

Widespread recent interest in the relationship between Mm and food properties was created primarily by Felix Franks [29] and the team of Louise Slade and Harry Levine [60–67, 112–118]. They showed that important basic principles underlying the behavior of synthetic, amorphous polymers, as developed by Ferry's group and others, apply to the behavior of glass-forming foods. Slade and Levine used the phrase "food polymer science approach" to describe the interrelationships just mentioned; however, the term "molecular mobility" (Mm) seems preferable because this term is simple and emphasizes the underlying aspect of importance. The work of Slade and Levine has been relied on heavily during preparation of this section on Mm.

The importance of the Mm concept, which lay unappreciated by food scientists for many years, lends support to an important principle: For those in applied sciences who aspire to engage in pioneering work, much more can be gained from scientific literature that underlies or is peripheral to their primary field of endeavor than from literature that is central to it. Evidence now suggests that Mm is causally related to diffusion-limited properties of foods that contain, besides water, substantial amounts of amorphous, primarily hydrophilic molecules, ranging in size from monomers to polymers. The key constituents with respect to Mm are water and the dominant solute or solutes (solute or solutes comprising the major fraction of the solute portion). Foods of this type include starch-containing foods, such as pasta, boiled confections, protein-based foods, intermediate-moisture foods, and dried, frozen, or freeze-dried foods. Some properties and behavioral characteristics of food that are dependent on Mm are shown in Table 7.

When a food is cooled and/or reduced in moisture content so that all or part of it is converted to a glassy state, Mm is greatly reduced and diffusion-limited properties become stable. It is important to note the qualifying term "diffusion-limited." Most physical properties/changes are diffusion-limited, but some chemical properties/reactions are controlled more by chemical reactivity than diffusion. Approaches to predicting whether rates of chemical reactions are limited by diffusion or chemical reactivity are available but calculations are not simple. This matter will be discussed further in Section 2.11.3.2. Even if future work establishes that the Mm approach applies to virtually all physical properties of foods but only to some chemical properties, the importance of this approach remains sufficient to justify careful study.

It is appropriate at this point to suggest that the RVP and Mm approaches to food stability are, for the most part, complementary rather than competitive. Although the RVP approach focuses on the "availability" of water in foods, such as the degree to which it can function as a solvent, and the Mm approach focuses on microviscosity and diffusibility of chemicals in foods, the latter, of course, is dependent on water and its properties [83].

Because several terms used in the following discussion will be unfamiliar to many readers of this book and/or have been poorly defined in other works, a glossary appears at the end of this chapter. Many readers will find it useful to study the glossary before proceeding further.

TABLE 7 Some Properties and Behavioral Characteristics of Foods That Are Governed by Molecular Mobility (Diffusion-Limited Changes in Products Containing Amorphous Regions)

Dry or semidry foods	Frozen foods
Flow properties and stickiness	Moisture migration (ice crystallization, formation of in-package ice)
Crystallization and recrystallization	Lactose crystallization ("sandiness" in frozen desserts)
Sugar bloom in chocolate	Enzymatic activity
Cracking of foods during drying	Structural collapse of amorphous phase during sublimation (primary) phase of freeze-drying
Texture of dry and intermediate moisture foods	Shrinkage (partial collapse of foam-like frozen desserts)
Collapse of structure during secondary (desorption) phase of freeze-drying	
Escape of volatiles encapsulated in a solid, amorphous matrix	
Enzymatic activity	
Maillard reaction	
Gelatinization of starch	
Staling of bakery products caused by retrogradation of starch	
Cracking of baked goods during cooling	
Thermal inactivation of microbial spores	

Source: Adapted from Ref. 114.

2.11.2 State Diagrams

Consideration of "state" diagrams is highly pertinent to the discussion of Mm and stability of foods that are frozen or have reduced moisture contents. State diagrams are much more suitable for this purpose than conventional phase diagrams. Phase diagrams pertain solely to equilibrium conditions. State diagrams contain equilibrium information as well as information on conditions of nonequilibrium and metastable equilibrium "states." State diagrams are, therefore, supplemented phase diagrams, and they are appropriate because foods that are dried, partially dried, or frozen do not exist in a state of thermodynamic equilibrium.

A simplified temperature–composition state diagram for a binary system is shown in Figure 25. Important additions to the standard phase diagram are the glass transition curve (T_g) and the line extending from T_E to T_g', with both lines representing metastable conditions. Samples located above the glass transition curve and not on any line exist, with few exceptions, in a state of nonequilibrium, as will be discussed subsequently. State diagrams of this format will be used several times during the following discussion of Mm in foods.

When using these diagrams, it is assumed that pressure is constant and the time dependency of the metastable states, although real, is of little or no commercial importance (not true of nonequilibrium states). It also should be recognized that each simple system will have its own characteristic state diagram that differs quantitatively, but not qualitatively, from that in Figure 25, and that most foods are so complex that they cannot be accurately or easily represented on a state diagram. For all complex foods, both dry and frozen, accurate determination of a glass transition curve (or more properly zone, as will be discussed later) is difficult, but estimates are essential if the Mm approach to food stability is to be used effectively. Although estimates of T_g for complex foods are not easy to obtain, this can be done with accuracy sufficient for commercial use.

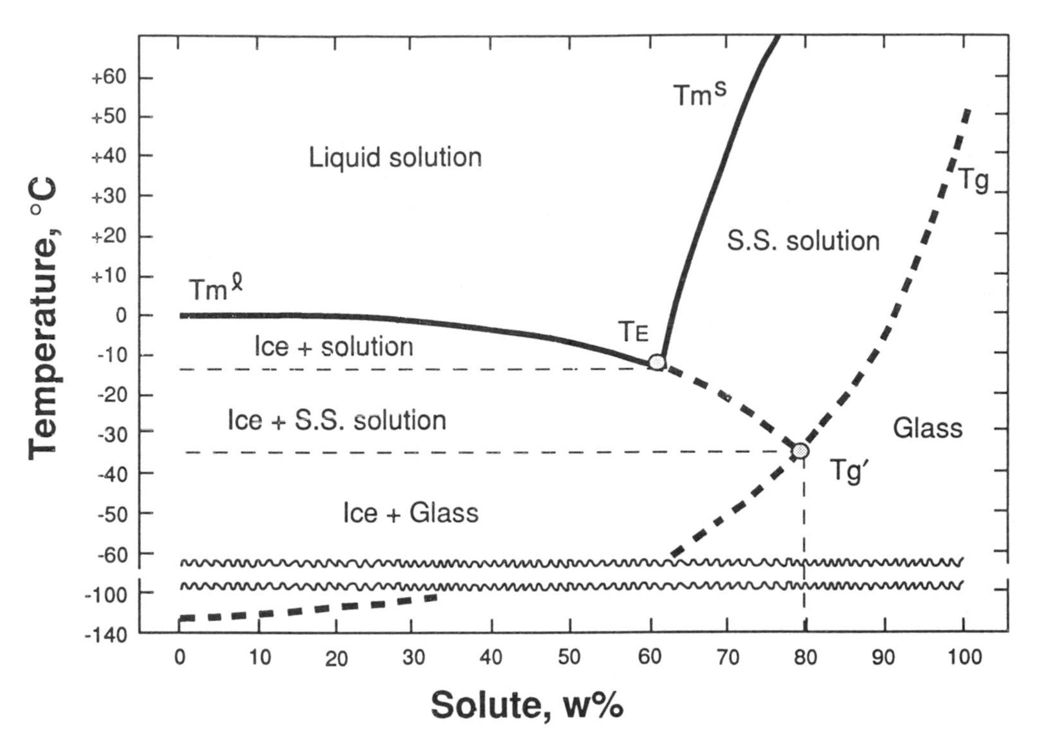

FIGURE 25 State diagram of a binary system. Assumptions: maximal freeze concentration, no solute crystallization, constant pressure, no time dependence. T_m^l is the melting point curve, T_E is the eutectic point, T_m^s is the solubility curve, T_g is the glass transition curve, and T_g' is the solute-specific glass transition temperature of a maximally freeze-concentrated solution. Heavy dashed lines represent conditions of metastable equilibrium. All other lines represent conditions of equilibrium.

Establishing the equilibrium curves (T_m^l and T_m^s; Fig. 25) for complex foods also can be difficult. For dry or semidry foods the T_m^s curve, the major equilibrium curve of importance, usually cannot be accurately depicted as a single line. A common approach is to base the state diagram on water and a food solute of dominating importance to the properties of the complex food, then deduce properties of the complex food from this diagram. For example, a state diagram for sucrose–water is useful for predicting the properties and behavior of cookies during baking and storage [67]. Determining T_m^s curves for dry or semidry complex foods that do not contain a dominating solute is a difficult matter that has not yet been satisfactorily resolved.

For frozen foods, the situation is somewhat better because the melting point curve (T_m^l), the major equilibrium curve of importance, is often known or easily determined. Thus, it is possible, with accuracy sufficient for commercial purposes, to prepare a state diagram for a complex frozen food.

For binary systems, the effect of different solutes on the glass transition curve is shown schematically in Figure 26. Note that the left end of the T_g curve is always fixed at –135°C, the T_g of water. Thus, differences in location of the curve depend on T_g' and on T_g of the dry solute.

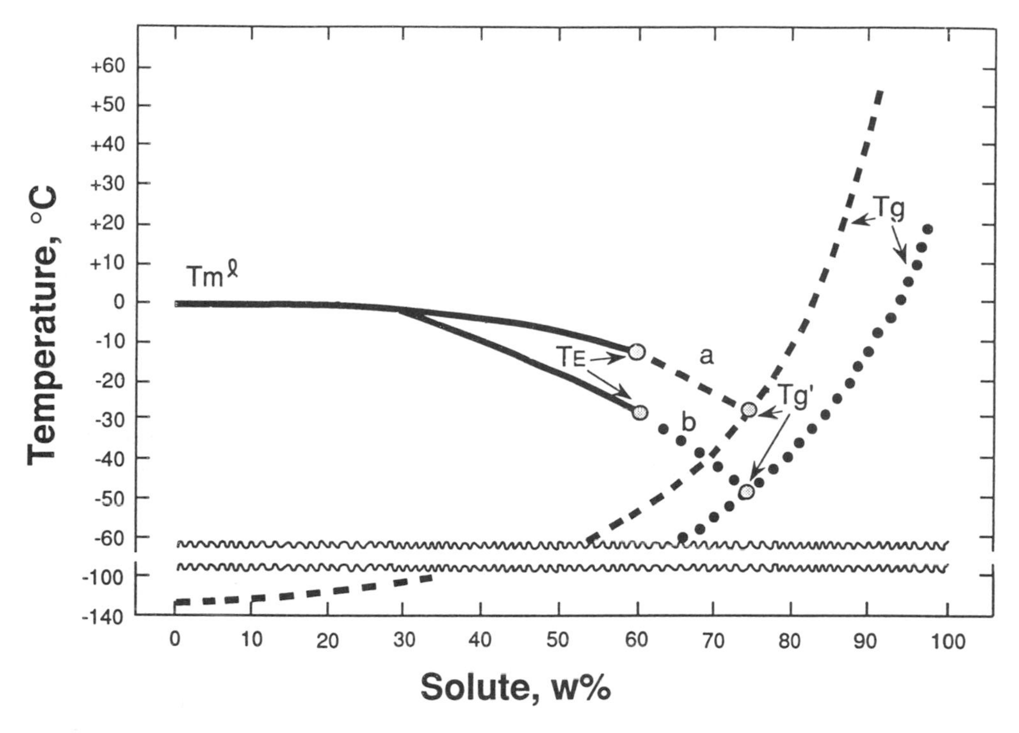

FIGURE 26 State diagram of a binary system showing the influence of solute type on the position of the glass transition curve. The extreme left position of the T_g curve is always fixed at the vitrification temperature of pure water (–135°C), the midpoint at the solute's T_g', and the extreme right position at the T_g of the pure solute; a and b are curves for different solutes. Assumptions stated in Figure 25 apply here.

2.11.3 Nine Key Concepts Underlying the Molecular Mobility Approach to Food Stability

2.11.3.1 Concept 1. Many Foods Contain Amorphous Components and Exist in a State of Metastable Equilibrium or Nonequilibrium

Complex foods frequently contain amorphous (noncrystalline solid or supersaturated liquid) regions. Biopolymers are typically amorphous or partly amorphous. Examples include proteins such as gelatin, elastin, and gluten, and carbohydrates such as amylopectin and amylose. Many small molecules such as sugars also can exist in an amorphous state, and all dried, partially dried, frozen, and freeze-dried foods contain amorphous regions.

Amorphous regions exist in metastable equilibrium or nonequilibrium. Attainment of thermodynamic equilibrium (minimum free energy) is not a goal during food processing, though this condition results in maximum stability. Thermodynamic equilibrium is incompatible with life, including that of fruits and vegetables postharvest, and is incompatible with satisfactory quality in foods. Thus, a major goal of food scientists/technologists, although they rarely view their duties in this manner, is to maximize the number of desirable food attributes that depend on metastable equilibrium states, and to achieve acceptable stability for those desirable attributes that must unavoidably depend on nonequilibrium states. Hard candy (amorphous solid) is a common example of a metastable food, and emulsions, small ice crystals, and unsaturated lipids are examples of food components that exist in a state of unstable nonequilibrium. Metastable states of food components often can be achieved by drying or freezing.

2.11.3.2 Concept 2. The Rates of Most Physical Events and Some Chemical Events Are Governed by Molecular Mobility (Mm)

Because most foods exist in metastable or nonequilibrium states, kinetic rather than thermodynamic approaches are often more appropriate for understanding, predicting, and controlling their properties. Molecular mobility (Mm) is a kinetic approach considered appropriate for this purpose because it is causatively related to rates of diffusion-limited events in foods. The WLF (Williams–Landel–Ferry) equation (see Section 2.11.3.5) provides the means for estimating Mm at temperatures above the glass transition temperature and below T_m^l or T_m^s. State diagrams indicate conditions of temperature and composition that permit metastable and nonequilibrium states to exist.

The utility of the Mm approach for predicting many kinds of physical changes has been reasonably well established. However, situations do exist where the Mm approach is of questionable value or is clearly unsuitable. Some examples are (a) chemical reactions whose rates are not strongly influenced by diffusion, (b) desirable or undesirable effects achieved through the action of specific chemicals (e.g., alteration of pH or oxygen tension), (c) situations in which sample Mm is estimated on the basis of a polymeric component (T_g of polymer) and where Mm of small molecules that can penetrate the polymer matrix is a primary determinant of the product attribute of interest [110,111], and (d) growth of vegetative cells of microorganisms (p/p_0 is a more reliable estimator than Mm) [12,13]. Point (a) deserves further attention because most authors of papers on Mm and its relevance to food stability are strangely silent on this important topic. Some useful references are Rice [95], Connors [14], Kopelman and Koo [50], Bell and Hageman [2a], and Haynes [40].

It is appropriate to first consider chemical reactions in a solution at ambient temperature. In this temperature range, some reactions are diffusion-limited but many are not. At constant temperature and pressure, three primary factors govern the rate at which a chemical reaction will occur: a diffusion factor, D (to sustain a reaction, reactants must first encounter each other), a frequency-of-collision factor, A (number of collisions per unit time following an encounter), and a chemical activation-energy factor, E_a (once a collision occurs between properly oriented reactants the energy available must be sufficient to cause a reaction, that is, the activation energy for the reaction must be exceeded). The latter two terms are incorporated in the Arrhenius relationship depicting the temperature dependence of the reaction rate constant. For a reaction to be diffusion-limited, it is clear that factors A and E_a must not be rate-limiting; that is, properly oriented reactants must collide with great frequency and the activation energy must be sufficiently low that collisions have a high probability of resulting in reaction. Diffusion-limited reactions typically have low activation energies (8–25 kJ/mol). In addition, most "fast reactions" (small E_a and large A) are diffusion-limited. Examples of diffusion-limited reactions are proton transfer reactions, radical recombination reactions, acid–base reactions involving transport of H^+ and OH^-, many enzyme-catalyzed reactions, protein folding reactions, polymer chain growth, and oxygenation/deoxygenation of hemoglobin and myoglobin [89,95]. These reactions may involve a variety of chemical entities including molecules, atoms, ions, and radicals. At room temperature, diffusion-limited reactions occur with bimolecular rate constants of about 10^{10} to 10^{11} $M^{-1}s^{-1}$; therefore a rate constant of this magnitude is regarded as presumptive evidence of a diffusion-limited reaction. It is also important to note that reactions in solution can go no faster than the diffusion-limited rate; that is, the diffusion-limited rate is the maximum rate possible (conventional reaction mechanisms are assumed). Thus, reactions occurring at rates significantly slower than the diffusion-limited maximum are limited by A or E_a or a combination of these two.

The theory describing diffusion-limited reactions was developed by Smoluchowski in 1917. For spherical, uncharged particles, the second-order diffusion-limited rate constant is

$$k_{dif} = \frac{4\pi N_A}{1000}(D_1 + D_2)r$$

where N_A is Avogadro's number, D_1 and D_2 are the diffusion constants for particles 1 and 2, respectively, and r is the sum of the radii of particles 1 and 2, that is, the distance of closest approach. This equation was subsequently modified by Debye to accommodate charged particles and by others to accommodate additional characteristics of real systems. Nonetheless, the Smoluchowski equation provides an order-of-magnitude estimate of k_{dif}.

Of considerable pertinence to the present discussion is the viscosity and temperature dependence of the diffusion constant. This relationship is represented by the Stokes–Einstein equation:

$$D = kT/\pi\beta\eta r_s$$

where k is the Boltzmann constant, T is absolute temperature, β is a numerical constant (about 6), η is viscosity, and r_s is the hydrodynamic radius of the diffusing species. This dependence of D (and therefore k_{dif}) on viscosity is a point of special interest because viscosity increases dramatically as temperature is reduced in the WLF region.

It appears likely that rates of some reactions in high moisture foods at ambient conditions are diffusion-limited and rates of others are not. Those that are would be expected to conform reasonably well to WLF kinetics as temperature is lowered or water content is reduced. Among those that are not diffusion-limited at ambient conditions (probably uncatalyzed, slow reactions), many probably become so as temperature is lowered below freezing or moisture is reduced to the point where solute saturation/supersaturation becomes common. This is likely because a temperature decrease would reduce the thermal energy available for activation and would increase viscosity dramatically, and/or because a reduction in water content would cause a pronounced increase in viscosity. Because the frequency-of-collision factor, A, is not strongly viscosity dependent, it probably is not an important determinant of the reaction-limiting mechanism under the circumstances described [9]. This likely conversion of some chemical reactions from non-diffusion-limited to diffusion-limited as temperature is lowered or water content is reduced should result, for those reactions behaving in this manner, in poor conformance to WLF kinetics in the upper part of the WLF region and much better conformance in the lower part of the WLF region.

2.11.3.3 Concept 3. Free Volume Is Mechanistically Related to Mm

As temperature is lowered, free volume decreases, making translational and rotational motion (Mm) more difficult. This has a direct bearing on movement of polymer segments, and hence on local viscosity in foods. Upon cooling to T_g, free volume becomes sufficiently small to stop translational motion of polymer segments. Thus, at temperatures $< T_g$, the stability of diffusion-limited properties of food is generally good. Increases in free volume (usually undesirable) can be achieved by addition of a low-molecular-weight solvent, such as water, or by increasing temperature, with both actions increasing translational motion of molecules. The free volume concept is useful in that it provides a mechanistic basis for the Mm versus stability relationship, but as yet, it has not been successfully applied as a quantitative tool for predicting food stability.

2.11.3.4 Concept 4. Most Foods Exhibit a Glass Transition Temperature (T_g or T_g') or Range

The T_g has an important relationship to the stability of diffusion-limited properties of food. Foods that normally contain amorphous regions or develop them during cooling or drying will exhibit a T_g or T_g range. In biological systems, solutes seldom crystallize during cooling or drying so amorphous regions and glass transitions are common. The stability of diffusion-limited properties of substances of this type often can be estimated from the relationship between Mm and T_g. Mm, and thus all diffusion-limited events, including many quality-deteriorating reactions, usually becomes severely restricted below T_g. Unfortunately, many foods are stored at $T > T_g$, resulting in much greater Mm and much poorer stability than that attainable at $T < T_g$.

To determine T_g with reasonable accuracy it must be measured. In simple systems this can be accomplished using a differential scanning calorimeter (DSC) equipped with a derivative plotting accessory, but great care must be exercised to achieve accurate results. In complex systems (most foods), accurate determination of T_g by DSC is very difficult, and dynamic mechanical analysis (DMA) and dynamic mechanical thermal analysis (DMTA) have been used as alternate techniques [77a]. All of these approaches are costly, and none is suited to in-plant measurements.

The observed T_g depends on the type of instrument used, the experimental conditions employed, and the interpretative skill of the operator [1,10,39,43,82,92,101,111]. A particularly good example concerns values for T_g' (T_g of frozen samples determined under conditions where maximal freeze concentration is presumably achieved). For a given substance, T_g' values reported after about 1990 tend to be lower than values determined earlier. These discrepancies are caused by differences in experimental procedures, clearly demonstrating the inappropriateness of reporting T_g values to an accuracy of greater than ±1°C. Whether the recent values more accurately reflect conditions prevailing in commercial foods is still a matter of controversy, and it should be noted that T_g' values reported by Slade and Levine pertain to samples with initial water contents of 80%. This is fairly typical of natural foods, whereas values reported more recently are based on initial water contents that are very low. Several criteria for ascertaining whether a glass transition has been accurately determined by DSC are given by Wolanczyk [131].

Various equations are available for calculating T_g of samples containing only a few components, and this approach can provide a useful estimate when sample composition and T_g values of the components are known. The oldest and simplest equation, as applied here to a binary system, is that of Gordon and Taylor [34].

$$T_g = \frac{w_1 T_{g_1} + k w_2 T_{g_2}}{w_1 + k w_2}$$

where T_{g_1} and T_{g_2} are the glass transition values (in K) of components 1 (water) and 2 of the sample, respectively; w_1 and w_2 are the weight fractions of components 1 and 2 of the sample; and k is an empirical constant. Refinements to this equation are available in the literature [38].

2.11.3.5 Concept 5. Molecular Mobility and Diffusion-Limited Food Stability Are Unusually Temperature Dependent Between T_m and T_g (Note: T_m Here Is Taken to Mean Either T_m^l or T_m^s, Whichever Is Appropriate)

Within this temperature range, which for foods may be as large as 100°C or as small as about 10°C, the vast array of products having amorphous regions will exhibit a temperature dependency for Mm and viscoelastic properties that is unusually large. The Mm is quite intense at T_m and is very subdued for most molecules at or below T_g. This temperature range encompasses product consistencies termed "rubbery" and "glassy" (although the "rubbery" term applies only when large polymers are present). Qualitative relationships between substance properties and substance temperature, including the range T_m–T_g, are depicted in Figure 27.

The temperature dependence of Mm, and that of food properties that depend strongly on Mm (most physical properties and some chemical properties), is far greater in zone T_m–T_g than it is at temperatures above or below this zone. Thus, one usually finds a change in slope (change in activation energy) of Arrhenius plots upon passage from temperatures outside zone T_m–T_g to temperatures inside. Estimating the temperature dependence of reaction rates in zone T_m–T_g has been the subject of considerable research, and an equation that applies accurately to all types of reactions and conditions has not been devised. In zone T_m–T_g, rates of many physical events conform more closely to the WLF equation [27,130], and to similar equations, than they do to the Arrhenius equation. Because the dependency of chemical reactions on Mm can vary greatly depending on reactant type, neither WLF nor the Arrhenius equation applies to all chemical reactions in zone T_m–T_g. Conformance of physical and chemical events to WLF or Arrhenius

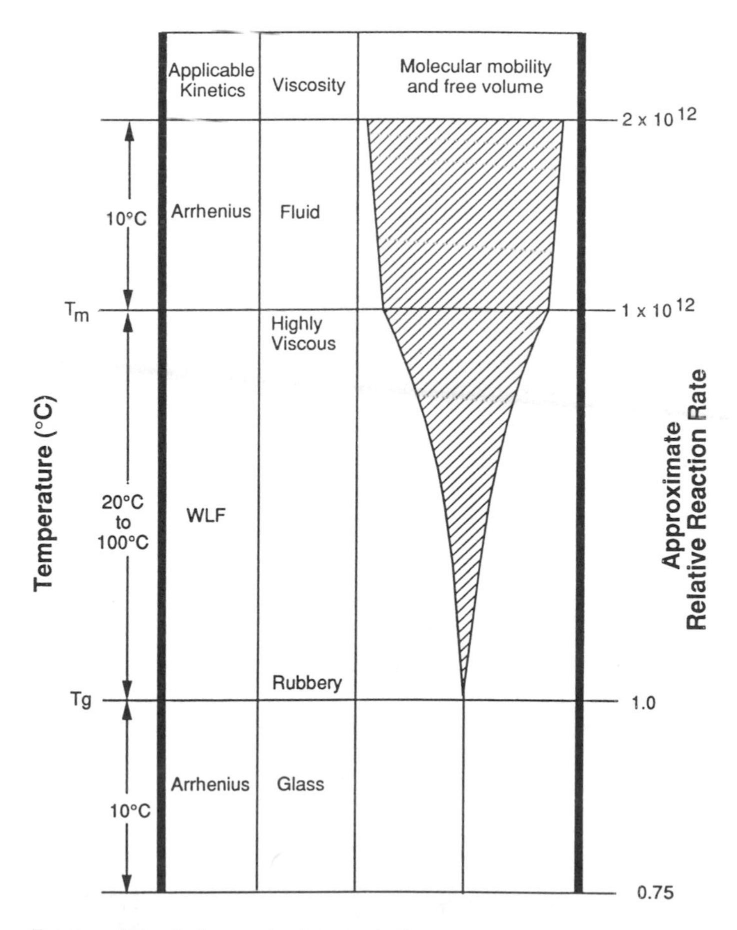

FIGURE 27 Schematic interrelations among temperature, appropriate type of kinetics, viscosity, molecular mobility, free volume, and relative rates of diffusion-dependent events. WLF kinetics based on mean constants. Other terms are defined in legend of Figure 25.

relationships is much less satisfactory in the presence of ice than in its absence because the concentrative effects of ice formation are not accommodated by either approach.

Because the WLF equation is a useful tool for estimating rates of physical events in zone T_m–T_g, it deserves further discussion. The WLF equation expressed in terms of viscosity is

$$\log(\eta/\eta_g) = \frac{-C_1(T - T_g)}{C_2 + (T - T_g)}$$

where η is viscosity at product temperature T(K) (Note: η can be replaced by 1/Mm, or any other diffusion-limited relaxation process), η_g is viscosity at product temperature T_g (K), and C_1 (dimensionless) and C_2(K) are constants [27,130]. The T_g is generally accepted as the reference temperature in the WLF equation when ice is not present. In the presence of ice, disagreement still exists as to whether T_g or T_g' is more appropriate [93,111]. This point will be discussed further in the section dealing with technological aspects of freezing preservation.

The terms C_1 and C_2 are substance-specific constants (i.e., not temperature dependent). They have mean (sometimes called "universal") values of 17.44 and 51.6, respectively, for many synthetic, totally amorphous polymers that are pure (no diluent). The numerical values of these constants vary substantially with water content and substance type, so values appropriate for foods often differ appreciably from the mean values. Constants specific for the food under study must be used if reasonable conformance of data to the equation is desired [49,111].

The WLF equation specifies a very large temperature dependence of substance properties in zone T_m–T_g. If one assumes conformance to WLF kinetics (mean constants), the absence of ice, and an initial substance temperature $< T_g$, then warming will result in the following sequence of changes in viscosity (or 1/Mm) [115,117]:

1. An approximate 10^3-fold decrease in viscosity and increase in Mm during an iso-temperature conversion from a glass to a supersaturated liquid.
2. An approximate 10^5-fold decrease in viscosity and increase in Mm with warming through the 20°C interval immediately above T_g. (In the unfrozen fraction of frozen samples, the change in these properties per °C is even greater.)
3. An approximate 10^{12}-fold decrease in viscosity and increase in Mm with warming through the entire range from T_g to T_m.

With respect to diffusion-limited food stability in the WLF region (T_m–T_g), two terms are of key importance: T–T_g (or T–T_g') and T_m/T_g. T–T_g, where T is product temperature, defines the location of the food in the WLF region. The $\log(\eta/\eta_g)$ (or any other viscosity-related property, such as Mm) varies curvilinearly with T–T_g. The term T_m/T_g (calculated on the basis of K) provides a rough estimate of product viscosity (inverse relationship with Mm) at T_g. Knowledge of product viscosity at T_g is important because viscosity at T_g is the reference value in the WLF equation, and this value can vary considerably with changes in product composition.

Valuable concepts pertaining to T_m–T_g, T–T_g, and T_m/T_g have been developed based largely on diffusion-limited properties of carbohydrates [62,112–114, 117]:

1. The magnitude of the T_m–T_g region can vary from about 10 to about 100°C, depending on product composition.
2. Product stability depends on product temperature, T, in the zone T_m–T_g, that is, it is inversely related to $\Delta T = T - T_g$.
3. At a given value of T_g and constant solids content, T_m/T_g varies inversely with Mm. Thus, T_m/T_g is directly related to both diffusion-limited product stability and product rigidity (viscosity) at both T_g and temperatures $> T_g$ in the WLF region. For example, at any given T in the WLF zone, substances with small T_m/T_g values (e.g., fructose)

result in larger values of Mm and greater rates of diffusion-limited events than do substances with large T_m/T_g values (e.g., glycerol [113]). Small differences in the value of T_m/T_g result in very large differences in both Mm and product stability [113].
4. T_m/T_g is highly dependent on solute type (Table 8).
5. For equal T_m/T_g at a given product temperature, an increase in solids content results in decreased Mm and increased product stability.

At this point it is appropriate to mention two approaches that have been used to study the interrelations between Mm, T_g, and food properties/stability. One approach, as just discussed, involves testing whether physical and chemical changes in foods over the temperature range T_m–T_g conform to WLF kinetics. Some studies of this kind have resulted in good conformity (physical properties) to WLF kinetics and others poor (some chemical reactions) (Fig. 28).

A second approach is simply to determine whether food stability differs markedly above and below T_g (or T_g'), with little attention given to characterization of kinetics. With this approach, the expectation is not so demanding and the results are often much more favorable; that is, it is found that desirable food properties, especially physical properties, are typically retained much better below T_g than they are above. T_g appears to be much less reliable for predicting stability of chemical properties than it is for physical properties. The temperature below which oxidation of ascorbic acid exhibits greatly reduced temperature dependency (practical termination temperature) appears to correspond fairly well to T_g', at least under the conditions used in Figure 29. However, the same is not true for nonenzymatic browning. In Figure 30, the practical termination temperature for nonenzymatic browning is well above sample T_g, whereas in Figure 31 the termination temperature is well below sample T_g. Differences in sample composition probably account for the dissimilar results in the two browning studies.

In closing this section, it is appropriate to note that an algebraic equation is available to interrelate food texture, temperature, and moisture content in the vicinity of T_g (or other critical temperatures) [88]. Although the equation cannot be used to predict textural changes, it can be used to create quantitative, three-dimensional graphs that effectively display these interrelationships near T_g. Large differences exist among products with respect to dependency of a given textural property on temperature and moisture, and these differences are clearly evident on graphs of this kind.

2.11.3.6 Concept 6. Water Is a Plasticizer of Great Effectiveness and It Greatly Affects T_g

This is especially true with regard to polymeric, oligomeric, and monomeric food substances that are hydrophilic and contain amorphous regions. This plasticizing action results in enhanced Mm, both above and below T_g. As water increases, T_g decreases and free volume increases (Fig. 32). This occurs because the average molecular weight of the mixture decreases. In general, T_g decreases about 5–10°C per wt% water added [118]. One should be aware, however, that the presence of water does not assure that plasticization has occurred; water must be absorbed in amorphous regions to be effective.

Water, because of its small molecular mass, can remain surprisingly mobile within a glassy matrix. This mobility no doubt accounts, as previously noted, for the ability of some chemical reactions involving small molecules to continue at measurable rates somewhat below the T_g of a polymer matrix, and for water to be desorbable during the secondary phase of freeze drying at temperatures $< T_g$.

TABLE 8 Glass Transition Values and Associated Properties of Pure Carbohydrates[a]

	Properties dry				Properties hydrated (water = W_g')			
Carbohydrate	MW	T_m (°C)	T_g (°C)[b]	T_m/T_g[c]	T_g' ($\approx T_c \approx T_r$) (°C)[b,d]	W_g' (wt%)[b,e]	MW_w[f]	MW_n[g]
Glycerol	92.1	18	−93	1.62	−65	46	58.0	31.9
Xylose	150.1	153	9–14	1.49 ± 0.01	−48	31	109.1	45.8
Ribose	150.1	87	−10 to −13	1.37 ± 0.01	−47	33	106.7	44.0
Glucose	180.2	158	31–39	1.39 ± 0.02	−43	29	133.0	49.8
Fructose	180.2	124	7–17[h]	1.39 ± 0.03	−42	49	100.8	33.3
Galactose	180.2	170	30–32[h]	1.45 ± 0.01	−41 to −42	29–45	107–151	35.6–50
Sorbitol	182.2	111	−2 to −4	1.45 ± 0.01	−43 to −44	19	151	66.7
Sucrose	342.3	192	52–70	1.40 ± 0.04	−32 to −46	20–36	225.9	45.8
Maltose	342.3	129	43–95	1.19 ± 0.1	−30 to −41	20	277.4	74.4
Trehalose	342.3	203	77–79	1.35 ± 0.01	−27 to −30	17	288.2	85.5
Lactose	342.3	214	101	1.37	−28	41	209.9	41.0
Maltotriose	504.5	134	76	1.17	−23 to −24	31	353.5	53.7
Maltopentose	828.9		125–165		−15 to −18	24–32	569.6	53.8
Maltohexose	990.9		134–175		−14 to −15	24–33	666.6	52.1
Maltoheptose	1153.0		139		−13 to −18	21–33	911.7	80.0

[a]Primarily from Levine and Slade [61, 64] and Slade and Levine [117].
[b]Most commonly reported value or a range of most commonly reported values.
[c]Calculated on basis of K.
[d]T_c = collapse temperature, T_r = temperature of incipient recrystallization.
[e]C_g', wt% solute at $T_g' = 100 - W_g'$.
[f]Weight-average molecular weight.
[g]Number-average molecular weight.
[h]Slade and Levine [117] report two T_g values for fructose (11 and 100°C) and galactose (30 and 110°C). They argue that the higher value is property-controlling.

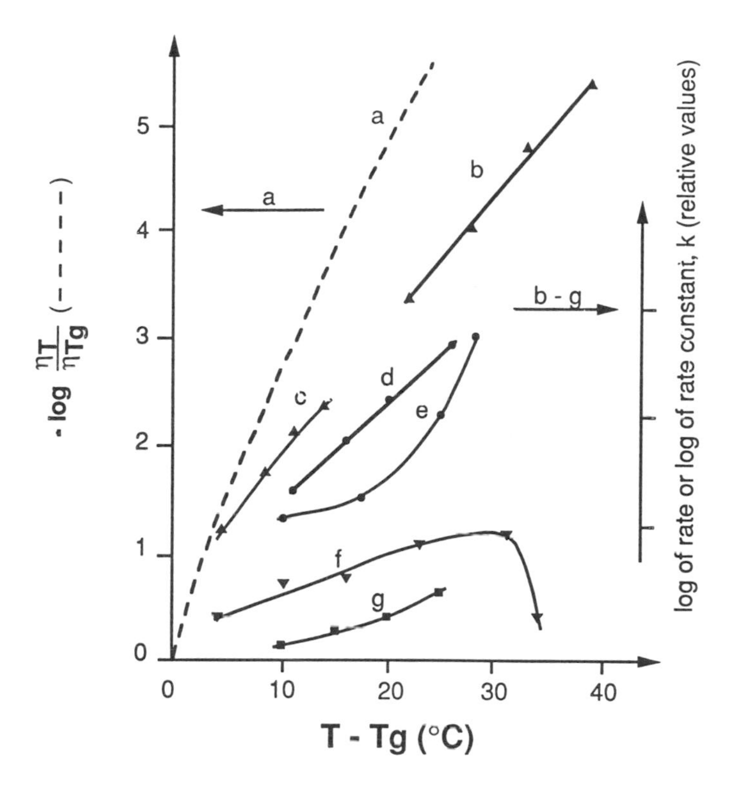

FIGURE 28 Comparison of the temperature dependence ($T - T_g$) of viscosity ($\log \eta_T/\eta_{T_g}$; inverse of molecular mobility) as estimated from the WLF equation using mean constants, with the temperature dependencies of the rates at which various foods deteriorate. Note: Curves for rate of deterioration of foods have been adjusted vertically to avoid overlap; thus the values shown are relative values and meaning should be attached only to the slopes of these curves. Curve a is WLF viscosity, which is usually assumed to be inversely proportional to the rates of diffusion-dependent reactions. Curve b is the pseudo-first-order rate constant for loss of ascorbic acid in frozen peas. Curve c is the rate of enzyme-catalyzed hydrolysis of disodium *p*-nitrophenyl phosphate in aqueous maltodextrin. Curve d is the rate constant for the decrease in protein solubility in frozen cod. Curve e is the rate constant for the increase in "Instron peak force" for frozen cod. Curve f is the mean rate of increase in the apparent viscosity of egg yolk (during cooling, the early stages of freezing account for the steep slope at the right end of the curve). Curve g is the "kinetic constant" for the growth of ice crystals in frozen beef. (Compiled from various sources by Simatos and Blond [110].)

2.11.3.7 Concept 7. Solute Type Greatly Affects T_g and T_g'

These relationships are important for predicting the behavior of compounds, but they are not simple. Thus, the discussion here is far from complete. For further information the reader is referred to Slade and Levine 113]. T_g is strongly dependent on both solute type and moisture content, except for observed T_g', which depends primarily on solute type and only slightly on water content (initial).

Attention will first be given to the relationship between solute molecular weight (MW) and T_g or T_g'. Figure 33 is a plot of T_g' versus MW for sugars, glycosides, and polyols with maximum MW of about 1200. T_g' (and T_g) increase proportionately with increases in solute MW

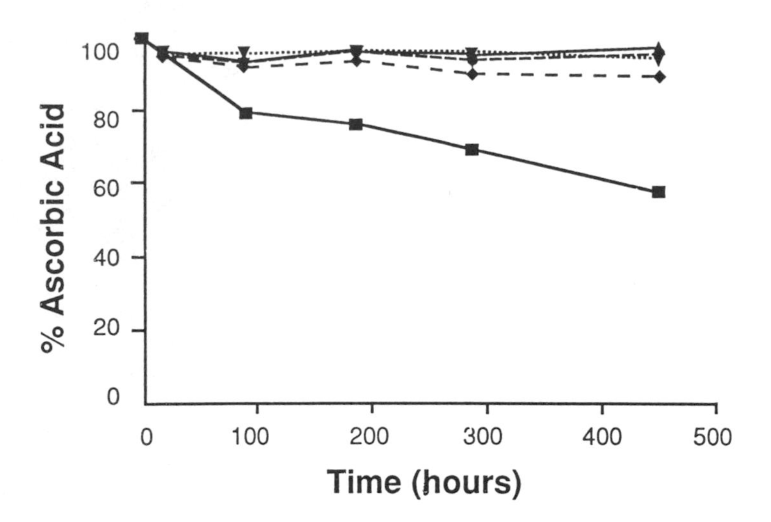

FIGURE 29 Loss of ascorbic acid with time at various temperatures. Initial samples contained L-ascorbic acid (40 mg/100 ml) and 10% w/w maltodextrin (M100) in degassed acetate buffer, pH 5.8. T_g' is −10°C. Upper three lines are data for temperatures of −11.5, −14.3, and −17.7°C, ◆ is −8.0°C, and ■ is −5.6°C. (From Ref. 70.)

over the range of MW shown. This is an expected relationship because translational mobility of molecules decreases with increasing size, so that a large molecule requires a higher temperature for movement than does a smaller one. However, with MW greater than 3000 (dextrose equivalent, DE, of < ~ 6 for starch hydrolysis products) T_g becomes independent of MW, as is shown in Figure 34. An exception occurs when time and large-molecule concentration are sufficient to allow "entanglement networks" (see later section) to form. In this instance, T_g continues to rise somewhat with increasing MW. A noteworthy aspect of Figure 34 is the relationship shown between functional properties of solutes and their DE (or MW_n). Compounds on the small-MW portion (vertical leg) of the curve serve, for example, as sweeteners,

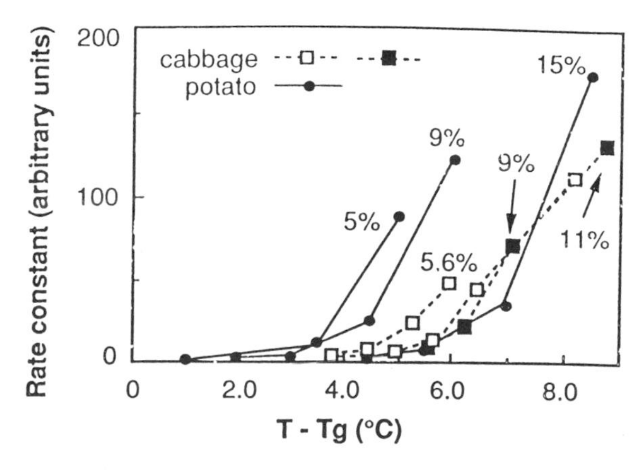

FIGURE 30 Rate of nonenzymatic browning in cabbage and potato as a function of product water content and $T - T_g$. (From Ref. 49.)

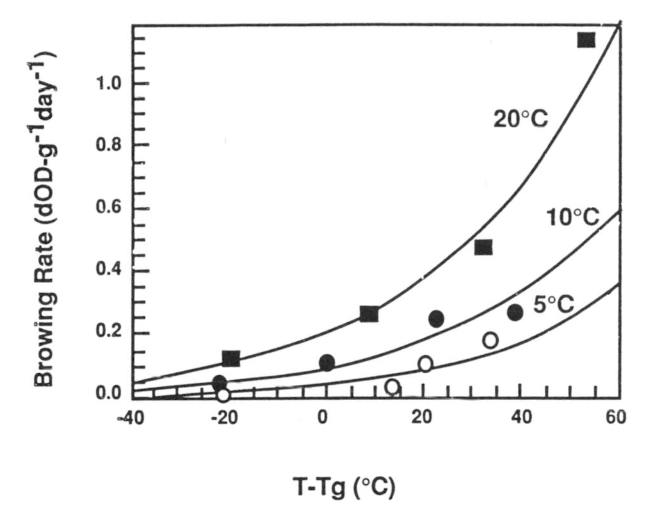

FIGURE 31 Rate of nonenzymatic browning in a model system as a function of $T - T_g$. Maltodextrin (DE 10), L-lysine, and D-xylose were used at ratios of 13:1:1. Storage temperature was held constant at each of the temperatures indicated and $T - T_g$ was altered by changing water content of the sample. (From Ref. 99.)

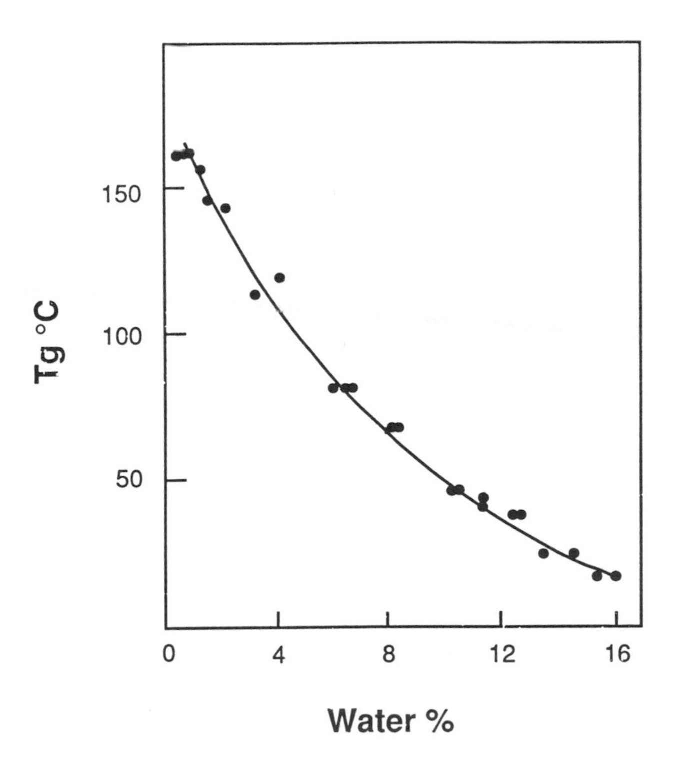

FIGURE 32 T_g of wheat gluten as a function of water content. (From Ref. 42.)

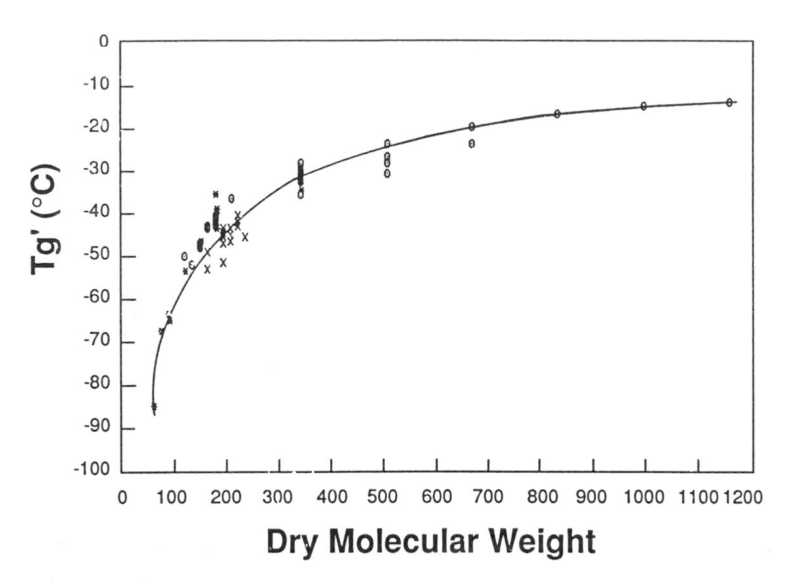

FIGURE 33 T_g' as influenced by solute molecular weight. T_g' values were determined from 20 wt% solutions of sugars (O), glycosides (x), and polyols (*) that were maximally freeze concentrated. (From Ref. 61.)

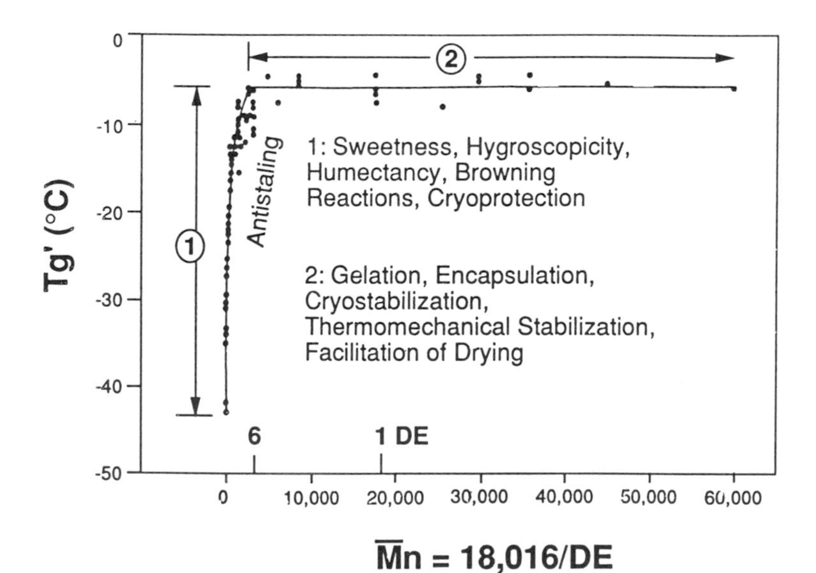

FIGURE 34 T_g' as influenced by number-average molecular weight and dextrose equivalent (DE) of commercial starch hydrolysis products. T_g' values were determined from maximally freeze-concentrated solutions that initially contained 80 wt% water. (Adapted from Ref. 60.)

humectants, participants in the Maillard reaction, and cryoprotectants. Compounds on the plateau region of the curve have entirely different functions, as indicated.

Plots of MW versus T_g that are linear and provide greater insight to the relationship between MW and T_g have been developed for solutes with MW of less than about 3000 [113]. For dry solutes, the appropriate plot is T_g versus $-1/MW_n$ of the solute. For samples containing significant amounts of water, the appropriate plot is T_g' versus $-1/MW_w$ of the water–solute solution that exists at T_g' of the sample. It should be noted that the good linearity of these plots deteriorates significantly when members of nonhomologous series are incorporated on the same plot. This indicates that solute attributes other than MW have an influence on T_g and T_g'. Thus, one cannot expect to obtain the same product properties (stability, processability) when chemicals of the same MW (or same DE) are interchanged. This failure to achieve uniform performance among compounds of the same MW is especially true when the interchange involves different chemical families. For example, for different sugars of equal MW, the difference in T_g' can be as much as 10°C. Different performance can also occur between molecules of the same type if fractions of different configurations are present [114]. T_g and T_g' values for selected pure carbohydrates, along with associated values for molecular weight and several other properties of importance, are listed in Table 8.

A final point should be mentioned before closing this section. Most, perhaps all, biopolymers of high MW have very similar glass curves and exhibit T_g' values near −10°C. Those with T_g' values in this range include polysaccharides such as starch, maltodextrin, cellulose, hemicellulose, carboxymethylcellulose, dextran, and xanthan; and proteins such as gluten, glutenin, gliadin, zein, collagen, gelatin, elastin, keratin, albumins, globulins, and casein [117].

2.11.3.8 Concept 8. Solute Type Has a Profound Effect on W_g

The term W_g is the water content of the sample at T_g, and W_g' is the unfrozen water content at T_g'. The terms C_g and C_g' are the solids content of the sample at T_g and T_g', respectively ($C_g' = 100 - W_g'$, and the same relationship applies to C_g and W_g at a given temperature). The following important relationships were derived for W_g' of carbohydrates but are believed to apply to W_g and to substances other than carbohydrates:

1. The value of W_g' varies directly with Mm and inversely with product stability at T_g'. Increases in product water content above W_g' or W_g cause product stability to decrease and Mm to increase.
2. The W_g' values vary greatly with solute type as is apparent from data in Table 8. The value of W_g' generally varies inversely with T_g' and MW_w, but these are qualitative relationships and are not suitable for predictive purposes, except for members of a homologous series of compounds. Generally, the most reliable values of solute or product W_g' are obtained by measurement. However, the experimental procedure chosen can greatly affect the value obtained, and the procedure most likely to give values of greatest relevance to food is still a matter of controversy [114].

Based on what has been said, one should not expect the same performance within the WLF zone when compounds are interchanged on an equal weight or equal p/p_0 basis, because these compounds may have different values for W_g' and T_g' (or W_g and T_g). Compare, for example, the W_g' values of glucose versus fructose and lactose versus trehalose (Table 8). Although each pair member has the same molecular weight, their effect on product stability usually is quite different, and this is at least partly attributable to the difference in W_g'.

2.11.3.9 Concept 9. Molecular Entanglement Can Greatly Affect the Properties of Food

Macromolecule entanglement (see glossary) can lead to the formation of entanglement networks (EN) when solute molecular size is sufficient (> ~3,000 MW, < DE of ~6 for carbohydrates), when solute concentration exceeds a critical value, and sufficient time elapses. Besides carbohydrates, proteins can form EN, examples being gluten in wheat flour dough and sodium caseinate in imitation mozzarella cheese.

ENs have a profound effect on the properties of food. For example, it has been suggested, based on some supporting evidence, that ENs can slow rates of crystallization in frozen foods, retard moisture migration in baked goods, help retain crispness of breakfast cereals, help reduce sogginess of pastries and pie crusts, and facilitate drying, gel formation, and encapsulation processes [114]. Once conditions result in an EN, further increases in MW will not only cause further increases in T_g or T_g', but will also result in firmer networks [64].

2.11.4 Technological Aspects: Freezing

Although freezing is regarded as the best method of long-term preservation for most kinds of foods, the benefits of this preservation technique derive primarily from low temperature as such, not from ice formation. The formation of ice in cellular foods and food gels has two important adverse consequences: (a) nonaqueous constituents become concentrated in the unfrozen phase (an unfrozen phase exists in foods at all storage temperatures used commercially), and (b) all water converted to ice increases 9% in volume. Both occurrences deserve further comment.

During freezing of aqueous solutions, cellular suspensions, or tissues, water from solution is transferred into ice crystals of variable but high degree of purity. Nearly all of the nonaqueous constituents are therefore concentrated in a diminishing quantity of unfrozen water. The net effect is similar to conventional dehydration except in this instance the temperature is lower and the separated water is deposited locally as ice. The degree of concentration is influenced mainly by the final temperature, and to a lesser degree by agitation, rate of cooling, and formation of eutectics (crystallization of solutes—uncommon).

Because of the freeze-concentration effect, the unfrozen phase changes significantly in properties such as pH, titratable acidity, ionic strength, viscosity, freezing point (and all other colligative properties), surface and interfacial tension, and oxidation–reduction potential. In addition, solutes sometimes crystallize, supersaturated oxygen and carbon dioxide may be expelled from solution, water structure and water–solute interactions may be drastically altered, and macromolecules will be forced closer together, making interactions more probable. These changes in concentration-related properties often favor increases in reaction rates. Thus, freezing can have two opposing effects on reaction rate: lowering temperature, as such, will always decrease reaction rates, and freeze-concentration, as such, will sometimes increase reaction rates. It should not be surprising, therefore, that reaction rates at subfreezing temperatures do not conform well to either Arrhenius or WLF kinetics, and that these deviations sometimes can be very large. In fact, it is not uncommon to find reactions that accelerate during freezing [22,23].

With this background, specific examples of freezing and the importance of Mm to the stability of frozen foods can be presented. Slow freezing of a complex food will be considered first. Very slow freezing results in close conformance to solid–liquid equilibrium and maximal freeze-concentration. Starting at A in Figure 35, removal of sensible heat moves the product to B, the initial freezing point of the sample. Because nucleation is difficult, further removal of heat results in undercooling and nucleation begins at point C. Nucleation is immediately followed by crystal growth, release of latent heat of crystallization and a rise in temperature to D. Further

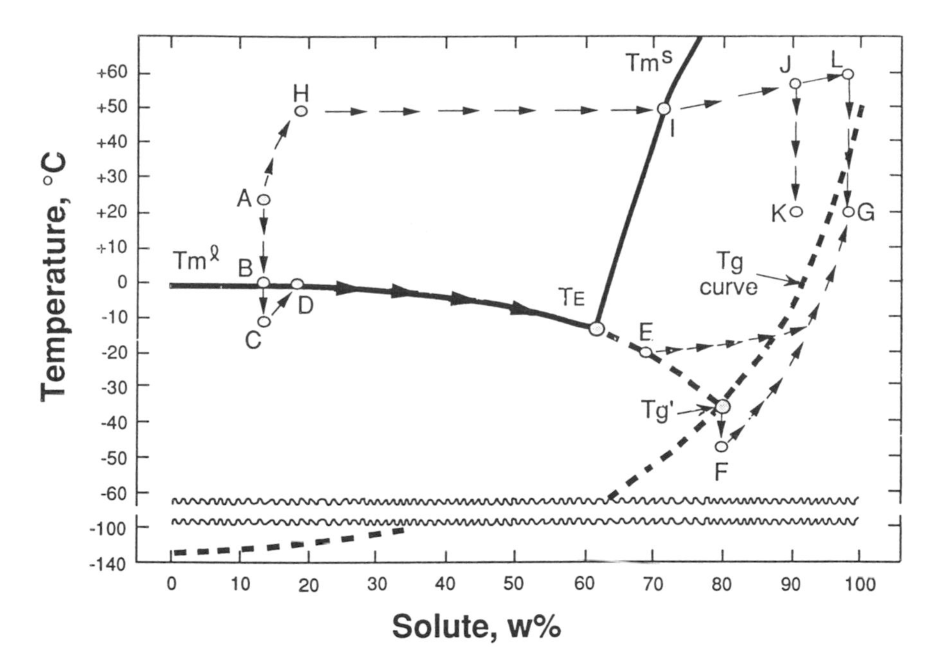

FIGURE 35 State diagram of a binary system showing possible paths for freezing (unstable sequence ABCDE; stable sequence ABCDET_g'F), drying (unstable sequence AHIJK; stable sequence AHIJLG) and freezing drying (unstable sequence ABCDEG; stable sequence ABCDET_g'FG). Temperatures during drying are lower than would occur in practice to facilitate entry of data on the graph. Assumptions stated in Figure 25 apply here.

removal of heat causes additional ice formation, concentration of the unfrozen phase, depression of its freezing point, and alteration of its composition in conformance with path D to T_E. The T_E in the complex food being considered represents T_{Emax} for the solute with the highest eutectic point (temperature at which saturation of the least soluble solute is achieved). Solutes in complex frozen foods seldom crystallize at or below their eutectic points. An occasional exception of commercial importance is formation of a lactose eutectic in frozen desserts. This results in a textural defect known as "sandiness."

Assuming eutectics do not form, further ice formation leads to metastable supersaturation of many solutes (an amorphous liquid phase) and compositional changes of the unfrozen fraction in accord with path T_E to E. Point E is the recommended storage temperature (–20°C) for most frozen foods. Unfortunately, point E is above the glass transition temperature of most foods, indicating that Mm will be moderately intense and the food's diffusion-limited physical and chemical properties will be relatively unstable and highly temperature-dependent. Exact conformance with WLF kinetics should not be expected because freeze-concentration effects during cooling, and melt-dilution effects during warming, are not accounted for in the WLF equation.

If cooling is continued below point E, additional ice formation and freeze concentration occur, causing composition of the unfrozen fraction to change from that at E to that at T_g' [77a,97,102,111]. At T_g', most of the supersaturated unfrozen phase converts to a glass, encompassing the ice crystals. The T_g' is a quasi-invariant T_g that applies only to the maximally freeze-concentrated unfrozen phase. The observed T_g' depends primarily on solute composition of the sample and secondarily on initial water content of the sample (T_g is strongly dependent on both solute composition and water content). Observed T_g' is not totally invari-

TABLE 9 Glass Transition (T_g') and DE Values for Commercial Starch Hydrolysis Products (SHP)

SHP	Manufacturer	Starch source	T_g'	DE
Staley 300	Staley[a]	Corn	–24	35
Maltrin M250	GPC[b] (1982)	Dent corn	–18	25
Maltrin M150	GPC	Dent corn	–14	15
Paselli SA-10	Avebe[c]	Potato (Ap)	–10	10
Star Dri 5	Staley (1984)	Dent corn	–8	5
Crystal gum	National[d]	Tapioca	–6	5
Stadex 9	Staley	Dent corn	–5	3.4
AB 7436	Anheuser-Busch	Waxy maize	–4	0.5

[a]A. E. Staley Manufacturing Co.
[b]Grain Processing Corp.
[c]Avebe America.
[d]National Starch and Chemical.
Source: From Ref. 114.

ant because maximum information is seldom obtained during procedures typically used for its determination.

Further cooling causes no further freeze-concentration, simply removal of sensible heat and alteration of product temperature in the direction of point F. Below T_g', Mm is greatly reduced and diffusion-limited properties usually exhibit excellent stability.

Some T_g' values for starch hydrolysis products, amino acids, proteins, and foods are listed in Tables 9–11. These values should be regarded as "observed" or "apparent" in T_g' values because maximum ice formation would be almost impossible under the measurement circumstances employed. These observed T_g' values are, however, probably of greater relevance to practical situations than the true (somewhat lower) T_g' values. The range of T_g' values for foods, and the variation of T_g' with location in a tissue, should be noted. More than one T_g' can occur in a product when, for example, the product contains a major chemical constituent that exists in two conformational forms, or when different domains in the product contain different ratios of macromolecules to small-solute molecules. In this instance the highest T_g' is usually considered most important.

Because most fruits have very low T_g' values, and storage temperature is typically $> T_g'$, texture stability during frozen storage is often poor. One would expect that vegetables, with T_g' values that are typically quite high, would exhibit storage lives that are longer than those of fruits. This is sometimes but not always true. The quality attribute that limits the storage life of vegetables (or any kind of food) can differ from one vegetable to another, and it is likely that some of these attributes are influenced less by Mm than others.

The T_g' values for fish (cod, mackerel) and beef in Table 11 were determined in 1996 and they differ markedly from earlier data (cod, –77°C [84]; beef, –60°C [91]). The earlier data are probably in error because the dominance of large protein polymers in muscle should result in T_g' values similar to those of other proteins (Table 10). Based on the muscle T_g' values in Table 11, one would be expect (as is generally observed) that all physical changes and all chemical changes that are diffusion limited would be effectively retarded during typical commercial frozen storage. Because storage lipids exist in domains separate from that of myofibrillar proteins, they probably are not protected by a glassy matrix during frozen storage and typically exhibit instability.

Values of W_g' for several solutes are shown in Tables 8, 10, and 11, but these values are subject to some uncertainty. W_g' values determined recently using altered techniques tend to be smaller than earlier values (mainly those of Slade and Levine). Consensus on what type of

TABLE 10 Glass Transition (T_g') Values and Associated Properties of Amino Acids and Proteins[a]

Substance	MW	pH (20wt%)	T_g' (°C)	W_g' (wt%)	W_g' (g UFW/g dry AA)[b]
Amino acids					
Glycine	75.1[c]	9.1	–58	63	1.7
DL-Alanine	89.1[d,e]	6.2	–51		
DL-Threonine	119.1[d]	6.0	–41	51	1.0
DL-Aspartic acid	133.1[c]	9.9	–50	66	2.0
DL-Glutamic acid · H_2O	147.1[c]	8.4	–48	61	1.6
DL-Lysine · HCl	182.7[d]	5.5	–48	55	1.2
DL-Arginine · HCl	210.7[d]	6.1	–44	43	0.7
Proteins					
Bovine serum albumin			–13	25–31	0.33–0.44
α-Casein			–13	38	0.6
Collagen (bovine, Sigma C9879)[f]			–6.6 ± 0.1		
Caseinate, sodium			–10	39	0.6
Gelatin (175 bloom, pigskin)			–12	34	0.5
Gelatin (300 bloom, pigskin)			–10	40	0.7
Gluten (Sigma, wheat)			–7	28	0.4
Gluten ("vital wheat gluten," commercial sample)			–5 to –10	7–29	0.07–0.4

[a] W_g' is unfrozen water existing in sample at T_g'.
[b] UFW is unfrozen water.
[c] Solubilized with NaOH.
[d] "As is" pH.
[e] Undergoes solute crystallization.
[f] From N. Brake and O. Fennema (unpublished). Determined by DSC; scanned from –60°C to 25°C at 5°C/min after tempering the sample for 1 hr at –10°C. Mean of 2 replicates ± SD.

Source: Ref. 118.

TABLE 11 Glass Transition (T_g') Values of Foods[a]

Food	T_g' (°C) [W_g' (wt%)]	Food	T_g' (°C) [W_g' (wt%)]
Fruit juices		Supermarket "fresh"	−8
Orange (various	−37.5 ± 1.0	Blanched	−10
samples)		Potato, Russet Burbank, fresh	−12
Pineapple	−37	Cauliflower, frozen stalk	−25
Pear	−40	Pea, frozen	−25
Apple	−40	Green bean, frozen	−27
Prune	−41	Broccoli, frozen	
White grape	−42	Stalk	−27
Lemon (various samples)	−43 ± 1.5	Head	−12
Fruits, fresh		Spinach, frozen	−17
Strawberry		**Frozen desserts**	
Sparkleberry, center	−41	Ice cream, vanilla	
Sparkleberry, edge	−39 and −33	Three commercial	−31 to −33
Sparkleberry,	−38.5 and −	brands	[32–37]
intermediate	33 [7]	Ice milk, vanilla,	−30 to −31
Other cultivars	−33 and −41	soft serve	[28–45]
Blueberry	[16–24]	**Cheese**	
Flesh	−41	Cheddar	−24
Skin	−41 and −32	Provolone	−13
Peach	−36	Cream cheese	−33
Banana	−35	**Fish**	
Apple		Cod muscle[b,c]	−11.7 ± 0.6
Red Delicious	−42	Cod muscle, water	−6.3 ± 0.1
Granny Smith	−41	insoluble fraction[b,d]	
Tomato, fresh, flesh	−41	Mackerel muscle[b,c]	−12.4 ± 0.2
Vegetables, fresh or frozen		Mackerel muscle, water	−7.5 ± 0.4
Sweet corn		insoluble fraction[b,d]	
Garden fresh, endosperm	−15	**Beef muscle**[b,c]	−12.0 ± 0.3

[a]From Levine and Slade [65, 66], unless otherwise indicated. W_g' is unfrozen water existing in sample at T_g'.
[b]From N. Brake and O. Fennema (unpublished). Determined by DSC; scanned from −60°C to 25°C at 5°C/min after annealing for 1 hr at −15°C.
[c]Means of 4 replicates ± SD.
[d]Means of 2 replicates ± SD.

measurement yields W_g' values most relevant to food stability has not yet been achieved [39,92,97]. It is important to note, however, that the Slade and Levine values reported here were determined using initial sample compositions close to those of high-moisture foods, and this has an important influence on the value obtained.

Two points need to be made about the term "unfrozen" as used in the definition of W_g'. First, unfrozen refers to a practical time scale. The unfrozen fraction will decrease somewhat over very long periods of time because water is not totally immobile at T_g' and equilibrium between the unfrozen phase and the glass phase is a metastable equilibrium, not a global one (not lowest free energy). Second, the term "unfrozen" has often been regarded as synonymous with "bound" water; however, bound water has been defined in so many other ways that the term has fallen into disrepute. A significant amount of W_g' water is engaged in interactions, mainly hydrogen bonds, that do not differ significantly in strength from water–water hydrogen bonds.

This water is unfrozen simply because local viscosity in the glass state is sufficiently great to preclude, over a practical time span, the translational and rotation motions required for further ice and solute crystallization (formation of eutectics). Thus, most of the W_g' water should be regarded as metastable and severely "hindered" in mobility.

Even though foods are frozen commercially at relatively slow rates (a few minutes to about 1 hr to attain –20°C) compared to typical rates for small samples of biological materials, maximal freeze concentration is unlikely. Increased rates of freezing affect the temperature–composition relationship, as shown schematically in Figure 36. This leads to the obvious question as to what is the appropriate reference temperature for foods frozen under commercial conditions—T_g or T_g'? This is another area of disagreement. Slade and Levine [114] argue that T_g' is the appropriate value. However, choice of this value is open to question because initial T_g (immediately following freezing) will always be $< T_g'$ and the approach of T_g to T_g' during frozen storage (caused by additional ice formation) will be slow and probably incomplete.

The choice of initial T_g as the reference T_g, as some have suggested, is also open to question because (a) initial T_g is influenced not only by product type but also by freezing rate [43], and (b) initial T_g does not remain constant with time of frozen storage, but rather it increases at a commercially important rate at storage temperatures in the T_m–T_g zone, and more slowly, but at a significant rate, at storage temperatures $< T_g$ [10,93,101]. The same considerations apply to W_g and W_g'.

Unfortunately, the important matter of selecting an appropriate reference T_g for frozen foods cannot be resolved unambiguously because appropriate data are not available. In the meantime, the best that can be done is to suggest that T_g' be regarded as a temperature zone rather than as a specific temperature. The lower boundary of the zone will depend on freezing rate and time/temperature of storage, but in commercially important situations it is reasonable

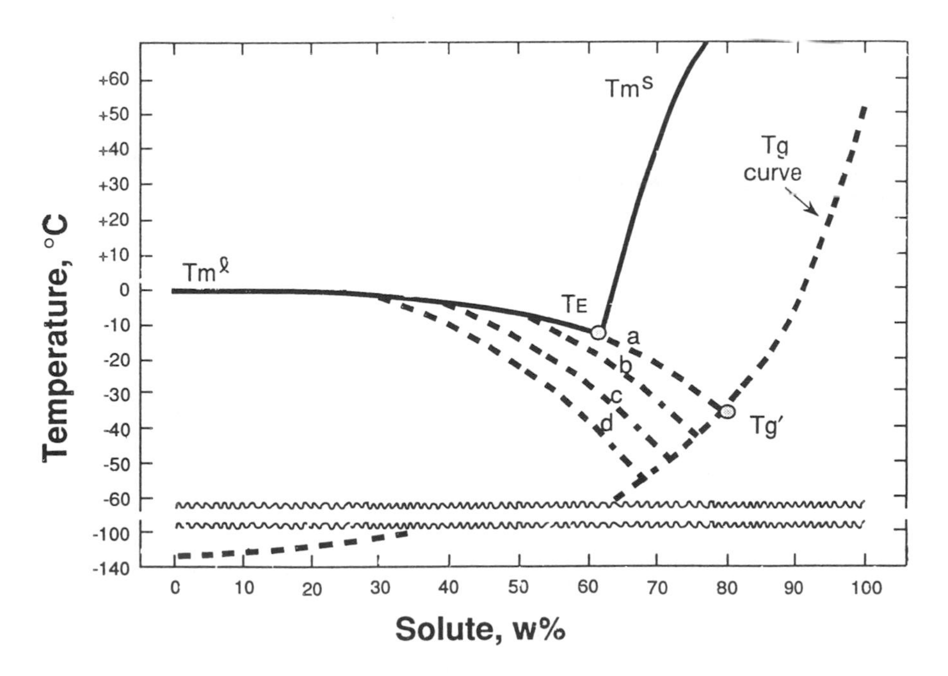

FIGURE 36 State diagram of a binary system showing the effects of increasing rates of freezing (rate a < b < c < d) on T_g. The T_m^l–T_E curve is the only one where maximal freeze-concentration occurs. Assumptions stated in Figure 25 apply here.

to suggest that this boundary (initial T_g) probably will not extend below $T_g' - 10°C$. The mean T_g for a food product marketed through retail channels, because of the relatively high mean storage temperature, is likely to be closer to T_g' than to initial T_g [43,101]. The term T_g' will continue to be used here with the understanding that it should be regarded as a temperature zone.

Several additional points regarding rates of diffusion-limited events in frozen foods deserve mentioning:

1. Product stability (diffusion-limited) can be increased by (a) lowering the storage temperature closer to, or preferably below, T_g', and/or (b) raising T_g' by incorporating high-molecular weight-solutes in the product. The latter is beneficial because it increases the probability that the storage temperature will be below T_g', and it reduces Mm at any given product temperature above T_g'.
2. The rate of recrystallization is explicitly related to T_g' (recrystallization refers to an increase in the mean size of ice crystals and a simultaneous decrease in their number). Provided maximum ice crystallization has occurred, the critical temperature of recrystallization (T_r) is the highest temperature at which ice recrystallization can be avoided (often $T_r \sim T_g'$; see Fig. 37). Rates of recrystallization in zone T_m to T_g' sometimes conform reasonably well to WLF kinetics (Fig. 37). If ice crystallization is not maximal, then the highest temperature at which ice recrystallization can be avoided is approximately equal to T_g. Generally, T must be somewhat greater than T_g' or T_g for rates to become of practical significance.
3. In general, high T_g' (achieved by adding water-soluble macromolecules) and low W_g' values are associated with firm frozen texture and good storage stability at a given

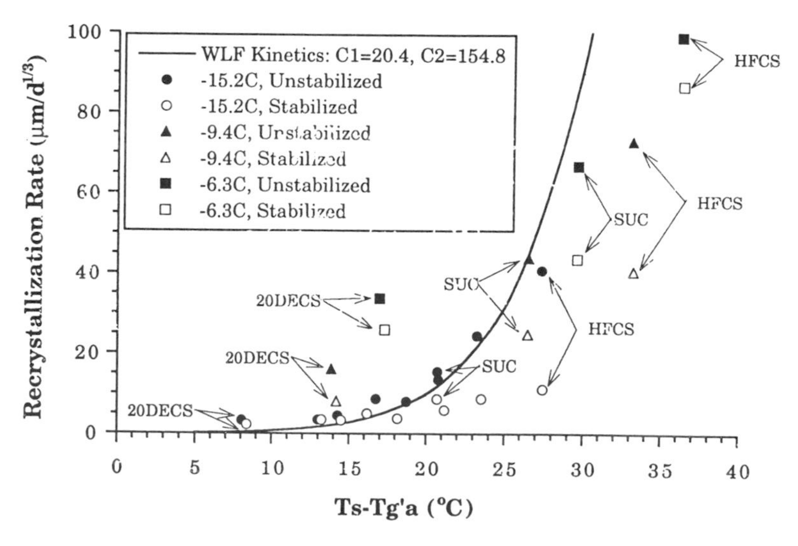

FIGURE 37 Rate of ice recrystallization in ice cream as a function of storage temperature (T_s), type of sweetener, and presence or absence of stabilizer. The solid curve is plotted from the WLF equation based on a nominal recrystallization rate of 30 μm/day$^{1/3}$ at $\Delta T = 25°C$. T_g' is "apparent" T_g'. HFCS is high-fructose corn syrup, SUC is sucrose, CS is corn syrup, DE is dextrose equivalent, d is days, and numerals in the body of the figure refer to dextrose equivalent. When single sweeteners were used they are indicated by arrows. Symbols without arrows represent ice creams made from 1:1 mixtures of two sweeteners selected from the sweeteners shown and 42DECS. Stabilizer, when used, was a 20–80 blend of carrageenan and locust bean gum at a combined concentration of 0.1 wt%. (From Ref. 37.)

subfreezing storage temperature. Conversely, low T_g' and high W_g' values (achieved by adding monomeric substances) are associated with soft frozen texture and relatively poor storage stability [65].

2.11.5 Technological Aspects: Air Drying

The paths (temperature–composition) of a product during air dehydration at a constant air temperature can also be followed on Figure 35. Indicated temperatures are lower than those used commercially simply to allow entry of data on the standardized state diagram used here. The example chosen for discussion is a complex food, and its $T_m{}^s$ curve is based on a food component that has a dominating influence on the location of this curve. Starting at point A, air drying will elevate product temperature and remove moisture, causing the product soon to acquire properties commensurate with point H (wet bulb temperature of air). Further moisture removal causes the product to arrive at and pass through I, the saturation point of the dominating solute (DS) present, with little or no solute crystallization (none is assumed). This sequence results in creation of major regions of liquid amorphous DS, in addition to smaller regions of liquid amorphous substances that may already have been present because of minor solutes with saturation temperatures higher than that of the DS. As drying continues to point J, product temperature approaches the dry bulb temperature of the air. If drying is terminated at point J and the product is cooled to point K then the product will be above the glass transition curve, Mm will be comparatively intense, and stability of diffusion limited properties will be relatively poor and strongly temperature-dependent (WLF kinetics). Alternatively, if drying is continued from point J to L and the product is then cooled to G, it will be below the T_g curve, Mm will be greatly subdued, and diffusion-limited properties will be stable and only weakly temperature-dependent.

2.11.6 Technological Aspects: Vacuum Freeze-Drying (Lyophilization)

Product paths during vacuum freeze-drying can also be followed on Figure 35. The first stage of freeze-drying coincides fairly closely with the path for slow freezing, ABCDE. If product temperature is not allowed to go below temperature E during ice sublimation (primary freeze drying), path EG would be a typical path. The early part of EG would involve ice sublimation (primary drying), during which collapse of the product cannot occur because of the presence of ice crystals. However, at some point along path E to G, ice sublimation is completed and the desorption period (secondary phase) ensues. This can (and often does) occur before the product passes the glass transition curve. Collapse during this phase of freeze drying is likely, not only for products that were initially fluid, but also to a lesser degree for food tissues. Collapse is likely because no ice is present to provide structural support and product T is $> T_g$ so Mm is sufficient to preclude rigidity. This scenario is not uncommon during freeze-drying of food tissues and it results in less than optimum product quality. Product collapse results in decreased product porosity (slower drying) and poorer rehydration characteristics. To prevent collapse, path ABCDEFG must be followed.

Provided maximum ice crystallization has occurred, the critical temperature for structural collapse (T_c) is the highest temperature at which collapse can be avoided during the primary stage of freeze drying ($T_c \sim T_g'$). The T_c values for some carbohydrates are shown in Table 8 [49,87,102]. If ice crystallization is not maximal then the highest temperature at which collapse can be avoided during primary freeze drying is approximately equal to T_g. Generally, T must be somewhat greater than T_g' or T_g for rates to become of practical significance.

If a product's composition can be altered, it is desirable to raise T_g' (T_c) and lower W_g' as much as possible. This can be accomplished by the addition of high-molecular-weight polymers and will enable higher freeze-drying temperatures to be used (less energy, greater rate of drying) without danger of product collapse [31].

2.11.7 Technological Aspects: Other Applications of the Mm Approach (Partial Listing)

1. Crystallization. Knowledge of T_g and T_m/T_g ratios allows accurate predictions of whether nucleation of solutes will occur, and if it does, the rate at which subsequent crystal growth will occur over the zone of T_m to T_g (or T_g') [97,102,104,114].
2. Enzyme activity [114]. The rates of many enzyme-catalyzed interactions are diffusion-limited, and in these instances, reaction rates are greatly slowed at $T < T_g$ or $T < T_g'$.
3. Thermal resistance of bacterial spores. Thermal inactivation of spores of *Bacillus stearothermophilus* occurs in accord with WLF kinetics, and higher T_g values are associated with greater resistance [105].
4. Color of gluten. For heat-treated gluten, plots of T_g versus moisture content (0–15%) and "critical temperature for color change" versus moisture content are virtually identical [32].
5. Other miscellaneous properties. Valuable information about gelation [97,114], caking [87], sticky point [87,97], and texture, including texture of bakery goods [67,82,114], can be obtained from knowledge of T_g and state diagrams.

2.11.8 Technological Aspects: Estimation of Relative Shelf Life

This topic will be dealt with in only a cursory manner. Shown on the temperature–composition state diagram in Figure 38 are zones of differing product stability. The reference line is derived from the T_g curve in the absence of ice and the T_g' zone in the presence of ice. Below this line (zone) physical properties are generally quite stable, and the same is true of those chemical properties for which stability is diffusion-limited. Above this line (zone) and below the intersecting curves for T_m^l and T_m^s, physical changes often conform to WLF kinetics. Product stability declines greatly as product conditions move upward or to the left in the WLF zone. This occurs with increases in product temperature and/or increases in moisture content. Above the T_m curves, attributes that depend on diffusion (Mm) are relatively unstable and become more so with movement to the upper left corner of the graph.

It should be reemphasized that storage below T_g or T_g' is highly desirable because this will stabilize food properties that are diffusion-limited. However, this is not an all-or-none situation. When storage temperatures $< T_g$ or $< T_g'$ are not feasible, minimizing temperature deviations above T_g or T_g' will greatly help.

2.11.9 Technological Aspects: Relationship of T_g and Mm to Relative Vapor Pressure (p/p_o) and Moisture Sorption Isotherms

It should come as no surprise that a product-specific, consistent relationship exists between T_g and p/p_o (RVP). This arises logically from two other well-established relationships: that of water content versus T_g, and water content versus p/p_o (moisture sorption isotherm, MSI). Plots of T_g versus p/p_o exhibit a large central zone of linearity and curvilinear tails. Examples for several maltodextrins are shown in Figure 39 [97].

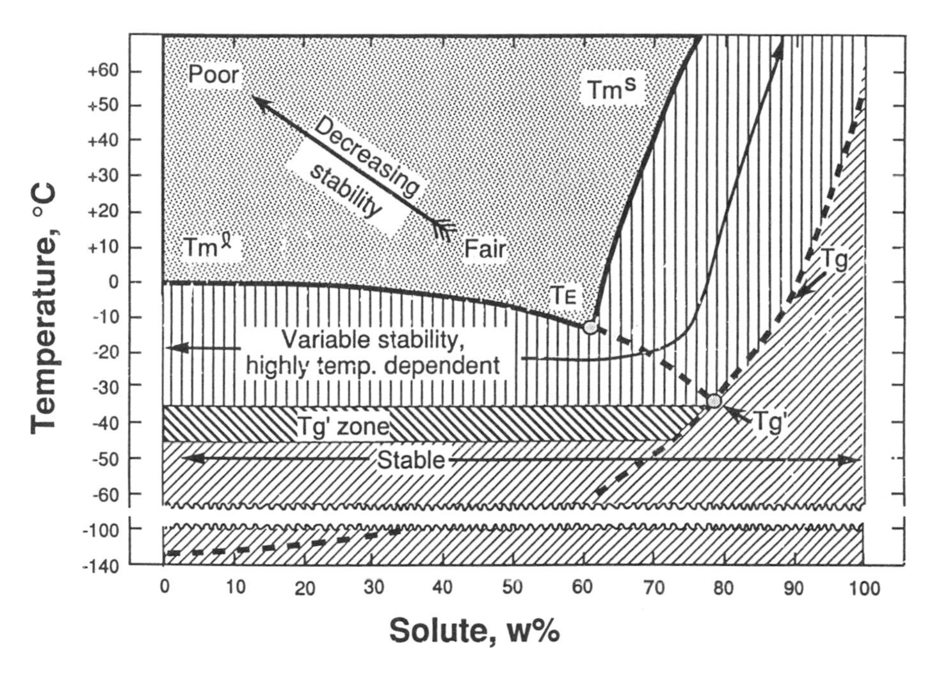

FIGURE 38 State diagram of a binary system showing stabilities of diffusion-dependent properties of food. It is assumed that the "T_g' zone" is the appropriate T_g for frozen foods, but this is still a matter of some disagreement. Other assumptions stated in Figure 25 apply here.

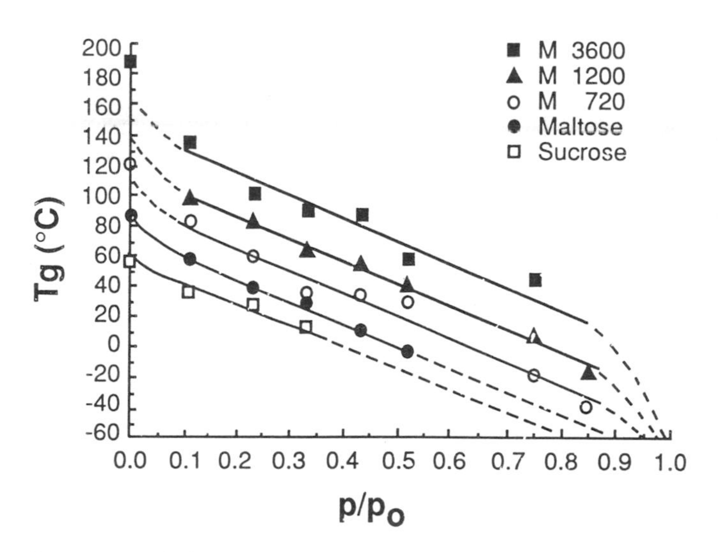

FIGURE 39 Relationship between calculated glass transition temperature and p/p_o (25°C) for several carbohydrates of differing molecular weight. M is maltodextrin and numerals following M are molecular weight. T_g values in the linear range are calculated from the equation of Fox and Flora $T_g = T_g(\infty) - 1/M$, where $T_g(\infty)$ is the T_g value of highest molecular weight substance. (From Ref. 100.)

Of greater interest is an isotemperature comparison between the p/p_0 value required to produce T_g (RVP_{Tg}) and the p/p_0 at the BET monolayer value (RVP_{mono}). Some data relating RVP_{Tg} and RVP_{mono}, both at 25°C, are shown in Figure 40 (note: maltodextrins were prepared using ultrapure water). In these examples, RVP_{Tg} and RVP_{mono} are never the same (note, however, that values for T_g and monolayer moisture are zones rather than precise values). Assuming that future studies establish the general applicability of these results, three conclusions can be suggested: (a) RVP_{Tg} tends to be greater than RVP_{mono} in samples containing primarily water and macromolecules, whereas the reverse tends to be true in samples containing water and solutes with molecular weights that are small and/or of mixed size; (b) because RVP_{mono} is often a good predictor of the RVP at which many chemical reactions attain a practical minimum rate during drying (see Fig. 23) and because RVP_{Tg} differs from the monolayer RVP, it follows that T_g is not a reliable indicator of this practical rate minimum for many chemical reactions (e.g., see Fig. 31); and (c) because T_g is often a good indicator of the point at which rates of diffusion-limited events become highly sensitive to temperature during sample warming or hydration, and because RVP_{Tg} and RVP_{mono} often differ, it follows that RVP_{mono} is not a reliable indicator of the critical point for diffusion-limited events.

2.11.10 Summary Statements Regarding the Mm Approach to Food Stability

1. Molecular mobility, Mm, as reflected by T_g, the magnitude of T_m–T_g, and the product location in the WLF zone T_m–T_g (i.e., deviations in product T above T_g, and W above W_g), provides a promising and potentially powerful tool for assessing the stability of important properties of food that are diffusion-dependent.
2. In general, the original diffusion-dependent properties of food are well retained during storage at $T < T_g$ (or T_g').

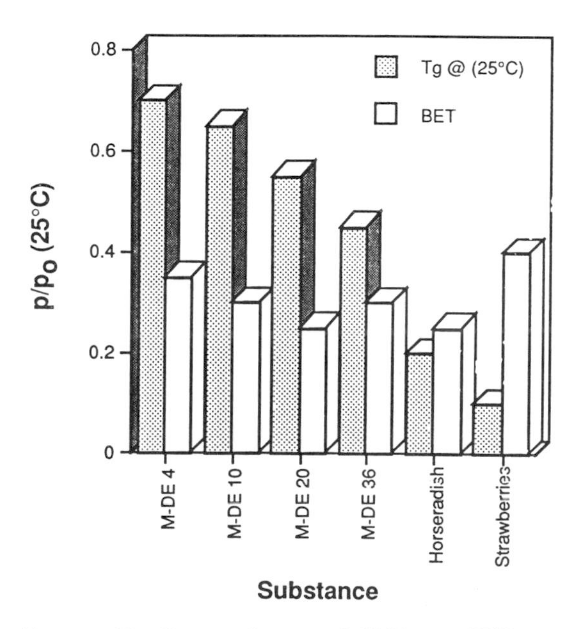

FIGURE 40 Comparisons of RVPs at 25°C required to produce a T_g of 25°C or a BET monolayer moisture value. M-DE is maltodextrin-dextrose equivalent. (Data from Ref. 98.)

3. The two key constituents with respect to Mm are water and the dominant solute or solutes (solute or solutes comprising the major fraction of the solute portion).
4. A solute's ability to reduce Mm usually varies inversely with (a) molecular weight (macromolecules effectively decrease Mm and raise T_g; water has the reverse effect) and (b) the amount of water, W_g', associated with the solute at T_g' (or W_g at T_g).
5. As compared to the RVP approach, the Mm approach appears (a) to be significantly more useful for estimating rates of occurrences that are diffusion-limited, such as the physical properties of frozen foods (RVP is not useful for predicting either physical or chemical properties of frozen foods), optimum conditions for freeze drying, and physical changes involving crystallization, recrystallization, gelatinization and starch retrogradation, (b) to be about equally useful for estimating conditions causing caking, stickiness, and crispness in products stored at near-ambient temperatures, and (c) to be significantly less useful or unreliable, in products free of ice, for estimating growth of vegetative microorganisms [12,13] and for estimating rates of chemical reactions that are not diffusion-limited (inherently slow reactions, e.g., those with high activation energies, and those occurring in a relatively low-viscosity medium such as high-moisture foods).
6. The Mm approach will not be brought to a level of usefulness that equals or exceeds that of the p/p_0 approach until methods that are rapid, accurate, and economical are developed for determining Mm and glass transitions in foods.

2.12 COMBINED METHODS APPROACH TO FOOD STABILITY

At this point, it is hoped that the reader will have realized that RVP and Mm are useful estimators of food stability but that neither is totally sufficient by itself. It has been mentioned several times that factors not accommodated by either of these approaches can have important influences on food stability and safety. The "combined methods approach" to controlling microbial growth in foods was developed specifically to deal with this reality. This approach is mentioned in a book on food chemistry because it demonstrates convincingly that (a) the RVP approach to controlling microbial growth is often inadequate when used alone, (b) the RVP approach, because it is based on a single parameter, is not a totally reliable predictor of chemical stability, and (c) the Mm approach, because it too is based on a single parameter, is unlikely to be a totally reliable predictor of chemical stability.

The combined methods approach (originally called "hurdle approach") was developed by Professor L. Leistner and others to determine conditions needed to limit the growth of microorganisms in nonsterile foods [57,58,59]. This approach involves manipulating various growth-controlling parameters in a manner such that growth will not occur; each parameter is a "hurdle" to microbial growth. The approach is best illustrated by several examples in Figure 41. The dashed, undulating line is meant to represent the progress of a microorganism in its attempt to overcome the inhibitory hurdles, with growth occurring only after all hurdles have been overcome. Size of the hurdle indicates relative inhibitory effectiveness. Obviously, a microorganism in real life would confront all hurdles simultaneously rather than in sequence as is shown, and some of the factors would act synergistically.

In example 1, six hurdles are present and growth is satisfactorily controlled because the microorganisms are unable to overcome all hurdles. Example 2 is a more realistic one in which a typical microbial population is present and the various hurdles differ in their inhibitory effectiveness, with RVP and preservatives being the most potent ones. This example also results

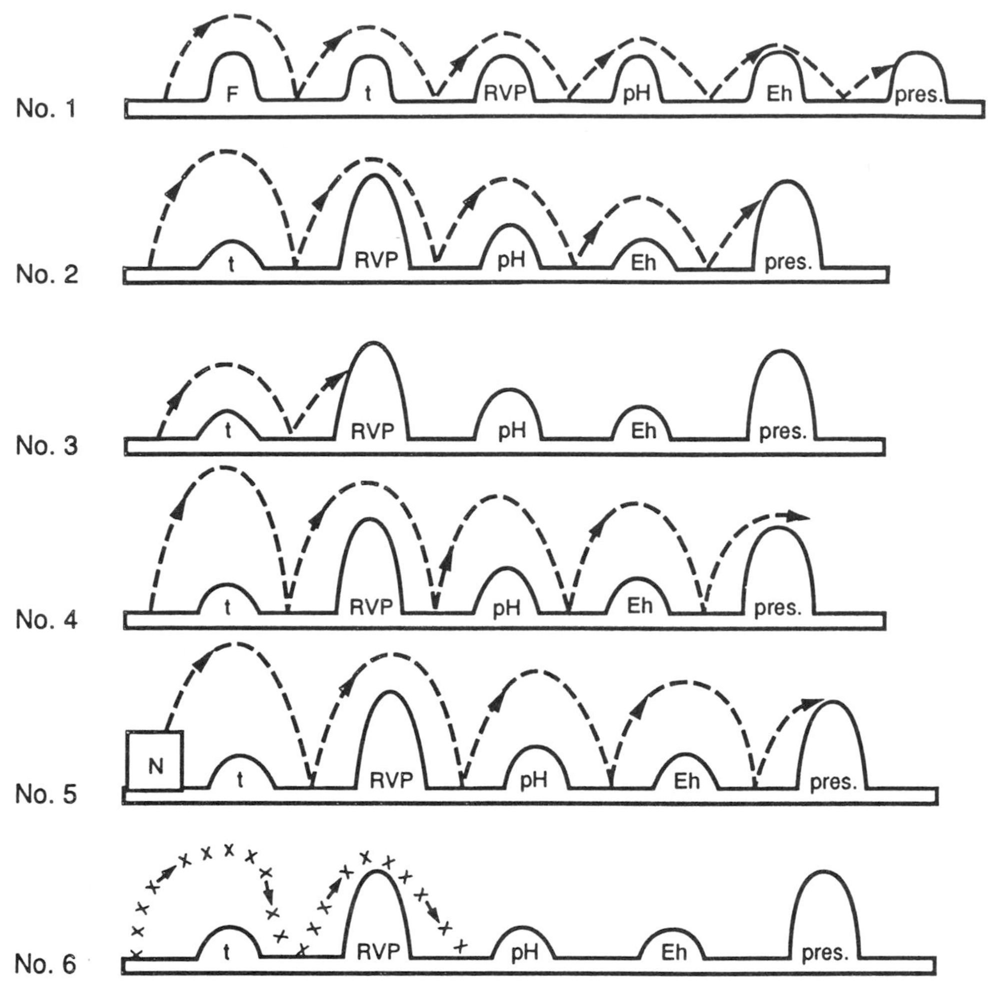

Figure 41 The combined methods approach to controlling growth of microorganisms in nonsterilized food. F is heating, t is chilling, RVP is relative vapor pressure, pH is acidification, E_h is redox potential, pres. is chemical preservative, and N is nutrients. (Adapted from Ref. 58.)

in satisfactory control of microbial growth. Example 3 represents the same product and the same hurdles with a small starting population of microorganisms that result from good sanitary practices. In this example, fewer hurdles suffice. Example 4 represents the same product and the same hurdles with a large starting population of microorganisms that result from poor sanitary practices. Here, the hurdles are insufficient to provide satisfactory control of microbial growth. Example 5 represents the same hurdles and population of organisms as in Example 2, but in this instance the sample is rich in nutrients. Because of the nutrients, the hurdles that were adequate in Example 2 now are inadequate. In example 6, the product and hurdles remain unchanged but the microorganisms are given a substerilizing treatment before storage. The surviving but damaged organisms are less able to overcome the hurdles and fewer hurdles suffice.

The lesson to be learned is this: Although RVP and Mm are powerful tools for predicting and controlling the properties and stability of food, there are many occasions when neither is sufficient alone and other factors, such as chemical properties of the solute, pH, and oxidation–reduction potential, must be also considered.

2.13 CONCLUDING COMMENTS ABOUT WATER

Water is typically the most abundant constituent in food, is of critical importance to the desirable qualities of food, is the cause of food's perishable nature, is a rate determinant of many chemical reactions, both desirable and undesirable, is a strong causative agent of undesirable side effects during freezing, is associated with nonaqueous food constituents in ways so complex that once these associations are disturbed by drying or freezing they can never again be completely restored, and, above all else, is frustratingly complex in behavior, inadequately studied, and poorly understood.

GLOSSARY: MOLECULAR MOBILITY AND FOOD STABILITY

Amorphous This refers to a nonequilibrium, noncrystalline state of a substance. When saturation conditions prevail and a solute remains noncrystalline the supersaturated solution can be regarded as amorphous. An amorphous solid is generally called a glass and is characterized by a viscosity of greater than about 10^{12} Pa sec.

Collapse This refers to both visibly evident collapse, such as the collapse of foods during freeze drying, and to collapse at the molecular level that consists of conversion from a nonequilibrium state to a lower state of free energy (relaxation). The kinetics of collapse is governed by molecular mobility, Mm, of the system.

Eutectic temperature (T_E) An invariant point on a temperature–composition phase diagram of a binary solution where solution can exist in equilibrium with both crystalline solute and crystalline solvent. Under equilibrium conditions, cooling at T_E results in simultaneous crystallization of solvent and solute in constant proportion and at constant temperature until maximum solidification has occurred. The T_E is, therefore, the highest temperature at which maximum crystallization can (but usually does not) occur.

Free volume Free volume is space not occupied by molecules. A useful mental image can be created by imagining a container filled with the maximum number of bees that can be accommodated in hovering flight. "Filled" means no more bees can be accommodated in a hovering state, not that all space is occupied. The analogous space in liquids is called "free volume" and it accounts for the fact that liquids can be compressed if the pressure is great enough. Free volume can also be thought of as the "elbow room" molecules require to undergo vibrational, rotational and translational motions [41]. Both free volume and molecular mobility increase with increasing temperature. Temperature dependence of free volume is small below T_g, and large between T_m and T_g.

Glass state (glassy) A substance existing as an amorphous (noncrystalline) solid is said to be in a glassy state. The Stokes viscosity (local viscosity, not bulk viscosity) is appropriate for characterizing glasses, and this value, at the temperature of incipient glass formation (T_g), ranges from 10^{10} to 10^{14} Pa sec, depending on the solute. This viscosity is sufficient to reduce translational and rotational mobility of most large molecules to a point of practical insignificance. In complex, polymer-dominated systems, very small molecules, most notably water, retain significant translational and rotational mobility at temperatures well below sample T_g. Vibrational mobility, of course, does not cease until the temperature is reduced to absolute zero.

Glass transition temperature (T_g, T_g') The glass transition temperature, T_g, is the temperature at which a supersaturated solution (amorphous liquid) converts to a glass. This is a second-order transition involving a step change in specific heat at the transition temperature, allowing this transition to be measured by differential scanning calorimetry (first-order transitions involve changes in physical state among gases, liquids, and crystalline solids). T_g values are observed in substances that contain sizeable regions that are amorphous or partially amorphous (all food tissues and many other foods), regardless of whether they contain ice. For partially crystalline polymeric substances, only the amorphous regions exhibit a glass transition. The T_g is dependent on solute type and water content. T_g' is a special T_g that applies only to samples containing ice, and only when ice has been formed so maximum freeze-concentration occurs (very slow cooling). For a given solute, the observed T_g' is a quasi-invariant point on a temperature-composition state diagram (not invariant because maximum ice formation is extremely difficult to obtain during typical measurement techniques, thus causing the observed T_g' to drift slightly lower with extended storage time). Below T_g or T_g' of a complex sample, all but small molecules lose their translational mobility while retaining limited rotational and vibrational mobility.

Macromolecular entanglement This refers to the interaction of large polymers in a random fashion without chemical bonding and with or without hydrogen bonding. When entanglement of macromolecules is sufficiently extensive (requires a minimum critical concentration of the macromolecule and time), a visoelastic entanglement network forms. This type of amorphous network can be dispersed by dilution, and can exist in conjunction with microcrystalline gels, which cannot be dispersed by dilution.

Metastable state This refers to a state of pseudo-equilibrium or apparent equilibrium that is stable over practical times. A metastable state is not, however, the most stable equilibrium because it possesses free energy that is greater than that of the global equilibrium state under the same conditions of pressure, temperature and composition. A metastable state can exist—that is, conversion to a more stable equilibrium state of lower free energy will not occur—if the activation energy is sufficiently high to prevent conversion to an equilibrium state of lower free energy during the period of interest.

Molecular mobility (Mm) This refers to either translational or rotational motion of molecules (vibrational mobility is not of concern in the context of food stability).

Molecular weight, number average (MW_n)

$$MW_n = \frac{\Sigma n_i M_i}{\Sigma n_i}$$

where n is the number of molecules of a given molecular species and M is the molecular weight of the same species, with i kinds of molecules present.

Molecular weight, weight average (MW_w)

$$MW_w = \frac{\Sigma n_i M_i^2}{\Sigma n_i M_i}$$

Plasticizer A substance incorporated into a polymeric material to increase its deformability. A true solvent is always a plasticizer, but a plasticizer is not always a true solvent [106,117]. A plasticizer decreases the T_g of a polymer. Water is a highly effective plasticizer of hydrophilic, amorphous polymers. Its low molecular weight leads to increased free volume, decreased local viscosity, and increased Mm.

Relaxation Relaxation refers to passage from a nonequilibrium state to a more stable (lower free energy) state. This term is also used to indicate the decay of a stress.

Rubbery state A term used to describe the viscoelastic nature of large polymers in the temperature range T_m to T_g, that is, when the substance or a part of the substance is between the glassy and liquid states. Small polymers and solutes of low molecular weight that exist in this temperature zone are highly viscous but not elastic, and are not called rubbery [27].

Sate diagram A phase diagram supplemented with lines depicting boundaries of various nonequilibrium and metastable states. Such a diagram is sometimes called a "supplemented phase diagram."

Temperature of solute crystallization/dissolution (T_m^s) for simple and complex aqueous samples T_m^s is the highest temperature at which a crystalline solute can exist in equilibrium with an aqueous solution of a given composition. The highest temperature at which this can occur for a given sample is the sample's initial crystallization temperature. The T_m^s also can be thought of as the incipient saturation temperature during cooling, or the temperature at which the last of a crystalline solute, present in a saturated solution, melts or dissolves upon warming. On a temperature–composition phase diagram for a binary aqueous solution, T_m^s values for samples with differing temperatures constitute the T_m^s curve.

Temperature of ice melting/crystallization (T_m^l) for simple and complex aqueous samples This is the temperature at which ice can exist in equilibrium with an aqueous solution of a given composition. The highest temperature at which this can occur is the initial freezing point. On a temperature–composition phase diagram for a binary aqueous solution, T_m^l values for samples with differing initial ratios of solute to water constitute the T_m^l curve.

Vitrification Solidification of an entire sample as a glass, that is, no crystallization of solvents or solutes.

ACKNOWLEDGMENTS

This chapter was reviewed totally or in part by Larry Beuchat, Nicole Brake, Kenneth Connors, Theodore Labuza, Marcus Karel, George Zografi, David Reid, Pieter Walstra, Harry Levine, and Richard Hartel. The helpful suggestions of these respected colleagues and good friends are acknowledged with deep appreciation.

ABBREVIATIONS AND SYMBOLS

C_1C_2	Constants for the WLF equation
C_g	Solute concentration (wt%) existing in sample at T_g
C_g'	Solute concentration (wt%) existing in sample at T_g'
DE	Dextrose equivalent
DS	Dominating solute
ERH	Percent equilibrium relative humidity
Mm	Molecular mobility
MSI	Moisture sorption isotherm
MW_n	Number-average molecular weight (see glossary)
MW_w	Weight-average molecular weight (see glossary)
η	Viscosity of sample at temperature T

η_g	Viscosity of sample at T_g (or T_g')
p	Vapor pressure of the sample
p_o	Vapor pressure of pure water
T	Sample temperature
T_c	Collapse temperature
T_E	Eutectic temperature
T_g	Glass transition temperature of the sample
T_g'	Glass transition temperature of a maximally freeze-concentrated sample
T_m	Either T_m^l or T_m^s
T_m^l	Melting or freezing temperature of water in a solution
T_m^s	Crystallization or melting (dissolution) temperature of solute in a solution
T_r	Recrystallization temperature
W	Water content of sample (wt%)
W_g	Water (wt%) existing in sample at T_g
W_g'	Unfrozen water (wt%) existing in sample at T_g'
W_m (or m_1)	"Monolayer" water content

BIBLIOGRAPHY

Blanshard, J. M. V., and P. J. Lillford (1933). *The Glassy State in Foods,* Nottingham University Press, Loughborough, U.K.

Franks, F., ed. (1972–1982). *Water—A Comprehensive Treatise,* 7 vols., Plenum Press, New York.

Iglesias, H. A., and J. Chirife (1982). *Handbook of Food Isotherms: Water Sorption Parameters for Food and Food Components,* Academic Press, Boca Raton, FL.

Rockland, L. B., and L. R. Beuchat, eds. (1987). *Water Activity: Theory and Applications to Food,* Marcel Dekker, New York.

Rupley, J. A., and G. Careri (1991). Protein hydration and function. *Adv. Protein Chem. 41*:37–172.

Simatos, D., and J. L. Multon (1985). *Properties of Water in Foods in Relation to Quality and Stability,* Martinus Nijhoff, Dordrecht, The Netherlands.

Slade, L., and H. Levine (1991). Beyond water activity: Recent advances based on an alternative approach to the assessment of food quality and safety. *Crit. Rev. Food Sci. Nutr. 30*:115–360.

Troller, J. A., and J. H. B. Christian (1978). *Water Activity and Food,* Academic Press, New York.

REFERENCES

1. Ablett, S., M. J. Izzard, and P. J. Lillford (1992). Differential scanning calorimetric study of frozen sucrose and glycerol solutions. *J. Chem. Soc. Faraday Trans. 88*(6):789–794.
2. Anderson, R. B. (1946). Modifications of the Brunauer, Emmett and Teller equation. *J. Amer. Chem. Soc. 68*:686–691.

2a. Bell, L. N., and M. J. Hageman (1994). Differentiating between the effect of water activity and glass transition dependent mobility on a solid state chemical reaction: Aspartame degradation. *J. Agric. Food Chem. 42*:2398–2401.

3. Berendsen, H. J. C. (1971). The molecular dynamics of water in biological systems. *Proc. First European Biophysics Congress* 1:483–488.
4. Berendsen, H. J. C. (1975). Specific interactions of water with biopolymers, in *Water—A Comprehensive Treatise,* vol. 5 (F. Franks, ed.), Plenum Press, New York, pp. 293–349.
5. Bertoluzza, A., C. Fagnano, M. A. Morelli, A. Tinti, and M. R. Tosi (1993). The role of water in biological systems. *J. Molec. Struct. 297*:425–437.
6. Beuchat, L. R. (1981). Microbial stability as affected by water activity. *Cereal Foods World 26*(7):345–349.

7. Brady, G. W., and W. J. Romanov (1960). Structure of water. *J. Chem. Phys. 32*:106.
8. Brunauer, S., H. P. Emmett, and E. Teller (1938). Adsorption of gases in multi-molecular layers. *J. Am. Chem. Soc. 60*:309–319.
9. Bull, H. B. (1964). *An Introduction to Physical Biochemistry,* F. A. Davis, Philadelphia.
10. Chang, Z. H., and J. G. Baust (1991). Physical aging of glassy state: DSC study of vitrified glycerol systems. *Cryobiology 28*:87–95.
11. Chirife, J. (1994). Specific solute effects with special reference to *Staphylococcus aureus. J. Food Eng. 22*:409–419.
12. Chirife, J., and M. P. Buera (1994). Water activity, glass transition and microbial stability in concentrated and semimoist food systems. *J. Food Sci. 59*:921–927.
13. Chirife, J., and M. P. Buera (1995). A critical review of the effect of some nonequilibrium situations and glass transitions on water activity values of foods in the microbiological growth range. *J. Food Eng. 25*:531–552.
14. Connors, K. A. (1990). *Chemical Kinetics: The Study of Reaction Rates in Solution,* VCH, New York.
15. Davidson, D. W. (1973). Clathrate hydrates, in *Water—A Comprehensive Treatise.* vol. 2 (F. Franks, ed.), Plenum Press, New York, pp. 115–234.
16. DeBoer, J. H., ed. (1968). *The Dynamical Character of Adsorption,* 2nd ed., Clarendon Press, Oxford, UK, pp. 200–219.
17. Duckworth, R. B., J. Y. Allison, and H. A. A. Clapperton (1976). The aqueous environment for chemical change in intermediate moisture foods, in *Intermediate Moisture Foods* (R. Davies, G. G. Birch, and K. J. Parker, eds.), Applied Science, London, pp. 89–99.
18. Eagland, D. (1975). Nucleic acids, peptides and proteins, in *Water—A Comprehensive Treatise,* vol. 5 (F. Franks, ed.), Plenum Press, New York, pp. 305–518.
19. Edelhoch, H., and J. C. Osborne, Jr. (1976). The thermodynamic basis of the stability of proteins, nucleic acids, and membranes. *Adv. Protein Chem. 30*:183–250.
20. Eichner, K. (1975). The influence of water content on non-enzymic browning reactions in dehydrated foods and model systems and the inhibition of fat oxidation by browning intermediates, in *Water Relations of Foods* (R. B. Duckworth, ed.), Academic Press, London, pp. 417–434.
21. Fennema, O. (1973). Water and ice, in *Low-Temperature Preservation of Foods and Living Matter* (O. Fennema, W. D. Powrie, and E. H. Marth, eds.), Marcel Dekker, New York, pp. 1–77.
22. Fennema, O. (1975). Activity of enzymes in partially frozen aqueous systems, in *Water Relations of Foods* (R. B. Duckworth, ed.), Academic Press, London, pp. 397–413.
23. Fennema, O. (1975). Reaction kinetics in partially frozen aqueous systems, in *Water Relations of Foods* (R. B. Duckworth, ed.), Academic Press, London, pp. 539–556.
24. Fennema, O. (1978). Enzyme kinetics at low temperature and reduced water activity, in *Dry Biological Systems* (J. H. Crowe and J. S. Clegg, eds.), Academic Press, New York, pp. 297–322.
25. Fennema, O., and L. A. Berny (1974). Equilibrium vapor pressure and water activity of food at subfreezing temperatures, in *Proc. IVth Int. Congress of Food Science and Technology,* Madrid, Spain, *2*:27–35.
26. Ferro Fontan, C., and J. Chirife (1981). The evaluation of water activity in aqueous solutions from freezing point depression. *J. Food Technol. 16*:21–30.
27. Ferry, J. D. (1980). *Viscoelastic Properties of Polymers,* 3rd ed., John Wiley & Sons, New York.
28. Franks, F. (1975). The hydrophobic interaction, in *Water—A Comprehensive Treatise,* vol. 4 (F. Franks, ed.), Plenum Press, New York, pp. 1–94.
29. Franks, F. (1982). The properties of aqueous solutions at subzero temperatures, in *Water—A Comprehensive Treatise,* vol. 7 (F. Franks, ed), Plenum Press, New York, pp. 215–338.
30. Franks, F. (1988). Protein hydration, in *Characterization of Proteins* (F. Franks, ed.), Humana Press, Clifton, NJ, pp. 127–154.
31. Franks, F. (1989). Improved freeze-drying. An analysis of the basic scientific principles. *Proc. Biochem. 24*(1):iii–vii.
32. Fujio, Y., and J. -K. Lim (1989). Correlation between the glass transition point and color change of heat-treated gluten. *Cereal Chem. 66*:268–270.

33. Gal, S. (1981). Recent developments in techniques for obtaining complete sorption isotherms, in *Water Activity: Influence on Food Quality* (L. B. Rockland and G. F. Steward, eds.), Academic Press, New York, pp. 89–110.
34. Gordon, M., and J. S. Taylor (1952). Ideal copolymers and the second-order transitions of synthetic rubbers. I. Non-crystalline copolymers. *J. Appl. Chem. 2*:493–500.
35. Görling, P. (1958). Physical phenomena during the drying of foodstuffs, in *Fundamental Aspects of the Dehydration of Foodstuffs,* Society of Chemical Industry, London, pp. 42–53.
36. Guggenheim, E. A., ed. (1966). *Applications of Statistical Mechanics,* Clarendon Press, Oxford, UK, pp. 186–206.
37. Hagiwara, T., and R. W. Hartel (1996). *J. Dairy Sci.* (in press).
38. Hancock, B. C., and G. Zografi (1994). The relationship between the glass transition temperature and water content of amorphous pharmaceutical solids. *Pharm. Res. 11*:471–477.
39. Hatley, R. H. M., and A. Mant (1993). Determination of the unfrozen water content of maximally freeze-concentrated carbohydrate solutions. *Int. J. Macromol. 15*:227–232.
40. Haynes, J. T. (1985). The theory of reactions in solution, in *Theory of Chemical Reaction Dynamics,* vol. IV (M. Baer, ed.), CRC Press, Boca Raton, FL, p. 172.
41. Hiemenz, P. (1984). *Polymer Chemistry—The Basic Concepts,* Marcel Dekker, New York, NY.
42. Hoseney, R. C., K. Zeleznak, and C. S. Lai (1986). Wheat gluten: A glassy polymer. *Cereal Chem. 63*:285–286.
43. Izzard, M. J., S. Ablett, and P. J. Lillford (1991). Calorimetric study of the glass transition occurring in sucrose solutions, in *Food Polymers, Gels, and Colloids* (E. Dickinson, ed.), Royal Society of Chemistry, London, pp. 289–300.
44. Johnson, M. R., and R. C. Lin (1987). FDA views on the importance of a_w in good manufacturing practice, in *Water Activity: Theory and Applications to Food* (L. B. Rockland and L. R. Beuchat, eds.), Marcel Dekker, New York, pp. 287–294.
45. Kapsalis, J. G. (1981). Moisture sorption hysteresis, in *Water Activity: Influences on Food Quality* (L. B. Rockland and G. F. Steward, eds.), Academic Press, New York, pp. 143–177.
46. Kapsalis, J. G. (1987). Influences of hysteresis and temperature on moisture sorption isotherms, in *Water Activity: Theory and Applications to Food* (L. B. Rockland and L. R. Beuchat, eds.), Marcel Dekker, New York, pp. 173–213.
47. Karel, M. (1988). Role of water activity, in *Food Properties and Computer-Aided Engineering of Food Processing Systems* (R. P. Singh and A. G. Medina, eds.), Kluwer Academic, Dordrecht, The Netherlands, pp. 135–155.
48. Karel, M., and S. Yong (1981). Autoxidation-initiated reactions in food, in *Water Activity: Influences on Food Quality* (L. B. Rockland and G. F. Stewart, eds.), Academic Press, New York, pp. 511–529.
49. Karel, M., M. P. Buera, and Y. Roos (1993). Effects of glass transitions on processing and storage, in *The Glassy State in Foods* (J. M. V. Blanshard and P. J. Lillford, eds.), Nottingham University Press, Loughborough, pp. 13–34.
50. Kopelman, R., and Y. -E. Koo (1992). Diffusion controlled elementary reactions in low dimensions. *Adv. Chem. Kinet. Dynam. 1*:113–138.
51. Kuntz, I. D., Jr., and W. Kauzmann (1974). Hydration of proteins and polypeptides. *Adv. Protein Chem. 28*:239–345.
52. Labuza, T. P. (1968). Sorption phenomena in foods. *Food Technol. 22*(3):15–24.
53. Labuza, T. P., and R. Contreras-Medellin (1981). Prediction of moisture protection requirements for foods. *Cereal Foods World 26*(7):335–344.
54. Labuza, T. P., and M. Saltmarch (1981). The nonenzymatic browning reaction as affected by water in foods, in *Water Activity: Influences on Food Quality* (L. B. Rockland and G. F. Stewart, eds.), Academic Press, New York, pp. 605–650.
55. Labuza, T. P., L. McNally, D. Gallagher, J. Hawkes, and F. Hurtado (1972). Stability of intermediate moisture foods. I. Lipid oxidation. *J. Food Sci. 37*:154–159.
56. Labuza, T. P., S. Cassil, and A. J. Sinskey (1972). Stability of intermediate moisture foods. 2. Microbiology. *J. Food Sci. 37*:160–162.

57. Leistner, L. (1978). Hurdle effect and energy saving, in *Food Quality and Nutrition* (W. K. Downey, ed.), Elsevier, Essex, UK, pp. 553–557.
58. Leistner, L. (1987). Shelf-stable products and intermediate moisture foods based on meat, in *Water Activity: Theory and Applications to Food* (L. B. Rockland and L. R. Beuchat, eds.), Marcel Dekker, New York, pp. 295–327.
59. Leistner, L., W. Rödel, and K. Krispien (1981). Microbiology of meat and meat products in high- and intermediate-moisture ranges, in *Water Activity: Influences on Food Quality* (L. B. Rockland and G. F. Stewart, eds.), Academic Press, Orlando, FL, pp. 855–916.
60. Levine, H., and L. Slade (1986). A polymer physico-chemical approach to the study of commercial starch hydrolysis products (SHPs). *Carbohydrate Polymers 6*:213–244.
61. Levine, H., and L. Slade (1988). "Collapse" phenomena—a unifying concept for interpreting the behaviour of low moisture foods, in *Food Structure—Its Creation and Evaluation* (J. M. V. Blanshard and J. R. Mitchell, eds.), Butterworths, London, pp. 149–180.
62. Levine, H., and L. Slade (1988). Water as a plasticizer: Physico-chemical aspects of low-moisture polymeric systems. *Water Sci. Rev. 3*:79–185.
63. Levine, H., and L. Slade (1988). Principles of "cryostabilization" technology from structure/property relationships of carbohydrate/water systems—A review. *Cryo-Letters 9*:21–45.
64. Levine, H., and L. Slade (1989). Interpreting the behavior of low-moisture foods, in *Water and Food Quality* (T. M. Hardman, ed.), Elsevier, London, pp. 71–134.
65. Levine, H. and L. Slade (1989). A food polymer science approach to the practice of cryostabilization technology. *Comm. Agric. Food Chem. 1*:315–396.
66. Levine, H., and L. Slade (1990). Cryostabilization technology: Thermoanalytical evaluation of food ingredients and systems, in *Thermal Analysis of Foods* (V. R. Harwalkar and C. -Y. Ma, eds.), Elsevier Applied Science, London, pp. 221–305.
67. Levine, H., and L. Slade (1993). The glassy state in applications for the food industry, with an emphasis on cookie and cracker production, in *The Glassy State in Foods* (J. M. V. Blanshard and P. J. Lillford, eds.), Nottingham University Press, Loughborough, pp. 333–373.
68. Lewin, S. (1974). *Displacement of Water and Its Control of Biochemical Reactions*, Academic Press, London.
69. Lide, D. R., ed. (1993/1994). *Handbook of Chemistry and Physics*, 74th ed., CRC Press, Boca Raton, FL.
70. Lim, M. H., and D. S. Reid (1991). Studies of reaction kinetics in relation to the T_g' of polymers in frozen model systems, in *Water Relationships in Foods* (H. Levine and L. Slade, eds.), Plenum Press, New York, pp. 103–122.
71. Lounnas, V., and B. M. Pettitt (1994). A connected-cluster of hydration around myoglobin: Correlation between molecular dynamics simulations and experiment. *Proteins: Structure, Function Genet. 18*:133–147.
72. Lounnas, V., and B. M. Pettitt (1994). Distribution function implied dynamics versus residence times and correlations: Solvation shells of myoglobin. *Proteins: Structure, Function Gent. 18*:148–160.
73. Luck, W. A. P. (1981). Structures of water in aqueous systems, in *Water Activity: Influences on Food Quality* (L. B. Rockland and G. F. Stewart, eds.), Academic Press, New York, pp. 407–434.
74. Luyet, B. (1960). On various phase transitions occurring in aqueous solutions at low temperatures. *Ann. NY Acad. Sci. 85*:549–569.
75. Luyet, B. J. (1966). Anatomy of the freezing process in physical systems, in *Cryobiology* (H. T. Meryman, ed.), Academic Press, New York, pp. 115–138.
76. Luyet, B. J., and P. M. Gehenio (1940). *Life and Death at Low Temperatures*, Biodynamica, Normandy, MO.
77. Luyet, B. J., and G. Rapatz (1958). Patterns of ice formation in some aqueous solutions. *Biodynamica 8*:1–68.
77a. MacInnes, W. M. (1993). Dynamic mechanical thermal analysis of sucrose solutions, in *The Glassy State of Foods* (J. M. V. Blanshard and P. J. Lillford, eds.), Nottingham University Press, Loughborough, pp. 223–248.

78. MacKenzie, A. P. (1966). Basic principles of freeze-drying for pharmaceuticals. *Bull. Parenteral Drug Assoc. 20*:101–129.
79. Mason, B. J. (1957). *The Physics of Clouds,* Clarendon, Oxford, p. 445.
80. Morgan, J., and B. E. Warren (1938). X-ray analysis of the structure of water. *J. Chem. Phys. 6*:666–673.
81. Nelson, K. A. (1993). *Reactions Kinetics of Food Stability: Comparison of Glass Transition and Classical Models for Temperature and Moisture Dependence,* PhD thesis, University of Minnesota, Minneapolis, MN.
82. Nelson, K. A., and T. P. Labuza (1993). Glass transition theory and the texture of cereal foods, in *The Glassy State in Foods* (J. M. V. Blanshard and P. J. Lillford, eds.), Nottingham University Press, Loughborough, pp. 513–517.
83. Nelson, K. A., and T. P. Labuza (1994). Water activity and food polymer science: Implications of state on Arrhenius and WLF models in predicting shelf life. *J. Food Eng. 22*:271–289.
84. Nesvadba, P. (1993). Glass transitions in aqueous solutions and foodstuffs, in *The Glassy State in Foods* (J. M. V. Blanshard and P. J. Lillford, eds.), Nottingham University Press, Loughborough, pp. 523–526.
85. Oakenfull, D., and D. E. Fenwick (1977). Thermodynamics and mechanism of hydrophobic interaction. *Aust. J. Chem. 30*:741–752.
86. Otting, G., E. Liepinsh, and K. Wuthrich (1991). Protein hydration in aqueous solution. *Science 254*:974–980.
87. Peleg, M. (1993). Mapping the stiffness-temperature-moisture relationship of solid biomaterials at and around their glass transition. *Rheol. Acta 32*:575–580.
88. Peleg, M. (1994). Mathematical characterization and graphical presentation of the stiffness-temperature-moisture relationship of gliadin. *Biotechnol. Prog. 10*:652–654.
89. Pilling, M. J. (1975). *Reaction Kinetics,* Clarendon Press, Oxford.
90. Prior, B. A. (1979). Measurement of water activity in foods: A review. *J. Food Protect. 42*:668–674.
91. Rasmussen, D. (1969). A note about "phase diagrams" of frozen tissue. *Biodynamica 10*:333–339.
92. Reid, D. S., J. Hsu, and W. Kerr (1993). Calorimetry, in *The Glassy State in Foods* (J. M. V. Blanshard and P. J. Lillford, eds.), Nottingham University Press, Loughborough, pp. 123–132.
93. Reid, D. S., W. Kerr, and J. Hsu (1994). The glass transition in the freezing process, in *Water in Foods* (P. Fito, A. Mulet, and B. McKenna, eds.), Elsevier Applied Science, London, pp. 483–494.
94. Rey, L. R. (1958). *Etude Physiologique et Physico-chimique de l-Action des Basses Température sur les Tissues Animaux Vivants,* PhD thesis, Université de Paris, France.
95. Rice, S. A. (1985). *Diffusion-Limited Reactions.* Vol. 25 of *Chemical Kinetics* (C. H. Bamford, C. F. H. Tipper, and R. G. Compton, eds.), Elsevier, Amsterdam, The Netherlands.
96. Röntgen, W. C. (1882). VIII. Ueber die Constitution des flüssigen Wassers. *Ann. Phy. Chem. 281*:91–97.
97. Roos, Y. H. (1992). Phase transitions and transformations in food systems, in *Handbook of Food Engineering* (D. R. Heldman and D. B. Lund, eds.), Marcel Dekker, New York, pp. 145–197.
98. Roos, Y. H. (1993). Water activity and physical state effects on amorphous food stability. *J. Food Proc. Preserv. 16*;433–447.
99. Roos, Y. H., and M. -J. Himberg (1994). Nonenzymatic browning behavior, as related to glass transition of a food model at chilling temperatures. *J. Agric. Food Chem. 42*:893–898.
100. Roos, J. H., and M. Karel (1991). Phase transitions of mixtures of amorphous polysaccharides and sugars. *Biotechnol. Prog. 7*:49–53.
101. Roos, J. H., and M. Karel (1991). Amorphous state and delayed ice formation in sucrose solutions. *Int. J. Food Sci. Technol. 26*:553–566.
102. Roos, Y., and M. Karel (1993). Effects of glass transitions on dynamic phenomena in sugar containing food systems, in *The Glassy State in Foods* (J. M. V. Blanshard and P. J. Lillford, eds.), Nottingham University Press, Loughborough, pp. 207–222.
103. Rupley, J. A., and G. Careri (1991). Protein hydration and function. *Adv. Protein Chem. 41 :37–172.*

104. Saleki-Gerhardt, A., and G. Zografi (1994). Non-thermal and isothermal crystallization of sucrose from the amorphous state. *Pharm. Res. 11*:1166–1173.
105. Sapru, V., and R. P. Labuza (1993). Temperature dependence of thermal inactivation rate constants of *Bacillus stearothermophilus* spores, in *The Glassy State in Foods* (J. M. V. Blanshard and P. J. Lillford, eds.), Nottingham University Press, Loughborough, pp. 499–505.
106. Sears, J. K., and J. R. Darby (1982). *The Technology of Plasticizers,* Wiley-Interscience, New York.
107. Sceats, M. G., and S. A. Rice (1982). Amorphous solid water and its relationship to liquid water: A random network model for water, in *Water—A Comprehensive Treatise,* vol. 7 (F. Franks, ed.), Plenum Press, New York, pp. 83–214.
108. Scott, W. J. (1953). Water relations of *Staphylococcus aureus* at 30°C. *Aust. J. Biol. Sci.* 6:549–556.
109. Scott, W. J. (1957). Water relations of food spoilage microorganisms. *Adv. Food Res. 7*:83–127.
110. Simatos, D., and G. Blond (1991). DSC studies and stability of frozen foods, in *Water Relationships in Foods* (H. Levine and L. Slade, eds.), Plenum Press, New York, pp. 139–155.
111. Simatos, D., and G. Blond (1993). Some aspects of the glass transition in frozen foods, in *The Glassy State in Foods* (J. M. V. Blanshard and P. J. Lillford, eds.). Nottingham University Press, Loughborough, pp. 395–415.
112. Slade, L., and H. Levine (1988). Structural stability of intermediate moisture foods—a new understanding, in *Food Structure—Its Creation and Evaluation* (J. M. V. Blanshard and J. R. Mitchell, eds.), Buttersworths, London, pp. 115–147.
113. Slade, L., and H. Levine (1988). Non-equilibrium behavior of small carbohydrate-water systems. *Pure Appl. Chem. 60*:1841–1864.
114. Slade, L., and H. Levine (1991). Beyond water activity: Recent advances based on an alternate approach to the assessment of food quality and safety. *Crit. Rev. Food Sci. Nutr. 30*:115–360.
115. Slade, L., and H. Levine (1993). Water relationships in starch transitions. *Carbohydrate Polymers 21*:105–131.
116. Slade, L., and H. Levine (1993). The glassy state phenomenon in food molecules, in *The Glassy State in Foods* (J. M. V. Blanshard and P. J. Lillford, eds.), Nottingham University Press, Loughborough, pp. 35–101.
117. Slade, L., and H. Levine (1995). Glass transitions and water-food structure interactions. *Adv. Food Nutr. Res. 38*:103–269.
118. Slade, L., H. Levine, and J. W. Finley (1989). Protein-water interactions: Water as a plasticizer of gluten and other protein polymers, in *Protein Quality and the Effects of Processing* (R. D. Phillips and J. W. Finley, eds.), Marcel Dekker, New York, pp. 9–124.
119. Spiess, W. E. L., and W. Wolf (1987). Critical evaluation of methods to determine moisture sorption isotherms, in *Water Activity: Theory and Applications to Food* (L. B. Rockland and L. R. Beuchat, eds.), Marcel Dekker, New York, pp. 215–233.
120. Stillinger, F. H. (1980). Water revisited. *Science 209*:451–457.
121. Storey, R. M., and G. Stainsby (1970). The equilibrium water vapour pressure of frozen cod. *J. Food Technol. 5*:157–163.
122. Suggett, A. (1976). Molecular motion and interactions in aqueous carbohydrate solutions. III. A combined nuclear magnetic and dielectric-relaxation strategy. *J. Solut. Chem. 5*:33–46.
123. Taborsky, G. (1979). Protein alterations at low temperatures: An overview, in *Proteins at Low Temperatures* (O. Fennema, ed.), Advances in Chemistry Series 180, American Chemical Society, Washington, DC, pp. 1–26.
124. Teeter, M. M. (1991). Water-protein interactions: Theory and experiment. *Annu. Rev. Biophys. Biophys. Chem. 20*:577–600.
125. van den Berg, C. (1981). *Vapour Sorption Equilibria and Other Water-Starch Interactions; A Physico-Chemical Approach,* PhD thesis, Wageningen Agricultural University, Wageningen, The Netherlands.
126. van den Berg, C. (1986). Water activity, in *Concentration and Drying of Foods* (D. MacCarthy, ed.), Elsevier, London, pp. 11–36.
127. van den Berg, C., and S. Bruin (1981). Water activity and its estimation in food systems: Theoretical

aspects, in *Water Activity: Influences on Food Quality* (L. B. Rockland and G. F. Stewart, eds.), Academic Press, New York, pp. 1–61.

128. van den Berg, C., and H. A. Leniger (1978). The water activity of foods, in *Miscellaneous Papers 15,* Wageningen Agricultural University, Wageningen, The Netherlands, pp. 231–244.
129. White, G. W., and S. H. Cakebread (1966). The glassy state in certain sugar-containing food products. *J. Food Technol. 1*:73–82.
130. Williams, M. L., R. F. Landel, and J. D. Ferry (1955). The temperature dependence of relaxation mechanisms in amorphous polymers and other glass-forming liquids. *J. Am. Chem. Soc. 77*:3701–3707.
131. Wolanczyk, J. P. (1989). Differential scanning calorimetry analysis of glass transitions. *Cryo-Letters 10*;73–76.

3

Dispersed Systems: Basic Considerations

PIETER WALSTRA
Wageningen Agricultural University, Wageningen, The Netherlands

3.1 INTRODUCTION

The subjects discussed in this chapter are rather different from most of the material in this book, in the sense that true chemistry, that is, reactions involving electron transfer, is hardly involved. Nevertheless, many aspects of dispersed systems are important to an understanding of the properties of most foods and the manufacture of "fabricated foods."

3.1.1 Foods as Dispersed Systems

Most foods are dispersed systems. A few are homogeneous solutions, like cooking oil and some drinks, but even beer—as consumed—has a foam layer. The properties of a dispersed system cannot be fully derived from its chemical composition, since they also depend on physical structure. The structure can be very intricate, as is the case with foods derived from animal or vegetable tissues; these are discussed in Chapters 15 and 16. Manufactured foods, as well as some natural foods, may have a somewhat simpler structure: Beer foam is a solution containing gas bubbles; milk is a solution containing fat droplets and protein aggregates (casein micelles); plastic fats consist of an oil containing aggregated triacylglycerol crystals; a salad dressing may be just an emulsion; several gels consist of a network of polysaccharide molecules that immobilize a solution. But other manufactured foods are structurally complicated in that they contain several different structural elements of widely varying size and state of aggregation: filled gels, gelled foams, materials obtained by extrusion or spinning, powders, margarine, doughs, bread, and so forth.

The existence of a dispersed state has some important consequences:

1. Since different components are in different compartments, there is no thermodynamic equilibrium. To be sure, even a homogeneous food may not be in equilibrium, but for dispersed systems this is a much more important aspect. It may have significant consequences for chemical reactions, as is briefly discussed in Section 3.1.3.
2. Flavor components may be in separate compartments, which may slow down their release during eating. Probably more important, compartmentalization of flavor components may lead to fluctuations in flavor release during eating, thereby enhancing flavor, because it offsets to some extent adaptation of the senses to flavor components. Thus, most "compartmentalized" foods taste quite different from the same food that has been homogenized before eating.

3. If, as is often the case, attractive forces act between structural elements, the system has a certain consistency, which is defined as its resistance against permanent deformation. This may be an important functional property as it is related to attributes such as stand-up, spreadability, or ease of cutting. Moreover, consistency affects mouth feel, as does any physical inhomogeneity of the food; food scientists often lump these properties under the word texture.
4. If the product has significant consistency, any solvent present—in most foods, water—will be immobilized against bulk flow. Transport of mass (and mostly of heat also) then has to occur by diffusion rather than convection. This may have a considerable effect on reaction rates.
5. The visual appearance of the system may be greatly affected. This is due to the scattering of light by structural elements, provided they are larger than about 50 nm. Large inhomogeneities are visible as such and give rise to what is the dictionary meaning of texture.
6. Since the system is physically inhomogeneous, at least at a microscopic scale, it may be physically unstable. Several kinds of change may occur during storage, which may be perceived as the development of macroscopic inhomogeneity, such as separation into layers. Moreover, during processing or usage, changes in the dispersed state may occur, which may be desirable—as in the whipping of cream—or undesirable—as in overwhipping of cream, where butter granules are formed.

In this chapter, some of the foregoing aspects will be discussed. Large-scale mechanical properties will be largely left out, and so will aspects of hydrodynamics and process engineering. Of course, most foods show highly specific behavior, but treating them all would take too much space and provide little understanding. Therefore, some general aspects of fairly simple model dispersions will be emphasized.

3.1.2 Characterization of Dispersions

A dispersion is a system of discrete particles in a continuous fluid phase. Various types exist and these are given in Table 1. Foods never exist as fog, and aerosols and powders will not be discussed. In addition to the dispersions listed in the table, solid foams, emulsions, or suspensions may exist: After the liquid systems have been made, the continuous phase may in some way solidify. In a foam omelette, the continuous protein solution has gelled; in margarine, the continuous oil phase now contains a continuous network of aggregated crystals; in chocolate much the same has happened and it thus contains solid particles (sugar, cacao) in a largely crystallized fat matrix.

There are two types of food emulsions, oil-in-water (o/w) and water-in-oil (w/o). For emulsions, as well as for other dispersions, the nature of the *continuous phase* determines some

TABLE 1 Various Types of Dispersions

Dispersed phase	Continuous phase	Dispersion type
Gas	Liquid	Foam
Liquid	Gas	Fog, aerosol
Liquid	Liquid	Emulsion
Solid	Gas	Smoke, powder
Solid	Liquid	Suspension, sol

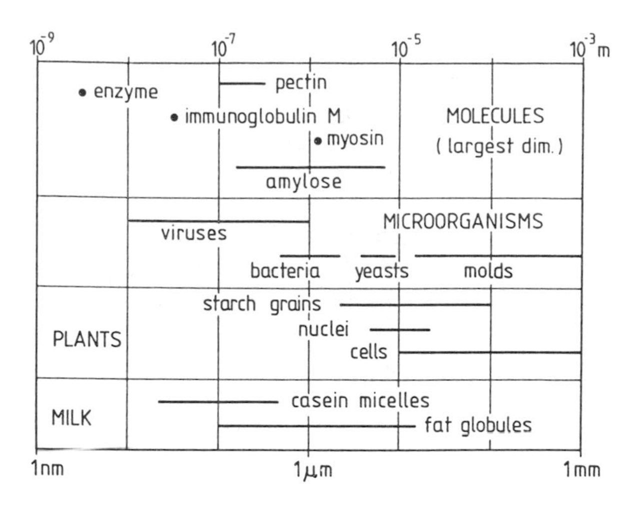

FIGURE 1 Approximate size of some structural elements in foods.

important properties of the dispersion, for instance, the type of liquid (aqueous or apolar) that can mix with the dispersion. In principle, there may be more than one continuous phase. The prime example is a wet sponge, where matrix and water both are continuous. Several foods are bicontinuous; for instance, in bread both the gas and solid phases are continuous. If not, the bread would lose most of its volume after baking, since the gas cells would shrink considerably on cooling, the more so as they partly consist of water vapor.

The term *phase* needs some clarification. A phase is commonly defined as a domain bounded by a closed surface, at which surface at least some properties (e.g., pressure, refractive index, density, heat capacity, chemical composition) change abruptly. The surface contains free energy and it thus resists enlargement; in other words, it requires energy to make the surface (or interface). Since changes in properties mostly occur over a distance of a few molecular diameters, the criterion of abruptness implies that the dimensions of structural elements constituting a phase must be far larger than a single molecule in any direction. For instance, protein aggregates that float in a solution do not constitute a phase, since the aggregates are small and contain solvent. The particles may even sediment (certainly in a centrifuge) and form separate layers, but even these layers are not phases, since there is no phase boundary. The constituents water and polysaccharide that form a gel do not constitute phases: The polysaccharide strands in the gel are only about 1 nm in thickness.

This brings us to the term *colloid*. A colloid is usually defined as a dispersion containing particles that are clearly larger than small molecules (say, solvent molecules), yet too small to be visible. This would imply a size range from about 10 nm to almost 1 mm. Two types of colloids are usually distinguished: reversible (or lyophilic) and irreversible (or lyophobic). The latter type consists of two (or more) phases of the types shown in Table 1 and these do not form spontaneously.

A reversible colloid forms by "dissolving" a material in a suitable solvent. The main examples are macromolecules (polysaccharides, proteins, etc.) and association colloids. The latter are formed from amphiphilic molecules, like soaps. These have a fairly long hydrophobic "tail" and a smaller, very polar (i.e., hydrophilic) "head." In an aqueous environment, the molecules tend to associate in such a way that the tails are close to each other and the heads are in contact with water. In this way, micelles or liquid crystalline structures are formed.

Micelles will be briefly discussed in Section 3.2.2; liquid crystalline phases [33] are not very prominent in foods.

In foods, the difference between reversible and irreversible colloids is not always clear. For instance, an oil-in-water emulsion is certainly irreversible in the sense that it will never form spontaneously. But if the oil droplets are covered by a protein layer, the interactions between droplets may be like that of protein particles in solution, that is, like a reversible colloid. This means that for some properties of the emulsion macroscopic considerations are appropriate, whereas for other properties molecular considerations may be more suitable.

The *size scale* of structural elements in foods can vary widely, spanning a range of six decades (Fig. 1). A water molecule has a diameter of about 0.3 nm. Also the shape of the particles is important, as is their volume fraction φ (i.e., the proportion of the volume of the system that is taken up by the particles). All these variables affect product properties. Some effects of size or scale are:

1. Visual appearance. An o/w emulsion, for example, will be almost transparent if the droplets have a diameter of 0.03 μm; blueish white if 0.3 μm; white if 3 μm; and the color of the oil (usually yellow) will be discernable for 30-μm droplets.
2. Surface area. For a collection of spheres each with a diameter d (in m), the specific surface area is given by

$$A = 6\varphi/d \tag{1}$$

   in square meters per cubic meter. The area can thus be large. For an emulsion of $\varphi = 0.1$ and $d = 0.3$ μm, $A = 2\ m^2$ per ml emulsion; if 5 mg protein is adsorbed per square meter of oil surface, the quantity of adsorbed protein would amount to 1% of the emulsion.
3. Pore size. Between particles, regions of continuous phase exist and their size is proportional to particle size and smaller for a larger φ. If the dispersed phase forms a continuous network, pores in this network follow the same rules. The permeability, that is, the ease with which solvent can flow through the pores, is inversely proportional to pore size squared. This is why a polymer gel is far less permeable than a gel made of fairly large particles (Sec. 3.5.2).
4. Time scales involved. (Note: Time scale is defined as the characteristic time needed for an event to occur—for instance, for two molecules to react, for a particle to rotate, for a bread to be baked.) The larger the particles, the longer are the time scales involved. For example, the root mean square value of the diffusion distance (z) of a particle of diameter d as a function of time t is

$$\langle z^2 \rangle^{0.5} \propto (t/d)^{0.5} \tag{2}$$

   In water, a particle of 10 nm diameter will diffuse over a distance equal to its diameter in 1 μs; a particle of 1 μm in 1 sec; and one of 0.1 mm in 12 days. Considering diffusion of a material into a structural element, the relation between diffusion coefficient D, distance l, and time $t_{0.5}$ needed to halve a difference in concentration (or temperature) is

$$l^2 = Dt_{0.5} \tag{3}$$

   The D of small molecules in water is approximately $10^{-9}\ m^2 \cdot sec^{-1}$ and in most cases (larger molecules, more viscous solution) it is smaller; D for evening out a temperature difference (by heat conduction) equals approximately $10^{-7}\ m^2 \cdot sec^{-1}$. Taking, for instance, $D = 10^{-10}$, we find a halving time of about 0.01 sec for a distance of 1 μm, and of 3 hr for a distance of 1 mm.

5. Physical stability. Most interaction forces between particles are roughly proportional to their diameter, and sedimentation rate to diameter squared. This implies that almost all dispersions become inhomogeneous more readily if the particles are larger.
6. Effect of external forces. Most external forces acting on particles are proportional to diameter squared, and most internal forces within or between particles are proportional to diameter. This implies that small particles are virtually impervious to external influences, like shearing forces or gravity. Large particles often can be deformed or even be disrupted by external forces.
7. Ease of separation. Some of the points raised earlier imply that it is much more difficult to separate small particles from a liquid than large ones.

Particles rarely are all of the same size. The subject of *size distributions* is a complicated one [2,58] and it will not be discussed here. Suffice it to say that mostly a size range may be used to characterize the size distribution and that the volume/surface average diameter d_{vs} or d_{32} is often a suitable estimate of the center of the distribution. However, different properties may need different types of averages. The wider the size distribution—width being defined as standard deviation divided by average—the greater the differences between average types (an order of magnitude is not exceptional). It is often very difficult to accurately determine a size distribution [2]. Difficulties in determination and interpretation increase with particles that are more anisometric or otherwise different in properties.

3.1.3 Effects on Reaction Rates

As already mentioned, components in a dispersed food may be compartmentalized, and this can greatly affect reaction rates. In a system containing an aqueous (α) and an oil phase (β), a component often is soluble in both. Nernst's distribution or partitioning law then states that the ratio of concentrations (c) in both phases is constant:

$$c_\alpha / c_\beta = \text{constant} \tag{4}$$

The constant will depend on temperature and possibly other conditions. For instance, pH has a strong effect on the partitioning of carboxylic acids, since these acids are oil soluble only when they are in a neutralized state. At high pH, where acids are fully ionized, almost all acid will be in the aqueous phase, whereas at low pH the concentration in the oil phase may be considerable. Note that the quantity of a reactant in a phase also depends on the phase volume fraction.

When a reaction occurs in one of the phases present, it is not the overall concentration of a reactant but its concentration in that phase that affects the reaction rate [87]. This concentration may be higher or lower than the overall concentration, depending on the magnitude of the partitioning constant (Eq. 4). Since many reactions in foods actually are cascades of several different reactions, the overall reaction pattern, and thereby the mix of components formed, may also depend on partitioning. Chemical reactions will often involve transport between compartments and will then depend on distances and molecular mobility. Applying Equation 3, it follows that diffusion times for transport into or out of fairly small structural elements, say emulsion droplets, would mostly be very short. However, if the solvent is immobilized in a network of structural elements, this may greatly slow down reactions, especially if reactants, say O_2, have to diffuse in from outside. Moreover, reactions may have to take place at the boundary between phases. An example is fat autoxidation, where the oxidizable material (unsaturated oil) is in oil droplets, and a catalyst, say Cu ions, is in the aqueous phase. Another example is that of an enzyme present in one structural element and the component on which it acts in another one. In such cases, the specific surface area may be rate-determining.

Adsorption of reactive substances onto interfaces between structural elements may diminish their effective concentration and thereby reactivity. On the other hand, if two substances that may react with each other are both adsorbed, their effective concentrations in the adsorption layer may be much greater, thereby enhancing reaction rate.

Thus, rates of chemical reactions and the mix of reaction products may be quite different in a dispersed system than in a homogeneous one. Examples in vegetable and animal tissues are well known, but other cases have not been studied in great detail, except for the activity of some additives [87] and, of course, for enzymatic lipolysis.

3.2 SURFACE PHENOMENA

As mentioned earlier, most foods have a large phase boundary or interfacial area. Often, substances adsorb onto interfaces, and this has a considerable effect on static and dynamic properties of the system. In this section basic aspects are discussed; applications are discussed later [see Refs. 1 and 51 for general literature].

3.2.1 Interfacial Tension and Adsorption

An excess of free energy is present at a phase boundary and the amount is conveniently expressed in joules per square meter. The interface can be solid or fluid; the latter is the case if both phases are fluid. A fluid interface is deformable and the interfacial free energy then becomes manifest as a two-dimensional interfacial tension, expressed in newtons per meter; it is usually denoted γ. Note that γ numerically equals the interfacial free energy ($1\ \mathrm{J} = 1\ \mathrm{N} \cdot \mathrm{m}$). In foods, we may have air (A), aqueous (W), oil (O), or solid (S) phases; if one of the phases is air, γ (e.g., γ_{AW}) is commonly called surface tension. The interfacial tension acts in the direction of the interface and resists enlargement of the interface. This provides a method to measure γ. For example, one may measure the force to pull a plate out of the interface and divide the force by the contour length of the interface–plate contact (Fig. 2).

Substances that lower γ adsorb onto the interface because this leads to a lower total free energy. Such substances are called surfactants. Reference 36 gives a detailed treatment of adsorption to fluid interfaces and the consequences thereof. The amount adsorbed is expressed as the surface excess Γ ($\mathrm{mol} \cdot \mathrm{m}^{-2}$ or $\mathrm{mg} \cdot \mathrm{m}^{-2}$), loosely called surface load. According to Gibbs, for one adsorbing solute in one of the phases and at equilibrium conditions

$$d\gamma = -RT\ \Gamma\ d(\ln a) \tag{5}$$

where R and T have their usual meaning and a is the activity of the solute. The latter is not equal to the total concentration of the solute because (a) part of the solute is adsorbed and a refers only to the surfactant in solution (usually called "the bulk"), and (b) the activity may be smaller than the bulk concentration, especially above the critical micelle concentration (see later). Whether adsorption does occur and γ is thus lowered depends on the properties of the solute (Sec. 3.2.2). The lowering is expressed in the surface pressure $\Pi \equiv \gamma_0 - \gamma$, which can be envisaged as a two-dimensional pressure exerted by the adsorbed molecules onto any boundary that confines a certain interfacial area (Fig. 2). Interfacial tensions of some systems are given in Table 2.

The Gibbs equation shows that γ will be lower for a larger bulk concentration of an adsorbing solute, but it does not give the relation between γ or Π and Γ, nor the relation between Γ and bulk concentration. The former relation, a surface equation of state, will not be discussed here. The latter relation may be seen as an adsorption isotherm, if equilibrium is attained. Examples are shown later, in Fig. 5. Usually a plateau value of Γ is reached, at which the

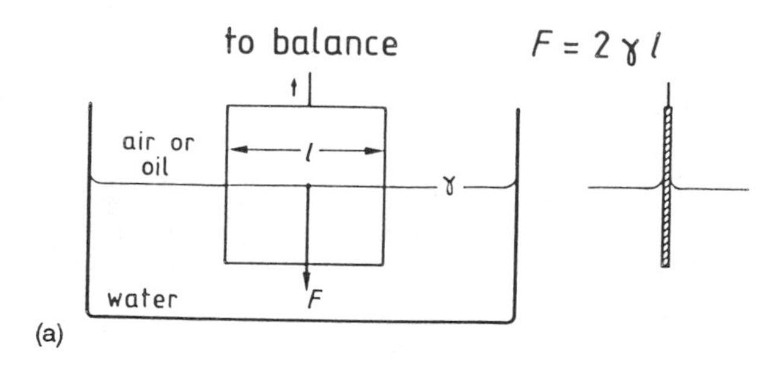

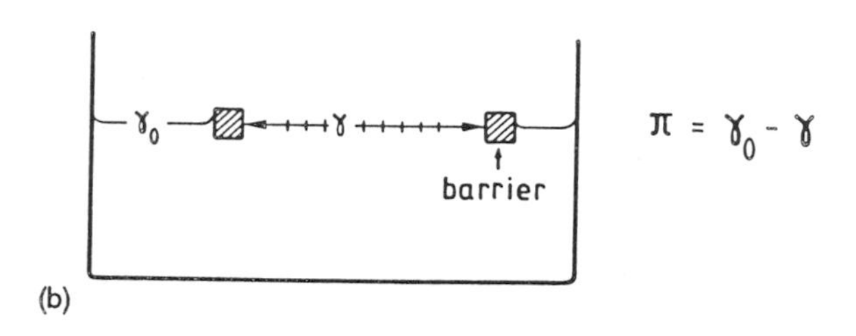

FIGURE 2 Surface (interfacial) tension measurements. (a) General principle: The surface tension pulls the Wilhelmy plate downward. (b) Surfactant molecules in the surface exert a two-dimensional pressure $\Pi = \gamma_0 - \gamma$ on the movable barriers confining the surfactant-containing surface.

interface is more or less fully packed with surfactant molecules. For most surfactants, the plateau value is a few milligrams per square meter.

Adsorption naturally takes time, since adsorbing molecules have to be transported, often by diffusion, to the interface. If the bulk concentration of surfactant is c, the layer of solution needed to provide the surfactant is given by Γ/c. Using Equation 3, the time needed for adsorption would be approximately

$$t_{ads} \approx 5(\Gamma/c)^2/D \tag{6}$$

This means that the time needed for adsorption would typically be < 1 sec for a surfactant concentration of 0.1%.

TABLE 2 Some Interfacial Tensions, Approximate Values ($mN \cdot m^{-1}$) at Room Temperature

Material	Against air	Against water
Water	72	0
Saturated NaCl solution	82	0
Ethanol	22	0
Paraffin oil	30	50[a]
Triacylglycerol oil	34	27[a]
Mercury	484	415

[a]Some buffer solutions give a lower interfacial tension than does pure water.

3.2.2 Surfactants

Surfactants come in two types, polymers and soap-like substances. The latter are fairly small amphiphilic molecules, the hydrophobic (lipophilic) part being typically an aliphatic chain. The hydrophilic part can vary widely; in the classical surfactant, common soap, it is an ionized carboxyl group. Most amphiphilic substances are not highly soluble either in water or oil, and they feel the least repulsive interaction from these solvents when they are partly in a hydrophilic environment (water, a hydrophilic substance) and partly in a hydrophobic one (oil, air, a hydrophobic substance), that is, at an interface. In solution, they tend to form micelles to lessen repulsive interaction with solvent. (Note on terminology: Some workers use the word surfactant for small-molecule surfactants only. Also, surfactants are often called emulsifiers, even when the surfactant is not involved in making an emulsion.)

Some *small-molecule surfactants* of importance to the food scientist are listed in Table 3 [32,53]. They are categorized as nonionic, anionic, and cationic, according to the nature of the hydrophilic part. Also, distinction is made between natural surfactants (e.g., soaps, monoacylglycerols, phospholipids) and synthetic ones. The Tweens are somewhat different from the other ones in that the hydrophilic part contains three or four polyoxyethylene chains of about five monomers in length. Phospholipids come in a wide range of composition and properties.

An important characteristic of a small-molecule surfactant is its *HLB value*, where HLB stands for hydrophile–lipophile balance. It is defined so that a value of 7 means that the substance has about equal solubility in water and oil. Lower values imply greater solubility in oil. Surfactants range in HLB value from about 1 to 40. The relation between HLB value and solubility is in itself useful, but it also relates to the suitability of the surfactant as an emulsifier: Surfactants with HLB > 7 are generally suitable for making o/w emulsions, and those with HLB < 7 for w/o emulsions (see also Sec. 3.6.2 about Bancroft's rule). Surfactants suitable as cleaning agents (detergents) in aqueous solutions have a high HLB number. Several other relations with HLB values have been claimed, but most of these are questionable.

In general, a longer aliphatic chain yields a lower HLB and a more polar (especially an

TABLE 3 Some Small-Molecule Surfactants and Their Hydrophile–Lipophile Balance (HLB) Values

Type	Example of surfactant	HLB value
Nonionics		
Aliphatic alcohol	Hexadecanol	1
Monoacylglycerol	Glycerol monostearate	3.8
Esters of monoacylglycerols	Lactoyl monopalmitate	8
Spans	Sorbitan monostearate	4.7
	Sorbitan monooleate	7
	Sorbitan monolaurate	8.6
Tween 80	Polyoxyethylene sorbitan monooleate	16
Anionics		
Soap	Na oleate	18
Lactic acid esters	Na stearoyl-2-lactoyl lactate	21
Phospholipids	Lecithin	Fairly large
Teepol[a]	Na lauryl sulfate	40
Cationics[a]		Large

[a]Not used in foods but as detergents.

ionized) or a larger polar group a higher HLB. For most surfactants, the HLB number decreases with increasing temperature. This implies that some surfactants exhibit a HLB temperature or phase inversion temperature (PIT), at which a value of 7 is reached. Above the PIT the surfactant tends to make a w/o emulsion, and below it an o/w emulsion. Near the PIT, γ_{OW} is generally very small.

Figure 3 shows the effect of concentration of surfactant on surface tension. Breaks in the curves occur at the *critical micelle concentration* (CMC). Beyond that concentration the surfactant molecules form micelles and their activity barely increases; hence, γ becomes essentially independent of concentration (see Eq. 5). At a concentration slightly below the CMC the surface load reaches a plateau value. In a homologous series of surfactants, a longer chain length results in a lower CMC and a lower concentration to obtain a given lowering of γ. This means that larger surfactant molecules are more surface active. Also, smaller surfactant molecules give a somewhat less steep slope near the CMC. This implies, according to Equation 5, that the plateau value of the surface load is somewhat smaller than it is for larger surfactant molecules. For ionic surfactants, the CMC markedly decreases and surface activity increases with increasing ionic strength. Both of these properties can also depend on pH.

At the oil–water interface much the same pattern is observed, but since γ_0 is lower and Π is roughly the same, γ is much smaller. The smallest value of γ obtained at the air–water interface is about 35 $mN \cdot m^{-1}$, whereas at the triacylglycerol oil–water interface it varies from < 1 to about 5 $mN \cdot m^{-1}$ for most small-molecule surfactants. The mode of adsorption of various surfactants is depicted in Figure 4.

It should be realized that commercially available surfactants generally are mixtures of several components, varying in chain length and possibly in other properties. These components may differ, for example, in the plateau value of γ (Fig. 3). In these instances, Equation 5 is no longer applicable and plots of the type in Figure 3 have a different shape, often with a local minimum in γ. Especially, some trace components may be present that give a lower γ than the main components, and at equilibrium the surfactants yielding the smallest γ will dominate in the interface. Because of their small concentration, it will, however, take a long time for them to reach the interface; see Equation 6. This implies that it will take a long time before an equilibrium composition, and thus a steady γ, is reached. Another complication is that in actual

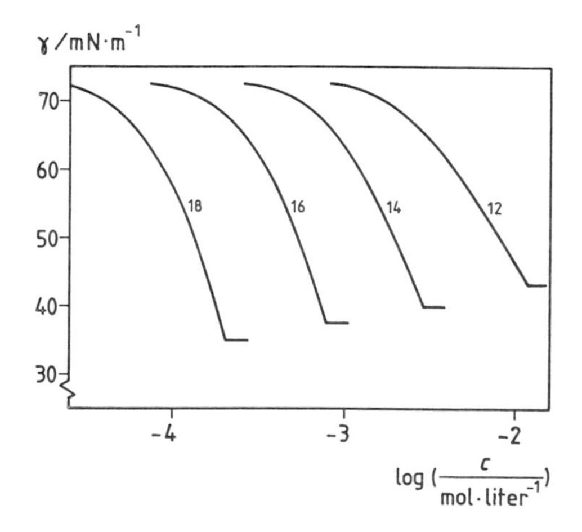

FIGURE 3 Surface tension γ (at the air–water interface) against bulk concentration c of sodium soaps of normal fatty acids of various chain length. Approximate values; various sources.

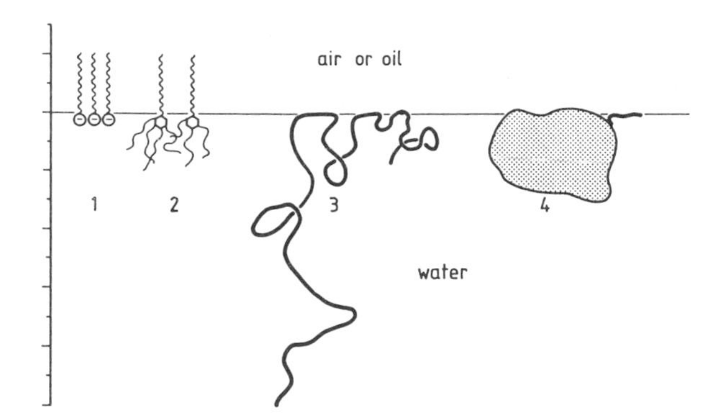

FIGURE 4 Mode of adsorption of various surfactants at an oil–water or air–water interface: 1, soap; 2, Tween; 3, a fairly small polymer molecule; 4, a globular protein. To the left is a scale of nanometers. Highly schematic.

dispersions the surface to volume ratio is very large, whereas this ratio is quite small in situations where γ is commonly measured, that is, at a macroscopic interface between the phases. This means that the result of such measurements of γ may not be representative for the actual values in a foam or emulsion. Finally, the CMC may not be clearly evident in mixtures of surfactants, especially for nonionics.

Macromolecules can be very surface active. Several synthetic polymers are used as surfactants, and a mass of experimental evidence as well as theory is available [22]. Copolymers, where part of the segments are fairly hydrophobic and others hydrophilic, are especially suitable. They tend to adsorb with "trains," "loops," and "tails" (Fig. 4). Actually, at an oil–water interface, parts of the molecule can also protrude for some instances into the oil phase (not into air). There are few natural polymers that adsorb in this way. Surface activity of polysaccharides still is controversial. Most workers assume that surface-active polysaccharides contain a protein moiety that is responsible for this attribute, and this is clearly the case for gum arabic [15]. On the other hand, it is believed that at least some galactomannans adsorb as such onto the o/w interface [24].

Proteins often are the surfactants of choice, especially for foams and o/w emulsions (because of their water solubility they are not suitable for w/o emulsions) [16,44,85]. The mode of adsorption of proteins varies. There always is a change of conformation, often considerably so. For instance, most enzymes completely lose their activity after adsorption at an oil–water interface due to conformational change. Some enzymes retain part of their activity after adsorption at an air–water interface. Most globular proteins appear to retain an approximately globular conformation at interfaces (Fig. 4), though not the native one. Proteins with little secondary structure, like gelatin and caseins, tend to adsorb more like a linear polymer (Fig. 4). Forms intermediate between those mentioned also occur. At high bulk protein concentration, multilayer adsorption may occur, but the second and more remote layers are only weakly adsorbed.

In Figure 5 adsorption of a protein and adsorption of a soap-like surfactant are compared. It is apparent that proteins (like synthetic high polymers) are much more surface active than soap-like surfactants. Even if the concentration scale in Figure 5 were in kilograms per cubic meter, the concentration needed for a certain adsorption to occur would differ by two decades. However, the "soap" yields a greater surface pressure (lower interfacial tension) than the protein at the plateau adsorption. For small-molecule surfactants, the Gibbs relation (Eq. 5) holds, but for most proteins (and most high polymers) this is not so. Desorption of adsorbed protein cannot

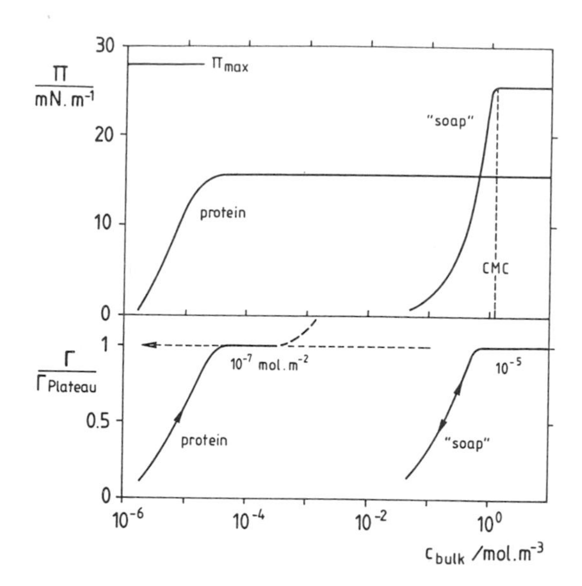

FIGURE 5 Surface pressure (Π) and surface load (Γ) as a function of the bulk concentration of a protein and a small-molecule surfactant ("soap"). CMC is critical micelle concentration. Typical examples. (After Ref. 85.)

be or barely can be achieved by dilution or "washing," at least over time scales of interest. The difficulty of desorption may be enhanced by cross-linking reactions between adsorbed protein molecules; this has been clearly shown for proteins containing a free thiol group, where cysteine–cystine interchange can occur [20].

As exemplified in Figure 5, small-molecule surfactants generally yield a greater surface pressure, that is, a lower interfacial tension, than proteins, if the former are present in sufficient

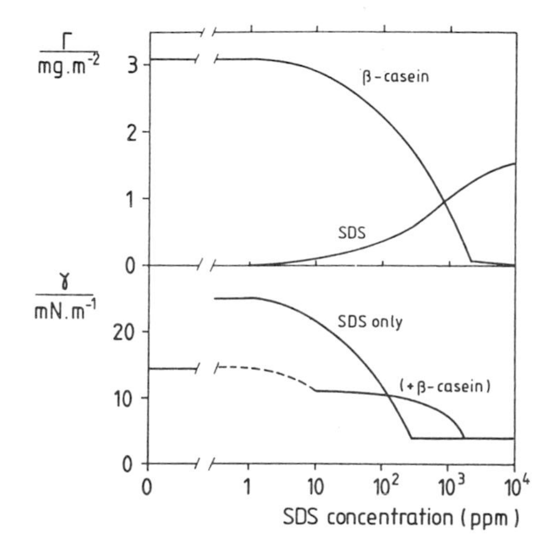

FIGURE 6 Surface load (Γ) in an o/w emulsion and interfacial tension (γ) at the o/w interface for β-casein in the presence of increasing concentration of Na lauryl sulfate (SDS); γ is also given for SDS only. (From Ref. 85, after Ref. 13.)

concentration. This implies that they will displace proteins from the interface [13]. This is illustrated in Figure 6. Proteins may also adsorb on top of an adsorbed monolayer of a displacing surfactant, especially if the monolayer is phospholipid. These are important aspects in relation to foam and emulsion stability. Many foods naturally contain some surfactants (fatty acids, monoacylglycerols, phospholipids), and these can modify the properties of adsorbed protein layers. To some extent, proteins can also displace each other in a surface layer, depending on concentration, surface activity, molar mass, molecular flexibility, etc. Although protein adsorption is irreversible in the sense that it is mostly not possible to lower Γ by diluting the system, the occurrence of mutual displacement nevertheless implies that individual protein molecules in the interfacial layer may interchange with those in solution, albeit slowly.

3.2.3 Contact Angles

When two fluids are in contact with a solid and with each other, there is a contact line between the three phases [1]. An example is given in Figure 7a for the system air–water–solid. There must be a balance between the surface forces acting in the plane of the solid surface and this leads to Equation 7, called the Young equation:

$$\gamma_{SA} = \gamma_{SW} + \gamma_{AW} \cos \theta \tag{7}$$

The contact angle θ is conventionally taken in the densest fluid phase. It depends on the three interfacial tensions. However, γ_{SA} and γ_{SW} cannot be measured, but their difference can be derived from the contact angle. If $(\gamma_{SA} - \gamma_{SW})/\gamma_{AW} > 1$, Equation 7 has no solution, $\theta = 0$, and the solid is completely wetted by the liquid; an example is water on clean glass. If the quotient

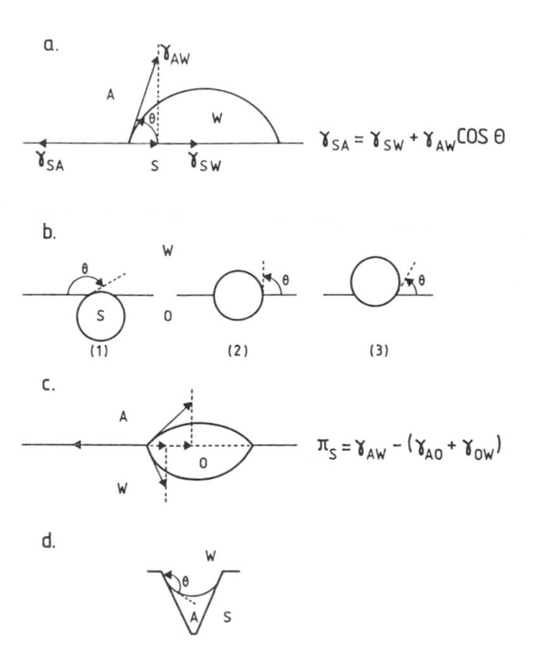

FIGURE 7 Contact angles (θ), shown as cross sections through three-phase systems. A = air, W = water, O = oil, S = solid. (a) Water drop on a solid substrate; (b) solid particle located in an oil–water interface; (c) oil droplet in an air–water interface; (d) air pocket in a crevice in a solid submerged in water. See text.

mentioned is < -1, there is no wetting at all; an example is water on Teflon or other strongly hydrophobic materials.

Figure 7b depicts, in fact, the same situation. In case b1 the contact angle is about 150 degrees, and this would be fairly typical for a triacylglycerol crystal in a triacylglycerol oil–water interface. The contact angle can in such a case be lowered by adding a suitable surfactant, such as Na lauryl sulfate, to the water phase. Addition of a large quantity of surfactant can even lead to $\theta = 0$, and thus to complete wetting of the crystal by the aqueous phase. This is accomplished in some processes to separate fat crystals from oil. Adherence of crystals to the w/o interface and the associated contact angle may be of importance for emulsion stability (e.g., Section 3.6.5).

In Figure 7c the more complicated situation of contact between three fluids is shown. Now there has to be a balance of surface forces in the horizontal as well as in the vertical plane, giving two contact angles. Equation 8 defines the spreading pressures:

$$\Pi_S = \gamma_{AW} - (\gamma_{AO} + \gamma_{OW}) \tag{8}$$

In Figure 7c, $\Pi_S < 0$. If it is greater than zero, the sum of the surface free energies of the a/o and the o/w interfaces is smaller than that of the a/w interface alone, and the oil will spread over the water surface. Use of the values in Table 2 leads to the conclusion that for paraffin oil $\Pi_S = -8\ \text{mN} \cdot \text{m}^{-1}$, implying that the droplet will not spread (but it does adhere to the a/w interface). For triacylglycerol oil, it follows that $\Pi_S = 11\ \text{mN} \cdot \text{m}^{-1}$, and spreading does occur. These aspects are of importance for the interactions between emulsion droplets and foam bubbles. The spreading pressures can, of course, be altered by surfactants. However, most proteins lower γ_{AW} and γ_{OW} by roughly the same amount, and the spreading pressure then is not greatly altered.

It should be noted that the action of gravity can alter the situations depicted in Figure 7. If the droplets are smaller than about 1 mm, the effect of gravity is small.

3.2.4 Curved Interfaces [1]

The pressure at the concave side of a curved phase boundary (interface) always is greater than that at the convex side. The difference is called the *Laplace pressure*. For a spherical surface of radius r, the Laplace pressure p_L is

$$p_L = 2\gamma/r \tag{9}$$

An important consequence is that drops and bubbles tend to be spherical and that it is difficult to deform them, the more so when they are smaller. If a drop is not spherical, the radius of curvature differs with location, which implies a pressure difference within the drop. This causes material in the drop to move from regions with a high pressure to those with a lower one, until a spherical shape is obtained. Only if an outside stress is applied can the drop (or bubble) be deformed from the spherical shape. Some examples may be enlightening. For an emulsion droplet of radius 0.5 μm and interfacial tension $0.01\ \text{N} \cdot \text{m}^{-1}$, the Laplace pressure would be 4×10^4 Pa and a considerable external pressure would be needed to cause substantial deformation. For an air bubble of 1 mm radius and $\gamma = 0.05\ \text{N} \cdot \text{m}^{-1}$, p_L would be 100 Pa, allowing deformation to occur more easily. These aspects will be discussed further in Sections 3.6.2, 3.6.4, and 3.7.1.

Another consequence of the Laplace pressure is *capillary rise*. If a vertical capillary contains a liquid that gives zero contact angle, such as water in a glass tube, a concave meniscus is formed, implying a pressure difference between the water just below the meniscus and that outside the tube at the same height. The liquid in the tube will then rise, until the pressure due to gravity (g times density times difference in height) balances the capillary pressure. For example, pure water in a cylindrical capillary of 0.1 mm internal diameter would rise 29 cm. If the contact angle is larger, the rise will be less; if it is > 90°, capillary depression occurs.

These aspects are relevant to the dispersion of powders in water. If a heap of powder is placed on water, capillary rise of the water through the pores (voids) between the powder particles must occur for wetting of the particles to occur, and this is prerequisite for dispersion. It requires a contact angle (between powder material, water, and air) < 90°. The effective contact angle in the powder is quite a bit smaller than that at a smooth surface of the powder material, so the angle must be distinctly smaller than 90° [see further in Ref. 65].

A third consequence of Laplace pressure is that the solubility of the gas in a bubble in the liquid around it is enhanced. According to Laplace (Eq. 9), the pressure of a gas in a (small) bubble is enhanced and, according to Henry's law, the solubility of a gas is proportional to its pressure. The effect of curvature of a particle on the solubility of the material in the particle is not restricted to gas bubbles and is in general given by the *Kelvin equation*:

$$RT \ln \frac{s}{s_\infty} = \frac{2\gamma M}{\rho r} \tag{10}$$

where s is solubility, s_∞ solubility at a plane surface (i.e., "normal" solubility), and M and ρ are molar mass and mass density, respectively, of the material in the particle. Examples of calculations according to Equation 10 are in Table 4. It is seen that for most systems, particle radius has to be very small (e.g., < 0.1 μm) for a significant effect. However, gas in bubbles of 1 mm has perceptibly enhanced solubility. If the surface is concave rather than convex, the solubility is, of course, decreased (Fig. 7d).

The increased solubility gives rise to *Ostwald ripening*, that is, the growth of large particles in a dispersion at the expense of small ones, and thus the eventual disappearance of the smallest particles. However, this only occurs if the material of the particles is at least somewhat soluble in the continuous phase. It may thus occur in foams and in water-in-oil emulsions, but not in triacylglycerol oil-in-water emulsions. The rate of Ostwald ripening is governed by several factors (see, e.g., Sec. 3.7.2).

Ostwald ripening will always occur with crystals in a saturated solution, albeit slowly if the crystals are large. Another effect is that it causes "rounding" of small crystals. At the edge of a crystal, the radius of curvature may be very small, say some nanometers, and this will lead to a greatly enhanced solubility (Table 4, fat crystal). The material near the edge will thus

TABLE 4 Examples of the Increase in Solubility of the Material in a Particle Due to Curvature, Calculated According to Equation 10 for Some Arbitrary Particle Radii and Some Reasonable Values of the Interfacial Tension (Temperature 300 K)

Variable	Water in oil	Air in water	Fat crystal in oil	Sucrose crystal in saturated solution
r (m)	10^{-6}	10^{-4}	10^{-8}	10^{-8}
γ ($N \cdot m^{-1}$)	0.005	0.05	0.005	0.005
ρ ($kg \cdot m^{-3}$)	990	1.2	1075	1580
M ($kg \cdot mol^{-1}$)	0.018	0.029	0.70	0.342
s_r/s_∞	1.000073	1.010	1.30	1.091

dissolve and be deposited somewhere else. In some systems, this leads to sintering of aggregated crystals (Section 3.4.4).

3.2.5 Interfacial Rheology [37,85]

If an interface contains surfactant, it has rheological properties. Two kinds of surface rheology can be distinguished, in shear and in dilation (Fig. 8). When the interface is sheared (leaving both area and amount of surfactant in the interface constant), one can measure the force in the plane of the interface needed to do this. Often this is done as a function of the shear rate, and a surface shear viscosity η_{ss} (units $N \cdot sec \cdot m^{-1}$) is obtained. For most surfactants η_{ss} is negligibly small, but not for several macromolecular surfactants. For most systems, shear rate thinning occurs and the observed viscosity is an apparent viscosity.

If the interfacial area is enlarged, leaving its shape unaltered, one measures an increase of interfacial tension, because now Γ becomes smaller. This is usually expressed in the surface dilational modulus, defined as

$$E_{sd} \equiv d\gamma/d \ln A \tag{11}$$

where A is the interfacial area. E_{sd} is finite for all surfactants, although it will be very small if surfactant activity is high and the rate of surface enlargement is small. In such a case surfactant rapidly diffuses to the enlarged surface, thereby increasing Γ and thus lowering γ. In other words, the Gibbs equilibrium (Eq. 5) will be rapidly restored. E_{sd} therefore strongly decreases with decreasing rate of deformation. For proteins E_{sd} may be large and less dependent on time scale, because proteins adsorb more or less irreversibly. Besides the interfacial concentration of protein, changes in its conformation can affect the modulus.

Surface rheological properties in dilation and in shear are fundamentally different. Essentially, η_{ss} is due to the viscosity of the material in the interfacial layer, which implies that it is a bulk property, but of a layer of unknown thickness; hence it is expressed as a two-dimensional quantity. On the other hand, E_{sd} is a real interfacial property, which appears in several equations relating to interfacial phenomena. A problem is, however, that measurement of E_{sd} is difficult or even impossible, except at fairly long time scales and/or at small deformation. By and large, for globular proteins at the a/w interface, values of about 30–100 $mN \cdot m^{-1}$ have been observed and for β-casein about 10–20 $mN \cdot m^{-1}$ [25,63]. Surface rheological parameters of protein layers naturally depend on pH, ionic strength, solvent quality, temperature, etc. Often moduli and viscosities are at maximum near the isoelectric pH. It should further be noted that one can also measure a surface dilational viscosity and a surface shear modulus.

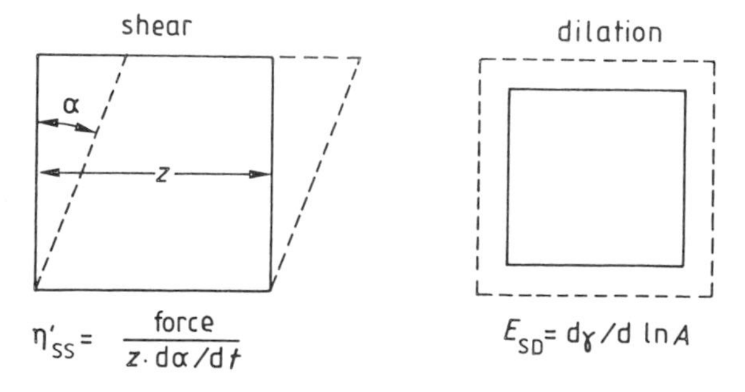

FIGURE 8 Illustration of the geometrical changes brought about in an interfacial element when performing interfacial rheology in simple shear and in dilation. (From Ref. 85.)

3.2.6 Surface-Tension Gradients

If a fluid interface contains a surfactant, surface–tension gradients can be created. This is illustrated for the case of an a/w interface in Figure 9. In Figure 9A a velocity gradient ($G = dv_y/dx$) sweeps surfactant molecules downstream (one may also say moves the surface), thereby producing a surface tension gradient; γ will now be smaller downstream. This implies a stress $\Delta\gamma/\Delta x$, which must be equal and opposite to the shearing stress ηG (η = viscosity of the liquid). If there were no surfactant, the surface would move with the flowing liquid; in the case of an o/w interface, the flow velocity would be continuous across the interface. This has important consequences, especially for foams, as is seen by comparing Figure 9C with D. In the absence of surfactant, the liquid between two foam bubbles rapidly streams downward, like a falling drop. In the presence of surfactant, flow is very much slower. In other words, development

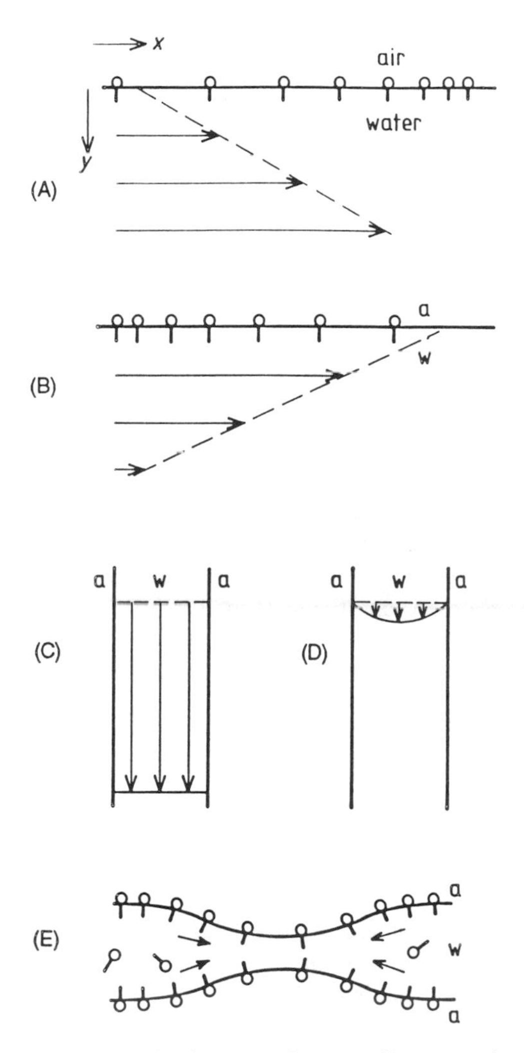

Figure 9 Surface-tension gradients at the a/w interface. (A) Streaming of liquid along a surface causes a surface-tension gradient. (B) A surface-tension gradient causes streaming of the adjacent liquid: Marangoni effect. (C) Drainage of liquid from a vertical film in the absence or (D) presence of surfactant. (E) Gibbs mechanism of film stability. (From Ref. 79.)

of surface-tension gradients is essential to formation of a foam. It also means that an air bubble or emulsion droplet moving through the surrounding liquid in virtually all cases has an immobile surface; that is, it behaves like a rigid particle.

Figure 9B illustrates that liquid adjacent to an interface will move with the interface when the latter exhibits (for some reason, say local adsorption of surfactant) an interfacial tension gradient. This is called the Marangoni effect. It is seen in a glass of wine, where wine drops above the liquid level tend to move upward; here evaporation of ethanol produces the γ gradient. An important consequence is the mechanism for stability of a thin film, illustrated in Figure 9E. If the film somehow acquires a thin spot, the surface area of the film is locally increased; hence Γ is lowered, γ is increased, and a γ gradient is established. The Marangoni effect ensues, causing liquid to flow to the thin spot, thereby restoring film thicknesses. This Gibbs mechanism provides the basis for the stability of films, as in a foam.

Another consequence is the Gibbs–Marangoni effect, which allows formation of emulsions (see Section 3.6.2). In all these situations, the effects depend on film or Gibbs elasticity, which is defined as twice the surface dilational modulus (twice because a film has two surfaces). Thin films typically have a large elasticity, because of the scarcity of dissolved surfactant. In a thick film containing a fairly high concentration of surfactant, surfactant molecules can rapidly diffuse toward a spot with a low surface load and restore the original surface tension. This cannot occur or occurs only very slowly, in a thin film, implying a large elasticity, except at very long time scale.

3.2.7 Functions of Surfactants

Surfactants in a food, whether small-molecule amphiphiles or proteins, can produce several effects, and these are briefly summarized next.

1. Due to the lowering of γ, the Laplace pressure is lowered and the interface can be more easily deformed. This is important for emulsion and foam formation (Sec. 3.6.2) and for the avoidance of coalescence (Sec. 3.6.4).
2. Contact angles are affected, which is important for wetting and dispersion events. The contact angle determines whether a particle can adsorb on a fluid interface and to what extent it then sticks out in either fluid phase. These aspects have an important bearing on stability of some emulsions (Sec. 3.6.5) and foams (Sec. 3.7.2).
3. A decrease in interfacial free energy will proportionally slow Ostwald ripening. The rate of Ostwald ripening may also be affected by the surface dilational modulus (Sec. 3.7.2).
4. The presence of surfactants allows the creation of surface-tension gradients and this may be their most important function. It is essential for formation and stability of emulsions and foams (Secs. 3.6.2, 3.6.4, 3.7.1, and 3.7.2).
5. Adsorption of surfactants onto particles may greatly modify (colloidal) interparticle forces, mostly enhancing repulsion and thereby stability. This is discussed in Section 3.3.
6. Small-molecule surfactants may undergo specific interactions with macromolecules. They often associate with proteins, thereby materially altering protein properties. Another example is the interaction of some polar lipids with amylose.

3.3 COLLOIDAL INTERACTIONS

In Section 3.1.2 colloids were defined and classified. Generally, between particles forces act that originate from material properties of the particles and the interstitial fluid. These colloidal interaction forces act in a direction perpendicular to the particle surface, contrary to the surface forces discussed in Section 3.2, which act in the direction of the surface. Colloidal interaction has important consequences:

1. It determines whether particles will aggregate (Sec. 3.4.3), which, in turn, may determine further physical instability, for instance, sedimentation rate. (Note on terminology: The terms flocculation and coagulation are also used, often with a more specific connotation; the former would, for instance refer to reversible aggregation, and the latter to irreversible.)
2. Aggregating particles may form a network (Sec. 3.5), and the rheological properties and the stability of systems containing networks strongly depend on colloidal interaction.
3. It may greatly affect susceptibility of emulsion droplets to (partial) coalescence (Secs. 3.6.4–5).

Literature on colloid science can be found in textbooks mentioned in the Bibliography.

3.3.1 Van der Waals Attraction

Van der Waals forces between molecules are ubiquitous, and they also act between larger entities such as colloidal particles. Since these forces are additive, it turns out that, within certain limits, the dependence of the interaction force on interparticle distance (as measured between the outer surfaces) is much weaker between particles than between molecules. For two identical spherical particles the van der Waals interaction free energy is

$$V_A \approx -Ar/12h \qquad h < \sim 10\ \text{nm} \tag{12}$$

where r is particle radius, h is interparticle distance, and A is the Hamaker constant. The last depends on the material of the particles and the fluid in between, and it increases in magnitude as the differences in the properties of the two materials increase. For most particles in water, A is between 1 and 1.5 times kT ($4–6 \times 10^{-21}$ J). Tabulated values are available [27,72].

If both particles are of the same material and the fluid in between is different, A always is positive and the particles attract each other. If the two particles are of different materials, A may be negative and there would be van der Waals repulsion, but this is fairly uncommon.

3.3.2 Electric Double Layers

Most particles in an aqueous solution exhibit an electric charge, because of adsorbed ions or ionic surfactants. In most foods, charges predominantly are negative. Since the system must be electroneutral, the particles are accompanied by a cloud of oppositely charged ions, called counterions. An example of the distribution of counterions and coions is given in Figure 10a. It is apparent that at a certain distance from the surface, the concentrations of positive and negative charges in the solution become equal. Beyond that distance, the charge on the particle is neutralized, due to an excess of counterions in the electric double layer. The latter is defined as the zone between the particle surface and the plane at which neutralization is achieved. The double layer should not be envisaged as being immobilized, because solvent molecules and ions diffuse in and out of the layer.

The electrical effects are usually expressed in the electric potential ψ. Its value, as a function of the distance h, from the surface, is given by

$$\psi = \psi_0 e^{-\kappa h} \tag{13}$$

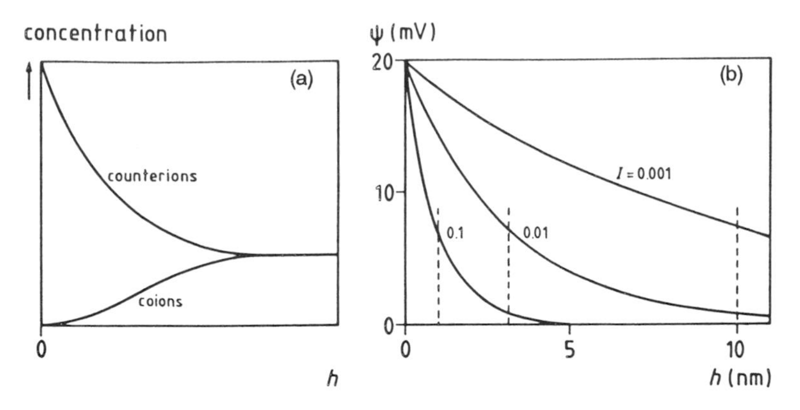

FIGURE 10 The electric double layer: (a) the distribution of counterions and coions as a function of the distance from the charged surface, and (b) the potential ψ as a function of distance for three values of the ionic strength I; the broken lines indicate the Debye length.

where ψ_0 is the potential at the surface and the nominal thickness of the electric double layer or Debye length $1/\kappa$ is given by

$$\kappa \approx 3.2 I^{0.5} \qquad (\text{nm}^{-1}) \tag{13a}$$

for dilute aqueous solutions at room temperature. The ionic strength I is defined as

$$I \equiv \tfrac{1}{2} \sum m_i z_i^2 \tag{14}$$

where m is molar concentration and z is valence of each of the ionic species present. Note that for a salt like NaCl, I equals molarity of the solution, but this is not so if ions of higher valence are present. For $CaCl_2$, I is three times the molarity.

Calculations of the potential as a function of distance are given in Figure 10b. Ionic strengths in aqueous foods vary from 1 mM (a typical tap water) to more than 1 M (pickled materials). The I of milk is about 0.075 M, and of blood about 0.14. Consequently, the thickness of the double layer is often only about 1 nm or less. This has important consequences when both negative and positive charges occur on one particle. If the distance between charged groups (or clusters of charged groups) on a surface (or on a macromolecule) is less than, say, $2/\kappa$, only the average potential is sensed from a distance. If, on the other hand, the distance between charges is greater than $2/\kappa$, the charges can be sensed separate from each other, enabling positive groups on one particle or macromolecule to react with negative groups on another one, forming salt bridges. This is, of course, relevant for proteins, which often are amphipolar.

Electrical interactions depend on the surface potential, and this, in turn, is often influenced by pH. For most food systems, values of $|\psi_0|$ are below 30 mV. At a high concentration of counterions (especially if these are divalent), ion pairs can be formed between counterions and charged groups on the particle surface thereby lowering $|\psi_0|$.

In a nonaqueous phase, the dielectric constant generally is much smaller than in water, and Equation 13a is no longer valid. Moreover, the ionic strength in this situation generally will be negligible. This means that even if there is a charged surface (as may be the case for aqueous droplets floating in oil), electrical interaction forces will usually be unimportant.

3.3.3 DLVO Theory

If electrically charged particles having the same sign come very close to each other, their electric double layers overlap. This is sensed and the particles repulse each other. The repulsive electric interaction free energy V_E can be calculated. For spheres of equal size it is

$$V_E \approx (4.3 \times 10^{-9}) r \psi_0^2 \ln(1 + e^{-\kappa h}) \tag{15}$$

which would be valid for $h < 10$ nm, $|\psi_0| < 40$ mV, $\kappa r >> 1$, in water at room temperature.

The interaction energies V_A (due to van der Waals attraction) and V_E can be added, and this has led to the first useful theory for colloid stability, the DLVO (Deryagin–Landau, Verwey–Overbeek) theory. This theory enables calculation of the total free energy V needed to bring two particles from infinite distance to a distance h. Particle radius can be determined, several Hamaker constants are known, κ is derived from the ionic strength and the electrokinetic (or zeta) potential, which often is assumed to be equal to ψ_0, can be experimentally determined by electrophoresis. The total interaction energy is usually divided by kT, that is, the average free energy involved in an encounter between two particles by Brownian (heat) motion.

Some plots of V/kT versus h are presented in Figure 11. If V is negative for all h, the particles will aggregate. If there is a maximum in the curve (as in Fig. 11a near x) that is much larger than kT, particles may never overcome this free energy barrier; if, on the other hand the maximum is somewhat lower, say $10kT$, two particles may occasionally reach the "primary minimum" (as in Fig. 11a near y) and become irreversibly aggregated. If there is a sufficiently deep "secondary minimum" (Fig. 11a near z), the particles readily become aggregated, but not fully irreversibly. It is seen that ψ_0 (Fig. 11b) and ionic strength (Fig. 11c) have large effects on stability. Also the effect of particle radius seems to be large (Fig. 11d), but the DLVO theory appears to be incorrect in predicting the effect of size.

Quite generally, the DLVO theory—though very successful for many inorganic systems—seems to be inadequate for predicting stability of biogenic systems. Milk fat globules, for example, are stable against aggregation at their isoelectric pH (3.8), where they have zero surface

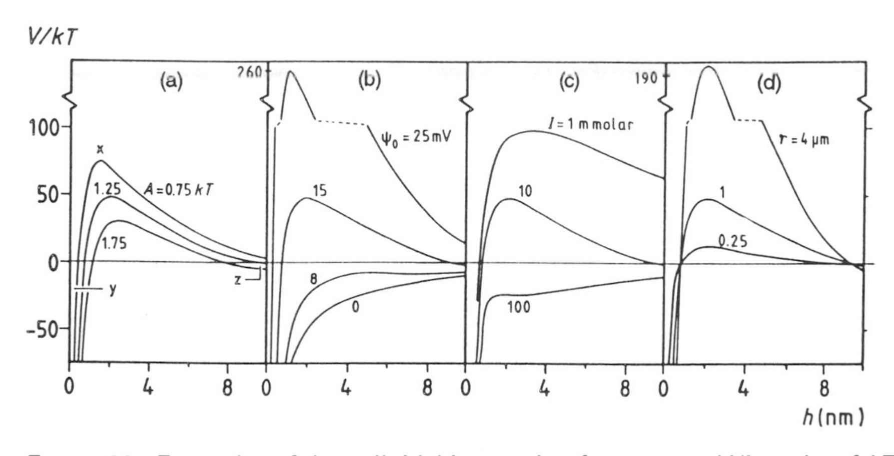

FIGURE 11 Examples of the colloidal interaction free energy V (in units of kT) as a function of interparticle distance h according to the DLVO theory for spherical particles in water. Illustrated are effects of the magnitude of (a) the Hamaker constant A; (b) surface potential ψ_0; (c) ionic strength I; and (d) particle radius r. Unless indicated otherwise, $A = 1.25\ kT$; $\psi_0 = 15$ mV; $I = 10$ mM; $r = 1\ \mu$m.

potential, so that the DLVO theory would predict zero repulsion [84]. Consequently, interaction forces other than those considered in this theory must be important.

3.3.4 Steric Repulsion

As depicted in Figure 4, some adsorbed molecules (polymers, Tweens©, etc.) have flexible molecular chains ("hairs") that protrude into the continuous phase. These may cause steric repulsion. Two mechanisms can be distinguished. First, if the surface of another particle comes close, the hairs are restricted in the conformations they can assume, which implies loss of entropy, that is, increase of free energy, and repulsion occurs. This volume-restriction effect can be very large, but it can be of importance only if the surfaces have a very low hair density (number of hairs per unit area). This is because the hairy layers start to overlap on approach of the particles and then a second mechanism will act before the first one comes into play. The overlap causes an increased concentration of protruding hairs and thereby an increased osmotic pressure; this would lead to water moving to the overlap region, which results in repulsion. However, this is true only if the continuous phase is a good solvent for the hairs; if it is not, attraction may result. For example, emulsion droplets covered by casein have protruding hairs, providing stability to the droplets. If ethanol is added to the emulsion, the solvent quality is strongly decreased and the droplets aggregate [16].

In some cases, steric repulsion free energy can be calculated with reasonable accuracy [22]. If these values are added to the van der Waals attraction, curves for total interaction versus interparticle distance are obtained. Schematic examples are shown in Figure 12. The solvent quality usually is of overriding importance and if it is good, repulsion can be very strong; see curve c versus d. Only for large particles (strong van der Waals attraction) and a fairly thin adsorbed layer (short hairs) will aggregation occur in a good solvent (curve a, which shows a minimum free energy of about –12 kT).

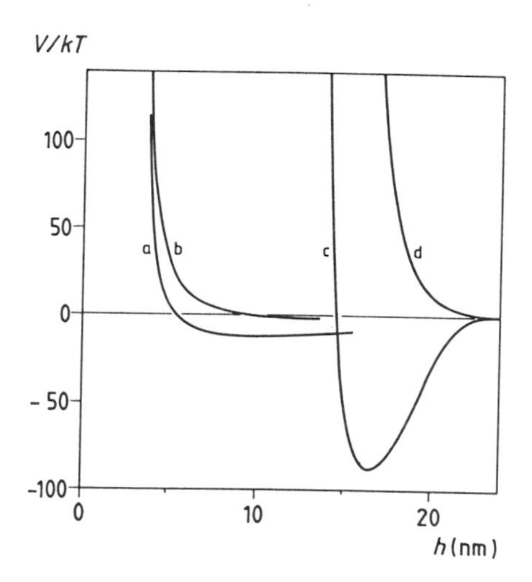

FIGURE 12 Hypothetical examples of the effects of particle size, thickness of the layer of protruding molecular chains, and the quality of the solvent for those chains on the colloidal interaction free energy (V in units of kT) between identical spherical particles as a function of the interparticle distance h. The interaction is supposed to be due to van der Waals attraction and steric repulsion. Curves: (a) good solvent, thin layer, particle diameter 2.5 μm; (b) same, but diameter 0.5 μm; (c) relatively poor solvent, thick layer, diameter 2.5 μm; (d) same, but good solvent.

In practical food systems calculation of steric repulsion usually is not possible because the situation can be complex. For example, the nature of the adsorbing molecules can vary greatly [18,22,81].

It should also be mentioned that adsorbing polymers may cause bridging aggregation, by becoming simultaneously adsorbed onto two particles [22,81]. This may happen if too little polymer is present to fully cover the particle surface area or with certain methods of processing.

3.3.5 Depletion Interaction

Nonadsorbing polymers are depleted near an interface, because the center of a molecule cannot come closer to the interface than the effective radius of the (random coil) molecule. Close to the interface, there is thus a layer with a polymer concentration lower than in the bulk. This depletion layer has a thickness (δ) that is about equal to the radius of the gyration (R_g) of the polymer molecules (Fig. 13). If the particles come close to each other, the depletion layers overlap, implying that the polymer concentration in the bulk becomes lower. This leads to a larger mixing entropy, that is, to a lower free energy and to a driving force for particle aggregation. An estimate of the depletion free energy can be obtained from

$$V_D(h) \approx -2\pi r \Pi_{osm} (2\delta - h)^2 \tag{16}$$

where Π_{osm} is the osmotic pressure caused by the dissolved polymer. It is, in first approximation, proportional to the molar concentration of the polymer, but it will be more concentration dependent in a good solvent. Thus polysaccharides at low concentrations can cause depletion aggregation in foods; for example, 0.03% xanthan may be sufficient [16]. Fairly high concentrations of polymer often lead to immobilization of the particles (Section 3.4.2), thereby causing the system to be fairly stable.

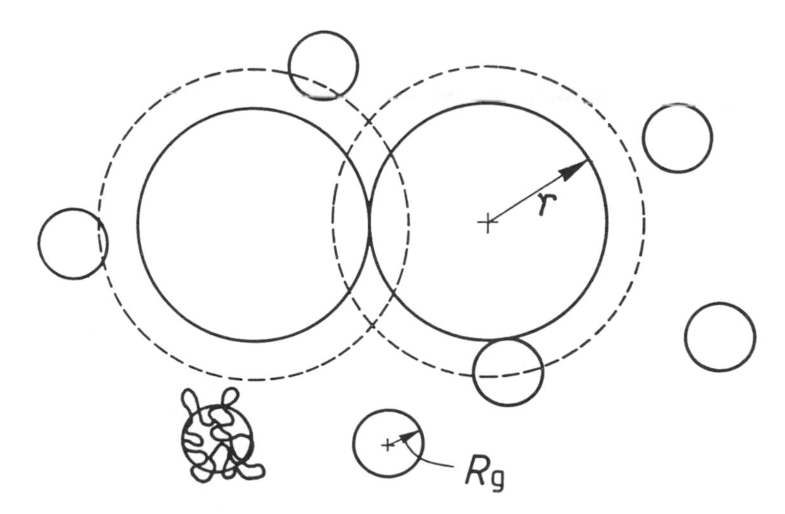

FIGURE 13 Schematic depiction of the depletion of non adsorbing macromolecules (radius of gyration R_g, depicted—except for one example of a random coil—by small open circles) from the surface of colloidal particles (radius r, depicted by large circles), and of the overlap of the depletion zones (bounded by broken lines) if the particles are aggregated. (From Ref. 81).

TABLE 5 Factors Affecting the Magnitude of the Contribution to the Interaction Free Energy Between Particles by Van der Waals Forces (V_A), Electrostatic Forces (V_E), and Steric Interactions (V_S): +, Effect; –, No Effect; (+), Effect Under Some Conditions

Variable	V_A	V_E	V_S
Particle size	+	+	(+)
Particle material	+	–	–
Adsorbed layer	(+)	+	+
pH	–	+	–[a]
Ionic strength	–	+	–[a]
Solvent quality	–	–	+

[a]In the absence of electrical charges.

3.3.6 Other Aspects

It should now be clear that several kinds of colloidal interactions can occur in foods and that the kind and concentration of surfactants present strongly influence these interactions. Even in the simplest cases, several variables are important (Table 5).

Several additional complications may be mentioned. The DLVO theory does not apply at very small distances. One cause may be surface roughness. Another may be *solvation repulsion*. If solvent molecules are attracted by the material on the outside of the particle, this may cause some repulsion at short distance [28]. This may affect the depth of the primary minimum (in plots of V vs. h) and thereby the irreversibility of aggregation.

Hydrophobic interactions also may occur, and they generally cause attraction. The effect is the result of poor solvent quality (Sec. 3.3.4 and Chap. 2). This type of interaction has a strong temperature dependence, being very weak near 0°C and increasing with increasing temperature.

Such hydrophobic interactions may, in principle, occur if a *protein* is the surfactant. However, repulsion is more likely in this case. Usually this is caused by a combination of steric and electrostatic repulsion, but calculation of the interaction free energy generally is not possible.

Another aspect is *double adsorption*, that is, adsorption of a substance onto a layer of surfactant. There are many examples of polysaccharides adsorbing onto particles that are already covered with small-molecule surfactants or proteins [3]. General rules cannot be given, but a kind of bridging aggregation occurs in many cases. It thus turns out that added polysaccharides can act in various ways. If they adsorb they can provide steric repulsion, but under other conditions they can cause bridging aggregation. If they do not adsorb they may cause depletion aggregation, but at high concentration they may stabilize the system.

3.4 LIQUID DISPERSIONS

3.4.1 Description

Several types of liquid dispersions exist. The discussions here will be limited to suspensions (solid particles in a liquid) and to those aspects of emulsions that follow the same rules. Foods that are suspensions include skim milk (casein micelles in serum); fat crystals in oil; many fruit and vegetable juices (cells, cell clusters, and cell fragments in an aqueous solution); and some fabricated foods, such as soups. During processing (food fabrication), suspensions are also encountered, such as starch granules in water, sugar crystals in a saturated solution, and protein aggregates in an aqueous phase.

Dispersions are subject to several kinds of instability, and these are schematically illustrated in Figure 14. Changes in particle size and in their arrangement are distinguished. Formation of small aggregates of particles may be considered to belong to both categories. Dissolution and growth of particles depend on concentration of the material, on its solubility, and on diffusion. In a supersaturated solution nucleation must occur before particles can be formed. Dissolution, nucleation, and growth will not be discussed further. Ostwald ripening is discussed in Sections 3.2.4 and 3.7.2 and coalescence in Section 3.6.4. The other changes are discussed next.

The various changes may affect each other, as is illustrated in Figure 14. Moreover, sedimentation is enhanced by any growth in particle size, and sedimentation will enhance the rate of aggregation if the particles tend to aggregate. Agitation of the liquid may enhance the rate of some changes, but it can also disturb sedimentation and disrupt large aggregates.

3.4.2 Sedimentation

If there is a difference in density (ρ) between the dispersed phase (subscript D) and the continuous phase (subscript C), there is a buoyancy force acting on the particles. According to Archimedes, this is for spheres $a\pi d^3(\rho_D - \rho_C)/6$, where a is acceleration. The sphere now

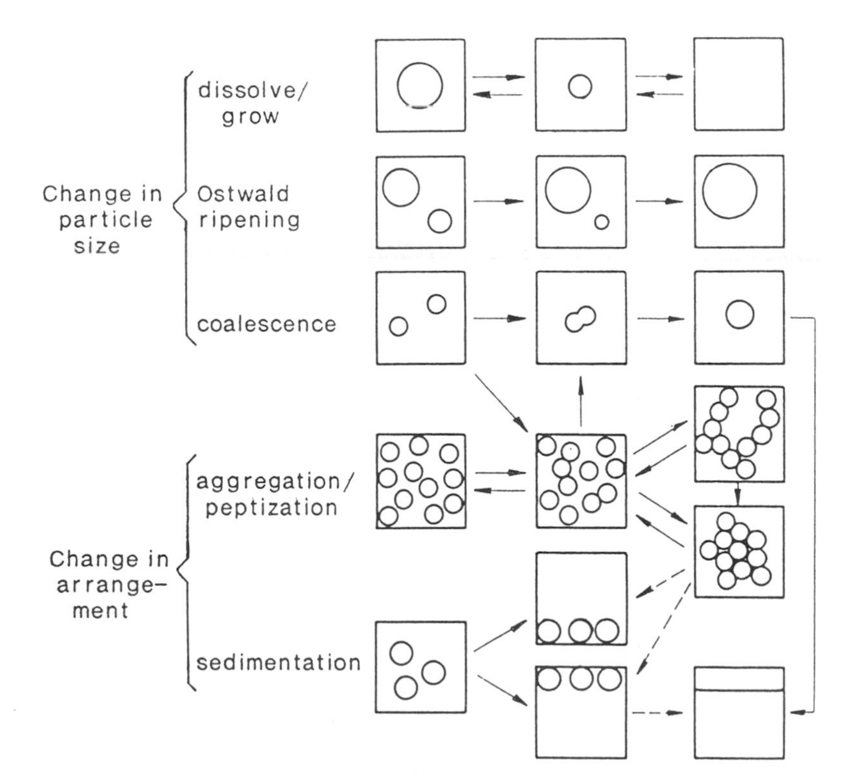

FIGURE 14 Illustration of the various changes in dispersity. Highly schematic.

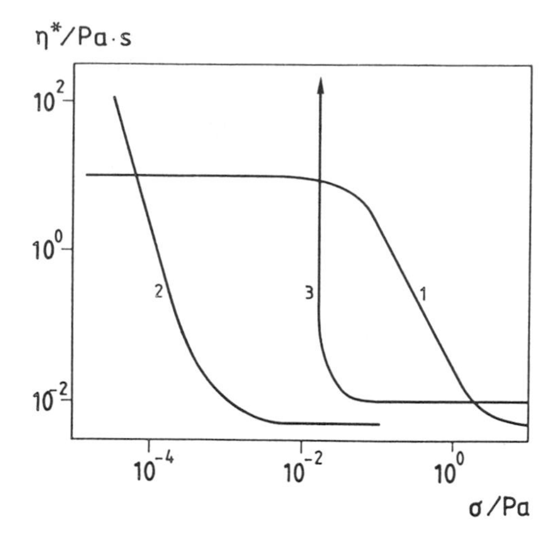

FIGURE 15 Schematic examples of non-Newtonian flow behavior of liquids: apparent viscosity η^* as a function of shearing stress σ. Curve 1 is typical for a polymer solution. Curve 2 is typical for a slightly aggregating dispersion of very small particles. Curve 3 is typical for a system exhibiting a yield stress.

sediments at a linear velocity v and feels a friction force, which is, according to Stokes, $3\pi d\eta_C v$, where η_C is the viscosity of the continuous phase. By putting both forces equal, the equilibrium or Stokes sedimentation velocity is obtained:

$$v_S = a(\rho_D - \rho_C)d^2/18\eta_C \tag{17}$$

If the particles show a size distribution, d^2 should be replaced by $\Sigma n_i d_i^5/\Sigma n_i d_i^3$, where n_i is the number of particles per unit volume in class i with diameter d_i.

For gravity sedimentation $a = g = 9.81\ \text{m} \cdot \text{sec}^{-2}$, and for centrifugal sedimentation $a = R\omega^2$, where R is the effective radius of the centrifuge and ω its rotation rate in radians per second. To give an example: If the sphere diameter is 1 μm, the density difference is 100 kg · m^{-3}, and the viscosity of the continuous phase is 1 mPa · sec (i.e., water), the spheres would sediment under gravity at a rate of 55 nm · sec^{-1} or 4.7 mm per day. Sedimentation greatly depends on particle size, and spheres of 10 μm would move 47 cm in a day. Normally, viscosity decreases and sedimentation rate increases with increasing temperature. If the density difference in Equation 17 is negative, sedimentation is upward, and one commonly speaks of creaming.

Equation 17 is very useful to predict trends, but it is almost never truly valid. Among the many factors causing deviation from Equation 17 [83], the following are the most important for foods:

1. The particles are not homogenous spheres. Any anisometric particle tends to sediment more slowly, because it orients itself during sedimentation in such a way as to maximize friction; that is, a plate-shaped particle will adopt a "horizontal" orientation. An aggregate of particles, even if spherical, sediments more slowly than a homogeneous sphere of the same size, since the interstitial liquid in the aggregate causes the effective density difference to be smaller.
2. Convection currents in the dispersion, caused, for instance, by slight temperature fluctuations, may strongly disturb sedimentation of small particles ($\leq \sim 1$ μm).

3. If the volume fraction of particles φ is not very small, sedimentation is hindered. For $\varphi = 0.1$, the sedimentation rate is already reduced by about 60%.
4. If particles aggregate, sedimentation velocity increases: The increase in d^2 is always larger than the decrease in $\Delta\rho$. Moreover, as larger aggregates sediment faster they overtake smaller ones, and thus become ever larger, leading to an even greater acceleration of sedimentation rate. This may enhance sedimentation by orders of magnitude. A good example is rapid creaming in cold raw milk, where fat globules aggregate due to the presence of cryoglobulins [84].
5. An assumption implicit in Equation 17 is that viscosity is Newtonian, that is, independent of shear rate (or shear stress), and this is not true for many liquid foods. Figure 15 gives some examples of the dependence of apparent viscosity on shear stress. The stress caused by a particle is given by the buoyancy force over the particle cross section, that is, $\frac{2}{3} g\ \Delta\rho d$ for spheres under gravity. The stress is on the order of a millipascal for many particles. This then is the stress that the particles sense during sedimentation. Viscosity should be measured at that stress (or the corresponding shear rate, given by η^*/σ), whereas most viscometers apply stresses of well over 1 Pa. Figure 15 shows that the apparent viscosity can differ by orders of magnitude, according to the shear stress applied.

   Also shown in Figure 15 is an example of a liquid exhibiting a small yield stress. Below that stress the liquid will not flow. However, this is never noticed during handling, because the yield stress is so very small (a stress of 1 Pa corresponds to a water "column" of 0.1 mm height). Nevertheless, such a small stress often is sufficient to prevent sedimentation (or creaming), as well as aggregation. Among liquid foods exhibiting a yield stress arc soya milk, many fruit juices, chocolate milk, and several dressings. These aspects are discussed further in Section 3.5.2 and in Ref. 71.

3.4.3 Aggregation Kinetics

Particles in a liquid exhibit Brownian motion and thereby frequently encounter each other. Such encounters may lead to aggregation, defined as a state in which the particles stay close together for a much longer time than they would in the absence of attractive colloidal interaction. The rate of aggregation is usually calculated according to Smoluchowski's theory of perikinetic aggregation [54]. The initial aggregation rate in a dilute dispersion of spheres of equal size is

$$-dN/dt = 4kTN^2/3\eta W \tag{18}$$

where N is the number of particles, that is, unaggregated particles plus aggregates, per unit volume. The stability factor W was assumed to equal unity by Smoluchowski. The time needed to halve the number of particles then is

$$t_{0.5} = \pi\eta/8kT\varphi \tag{18a}$$

where φ is the particle volume fraction. This results in $d^3/10\varphi$ seconds for particles in water at room temperature where d is in micrometers. For $d = 1$ μm and $\varphi = 0.1$, this would be 1 sec, implying that aggregation is very rapid.

In most practical situations, aggregation is much slower, because W often has a large value. If it is desired to increase the halving time from 1 sec to 4 months, this would need a W of 10^7. The magnitude of the stability factor is primarily determined by the colloidal repulsion between the particles (Sec. 3.3).

Direct use of Equation 18 to predict stability is rarely possible in food systems. There are numerous complications. Some of the more important ones are that (a) it is mostly impossible to establish the value of W, (b) the stability factor may change with time, (c) there are other encounter mechanisms, due to streaming (agitation) or to sedimentation, and (d) aggregation may take various forms, leading to coalescence or to aggregates, which may be compact or tenuous. There are even more complications. Nevertheless, application of aggregation theory often is possible and useful, but it is far more intricate than can be discussed here [8,81,83].

3.4.4 Reversibility of Aggregation

According to the nature of the interaction forces between aggregated particles (Sec. 3.3), agents can be added to cause *deaggregation*. This can occur during processing, or can be done in the laboratory to establish the nature of the forces. It should be realized that often more than one type of force is acting. Diluting with water may cause deaggregation (a) due to lowering of osmotic pressure (if depletion interaction was the main cause of aggregation), (b) due to lowering ionic strength (which enhances electrostatic repulsion), or (c) due to enhancing solvent quality (which can increase steric repulsion). Electric forces can also be manipulated by altering pH. Bridging by divalent cations can often be undone by addition of a chelating agent, say ethylenediamine tetraacetic acid (EDTA). Bridging by adsorbed polymers or proteins can mostly be undone by addition of a suitable small-molecule surfactant (Sec. 3.2.2). Reversal of specific interactions, such as -S-S- bridges, requires specific reagents. Also, a change in temperature can affect aggregate stability, by altering solvent quality.

If the forces between the particles in an aggregate are not very strong, deaggregation can be achieved by *shear forces*. These exert a stress ηG, where G is the shear rate. In water, $G = 10^3\ \text{sec}^{-1}$ would be needed to achieve a shear stress of 1 Pa, which does not seem very large. However, the shear force acting on an aggregate is proportional to aggregate diameter squared, and large aggregates often have weak spots, because of their inhomogeneity. This means that aggregates often are degraded to a certain size by a given shear rate.

Another aspect is that *bonds* may *strengthen* after aggregation. It may be better to speak of junctions between particles, since any such junction may represent many (often hundreds of) separate bonds. The strengthening may occur by several mechanisms [81].

3.5 GELS

Many foods are "soft solids," and these can often be said to be gels or gel-like. Some general aspects of idealized model systems will be discussed here, with principles being emphasized. Especially the mechanical (i.e., rheological and fracture) properties [e.g., 47] cannot be discussed in depth, because it would take far too much space and is outside the realm of this book. Nevertheless, some basic facts involving rheology are needed to understand gel properties.

3.5.1 Description

From a rheological viewpoint a typical gel is a material that exhibits a yield stress, has viscoelastic properties and has a moderate modulus (say, $< 10^6$ Pa). This is illustrated in Figure 16a. When a small stress ($\sigma \equiv$ force/area) acts on the material, it behaves elastically: It instantaneously deforms, keeps the shape obtained as long as the stress acts, and returns instantaneously to its original shape as soon as the stress is removed. For a greater stress, the material may show viscoelastic behavior: It first deforms elastically but then starts to flow; after removal of the stress it only partly regains its original shape. Figure 16b illustrates the differences between a liquid, a solid, and a viscoelastic material. Strain (ε) means relative deformation, and strain rate is its change with time ($d\varepsilon/dt$) (the latter may be equal to the shear rate or to some other parameter, according to the type of deformation). Figure 16 also explains what yield stress

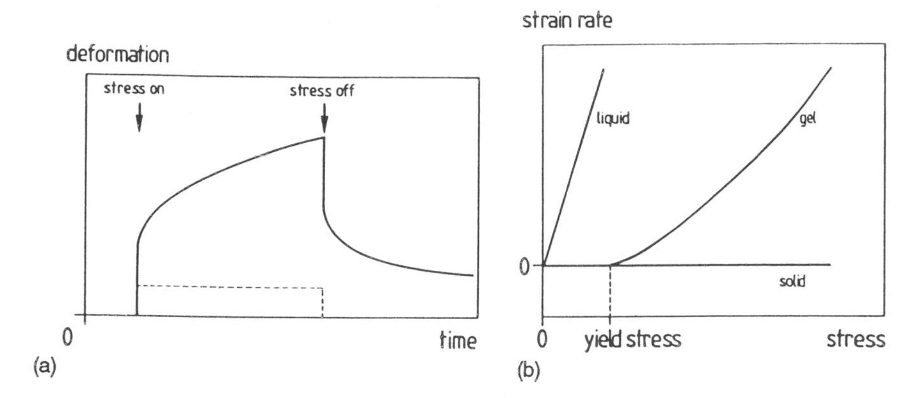

Figure 16 Characteristics of a visco-elastic material: (a) Example of the relation between deformation (strain) and time, when a visco-elastic material is suddenly brought under a certain stress, as well as after removal of the stress. The broken line is for a stress below the yield stress. (b) Strain rate as a function of stress for a Newtonian liquid, a viscoelastic material, and an elastic solid.

(σ_y) means. The behavior of a viscoelastic material greatly depends on the time scale of the deformation. Rapid deformation—for distance, achieved by applying a fluctuating stress at high frequency—implies a short time scale. At very short time scales a gel is almost purely elastic, and at very long time scales almost purely viscous. Also the yield stress mostly is time dependent, being smaller at longer time scales. All these rheological parameters may vary greatly in magnitude among gels. For instance, yield stresses from 10^{-5} to 10^4 Pa have been observed.

From a structural point of view, a gel has a continuous matrix of interconnected material with much interstitial solvent. Figure 17 illustrates three main types of gels occurring in food systems; moreover, intermediate and combined types occur. The various types have different properties; for instance, particle gels are much coarser than polymer gels, thus having a much greater permeability (Sec. 3.5.2).

The *mechanical properties* of various gels differ greatly. To explain this, the behavior at large deformations should be considered [47,55,66,67]. Figure 18a gives a hypothetical stress–strain curve, ending at the point where fracture occurs. Fracture implies that the stressed specimen breaks, mostly into many pieces; if the material contains a large proportion of solvent, the space between the pieces may immediately become filled with solvent, rather than air. The modulus of the material (G), also called stiffness, is stress divided by strain, provided this quotient is constant. For most gels, proportionality of stress and strain is only observed at very small strains, and at larger strains the quotient may be called an apparent modulus. The strength of the material is the stress at fracture (σ_{fr}). Terms like firmness, hardness, and strength are often used rather indiscriminately; sensoric firmness or hardness often correlates with fracture stress. Modulus and fracture stress need not be closely correlated, as is clear in Figure 18b. These parameters also vary greatly with concentration of the gelling material. It is frequently observed that addition of inert particles ("fillers") to a gelling material increases the modulus but decreases the fracture stress [38]. Part of the explanation of these divergencies is that a modulus is predominantly determined by number and strength of the bonds in the gel, whereas fracture properties highly depend on large-scale inhomogeneities.

The strain at fracture (ε_{fr}) may be called longness, but this term is rarely used. The terms shortness and brittleness are used, and they are closely related to $1/\varepsilon_{fr}$. The strain at fracture may vary widely; for gelatin ε_{fr} may be 3, and for some polysaccharide gels only 0.1. For gels as

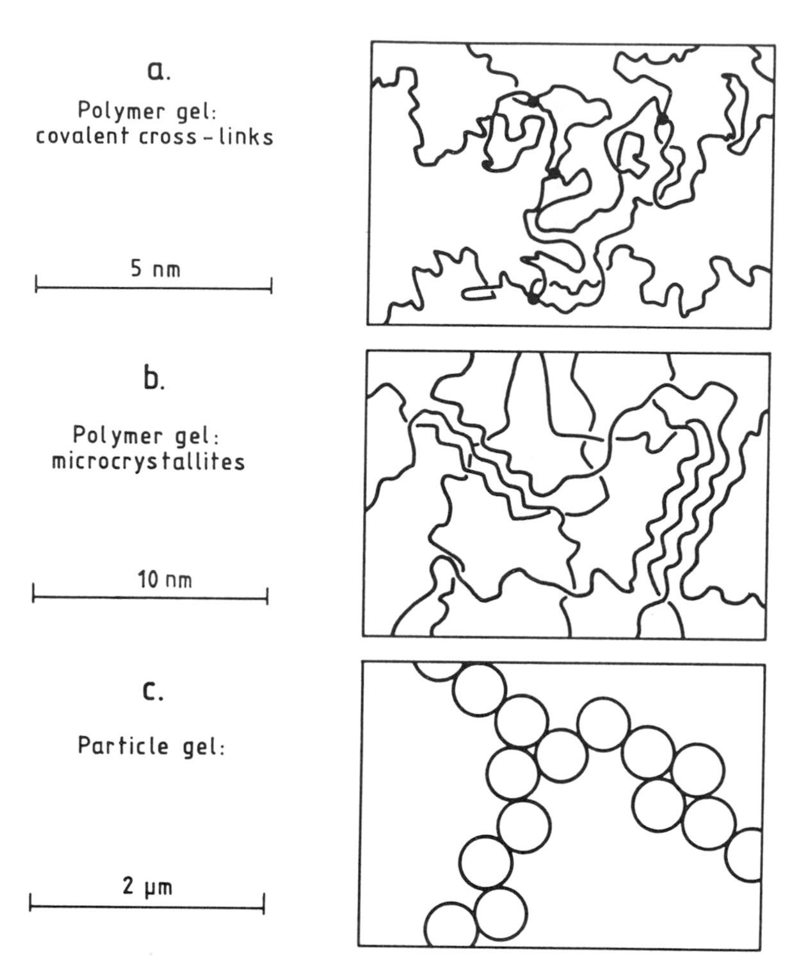

FIGURE 17 Schematic illustrations of three main types of gel. The dots in frame (a) denote cross-links. Note the differences in approximate scale.

depicted in Figure 17a and b, ε_{fr} greatly depends on length and stiffness of the polymer chains between cross links.

Another parameter is the toughness or work of fracture E_{fr}. This is derived from the area under the curve in Figure 18a and expressed in joules per cubic meter.

All these parameters depend on time scale or rate of deformation, but in various manners. Several gels do not fracture at all when deformed slowly, but do so when deformation is rapid. In some cases, it is even difficult to distinguish between yielding and fracture. For some gels, fracture itself takes a long time.

Gels may be formed in various ways, according to the kind of gelling material. *Polymer* molecules in solution [e.g., 4] behave more or less as random coils, effectively immobilizing a large amount of solvent (water), thereby increasing viscosity considerably. If the polymer concentration is not very low, the individual molecules tend to interpenetrate and form entanglements. This gives the solution some elasticity, but no yield stress. Gelation is caused by formation of intermolecular cross-links. These can be covalent bonds, salt bridges, or microcrystalline regions. Covalent bonds are provoked by added reagents or by increased temperature. The latter can occur with some proteins, where -S–S- bridges form during heating. For polyelectrolytes,

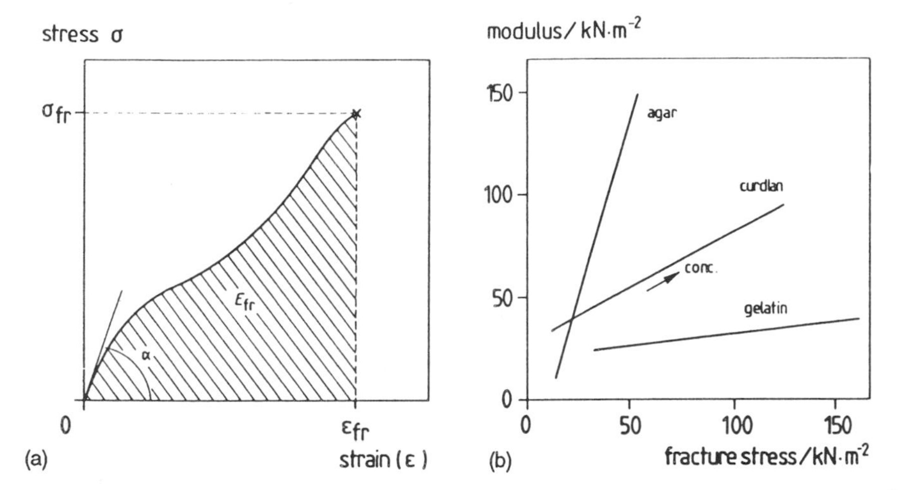

FIGURE 18 (a) Hypothetical example of the relation between stress and strain when deforming a viscoelastic material until it fractures. The modulus equals tan α. See text. (b) Relation between modulus and fracture stress for gels of various materials (curdlan is a bacterial β-1,3-glucan) at varying concentration. (After Ref. 31.)

that is, charged polymers such as proteins, formation of salt bridges may induce gelation. Other polymers gel by forming microcrystalline regions that act as cross-links. These are commonly formed on cooling, because then a reversible transition occurs that involves stiffening and localized crystallization of polymer chains.

Polymer gels are often discussed in terms of the so-called rubber theory, where it is assumed that the polymer chains between cross-links are very long and can assume a large number of conformations. Deforming the gel leads to a decrease in the possible number of conformations and thereby to a decrease in entropy. In such an ideal entropic gel, the elastic modulus is simply

$$G = \nu kT \tag{19}$$

where ν is the number of effective cross-links per unit volume. As will be seen later, only a few food polymer gels behave like this.

Particle gels may form because of aggregation, induced by a change in pH, ionic strength, solvent quality, etc. (Section 3.3). In these gels, the entropy loss on deformation is negligible and they derive their elastic modulus from deformation of bonds, an enthalpic effect. Particle gels are often *fractal* in nature [86]. If attractive particles encounter each other at random they form aggregates, and as these aggregates encounter other aggregates, larger aggregates form. This is called cluster–cluster aggregation. The relation between the average number of particles in an aggregate N_p and the radius of the aggregate R then is

$$N_p = (R/r)^D \tag{20}$$

where r is the radius of the primary (component) particles and D is a constant < 3 and is called the fractal dimensionality. Because it is smaller than 3, aggregates become more tenuous (rarefied) when they increase in size. The average volume fraction of particles in an aggregate is given by

$$\varphi_{ag} \equiv \frac{N_p}{N_m} = \frac{(R/r)^D}{(R/r)^3} = (R/r)^{D-3} \tag{21}$$

where N_m is the number of primary particles that a sphere of the same radius can contain assuming close packing. It is apparent that φ_{ag} decreases as R increases. When φ_{ag} becomes equal to the volume fraction of primary particles in the system, φ, the aggregates touch each other and a gel forms. This implies an unequivocal gel point. It also implies that a gel is formed for any φ, however small, although the gel may be extremely weak. The critical radius of the aggregates at gelation is given by

$$R_{cr} = r\varphi^{1/(D-3)} \tag{22}$$

The size of the largest holes in the gel network equals about R_{cr}. Often, it is found that $D \approx 2.2$ and it then follows that for $\varphi = 0.01$, $R_{cr}/r = 316$. For $\varphi = 0.1$ the ratio is 18, and for $\varphi = 0.4$ it is about 3.

Although particles gels may seem to be rather disordered structures, simple scaling laws hold. Such relations can be given for rheological properties and permeability (see later). Such simple relations do not exist for most polymer gels.

3.5.2 Functional Properties

Food technologists make gels for a purpose, often to obtain a certain consistency or to provide physical stability. The properties desired and the means of achieving those are summarized in Tables 6 and 7. *Consistency* has already been briefly discussed, but the message of Table 6 is an important one: According to the purpose in mind, the rheological measurements must be of a relevant type and conducted at the relevant time scale or strain rate. This need not be difficult. For instance, to evaluate stand-up (i.e., the propensity of a piece of gel, say, a pudding, to keep its shape under its own weight), measurement of a modulus makes no sense. The proper experiment is simply watching the piece, and possibly measuring the height of a specimen that will just start yielding. To ensure stand-up, the yield stress must be greater than $g \times \rho \times H$, where H is specimen height. For a piece 10 cm tall, this would be about $10 \times 10^3 \times 0.1 = 10^3$ Pa. It should be realized that yield stress often is smaller when time scale is longer.

Very *weak gels* were briefly discussed in Section 3.4.2. In daily life, such a system appears to be liquid—that is, it will readily flow out of a bottle if its yield stress is < 10 Pa or so; but it nevertheless has elastic properties at extremely small stresses. This small yield stress may be sufficient to prevent sedimentation. A good example is soya milk (Fig. 19). Soya milk contains small particles, consisting of cell fragments and organelles. These particles aggregate, forming

TABLE 6 Consistency of Gels: Specification of the Mechanical Characteristics Desired of Gels Made for a Given Purpose

Property desired	Relevant parameters	Relevant conditions
"Stand-up"	Yield stress	Time scale
Firmness	Modulus or fracture stress or yield stress	Time scale, strain
Shaping	Yield stress + restoration time	Several
Handling, slicing	Fracture stress and work of fracture	Strain rate
Eating characteristics	Yield or fracture properties, or both	Strain rate
Strength (e.g., of film)	Fracture properties	Stress, time scale

TABLE 7 Gel Properties Needed to Provide Physical Stability

Prevent or impede	Gel property needed
Motion of particles	
Sedimentation	High viscosity; or significant yield stress + short restoration time
Aggregation	High viscosity or significant yield stress
Local volume changes	
Ostwald ripening	Very high yield stress
Coalescence	Very high yield stress
Motion of liquid	
Leakage	Low permeability and significant yield stress
Convection	High viscosity or significant yield stress
Motion of solute	
Diffusion	Small diffusivity

a weak reversible gel. If processing conditions are appropriate, the yield stress is sufficient to prevent these particles, and even larger particles, from sedimenting.

Also, some mixtures of polysaccharides, such as solutions of xanthan gum and locust bean gum, even if very dilute, can exhibit a small yield stress (Fig. 15, curve 3). This yield stress can prevent sedimentation of any particles present [39].

Sometimes it is desired to arrest the motion of liquid, in which a case *permeability* is an essential parameter. According to Darcy's law, the superficial velocity v of a liquid through a porous matrix, for instance a gel, is

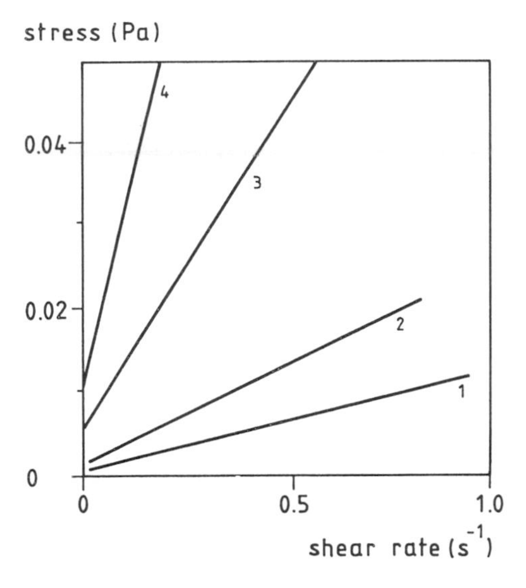

FIGURE 19 Flow curves (shear stress vs. shear rate) of soya milk. The yield stress is given by the *y* intercept of the lines. Curves 1 and 2 are for milk made from dehulled soya beans, about 6% dry matter; 3 and 4 from whole beans, about 7% dry matter. Curves 1 and 3, soaking of the beans overnight at room temperature; curves 2 and 4, soaking 4 h at 60°C. (From Ref. 45.)

$$v \equiv \frac{Q}{A} = \frac{B}{\eta}\frac{\Delta p}{x} \tag{23}$$

where Q is the volume flow rate ($m^3 \cdot sec^{-1}$) through a cross-sectional area A and Δp is the pressure difference over distance x. The permeability $B(m^2)$ is a material constant that varies greatly among gels. A particle gel like renneted milk (built of paracasein micelles) has a permeability of the order of 10^{-12} m^2, whereas a polymer gel (e.g., gelatin) typically would have a B of 10^{-17}. In the latter case, leakage of liquid from the gel would be negligibly slow.

Swelling and *syneresis* are additional properties of gels. Syneresis refers to expulsion of liquid from the gel, and it is the opposite of swelling. There are no general rules governing their occurrence. In a polymer gel, lowering of solvent quality (e.g., by changing temperature), adding salt (in the case of polyelectrolytes), or increasing the number of cross-links or junctions may cause syneresis. However, because both the pressure difference in Equation 23 and B are usually very small, syneresis (or swelling) often is very slow. In particle gels, syneresis may occur much faster, due to the far greater permeability. It is well known that renneted milk is prone to syneresis, an essential step in cheese making. The combination of variables influencing syneresis is intricate [68].

Transport of a solute through the liquid in a gel has to occur by *diffusion*, because convection generally is not possible. The diffusion coefficient D is not greatly different from that in solution, at least for small molecules in a not very concentrated gel. The Stokes relation for diffusivity $D = kT/6\pi\eta r$, where r is molecule radius, cannot be applied. The macroscopic viscosity of the system is irrelevant, since it concerns here the viscosity as sensed by the diffusing molecules, which would be the viscosity of the solvent. On the other hand, the solute has to diffuse around the strands of the gel matrix, and the hindrance will be greater for larger molecules and smaller pores between strands in the gel. These aspects are illustrated in Figure 20.

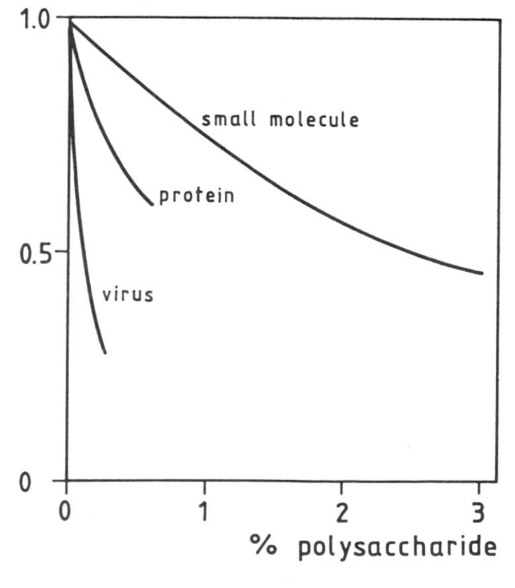

FIGURE 20 Diffusion of solutes in polysaccharide gels. Highly schematic examples after results by Muhr and Blanshard [42].

3.5.3 Some Food Gels

The fairly theoretical points discussed earlier will now be illustrated by discussing some food gels.

3.5.3.1 Plastic Fats [60,77]

If a triacylglycerol oil is cooled, fat crystals can form and these will aggregate. There is van der Waals attraction between the crystals and no repulsion (except at very small distance). The crystals form a fractal gel, at least under some conditions [75]. Subsequently, the aggregated crystals may to some extent grow together (fuse) by local accretion of solid material, because it may take a long time before fat crystallization has reached an equilibrium state. This results in a rigid and brittle network. The modulus is fairly high (e.g., 10^6 Pa), the proportionality between stress and strain extends only for a strain of at most 0.01, and yielding occurs at a strain of about 0.1; the yield stress in this case is about 10^5 Pa. These values greatly increase as the fraction solid increases. Permeability ranges from about 10^{-16} to 10^{-13} m^2, according to crystal size and fraction solid. These properties are characteristic of what is known as a plastic fat.

3.5.3.2 Caseinate Gels [70,82,86]

Milk contains casein micelles, proteinaceous aggregates of about 120 nm average diameter, each containing some 10^4 casein molecules (see Chap. 14). The micelles can be made to aggregate, by lowering the pH to about 4.6 (thereby decreasing electric repulsion) or by adding a proteolytic enzyme that removes the parts of κ-casein molecules that protrude into the solvent (thereby decreasing steric repulsion). Fractal gels are formed with a fractal dimensionality of about 2.3. The permeability, which is about 2×10^{-13} m^2 for average casein concentration (c), is thus strongly dependent on the latter, being about proportional to c^{-4}. For casein gels (Fig. 21a) a linear relation exists between log modulus and log casein concentration, in accordance with their fractal nature. The different slopes imply a difference in structure. Apparently, the initially tortuous strands in rennet gels become straightened soon after formation, whereas the strands in acid gels remain tortuous.

The building blocks of the gel, that is, the casein micelles, are themselves deformable, and

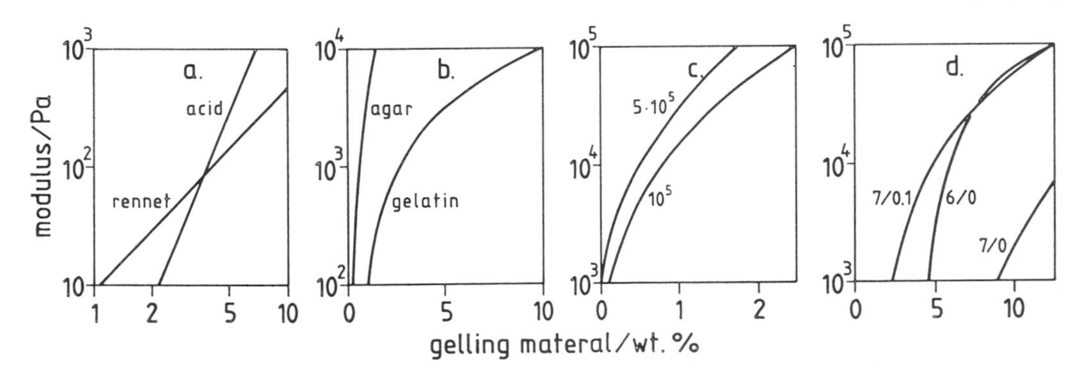

FIGURE 21 Dependence of the modulus of various gels on mass concentration of gelling material. Note that some of the abscissas are logarithmic, others linear. (a) Casein gels, made by slow acidification or by renneting. (b) Agar and gelatin gels. (c) Gels of κ-carrageenan of two molar masses in 0.1 molar KCl. (d) Heat-set gels of bovine serum albumin; the figures near the curves denote pH/added NaCl (molar). Approximate results, only meant to illustrate trends. From various sources.

the junctions between them are flexible. Thus, the gel is rather weak and deformable. For acid casein gels, fracture stress is about 100 Pa, and fracture strain about 1.1; for rennet gels, these values are about 10 Pa and 3, respectively. The acid gel is thus shorter (more brittle). These results apply to long time scales, say 15 min; at shorter time scales, the fracture stress is much greater. Applying a stress of slightly over 10 Pa to a rennet gel will cause flow (there is no detectable yield stress), and after considerable time fracture will occur. Applying a stress of 100 Pa leads to fracture within 10 sec. Similar behavior is found for some other types of particle gels but is by no means universal.

All these values depend on conditions applied, especially temperature. It is seen that the modulus of a casein gel is larger at lower temperature (Fig. 22a). This may appear strange, since hydrophobic bonds between casein molecules are considered to play a major part in keeping the gel together, and these bonds decrease in strength with decreasing temperature. Presumably, a decrease of hydrophobic bond strength (low T) leads to swelling of the micelles, to a large contact area between adjoining micelles, and to a greater number of bonds per junction. Conversely, a higher temperature of gel formation leads to a larger modulus (at least for acid gels; Fig. 22a), and this is due to a somewhat different network geometry, not to a difference in type of bonds.

At temperatures above about 20°C, rennet gels show syneresis. Syneresis accompanies rearrangement of the network of particles, which implies that some deaggregation occurs. In a region where no liquid can be expelled, that is, internally in the gel, rearrangement occurs as well, leading to both denser and less dense regions. This is called microsyneresis; it goes along with an increase in permeability and causes the straightening of the network strands mentioned above.

3.5.3.3 Gelatin [10,34]

Of all food gels, gelatin is closest to an ideal entropic gel. The flexible molecular strands between cross-links are long and this causes the gel to be very extensible. It is also predominantly elastic because the cross-links are fairly permanent (at least at low temperature). The dependence of modulus on concentration (Figure 21b) is in reasonable agreement with Equation 19 (data not

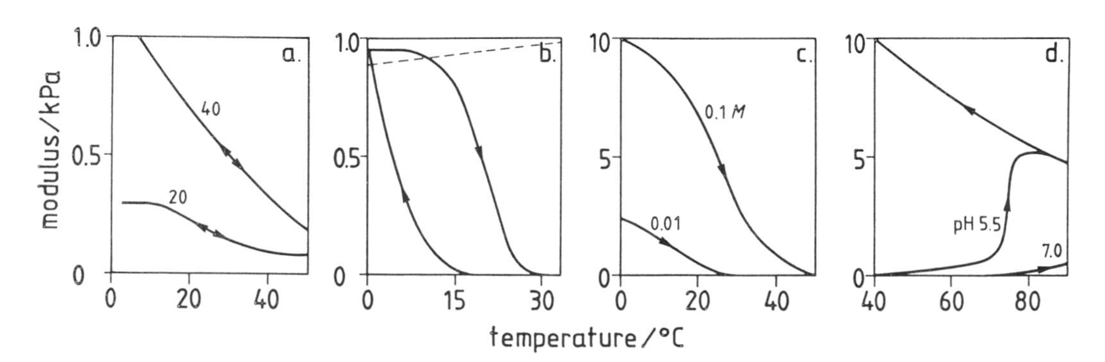

FIGURE 22 Dependence of the modulus of various gels on temperature of measurement. Arrows indicate the temperature sequence. (a) Acid casein gels (2.5%); figures on curves denote temperature of formation and ageing. (b) Gelatin (2.5%); the broken line indicates the relation according to Equation 19. (c) κ-carrageenan (1%); figures on curves denote concentration of $CaCl_2$. (d) β-Lactoglobulin (10%) at two pH values. Data represent trends; results may vary greatly depending on heating or cooling rate and other conditions. After various sources.

shown), but the temperature dependence is not (see Fig. 22b). This discrepancy stems from the mechanism of cross-linking. Despite the severe treatment of the collagen during preparation of gelatin, the molecules retain much of their length and produce highly viscous aqueous solutions. Upon cooling, the molecules tend to form triple helices (broadly speaking proline helices) like those in collagen. This applies to only part of the gelatin, and the helical regions are relatively short. Presumably, a gelatin molecule sharply bends at a so-called β-turn, and then forms a short double helix. Subsequently, a third strand may wind around this helix, thereby completing it. If the third strand is part of another molecule, a cross-link is formed; if not, there is no cross-link. Regardless of how this develops, groups of triple helices tend to align themselves in a parallel array, forming microcrystalline regions, and these constitute cross-links (Fig. 23a). It would follow from this mechanism that the kinetics of gelation are intricate, which is indeed the case. Also, rheological properties can be strongly dependent on temperature history.

3.5.3.4 Polysaccharides [40,49]

Despite the wide range of polysaccharide types (see Chap. 4), some general rules governing their gelling behavior exist. Unlike gelatin, most polysaccharide chains are fairly stiff; appreciable bending can occur only if chain length exceeds about 10 monomers (monosaccharide residues). This characteristic causes polysaccharides to produce highly viscous solutions; for example, 0.1% xanthan will increase the viscosity of water by at least a factor of 10. Some polysaccharides can form gels. Broadly speaking, the gel cross-links consist of microcrystalline regions (see Fig. 17b), involving an appreciable portion of the material. This means that the length of strands between cross-links is not very long. Combined with chain stiffness, this leads to rather short gels, very unlike rubber gels. Actually, they are intermediate between entropic and enthalpic gels. Of course, there is considerable variation among polysaccharides in this regard.

Cross-links among polysaccharide molecules can be any of the following three types, each illustrated by one example:

Type 1. Single helices. These helices, as found in amylose, arrange themselves in microcrystalline regions and, if the concentration is sufficiently great, gelation will occur. With amylopectin, similar behavior is observed.

Type 2. Double helices. These occur in κ-carrageenan below a sharply defined temperature. Each helix generally involves one molecule. Double helix formation between two molecules is very unlikely for geometric reasons. As soon as even a very short helix would have formed, the rotational freedom of the remaining part of each molecule involved would be very much reduced. Consequently, double helices involving

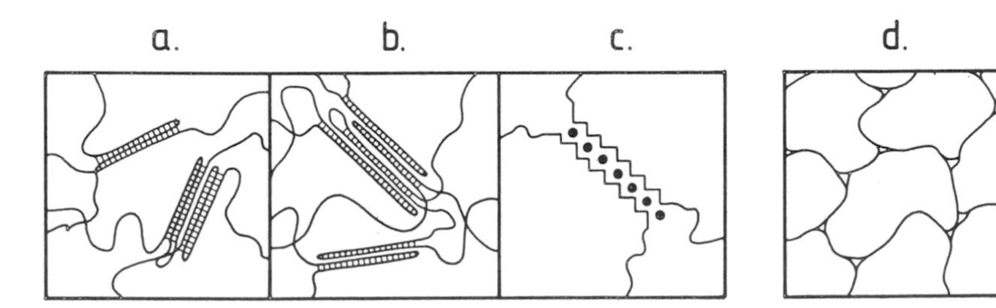

FIGURE 23 Various types of junctions in polymer gels. (a) Partly stacked triple helices as in gelatin. (b) Stacked double helices as in carrageenans. (c) "Egg box" junction as in alginate, dots denote calcium ions. (d) Swollen starch grains in a concentrated starch gel. Helices are schematically indicated by crosslines. The scales of a–c and d are very different.

two molecules will be rare. Cross-links arise because the stiff helices form microcrystalline regions (Fig. 23b). Helix formation is very rapid (milliseconds), whereas gelation takes longer (several seconds). As soon as the helices "melt," so does the gel.

Type 3. "Egg-box" junctions. These occur with some charged polysaccharides, such as alginate, when divalent cations are present (Fig. 23c). Alginate has negative charges, often spaced at regular distances, allowing divalent cations, such as Ca^{2+}, to establish bridges between two parallel polymer molecules. In this way, fairly rigid junctions are formed. It is likely that the junctions further arrange themselves in microcrystalline regions. The junctions do not readily "melt" unless the temperature is near 100°C.

Many factors may affect gelation and gel properties of polysaccharides. These include molecular structure, molar mass (Fig. 21c), temperature (Fig. 22c), solvent quality, and, for polyelectrolytes, pH and ionic strength (Fig. 22c).

3.5.3.5 Globular Proteins [11,56,57]

Several globular proteins form irreversible heat-setting gels (Fig. 22d). The proteins denature, that is , unfold, and then mutually react. The gel bonds formed may be -S-S- linkages (or possibly other covalent bonds), salt bridges, and/or hydrophobic bonds. A fairly large concentration of protein is needed to obtain stiff gels (Fig. 21d). The structures, and thereby rheological properties of the gels, vary greatly with type of protein, pH, ionic strength and rate of heating. Although proteins are polymers, the gels are not typical polymer gels, nor are they quite like casein gels. At pH values far from the isoelectric pH, fairly weak fine-stranded gels are typically formed. The strands of these gels have a thickness of ~ 10 nm and their structure is somewhat irregular. Near the isoelectric pH, stiffer gels form and they look like particle gels. The particles are about 1 μm in diameter or more, and thus contain numerous molecules, at least 10^7. These gels are often fractal-like [74]. The two types should differ greatly in permeability, though substantiating determinations appear not to have been made.

Structure formation during extrusion is comparable to heat-setting of globular proteins. An important example involves protein-rich products from soya [35]. Spinning of proteins, however, may involve different mechanisms [73].

3.5.3.6 Concentrated Starch Gels [30]

Starch (see Chap. 4) occurs naturally in rigid granules, mostly between 5 and 100 μm in diameter, that are insoluble in water. Part of the amylopectin is present in microcrystalline regions, which gives the granules considerable rigidity. If starch granules are heated in excess water, they gelatinize. This involves swelling, as water up to several times their own weight is taken up; melting of microcrystallites; and leaching of some amylose. The swollen granules remain intact upon cooling, unless vigorous agitation is applied. Dilute gelatinized starch suspensions gel in roughly the same way as does amylose; that is, leached amylose forms a gel network in which the swollen granules are trapped.

A concentrated, gelatinized starch solution—above 5–15%, depending on starch type—forms a very different kind of gel. The granules swell until they virtually fill the whole volume, deforming each other in the process (Fig. 23d). Especially for potato starch, the swollen granules interlock, as in a jig-saw puzzle. A very thin layer of gelled amylose solution is present between the granules and may act as intergranule glue. The rheological properties of the gel are dominated by those of the granules. A freshly made gel can be deformed very much and it is purely elastic (like a rubber ball) if deformation is very fast. Slower deformation may lead to fracture. Presumably, the individual granules are rubber-like.

During storage of concentrated starch gels, considerable rearrangement of the molecules occurs. This is like partial crystallization and happens faster at a lower nonfreezing temperature. This leads to an increase in number and particularly strength of microcrystalline junctions. The granules and the gel become stiffer (the modulus increases). It also results in an increase of the fracture stress and a decrease of the fracture strain of the gel upon aging. These "retrogradation" phenomena are of great importance in the staling of bread.

3.5.3.7 Mixed Gels

It should now be clear that the structure and properties of gels can vary greatly. The modulus of a 1% gel can vary by almost 5 orders of magnitude and the strain at fracture by a factor of 100. The preceding discussion of starch gels gives one example of how the various rheological and fracture properties can be related to each other. In that case most relations can be explained, even semiquantitatively. However, nearly every system exhibits specific relations that are often poorly understood.

The situation becomes even more complicated when mixed gels are considered. Some mixtures of polysaccharides may show enhanced gelation at a lower concentration [9,41]. For instance, dilute xanthan or locust bean gum solutions do not show an appreciable yield stress, but dilute mixed solutions do. Filled systems may also have greatly altered properties from unfilled [9,38]. Phase separation can also occur if the polymers are thermodynamically incompatible. For instance, mixed solutions of a highly soluble polysaccharide and a protein can separate into two phases, one rich in protein and the other rich in polysaccharide [61,62]. In other cases, a mixed polymer solution forms a coacervate, that is, a phase containing a high concentration of both, leaving another phase that is depleted of polymers. This can occur if the polymers have opposite charge.

Complete foods are still more complicated than the systems so far discussed [26]. Nevertheless, knowledge of the principles provided can be of great help in understanding food behavior and for designing experiments to study it.

3.6 EMULSIONS

3.6.1 Description

Emulsions are dispersions of one liquid in another. The most important variables determining emulsion properties are:

1. Type, that is, o/w or w/o. This determines, among other things, with what liquid the emulsion can be diluted (Section 3.1.2). The o/w emulsions are most common, examples being milk and milk products, sauces, dressings, and soups. Butter and margarine are w/o emulsions, but they contain other structural elements as well.
2. Droplet size distribution. This has an important bearing on physical stability, smaller drops generally giving more stable emulsions. Also the energy and the amount of emulsifier needed to produce the emulsion depend on the droplet size desired. A typical mean droplet diameter is 1 μm, but it can range between 0.2 and several micrometers. Because of the great dependence of stability on droplet size, the width of the size distribution is also important.
3. Volume fraction of dispersed phase (φ). In most foods, φ is between 0.01 and 0.4. For mayonnaise it may be 0.8, which is above the value for maximum packing of rigid spheres, roughly 0.7; this means that the oil droplets are somewhat distorted.
4. Composition and thickness of the surface layer around the droplets. This determines interfacial tension, colloidal interaction forces, etc. (Section 3.2.7).

5. Composition of the continuous phase. This determines solvent conditions for the surfactant and thereby colloidal interactions. The viscosity of the continuous phase has a pronounced effect on creaming.

Unlike the solid particles in a suspension, emulsion droplets are spherical (greatly simplifying many predictive calculations) and deformable (allowing droplet disruption and coalescence), and their interface is fluid (allowing interfacial tension gradients to develop). Nevertheless, in most conditions, emulsion droplets behave like solid particles. From Equation 9, the Laplace pressure of a droplet of 1 μm radius and interfacial tension $\gamma = 5\ \mathrm{mN \cdot m^{-1}}$ is 10^4 Pa. For a liquid viscosity of $\eta = 10^{-3}\ \mathrm{Pa \cdot sec}$ (water) and a velocity gradient achieved by stirring of $G = 10^6\ \mathrm{sec^{-1}}$ (this is extremely vigorous), the shear stress ηG acting on the droplet would be 10^3 Pa. This implies that the droplet would be deformed only slightly. Also, the surfactant at the droplet surface allows this surface to withstand a shear stress (Section 3.2.6). For the conditions mentioned, an interfacial tension difference between two sides of the droplet of $1\ \mathrm{mN \cdot m^{-1}}$ would suffice to prevent lateral motion of the interface, and a difference of this magnitude can be achieved readily. It can be concluded that emulsion droplets behave like solid spheres, unless stirring is extremely vigorous or droplets are very large.

3.6.2 Emulsion Formation [76,80]

In this section the size of the droplets and the surface load obtained during making of emulsions are discussed, especially when protein is used as a surfactant.

To make an emulsion, one needs oil, water, an emulsifier (i.e., a surfactant), and energy (generally mechanical energy). Making drops is easy, but to break them up into small droplets mostly is difficult. Drops resist deformation and thereby break up because of their Laplace pressure, which becomes larger as droplet size decreases. This necessitates a large input of energy. The energy needed can be reduced if the interfacial tension, hence the Laplace pressure, is reduced by adding an emulsifier, though this is not the latter's main role.

The energy needed to deform and break up droplets is generally provided by intense agitation. Agitation can cause sufficiently strong shear forces if the continuous phase is very viscous. This is common when making w/o emulsions, resulting in droplets with diameters down to a few micrometers (which is not very small). In o/w emulsions the viscosity of the continuous phase tends to be low, and to break up droplets inertial forces are needed. These are produced by the rapid, intensive pressure fluctuations occurring in turbulent flow. The machine of choice is the high-pressure homogenizer, which can produce droplets as small as 0.2 μm. The average droplet size obtained is about proportional to the homogenization pressure to the power –0.6. When using high-speed stirrers, faster stirring, longer stirring, or stirring in a smaller volume result in smaller droplets; however, average droplet diameters below 1 or 2 μm usually cannot be obtained.

However, this is not the whole story. Figure 24 shows the various processes that occur during emulsification. Besides disruption of droplets (Fig. 24a), emulsifier has to be transported to the newly created interface (Fig. 24b). The emulsifier is not transported by diffusion but by convection, and this occurs extremely quickly. The intense turbulence (or the high shear rate, if that is the situation) also leads to frequent encounters of droplets (Fig. 24c and d). If these are as yet insufficiently covered by surfactant, they may coalesce again (Fig. 24c). All these processes have their own time scales, which depend on several conditions, but a microsecond is fairly characteristic. This means that all processes occur numerous times, even during one passage through a homogenizer valve, and that a steady state—where break-up and coalescence balance each other—is more or less obtained.

The role of the emulsifier is not yet fully clear, but it is fairly certain that the Gibbs–

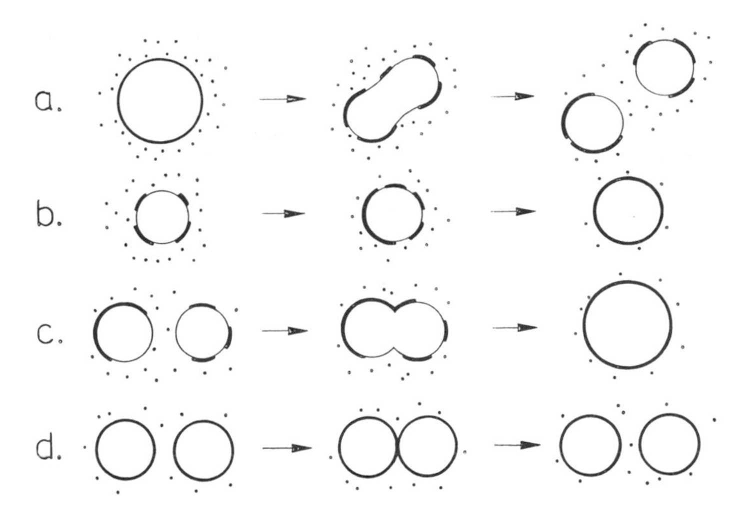

FIGURE 24 Important processes occurring during emulsification. The drops are depicted by thin lines and the emulsifier by heavy lines and dots. Highly schematic and not to scale.

Marangoni effect is crucial (Fig. 25). If two drops move toward each other (Fig. 25a), which they frequently do at great speed during emulsification, and if they are insufficiently covered by emulsifier, they will acquire more emulsifier during their approach, but the least amount of emulsifier will be available where the film between droplets is thinnest. This lead to an interfacial tension gradient, with interfacial tension (γ) being largest where the film is thinnest (Fig. 25b).

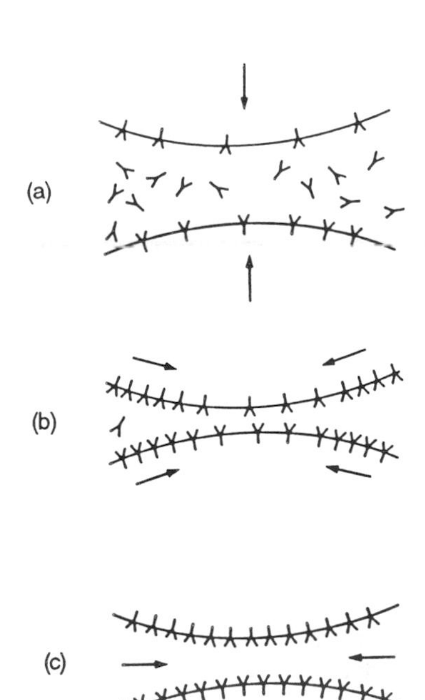

FIGURE 25 Diagram of the Gibbs–Marangoni effect acting on two approaching emulsion droplets during emulsification. Emulsifier molecules indicated by Y. See text for further discussion. (After Ref. 80.)

The gradient causes the emulsifier, and thereby the droplet surfaces, to move in the direction of large γ, carrying liquid along with it (Fig. 25b). This is the Marangoni effect (see also Fig. 9). The streaming liquid will thus drive the droplets away from each other (Fig. 25c), thereby providing a self-stabilizing mechanism. The magnitude of the effect depends on the Gibbs elasticity of the film, which is twice the surface dilational modulus (Sec. 3.2.5). The Gibbs elasticity generally increases with increasing molar concentration of surfactant in the continuous phase. It therefore decreases during emulsification, because the interfacial area increases, whereby the adsorbed amount of surfactant increases.

The Gibbs–Marangoni effect occurs only if the emulsifier is in the continuous phase, since otherwise a γ gradient cannot develop. This is the basis of Bancroft's rule: The phase in which the emulsifier is (most) soluble becomes the continuous phase. Hence the need for a surfactant with high HLB number for o/w emulsions and a low HLB for w/o emulsions (Sec. 3.2.2).

Proteins are the emulsifiers of choice for o/w food emulsions because they are edible, surface active, and provide superior resistance to coalescence [78]. They cannot be used for w/o emulsions because of their insolubility in oil. Proteins do not give a very low interfacial tension (Section 3.2.2, Fig. 5) and their Gibbs–Marangoni effect is not very strong, presumably because of their low molar concentration. Therefore, the droplets obtained are typically not very small for given conditions. However, droplets can be made smaller by applying more intense emulsification, such as a greater homogenization pressure. Some examples of average droplet size (d_{vs}) are given in Figure 26. At large emulsifier concentration, d_{vs} reaches a plateau value. This value is smaller for the nonionic surfactant than for the proteins, because the former produces a lower interfacial tension. The nonionic is also more effective at low concentration than the proteins, presumably because of a stronger Gibbs–Marangoni effect.

It is also seen that the various proteins give about the same plateau value for d_{vs}. This is not strange because they produce comparable values for the interfacial tension (about 10 mN · m^{-1}). However, at low concentrations large differences in d_{vs} are apparent. Several tests

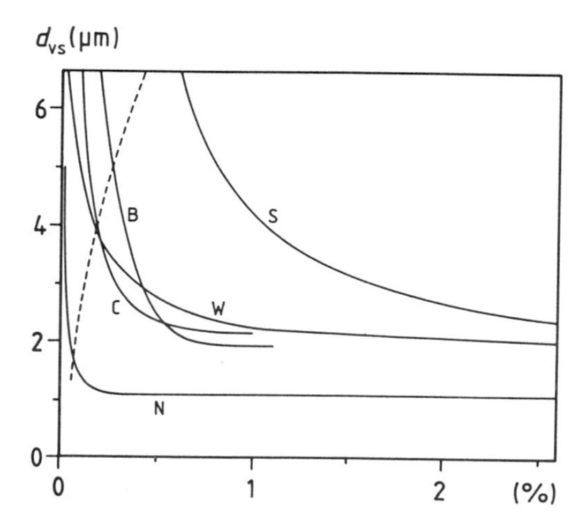

FIGURE 26 Influence of emulsifier concentration (%w/w) on volume/surface average droplet diameter d_{vs} for various emulsifiers. Emulsions contained about 20% triacylglycerol oil in water and were prepared under constant conditions (moderate emulsification intensity). B, blood protein; C, sodium caseinate; N, nonionic small-molecule surfactant; S, soya protein; W, whey protein. Approximate results from various sources. The broken line indicates approximately the conditions during determination of the so-called emulsifying activity index.

have been developed to evaluate the suitability of proteins as emulsifiers. The well-known emulsifying activity index (EAI) involves emulsifying a large quantity of oil in a dilute protein solution [46]. This test corresponds roughly to the conditions indicated by the broken line in Figure 26. This is not realistic for most practical situations, because the protein/oil ratio normally would be much larger. Consequently, EAI values are often irrelevant.

Attempts have been made to explain differences in EAI of various proteins by differences in their surface hydrophobicity [29]. However, the correlation is poor and other workers have refuted this concept [e.g., 52]. In the author's view, proteins differ primarily in emulsifying efficiency because they differ in molar mass. For a larger molar mass and the same mass concentration, the molar concentration is smaller, and the latter is presumably the most important variable in providing a strong Gibbs–Marangoni effect. Proteins of a smaller molar mass would thus be more efficient emulsifiers. (However, a very small molar mass, as obtained, for instance, by partial hydrolysis of the protein, would not be desirable, because emulsions obtained with fairly small peptides usually show rapid coalescence.) It should be realized that several protein preparations, especially industrial products, contain molecular aggregates of various size, thereby greatly increasing effective molar mass and decreasing emulsifying efficiency. By and large, protein preparations that are poorly soluble are poor emulsifiers. As a rule, well-soluble proteins are about equally suitable to make an emulsion (i.e., obtain small droplcts), if present at not too low a concentration.

Another important variable is the *surface load* (Γ). If an emulsifier tends to give a high Γ, relatively much of it is needed to produce an emulsion. Compare, for instance, whey protein and soya protein in Figures 26 and 27. Moreover, a fairly high Γ is usually needed to obtain a stable emulsion. In the case of small-molecule surfactants, equilibrium is reached between Γ and bulk concentration of surfactant according to the Gibbs relation (Eq. 5). Consequently, knowledge of total surfactant concentration, of o/w interfacial area, and of the adsorption isotherm (e.g., Fig. 5, lower part) allows calculation of Γ, irrespective of the manner of formation of the emulsion. This

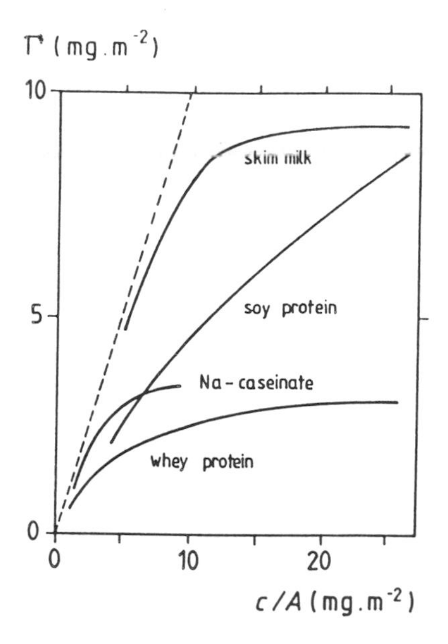

FIGURE 27 Protein surface load (Γ) as a function of protein concentration (c) per unit oil surface area (A) created by emulsification. The broken line indicates 100% adsorption. (From Ref. 85.)

is not the case when a protein (or other polymer) is the emulsifier, because thermodynamic equilibrium is not reached (Sec. 3.2.2). The surface load of a protein can depend on the way of making the emulsion, in addition to the variables mentioned. It has been observed that plots as given in Figure 27 are more suitable to relate Γ to protein concentration.

Figure 27 gives some examples of results obtained for various proteins. If c/A is very small, some proteins presumably almost fully unfold at the o/w interface, and form a stretched polypeptide layer, with a Γ of about 1 mg · m^{-2}. Many highly soluble proteins give a plateau value of about 3 mg · m^{-2}. Aggregated proteins can yield much larger values. It may also be noted that any large protein aggregates present tend to become preferentially adsorbed during emulsification, thereby further increasing Γ.

An emulsifier is needed not only for the formation of an emulsion, but also for providing stability of the emulsion once made. It is of importance to clearly distinguish between these main functions, since they often are not related. An emulsifier may be very suitable for making small droplets, but may not provide long-term stability against coalescence, or vice versa. Evaluation of proteins merely for their ability to produce small droplets is therefore not very useful. Another, usually desirable, feature of a surfactant is to prevent aggregation under various conditions (pH near the isoelectric pH, high ionic strength, poor solvent quality, high temperature). Types of emulsion instability and means of prevention are discussed next.

3.6.3 Types of Instability [17,43,59,78,83]

Emulsions can undergo several types of physical change as illustrated in Figure 28. The figure pertains to o/w emulsions; the difference with w/o emulsions is that downward sedimentation rather than creaming would occur.

Ostwald ripening (Fig. 28a) does not normally occur in o/w emulsions, because triacylglycerol oils are commonly used and they are insoluble in water. When essential oils are present (e.g., citrus juices), some have sufficient solubility so that smaller droplets gradually disap-

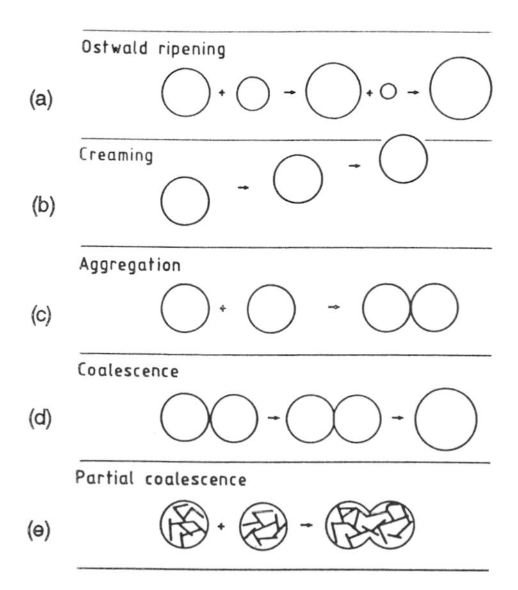

FIGURE 28 Types of physical instability for oil-in-water emulsions. Highly schematic. The size of the contact area denoted in (d) may be greatly exaggerated; the short heavy lines in (e) denote triacylglycerol crystals.

pear [19]. W/o emulsions may exhibit Ostwald ripening. Data in Table 4 show only a very small solubility excess for a 1-μm droplet, but it would be sufficient to produce marked Ostwald ripening during prolonged storage. This can easily be prevented by adding a suitable solute to the water phase, that is, one that is insoluble in oil. A low concentration of salt (say, NaCl) will do: As soon as a small droplet shrinks, its salt concentration and osmotic pressure increase, thereby producing a driving force for water transport in the opposite direction. The net result is a stable droplet size distribution.

The other instabilities are discussed in other sections: creaming in 3.4.2, aggregation in 3.3 and 3.4.3, coalescence in 3.6.4, and partial coalescence in 3.6.5.

The various changes may affect each other. Aggregation greatly enhances creaming, and if this occurs, creaming further enhances aggregation rate, and so on. Coalescence can only occur if the droplets are close to each other, that is, in an aggregate or in a cream layer. If the cream layer is more compact, which may occur if individual, fairly large droplets cream, coalescence will be faster. If partial coalescence occurs in a cream layer, the layer may assume characteristics of a solid plug.

It is often desirable to establish what kind of instability has occurred in an emulsion. Coalescence leads to large drops, not to irregular aggregates or clumps. Clumps due to partial coalescence will coalesce into large droplets when heated sufficiently to melt the triacylglycerol crystals. A light microscope can be used to establish whether aggregation, coalescence, or partial coalescence has occurred. Section 3.4.4 gives some hints for distinguishing among various causes of aggregation. It is fairly usual that coalescence or partial coalescence leads to broad size distributions, and then the larger droplets or clumps cream very rapidly.

Agitation can disturb creaming and may disrupt aggregates of weakly held droplets, but not clumps formed by partial coalescence. Slow agitation tends to prevent any true coalescence.

If air is beaten in an o/w emulsion, this may lead to adsorption of droplets onto air bubbles. The droplets may then be disrupted into smaller ones, due to spreading of oil over the a/w interface (Sec. 3.2.3). If the droplets contain crystalline fat, clumping may occur; beating in of air thus promotes partial coalescence. This is what happens during churning of cream to make butter and also during whipping of cream. In the latter case, the clumped, partially solid droplets form a continuous network that encapsulates and stabilizes the air bubbles and lends stiffness to the foam.

A way to prevent or retard all changes except Ostwald ripening is to cause the continuous phase to gel (Sec. 3.5). Examples are butter and margarine. Here, the water droplets are immobilized by a network of fat crystals. Moreover, some crystals become oriented at the oil/water interface because of the favorable contact angle (Sec. 3.2.3). In this way, the droplets cannot closely encounter each other. If the product is heated to melt the crystals, the aqueous droplets readily coalesce. Often, a suitable surfactant is added to margarine to prevent rapid coalescence during heating, since this would cause undesirable spattering.

3.6.4 Coalescence [21,78,83]

This discussion will focus on o/w emulsions. The theory is still in a state of some confusion. As a preliminary remark it should be noted that factors governing coalescence during emulsion formation (Sec. 3.6.2) differ completely from those that are important in "finished" emulsions. The latter are considered here.

Coalescence is induced by rupture of the thin film (lamella) between close droplets. If a small hole is somehow formed in the film, the droplets will immediately flow together. Film rupture is a chance event and this has important consequences: (a) The probability of coalescence, if possible, will be proportional to the time that the droplets are close to each other. Hence, it is especially likely in a cream layer or in aggregates. (b) Coalescence is a first-order rate process, unlike aggregation, which is in principle second order with respect to time and concentration. (c) The probability that rupture of a film occurs will be proportional to its area. This implies that flattening of the droplets on approach leading to formation of a greater film area, will promote coalescence. Oil droplets normally found in food emulsions do not show such flattening, because their Laplace pressures are too high (Fig. 28d is thus misleading in this respect).

There is no generally accepted coalescence theory, but the author feels that the following principles are valid. Coalescence is less likely for:

1. Smaller droplets. They lead to a smaller film area between droplets, and hence a lower probability of rupture of the film; more coalescence events are needed to obtain droplets of a certain size; the rate of creaming is decreased. In practice, average droplet size often is the overriding variable.
2. A thicker film between droplets. This implies that strong or far-reaching repulsive forces between droplets (Sec. 3.3) provide stability against coalescence. For DLVO-type interactions, it turns out that coalescence will always occur if the droplets are aggregated in the primary minimum (Fig. 11). Steric repulsion is especially effective against coalescence, because it keeps the droplets relatively far apart (Fig. 12).
3. A greater interfacial tension. This may appear strange, because a surfactant is needed to make an emulsion and a surfactant decreases γ. Moreover, a smaller γ implies a smaller surface free energy of the system. However, it is the activation free energy for film rupture that counts and that is larger for a larger γ. This is because a larger γ makes it more difficult to deform this film (bulging, development of a wave on it), and deformation is needed to induce rupture.

Based on these principles, proteins appear to be very suitable for preventing coalescence, and this agrees with observation. Proteins do not produce a very small γ, and they often provide considerable repulsion, both electric and steric. Figure 29 shows results of experiments in which

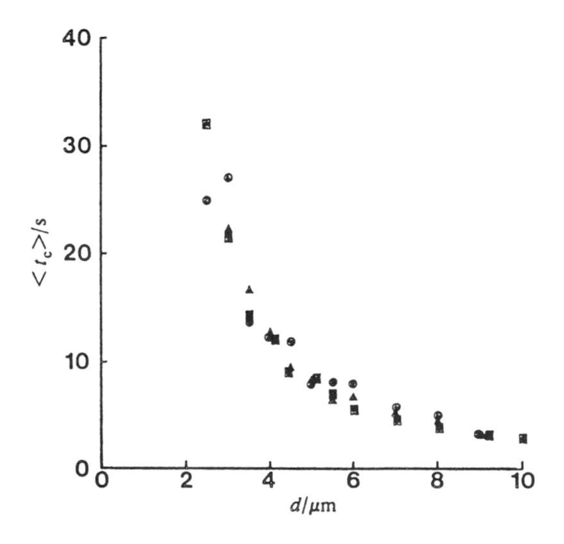

FIGURE 29 Time (t_c) for coalescence of droplets of various diameter (d) with a planar o/w interface, in 20-min-old 1 ppm protein solutions. ▲, β-Casein; ●, κ-casein; ■, lysozyme. (From Ref. 21.)

small droplets in a dilute protein solution were allowed to cream to planar o/w interfaces of various age and the time needed for coalescence was observed. A strong effect of droplet size is apparent. The results were obtained under conditions (protein concentration and adsorption time) where the surface load of the proteins would have been only about 0.5 $mg \cdot m^{-2}$, implying very weak repulsion. In cases where a thicker adsorbed layer was allowed to form, the authors found virtually no coalescence of small droplets.

Figure 29 shows no significant difference between proteins in their ability to prevent coalescence. This is also the general experience in practice, except for gelatin, which is less effective than most other proteins. Under severe conditions (see later), differences among proteins may be observed, with caseinates tending to be superior. Partial hydrolysis of proteins can severely impair their ability to prevent coalescence. Attempts have been made to relate the coalescence inhibiting ability of proteins (and of other surfactants) to various properties, particularly surface shear viscosity of the adsorbed protein layer (Section 3.2.5). Unfortunately, this quantity often is named "film strength," a misleading term. In some cases a positive correlation between surface shear viscosity and coalescence stability is observed, but there are several instances where deviation from this relationship are very large. Some positive correlations may have been due to a relation between surface shear viscosity and surface load.

Most small-molecule surfactants yield a small interfacial tension. Because a small γ favors coalescence, surfactants that provide considerable steric repulsion, such as the Tweens©, are among the most effective. Ionic surfactants are effective against coalescence only at very low ionic strength.

Small-molecule surfactants present in (added to) protein-stabilized emulsions tend to displace protein from the droplet surface (Sec. 3.2.2, Fig. 6), and this generally leads to less resistance to coalescence. If coalescence is desired, this provides a method to achieve it; for example, add sodium lauryl sulfate and some salt (to lower double layer thickness), and rapid coalescence will usually occur.

Food emulsions may exhibit coalescence under extreme conditions. For example, during freezing, formation of ice crystals will force the emulsion droplets closer together, often causing copious coalescence on thawing. Something similar happens upon drying and subsequent redispersion; here coalescence is alleviated by a relatively high concentration of "solids not fat." In such cases, best stability is obtained by having small droplets and a thick protein layer, such as of Na caseinate.

Another extreme condition is centrifuging. This causes a cream layer to form rapidly, a pressing of droplets together with sufficient force to cause considerable flattening even of small droplets, and likely coalescence. This implies that centrifugation tests to predict coalescence stability of emulsions are usually not valid, since the conditions during centrifugation differ so greatly from those during handling of emulsions. (This does not mean that centrifugation tests to predict creaming are useless. They can be quite helpful, provided that the complications discussed in Sec. 3.4.2 are taken into account.)

Predicting coalescence rate is always very difficult. The best approach is to use a sensitive method to estimate average droplet size (e.g., turbidity at a suitable wavelength) and establish the change over time (perhaps 1 day). Even then, extrapolation may result in error: protein layers around droplets tend to age (Sec. 3.2.2) and become better inhibitors of coalescence.

3.6.5 Partial Coalescence [5–7,12,64]

In many o/w food emulsions part of the oil in the droplets can crystallize. The proportion of fat solid, ψ, depends on the composition of the triacylglycerol mixture and on temperature (Chap. 5). In emulsion droplets, ψ can also depend on temperature history, since a finely emulsified oil can show considerable and long-lasting supercooling [77]. If an emulsion droplet contains fat crystals, they usually form a continuous network (Sec. 3.5.3). These phenomena greatly affect emulsion stability.

Oil droplets containing a network of fat crystals cannot fully coalesce (Fig. 28e). If the film between them ruptures, they form an irregular clump, held together by a "neck" of liquid oil. True and partial coalescence thus have rather different consequences. Partial coalescence causes an increase in the apparent volume fraction of dispersed material, and if the original volume fraction is moderate or high and shear rate is fairly small, a solid or gel-like network of partially coalesced clumps can form. This is desired, for instance, in ice cream, where it gives the product a desirable texture ("dryness," lack of stickiness, melt-down resistance).

Rupture of the film between close droplets containing some solid fat can be triggered by a crystal that protrudes from the droplet surface and pierces the film. This happens particularly in a shear field, and then it may occur very much faster, for instance, about six orders of magnitude faster than true coalescence (same emulsion, no fat crystals). This implies that partial coalescence is far more important than true coalescence in o/w emulsions that are subject to fat crystallization.

The kinetics of particle coalescence are complicated and variable. This is due to the numerous variables affecting it. In many emulsions, large particles (original droplets or clumps already formed) are more prone to partial coalescence than small ones, leading to a self-accelerating process, which soon leads to very large clumps that cream very rapidly. The remaining layer may then exhibit a decreased average droplet size. Other emulsions may simply show a gradual increase in average particle size with time.

The most important factors affecting the rate of partial coalescence generally are (Fig. 30):

1. Shear rate. This has various effects. (a) Because of the shear, droplets encountering each other tend to roll over each other, thereby greatly enhancing the probability that

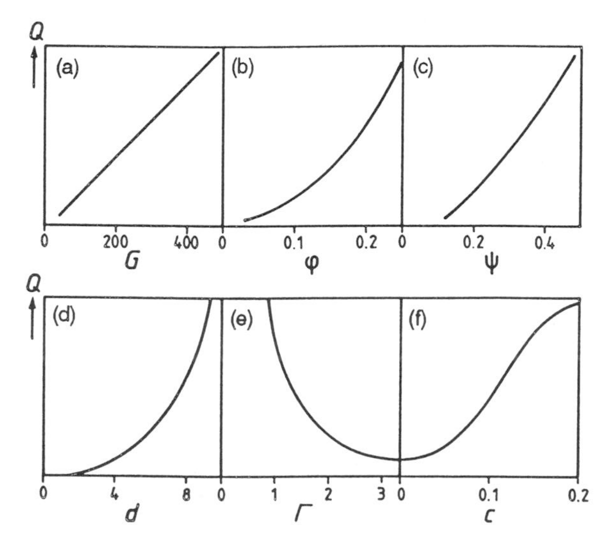

FIGURE 30 Schematic depiction of rate of partial coalescence (Q) in protein-stabilized emulsions as influenced by: (a) shear rate G (sec^{-1}); (b) volume fraction φ; (c) proportion of fat solid ψ; (d) mean droplet diameter d (μm); (e) protein surface load Γ (mg · m^{-2}); and (f) concentration of added small-molecule surfactant c (%), on partial coalescence rate Q. Various sources.

a protruding crystal can pierce the film between them. (b) The encounter rate between particles is proportional to shear rate (Section 3.4.3). (c) The shear force tends to press approaching droplets closer together, thereby enhancing the possibility that a protruding crystal can pierce the film. Thus, shear rate has a very large effect on the rate of partial coalescence, and this influence is accentuated when flow is turbulent.

2. Volume fraction of droplets. For a higher φ, the rate of partial coalescence is obviously greater, being about second order with respect to φ.
3. Fat crystallization. If the fraction solid, ψ, is 0 partial coalescence is impossible, and it also can not occur for $\psi = 1$. For fairly low ψ, partial coalescence rate generally increases with increasing ψ, as this causes more crystals to protrude. However, the relation between ψ and rate is variable, and the curve in Figure 30c only provides an example. The variability is largely due to variation in crystal size and arrangement. An important aspect is that the crystals must form a network throughout the droplet to support protruding crystals. The minimum ψ needed for such a network often is of the order 0.1. If most of the oil is crystallized and the crystals are very small, the crystal network may tenaciously hold the remaining oil, thereby preventing partial coalescence, even if the film is pierced. The protrusion distance may also depend on ψ, temperature history, crystal size, and crystal shape.
4. Droplet diameter. A relation as depicted in Figure 30d is usually observed, but the scale of droplet size may vary among emulsions. The effect of *d* presumably is due to (a) larger droplets sensing a larger shearing force; (b) larger droplets exhibiting a larger film area between two droplets; and (c) a positive correlation between droplet size and crystal size.
5. Surfactant type and concentration. Two effects are of major importance. First, these variables will determine the oil–crystal–water contact angle (Sec. 3.2.3) and thereby affect the distance a given crystal can protrude. Second, these variables determine repulsion (strength and range) between the droplets. The weaker the repulsion, the easier it is for two droplets to closely approach each other, thus increasing the likelihood that a protruding crystal will pierce the film between them. The repulsion, together with the droplet size, will thus determine what minimum shear rate is needed for partial coalescence to occur; values between 5 and 120 sec^{-1} have been observed. Some emulsions show no partial coalescence at all at the shear rates studied. The best type of surfactant to achieve this is, again, a protein, if the surface load is large enough (Fig. 30e). Addition of a small-molecule surfactant generally leads to displacement of the protein from the interface (Sec. 3.2.2), thereby greatly enhancing partial coalescence (Fig. 30f). This approach is commonly used in ice cream mixes.

3.7 FOAMS

In a sense, foams are much like o/w emulsions; both are dispersions of a hydrophobic fluid in a hydrophilic liquid. However, because of considerable quantitative differences, their properties also are qualitatively different. Quantitative information is given in Table 8. It is evident that bubble diameter is so large as to exclude foam bubbles from the realm of colloids. Large diameters, combined with large density difference, cause foam bubbles to cream faster than emulsion droplets by several orders of magnitude. The relatively high solubility of air in water causes rapid Ostwald ripening (often called disproportionation in foams). If the gas phase is CO_2, as it is in some foods (bread, carbonated beverages), the solubility is even higher, by a factor of about 50. The characteristic time scales during formation are two or three orders of magnitude longer for

TABLE 8 Comparison of Foams and Emulsions: Order of Magnitude of Some Quantities

Property	Foam	Emulsion o/w	Emulsion w/o	Units
Drop/bubble diameter	10^{-3}	10^{-6}	10^{-6}	m
Volume fraction	0.9	0.1	0.1	—
Drop/bubble number	10^{9}	10^{17}	10^{17}	m^{-3}
Interfacial tension	0.04	0.006	0.006	$N \cdot m^{-1}$
Laplace pressure	10^{2}	10^{4}	10^{4}	Pa
Solubility D in C	2.2	0	0.15	vol%
Density difference D – C	-10^{3}	-10^{2}	10^{2}	$kg \cdot m^{-3}$
Viscosity ratio D/C	10^{-4}	10^{2}	10^{-2}	—
Time scale[a]	10^{-3}	10^{-6}	10^{-5}	sec

Key: D, dispersed phase (air in the case of foam); C, continuous phase.
[a]Characteristic times during formation.

foams than for emulsions. Because creaming and Ostwald ripening occur so fast, formation and instabilities often cannot be separated, which makes the study of foams rather difficult.

Several aspects of foams will be briefly discussed here. A few have already been mentioned in Section 3.1.2. Surface phenomena are of overriding importance to foam production and properties, and background information is provided in Section 3.2. Further sources of literature are Refs. 23, 48, and 79. For churning and whipping of cream see Ref. 43. Also, some of the books mentioned in the bibliography, especially Dickinson's, have chapters on foam.

3.7.1 Formation and Description

In principle, foams can be made in two ways, by supersaturation or mechanically.

3.7.1.1 Via Supersaturation

A gas, usually CO_2 or N_2O because of their high solubility, is dissolved in the liquid at high pressure (a few bars). When the pressure is released, gas bubbles form. They do not form by nucleation; a spontaneously formed gas bubble would have a radius of, say, 2 nm, which would imply a Laplace pressure (Eq. 9) of about 10^8 Pa, or 10^3 bar. To achieve this, the gas would have to be brought to this pressure, which is, of course, impractical. Instead, gas bubbles always grow from small air pockets that are already present at the wall of the vessel or on boiling chips. The contact angle gas/water/solid may be as high as 150 degrees for a fairly hydrophobic solid, and this allows small air pockets to remain in crevices or sharp dents in the solid (Fig. 7d). For a negative curvature, air could even remain there if less than saturated.

To give an example, if a pressurized bottle of a carbonated liquid is opened, the overpressure is released, CO_2 becomes supersaturated, and it diffuses toward small air pockets in the bottle wall. These grow, and become dislodged when large enough, leaving a remnant from which another bubble can grow. The bubbles rise while growing further, and a creamed layer of bubbles is formed, that is, a foam. These bubbles always are fairly large, say 1 mm.

Another example is the formation of CO_2 in a leavened dough. Excess CO_2 collects at sites of tiny entrapped small air bubbles, and these sites grow in size. Some of them grow to form visible gas cells, creating a macroscopic foam structure.

3.7.1.2 By Mechanical Forces

A gas stream can be led through narrow openings into the aqueous phase (sparging); this causes bubbles to form, but they are fairly large. Smaller bubbles can be made by beating air into the liquid. At first, large bubbles form, and these are broken up into progressively smaller ones. Shear forces are typically too weak to obtain small bubbles and the breakup mechanism presumably involves pressure fluctuations in a turbulent field, as is true during formation of o/w emulsions (Sec. 3.6.2). Bubbles of about 0.1 mm can be obtained in this way, which is the method of choice in industrial processing. The method also enables the amount of air incorporated to be controlled. This is often expressed as percent overrun, that is, the relative increase in volume, which is equal to $100\varphi/(1-\varphi)$, where φ is the volume fraction of air (or gas) in the system.

3.7.1.3 Foam Formation and Structure

To make a foam, a surfactant is needed. Almost any type will do, since the only criterion for its functionality is that a certain γ gradient be created. This does not mean that any surfactant is suitable to make a stable foam, as will be discussed later. A fairly low concentration of surfactant suffices. During beating, a surface area of, say 10^5 m^2 per m^3 aqueous phase is produced. Assuming a surface load of 3 mg surfactant per m^2, about 0.1% of surfactant would be more than sufficient.

Figure 31 illustrates the stages in foam formation after creation of initial bubbles. For most surfactants, Ostwald ripening will be substantial even during foam formation, implying that very small bubbles will soon disappear and the bubble size distribution will become fairly narrow. As soon as beating stops, bubbles rise rapidly and form a foam layer (unless liquid viscosity is quite high). The buoyancy force soon is sufficient to cause mutual deformation of bubbles, causing the formation of flat lamellae between them. The stress due to buoyancy is roughly equal to $\rho_{water}gH$, where H is the height in the foam layer, about 100 Pa for $H = 1$ cm.

However, there is marked stress concentration as spherical bubbles come into contact, and this means that bubbles with a Laplace pressure of 10^3 Pa would become significantly flattened. Further drainage of interstitial liquid causes the bubbles to attain a polyhedral shape. Where three lamellae meet (never > 3, because that would yield an unstable conformation), a prism-shaped water volume, bounded by cylindrical surfaces, is formed. This structural element is called a Plateau border. Residual small bubbles usually disappear by Ostwald ripening. In this way, a fairly regular polyhedral foam is formed, not unlike a honeycomb structure. In the lower part of a foam layer, bubbles remain more or less spherical.

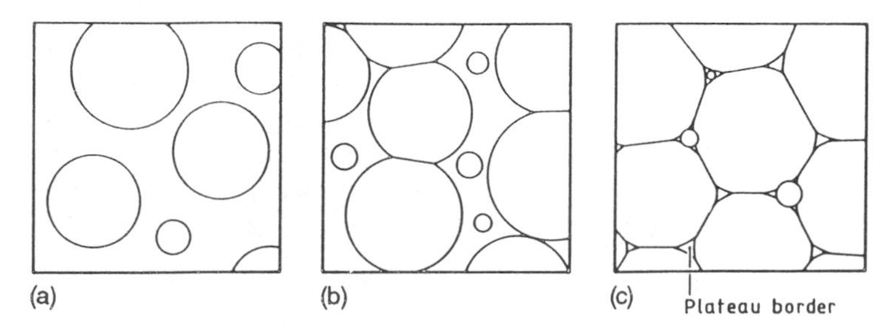

FIGURE 31 Subsequent stages (a, b, and c) in the formation of a foam, once bubbles have been created. Thickness of the lamellae between bubbles is too small to be shown on this scale (bubble diameter < 1 mm).

As the foam keeps draining (see also the following section), the volume fraction of air increases. The Laplace pressure in the Plateau borders is lower that in the lamellae, and this causes liquid to flow to the Plateau borders. Because the latter are interconnected, they provide pathways through which the liquid can drain. As drainage continues, a φ value of 0.95 can readily be reached, which corresponds to an "overrun" of 2000%. Such a foam is not very substantial as a food. To avoid excessive drainage, small filler particles may be incorporated, but they should be hydrophilic; otherwise, considerable coalescence of bubbles can occur (Sec. 3.7.2). Small protein-coated emulsion droplets will function well, and they are incorporated in several whipped toppings. Another approach is gelation of the aqueous phase. This is employed in many aerated food products, such as meringues, foam omelets, bavarois, bread, and cakes. By letting the system gel in an early stage, it is also possible to make a foam with spherical rather than polyhedral bubbles ("Kugelschaum"). Otherwise, only small bubbles in a fairly thin layer of foam can retain a spherical shape.

A polyhedral foam itself may be considered a gel. Deformation of the foam causes an increase in curvature of bubbles, a corresponding increase in Laplace pressure, and elastic behavior at small deformation. Then, at greater stress, bubbles slip past each other and viscoelastic deformation occurs. There is thus a yield stress, which is readily observed, since even fairly large portions of foam retain shape under their own weight. The yield stress usually exceeds 100 Pa.

3.7.2 Stability

Foams are subject to three main types of instability:

1. Ostwald ripening (disproportionation), which is the diffusion of gas from small to larger bubbles (or to the atmosphere). This occurs because the pressure in a small bubble is greater than in larger ones.
2. Drainage of liquid from and through the foam layer, due to gravity.
3. Coalescence of bubbles due to instability of the film between them.

These changes are to some extent interdependent: Drainage may promote coalescence, and Ostwald ripening or coalescence may affect drainage rate.

These instabilities are governed by fundamentally different factors, as will be made clear next. Unfortunately, most studies have failed to distinguish between the three types of instability. One reason for this may be the lack of suitable methods to follow bubble size distribution.

3.7.2.1 Ostwald Ripening*

This is often by far the most important type of instability. Within minutes after formation, noticeable coarsening of the bubble size distribution often occurs. It happens most rapidly at the top of a foam layer, because the air can diffuse directly to the atmosphere and the layer of water between bubbles and atmosphere is very thin. Ostwald ripening also occurs inside a foam at a significant rate.

The classical treatment of Ostwald ripening rate, based on Equation 10 and the diffusion laws, is by de Vries [14]. He considered a small bubble of radius r_0 surrounded by very much larger bubbles at an average distance δ. The change in radius with time t then would be given by

$$r^2(t) = r_0^2 - (RTDs_\infty\gamma/p\delta)t \tag{24}$$

where D is the diffusion coefficient of the gas in water, s_∞ is solubility for $r = \infty$ ($\text{mol} \cdot \text{m}^{-3} \cdot \text{Pa}^{-1}$), γ is interfacial tension (mostly about 0.05 $\text{N} \cdot \text{m}^{-1}$) and p is ambient pressure (Pa). It follows

*See Sec. 3.2 for fundamental aspects.

from Equation 24 that a bubble will shrink ever faster as it becomes smaller. Furthermore, since γ and solubility for most gases in water are high, shrinkage is fast as illustrated by the following examples. A nitrogen bubble of radius of 0.1 mm at $\delta = 1$ mm in water will disappear in about 3 min, and a CO_2 bubble in about 4 sec. This is not quite realistic, since the geometric assumptions underlying Equation 24 are not fully met in practice, and also because the process is somewhat slower if a mixture of gases, like air, is present. Moreover, as the remaining bubbles become larger, the rate of change decreases with time. Nevertheless, Ostwald ripening can occur very fast.

Can Ostwald ripening be stopped or retarded? If a bubble shrinks, its area decreases and its surface load (Γ) increases, provided the surfactant does not desorb. If no desorption occurs, γ is lowered, the Laplace pressure (Eq. 9) is lowered, and the driving force for Ostwald ripening decreases. It will even stop as soon as the surface dilational modulus, E_{sd}, which is a measure of the change in γ with change in area (see Eq. 11), becomes equal to $\gamma/2$. However, surfactant normally desorbs and E_{sd} therefore decreases, at a rate that depends on several factors, especially surfactant type. For a foam made with small-molecule surfactants, desorption occurs readily and retardation of Ostwald ripening is never great. Proteins, however, desorb very sluggishly (see Sec. 3.2.2), E_{sd} remains high (Section 3.2.6), and Ostwald ripening will be greatly retarded [50].

Some proteins produce tenacious layers at the a/w interface because of cross-linking reactions between adsorbed molecules (Section 3.2.2). Egg white especially is a good foam stabilizer. During foaming, strong surface denaturation occurs, leading to fairly large protein aggregates. These remain irreversibly adsorbed, resulting in a very strong resistance to Ostwald ripening. Something similar can be achieved with solid particles, if they have a suitable contact angle (Fig. 7). An example is provided by the partially solid fat globules in whipped cream, which completely coat the air bubbles, and also form a network throughout the system (see Sec. 3.6.5).

Many complex systems contain at least some solid particles that act in this manner (being small and fairly hydrophilic). Bubble shrinkage occurs until the adsorbed solid particles touch each other. Then a small but stable bubble remains. This is presumably the cause of many undesirable persistent foams. Another example is the gas cells in bread dough [69]. They show extensive Ostwald ripening, and the number of visible cells in the final product is less than 1% of those originally present. This does not mean that all the others have disappeared. In fact, many tiny cells remain, presumably stabilized by solid particles. They are not visible but scatter light sufficiently to give the bread its white appearance.

It should be mentioned that Ostwald ripening can be prevented by a yield stress in the aqueous phase, but it would need to be high, at least about 10^4 Pa.

3.7.2.2 Drainage

As mentioned in Sec. 3.2.6, immobilization of the a/w interface by means of a γ gradient is essential to prevent almost instantaneous drainage (Figure 9C and D). The maximum height that a vertical film (lamella) between two bubbles can have while preventing motion of the film surfaces is given by

$$H_{max} = 2\,\Delta\gamma/\rho g \delta \tag{25}$$

The maximum value that $\Delta\gamma$ (between top and bottom of a vertical film) can assume equals the surface pressure Π, which would be about 0.03 $N \cdot m^{-1}$. For an aqueous film of thickness $\delta = 0.1$ mm, H_{max} would be 6 cm, far more than needed in food foams (6 cm is, indeed, about the height of the largest foam bubbles floating on a detergent solution).

The drainage time of a single vertical film with immobilized surfaces is given by

$$t(\delta) \approx 6\eta H/\rho g \delta^2 \qquad (26)$$

where $t(\delta)$ is the time needed for the film to drain to a given thickness δ. For a water film of a 1 mm height, only 6 sec of drainage is required to achieve a thickness of 10 μm. However, the drainage rate diminishes with decreasing thickness, and it would take 17 days of drainage to achieve $\delta = 20$ nm. The latter is the approximate thickness at which colloidal interaction forces between the two film surfaces come into play, and it appears that this point is usually not reached within the lifetime of a foam.

Predicting the drainage rate in a real foam is far more difficult, and accurate calculations cannot be made. Equation 26 will serve to provide approximate (order of magnitude) values. Drainage can, of course, be slowed down considerably by increasing the viscosity. For this purpose, viscosity should be measured at fairly low shear stress. A yield stress of about $gH\rho_{water}$ (where H is the height of the foam layer) will also arrest drainage.

3.7.2.3 Coalescence

The mechanism differs with circumstances. Three cases are important.

1. Thick films. This refers to films thick enough so that colloidal interaction between the two surfaces is negligible. In this section the Gibbs stabilizing mechanism is essential (Sec. 3.2.6, especially Fig. 9E). Film rupture, and thereby bubble coalescence, will occur only when surfactant concentration is very low, and only during foam formation. If a film is rapidly stretched, as will always occur during whipping, rupture occurs more readily (it causes E_{sd} to decrease). Thus, it is observed that there is an optimum whipping speed for foam formation, that is, one that achieves greatest air incorporation.
2. Thin films. As mentioned earlier, drainage will almost never lead to films thin enough for colloidal interactions to become important. However, water may evaporate from the film, especially at the top of a foam. If so, film rupture may, and often will, occur. The considerations given in Section 3.6.4 roughly apply. Compared to emulsions, γ is large (more stable), but film area is very large (less stable). Again, proteins may yield the most stable films, especially if they form thick adsorbed layers.
3. Films containing extraneous particles. It is often observed that the presence of extraneous particles, especially lipids, is very detrimental to foam stability. Such particles can cause rupture of relatively thick films, and possible mechanisms are indicated in Figure 32. In cases a and b, a hydrophobic particle becomes trapped in a thinning film, coming into contact with both air bubbles. Because of the obtuse contact angle (Fig. 7), the curvature of the film surface where it reaches the particle results in a high Laplace pressure. Hence, the water flows away from this high pressure region and the film breaks. Hydrophilic particles, on the other hand, may adsorb and not induce film rupture.

In cases c and d, contact with one air bubble would suffice. An oil droplet reaching the a/w interface (case c) will suddenly alter its shape. It may greatly flatten, depending on the three interfacial tensions (Fig. 7). This induces water to flow away from the flattening droplet, and if the film is fairly thin, it can lead to film rupture. In case d the contact angle as measured in the oil is zero. This is equivalent to the existence of a positive spreading pressure Π_S. Oil then spreads over the a/w interface, dragging water along with it, which may, again, lead to film rupture.

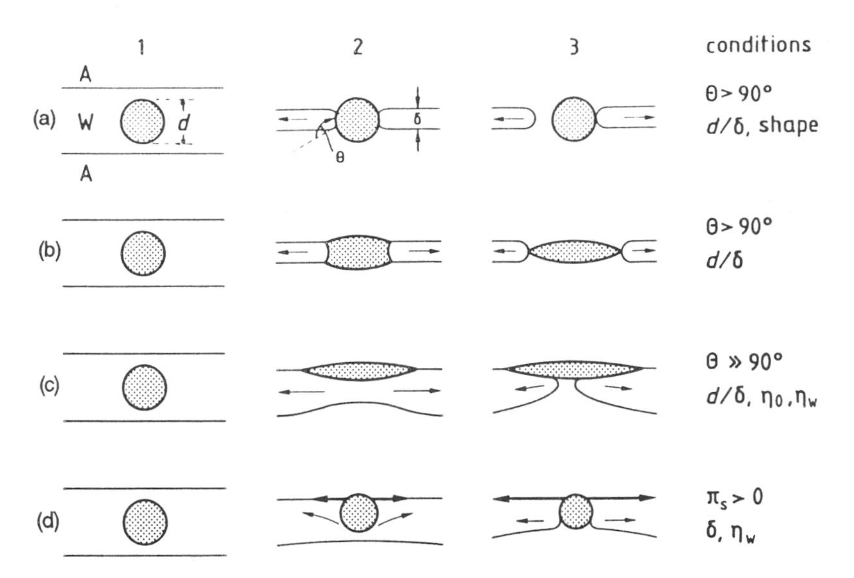

FIGURE 32 Mechanisms of rupture of an aqueous film (lamella) between air bubbles, as induced by extraneous particles; 1, 2, and 3 denote subsequent stages. (a) Solid particle; (b and c) oil droplets; (d) oil droplet or composite particle. A is air; W is water; θ is contact angle (measured in the water phase); η is viscosity; Π_S is spreading pressure. See text for further discussion.

The surface pressure usually is negative if the foaming agent is a small-molecule surfactant, but for many proteins it can be positive—hence the detrimental effect of fat on the stability of beer foam, which is primarily stabilized by proteins and other polymers.

The situation can even be more complicated. For a droplet to come into contact with air, the thin aqueous film between droplet and air bubble has to break. Since both bubble and droplet are covered by surfactants that may provide colloidal repulsion, rupture of this film may not readily occur. If the droplet contains fat crystals, something similar to the induction of partial coalescence may happen (Sec. 3.6.5)—that is, a protruding crystal can pierce the film. It is indeed observed, that partial crystallization of the oil in the droplets may greatly enhance their destabilizing activity.

Finally, the number concentration of particles should be considered. A typical foam (Table 8) contains about 10^{11} lamellae per cubic meter of liquid phase. Presumably, 10^{10} particles per cubic meter would suffice to cause overwhelming coalescence, provided the particles can induce film rupture. If the concentration of extraneous particles is much smaller, the foam may remain fairly stable.

In a typical whipping cream, the number of particles, that is, partially solid fat globules, is very large, about $10^{16}/m^3$. These globules can induce film rupture according to mechanism d (Fig. 32). However, their large concentration causes almost simultaneous adsorption of many globules very close to each other. Spreading of liquid oil over any distance then is not possible, and film rupture will rarely occur. However, if whipping goes on, the fat globules undergo extensive partial coalescence, large clumps are formed, and eventually their number becomes so small that film rupture can occur. In other words, overwhipping destroys the foam made at an earlier stage.

ACKNOWLEDGMENTS

The author gratefully acknowledges the cooperation on subjects as discussed in this chapter with his Wageningen colleagues Drs. A. Prins and T. van Vliet.

FREQUENTLY USED SYMBOLS

A	(Specific) surface area	(m^{-1}, m^2)
	Hamaker constant	(J)
a	Thermodynamic activity	(mole fraction)
	Acceleration	($m \cdot sec^{-2}$)
B	Permeability	(m^2)
c	Concentration	($kg \cdot m^{-3}$; $mol \cdot m^{-3}$; moles per liter)
D	Diffusion coefficient	($m^2 \cdot sec^{-1}$)
	Fractal dimensionality	(—)
d	Particle diameter	(m)
E_{sd}	Surface dilational modulus	($N \cdot m^{-1}$)
F	Force	(N)
G	Velocity gradient, shear rate	(sec^{-1})
	Elastic shear modulus	(Pa)
g	Acceleration due to gravity	($9.81\ m \cdot sec^{-2}$)
H	Height	(m)
h	Interparticle distance	(m)
I	Ionic strength	(moles per liter)
k	Boltzmann constant	($1.38 \times 10^{-23}\ J \cdot K^{-1}$)
l	Distance, length	(m)
m	Concentration	(moles per liter)
N	Total number concentration	(m^{-3})
n_i	Number of particles in class i	(m^{-3})
p	Pressure	(Pa)
p_L	Laplace pressure	(Pa)
Q	Volume flow rate	($m^3 \cdot sec^{-1}$)
R	Universal gas constant	($8.314\ J \cdot mol^{-1} \cdot K^{-1}$)
	Radius of aggregate (floc)	(m)
R_{cr}	Critical radius	(m)
R_g	Radius of gyration	(m)
r	Particle radius	(m)
s	Solubility of gas	($mol \cdot m^{-3} \cdot Pa^{-1}$)
T	Temperature (absolute)	(K)
t	Time	(sec)
$t_{0.5}$	Halving time	(sec)
V	Interaction free energy	(J)
V_A	Part of V due to van der Waals attraction	
V_E	Part of V due to electrostatic repulsion	
V_R	Part of V due to steric repulsion	
v	Velocity	($m \cdot sec^{-1}$)
v_s	Stokes velocity of particle	($m \cdot sec^{-1}$)
W	Stability ratio	(—)
x	Distance	(m)
z	Valence	(—)

Greek Symbols

Γ	Surface excess (load)	($mol \cdot m^{-2}$, $kg \cdot m^{-2}$)
γ	Surface (interfacial) tension	($N \cdot m^{-1}$)
δ	Layer (film) thickness	(m)
ε	Strain (relative deformation)	(—)
ε_{fr}	Strain at fracture	(—)
θ	Contact angle	(rad)
κ	Reciprocal Debye length	(m^{-1})
η	Viscosity	($Pa \cdot sec$)
η_{ss}	Surface shear viscosity	($N \cdot sec \cdot m^{-1}$)
Π	Surface pressure	($N \cdot m^{-1}$)
Π_s	Spreading pressure	($N \cdot m^{-1}$)
Π_{osm}	Osmotic pressure	(Pa)
ρ	Mass density	($kg \cdot m^{-3}$)
σ	Stress	(Pa)
σ_{fr}	Fracture stress	(Pa)
σ_y	Yield stress	(Pa)
φ	Volume fraction	(—)
ψ	Fraction solid	(—)

Subscripts

A	Air
C	Continuous phase
D	Dispersed phase
O	Oil
S	Solid
W	Water (aqueous phase)

BIBLIOGRAPHY

Blanshard, J. M. V. and P. Lillford, eds. (1987). *Food Structure and Behaviour*, Academic Press, London.

Blanshard, J. M. V., and J. R. Mitchell, eds. (1988). *Food Structure—Its Creation and Evaluation*, Butterworth, London.

Dickinson, E. (1992). *An Introduction into Food Colloids*, Oxford Science, Oxford.

Dickinson, E., and G. Stainsby, (1982). *Colloids in Foods*, Applied Science Publishers, London.

Larsson, K., and S. E. Friberg, eds. (1990). *Food Emulsions*, 2nd. ed., Marcel Dekker, New York.

Mitchell, J. R., and D. A. Ledward, eds. (1986). *Functional Properties of Food Macromolecules*, Elsevier, London.

P. Walstra (In preparation). *Physical Chemistry of Foods*, Marcel Dekker, New York.

REFERENCES

1. Adamson, A. W. (1976). *Physical Chemistry of Surfaces*, 3rd ed., Interscience, New York.
2. Allen, T. (1981). *Particle Size Measurement*, 3rd. ed., Chapman & Hall, London.
3. Bergenståhl, B. (1988). Gums as stabilizers of emulsifier covered emulsion droplets, in *Gums and Stabilizers for the Food Industry*, Vol. 4 (G. O. Phillips, P. A. Williams, and D. J. Wedlock, eds.), IRL Press, Oxford, pp. 363–372.
4. Blanshard, J. M. V. (1982). Hydrocolloid water interactions. *Progr. Food Nutr. Sci. 6*:3–20.
5. Boode, K., and P. Walstra (1993). Kinetics of partial coalescence in oil-in-water emulsions, in *Food*

Colloids and Polymers: Stability and Mechanical Properties (eds. E. Dickinson and P. Walstra), Royal Society Chemistry, Cambridge, pp. 23–30.

6. Boode, K., and P. Walstra (1993). Partial coalescence in oil-in-water emulsions 1. Nature of the aggregation. *Colloids Surf. 81*:121–137.
7. Boode, K., P. Walstra, and A. E. A. de Groot-Mostert (1993). Partial coalescence in oil-in-water emulsions 2. Influence of the properties of the fat. *Colloids Surf 81*:139—151.
8. Bremer, L. G. B., P. Walstra, and T. van Vliet (1995). Aggregation kinetics related to the observation of instability of colloidal systems. *Colloids Surf. A99*:121–127.
9. Brownsey, G. J., and V. J. Morris (1988). Mixed and filled gels—Models for foods, in *Functional Properties of Food Macromolecules* (J. R. Mitchell and D. A. Ledward, eds.), Elsevier, London, pp. 7–23.
10. Busnell, J. P., S. M. Clegg, and E. R. Morris (1988). Melting behaviour of gelatin: Origin and control, in *Gums and Stabilizers for the Food Industry*, Vol. 4 (G. O. Phillips, D. J. Wedlock, and P. A. Williams, eds.), IRL Press, Oxford, pp. 105–115.
11. Clark, A. H., and C. D. Lee-Tufnell (1986). Gelation of globular proteins, in *Functional Properties of Food Macromolecules* (J. R. Mitchell and D. A. Ledward, eds.), Elsevier, London, pp. 203–272.
12. Darling, D. F. (1982). Recent advances in the destabilization of dairy emulsions. *J. Dairy Res. 49*:695–712.
13. De Feijter, J. A., J. Benjamins, and M. Tamboer (1987). Adsorption displacement of proteins by surfactants in oil-in-water emulsions. *Colloids Surf. 27*:243–266.
14. De Vries, A. J. (1958). Foam stability. II. Gas diffusion in foams. *Rec. Trav. Chim. 77*:209–225.
15. Dickinson, E. (1988). The role of hydrocolloids in stabilising particulate dispersions and emulsions, in *Gums and Stabilizers for the Food Industry*, Vol. 4 (G. O. Phillips, D. J. Wedlock, and P. A. Williams, eds.), IRL Press, Oxford, pp. 249–263.
16. Dickinson, E. (1992). Structure and composition of adsorbed protein layers and the relation to emulsion stability. *J. Chem. Soc. Faraday Trans. 88*:2973–2983.
17. Dickinson, E. (1994). Protein-stabilized emulsions. *J. Food Eng. 22*:59–74.
18. Dickinson, E., and L. Eriksson (1991). Particle flocculation by adsorbing polymers. *Adv. Colloid Interf. Sci. 34*:1–29.
19. Dickinson, E., V. B. Galazka, and D. M. W. Anderson (1991). Emulsifying behaviour of gum arabic. Part 1: Effect of the nature of the oil phase on the emulsion droplet-size distribution. *Carbohydrate Polym. 14*:373–383.
20. Dickinson, E., and Y. Matsumura (1991). Time-dependent polymerization of β-lactoglobulin through disulphide bonds at the oil-water interface in emulsions. *Int. J. Biol. Macromol. 13*:26–30.
21. Dickinson, E., B. S. Murray, and G. Stainsby (1988). Coalescence stability of emulsion-sized droplets at a planar oil-water interface and the relation to protein film surface rheology. *J. Chem. Soc., Faraday Trans. 1 84*:871–883.
22. Fleer, G. J., M. A. Cohen Stuart, J. M. H. M. Scheutjens, T. Cosgrove, and B. Vincent (1993). *Polymers at Interfaces*, Chapman & Hall, London.
23. Garrett, P. R. (1993). Recent developments in the understanding of foam generation and stability. *Chem. Eng. Sci. 48*:367–392.
24. Garti, N., and D. Reichman (1994). Surface properties and emulsification activity of galactomannans. *Food Hydrocolloids 8*:155–173.
25. Graham, D. E., and M. C. Phillips (1980). Proteins at liquid interfaces IV. Dilational properties. *J. Colloid Interface Sci. 76*:227–239.
26. Hermansson, A.-M. (1988). Gel structure of food biopolymers, in *Functional Properties of Food Macromolecules* (J. R. Mitchell and D. A. Ledward, eds.), Elsevier, London, pp. 25–40.
27. Hough, D. B., and L. R. White (1980). The calculation of Hamaker constants from Lifshitz theory with application to wetting phenomena. *Adv. Colloid Interface Sci. 14*:3–41.
28. Israelachvili, J. N. (1992). *Intermolecular and Surface Forces*, 2nd ed., Academic Press, London.
29. Kato, A., and S. Nakai (1980). Hydrophobicity determined by a fluorescent probe method and its correlation with surface properties of proteins. *Biochim. Biophys. Acta 624*:13–20.
30. Keetels, C., T. van Vliet, and H. Luyten (1995). The effect of retrogradation on the structure and

mechanics of concentrated starch gels, in *Food Macromolecules and Colloids* (E. Dickinson and D. Lorient, eds.), Roy. Soc. Chem., Cambridge, pp. 472–479.

31. Kimura, H., S. Morikata, and M. Misaki (1973). Polysaccharide 13140: A new thermo-gelable polysaccharide. *J. Food Sci. 38*:668–670.
32. Krog, N. J. (1990). Food emulsifiers and their chemical and physical properties, in *Food Emulsions* (K. Larsson and S. E. Friberg, eds.), 2nd. ed., Marcel Dekker, New York, pp. 127–180.
33. Larsson, K., and P. Dejmek (1990). Crystal and liquid crystal structure of lipids, in *Food Emulsions* (K. Larsson and S. E. Friberg, eds.), 2nd ed., Marcel Dekker, New York, pp. 97–125.
34. Ledward, D. A. (1986). Gelation of gelatin, in *Functional Properties of Food Macromolecules* (J. R. Mitchell and D. A. Ledward, eds.), Elsevier, London, pp. 171–201.
35. Ledward, D. A. and J. R. Mitchell (1988). Protein extrusion—More questions than answers?, in *Food Structure—Its Creation and Evaluation* (J. M. V. Blanshard and J. R. Mitchell, eds.), Butterworths, London, pp. 219–229.
36. Lucassen-Reynders, E. H. (1981). Adsorption at fluid interfaces, in *Anionic Surfactants: Physical Chemistry of Surfactant Action* (E. H. Lucassen-Reynders, ed.), Marcel Dekker, New York, pp.1–54.
37. Lucassen-Reynders, E. H. (1981). Surface elasticity and viscosity in compression/dilation, in *Anionic Surfactants: Physical Chemistry of Surfactant Action* (E. H. Lucassen-Reynders, ed.), Marcel Dekker, New York, pp. 173–216.
38. Luyten, H., and T. van Vliet (1990). Influence of a filler on the rheological and fracture properties of food materials, in *Rheology of Foods, Pharmaceutical and Biological Materials with General Rheology* (R. E. Carter, ed.), Elsevier, London, pp. 43–56.
39. Luyten, H., T. van Vliet, and W. Kloek (1991). Sedimentation in aqueous xanthan + galactomannan mixtures, in *Food Polymers, Gels and Colloids* (E. Dickinson, ed.), Royal Society of Chemistry, Cambridge, pp. 527–530.
40. Morris, V. J. (1986). Gelation of polysaccharides, in *Functional Properties of Food Macromolecules* (J. R. Mitchell, and D. A. Ledward, eds.), Elsevier, London, pp. 121–170.
41. Morris, V. J. (1992). Designing polysaccharides for synergistic interactions, in *Gums and Stabilizers for the Food Industry,* Vol. 6 (G. O. Phillips, P. A. Williams and D. J. Ledward, eds.) IRL Press, Oxford, pp. 161–172.
42. Muhr, A. H., and J. M. V. Blanshard (1982). Diffusion in gels. *Polymer 23* (suppl.):1012–1026.
43. Mulder, H., and P. Walstra (1974). *The Milk Fat Globule: Emulsion Science as Applied to Milk Products and Comparable Foods*, PUDOC, Wageningen, The Netherlands, and CAB, Farnham Royal.
44. Norde, W., and J. Lyklema (1991). Why proteins prefer interfaces. *J. Biomater. Sci. Polymer Ed. 2*:183–202.
45. Oguntunde, A. O., P. Walstra, and T. van Vliet (1989). Physical characterization of soymilk, in *Trends in Food Biotechnology* (Ang How Ghee, ed.), Proc. 7th World Congr. Food Sci. Technol., 1987, pp. 307–308.
46. Pearce, K. N., and J. E. Kinsella (1978). Emulsifying properties of proteins: Evaluation of a turbidimetric technique. *J. Agric. Food Chem. 26*:716–723.
47. Peleg, M. (1987). The basics of solid food rheology, in *Food Texture* (H. R. Moskowitz, ed.), Marcel Dekker, New York, pp.3–33.
48. Prins, A. (1988). Principles of foam stability, in *Advances in Food Emulsions and Foams* (E. Dickinson and G. Stainsby, eds.), Elsevier, London, pp. 91—122.
49. Rinaudo, M. (1992). The relation between the chemical structure of polysaccharides and their physical properties, in *Gums and Stabilizers for the Food Industry*, vol. 6 (G. O. Phillips, P. A. Williams, and D. J. Wedlock, eds.), IRL Press, Oxford, pp. 51–61.
50. Ronteltap, A. D., and A. Prins (1990). The role of surface viscosity in gas diffusion in aqueous foams. II. Experimental. *Colloids Surf. 47*:285–298.
51. Shaw, D. J. (1970). *Introduction to Colloid and Surface Chemistry*, 2nd. ed., Butterworth, London.
52. Shimizu, M., M. Saito, and K. Yamauchi (1986). Hydrophobicity and emulsifying activity of milk proteins. *Agric. Biol. Chem. 50*:791–792.
53. Shinoda, K., and H. Kunieda (1983). Phase properties of emulsions: PIT and HLB, in *Encyclopedia of Emulsion Technology*, Vol. 1, *Basic Theory* (P. Becher, ed.), Marcel Dekker, New York, pp. 337–367.

54. Spielman, L. A. (1978). Hydrodynamic aspects of flocculation, in *The Scientific Basis of Flocculation* (K. J. Ives, ed.), Sijthoff & Noordhoff, Alphen aan den Rijn, pp. 63–88.
55. Stading, M., and A.-M. Hermansson (1991). Large deformation properties of β-lactoglobulin gel structures. *Food Hydrocolloids 5*:339–352.
56. Stading, M., M. Langton, and A.-M. Hermansson (1992). Inhomogeneous fine-stranded β-lactoglobulin gels. *Food Hydrocolloids 6*:455–470.
57. Stading, M., M. Langton, and A.-M. Hermansson (1993). Microstructure and rheological behaviour of particulate β-lactoglobulin gels. *Food Hydrocolloids 7*:195–212.
58. Stockham, J. D., and E. G. Fochtman (1977). *Particle Size Analysis*, Ann Arbor Science, Ann Arbor, MI.
59. Tadros, T. F., and B. Vincent (1983). Emulsion stability, in *Encyclopedia of Emulsion Technology*, vol. 1, *Basic Theory* (P. Becher, ed.), Marcel Dekker, New York, pp. 129–286.
60. Timms, R. E. (1994). Physical chemistry of fats, in *Fats in Food Products* (D. P. J. Morgan and K. K. Rajah, eds.), Blackie, London, pp. 1–27.
61. Tolstoguzov, V. B. (1986). Functional properties of protein–polysaccharide mixtures, in *Functional Properties of Food Macromolecules* (J. R. Mitchell and D. A. Ledward, eds.), Elsevier, London, pp. 385–415.
62. Tolstoguzov, V. B. (1993). Thermodynamic incompatibility of food macromolecules, in *Food Colloids and Polymers: Stability and Mechanical Properties* (eds. E. Dickinson and P. Walstra) Royal Society of Chemistry, Cambridge, pp. 94–102.
63. Van Aken, G. A., and M. T. E. Merks, (1993). [Surface-active properties of milk proteins], *Voedingsmiddelentechnologie 27* (9):11–13.
64. Van Boekel, M. A. J. S., and P. Walstra (1981). Stability of oil-in-water emulsions with crystals in the disperse phase. *Colloids Surf. 3*:109–118.
65. Van Kreveld, A. (1974). Studies on the wetting of milk powder. *Neth. Milk Dairy J. 28*:23–45.
66. Van Vliet, T., H. Luyten, and P. Walstra (1991). Fracture and yielding of gels, in *Food Polymers, Gels and Colloids* (E. Dickinson, ed.), Royal Society of Chemistry, Cambridge, pp. 392–403.
67. Van Vliet, T., H. Luyten, and P. Walstra (1993). Time dependent fracture behavior of food, in *Food Colloids and Polymers: Stability and Mechanical Properties* (eds. E. Dickinson and P. Walstra), Royal Society of Chemistry, Cambridge, pp. 175–190.
68. Van Vliet, T., H. J. M. van Dijk, P. Zoon, and P. Walstra (1991). Relation between syneresis and rheological properties of particle gels. *Colloid Polym. Sci. 269*:620–627.
69. Van Vliet, T., A. M. Janssen, A. H. Bloksma, and P. Walstra (1992). Strain hardening of dough as a requirement for gas retention. *J. Texture Studies 23*:439–460.
70. Van Vliet, T., S. P. F. M. Roefs, P. Zoon, and P. Walstra (1989). Rheological properties of casein gels. *J. Dairy Res. 56*:529–534.
71. Van Vliet, T., and P. Walstra (1989). Weak particle networks, in *Food Colloids* (R. D. Bee, P. Richmond, and J. Mingins, eds.), Royal Society of Chemistry, Cambridge, pp. 206–217.
72. Visser, J. (1972). On Hamaker constants: A comparison between Hamaker constants and Lifshitz-van der Waals constants. *Adv. Colloid Interface Sci. 3*:331–363.
73. Visser, J. (1988). Dry spinning of milk proteins, in *Food Structure—Its Creation and Evaluation* (J. M. V. Blanshard and J. R. Mitchell, eds.) Butterworths, London, pp. 197–218.
74. Vreeker, R., L. L. Hoekstra, D. C. den Boer, and W. G. M. Agterof (1992). Fractal aggregation of whey proteins. *Food Hydrocolloids 6*:423–435.
75. Vreeker, R., L. L. Hoekstra, D. C. den Boer, and W. G. M. Agterof (1993). The fractal nature of fat networks, in *Food Colloids and Polymers: Stability and Mechanical Properties* (E. Dickinson and P. Walstra, eds.), Royal Society of Chemistry, Cambridge, pp. 16–23.
76. Walstra, P. (1983). Formation of emulsions, in *Encyclopedia of Emulsion Technology*, Vol. 1, *Basic Theory* (P. Becher, ed.), Marcel Dekker, New York, pp. 57–127.
77. Walstra, P. (1987). Fat crystallization, in *Food Structure and Behaviour* (J. M. V. Blanshard and P. Lillford, eds.), Academic Press, London, pp. 67–85.
78. Walstra, P. (1988). The role of proteins in the stabilization of emulsions, in *Gums and Stabilizers for*

the Food Industry, Vol. 4 (G. O. Phillips, D. J. Wedlock, and P. A. Williams, eds.), IRL Press, Oxford, pp. 323–336.
79. Walstra, P. (1989). Principles of foam formation and stability, in *Foams: Physics, Chemistry and Structure* (A. J. Wilson, ed.), Springer, London, pp. 1–15.
80. Walstra, P. (1993). Principles of emulsion formation. *Chem. Eng. Sci. 48*:333–349.
81. Walstra, P. (1993). Introduction to aggregation phenomena in food colloids, in *Food Colloids and Polymers: Stability and Mechanical Properties* (eds. E. Dickinson and P. Walstra), Royal Society of Chemistry, Cambridge, pp. 1–15.
82. Walstra, P. (1993). Syneresis of curd, in *Cheese: Chemistry, Physics and Microbiology*, Vol. 1, *General Aspects* (P. F. Fox, ed.), Chapman & Hall, London, pp. 141–191.
83. Walstra, P. (1996). Emulsion stability, in *Encyclopedia of Emulsion Technology*, Vol. 4 (ed. P. Becher), Marcel Dekker, New York, in press.
84. Walstra, P., and R. Jenness (1984). *Dairy Chemistry and Physics*, Wiley, New York.
85. Walstra, P., and A. L. de Roos (1993). Proteins at air-water and oil-water interfaces: Static and dynamic aspects. *Food Rev. Int. 9*:503–525.
86. Walstra, P., T. van Vliet, and L. G. B. Bremer (1991). On the fractal nature of particle gels, in *Food Polymers, Gels and Colloids* (E. Dickinson, ed.), Royal Society of Chemistry, Cambridge, pp. 369–382.
87. Wedzicha, B. L. (1988). Distribution of low-molecular-weight food additives in dispersed systems, in *Advances in Food Emulsions and Foams* (E. Dickinson and G. Stainsby, eds.), Elsevier, London, pp. 329–371.

4

Carbohydrates

JAMES N. BEMILLER AND ROY L. WHISTLER
Purdue University, West Lafayette, Indiana

Carbohydrates comprise more than 90% of the dry matter of plants. They are abundant, widely available, and inexpensive. They are common components of foods, both as natural components and as added ingredients. Their use is large in terms of both the quantities consumed and the variety of products in which they are found. They have many different molecular structures, sizes, and shapes and exhibit a variety of chemical and physical properties. They are amenable to both chemical and biochemical modification, and both modifications are employed commercially in improving their properties and extending their use. They are also safe (nontoxic).

Starch, lactose, and sucrose are digestible by humans, and they, along with D-glucose and D-fructose, are human energy sources, providing 70–80% of the calories in the human diet worldwide. In the United States, they supply less than that percentage, in fact only about 50%, of the human caloric intake. Health organizations recommend that the percentage of the total calories the average American consumes in the form of fat (about 37%) be reduced to no more than 30% and that the difference be made up with carbohydrate, especially starch.

The term carbohydrate suggests a general elemental composition, namely $C_x(H_2O)_y$, which signifies molecules containing carbon atoms along with hydrogen and oxygen atoms in the same ratio as they occur in water. However, the great majority of natural carbohydrate compounds produced by living organisms do not have this simple empirical formula. Rather, most of the natural carbohydrate is in the form of oligomers (oligosaccharides) or polymers (polysaccharides) of simple and modified sugars.* Lower molecular weight carbohydrates are often obtained by depolymerization of the natural polymers. This chapter begins with a presentation of the simple sugars and builds from there to larger and more complex structures.

4.1 MONOSACCHARIDES [8,27]

Carbohydrates contain chiral carbon atoms. A chiral carbon atom is one that can exist in two different spatial arrangements (configurations). Chiral carbon atoms are those that have four different groups attached to them. The two different arrangements of the four groups in space (configurations) are nonsuperimposable mirror images of each other. In other words, one is the

*Carbohydrates that cannot be broken down to lower molecular weight carbohydrates by hydrolysis are monosaccharides, a term that indicates that they are the monomeric building units of the oligo- and polysaccharides. Monosaccharides are commonly referred to simply as sugars. However, as will be explained later, table sugar (sucrose) is not a monosaccharide.

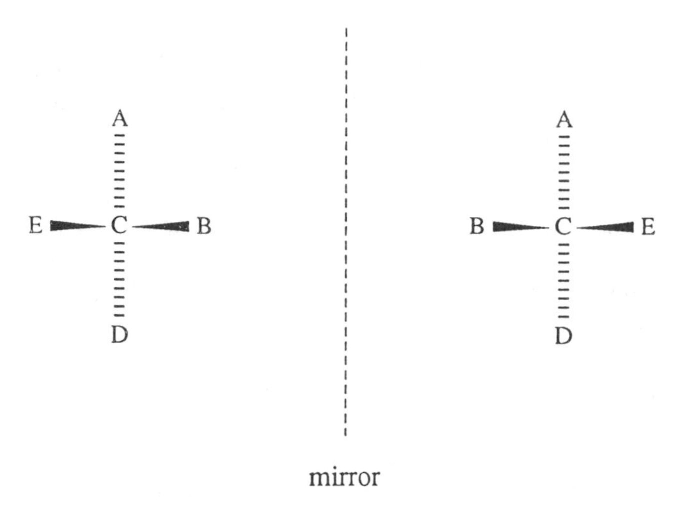

FIGURE 1 Chiral carbon atom. A, B, D, and E represent different atoms, functional groups, or other groups of atoms attached to carbon atom C. Wedges indicate chemical bonds projecting outward from the plane of the page; dashes indicate chemical bonds projecting into or below the plane of the page.

reflection of the other that we would see in a mirror, with everything that is on the right in one configuration on the left in the other and vice versa (Fig. 1).

D-Glucose, the most abundant carbohydrate and the most abundant organic compound (if all its combined forms are considered), belongs to the class of carbohydrates called monosaccharides. Monosaccharides are carbohydrate molecules that cannot be broken down to simpler carbohydrate molecules by hydrolysis, so they are sometimes referred to as simple sugars. They can be joined together to form larger structures, namely, oligosaccharides and polysaccharides (see Secs. 4.2 and 4.3), that can be converted into monosaccharides by hydrolysis.

D-Glucose is both a polyalcohol and an aldehyde. It is classified as an aldose, a designation for sugars containing an aldehyde group (Table 1). The ending -ose signifies a sugar; ald- signifies an aldehyde group. When D-glucose is written in an open or vertical, straight-chain fashion, known as an acyclic structure to organic chemists, with the aldehyde group (position 1) at the top and the primary hydroxyl group (position 6) at the bottom, it is seen that all secondary hydroxyl groups are on carbon atoms having four different substituents attached to them.

TABLE 1 Classification of Monosaccharides

Number of carbon atoms	Kind of carbonyl group	
	Aldehyde	Ketone
3	Triose	Triulose
4	Tetrose	Tetrulose
5	Pentose	Pentulose
6	Hexose	Hexulose
7	Heptose	Heptulose
8	Octose	Octulose
9	Nonose	Nonulose

These carbon atoms are therefore chiral. Glucose has four chiral carbon atoms: C-2, C-3, C-4, and C-5. Naturally occurring glucose is designated as the D form, specifically D-glucose. It has a molecular mirror image, termed the L form, specifically L-glucose. Since each chiral carbon atom has a mirror image, there are 2^n arrangements for these atoms. Therefore, in a six-carbon aldose, there are 2^4 or 16 different arrangements of the carbon atoms containing secondary hydroxyl groups, allowing formation of 16 different six-carbon sugars with an aldehyde end. Eight of these belong to the D-series (see Fig. 3); eight are their mirror images and belong to the L-series. All sugars that have the hydroxyl group on the highest numbered chiral carbon atom (C-5 in this case) positioned on the right-hand side are arbitrarily called D-sugars, and all with a left-hand positioned hydroxyl group on the highest numbered chiral carbon atom are designed L-sugars. Two structures of D-glucose in its open-chain, acyclic form (called the Fischer projection) with the carbon atoms numbered in the conventional manner are given in Figure 2. In this convention, each horizontal bond projects outward from the plane of the page and each vertical bond projects into or below the plane of the page. It is customary to omit the horizontal lines for covalent chemical bonds to the hydrogen atoms and hydroxyl groups as in the structure on the right. Because the lowermost carbon atom is nonchiral, it is meaningless to designate the relative positions of the atoms and groups attached to it. Thus, it is usually written as $-CH_2OH$.

D-Glucose and all other sugars containing six carbon atoms are called hexoses, the most common group of aldoses. The categorical names are often combined, with a six-carbon-atom aldehyde sugar being termed an aldohexose.

There are two aldoses containing three carbon atoms. They are D-glycerose (D-glyceraldehyde) and L-glycerose (L-glyceraldehyde), each possessing only one chiral carbon atom. Aldoses with four carbon atoms, the tetroses, have two chiral carbon atoms; aldoses with five carbon atoms, the pentoses, have three chiral carbon atoms and comprise the second most common group of aldoses. Extending the series above six carbon atoms gives heptoses, octoses, and nonoses, which is the practical limit of naturally occurring sugars. Development of the eight D-hexoses from D-glycerose is shown in Figure 3. In this figure, the circle represents the aldehyde group; the horizontal lines designate the location of each hydroxyl group on its chiral carbon atom, and at the bottoms of the vertical lines are the terminal, nonchiral primary hydroxyl groups. This

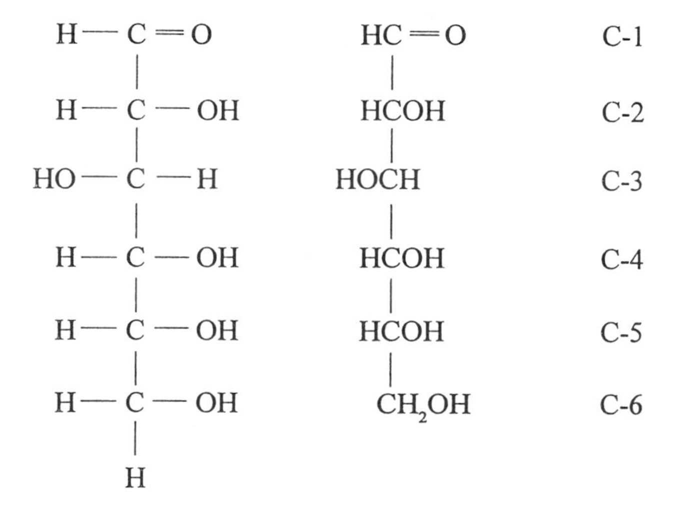

FIGURE 2 D-Glucose (open-chain or acyclic structure).

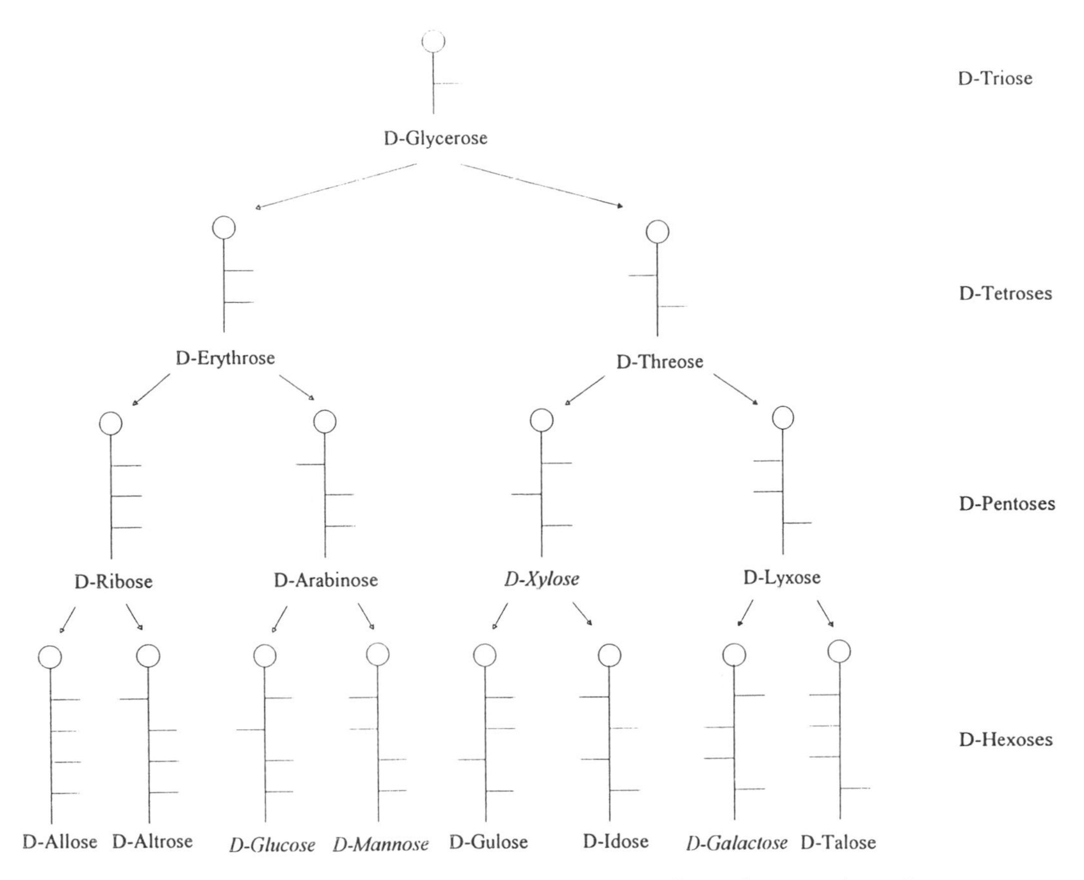

FIGURE 3 Rosanoff structure of the D-aldoses containing from three to six carbon atoms.

shorthand way of indicating monosaccharide structures is called the Rosanoff method. Sugars whose names are in italics in Figure 3 are commonly found in plants, almost exclusively in combined forms. They therefore are present in our diets in combined forms. D-Glucose is the only free aldose usually present in natural foods, and then only in small amounts.

L-Sugars are less numerous and less abundant in nature than are the D forms but nevertheless have important biochemical roles. Two L-sugars found in foods are L-arabinose and L-galactose, both of which occur as units in carbohydrate polymers (polysaccharides).

In the other type of monosaccharide, the carbonyl function is a ketone group. These sugars are termed ketoses. (Ket- signifies the ketone group.) The suffix designating a ketose in systematic carbohydrate nomenclature is -ulose (Table 1). D-Fructose (Fig. 4) is the prime example of this sugar group. It is one of the two monosaccharide units of the disaccharide sucrose (see Sec. 4.2.3) and makes up about 55% of high-fructose corn syrup and about 40% of honey. D-Fructose has only three chiral carbon atoms, C-3, C-4, and C-5. Thus, there are only 2^3 or 8 D-ketohexoses. D-Fructose is the principal commercial ketose and the only one found free in natural foods, but, like D-glucose, only in small amounts.

4.1.1 Monosaccharide Isomerization

Simple aldoses and ketoses containing the same number of carbon atoms are isomers of each other; that is, a hexose and a hexulose both have the empirical formula $C_6H_{12}O_6$ and can

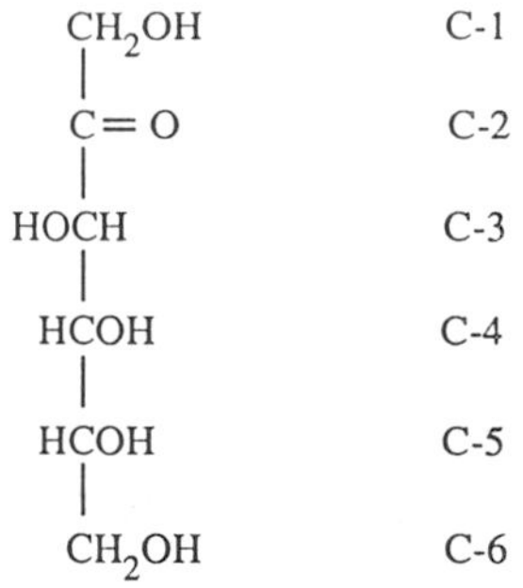

```
   CH2OH        C-1
   |
   C=O          C-2
   |
HOCH            C-3
   |
 HCOH           C-4
   |
 HCOH           C-5
   |
   CH2OH        C-6
```

FIGURE 4 D-Fructose (open-chain or acyclic structure).

be interconverted by isomerization. Isomerization of monosaccharides involves both the carbonyl group and the adjacent hydroxyl group. By this reaction, an aldose is converted into another aldose (with opposite configuration of C-2) and the corresponding ketose, and a ketose is converted into the corresponding two aldoses. Therefore, by isomerization, D-glucose, D-mannose, and D-fructose can be interconverted (Fig. 5). Isomerization can be catalyzed by either a base or an enzyme.

4.1.2 Monosaccharide Ring Forms

Carbonyl groups of aldehydes are reactive and easily undergo nucleophilic attach by the oxygen atom of a hydroxyl group to produce a hemiacetal. The hydroxyl group of a hemiacetal can react further (by condensation) with a hydroxyl group of an alcohol to produce an acetal (Fig. 6). The carbonyl group of a ketone reacts similarly.

Hemiacetal formation can occur within the same aldose or ketose sugar molecule wherein the carbonyl function reacts with one of its own properly positioned hydroxyl groups, as illustrated in Figure 7 with D-glucose laid coiled on its side. The resulting six-membered sugar ring is called a pyranose ring. Notice that for the oxygen atom of the hydroxyl group at C-5 to react to form the ring, C-5 must rotate to bring its oxygen atom upward. This rotation brings the hydroxymethyl group (C-6) to a position above the ring. The representation of the D-glucopyranose ring used in Figure 7 is termed a Haworth projection.

Sugars occur less frequently in five-membered (furanose) rings (Fig. 8).

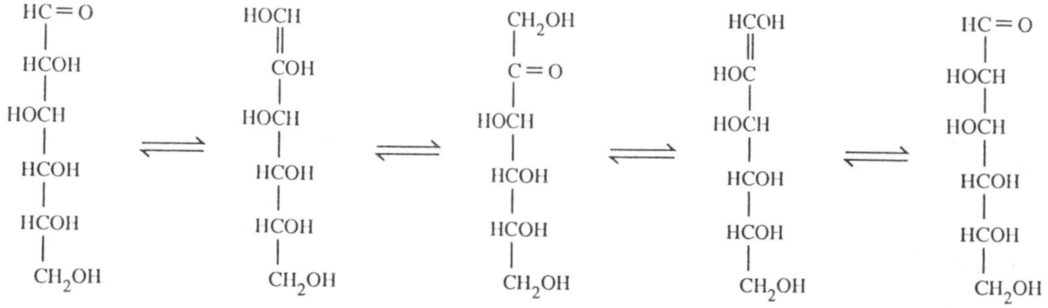

```
 HC=O          HOCH          CH2OH         HCOH          HC=O
 |             ||            |             ||            |
 HCOH          COH           C=O           HOC           HOCH
 |             |             |             |             |
HOCH    ⇌     HOCH    ⇌     HOCH    ⇌     HOCH    ⇌     HOCH
 |             |             |             |             |
 HCOH          HCOH          HCOH          HCOH          HCOH
 |             |             |             |             |
 HCOH          HCOH          HCOH          HCOH          HCOH
 |             |             |             |             |
 CH2OH         CH2OH         CH2OH         CH2OH         CH2OH

D-Glucose   trans-enediol   D-Fructose   cis-enediol   D-Mannose
```

FIGURE 5 Interrelationship of D-glucose, D-mannose, and D-fructose via isomerization.

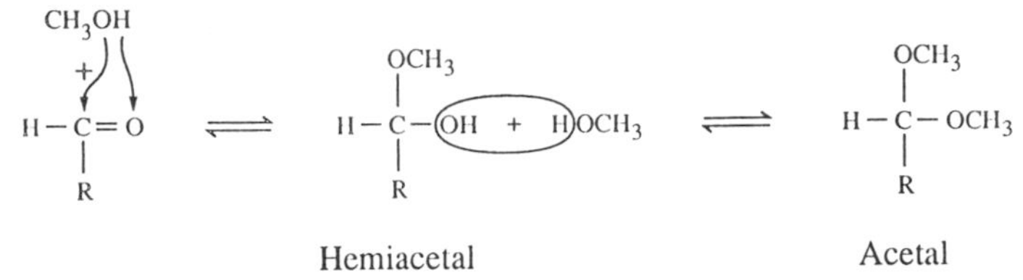

Hemiacetal

Acetal

FIGURE 6 Formation of an acetal by reaction of an aldehyde with methanol.

To avoid clutter in writing the ring structures, common conventions are adopted wherein ring carbon atoms are indicated by angles in the ring and hydrogen atoms attached to carbon atoms are eliminated altogether. A mixture of chiral forms is indicated by a wavy line (Fig. 9).

When the carbon atom of the carbonyl group is involved in ring formation, leading to hemiacetal (pyranose or furanose ring) development, it becomes chiral. With D-sugars, the configuration that has the hydroxyl group located below the ring is the alpha form. For example, therefore, α-D-glucopyranose is D-glucose in the pyranose (six-membered) ring form with the configuration of the new chiral carbon atom, C-1, termed the anomeric carbon atom, alpha. When the newly formed hydroxyl group at C-1 is above the ring, it is in the beta position, and the structure is named β-D-glucopyranose. This designation holds for all D-sugars. For sugars in the L-series, the opposite is true—that is, the anomeric hydroxyl group is up in the alpha anomer and down in the beta anomer.* (See, for example, Fig. 8.) This is so because, for example, α-D-glucopyranose and α-L-glucopyranose are mirror images of one another.

However, pyranose rings are not flat with the attached groups sticking straight up and straight down as the Haworth structure suggests. Rather, they occur in a variety of shapes (conformations), most commonly in one of two chair conformations, so called because they are shaped somewhat like a chair. In this shape, one bond on each carbon atom does project either up or down from the ring; these are called axial bonds or axial positions. The other bond not involved in ring formation is either up or down with respect to the axial bonds but with respect to the ring projects out around the perimeter in what is called an equatorial position (Fig. 10).

Using β-D-glucopyranose as an example, C-2, C-3, C-5, and the ring oxygen atom remain in a plane, but C-4 is raised slightly above the plane and C-1 is positioned slightly below the plane as in Figures 10 and 11. This conformation is designated $^4C_1$. The notation C indicates that the ring is chair-shaped; the superscript and subscript numbers indicate that C-4 is above the plane of the ring and C-1 below the plane. There are two chair forms. The second, $^1C_4$, has

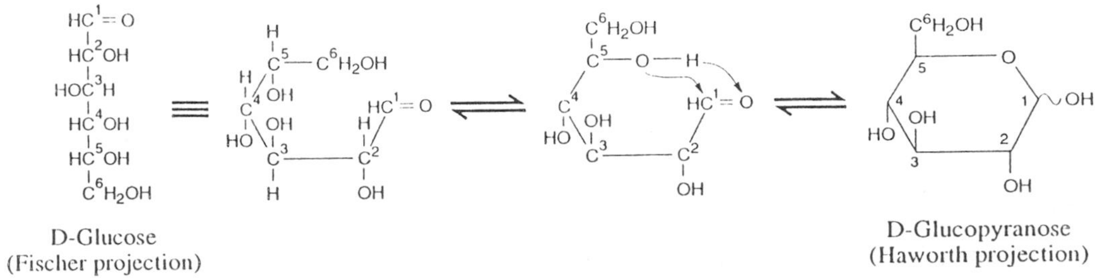

D-Glucose
(Fischer projection)

D-Glucopyranose
(Haworth projection)

FIGURE 7 Formation of a pyranose hemiacetal ring from D-glucose.

*The α and β ring forms of a sugar are known as anomers. The two anomers comprise an anomeric pair.

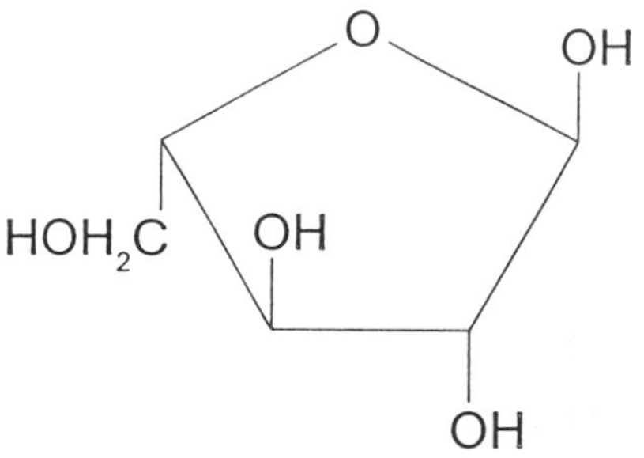

FIGURE 8 L-Arabinose in the furanose ring form and α-L-configuration.

all the axial and equatorial groups reversed. The six-membered ring distorts the normal carbon and oxygen atom bond angles less than do rings of other sizes. The strain is further lessened when the bulky hydroxyl groups are separated maximally from each other by the ring conformation that arranges the greatest number of them in equatorial rather than axial positions. The equatorial position is energetically favored, and rotation of carbon atoms takes place on their connecting bonds to swivel the bulky groups to equatorial positions in so far as possible.

As noted, β-D-glucopyranose has all its hydroxyl groups in the equatorial arrangement, but each is bent either slightly above or slightly below the true equatorial position. In β-D-glucopyranose, the hydroxyl groups, all of which are in an equatorial position, alternate in an up-and-down arrangement, with that at C-1 positioned slightly up, that on C-2 slightly down, and continuing with an up-and-down arrangement. The bulky hydroxymethyl group, C-6 in hexoses, is almost always in a sterically free equatorial position. If D-glucopyranose were in a

D-Glucopyranose

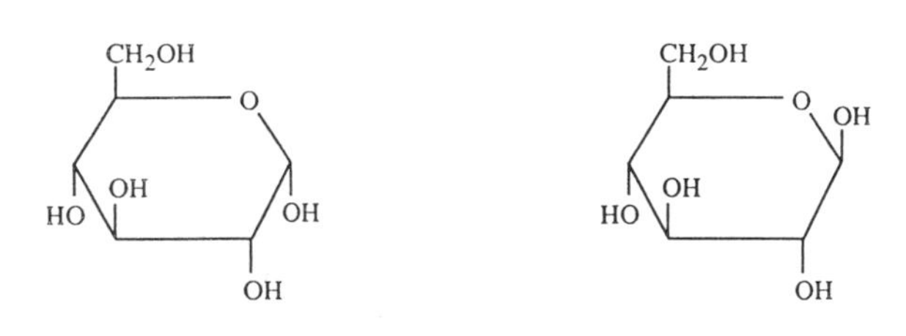

α-D-Glucopyranose β-D-Glucopyranose

FIGURE 9 D-Glucopyranose as a mixture of two chiral forms.

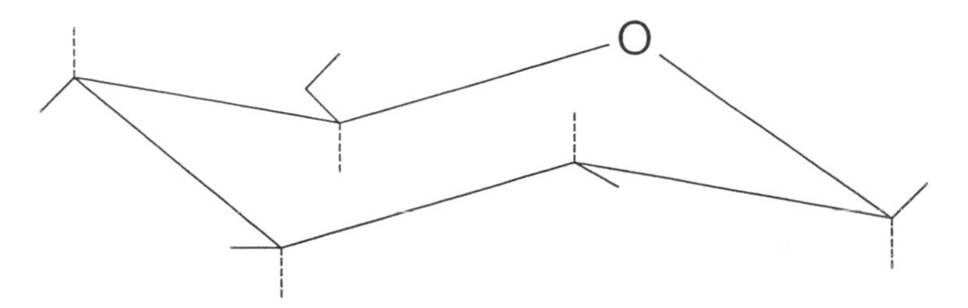

FIGURE 10 A pyranose ring showing the equatorial (solid line) and axial (dashed line) bond positions.

$^1C_4$ conformation, all the bulky groups would be axial; so very little D-glucopyranose exists in the $^1C_4$ conformation, a much higher energy form.

Six-membered sugar rings are then quite stable if bulky groups, such as hydroxyl groups and the hydroxymethyl group, are in equatorial positions. Thus, β-D-glucopyranose dissolves in water to give a rapidly equilibrating mixture containing the open chain form and its five-, six-, and seven-membered ring forms. At room temperature, the six-membered (pyranose) ring forms predominate, followed by the five-membered (furanose) ring forms and only a trace of the seven-membered ring forms. The anomeric arrangement of each ring may be alpha or beta. The open-chain, aldehydo form constitutes only about 0.003% of the total forms (Fig. 12).

4.1.3 Glycosides

The hemiacetal form of sugars can react with an alcohol to produce a full acetal, called a glycoside. The acetal linkage at the anomeric carbon atom is indicated by the -ide ending. In the case of D-glucose reacting with methanol, the product is mainly methyl α-D-glucopyranoside, with less methyl β-D-glucopyranoside (Fig. 13).

The two anomeric forms of the five-membered-ring furanosides are also formed, but being higher energy structures, they reorganize into more stable forms and are present at equilibrium in comparatively low quantities. The methyl group in this case, and any other group bonded to a sugar to make a glycoside, is termed an aglycon. Glycosides undergo hydrolysis to yield a reducing sugar (see Sec. 4.1.4.1) and a hydroxylated compound in the presence of warm or hot aqueous acid.

4.1.4 Monosaccharide Reactions

All carbohydrate molecules have hydroxyl groups available for reaction. Simple monosaccharide and most other low-molecular-weight carbohydrate molecules also have carbonyl groups available for reaction. Formation of pyranose and furanose rings (cyclic hemiacetals) and of glycosides (acetals) of monosaccharides has already been presented.

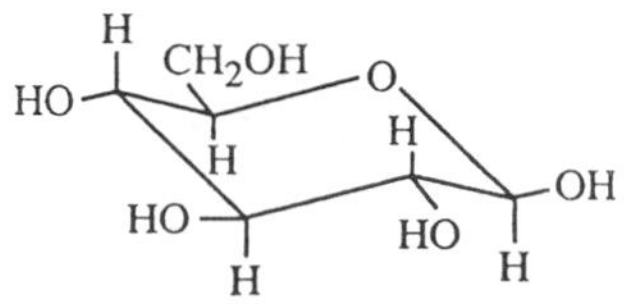

FIGURE 11 β-D Glucopyranose in the $^4C_1$ conformation. All bulky groups are in equatorial positions, and all hydrogen atoms are in axial positions.

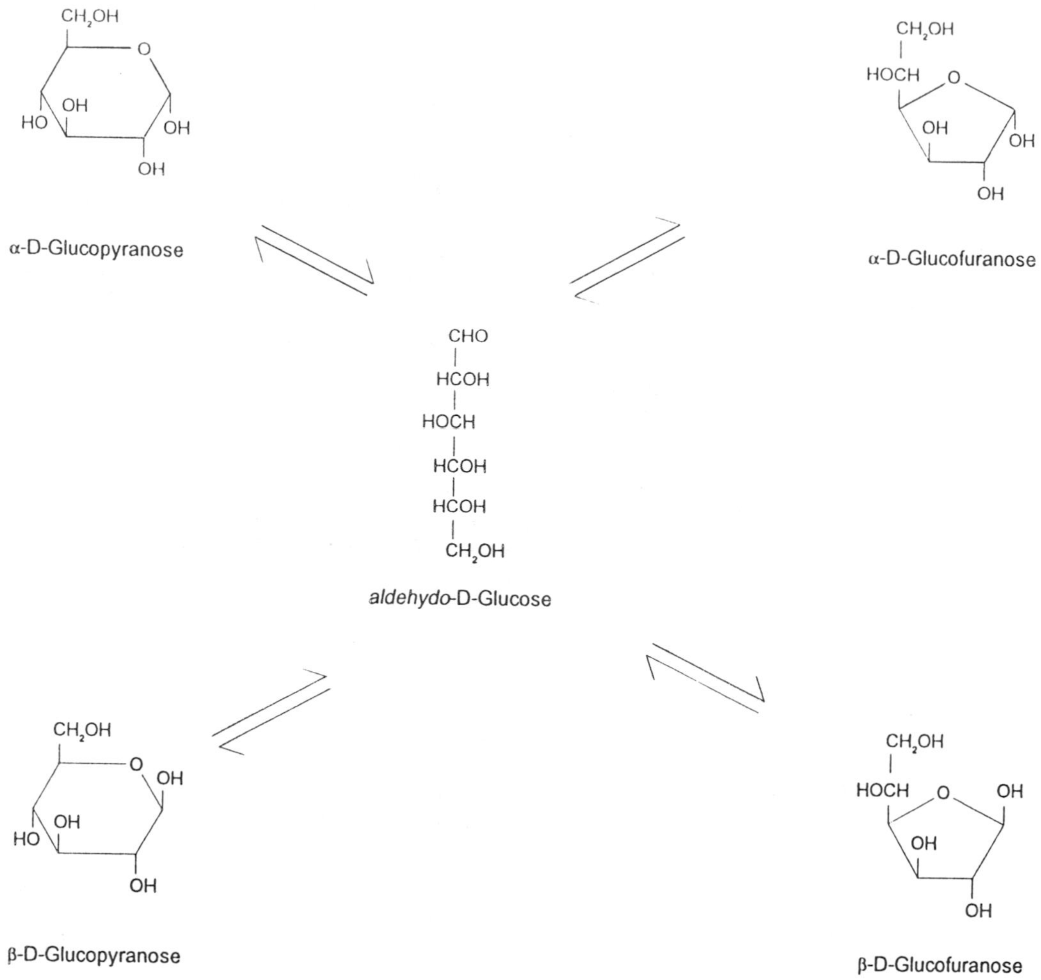

α-D-Glucopyranose

α-D-Glucofuranose

aldehydo-D-Glucose

β-D-Glucopyranose

β-D-Glucofuranose

FIGURE 12 Interconversion of the acyclic and cyclic forms of D-glucose.

4.1.4.1 Oxidation to Aldonic Acids and Aldonolactones

Aldoses are readily oxidized to aldonic acids. The reaction is commonly used for quantitative determination of sugars. One of the earliest methods for quantitative measurement of sugars employed Fehling solution. Fehling solution is an alkaline solution of copper(II) that oxidizes an aldose to an aldonate and in the process is reduced to copper(I), which precipitates as brick-red Cu_2O. Variations, the Nelson–Somogyi and Benedict reagents, are still used for determining amounts of reducing sugars in foods and other biological materials.

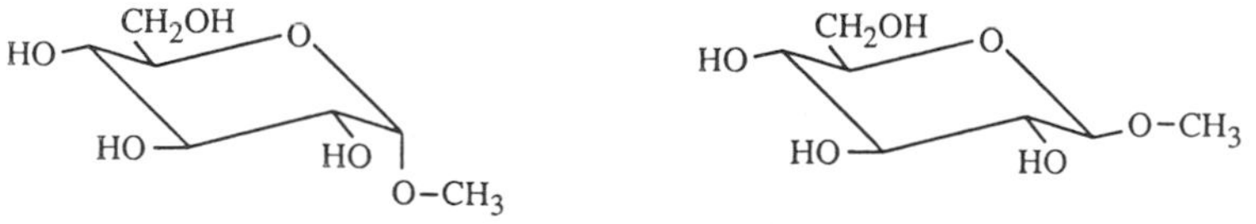

FIGURE 13 Methyl α-D-glucopyranoside (left) and methyl β-D-glucopyranoside (right).

$$2Cu(OH)_2 + R\text{-}\overset{\displaystyle H}{\overset{|}{C}}{=}O \longrightarrow R\text{-}\overset{\displaystyle O}{\overset{\|}{C}}\text{-}OH + Cu_2O + H_2O \qquad (1)$$

Because, in the process of oxidizing the aldehyde group of an aldose to the salt of a carboxylic acid group, the oxidizing agent is reduced, aldoses are called reducing sugars. Ketoses are also termed reducing sugars because, under the alkaline conditions of the Fehling test, ketoses are isomerized to aldoses. Benedict reagent, which is not alkaline, will react with aldoses but not with ketoses.

A simple and specific method for quantitative oxidation of D-glucose to D-gluconic acid uses the enzyme glucose oxidase, with the initial product being the 1,5-lactone (an intramolecular ester) of the acid (Fig. 14). The reaction is commonly employed to measure the amount of D-glucose in foods and other biological materials, including the D-glucose level of blood. D-Gluconic acid is a natural constituent of fruit juices and honey.

The reaction given in Figure 14 is also used for the manufacture of commercial D-gluconic acid and its lactone. D-Glucono-delta-lactone (GDL), D-glucono-1,5-lactone according to systematic nomenclature, hydrolyzes to completion in water in about 3 hr at room temperature, effecting a decrease in pH. Its slow hydrolysis, slow acidification, and mild taste set GDL apart from other food acidulants. It is used in meats and dairy products, but particularly in baked goods as a component of chemical leavening agents for preleavened products.

4.1.4.2 Reduction of Carbonyl Groups [9]

Hydrogenation is the addition of hydrogen to a double bond. When applied to carbohydrates, it most often entails addition of hydrogen to the double bond between the oxygen atom and the carbon atom of the carbonyl group of an aldose or ketose. Hydrogenation of D-glucose is easily accomplished with hydrogen gas under pressure in the presence of Raney nickel. The product is D-glucitol, commonly known as sorbitol, where the -itol suffix denotes a sugar alcohol (an alditol) (Fig. 15). Alditols are also known as polyhydroxy alcohols and as polyols. Because it is derived from a hexose, D-glucitol (sorbitol) is specifically a hexitol. It is found widely distributed throughout the plant world, ranging from algae to higher orders where it is found in fruits and berries, but the amounts present are generally small. It is 50% as sweet as sucrose, is sold both as a syrup and as crystals, and is used as a general humectant.

D-Mannitol can be obtained by hydrogenation of D-mannose. Commercially, it is obtained along with sorbitol from hydrogenolysis of sucrose. It develops from hydrogenation of the D-fructose component of sucrose and from isomerization of D-glucose, which can be controlled

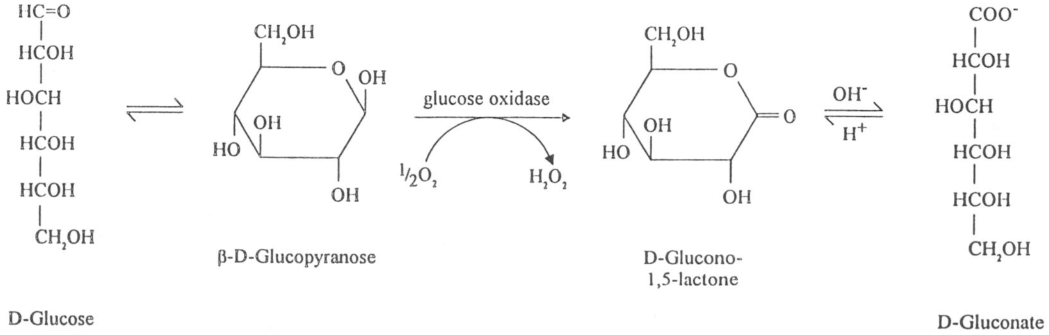

FIGURE 14 Oxidation of D-glucose catalyzed by glucose oxidase.

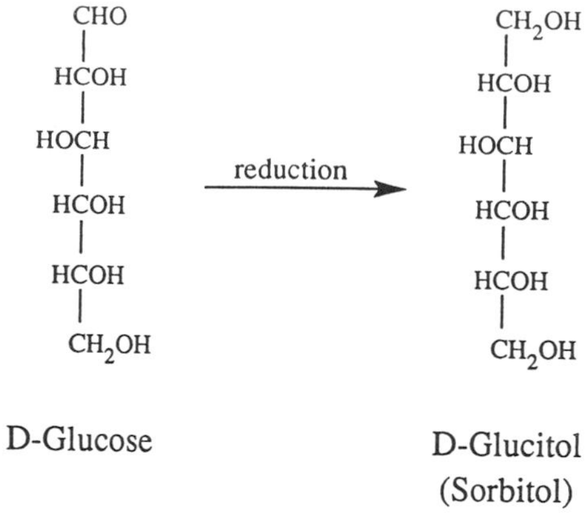

$$\begin{array}{c} CHO \\ | \\ HCOH \\ | \\ HOCH \\ | \\ HCOH \\ | \\ HCOH \\ | \\ CH_2OH \end{array} \xrightarrow{\text{reduction}} \begin{array}{c} CH_2OH \\ | \\ HCOH \\ | \\ HOCH \\ | \\ HCOH \\ | \\ HCOH \\ | \\ CH_2OH \end{array}$$

D-Glucose D-Glucitol (Sorbitol)

FIGURE 15 Reduction of D-glucose.

by the alkalinity of the solution undergoing catalytic hydrogenation (Fig. 16). D-Mannitol, unlike sorbitol, is not a humectant. Rather, it crystallizes easily and is only moderately soluble. It has been used as a nonsticky coating on candies. It is 65% as sweet as sucrose and is used in sugar-free chocolates, pressed mints, cough drops, and hard and soft candies.

Xylitol (Fig. 17) is produced from hydrogenation of D-xylose obtained from hemicelluloses, especially from birch trees. Its crystals have an endothermic heat of solution and give a cool feel when placed in the mouth. It is used in dry hard candies and in sugarless chewing gum. It is about 70% as sweet as sucrose. When xylitol is used in place of sucrose, there is a reduction in dental caries because xylitol is not metabolized by the microflora of the mouth to produce plaques.

4.1.4.3 Uronic Acids

The terminal carbon atom (at the other end of the carbon chain from the aldehyde group) of a monosaccharide unit of an oligo- or polysaccharide may occur in an oxidized (carboxylic acid) form. Such an aldohexose with C-6 in the form of a carboxylic acid group is called a uronic acid. When the chiral carbon atoms of a uronic acid are in the same configuration as they are in D-galactose, for example, the compound is D-galacturonic acid (Fig. 18), the principal component of pectin (see Sec. 4.10).

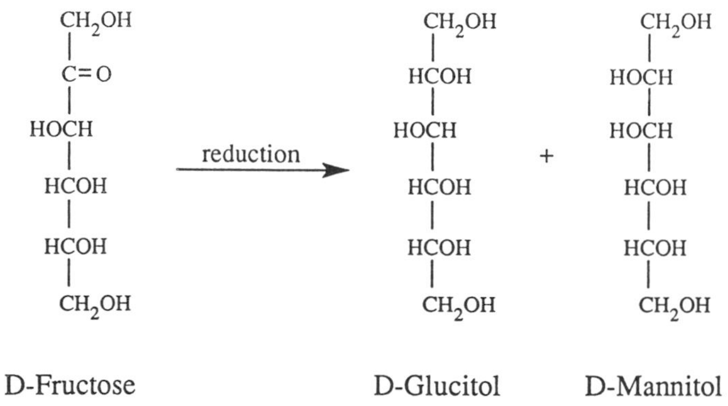

$$\begin{array}{c} CH_2OH \\ | \\ C{=}O \\ | \\ HOCH \\ | \\ HCOH \\ | \\ HCOH \\ | \\ CH_2OH \end{array} \xrightarrow{\text{reduction}} \begin{array}{c} CH_2OH \\ | \\ HCOH \\ | \\ HOCH \\ | \\ HCOH \\ | \\ HCOH \\ | \\ CH_2OH \end{array} + \begin{array}{c} CH_2OH \\ | \\ HOCH \\ | \\ HOCH \\ | \\ HCOH \\ | \\ HCOH \\ | \\ CH_2OH \end{array}$$

D-Fructose D-Glucitol D-Mannitol

FIGURE 16 Reduction of D-fructose.

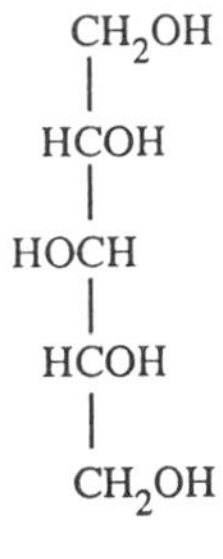

CH_2OH
HCOH
HOCH
HCOH
CH_2OH

FIGURE 17 Xylitol.

4.1.4.4 Hydroxyl Group Esters

The hydroxyl groups of carbohydrates, like the hydroxyl groups of simple alcohols, form esters with organic and some inorganic acids. Reaction of hydroxyl groups with a carboxylic acid anhydride or chloride (an acyl chloride) in the presence of a suitable base produces an ester.

$$\text{ROH} + \text{R'-}\overset{\text{O}}{\overset{\|}{\text{C}}}\text{-O-}\overset{\text{O}}{\overset{\|}{\text{C}}}\text{-R'} \textit{ or } \text{R'-}\overset{\text{O}}{\overset{\|}{\text{C}}}\text{-Cl} \longrightarrow \text{R-O-}\overset{\text{O}}{\overset{\|}{\text{C}}}\text{-R'} + \text{HO-}\overset{\text{O}}{\overset{\|}{\text{C}}}\text{-R'} \textit{ or } \text{HCl} \qquad (2)$$

Acetates, succinate half-esters, and other carboxylic acid esters of carbohydrates occur in nature. They are found especially as components of polysaccharides. Sugar phosphates are common metabolic intermediates (Fig. 19).

Monoesters of phosphoric acid are also found as constituents of polysaccharides. For example, potato starch contains a small percentage of phosphate ester groups. Corn starch contains even less. In producing modified food starch, corn starch is often derivatized with one or the other or both of mono- and distarch ester groups (see Sec. 4.4.9). Other esters of starch, most notably the acetate, succinate and substituted succinate half-esters, and distarch adipates, are in the class of modified food starches (see Sec. 4.4.9). Sucrose (see Sec. 4.2.3) fatty acid esters are produced commercially as emulsifiers. Members of the family of red seaweed polysaccharides, which includes the carrageenans (see Sec. 4.8), contain sulfate groups (half-esters of sulfuric acid, $R\text{-}OSO_3^-$).

4.1.4.5 Hydroxyl Group Ethers

The hydroxyl groups of carbohydrates, like the hydroxyl groups of simple alcohols, can form ethers as well as esters. Ethers of carbohydrates are not as common in nature as are esters. However, polysaccharides are etherified commercially to modify their properties and make them more useful. Examples are the production of methyl ($\text{-O-}CH_3$), sodium carboxy-

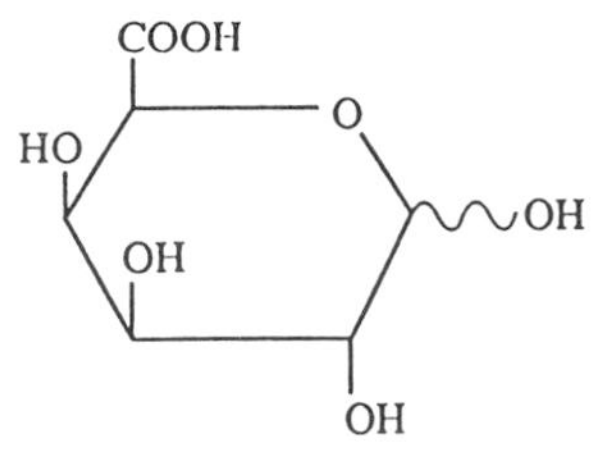

COOH
O
HO
OH
OH
OH

FIGURE 18 D-Galacturonic acid.

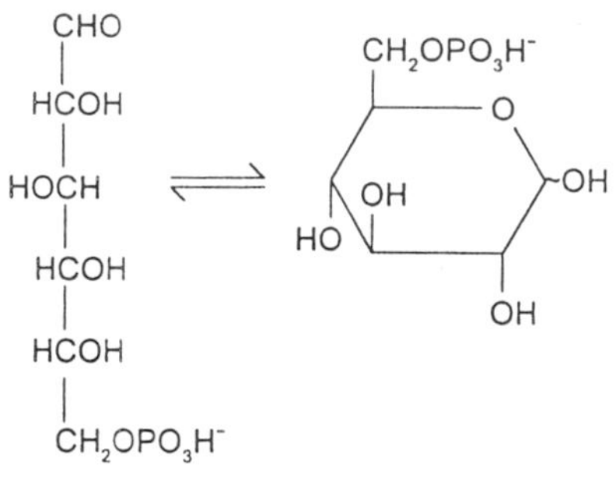

CHO
HCOH
HOCH
HCOH
HCOH
$CH_2OPO_3H^-$

$CH_2OPO_3H^-$
O
OH
OH
HO
OH

D-Glucose 6-phosphate

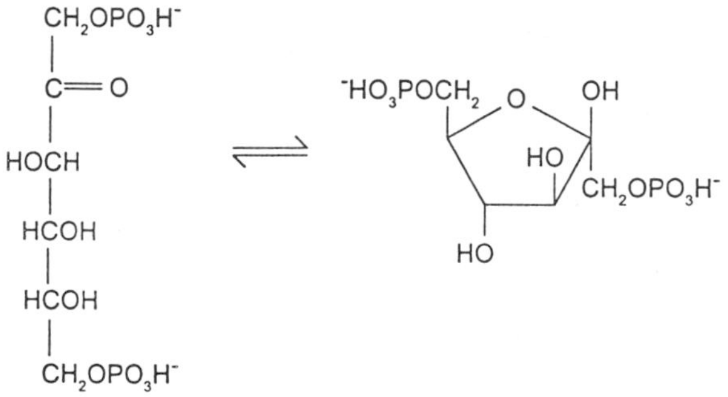

$CH_2OPO_3H^-$
C=O
HOCH
HCOH
HCOH
$CH_2OPO_3H^-$

$^-HO_3POCH_2$
O
OH
HO
$CH_2OPO_3H^-$
HO

D-Fructose 1,6-bisphosphate

FIGURE 19 Examples of sugar phosphate metabolic intermediates.

methyl ($-O-CH_2-CO_2^-Na^+$), and hydroxypropyl ($-O-CH_2-CHOH-CH_3$) ethers of cellulose and hydroxypropyl ethers of starch, all of which are approved for food use.

A special type of ether, an internal ether formed between carbon atoms 3 and 6 of a D-galactosyl unit, is found in the red seaweed polysaccharides, specifically agar, furcellaran, kappa-carrageenan, and iota-carrageenan (see Sec. 4.8) (Fig. 20). Such an internal ether is known

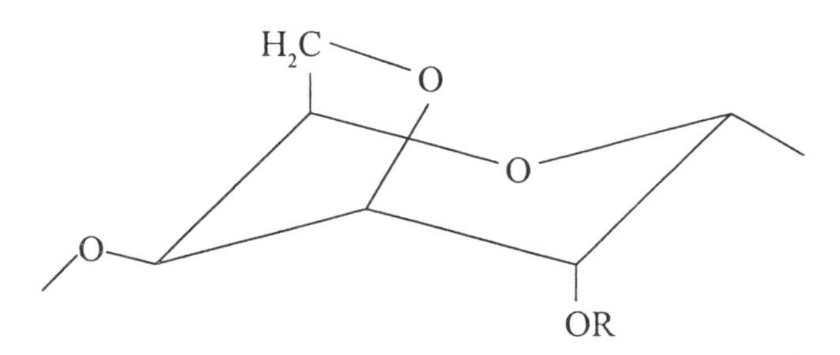

FIGURE 20 A 3,6-anhydro-α-D-galactopyranosyl unit found in red seaweed polysaccharides.

as a 3,6-anhydro ring, whose name derives from the fact that the elements of water (HOH) are removed during its formation.

A series of nonionic surfactants based on sorbitol (D-glucitol) is used in foods as water-in-oil emulsifiers and as defoamers. They are produced by esterification of sorbitol with fatty acids. Cyclic dehydration accompanies esterification (primarily at a primary hydroxyl group, i.e., C-1 or C-6) so that the carbohydrate (hydrophilic) portion is not only sorbitol but also its mono- and dianhydrides (cyclic ethers). The products are known as sorbitan esters (Fig. 21). The product called sorbitan monostearate is actually a mixture of partial stearic (C_{18}) and palmitic (C_{16}) acid esters of sorbitol (D-glucitol), 1,5-anydro-D-glucitol (1,5-sorbitan), **1,4-anhydro-D-glucitol** (1,4-sorbitan), both internal (cyclic) ethers, and 1,4:3,6-dianhydro-D-glucitol (isosorbide), an internal dicyclic ether. Sorbitan fatty acid esters, such as sorbitan monostearate, sorbitan monolaurate, and sorbitan monooleate, are sometimes modified by reaction with ethylene oxide to produce so-called ethoxylated sorbitan esters, also nonionic detergents approved by the FDA for food use.

4.1.4.6 Nonenzymic Browning [10,12,30,59]

Under some conditions, reducing sugars produce brown colors that are desirable and important in some foods. Other brown colors obtained upon heating or during long-term storage of foods containing reducing sugars are undesirable. Common browning of foods on heating or on storage is usually due to a chemical reaction between reducing sugars, mainly D-glucose, and a free amino acid or a free amino group of an amino acid that is part of a protein chain. This reaction is called the Maillard reaction. It is also called nonenzymic browning to differentiate it from the often rapid, enzyme-catalyzed browning commonly observed in freshly cut fruits and vegetables, such as apples and potatoes.

When aldoses or ketoses are heated in solution with amines, a variety of reactions ensue, producing numerous compounds, some of which are flavors, aromas, and dark-colored polymeric materials, but both reactants disappear only slowly. The flavors, aromas, and colors may be either desirable or undesirable. They may be produced by frying, roasting, baking, or storage.

The reducing sugar reacts reversibly with the amine to produce a glycosylamine, as illustrates with D-glucose (Fig. 22). This undergoes a reaction called the Amadori rearrangement to give, in the case of D-glucose, a derivative of 1-amino-1-deoxy-D-fructose. Reaction continues, especially at pH 5 or lower, to give an intermediate that dehydrates. Eventually a furan derivative is formed; that from a hexose is 5-hydroxymethyl-2-furaldehyde (HMF) (Fig. 23). Under less acidic conditions (higher than pH 5), the reactive cyclic compounds (HMF and others) polymerize quickly to a dark-colored, insoluble material containing nitrogen.

Maillard browning products, including soluble and insoluble polymers, are found where reducing sugars and amino acids, proteins, and/or other nitrogen-containing compounds are heated together, such as in soy sauce and bread crusts. Maillard reaction products are important contributors to the flavor of milk chocolate. The Maillard reaction is also important in the production of caramels, toffees, and fudges, during which reducing sugars also react with milk

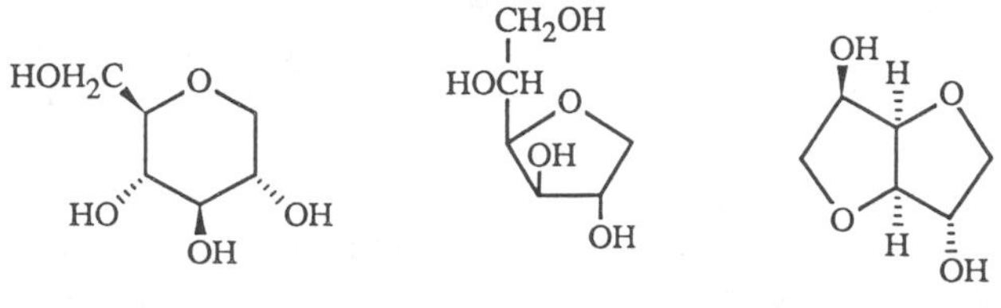

HOH_2C O HO OH OH

CH_2OH HOCH O OH OH

OH H O O H OH

FIGURE 21 Anhydro-D-glucitols (sorbitans).

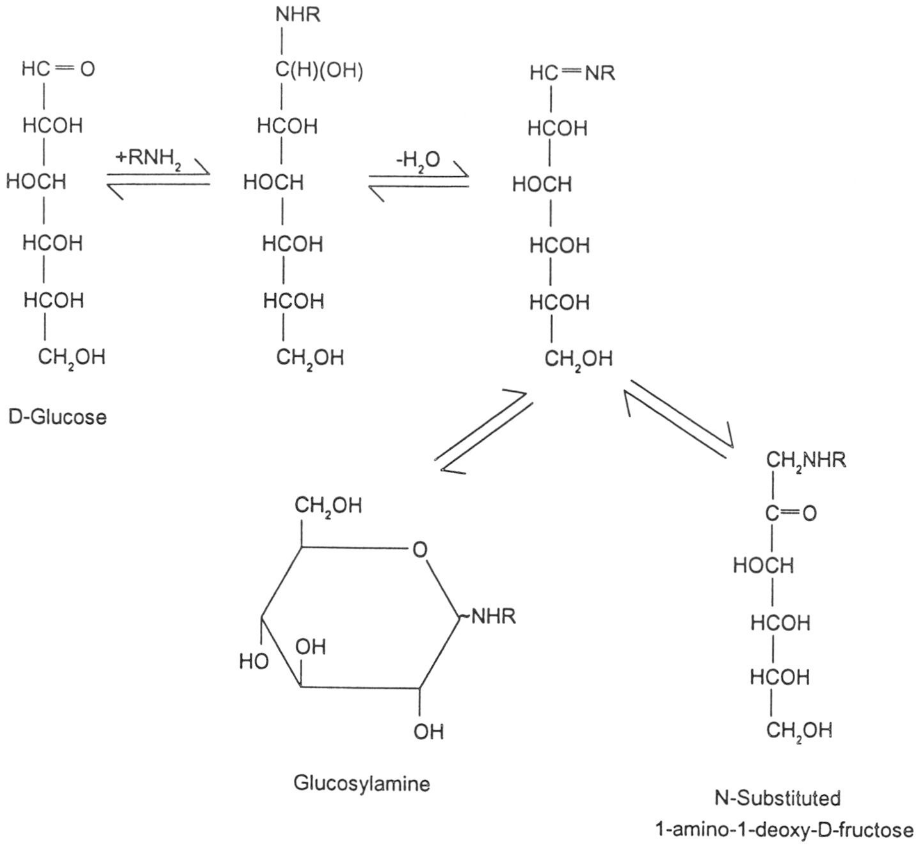

D-Glucose $\xrightleftharpoons{+RNH_2}$ $\xrightleftharpoons{-H_2O}$

Glucosylamine

N-Substituted 1-amino-1-deoxy-D-fructose

FIGURE 22 Products of reaction of D-glucose with an amine (RNH_2).

proteins. D-Glucose undergoes the browning reaction faster than does D-fructose. Application of heat is generally required for nonenzymic browning. While Maillard reactions are useful, they also have a negative side. Reaction of reducing sugars with amino acids destroys the amino acid. This is of particular importance with L-lysine, an essential amino acid whose ε-amino group can react when the amino acid is part of a protein molecule. Also, a relationship has been found between formation of mutagenic compounds and cooking of protein-rich foods. Mutagenic heterocyclic amines have been isolated from broiled and fried meat and fish, and from beef extracts.

Heating of carbohydrates, in particular sucrose (see Sec. 4.2.3) and reducing sugars, without nitrogen-containing compounds effects a complex group of reactions termed caramelization. Reaction is facilitated by small amounts of acids and certain salts. Mostly thermolysis causes dehydration of the sugar molecule with introduction of double bonds or formation of anhydro rings. Introduction of double bonds leads to unsaturated rings such as furans. Conjugated double bonds absorb light and produce color. Often unsaturated rings will condense to polymers yielding useful colors. Catalysts increase the reaction rate and are often used to direct the reaction to specific types of caramel colors, solubilities, and acidities.

Brown caramel color made by heating a sucrose (see Sec. 4.2.3) solution with ammonium bisulfite is used in cola soft drinks, other acidic beverages, baked goods, syrups, candies, pet

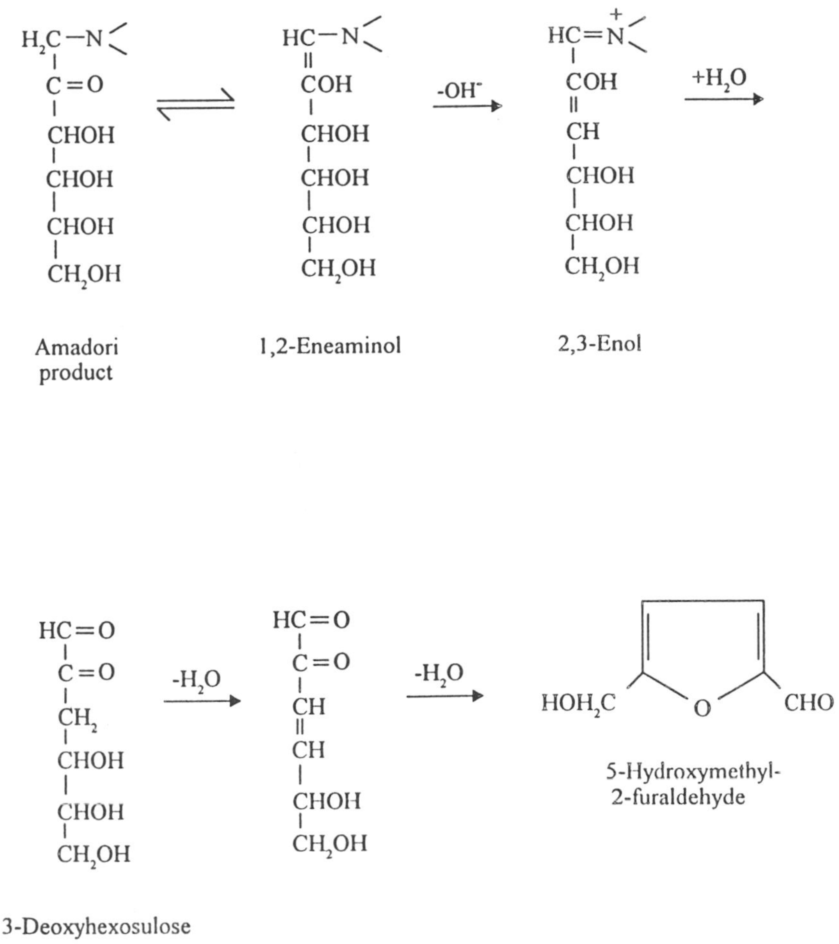

$H_2C-N<$ / $C=O$ / $CHOH$ / $CHOH$ / $CHOH$ / CH_2OH ⇌ $HC-N<$ / $\|$ / COH / $CHOH$ / $CHOH$ / $CHOH$ / CH_2OH $\xrightarrow{-OH^-}$ $HC=\overset{+}{N}<$ / COH / $\|$ / CH / $CHOH$ / $CHOH$ / CH_2OH $\xrightarrow{+H_2O}$

Amadori product — 1,2-Eneaminol — 2,3-Enol

$HC=O$ / $C=O$ / CH_2 / $CHOH$ / $CHOH$ / CH_2OH $\xrightarrow{-H_2O}$ $HC=O$ / $C=O$ / CH / $\|$ / CH / $CHOH$ / CH_2OH $\xrightarrow{-H_2O}$ HOH_2C (furan ring, O) CHO

3-Deoxyhexosulose — 5-Hydroxymethyl-2-furaldehyde

FIGURE 23 Conversion of the Amadori product into HMF.

foods, and dry seasonings. Its solutions are acidic (pH 2–4.5) and contain colloidal particles with negative charges. The acidic salt catalyzes cleavage of the glycosidic bond of sucrose; the ammonium ion participates in the Amadori rearrangement. Another caramel color, also made by heating sugar with ammonium salts, is reddish brown, imparts pH values of 4.2–4.8 to water, contains colloidal particles with positive charges, and is used in baked goods, syrups, and puddings. Caramel color made by heating sugar without an ammonium salt is also reddish brown, but contains colloidal particles with slightly negative charges and has a solution pH of 3–4. It is used in beer and other alcoholic beverages. The nonenzymic browning caramel pigments are large polymeric molecules with complex, variable, and unknown structures. It is these polymers that form the colloidal particles. Their rate of formation increases with increasing temperature and pH.

Certain pyrolytic reactions of sugars (Fig. 24) produce unsaturated ring systems that have unique flavors and fragrances in addition to the coloring materials. Maltol (3-hydroxy-2-methylpyran-4-one) and isomaltol (3-hydroxy-2-acetylfuran) contribute to the flavor of bread. 2*H*-4-Hydroxy-5-methylfuran-3-one can be used to enhance various flavors and sweeteners.

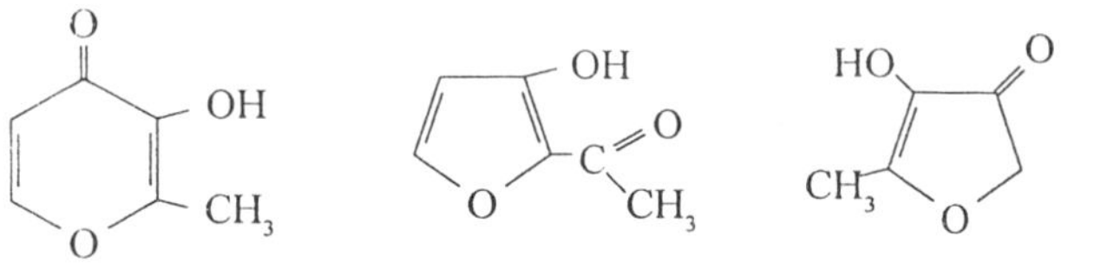

Maltol Isomaltol 2H-4-Hydroxy-5-methylfuran-3-one

FIGURE 24 Some pyrolytic reaction products of sugars.

4.2 OLIGOSACCHARIDES

An oligosaccharide contains 2 to 20 sugar units joined by glycosidic bonds. When a molecule contains more than 20 units, it is a polysaccharide.

Disaccharides are glycosides in which the aglycon is a monosaccharide unit. A compound containing three monosaccharide units is a trisaccharide. Structures containing from 4 to 10 glycosyl units, whether linear or branched, are tetra-, penta-, hexa-, octa-, nona-, and decasaccharides, and so on.

Because glycosidic bonds are part of acetal structures, they undergo acid-catalyzed hydrolysis, that is, cleavage in the presence of aqueous acid and heat. Only a few oligosaccharides occur in nature. Most are produced by hydrolysis of polysaccharides into smaller units.

4.2.1 Maltose

Maltose (Fig. 25), which can be obtained by hydrolysis of starch, is an example of a disaccharide. The end unit on the right (as customarily written) has a potentially free aldehyde group; therefore, it is the reducing end and will be in equilibrium with alpha and beta six-membered ring forms, as described earlier for monosaccharides. Since O-4 is blocked by attachment of the second D-glucopyranosyl unit, a furanose ring cannot form. Maltose is a reducing sugar because its aldehyde group is free to react with oxidants and, in fact, to undergo almost all reactions as though it were present as a free aldose.

Maltose is readily produced by hydrolysis of starch using the enzyme β-amylase (see Sec. 4.4.8). It occurs only rarely in plants, and even then, it results from partial hydrolysis of starch. Maltose is produced during malting of grains, especially barley, and commercially by the specific enzyme-catalyzed hydrolysis of starch using β-amylase from *Bacillus* bacteria, although the β-amylases from barley seed, soybeans, and sweet potatoes may be used. Maltose is used sparingly as a mild sweetener for foods.

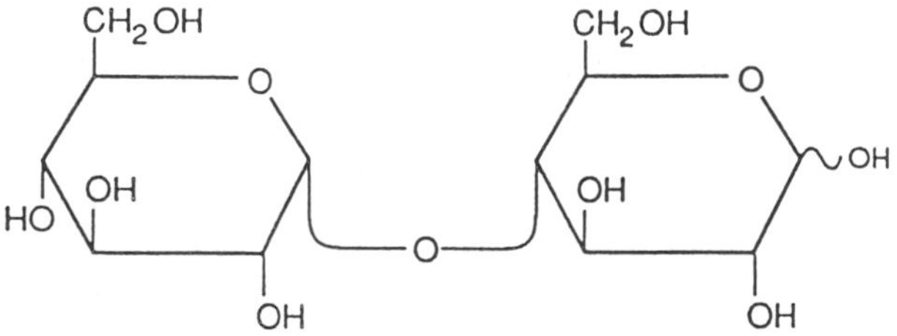

FIGURE 25 Maltose.

4.2.2 Lactose

The disaccharide lactose (Fig. 26) occurs in milk, mainly free, but to a small extent as a component of higher oligosaccharides. The concentration of lactose in milk varies with the mammalian source from 2.0 to 8.5%. Cow and goat milks contain 4.5–4.8%, and human milk about 7%. Lactose is the primary carbohydrate source for developing mammals. In humans, lactose constitutes 40% of the energy consumed during nursing. Utilization of lactose for energy must be preceded by hydrolysis to the constituent monosaccharides, D-glucose and D-galactose. Milk also contains 0.3–0.6% of lactose-containing oligosaccharides, many of which are important as energy sources for growth of a specific variant of *Lactobacillus bifidus*, which, as a result, is the predominant microorganism of the intestinal flora of breast-fed infants.

Lactose is ingested in milk and other unfermented dairy products, such as ice cream. Fermented dairy products, such as most yogurt and cheese, contain less lactose because during fermentation some of the lactose is converted into lactic acid. Lactose stimulates intestinal adsorption and retention of calcium. Lactose is not digested until it reaches the small intestine, where the hydrolytic enzyme lactase is located. Lactase (a β-galactosidase) is a membrane-bound enzyme located in the brush border epithelial cells of the small intestine. It catalyzes the hydrolysis of lactose into its constituent monosaccharides, D-glucose and D-galactose. Of the carbohydrates, only monosaccharides are absorbed from the intestines. Both D-glucose and D-galactose are rapidly absorbed and enter the blood stream.

$$\text{lactose} \xrightarrow{\text{lactase}} \text{D-glucose} + \text{D-galactose} \qquad (3)$$

If for some reason the ingested lactose is only partially hydrolyzed, that is, only partially digested, or is not hydrolyzed at all, a clinical syndrome called lactose intolerance results. If there is a deficiency of lactase, some lactose remains in the lumen of the small intestine. The presence of lactose tends to draw fluid into the lumen by osmosis. This fluid produces abdominal distention and cramps. From the small intestine, the lactose passes into the large intestine (colon) where it undergoes anaerobic bacterial fermentation to lactic acid (present as the lactate anion) and other short-chain acids (Fig. 27). The increase in the concentration of molecules, that is, the increase in osmolality, results in still greater retention of fluid. In addition, the acidic products of fermentation lower the pH and irritate the lining of the colon, leading to an increased movement of the contents. Diarrhea is caused both by the retention of fluid and the increased movement of the intestinal contents. The gaseous products of fermentation cause bloating and cramping.

Lactose intolerance is not usually seen in children until after about 6 years of age. At this

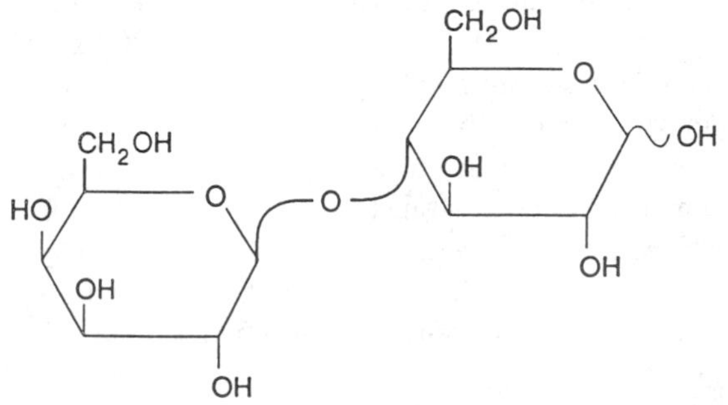

FIGURE 26 Lactose.

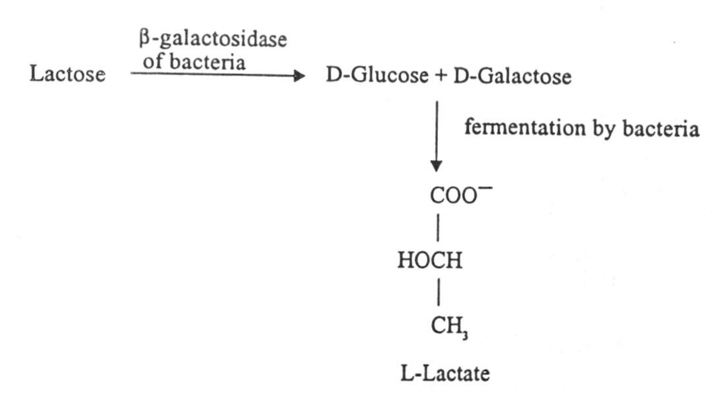

FIGURE 27 The fate of lactose in the large intestine of persons with lactase deficiency.

point, the incidence of lactose-intolerant individuals begins to rise and increases throughout the life span, with the greatest incidence in the elderly. Both the incidence and the degree of lactose intolerance vary by ethnic group, indicating that the presence or absence of lactase is under genetic control.

There are two ways to overcome the effects of lactase deficiency. One is to remove the lactose by fermentation; that produces yogurt and buttermilk products. Another is to produce reduced-lactose milk by adding lactase to it. However, both products of hydrolysis, D-glucose and D-galactose, are sweeter than lactose, and at about 80% hydrolysis, the taste change becomes too evident. Therefore, most reduced-lactose milk has the lactose reduced as close as possible to the 70% government-mandated limit for a claim. In a technology under development, live yogurt cultures are added to refrigerated milk. The bacteria remain dormant in the cold and do not change the flavor of the milk, but upon reaching the small intestine, release lactase.

Other carbohydrates that are not completely broken down into monosaccharides by intestinal enzymes and are not absorbed also pass into the colon. There they also are metabolized by microorganisms, producing lactate and gas. Again diarrhea and bloating result. This problem can occur from eating beans, because beans contain a trisaccharide (raffinose) and a tetrasaccharide (stachyose) (see Sec. 4.2.3) that are not hydrolyzed to monosaccharides by intestinal enzymes and thus pass into the colon, where they are fermented.

4.2.3 Sucrose [35,39]

The per person daily utilization of sucrose, usually called simply sugar or table sugar, in the United States averages about 160 g, but sucrose is also used extensively in fermentations, in bakery products where it is also largely used up in fermentation, and in pet food, so the actual average daily amount consumed by individuals in foods and beverages is about 55 g (20 kg or 43 lb/year). Sucrose is composed of an α-D-glucopyranosyl unit and a β-D-fructofuranosyl unit linked head to head (reducing end to reducing end) rather than by the usual head-to-tail linkage (Fig. 28). Since it has no reducing end, it is a nonreducing sugar.

There are two principal sources of commercial sucrose—sugar cane and sugar beets. Also present in sugar beet extract are a trisaccharide, raffinose, which has a D-galactopyranosyl unit attached to sucrose, and a tetrasaccharide, stachyose, which contains another D-galactosyl unit (Fig. 29). These oligosaccharides are also found in beans, are nondigestible, and are the source of the flatulence associated with eating beans.

Commercial brown sugar is made by treating white sugar crystals with molasses to leave a coating of desired thickness. Grades range from light yellow to dark brown. Confection, or

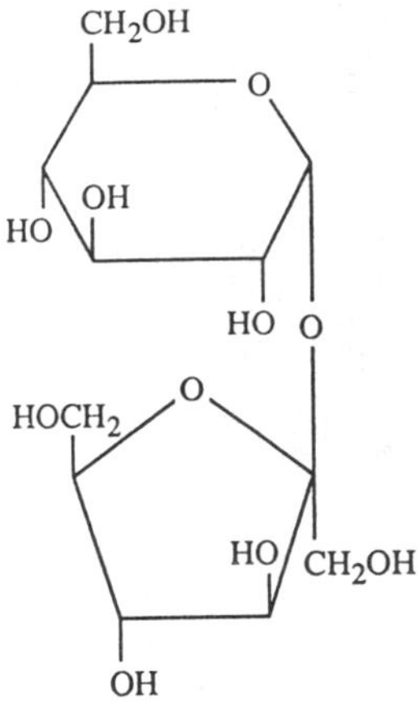

FIGURE 28 Sucrose.

powdered, sugar is pulverized sucrose. It usually contains 3% corn starch as an anticaking agent. To make fondant sugar, which is used in icings and confections, very fine sucrose crystals are surrounded with a saturated solution of invert sugar, corn syrup, or maltodextrin.*

For many food product applications, sucrose is not crystallized; rather, it is shipped as a refined aqueous solution known as liquid sugar. Sucrose and most other low-molecular-weight carbohydrates (for example, monosaccharides, alditols, disaccharides, and other low-molecular-weight oligosaccharides), because of their great hydrophilicity and solubility, can form highly concentrated solutions of high osmolality. Such solutions, as exemplified by pancake and waffle syrups and honey, need no preservatives themselves and can be used not only as sweeteners (although not all such carbohydrate syrups need have much sweetness) but also as preservatives and humectants.

A portion of the water in any carbohydrate solution is nonfreezable. When the freezable water crystallizes, that is, forms ice, the concentration of solute in the remaining liquid phase increases, and the freezing point decreases. There is a consequential increase in viscosity of the remaining solution. Eventually, the liquid phase solidifies as a glass in which the mobility of all molecules becomes greatly restricted and diffusion-dependent reactions become very slow

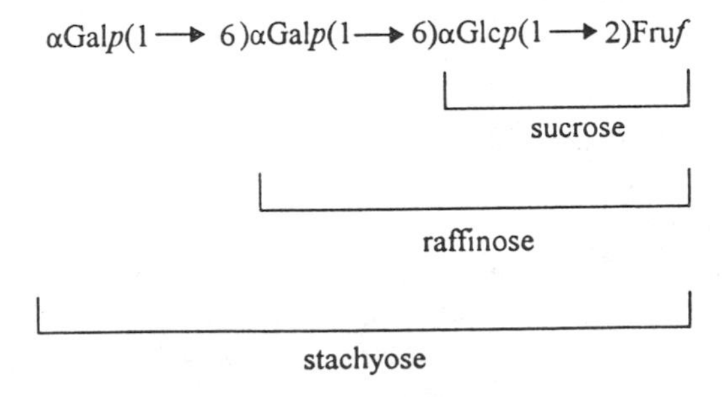

FIGURE 29 Sucrose, raffinose, and stachyose. (For explanation of the shorthand designations of structures, see Sec. 4.3.)

*See Section 4.4.8 for a description of the latter two products. Invert sugar is the equimolar mixture of D-glucose and D-fructose formed by hydrolysis of sucrose.

(see Chap. 2); because of the restricted motion, these glass-state water molecules cannot crystallize. In this way, carbohydrates function as cryoprotectants and protect against the dehydration that destroys structure and texture caused by freezing.

The sucrase of the human intestinal tract catalyzes hydrolysis of sucrose into D-glucose and D-fructose, making sucrose one of the three carbohydrates (other than monosaccharides) humans can utilize for energy, the other two being lactose and starch. Monosaccharides (D-glucose and D-fructose being the only significant ones in our diets) do not need to undergo digestion before absorption.

4.3 POLYSACCHARIDES [42,47,55]

Polysaccharides are polymers of monosaccharides. Like the oligosaccharides, they are composed of glycosyl units in linear or branched arrangements, but most are much larger than the 20-unit limit of oligosaccharides. The number of monosaccharide units in a polysaccharide, termed its degree of polymerization (DP), varies. Only a few polysaccharides have a DP less than 100; most have DPs in the range 200–3000. The larger ones, like cellulose, have a DP of 7000–15,000. It is estimated that more than 90% of the considerable carbohydrate mass in nature is in the form of polysaccharides. Polysaccharides can be either linear or branched. The general scientific term for polysaccharides is glycans.

If all the glycosyl units are of the same sugar type, they are homogeneous as to monomer units and are called homoglycans. Examples of homoglycans are cellulose (see Sec. 4.5) and starch amylose (see Sec. 4.4.1), which are linear, and starch amylopectin (see Sec. 4.4.2), which is branched. All three are composed only of D-glucopyranosyl units.

When a polysaccharide is composed of two or more different monosaccharide units, it is a heteroglycan. A polysaccharide that contains two different monosaccharide units is a diheteroglycan, a polysaccharide that contains three different monosaccharide units is a triheteroglycan, and so on. Diheteroglycans generally are either linear polymers of blocks of similar units alternating along the chain, or consist of a linear chain of one type of glycosyl unit with a second present as single-unit branches. Examples of the former type are algins (see Section 4.9) and of the latter guar and locust bean gums (see Sec. 4.6).

In the shorthand notations of oligo- and polysaccharides, the glycosyl units are designated by the first three letters of their names with the first letter being capitalized, except for glucose, which is Glc. If the monosaccharide unit is that of a D-sugar, the D is omitted; only L sugars are so designated, such as LAra. The size of the ring is designated by an italicized *p* for pyranosyl or *f* for furanosyl. The anomeric configuration is designated with α or β as appropriate; for example, an α-D-glucopyranosyl unit is indicated as αGlc*p*. Uronic acids are designated with a capital A; for example, an L-gulopyranosyluronic acid unit (see Sec. 4.9) is indicated as LGul*p*A. The position of linkages are designated either as, for example, 1 → 3 or 1,3, with the latter being more commonly used by biochemists and the former more commonly used by carbohydrate chemists. Using the shorthand notation, the structure of lactose is represented as βGal*p*(1 → 4)Glc or βGal*p*1,4Glc and maltose as αGlc*p*(1 → 4)Glc or αGlc*p*1,4Glc. (Note that the reducing end cannot be designated as α or β or as pyranose or furanose because the ring can open and close; that is, in solutions of lactose and maltose and other oligo- and polysaccharides, the reducing end unit will occur as a mixture of α- and β-pyranose ring forms and the acyclic form, with rapid interconversion between them; see Fig. 12.)

4.3.1 Polysaccharide Solubility

Most polysaccharides contain glycosyl units that, on average, have three hydroxyl groups. Polysaccharides are thus polyols in which each hydroxyl group has the possibility of hydrogen bonding to one or more water molecules. Also, the ring oxygen atom and the glycosidic oxygen atom connecting one sugar ring to another can form hydrogen bonds with water. With every sugar unit in the chain having the capacity to hold water molecules avidly, glycans possess a strong affinity for water and readily hydrate when water is available. In aqueous systems, polysaccharide particles can take up water, swell, and usually undergo partial or complete dissolution.

Polysaccharides modify and control the mobility of water in food systems, and water plays an important role in influencing the physical and functional properties of polysaccharides. Together polysaccharides and water control many functional properties of foods, including texture.

The water of hydration that is naturally hydrogen-bonded to and thus solvates polysaccharide molecules is often described as water whose structure has been sufficiently modified by the presence of the polymer molecule so that it will not freeze. This water has also been referred to as plasticizing water. The molecules that make up this water are not energetically bound in a chemical sense. While their motions are retarded, they are able to exchange freely and rapidly with other water molecules. This water of hydration makes up only a small part of the total water in gels and fresh tissue foods. Water in excess of the hydration water is held in capillaries and cavities of various sizes in the gel or tissue.

Polysaccharides are cryostabilizers rather than cryoprotectants, because they do not increase the osmolality or depress the freezing point of water significantly, since they are large, high-molecular-weight molecules and these are colligative properties. As an example, when a starch solution is frozen, a two-phase system of crystalline water (ice) and a glass consisting of about 70% starch molecules and 30% nonfreezable water is formed. As in the case of solutions of low-molecular-weight carbohydrates, the nonfreezable water is part of a highly concentrated polysaccharide solution in which the mobility of the water molecules is restricted by the extremely high viscosity. However, while most polysaccharides provide cryostabilization by producing this freeze-concentrated matrix, which severely limits molecular mobility, there is evidence that others provide cryostabilization by restricting ice crystal growth by adsorption to nuclei or active crystal growth sites. Other polysaccharides may be ice nucleators.

Thus both high- and low-molecular-weight carbohydrates are generally effective in protecting food products stored at freezer temperatures (typically –18°C) from destructive changes in texture and structure. In both cases, the improvement in product quality and storage stability is a result of controlling both the amount (particularly in the case of low-molecular-weight carbohydrates) and the structural state (particularly in the case of polymeric carbohydrates) of the freeze-concentrated, amorphous matrix surrounding ice crystals.

Most, if not all, polysaccharides, except those with very bushlike, branch-on-branch structures, exist in some sort of helical shape. Certain linear homoglycans, like cellulose (see Sec. 4.5), have flat, ribbon-like structures. Such uniform linear chains undergo hydrogen bonding with each other so as to form crystallites separated by amorphous regions. Crystalline arrangements of this sort are called fringed micelles (Fig. 30). It is these crystallites of linear chains that give cellulose fibers, like wood and cotton fibers, their great strength, insolubility, and resistance to breakdown, the latter because the crystalline regions are nearly inaccessible to

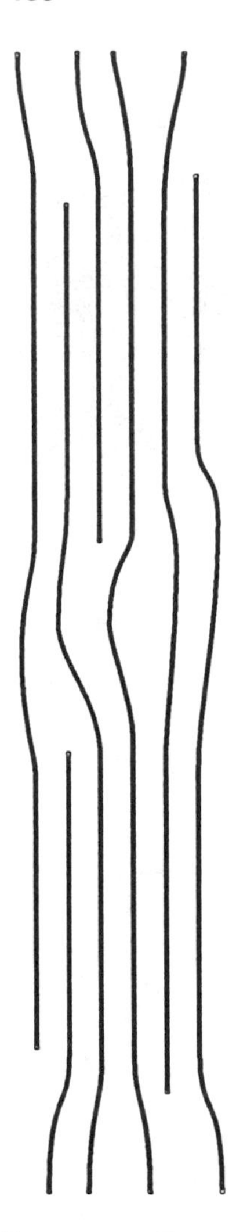

FIGURE 30 Fringed micelles. The crystalline regions are those in which the chains are parallel and ordered.

enzyme penetration. These highly ordered polysaccharides with orientation and crystallinity comprise the exception, rather than the rule. Most polysaccharides are not so crystalline as to impart water insolubility, but are readily hydrated and dissolved.

Most unbranched diheteroglycans containing nonuniform blocks of glycosyl units and most branched glycans cannot form micelles because chain segments are prevented from becoming closely packed over lengths necessary to form strong intermolecular bonding. Hence, these chains have a degree of solubility that increases as chains become less able to fit closely together. In general, polysaccharides become more soluble in proportion to the degree of irregularity of the molecular chains, which is another way of saying that, as the ease with which molecules fit together decreases, the solubility of the molecules increases.

Water-soluble polysaccharides and modified polysaccharides used in food and other industrial applications are known as gums or hydrocolloids. Gums are sold as powders of varying particle size.

4.3.2 Polysaccharide Solution Viscosity and Stability [13]

Polysaccharides (gums, hydrocolloids) are used primarily to thicken and/or gel aqueous solutions and otherwise to modify and/or control the flow properties and textures of liquid food and beverage products and the deformation properties of semisolid foods. They are generally used in food products at concentrations of 0.25–0.50%, indicating their great ability to produce viscosity and to form gels.

The viscosity of a polymer solution is a function of the size and shape of its molecules and the conformations they adopt in the solvent. In foods and beverages, the solvent is an aqueous solution of other solutes. The shapes of polysaccharide molecules in solution are a function of oscillations around the bonds of the glycosidic linkages. The greater the internal freedom at each glycosidic linkage, the greater the number of conformations available to each individual segment. Chain flexibility thus provides a strong entropic drive, which generally overcomes energy considerations and induces the chain to approach disordered or random coil states in solution (Fig. 31). However, most polysaccharides exhibit deviations from strictly random coil states, forming stiff coils, with the specific nature of the coils being a function of the monosaccharide composition and linkages, some being compact, some expanded.

Linear polymer molecules in solution gyrate and flex, sweeping out a large space. They frequently collide with each other, creating friction, consuming energy, and thereby producing viscosity. Linear polysaccharides produce highly viscous solutions, even at low concentrations.Viscosity depends both on the DP (molecular weight) and the extension and rigidity, that is, the shape and flexibility, of the solvated polymer chain.

A highly branched polysaccharide molecule will sweep out much less space than a linear polysaccharide of the same molecular weight (Fig. 32). As a result, highly branched molecules will collide less frequently and will produce a much lower viscosity than will linear molecules of the same DP. This also implies that highly branched polysaccharide molecules must be significantly larger than linear polysaccharide molecules to produce the same viscosity at the same concentration.

Likewise, linear polysaccharide chains bearing only one type of ionic charge (always a negative charge derived from ionization of carboxyl or sulfate half-ester groups) assume an extended configuration due to repulsion of the like charges, increasing the end-to-end chain length and thus the volume swept out by the polymer. Therefore, these polymers tend to produce solutions of high viscosity.

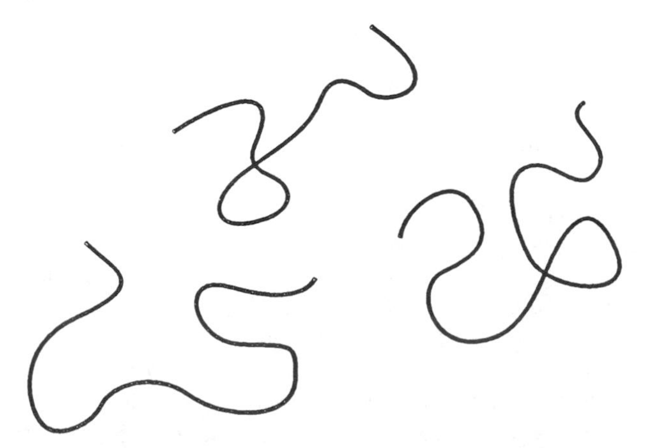

FIGURE 31 Randomly coiled polysaccharide molecules.

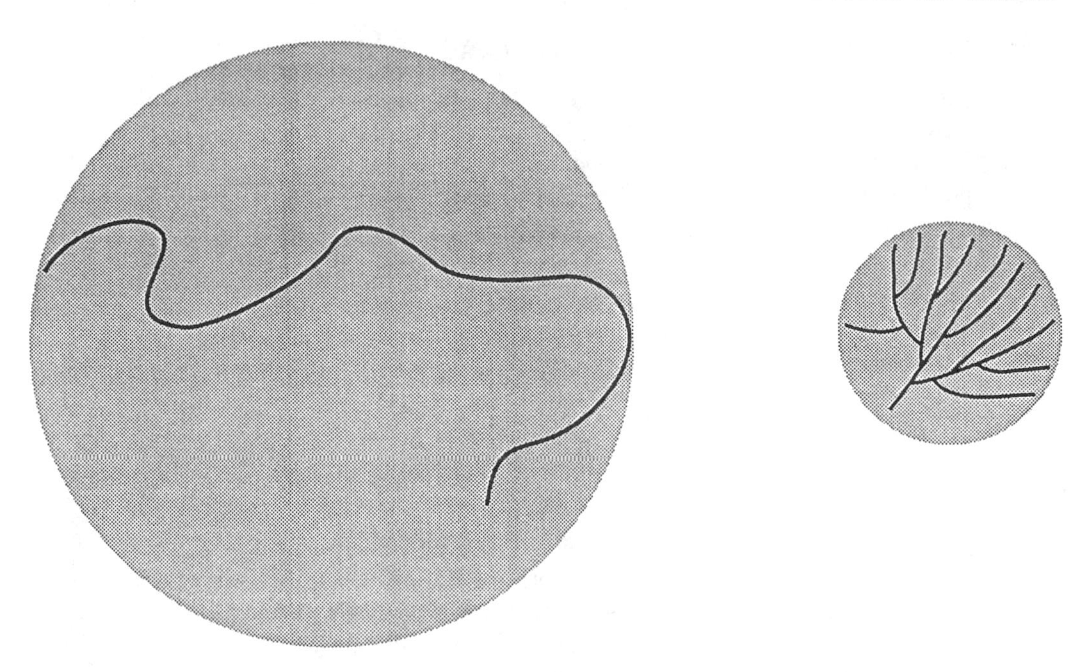

FIGURE 32 Relative volumes occupied by a linear polysaccharide and a highly branched polysaccharide of the same molecular weight.

Unbranched, regular glycans, which dissolve in water by heating, form unstable molecular dispersions that precipitate or gel rapidly. This occurs as segments of the long molecules collide and form intermolecular bonds over the distance of a few units. Initial short alignments then extend in a zipper-like fashion to greatly strengthen intermolecular associations. Other segments of other chains colliding with this organized nucleus bind to it, increasing the size of the ordered, crystalline phase. Linear molecules continue to bind to fashion a fringed micelle that may reach a size where gravitational forces cause precipitation. For example, starch amylose, when dissolved in water with the aid of heat followed by cooling the solution, undergoes molecular aggregation and precipitates, a process called retrogradation. During cooling of bread and other baked products, amylose molecules associate to produce a firming. Over a longer storage time, the branches of amylopectin associate to produce staling (see Sec. 4.4.6).

In general, molecules of all unbranched, neutral homoglycans have an inherent tendency to associate and partially crystallize. However, if linear glycans are derivatized, or occur naturally derivatized, as does guar gum (see Sec. 4.6), which has single-unit glycosyl branches along a backbone chain, their segments are prevented from association and stable solutions result.

Stable solutions are also formed if the linear chains contain charged groups where coulombic repulsions prevent segments from approaching each other. As already mentioned, charge repulsion also causes chains to extend, which provides high viscosities. Such highly viscous, stable solutions are seen with sodium alginate (see Sec. 4.9), where each glycosyl unit is a uronic acid unit having a carboxylic acid group in the salt form, and in xanthan (see Sec. 4.7), where one out of five glycosyl units is a uronic acid unit and another carboxylate group may be present. But if the pH of an alginate solution is lowered to 3, where ionization of carboxylic acid groups is somewhat repressed because the pK_a values of the constituent monomers are 3.38 and 3.65, the resulting less ionic molecules can associate and precipitate or form a gel as expected for unbranched, neutral glycans.

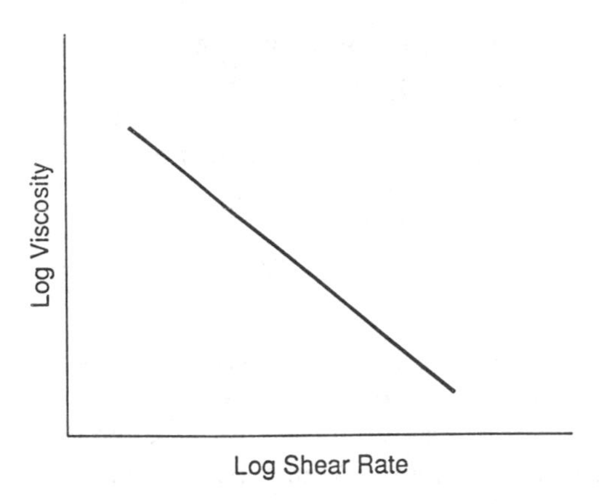

FIGURE 33 The logarithm of viscosity as a function of the shear rate for a pseudoplastic shear-thinning fluid.

Carrageenans are mixtures of linear chains that have a negative charge due to numerous ionized sulfate half-ester groups along the chain (see Sec. 4.8). These molecules do not precipitate at low pH because the sulfate group remains ionized at all practical pH values.

Solutions of gums are dispersions of hydrated molecules and/or aggregates of hydrated molecules. Their flow behavior is determined by the size, shape, ease of deformation (flexibility), and presence and magnitude of charges on these hydrated molecules and/or aggregates. There are two general kinds of flow exhibited by polysaccharide solutions: pseudoplastic (by far the most common) and thixotropic.

Pseudoplastic fluids are shear-thinning. In pseudoplastic flow, a more rapid increase in flow results from an increase in shear rate; that is, the faster a fluid flows, the less viscous it becomes (Fig. 33). The flow rate can be increased by increasing the applied force by pouring, chewing, swallowing, pumping, mixing, etc. The change in viscosity is independent of time; that is, the rate of flow changes instantaneously as the shear rate is changed. Linear polymer molecules form shear-thinning, usually pesudoplastic, solutions. In general, higher molecular weight gums are more pseudoplastic.

Gum solutions that are less pseudoplastic are said to give long flow;* such solutions are generally perceived as being slimy. More pseudoplastic solutions are described as having short flow and are generally perceived as being nonslimy. In food science, a slimy material is one that is thick, coats the mouth, and is difficult to swallow. Sliminess is inversely related to pseudoplasticity; that is, to be perceived as being nonslimy, there must be marked thinning at the low shear rates produced by chewing and swallowing.

Thixotropic flow is a second type of shear-thinning flow. In this case, the viscosity reduction that results from an increase in the rate of flow does not occur instantaneously. The viscosity of thixotropic solutions decreases under a constant rate of shear in a time-dependent manner and regains the original viscosity after cessation of shear, but again only after a clearly

*"Short flow" is exhibited by shear-thinning, primarily pseudoplastic, viscous solutions and "long flow" by viscous solutions that exhibit little or no shear thinning. These terms were applied long before there were instruments to determine and measure rheological phenomena. They were arrived at in this way. When a gum or starch solution is allowed to drain from a pipette or a funnel, those that are not shear-thinning come out in long strings, while those that shear-thin form short drops. The latter occurs because, as more and more fluid exits the orifice, the weight of the string becomes greater and greater, which causes it to flow faster and faster, which causes it to shear-thin to the point that the string breaks into drops.

defined and measured time interval. This behavior is due to a gel → solution → gel transition. In other words, a thixotropic solution is at rest a weak (pourable) gel (see Sec. 4.3.3).

For solutions of most gums, an increase in temperature results in a decrease in viscosity. (Xanthan gum is an exception between 0 and 100°C; see Sec. 4.7.) This is often an important property, for it means that higher solids can be put into solution at a higher temperature; then the solution can be cooled for thickening.

4.3.3 Gels [6,18]

A gel is a continuous, three-dimensional network of connected molecules or particles (such as crystals, emulsion droplets, or molecular aggregates/fibrils) entrapping a large volume of a continuous liquid phase, much as does a sponge. In many food products, the gel network consists of polymer (polysaccharide and/or protein) molecules or fibrils formed from polymer molecules joined in junction zones by hydrogen bonding, hydrophobic associations (van der Waals attractions), ionic cross bridges, entanglements, or covalent bonds, and the liquid phase is an aqueous solution of low-molecular-weight solutes and portions of the polymer chains.

Gels have some characteristics of solids and some of liquids. When polymer molecules or fibrils formed from polymer molecules interact over portions of their lengths to form junction zones and a three-dimensional network (Fig. 34), a fluid solution is changed into a material that

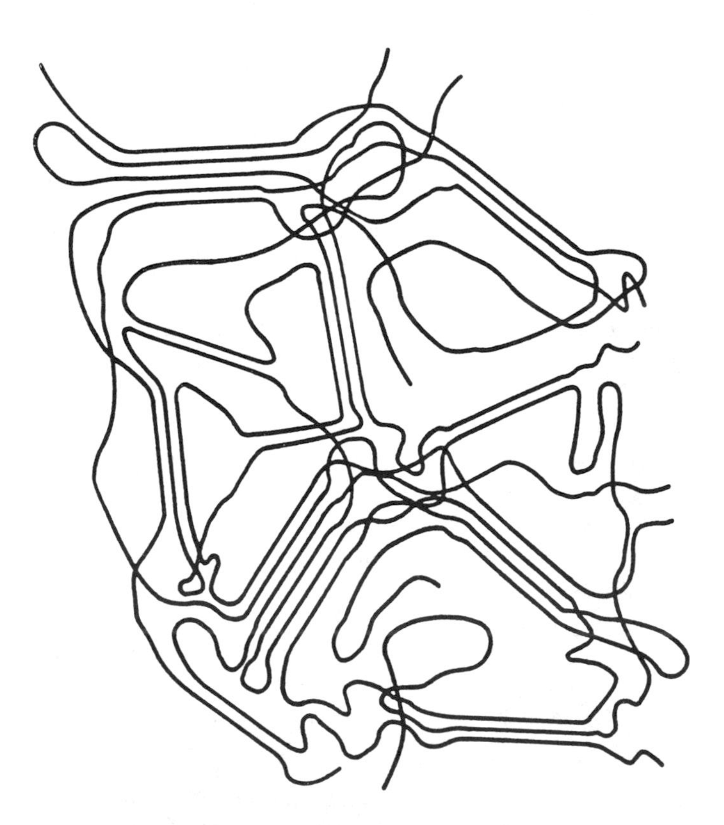

FIGURE 34 A diagrammatic representation of the type of three-dimensional network structure found in gels. This type of structure is known as a fringed micelle structure. Parallel side-by-side chains indicate the ordered, crystalline structure of a junction zone. The holes between junction zones contain an aqueous solution of dissolved segments of polymer chains and other solutes.

has a sponge-like structure and can retain its shape. The three-dimensional network structure offers significant resistance to an applied stress causing it to behave in some respects as an elastic solid. However, the continuous liquid phase, in which molecules are completely mobile, makes a gel less stiff than an ordinary solid, causing it to behave in some respects as a viscous liquid. Therefore, a gel is a viscoelastic semisolid; that is, the response of a gel to stress is partly characteristic of an elastic solid and partly characteristic of a viscous liquid.

Although gel-like or salve-like materials can be formed by high concentrations of particles (much like tomato paste), to form a true gel, the polymer molecules or aggregates of molecules must first be in solution, then partially come out of solution in junction zone regions to form the three-dimensional gel network structure. In general, if junction zones grow larger after formation of the gel, the network becomes more compact, the structure contracts, and syneresis results. (The appearance of fluid droplets on the gel surface is called syneresis.)

Although polysaccharide gels generally contain only about 1% polymer—that is, they may contain as much as 99% water—they can be quite strong. Examples of polysaccharide gels are dessert gels, aspics, structured fruit pieces, structured onion rings, meat-analog pet foods, and icings.

Choice of a specific gum for a particular application depends on the viscosity or gel strength desired, desired rheology, pH of the system, temperatures during processing, interactions with other ingredients, desired texture, and cost of the amount needed to impart the desired properties. In addition, consideration is also given to desired functional characteristics. These include a gum's ability to function as a binder, bodying agent, bulking agent, crystallization inhibitor, clarifying agent, cloud agent, coating agent/film former, fat mimetic, flocculating agent, foam stabilizer, mold release agent, suspension stabilizer, swelling agent, syneresis inhibitor, and whipping agent and as an agent for water absorption and binding (water retention and migration control), adhesion, emulsification, emulsion stabilization, and encapsulation. Each food gum tends to have one or more outstanding unique property related to this list, and this property is often the basis for its choice (Table 2).

4.3.4 Polysaccharide Hydrolysis

Polysaccharides are relatively less stable to hydrolytic cleavage than are proteins and may, at times, undergo depolymerization during food processing and/or storage of foods. Often, food gums are deliberately depolymerized so a relatively high concentration can be used to provide body without producing undesirable viscosity.

Hydrolysis of glycosidic bonds joining monosaccharide (glycosyl) units in oligo- and polysaccharides can be catalyzed by either acids (H^+) or enzymes. The extent of depolymerization, which has the effect of reducing viscosity, is determined by the acid strength, time, temperature, and structure of the polysaccharide. Generally, hydrolysis occurs most readily during thermal processing, because many foods are somewhat acidic. Defects associated with depolymerization during processing can usually be overcome by using more of the polysaccharide (gum) in the formulation to compensate for breakdown, using a higher viscosity grade of the gum, again to compensate for any depolymerization, or using a more acid-stable gum. Depolymerization can also be an important determinant of shelf life.

Polysaccharides are also subject to enzyme-catalyzed hydrolysis. The rate and end products of this process are controlled by the specificity of the enzyme, pH, time, and temperature. Polysaccharides, like any and all other carbohydrates, are subject to microbial attack because of their susceptibility to enzyme-catalyzed hydrolysis. Furthermore, gum products are very seldom delivered sterile, and this fact must be considered when using them as ingredients.

TABLE 2 Predominantly Used, Water-Soluble, Nonstarch Food Polysaccharides

Gum	Source	Class	General shape	Monomer units and linkages (approx. ratios)	Noncarbohydrate substituent groups	Water solubility	Key general characteristics	Major food applications
Carboxymethylcellulose (CMC)	Derived from cellulose	Modified cellulose	Linear	→4)-β-D-Glc*p*	Carboxymethyl ether (DS 0.4–0.8)*	High	Clear, stable solutions that can be either pseudoplastic or thixotropic	Retarder of ice crystal growth in frozen dessert products Thickener, suspending aid, protective colloid, and improver of mouthfeel, body, and texture in a variety of dressings, sauces and spreads Lubricant, film former, and processing aid for extruded products Batter thickener and humectant in cake and related mixes Moisture binder and retarder of crystallization and/or syneresis in icings, frostings, toppings, fillings, and puddings Syrup thickener Suspending aid and thickener in dry powder, hot and cold drink mixes Gravy maker in dry pet food

Methylcelluloses (MC) and hydroxypropylmethylcelluloses (HPMC)	Derived from cellulose	Modified cellulose	Linear	→4)-β-D-Glc*p*	Hydroxypropyl (MS 0.02–0.3)* and methyl (DS 1.1–2.2)* ether groups	Soluble in cold water; insoluble in hot water	Clear solutions that are thermal gelling; surface active	MC: Provides fat-like characteristics Reduces fat absorption in fried products Imparts creaminess through film and viscosity formation Provides lubricity Gas retention during baking Moisture retention and control of moisture distribution in bakery products (increases shelf life and imparts tenderness) HPMC: Nondairy whipped toppings, where it stabilizes foams, improves whipping characteristics, prevents phase separation, and provides freeze–thaw stability
Guar gum	Guar seed	Seed galactomannan	Linear with single-unit branches (behaves as a linear polymer)	→4)-β-D-Man*p* (~0.56) α-D-Gal*p* 1 ↓ 6 →4)-β-D-Man*p* (~1.0) (Man:Gal = 1.56:1)		High	Stable, opaque, very viscous, moderately pseudoplastic solutions Economical thickening	Binds water, prevents ice crystal growth, improves mouthfeel, softens texture produced by carrageenan + LBG, and slows meltdown in ice cream and ices Dairy products, prepared meals, bakery products, sauces, pet food
Locust bean gum (carob gum, LBG)	Locust bean (carob) seed	Seed galactomannan	Linear with single-unit branches (behaves as a linear polymer)	→4)-β-D-Man*p* (~2.5) α-D-Gal*p* 1 ↓ 6 →4)-β-D-Man*p* (~1.0) (Man:Gal = 3.5:1)		Soluble only in hot water; requires 90°C for complete solubilization	Interacts with xanthan and carrageenan to form rigid gels; rarely used alone	Provides excellent heat shock resistance, smooth meltdown, and desirable texture and chewiness in ice cream and other frozen dessert products

TABLE 2 Continued

Gum	Source	Class	General shape	Monomer units and linkages (approx. ratios)	Noncarbohydrate substituent groups	Water solubility	Key general characteristics	Major food applications
Xanthan gum	Fermentation medium	Microbial polysaccharide	Linear with trisaccharide unit branches on every other main chain unit (behaves as a linear polymer)	→4)-β-D-Glc*p* (1) β-D-Man*p* 1 ↓ 4 β-D-Glc*p* A 1 ↓ 2 β-D-Man*p* 6-Ac 1 ↓ 6 →4)-β-D-Glc*p* (1)	Acetyl ester Pyruvyl cyclic acetal	High	Very pseudoplastic, high viscosity solutions; excellent emulsion and suspension stabilizer; solution viscosity unaffected by temperature; solution viscosity unaffected by pH; excellent salt compatibility; synergistic increase in viscosity upon interaction with guar gum; heat reversible gelation with LBG	Stabilization of dispersions, suspensions, and emulsions Thickener
Carrageenans	Red algae	Seaweed (algal) extracts Sulfated galactans			Sulfate half-ester			Secondary stabilizer in ice cream and related products Preparation of evaporated milk, infant formulas, freeze–thaw stable whipped cream, dairy desserts, and chocolate milk Meat coating Improves adhesion and increases water-holding capacity of meat emulsion products Improves texture and quality of low-fat meat products

Kappa types: →3)-β-D-Gal*p* 4-SO_3^- (1) →4)-3,6-An-α-D-Gal*p* (1)	Kappa types: Na^+ salt soluble in cold water, K^+ and Ca^{2+} salts insoluble; all salts soluble at temperatures > 65°C; soluble in hot milk, insoluble in cold milk	Forms stiff, brittle, thermoreversible gels with $K^+ > Ca^{2+}$; thickens and gels milk at low concentration; synergistic gelation with LBG
Iota types: →3)-β-D-Gal*p* 4-SO_3^- (1) →4)-3,6-An-α-D-Gal*p* 2-SO_3^- (1)	Iota types: Na^+ salt soluble in cold water, K^+ and Ca^{2+} salts insoluble: all salts soluble at temperatures > 55°C; soluble in hot milk, insoluble in cold milk	Forms soft, resilient, thermoreversible gels with $Ca^{2+} > K^+$; gels do not synerese and have good freeze–thaw stability
Lambda types: →3)-β-D-Gal*p* 2-SO_3^- (1) →4)-α-D-Gal*p* 2,6-diSO_3^- (1)	Lambda types: all salts soluble in hot and cold water and milk	Thickens cold milk

TABLE 2 Continued

Gum	Source	Class	General shape	Monomer units and linkages (approx. ratios)	Noncarbohydrate substituent groups	Water solubility	Key general characteristics	Major food applications
Algins (alginates) (generally sodium alginate)	Brown algae	Seaweed (algal) extract Poly(uronic acid)	Linear	Block copolymer of the following units: →4)-β-D-ManpA (1.0) →4)-α-L-GulpA (0.5–2.5)		Sodium alginate soluble Alginic acid insoluble	Gels with Ca^{2+} Viscous, not very pseudoplastic solutions	Forms nonmelting gels (dessert gels, fruit analogs, other structured foods) Meat analogs Alginic acid forms soft, thixotropic, nonmelting gels (tomato aspic, jelly-type bakery fillings, filled fruit-containing breakfast cereal products)
					Hydroxypropyl ester groups in propylene glycol alginate	Soluble	Surface active Solutions stable to acids and Ca^{2+}	Emulsion stabilization in creamy salad dressings Thickener in low-calorie salad dressings
Pectins	Citrus peel Apple pomace	Plant extract Poly(uronic acid)	Linear	Primarily composed of →4)-α-GalpA units	Methyl ester groups May contain amide groups	Soluble	Forms jelly- and jam-type gels in presence of sugar and acid or with Ca^{2+}	HM pectin: high-sugar jellies, jams, preserves, and marmalades Acid milk drinks LM pectin: dietetic jellies, jams, preserves, and marmalades
Gum arabic (gum acacia)	Acacia tree	Exudate gum	Branch-on-branch, highly branched	Complex, variable structure Contains polypeptide		Very high	Emulsifier and emulsion stabilizer Compatible with high concentrations of sugar Very low viscosity at high concentrations	Prevents sucrose crystallization in confections Emulsifies and distributes fatty components in confections Preparation of flavor oil-in-water emulsions Component of coating of pan-coated candies Preparation of flavor powders

*For definitions of DS and MS see Sec. 4.4.9 and 4.5.3.

4.4 STARCH [32,33,38,56,61]

Starch's unique chemical and physical characteristics and nutritional quality set it apart from all other carbohydrates. Starch is the predominant food reserve substance in plants and provides 70–80% of the calories consumed by humans worldwide. Starch and starch hydrolysis products constitute most of the digestible carbohydrate in the human diet. Also, the amount of starch used in the preparation of food products, without counting that present in flours used to make bread and other bakery products, that naturally occurring in grains used to make breakfast cereals, or that naturally consumed in fruits and vegetables, greatly exceeds the combined use of all other food hydrocolloids.

Commercial starches are obtained from cereal grain seeds, particularly from corn, waxy corn (waxy maize), high-amylose corn, wheat, and various rices, and from tubers and roots, particularly potato, sweet potato, and tapioca (cassava). Starches and modified starches have an enormous number of food uses, including adhesive, binding, clouding, dusting, film forming, foam strengthening, antistaling, gelling, glazing, moisture retaining, stabilizing, texturizing, and thickening applications.

Starch is unique among carbohydrates because it occurs naturally as discrete particles (granules). Starch granules are relatively dense and insoluble and hydrate only slightly in cold water. They can be dispersed in water, producing low-viscosity slurries that can be easily mixed and pumped, even at concentrations of greater than 35%. The viscosity-building (thickening) power of starch is realized only when a slurry of granules is cooked. Heating a 5% slurry of unmodified starch granules to about 80°C (175°F) with stirring produces very high viscosity. A second uniqueness is that most starch granules are composed of a mixture of two polymers: an essentially linear polysaccharide called amylose, and a highly branched polysaccharide called amylopectin.

4.4.1 Amylose

While amylose is essentially a linear chain of $(1 \rightarrow 4)$-linked α-D-glucopyranosyl units, many amylose molecules have a few α-D-$(1 \rightarrow 6)$ branches, perhaps 1 in 180–320 units, or 0.3–0.5% of the linkages [49]. The branches in branched amylose molecules are either very long or very short, and the branch points are separated by large distances so that the physical properties of amylose molecules are essentially those of linear molecules. Amylose molecules have molecular weights of about 10^6.

The axial $\rightarrow$ equatorial position coupling of the $(1 \rightarrow 4)$-linked α-D-glucopyranosyl units in amylose chains gives the molecules a right-handed spiral or helical shape (Fig. 35). The interior of the helix contains only hydrogen atoms and is lipophilic, while the hydroxyl groups are positioned on the exterior of the coil.

Most starches contain about 25% amylose (Table 3). The two so-called high-amylose corn starches that are commercially available have apparent amylose contents of about 52% and 70–75%.

4.4.2 Amylopectin [21,22,34]

Amylopectin is a very large, very highly branched molecule, with branch-point linkages constituting 4–5% of the total linkages. Amylopectin consists of a chain containing the only reducing end-group, called a C-chain, which has numerous branches, termed B-chains, to which one to several third-layer A-chains are attached.* The branches of amylopectin mole-

*A-chains are unbranched. B-chains are branched with A-chains or other B-chains.

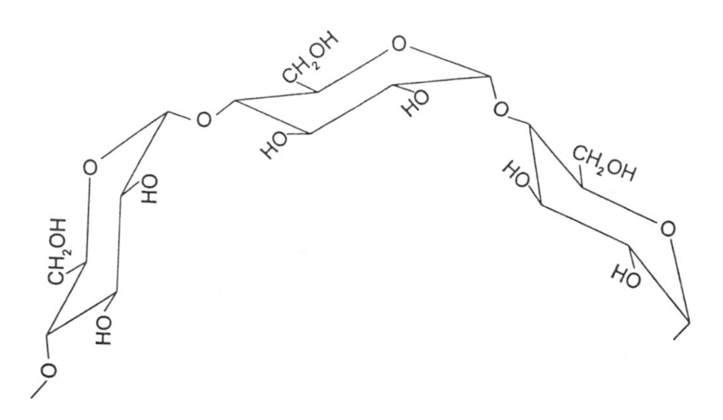

Figure 35 A trisaccharide segment of an unbranched portion of an amylose or amylopectin molecule.

cules are clustered (Fig. 36) and occur as double helices. Molecular weights of from 10^7 to 5×10^8 make amylopectin molecules among the largest, if not the largest, molecules found in nature.

Amylopectin is present in all starches, constituting about 75% of most common starches (Table 3). Some starches consist entirely of amylopectin and are called waxy starches. Waxy corn (waxy maize), the first grain recognized as one in which the starch consists only of amylopectin, is so termed because when the kernel is cut the new surface appears vitreous or waxy. Other all-amylopectin starches are also called waxy although, as in corn, there is no wax content.

Potato amylopectin is unique in having phosphate ester groups, attached most often (60–70%) at an O-6 position, with the other third at O-3 positions. These phosphate ester groups occur about once in every 215–560 α-D-glucopyranosyl units, and about 88% of them are on B chains.

4.4.3 Starch Granules [2,21,24,37,44]

Starch granules are made up of amylose and/or amylopectin molecules arranged radially. They contain both crystalline and noncrystalline regions in alternating layers.* The clustered branches of amylopectin occur as packed double helices. It is the packing together of these double-helical structures that forms the many small crystalline areas comprising the dense layers of starch granules that alternate with less dense amorphous layers. Because the crystallinity is produced by ordering of the amylopectin chains, waxy starch granules, that is, granules without amylose, have about the same amount of crystallinity as do normal starches. Amylose molecules occur among the amylopectin molecules, and some diffuse from partially water-swollen granules. The radial, ordered arrangement of starch molecules in a granule is evident from the polarization cross (white cross on a black background) seen in a polarizing microscope with the polarizers set 90° to each other. The center of the cross is at the hilum, the origin of growth of the granule.

Corn starch granules, even from a single source, have mixed shapes, with some being almost spherical, some angular, and some idented. (For the size, see Table 3.) Wheat starch granules are lenticular and have a bimodal or trimodal size distribution (> 14 μm, 5–14 μm, 1–5 μm).

*Starch granules are composed of layers somewhat like the layers of an onion, except that the layers cannot be peeled off.

TABLE 3 General Properties of Some Starch Granules and Their Pastes

	Common corn starch	Waxy maize starch	High-amylose corn starch	Potato starch	Tapioca starch	Wheat starch
Granule size (major axis, μm)	2–30	2–30	2–24	5–100	4–35	2–55
Amylose (%)	28	< 2	50–70	21	17	28
Gelatinization/pasting temperature (°C)[a]	62–80	63–72	66–170[b]	58–65	52–65	52–85
Relative viscosity	Medium	Medium high	Very low[b]	Very high	High	Low
Paste rheology[c]	Short	Long (cohesive)	Short	Very long	Long (cohesive)	Short
Paste clarity	Opaque	Very slightly cloudy	Opaque	Clear	Clear	Opaque
Tendency to gel/retrograde	High	Very low	Very high	Medium to low	Medium	High
Lipid (%DS)	0.8	0.2	—	0.1	0.1	0.9
Protein (%DS)	0.35	0.25	0.5	0.1	0.1	0.4
Phosphorus (%DS)	0.00	0.00	0.00	0.08	0.00	0.00
Flavor	Cereal (slight)	"Clean"		Slight	Bland	Cereal (slight)

[a]From the initial temperature of gelatinization to complete pasting.
[b]Under ordinary cooking conditions, where the slurry is heated to 95–100°C, high-amylose corn starch produces essentially no viscosity. Pasting does not occur until the temperature reaches 160–170°C (320–340°F).
[c]For a description of long and short flow, see Section 4.3.2.

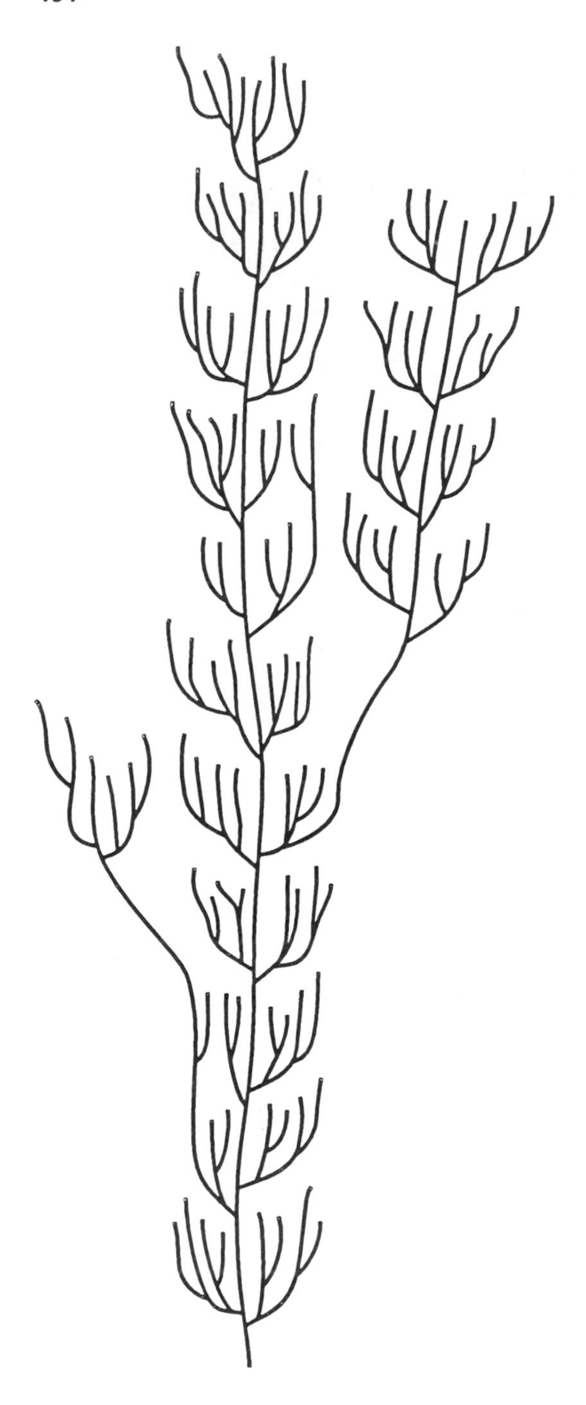

FIGURE 36 A diagrammatic representation of a portion of an amylopectin molecule.

Rice granules, on average, are the smallest of the commercial starch granules (1.5–9 μm), although the small granules of wheat starch are almost the same size. Many of the granules in tuber and root starches, such as potato and tapioca starches, tend to be larger than those of seed starches and are generally less dense and easier to cook. Potato starch granules may be as large as 100 *μm* along the major axis.

All starches retain small amounts of ash, lipid, and protein (Table 3). The phosphorus content of potato starch (0.06–0.1%) is due to the presence of the phosphate ester groups on amylopectin molecules. The phosphate ester groups give potato starch amylopectin a slight negative charge, resulting in some coulombic repulsion that may contribute to the rapid swelling of potato starch granules in warm water and to several properties of potato starch pastes, namely, their high viscosities, good clarity (Table 3), and low rate of retrogradation (see Sec. 4.4.6). Cereal starch molecules either do not have phosphate ester groups or have very much smaller amounts than occurs in potato starch. Only the cereal starches contain endogenous lipids in the granules. These internal lipids are primarily free fatty acids (FFA) and lysophospholipid (LPL), largely lysophosphatidyl choline (89% in corn starch), with the ratio of FFA to LPL varying from one cereal starch to another.

4.4.4 Granule Gelatinization and Pasting

Undamaged starch granules are insoluble in cold water, but can imbibe water reversibly; that is, they can swell slightly, and then return to their original size on drying. When heated in water, starch granules undergo a process called gelatinization. Gelatinization is the disruption of molecular order within granules. Evidence for the loss of order includes irreversible granule swelling, loss of birefringence, and loss of crystallinity. Leaching of amylose occurs during gelatinization, but some leaching of amylose can also occur prior to gelatinization. Total gelatinization usually occurs over a temperature range (Table 3), with larger granules generally gelatinizing first. The apparent temperature of initial gelatinization and the range over which gelatinization occurs depend on the method of measurement and on the starch:water ratio, granule type, and heterogeneities within the granule population under observation. Several stages of gelatinization can be determined using a polarizing microscope equipped with a hot stage. These are the initiation temperature (first observed loss in birefringence), the midpoint temperature, the completion or birefringence endpoint temperature (BEPT, the temperature at which the last granule in the field under observation loses its birefringence), and the gelatinization temperature range.

Continued heating of starch granules in excess water results in further granule swelling, additional leaching of soluble components (primarily amylose), and eventually, especially with the application of shear forces, total disruption of granules. This phenomenon results in the formation of a starch paste. (In starch technology, what is called a paste is what results from heating a starch slurry.) Granule swelling and disruption produce a viscous mass (the paste) consisting of a continuous phase of solubilized amylose and/or amylopectin and a discontinuous phase of granule remnants (granule ghosts* and fragments). Complete molecular dispersion is not accomplished except, perhaps, under conditions of high temperature, high shear, and excess water, conditions seldom, if ever, encountered in the preparation of food products. Cooling of a hot corn-starch paste results in a viscoelastic, firm, rigid gel.

Because gelatinization of starch is an endothermic process, differential scanning calorimetry (DSC), which measures both the temperature and the enthalpies of gelaninization, is

*The granule ghost is the remnant remaining after cooking under no to moderate shear. It consists of the outer portion of the granule. It appears as an insoluble, outer envelope, but is not a membrane.

widely used. Although there is not complete agreement on the interpretation of DSC data and the events that take place during gelatinization of starch granules, the following general picture is widely accepted. Water acts as a plasticizer. Its mobility-enhancing effect is first realized in the amorphous regions, which physically have the nature of a glass. When starch granules are heated in the presence of sufficient water (at least 60%), and a specific temperature (T_g, the glass transition temperature) is reached, the plasticized amorphous regions of the granule undergo a phase transition from a glassy state to a rubbery state.* However, the peak for absorption of energy associated with this transition is not often seen by DSC because the regions of crystallinity, that is, the ordered, packed, double-helical branches of amylopectin, are contiguous and connected by covalent bonds to the amorphous regions and melting of the crystallites immediately follows the glass transition. Because the enthalpy of initial melting (T_m) is so much larger than that of the glass transition, the latter is usually not evident.

Melting of lipid–amylose complexes occurs at much higher temperatures (100–120°C in excess water) than does melting of the amylopectin double-helical branches packed in crystalline order. These complexes are made with single-helical segments of amylose molecules when a starch paste containing monoacyl lipids is cooled. A DSC peak for this event is absent in waxy starches (without amylose).

Under normal food processing conditions (heat and moisture, although many food systems contain limited water as far as starch cooking is concerned), starch granules quickly swell beyond the reversible point. Water molecules enter between chains, break interchain bonds, and establish hydration layers around the separated molecules. This plasticizes (lubricates) chains so they become more fully separated and solvated. Entrance of large amounts of water causes granules to swell to several times their original size. When a 5% starch suspension is gently stirred and heated, granules imbibe water until much of the water is absorbed by them, forcing them to swell, press against each other, and fill the container with a highly viscous starch paste. Such highly swollen granules are easily broken and disintegrated by stirring, resulting in a decrease in viscosity. As starch granules swell, hydrated amylose molecules diffuse through the mass to the external phase (water), a phenomenon responsible for some aspects of paste behavior. Results of starch swelling can be recorded using a Brabender Visco/amylo/graph, which records the viscosity continuously as the temperature is increased, held constant for a time, and then decreased (Fig. 37).

By the time peak viscosity is reached, some granules have been broken by stirring. With continued stirring, more granules rupture and fragment, causing a further decrease in viscosity. On cooling, some starch molecules partially reassociate to form a precipitate or gel. This process is called retrogradation (see Sec. 4.4.6). The firmness of the gel depends on the extent of junction zone formation (see Sec. 4.3.3). Junction zone formation is influenced (either facilitated or hindered) by the presence of other ingredients such as fats, proteins, sugars, and acids and the amount of water present.

4.4.5 Uses of Unmodified Starches

Starches serve a variety of roles in food production. Principally they are used to take up water and to produce viscous fluids/pastes and gels and to give desired textural qualities (see also Sec. 4.4.9). The extent of starch gelatinization in baked goods strongly affects product properties, including storage behavior and rate of digestion. In some baked products, many starch granules remain ungelatinized. In certain cookies and pie crust that are high in fat and low in

*A glass is a mechanical solid capable of supporting its own weight against flow. A rubber is an undercooled liquid that can exhibit viscous flow. (See Chap. 2 for further details.)

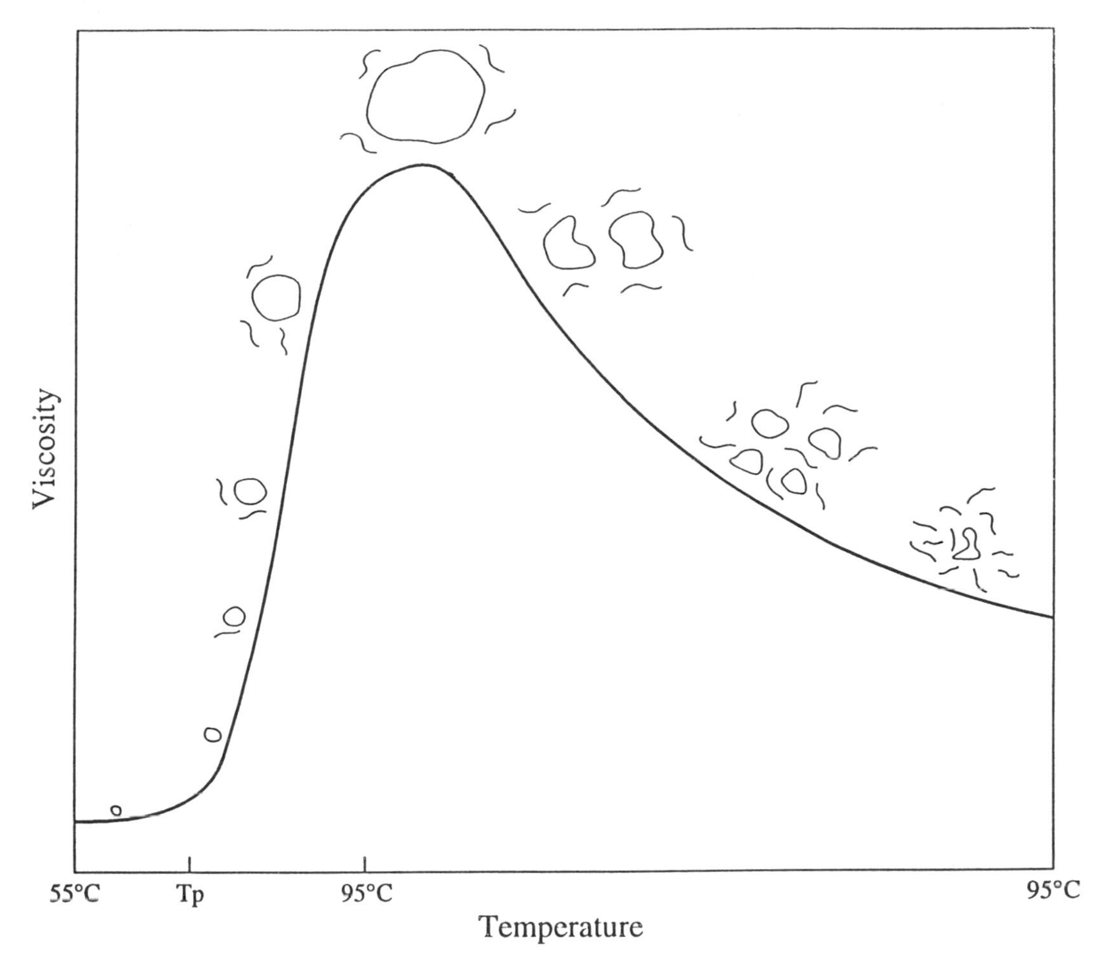

FIGURE 37 Representative Brabender Visco/amylo/graph curve showing viscosity changes related to typical starch granule swelling and disintegration as a granule suspension is heated to 95°C and then held at that temperature. (The instrument imparts moderate shear to the system.) Tp is the pasting temperature, that is, the temperature at which a viscosity increase is recorded by the instrument.

water, about 90% of the wheat starch granules remain ungelatinized (as observed microscopically and as evidenced by their slow rate of attack by amylases). In other products, such as angel food cake and white bread, which are higher in moisture, about 96% of the wheat starch granules are gelatinized and many become deformed.

Food companies value the clear, cohesive pastes produced from waxy maize starch. Potato starch is used in extruded cereal and snack food products and in dry mixes for soups and cakes. Rice starch produces opaque gels useful for baby food. Waxy rice starch gels are clear and cohesive. Wheat starch gels are weak; their slight flavor may be due to residual flour components. Tuber (potato) and root (tapioca) starches have weak intermolecular bonding and swell greatly to give high-viscosity pastes. Because the highly swollen granules break easily, the viscosity quickly decreases with only moderate shear. Starches are often modified (see Sec. 4.4.9) before use in foods.

4.4.6 Retrogradation and Staling [36,37,44]

As already pointed out, cooling a hot starch paste generally produces a viscoelastic, firm, rigid gel. The formation of the junction zones of a gel can be considered to be the first stage of an attempt by starch molecules to crystallize. As starch pastes are cooled and stored, the starch becomes progressively less soluble. In dilute solution, starch molecules will precipitate, with the insoluble material being difficult to redissolve by heating. The collective processes of dissolved starch becoming less soluble are called retrogradation. Retrogradation of cooked starch involves both the two constituent polymers, amylose and amylopectin, with amylose undergoing retrogradation at a much more rapid rate than does amylopectin. The rate of retrogradation depends on several variables, including the molecular ratio of amylose to amylopectin; structures of the amylose and amylopectin molecules, which are determined by the botanical source of the starch; temperature; starch concentration; and presence and concentration of other ingredients, such as surfactants and salts. Many quality defects in food products, such as bread staling and loss of viscosity and precipitation in soups and sauces, are due, at least in part, to starch retrogradation.

Staling of baked goods is noted by an increase in crumb firmness and a loss in product freshness. Staling begins as soon as baking is complete and the product begins to cool. The rate of staling is dependent on the product formulation, the baking process, and storage conditions. Staling is due, at least in part, to the gradual transition of amorphous starch to a partially crystalline, retrograded state. In baked goods, where there is just enough moisture to gelatinize starch granules (while retaining a granule identity), amylose retrogradation (insolubilization) may be largely complete by the time the product has cooled to room temperature. Retrogradation of amylopectin is believed to involve primarily association of its outer branches and requires a much longer time than amylose retrogradation, giving it prominence in the staling process that occurs with time after the product has cooled.

Most polar lipids with surfactant properties retard crumb firming. Compounds such as glyceryl monopalmitate (GMP), other monoglycerides and their derivatives, and sodium stearoyl 2-lactylate (SSL) are incorporated into doughs of bread and other baked goods, in part to increase shelf life.

4.4.7 Starch Complexes [3]

Amylose chains are helical with hydrophobic (lipophilic) interiors and are able to form complexes with linear hydrophobic portions of molecules that can fit in the hydrophobic tube. Iodine (as I_3^-) complexes with both amylose and amylopectin molecules. Again, the complexing occurs within the hydrophobic interior of helical segments. With amylose, the long helical segments allow long chains of poly(I_3^-) to form and produce the blue color used as a diagnostic test for starch. The amylose–iodine complex contains 19% iodine, and determination of the amount of complexing is used to measure the amount of apparent amylose present in a starch. Amylopectin is colored a reddish-purple by iodine because the branches of amylopectin are too short for formation of long chains of poly(I_3^-).

Polar lipids (surfactants/emulsifiers and fatty acids) can affect starch pastes and starch-based foods in one or more of three ways as a result of complex formation: (a) by affecting the processes associated with starch gelatinization and pasting (that is, the loss of birefringence, granular swelling, leaching of amylose, melting of the crystalline regions of starch granules, and viscosity increases during cooking), (b) by modifying the rheological behavior of the resulting pastes, and (c) by inhibiting the crystallization of starch molecules associated with the retrogradation process. The specific changes observed upon addition of lipid depend on its structure, the starch employed, and the product to which it is added.

4.4.8 Hydrolysis of Starch [43,48,52]

Starch molecules, like all other polysaccharide molecules, are depolymerized by hot acids. Hydrolysis of the glycosidic bonds occurs more or less randomly to produce, initially, very large fragments. Commercially, hydrochloric acid is sprayed onto well-mixed starch, or stirred moist starch is treated with hydrogen chloride gas; the mixture is then heated until the desired degree of depolymerization is obtained. The acid is then neutralized, and the product is recovered, washed, and dried. The products are still granular, but break up (cook out) easily. They are called acid-modified or thin-boiling starches, and the process of making them is called thinning. Even though only a few glycosidic bonds are hydrolyzed, the starch granules disintegrate much more easily during heating in water. Acid-modified starches form gels with improved clarity and increased strength, even though they provide less solution viscosity. Thin-boiling starches are used as film formers and adhesives in products such as pan-coated nuts and candies and whenever a strong gel is desired, such as in gum candies like jelly beans, jujubes, orange slices, and spearmint leaves and in processed cheese loaves. To prepare especially strong and fast-setting gels, a high-amylose corn starch is used as the base starch.

More extensive modification with acid produces dextrins. Low-viscosity dextrins can be used at high concentrations in food processing. They have film-forming and adhesive properties and are used in products such as pan-coated roasted nuts and candy. They are also used as fillers, encapsulating agents, and carriers of flavors, especially spray-dried flavors. They are classified by their cold water solubility and color. Dextrins that retain large amounts of linear chains or long chain fragments form strong gels.

Hydrolysis of starch dispersions with either an acid or an enzyme produces first malto-dextrins. Maltodextrins are usually described by their dextrose equivalency (DE). The DE is related to the degree of polymerization (DP) through the following equation: $DE = 100/DP$. (Both DE and DP are average values for populations of molecules.) Therefore, the DE of a product of hydrolysis is its reducing power as a percentage of the reducing power of pure dextrose (D-glucose); thus, DE is inversely related to average molecular weight. Maltodextrins are defined as products with DE values that are measurable, but less than 20. Maltodextrins of lowest DE are nonhygroscopic, while those of highest DE (that is, lowest average molecule weight) tend to absorb moisture. Maltodextrins are bland with virtually no sweetness and are excellent for contributing body or bulk to food systems. Hydrolysis to DE values of 20–60 gives mixtures of molecules that, when dried, are called corn syrup solids. Corn syrup solids dissolve rapidly and are mildly sweet. Table 4 lists functional properties of starch hydrolysis products.

Continued hydrolysis of starch produces a mixture of D-glucose, maltose, and other

TABLE 4 Functional Properties of Starch Hydrolysis Products

Properties enhanced by greater hydrolysis[a]	Properties enhanced in products of less conversion[b]
Sweetness	Viscosity production
Hygroscopicity and humectancy	Body formation
Freezing point depression	Foam stabilization
Flavor enhancement	Sugar crystallization prevention
Fermentability	Ice crystal growth prevention
Browning reaction	

[a]High-conversion (high-DE) syrups.
[b]Low-DE syrups and maltodextrins.

malto-oligosaccharides. Syrups of this composition are produced in enormous quantities. One of the most common has a DE of 42. These syrups are stable because crystallization does not occur easily in such complex mixtures. They are sold in concentrations of high osmolality, high enough that ordinary organisms cannot grow in them. An example is waffle and pancake syrup, which is colored with caramel coloring and flavored with maple flavoring.

Three to four enzymes are used for the industrial hydrolysis of starch to D-glucose. α-Amylase is an endo-enzyme that cleaves both amylose and amylopectin molecules internally, producing oligosaccharides. The larger oligosaccharides may be singly, doubly, or triply branched via $(1 \rightarrow 6)$ linkages, since α-amylase acts only on the $(1 \rightarrow 4)$ linkages of starch. α-Amylase does not attack double-helical starch polymer segments or polymer segments complexed with a polar lipid (stabilized single helical segments).

Glucoamylase (amyloglucosidase) is used commercially, in combination with an α-amylase, for producing D-glucose (dextrose) syrups and crystalline D-glucose. The enzyme acts upon fully gelatinized starch as an exo-enzyme, sequentially releasing single D-glucosyl units from the nonreducing ends of amylose and amylopectin molecules, even those joined through $(1 \rightarrow 6)$ bonds. Consequently, the enzyme can completely hydrolyze starch to D-glucose, but is used on starch that has been previously depolymerized with α-amylase to generate small fragments and more nonreducing ends.

β-Amylase releases the disaccharide maltose sequentially from the nonreducing end of amylose. It also attacks the nonreducing ends of amylopectin, sequentially releasing maltose, but it cannot cleave the $(1 \rightarrow 6)$ linkages at branch points, so it leaves a pruned amylopectin residue termed a limit dextrin, specifically a beta-limit dextrin.

There are several debranching enzymes that specifically catalyze hydrolysis of $(1 \rightarrow 6)$-linkages in amylopectin, producing numerous linear but low-molecular-weight molecules. One such enzyme is isoamylase; another is pullulanase.

Cyclodextrin glucanotransferase is a unique *Bacillus* enzyme that forms rings of $(1 \rightarrow 4)$-linked α-D-glucopyranosyl units from starch polymers. The enzyme can form six-, seven-, and eight-membered rings, which are respectively alpha-, beta-, and gamma-cyclodextrins. Since the normal helical conformation of a linear portion of a starch molecule contains six to seven glucosyl units per turn of the helix, this transfer of a glycosidic bond from one that joins adjacent segments of a spiral to one that forms a doughnut-like circular structure is easy to picture. These products, originally called Schardinger dextrins after their discoverer, are now known as cyclodextrins or cycloamyloses. They have the ability to complex with hydrophobic substances that are held in the center of the ring. Through such complexing of guest molecules, volatile essential oils can be converted into dry powders in which the flavoring or aromatic substance is protected from light and oxygen but is readily released when the complex is added to an aqueous system because of the water solubility of the cyclodextrin. Cyclodextrins are not yet approved for food use in the United States. However, chiral supports that are useful for chromatographic separations are made by converting cyclodextrins into insoluble polymeric materials. In a similar application, insoluble polymeric beads of cyclodextrins have been shown to be useful for removal of bitter components of citrus juices.

Corn syrup is the major source of D-glucose and D-fructose. To make a corn syrup, a slurry of starch in water is mixed with a thermally stable α-amylase and put through a cooker where rapid gelatinization and enzyme-catalyzed hydrolysis (liquefaction) takes place. After cooling to 55–60°C (130–140°F), hydrolysis is continued with glucoamylase, whereupon the syrup is clarified, concentrated, carbon-refined, and ion-exchanged. If the syrup is properly refined and combined with seed crystals, crystalline D-glucose (dextrose) or its monohydrate can be obtained.

For production of D-fructose, a solution of D-glucose is passed through a column containing bound (immobilized) glucose isomerase. The enzyme catalyzes the isomerization of D-glucose to D-fructose (see Fig. 5) to an equilibrium mixture of approximately 58% D-glucose and 42% D-fructose. Higher concentrations of D-fructose are usually desired. [The high-fructose corn syrup (HFCS) most often used as a soft drink sweetener is approximately 55% D-fructose.] So the isomerized syrup is passed through a bed of cation-exchange resin in the calcium salt form. The resin binds D-fructose, which can be recovered to provide an enriched syrup fraction.

4.4.9 Modified Food Starch [1,56,58]

Food processors generally prefer starches with better behavioral characteristics than provided by native starches. Native starches produce particularly weak-bodied, cohesive, rubbery pastes when cooked and undesirable gels when the pastes are cooled. The properties of starches can be improved by modification. Modification is done so that resultant pastes can withstand the conditions of heat, shear, and acid associated with particular processing conditions and to introduce specific functionalities. Modified food starches are functional, useful, and abundant food macroingredients and additives.

Types of modifications that are most often made, sometimes singly, but often in combinations, are crosslinking of polymer chains, non-crosslinking derivatization, depolymerization (see Sec. 4.4.8), and pregelatinization (see Sec. 4.4.10). Specific property improvements that can be obtained by proper combinations of modifications are reduction in the energy required to cook (improved gelatinization and pasting), modification of cooking characteristics, increased solubility, either increased or decreased paste viscosity, increased freeze–thaw stability of pastes, enhancement of paste clarity, increased paste sheen, inhibition of gel formation, enhancement of gel formation and gel strength, reduction of gel syneresis, improvement of interaction with other substances, improvement in stabilizing properties, enhancement of film formation, improvement in water resistance of films, reduction in paste cohesiveness, and improvement of stability to acid, heat, and shear.

Starch, like all carbohydrates, can undergo reactions at its various hydroxyl groups. In modified food starches, only a very few of the hydroxyl groups are modified. Normally ester or ether groups are attached at very low DS values (degrees of substitution).* DS values are usually < 0.1 and generally in the range 0.002–0.2. Thus, there is, on average, one substituent group on every 500–5 D-glucopyranosyl units. Small levels of derivatization change the properties of starches dramatically and greatly extend their usefulness. Starch products that are esterified or etherified with monofunctional reagents resist interchain associations. Derivatization of starches with monofunctional reagents reduces intermolecular associations, the tendency of the starch paste to gel, and the tendency for precipitation to occur. Hence this modification is often called stabilization, and the products are called stabilized starches. Use of difunctional reagents produces crosslinked starches. Modified food starches are often both cross-linked and stabilized.

Chemical reactions currently both allowed and used to produce modified food starches in the United States are as follows: esterification with acetic anhydride, succinic anhydride, the mixed anhydride of acetic and adipic acids, 1-octenylsuccinic anhydride, phosphoryl chloride, sodium trimetaphosphate, sodium tripolyphosphate, and monosodium orthophosphate; etherification with propylene oxide; acid modification with hydrochloric and sulfuric acids; bleaching

---

*The degree of substitution (DS) is defined as the average number of esterified or etherified hydroxyl groups per monosaccharide unit. Both branched and unbranched polysaccharides composed of hexopyranosyl units have an average of three hydroxyl groups per monomeric unit. Therefore, the maximum DS for a polysaccharide is 3.0.

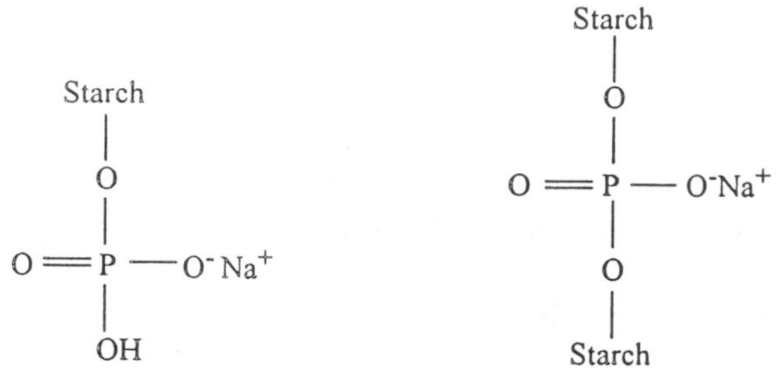

Starch–O–P(=O)(–O^-Na^+)–OH   Starch–O–P(=O)(–O^-Na^+)–O–Starch

FIGURE 38 Structures of starch monoester phosphate (left) and diester phosphate (right). The diester joins two molecules together, resulting in crosslinked starch granules.

with hydrogen peroxide, peracetic acid, potassium permanganate, and sodium hypochlorite; oxidation with sodium hypochlorite; and various combinations of these reactions.

Modified waxy maize starches are especially popular in the U.S. food industry. Pastes of unmodified common corn starch will gel, and the gels will generally be cohesive, rubbery, long textured, and prone to syneresis (that is, to weep or exude moisture). However, pastes of waxy maize starch show little tendency to gel at room temperature, which is why waxy maize starch is generally preferred as the base starch for food starches. But pastes of waxy maize starch will become cloudy and chunky and exhibit syneresis when stored under refrigerator or freezing conditions, so even waxy maize starch is usually modified to increase the stability of its pastes. The most common and useful derivatives employed for starch stabilization are the hydroxypropyl ether, the monostarch phosphate ester, and the acetate ester.

Acetylation of starch to the maximum allowed in foods (DS 0.09) lowers the gelatinization temperature, improves paste clarity, and provides stability to retrogradation and freeze-thaw stability (but not as well as hydroxypropylation). Starch phosphate monoesters (Fig. 38) made by drying starch in the presence of sodium tripolyphosphate or monosodium orthophosphate can be used to make pastes that are clear and stable, have emulsifying properties, and have freeze-thaw stability. Monostarch phosphates have a long, cohesive texture. Paste viscosity is generally high and can be controlled by varying the concentration of reagent, time of reaction, temperature, and pH. Phosphate esterification lowers the gelatinization temperature. In the United States, the maximum allowable DS with phosphate groups is 0.002.

Preparation of an alkenylsuccinate ester of starch attaches a hydrocarbon chain to its polymer molecules (Fig. 39). Even at very low degrees of substitution, starch 1-octenylsuccinate molecules concentrate at the interface of an oil-in-water emulsion because of the hydrophobicity of the alkenyl group. This characteristic makes them useful as emulsion stabilizers. Starch 1-octenylsuccinate products can be used in a variety of food applications where emulsion

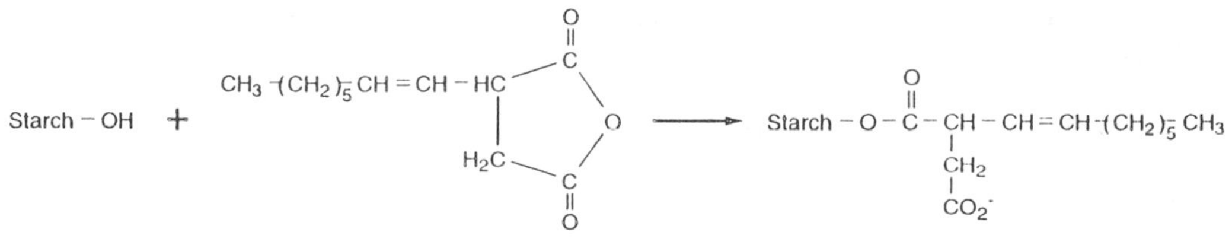

Starch – OH + $CH_3(CH_2)_5CH=CH-HC$ (succinic anhydride ring: HC–C(=O)–O–C(=O)–H_2C) ⟶ Starch – O – C(=O) – CH($CH_2CO_2^-$) – CH=CH$(CH_2)_5CH_3$

FIGURE 39 Preparation of starch 2-(1-octenyl)succinyl ester.

stability is needed, such as in flavored beverages. The presence of the aliphatic chain tends to give the starch derivative a sensory perception of fattiness, so it is possible to use the derivatives as a partial replacement for fat in certain foods.

Hydroxypropylstarch (starch-O-CH_2-CHOH-CH_3), prepared by reacting starch with propylene oxide to produce a low level of etherification (DS 0.02–0.2, 0.2 being the maximum allowed) is a product with properties similar to those of starch acetate because it similarly has "bumps" along the starch polymer chains—that is, it is a stabilized starch. Hydroxypropylation reduces the gelatinization temperature. Hydroxypropylstarches form clear pastes that do not retrograde and withstand freezing and thawing. They are used as thickeners and extenders. To improve viscosity, particularly under acidic conditions, acetylated and hydroxypropylated starches are often also crosslinked with phosphate groups.

The majority of modified food starch is crosslinked. Crosslinking occurs when starch granules are reacted with difunctional reagents that react with hydroxyl groups on two different molecules within the granule. Crosslinking is accomplished most often by producing distarch phosphate esters. Starch is reacted with either phosphoryl chloride, PO_3Cl_2, or sodium trimetaphosphate in an alkaline slurry, then dried. Linking together of starch chains with phosphate diester or other crosslinks reinforces the granule and reduces both the rate and the degree of granule swelling and subsequent disintegration. Thus, granules exhibit reduced sensitivity to processing conditions (high temperature; extended cooking times; low pH; high shear during mixing, milling, homogenization, and/or pumping). Cooked pastes of crosslinked starches are more viscous,* heavier bodied, shorter textured, and less likely to break down during extended cooking or during exposure to low pH and/or severe agitation than are pastes of native starches from which they are prepared. Only a small amount of crosslinking is required to produce a noticeable effect; with lower levels of crosslinking, granules exhibit hydration swelling in inverse proportion to DS. As crosslinking is increased, the granules become more and more tolerant to physical conditions and acidity, but less and less dispersible by cooking. Energy requirements to reach maximum swelling and viscosity are also increased. For example, treatment of a starch with only 0.0025% of sodium trimetaphosphate greatly reduces both the rate and the degree of granule swelling, greatly increases paste stability, and changes dramatically the Brabender Visco/amylo/graph viscosity profile and textural characteristics of its paste. Treatment with 0.08% of trimetaphosphate produces a product in which granule swelling is restricted to the point that a peak viscosity is never reached during the hot holding period in a Brabender Visco/amylo/graph. As the degree of crosslinking increases, the starch also becomes more acid stable. Though hydrolysis of glycosidic bonds occurs during heating in aqueous acid, chains tied to each other through phosphate crosslinks continue to provide large molecules and an elevated viscosity. The only other crosslink permitted in a food starch is the distarch ester of adipic acid.

Most crosslinked food starches contain less than one crosslink per 1000 α-D-glucopyranosyl units. Trends toward continuous cooking require increased shear resistance and stability to hot surfaces. Storage-stable thickening is also provided by crosslinked starches. In retort sterilization of canned foods, crosslinked starches, because of their reduced rate of gelatinization and swelling, maintain a low initial viscosity long enough to facilitate the rapid heat transfer and temperature rise that is needed to provide uniform sterilization before granule swelling brings about the ultimately desired viscosity, texture, and suspending characteristics. Crosslinked

---

*Note in Figure 37 that maximum viscosity is reached when the system contains highly swollen granules. Crosslinked granules hold together. Thus, there is minimal loss of viscosity after the peak is reached.

starches are used in canned soups, gravies, and puddings and in batter mixes. Crosslinking of waxy maize starch gives the clear paste sufficient rigidity so that, when used in pie fillings, cut sections of pie hold their shape.

Starches that are both crosslinked and stabilized are used in canned, frozen, baked, and dry foods. In baby foods and fruit and pie fillings in cans and jars, they provide long shelf life. They also allow frozen fruit pies, pot pies, and gravies to remain stable under long-term storage.

Modified food starches are tailor-made for specific applications. Properties that can be controlled by combinations of crosslinking, stabilization, and thinning of corn, waxy maize, potato, and other starches include, but are not limited to, the following: adhesion, clarity of solutions/pastes, color, emulsion stabilization ability, film-forming ability, flavor release, hydration rate, moisture holding capacity, stability to acids, stability to heat and cold, stability to shear, temperature required to cook, and viscosity (hot paste and cold paste). Some characteristics imparted to the food product include, but are not limited to, the following: mouth feel, reduction of oil migration, texture, sheen, stability, and tackiness.

4.4.10 Cold-Water-Soluble (Pregelatinized) Starch

Once starch has been pasted and dried without excessive retrogradation, it can be redissolved in cold water. The largest commercial quantity of such starch is made by flowing a starch–water slurry into the nip between two nearly touching and counterrotating, steam-heated rolls. The starch slurry is gelatinized and pasted almost instantaneously, and the paste coats the rolls where it dries quickly. The dry film is scraped from the roll and ground. The resulting products, known as pregelatinized starches or instant starches, are precooked starches. They are also prepared using extruders.

Both chemically modified and unmodified starches can be used to make pregelatinized starches. If chemically modified starches are used, properties introduced by the modification(s) carry through to the pregelatinized products; thus, paste properties such as stability to freeze–thaw cycling can also be characteristics of pregelatinized starches. Pregelatinized, slightly crosslinked starch is useful in instant soup, pizza topping, and extruded snacks and in breakfast cereals.

Pregelatinized starches can be used without cooking. Finely ground pregelatinized starch forms small gel particles like a water-soluble gum, but when properly dissolved gives solutions of high viscosity. Coarsely ground products "dissolve" much more readily and produce dispersions of lower viscosity and with graininess or pulpiness that is desirable in some products. Many pregelatinized starches are used in dry mixes such as instant pudding mixes; they disperse readily with high-shear stirring or when mixed with sugar or other dry ingredients.

4.4.11 Cold-Water-Swelling Starch

Granular starch that swells extensively in cold water is made by heating common corn starch in 75–90% ethanol or by a special spray-drying process. The product is dispersible in sugar solutions or corn syrups by rapid stirring; the resulting dispersion can be poured into molds, where it sets to a rigid gel that can be sliced easily. The result is a gum candy. Cold-water-swelling starch is also useful in making desserts and in muffin batters containing particles, such as blueberries, that otherwise would settle to the bottom before the batter is thickened by heating during baking.

4.5 CELLULOSE: MODIFICATIONS AND DERIVATIVES [60]

Cellulose is the most abundant organic compound, and therefore the most abundant carbohydrate, on earth because it is the principal cell-wall component of higher plants. (It can be argued that D-glucose is the most abundant carbohydrate and organic compound if we consider cellulose as a combined form of this monomeric building block.) Cellulose is a high-molecular-weight, linear, insoluble homopolymer of repeating β-D-glucopyranosyl units joined by $(1 \rightarrow 4)$ glycosidic linkages (Fig. 40). Because of their linearity and stereoregular nature, cellulose molecules associate over extended regions, forming polycrystalline, fibrous bundles. Crystalline regions are held together by large numbers of hydrogen bonds. They are separated by, and connected to, amorphous regions. Cellulose is insoluble because, in order for it to dissolve, most of these hydrogen bonds would have to be released at once. Cellulose can, however, through substitution, be converted into water-soluble gums.

Cellulose and its modified forms serve as dietary fiber because they do not contribute significant nourishment or calories as they pass through the human digestive system. Dietary fiber does, however, serve important functions (see Sec. 4.12).

A purified cellulose powder is available as a food ingredient. High-quality cellulose can be obtained from wood through pulping and subsequent purification. Chemical purity is not required for food use because cellulosic cell-wall materials are components of all fruits and vegetables and many of their products. The powdered cellulose used in foods has negligible flavor, color, and microbial contamination. Powdered cellulose is most often added to bread to provide noncaloric bulk. Reduced-calorie baked goods made with powdered cellulose, not only have an increased content of dietary fiber, but also stay moist and fresh longer.

4.5.1 Microcrystalline Cellulose [51]

A purified, insoluble cellulose termed microcrystalline cellulose (MCC) is useful in the food industry. It is made by hydrolysis of purified wood pulp, followed by separation of the constituent microcrystals of cellulose. Cellulose molecules are fairly rigid, completely linear chains of about 3000 β-D-glucopyranosyl units and associate easily in long junction zones. However, the long and unwieldy chains do not align over their entire lengths. The end of the crystalline region is simply the divergence of cellulose chains away from order into a more random arrangement. When purified wood pulp is hydrolyzed with acid, the acid penetrates the lower density, amorphous regions, effects hydrolytic cleavage of chains in these regions, and releases individual, fringed crystallites. The released crystallites grow larger because the chains that constitute the fringes now have greater freedom of motion and can order themselves.

Two types of microcrystalline cellulose are produced, both of which are stable to both heat

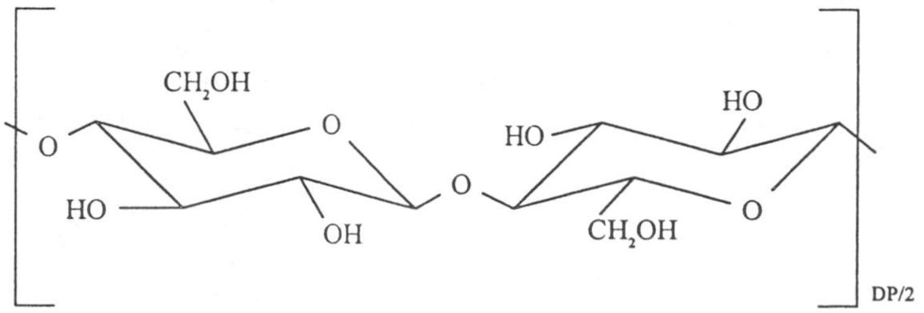

FIGURE 40 Cellulose (repeating unit).

and acids. Powdered MCC is a spray-dried product. Spray-drying produces agglomerated aggregates of microcrystals that are porous and sponge-like. Powdered MCC is used primarily as a flavor carrier and as an anticaking agent for shredded cheese. The second type, colloidal MCC, is water dispersible and has functional properties similar to those of water-soluble gums. To make colloidal MCC, considerable mechanical energy is applied after hydrolysis to tear apart the weakened microfibrils and provide a major proportion of colloidal-sized aggregates (< 0.2 μm in diameter). To prevent rebonding of the aggregates during drying, sodium carboxymethylcellulose (CMC) is added (see Sec. 4.5.2). CMC aids in redispersion and acts as a barrier to reassociation by giving the particles a stabilizing negative charge.

The major functions of colloidal MCC are to stabilize foams and emulsions, especially during high-temperature processing; to form gels with salve-like textures (MCC does not dissolve, nor does it form intermolecular junction zones; rather it forms a network of hydrated microcrystals); to stabilize pectin and starch gels to heat; to improve adhesion; to replace fat and oil, and to control ice crystal growth. MCC stabilizes emulsions and foams by adsorbing at interfaces and strengthening interfacial films. It is a common ingredient of reduced-fat ice cream and other frozen dessert products.

4.5.2 Carboxymethylcellulose [11,26]

Carboxymethylcellulose (CMC) is widely and extensively used as a food gum. Treatment of purified wood pulp with 18% sodium hydroxide solution produces alkali cellulose. When alkali cellulose is reacted with the sodium salt of chloroacetic acid, the sodium salt of the carboxymethyl ether (cellulose-O-CH_2-$CO_2^-Na^+$) is formed (Table 2). Most commercial sodium carboxymethylcellulose (CMC) products have a degree of substitution (DS) in the range 0.4–0.8. The most widely sold type for use as a food ingredient has a DS of 0.7.

Since CMC consists of long, fairly rigid molecules that bear a negative charge due to numerous ionized carboxyl groups, electrostatic repulsion causes its molecules in solution to be extended. Also, adjacent chains repel each other. Consequently, CMC solutions tend to be both highly viscous and stable. CMC is available in a wide range of viscosity types.

CMC stabilizes protein dispersions, especially near their isoelectric pH value. Thus, egg white is stabilized with CMC for co-drying or freezing, and milk products are given improved stability against casein precipitation.

4.5.3 Methylcelluloses and Hydroxypropylmethylcelluloses [16,17]

Alkali cellulose is treated with methyl chloride to introduce methyl ether groups (cellulose-O-CH_3). Many members of this class of gums also contain hydroxypropyl ether groups (cellulose-O-CH_2-CHOH-CH_3). Hydroxypropylmethylcelluloses (HPMC) are made by reacting alkali cellulose with both propylene oxide and methyl chloride. The degree of substitution with methyl ether groups of commercial methylcelluloses (MC) ranges from 1.1 to 2.2. The moles of substitution (MS)* values with hydroxypropyl ether groups in commercial hydroxypropylmethylcelluloses range from 0.02 to 0.3. (Both the methylcellulose and hydroxypropylmethyl-

*The moles of substitution or molar substitution (MS) is the average number of moles of substituent attached to a glycosyl unit of a polysaccharide. Because reaction of a hydroxyl group with propylene oxide creates a new hydroxyl group with which propylene oxide can react further, poly(propylene oxide) chains, each terminated with a free hydroxyl group, can form. Thus, because more than three moles of propylene oxide can react with a single hexopyranosyl unit, MS rather than DS must be used.

cellulose members of this gum family are generally referred to simply as methylcelluloses.) Both products are cold-water soluble because the methyl and hydroxypropyl ether group protrusions along the chains prevent the intermolecular association characteristic of cellulose.

While a few added ether groups spread along the chains enhance water solubility of cellulose (by decreasing internal hydrogen bonding), they also decrease chain hydration by replacing water-binding hydroxyl groups with less polar ether groups, giving members of this family unique characteristics. The ether groups restrict solvation of the chains to the point that they are on the borderline of water solubility. Hence, when an aqueous solution is heated, the water molecules of polymer solvation dissociate from the chain and hydration is decreased sufficiently that intermolecular associations increase and gelation occurs. Reducing the temperature once again brings about solubility, so the gelation is reversible.

Because of the ether groups, the gum chains are somewhat surface active and absorb at interfaces. This helps stabilize emulsions and foams. Methylcelluloses also can be used to reduce the amount of fat in food products through two mechanisms: (a) they provide fat-like properties so that the fat content of a product can be reduced, and (b) they reduce adsorption of fat in products being fried. The gel structure produced by thermogelation provides a barrier to oil, holds moisture, and acts as a binder.

4.6 GUAR AND LOCUST BEAN GUMS [19,20,31]

Guar and locust bean gums are important thickening polysaccharides for both food and nonfood uses (Table 2). Guar gum produces the highest viscosity of any natural, commercial gum. Both gums are the ground endosperm of seeds. The main component of both endosperms is a galactomannan. Galactomannans consist of a main chain of β-D-mannopyranosyl units joined by $(1 \rightarrow 4)$ bonds with single-unit α-D-galactopyranosyl branches attached at O-6 (Fig. 41). The specific polysaccharide component of guar gum is guaran. In guaran, about one-half of the D-mannopyranosyl main-chain units contain a D-galactopyranosyl side chain.

The galactomannan of locust bean gum (LBG, also called carob gum) has fewer branch units than does guaran and its structure is more irregular, with long stretches of about 80 underivatized D-mannosyl units alternating with sections of about 50 units in which almost every main chain unit has an α-D-galactopyranosyl group glycosidically connected to its O-6 position.

Because of the difference in structures, guar gum and LBG have different physical

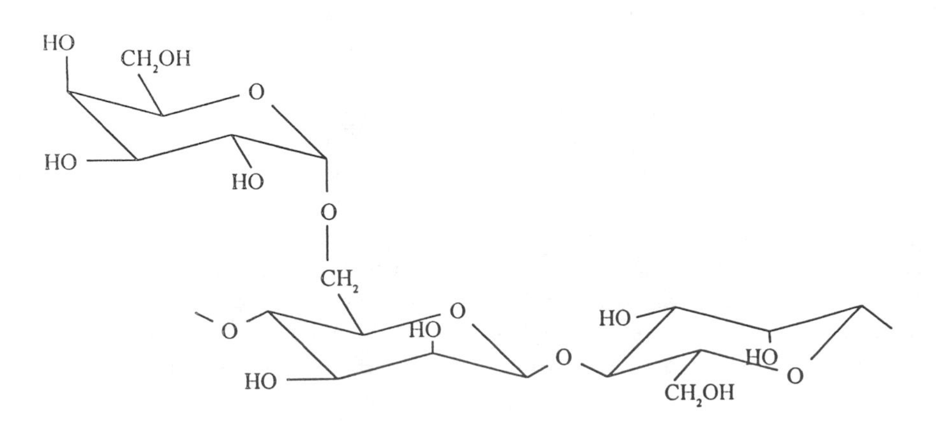

FIGURE 41 A representative segment of a galactomannan molecule.

properties, even though both are composed of long, rather rigid chains that provide high solution viscosity. Because guaran has its galactosyl units fairly evenly placed along the chain, there are few locations on the chains that are suitable for formation of junction zones. However, LBG, with its long "naked chain" sections, can form junction zones. LBG molecules can interact with bare portions of cellulose derivatives to form junctions. This produces an increase in viscosity. LBG also interacts with xanthan (see Sec. 4.7) and carrageenan (see Sec. 4.8) helices to form rigid gels.

Guar gum provides economical thickening to numerous food products. It is used frequently in combination with other food gums, such as in ice cream, where it is used in combination with carboxymethylcellulose (see Sec. 4.5.2), carrageenan (see Sec. 4.8), xanthan (see Sec. 4.7), and LBG.

Typical products in which LBG is found are the same as those for guar gum. About 85% of LBG is used in dairy and frozen dessert products. It is rarely used alone, but in combination with other gums such as CMC, carrageenan, xanthan, and guar gum. A typical use level is 0.05–0.25%.

4.7 XANTHAN [25,40]

Xanthomonas campestris, a bacterium commonly found on leaves of plants of the cabbage family, produces a polysaccharide, termed xanthan, that is widely used as a food gum. The polysaccharide is known commercially as xanthan gum (Table 2).

Xanthan has a backbone chain identical to that of cellulose (Fig. 43; compare with Fig. 40). In the xanthan molecule, every other β-D-glucopyranosyl unit in the cellulose backbone has attached, at the O-3 position, a β-D-mannopyranosyl-(1 → 4)-β-D-glucuronopyranosyl-(1 → 2)-6-*O*-acetyl-β-D-mannopyranosyl trisaccharide unit (Fig. 42).* About half of the terminal β-D-mannopyranosyl units have pyruvic acid attached as a 4,6-cyclic acetal. The trisaccharide side chains interact, by secondary bonding forces, with the main chain and make the molecule rather stiff. The molecular weight is probably in the order of 2×10^6, although much larger values, presumably due to aggregation, have been reported.

Xanthan interacts with guar gum, giving a synergistic increase in solution viscosity. The interaction with LBG produces a heat-reversible gel (Fig. 42).

Xanthan is widely used as a food gum because of the following important characteristics: solubility in hot or cold water; high solution viscosity at low concentrations; no discernible change in solution viscosity in the temperature range from 0 to 100°C, which makes it unique among food gums; solubility and stability in acidic systems; excellent compatibility with salt; interaction with other gums such as LBG; ability to stabilize suspensions and emulsions; and good solution stability when exposed to freezing and thawing. The unusual and very useful properties of xanthan undoubtedly result from the structural rigidity and extended nature of its molecules, which in turn result from its linear, cellulosic backbone, which is stiffened and shielded by the anionic trisaccharide side chains.

Xanthan is ideal for stabilizing aqueous dispersions, suspensions, and emulsions. The fact that the viscosity of its solutions changes very little with temperature—that is, its solutions do not thicken upon cooling—makes it irreplaceable for thickening and stabilizing such products as pourable salad dressings and chocolate syrup, which need to pour as easily when taken from the refrigerator as they do at room temperature, and gravies, which should

*Bacterial heteroglycans, unlike plant heteroglycans, usually have regular, repeating-unit structures.

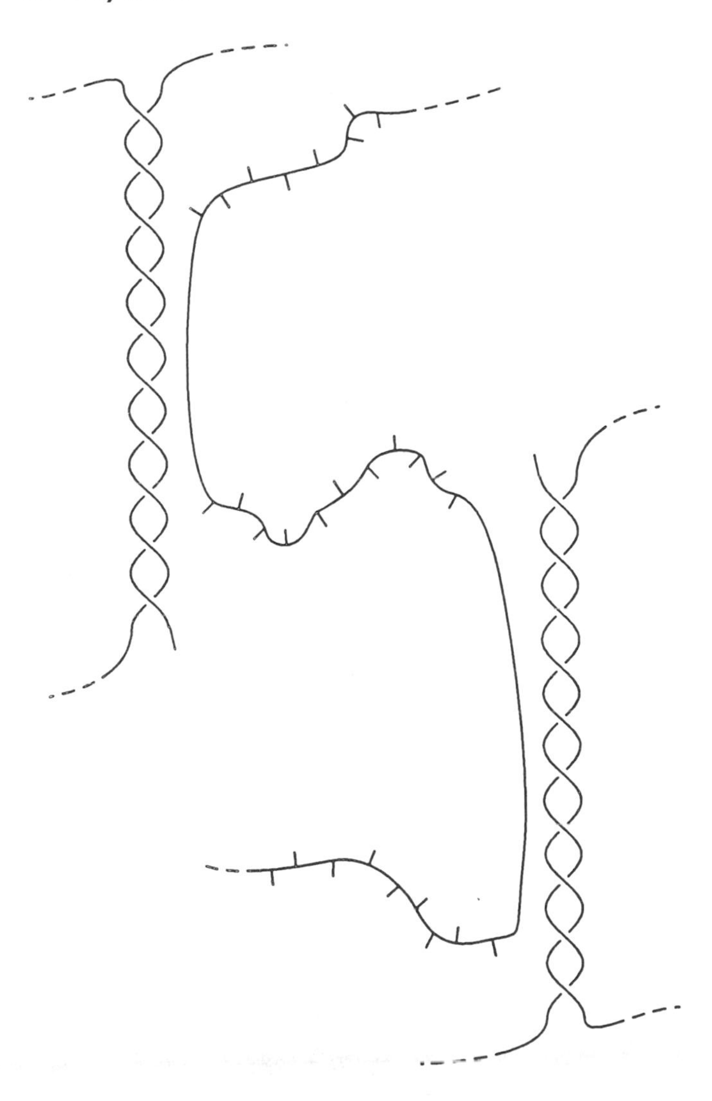

FIGURE 42 Representation of the hypothesized interaction of a locust bean gum molecule with double-helical portions of xanthan or carrageenan molecules to form a three-dimensional network and a gel.

neither thicken appreciably as they cool nor thin too much when hot. In regular salad dressings, xanthan is at the same time a thickener and a stabilizer for both the suspension of particulate materials and the oil-in-water emulsion. It is also a thickener and suspending agent in no-oil (reduced-calorie) dressings. In both oil-containing and no-oil salad dressings, xanthan is almost always used in combination with propylene glycol alginate (PGA) (see Sec. 4.9). PGA decreases solution viscosity and pseudoplasticity. Together they give the desired pourability associated with the pseudoplastic xanthan and the creaminess sensation associated with a nonpseudoplastic solution.

Blends of xanthan and LBG and/or guar gum are excellent ice cream stabilizers. Carrageenan is added to the mix to prevent whey separation.

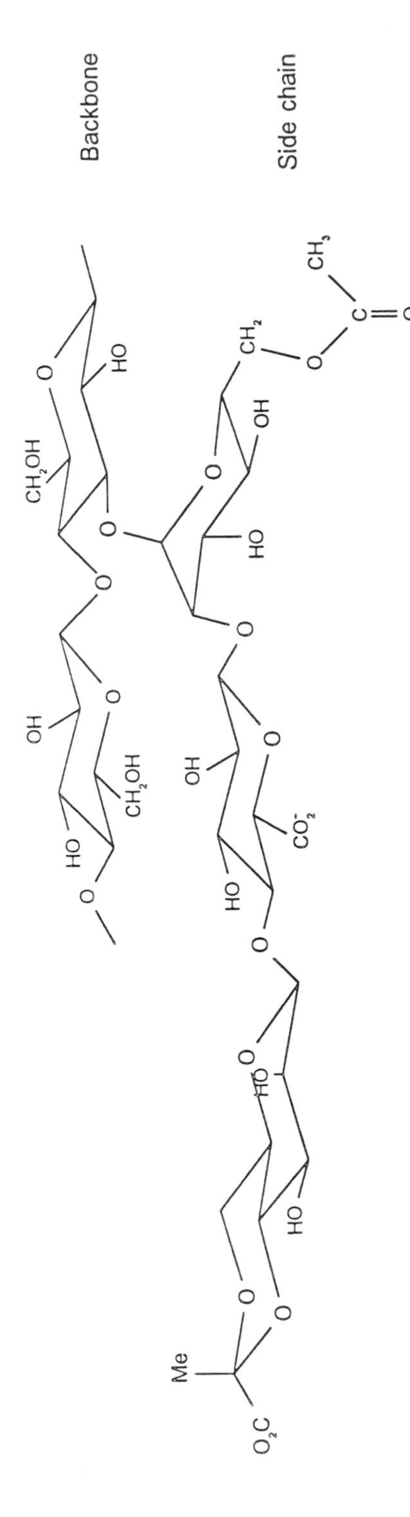

FIGURE 43 Structure of the pentasaccharide repeating unit of xanthan. Note the 4,6-*O*-pyruvyl-D-mannopyranosyl nonreducing end unit of the trisaccharide side chain. About half of the side chains are normally pyruvylated.

4.8 CARRAGEENANS [15,50]

The carrageenans are mixtures of several related galactans having sulfate half-ester groups attached to the sugar units (Table 2). They are extracted from red seaweeds with a dilute alkaline solution; the sodium salt of a carrageenan is normally produced. Also prepared and used is an alkali-modified seaweed flour called processed Euchema seaweed (PES) or Phillippine natural grade (PNG) carrageenan. To prepare PES/PNG carrageenan, red seaweed is treated with a potassium hydroxide solution. Because the potassium salts of the types of carrageenans found in these seaweeds are insoluble, they are not extracted out. Primarily low-molecular-weight soluble components are removed from the plants during this treatment. The remaining seaweed is dried and ground to a powder. PES/PNG carrageenan is, therefore, a composite material that contains not only the molecules of carrageenan that would be extracted with dilute sodium hydroxide, but also other cell wall materials.

The term carrageenan denotes a group or family of sulfated galactans extracted from red seaweeds. Carrageenans are linear chains of D-galactopyranosyl units joined with alternating $(1 \rightarrow 3)$-α-D- and $(1 \rightarrow 4)$-β-D-glycosidic linkages, with most sugar units having one or two sulfate groups esterified to a hydroxyl group at carbon atoms C-2 or C-6. This gives a sulfate content ranging from 15 to 40%. Units often contain a 3,6-anhydro ring. The principal structures are termed kappa (κ), iota, (ι), and lambda (λ) (Fig. 44). The disaccharide units shown in Figure 44 represent the predominant building block of each type but are not repeating unit structures. Carrageenans, as extracted, are mixtures of nonhomogeneous polysaccharides. Carrageenan products, of which there may be well more than 100 for different specific applications from a single supplier, contain different proportions of the three main behavioral types: kappa, iota, and lambda.

Carrageenan products dissolve in water to form highly viscous solutions. The viscosity is quite stable over a wide range of pH values because the sulfate half-ester groups are always ionized, even under strongly acidic conditions, giving the molecules a negative charge.

Segments of molecules of kappa- and iota-type carrageenans exist as double helices of parallel chains. In the presence of potassium or calcium ions, thermoreversible gels form upon cooling a hot solution containing such double-helical segments. Gelation can occur in water at concentrations as low as 0.5%. When kappa-type carrageenan solutions are cooled in the presence of potassium ions, a stiff, brittle gel results. Calcium ions are less effective in causing gelation. Potassium and calcium ions together produce a high gel strength. Kappa-type gels are the strongest of the carrageenan gels. These gels tend to synerese as junction zones extend within the structure. The presence of other gums retards syneresis.

Iota types are a little more soluble than are kappa-type carrageenans, but again only the sodium salt form is soluble in cold water. Iota types gel best with calcium ions. The resulting gel is soft and resilient, has good freeze–thaw stability, and does not synerese, presumably because iota-type carrageenans are more hydrophilic and form fewer junction zones than do kappa-type carrageenans.

During cooling of solutions of kappa- or iota-type carrageenans, gelation occurs because the linear molecules are unable to form continuous double helices due to the presence of structural irregularities. The linear helical portions then associate to form a rather firm three-dimensional, stable gel in the presence of the appropriate cation (Fig. 45). All salts of lambda-type carrageenans are soluble and nongelling.

Under conditions in which double helical segments occur, carrageenan molecules, particularly those of the kappa type, form junction zones with the naked segments of LBG to produce rigid, brittle, syneresing gels. This gelation occurs at a concentration one-third that needed to form a pure kappa-type carrageenan gel.

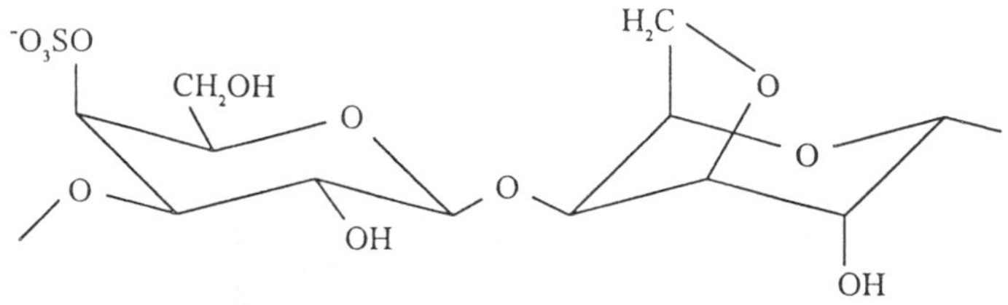

κ-Carrageenan

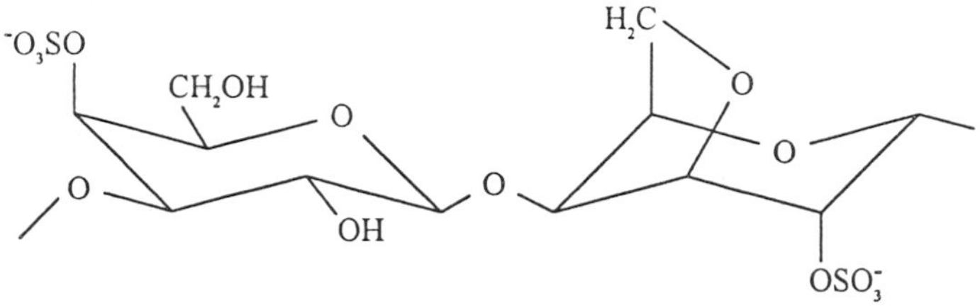

ι-Carrageenan

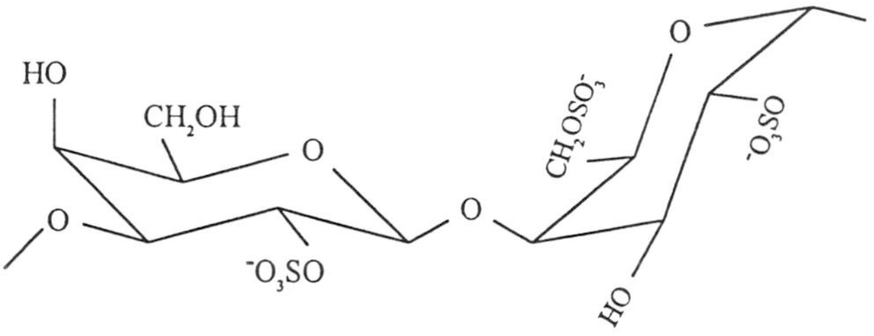

λ–Carrageenan

FIGURE 44 Idealized unit structures of kappa-, iota-, and lambda-type carrageenans.

Carrageenans are most often used because of their ability to form gels with milk and water. Blending provides a wide range of products that are standardized with various amounts of sucrose, glucose (dextrose), buffer salts, or gelling aids, such as potassium chloride. The available commercial products form a variety of gels: gels that are clear or turbid, rigid or elastic, tough or tender, heat-stable or thermally reversible, and do or do not undergo syneresis. Carrageenan gels do not require refrigeration because they do not melt at room temperature. They are freeze–thaw stable.

A useful property of carrageenans is their reactivity with proteins, particularly those of milk. Kappa-type carrageenans complex with kappa-casein micelles of milk, forming a weak, thixotropic, pourable gel. The thickening effect of kappa-carrageenans in milk is 5–10 times greater than it is in water. This property is used in the preparation of chocolate milk where the thixotropic gel structure prevents settling of cocoa particles. Such stabilization requires only about 0.025% gum. This property is also utilized in the preparation of ice cream, evaporated

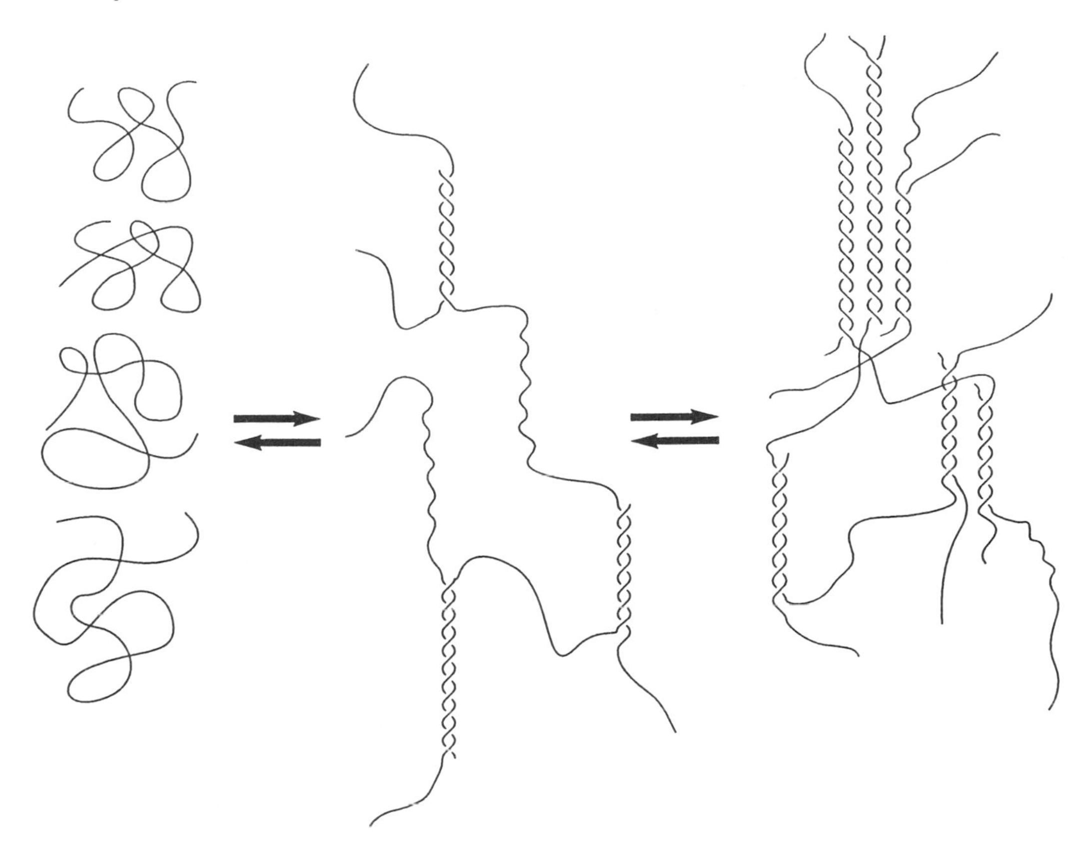

FIGURE 45 A representation of the hypothesized mechanism of gelation of kappa- and iota-type carrageenans. In a hot solution, the polymer molecules are in a coiled state. As the solution is cooled, they intertwine in double-helical structures. As the solution is cooled further, the double helices are believed to nest together with the aid of potassium or calcium ions.

milk, infant formulas, freeze–thaw stable whipped cream, and emulsions in which milk fat is replaced with a vegetable oil.

The synergistic effect between kappa-carrageenan and locust bean gum (LBG) (Fig. 42) produces gels with greater elasticity and gel strength, and less syneresis than gels of potassium kappa-carrageenan alone. As compared to kappa-type carrageenan alone, the kappa-type carrageenan-LBG combination provides greater stabilization and air bubble retention (overrun) in ice cream, but also a little too much chewiness, so guar gum is added to soften the gel structure.

Cold hams and poultry rolls take up 20–80% more brine when they contain 1–2% of a kappa-type carrageenan. Improved slicing also results. Carrageenan coatings on meats can serve as a mechanical protection and a carrier for seasonings and flavors. Carrageenan is sometimes added to meat analogs made from casein and vegetable proteins. A growing use of carrageenan is to hold water and maintain water content, and therefore softness of meat products, such as wieners and sausages, during the cooking operation. Addition of a kappa- or iota-type carrageenan in the Na^+ form or PES/PNG carrageenan to low-fat ground beef improves texture and hamburger quality. Normally, fat serves the purpose of maintaining softness, but because of the binding power of carrageenan for protein and its high affinity for water, carrageenans can be used to replace in part this function of natural animal fat in lean products.

Two other food gums, agar and furcellaran (also called Danish agar), also come from red seaweeds and have structures and properties that are closely related to those of the carrageenans.

4.9 ALGINS [5,28]

Commercial algin is a salt, most often the sodium salt, of a linear poly(uronic acid), alginic acid, obtained from brown seaweeds (Table 2). Alginic acid is composed of two monomeric units, β-D-mannopyranosyluronic acid and α-L-gulopyranosyluronic acid units. These two monomers occur in homogeneous regions (composed exclusively of one unit or the other) and in regions of mixed units. Segments containing only D-mannuronopyranosyl units are referred to as M blocks and those containing only L-guluronopyranosyl units as G blocks. D-Mannuronopyranosyl units are in the $^4C_1$ conformation, while L-guluronopyranosyl units are in the $^1C_4$ conformation (see Sec. 4.1.2 and Fig. 46), which gives the different blocks quite different chain conformations. M-Block regions are flat and ribbon-like, similar to the conformation of cellulose (see Sec. 4.5), because of the equatorial → equatorial bonding. G-block regions have a pleated (corrugated) conformation as a result of its axial → axial glycosidic bonds. Different percentages of the different block segments cause algins (alginates) from different seaweeds to have different properties. Algins with high G-block contents produce gels of high strength.

Solutions of sodium alginates are highly viscous. The calcium salt of alginates is insoluble. The insoluble salt results from the autocooperative reaction between calcium ions and the G-block regions of the chain. The holes formed between two G-block chains are cavities that bind calcium ions. The result is a junction zone that has been called an "egg box" arrangement

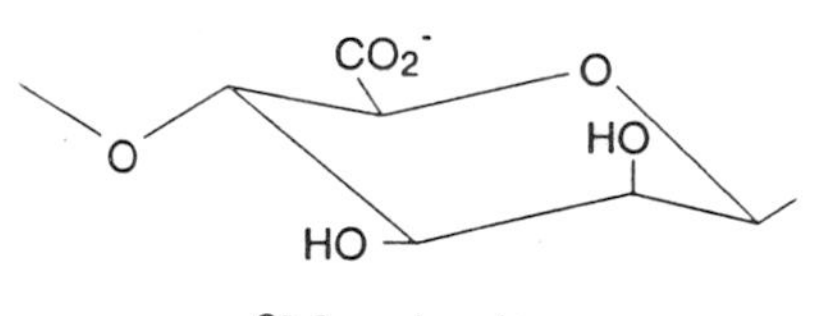

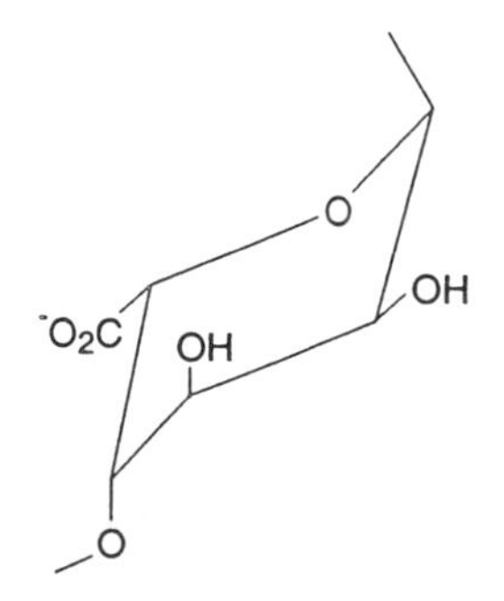

FIGURE 46 Units of β-D-mannopyranosyluronic acid (βMan*p*A) in the $^4C_1$ conformation and α-L-gulopyranosyluronic acid (αLGul*p*A) in the $^1C_4$ conformation.

with the calcium ions being likened to eggs in the pockets of an egg carton (Fig. 47). The strength of the gel depends on the content of G blocks in the alginate used and the concentration of calcium ions.

Propylene glycol alginates are made by reacting moist alginic acid with propylene oxide to produce a partial ester with 50–85% of the carboxyl groups esterified. Solutions of propylene alginates (PGA) are much less sensitive to low pH values and polyvalent cations, including calcium ions and proteins, than are solutions of nonesterified alginates, because esterified carboxyl groups cannot ionize. Also, the propylene glycol group introduces a bump on the chain that prevents close association of chains. Therefore, PGA solutions are quite stable. Because of its tolerance to calcium ions, propylene glycol alginates can be used in dairy products. The hydrophobic propylene glycol groups also give the molecule mild interfacial activity, that is, foaming, emulsifying, and emulsion-stabilizing properties.

Algin is widely used to provide high viscosity at low concentrations. Even greater viscosity under low-shear conditions can be achieved by introducing a small amount of calcium ions. If PGA is used, some calcium ion cross-linking of chains still occurs through the remaining carboxylate groups, and results in thickening of solutions (rather than gelling).

Calcium alginate gels are obtained by diffusion setting, internal setting, and setting by cooling. Diffusion setting can be used to prepare structured foods. Perhaps the best example is the structured pimento strip. In the production of pimento strips for stuffing green olives, pimento puree is first mixed with water containing a small amount of guar gum as an immediate thickener, and then with sodium alginate. The mixture is pumped onto a conveyor belt and gelled by the addition of calcium ions. The set sheet is cut into thin strips and stuffed into olives. Internal setting for fruit mixes, purees, and fruit analogs involves slow release of calcium ions within the mixture. Setting by cooling involves dissolving a calcium salt, a slightly soluble acid, and a sequestrant in hot water and allowing the mixture to set on cooling. Gels produced in this way are quite stable. Alginate gels are reasonably heat stable and show little or no syneresis; those containing fruit can be used for fillings that remain stable through pasteurization and cooking. Unlike gelatin gels, alginate gels are not thermoreversible and can be used as dessert gels that do not require refrigeration.

Alginic acid, an alginate solution whose pH has been lowered, with and without addi-

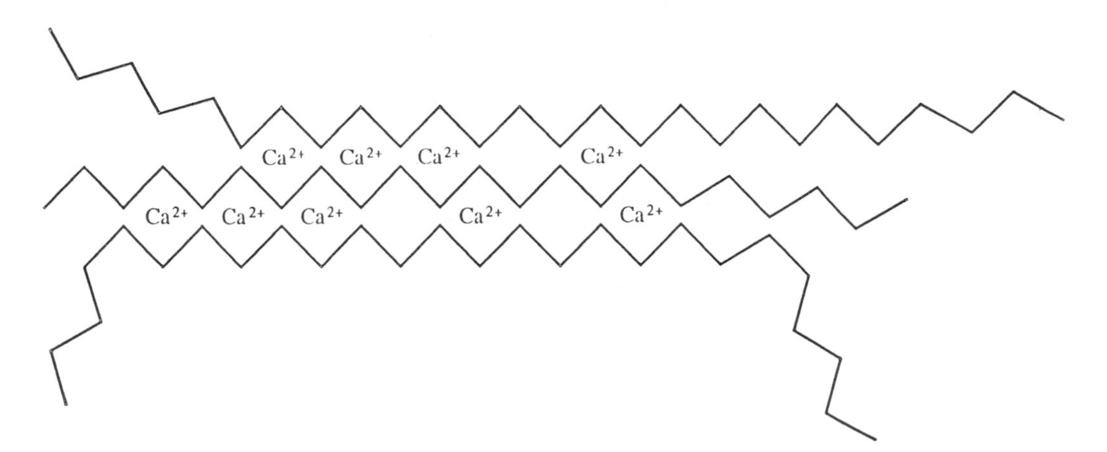

FIGURE 47 A representation of the proposed formation of a junction between G-block regions of three alginate molecules promoted by calcium ions.

tion of some calcium ions, is employed in the preparation of soft, thixotropic, nonmelting gels (Table 2).

Propylene glycol alginate is used when stability to acid, nonreactivity with calcium ions (for example, in milk products), or its surface-active property is desired. Accordingly, it finds use as a thickener in salad dressings (Table 2). In low-calorie dressings, it is often used in conjunction with xanthan (see Sec. 4.7).

4.10 PECTINS [4,41]

Commercial pectins are galacturonoglycans [poly(α-D-galactopyranosyluronic acids)] with various contents of methyl ester groups (Table 2). Native pectins found in the cell walls and intercellular layers of all land plants are more complex molecules that are converted into commercial products during extraction with acid. Commercial pectin is obtained from citrus peel and apple pomace. Pectin from lemon and lime peel is the easiest to isolate and is of the highest quality. Pectins have an unique ability to form spreadable gels in the presence of sugar and acid or in the presence of calcium ions and are used almost exclusively in these types of applications.

The compositions and properties of pectins vary with source, the processes used during preparation, and subsequent treatments. During extraction with mild acid, some hydrolytic depolymerization and hydrolysis of methyl ester groups occurs. Therefore, the term pectin denotes a family of compounds, and this family is part of a still larger family called pectic substances. The term pectin is usually used in a generic sense to designate those water-soluble galacturonoglycan preparations of varying methyl ester contents and degrees of neutralization that are capable of forming gels. In all natural pectins, some of the carboxyl groups are in the methyl ester form. Depending on the isolation conditions, the remaining free carboxylic acid groups may be partly or fully neutralized, that is, partly or fully present as sodium, potassium, or ammonium carboxylate groups. Typically, they are present in the sodium salt form.

By definition, preparations in which more than half of the carboxyl groups are in the methyl ester form ($-COOCH_3$) are classified as high-methoxyl (HM) pectins (Fig. 48); the remainder of the carboxyl groups will be present as a mixture of free acid (-COOH) and salt ($-COO^-Na^+$) forms. Preparations in which less than half of the carboxyl groups are in the methyl ester form are called low-methoxyl (LM) pectins. The percentage of carboxyl groups esterified with methanol is the degree of esterification (DE). Treatment of a pectin preparation with ammonia dissolved in methanol converts some of the methyl ester groups into carboxamide groups (15–25%). In the process, a low-methoxyl (LM) pectin (by definition) is formed. These products are known as amidated LM pectins.

The principal and key feature of all pectin molecules is a linear chain of (1 → 4)-linked α-D-galactopyranosyluronic acid units. Neutral sugars, primarily L-rhamnose, are also present. In citrus and apple pectins, the α-L-rhamnopyranosyl units seem to be inserted into the polysaccharide chain at rather regular intervals. The inserted L-rhamnopyranosyl units may

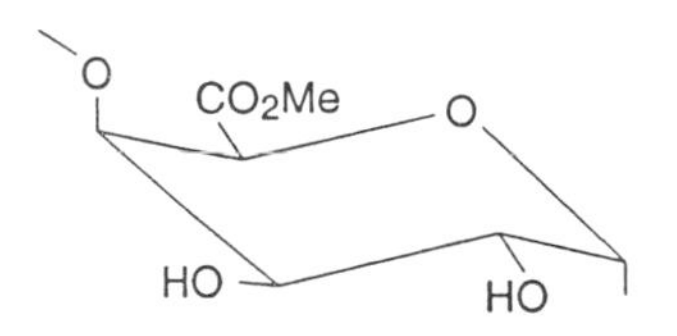

FIGURE 48 The most prevalent monomeric unit of a high-methoxyl pectin.

provide the necessary irregularities in the structure required to limit the size of the junction zones and effect gelation. At least some pectins contain covalently attached, highly branched arabinogalactan chains and/or short side chains composed of D-xylosyl units. The presence of side chains may also be a factor that limits the extent of chain association. Junction zones are formed between regular, unbranched pectin chains when the negative charges on the carboxylate groups are removed (addition of acid), when hydration of the molecules is reduced (by addition of a cosolute, almost always sugar, to a solution of HM pectin), and/or when pectinic acid polymer chains are bridged by calcium cations.

HM pectin solutions gel when sufficient acid and sugar are present. As the pH of a pectin solution is lowered, the highly hydrated and charged carboxylate groups are converted into uncharged, only slightly hydrated carboxylic acid groups. As a result of losing some of their charge and hydration, the polymer molecules can now associate over a portion of their length, forming junctions and a network of polymer chains that entraps the aqueous solution of solute molecules. Junction zone formation is assisted by the presence of a high concentration (~65%, at least 55%) of sugar, which competes for water of hydration and reduces solvation of the chains, allowing them to interact with one another.

LM pectin solutions gel only in the presence of divalent cations, which provide cross-bridges. Increasing the concentration of divalent cations (only calcium ion is used in food applications) increases the gelling temperature and gel strength. The same general egg-box model used to describe the formation of calcium alginate gels (see Sec. 4.9) is used to explain gelation of solutions of LM and amidated LM pectins upon addition of calcium ions. LM pectin, since it does not require sugar for gelation, is used to make dietetic jams, jellies, and marmalades.

4.11 GUM ARABIC [14,53]

When the bark of some trees and shrubs is injured, the plants exude a sticky material that hardens to seal the wound and give protection from infection and desiccation. Such exudates are commonly found on plants that grow in semiarid climates. All require cleaning and pasteurization, since they are sticky when freshly exuded and trap dust, sand particles, insects, bacteria, and/or pieces of bark. Gum arabic (gum acacia), gum karaya, and gum ghatti are exudates of trees; gum tragacanth is the exudate of a shrub. Their overall use has diminished and continues to diminish because of their uncertain and restricted availability and their increasing cost. Only gum arabic still has a good market in its traditional food applications.

Gum arabic (gum acacia) is an exudate of acacia trees, of which there are many species distributed over tropical and subtropical regions (Table 2). The most important growing areas for species that give the best gum are the Sudan and Nigeria.

Gum arabic is a heterogeneous material, but generally consists of two fractions. One, which accounts for about 70% of the gum, is composed of polysaccharide chains with little or no nitrogenous material. The other fraction contains molecules of higher molecular weight that have protein as an integral part of their structures. The protein–polysaccharide fraction is itself heterogeneous with respect to protein content. The polysaccharide structures are covalently attached to the protein component by linkage to hydroxyproline and, perhaps, serine units, the two predominant amino acids in the polypeptide. The overall protein content is about 2 wt%, but specific fractions may contain as much as 25 wt% protein.

The polysaccharide structures, both those attached to protein and those that are not, are highly branched, acidic arabinogalactans with the following approximate composition: D-galactose, 44%; L-arabinose, 24%; D-glucuronic acid, 14.5%; L-rhamnose, 13%; 4-*O*-methyl-D-glucuronic acid, 1.5%. They contain main chains of (1 → 3)-linked β-D-galactopyranosyl units having two- to four-unit side chains consisting of (1 → 3)-β-D-galactopyranosyl units joined to it by (1 → 6)-linkages. Both the main chain and the numerous side chains have attached α-L-arabinofuranosyl, α-L-rhamnopyranosyl, β-D-glucuronopyranosyl, and 4-*O*-methyl-β-D-glucuronopyranosyl units. The two uronic acid units occur most often as ends of chains.

Gum arabic dissolves easily when stirred in water. It is unique among the food gums because of its high solubility and the low viscosity of its solutions. Solutions of 50% concentration can be made. At this concentration, the dispersion is somewhat gel-like.

Gum arabic is both a fair emulsifying agent and a very good emulsion stabilizer for flavor oil-in-water emulsions. It is the gum of choice for emulsification of citrus, other essential oils, and imitation flavors used as baker's emulsions and concentrations for soft drinks. The soft drink industry consumes about 30% of the gum supply as an emulsifier and stabilizer. For a gum to have an emulsion-stabilizing effect, it must have anchoring groups with a strong affinity for the surface of the oil and a molecular size large enough to cover the surfaces of dispersed droplets. Gum arabic has surface activity and forms a thick, sterically stabilizing, macromolecular layer around oil droplets. Emulsions made with flavor oils and gum arabic can be spray-dried to produce dry flavor powders that are nonhygroscopic and in which the flavor oil is protected from oxidation and volatization. Rapid dispersion and release of flavor without affecting product viscosity are other attributes. These stable flavor powders are used in dry package products such as beverage, cake, dessert, pudding, and soup mixes.

Another important characteristic is its compatibility with high concentrations of sugar. Therefore, it finds widespread use in confections with a high sugar content and a low water content. More than half the world's supply of gum arabic is used in confections such as caramels, toffees, jujubes, and pastilles. Its functions in confections are to prevent sucrose crystallization and to emulsify and distribute fatty components. Avoidance of surface accumulation of lipids is important because this occurrence results in a greasy surface whitening, called bloom. Another use is as a component of the glaze or coating of pan-coated candies.

4.12 DIETARY FIBER AND CARBOHYDRATE DIGESTIBILITY [7,23,29,45,46,54,57]

Dietary fiber is not necessarily fibrous in nature. Dietary fiber is a nutritional term that has nothing to do with its physical or chemical nature, although both chemical and physical properties are involved in its determination. Dietary fiber is actually defined by the method used to measure it, of which there are several. Both insoluble plant cell-wall materials, primarily cellulose and lignin, and nonstarch, water-soluble polysaccharides are components of dietary fiber. The only common feature of these substances is that they are nondigestible polymers. Therefore, not only do natural components of foods contribute dietary fiber, but so also do gums that are added to modify rheological properties, to provide bulk, and/or to provide other functionalities as already described.

One natural component of dietary fiber is a water-soluble polysaccharide, commonly known as β-glucan, but more properly β-D-glucan, that is present in oat and barley brans. Oat β-glucan has become a commercial food ingredient because it has been shown to be effective in reducing the level of serum cholesterol. Oat β-glucan is a linear chain of β-D-glucopyranosyl units. About 70% are linked (1 → 4) and about 30% (1 → 3). The (1 → 3) linkages occur singly and are separated by sequences of two or three (1 → 4) linkages. Thus, the molecule is composed of (1 → 3)-linked β-cellotriosyl [→ 3)-βGl*cp*-(1 → 4)-βGl*cp*-(1 → 4)-

$$\longrightarrow 3)\text{—}\beta\text{Glc}p\text{—}(1 \left[\rightarrow 4)\text{—}\beta\text{Glc}p\text{—}(1 \right]_n \rightarrow$$

FIGURE 49 Representative structure of a segment of oat and barley β-glucans where *n* usually is 1 or 2, but occasionally may be larger (shorthand notation).

βGl*cp*-(1 →] and β-cellotetraosyl units (Fig. 49). Such (1 → 4,1 → 3)-β-glucans are often called mixed-linkage β-glucans.

When taken orally in foods, β-glucans reduce postprandial serum glucose levels and the insulin response—that is, they moderate the glycemic response—in both normal and diabetic human subjects. This effect seems to be correlated with viscosity. They also reduce serum cholesterol concentrations in rats, chicks, and humans. These physiological effects are typical of those of soluble dietary fiber. Other soluble polysaccharides have similar effects but to differing degrees. The mechanism(s) of action remains to be determined.

Carbohydrates have always been the principal source of metabolic energy for humans and the means for maintaining health of the human gastrointestinal tract. Carbohydrates are the principal providers of the bulk and body of food products.

The higher saccharides may be digestible (most starch-based products), partially digestible (retrograded amylose, the so-called resistant starch), or nondigestible (essentially all other polysaccharides). When digestive hydrolysis to monosaccharides occurs, the products of digestion are absorbed and catabolized. [Only D-glucose is produced by digestion of polysaccharides (starch) in humans.] Those carbohydrates not digested to monosaccharides by human enzymes in the small intestine (all others except sucrose, lactose, and those related to starch) may be metabolized by microorganisms in the large intestine, producing substances that are absorbed and catabolized for energy. Therefore, carbohydrates may be caloric, partially caloric, or essentially noncaloric. They may be soluble or insoluble, and they may produce high or low viscosities. Naturally occurring plant carbohydrates are nontoxic.

The most common bulking agents in natural food are remnants of plant cells resistant to hydrolysis by enzymes in the alimentary tract. This material includes cellulose, hemicelluloses, pectin, and lignin. Dietary fiber bulking agents are important in human nutrition because they maintain normal functioning of the gastrointestinal tract. They increase intestinal and fecal bulk, which lowers intestinal transit time and helps prevent constipation. Their presence in foods induces satiety at meal time. Nutritionists set requirements of dietary fiber at 25–50 g/day. Insoluble fiber bulking agents are claimed to decrease blood cholesterol levels, lessening the chance of heart disease. They also reduce the chances of colonic cancer, probably due to their sweeping action.

Soluble gums other than β-glucans have similar effects in the gastrointestinal tract and on the level of cholesterol in blood, but to different extents. Some gums that have been specifically examined in this regard are pectin, guar gum, xanthan, and hemicelluloses. [For example, guar gum ingested at a rate of 5 g/day results in an improved glycemic index, a 13% lowering of serum cholesterol, and no decrease in the high-density lipoprotein (HDL) fraction, the beneficial cholesterol carrier.] In addition to cereal brans, kidney and navy beans are especially good sources of dietary fiber. A product based on psyllium seed hulls has high water-binding properties, leading to rapid transit time in the gastrointestinal tract, and is used to prevent constipation. A product with a methylcellulose base is sold for the same purpose.

(= metamucil)

The starch polysaccharides are the only polysaccharides that can be hydrolyzed by human digestive enzymes. They, of course, provide D-glucose, which is absorbed by microvilli of the small intestine to supply the principal metabolic energy of humans. Other polysaccharides consumed normally as natural components of edible vegetables, fruits, and other plant materials, and those food gums added to prepared food products, are not digested in the upper digestive tract of humans, but pass into the large intestine (colon) with little or no change. (The acidity of the stomach is not strong enough, nor is the residence time of polysaccharides in the stomach sufficiently long, to cause significant chemical cleavage.) When the undigested polysaccharides reach the large intestine, they come into contact with normal intestinal microorganisms, some of which produce enzymes that catalyze hydrolysis of certain polysaccharides or certain parts of polysaccharide molecules. The consequence of this is that polysaccharides not cleaved in the upper intestinal tract may undergo cleavage and microbial metabolism within the large intestine.

Sugars that are split from the polysaccharide chain are used by the microorganisms of the large intestine as energy sources in anaerobic fermentation pathways that produce lactic, propionic, butyric, and valeric acids. These short-chain acids can be absorbed through the intestinal wall and metabolized, primarily in the liver. In addition, a small, though significant in some cases, fraction of the released sugars can be taken up by the intestinal wall and transported to the portal bloodstream where they are conveyed to the liver and metabolized. It is calculated that, on average, 7% of human energy is derived from sugars split from polysaccharides by microorganisms in the large intestine and/or from the acid by-products produced from them by these microorganisms via anaerobic fermentation. The extent of polysaccharide cleavage depends on the abundance of the particular organism(s) producing the specific enzymes required. Thus, when changes occur in the type of polysaccharide consumed, utilization of the polysaccharide by colonic microorganisms may be temporarily reduced until organisms capable of splitting the new polysaccharide proliferate.

Some polysaccharides survive almost intact during their transit through the entire gastrointestinal tract. These, plus larger segments of other polysaccharides, give bulk to the intestinal contents and lower transit time. They can be a positive factor in health through a lowering of blood cholesterol concentration, perhaps by sweeping out bile salts and reducing their chances for reabsorption from the intestine. In addition, the presence of large amounts of hydrophilic molecules maintains a water content of the intestinal contents that results in softness and consequent easier passage through the large intestine.

BIBLIOGRAPHY

Dey, P. M., and R. A. Dixon, eds. (1985). *Biochemistry of Storage Carbohydrates in Green Plants,* Academic Press, London.

Eliason, A.-C., ed. (1996). *Carbohydrates in Food,* Marcel Dekker, New York.

El Khadem, H. S. (1988). *Carbohydrate Chemistry,* Academic Press, San Diego.

Glicksman, M., ed. (1982–1984). *Food Hydrocolloids,* vols. I–III, CRC Press, Boca Raton, FL.

Kennedy, J. F., ed. (1988). *Carbohydrate Chemistry,* University Press, New York.

Lineback, D. R., and G. E. Inglett, eds. (1982). *Food Carbohydrates,* AVI, Westport, CT.

Mathlouthi, M., and P. Reiser, eds. (1995). *Sucrose: Properties and Applications,* Blackie, Glasgow.

Stephen, A. M., ed. (1995). *Food Polysaccharides and Their Applications,* Marcel Dekker, New York.

Whistler, R. L., and J. N. BeMiller, eds. (1993). *Industrial Gums,* 3rd ed., Academic Press, San Diego.

Whistler, R. L., J. N. BeMiller, and E. F. Paschall, eds. (1994). *Starch: Chemistry and Technology,* 2nd ed., Academic Press, New York.

Wurzburg, O. B., ed. (1986). *Modified Starches: Properties and Uses,* CRC Press, Boca Raton, FL.

REFERENCES

1. BeMiller, J. N. (1993). Starch-based gums, in *Industrial Gums* (R. L. Whistler and J. N. BeMiller, eds.), Academic Press, San Diego, pp. 579–600.
2. Biliaderis, C. G. (1990). Thermal analysis of food carbohydrates, in *Thermal Analysis of Foods* (V. R. Harwalkar and C. -Y. Ma, eds.), Elsevier Science, New York, pp. 168–220.
3. Biliaderis, C. G. (1991). The structure and interactions of starch with food constituents. *Can. J. Physiol. Pharmacol. 69*:60–78.
4. Christensen, S. H. (1984). Pectins, in *Food Hydrocolloids,* vol. III (M. Glicksman, ed.), CRC Press, Boca Raton, FL, pp. 205–230.
5. Clare, K. (1993). Algin, in *Industrial Gums* (R. L. Whistler and J. N. BeMiller, eds.), Academic Press, San Diego, pp. 105–143.
6. Dickinson, E., ed. (1991). *Food Polymers, Gels, and Colloids,* Royal Society of Chemistry, London.
7. Dixon, R. A. (1985). β-(1 → 3)-Linked glucans from higher plants, in *Biochemistry of Storage Carbohydrates in Green Plants* (P. M. Dey and R. A. Dixon, eds.), Academic Press, London, pp. 229–264.
8. El Khadem, H. S. (1988). *Carbohydrate Chemistry,* Academic Press, San Diego.
9. Emodi, A. (1982). Polyols: chemistry and application, in *Food Carbohydrates* (D. R. Lineback and G. E. Inglett, eds.), AVI, Westport, CT, pp. 49–61.
10. Feather, M. S. (1982). Sugar dehydration reactions, in *Food Carbohydrates* (D. R. Lineback and G. E. Inglett, eds.), AVI, Westport, CT, pp. 113–133.
11. Feddersen, R. L., and S. N. Thorp (1993). Sodium carboxymethylcellulose, in *Industrial Gums* (R. L. Whistler and J. N. BeMiller, eds.), Academic Press, San Diego, pp. 537–578.
12. Fujimaki, M., M. Namiki, and H. Kato, eds. (1986). *Amino-Carbonyl Reactions in Food and Biological Systems,* Elsevier, Amsterdam.
13. Glicksman, M. (1982). Functional properties of hydrocolloids, in *Food Hydrocolloids,* Vol. I (M. Glicksman, ed.), CRC Press, Boca Raton, FL, pp. 47–99.
14. Glicksman, M. (1983). Gum arabic (gum acacia), in *Food Hydrocolloids,* vol. II (M. Glicksman, ed.), CRC Press, Boca Raton, FL, pp. 7–29.
15. Glicksman, M. (1983). Red seaweed extracts (agar, carrageenan, furcellaran), in *Food Hydrocolloids,* vol. II (M. Glicksman, ed.), CRC Press, Boca Raton, FL, pp. 73–113.
16. Grover, J. A. (1984). Methylcellulose (MC) and hydroxypropylmethylcellulose (HPMC), in *Food Hydrocolloids,* vol. III (M. Glicksman, ed.), CRC Press, Boca Raton, FL, pp. 121–154.
17. Grover, J. A. (1993). Methylcellulose and its derivatives, in *Industrial Gums* (R. L. Whistler and J. N. BeMiller, eds.), Academic Press, San Diego, pp. 475–504.
18. Harris, P., ed. (1990). *Food Gels,* Elsevier Applied Science, London.
19. Herald, C. T. (1984). Locust/carob bean gum, in *Food Hydrocolloids,* vol. III (M. Glicksman, ed.), CRC Press, Boca Raton, FL, pp. 161–170.
20. Herald, C. T. (1984). Guar gum, in *Food Hydrocolloids,* vol. III (M. Glicksman, ed.), CRC Press, Boca Raton, FL, pp. 171–184.
21. Hizukuri, S. (1985). Relationship between the distribution of the chain length of amylopectin and the crystalline structure of starch granules. *Carbohydr. Res. 141*:295–306.
22. Huzukuri, S. (1986). Polymodal distribution of the chain lengths of amylopectins and its significance. *Carbohydr. Res. 147*:342–347.
23. Johnson, I. T., and Southgate, D. A. T. (1994). *Dietary Fibre and Related Substances,* Chapman & Hall, London.
24. Kainuma, K. (1988). Structure and chemistry of the starch granule, in *The Biochemistry of Plants* (J. Preiss, ed.), Academic Press, San Diego, pp. 141–180.
25. Kang, K. S., and D. J. Pettitt (1993). Xanthan, gellan, welan, and rhamsan, in *Industrial Gums* (R. L. Whistler and J. N. BeMiller, eds.), Academic Press, San Diego, pp. 341–397.
26. Keller, J. D. (1984). Sodium carboxymethylcellulose (CMC), in *Food Hydrocolloids,* vol. III (M. Glicksman, ed.), CRC Press, Boca Raton, FL, pp. 43–109.

27. Kennedy, J. F., ed. (1988). *Carbohydrate Chemistry,* University Press, New York.
28. King, A. H. (1983). Brown seaweed extracts (alginates), in *Food Hydrocolloids,* vol. II (M. Glicksman, ed.), CRC Press, Boca Raton, FL, pp. 115–188.
29. Kritchevsky, D., and Bonfield, C., eds. (1995). *Dietary Fiber in Health and Disease,* Eagan Press, St. Paul, MN.
30. Labuza, T. P., Reineccius, G. A., Monnier, V., O'Brien, J., and Baynes, J., eds. (1995). *Maillard Reactions in Chemistry, Food, and Health,* CRC Press, Boca Raton, FL.
31. Maier, H., M. Anderson, C. Karl, K. Magnuson, and R. L. Whistler (1993). Guar, locust bean, tara and fenugreek gums, in *Industrial Gums* (R. L. Whistler and J. N. BeMiller, eds.), Academic Press, San Diego, pp. 181–226.
32. Manners, D. J. (1985). Some aspects of the structure of starch. *Cereal Foods World 30*:461–467.
33. Manners, D. J. (1985). Starch, in *Biochemistry of Storage Carbohydrates in Green Plants* (P. M. Dey and R. A. Dixon, eds.), Academic Press, London, pp. 149–203.
34. Manners, D. J. (1989). Recent developments in our understanding of amylopectin structure. *Carbohydr. Polym. 11*:87–112.
35. Mathlouthi, M., and Reiser, P., eds. (1995). *Sucrose: Properties and Applications,* Blackie, Glasgow.
36. Miles, M. J., V. J. Morris, P. D. Orford, and S. G. Ring (1985). The roles of amylose and amylopectin in the gelation and retrogradation of starch. *Carbohydr. Res. 135*:271–281.
37. Morris, V. J. (1994). Starch gelation and retrogradation. *Trends Food Sci. Technol. 1*:2–6.
38. Morrison, W. R., and J. Karkaas (1990). Starch, in *Methods in Plant Biochemistry,* vol. 2, *Carbohydrates* (P. M. Dey, ed.), Academic Press, San Diego, pp. 323–352.
39. Pennington, N. L., and C. W. Baker, eds. (1990). *Sugar, A User's Guide to Sucrose,* Van Nostrand Reinhold, New York.
40. Pettitt, D. (1982). Xanthan gum, in *Food Hydrocolloids,* vol. I (M. Glicksman, ed.), CRC Press, Boca Raton, FL, pp. 127–149.
41. Rolin, C. (1993). Pectin, in *Industrial Gums* (R. L. Whistler and J. N. BeMiller, eds.), Academic Press, San Diego, pp. 257–293.
42. Sand, R. E. (1982). Nomenclature and structure of carbohydrate hydrocolloids, in *Food Hydrocolloids,* vol. I (M. Glicksman, ed.), CRC Press, Boca Raton, FL, pp. 19–46.
43. Schenck, F. W., and R. E. Hebeda, eds. (1992). *Starch Hydrolysis Products,* VCH, New York.
44. Slade, L., and H. Levine (1989). A food polymer science approach to selected aspects of starch gelatinization and retrogradation, in *Frontiers in Carbohydrate Research—1,* (R. P. Millane, J. N. BeMiller, and R. Chandrasekaran, eds.), Elsevier Applied Science, London, pp. 215–270.
45. Spiller, G. A. (1993). *CRC Handbook of Dietary Fiber in Human Nutrition,* 2nd ed., CRC Press, Boca Raton, FL.
46. Staub, H. W., and R. Ali (1982). Nutritional and physiological value of gums, in *Food Hydrocolloids,* vol. I (M. Glicksman, ed.), CRC Press, Boca Raton, FL, pp. 101–121.
47. Stephen, A. M., ed. (1995). *Food Polysaccharides and Their Applications,* Marcel Dekker, New York.
48. Strickler, A. J. (1982). Corn syrup selections in food applications, in *Food Carbohydrates* (D. R. Lineback and G. E. Inglett, eds.), AVI, Westport, CT, pp. 12–24.
49. Takeda, Y., T. Shitaozono, and S. Hizukuri (1990). Structures of subfractions of corn amylose. *Carbohydr. Res. 199*:207–214.
50. Therkelsen, G. H. (1993). Carrageenan, in *Industrial Gums* (R. L. Whistler and J. N. BeMiller, eds.), Academic Press, San Diego, pp. 145–180.
51. Thomas, W. R. (1984). Microcrystalline cellulose (MCC or cellulose gel), in *Food Hydrocolloids,* vol. III (M. Glicksman, ed.), CRC Press, Boca Raton, FL, pp. 9–42.
52. Van Beynum, G. M. A., and J. A. Roels, eds. (1985). *Starch Conversion Technology,* Marcel Dekker, New York.
53. Whistler, R. L. (1993). Exudate gums, in *Industrial Gums* (R. L. Whistler and J. N. BeMiller, eds.), Academic Press, San Diego, pp. 309–339.
54. Whistler, R. L. (1993). Hemicelluloses, in *Industrial Gums* (R. L. Whistler and J. N. BeMiller, eds.), Academic Press, San Diego, pp. 295–308.

55. Whistler, R. L., and J. N. BeMiller, eds. (1993). *Industrial Gums,* 3rd ed., Academic Press, San Diego.
56. Whistler, R. L., J. N. BeMiller, and E. F. Paschall, eds. (1984). *Starch: Chemistry and Technology,* 2nd ed., Academic Press, New York.
57. Wood, P. J., ed. (1993). *Oat Bran,* American Association of Cereal Chemists, St. Paul, MN.
58. Wurzburg, O. B., ed. (1986). *Modified Starches: Properties and Uses,* CRC Press, Boca Raton, FL.
59. Yaylayan, V. A., and A. Huyghues-Despointes (1994). Chemistry of Amadori rearrangement products: Analysis, synthesis, kinetics, reactions, and spectroscopic properties. *Crit. Rev. Food Sci. Nutr. 34*:321–369.
60. Young, R. A., and R. M. Rowell, eds. (1986). *Cellulose,* John Wiley, New York.
61. Zobel, H. F. (1988). Molecules to granules: A comprehensive starch review. *Starch/Stärke 40*:44–50.

5

Lipids

Wassef W. Nawar
University of Massachusetts, Amherst, Massachusetts

5.1 INTRODUCTION

Lipids consist of a broad group of compounds that are generally soluble in organic solvents but only sparingly soluble in water. They are major components of adipose tissue, and together with proteins and carbohydrates, they constitute the principal structural components of all living cells. Glycerol esters of fatty acids, which make up to 99% of the lipids of plant and animal origin, have been traditionally called fats and oils. This distinction, based solely on whether the material is solid or liquid at room temperature, is of little practical importance and the two terms are often used interchangeably.

Food lipids are either consumed in the form of "visible" fats, which have been separated from the original plant or animal sources, such as butter, lard, and shortening, or as constituents of basic foods, such as milk, cheese, and meat. The largest supply of vegetable oil comes from the seeds of soybean, cottonseed, and peanut, and the oil-bearing trees of palm, coconut, and olive.

Lipids in food exhibit unique physical and chemical properties. Their composition, crystalline structure, melting properties, and ability to associate with water and other non-lipid molecules are especially important to their functional properties in many foods. During the processing, storage, and handling of foods, lipids undergo complex chemical changes and react with other food constituents, producing numerous compounds both desirable and deleterious to food quality.

Dietary lipids play an important role in nutrition. They supply calories and essential fatty acids, act as vitamin carriers, and increase the palatability of food, but for decades they have been at the center of controversy with respect to toxicity, obesity, and disease.

5.2 NOMENCLATURE

Lipid nomenclature can be understood more readily if simple nomenclature of the various classes of organic compounds is reviewed first. International recommendations for the nomenclature of lipids have been published [66].

5.2.1 Fatty Acids

This term refers to any aliphatic monocarboxylic acid that can be liberated by hydrolysis from naturally occurring fats.

5.2.1.1 Saturated Fatty Acids

Five accepted naming systems exist for fatty acids.

1. The acids can be named after hydrocarbons with the same number of carbon atoms (CH_3 replaced by COOH). The terminal letter *e* in the name of the parent hydrocarbon is replaced with *oic*. If the acid contains two carboxyl groups the suffix becomes *dioic* (e.g., hexanedioic). When a single carboxyl group is present it is regarded as carbon number 1, as shown here.

Hydrocarbon	Fatty acid
$CH_3CH_2CH_2CH_2CH_2CH_3$	$\overset{6}{C}H_3\overset{5}{C}H_2\overset{4}{C}H_2\overset{3}{C}H_2\overset{2}{C}H_2\overset{1}{C}OOH$
alkane	alkanoic
hexane	hexanoic

2. The acids can be called carboxylic acids with the prefix being the hydrocarbon to which the carboxyl group is attached. Thus, when a carboxyl group is attached to pentane, the fatty acid is 1-pentanecarboxylic acid. In this system carbon number 1 is the carbon atom adjacent to the terminal carboxyl group. This convention corresponds to the long-practiced use of Greek letters α, β, γ, δ, and so on, in which the α-carbon atom is that adjacent to the carboxyl carbon.

$$CH_3CH_2CH_2CH_2\overset{H}{\underset{H}{C}}-H \qquad \overset{5}{C}H_3\overset{4}{C}H_2\overset{3}{C}H_2\overset{2}{C}H_2\overset{1}{C}H_2COOH$$

Pentane — 1-Pentanecarboxylic acid

3. A common name can be used, such as butyric, stearic, or oleic.
4. Fatty acids can be represented by a simple numerical expression consisting of two terms separated by a colon, with the first term depicting the number of carbon atoms and the second the number of double bonds; for example, 4:0, 18:1, and 18:3 represent butyric, oleic, and linolenic acids, respectively.
5. In abbreviated designations for triacylglycerols, each acid can be given a standard letter abbreviation such as P for palmitic and L for linoleic.

Thus, the fatty acid $CH_3CH_2CH_2COOH$ can be referred to as 4:0, *n*-butanoic, 1-propanecarboxylic, or butyric acid. Similarly, the compound here is named 3-methylbutanoic, 2-methyl-1-propanecarboxylic, or β-methylbutyric acid.

$$CH_3-\underset{\displaystyle CH_3}{\underset{|}{C}H}-CH_2-COOH$$

5.2.1.2 Unsaturated Fatty Acids

As in the case of the saturated fatty acids, unsaturated acids can be named after the parent unsaturated hydrocarbons. Replacement of the terminal *anoic* by *enoic* indicates unsaturation, and the *di, tri,* and so on represent the number of double bonds present. Hence, we have hexadecenoic for 16:1, octadecatrienoic for 18:3, and so on.

The simplest way to specify the location of double bonds is to put, before the name of the acid, one number for each unsaturated linkage. Oleic acid, for example, with one double bond between carbons 9 and 10, is named 9-octadecenoic acid. In certain cases it is convenient to distinguish unsaturated fatty acids by the location of the first double bond from the methyl end of the molecule, that is, the omega carbon. Linoleic acid (9,12-octadecadienoic acid) is therefore an 18:2ω6 (or n-6) acid.

The geometric configuration of double bonds is usually designated by the use of cis (Latin, on this side), and trans (Latin, across), indicating whether the alkyl groups are on the same or opposite sides of the molecule.

```
R1   H          R1   H
  \ /             \ /
   C               C
   ||              ||
   C               C
  / \             / \
R2   H           H   R2

 cis-            trans-
```

The cis configuration is the naturally occurring form, but the trans configuration is thermodynamically favored. Linoleic acid, with both double bonds in the cis configuration, is named *cis*-9, *cis*-12-octadecadienoic acid. Difficulty arises, however, if the four attached groups are all different as in this compound:

```
R1   H
  \ /
   C
   ||
   C
  / \
R2   R3
```

In this case the two atoms or groups attached to each carbon are assigned "priorities" in accord with the Cahn–Ingold–Prelog procedure (described in Sec. 5.2.2.1, under *R/S* System). If the high-priority groups (greater atomic number) lie on the same side of both carbons, the letter *Z* (German, *zusammen*) is used to designate the configuration. If the two high-priority groups are on opposite sides, the letter *E* (German, *entgegen*) is used.

Table 1 gives a list of some of the fatty acids commonly found in natural fats, with various designations for each. Table 2 lists some uncommon, naturally occurring, polyunsaturated fatty acids that are claimed to offer attractive applications in nutrition, cosmetics, and pharmacology.

5.2.2 Acylglycerols

Neutral fats are mono-, di-, and triesters of glycerol with fatty acids, and are termed monoacylglycerols, diacylglycerols, and triacylglycerols, respectively. Use of the old terms mono-, di-, and triglycerides is discouraged. The compound shown here can be named any of the following: tristearoylglycerol, glycerol tristearate, tristearin, or StStSt.

$$
\begin{array}{r}
CH_2OOC(CH_2)_{16}CH_3 \\
| \\
CH_3(CH_2)_{16}COOCH \\
| \\
CH_2OOC(CH_2)_{16}CH_3
\end{array}
$$

Although glycerol by itself is a completely symmetrical molecule, the central carbon atom acquires chirality (asymmetry) if one of the primary hydroxyl groups (on carbons 1 and 3) is esterified, or if the two primary hydroxyls) are esterified to different acids. Several methods have been used to specify the absolute configuration of glycerol derivatives.

TABLE 1 Nomenclature of Some Common Fatty Acids

Abbreviation	Systematic name	Common name	Symbol
4:0	Butanoic	Butyric	B
6:0	Hexanoic	Caproic	H
8:0	Octanoic	Caprylic	Oc
10:0	Decanoic	Capric	D
12:0	Dodecanoic	Lauric	La
14:0	Tetradecanoic	Myristic	M
16:0	Hexadecanoic	Palmitic	P
16:1 (n-7)	9-Hexadecenoic	Palmitoleic	Po
18:0	Octadecanoic	Stearic	St[a]
18:1 (n-9)	9-Octadecenoic	Oleic	O
18:2 (n-6)	9,12-Octadecadienoic	Linoleic	L
18:3 (n-3)	9,12,15-Octadecatrienoic	Linolenic	Ln
20:0	Arachidic	Eicosanoic	Ad
20:4 (n-6)	5,8,11,14-Eicosatetraenoic	Arachidonic	An
20:5 (n-3)	5,8,11,14,17-Eicosapentaenoic	EPA	
22:1 (n-9)	13-Docosenoic	Erucic	E
22:5 (n-3)	7,10,13,16,19-Docosapentaenoic		
22:6 (n-6)	4,7,10,13,16,19-Docosahexaenoic	DHA	

[a]Some authors use S for stearic, but this can be confusing, since S is also used for "saturated" whenever triaclycerol composition is expressed in terms of saturated (S) and unsaturated (U) fatty acids. For example, S_3 or SSS = all three fatty acids saturated, SU_2 or SUU = diunsaturated–monosaturated, and so on.

5.2.2.1 *R/S* System

Use of the prefixes *R* and *S* was proposed by Cahn et al. [17]. A sequence of priority is assigned to the four atoms or groups of atoms attached to a chiral carbon, with the atom of greatest atomic number assigned the highest priority. The molecule is oriented so that the group of lowest priority is directed straight away from the viewer, and the remaining groups are directed toward the viewer in a tripodal fashion. If the direction of decrease in order of priority is clockwise, the configuration is *R* (Latin, *rectus*); if counterclockwise, it is *S* (Latin, *sinister*). To apply this system to acylglycerols, it is helpful to consider the structures in Figure 1. In both instances, the H atom on asymmetric carbon 2 is the substituent of lowest priority and thus is depicted as being beneath the plane of the paper. Among the substituents of carbon 2, oxygen has the highest rank. The remaining two substituents are -CO- (hydrogens on carbon 1 and 3 are disregarded since O

TABLE 2 Polyunsaturated Fatty Acids of Biological Significance in Oils from Various Plant Sources (%)

Fatty acid	Black currant	Borage	Evening primrose
18:2 n-6 (LA)	48	38	72
18:3 n-6 (GLA)	17	23	9
18:3 n-3 (α)	13	—	—
18:4 n-3 (SA)	4	—	—

Note: LA, linoleic; GLA, gamma-linolenic; α, alpha-linolenic; SA, stearidonic.

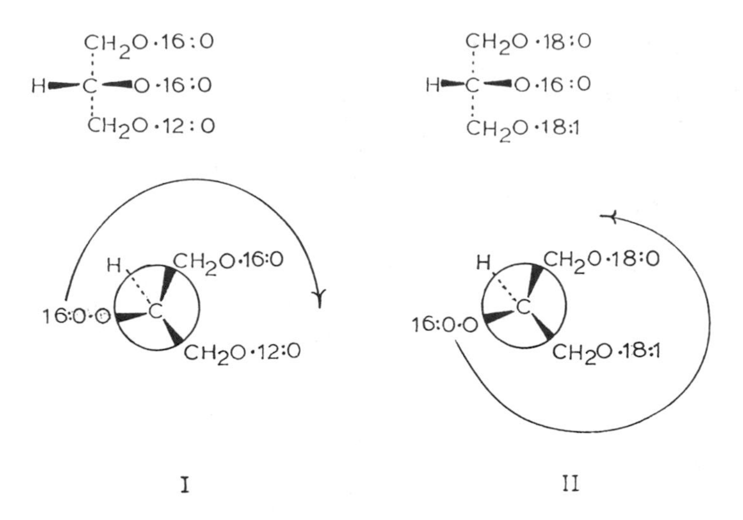

FIGURE 1 Application of the *R/S* system of nomenclature for triacylglycerols.

has a higher priority). Thus, a comparison must be made of the atoms attached to these -CO- groups. Long saturated acyls have higher priority than short. Unsaturated chains outrank saturated chains, a double bond outranks single branching, two double bonds outrank one, cis outranks trans, and a branched chain outranks an unbranched chain. Accordingly, the groupings in configuration I have a clockwise arrangement in order of decreasing priority, and thus configuration I is regarded as *R*. Similarly, configuration II is *S*.

Although the *R/S* system does convey the stereochemical configuration of acylglycerols, it is obvious that the designated structure depends on the nature of the acyl groups in positions 1 and 3 of the triacylglycerol molecule. Consequently, this system cannot be applied in situations in which these positions contain mixtures of fatty acids, as in the case of natural fats, unless individual triacylglycerols are separated.

5.2.2.2 Stereospecific Numbering

The *sn* system, as proposed by Hirschmann [60], is simple, is applicable to both synthetic and natural fats and has now been universally adopted. The usual Fischer planar projection of glycerol is utilized with the middle hydroxyl group positioned on the left side of the central carbon. The carbon atoms are numbered 1 to 3 in the conventional top-to-bottom sequence, as shown:

$$
\begin{array}{ccccl}
 & & CH_2\text{–}OH & & \quad sn\text{-}1 \\
 & & | & & \\
HO & \text{—} & C & \text{—} H & \quad sn\text{-}2 \\
 & & | & & \\
 & & CH_2\text{–}OH & & \quad sn\text{-}3
\end{array}
$$

If, for example, stearic acid is esterified at the *sn*-1 position, oleic at *sn*-2, and myristic at *sn*-3, the triacylglycerol would appear as:

$$
\begin{array}{rl}
 & CH_2OOC(CH_2)_{16}CH_3 \\
 & | \\
CH_3(CH_2)_7CH{=}CH(CH_2)_7COO & CH \\
 & | \\
 & CH_2OOC(CH_2)_{12}CH_3
\end{array}
$$

and can be designated any one of the following: 1-stearoyl-2-oleoyl-3-myristoyl-*sn*-glycerol, *sn*-glycerol-1-stearate-2-oleate-3-myristate, *sn*-StOM, or *sn*-18:0-18:1-16:0.

The following prefixes are now widely used with abbreviations to designate the positional distribution of fatty acids within triacylglycerol molecules.

sn: Used immediately preceding the term "glycerol," indicates that the *sn*-1, *sn*-2, and *sn*-3 positions are listed in that order.

rac: Racemic mixture of two enantiomers. This specifies that the middle acid in the abbreviation is attached at the *sn*-2 position, and that the remaining two acids are equally divided between *sn*-1 and *sn*-3 (e.g., rac-StOM indicates equal amounts of *sn*-StOM and *sn*-MOSt).

β-: Middle acid in the abbreviation is at the *sn*-2 position but positioning of the remaining two is unknown (e.g., β-StOM indicates a mixture of *sn*-StOM and *sn*-MOSt in any proportion).

No prefix is given in case of monoacid acylglycerols (e.g., MMM) or if the positional distribution of the acids is unknown, and hence any mixture of isomers is possible (e.g., StOM indicates a possible mixture of *sn*-StOM, *sn*-MOSt, *sn*-OStM, *sn*-MStO, *sn*-StMO, and *sn*-OMST in any proportion).

5.2.3 Phospholipids

The term "phospholipid" may be used for any lipid containing phosphoric acid as a mono- or diester. "Glycerophospholipid" signifies any derivative of glycerophosphoric acid that contains an *O*-acyl, *O*-alkyl, or *O*-alkenyl group attached to the glycerol residue. Thus all phosphoglycerols contain a polar head (hence the term polar lipids) and two hydrocarbon tails. These compounds differ from one another in the size, shape, and polarity of the alcohol component of their polar head. The two fatty acid substituents also vary. Usually one is saturated while the other unsaturated and mainly located in the *sn*-2 position.

The common glycerophospholipids are named as derivatives of phosphatidic acid, such as 3-*sn*-phosphatidylcholine (common name, lecithin), or by their systematic name, similar to the system for triacylglycerols. The term "phospho" is used to indicate the phosphodiester bridge; for example, 1-stearoyl-2-linoleoyl-*sn*-glycero-3-phosphocholine is the designation for this compound:

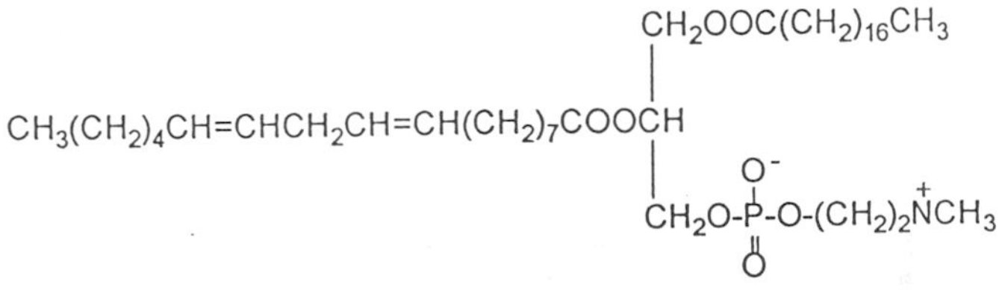

CH2OOC(CH2)16CH3
CH3(CH2)4CH=CHCH2CH=CH(CH2)7COOCH
O-
CH2O-P-O-(CH2)2NCH3
O

In addition to phosphatidycholine, phosphatidylethanolamine, phosphatidylserine, and phosphatidylinositol are commonly found in food.

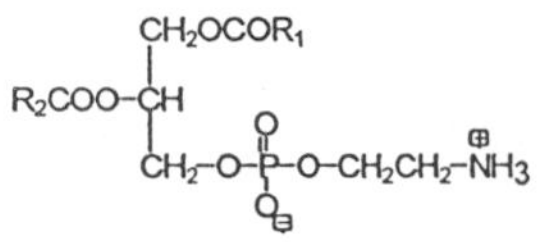

CH2OCOR1
R2COO-CH
CH2-O-P-O-CH2CH2-NH3

3-sn-Phosphatidylethanolamine
(PE)

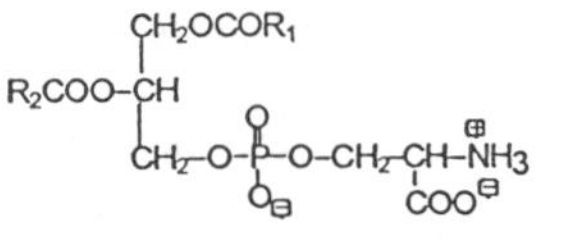

CH2OCOR1
R2COO-CH
CH2-O-P-O-CH2-CH-NH3
COO

3-sn-Phosphatidylserine
(PS)

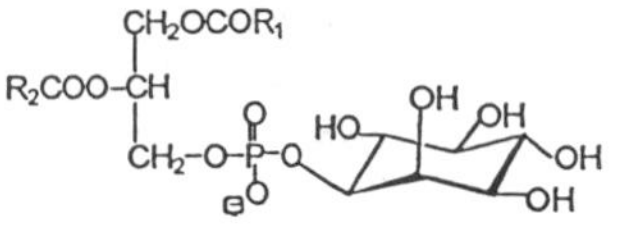

CH2OCOR1
R2COO-CH
CH2-O-P-O
HO
OH
OH
OH
OH

3-sn-Phosphatidylinositol
(PI)

5.3 CLASSIFICATION

A general classification of lipids based on their structural components is presented in Table 3. Such a classification, however, is possibly too rigid for a group of compounds so diverse as lipids, and should be used only as a guide. It should also be recognized that other classifications may sometimes be more useful. For example, the sphingomyelins can be classed as phospholipids because of the presence of phosphate. The cerebrosides and the gangliosides can also be classed as glycolipids because of the presence of carbohydrate, and the sphingomyelins and the glycolipids can be classed as sphingolipids because of the presence of sphingosine.

The most abundant class of food lipids is the acylglycerols, which dominate the composition of depot fats in animals and plants. The polar lipids are found almost entirely in cellular membranes (phospholipids being the main components of the bilayer) with only very small amounts in depot fats. In some plants, glycolipids constitute the major polar lipids in cell membranes. Waxes are found as protective coatings on skin, leaves, and fruits. Edible fats are traditionally classified into the following subgroups.

5.3.1 Milk Fats (See also Chap. 14)

Fats of this group are derived from the milk of ruminants, particularly dairy cows. Although the major fatty acids of milk fat are palmitic, oleic, and stearic, this fat is unique among animal fats in that it contains appreciable amounts of the shorter chain acids C4 to C12, small amounts of branched and odd-numbered acids, and trans-double bonds.

5.3.2 Lauric Acids (See also Chap. 16)

Fats of this group are derived from certain species of palm, such as coconut and babasu. The fats are characterized by their high content of lauric acid (40–50%), moderate amounts of C6, C8, and C10 fatty acids, low content of unsaturated acids, and low melting points.

5.3.3 Vegetable Butters (See also Chap. 16)

Fats of this group are derived from the seeds of various tropical trees and are distinguished by their narrow melting range, which is due mainly to the arrangement of fatty acids in the

TABLE 3 Classification of Lipids

Major class	Subclass	Description
Simple lipids	Acylglycerols	Glycerol + fatty acids
	Waxes	Long-chain alcohol + long-chain fatty acid
Compound lipids	Phosphoacylglycerols (or glycerophospholipids)	Glycerol + fatty acids + phosphate + another group usually containing nitrogen
	Sphingomyelins	Sphingosine + fatty acid + phosphate + choline
	Cerebrosides	Sphingosine + fatty acid + simple sugar
	Gangliosides	Sphingosine + fatty acid + complex carbohydrate moiety that includes sialic acid
Derived lipids	Materials that meet the definition of a lipid but are not simple or compound lipids	Examples: carotenoids, steroids, fat-soluble vitamins

triacylglycerol molecules. In spite of their large ratio of saturated to unsaturated fatty acids, trisaturated acylglycerols are not present. The vegetable butters are extensively used in the manufacture of confections, with cocoa butter being the most important member of the group.

5.3.4 **Oleic–Linoleic Acids** (See also Chap. 16)

Fats in this group are the most abundant. The oils are all of vegetable origin and contain large amounts of oleic and linoleic acids, and less than 20% saturated fatty acids. The most important members of this group are cottonseed, corn, peanut, sunflower, safflower, olive, palm, and sesame oils.

5.3.5 **Linolenic Acids** (See also Chap. 16)

Fats in this group contain substantial amounts of linolenic acid. Examples are soybean, rapeseed and flaxseed, wheat germ, hempseed, and perilla oils, with soybean being the most important. The abundance of linolenic acid in soybean oil is responsible for the development of an off-flavor problem known as flavor reversion.

5.3.6 **Animal Fats** (See also Chap. 15)

This group consists of depot fats from domestic land animals (e.g., lard and tallow), all containing large amounts of C16 and C18 fatty acids, medium amounts of unsaturated acids, mostly oleic and linoleic, and small amounts of odd-numbered acids. These fats also contain appreciable amounts of fully saturated triacylglycerols and exhibit relatively high melting points. Egg lipids are of particular importance because of their emulsifying properties and their high content of cholesterol.

The lipid content of whole eggs is approximately 12%, almost exclusively present in the yolk, which contains 32–36% lipid. The major fatty acids in egg yolks are 18:1 (38%), 16:0 (23%), and 18:2 (16%). Yolk lipids consist of about 66% triacylglycerols, 28% phospholipids, and 5% cholesterol. The major phopholipids of egg yolk are phosphatidylcholine (73%) and phosphatidylethanolamine (15%).

5.3.7 **Marine Oils** (See also Chap. 15)

These oils typically contain large amounts of long-chain omega-3-polyunsaturated fatty acids, with up to six double bonds, and they are usually rich in vitamins A and D. Because of their high degree of unsaturation, they are less resistant to oxidation than other animal or vegetable oils.

5.4 PHYSICAL ASPECTS

5.4.1 Theories of Triacylglycerol Distribution Patterns

5.4.1.1 Even or Widest Distribution

This theory resulted from a series of systematic studies initiated by Hilditch and Williams on the triacylglycerol composition of natural fats [59]. Component fatty acids in the triacylglycerol molecules of natural fats tend to be distributed as broadly as possible. If this is conformed to, when an acid, S, forms less than one-third of the total fatty acids present, it should not appear more than once in any triacylglycerol. If X refers to the other acids present only XXX and SXX species should be found. If an acid forms between one-third and two-thirds of the total fatty acids, it should, according to this theory, occur at least once but never three times in any one molecule; that is, only SXX and SSX should be present. If an acid forms more than two-thirds of the total acids, it should occur at least twice in every molecule; that is, only SSX and SSS should be present.

Shortcomings of the even distribution theory were soon realized. Analysis of many natural fats, especially those of animal origin, revealed marked deviation from theory. Trisaturated acylglycerols were found in fats containing less than 67% saturated fatty acids. Also, the theory can be applied only to two component systems and does not take into account positional isomers. Thus, this theory is no longer considered valid.

5.4.1.2 Random (21,2,3-Random) Distribution

According to this theory, fatty acids are distributed randomly both within each triacylglycerol molecule and among all the triacylglycerols. Thus, the fatty acid composition of all three positions should be the same and also equivalent to the fatty acid composition of the total fat. The proportion of any given fatty acid species expected on the basis of this theory can be calculated according to the equation.

$$\begin{aligned}\%sn\text{-XYZ} &= (\text{mol\% X in total fat}) \times (\text{mol\% Y in total fat}) \\ &\quad \times (\text{mol\% Z in total fat}) \times 10^{-4}\end{aligned}$$

where X, Y, and Z are the component fatty acids at positions 1, 2, and 3, respectively, of the acylglycerol. For example, if a fat contains 8% palmitic acid, 2% stearic, 30% oleic, and 60% linoleic, 64% triacylglycerol species ($n = 4$, $n^3 = 64$) would be predicted. The following examples illustrate calculations for such species.

$$\begin{aligned}\%sn\text{-OOO} &= 30 \times 30 \times 30 \times 10^{-4} = 2.7 \\ \%sn\text{-PLSt} &= 8 \times 60 \times 2 \times 10^{-4} = 0.096 \\ \%sn\text{-LOL} &= 60 \times 30 \times 60 \times 10^{-4} = 10.8\end{aligned}$$

Most fats do not conform to a completely random distribution pattern. For example, the proportion of fully saturated triacylycerols expected on the basis of random distribution exceeds, in many cases, that found experimentally. Modern techniques of analysis have revealed that in natural fats the fatty acid composition of the *sn*-2 position is always different from that of the combined 1,3 positions. Calculations of random distribution are, however, useful for understanding other hypotheses and for predicting distribution patterns of fatty acids in fats randomized by interesterification.

5.4.1.3 Restricted Random

According to this hypothesis, first proposed by Kartha [77], saturated and unsaturated fatty acids in animal fats are distributed randomly. However, fully saturated triacylglycerols (SSS) can be present, but only to the extent that the fat can remain fluid in vivo. Excess SSS, according to this theory, can be exchanged with UUS and UUU to form SSU and SUU. Kartha's calculations do not account for positional isomers or positioning of individual acids.

5.4.1.4 1,3-Random-2-Random

The fatty acid composition at the 2 position is known to be different from that of the 1 or 3 positions. This theory assumes that two different pools of fatty acids are separately and randomly esterified to the 2 and 1,3 positions. Thus, composition at positions 1 and 3 will, presumably, be identical.

On the basis of this hypothesis, the amount of a given triacylglycerol can be computed as

$$\%sn\text{-XYZ} = (\text{mol\% X at 1,3}) \times (\text{mol\% Y at 2}) \times (\text{mol\% Z at 1,3}) \times 10^{-4}$$

Composition of the *sn*-2 and/or the combined 1,3 positions can be obtained by analysis of the mono- or diacylglycerols derived from partial deacylation by chemical or enzymic methods [27,98].

Chemical Deacylation

Representative diacylglycerols are prepared by deacylation with a Grignard reagent (e.g., ethyl magnesium bromide). The reagent reacts randomly with one of the ester linkages of the triacylglycerol molecules to produce a diacylglycerol and a tertiary alcohol:

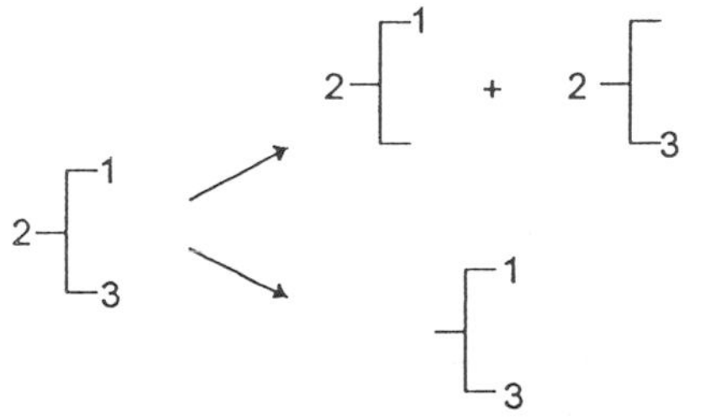

2-[1, 3] → 2-[1] + 2-[3]
2-[1, 3] → [1, 3]

The reaction is stopped with acetic acid at a point of maximum diacylglycerol yield, and the *sn*-1,2(2,3)- and *sn*-1,3-diacylglycerols are separated by thin-layer chromatography. Fatty acid composition of the combined 1 and 3 positions can be obtained by direct analysis of the 1,3-diacylglycerols, while that of the *sn*-2 position can be deduced by difference.

$$\%\text{ X at sn-2-position} = 3\left[\begin{matrix}\%\text{ X in origional}\\ \text{fat}\end{matrix}\right] - 2\left[\begin{matrix}\%\text{ X in sn-1,3-}\\ \text{diacylglycerols}\end{matrix}\right]$$

Alternatively, the compositions of the *sn*-1,3 position can be calculated as follows:

$$\%\text{ X at sn-2-position} = 4\left[\begin{matrix}\%\text{ X in sn-1,2(2,3)}\\ \text{diacylglycerols}\end{matrix}\right] - 3\left[\begin{matrix}\%\text{ X in original}\\ \text{triacylglycerols}\end{matrix}\right]$$

$$\%\text{ X at sn-1,3-position} = 3\left[\begin{matrix}\%\text{ X in original}\\ \text{triacylglycerols}\end{matrix}\right] - 2\left[\begin{matrix}\%\text{ X in sn-1,2(2,3)}\\ \text{diacylglycerols}\end{matrix}\right]$$

Enzymatic Deacylation

Pancreatic lipase selectively hydrolyzes the primary ester bonds of triacylglycerols.

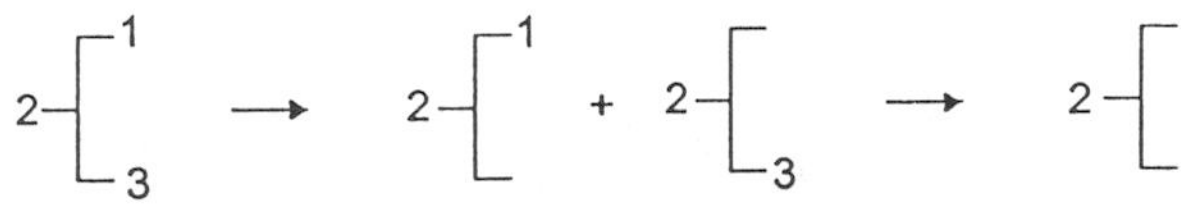

2-[1, 3] → 2-[1] + 2-[3] → 2-[

To achieve rapid hydrolysis and minimum acyl migration, the reaction is carried out at 37°C and pH 8 with vigorous agitation. Compositional requirements include the presence of calcium, an emulsifier, and a large enzyme:sample ratio [98].

Fatty acid composition of the monoacylglycerols produced gives the composition at the *sn*-2 position, allowing composition at the combined 1,3 position to be calculated.

$$\%\text{ X at 1,3-positions} = \frac{3\left[\begin{matrix}\%\text{ X in original}\\ \text{triacylglycerols}\end{matrix}\right] - \left[\%\text{ X at 2-position}\right]}{2}$$

5.4.1.5 1-Random-2-Random-3-Random

According to this theory, three different pools of fatty acids are separately but randomly distributed at each of the three positions of the triacylglycerol molecules of a natural fat. Thus, the possibility of a given fatty acid appearing in each *sn* position would likely be different. The content of a given triacylglycerol species can be calculated as

$$\%sn\text{-XYZ} = (\text{mol\% X at } sn\text{-1}) \times (\text{mol\% Y at } sn\text{-2}) \times (\text{mol\% Z at } sn\text{-3}) \times 10^{-4}$$

To calculate the range of molecular species expected to be present in a natural fat on the basis of this theory, the fatty acid compositions of the *sn*-1 and the *sn*-3 position must be distinguished [27]. Two techniques are available:

The Method of Brockeroff [15]

The *sn*-1,2(2,3) diacylglycerols, prepared using pancreatic lipase, are reacted with phenyl dichlorophenol to produce a mixture of *sn*-1,2-diacyl-3-phosphatidylphenol and *sn*-2,3-diacyl-1-phosphatidylphenol. Incubation with phospholipase A liberates fatty acids from the 2 position of the *sn*-3-phosphatide but leaves the 2 position of the *sn*-1-phosphatide intact.

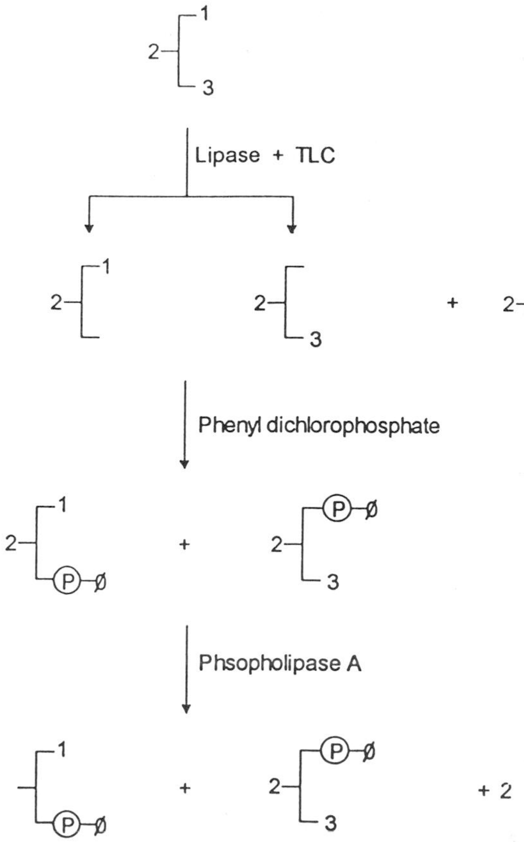

Lipase + TLC

Phenyl dichlorophosphate

Phsopholipase A

Fatty acid analysis of the lysophosphatide gives composition of the *sn*-1-position. Analysis of the free acids liberated from phospholipase hydrolysis gives composition of the *sn*-2 position. Composition of the *sn*-3 position can be computed:

$$sn\text{-3} = 2(\text{unhydrolyzed phosphatide}) - \text{monoglyceride}$$

The Method of Lands [93]

The *sn*-1,2(2,3)-diacylglycerols, prepared by reaction with pancreatic lipase or a Grignard reagent, are incubated with diacylglycerol kinase, which phosphorylates *sn*-1,2-diacylglycerols but not *sn*-2,3-diacylglycerols.

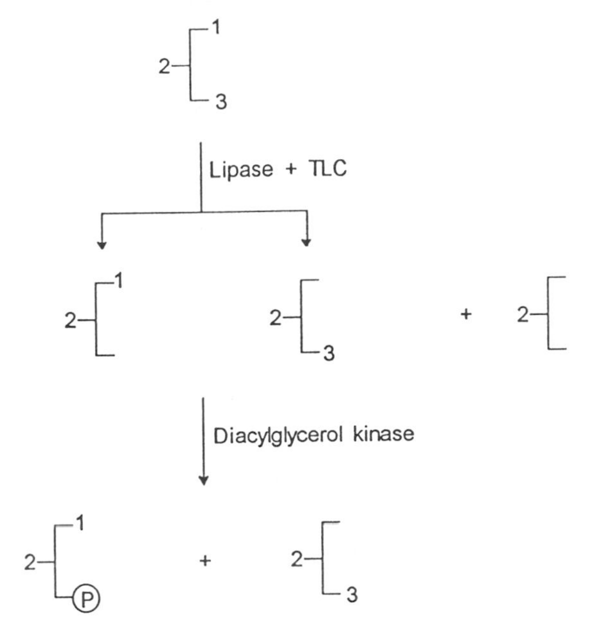

Fatty acid analysis of each of the three triacylglycerol positions can be computed from analysis of the resulting phosphatides, the monoacylglycerols (from lipase hydrolysis), and the total fat, according to the following equations:

sn-1 = 2(phosphatide) − (monoglyceride)
sn-2 = monoglyceride
sn-3 = 3(total fat) − 2(phosphatide)

5.4.2 Positional Distribution of Fatty Acids in Natural Fats

In earlier studies, major classes of triacylglycerols were separated on the basis of unsaturation (i.e., trisaturated, disaturated, diunsaturated, and triunsaturated) via fractional crystallization and oxidation-isolation methods. More recently, the techniques of stereospecific analysis made possible the detailed determinations of individual fatty acid distribution in each of the three positions of the triacylglycerols of many fats. The data listed in Table 4 clearly indicate the differences in distribution patterns among plant and animal fats.

5.4.2.1 Plant Triacylglycerols

In general, seed oils containing common fatty acids show preferential placement of unsaturated fatty acids at the *sn*-2 position. Linoleic acid is especially concentrated at this position. The saturated acids occur almost exclusively at the 1,3 positions. In most cases, the individual saturated or unsaturated acids are distributed in approximately equal quantities between the *sn*-1 and the *sn*-3 positions.

The more saturated fats of plant origin show a different distribution pattern. Approximately 80% of the triacylglycerols in cocoa butter are disaturated, with 18:1 concentrated in the

TABLE 4 Positional Distribution of Individual Fatty Acids in Triacylglycerols of Some Natural Fats

		Fatty acid (mol%)														
Source	Position	4:0	6:0	8:0	10:0	12:0	14:0	16:0	18:0	18:1	18:2	18:3	20:0	20:1	22:0	24:0
Cow's milk	1	5	3	1	3	3	11	36	15	21	1					
	2	3	5	2	6	6	20	33	6	14	3					
	3	43	11	2	4	3	7	10	4	15	0.5					
Coconut	1		1	4	4	39	29	16	3	4						
	2		0.3	2	5	78	8	1	0.5	3	2					
	3		3	32	13	38	8	1	0.5	3	2					
Cocoa butter	1							34	50	12	1					
	2							2	2	87	9					
	3							37	53	9						
Corn	1							18	3	28	50					
	2							2		27	70					
	3							14	31	52	1					
Soybean	1							14	6	23	48	9				
	2							1		22	70	7				
	3							13	6	28	45	8				
Olive	1							13	3	72	10	0.6				
	2							1		83	14	0.8				
	3							17	4	74	5	1				
Peanut	1							14	5	59	19		1	1		1
	2							2		59	39					0.5
	3							11	5	57	10		4	3	6	3
Beef (depot)	1						4	41	17	20	4	1				
	2						9	17	9	41	5	1				
	3						1	22	24	37	5	1				
Pig (outer back)	1						1	10	30	51	6					
	2						4	72	2	13	3					
	3								7	73	18					

2 position and saturated acids almost exclusively located in the primary positions (β-POSt constitutes the major species). There is approximately 1.5 times more oleic in the *sn*-1 than in the *sn*-2 position.

Approximately 80% of the triacylglycerols in coconut oil are trisaturated, with lauric acid concentrated at the *sn*-2 positions, octanoic at the *sn*-3, and myristic and palmitic at the *sn*-1 positions.

Plants containing erucic acid, such as rapeseed oil, show considerable positional selectivity in placement of their fatty acids. Erucic acid is preferentially located at the 1,3 position, but more of it is present at the *sn*-3 position than at the *sn*-1 position.

5.4.2.2 Animal Triacylglycerols

Distribution patterns of fatty acids in triacylglycerols differ among animals and vary among parts of the same animal. Depot fat can be altered by changing dietary fat. In general, however, the saturated acid content of the *sn*-2 position in animal fats is greater than that in plant fats, and the difference in composition between the *sn*-1 and *sn*-2 positions is also greater. In most animal fats the 16:0 acid is preferentially esterified at the *sn*-1 position and the 14:0 at the *sn*-2 position. Short-chain acids in milk fat are selectively associated with the *sn*-3 position. The major triacylglycerols of beef fat are of the SUS type.

Pig fat is unique among animal fats. The 16:0 acid is significantly concentrated at the central position, the 18:0 acid is primarily located at the *sn*-1 position, 18:2 at the *sn*-3 position, and a large amount of oleic acid occurs at positions 3 and 1. Major triacylglycerol species in lard are *sn*-StPSt, OPO, and POSt.

The long polyunsaturated fatty acids, characteristic of marine oils, are preferentially located at the *sn*-2 position.

5.4.3 Crystallization and Consistency

5.4.3.1 Nucleation and Crystal Growth

The formation of a solid from a solution or a melt is a complicated process in which molecules must first come in contact, orient, and then interact to form highly ordered structures known as nuclei. Nucleation can be encouraged by stirring, or the nucleation process can be circumvented by seeding the supercooled liquid with tiny crystals of the type ultimately desired. As with chemical reactions, an energy barrier exists to hinder nucleation. The more complex and more stable polymorphic forms (i.e., highly ordered, compact, of high melting point) are more difficult to nucleate. Thus, stable crystals generally do not form at temperatures just below their melting points; rather, the liquid persists for some time in a supercooled state. On the other hand, the least stable, least ordered form (α) nucleates and crystallizes readily at a temperature just below its melting point.

Although the magnitude of the energy barrier to nucleation decreases as the temperature is lowered, the rate of nucleation does not increase indefinitely with decrease in temperature. At some point, decreasing the temperature increases the viscosity of the lipid sufficiently to slow the rate of nucleation.

Following nucleation, enlargement of these nuclei (crystal growth) progresses at a rate dependent mainly on the temperature.

5.4.3.2 Crystal Structure [8, 99, 131]

Most of our present knowledge regarding the crystalline structure and behavior of fats has resulted from x-ray diffraction studies. However, significant insights also have been gained from the application of other techniques, such as nuclear magnetic resonance, infrared spectroscopy, calorimetry, dilatometry, microscopy, and differential thermal analysis. Indeed recent advances in magnetic resonance imaging (MRI) will allow greater insight into the actual dynamics of the crystallization process [130].

In the crystalline state, atoms or molecules assume rigid positions forming a repeatable, highly ordered, three-dimensional pattern. If reference points representing this regularity of structure are chosen (e.g., the center of a certain atom or a convenient point in a molecule), the resulting three-dimensional arrangement in space is known as a *space lattice*. This network of points embodies all the symmetry properties of the crystal. If the points of a space lattice are joined, a series of parallel-sided "unit cells" is produced, each of which contains all elements of the lattice. A complete crystal can thus be regarded as unit cells packed side by side in space. In the example of the simple space lattice given in Figure 2, each of the 18 unit cells has an atom or molecule at each of its corners. However, since each corner is shared by eight other adjacent cells, there is only one atom (or molecule) per unit cell. It can be seen that each point in the space lattice is equivalent in its environment to all other points. The ratios a:b:c (called axial ratios), as well as the angles between the crystallographic axes OX, OY, and OZ, are constant values used to distinguish different lattice arrangements.

Organic compounds having long chains pack side by side in the crystal to obtain maximum van der Waals interaction. In the unit cell, three spacings can be identified, two short and one long. Thus, the long spacing of normal alkanes increases steadily with an increase in carbon number, but the short spacings remain constant. The molecular end groups (e.g., methyl or carboxyl) associate with each other to form planes. If the chains are tilted with respect to the base of the unit cell, the long spacing will be smaller to an extent dependent on the angle of tilt. Fatty acids tend to form double molecules oriented head to head by sharing hydrogen bonds between carboxyl groups. Consequently, the long spacings of fatty acids are almost twice as great as those of hydrocarbons of equal carbon number (Fig. 3).

When lipid mixtures of different but similar compounds are present, crystals containing more than one kind of molecule can be formed. In the case of medium- or low-molecular-weight fatty acids differing in chain length by one carbon atom, compound crystals are formed in which dissimilar pairs are bound carboxyl to carboxyl but are otherwise arranged as in crystals containing only one acid. Also common are solid solutions in which component molecules of one type are distributed at random into the crystal lattice of another. Under certain conditions slow cooling can result in formation of layer crystals in which layers of one type of crystal deposit on the surfaces of other crystals.

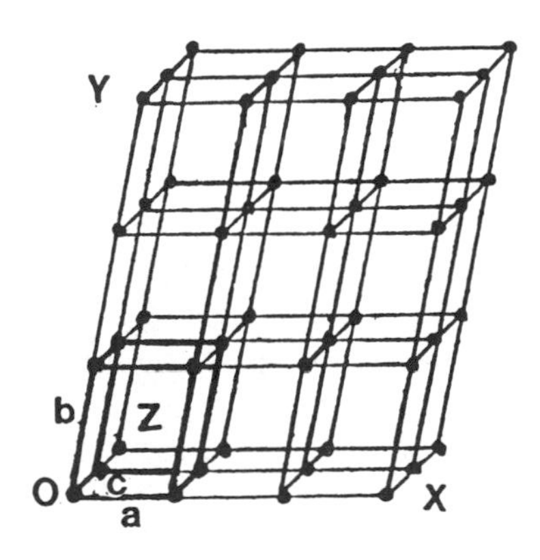

FIGURE 2 Crystal lattice.

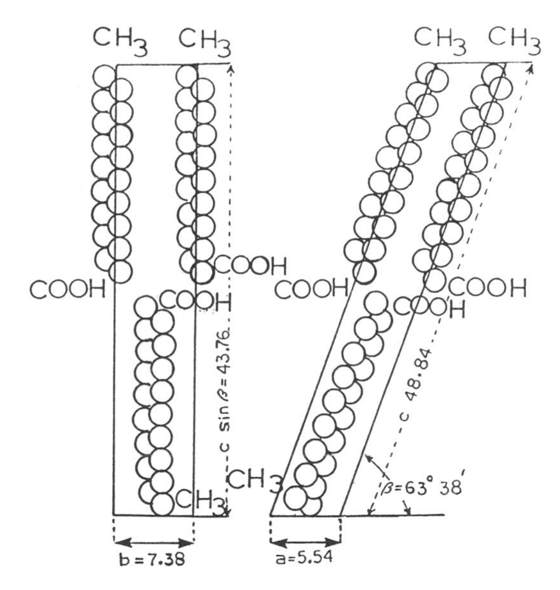

FIGURE 3 Unit cell of stearic acid. Long spacing 43.76 Å, short spacings 7.38 Å, 5.54 Å. (From Ref. 8.)

5.4.3.3 Polymorphism [8, 33, 53]

Polymorphic forms are crystalline phases of the same chemical composition that differ among themselves in structure but yield identical liquid phases upon melting. Each polymorphic form, sometimes termed polymorphic modification, is characterized by specific properties, such as x-ray spacings, specific volume, and melting point. Several factors determine the polymorphic form assumed upon crystallization of a given compound. These include purity, temperature, rate of cooling, presence of crystalline nuclei, and type of solvent.

Depending on their particular stabilities, transformation of one polymorphic form into another can take place in the solid state without melting. Two crystalline forms are said to be "monotropic" if one is stable and the other metastable throughout their existence and regardless of temperature change. Transformation will take place only in the direction of the more stable form. Two crystalline forms are "enantiotropic" when each has a definite range of stability. Either modification may be the stable one, and transformation in the solid state can go in either direction depending upon the temperature. The temperature at which their relative stability changes is known as the *transition point*. Natural fats are invariably monotropic, although enantiotropism is known to occur with some fatty acid derivatives.

With long-chain compounds, polymorphism is associated with different packing arrangements of the hydrocarbon chains or different angles of tilt. The mode of packing can be described using the subcell concept.

Subcells

A subcell is the smallest spatial unit of repetition along the chain axis within the unit cell. The schematic in Figure 4 represents a "subcell lattice" in a fatty acid crystal. In this case, each subcell contains one ethylene group and the height of the subcell is equivalent to the distance between alternate carbon atoms in the hydrocarbon chain, that is, 2.54 Å. The methyl and acid groups are not part of the subcell lattice.

Seven packing types (crystal systems) for hydrocarbon subcells are known to exist.

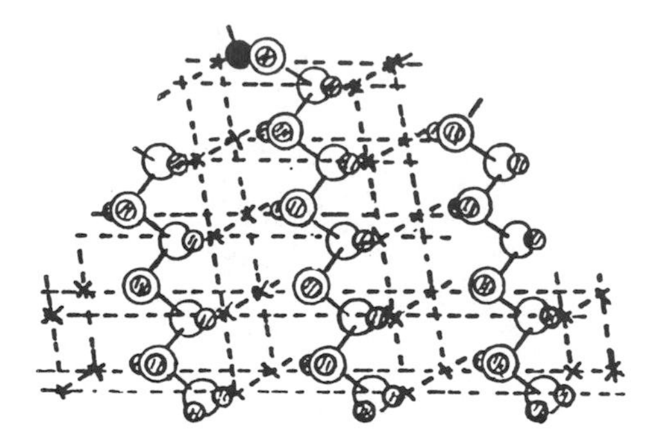

FIGURE 4 Subcell lattice in a fatty acid crystal. (From Ref. 131.)

The most common types are the three shown in Figure 5. In the triclinic (T//) system, called β, the two methylene units together make up the ethylene repeat unit of which there is one per subcell, and all zigzag planes are parallel. This subcell packing occurs in *n*-hydrocarbons, fatty acids, and triacylglycerols. It is the most stable polymorphic form.

In the common orthorhombic (O⊥) system, also called β′, there are two ethylene units in each subcell, and alternate chain planes are perpendicular to their adjacent planes. This packing occurs with *n*-paraffins and with fatty acids and their esters. The β′ form has intermediate stability.

The hexagonal (H) system, generally called α, often occurs when hydrocarbons are rapidly cooled to just below their melting points. The chains are randomly oriented and exhibit rotation about their long vertical axes. This type of packing is observed with hydrocarbons, alcohols, and ethyl esters. The α form is the least stable polymorphic form.

Fatty Acids

Even-numbered saturated fatty acids can be crystallized in any of the polymorphic forms. In decreasing length of long spacings (or increasing tilt of the chains), these forms are designated A, B, and C. Similarly, the acids with odd numbers of carbon atoms are designated A′, B′, and

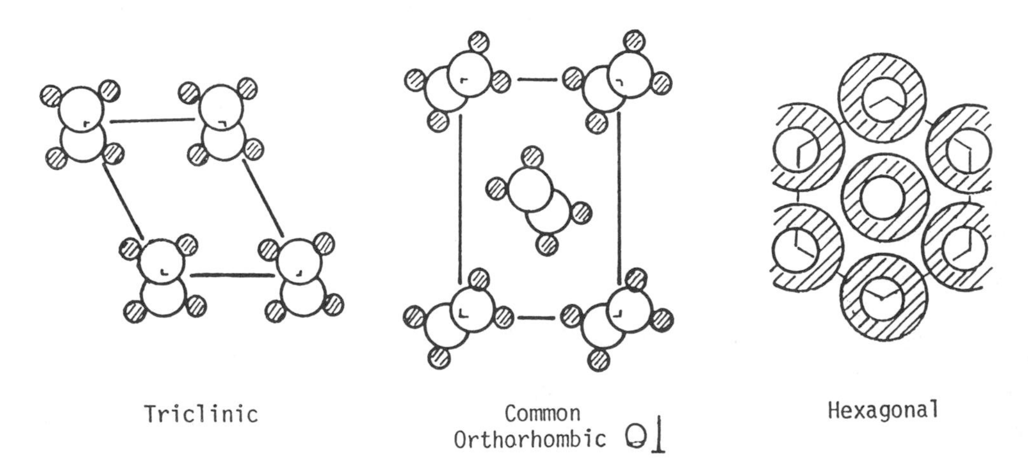

FIGURE 5 Common types of hydrocarbon subcell packing. (From Ref. 131.)

C′. The A and A′ forms have triclinic subcell chain packing (T//); the remaining forms are packed in the common orthorhombic (O⊥) manner.

The β form of stearic acid has been studied in detail. The unit cell is monoclinic and contains four molecules. Its axial dimensions are $a = 5.54$ Å, $b = 7.38$ Å, and $c = 48.84$ Å. However, the c axis is inclined at an angle of 63°38′ from the a axis, which results in a long spacing of 43.76 Å (Fig. 3).

In the case of oleic acid, the low-melting form has double molecules per unit cell length, and the hydrocarbon portions around the cis double bond are tilted in opposite directions in the plane of molecules (Fig. 6).

Triacylglycerols

The nomenclature used in earlier literature to designate the different polymorphic forms of triacylglycerols is extremely confusing. Different authors used different criteria, such as melting points or x-ray spacings, as the basis for their nomenclature. Often the same symbols were used by different investigators to designate different polymorphic forms, and much debate took place regarding the number of forms exhibited by certain triacylglycerol species. Much of the disagreement was finally resolved based on results from infrared spectroscopy. In general, triacylglycerols, due to their relatively long chains, take on many of the features of hydrocarbons. They exhibit, with some exceptions, three principal polymorphic forms: α, β′, and β. Characteristics typical of each form are summarized in Table 5.

If a monoacid triacylglycerol, such as StStSt, is cooled from the melt, it crystallizes in the least dense, lowest melting form (α). On further cooling of the α form, the chains associate more compactly, and gradual transition into the β form takes place. If the α form is heated to its melting point, a transformation into the most stable β form occurs rapidly. The β′ form can also be obtained directly by cooling the melt and maintaining the temperature a few degrees above the melting point of the α form. On heating the β′ form to its melting point, some melting takes place and transition to the stable β form occurs.

The general molecular arrangement in the lattice of monoacid triacylglycerols is a double-chainlength modified tuning fork, or chair structure, as shown in Figure 7 for trilaurin. The chains in the 1 and 3 positions are oriented in a direction opposite that of the chain in the 2 position. Because natural triacylglycerols contain a variety of fatty acids, departure from the simple polymorphic classification outlined earlier is bound to occur. In the case of triacylglycerols containing different fatty acids, some polymorphic forms are more difficult to obtain than others; for example, β′ forms have been observed that have higher melting points than the β forms. PStP glycerides, as found in cottonseed hardstock, tend to crystallize in a β′ form of relatively high density. This form results in greater stiffening power when added to oils than does the expanded (snowlike) β form commonly obtained from the StStSt glycerides of soybean hardstock.

The polymorphic structure of mixed triacylglycerols is further complicated by a tendency for carbon chains to segregate according to length or degree of unsaturation to form structures in which the long spacing is made up of triple chain lengths. Various tuning-fork structures have been proposed for mixed triacylglycerols containing fatty acids of different chain lengths. If the middle chain is shorter or longer than the other two by four or more carbons, there may be a segregation of chains, as in Figure 8a. In the case of unsymmetrical triacylglycerols, a chair-type arrangement similar to Figure 8b may result. Sorting of chains may also arise on the basis of unsaturation, as in Figure 8c. Such structures are designated by a number following the Greek letter. For example, β-3 would indicate the β modification with a triple chain length (Fig. 9). In the liquid state lamellar units exist in which the triacylglycerols are arranged in a chair conformation with their hydrocarbon chains disordered. A proposed model based on x-ray and

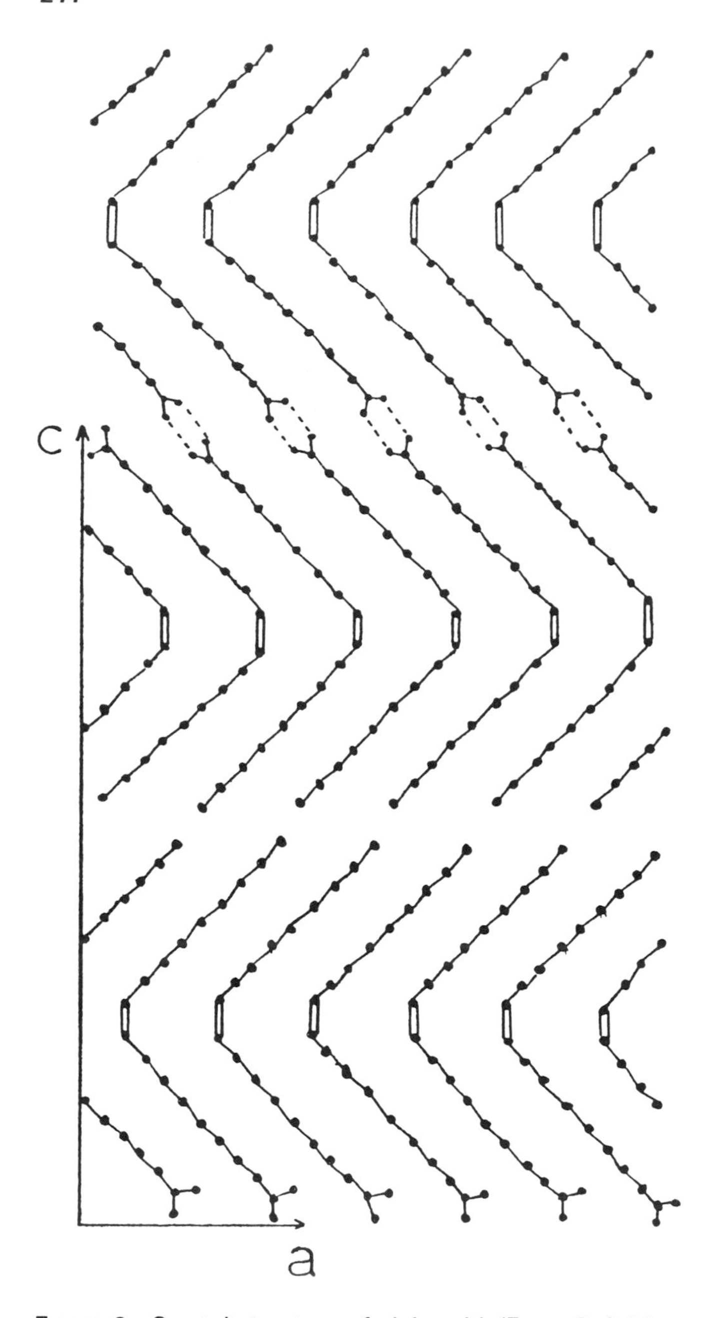

FIGURE 6 Crystal structure of oleic acid. (From Ref. 1.)

Raman spectroscopy [58] is shown in Figure 10. The size of the lamellar units decreases with increase in temperature. Upon cooling of a melt, lamellar units increase in size until crystallization occurs.

Polymorphic Behavior in Commercial Fats [26, 31, 49, 124]

It is evident from the foregoing that the polymorphic behavior of a fat is largely influenced by the composition of its fatty acids and their positional distribution in the acylglycerols. In general,

Table 5 Characteristics of the Polymorphic Forms of Monoacid Triacylglycerols

Characteristic	α Form	β′ Form	β Form
Chain packing	Hexagonal	Orthorhombic	Triclinic
Short spacing (Å)	4.15	4.2 and 3.8	4.6, 3.9, 3.7
Characteristic infrared spectrum	Single band at 720 cm^{-1}	Doublet at 727 and 719 cm^{-1}	Single band at and 717 cm^{-1}
Density	Least dense	Intermediate	Most dense
Melting point	Lowest	Medium	Highest

fats that consist of relatively few closely related triacylglycerol species tend to transform rapidly to stable β forms. Conversely, heterogeneous fats tend to transform more slowly to stable forms. For example, highly randomized fats exhibit β′ forms that transform slowly.

Fats that tend to crystallize in β forms include soybean, peanut, corn, olive, coconut, and safflower oils, as well as cocoa butter and lard. On the other hand, cottonseed, palm, and rapeseed oils, milk fat, tallow, and modified lard tend to produce β′ crystals that tend to persist for long periods. The β′ crystals are desirable in the preparation of shortenings, margarine, and baked products since they aid in the incorporation of a large amount of air in the form of small air bubbles, giving rise to products of better plastic and creaming properties.

In the food industry, emulsifiers are frequently added to stabilize certain metastable forms. It is believed that the action of such agents, such as sorbitan esters, is not due to their surface activity, but rather to their unique chemical structure, which allows them to fit into the crystallographic structure with the hydrophobic parts of the surfactant molecules parallel to the triacylglycerol hydrocarbon chains, with their hydrophilic moieties forming hydrogen bonds with the carbonyl groups of the triacylglycerols—a phenomenon called the "Button syndrome" [6]. Transformation of the fat crystals into other polymorphic forms is thus prevented or delayed without altering the crystal lattice.

In the case of cocoa butter, in which the three main glycerides are POSt (40%), STOSt (30%), and POP (15%), six polymorphic forms (I–VI) have been recognized [102, 106, 124, 145]. Form I is the least stable and has the lowest melting point. Form V is the most stable form that can be crystallized from the melted fat and is the desired structure since it produces the bright glossy appearance of chocolate coatings. Form VI has a higher melting point than form V but cannot be crystallized from the melt; it only forms by very slow transformation of form V. The V–VI transformation that occurs during storage is of particular importance since it usually coincides with the appearance of a defect called "chocolate bloom." This defect usually involves a loss of desirable chocolate gloss and the development of a dull appearance with white or grayish spots on the surface.

In addition to theories involving polymorphic transformation as a cause of chocolate bloom, it is believed that melted chocolate fat migrates to the surface and recrystallizes upon recooling, causing the undesirable appearance. Although it is clear that bloom does not depend on the appearance of a specific polymorphic form, polymorphic behavior of the cocoa fat is believed to play an important role in bloom. To delay the appearance of bloom, proper solidification of the chocolate is necessary. This is achieved by a tempering process that involves heating the cocoa butter–sugar–cocoa powder mixture to 50°C, seeding with the stable crystal "seeds," slowly crystallizing it with continuous stirring as the temperature is reduced to 26–29°C, and then slowly heating it to 32°C. Without the stable seeds, unstable crystal forms will form initially and these are likely to melt, migrate, and transform to more stable forms (bloom).

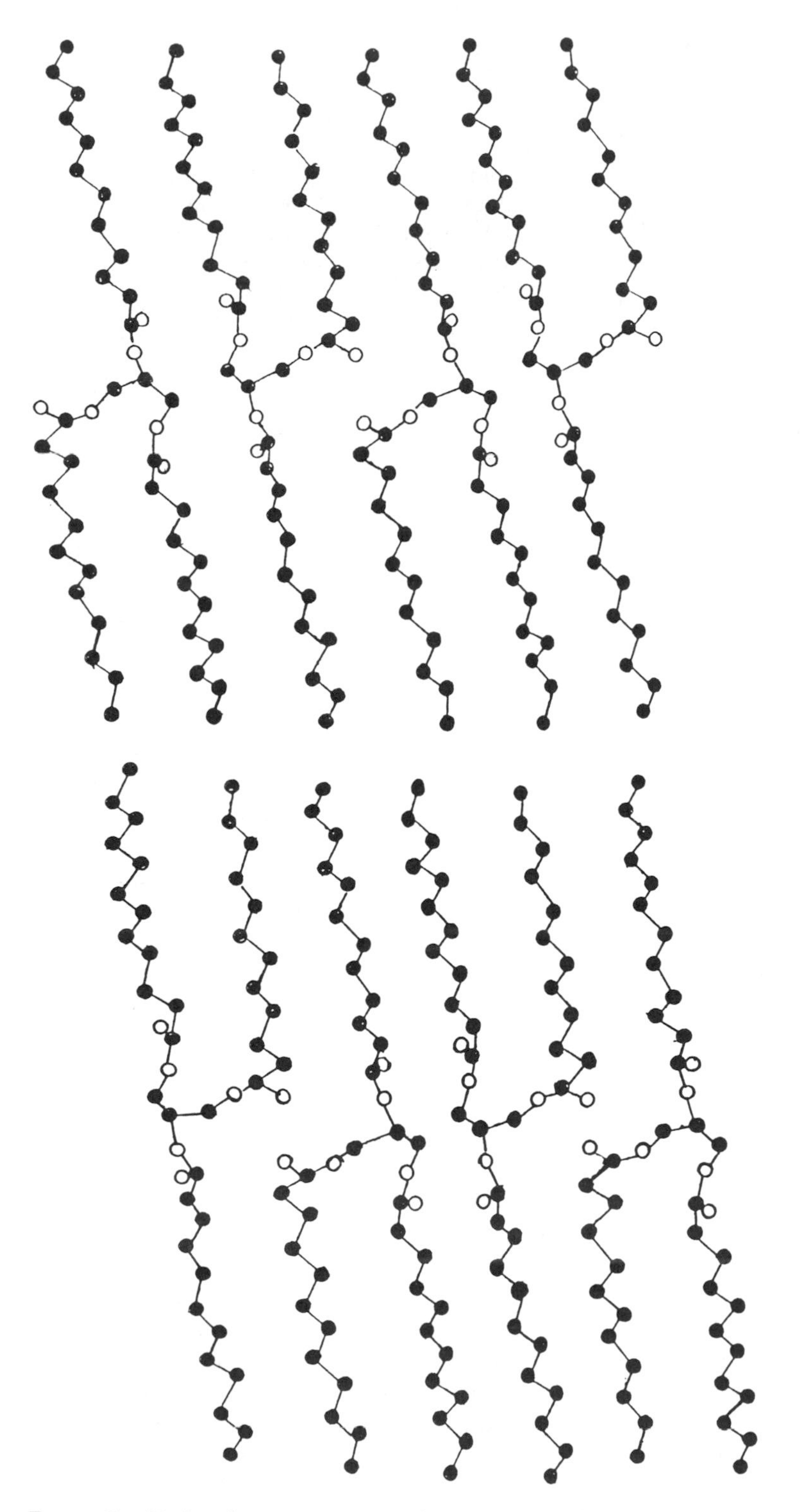

FIGURE 7 Molecular arrangement in trilaurin lattice. (From Ref. 95.)

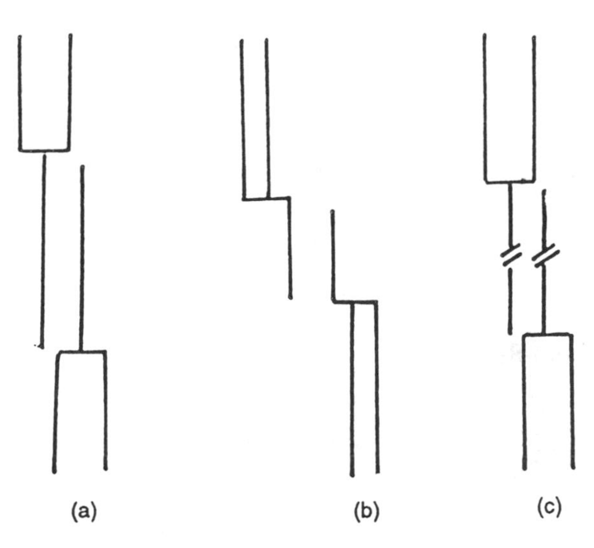

FIGURE 8 Arrangement of molecules in triacylglycerol crystals.

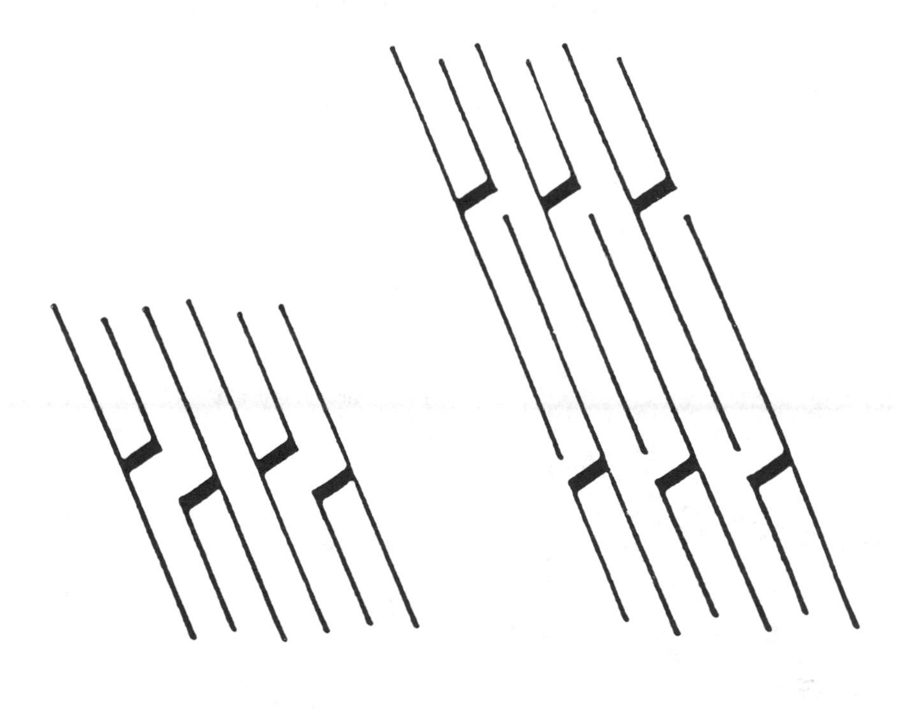

FIGURE 9 Double and triple chair arrangements of β form.

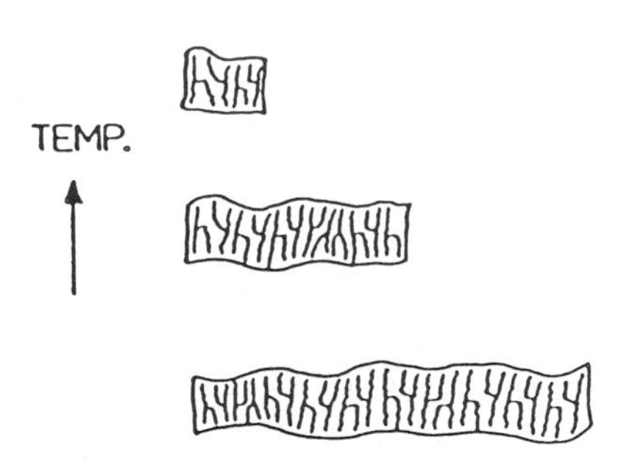

FIGURE 10 Lamellar units in the liquid. (From Ref. 58.)

Furthermore, emulsifiers have been successfully used to retard undesirable polymorphic transformation and/or migration of melted fat to the surface.

5.4.3.4 Melting [8, 53]

Shown in Figure 11 are schematic heat content (enthalpy) curves for the stable β form and the metastable α form of a simple triacylglycerol. Heat is absorbed during melting. Curve ABC represents the increase in heat content of the β form with increasing temperature. At the melting point, heat is absorbed with no rise in temperature (heat of fusion) until all solid is transformed

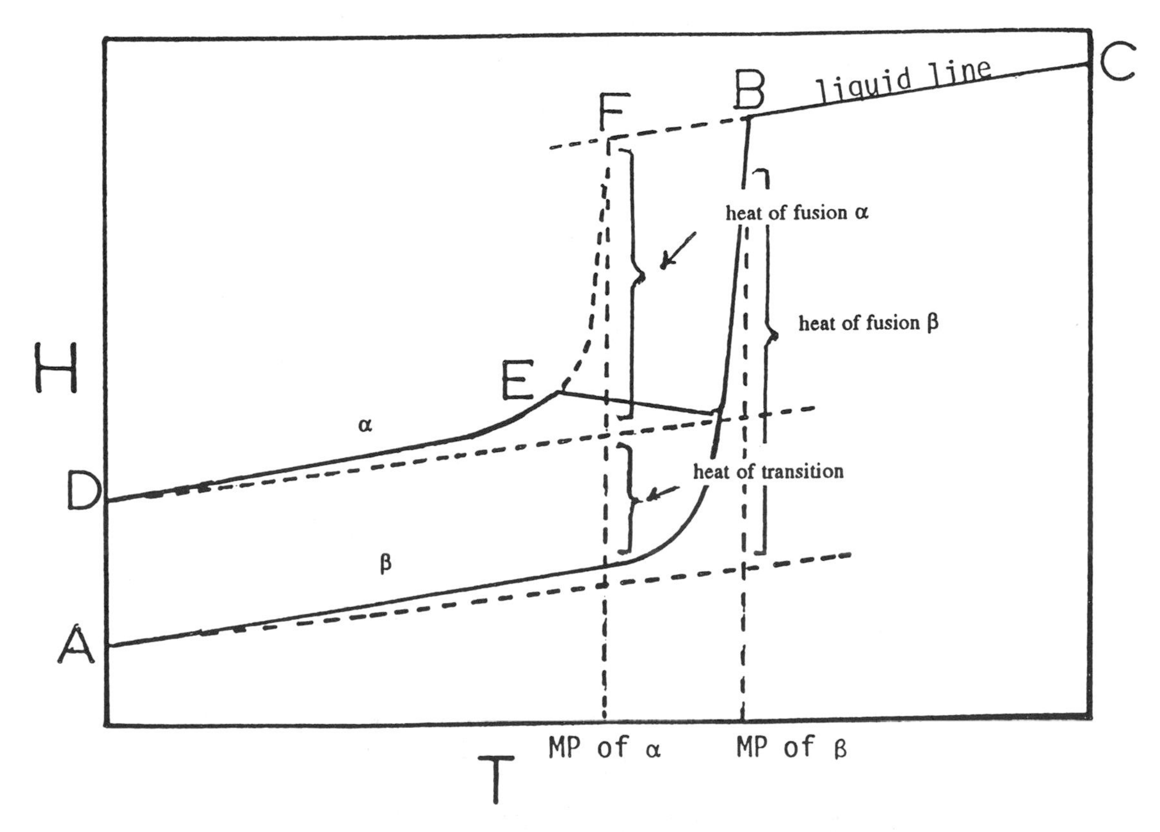

FIGURE 11 Heat content (*H*) melting curves of stable (β) and unstable (α) polymorphic forms.

into liquid (final melting point, B). On the other hand, transformation from an unstable to a stable polymorphic form (beginning at point E in Fig. 11 and intersecting with curve AB) involves evolution of heat.

Fats expand upon melting and contract upon polymorphic transformation. Consequently, if the change in specific volume (dilation) is plotted against temperature, dilatometric curves very similar to calorimetric curves are obtained, with melting dilation corresponding to specific heat. Since dilatometric measurements involve very simple instruments, they are often used in preference to calorimetric methods. Dilatometry has been used widely to determine the melting behavior of fats. If several components of different melting points are present, melting takes place over a wide range of temperature, and dilatometric or calorimetric curves similar to the schematic in Figure 12 are obtained.

Point X represents the beginning of melting; below this point the system is completely solid. Point Y represents the end of melting; above this point the fat is completely liquid. Curve XY represents the gradual melting of the solid components in the system. If the fat melts over a narrow range of temperature, the slope of the melting curve is steep. Conversely, a fat is said to have a "wide plastic range" if the difference in temperature between the beginning and the end of melting is large. Thus, the plastic range of a fat can be extended by adding a relatively high-melting and/or a low-melting component.

Solid Fat Index [67, 96, 100]

The proportion of solids and liquids in a plastic fat at different temperatures can be estimated by constructing calorimetric or melting dilation curves similar to that of Figure 12, or preferably by using nuclear magnetic resonance. Dilatometric measurements can be made starting at temperatures sufficiently low to establish a solid line, at other temperatures high enough to establish a liquid line, and at intervals between these to determine the melting curve. The solid and liquid lines can then be extrapolated, and the solid or liquid fraction at any temperature can be calculated. As shown in Figure 12, ab/ac represents the solid fraction, and bc/ac is the liquid fraction at temperature *t*. The ratio of solid to liquid is known as the solid fat index (SFI), and this has relevance to the functional properties of fats in foods.

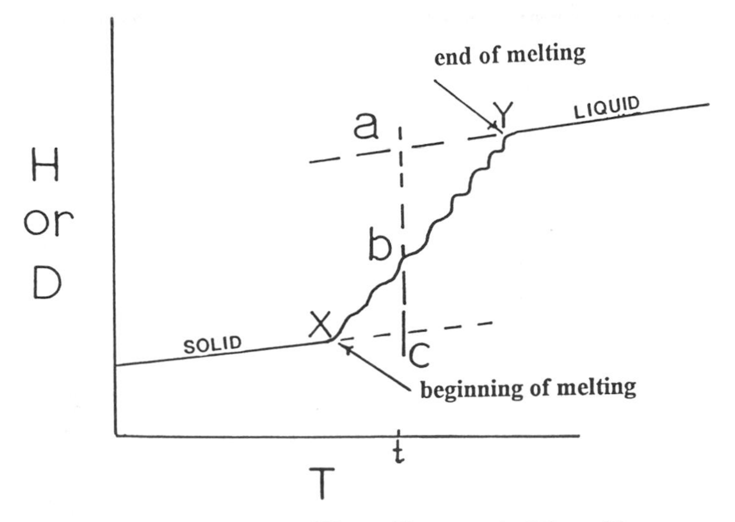

FIGURE 12 Heat content (*H*) or dilatometric (*D*) melting curve of a glyceride mixture.

Dilatometric methods for SFI determination are relatively accurate and precise but are time-consuming and applicable only below 50% SFI. Dilatometry has been largely replaced by wide-line nuclear magnetic resonance (NMR) methods, which presumably measure the fat solids directly since they involve measurement of the ratio of the number of hydrogen nuclei in the solid (which produce faster decaying signals than do those in the liquid) to the total number of hydrogen nuclei in the sample. This ratio is expressed as NMR solids percent. At present, automated pulsed NMR procedures are commonly used and are believed to offer accuracy and precision superior to wide-line NMR techniques. However, low-resolution NMR is even more appropriate for industrial use in view of its reliability and lower cost. An ultrasonic technique has also been recently proposed as an alternative or adjunct to pulsed NMR. It is based on the observation that the velocity of ultrasound is greater in solid fat than in liquid oil [64, 103, 104]. It has the advantages of lower capital cost, faster measurements, higher sensitivity to low concentrations of solid fat, and easy adaptability for online measurements.

5.4.3.5 Consistency of Commercial Fats

Although natural fats and products derived from them contain exceedingly complex mixtures of large numbers of individual acylglycerols of different composition and structure, they show a remarkable tendency to behave as simple mixtures of only a few components. Each group of similar compounds appears to act as a single component, so only the distinctly different groups are apparent in melting behavior. This tendency for simplification in the melting behavior of complex mixtures is indeed fortunate since it permits the application of rules governing simple mixtures to the more complex natural or processed fats. An initial reduction in the solidification point of certain mixtures of commercial fats has been observed upon hydrogenation or the addition of a high-melting component. Such systems give rise to eutectic-

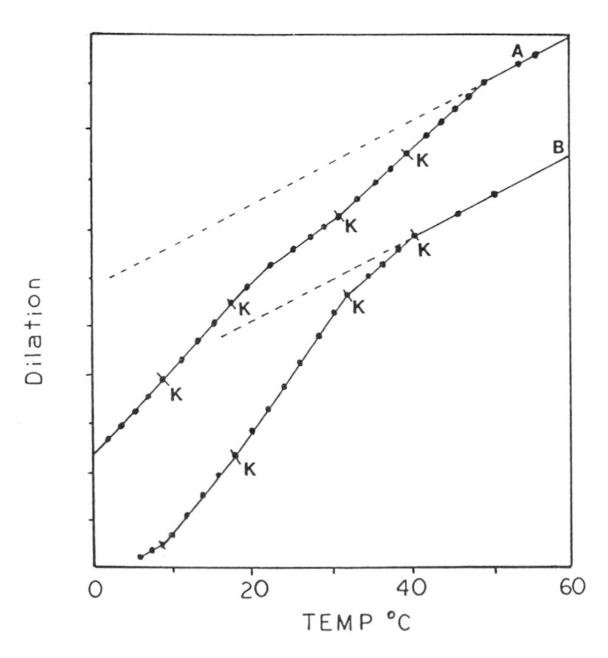

FIGURE 13 Dilatometric curves of (a) a typical all-hydrogenated shortening and (b) a typical American margarine. (From Ref. 8.)

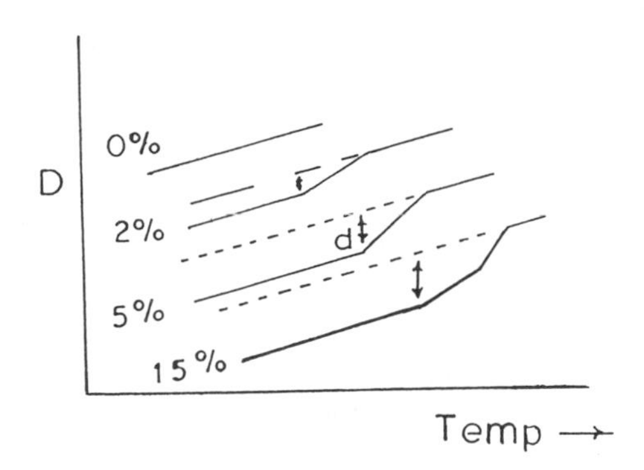

FIGURE 14 Upper portions of dilatomertric curves of cottonseed oil mixed with 0–15% highly hardened cottonseed oil. (From Ref. 8.)

type behavior and solid–liquid equilibrium curves of striking resemblance to those typical of simple binary systems.

The basic concepts involving the melting behavior of commercial fats can be simply illustrated by dilatometric curves such as those in Figures 13–15. However, as already indicated, NMR techniques give comparable information more quickly and are now the preferred methods. The dilatometric curves of plastic fats do not show a smooth melting line, but rather they exhibit a series of somewhat linear segments with visible changes in slope (Fig. 13). The inflection points between the segments are sometimes designated K points, with K being the temperature of final melting. The K points obviously correspond to specific boundaries in the phase diagrams of the complex fat; that commercial fats give rise to only a few K points

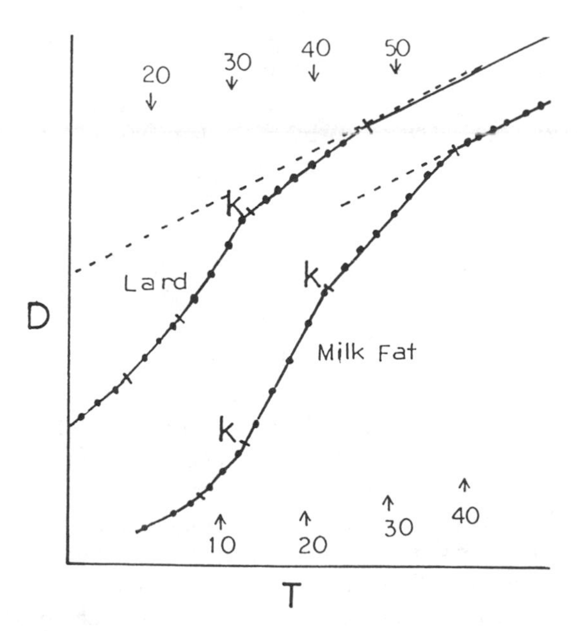

FIGURE 15 Dilatometric curves of lard and milk fat. (From Ref. 8.)

emphasizes the tendency for different individual components in a narrow melting range to behave like a single component.

If a high-melting fraction is added to a natural fat, melting of the "hard" component is clearly reflected in the dilatometric curve. The vertical distance (*d* in Fig. 14) provides a relative measure of the amount of the hard component. It can be seen from Figures 13, 14, and 15 that much useful information regarding the melting characteristics of plastic fats can be obtained from dilatometric curves.

Hard "butters" melt over a relatively narrow range of temperature because their triacylglycerols are mainly POSt, StOSt, and POP. This abrupt melting, which occurs at the temperature of the mouth, makes such fats particularly suitable for confectionary coatings.

A typical dilatometric curve for milk fat, showing almost complete melting at the temperature of the mouth, is quite different from that of lard, which exhibits a more gradual course of melting (Fig. 15). The difference in solid content of the two fats at any given temperature is clearly evident.

5.4.3.6 Mesomorphic Phase (Liquid Crystals)

As discussed earlier, crystalline lipids are highly ordered structures with regular three-dimensional arrangements of their molecules. In the liquid state, the intermolecular forces are weakened and the molecules acquire freedom of movement and assume a state of disorder. In addition, phases with properties intermediate between those of the liquid and the crystalline states are known to occur, and these mesomorphic phases consist of so-called liquid crystals. Typically, amphiphilic compounds give rise to mesomorphic phases. For example, when an amphiphilic crystalline compound is heated, the hydrocarbon region may melt before the final melting point is reached. This occurs because relatively weak van der Waals forces exist among the hydrocarbon chains as compared with the somewhat stronger hydrogen bonding that exists among polar groups. Pure crystals that form liquid crystals upon heating are said to be *thermotropic*. In the presence of water, at temperatures above the melting point of the hydrocarbon region (the so-called Krafft temperature), hydrocarbon chains of triacylglycerols transform into a disordered state and water penetrates among the ordered polar groups. Liquid crystals formed in this manner, that is, with the aid of a solvent, are said to be *lyotropic*.

Mesomorphic structure depends on factors such as concentration and chemical structure of the amphiphilic compound, water content, temperature, and the presence of other components in the mixture. The principal kinds of mesomorphic structures are lamellar, hexagonal, and cubic.

Lamellar or Neat

This structure corresponds to that existing in biological bilayer membranes. It is made up of double layers of lipid molecules separated by water (Fig. 16a). Structures of this kind are usually less viscous and less transparent than the other mesomorphic structures.

The capacity of a lamellar phase to retain water depends on the nature of the lipid constituents. Lamellar liquid crystals of monoacylglycerols, for example, can accommodate up to approximately 30% water, corresponding to a water layer thickness of about 16 Å between the lipid bilayers. However, almost infinite swelling of the lamellar phase can occur if a small amount of an ionic surface-active substance is added to the water. If the water content is increased above thc swelling limit of the lamellar phase, a dispersion of spherical aggregates consisting of concentric, alternating layers of lipid and water gradually forms.

In general, a lamellar liquid crystalline phase tends to transform upon heating into hexagonal II or cubic mesophases. On the other hand, if the lamellar liquid-crystal phase is cooled below the Krafft temperature, a metastable "gel" will form in which water remains between the lipid bilayers and the hydrocarbon chains recrystallize. During extended holding

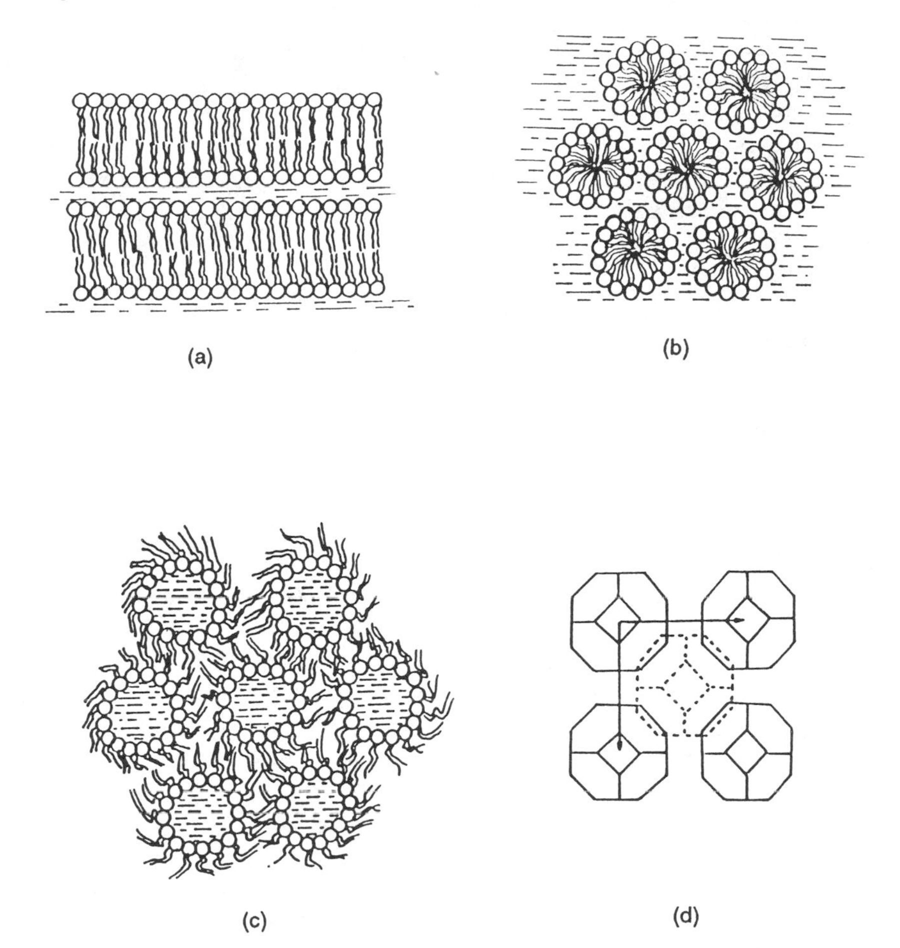

FIGURE 16 Mesomorphic structures of lipids: (a) lamellar, (b) hexagonal I, (c) hexagonal II, (d) cubic. (From Ref. 95.)

the water is expelled and the gel phase transforms into a microcrystalline suspension in water, called a coagel.

Hexagonal

In this structure the lipids form cylinders arranged in a hexagonal array. The liquid hydrocarbon chains fill the interior of the cylinders, and the space between the cylinders is taken up by water (Fig. 16b). This type of liquid crystal is termed "hexagonal I" or "middle." A reversed hexagonal structure, "hexagonal II," is also possible in which water fills the interior of the cylinders and is surrounded by the polar groups of the amphiphile. The hydrocarbon chains extend outward making up the continuous phase between the cylinders (Fig. 16c). If hexagonal I liquid crystals are diluted with water, spherical micelles form. However, dilution of hexagonal II liquid crystals with water is not possible.

Cubic or Viscous Isotropic

Although this state is encountered with many long-chain compounds, it is not as well characterized as are the lamellar and hexagonal liquid crystals. Cubic-phase structures were studied in monoacylglycerol–water systems by Larsson [94, 95], who observed the existence of closed water regions in these systems and proposed a cubic-phase model based on space-filling polyhedra packed in a body-centered cubic lattice (Fig. 16d). Cubic phases are usually very viscous and completely transparent.

Mesomorphic states are of great importance in many physiological processes; for example, they influence the permeability of cell membranes [148]. Liquid crystals also play a significant role with regard to the stability of emulsions.

5.4.3.7 Factors Influencing Consistency [8]

Several factors have important influences on the consistency of commercial fats:

Proportion of solids in the fat. In general, the greater the solids content, the firmer the fat. It has been estimated that plastic commercial fats, at workable temperatures, increase in either firmness or viscosity by about 10% with each increment of crystals [8].

Number, size, and kind of crystals. At a given solids content, a large number of small crystals produces a harder fat than does a small number of large crystals. Larger, soft crystals are typically produced by slow cooling. Crystals composed of high-melting acylglycerols provide greater stiffening power than do those of lower melting acylglycerols.

Viscosity of the liquid. Oils differ in viscosity at a given temperature and this will influence viscosity of the melt, as well as consistency of a solid–liquid lipid mixture.

Temperature treatment. If a fat tends to supercool excessively, this can be overcome by melting the crystalline fat at the lowest possible temperature, holding it for an extended period of time at a temperature just above its melting point, and then cooling it. This facilitates formation of numerous crystal nuclei, numerous small crystals, and a firm consistency.

Mechanical working. Crystallized fats are generally thixotropic; that is, they become reversibly softer after vigorous agitation and only gradually regain their original firmness. If a melted fat is mechanically agitated during solidification, it will be much softer than if allowed to solidify in a quiescent condition. In the quiescent state the growing crystals form structures of relatively great strength. These structures can be deformed by mechanical working.

5.5 CHEMICAL ASPECTS

5.5.1 Lipolysis

Hydrolysis of ester bonds in lipids (lipolysis) may occur by enzyme action [149] or by heat and moisture, resulting in the liberation of free fatty acids. Free fatty acids are virtually absent in fat of living animal tissue. They can form, however, by enzyme action after the animal is killed. Since edible animal fats are not usually refined, prompt rendering is of particular importance. The temperatures commonly used in the rendering process are capable of inactivating the enzymes responsible for hydrolysis.

The release of short-chain fatty acids by hydrolysis is responsible for the development of an undesirable rancid flavor (hydrolytic rancidity) in raw milk. On the other hand, certain typical cheese flavors are produced by deliberate addition of microbial and milk lipases. Controlled and selective lipolysis is also used in the manufacture of other food items, such as yogurt and bread.

In contrast to animal fats, oils in mature oil seeds may have undergone substantial hydrolysis by the time they are harvested, giving rise to significant amounts of free fatty acids. Neutralization with alkali is thus required for most vegetable oils after they are extracted.

Lipolysis is a major reaction occurring during deep-fat frying due to the large amounts of water introduced from the food, and the relatively high temperatures used. Development of high levels of free fatty acids during frying is usually associated with decreases in smoke point and surface tension of the oil and a reduction in quality of the fried food. Furthermore, free fatty acids are more susceptible to oxidation than are fatty acids esterified to glycerol.

Phospholipase A and lipases from muscle tissues have been described. Significant phospholipid hydrolysis occurs in most species of fish during frozen storage and is usually associated with deterioration in quality. A number of studies appear to indicate that while the hydrolysis of triacylglycerols leads to increased lipid oxidation, phospholipid hydrolysis inhibits the oxidation [123].

Enzymic lipolysis is used extensively as an analytical tool in lipid research. As discussed earlier in this chapter, pancreatic lipase and snake venom phospholipase are used to determine the positional distribution of fatty acids in acylglycerol molecules. The specificities of these and numerous other enzymes make them particularly useful in the preparation of intermediates in the chemical synthesis of certain lipids, and in the manufacture of "structured fats" useful for specific nutritional, pharmaceutical, and cosmetic applications [2, 16, 141].

5.5.2 Autoxidation

Lipid oxidation is one of the major causes of food spoilage. It is of great economic concern to the food industry because it leads to the development, in edible oils and fat-containing foods, of various off flavors and off odors generally called rancid (oxidative rancidity), which render these foods less acceptable. In addition, oxidative reactions can decrease the nutritional quality of food, and certain oxidation products are potentially toxic. On the other hand, under certain conditions, a limited degree of lipid oxidation is sometimes desirable, as in aged cheeses and some fried foods.

For these reasons, extensive research has been done not only to identify the products of lipid oxidation and the conditions that influence their production, but also to study the mechanisms involved. Since oxidative reactions in food lipids are exceedingly complex, simpler model systems, such as oleate, linoleate, and linolenate, have been used to ascertain mechanistic pathways. Although this approach has provided useful information, extrapolation to the more complex food lipid systems must be done with caution.

It is generally agreed that "autoxidation," that is, the reaction with molecular oxygen via a self-catalytic mechanism, is the main reaction involved in oxidative deterioration of lipids. Although photochemical reactions have been known for a long time, only recently has the role of photosensitized oxidation and its interaction with autoxidation emerged. In foods, the lipids can be oxidized by both enzymic and nonenzymic mechanisms.

5.5.2.1 General Characteristics of the Autoxidation Reaction

Our present knowledge regarding the fundamental mechanisms of lipid oxidation resulted largely from the pioneering work of Farmer and his coworkers [36], Bolland and Gee [12], and Bateman et al. [10].

Autoxidation of fats proceeds via typical free radical mechanisms as characterized by (a) marked inhibition in rate by chemicals known to interfere with other well-established free radical reactions, (b) catalytic effects of light and free radical-producing substances, (c) high yields of the hydroperoxide, ROOH, (d) quantum yields exceeding unity when the oxidation reaction is initiated by light, and (e) a relatively long induction period observed when starting with a pure substrate.

Based on experimental results, mostly with ethyl linoleate, the rate of oxygen absorption can be expressed as shown here,

$$\text{Rate} = -\frac{d[O_2]}{dt} = \frac{K_a[RH][ROOH]}{1 + \lambda[RH]/p}$$

where RH is the substrate fatty acid, H is an α-methylenic hydrogen atom easily detachable due to the influence of a neighboring double bond or bonds, ROOH is the hydroperoxide formed, p is oxygen pressure, and λ and K_a are empirical constants.

To explain the experimental results, a three-step simplified, free-radical scheme has been postulated:

$$\text{Initiator} \xrightarrow{k_1} \text{free radicals } (R^{\cdot}, ROO^{\cdot}) \qquad \text{INITIATION} \qquad (1)$$

$$R^{\cdot} + O_2 \xrightarrow{k_2} ROO^{\cdot} \qquad (2)$$

$$ROO^{\cdot} + RH \xrightarrow{k_3} ROOH + R^{\cdot} \qquad (3)$$

PROPAGATION (2)–(3)

$$R^{\cdot} + R^{\cdot} \xrightarrow{K_4} \qquad (4)$$

$$R^{\cdot} + ROO^{\cdot} \xrightarrow{K_5} \text{Nonradical products} \qquad \text{TERMINATION} \qquad (5)$$

$$ROO^{\cdot} + ROO^{\cdot} \xrightarrow{K_6} \qquad (6)$$

At high oxygen pressure ($\lambda[RH]/p << 1$), reactions (4) and (5) can be neglected to give this form:

$$\text{Rate} = k_3\left(\frac{k_1}{k_6}\right)^{1/2}[ROOH][RH]$$

In this situation, the rate of oxidation is independent of oxygen pressure.

At low oxygen pressure ($\lambda[RH]/p > 1$), steps (5) and (6) can be neglected to give this form:

$$\text{Rate} = k_2\left(\frac{k_1}{k_4}\right)^{1/2}[ROOH][O_2]$$

Since the reaction $RH + O_2 \rightarrow$ free radicals is thermodynamically difficult (activation energy of about 35 kcal, or 146 kJ/mol), production of the first few radicals (initiation) necessary to start the propagation reaction must be catalyzed. It has been proposed that initiation of oxidation may take place by hydroperoxide decomposition, by metal catalysis, or by exposure to light. More recently, it has been postulated that singlet oxygen is the active species involved, with tissue pigments such as chlorophyll and myoglobin acting as sensitizers.

After initiation, oxidation is propagated by abstraction of hydrogen atoms at positions α to fatty acid double bonds, producing free radical species $R^{\cdot}$. Oxygen addition then occurs at these locations, resulting in the production of peroxy radicals ROO˙, and these in turn abstract hydrogen from α-methylenic groups of other molecules (RH) to yield hydroperoxides (ROOH) and new free radicals (R˙). The new R˙ groups react with oxygen, and the sequence of reactions just described is repeated.

Due to resonance stabilization of these R˙ species, the reaction sequence is usually accompanied by a shift in position of the double bonds, resulting in the formation of isomeric hydroperoxides that often contain conjugated diene groups (atypical of unoxidized, natural acylglycerols).

Hydroperoxides, the primary initial products of lipid autoxidation, are relatively unstable. They enter into numerous complex reactions involving substrate degradation and interaction, resulting in a myriad of compounds of various molecular weights, flavor thresholds, and biological significance.

A general scheme summarizing the overall picture of lipid autoxidation is given in Figure 17, and some aspects of the reaction sequence are discussed in more detail later.

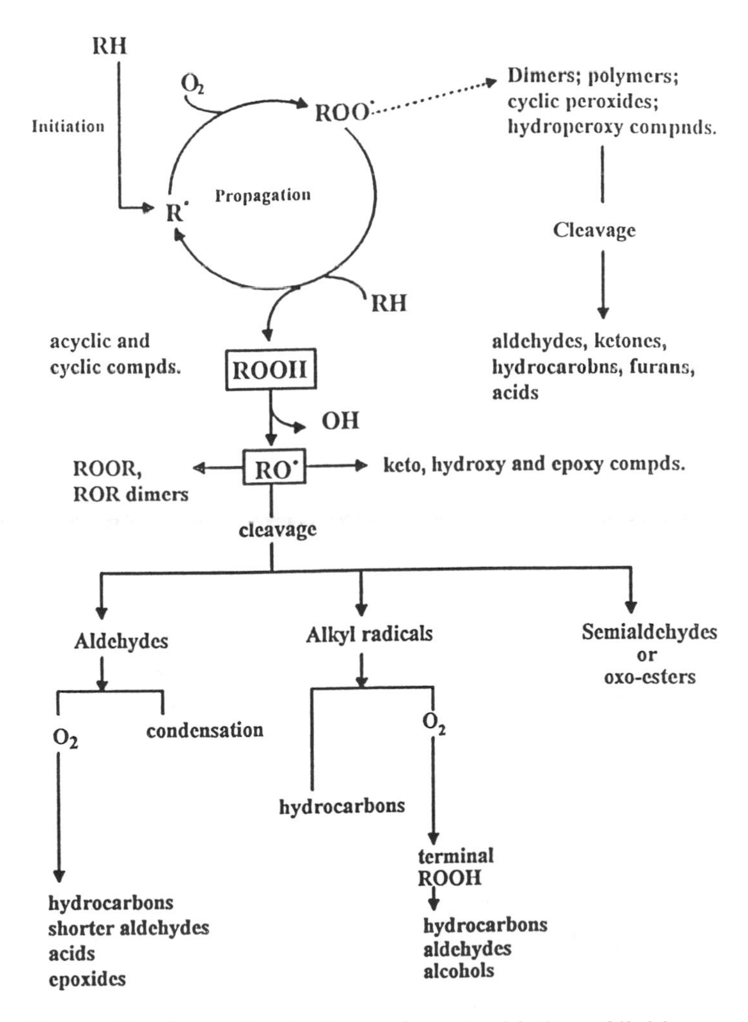

FIGURE 17 Generalized scheme for autoxidation of lipids.

5.5.2.2 Formation of Hydroperoxides

Qualitative and quantitative analyses of the isomeric hydroperoxides from oleate, linoleate, and linolenate have been conducted [22, 23, 39].

Oleate

Hydrogen abstraction at carbons 8 and 11 of oleate results in formation of two allylic radical intermediates. Oxygen attack at the end carbons of each radical produces an isomeric mixture of 8-, 9-, 10-, and 11-allylic hydroperoxides.

```
              11  10   9   8
              C - C = C - C

 11 10 9          11 10 9            10 9 8          10 9 8
-C=C-C·    ↔    -C·-C=C-          -C=C-C·     ↔    -C·-C=C-
    ↓ O2            ↓ O2              ↓ O2             ↓ O2
-C=C-C-OO·     ·OO-C-C=C-         -C=C-C-OO·      ·OO-C-C=C-
    ↓               ↓                 ↓                ↓
      9          11                     8          10
-C=C-C-OOH     HOO-C-C=C-         -C=C-C-OOH      HOO-C-C=C-
```

The amounts of 8- and 11-hydroperoxides formed are slightly greater than those of the 9- and 10-isomers. At 25°C the amounts of *cis* and *trans* 8- and 11-hydroperoxides are similar, but the 9- and 10-isomers are mainly *trans*.

Linoleate

The 1,4-pentadiene structure in linoleates makes them much more susceptible (by a factor of about 20) to oxidation than the propene system of oleate. The methylene group at position 11 is doubly activated by the two adjacent double bonds.

```
 13 12 11 10 9
-C=C-C-C=C-
      ↓
-C=C-C·-C=C-

[ -C=C-C=C-C·- ]
[ -C·-C=C-C=C- ]
      ↓
            9
-C=C-C=C-C-OOH
      +
 13
HOO-C-C=C-C=C-
```

Hydrogen abstraction at this position produces a pentadienyl radical intermediate, which upon reaction with molecular oxygen produces an equal mixture of conjugated 9- and 13-diene hydroperoxides. Evidence reported in the literature indicates that the 9- and 13-*cis, trans*-hydroperoxides undergo interconversion, along with some geometric isomerization, forming *trans,trans*-isomers. Thus, each of the two hydroperoxides (9- and 13-) is found in both the *cis,trans* and the *trans,trans* forms.

Linolenate

In linolenate, two 1,4-pentadiene structures are present. Hydrogen abstraction at the two active methylene groups of carbons 11 and 14 produces two pentadienyl radicals.

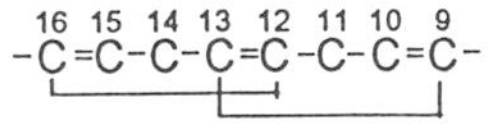

```
16 15 14 13 12 11 10  9
-C=C-C-C=C-C-C=C-
```

Oxygen attack at the end carbon of each radical results in the formation of a mixture of 9-, 12-, 13-, and 16-hydroperoxides. For each of the hydroperoxides, geometric isomers exist, each having a conjugated diene system in either the *cis,trans* or the *trans,trans* configuration, and an isolated double bond that is always *cis*.

The 9- and 16-hydroperoxides are formed in significantly greater amounts than the 12- and 13-isomers. This has been attributed to (a) a preference of oxygen to react with carbons 9 and 16; (b) faster decomposition of the 12- and 13-hydroperoxides, or (c) a tendency of the 12- and 13-hydroperoxides to form six-membered peroxide hydroperoxides via 1,4-cyclization, as shown here, or prostaglandin-like endoperoxides via 1,3-cyclization. The cyclization mechanisms are thought to be the most likely explanations.

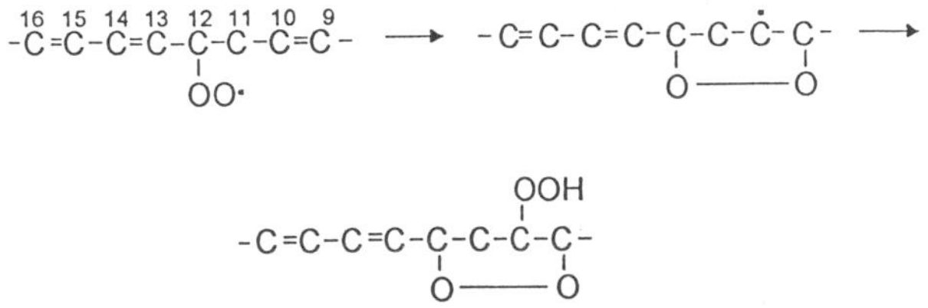

```
16 15 14 13 12 11 10  9
-C=C-C=C-C-C-C=C-   ⟶   -C=C-C=C-C-C-Ċ-C-   ⟶
          |                      |     |
         OO·                     O-----O

            OOH
             |
-C=C-C=C-C-C-C-C-
         |     |
         O-----O
```

5.5.2.3 Oxidation with Singlet Oxygen

As just discussed, the major pathway for oxidation of unsaturated fatty acids involves a self-catalytic free-radical mechanism (autoxidation) that accounts for the chain reaction of hydroperoxide (ROOH) formation and decomposition. However, the origin of the initial free radicals necessary to begin the process has been difficult to explain. It is unlikely that initiation occurs by direct attack of oxygen in its most stable form (triplet state) on double bonds of fatty acids (RH). This is because the C=C bonds in RH and ROOH are in singlet states and thus such reaction does not obey the rule of spin conservation. A more satisfactory explanation is that singlet oxygen ($^1O_2$), believed to be the active species in photooxidative deterioration, is responsible for initiation.

Since electrons are charged, they act like magnets that can exist in two different orientations, but with equal magnitude of spin, +1 and −1. The total angular momentum of the electrons in an atom is described by the expression $2S + 1$ where S is total spin. If an atom such as oxygen has two unpaired electrons in its outer orbitals, they can align their spins parallel or antiparallel with each other, giving rise to two different multiplicities of state, that is, $2(½ + ½) + 1 = 3$ and $2(½ - ½) + 1 = 1$. These are called the triplet ($^3O_2$) and singlet state ($^1O_2$), respectively. In the triplet state, the two electrons in the antibonding $2p$ orbitals have the same spin but are in different orbitals. These electrons are kept apart via "Pauli exclusion" and therefore have only a small repulsive electrostatic energy.

In the singlet state, the two electrons have opposite spins, and therefore electrostatic repulsion will be great, resulting in an excited state. Singlet oxygen can be found in two energy states, $^1\delta$, with an energy of 22 kcal (94 kJ) above ground state, and $^1\varepsilon$, with an energy of 37 kcal (157 kJ) above ground state.

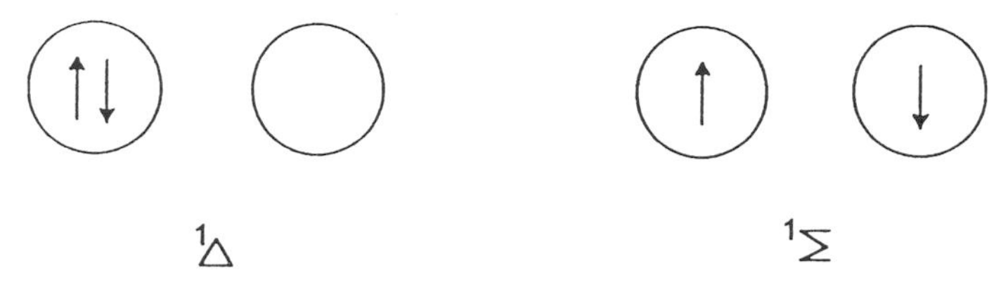

Singlet-state oxygen is more electrophilic than triplet state oxygen. It can thus react rapidly (1500 times faster than $^3O_2$) with moieties of high electron density, such as C=C bonds. The resulting hydroperoxides can then cleave to initiate a conventional, free-radical chain reaction.

Singlet oxygen can be generated in a variety of ways; probably the most important is via photosensitization by natural pigments in foods. Two pathways have been proposed for photosensitized oxidation [21]. In the type 1 pathway, the sensitizer presumably reacts, after light absorption, with substrate (A) to form intermediates, which then react with ground-state (triplet) oxygen to yield the oxidation products.

Sens + A + $h\nu$ $\rightarrow$ intermediates-I
Intermediates-I + O_2 $\rightarrow$ products + sens

In the type 2 pathway, molecular oxygen rather than the substrate is presumably the species that reacts with the sensitizer upon light absorption.

Sens + O_2 + $h\nu$ $\rightarrow$ intermediates-II
Intermediates-II + A $\rightarrow$ products + sens

Several substances are commonly found in fat-containing foods that can act as photosensitizers to produce $^1O_2$. These include natural pigments, such as chlorophyll-a, pheophytin-a, and hematoporphyrin, the pigment portion of hemoglobin, and myoglobin. The synthetic colorant erythrosine also acts as an active photosensitizer.

β-Carotene is a very effective $^1O_2$ quencher, and tocopherols are also somewhat effective. Synthetic quenchers such as butylated hydroxyanisole (BHA) and butylated hydroxytoluene (BHT) are also effective and permissible in foods.

The formation of hydroperoxides by singlet oxygen proceeds via mechanisms different from those of free-radical autoxidation. The most important of these is the "ene" reaction, which involves the formation of a six-membered ring transition state. Oxygen is thus inserted at the ends of the double bond, which then shifts to yield an allylic hydroperoxide in the trans configuration. Accordingly, oleate produces the 9- and 10-hydroperoxides (instead of the 8-, 9-, 10-, and 11-hydroperoxides occurring by free radical autoxidation), and linoleate produces a mixture of 9-, 10-, 12-, 13-, 15-, and 16-hydroperoxides (instead of 9-, 12-, 13-, and 16-).

In addition to the role of singlet oxygen as an active initiator of free radicals, it has been suggested that certain products of lipid oxidation can be explained only on the basis of decomposition of hydroperoxides typical of those produced by singlet oxygen. However, there is general agreement that, once the initial hydroperoxides are formed, the free-radical chain reaction prevails as the main mechanism. Further research is needed to clarify the extent to which singlet-oxygen oxidation may be involved in the production of oxidative decomposition products.

5.5.2.4 Decomposition of Hydroperoxides

Hydroperoxides break down in several steps, yielding a wide variety of decomposition products. Each hydroperoxide produces a set of initial breakdown products that are typical of the specific hydroperoxide and depend on the position of the peroxide group in the parent molecule [40]. The decomposition products can themselves undergo further oxidation and decomposition, thus contributing to a large and varied free-radical pool. The multiplicity of the many possible reaction pathways results in a pattern of autoxidation products so complex that in most cases their hydroperoxide origin is completely obliterated.

It should also be mentioned that hydroperoxides begin to decompose as soon as they are formed. In the first stages of autoxidation, their rate of formation exceeds their rate of decomposition. The reverse takes place at later stages.

The first step in hydroperoxide decomposition is scission at the oxygen–oxygen bond of the hydroperoxide group, giving rise to an alkoxy radical and a hydroxy radical, as shown. Carbon–carbon bond cleavage on either side of the alkoxy group (homolytic cleavage) is the second step in decomposition of the hydroperoxides:

$$R_1\text{-}CH(\text{O-OH})\text{-}R_2 \longrightarrow R_1\text{-}CH(\text{O}\cdot)\text{-}R_2 + \dot{O}H$$

alkoxy radical

HOMOLYTIC CLEAVAGE

$$R\text{-}CH{=}CH\text{-}CH(\text{O-OH})\text{-}CH_2\text{-}R'$$

$$\downarrow$$

$$R\text{-}CH{=}CH\text{-}CH(\text{O}\cdot)\text{-}CH_2\text{-}R'$$

$$R\text{-}CH{=}CH\cdot + OHC\text{-}CH_2\text{-}R' \qquad R\text{-}CH{=}CH\text{-}CHO + \cdot CH_2\text{-}R'$$

In general, cleavage on the acid side (i.e., the carboxyl or ester side) results in formation of an aldehyde and an acid (or ester), while scission on the hydrocarbon (or methyl) side produces a hydrocarbon and an oxoacid (or oxoester). If, however, a vinylic radical results from such cleavage, an aldehydic functional group is formed:

$$R_1\text{-}CH{=}CH\cdot \xrightarrow{\cdot OH} R_1\text{-}CH{=}CH\text{-}OH \updownarrow R_1\text{-}CH_2\text{-}CHO$$

For example, with the 8-hydroperoxide isomer from methyl oleate, cleavage on the hydrocarbon side (at a) yields decanal and methyl-8-oxooctanoate, and scission on the ester side (at b) forms 2-undecenal and methyl heptanoate:

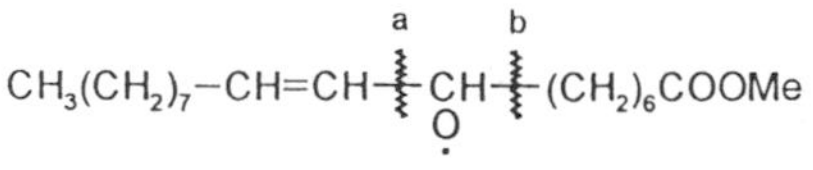

$$CH_3(CH_2)_7-CH=CH \overset{a}{\wr} \underset{\underset{\bullet}{O}}{CH} \overset{b}{\wr} (CH_2)_6COOMe$$

In the same manner, each of the remaining three oleate hydroperoxides would be expected to produce four typical products; that is, the 9-hydroperoxide, shown here,

$$CH_3(CH_2)_6-CH=CH \wr \underset{\underset{\bullet}{O}}{CH} \wr (CH_2)_7COOMe$$

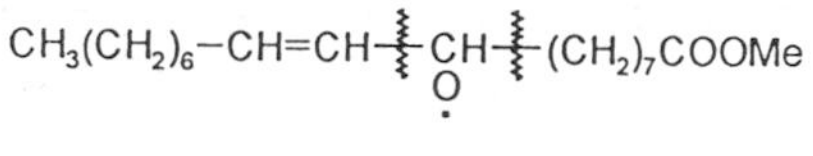

would produce nonanal, methyl-9-oxononaoate, 2-decenal, and methyl octanoate; the 10-hydroperoxide, shown here,

$$CH_3(CH_2)_7 \wr \underset{\underset{\bullet}{O}}{CH} \wr CH=CH-(CH_2)_6COOMe$$

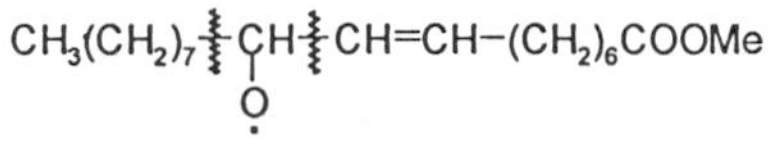

would produce octane, methyl-10-oxo-8-decenoate, nonanal, and methyl-9-oxononanoate; and the 11-hydroperoxide, shown here,

$$CH_3(CH_2)_6 \wr \underset{\underset{\bullet}{O}}{CH} \wr CH=CH-(CH_2)_7COOMe$$

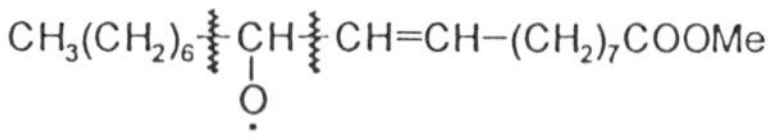

would produce heptane, methyl-11-oxo-9-undecenoate, octanal, and methyl-10-oxo-decanoate.

As mentioned earlier, autoxidation of linoleate produces two conjugated hydroperoxides, the 9- and the 13-hydroperoxides. Figure 18 shows the typical cleavage pattern of the 9-alkoxy radical.

In addition to the compounds resulting from cleavage on either side of the alkoxy group, several other compounds have been observed. For example, in studies on heated ethyl linoleate, several esters and oxoesters with chain lengths $< C_9$ were observed (Fig. 19). The C_8 aldehyde ester, a compound not predictable from any of the "confirmed" hydroperoxides of linoleate but found in relatively large amounts, may be produced by cleavage of the 9-alkoxy radical followed by terminal hydroperoxidation. Further oxidation of the C_8 aldehyde ester would produce the corresponding acid $HOOC(CH_2)_6COOH_2H_5$, which, upon decarboxylation, would give rise to ethyl heptanoate. Both the C_8 dicarboxylic acid and the C_7 ethyl ester are major products of linoleate thermal oxidation. The 8-alkoxy radical may also decompose to give formaldehyde and the C_7 alkyl radical, which in turn would produce ethyl heptanoate or form a terminal hydroperoxide, and so on. This mechanism explains the formation of C_8, C_7, C_6 aldehyde esters, and dicarboxylic acid series, and the C_7, C_6, C_5 ethyl ester series. Concentrations of the latter decrease with decreasing chain length.

An alternative mechanism for formation of the C_8 aldehyde ester and the C_7 ethyl ester involves formation and decomposition of the unconjugated C_8 hydroperoxide. Oxygen attack outside the 1,4-pentadiene system of linoleate is considered unlikely.

Although C_8 or C_{14} linoleate hydroperoxides have never been isolated, all studies regarding the identity and relative distribution of hydroperoxide intermediates were conducted at 80°C or lower. It is conceivable that formation of C_8 or C_{14} hydroperoxides may occur at higher temperatures. Indeed, many authors have speculated that hydrogen abstraction at allylic positions outside the pentadiene systems is responsible for many of the "unpredicted" decomposition products found after high-temperature treatment. Such a mechanism has been proposed to rationalize the formation of aromatic compounds and cyclic monomers from methyl linoleate,

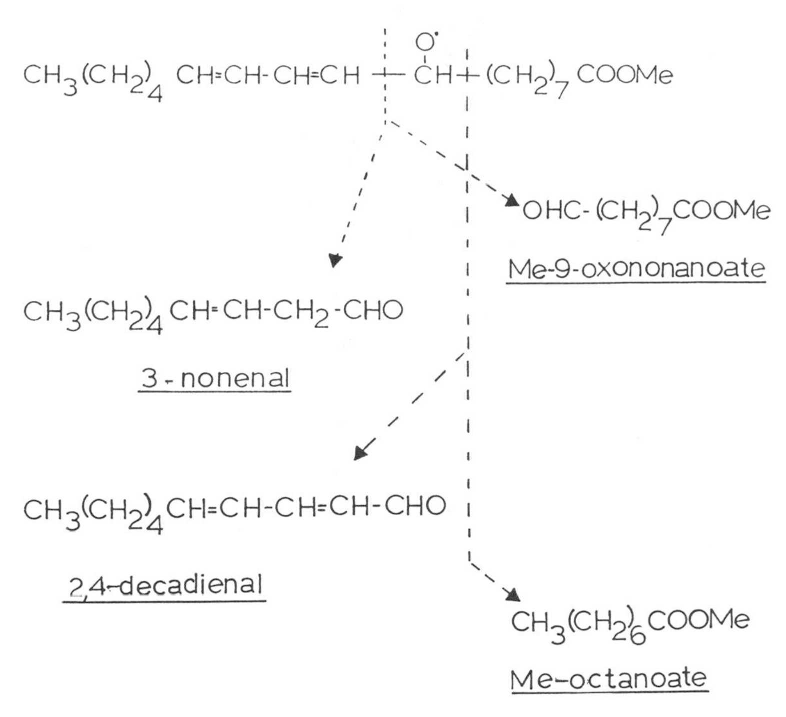

FIGURE 18 Decomposition of methyl-9-hydroperoxy-10,12-octadecadienoate.

and to explain the production of 2-cyclohexenyl acetate and 5-cyclohexenyl pentenoate from methyl docosahexaenoate.

In contrast to the homolytic cleavage described earlier, specific conditions will cause heterolytic cleavage to occur between the hydroperoxide group and the allylic double bond. Such cleavage results in a greater yield and more selective distribution of carbonyl compounds than that occurring with homolytic cleavage [84]. Heterolytic cleavage of methyl linoleate hydroperoxides, for example, would give rise mainly to hexanal, nonanal, methyl 9-oxononanoate and methyl 12-oxododecenoate [48, 56]. A mechanism proposed for heterolytic cleavage is shown in Figure 20.

It was also pointed out that in the case of linoleate, oxidation by singlet oxygen leads to the formation of four isomeric hydroperoxides, two of which are not typical of autoxidation. The 10-hydroperoxide would be expected to yield 2-octene, methyl-10-oxo-8-decenoate, 3-nonenal, and methyl-9-oxononanoate; the 12-hydroperoxide would give hexanal, methyl-12-oxo-9-dodecenoate, 2-heptenal, and methyl-9-undecenoate.

For linolenates, some of the cleavage products expected from classic scisson of the 9-, 12-, 13-, and 16-hydroperoxides have been reported; however, several others, particularly the longer chain and the polyunsaturated oxoesters, have not been detected. It is possible that these types of compounds rapidly undergo further decomposition.

The cyclic peroxides or the hydroperoxy cyclic peroxides, which are commonly formed during oxidation of polyunsaturated fatty acids, also decompose, yielding a variety of compounds. Formation of 3,5-octadiene-2-one is shown here:

$$\text{C-C-C=C-C=C-C-C-C(OOH)-C-(C)}_7\text{-COOMe (with O–O bridge)} \longrightarrow \text{C-C-C=C-C=C-C(=O)-C}$$

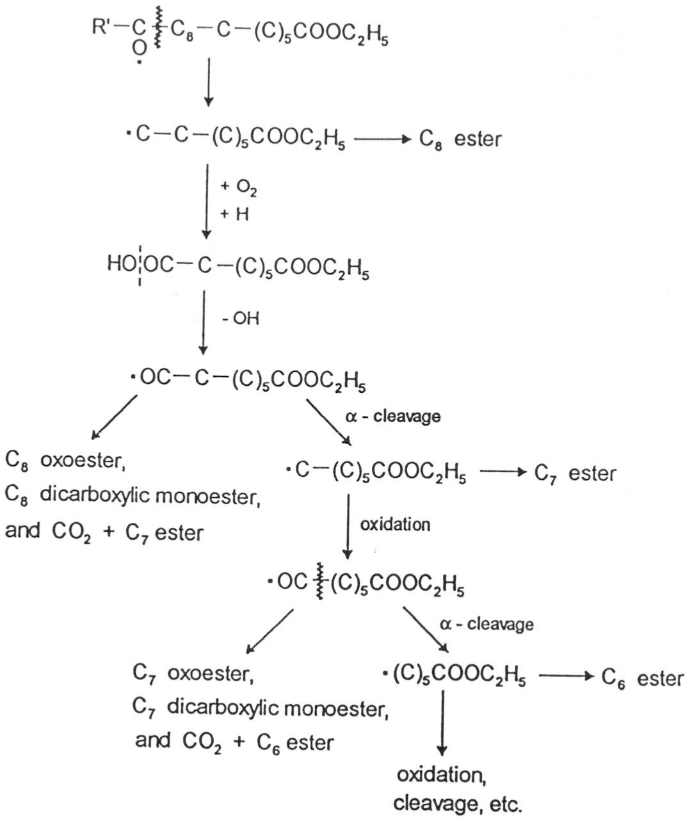

$$R'-\underset{\dot{O}}{C} \wr C_8-C-(C)_5COOC_2H_5$$

$$\downarrow$$

$$\cdot C-C-(C)_5COOC_2H_5 \longrightarrow C_8 \text{ ester}$$

$$\downarrow + O_2, + H$$

$$HO \vdots OC-C-(C)_5COOC_2H_5$$

$$\downarrow - OH$$

$$\cdot OC-C-(C)_5COOC_2H_5$$

↙ C_8 oxoester, C_8 dicarboxylic monoester, and CO_2 + C_7 ester

↘ α - cleavage

$$\cdot C-(C)_5COOC_2H_5 \longrightarrow C_7 \text{ ester}$$

$$\downarrow \text{oxidation}$$

$$\cdot OC \wr (C)_5COOC_2H_5$$

↙ C_7 oxoester, C_7 dicarboxylic monoester, and CO_2 + C_6 ester

↘ α - cleavage

$$\cdot (C)_5COOC_2H_5 \longrightarrow C_6 \text{ ester}$$

$$\downarrow$$

oxidation, cleavage, etc.

FIGURE 19 Possible mechanism for the formation of the short-chain esters, oxoesters, and dicarboxylic acids.

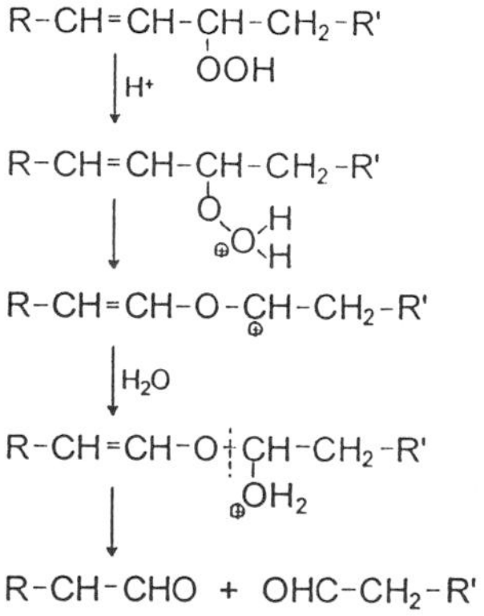

$$R-CH=CH-\underset{OOH}{CH}-CH_2-R'$$

$$\downarrow H^+$$

$$R-CH=CH-\underset{O-\overset{\oplus}{O}H_2}{CH}-CH_2-R'$$

$$\downarrow$$

$$R-CH=CH-O-\overset{\oplus}{C}H-CH_2-R'$$

$$\downarrow H_2O$$

$$R-CH=CH-O \vdots \underset{\oplus OH_2}{CH}-CH_2-R'$$

$$\downarrow$$

$$R-CH-CHO + OHC-CH_2-R'$$

FIGURE 20 Heterolytic hydroperoxide cleavage.

5.5.2.5 Further Decomposition of the Aldehydes

As demonstrated earlier, aldehydes are a major class of compounds typically arising from fat oxidation. Many of the aldehydes found in oxidized fats, however, cannot be explained solely on the basis of the classic hydroperoxide cleavage just discussed. This is not surprising in view of the ease and variety of reactions that aldehydes can undergo.

Saturated aldehydes can easily oxidize to form the corresponding acids, and they can also participate in dimerization and condensation reactions. Three molecules of hexanal, for example, can combine to form tripentyltrioxane:

$$3\ C_5H_{11}CHO \longrightarrow$$

$H_{11}C_5$ C_5H_{11} C_5H_{11} (trioxane ring: O, O, O)

Trialkyltrioxanes possessing relatively strong odors have been reported as secondary oxidation products of linoleate.

In a study of oleic acid and nonanal, a speculative but interesting mechanism was proposed to account for the aldehydes, alcohols, alkyl formates, and hydrocarbons produced from autoxidation of oleic acid [97]. Abstraction of hydrogen from nonanal, the aldehyde formed from the 10-hydroperoxide of oleate, results in formation of a resonance equilibrium between two forms of the carbonyl free radical (Fig. 21). This leads to formation of a peracid and an α-hydroperoxyaldehyde. Carbon–carbon and oxygen–oxygen cleavages can produce a variety of free radicals that can initiate chain reactions or combine to form oxidation products.

Unsaturated aldehydes can undergo classic autoxidation with oxygen attack at α-methylenic positions giving rise to short-chain hydrocarbons, alehydes, and dialdehydes:

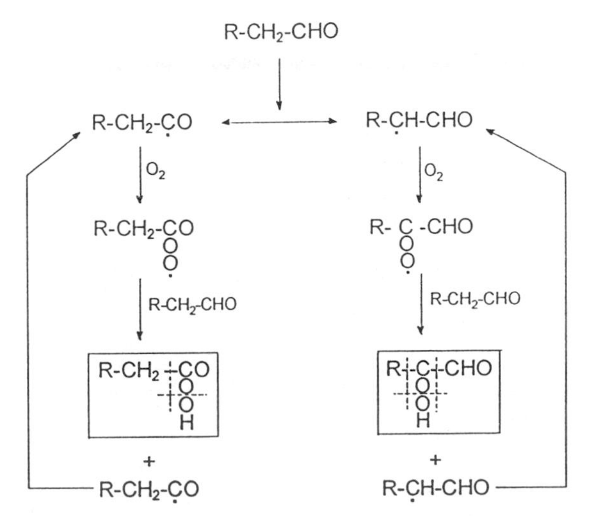

FIGURE 21 Speculative mechanism for the autoxidation of nonanal. (From Ref. 97.)

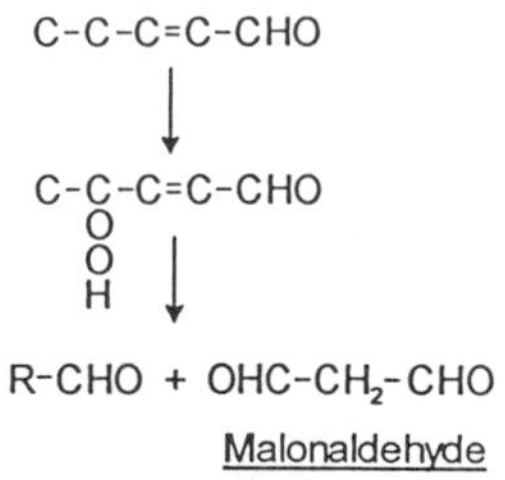

$$C\text{-}C\text{-}C{=}C\text{-}CHO$$
$$\downarrow$$
$$C\text{-}\underset{\substack{O\\O\\H}}{C}\text{-}C{=}C\text{-}CHO$$
$$\downarrow$$
$$R\text{-}CHO + OHC\text{-}CH_2\text{-}CHO$$

Malonaldehyde

Formation of malonaldehyde is the basis for the well-known TBA method used for measuring fat oxidation, as is discussed later.

For oxidation of aldehydes with conjugated double bonds, the proposed mechanism involves formation of epoxides by oxygen attack at the olefinic centers [102]. In the case of 2,4-decadienal, the major aldehyde from the 9-hydroperoxide of linoleate, either the 2,3-epoxy or the 4,5-epoxy derivative, can be produced as an intermediate. Figure 22 shows the formation and decomposition of the 2,3-epoxide. Similarly, hexanal, 2-butenal, hexane, and 2-butene-1,4-dial can be formed from decomposition of the 4,5-epoxide.

5.5.2.6 Cholesterol Oxidation [133]

Oxidation of cholesterol is of major concern because certain oxidation products have been reported to produce cytotoxic, angiotoxic, and carcinogenic effects. The products of choles-

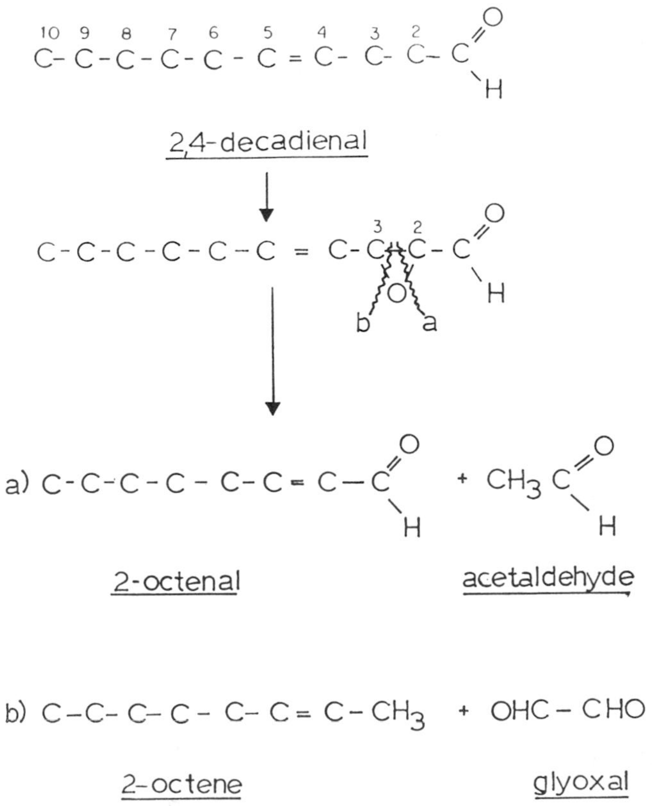

$$\overset{10}{C}\text{-}\overset{9}{C}\text{-}\overset{8}{C}\text{-}\overset{7}{C}\text{-}\overset{6}{C}\text{-}\overset{5}{C}{=}\overset{4}{C}\text{-}\overset{3}{C}\text{-}\overset{2}{C}\text{-}C({=}O)H$$

2,4-decadienal

$$\downarrow$$

$$C\text{-}C\text{-}C\text{-}C\text{-}C\text{-}C{=}C\text{-}\overset{3}{C}\text{-}\overset{2}{C}\text{-}C({=}O)H \quad \text{(2,3-epoxide, O bridging C3–C2; cleavage paths b, a)}$$

$$\downarrow$$

a) $C\text{-}C\text{-}C\text{-}C\text{-}C\text{-}C{=}C\text{-}C({=}O)H \;+\; CH_3C({=}O)H$

2-octenal acetaldehyde

b) $C\text{-}C\text{-}C\text{-}C\text{-}C\text{-}C{=}C\text{-}CH_3 \;+\; OHC\text{-}CHO$

2-octene glyoxal

FIGURE 22 Formation and decomposition of the 2,3-epoxide from 2,4-decadienal. (From Ref. 102.)

terol oxidation and the mechanisms for their formation have been reviewed [133]. In general, the epimeric 7α- and 7β-hydroperoxides are recognized as the initial products, with the 7β-hydroperoxides being more abundant than the α-isomers. Decomposition of the hydroperoxides gives rise to the isomeric 7α- and 7β-hydroxycholesterols, cholesterol α- and β-epoxides, and 7-ketocholesterol, with the latter being a major product (Fig. 23). In addition, the 20- and 25-hydroxycholesterols may arise from side-chain derivatizations, whereas cholesta-3,5-diene, cholesta-3,5-dien-7-one, and several other ketones arise from elimination reactions. Structures of some of these compounds are given here.

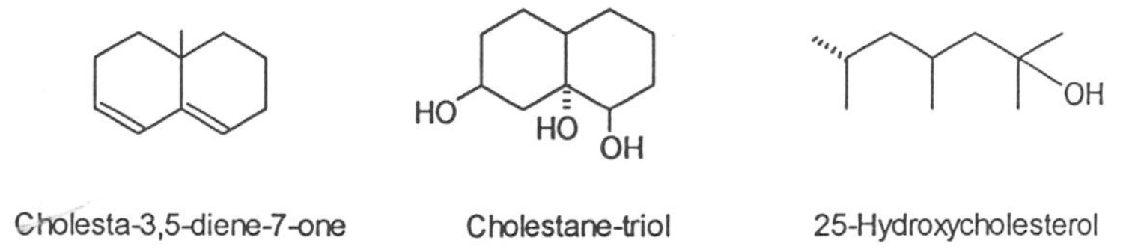

Cholesta-3,5-diene-7-one — Cholestane-triol — 25-Hydroxycholesterol

Other products, such as volatiles, relatively high-molecular-weight compounds, and numerous other minor oxides, have been reported.

Cholesterol oxidation products have been identified in several processed foods including dried eggs, meat and dairy products, fried foods, and heated fats.

5.5.2.7 Other Reactions of the Alkyl and Alkoxy Radicals

The alkyl radical arising from cleavage on the methyl side of an alkoxy group can enter into a variety of reactions (Fig. 24). It can combine with a hydroxyl radical to give an alcohol, abstract a hydrogen to form 1-alkene, or peroxidize to form a terminal hydroperoxide. The hydroperoxide

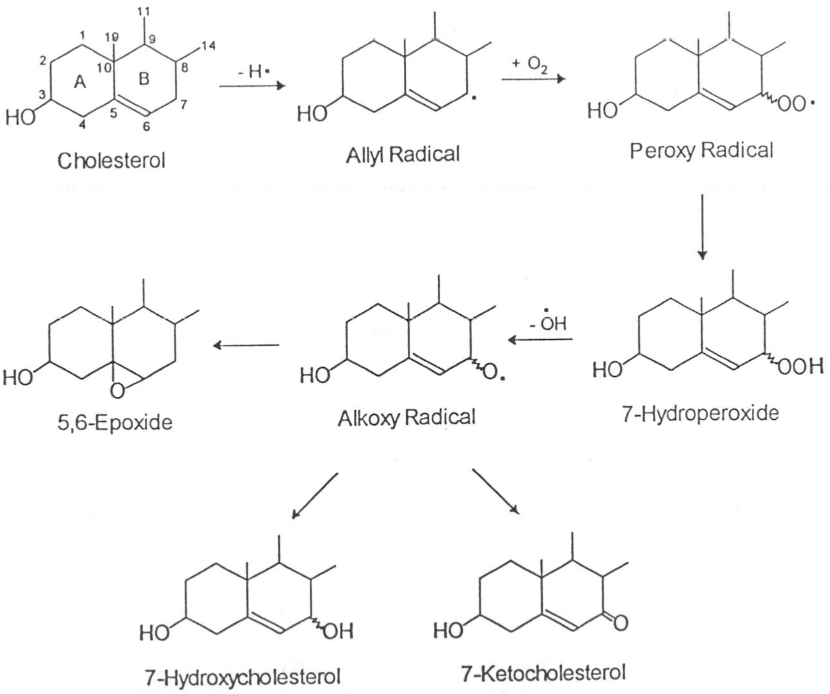

Cholesterol — $-H\cdot$ → Allyl Radical — $+O_2$ → Peroxy Radical → 7-Hydroperoxide — $-\dot{O}H$ → Alkoxy Radical → 5,6-Epoxide; Alkoxy Radical → 7-Hydroxycholesterol; Alkoxy Radical → 7-Ketocholesterol

FIGURE 23 Mechanism of cholesterol autoxidation.

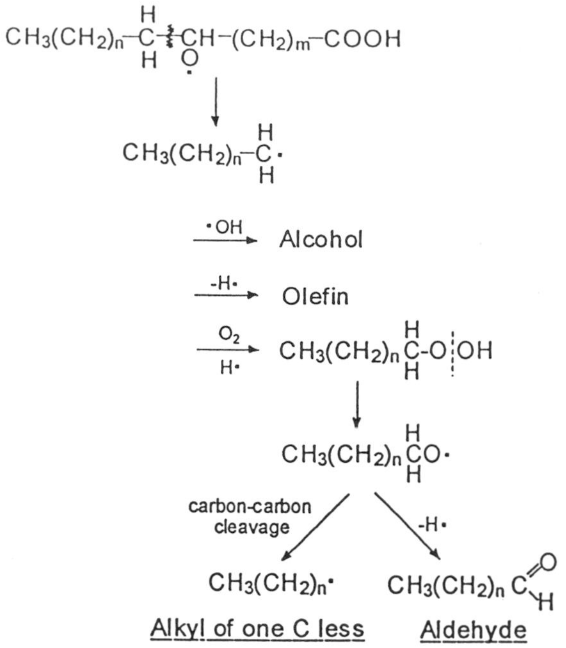

$$CH_3(CH_2)_n-\underset{H}{\overset{H}{C}}\,\wr\,\underset{O\cdot}{CH}-(CH_2)_m-COOH$$

$$\downarrow$$

$$CH_3(CH_2)_n-\underset{H}{\overset{H}{C}}\cdot$$

$$\xrightarrow{\cdot OH} \text{Alcohol}$$

$$\xrightarrow{-H\cdot} \text{Olefin}$$

$$\xrightarrow[H\cdot]{O_2} CH_3(CH_2)_n\underset{H}{\overset{H}{C}}-O\,\vdots\,OH$$

$$\downarrow$$

$$CH_3(CH_2)_n\underset{H}{\overset{H}{C}}O\cdot$$

carbon-carbon cleavage ↙ ↘ -H·

$$CH_3(CH_2)_n\cdot \qquad CH_3(CH_2)_nC\overset{\displaystyle O}{\underset{\displaystyle H}{\lessgtr}}$$

Alkyl of one C less   Aldehyde

FIGURE 24 Some reactions of the alkyl radical.

can decompose to the corresponding alkoxyradical, which in turn may form an aldehyde, or it can further cleave to yield another alkyl radical.

The alkoxy radical can also abstract a hydrogen atom from the α-methylene group of another molecule, producing a hydroxy acid, or it can lose a hydrogen to give a keto acid:

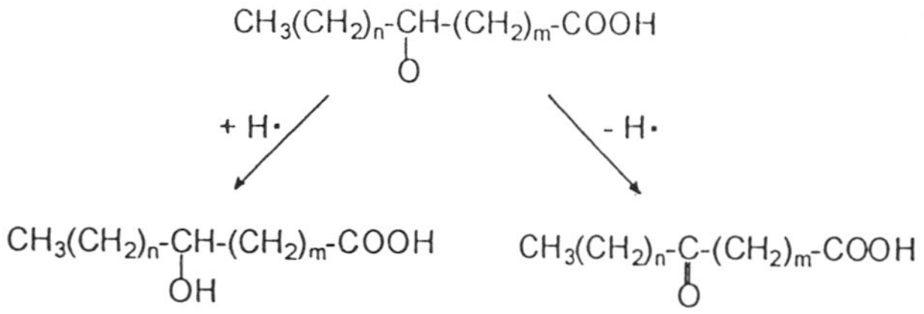

$$CH_3(CH_2)_n-\underset{O\cdot}{CH}-(CH_2)_m-COOH$$

+ H· ↙ ↘ - H·

$$CH_3(CH_2)_n-\underset{OH}{CH}-(CH_2)_m-COOH \qquad CH_3(CH_2)_n-\underset{O}{\overset{\|}{C}}-(CH_2)_m-COOH$$

Reaction of alkoxy and peroxy radicals with double bonds can produce epoxides as follows:

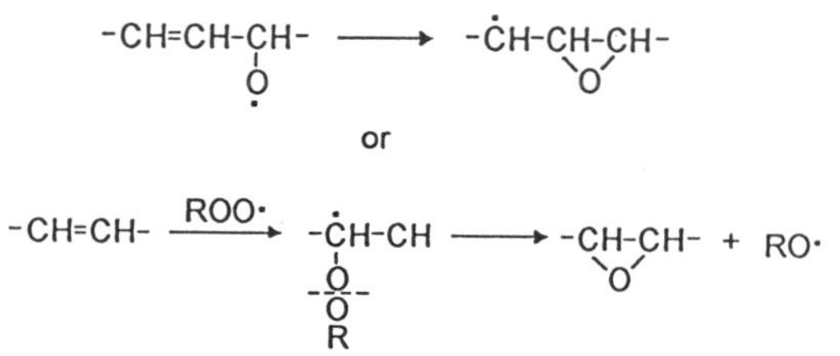

$$-CH{=}CH-\underset{O\cdot}{CH}- \longrightarrow -\dot{C}H-\underbrace{CH-CH}_{O}-$$

or

$$-CH{=}CH- \xrightarrow{ROO\cdot} -\underset{\underset{\underset{R}{O}}{O}}{\dot{C}H}-CH \longrightarrow -\underbrace{CH-CH}_{O}- + RO\cdot$$

5.5.2.8 Formation of Dimers and Polymers

Dimerization and polymerization are major reactions that occur in lipids during heating and/or oxidation. Such changes are usually accompanied by a decrease in iodine value and increases in molecular weight, viscosity, and refractive index.

Diels–Alder Reactions Between a Double Bond and a Conjugated Diene to Produce a Tetrasubstituted Cyclohexene

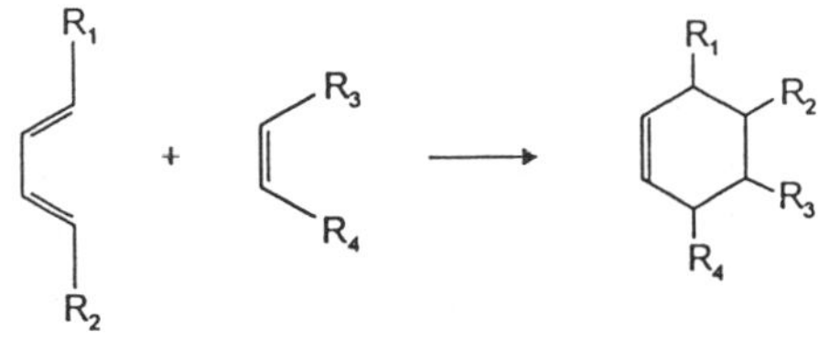

R_1 R_3 R_4 R_2 R_1 R_2 R_3 R_4

1,4 Diels - Alder Reaction

Linoleate, for example, can develop a conjugated double-bond system during thermal oxidation and then react with another molecule of linoleate (or with oleate) to produce a cyclic dimer:

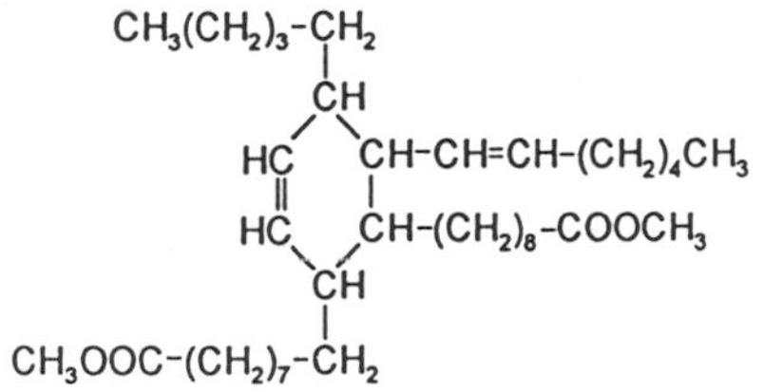

$CH_3(CH_2)_3$-CH_2 CH HC CH-CH=CH-$(CH_2)_4CH_3$ HC CH-$(CH_2)_8$-$COOCH_3$ CH CH_3OOC-$(CH_2)_7$-CH_2

In the case of acylglycerols, dimerization can take place between acyl groups in two triacylglycerol molecules or between two acyl groups in the same molecule.

$CH_2OO(CH_2)_x$ R $CHOO(CH_2)_x$ R $CH_2OO(CH_2)_yCH_3$ → $CH_2OO(CH_2)_x$ R $CHOO(CH_2)_x$ R $CH_2OO(CH_2)_yCH_3$

Combination of Free Radicals to Give Noncyclic Dimers

Under conditions of low oxygen tension, carbon–carbon bond formation between two acyl groups can occur in several ways: Oleate, for example, can give rise to a mixture of dimers (dehydrodimers) coupled at the 8, 9, 10, or 11 positions.

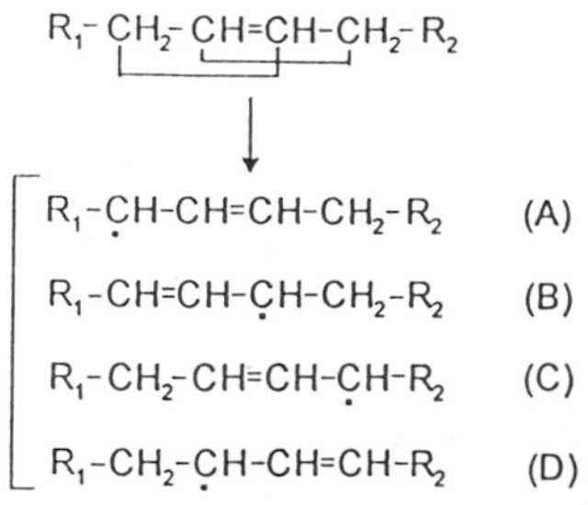

R_1-CH_2-CH=CH-CH_2-R_2

↓

R_1-ĊH-CH=CH-CH_2-R_2 (A)

R_1-CH=CH-ĊH-CH_2-R_2 (B)

R_1-CH_2-CH=CH-ĊH-R_2 (C)

R_1-CH_2-ĊH-CH=CH-R_2 (D)

AA, AB, AC, AD, BB, BC, BD, CC, CD, or DD

Addition of Free Radical to Double Bonds

This yields a dimeric radical, which may abstract hydrogen from another molecule or attack other double bonds to produce acyclic or cyclic compounds. Figures 25 and 26 show dimerization of

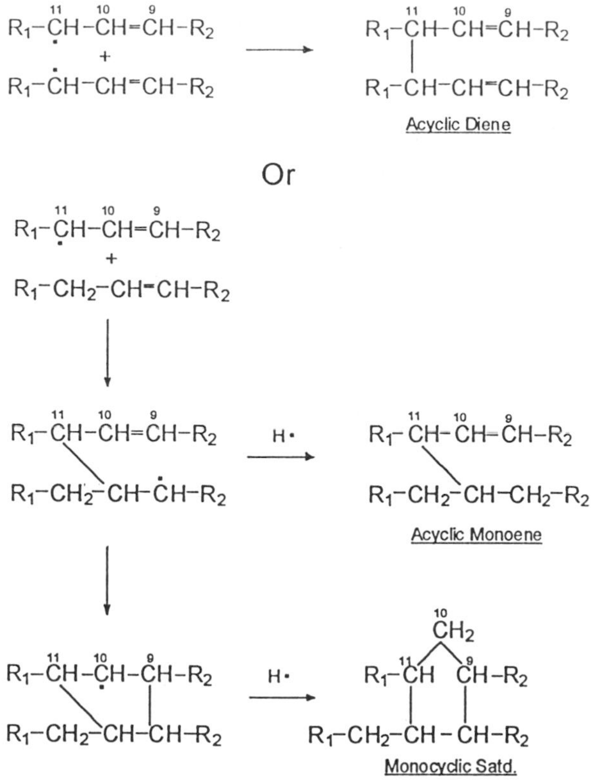

11 10 9
R_1–CH–CH=CH–R_2 + R_1–CH–CH=CH–R_2 → R_1–CH–CH=CH–R_2 | R_1–CH–CH=CH–R_2
Acyclic Diene

Or

11 10 9
R_1–CH–CH=CH–R_2 + R_1–CH_2–CH=CH–R_2
↓
R_1–CH–CH=CH–R_2 / R_1–CH_2–CH–CH–R_2 —H•→ R_1–CH–CH=CH–R_2 / R_1–CH_2–CH–CH_2–R_2
Acyclic Monoene
↓
R_1–CH–CH–CH–R_2 / R_1–CH_2–CH–CH–R_2 —H•→ CH_2 (10) / R_1–CH (11) CH–R_2 (9) / R_1–CH_2–CH – CH–R_2
Monocyclic Satd.

FIGURE 25 Dimerization of oleate.

oleate and linoleate, respectively. Similar reactions can occur between acyl groups in different acylglycerols to produce dimeric and trimeric triacylglycerols.

In the presence of abundant oxygen, combinations between alkyl, alkoxy, and peroxy free radicals can result in a variety of dimeric and polymeric acids and acylglycerols with carbon–oxygen–carbon or carbon–oxygen–oxygen–carbon cross-links (Fig. 27).

The addition of a free radical to a double bond may take place in the same molecule, giving rise to cyclic monomers. Cyclization occurs more readily in case of the longer chain polyunsaturated acids, as shown for arachidonic acid (Fig. 28).

5.5.2.9 Oxidation in Biological Systems

General Conditions

The preceding discussion dealt mainly with oxidation of pure lipid substrates in liquid bulk phase. In biological systems, including food, the lipid molecules often exist in a highly ordered state, where their mobility is restricted, and they are often closely associated with neighboring nonlipid material. For example, in addition to the phospholipids, proteins, cholesterol, tocopherols, enzymes, and trace metals are present within the membrane bilayer. Also, the lipids in the membrane have a very large surface area that is in contact with the components in the aqueous phase [63].

These characteristics vary greatly with the plant or animal species and with location of the lipid within the individual organism. Obviously, the consequences and mechanisms of oxidative reactions in such natural systems are often quite different from those in pure bulk-lipid models.

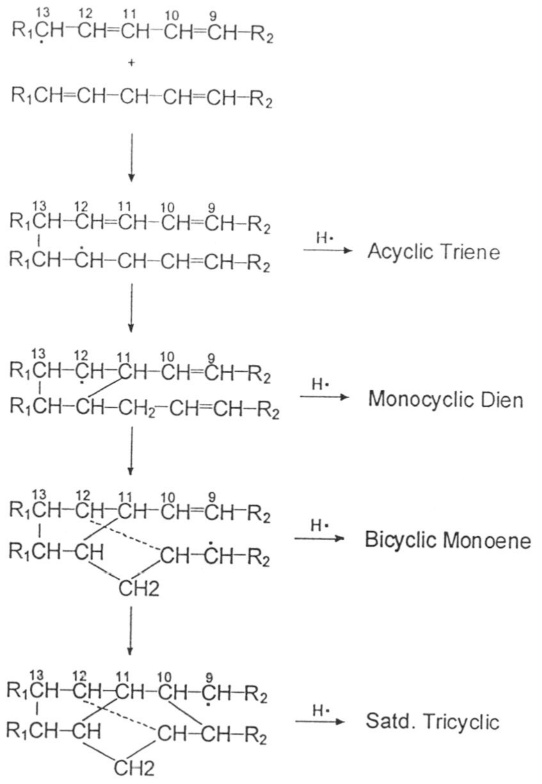

$R_1\dot{C}H-CH=CH-CH=CH-R_2$ (positions 13, 12, 11, 10, 9)

+

$R_1CH=CH-CH-CH=CH-R_2$

↓

$R_1CH-CH=CH-CH=CH-R_2$ / $R_1CH-\dot{C}H-CH-CH=CH-R_2$ —H•→ Acyclic Triene

↓

$R_1CH-\dot{C}H-CH-CH=CH-R_2$ / $R_1CH-CH-CH_2-CH=CH-R_2$ —H•→ Monocyclic Dien

↓

$R_1CH-CH-CH-CH=CH-R_2$ / R_1CH-CH $CH-\dot{C}H-R_2$ / CH2 —H•→ Bicyclic Monoene

↓

$R_1CH-CH-CH-CH-\dot{C}H-R_2$ / R_1CH-CH $CH-CH-R_2$ / CH2 —H•→ Satd. Tricyclic

FIGURE 26 Dimerization of linoleate.

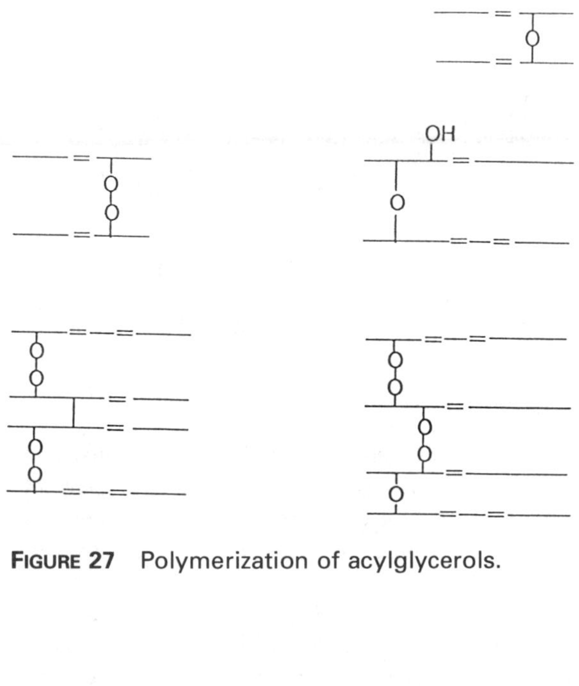

FIGURE 27 Polymerization of acylglycerols.

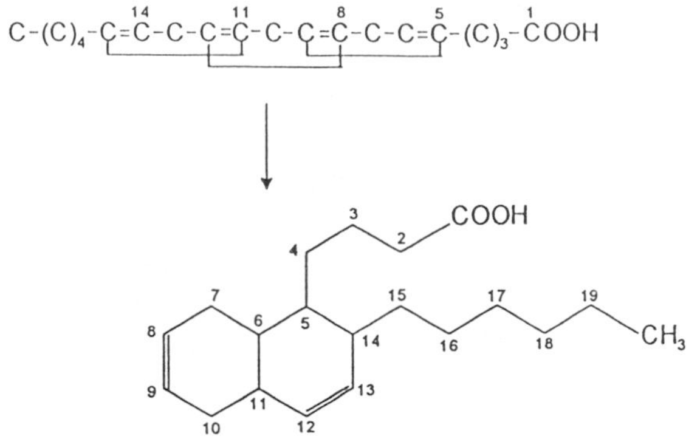

FIGURE 28 Cyclization of arachidonic acid.

Lipid Oxidation in Simple Nonbulk Lipid Systems

Because the mechanisms of lipid oxidation in intact biological systems are exceedingly difficult to determine, interest has focused on simpler nonbulk lipid systems with the hope that the results obtained can be extrapolated to more complex biological systems. Such nonbulk lipid models include monolayers (on the surface of an aqueous phase or adsorbed on silica), artificial lipid vesicles, and phospholipid bilayers. In a study on the oxidation of monolayers of linoleic acid esters, it was concluded that *cis* and *trans* epoxy compounds are the primary intermediates in oxidized bulk-phase lipids, rather than the hydroperoxides. Kinetically, the monolayer reaction is first order, whereas bulk-phase oxidation exhibits more complex kinetics. When the resistant saturated chains are interposed between reactive unsaturated acids in the monolayer, the rate of peroxidation decreases provided the chains are long enough to interfere with oxygen transfer. Little is known about the influence of molecular orientation on the mechanisms and products of lipid oxidation.

Enzyme-Catalyzed Oxidation of Lipids

Gardner [45–47] reviewed the enzymic oxidation of lipids. Sequential enzyme action starts with lipolysis. Released polyunsaturated fatty acids are then oxidized by either lipoxygenase or cyclo-oxygenase to form hydroperoxides or endoperoxides, respectively. Until recently, lipoxygenases were thought to be restricted to plants. However, animal lipoxygenase systems are now known to exist. Plant and animal lipoxygenases are both regiospecific (catalyze oxygenation at specific sites) and stereospecific (produce enantiomeric hydroperoxides). The next sequence of events involves enzymic cleavage of the hydroperoxides and endoperoxides to yield a variety of breakdown products, which are often responsible for the characteristic flavors of natural products.

Lipid Oxidation in Mixed-Lipid Complex Systems

In biological systems, various oxidizable lipids (e.g., triacylglycerols with different fatty acid constituents, phospholipids, cholesterol) often exist in proximity, and this makes the pattern of oxidation even more complex. These lipids also can exist in conjunction with, or sometimes bound to, nonlipid components such as proteins, carbohydrates, and water. This enables interactions to take place not only among lipids and the other nonlipid components, but also among the many oxidation products.

Much less clear is how mixtures of different oxidizable components affect oxidation of each other. In studies on oxidative interactions among cholesterol, triacylglycerols, and phospholipids in model systems, and among these same substances in membranes of the milk-fat globule and in red blood cells, it has been shown that two neighboring lipids can accelerate and/or inhibit the oxidation of each other, depending on conditions that prevail [81–83, 109]. Important conditions include chemical nature of the lipids, their concentrations, temperature and pH, and especially their physical state. Furthermore, during oxidation, one component may switch from being an accelerator of oxidation to an inhibitor, or vice versa.

Lipid Oxidation as Influenced by the Presence of Nonlipid Substances

In the presence of other compounds (e.g., proteins, pro- and antioxidants), oxidative reactions (both enzymic and nonenzymic) can be terminated by reactions with compounds other than those originating from oxidation of the lipid substrate, and this can influence reaction rates. One example of importance involves the interaction between free amino groups, for example, in amino acids, primary amines, or phosphatidylethanolamine, with aldehydes produced from oxidation of lipids. This can result in browning, the creation of antioxidant properties, and/or toxic effects. Also, the basic groups in proteins can catalyze aldol condensation of carbonyls produced from lipid oxidation, resulting in formation of brown pigments. Furthermore, lipid hydroperoxides or secondary products of lipid oxidation can damage the nutritive value of proteins by oxidizing sulfur-containing proteins [76], by free radical reactions with proteins, or by formation of Schiff-base addition products with the ε-amino groups of lysine residues. It has also been reported that free carbohydrates can increase the rate of oxidation of lipids in emulsions.

Biological systems also contain a multitude of nonlipid pro- and antioxidants that are present at small concentrations. The overall effect of these substances (pro- or antioxidant) on lipid oxidation is precarious, complex, and dictated by factors that are poorly understood. As is true of some of the major oxidizable components, some of these substances can switch roles (pro, anti) as conditions change.

5.5.2.10 Factors Influencing Rate of Lipid Oxidation in Foods

Food lipids contain a variety of fatty acids that differ in chemical and physical properties and also in their susceptibility to oxidation. In addition, foods contain numerous nonlipid components that may co-oxidize and/or interact with the oxidizing lipids and their oxidation products. Thus, oxidation of lipids in food is a dynamic, multifaceted series of events, often overlapping and continually interacting, which makes accurate description of oxidation kinetics almost impossible. Understandably, reviews on the kinetics of lipid oxidation have not been able to go beyond the treatment of simple model systems [28, 75, 88–90].

Fatty Acid Composition

The number, position, and geometry of double bonds affect the rate of oxidation. Relative rates of oxidation for arachidonic, linolenic, linoleic, and oleic acids are approximately 40:20:10:1, respectively. *Cis* acids oxidize more readily than their *trans* isomers, and conjugated double bonds are more reactive than nonconjugated. Autoxidation of saturated fatty acids is extremely slow; at room temperature, they remain practically unchanged when oxidative rancidity of unsaturates becomes detectable. At high temperatures, however, saturated acids can undergo oxidation at significant rates (see Temperature).

Free Fatty Acids Versus the Corresponding Acylglycerols

Fatty acids oxidize at a slightly greater rate when free than when esterified to glycerol. Randomizing the fatty acid distribution of a natural fat reduces the rate of oxidation.

The existence in a fat or oil of a small amount of free fatty acids does not have a marked effect on oxidative stability. In some commercial oils, however, the presence of relatively large amounts of free acids can facilitate incorporation of catalytic trace metals from equipment or storage tanks and thereby increase the rate of lipid oxidation.

Oxygen Concentration

When oxygen is abundant, the rate of oxidation is independent of oxygen concentration, but at very low oxygen concentration the rate is approximately proportional to oxygen concentration. However, the effect of oxygen concentration on rate is also influenced by other factors, such as temperature and surface area.

Temperature

In general, the rate of oxidation increases as the temperature is increased. Temperature also influences the relationship between rate and oxygen partial pressure. As temperature is increased, changes in oxygen partial pressure have a smaller influence on rate because oxygen becomes less soluble in lipids and water as the temperature is raised.

Surface Area

The rate of oxidation increases in direct proportion to the surface area of the lipid exposed to air. Furthermore, as surface–volume ratio is increased, a given reduction in oxygen partial pressure becomes less effective in decreasing the rate of oxidation. In oil-in-water emulsions the rate of oxidation is governed by the rate at which oxygen diffuses into the oil phase.

Moisture

In model lipid systems and various fat-containing foods, the rate of oxidation depends strongly on water activity (see also Chap. 2) [50, 72, 74]. In dried foods with very low moisture contents (a_W values of less than about 0.1), oxidation proceeds very rapidly. Increasing the a_W to about 0.3 retards lipid oxidation and often produces a minimum rate. This protective effect of small amounts of water is believed to occur by reducing the catalytic activity of metal catalysts, by quenching free radicals, and/or by impeding access of oxygen to the lipid.

At somewhat higher water activities ($a_W = 0.55–0.85$), the rate of oxidation increases again, presumably as a result of increased mobilization of catalysts and oxygen.

Molecular Orientation

Molecular orientation of substrates has an important influence on the rate of lipid oxidation. For example, when oxidative stability of polyunsaturated fatty acids (PUFA) was studied in aqueous solutions at 37°C and pH 7.4 with Fe^{2+}-ascorbic acid present as a catalyst, stability increased with increasing degree of unsaturation, contrary to what would be expected. This outcome was attributed to the conformations of the PUFAs existing in the aqueous medium [107].

Further insight regarding the effect of molecular orientation of lipids on the rate of oxidation was gained from a study involving ethyl linoleate [56]. In this study, linoleate was present either in an ordered state (adsorbed as a monolayer on silica) or in a bulk phase. At 60°C, oxidation of ethyl linoleate occurred more rapidly in the monolayer than it did in the bulk phase (expected result) because of greater access of the monolayer linoleate to oxygen. Surprisingly, at 180°C, the reverse was true. Apparently, the molecules of bulk-phase linoleate and associated free radicals were sufficiently more mobile than those in the monolayer sample to offset the effect of differences in access to oxygen. For both bulk and monolayer samples, the major decomposition products, although not identical, were those typically arising from classical decomposition of hydroperoxides.

Physical State

A recent study of cholesterol oxidation demonstrates the importance of physical state on rate of lipid oxidation. Fragments of microcrystalline cholesterol films, both solid and liquid, were suspended in aqueous media, and oxidation was studied under various conditions of temperature, time, pH and buffer composition. The types and ratios of oxides differed depending on conditions, with physical state of the cholesterol having the most pronounced influence [83]. Indeed, many of the factors considered in this section exert their effects on oxidation rates by influencing the physical state of both the substrate and/or the medium in which it exists [42, 129, 132].

Emulsification

In oil-in-water emulsions, or in foods where oil droplets are dispersed into an aqueous matrix, oxygen must gain access to the lipid by diffusion into the aqueous phase and passage through the oil–water interface. The rate of oxidation will depend on the interplay between a number of factors including type and concentration of emulsifier, size of oil droplets, surface area of interface, viscosity of the aqueous phase, composition and porosity of the aqueous matrix, and pH.

Molecular Mobility and the Glass Transition

If the rate of lipid oxidation is diffusion limited (see Chap. 2), a low rate of oxidation would be expected below the glass transition and the rate would exhibit a very large temperature dependence at temperatures above the glass transition [42].

Pro-Oxidants

Transition metals, particularly those possessing two or more valency states and a suitable oxidation–reduction potential between them (e.g., cobalt, copper, iron, manganese, and nickel), are effective pro-oxidants. If present, even at concentrations as low as 0.1 ppm, they can decrease the induction period and increase the rate of oxidation. Trace amounts of heavy metals are commonly encountered in edible oils, and they originate from the soil in which the oil-bearing plant was grown, from the animal, or from metallic equipment used in processing or storage. Trace metals are also naturally occurring components of all food tissues and of all fluid foods of biological origin (eggs, milk, and fruit juices), and they are present in both free and bound forms.

Several mechanisms for metal catalysis of oxidation have been postulated:

1. Acceleration of hydroperoxide decomposition:

$$M^{n+} + ROOH \longrightarrow M^{(n+1)+} + OH^- + RO^\bullet$$

$$M^{n+} + ROOH \longrightarrow M^{(n-1)+} + H^+ + ROO^\bullet$$

2. Direct reaction with the unoxidized substrate:

$$M^{n+} + RH \longrightarrow M^{(n-1)+} + H^+ + R^\bullet$$

3. Activation of molecular oxygen to give singlet oxygen and peroxy radical:

$$M^{n+} + O_2 \longrightarrow M^{(n+1)+} + O_2^{-\bullet} \quad \begin{cases} \xrightarrow{-e^-} {}^1O_2 \\ \xrightarrow{+H^+} HO_2^\bullet \end{cases}$$

Hematin compounds, present in many food tissues, are also important pro-oxidants.

Radiant Energy

Visible, ultraviolet, and γ-radiation are effective promoters of oxidation.

Antioxidants

Because lipid antioxidants are of such great importance in food chemistry, this subject is treated separately in a later section.

5.5.2.11 Methods for Measuring Lipid Oxidation

It is obvious from the preceding discussion that lipid oxidation is an exceedingly complex process involving numerous reactions that cause a variety of chemical and physical changes. Although these reactions appear to follow recognized stepwise pathways, they often occur simultaneously and competitively. The nature and extent of each change is influenced by many variables that have been mentioned previously. Since oxidative decomposition is of major significance in regard to both the acceptability and nutritional quality of food products, many methods have been devised for assessing this occurrence. However, a single test cannot possibly measure all oxidative events at once, nor can it be equally useful at all stages of the oxidative process, and for all fats, all foods, and all conditions of processing. At best, a test can monitor only a few changes as they apply to specific systems under specific conditions. For many purposes, a combination of tests is needed. These considerations should be kept in mind, as the following commonly used methods are reviewed.

Peroxide Value

Peroxides are the main initial products of autoxidation. They can be measured by techniques based on their ability to liberate iodine from potassium iodide (iodimetry):

$$ROOH + 2KI \rightarrow ROH + I_2 + K_2O$$

or to oxidize ferrous to ferric ions (thiocyanate method):

$$ROOH + Fe^{2+} \rightarrow ROH + HO + Fe^{3+}$$

The peroxide value is usually expressed in terms of milliequivalents of oxygen per kilogram of fat [113]. Various other colorimetric techniques are available. Although the peroxide value is applicable for following peroxide formation at the early stages of oxidation, it is nevertheless highly empirical. Its accuracy is questionable, and the results vary with the specific procedure used, and with test temperature. During the course of oxidation, peroxide values reach a peak and then decline.

Various attempts have been made to correlate peroxide values with development of oxidative off flavors. Good correlations are sometimes obtained, but often not. It should be pointed out that the amount of oxygen that must be absorbed, or peroxides that must be formed, to produce noticeable oxidative rancidity varies with composition of the oil (those that are more saturated require less oxygen absorption to become rancid), the presence of antioxidants and trace metals, and the conditions of oxidation.

Thiobarbituric Acid Test

This is one of the most widely used tests for evaluating the extent of lipid oxidation. Oxidation products of unsaturated systems produce a color reaction with thiobarbituric acid (TBA). It is believed that the chromagen results from condensation of two molecules of TBA with one molecule of malonaldehyde. However, malonaldehyde is not always present in all oxidized systems. Many alkanals, alkenals, and 2,4-dienals produce a yellow pigment (at 450 nm) in conjunction with TBA, but only dienals produce a red pigment (at 530 nm). It has been suggested that measurement at both absorption maxima is desirable.

In general, TBA-reactive material is produced in significant amounts only from fatty acids containing three or more double bonds. The reaction mechanism shown in Figure 29 was offered by Dahle and coworkers [30]. They suggested that radicals with a double bond β to the carbon bearing the peroxy groups (which can only arise from acids containing more than two double bonds) cyclize to form peroxides with five-membered rings. These then decompose to give malonaldehyde. More recently, however, Pryor et al. [120] concluded that malonaldehyde arises, at least in part, from decomposition of prostaglandin-like endoperoxides produced during autoxidation of polyunsaturated fatty acids.

Various compounds, other than those found in oxidized lipids, have been found to react with TBA to yield the characteristic red pigment. For example, sucrose and some compounds in wood smoke have been reported to give a red color upon reaction with TBA, and in these situations the observed results must be corrected for these interfering compounds. On the other hand, abnormally low TBA values can result if some of the malonaldehyde reacts with proteins in an oxidizing system. In addition, flavor scores for different systems cannot be consistently estimated from TBA values because the amount of TBA products obtained from a given amount of oxidation varies from product to product. The TBA test is often useful for comparing samples of a single material at different stages of oxidation.

Total and Volatile Carbonyl Compounds

Methods for determining total carbonyl compounds are usually based on measurement of hydrazones that arise from reaction of aldehydes or ketones (oxidation products) with 2,4-

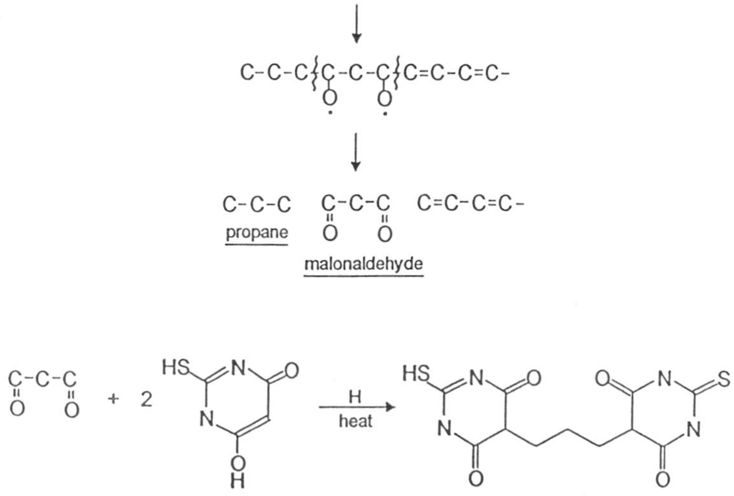

FIGURE 29 Speculative mechanism for malonaldehyde formation. (From Ref. 120.)

dinitrophenylhydrazine. However, under the experimental conditions used for these tests, carbonyl compounds may be generated by decomposition of unstable intermediates, such as hydroperoxides, thus detracting from accuracy of the results. Attempts to minimize such interference have involved reduction of hydroperoxides to noncarbonyl compounds prior to determination of carbonyls, or conducting the reaction at a low temperature.

Because the carbonyl compounds in oxidized fats are of relatively high molecular weight, they can be separated by a variety of techniques from lower molecular weight volatile carbonyl compounds. The lower molecular weight volatile carbonyl compounds are of interest because of their influence on flavor. The volatile carbonyls are usually recovered by distillation at atmospheric or reduced pressure and then determined by reaction of the distillate with appropriate reagents or by chromatographic methods. Quantitative measurement of hexanal by headspace analysis is a common technique [43].

Anisidine Value [136]

In the presence of acetic acid, *p*-anisidine reacts with aldehydes producing a yellowish color. The molar absorbance at 350 nm increases if the aldehyde contains a double bond. Thus, the anisidine value is mainly a measure of 2-alkenals. An expression termed the Totox or oxidation value (OV), which is equivalent to 2 × peroxide value + anisidine value, has been suggested for the assessment of oxidation in oils.

Kreis Test

This was one of the first tests used commercially to evaluate oxidation of fats. The procedure involves measurement of a red color that is believed to result from reaction with phloroglucinol (Kreis reagent). Deficiencies of this test include: (a) fresh samples free of oxidized flavor frequently develop some color upon reaction with the Kreis reagent, and (b) consistent results among different laboratories are difficult to obtain.

Ultraviolet Spectrophotometry

Measurement of absorbance at 234 nm (conjugated dienes) and 268 nm (conjugated trienes) is sometimes used to monitor oxidation. However, the magnitude of the absorbance does not correlate well with the degree of oxidation except in the early stages.

Oxirane Test

This method, which provides a measure of epoxide content, is based on the addition of hydrogen halides to the oxirane group. Epoxide content is determined by dissolving the sample in aqueous acetic acid in the presence of crystal violet, and titrating with HBr to a bluish green endpoint [113]. The test, however, has poor sensitivity, lacks specificity, and does not provide an accurate quantitative result with some *trans*-epoxides. Hydrogen halides also react with β-unsaturated carbonyls and conjugated dienols. A colorimetric method for epoxides, based on the reaction of the oxirane group with picric acid, is reportedly more sensitive than the oxirane test and is not subject to the shortcomings mentioned [37].

Iodine Value

This test is a measure of the unsaturated linkages in a fat and is determined by reacting the fat with a solution of iodine monochloride in a mixture of acetic acid and CCl_4, liberating excess iodine with KI, and titrating with $Na_2S_2O_3$ (Wijs method). The results are expressed as grams of iodine absorbed by a 100-g sample [113]. In the Hanus method, an iodine bromide reagent is used instead of iodine monochloride. A decline in iodine value is sometimes used to measure the reduction of dienoic acids during the course of autoxidation.

Fluorescence

Fluorescent compounds may develop from interaction of carbonyl compounds (produced by lipid oxidation) with constituents possessing free amino groups [32]. Fluorescence methods provide a relatively sensitive measure of oxidation products in biological tissues.

Chromatographic Methods

Various chromatographic techniques, including liquid, thin-layer, high-performance liquid, size exclusion, and gas, have been used to determine oxidation in oils or lipid-containing foods. This approach is based on the separation and quantitative measurement of specific fractions, such as volatile [34, 122], polar [11], or polymeric compounds [142, 143] or individual components, such as pentane or hexanal [43], that are known to be typically produced during autoxidation.

Sensory Evaluation

The ultimate test for oxidized flavor in foods is a sensory one. The value of any objective chemical or physical method is judged primarily on how well it correlates with results from sensory evaluation. The testing of flavor is usually conducted by trained or semitrained taste panels using highly specific procedures [69].

Several "accelerated" tests are also available to estimate resistance of a lipid to oxidation.

Schaal Oven Test

The sample is stored at about 65°C and periodically tested until oxidative rancidity is detected. Detection can be done organoleptically or by measuring the peroxide value.

Active Oxygen Method (AOM)

This test is widely used. It involves maintenance of the sample at 98°C while air is continuously bubbled through it at a constant rate. The time required to obtain a specific peroxide value is then determined [113].

Rancimat Method [55]

Air is bubbled through the oil as in the AOM test, and the increase in electrical conductivity due to generation of oxidation products is measured, usually at 100°C, and expressed in terms of "induction time."

Oxygen Absorption

The sample is placed in a closed chamber and the amount of oxygen absorbed is determined and used as a measure of stability. This is done by measuring the time to produce a specific pressure decline, or by the time to absorb a preestablished quantity of oxygen under specific oxidizing conditions. This test has been particularly useful in studies of antioxidant activity.

5.5.2.12 Antioxidants [14, 24, 62, 127]

Antioxidants are substances that can delay onset, or slow the rate, of oxidation of autoxidizable materials. Literally hundreds of compounds, both natural and synthesized, have been reported to possess antioxidant properties. Their use in foods, however, is limited by certain obvious requirements not the least of which is adequate proof of safety. The main lipid-soluble antioxidants currently used in food are monohydric or polyhydric phenols with various ring substitutions (Fig. 30). For maximum efficiency, primary antioxidants are often used in combination with other phenolic antioxidants or with various metal sequestering agents (see Table 6).

Although the mechanisms by which many antioxidants impart stability to pure oils are relatively well known, much remains to be learned about their action in complex foods.

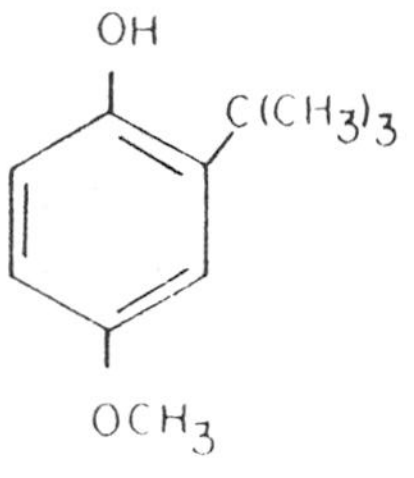

OH
$C(CH_3)_3$
OCH_3

OH
$C(CH_3)_3$
OCH_3

OH
$(CH_3)_3C$ $C(CH_3)_3$
CH_3

2-BHA 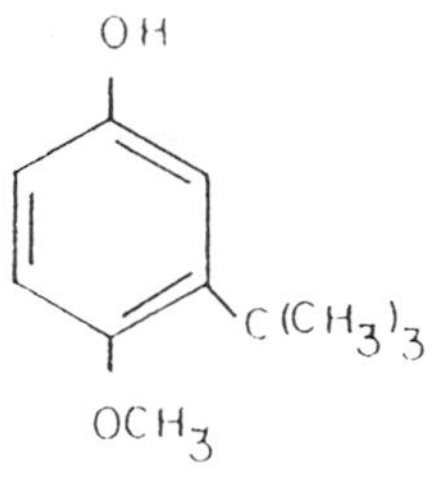 3-BHA
(Butylated hydroxyanisole)

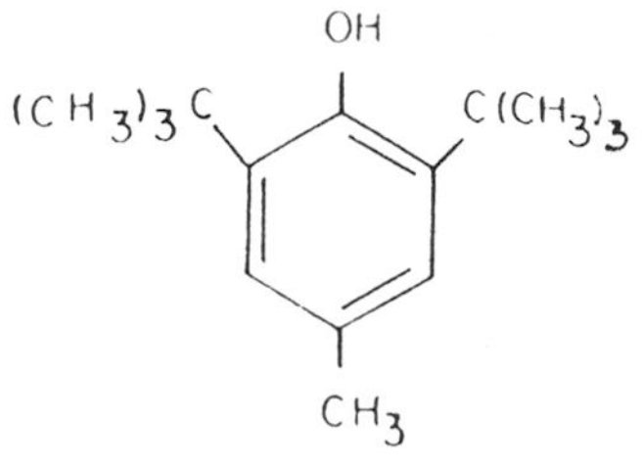

BHT
(Butylated hydroxytoluene)

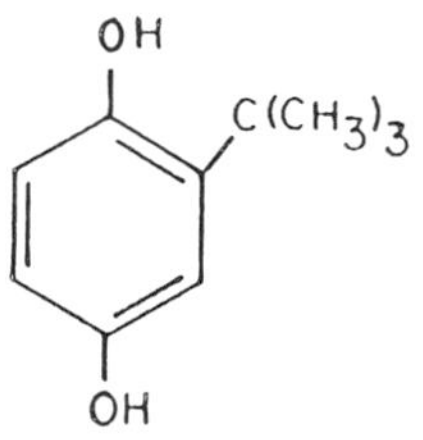

OH
$C(CH_3)_3$
OH

OH
HO OH
$COOC_3H_7$

TBHQ
(Tertiary butylhydroquinone)

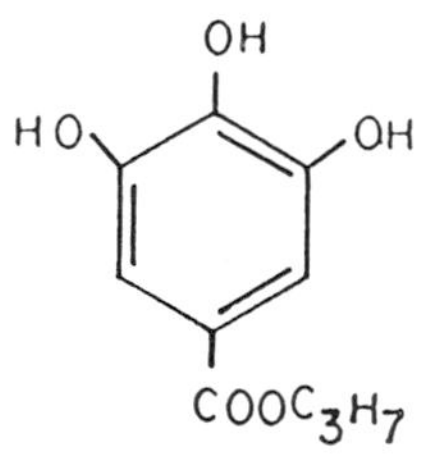

PG
(Propyl gallate)

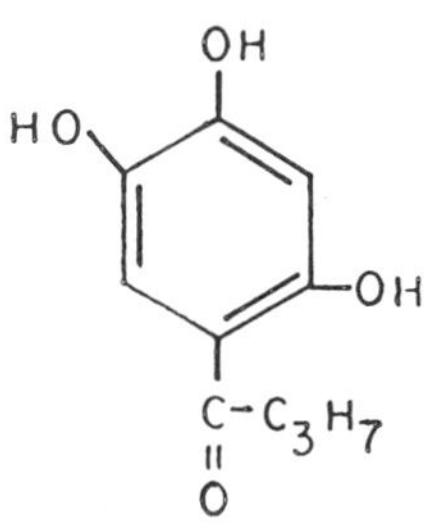

OH
HO
OH
$C-C_3H_7$
O

OH
$(CH_3)_3C$ $C(CH_3)_3$
CH_2OH

THBP
(2,4,5-Trihydroxy-butyrophenone)

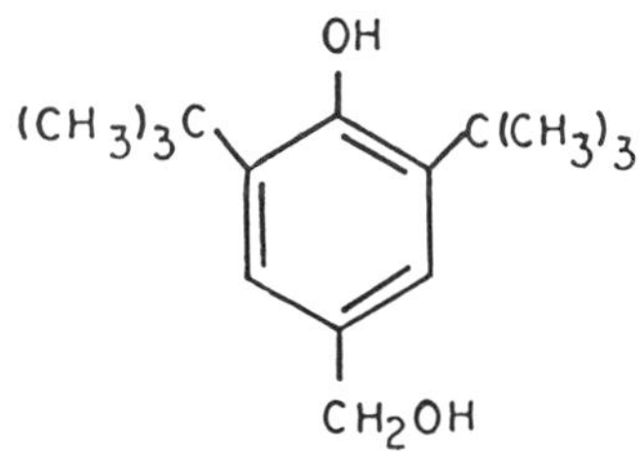

4-Hydroxymethyl-2,6-ditertiarybutylphenol

FIGURE 30 Major antioxidants used in food.

Effectiveness and Mechanisms of Action

Several reviews of antioxidant kinetics and mechanism of action have been published [118, 125, 128, 139]. As should be clear from the autoxidative mechanisms outlined earlier, a substance delays the autoxidation reaction if it inhibits the formation of free radicals in the initiation step, or if it interrupts propagation of the free radical chain. The initiation of free radicals can be retarded by using peroxide decomposers, metal chelating agents, or singlet-oxygen inhibitors. Because it is very difficult to eliminate traces of peroxides and metal initiators, most studies have focused on the use of free radical acceptors.

The first detailed kinetic study of antioxidant action was conducted in 1947 by Bolland and ten Have using a model system of autoxidizing ethyl linoleate containing hydroquinone as an inhibitor [13]. They postulated that antioxidants inhibit the chain reaction by acting as hydrogen donors or free radical acceptors and concluded that the free-radical acceptor (AH) reacts primarily with ROO˙ and not with R˙ radicals.

$$ROO^{\cdot} + AH \rightarrow ROOH + A^{\cdot}$$

These authors also concluded that the most probable number of oxidation chains terminated by one inhibitor molecule is two, and that the reaction can, therefore, be considered as occurring in two stages. Different versions of antioxidation mechanisms were suggested by other workers. It was proposed, for example, that the free-radical intermediate, AH˙, forms stable products by reacting with an $RO_2^{\cdot}$ radical or that a complex between the RO_2 radical and the inhibitor is formed followed by reaction of the complex with another $RO_2^{\cdot}$ radical to yield stable products:

$$RO_2^{\cdot} + inh^{\cdot} \rightarrow [RO_2\text{-}inh^{\cdot}]$$

$$[RO_2\text{-}inh^{\cdot}] + RO_2^{\cdot} \rightarrow \text{stable products}$$

Although all of the preceding reactions may be taking place, the original concept of Bolland and ten Have is considered to be the most important, and the basic mechanism can thus be visualized as a competition between the "inhibitor reaction,"

$$RO_2^{\cdot} + AH \rightarrow ROOH + A^{\cdot}$$

and the chain propagation reaction,

$$RO_2^{\cdot} + RH \rightarrow ROOH + R^{\cdot}$$

The effectiveness of an antioxidant is related to many factors, including activation energy, rate constants, oxidation reduction potential, ease of antioxidant loss or destruction, and solubility properties. For the two competing reactions discussed earlier (i.e., the inhibitor reaction and the chain propagation reaction), both of which are exothermic, the activation energy increases with increasing A-H and R-H bond dissociation energies, and therefore the efficiency of the antioxidant (AH) increases with decreasing A-H bond strength. Ideally, however, the resulting antioxidant free radical must not itself initiate new free radicals or be subject to rapid oxidation by a chain reaction. In this regard phenolic antioxidants occupy a favored status. They are excellent hydrogen or electron donors and, in addition, their radical intermediates are relatively stable due to resonance delocalization and to a lack of positions suitable for attack by molecular oxygen. Hydroquinones, for example, react with hydroperoxy radicals, forming stable semiquinone resonance hybrids.

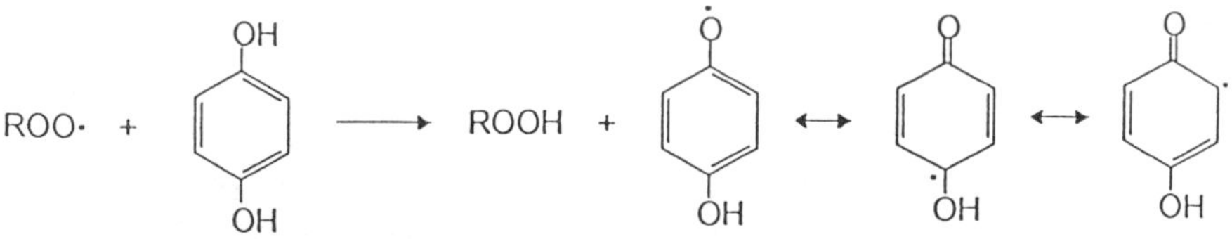

$$ROO\cdot + \text{hydroquinone (HO–C}_6\text{H}_4\text{–OH)} \longrightarrow ROOH + \text{semiquinone radical} \leftrightarrow \leftrightarrow$$

The semiquinone radical intermediates may undergo a variety of reactions forming more stable products. They can react with one another to form dimers, or dismutate to yield quinones with renewed formation of the original inhibitor molecule:

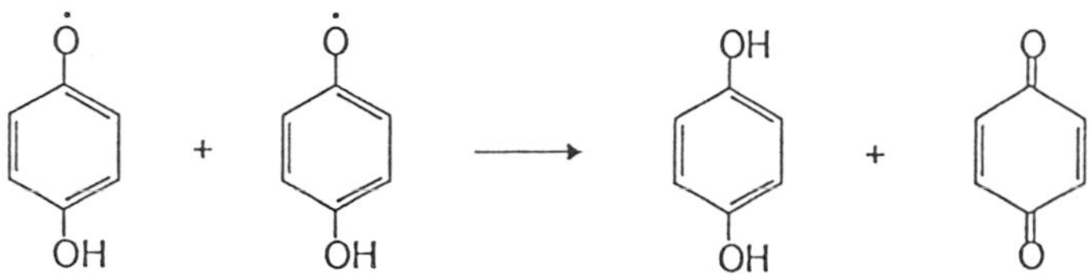

$$\text{semiquinone} + \text{semiquinone} \longrightarrow \text{hydroquinone} + \text{quinone}$$

or they can react with another $RO_2^{\cdot}$ radical.

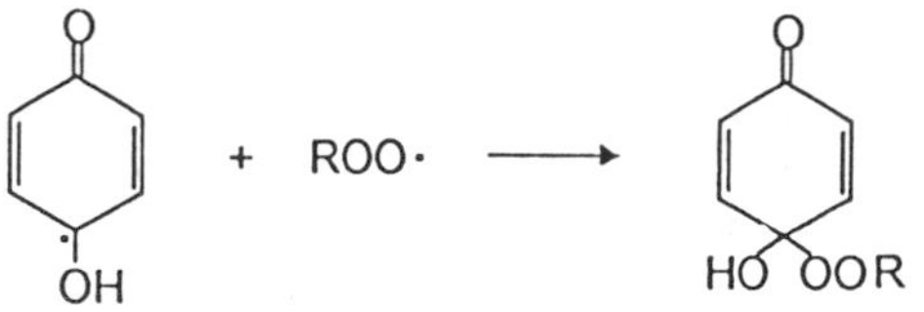

$$\text{semiquinone} + ROO\cdot \longrightarrow \text{(HO, OOR adduct)}$$

The monohydric phenols cannot form semiquinones or quinones, as such, but they give rise to radical intermediates with moderate resonance delocalization, and they can be sterically hindered, as, for example, by the two *t*-butyl groups in BHT. The *t*-butyl groups reduce further chain initiation (by the alkoxy radical) after the initial hydrogen donation. The net reaction between linoleate and BHA can be represented as:

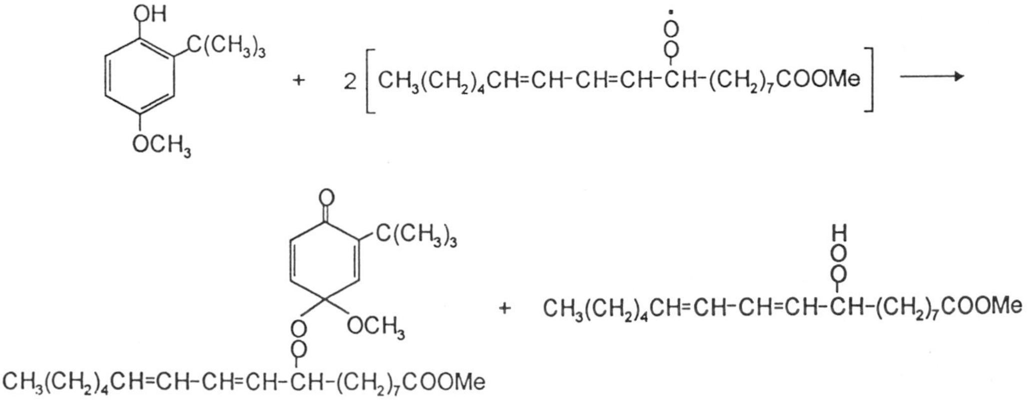

$$\text{BHA (OH, C(CH}_3)_3\text{, OCH}_3) + 2\left[CH_3(CH_2)_4CH{=}CH{-}CH{=}CH{-}CH(OO\cdot){-}(CH_2)_7COOMe\right] \longrightarrow$$

$$\text{adduct: }CH_3(CH_2)_4CH{=}CH{-}CH{=}CH{-}CH(OO{-}){-}(CH_2)_7COOMe \text{ bonded to ring (O, C(CH}_3)_3\text{, OCH}_3) + CH_3(CH_2)_4CH{=}CH{-}CH{=}CH{-}CH(OOH){-}(CH_2)_7COOMe$$

In addition to the innate chemical potency of an antioxidant, its solubility in oil and its volatility influence its effectiveness. Its solubility affects accessibility to peroxy radical sites, and its volatility affects its persistence during storage or heating.

In the last few years, emphasis has shifted to the study of antioxidant action at surfaces, especially in such systems as membranes, micelles, and emulsions. The importance of the amphiphilic character of antioxidant molecules to their effectiveness in biphasic and multiphasic systems is demonstrated in a provocative discussion by Porter [118].

Synergism

Synergism occurs when a mixture of antioxidants produces greater activity than the sum of the activities of the antioxidants when tested individually. Two categories of synergism are recognized. One involves the action of mixed free-radical acceptors; the other involves the combined action of a free-radical acceptor and a metal chelating agent. Frequently, however, a so-called synergist may play more than one role. Ascorbic acid, for example, may function as an electron donor, a metal chelator, an oxygen scavenger, and a contributor to the formation of browning products that have antioxidant activity.

1. Mixed free radical acceptors. Uri [139] introduced a hypothesis to explain the action of two mixed free-radical acceptors, designated AH and BH, where it is assumed that the bond dissociation energy of B-H is smaller than that of A-H. It is further assumed that BH reacts slowly with $RO_2^{\cdot}$ because of steric hindrance. The following reaction would then take place:

$$RO^{\cdot}_2 + AH \rightarrow ROOH + A^{\cdot}$$

$$A^{\cdot} + BH \rightarrow B^{\cdot} + AH$$

Thus, the presence of BH will have a sparing effect since it results in regeneration of the primary antioxidant. In addition, the tendency of $A^{\cdot}$ to disappear via a chain reaction is substantially diminished. An example of such a system is a phenolic antioxidant and ascorbic acid. Since the phenolic compound, if used by itself, is the more effective of the two, it is customarily termed the primary antioxidant while ascorbic acid is called the synergist. In a similar way, it is also possible for two phenolic antioxidants to exhibit synergism.

2. Metal chelating agents. In general, metal chelating agents partly deactivate trace metals, which are often present as salts of fatty acids. When the antioxidant property of a free radical acceptor is enhanced greatly by the presence of a metal complexing agent, a synergistic effect is said to occur. Citric and phosphoric acids, polyphosphates, and ascorbic acid are typical examples of metal chelating agents.

Choice of Antioxidant

Due to differences in their molecular structure, various antioxidants exhibit substantial differences in effectiveness when used with different types of oils or fat-containing foods, and when used under different processing and handling conditions. In addition to the potency of an antioxidant(s) in a particular application, due consideration must be given to other factors, such as ease of incorporation in the food, carry-through characteristics, sensitivity to pH, tendency to discolor or produce off flavors, availability, and cost. The problem of selecting the optimum antioxidant or combination of antioxidants is further complicated by the difficulty of predicting how the added antioxidant(s) will function in the presence of pro-oxidants and antioxidants already present in the food or produced in the course of processing.

The hydrophilic–lipophilic properties of various antioxidants also have a bearing on their effectiveness in different applications. Porter [118] described two basic situations that call for two different types of antioxidants. The first involves lipids with a small surface–volume ratio, such as occurs with bulk oils. In this case antioxidants with relatively large values for hydrophilic–lipophilic balance [e.g., propyl gallate (PG) or *tert*-butylhydroquinone) TBHQ)] are most effective since they concentrate at the surface of the oil where reaction of fat with molecular oxygen is most prevalent. The second situation involves lipids with a large surface–volume ratio, such as occurs with polar lipid membranes in intact tissue foods, with intracellular micelles of neutral lipids, and with oil-in-water emulsions (e.g., salad dressing). These are multiphasic systems in which the concentration of water is large and the lipid is often in a mesophasic state.

For these situations, the more lipophilic antioxidants, such as, BHA, BHT, higher alkyl gallates, and tocopherols, are most effective.

Characteristics of Some Commonly Used Primarily Antioxidants

Tocopherols: These are the most widely distributed antioxidants in nature, and they constitute the principal antioxidants in vegetable oils. The small amounts of tocopherols present in animal fats originate from vegetable components in the animal diet. Eight tocopherol structures, all methyl-substituted forms of tocol, are known. Of these tocepherols, 5,7,8-trimethyl-, 7,8-dimethyl-, and 8-methyltocol, respectively, predominate in vegetable oils.

Tocol

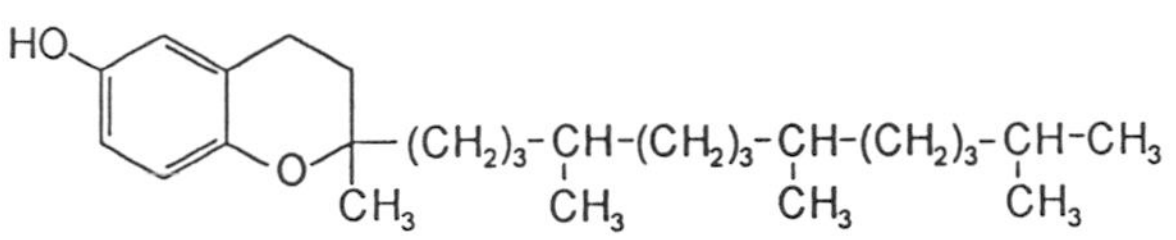

HO

$-(CH_2)_3-CH(CH_3)-(CH_2)_3-CH(CH_3)-(CH_2)_3-CH(CH_3)-CH_3$ (ring carbon bearing CH_3)

A relatively high proportion of the tocopherols present in crude vegetable oils survives the oil processing steps and remains in sufficient quantities to provide oxidative stability in the finished product. As antioxidants, tocopherols exert their maximum effectiveness at relatively low levels, approximately equal to their concentration in vegetable oils. If used at a very high concentration, they may actually act as pro-oxidants.

Mechanism of Action: Like all phenolic antioxidants, tocepherol exerts its antioxidant effect by competing with the reaction:

$$ROO^{\cdot} + RH \rightarrow ROOH + R^{\cdot}$$

Tocopherol (TH_2) reacts with peroxy radicals in the following manner:

$$ROO^{\cdot} + TH_2 \rightarrow ROOH + TH^{\cdot}$$

The α-tocopherol radical ($TH^{\cdot}$) is relatively stable due to delocalization of the unpaired electron and therefore is less reactive than the peroxyl radical. This is why α-tocopherol is an effective antioxidant. Figure 31 shows five resonance structures of the tocopherol radical. Resonance structure V is energetically favorable [9]. The α-tocopherol radical may quench another peroxyl radical to yield methyltocopherylquinone (T), as shown,

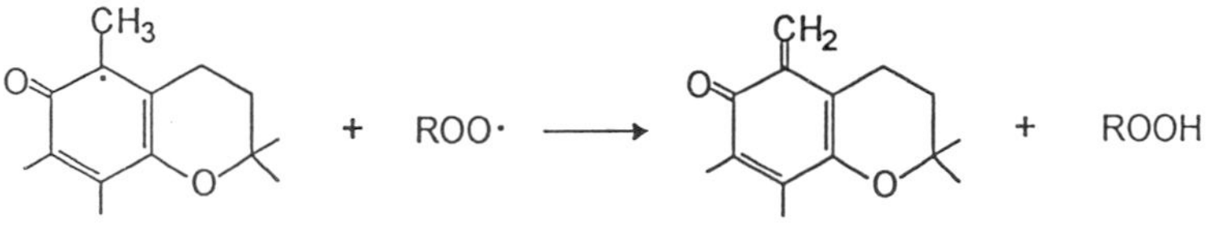

$TH^{\cdot}$ + $ROO^{\cdot}$ ⟶ T + ROOH

TH·

T
(Methyl tocopherylquinone)

or may react with another α-tocopheryl radical to give methyl tocopherylquinone and a regenerated molecule of tocopherol:

$$TH^{\cdot} + TH^{\cdot} \rightarrow T + TH_2$$

In general, tocopherols with high vitamin E activity are less effective antioxidants than those with low vitamin E activity. The order of antioxidant activity is thus $\delta > \gamma > \beta > \alpha$. However, the relative activity of these compounds is significantly influenced by temperature and light.

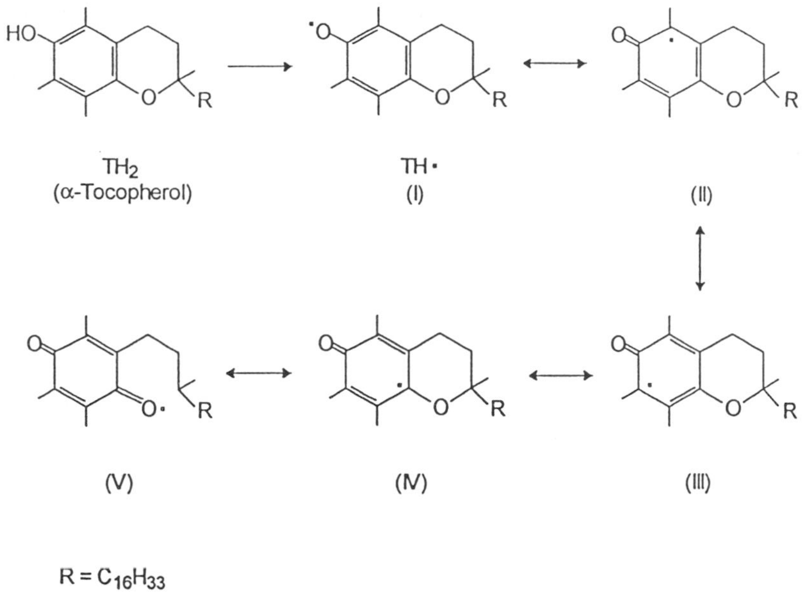

TH_2
(α-Tocopherol)

$TH\cdot$
(I)

(II)

(V)

(IV)

(III)

$R = C_{16}H_{33}$

FIGURE 31 Resonance structures of the α-tocopherol radical. (From Ref. 9.)

In biological systems, ascorbate is capable of "recycling" α-tocopherol by reducing the α-tocopherol radical [9].

Tocopherol as a Pro-Oxidant: It has been observed that under certain conditions tocopherol may act as a pro-oxidant. Under normal circumstances, where lipid concentration greatly exceeds the concentration of tocopherol, progressive oxidation depletes tocopherol, leaves the pool of lipid relatively unchanged and causes accumulation of ROOH. Accumulation of ROOH reverses the equilibrium reaction,

$$ROO^{\cdot} + TH_2 \rightarrow ROOH + TH^{\cdot}$$

thus stimulating the propagation reaction

$$RH + ROO^{\cdot} \rightarrow ROOH + R^{\cdot}$$

When α-tocopherol is present at a relatively high concentration, a pro-oxidant effect can occur via radical formation according to the reaction

$$ROOH + TH_2 \rightarrow RO^{\cdot} + TH^{\cdot} + H_2O$$

Gum guaiac is a resinous exudate from a tropical tree. Its antioxidant effectiveness, due mainly to its appreciable content of phenolic acids, is more pronounced in animal fats than in vegetable oils. Gum guaic has a reddish brown color, is very slightly soluble in oil, and gives rise to some off flavor.

Both *butylated hydroxyanisole* (BHA) which is commercially available as a mixture of two isomers (Fig. 27), and *butylated hydroxytoluene* (BHT) have found wide commercial use in the food industry. Both are highly soluble in oil and exhibit weak antioxidant activity in vegetable oils, particularly those rich in natural antioxidants. BHT and BHA are relatively effective when used in combination with other primary antioxidants. BHA has a typical phenolic odor that may become noticeable if the oil is subjected to high heat.

Nordihydroquaiaretic acid (NDGA) is extracted from a desert plant, *Larrea divaricata*. Its solubility in oils is limited (0.5–1%), but greater amounts can be dissolved if the oil is heated. NDGA has poor carry-through properties and tends to darken slightly on storage, in the presence of iron, or when subjected to high temperatures. The antioxidant activity of NDGA is markedly influenced by pH, and it is readily destroyed in highly alkaline conditions. NDGA is reported to effectively retard hematin-catalyzed oxidation in fat–aqueous systems and in certain meats. In the United States NDGA is not permitted for use as a food additive. However, it can be used in food packaging material.

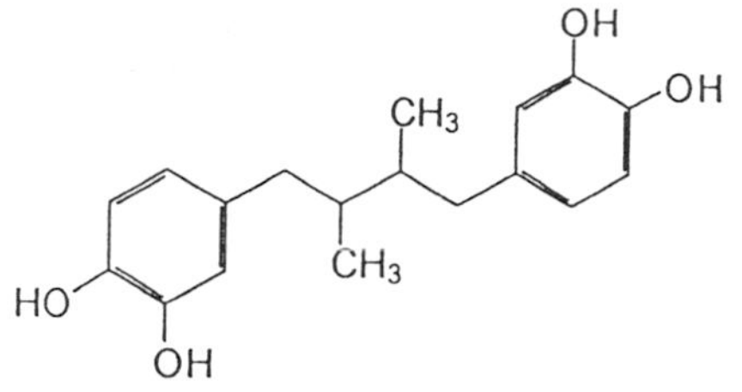

OH OH CH3 CH3 HO OH

Nordihydroguaiaretic acid (NDGA)

As would be expected from their phenolic structure and three hydroxyl groups, *alkyl gallates* and *gallic acid* exhibit considerable antioxidant activity. Gallic acid is soluble in water but nearly insoluble in oil. Esterification of the carboxyl group with alcohols of varying chain length produces alkyl gallates with increased oil solubility. Of these, propyl gallate is widely used in the United States. This compound reportedly is effective for retarding lipoxygenase-catalyzed oxidation of linoleate. In the presence of traces of iron and alkaline conditions, gallates will cause a blue-black discoloration. They are unstable during baking or frying.

Tertiary Butylhydroquinone (TBHQ): TBHQ is moderately soluble in oil and slightly soluble in water. In many cases, TBHQ is more effective than other common antioxidants in providing oxidative stability to crude and refined polyunsaturated oils, without problems of color or flavor stability. TBHQ is also reported to exhibit good carry-through characteristics in the frying of potato chips.

As would be expected from their similarity in structure, *2,4,5-trihydroxybutyrophenone* (THBP) and the gallates exhibit similar antioxidant properties. THBP is not widely used in the United States.

4-Hydroxymethyl-2,6-di-tertiary-butylphenol is produced by substituting a hydroxyl for one hydrogen in the methyl group of BHT. It is therefore less volatile than BHT but otherwise behaves similarly as an antioxidant.

Decomposition of Antioxidants

Antioxidants may exhibit significant decomposition, particularly at elevated temperatures, and several degradation products can arise. The amounts of these products are very small because the concentrations of antioxidants allowed in food are small. However, some of these degradation products have antioxidative properties.

The stabilities of four phenolic antioxidants have been studied during heating for 1 hr at 185°C [54]. Apparent stability increased in the order TBHQ < BHA < PG < BHT. This tendency is based partly on heat stability and partly on loss by evaporation (volatility). PG is the least volatile while BHT and TBHQ are the most volatile.

Figure 32 shows reaction mechanisms proposed for thermal decomposition of BHA. The antioxidant first loses a hydrogen atom, giving rise to the free radical species a, b, and c.

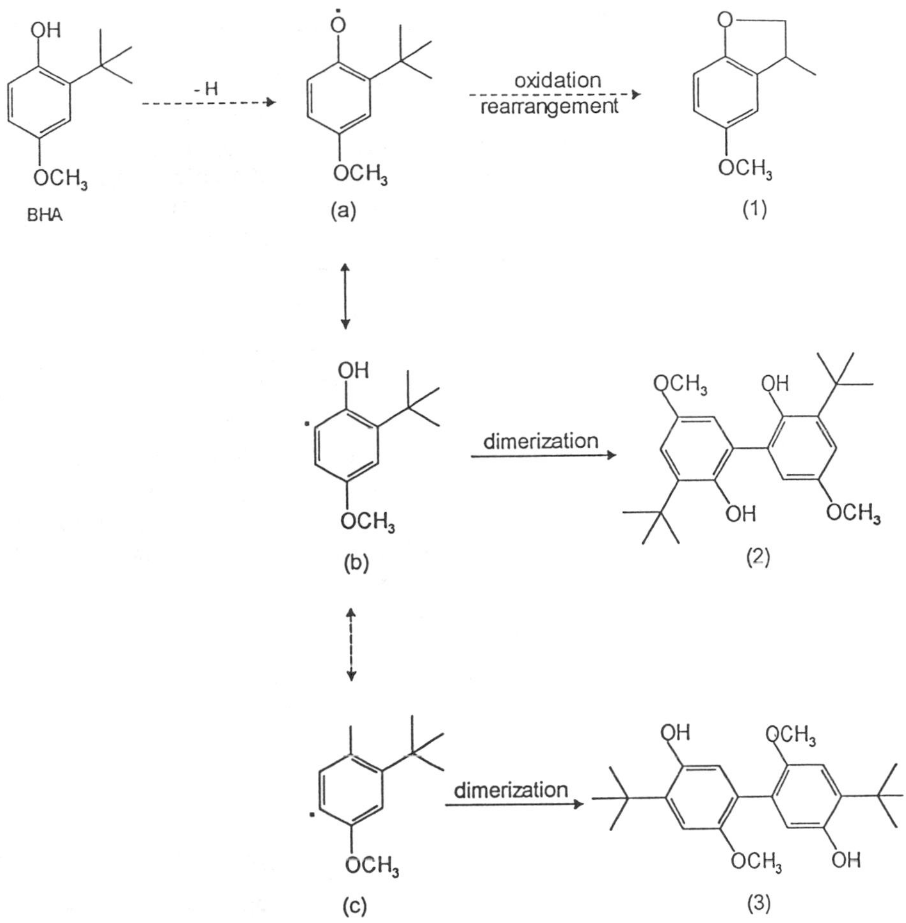

FIGURE 32 Thermal decomposition of BHA.

Oxidation of radical a produces compound 1, while various recombinations of the resonant hybrid result in formation of the dimeric products 2 and 3. The biological properties of these degradation products are not known.

Regulation

Antioxidants are food additives. In the United States their use is subject to regulation under the Federal Food Drug and Cosmetic Act. Antioxidants for food products are also regulated under the Meat Inspection Act, the Poultry Inspection Act, and various state laws. Antioxidants permitted for use in foods are listed in Table 6. In most instances the total concentration of authorized antioxidants, added singly or in combination, must not exceed 0.02% by weight based on the fat content of the food. Certain exceptions exist in the case of standardized foods and products covered by special regulations. Under the Meat Inspection Act, concentrations up to 0.01% are permitted for single antioxidants, based on fat content, with a combined total of no more than 0.02%. Tocopherols and the major acid synergists are virtually unregulated. For some foods, however, the tocopherols are not permitted above certain concentrations: 0.0002% in milk products, 0.002% in infant formulas, 0.03% (lipid) in shellfish, and 0.05% (lipid) in herbs, spices, or flavorings. BHT and BHA are not allowed in fish products.

Table 6 Antioxidants Permitted in Foods in the United States

Primary antioxidants	Synergists
Tocopherols	Citric acid and isopropyl citrate
Gum guaiac	Phosphoric acid
Propyl gallate	Thiodipropionic acid and its didodecyl, dilauryl, and dioctadecyl esters
Butylated hydroxyanisole (BHA)	Ascorbic acid and ascorbyl palmitate
Butylated hydroxytoluene (BHT)	Tartaric acid
2,4,5-Trihydroxybutyrophenone (THBP)	Lecithin
4-Hydroxymethyl-2,6-di-*tert*-butylphenol	
tert-Butylhydroquinone (TBHQ)	

The general public concern with the safety of chemical additives has stimulated a continuing search for new antioxidants that may occur naturally in food or may form inadvertently during processing. Compounds with antioxidant properties have been found in many spices, oil seeds, citrus pulp and peel, cocoa shells, oats, soybean, hydrolyzed plants, animal and microbial proteins, and in products that have been heated and/or have undergone nonenzymic browning.

Another result of the effort to achieve "antioxidant safety" is the development of nonabsorbable polymeric antioxidants. These are generally hydroxyaromatic polymers with various alkyl and alkoxyl substitutions. These compounds are usually large and stable so absorption in the intestinal tract is practically nil. In addition to their reportedly high antioxidant activity they are nonvolatile under deep-fat frying conditions, which results in nearly quantitative carry-through to the fried items. They have not yet, however, received FDA approval.

Modes of Application

Antioxidants can be added directly to vegetable oils or to melted animal fats after they are rendered. In some cases, however, better results are achieved when the antioxidant is administered in a diluent. Examples are mixtures of monoacylglycerols and glycerol in propylene glycol, monoacylglycerol–water emulsions, and mixtures of antioxidants in volatile solvents. Food products can also be sprayed with or dipped in solutions or suspensions of antioxidants, or they can be packaged in films that contain antioxidants.

5.5.3 Thermal Decomposition

Heating of food produces various chemical changes, some of which can be important to flavor, appearance, nutritive value, and toxicity. Not only do the different nutrients in food undergo decomposition reactions, but these nutrients also interact among themselves in extremely complex ways to form a very large number of new compounds.

The chemistry of lipid oxidation at high temperatures is complicated by the fact that both thermolytic and oxidative reactions are simultaneously involved. Both saturated and unsaturated fatty acids undergo chemical decomposition when exposed to heat in the presence of oxygen. A schematic summary of these mechanisms is shown in Fig. 33.

5.5.3.1 Thermal Nonoxidative Reactions of Saturated Fats

In general, very high temperatures of heating are required to produce substantial nonoxidative decomposition of saturated fatty acids. Thus, heating of saturated triacylglycerols and methyl esters of fatty acids at temperatures of 200–700°C yields detectable amounts of decomposition products consisting mostly of hydrocarbons, acids, and ketones. However, using very sensitive

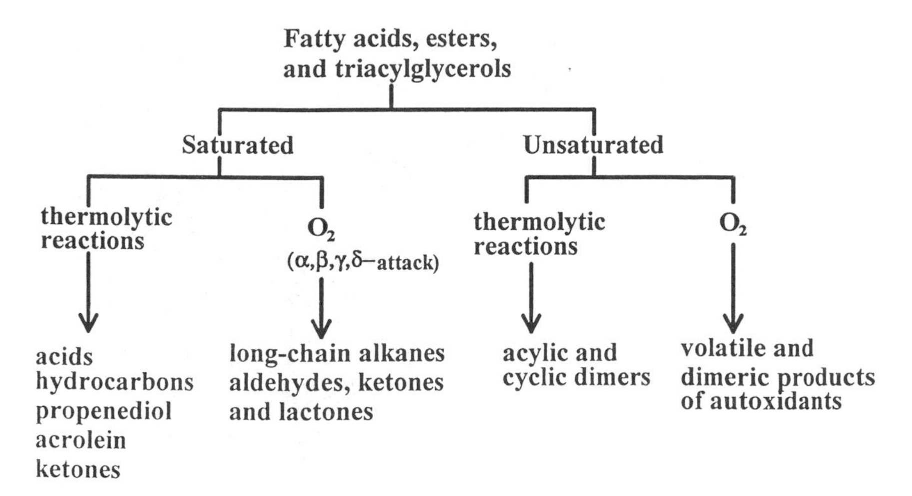

FIGURE 33 Generalized scheme for thermal decomposition of lipids.

measurement techniques, thermolytic products were detected in triacylglycerols after heating in vacuum for only 1 hr at 180°C [29].

Shown in Figure 34 are the products arising from anaerobic heating of tributyrin (Tri-4), tricaproin (Tri-6), and tricaprylin (Tri-8). Each triacylglycerol produces the following products (n is the number of carbons in the fatty acid molecule): a series of normal alkanes and alkenes with the C_{n-1} alkane predominating; a C_n fatty acid; a C_{2n-1} symmetrical ketone; a C_n oxopropyl ester; C_n propene and propanediol diesters; and C_n diacylglycerols. Acrolein, CO, and CO_2 are also formed. Quantitatively, the constituent fatty acids are the major compounds produced from

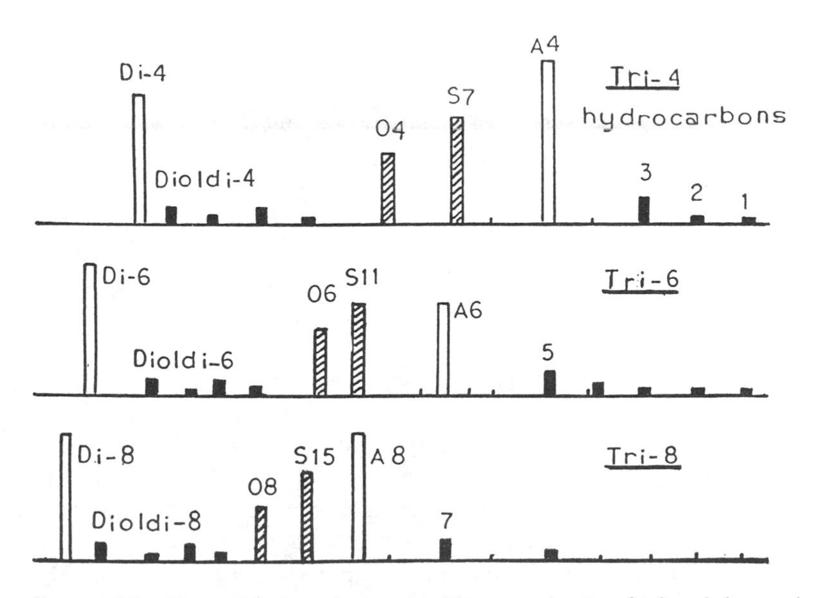

FIGURE 34 Nonoxidative decomposition products of triacylglycerols: A, free acid; S, symmetrical ketone; O, oxopropyl ester; Dioldi, propane- and propenediol diesters; Di, diglycerides. Numbers refer to hydrocarbon or fatty acid carbon chain.

thermolytic decomposition of triacylglycerols. In the absence of moisture the free fatty acid can be formed via a "six-atom-ring closure," as follows:

$$\begin{array}{c} H_2C{-}O{-}C(R'){=}O\cdots H \\ | \\ R''COOC \\ | \\ H_2COOCR'' \end{array} \longrightarrow \begin{array}{c} H_2C \\ \| \\ R''COOC \\ | \\ H_2COOCR'' \end{array} + \begin{array}{c} O{=}C{-}R' \\ | \\ O{-}H \end{array}$$

This mechanism also explains the formation of propenedioldiesters. Expulsion of the acid anhydride from the triacylglycerol molecule produces 1- or 2-oxopropyl esters and the acid anhydride:

$$\begin{array}{l} CH_2OOCR \\ | \\ CHOOCR \\ | \\ CH_2OOCR \end{array} \longrightarrow \begin{array}{l} CH_3 \\ | \\ CO \\ | \\ CH_2OOCR \end{array} + \begin{array}{l} O{=}C{-}R \\ \quad O \\ O{=}C{-}R \end{array}$$

Decomposition of the 1-oxopropyl gives rise to acrolein and the C_n fatty acid; decarboxylation of the acid anhydride intermediate produces the symmetrical ketone. Free-radical mechanisms similar to those proposed for radiolysis of triacylglycerols may also play a significant role in the formation of thermolytic products, particularly when relatively high temperatures are used.

5.5.3.2 Thermal Oxidative Reactions of Saturated Fats

Saturated fatty acids and their esters are considerably more stable than their unsaturated analogs. However, when heated in air at temperatures higher than 150°C, they undergo oxidation, giving rise to a complex decomposition pattern. The major oxidative products consist of a homologous series of carboxylic acids, 2-alkanones, *n*-alkanals, lactones, *n*-alkanes, and 1-alkenes.

In a systematic study, triacylglycerols in a series containing the even-numbered fatty acid chains C_6 to C_{16} were heated in air for 1 hr at either 180 or 250°C and their oxidation products were examined [29]. The hydrocarbon series was the same as that produced in the absence of oxygen, but the amounts formed were much greater under the oxidative conditions. In all cases the C_{n-1} alkane was the major hydrocarbon produced.

In general, the 2-alkanones were produced in larger quantities during oxidative heating than the alkanals, with the C_{n-1} methyl ketone being the most abundant carbonyl compound. All the lactones found were gamma-lactones, except for those with a carbon number equal to that of the parent fatty acid, in which case both C_n γ- and δ-lactones were produced in relatively large amounts.

It is generally accepted that the thermal oxidation of saturated fatty acids involves the formation of monohydroperoxides as the principal mechanism, and that oxygen attack can occur at all methylene groups of the fatty acid. There is, however, controversy regarding whether hydroperoxide formation is favored at certain locations along the alkyl chain. Since the dominant oxidative products of saturated fatty acids are those with chain lengths near or equal to the parent fatty acid, it is likely that oxidation occurs preferentially at the α, β, and γ positions.

Oxidative attack at the β-carbon of the fatty acid, for example, results in the formation of β-keto acids, which in turn yield C_{n-1} methyl ketones upon decarboxylation. Cleavage between the α and β carbons of the alkoxy radical intermediate gives rise to C_{n-2} alkanal; scission between the β and γ carbons produces C_{n-3} hydrocarbons:

$$R_2O-\overset{O}{\overset{\|}{C}}-C-\overset{\dot{O}}{\overset{|}{C}}-C-R_1$$

$\rightarrow C_{n-3}$ alkane
$\rightarrow C_{n-2}$ alkanal
$\rightarrow C_{n-1}$ methyl ketone

Oxygen attack at the γ position yields a $C_{n\text{-}4}$ hydrocarbon, a $C_{n\text{-}3}$ alkanal, and a $C_{n\text{-}2}$ methyl ketone.

$$R_2O-\overset{O}{\overset{\|}{C}}-C-C-\overset{\dot{O}}{\overset{|}{C}}-R_1$$

$\rightarrow C_{n-4}$ alkane
$\rightarrow C_{n-3}$ alkanal
$\rightarrow C_{n-2}$ methyl ketone

In addition, hydroperoxide formation at the γ position is believed to be responsible for the production of the C_n γ-lactones via cyclization of the resulting hydroxy acids.

Oxygen attack at the α-carbon accounts for the production of a $C_{n\text{-}1}$ fatty acid (via formation of the α-keto acid), a $C_{n\text{-}1}$ alkanal, and a $C_{n\text{-}2}$ hydrocarbon.

$$R_2O-\overset{O}{\overset{\|}{C}}-\overset{\dot{O}}{\overset{|}{C}}-C-C-R_1$$

$\rightarrow C_{n-3}$ alkane
$\rightarrow C_{n-1}$ alkanal

or

$$R_2O-\overset{O}{\overset{\|}{C}}-\overset{OOH}{\overset{|}{C}}-C-C-R_1 \longrightarrow R_2+O-\overset{O}{\overset{\|}{C}}-\overset{O}{\overset{\|}{C}}-C-C-R_1$$

$$CO + HO-\overset{O}{\overset{\|}{C}}-C-C-R_1 \longleftarrow HO-\overset{O}{\overset{\|}{C}}-\overset{O}{\overset{\|}{C}}-C-C-R_1$$

C_{n-1} acid

Further stepwise oxidations give rise to a series of smaller acids, which may themselves undergo oxidation, producing typical decomposition products of their own. This may explain the formation, in small amounts, of the shorter chain hydrocarbons, lactones, and carbonyl compounds.

5.5.3.3 Thermal Nonoxidative Reactions of Unsaturated Fatty Acid Esters

Formation of dimeric compounds appears to be the predominant reaction of unsaturated fatty acids when heated in the absence of oxygen. In addition, certain other substances of lower molecular weight are formed. Relatively severe heat treatment, however, is required for these reactions to take place. Thus, no significant decomposition of methyl oleate can be detected at temperatures below 220°C, but when samples are heated at 280°C for 65 hr under argon, hydrocarbons, short- and long-chain fatty acid esters, and straight-chain dicarboxylic dimethyl esters, as well as dimers, are produced. The occurrence of many of these compounds has been explained on the basis of the formation and/or combination of free radicals resulting from homolytic cleavage of C-C linkages near the double bond. The dimeric compounds, which include alicyclic monoene and diene dimers, as well as saturated dimers with cyclopentane structure, are believed to arise via allyl radicals resulting from hydrogen abstraction at methylene groups α to the double bond. Such radicals may undergo disproportionation into monoenoic and dienoic acids, or inter- and intramolecular addition to C-C double bonds (Fig. 25).

Methyl linoleate, heated under the same conditions, produces a more complex mixture of dimers consisting of saturated tricyclic, monounsaturated bicyclic, diunsaturated monocyclic, and triunsaturated acyclic dimers, as well as dehydrodimers containing one or two double bonds (Fig. 26). At a less severe heat treatment of 180°C for 1 hr under vacuum, ethyl linoleate yields only traces of volatile decomposition products. These compounds appear to be hydrocarbons, but the amounts produced are so small that positive identification has not been accomplished.

5.5.3.4 Thermal Oxidative Reactions of Unsaturated Fatty Acid Esters

Unsaturated fatty acids are much more susceptible to oxidation than their saturated analogs. At elevated temperatures their oxidative decomposition proceeds very rapidly. Although certain specific differences between high- and low-temperature oxidations have been observed, the evidence accumulated to date indicates that in both cases the principal reaction pathways are the same. The formation and decomposition of hydroperoxide intermediates, predictable according to location of double bonds, appear to occur over a wide temperature range. Many decomposition products have been isolated from heated fats. The major compounds produced at high temperature are qualitatively typical of those ordinarily produced at room-temperature autoxidation. However, at elevated temperatures, hydroperoxide decomposition and secondary oxidations occur at extremely rapid rates. The amount of a given decomposition product, at a given time during the autoxidation process, is determined by the net balance between the complex effects of many factors. Hydroperoxide structure, temperature, the degree of unsaturation, and the stability of the decomposition products themselves undoubtedly exert major influences on the final quantitative pattern. The picture is further complicated, since these factors influence not only C-C bond scission, but also a large number of other possible decomposition reactions that occur simultaneously and in competition with C-C cleavage. The latter reactions include C-O scission, which may lead to positional isomerization of hydroperoxides, epoxidation, formation of dihydroperoxides, intramolecular cyclization, and dimerization.

We have already discussed the basic mechanisms of thermal and oxidative polymerization of unsaturated fatty acids. Heating of unsaturated fatty acids in air at elevated temperatures leads to the formation of oxydimers or polymers with hydroperoxide, hydroxide, epoxide, and carbonyl groups, as well as ether and peroxide bridges. The precise structures of many of these compounds, as well as the effects of the various oxidation parameters on reactions leading to their formation, are still ambiguous. However, with the development of highly efficient chromatographic techniques for the separation of high-molecular-weight compounds, renewed interest has been generated in the study of oxidative polymerization.

5.5.4 Chemistry of Frying

Foods fried in fat make significant contributions to the calories in the average U.S. diet. In the course of deep-fat frying, food contacts oil at about 180°C and is partially exposed to air for various periods of time. Thus frying, more than any other standard food process or handling method, has the greatest potential for causing chemical changes in fat, and sizeable amounts of this fat are carried with the food (5–40% fat by weight is absorbed).

5.5.4.1 Behavior of the Frying Oil

The following classes of compounds are produced from the oil during frying.

Volatiles. During frying, oxidative reactions involving the formation and decomposition of hydroperoxides lead to such compounds as saturated and unsaturated aldehydes, ketones, hydrocarbons, lactones, alcohols, acids, and esters [25]. After oil is heated for 30 min at 180°C in the presence of air, the primary volatile oxidation products can be detected by gas chromatography. Although the amounts of volatiles produced vary widely, depending on oil type, food type, and the heat treatment, they generally reach plateau values, probably because a balance is achieved between formation of the volatiles and loss through evaporation or decomposition.

Nonpolymeric polar compounds of moderate volatility (e.g., *hydroxy and epoxy acids*). These compounds are produced according to the various oxidative pathways involving the alkoxy radical as discussed earlier.

Dimeric and polymeric acids, and dimeric and polymeric glycerides. These compounds occur, as expected, from thermal and oxidative combinations of free radicals. Polymerization results in a substantial increase in the viscosity of the frying oil.

Free fatty acids. These compounds arise from hydrolysis of triacylglycerols in the presence of heat and water.

The reactions just described are responsible for a variety of physical and chemical changes that can be observed in the oil during the course of frying. These changes include increases in viscosity and free fatty acid content, development of a dark color, decreases in iodine value and surface tension, changes in refractive index, and an increased tendency to foam.

5.5.4.2 Behavior of the Food During Frying

The following events occur during frying of food:

Water is continuously released from the food into the hot oil. This produces a steam-distillation effect, sweeping volatile oxidative products from the oil. The released moisture also agitates the oil and hastens hydrolysis. The blanket of steam formed above the surface of the oil tends to reduce the amount of oxygen available for oxidation.

Volatiles (e.g., sulfur compounds and pyrazine derivatives in potato) may develop in the food itself or from the interactions between the food and oil.

Food absorbs varying amounts of oil during deep-fat frying (potato chips have a final fat content of about 35%), resulting in the need for frequent or continuous addition of fresh oil. In continuous fryers, this results in rapid attainment of a steady-state condition for oil properties.

The food itself can release some of its endogenous lipids (e.g., fat from chicken) into the frying fat and consequently the oxidative stability of the new mixture may be different from that of the original frying fat. The presence of food causes the oil to darken at an accelerated rate.

5.5.4.3 Chemical and Physical Changes

The changes that occur in the oil and food during frying should not be automatically construed as undesirable or harmful. In fact, some of these changes are necessary to provide the sensory qualities typical of fried food. On the other hand, extensive decomposition, resulting from lack of adequate control of the frying operation, can be a potential source of damage not only to sensory quality of the fried food but also to nutritional value.

The chemical and physical changes in the frying fat are influenced by a number of frying parameters. Obviously, the compounds formed depend on the composition of both the oil and the food being fried. High temperature, long frying times, and metal contaminants favor extensive decomposition of the oil. The design and type of the fryer (pan deep-fat batch, or deep-fat continuous) is also important. Oxidation of the oil will occur faster at large surface–volume ratios. Other factors of concern are turnover rate of the oil, the heating pattern (continuous or intermittent), and whether antioxidants are present.

5.5.4.4 Tests for Assessing the Quality of Frying Oils

Several of the methods just discussed for determining fat oxidation are commonly used to monitor thermal and oxidative decomposition of oils during the frying process. Measures of viscosity, free fatty acids, sensory quality, smoke point, foaming, polymer formation, and specific degradation products are also used with various degrees of success. In addition, several methods have been developed specifically to assess the chemical status of used frying oil. Some of these require standard laboratory instruments, while others require special measurements.

Because changes occurring during frying are numerous and variable, one test may be adequate for one set of conditions but totally unsatisfactory for another.

Petroleum Ether Insolubles

This method was developed in Germany and later recommended by the German Society for Fat Research. It is suggested that a used frying fat be regarded as "deteriorated" if the petroleum ether insolubles are 0.7% and the smoke point is less than 170°C, or if the petroleum ether insolubles are 1.0% regardless of the smoke point. The method is tedious and not very accurate, since oxidation products are partially soluble in petroleum ether.

Polar Compounds [11]

The heated fat is fractionated on a silica-gel column and the nonpolar fraction is eluted with a petroleum ether–diethyl ether mixture. The percentage weight of the polar fraction is calculated by difference. A value of 27% polar components is suggested as the maximum tolerable for usable oil.

Dimer Esters [115]

This technique involves complete conversion of the oil to the corresponding methyl esters followed by separation and detection on a short column in a gas chromatograph. The increase in dimer esters is used as a measure of thermal decomposition.

Quick Tests

Changes in the dielectric constant of the oil can be measured quickly in an instrument known as the Foodoil Sensor [44]. The dielectric constant increases with an increase in polarity, and increased polarity is used as an indicator of deterioration. This technique has several advantages. The unit is compact and portable, it can be operated in the field by unskilled personnel, and analysis time is less than 5 min. However, caution must be exercised in interpreting the data obtained [114]. Dielectric-constant readings represent the net balance between polar and nonpolar components, both of which develop upon frying. The increase in the polar fraction usually predominates, but the net difference between the two fractions depends on a variety of factors, some of which may be unrelated to oil quality (e.g., moisture).

"Diagnostic kits" are available for assessing oil deterioration. Most involve simple addition of a reagent to the sample and matching the developed color with a color standard. Fritest©, based on the presence of carbonyl compounds, and Oxifrit©, based on the presence of oxidized compounds, are sold by Merck (Dramstadt, Germany). Verify-FAA.500©, based on free fatty acids, and Verify-TAM.150©, based on "alkaline material," are tests available from Libra Labs (Piscataway, NJ).

5.5.4.5 Control Measures

From the standpoints of food acceptability and wholesomeness, it is desirable to minimize extensive oxidative decomposition, and to avoid development of objectionable flavors and excessive formation of cyclic or polymeric material during frying. Good manufacturing practices include:

1. Choice of frying oil of good quality and consistent stability.
2. Use of properly designed equipment.
3. Selection of the lowest frying temperature consistent with producing a fried product of good quality.
4. Frequent filtering of the oil to remove food particles.
5. Frequent shutdown and cleaning of equipment.
6. Replacement of oil as needed to maintain high quality.
7. Consideration of antioxidant use. When used, the level of antioxidant will sometimes decrease rapidly, and this is important to monitor.
8. Adequate training of personnel.
9. Frequent testing of the oil throughout the frying process.

5.5.5 Effects of Ionizing Radiation on Fats

The objective of food irradiation is comparable to that of other methods of preservation. Its primary purpose is to destroy microorganisms and to prolong the shelf life of the food. The process is capable of sterilizing meats or meat products [at high doses, e.g., 10–50 kGy (kilogray = 100 kilorads)]; extending the shelf life of refrigerated fresh fish, chicken, fruits, and vegetables (at medium doses, e.g., 1–10 kGy); preventing the sprouting of potatoes and onions; delaying fruit ripening; and killing insects, in spices, cereals, peas, and beans (at low doses, e.g., up to 3 kGy). Radiation preservation of food is becoming increasingly attractive to industry from both food stability and economic points of view.

As with heating, chemical changes are induced in food as a result of the irradiation. Indeed, in both cases, the primary objective of the treatment is achieved as a result of these chemical changes. On the other hand, treatment conditions must be controlled such that the nature and extent of the changes do not compromise the wholesomeness or quality of the food.

Considerable effort has been devoted to determining the effects of irradiation on various food components. Most of the early work on irradiation of lipids was concerned with the study of radiation-induced oxidation in natural fats or in certain model lipid systems. More recently, strictly radiolytic (i.e., nonoxidative) changes in fats have been investigated.

5.5.5.1 Radiolytic Products

Listed in Table 7 are the classes of compounds found to result from irradiation of natural fats or model systems of fatty acids and their derivatives. This listing of compounds is derived from studies of beef and pork fats, mackerel oil, corn, soybean, olive, safflower, and cottonseed oils, and pure triacylglycerols, all treated under vacuum with radiation doses ranging from 0.5 to 600 kGy [108]. Hydrocarbons, aldehydes, methyl and ethyl esters, and free fatty acids are the major volatile compounds produced by irradiation. Found in all irradiated fats studied are hydrocarbons consisting of complete homologous series of C_1–C_{17} *n*-alkanes, C_2–C_{17} 1-alkenes, certain internally unsaturated monoenes, a series of alkadienes, and in some cases, several polyenes. The aldehydes and esters include normal alkanals containing 16 and 18 carbon atoms and the methyl and ethyl esters of C_{16} and C_{18} fatty acids. In these studies, unsaturated

TABLE 7 Classes of Compounds Identified in Irradiated Fats

n-Alkanes	Methyl and ethyl esters
l-Alkenes	Propane and propenediol diesters
Alkadienes	Ethanediol diesters
Alkynes	Oxopropanediol diesters
Aldehydes	Glyceryl ether diesters
Ketones	Long-chain hydrocarbons
Fatty acids	Long-chain alkyl esters
Lactones	Long-chain alkyldioldiesters
Monoacylglycerols	Long-chain ketones
Diacylglycerols	Dimers
Triacylglycerols	Trimers

Note: Approximately 150 compounds were identified in various studies in the author's laboratory. Natural fats and model systems were irradiated at various doses in the absence of oxygen.

hydrocarbons larger than C_{17} arose only in irradiated fish oil, and aldehydes of chain length shorter than C_{16} were found only in irradiated coconut oil.

The quantitative pattern of the radiolytic products is strictly dependent on the fatty acid composition of the original fat. Although most members of the hydrocarbon series are produced in only small amounts, those containing one or two carbon atoms less than the major component fatty acids of the original fat are formed in the greatest quantities. For example, in the case of pork fat, in which 18:1 is the major fatty acid, C_{17} and C_{16} hydrocarbons are the most abundant hydrocarbons; for coconut oil, in which lauric acid is the main fatty acid, undecane and decene are the hydrocarbons produce in the greatest quantities. Based on the foregoing, and on the observation that the two major hydrocarbons produced from each fatty acid increase linearly with dose, a method was developed for the detection of irradiated lipid-containing foods. The technique involves solvent extraction of the lipids, collection of volatiles, and quantitative measurement of the "key" hydrocarbon markers by gas chromatography [110].

The major aldehydes in the radiolytic mixture are those with the same chain length as those of the major component fatty acids of the original fat, and the methyl and ethyl esters produced in the largest quantity are those corresponding to the most abundant saturated acids in the original fat. If the fats are partially hydrogenated before irradiation, the change in their original fatty acid composition results in a corresponding change in the composition of the radiolytic compounds. The radiolytic compounds of relatively high molecular weight have been identified in a study with beef fat [142].

5.5.5.2 Mechanisms of Radiolysis

General Principles

Ions and excited molecules are the first species formed when ionizing radiation is absorbed by matter. Chemical breakdown is brought about by decomposition of the excited molecules and ions, or by their reaction with neighboring molecules. The fate of the excited molecules includes dissociation into free radicals. Among the reactions of ions is neutralization, producing excited molecules that may then dissociate into smaller molecules or free radicals. The free radicals formed by dissociation of excited molecules and by ion reactions may combine with each other in regions of high radical concentrations, or may diffuse into the bulk of the medium and react with other molecules.

Radiolysis of Fats

The reactions induced by irradiation are not the result of a statistical distribution of random cleavages of chemical bonds, but rather they follow preferred pathways largely influenced by molecular structure. In the case of saturated fats, as with all oxygen-containing compounds, the high localization of the electron deficiency on the oxygen atom directs the location of preferential cleavage to the vicinity of the carbonyl group.

Based on the results available to date, the mechanisms of radiolysis in triacylglycerols appear to proceed in accord with the concepts of Williams, regarding the primary ionization events in oxygen-containing compounds [147]. In a triacylglycerol molecule, as shown here,

$$\begin{array}{l} CH_2 \overset{a}{|} O \overset{b}{|} \overset{O}{\overset{\|}{C}} \overset{c}{|} CH_2 \overset{d}{|} CH_2 \vdots CH_2 \vdots CH_2 \vdots (CH_2) \vdots CH_3 \\ e - | \\ CHOOCR \\ | \\ CH_2OOCR \end{array}$$

radiolytic cleavage occurs preferentially at five locations in the vicinity of the carbonyl group (wavy solid lines), and randomly at all the remaining carbon–carbon bonds in the fatty acid moiety (dashed lines).

These cleavages result in the formation of alkyl, acyl, acyloxy, and acyloxymethylene free radicals, and in addition, free radicals representing the corresponding glycerol residues. Free radical termination may occur by hydrogen abstraction, and to a lesser degree by loss of hydrogen with the formation of an unsaturated linkage. Thus, scission of an acyloxy-methylene bond (location a) produces a free fatty acid and a propanediol ester or propenediol diester; cleavage at the acyl-oxygen bond (at b) produces an aldehyde of equal chain length to the parent fatty acid and a diacylglycerol; cleavage at c or d produces a hydrocarbon with one or two carbon atoms less than the parent fatty acid and a triacylglycerol; and cleavage between carbons of the glycerol skeleton (e.g., at e) produces a methyl ester of the parent fatty acid and an ethanediol diester.

Alternatively, the free radicals may recombine, giving rise to a variety of radiolytic products. Thus, combination of alkyl radicals with other alkyl radicals results in the production of longer chain or dimeric hydrocarbons:

$$R^{\cdot} + R'^{\cdot} \rightarrow R\text{-}R'$$

combination of an acyl radical with an alkyl radical produces a ketone:

$$R\overset{O}{\overset{\|}{C}}\cdot + R'\cdot \longrightarrow R\overset{O}{\overset{\|}{C}}R'$$

combination of an acyloxy radical with an alkyl radical produces an ester:

$$R\overset{O}{\overset{\|}{C}}O\cdot + R'\cdot \longrightarrow R\overset{O}{\overset{\|}{C}}OR'$$

and combination of alkyl radicals with various glyceryl residues produces alkyl diglycerides:

$$\begin{array}{l} CH_2OCR \\ | \\ CH\cdot \\ | \\ CH_2OCR \end{array} + R'\cdot \longrightarrow \begin{array}{l} CH_2OCR \\ | \\ CHR' \\ | \\ CH_2OCR \end{array}$$

(each OCR group bearing $=O$ on the carbonyl carbon)

and glyceryl ether diesters.

$$\begin{array}{l} CH_2OCR \\ | \\ CHO\cdot \\ | \\ CH_2OCR \end{array} + R'\cdot \longrightarrow \begin{array}{l} CH_2OCR \\ | \\ CHOR' \\ | \\ CH_2OCR \end{array}$$

It is indeed likely that other free-radical combinations take place to produce dimeric or high-molecular-weight compounds that may have escaped detection by current analytical techniques. Furthermore, mechanisms other than those presented here are probably involved in the radiolysis of fats. In particular, radiolytic reactions resulting from unsaturation of the glyceride molecule (e.g., cross-linking, dimerization, and cyclization) have not been thoroughly investigated.

In the presence of oxygen, irradiation will accelerate the autoxidation of fats via one or more of the following reactions: (a) formation of free radicals, which may combine with oxygen to form hydroperoxides, (b) breakdown of hydroperoxides, giving rise to various decomposition products, particularly carbonyl compounds, and (c) destruction of antioxidants.

Complex Foods Containing Fats

The irradiation of isolated fats does not have a practical value in terms of industrial application. Food lipids are normally exposed to irradiation as constituents coexisting with other major and minor components of complex food systems. The radiolytic products obtained from an isolated fat are also obtained when the complex food containing such a fat is irradiated. However, the concentration of these substances in the irradiated food will be considerably reduced by the diluting effect of the other substances present. Of course, additional changes can be anticipated from radiolysis of the nonlipid constituents and from interaction between these constituents and the lipids.

Some researchers concluded that irradiation reduces the stability of fatty foods by destroying antioxygenic factors, and suggested that irradiation be performed in the absence of air and that antioxidants be added after irradiation. On the other hand, others have found that, in certain cases, irradiation resulted in the formation of new protective factors that can improve product stability.

5.5.5.3 Comparison with Heat Effects

Although the mechanisms involved are different, many of the compounds produced from fats by irradiation are similar to those formed by heating. Far more decomposition products, however, have been identified from heated or thermally oxidized fats than from irradiated fats. Recent research in our laboratory with fatty acid esters and triacylglycerols indicates that even when irradiation is conducted at doses as high as 250 kGy, both the volatile and the nonvolatile patterns of products formed are much simpler than those obtained by heating at frying temperatures (180°C) for 1 hr.

5.5.6 Concluding Thoughts on Lipid Oxidation in Food

We are witnessing a revolution in the field of lipid oxidation caused by remarkable advances in the sciences, analytical methodology, and cross-disciplinary communication. Some of the long established theories, which in the past helped us to develop an understanding of the basic mechanisms of lipid oxidation, must now give way to new thinking and a new approach to the examination and control of oxidizing lipids in food.

It may not be too unreasonable to envision lipid oxidation in food in the same way one looks at the weather map in the United States. The atmospheric phenomena and weather activities vary from region to region and the weather map may change significantly from moment to moment:

1. Lipid oxidation is not a single process that proceeds according to a strict sequence. Rather, numerous events take place in different regions of the food matrix. The different reactions usually occur with considerable overlap, and the rate of each may vary depending on conditions in each region.
2. The "conditions" in each region (oxidation parameters) are not static. They change continuously.
3. In food, oxidation involves oxidizable lipid substrates differing in composition, chemical and physical properties, and in their sensitivity to oxidation (fatty acids, mixtures of acylglycerols, phospholipids).
4. The lipid substrates coexist in proximity with various other nonlipid, major (proteins, carbohydrates, water) and minor (trace metals, vitamins, enzymes, pro- and antioxidants) components. A multitude of interactions take place.
5. Nearly all oxidative events are not "strictly chemical." Instead, they are largely dictated by the physical state of both substrate and medium.
6. The minor components play a major role in a critical, delicate oxidative–antioxidative balance, the mechanism of which is not fully understood.
7. The same inhibitors (or accelerators) may play opposite roles depending on the oxidative event being considered. For example, a phenolic antioxidant may block the formation of volatile oxidation products while enhancing the accumulation of peroxides.

It is clear that a realistic evaluation of lipid oxidation in food can only be achieved if the foregoing scenario, in all of its detail, is considered.

5.6 CHEMISTRY OF FAT AND OIL PROCESSING

5.6.1 Refining

Crude oils and fats contain varying amounts of substances that may impart undesirable flavor, color, or keeping quality. These substances include free fatty acids, phospholipids, carbohydrates, proteins and their degradation products, water, pigments (mainly carotenoids and chlorophyll), and fat oxidation products. Crude oils are subjected to several commercial refining processes designed to remove these materials.

5.6.1.1 Settling and Degumming

Settling involves heating the fat and allowing it to stand until the aqueous phase separates and can be withdrawn. This rids the fat of water, proteinaceous material, phospholipids, and carbohydrates. In some cases, particularly with oils containing substantial amounts of phospholipids (e.g., soybean oil), a preliminary treatment known as degumming is applied by adding 2–3% water, agitating the mixture at about 50°C, and separating the hydrated phospholipids by settling or centrifugation.

5.6.1.2 Neutralization

To remove free fatty acids, caustic soda in the appropriate amounts and strength is mixed with the heated fat and the mixture is allowed to stand until the aqueous phase settles. The resulting aqueous solution, called foots or soapstock, is separated and used for making soap. Residual soapstock is removed from the neutral oil by washing it with hot water, followed by settling or centrifugation.

Although free fatty acid removal is the main purpose of the alkali treatment, this process also results in a significant reduction of phospholipids and coloring matter.

5.6.1.3 Bleaching

An almost complete removal of coloring materials can be accomplished by heating the oil to about 85°C and treating it with adsorbants, such as fuller's earth or activated carbon. Precautions should be taken to avoid oxidation during bleaching. Other materials, such as phospholipids, soaps, and some oxidation products, are also absorbed along with the pigments. The bleaching earth is then removed by filtration.

5.6.1.4 Deodorization

Volatile compounds with undesirable flavors, mostly arising from oxidation of the oil, are removed by steam distillation under reduced pressure. Citric acid is often added to sequester traces of pro-oxidant metals. It is believed that this treatment also results in thermal destruction of nonvolatile off-flavor substances, and that the resulting volatiles are distilled away.

Although the oxidative stability of oils is generally improved by refining, this is not always the case. Crude cottonseed oil, for example, has a greater resistance to oxidation than its refined counterparts due to the greater amounts of gossypol and tocopherols in the crude oil. On the other hand, there can be little doubt as to the remarkable quality benefits that accrue from refining edible oils. An impressive example is the upgrading of palm oil quality that has occurred in recent years. Furthermore, in addition to the obvious improvements in color, flavor, and stability, powerful toxicants (e.g., aflatoxins in peanut oil and gossypol in cottonseed oil) are effectively eliminated during the refining process.

5.6.2 Hydrogenation

Hydrogenation of fats involves the addition of hydrogen to double bonds in the fatty acid chains. The process is of major importance in the fats and oils industry. It accomplishes two major objectives. First, it allows the conversion of liquid oils into semisolid or plastic fats more suitable for specific applications, such as in shortenings and margarine, and second, it improves the oxidative stability of the oil.

In practice, the oil is first mixed with a suitable catalyst (usually nickel), heated to the desired temperature (140–225°C), then exposed, while stirred, to hydrogen at pressures up to 60 psig. Agitation is necessary to aid in dissolving the hydrogen, to achieve uniform mixing of the catalyst with oil, and to help dissipate the heat of the reaction. The starting oil must be refined, bleached, low in soap, and dry; the hydrogen gas must be dry and free of sulfur, CO_2, or ammonia; and the catalyst must exhibit long-term activity, function in the desired manner with respect to selectivity of hydrogenation and isomer formation, and be easily removable by filtration. The course of the hydrogenation reaction is usually monitored by determining the change in refractive index, which is related to the degree of saturation of the oil. When the desired end point is reached, the hydrogenated oil is cooled and the catalyst is removed by filtration.

5.6.2.1 Selectivity

During hydrogenation, not only are some of the double bonds saturated, but some may also be relocated and/or transformed from the usual *cis* to the *trans* configuration. The isomers produced are commonly called iso acids. Partial hydrogenation thus may result in the formation of a relatively complex mixture of reaction products, depending on which of the double bonds are hydrogenated, the type and degree of isomerization, and the relative rates of these various reactions. A simplistic scheme showing the possible reactions that linolenate can undergo during hydrogenation is shown here:

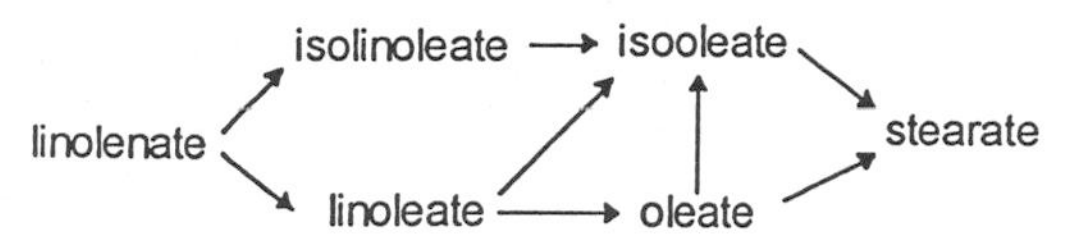

In the case of natural fats, the situation is further complicated by the fact that they already contain an extremely complex mixture of starting materials.

The term "selectivity" refers to the relative rate of hydrogenation of the more unsaturated fatty acids as compared with that of the less unsaturated acids. When expressed as a ratio (selectivity ratio), a quantitative measure of selectivity can be obtained in more absolute terms. The term "selectivity ratio," as defined by Albright [3], is simply the ratio (rate of hydrogenation of linoleic to oleic)/(rate of hydrogenation of oleic to stearic). Reaction rate constants can be calculated from the starting and ending fatty acid compositions and the hydrogenation time (Fig. 35). For the reactions just mentioned, the selectivity ratio (SR) is

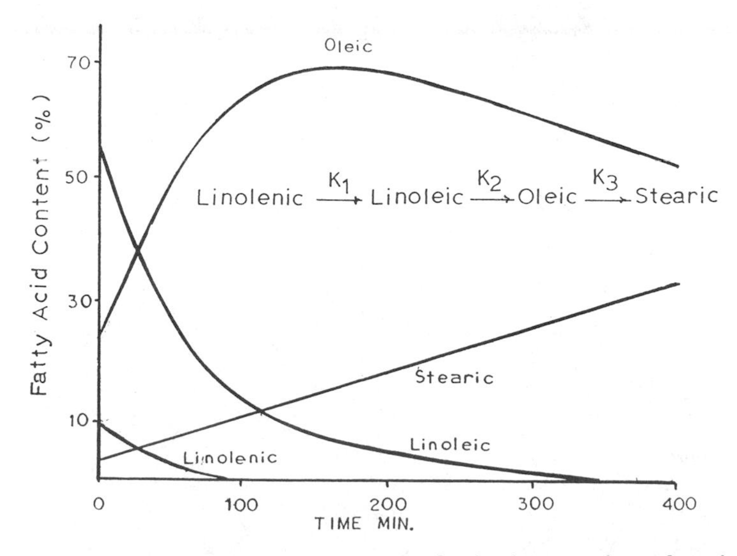

FIGURE 35 Reaction rate constants for hydrogenation of soybean oil. (From Ref. 5.)

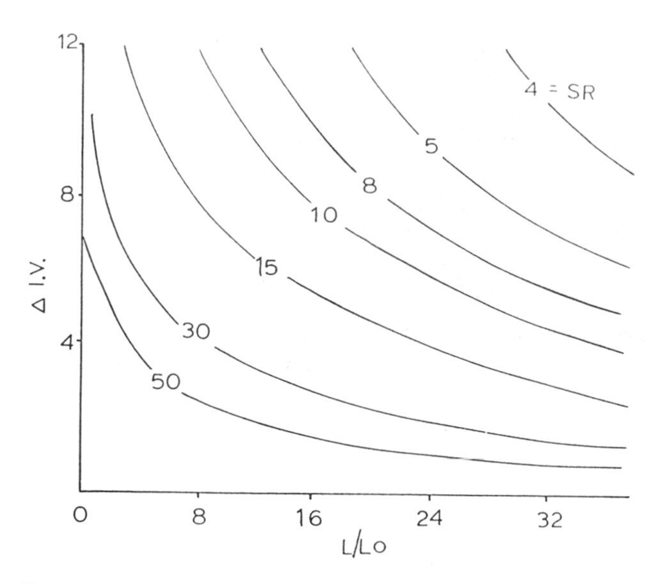

FIGURE 36 SR curves for soybean oil [3], ΔIV + decrease in iodine value; L/L_0 + fraction of linoleic acid unhydrogenated.

$K_2/K_3 = 0.159/0.013 = 12.2$, which means that linoleic acid is being hydrogenated 12.2 times faster than oleic acid.

Since calculations of SR for every oil hydrogenated would be quite tedious, Albright [3] prepared a series of graphs for various oils by calculation of the fatty acid compostion at constant SR. In these graphs the decrease in iodine value (ΔIV) is plotted against the fraction of linoleic acid that remains unhydrogenated (L/L_0). Although the curves are calculated with the assumption that the reaction rates are first order and that isooleic acid is hydrogenated at the same rate as oleic, they are nonetheless very useful for predicting selectivity. Selectivity ratio curves for soybean oil (K_2/K_3) are shown in Figure 36. Of course, linolenic acid selectivity can be similarly expressed; that is, $\text{LnSR} = K_1/K_2$ where K_1 and K_2 are as defined in Figure 30. This is relevant to the hydrogenation of soybean oil, since flavor reversion in this oil is believed to arise from its linolenate content.

Different catalysts result in different selectivities, and operating parameters also have a profound effect on selectivity. As shown in Table 8, larger SR values result from high temperatures, low pressures, high catalyst concentration, and low intensity of agitation. The effects of processing conditions on the rate of hydrogenation and on the formation of *trans* acids are also

TABLE 8 Effects of Processing Parameters on Selectivity and Rate of Hydrogenation

Processing parameter	SR	*Trans* acids	Rate
High temperature	High	High	High
High pressure	Low	Low	High
High catalyst concentration	High	High	High
High-intensity agitation	Low	Low	High

given. Several mechanistic speculations have been advanced to explain the observed influence of process conditions on selectivity and rate of hydrogenation, and these are discussed next.

5.6.2.2 Mechanism

The mechanism involved in fat hydrogenation is believed to be the reaction between unsaturated liquid oil and atomic hydrogen adsorbed on a metal catalyst. First, a carbon–metal complex is formed at either end of the olefinic bond (complex a in Fig. 37). This intermediate complex is then able to react with an atom of catalyst-adsorbed hydrogen to form an unstable half-hydrogenated state (b or c in Fig. 37) in which the olefin is attached to the catalyst by only one link and is thus free to rotate. The half-hydrogenated compound now may either (a) react with another hydrogen atom and dissociate from the catalyst to yield the saturated product (d in Fig. 37), or (b) lose a hydrogen atom to the nickel catalyst to restore the double bond. The regenerated double bond can be either in the same position as in the unhydrogenated compound, or a positional and/or geometric isomer of the original double bond (e and f in Fig. 37).

In general, evidence seems to indicate that the concentration of hydrogen adsorbed on the catalyst is the factor that determines selectivity and isomer formation [5]. If the catalyst is saturated with hydrogen, most of the active sites hold hydrogen atoms and the chance is greater that two atoms are in the appropriate position to react with any double bond upon approach. This results in low selectivity, since the tendency will be toward saturation of any double bond approaching the two hydrogens. On the other hand, if the hydrogen atoms on the catalyst are scarce, it is more likely that only one hydrogen atom reacts with the double bonds, leading to the half-hydrogenation–dehydrogenation sequence and a greater likelihood of isomerization. Thus, operating conditions (hydrogen pressure, intensity of agitation, temperature, and kind and concentration of catalyst) influence selectivity through their effect on the ratio of hydrogen to catalyst sites. An increase in temperature, for example, increases the speed of the reaction and causes faster removal of hydrogen from the catalyst, giving rise to increased selectivity.

The ability to change the SR by changing the processing conditions enables processors to exert considerable control over the properties of the final oil. A more selective hydrogenation, for example, allows linoleic acid to be decreased and stability to be improved while minimizing

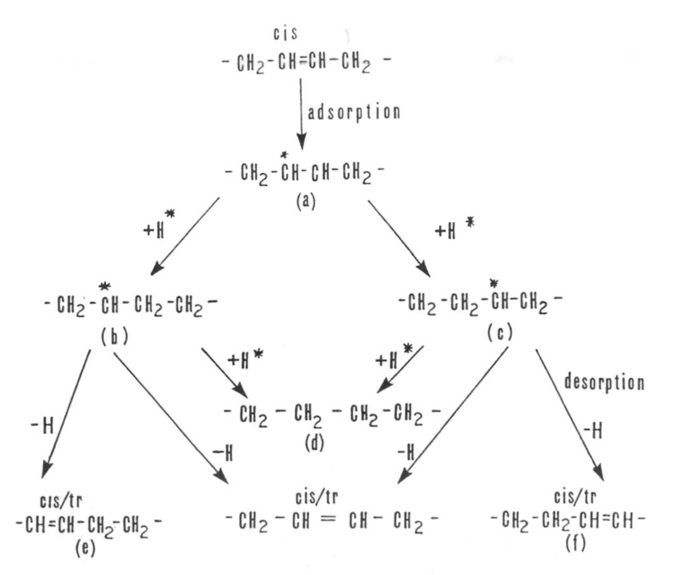

FIGURE 37 Half-hydrogenation–hydrogenation reaction scheme. Asterisk indicates metal link.

the formation of fully saturated compounds and avoiding excessive hardness. On the other hand, the more selective the reaction, the greater will be the formation of *trans* isomers, which are of concern from a nutritional standpoint. For many years manufacturers of food fats have been trying to devise hydrogenation processes that minimize isomerization while avoiding the formation of excessive amounts of fully saturated material.

5.6.2.3 Catalysts

As indicated earlier, catalysts vary with regard to the degree of selectivity they provide. Nickel on various supports is almost invariably used commercially to hydrogenate fats. Other catalysts, however, are available. These include copper, copper/chromium combinations, and platinum. Palladium has been found to be considerably more efficient (in terms of the amount of catalyst required) than nickel, although it produces a high proportion of *trans* isomers. The so-called homogeneous catalysts, which are soluble in oil, provide greater contact between oil and catalyst and better control of selectivity. A host of different compounds are capable of poisoning the catalyst used, and these compounds are often the major source of problems encountered during commercial hydrogenation. Poisons include phospholipids, water, sulfur compounds, soaps, certain glycerol esters, CO_2, and mineral acids.

5.6.3 Interesterification

It has been mentioned that natural fats do not contain a random distribution of fatty acids among the glyceride molecules. The tendency of certain acids to be more concentrated at specific *sn* positions varies from one species to another and is influenced by factors such as environment and location in the plant or animal. The physical characteristics of a fat are greatly affected not only by the nature of constituent fatty acids (i.e., chain length and unsaturation) but also by their distribution in the triacylglycerol molecules. Indeed, unique fatty acid distribution patterns of some natural fats limit their industrial applications. Interesterification is one of the processes that can be applied to improve the consistency of such fats and to improve their usefulness. This process involves rearranging the fatty acids so they become distributed randomly among the triacylglycerol molecules of the fat.

5.6.3.1 Principle

The term "interesterification" refers to the exchange of acyl radicals between an ester and an acid (acidolysis), or an ester and an alcohol (alcoholysis), or an ester and an ester (trans-esterfication). It is the latter reaction that is relevant to industrial interestification of fat, also known as randomization, since it involves ester interchange within a single triacylglycerol molecule (intraesterification) as well as ester exchange among different molecules.

If a fat contains only two fatty acids (A and B), eight possible triacylglycerol species n_3 are possible according to the rule of chance:

A	A	A	A	B	B	B	B
A	A	B	B	A	A	B	B
A	B	A	B	A	B	A	B

Regardless of the distribution of the two acids in the original fat (e.g., AAA and BBB or ABB, ABA, BBA), interesterification results in the "shuffling" of fatty acids within a single molecule and among triacylglycerol molecules until an equilibrium is achieved in which all possible combinations are formed. The quantitative proportions of the different species depend on the amount of each acid in the original fat and can be predicted by simple calculation, as already discussed under the 1,2,3-random distribution hypothesis.

5.6.3.2 Industrial Process

Interesterification can be accomplished by heating the fat at relatively high temperatures (< 200°C) for a long period. However, catalysts are commonly used that allow the reaction to be completed in a short time (e.g., 30 min) at temperatures as low as 50°C. Alkali metals and alkali metal alkylates are effective low-temperature catalysts, with sodium methoxide being the most popular. Approximately 0.1% catalyst is required. Higher concentrations may cause excessive losses of oil resulting from the formation of soap and methyl esters.

The oil to be esterified must be extremely dry and low in free fatty acids, peroxides, and any other material that may react with sodium methoxide. Minutes after the catalyst is added the oil acquires a reddish brown color due to the formation of a complex between the sodium and the glycerides. This complex is believed to be the "true catalyst." After esterification, the catalyst is inactivated by addition of water or acid and removed.

5.6.3.3 Mechanisms

Two mechanisms for interesterification have been proposed [135].

Enolate Ion Formation

According to this mechanism, an enolate ion (II), typical of the action of a base on an ester, is formed. The enolate ion reacts with another ester group in the triacylglycerol molecule to produce a β-keto ester (III), which in turn reacts further to give another β-keto ester (IV). Intermediate IV yields the intramolecularly esterified product V.

$$\begin{array}{l} \text{-O CO CH}_2\text{R}_1 \\ \text{-O CO CH}_2\text{R}_2 \\ \text{-O CO CH}_2\text{R}_3 \end{array} + \text{OCH}_3^- \rightleftharpoons \left[\begin{array}{l} \text{-O }\overset{\text{O}}{\text{C}}\text{=C(H)R}_1 \\ \text{-O CO CH}_2\text{R}_2 \\ \text{-O CO CH}_2\text{R}_3 \end{array} \right]^- \rightleftharpoons$$

I $\qquad\qquad$ II

$$\left[\begin{array}{l} \text{-O }\overset{\text{O}}{\text{C}}\text{-}\underset{\text{R}_1}{\overset{\text{H}}{\text{C}}}\text{-}\overset{\text{O}}{\text{C}}\text{-C-R}_1 \\ \text{-O} \\ \text{-O CO CH}_2\text{R}_3 \end{array} \right]^- \rightleftharpoons \left[\begin{array}{l} \text{-O} \\ \text{-O }\overset{\text{O}}{\text{C}}\text{-}\underset{\text{R}_1}{\overset{\text{H}}{\text{C}}}\text{-}\overset{\text{O}}{\text{C}}\text{-CH}_2\text{R}_2 \\ \text{-O CO CH}_2\text{R}_3 \end{array} \right]^- \rightleftharpoons \left[\begin{array}{l} \text{-O CO CH}_2\text{R}_2 \\ \text{-O }\overset{\text{O}}{\text{C}}\text{-}\underset{\text{H}}{\text{C}}\text{-R}_1 \\ \text{-O CO CH}_2\text{R}_3 \end{array} \right]^-$$

III $\qquad\qquad$ IV $\qquad\qquad$ V

The same mode of action applies to ester interchange between two or more triacylglycerol molecules. The intra-ester-ester interchange is believed to predominate in the initial stages of the reaction.

Carbonyl Addition

In this proposed mechanism, the alkylate ion adds onto a polarized ester carboxyl, producing a diglycerinate intermediate:

$$R_2\left[\begin{array}{l}-ONa\\ \\ -R_3\end{array}\right.$$

This intermediate reacts with another glyceride by abstracting a fatty acid, thus forming a new triacylglycerol and regenerating a diglycerinate for further reaction. Ester interchange between fully saturated S_3 and unsaturated U_3 molecules is shown here:

$$S_3 + U_2ONa \underset{K}{\overset{3K}{\rightleftharpoons}} SU_2 + S_2ONa$$

$$SU_2 + U_2ONa \underset{3K}{\overset{2K}{\rightleftharpoons}} U_3 + SUONa$$

$$U_3 + S_2ONa \underset{K}{\overset{3K}{\rightleftharpoons}} S_2U + U_2ONa$$

$$S_2U + S_2ONa \underset{3K}{\overset{2K}{\rightleftharpoons}} S_3 + SUONa$$

$$S_2U + U_2ONa \underset{K}{\overset{2K}{\rightleftharpoons}} SU_2 + SUONa$$

$$SU_2 + S_2ONa \underset{K}{\overset{2K}{\rightleftharpoons}} S_2U + SUONa$$

5.6.3.4 Directed Interesterification

A random distribution, such as that produced by interesterification, is not always the most desirable. Interesterification can be directed away from randomness if the fat is maintained at a temperature below its melting point. This results in selective crystallization of the trisaturated glycerides, which has the effect of removing them from the reaction mixture and changing the fatty acid equlibrium in the liquid phase. Interesterification proceeds with the formation of more trisaturated glycerides than would have otherwise occurred. The newly formed trisaturated acylglycerols crystallize and precipitate, thus allowing the formation of still more trisaturated glycerides, and the process continues until most of the saturated fatty acids in the fat have precipitated. If the original fat is a liquid oil containing a substantial amount of saturated acids, it is possible, by this method, to convert the oil into a product with the consistency of shortening without resorting to hydrogenation or blending with a hard fat. The procedure is relatively slow due to the low temperature used, the time required for crystallization, and the tendency of the catalyst to become coated. A dispersion of liquid sodium–potassium alloy is commonly used to slough off the coating as it forms.

Rearrangement can also be selectively controlled during interesterification by adding excess fatty acids and continuously distilling out the liberated acids that are highly volatile. This impoverishes the fat of its acids of lower molecular weight. The content of certain acids in a fat also can be reduced by using suitable solvents to extract appropriate acids during the interesterification process.

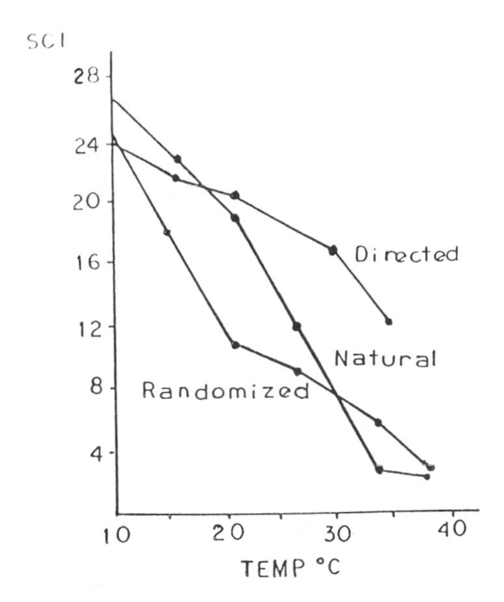

FIGURE 38 Effect of interesterification on solid content index. (From Ref. 135.)

5.6.3.5 Applications

Interesterification finds its greatest application in the manufacture of shortenings. Lard, due to its high proportion of disaturated triacylglycerols with palmitic acid in the 2 position, forms relatively large and coarse crystals, even when rapidly solidified in commercial chilling machines. Shortenings made from natural lard possess a grainy consistency and exhibit poor performance in baking. Randomization of lard improves its plastic range and makes it a better shortening. Directed interesterification, however, produces a product with a higher solids content at high temperatures (Fig. 38) and thus extends its plastic range.

Salad oil of a relatively low cloud point can be made from palm oil by fractionation after directed interesterification. The use of interesterification has also been applied to the production of high-stability margarine blends and hard butters that have highly desirable melting qualities.

Using a column with countercurrent flow of dimethylformamide, a process has been developed to selectively reduce the content of linolenic acid in soybean oil by direct interesterification.

5.7 ROLE OF FOOD LIPIDS IN FLAVOR

5.7.1 Physical Effects

Pure food lipids are nearly odorless. However, apart from their major contributions as precursors of flavor compounds, they modify the overall flavor of many foods through their effect on mouth feel (e.g., the richness of whole milk and the smooth or creamy nature of ice cream) and on the volatility and threshold value of the flavor components present.

5.7.2 Lipids as Flavor Precursors

In the preceding discussion we have seen that lipids can undergo a variety of reactions and give rise to a multitude of intermediates and decomposition end products. These compounds vary widely in their chemical and physical properties. They also vary in their impact on flavor. Some are responsible for pleasant aromas, as is typical in fresh fruits and vegetables; others produce offensive odors and flavors, often causing major problems in the storage and processing of foods. A review by Forss [38] provides detailed information regarding the characteristics of volatile flavor compounds derived from lipids. In the following, some typical off flavors of lipid origin are discussed (see also Chap. 11).

5.7.2.1 Rancidity

Hydrolytic rancidity results from the release of free fatty acids by lipolysis. Since only the shortchain fatty acids have unpleasant odors, the problem is typically encountered in milk and dairy products.

The word *rancid,* however, is a general term that also commonly refers to off flavors resulting from lipid oxidation. The qualitative nature of oxidation flavors, however, varies significantly from product to product and even in the same food item. For example, the rancid off flavors developed from oxidation of fat in meat, walnuts, or butter are not the same. Fresh milk develops various off flavors described as rancid, cardboard, fishy, metallic, stale, or chalky; all are believed to be oxidative. Although considerable advances have been made in understanding the basic mechanisms of lipid oxidation and the flavor significance of many individual compounds, progress in correlating specific descriptions of rancid flavors with individual compounds or combinations of selected compounds has been slow. This is not surprising if one considers (a) the vast number of oxidative decomposition products identified so far (literally in the thousands), (b) their wide range of concentrations, volatiles, and flavor potencies, (c) the many possible interactions among the lipid decomposition products or between these products and other nonlipid food components, (d) the different food environments in which these flavor compounds reside, (e) the subjective nature of flavor description, (f) the complex influence of oxidative conditions on the multitude of possible reaction pathways and reaction products, and (g) the persisting suspicion that some trace but important flavor components may still escape detection by the most elegant analysis. Unfortunately, much more extensive research has been done on identification of volatiles in oxidized fats than on correlating these volatiles with flavor. With the continuous improvements of sensitivity and resolution in analytical instrumentation the detection of whole series of new compounds is made possible, and interpretations regarding flavor become more complex. Perhaps new approaches are needed to establish more precise relationships between volatile (and pertinent nonvolatile) lipid-derived components and specific oxidized flavors. An interesting technique, aimed at identifying the most important volatile oxidation products, is one in which a flavor dilution factor (FD factor) is determined by gas chromatographic analysis and effluent sniffing of a dilution series of the original aroma extract [138].

5.7.2.2 Flavor Reversion

This problem is unique to soybean oil and other linolenate-containing oils. The off flavor has been described as beany and grassy and usually develops at low peroxide values (about 5 mEq/kg). Several compounds have been suggested as responsible for or contributing to reversion flavors.

Smouse and Chang [134] postulated that 2-*n*-pentylfuran, one of the compounds identified in reverted soybean oil and found to produce reversion-like flavors in other oils if added at 2 ppm levels, is formed from the autoxidation of linoleate by the following mechanism:

$$C-(C)_4-C=C-\overset{10}{\underset{OOH}{C}}-C=C-R'$$

↓

$$C-(C)_4-C=C-C-CHO$$

↓

$$C-(C)_4-\underset{O\,-}{C}-\underset{O}{C}-C-CHO$$

↓

$$C-(C)_4-\underset{O\,\vdots\,OH}{C}-C-C-CHO$$

↓

$$C-(C)_4-\underset{O}{\overset{\|}{C}}-C-C-CHO$$

↓

$$C-(C)_4-\underset{OH}{C}-C-C=\underset{OH}{C}$$

↓

$$C-(C)_4-\text{(furan ring)}$$ 2-Pentylfuran

The *cis*- and *trans*-2-(1-pentenyl) furans were also reported as possible contributors to the reversion flavor.

Presence of the 18:3 acid catalyzes this reaction. It should be pointed out that the hydroperoxide intermediate involved here is the 10-hydroperoxide, which is not typical of linoleate autoxidation. It can arise, however, from singlet oxygen oxidation. An alternative mechanism involving singlet oxygen and the formation of a hydroperoxy cyclic peroxide [40] is given:

$$C-(C)_4-CH=CH-CH=CH-\underset{OOH}{C}-(C)_7COOMe$$

$\downarrow\ ^1O_2$

$$C-(C)_4-\text{(cyclic peroxide, O–O)}-\underset{OOH}{C}-(C)_7COOMe$$

↓

$$C-(C)_4-\text{(furan ring)}-CHO$$

↓

$$C-(C)_4-\text{(furan ring)}$$

2-Pentylfuran

Other compounds reported by various workers to be significant in soybean reversion are 3-*cis*- and 3-*trans*-hexenals, phosphatides, and nonglyceride components.

5.7.2.3 Hardening Flavor

This off flavor develops in hydrogenated soybean and marine oils during storage. Compounds reported to contribute to this defect include 6-*cis*- and 6-*trans*-nonenal, 2-*trans*-6-*trans*-octadecadienal, ketones, alcohols, and lactones. These compounds are presumed to arise from autoxidation of the isomeric dienes, known as isolinoleates, formed during the course of hydrogenation.

5.8 PHYSIOLOGICAL EFFECTS OF LIPIDS

5.8.1 Composition and Trends

Intakes of lipids in the United States have been estimated in several national surveys. The 1985 and 1986 USDA Continuing Survey of Food Intakes of Individuals [140, 141] indicated that fat provided an average of 36–37% of total calories for men and women 19–50 years of age. A mean of 13% of total calories were consumed from saturated fatty acids, 14% from monounsaturated fatty acids, and 7% from polyunsaturated fatty acids.

The daily intake of cholesterol averaged 280 mg for women and 439 mg for men. Dietary cholesterol came mainly from meat, eggs, and dairy products. Recent trends in the United States include greater consumption of vegetable fats, especially salad and cooking oils; decreased consumption of animal fats; increased consumption of seafood; and a shift from lard to shortenings, from butter to margarine, and from whole milk to skim and low-fat milks. Low-fat and fat-free desserts and snack-food items are becoming more and more available.

5.8.2 Nutritional Functions

Fats are a concentrated source of energy. As compared with proteins and carbohydrates, they supply more than twice as many calories per gram, that is, 9 kcal/g (37.7 kJ/g). They confer a feeling of satiety and contribute to the palatability of food. The essential fatty acids linoleic and arachidonic, as well as fat-soluble vitamins A, D, E, and K, are obtained from the lipid fraction of the diet. Only recently have the biological functions of the n-3 fatty acids been recognized [85, 92]. Linoleic acid (18:2, n-6) and α-linolenic acid (18:3, n-3) are recognized as the parent fatty acids of the n-6 and the n-3 series, respectively. They undergo successive desaturations and elongations, giving rise to various metabolites with important biological functions (Fig. 39). A nutritional strategy to provide optimum amounts and ratios of the n-6 and n-3 fatty acid series is being vigorously pursued.

5.8.3 Safety of Lipids Exposed to Frying Conditions

The various changes induced in fats by heating and oxidation, and the factors that influence such changes were discussed earlier in this chapter. The possibility that consumption of heated and/or oxidized fats may produce adverse effects has been a major concern and has stimulated extensive research. Several reports and reviews on this subject are available [4, 7, 20, 68, 111, 117, 121, 137]. Nutritional and toxicity studies have been conducted with animals fed heated or oxidized fats, used frying oils, or certain fractions or pure compounds typical of those identified in such oils. Unfortunately, some information reported from laboratory experiments has been misleading because oils were heated continuously at abusive temperatures in the absence of food, or because fractions of heated oils were fed to animals at unreasonably high levels. For example, cyclic monomers that form in thermally abused oils cause acute toxicity when

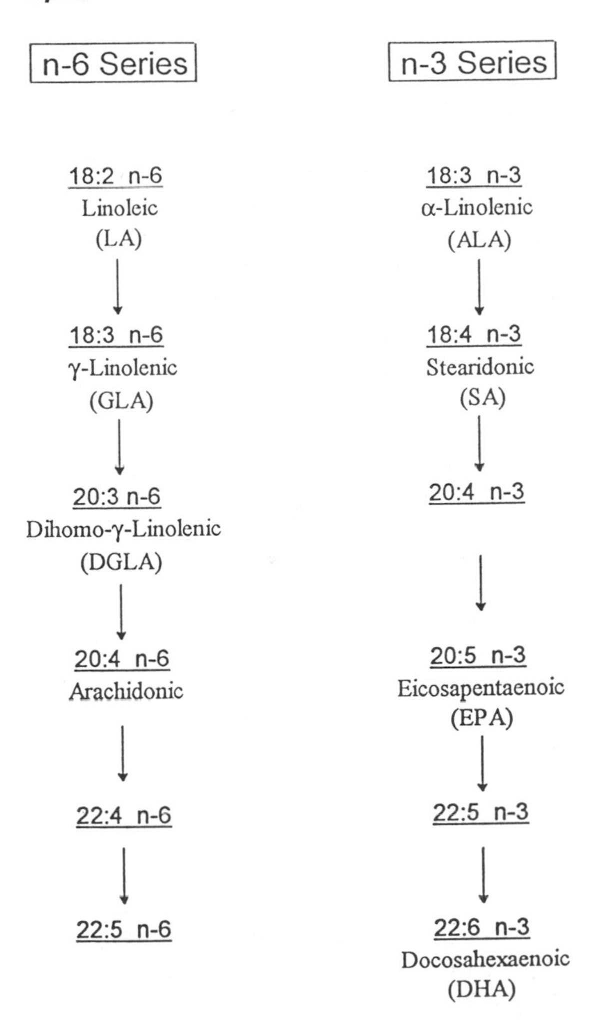

FIGURE 39 Metabolic pathways of the n-6 and n-3 fatty acids.

administered to rats in large doses. Results from studies of these kinds are inapplicable to "real-life" frying situations.

The current consensus is that toxic compounds can be generated in fat by abusive heating and/or oxidation, but that moderate ingestion of foods fried in high quality oils using recommended practices is unlikely to pose a significant hazard to health.

5.8.4 Safety of Hydrogenated Fats [35, 51, 65, 71, 86, 101, 119, 126, 146]

As pointed out earlier, some double-bond migration occurs during hydrogenation of oils, and this results in the formation of positional and geometric isomers. *Trans* fatty acids constitute 20–40% of the total acids of some margarines and shortenings. *Trans* fatty acids are not biologically equivalent to their *cis* isomers, and precise knowledge of their physiological properties, metabolism, and long-term effects on health remains controversial.

5.8.5 Safety of Irradiated Food

Irradiation results in partial destruction of fat-soluble vitamins, with tocopherol being particularly sensitive. Extensive research, conducted over several decades, indicates that radiation pasteurization of food, conducted under proper conditions, is safe and wholesome [144]. However, widespread acceptance of this novel process has been hampered by fears, misconceptions, politics, false association with nuclear reactors or weapons, and lack of education.

In November 1980 the Joint FAO/WHO/IAEA Expert Committee on Wholesomeness of Irradiated Food concluded that "the irradiation of any food commodity up to an overall average dose of 10 kGy causes no toxicological hazard and hence toxicological testing of foods so treated is no longer required." In 1986 the U.S. Food and Drug Administration approved irradiation of fresh food at up to a maximum dose of 1 kGy to inhibit growth and maturation. Also, irradiation of spices was approved to a maximum dose of 30 kGy for disinfestation. Irradiation of raw poultry was approved by the Food and Drug Administration to control foodborne pathogens in 1990, and by the U.S. Department of Agriculture to control salmonellosis in 1992.

5.8.6 Dietary Lipids, Coronary Heart Disease, and Cancer [18, 19, 52, 57, 61, 70, 73, 78–80, 87, 91, 92, 105, 112, 116, 140, 141]

The type and amount of lipids in the diet is one of several factors believed to have an influence on the incidence of coronary heart disease. Lipid intake can, in some individuals, have a moderately adverse influence on serum cholesterol concentration and on the ratios of low- and high-density lipoproteins, both factors affecting the likelihood of coronary heart disease (CHD). It is generally agreed that persons in the U.S. population consume greater amounts of lipids, especially saturated lipids and cholesterol, than is desirable. The fatty acid composition of lipids has thus received considerable attention with respect to its influence on risk of CHD. Fatty acid attributes that have received particular attention are chain length, degree and position (n-3, n-6, n-9) of unsaturation, geometric configuration (*cis, trans*), and *sn* position. The effects of these attributes on human health are frequently the subject of considerable controversy, and the reader is referred to the many sources of information on this subject.

With respect to cancer, epidemiologic and animal data support a relationship between total fat intake and risk [101]. Some studies on humans suggest that high intake of saturated fatty acids is associated with increased risk of colon, prostate, and breast cancers. Results of some animal studies indicated increased risk caused by injection of n-3 fatty acids, whereas others suggest that diets with a high content of n-3 fatty acids may decrease the risk of cancers of the mammary gland, intestine, and pancreas.

5.8.7 Concluding Thoughts on Dietary Lipids and Health

The relationship between dietary fats and health and disease is one of the most active areas of biochemical, nutritional, and medical research. This field is advancing rapidly, and the consumer is bombarded daily with information and advice that are often contradictory and confusing. Exceedingly complex metabolic interrelationships are involved, and therefore it is quite understandable that experimental work in this field is complicated and difficult. I do not, however, wish to leave the impression that published material should be dismissed, or that all recommendations should be ignored. One cannot quarrel with dietary modifications directed at correcting dietary excesses or nutritional imbalances, or with appropriate adjustments in total caloric intake. A proper balance among mono-unsaturated, n-3, and n-6 unsaturated fatty acids is justified.

But it should be understood that the role of dietary fat in health and disease remains one of considerable controversy. Thus, the general public would be advised not to engage in dietary practices without proper consideration of potential adverse consequences. Probably the best advice for the majority of human subjects remains as it has been for years: a well-balanced diet that includes ample amounts of all essential nutrients, fresh fruits and vegetables, and regular exercise. Finally, it is highly desirable that intensive, well-coordinated research efforts continue for the purpose of determining more fully the role of lipids in human health, and that appropriate nutrition education programs for all age groups be conducted simultaneously.

ACKNOWLEDGMENTS

The author thanks Michelle Nawar, Yinhuai Chen, and Wen Zhang for their kind assistance in the preparation of this chapter.

BIBLIOGRAPHY

Allen, R. R. (1978). Principles and catalysts for hydrogenation of fats and oils. *J. Am. Oil Chem. Soc. 55*:792–795.

Bailey, A. E. (1950). *Melting and Solidification of Fats,* Interscience, New York.

Brisson, G. J. (1981). *Lipids in Human Nutrition,* J. K. Burgess, Englewood Cliffs, NJ.

Frankel, E. N. (1979). Autoxidation, in *Fatty Acids* (E. H. Pryde, ed.), American Oil Chemists Society, Champaign, IL, pp. 353–390.

Garti, N., and K. Sato, eds. (1988). *Crystallization and Polymorphism of Fats,* Marcel Dekker, New York.

Hamilton, R. J., and J. B. Rossell (1986). *Analysis of Oils and Fats,* Elsvier, New York.

Hudson, B. J. D., ed. (1990). *Food Antioxidants,* Elsevier, Amsterdam.

Labuza, T. P. (1971). Kinetics of lipid oxidation. *CRC Crit. Rev. Food Technol.* October, 355–404.

Lichfield, C. (1972). *Analysis of Triglycerides,* Academic Press, New York.

Lundberg, W. O., ed. (1961, 1962). *Autoxidation and Antioxidants,* vols. I and II, John Wiley & Sons, New York.

Markley, K. S. (1960). *Fatty Acids: Their Chemistry, Properties, Production, and Uses,* Interscience, New York.

Pryde, E. H. (ed.) (1979). *Fatty Acids,* American Oil Chemists Society, Champaign, IL.

Richardson, T., and J. W. Finley (1985). *Chemical Changes in Food During Processing,* AVI, Westport, CT.

Schultz, H. W., E. A. Day, and R. O. Sinnhuber, eds. (1962). *Symposium on Foods: Lipids and Their Oxidation,* AVI, Westport, CT.

Sebedio, J. L., and E. G. Perkins, eds. (1995). *New Trends in Lipid and Lipoprotein Analysis,* AOCS Press, Champaign, IL.

Simic, M. G., and M. Karel (1980). *Autoxidation in Food and Biological Systems,* Plenum Press, New York, p. 659.

Small, D. M. (1986). The physical chemistry of lipids, in *Handbook of Lipid Research,* vol. 4, Plenum Press, New York.

Swern, D., ed. (1979). *Bailey's Industrial Oil and Fat Products,* John Wiley & Sons, New York.

U.S. Department of Agriculture (1986). Nationwide food consumption survey. Continuing Survey of Food Intakes of Individuals. 1985 report no. 85-3. Nutrition Monitoring Division, Human Nutrition Service, Hyattsville, MD.

U.S. Department of Agriculture (1987). Nationwide food consumption survey. Continuing Survey of Food Intakes of Individuals. 1985 report no. 85-4. Nutrition Monitoring Division, Human Nutrition Service, Hyattsville, MD.

Weiss, T. J. (1983). *Food Oils and Their Uses,* 2nd ed., AVI, Westport, CT.

REFERENCES

1. Abrahamsson, S., and I. Ryderstedt-Nahringbauer (1962). The crystal structure of the low-melting form of oleic acid. *Acta Crystallogr. 15*:1261–1268.
2. Alberghina, L., R. D. Schmid, and R. Verger, eds. (1991). *Lipases: Structure, Mechanism and Genetic Engineering. GBF Monographs,* vol. 16, VCH, Weinheim.
3. Albright, L. F. (1965). Quantitative measure of selectivity of hydrogenation of triglycerides. *J. Am. Oil Chem. Soc. 42*:250–253.
4. Alexander, J. C. (1978). Biological effects due to changes in fats during heating. *J. Am. Oil Chem. Soc. 55*:711–717.
5. Allen, R. R. (1978). Principles and catalysts for hydrogenation of fats and oils. *J. Am. Oil Chem. Soc. 55*:792–795.
6. Aronhime, J. S., S. Sarig, and N. Garti (1988). Dynamic control of polymorphic transformation in triglycerides by surfactants: The Button syndrome. *J. Am. Oil Chem. Soc. 65*:1144–1159.
7. Artman, N. R. (1969). The chemical and biological properties of heated and oxidized fats. *Adv. Lipid Res. 7*:245–330.
8. Bailey, A. E. (1950). *Melting and Solidification of Fats,* Interscience, New York.
9. Bast, A., G. Haenen, and C. Doelman (1991). Oxidants and antioxidants: State of the art. *Am. J. Med. 91*(suppl. 3C):2–13S
10. Bateman, L., H. Hughes, and A. L. Morris (1953). Hydroperoxide decomposition in relation to the initiation of radical chain reactions. *Disc. Faraday Soc. 14*:190–199.
11. Billek, G., G. Guhr, and J. Waibel (1978). Quality assessment of used frying fats: A comparison of four methods. *J. Amer. Oil Chem. Soc. 55*:728–733.
12. Bolland, J. L., and G. Gee (1946). Kinetic studies in the chemistry of rubber and related materials. *Trans. Faraday Soc. 42*:236–252.
13. Bolland, J. L., and P. ten Have (1947). Kinetic studies in the chemistry of rubber and related materials. IV. The inhibitory effect of hydroquinone on the thermal oxidation of ethyl linoleate. *Trans. Faraday Soc. 43*:201–210.
14. Bracco, U., J. Loliger, and J. L. Viret (1981). Production and use of natural antioxidants. *J. Am. Oil Chem. Soc. 68*:669–671.
15. Brockerhoff, H. (1965). A stereospecific analysis of triglycerides. *J. Lipid Res. 6*:10–15.
16. Brockerhoff, H., and R. G. Jensen (1974). *Lipolytic Enzymes,* Academic Press, New York, pp. 330.
17. Cahn, R. S., C. K. Ingold, and V. Prelog (1956). The specification of asymmetric configuration in organic chemistry. *Experientia 12*:81–94.
18. Carroll, K. K. (1991). Dietary fats and cancer. *Am. J. Clin. Nutr. 53*:1064S–1067S.
19. Carroll, K. K., and G. J. Hopkins (1978). Dietary polyunsaturated fat versus saturated fat in relation to mammary carcinogenesis. *Lipids 14*(2):155–158.
20. Causert, J. (1982). Chauffage des corps gras et risques de toxicité. (1). *Cahiers Nutr. Diet 17*:19–23.
21. Chan, H. W. S. (1977). Photo-sensitized oxidation of unsaturated fatty acid methyl esters. The identification of different pathways. *J. Am. Oil Chem. Soc. 54*:100–104.
22. Chan, H. W. S., and G. Levett (1977). Autoxidation of methyl linoleate: Separation and analysis of isomeric mixtures of methyl linoleate hydroperoxides and methyl hydroxylinoleates. *Lipids 12*:99–104.
23. Chan, H. W. S., and G. Levett (1977). Autoxidation of methyl linoleate: Analysis of methyl hydroxylinoleate isomers by high performance liquid chromatography. *Lipids 12*:837–840.
24. Chang, S. S., and B. Matijasevic, O. Hsieh, and C. Huang (1977). Natural antioxidants from rosemary and sage. *J. Food Sci. 42*:1102–1106.
25. Chang, S. S., R. J. Peterson, and C. T. Ho (1978). Chemistry of deep fat fried flavor, in *Lipids as a Source of Flavor* (M. K. Supton, ed.), ACS Symposium Series 75, ACS, Washington, DC, pp. 18–41.
26. Chapman, G. M., E. E. Akenhurst, and W. B. Wright (1971). Cocoa butter and confectionary fats. Studies using programmed temperature x-ray diffraction and differential scanning calorimetry. *J. Am. Oil Chem. Soc. 48*:824–830.

27. Christie, W. W., and J. H. Moore (1969). A semimicro method for the stereospecific analysis of triglycerides. *Biochim. Biophys. Acta 176*:445–452.
28. Cosgrove, J. P., D. F. Church, and W. A. Pryor (1987). The kinetics of the autoxidation of polyunsaturated fatty acids. *Lipids 22*:299–304.
29. Crnjar, E. D., A. Witchwoot, and W. W. Nawar (1981). Thermal oxidation of a series of saturated triacylglycerols. *J. Agric. Food Chem. 29*:39–42.
30. Dahle, L. K., E. G. Hill, and R. T. Holman (1962). The thiobarbituric acid reaction and the autoxidations of polyunsaturated fatty acid methyl esters. *Arch. Biochem. Biophys. 98*:253–267.
31. De Man, L., J. M. de Man, and B. Blackman (1989). Physical and textural evaluation of some shortenings and margarines. *J. Am. Oil Chem. Soc. 66*:128–132.
32. Dillard, C. J., and A. L. Tappel (1971). Fluorescent products of lipid peroxidation and mitochondria and microsomes. *Lipids 6*:715–721.
33. D'Souza, V., J. M. de Man, and L. de Man (1990). Short spacings and polymorphic forms of natural and commercial solid fats: A review. *J. Am. Oil Chem. Soc. 67*:835–843.
34. Dupuy, H. P., E. T. Rayner, J. I. Wadsworth, and M. L. Legendre (1977). Analysis of vegetable oils for flavor quality by direct gas chromatography. *J. Am. Oil Chem. Soc. 54*:445–449.
35. Enig, M. G. (1993). *Trans*-fatty acids—An update. *Nutr. Q. 17*:79–95.
36. Farmer, E. H., G. F. Bloomfield, A. Sundralingam, and D. A. Sutton (1942). The course and mechanism of autoxidation reactions in olefinic and polyolefinic substances, including rubber. *Trans. Faraday Soc. 38*:348–356.
37. Fioriti, J. A., A. P. Bentz, and R. J. Sims (1966). The reaction of picric acid with epoxides. II, The detection of epoxides in heated oils, *J. Am. Oil Chem. Soc. 43*:487–490.
38. Forss, D. A. (1972). Odor and flavor compounds from lipids, in *Progress in the Chemistry of Fats and Other Lipids,* vol. 13 (R. T. Holman, ed.), Pergamon Press, London, pp. 177–258.
39. Frankel, E. N. (1979). Autoxidation, in *Fatty Acids* (E. H. Pryde, ed.), American Oil Chemists Society, Champaign, IL, pp. 353–390.
40. Frankel, E. N. (1982). Volatile lipid oxidation products. *Prog. Lipid Res. 22*:1–33.
41. Frankel, E. N., W. E. Neff, and E. Selke (1984). Analysis of autoxidized fats by gas chromatography—mass spectrometry. IX. Homolytic cleavage vs. heterolytic cleavage of primary and secondary oxidation products. *Lipids 19*:790–800.
42. Fritsch, C. W. (1994). Lipid oxidation—The other dimensions. *Inform 5*:423–436.
43. Fritsch, C. W., and J. A. Gale (1977). Hexanal as a measure of rancidity in low fat foods. *J. Am. Oil Chem. Soc. 54*:225–228.
44. Fritsch, C. W., D. C. Egberg, and J. S. Magnuson (1979). Changes in dielectric constant as a measure of frying oil deterioration. *J. Am. Oil Chem. Soc. 56*:746–750.
45. Gardner, H. W. (1975). Decomposition of linoleic acid hydroperoxides. Enzymatic reactions compared with nonenzymic. *J. Agric. Food Chem. 23*:129–136.
46. Gardner, H. W.(1980). Lipid enzymes: Lipases, lipoxygenases and hydroperoxides, in *Autoxidation in Food and Biological Systems* (M. G. Simic and M. Karel, eds.), Plenum Press, New York, pp. 447–504.
47. Gardner, H. W. (1985). Oxidation of lipids in biological tissue and its significance, in *Chemical Changes in Food During Processing* (T. Richardson and J. W. Finley, eds.), AVI Publishing, Westport, CT, pp. 177–203.
48. Gardner, H. W., and R. D. Plattner (1984). Linoleate hydroperoxides are cleaved heterolytically into aldehydes by a Lewis acid in aprotic solvent. *Lipids 19*:294–299.
49. Garti, N., and Sato, K., eds. (1988). *Crystallization and Polymorphism of Fats,* Marcel Dekker, New York.
50. Gopala Krishna, A. G., and J. V. Prabhakar (1992). Effect of water activity on secondary products formation in autoxidizing methyl linoleate. *J. Am. Oil Chem. Soc. 69*:178–183.
51. Grundy, S. M. (1990). *Trans*-monounsaturated fatty acids and serum cholesterol levels. *N. Engl. J. Med. 323*:480–481.
52. Grundy, S. M., and G. V. Vega (1988). Plasma cholesterol responsiveness to saturated fatty acids. *Am. J. Clin. Nutr. 47*:822–824.

53. Hagemann, J. W. (1988). Thermal behavior of acylglycerides, in *Crystallization and Polymorphism of Fats and Fatty Acids* (N. Garti and K. Sato, eds.), Marcel Dekker, New York, pp. 9–95.
54. Hamama, A. A., and W. W. Nawar (1991). Thermal decomposition of some phenolic antioxidants. *J. Agric. Food Chem. 39*:1063–1069.
55. Hamilton, R. J., and J. B. Rossell (1986). *Analysis of Oils and Fats,* Elsevier, New York.
56. Hau, L. B., and W. W. Nawar (1988). Thermal oxidation of lipids in monolayers. Unsaturated fatty acid esters. *J. Am. Oil Chem. Soc. 65*:1307–1310.
57. Hegsted, D. M., R. B. McGandy, M. L. Myers, and F. J. Stare (1965). Quantitative effects of dietary fat on serum cholesterol in man. *Am. J. Clin. Nutr. 17*:281–295.
58. Hernqvist, L. (1984). Structure of triglycerides in the liquid state and fat crystallization. *Fette Seifen Anstrich 86*(8):297–300.
59. Hilditch, T. P., and P. N. Williams (1964). *The Chemical Constitution of Natural Fats,* 4th ed., Chapman & Hall, London.
60. Hirschmann, H. (1960). The nature of substrate asymmetry in stereoselective reactions. *J. Biol. Chem. 235*:2762–2767.
61. Holman, R. T., and S. B. Johnsen (1983). Essential fatty acid deficiencies in man, in *Dietary Fats and Health* (E. G. Perkings and W. J. Visek, eds.), American Oil Chemists' Society, Champaign, IL, pp. 247–266.
62. Hudson, B. J. D., ed. (1990). *Food Antioxidants,* Elsevier, Amsterdam.
63. Hultin, H. O. (1995). Role of membranes in fish quality. *Proc. Nordic Conf. Fish Quality—Role of Biological Membranes,* Hillerod, Denmark.
64. Hussein, A. B. B. H., and M. J. W. Povey (1984). Study of dilation and acoustic propagation in solidifying fats and oils: Experimental. *J. Am. Oil Chem. Soc. 61*:560–564.
65. Hunter, J. E., and T. H. Applewhite (1986). Isomeric fatty acids in the U.S. diet: Levels and health perspectives. *Am. J. Clin. Nutr. 44*:707–717.
66. IUPAC-IUB Commission on Biochemical Nomenclature (1977). The nomenclature of lipids. Recommendations (1976). *Lipids 12*:445–468.
67. IUPAC (1979). *Standard Methods for the Analysis of Oils, Fats and Derivatives,* Pergamon Press, Oxford.
68. Iwaoka, W. T., and E. G. Perkins (1978). Metabolism and lipogenic effects of the cyclic monomers of methyl linolenate in the rat. *J. Am. Oil Chem. Soc. 55*:734–738.
69. Jackson, H. W. (1981). Techniques for flavor and odor evaluation. *J. Am. Oil Chem. Soc. 58*:227–231.
70. Jackson, R. L., O. D. Taunton, J. D. Morrissett, and A. M. Gotts (1978). The role of dietary polyunsaturated fat in lowering blood cholesterol in man. *Circ. Res. 42*:447–453.
71. Judd, J. T. (1994). Dietary *trans*-fatty acids: Effects on plasma lipids and lipoproteins of healthy men and women. *Am. J. Clin. Nutr. 59*:861–868.
72. Kahl, J. L., W. E. Artz, and E. G. Schanus (1988). Effects of relative humidity on lipid autoxidation in a model system. *Lipids 23*:275–279.
73. Kannel, W. B., D. McGee, and T. Gordon (1976). A general cardiovascular risk profile: The Framingham study. *Am. J. Cardiol.* 38:46–51.
74. Karel, M. (1980). Lipid oxidation, secondary reactions, and water activity of foods, in *Autoxidation in Food and Biological Systems* (M. S. Simic and M. Karel, eds.), Plenum Press, New York, pp. 191–206.
75. Karel, M. (1985). Environmental effects on chemical changes in foods, in *Chemical Changes in Food During Processing* (T. Richardson and J. W. Finley, Eds.), AVI Publishing, Westport, CT, pp. 483–501.
76. Karel, M., K. Schaich, and B. R. Roy (1975). Interaction of peroxidizing methyl linoleate with some proteins and amino acids. *J. Agric. Food Chem. 23*:159–163.
77. Kartha, A. R. S. (1953). The glyceride structure of natural fats. II. The rule of glyceride type distribution of natural fats. *J. Am. Oil Chem. Soc. 30*:326–329.
78. Keys, A. (1974). Bias and misrepresentation revisited: Perspective on saturated fat. *Am. J. Clin. Nutr. 27*:188–212.
79. Keys, A., J. T. Anderson, and F. Grande (1965). Serum cholesterol response to changes in the diet. IV. Particular saturated fatty acids in the diet. *Metab. Clin. Exp. 14*:776–787.

80. Keys, A., J. T. Anderson, O. Mickelson, S. F. Adelson, and F. Fidanza (1956). Diet and serum cholesterol in man: Lack of effect of dietary cholesterol. *J. Nutr. 59*:39–56.
81. Kim, S. K., and W. W. Nawar (1991). Oxidation interactions of cholesterol with triacylglycerols. *J. Am. Oil Chem. Soc. 68*:931–934.
82. Kim, S. K., and W. W. Nawar (1992). Oxidative interactions of cholesterol in the milk fat globule membrane. *Lipids 27*:928–932.
83. Kim, S. K., and W. W. Nawar (1993). Parameters influencing cholesterol oxidation. *Lipids* 28:917–920.
84. Kimoto, W. I., and A. M. Gaddis (1969). Precursors of alk-2,4-dienals in autoxidized lard. *J. Am. Oil Chem. Soc. 46*:403–408.
85. Kinsella, J. E. (1987). *Seafoods and Fish Oils in Human Health and Disease.* Marcel Dekker, New York.
86. Kinsella, J. E., G. Bruckner, J. Mai, and J. Shimp (1981). Metabolism of trans fatty acids with emphasis on the effects of trans, trans, octadecadienoate on lipid compostion, essential fatty acid, and prostaglandins: An overview, *Am. J. Clin. Nutr. 34*:2307–2318.
87. Kritchersky, D. (1976). Diet and atherosclerosis. *Am. J. Pathol. 84*:615–632.
88. Labuza, T. P. (1971). Kinetics of lipid oxidation. In *CRC Crit. Rev. Food Technol*. October, 355–404.
89. Labuza, T. P., L. McNally, D. Gallagher, J. Hawekes, and F. Hurtado (1972). Stability of intermediate moisture foods. 1. Lipid oxidation. *J. Food Sci. 37*:154–159.
90. Labuza, T. P., H. Tsuyuki, and M. Karel (1969). Kinetics of linoleate oxidation in model systems. *J. Am. Oil Chem. Soc. 46*:409–416.
91. Lambert-Legace, L., and M. Laflame (1995). *Good Fat, Bad Fat,* Stoddart, Toronto.
92. Lands, W. E. M. (1986). *Fish and Human Health,* Academic Press, New York.
93. Lands, W. E. M., R. A. Pieringer, P. M. Slakey, and A. Zschoke (1966). A micro-method for the stereospecific determination of triglyceride structure. *Lipids J. 1*:444–448.
94. Larsson, K. (1972). On the structure of isotropic phases in lipid-water systems. *Chem. Phys. Lipids 9*:181–195.
95. Larsson, K. (1976). Crystal and liquid crystal structures of lipids, in *Food Emulsions* (S. Friberg, ed.), Marcel Dekker, New York, pp. 39–66.
96. Link, W. E., ed. (1974). *Official and Tentative Methods of the American Oil Chemists' Society,* AOCS, Champaign, IL.
97. Loury, M. (1972). Possible mechanisms of autoxidative rancidity. *Lipids 7*:671–675.
98. Luddy, F. E., R. A. Barford, S. F. Herb, P. Magidman, and R. Riemenschneider (1964). Pancreatic lipase hydrolysis of triglycerides by a semimicro technique. *J. Am. Oil Chem. Soc. 41*:693–696.
99. Luzatti, V. (1968). in *X-ray Diffraction Studies of Lipid–Water Systems in Biological Membranes* (D. Chapman, ed.), Academic Press, London, pp. 71–123.
100. Madison, B. L., and R. C. Hill (1978). Determination of solid fat content of commercial fats by pulsed nuclear magnetic resonance. *J. Am. Oil Chem. Soc. 55*:328–331.
101. Mann, G. (1994). Metabolic consequences of dietary *trans*-fatty acids. *Lancet 343*:1268–1271.
102. Matthews, R. F., R. A. Scanlon, and L. M. Libbey (1971). Autoxidation products of 2,4-decadienal. *J. Am. Oil Chem. Soc. 48*:745–747.
103. McClements, D. J., and M. J. W. Povey (1987). Solid fat content determination using ultrasonic velocity measurements. *Int. J. Food Sci. Technol. 22*:491–499.
104. McClements, D. J., and M. J. W. Povey (1988). Comparison of pulsed NMR and ultrasonic velocity techniques for determining solid fat contents. *Int. J. Food Sci. and Technol. 23*:159–170.
105. McGill, H. C. Jr. (1979). The relationship of dietary cholesterol to serum cholesterol concentration and to atherosclerosis in man. *Am. J. Clin. Nutr. 32*:2664–2702.
106. Minifie, B. W. (1980). *Chocolate, Cocoa and Confectionary,* 2d ed., AVI, Westport, CT.
107. Miyashita, K., E. Nara, and T. Ota (1993). Oxidation stability of polyunsaturated fatty acids in an aqueous solution. *Biosci. Biotechnol. Biochem. 57*(10):1638–1640.
108. Nawar, W. W. (1977). Radiation chemistry of lipids, in *Radiation Chemistry of Major Food Components* (P. S. Elias and A. J. Cohen, eds.), Elsevier, Amsterdam, pp. 26–61.
109. Nawar, W. W., S. K. Kim, and M. Vajdi (1991). Measurement of oxidative interactions of cholesterol. *J. Am. Oil Chem. Soc. 68*:496–498.

110. Nawar, W. W., Z. R. Zhu, and Y. J. Yoo (1990). Radiolytic products of lipids as markers for the detection of irradiated meats, in *Food Irradiation and the Chemist* (D. E. Johnston, M. H. Stevenson, eds.) Royal Society of Chemists, Cambridge, pp. 13–24.
111. Nolen, G. A., J. C. Alexander, and N. R. Artman (1967). Long-term rat feeding study with used frying fats. *J. Nutr. 93*:337–348.
112. NRC Committee on Diet and Health (1990). *Diet and Health: Implications for Reducing Chronic Disease Risk,* National Academy Press, Washington, DC.
113. Official and Tentative Methods of the American Oil Chemists Society (1980). Peroxide value, Cd 8-53; oxirane test, Cd 9-57; iodine value Cd 1-25; AOM, CD 12-57. *J. Am. Oil Chem. Soc.*
114. Paradis, A. J., and W. W. Nawar (1981). Evaluation of new methods for the assessment of used frying oils. *J. Food Sci. 46*:449–451.
115. Paradis, A. J., and W. W. Nawar (1981). A gas chromatographic method for the assessment of used frying oils: Comparison with other methods. *J. Am. Oil Chem. Soc. 58*:635–638.
116. Parker, R. S. (1989). Dietary and biochemical aspects of vitamin E. *Adv. Food Nutr. Res. 33*:157–232.
117. Perkins, E. G. (1976). Chemical, nutritional and metabolic studies of heated fats. II. Nutritional aspects. *Rev. Fr. Corps Gras 23*:313–322.
118. Porter, W. L. (1980). Recent trends in food applications of antioxidants, in *Autoxidation in Food in Biological Systems* (M. G. Simic and M. Karel, eds.), Plenum Press, New York, pp. 295–365.
119. Privett, O. S., F. Phillips, H. Shimasaki, T. Nazawa, and E. C. Nickell (1977). Studies of effects of trans fatty acids on the diet on lipid metabolism in essential fatty acid deficient rats. *Am. J. Clin. Nutr. 30*:1009–1017.
120. Pryor, W. A., J. P. Stanley, and E. Blair (1976). Autoxidation of polyunsaturated fatty acids. II. A suggested mechanism for the formation of TBA-reactive materials from prostaglandin-like endoperoxides. *Lipids 11*:370–379.
121. Scheutwinkel-Reich, M., G. Ingerowski, and H. J. Stan (1980). Microbiological studies investigating mutagenicity of deep frying fat fractions and some of their components. *Lipids 15*:849–852.
122. Selke, E., and E. N. Frankel (1987). Dynamic headspace capillary gas chromatographic analysis of soybean oil volatiles. *J. Am. Oil Chem. Soc. 64*:749–753.
123. Shewfelt, R. L. (1981). Fish muscle lipolysis—A review. *J. Food Biochem. 5*:79–100.
124. Schlicter-Aronhime, J., and N. Garti (1988). Solidification and polymorphism in cocoa butter and the blooming problems, in *Crystallization and Polymorphism of Fats and Fatty Acids*. (N. Garti and K. Sato, eds.), Marcel Dekker, New York.
125. Schuler, P. (1990). Natural antioxidants exploited commercially, in *Food Antioxidants* (B. J. F. Hudson, ed.), Elsevier, Amsterdam, pp. 99–170.
126. Senti, F. R., ed. (1985). *Health Aspects of Dietary Trans Fatty Acids,* Federation of American Societies for Experimental Biology, Bethesda, MD.
127. Shahidi, F., P. K. Janitha, and P. D. Wanasundara (1992). Phenolic antioxidants. *Crit. Rev. Food Sci. Nutr. 32*:67–103.
128. Sherwin, E. R. (1976). Antioxidants for vegetable oils. *J. Am. Oil Chem. Soc. 53*:430–436.
129. Shimada, Y., Y. Roos, and M. Karel (1991). Oxidation of methyl linoleate encapsulated in amorphous lactose-based food model. *J. Agric. Food Chem. 39*:637–641.
130. Simoneau, C., M. J. McCarthy, R. J. Kauten, and J. B. German (1991). Crystallization dynamics in model emulsions from magnetic resonance imaging. *J. Am. Oil Chemists' Soc. 68*:481–487.
131. Simpson, T. D. (1979). Crystallography, in *Fatty Acids* (E. H. Pryde, ed.), American Oil Chemists Society, Champaign, IL, pp. 157–172.
132. Sims, R. (1994). Oxidation of fats in food products. *Inform 5*:1020–1028.
133. Smith, L. L. (1981). *Cholesterol Autoxidation*. Plenum Press, New York.
134. Smouse, T. H., and S. S. Chang (1967). A systematic characterization of the reversion flavor of soybean oil. *J. Am. Oil Chem. Soc. 44*:509–514.
135. Sreenivasan, B. (1978). Interesterification of fats. *J. Am. Oil Chem. Soc. 55*:796–805.
136. *Standard Methods for the Analysis of Oils, Fats, and Derivatives* (1979). Determination of the *p*-anisidine value. Method 2.504, 6th ed. Pergamon Press, London, pp. 143–144.

137. Taylor, S. L., C. M. Berg, N. H. Shoptaugh, and E. Traisman (1983). Mutagen formation in deep-fat fried foods as a function of frying conditions. *J. Am. Oil Chem. Soc. 60*:576–580.
138. Ullrich, F., and W. Grosch (1987). Identification of the most intense volatile flavor compounds formed during autoxidation of linoleic acid. *Z. Lebensm. Unters. Fortsch. 184*:277–282.
139. Uri, N. (1961). Mechanism of antioxidation, in *Autoxidation and Antioxidants* (W. O. Lundberg, ed.), Interscience, New York, pp. 133–169.
140. U.S. Department of Agriculture (1986). Nationwide food consumption survey. Continuing Survey of Food Intakes of Individuals. 1985 report no. 85-3. Nutrition Monitoring Division, Human Nutrition Service, Hyattsville, MD.
141. U.S. Department of Agriculture (1987). Nationwide food consumption survey. Continuing Survey of Food Intakes of Individuals. 1985 report no. 85-4. Nutrition Monitoring Division, Human Nutrition Service, Hyattsville, MD.
142. Vajdi, M., and W. W. Nawar (1979). Identification of radiolytic compounds from beef. *J. Am. Oil Chem. Soc. 56*:611–615.
143. Waltking, A. E., W. E. Seery, and G. W. Bleffert (1975). Chemical analysis of polymerization products in abused fats and oils. *J. Am. Oil Chem. Soc. 52*:96–100.
144. WHO (1994). *Safety and Nutritional Adequacy of Irradiated Food,* World Health Organization, Geneva, pp. 161.
145. Wille, R. L., and E. S. Lutton (1966). Polymorphism of cocoa butter. *J. Am. Oil Chem. Soc. 43*:491–496.
146. Willet, W. C., and A. Ascherio (1994). *Trans*-fatty acids: are the effects only marginal? *Am. J. Public Health 84*(5):1–3.
147. Williams, T. F. (1962). Specific elementary processes in the radiation chemistry of organic compounds. *Nature (Lond.)* 194:348–351.
148. Williams, R. M., and D. Chapman (1970). in *Progress in the Chemistry of Fats and Other Lipids* (P. T. Holman, ed.), vol. 11, part 1, pp. 1–79, *Phosophlipids, Liquid Crystals, and Cell Membranes,* Pergamon Press, London.
149. Woolley, P., and S. Petersen, eds. (1994). *Lipases: Their Structure, Biochemistry, and Application,* Cambridge University Press, Cambridge.

6

Amino Acids, Peptides, and Proteins

SRINIVASAN DAMODARAN
University of Wisconsin—Madison, Madison, Wisconsin

6.1 INTRODUCTION

Proteins play a central role in biological systems. Although the information for evolution and biological organization of cells is contained in DNA, the chemical and biochemical processes that sustain the life of a cell/organism are performed exclusively by enzymes. Thousands of enzymes have been discovered. Each one of them catalyzes a highly specific biological reaction in cells. In addition to functioning as enzymes, proteins (such as collagen, keratin, elastin, etc.) also function as structural components of cells and complex organisms. The functional diversity of proteins essentially arises from their chemical makeup. Proteins are highly complex polymers, made up of 20 different amino acids. The constituents are linked via substituted amide bonds. Unlike the ether and phosphodiester bonds in polysaccharides and nucleic acids, the amide linkage in proteins is a partial double bond, which further underscores the structural complexity of protein polymers. The myriad of biological functions performed by proteins might not be possible but for the complexity in their composition, which gives rise to a multitude of three-dimensional structural forms with different biological functions. To signify their biological importance, these macromolecules were named proteins, derived from the Greek word *proteois*, which means of the first kind.

At the elemental level, proteins contain 50–55% carbon, 6–7% hydrogen, 20–23% oxygen, 12–19% nitrogen, and 0.2–3.0% sulfur. Protein synthesis occurs in ribosomes. After the synthesis, some amino acid constituents are modified by cytoplasmic enzymes. This changes the elemental composition of some proteins. Proteins that are not enzymatically modified in cells are called *homoproteins*, and those that are modified or complexed with nonprotein components are called *conjugated proteins* or *heteroproteins*. The nonprotein components are often referred to as *prosthetic groups*. Examples of conjugated proteins include *nucleoproteins* (ribosomes), *glycoproteins* (ovalbumin, κ-casein), *phosphoproteins* (α- and β-caseins, kinases, phosphorylases), *lipoproteins* (proteins of egg yolk, several plasma proteins), and *metalloproteins* (hemoglobin, myoglobin, and several enzymes). Glyco- and phosphoproteins contain covalently linked carbohydrate and phosphate groups, respectively, whereas the other conjugated proteins are noncovalent complexes containing nucleic acids, lipids, or metal ions. These complexes can be dissociated under appropriate conditions.

Proteins also can be classified according to their gross structural organization. Thus, *globular proteins* are those that exist in spherical or ellipsoidal shapes, resulting from folding of the polypeptide chain(s) on itself. On the other hand, *fibrous proteins* are rod-shaped molecules containing twisted linear polypeptide chains (e.g., tropomyosin, collagen, keratin, and elastin). Fibrous proteins also can be formed as a result of linear aggregation of small globular proteins, such as actin and fibrin. A majority of enzymes are globular proteins, and fibrous proteins invariably function as *structural proteins*.

The various biological functions of proteins can be categorgized as *enzyme catalysts, structural proteins, contractile proteins* (myosin, actin, tubulin), *hormones* (insulin, growth hormone), *transfer proteins* (serum albumin, transferrin, hemoglobin), *antibodies* (immunoglobulins), *storage proteins* (egg albumen, seed proteins), and protective proteins (toxins and allergens). Storage proteins are found mainly in eggs and plant seeds. These proteins act as sources of nitrogen and amino acids for germinating seeds and embryos. The protective proteins are a part of the defense mechanism for the survival of certain microorganisms and animals.

All proteins are essentially made up of the same primary 20 amino acids; however, some proteins may not contain one or a few of the 20 amino acids. The differences in structure and function of these thousands of proteins arise from the sequence in which the amino acids are linked together via amide bonds. Literally, billions of proteins with unique properties can be synthesized by changing the amino acid sequence, the type and ratio of amino acids, and the chain length of polypeptides.

All biologically produced proteins can be used as food proteins. However, for practical purposes, food proteins may be defined as those that are easily digestible, nontoxic, nutritionally adequate, functionally useable in food products, and available in abundance. Traditionally, milk, meats (including fish and poultry), eggs, cereals, legumes, and oilseeds have been the major sources of food proteins. However, because of the burgeoning world population, nontraditional sources of proteins for human nutrition need to be developed to meet the future demand. The suitability of such new protein sources for use in foods, however, depends on their cost and their ability to fulfill the normal role of protein ingredients in processed and home-cooked foods. The functional properties of proteins in foods are related to their structural and other physicochemical characteristics. A fundamental understanding of the physical, chemical, nutritional, and functional properties of proteins and the changes these properties undergo during processing is essential if the performance of proteins in foods is to be improved, and if new or less costly sources of proteins are to compete with traditional food proteins.

6.2 PHYSICOCHEMICAL PROPERTIES OF AMINO ACIDS

6.2.1 General Properties

6.2.1.1 Structure and Classification

α-Amino acids are the basic structural units of proteins. These amino acids consist of an α-carbon atom covalently attached to a hydrogen atom, an amino group, a carboxyl group, and a side-chain R group.

$$\begin{array}{c} \text{COOH} \\ | \\ \text{H}-\text{C}_{\alpha}-\text{NH}_2 \\ | \\ \text{R} \end{array} \qquad (1)$$

Natural proteins contain up to 20 different primary amino acids linked together via amide bonds. These amino acids differ only in the chemical nature of the side chain R group (Table 1). The physicochemical properties, such as net charge, solubility, chemical reactivity, and hydrogen bonding potential, of the amino acids are dependent on the chemical nature of the R group.

Amino acids can be classified into several categories based on the degree of interaction of the side chains with water. Amino acids with aliphatic (Ala, Ile, Leu, Met, Pro, and Val) and aromatic side chains (Phe, Trp, and Tyr) are hydrophobic, and hence they exhibit limited solubility in water (Table 2). Polar (hydrophilic) amino acids are quite soluble in water, and they are either charged (Arg, Asp, Glu, His, and Lys) or uncharged (Ser, Thr, Asn, Gln, and Cys). The side chains of Arg and Lys contain guanidino and amino groups, respectively, and thus are positively charged (basic) at neutral pH. The imidazole group of His is basic in nature. However,

TABLE 1 Primary α-Amino Acids That Occur in Proteins

Amino acid					
	Symbol				
Name	Three letters	One letter	Molecular weight	Genetic code	Structure at neutral pH
Alanine	Ala	A	89.1	GC(N)	$CH_3—CH(^+NH_3)—COO^-$
Arginine	Arg	R	174.2	AGA AGG CG(N)	$H_2N—C(=^+NH_2)—NH—(CH_2)_3—CH(^+NH_3)—COO^-$
Asparagine	Asn	N	132.1	AAU AAC	$H_2N—C(=O)—CH_2—CH(^+NH_3)—COO^-$
Aspartic acid	Asp	D	133.1	GAU GAC	$^-O—C(=O)—CH_2—CH(^+NH_3)—COO^-$
Cysteine	Cys	C	121.1	UGU UGC	$HS—CH_2—CH(^+NH_3)—COO^-$
Glutamine	Glu	Q	146.1	CAA CAG	$H_2N—C(=O)—(CH_2)_2—CH(^+NH_3)—COO^-$
Glutamic acid	Glu	E	147.1	GAA GAG	$^-O—C(=O)—(CH_2)_2—CH(^+NH_3)—COO^-$
Glycine	Gly	G	75.1	GG(N)	$H—CH(^+NH_3)—COO^-$
Histidine	His	H	155.2	CAU CAC	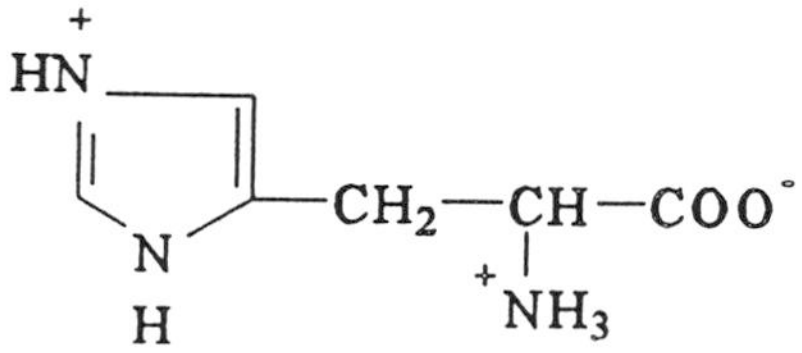 HN^+ / N / H (imidazole ring) $—CH_2—CH(^+NH_3)—COO^-$
Isoleucine	Ile	I	131.2	AUU AUC AUA	$CH_3—CH_2—CH(CH_3)—CH(^+NH_3)—COO^-$
Leucine	Leu	L	131.2	UUA UUG CU(N)	$CH_3—CH(CH_3)—CH_2—CH(^+NH_3)—COO^-$
Lysine	Lys	K	146.2	AAA AAG	$^+NH_3—(CH_2)_4—CH(^+NH_3)—COO^-$
Methionine	Met	M	149.2	AUG	$CH_3—S—(CH_2)_2—CH(^+NH_3)—COO^-$
Phenylalanine	Phe	F	165.2	UUU UUC	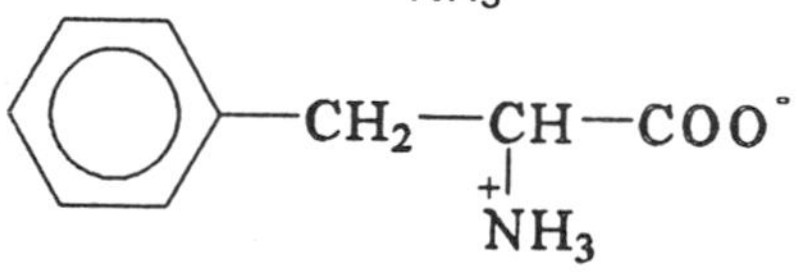 (benzene ring) $—CH_2—CH(^+NH_3)—COO^-$

TABLE 1 Continued

Amino acid: Name	Symbol: Three letters	Symbol: One letter	Molecular weight	Genetic code	Structure at neutral pH
Proline	Pro	P	115.1	CC(N)	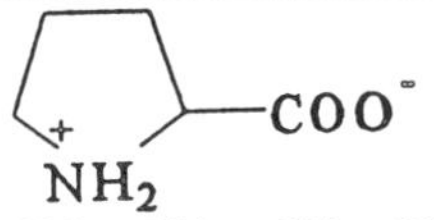 (pyrrolidine ring, $^{+}NH_2$)—COO^-
Serine	Ser	S	105.1	AGU AGC	$HO—CH_2—CH(^{+}NH_3)—COO^-$
Threonine	Thr	T	119.1	AC(N)	$CH_3—CH(OH)—CH(^{+}NH_3)—COO^-$
Trytophan	Trp	W	204.2	UGG	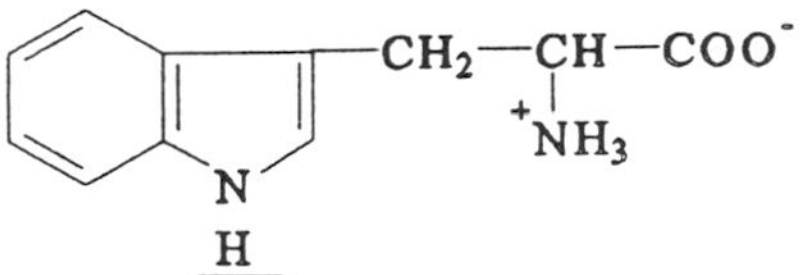 (indole ring, N—H)—$CH_2—CH(^{+}NH_3)—COO^-$
Tyrosine	Tyr	Y	181.2	AUA UAC	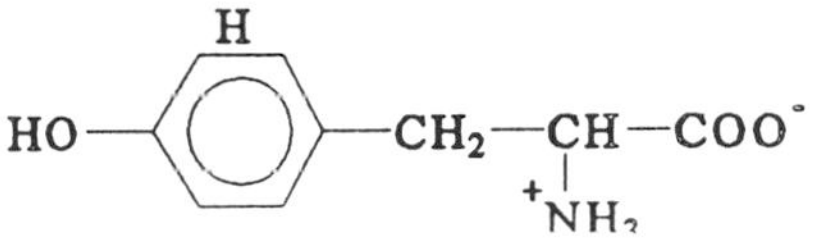 HO—(benzene ring)—$CH_2—CH(^{+}NH_3)—COO^-$
Valine	Val	V	117.1	GU(N)	$CH_3—CH(CH_3)—CH(^{+}NH_3)—COO^-$

at neutral pH its net charge is only slightly positive. The side chains of Asp and Glu acids contain a carboxyl group. These amino acids carry a net negative charge at neutral pH. Both the basic and acidic amino acids are strongly hydrophilic. The net charge of a protein at physiological conditions is dependent on the relative numbers of basic and acidic amino acids residues in the protein.

TABLE 2 Solubilities of Amino Acids in Water at 25°C

Amino acid	Solubility (g/L)	Amino acid	Solubility (g/L)
Alanine	167.2	Leucine	21.7
Arginine	855.6	Lysine	739.0
Asparagine	28.5	Methionine	56.2
Aspartic acid	5.0	Phenylalanine	27.6
Cysteine	—	Proline	1620.0
Glutamine	7.2 (37°C)	Serine	422.0
Glutamic acid	8.5	Threonine	13.2
Glycine	249.9	Tryptophan	13.6
Histidine	—	Tyrosine	0.4
Isoleucine	34.5	Valine	58.1

The polarities of uncharged neutral amino acids fall between those of hydrophobic and charged amino acids. The polar nature of Ser and Thr is attributed to the hydroxyl group, which is able to hydrogen bond with water. Since Tyr also contains an ionizable phenolic group that ionizes at alkaline pH, it is also considered to be a polar amino acid. However, based on its solubility characterisitcs at neutral pH, it should be regarded as a hydrophobic amino acid. The amide group of Asn and Gln is able to interact with water through hydrogen bonding. Upon acid or alkaline hydrolysis, the amide group of Asn and Gln is converted to a carboxyl group with release of ammonia. A majority of the Cys residues in proteins exists as cystine, which is a Cys dimer created by oxidation of thiol groups to form a disulfide cross-link.

Proline is a unique amino acid because it is the only imino acid in proteins. In proline, the propyl side chain is covalently linked to both the α-carbon atom and the α-amino group, forming a pyrrolidine ring structure.

The amino acids listed in Table 1 have genetic codes. That is, each one of these amino acids has a specific *t*-RNA that translates the genetic information on *m*-RNA into an amino acid sequence during protein synthesis. Apart from the 20 primary amino acids listed in Table 1, several proteins also contain other types of amino acids that are derivatives of the primary amino acids. These derived amino acids are either cross-linked amino acids or simple derivatives of single amino acids. Proteins that contain derived amino acids are called *conjugated* proteins. Cystine, which is found in most proteins, is a good example of a cross-linked amino acid. Other cross-linked amino acids, such as desmosine, isodesmosine, and di- and trityrosine, are found in structural proteins such as elastin and resilin. Several simple derivatives of amino acids are found in several proteins. For example, 4-hydroxyproline and 5-hydroxylysine are found in collagen. These are the result of posttranslational modification during maturation of collagen fiber. Phosphoserine and phosphothreonine are found in several proteins, including caseins. *N*-Methyllysine is found in myosin, and γ-carboxyglutamate is found in several blood clotting factors and calcium binding proteins.

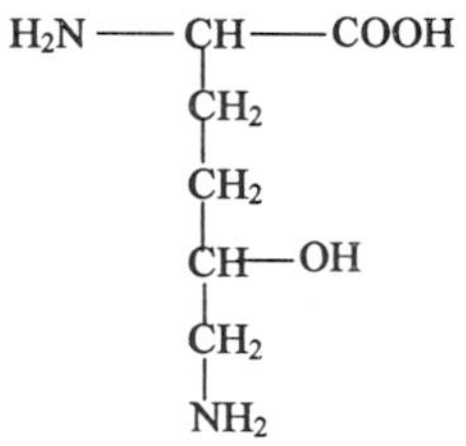

H_2N—CH—COOH
CH_2
CH_2
CH—OH
CH_2
NH_2

Hydroxylysine

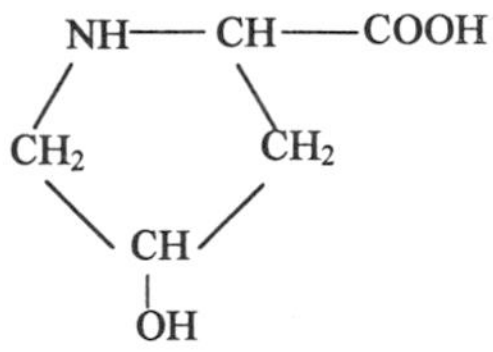

NH—CH—COOH
CH_2 CH_2
CH
OH

Hydroxyproline

(2)

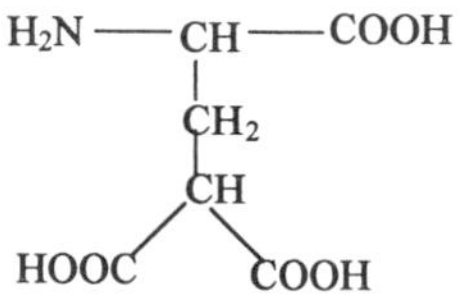

H_2N—CH—COOH
CH_2
CH
HOOC COOH

γ-Carboxyglutamate

H_2N—CH—COOH
CH_2
O—$PO_3^{=}$

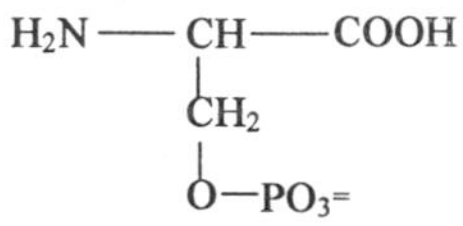

Phosphoserine

6.2.1.2 Stereochemistry of Amino Acids

With the exception of Gly, the α-carbon atom of all amino acids is asymmetric, meaning that four different groups are attached to it. Because of this asymmetric center, amino acids exhibit optical activity; that is, they rotate the plane of linearly polarized light. In addition to the asymmetric α-carbon atom, the β-carbon atoms of Ile and Thr are also asymmetric, and thus both Ile and Thr can exist in four enantiomeric forms. Among the derived amino acids, hydroxyproline and hydroxylysine also contain two asymmetric carbon centers. All proteins found in nature contain only L-amino acids. Conventionally, the L- and D-enantiomers are represented as

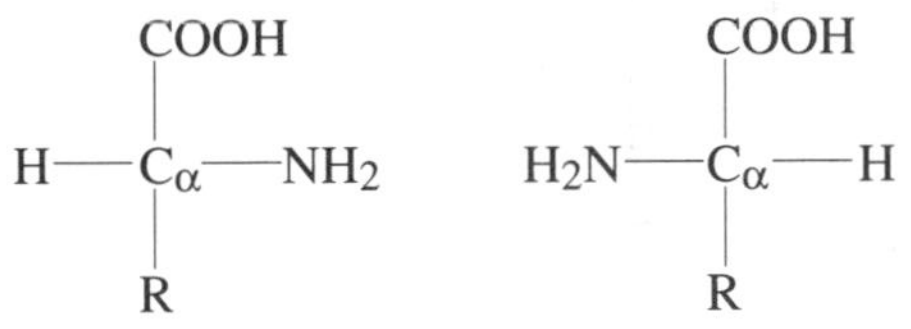

$$\begin{array}{ccc} & COOH & \\ & | & \\ H— & C_\alpha & —NH_2 \\ & | & \\ & R & \end{array} \qquad \begin{array}{ccc} & COOH & \\ & | & \\ H_2N— & C_\alpha & —H \\ & | & \\ & R & \end{array} \qquad (3)$$

D-Amino acid L-Amino acid

This nomenclature is based on D- and L-glyceraldehyde configurations, and not on the actual direction of rotation of linearly polarized light. That is, the L-configuration does not refer to levorotation as in the case of L-glyceraldehyde. In fact most of the L-amino acids are dextrorotatory, not levorotatory.

6.2.1.3 Acid-Base Properties of Amino Acids

Since amino acids contain a carboxyl group (acidic) and an amino group (basic), they behave both as acids and bases; that is, they are *ampholytes*. For example, Gly, the simplest of all amino acids, can exist in three different ionized states, depending on the pH of the solution.

$$H_3^+N—CH_2—COOH \underset{H^+}{\overset{K_1}{\rightleftharpoons}} H_3^+N—CH_2—COO^- \underset{-H^+}{\overset{K_2}{\rightleftharpoons}} H_2N—CH_2—COO^- \qquad (4)$$

Acidic Neutral Basic

At around neutral pH, both the α-amino and α carboxyl groups are ionized, and the molecule is a *dipolar ion* or a *zwitterion*. The pH at which the dipolar ion is electrically neutral is called the *isoelectric point* (pI). When the zwitterion is titrated with an acid, the COO^- group becomes protonated. The pH at which the concentrations of COO^- and COOH are equal is known as pK_{a1} (i.e., negative logarithm of the dissociation constant K_{a1}). Similarly, when the zwitterion is titrated with a base, the NH_3^+ group becomes deprotonated. As before, the pH at which $[NH_3^+] = [NH_2]$ is known as pK_{a2}. A typical electrometric titration curve for a dipolar ion is shown in Figure 1. In addition to the α-amino and α-carboxyl groups, the side chains of Lys, Arg, His, Asp, Glu, Cys, and Tyr also contain ionizable groups. The pK_a values of all the ionizable groups in amino acids are given in Table 3. The isoelectric points of amino acids can be estimated from their pK_{a1}, pK_{a2}, and pK_{a3} values, using the following expressions:

For amino acids with no charged side chain, $pI = (pK_{a1} + pK_{a2})/2$.
For acidic amino acids, $pI = (pK_{a1} + pK_{a3})/2$.
For basic amino acids, $pI = (pK_{a2} + pK_{a3})/2$.

The subscripts 1,2, and 3 refer to α-carboxyl, α-amino, and side-chain ionizable groups, respectively.

In proteins, the α-COOH of one amino acid is coupled to the α-NH_2 of the next amino acid through an amide bond; thus the only ionizable groups are the N-terminus amino group, the C-terminus carboxyl group, and ionizable groups on side chains. The pK_a of these ionizable groups in proteins are different from those of free amino acids (Table 3). In protein, the pK_{a3}

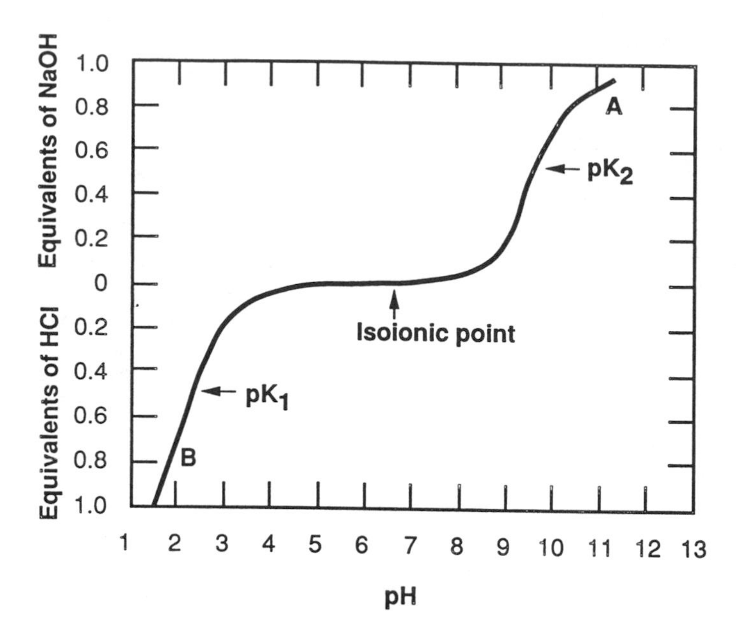

FIGURE 1 Titration curve of a typical amino acid.

TABLE 3 pK_a and pI Values of Ionizable Groups in Free Amino Acids and Proteins at 25°C

Amino acid	pK_{a1} (α-COOH)	pK_{a2} (α-NH_3^+)	pK_{aR} AA	pK_{aR} Side chain[a]	pI
Alanine	2.34	9.69	—		6.00
Arginine	2.17	9.04	12.48	> 12.00	10.76
Asparagine	2.02	8.80	—		5.41
Aspartic acid	1.88	9.60	3.65	4.60	2.77
Cysteine	1.96	10.28	8.18	8.80	5.07
Glutamine	2.17	9.13	—		5.65
Glutamic acid	2.19	9.67	4.25	4.60	3.22
Glycine	2.34	9.60			5.98
Histidine	1.82	9.17	6.00	7.00	7.59
Isoleucine	2.36	9.68	—		6.02
Leucine	2.30	9.60	—		5.98
Lysine	2.18	8.95	10.53	10.20	9.74
Methionine	2.28	9.21	—		5.74
Phenylalanine	1.83	9.13	—		5.48
Proline	1.94	10.60	—		6.30
Serine	2.20	9.15	—		5.68
Threonine	2.21	9.15	—		5.68
Tryptophan	2.38	9.39	—		5.89
Tyrosine	2.20	9.11	10.07	9.60	5.66
Valine	2.32	9.62	—		5.96

[a]pK_a values in proteins.

values of acidic side chains (Glu and Asp) are larger and those of basic side chains are lower than those of the corresponding free amino acids.

The degree of ionization of a group at any given solution pH can be determined by using the Henderson–Hasselbach equation:

$$pH = pK_a + \log \frac{[\text{conjugated base}]}{[\text{conjugated acid}]} \tag{5}$$

The net charge of a protein at a given pH can be estimated by determining the degree of ionization of individual ionizable groups using this equation, and then adding up the total number of negative and positive charges.

6.2.1.4 Hydrophobic Properties of Amino Acids

One of the major factors affecting physicochemical properties, such as structure, solubility, fat-binding properties, etc., of proteins and peptides is the hydrophobicity of the constituent amino acid residues. Hydrophobicity can be defined as the excess free energy of a solute dissolved in water compared to that in an organic solvent under similar conditions. The most direct and simplest way to estimate relative hydrophobicities of amino acid side chains involves experimental determination of free energy changes for dissolution of amino acid side chains in water and in an organic solvent, such as ethanol. The chemical potential of an amino acid dissolved in water can be expressed by the equation

$$\mu_{AA,w} = \mathring{\mu}_{AA,w} + RT \ln \gamma_{AA,w} C_{AA,w} \tag{6}$$

where $\mathring{\mu}_{AA,w}$ is the standard chemical potential of the amino acid, γ_{AA} is the activity coefficient, C_{AA} is concentration, T is absolute temperature, and R is the gas constant. Similarly, the chemical potential of an amino acid dissolved in ethanol can be expressed as

$$\mu_{AA,Et} = \mathring{\mu}_{AA,Et} + RT \ln \gamma_{AA,Et} C_{AA,Et} \tag{7}$$

In saturated solutions, in which $C_{AA,W}$ and $C_{AA,Et}$ represent solubilities in water and ethanol, respectively, the chemical potentials of the amino acid in water and in ethanol are the same, that is,

$$\mu_{AA,w} = \mu_{AA,Et} \tag{8}$$

Thus

$$\mathring{\mu}_{AA,Et} + RT \ln \gamma_{AA,Et} C_{AA,Et} = \mathring{\mu}_{AA,w} + RT \ln \gamma_{AA,w} C_{AA,w} \tag{9}$$

The quantity $(\mathring{\mu}_{AA,Et} - \mathring{\mu}_{AA,w})$, which represents the difference between the chemical potentials arising from the interaction of the amino acid with ethanol and with water, can be defined as the free energy change $(\Delta G_{t,Et \rightarrow w})$ of transfer of the amino acid from ethanol to water. Thus, assuming that the ratio of activity coefficients is one, the preceding equation can be expressed as

$$\Delta G_{t,(Et \rightarrow w)} = -RT \ln(S_{AA,Et}/S_{AA,W}) \tag{10}$$

where $S_{AA,Et}$ and $S_{AA,w}$ represent solubilities of the amino acid in ethanol and water, respectively.

As is true of all other thermodynamic parameters, ΔG_t is an additive function. That is, if a molecule has two groups, A and B, covalently attached, the ΔG_t for transfer from one solvent to another solvent is the sum of the free energy changes for transfer of group A and group B. That is,

$$\Delta G_{t,AB} = \Delta G_{t,A} + \Delta G_{t,B} \tag{11}$$

The same logic can be applied to the transfer of an amino acid from water to ethanol. For example, Val can be considered as a derivative of Gly with an isopropyl side chain at the α-carbon atom.

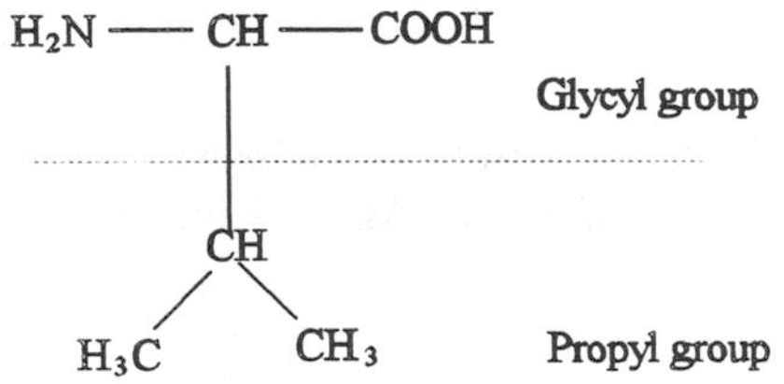

$$\begin{array}{l} H_2N - CH - COOH \quad \text{Glycyl group} \\ \qquad\quad | \\ \qquad\quad CH \\ \quad H_3C \quad CH_3 \quad \text{Propyl group} \end{array} \tag{12}$$

The free energy change of transfer of valine from ethanol to water can then be considered as

$$\Delta G_{t,val} = \Delta G_{t,glycine} + \Delta G_{t,side\ chain} \tag{13}$$

or

$$\Delta G_{t,side\ chain} = \Delta G_{t,val} - \Delta G_{t,glycine} \tag{14}$$

In other words, the hydrophobicities of amino acid side chains can be determined by subtracting $\Delta G_{t,gly}$ from $\Delta G_{t,AA}$.

The hydrophobicity values of amino acid side chains obtained in this manner are given in Table 4. Amino acid side chains with large positive ΔG_t values are hydrophobic; they would prefer to be in an organic phase rather than in an aqueous phase. In proteins, these residues tend to locate themselves in the protein interior. Amino acid residues with negative ΔG_t values are hydrophilic, and these residues tend to locate themselves on the surface of protein molecules. It should be noted that although Lys is considered to be a hydrophilic amino acid residue, it has a positive ΔG_t; this is due to the four $-CH_2-$ groups, which prefer to be in an organic environment. In fact, with proteins, part of this chain is usually buried with the ε-amino group protruding at the protein surface.

TABLE 4 Hydrophobicity of Amino Acid Side Chains at 25°C

Amino acid	ΔG_t (ethanol → water) (kJ/mol)	Amino acid	ΔG_t (ethanol → water) (kJ/mol)
Alanine	2.09	Leucine	9.61
Arginine	—	Lysine	—
Asparagine	0	Methionine	5.43
Aspartic acid	2.09	Phenylalanine	10.45
Cysteine	4.18	Proline	10.87
Glutamine	–0.42	Serine	–1.25
Glutamic acid	2.09	Threonine	1.67
Glycine	0	Tryptophan	14.21
Histidine	2.09	Tyrosine	9.61
Isoleucine	12.54	Valine	6.27

Source: Refs. 80 and 104.

Table 5 Ultraviolet Absorbance and Fluorescence of Aromatic Amino Acids

Amino acid	λ_{max} of absorbance (nm)	Molar extinction coefficient ($1\ cm^{-1}\ mol^{-1}$)	λ_{max} of fluorescence (nm)
Phenylalanine	260	190	282[a]
Tryptophan	278	5500	348[b]
Tyrosine	275	1340	304[b]

[a]Excitation at 260 nm.
[b]Excitation at 280 nm.

6.2.1.5 Optical Properties of Amino Acids

The aromatic amino acids Trp, Tyr, and Phe absorb light in the near-ultraviolet region (250–300 nm). In addition, Trp and Tyr also exhibit fluorescence in the ultraviolet region. The maximum wavelengths of absorption and fluorescence emission of the aromatic amino acids are given in Table 5. Since both absorption and fluorescence properties of these amino acids are influenced by the polarity of their environment, changes in their optical properties are often used as a means to monitor conformational changes in proteins.

6.2.2 Chemical Reactivity of Amino Acids

The reactive groups, such as amino, carboxyl, sulfhydryl, phenolic, hydroxyl, thioether (Met), imidazole, and guanyl, in free amino acids and proteins are capable of undergoing chemical reactions that are similar to those that would occur if they were attached to other small organic molecules. Typical reactions for various side-chain groups are presented in Table 6. Several of these reactions can be used to alter the hydrophilic and hydrophobic properties and the functional properties of proteins and peptides. Some of these reactions also can be used to quantify amino acids and specific amino acid residues in proteins. For example, reaction of amino acids with ninhydrin, *O*-phthaldialdehyde, or fluorescamine is regularly used in the quantification of amino acids.

Reaction with Ninhydrin

The ninhydrin reaction is often used to quantify free amino acids. When an amino acid is reacted with an excess amount of ninhydrin, the following products are formed. For every mole of amino acid reacted with ninhydrin, one mole each of ammonia, aldehyde, CO_2, and hydrindantin are formed as intermediates. The liberated ammonia subsequently reacts with one mole of ninhydrin and one mole of hydrindantin, forming a purple product known as Ruhemann's purple, which has maximum absorbance at 570 nm. Proline and hydroxyproline give a yellow product that has a maximum absorbance at 440 nm. These color reactions are the bases for colorimetric determination of amino acids.

Ninhydrin + R—CH(NH₂)—COOH → Ruhemann's purple + R-CHO + CO_2 + $3H_2O$ (15)

TABLE 6 Chemical Reactions of Functional Groups in Amino Acids and Proteins

Type of reaction	Reagent and conditions	Product	Remarks
A. Amino groups			
1. Reductive alkylation	HCHO, $NaBH_4$ (formaldehyde)	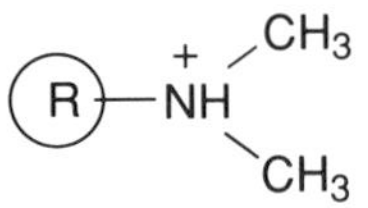 (R)—$\overset{+}{N}H(CH_3)_2$	Useful for radiolabeling proteins
2. Guanidation	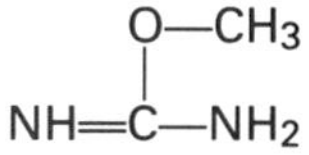 $NH{=}C(O{-}CH_3){-}NH_2$ (*O*-methylisourea) pH 10.6, 4°C for 4 days	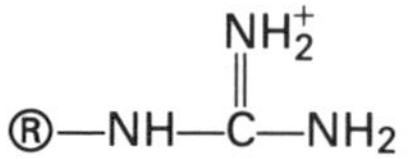 (R)—NH—$C(=NH_2^+)$—NH_2	Converts lysyl side chain to homoarginine
3. Acetylation	Acetic anhydride	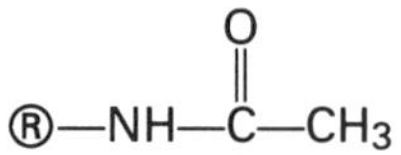 (R)—NH—C(=O)—CH_3	Eliminates the positive charge
4. Succinylation	Succinic anhydride	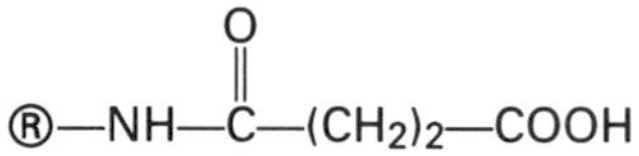 (R)—NH—C(=O)—$(CH_2)_2$—COOH	Introduces a negative charge at lysyl residues
5. Thiolation	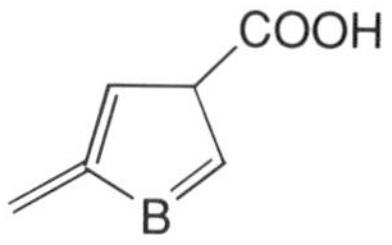 (Thioparaconic acid)	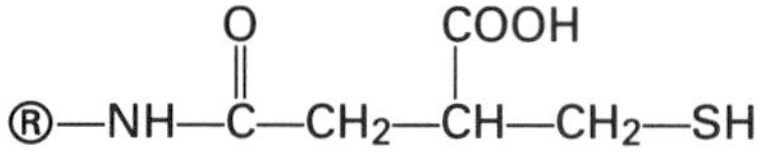 (R)—NH—C(=O)—CH_2—CH(COOH)—CH_2—SH	Eliminates positive charge, and initiates thiol group at lysyl residues
6. Arylation	1-Fluoro-2,4-dinitrobenzene (FDNB)	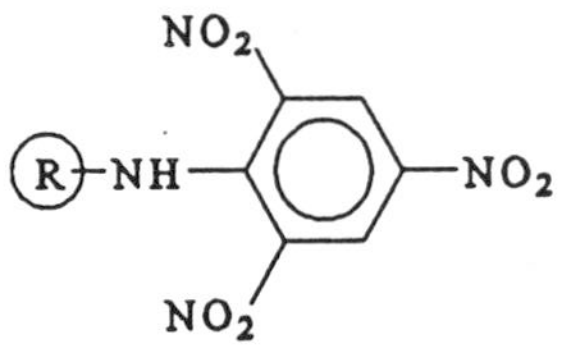 (R)—NH—$C_6H_2(NO_2)_3$	Used for the determination of amino groups
	2,4,6-Trinitrobenzene sulfonic acid (TNBS)	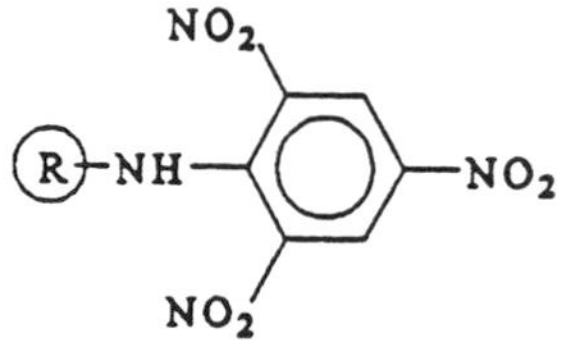 (R)—NH—$C_6H_2(NO_2)_3$	The extinction coefficient is 1.1×10^4 M^{-1} cm^{-1} at 367 nm; used to determine reactive lysyl residues in proteins

Reaction	Reagent	Product	Comments
7. Deamination	1.5 *M* $NaNO_2$ in acetic acid, 0°C	$R—OH + N_2 + H_2O$	
B. Carboxyl groups			
1. Esterification	Acidic methanol	Ⓡ—$COOCH_3 + H_2O$	Hydrolysis of the ester occurs at pH > 6.0
2. Reduction	Borohydride in tetrahydrofuran, trifluoracetic acid	Ⓡ—CH_2OH	
3. Decarboxylation	Acid, alkali, heat treatment	$R—CH_2—NH_2$	Occurs only with amino acid, not with proteins
C. Sulfhydryl group			
1. Oxidation	Performic acid	Ⓡ—$CH_2—SO_3H$	
2. Blocking	$CH_2—CH_2$ (NH bridged) (ethyleneimine)	Ⓡ—$CH_2—S—(CH_2)_2—NH_3^+$	Introduces amino group
	Iodoacetic acid	Ⓡ—$CH_2—S—CH_2—COOH$	Introduces one amino group
	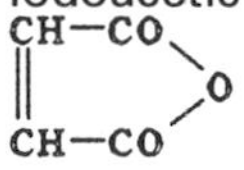 CH—CO—O—CO—CH (Maleic anhydride)	Ⓡ—$CH_2—S—CH(COOH)—CH_2—COOH$	Introduces two negative charges for each SH group blocked
	p-Mercuribenzoate	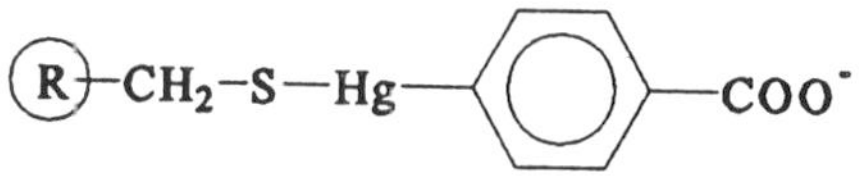 Ⓡ—CH_2—S—Hg—C_6H_4—COO^-	The extinction coefficient of this derivative at 250 nm (pH 7) is 7500 M^{-1} cm^{-1}; this reaction is used to determine SH content of proteins
	N-Ethylmaleimide	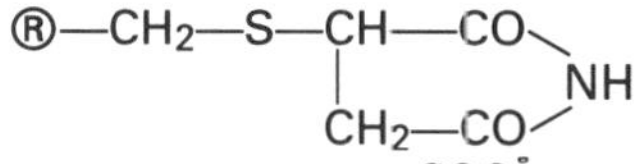 Ⓡ—CH_2—S—CH—CO—NH—CO—CH_2 (ring)	Used for blocking SH groups
	5,5′-Dithiobis (2-nitrobenzoic acid) (DTNB)	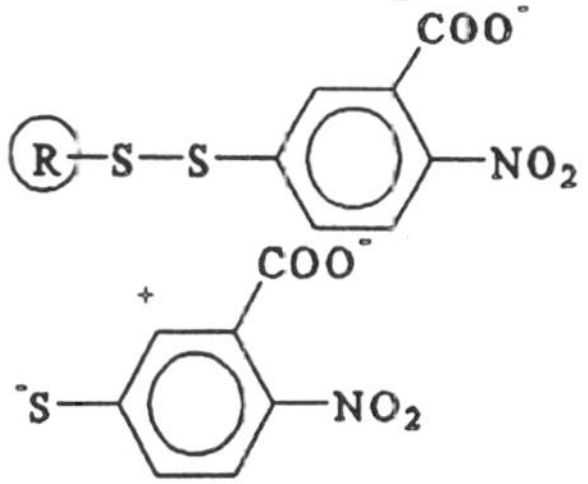 Ⓡ—S—S—$C_6H_3(COO^-)(NO_2)$ + $^-S—C_6H_3(COO^-)(NO_2)$ (Thionitrobenzoate)	One mole of thionitrobenzoate is released; the ϵ_{412} of thionitrobenzoate is 13,600 M^{-1} cm^{-1}; this reaction is used to determine SH groups in proteins

TABLE 6 Continued

Type of reaction	Reagent and conditions	Product	Remarks
D. Serine and threonine			
1. Esterification	$CH_3—COCl$	$Ⓡ—O—C(=O)—CH_3$	
E. Methionine			
1. Alkyl halides	CH_3I	$Ⓡ—CH_2—\overset{\oplus}{S}(CH_3)—CH_3$	
2. β-Propiolactone	$CH_2—CH_2—CO$ (ring closed through —O—)	$Ⓡ—CH_2—\overset{\oplus}{S}(—CH—CH_2—COOH)—CH_3$	

The ninhydrin reaction is usually used to help determine the amino acid composition of proteins. In this case, the protein is first acid hydrolyzed to the amino acid level. The freed amino acids are then separated and identified using ion exchange/hydrophobic chromatography. The column eluates are reacted with ninhydrin and quantified by measuring absorbance at 570 and 440 nm.

Reaction with *O*-phthaldialdehyde

Reaction of amino acids with *O*-phthaldialdehyde (1,2-benzene dicarbonal) in the presence of 2-mercaptoethanol yields a highly fluorescent derivative that has an excitation maximum at 380 nm and a fluorescence emission maximum at 450 nm.

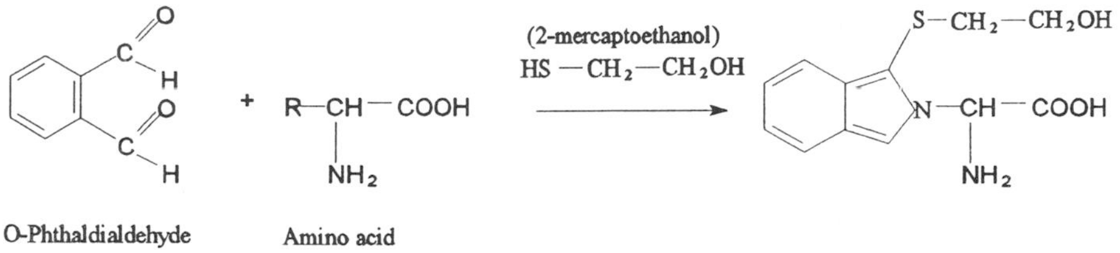

O-Phthaldialdehyde + Amino acid $\xrightarrow{\text{(2-mercaptoethanol) } HS-CH_2-CH_2OH}$ (16)

Reaction with Fluorescamine

Reaction of amino acids, peptides, and proteins containing primary amines with fluorescamine yields a highly fluorescent derivative with fluorescence emission maximum at 475 nm when excited at 390 nm. This method can be used to quantify amino acids as well as proteins and peptides.

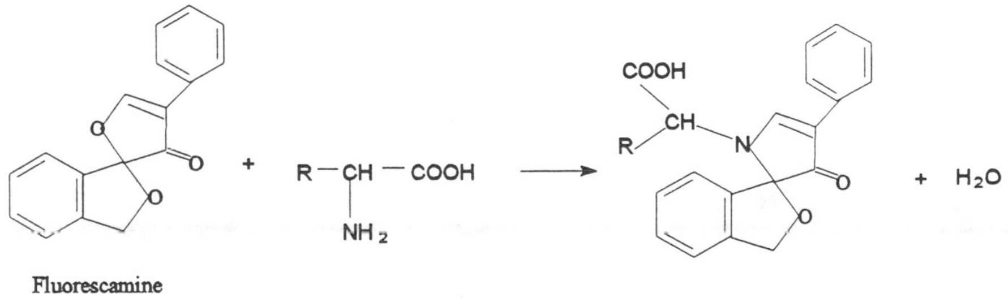

Fluorescamine + R–CH(NH$_2$)–COOH ⟶ + H_2O (17)

6.3 PROTEIN STRUCTURE

6.3.1 Structural Hierarchy in Proteins

Four levels of protein structure exist: primary, secondary, tertiary, and quaternary.

6.3.1.1 Primary Structure

The primary structure of a protein refers to the linear sequence in which the constituent amino acids are covalently linked through amide bonds, also known as peptide bonds. The peptide linkage results from condensation of the α-carboxyl group of the ith amino acid and the α-amino group of the $i+1$th amino acid with removal of a water molecule. In this linear sequence, all the amino acid residues are in the L-configuration. A protein with n amino acid residues

$$
\begin{array}{c}
\text{—NH—}\underset{\displaystyle R_i}{\underset{|}{\text{CH}}}\text{—COOH} \quad + \quad \text{H}_2\text{N—}\underset{\displaystyle R_{i+1}}{\underset{|}{\text{CH}}}\text{—COOH} \\
\downarrow \\
\text{—NH—}\underset{\displaystyle R_i}{\underset{|}{\text{CH}}}\text{—CO—NH—}\underset{\displaystyle R_{i+1}}{\underset{|}{\text{CH}}}\text{—COOH} + \text{H}_2\text{O}
\end{array}
\tag{18}
$$

contains n–1 peptide linkages. The terminus with the free α-amino group is known as the N-terminal, and that with the free α-COOH group is known as the C-terminal. By convention, N represents the beginning and C the end of the polypeptide chain.

The chain length (n) and the sequence in which the n residues are linked determine the physicochemical, structural, and biological properties and functions of a protein. The amino acid sequence acts as the code for formation of secondary and tertiary structures, and ultimately determines the protein's biological functionality. The molecular mass of proteins ranges from a few thousand daltons to over a million daltons. For example, titin, which is a single-chain protein found in muscle, has a molecular weight of over one million, whereas secretin has a molecular weight of about 2300. Many proteins have molecular masses in the range of 20,000 to 100,000 daltons.

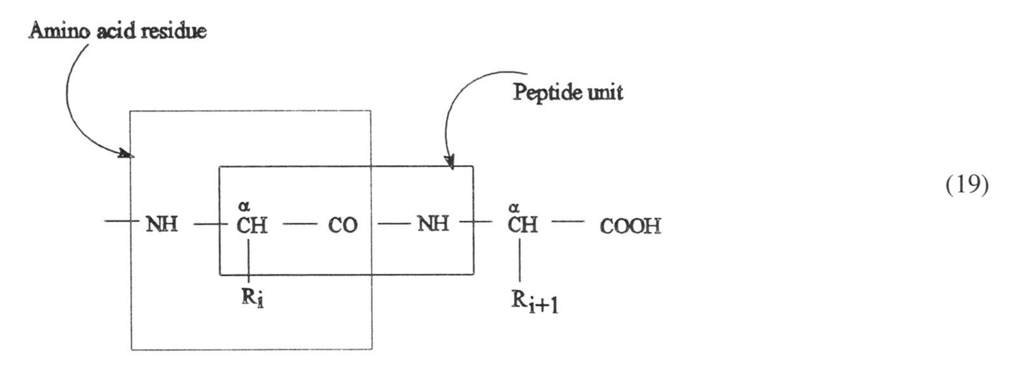

(19)

The backbone of polypeptides can be depicted as repeating units of -N-C-C- or -C-C-N-. The expression -NH-CHR-CO- relates to an amino acid residue, whereas -CHR-CO-NH- represents a peptide unit. Although the CO-NH bond is depicted as a single covalent bond, in reality it has a partial double bond character because of the resonance structure caused by delocalization of electrons.

$$
\text{—}\overset{\displaystyle O}{\overset{\|}{C}}\text{—}\underset{\displaystyle H}{\underset{|}{\ddot{N}}}\text{—} \rightleftharpoons \text{—}\overset{\displaystyle O^{\ominus}}{\overset{|}{C}}\text{=}\underset{\displaystyle H^{\oplus}}{\underset{|}{N}}\text{—}
\tag{20}
$$

This has several important structural implications in proteins. First, the resonance structure precludes protonation of the peptide N-H group. Second, because of the partial double bond character, the rotation of the CO-NH bond is restricted to a maximum of 6°, known as the ω angle. Because of this restriction, each six-atom segment (-C_α-CO-NH-C_α-) of the peptide backbone lies in a single plane. The polypeptide backbone, in essence, can be depicted as a series of -C_α-CO-NH-C_α- planes connected at the C_α atoms (Figure 2). Since peptide bonds constitute about one-third of the total covalent bonds of the backbone, their restricted rotational freedom drastically reduces backbone flexibility. Only the N-C_α and the C_α-C bonds have rotational freedoms, and these are termed ϕ (phi) and ψ (psi) dihedral angles, respectively. These are also known as main-chain torsion angles. Third, delocalization of electrons also imparts a partial negative charge to the carbonyl oxygen atom and a partial positive charge to the hydrogen atom of the N-H group. Because of this, hydrogen bonding (dipole–dipole interaction) between the C=O and N-H groups of the peptide backbone is possible under appropriate conditions.

Another consequence of the partial-bond nature of the peptide bond is that the four atoms attached to the peptide bond can exist either in *cis* or *trans* configuration.

$\delta\ominus$ O, C, N, $C_{\alpha,\,i+1}$, $C_{\alpha,i}$, H $\delta\oplus$ — *trans*

$\delta\ominus$ O, C, N, H $\delta\oplus$, $C_{\alpha,i}$, $C_{\alpha,\,i+1}$ — *cis*

(21)

However, almost all protein peptide bonds exist in the *trans* configuration. This is due to the fact that the trans configuration is thermodynamically more stable than the *cis* configuration. Since *trans* $\rightarrow$ *cis* transformation increases the free energy of the peptide bond by 34.8 kJ/mol, isomerization of peptide bonds does not occur in proteins. One exception to this is peptide bonds

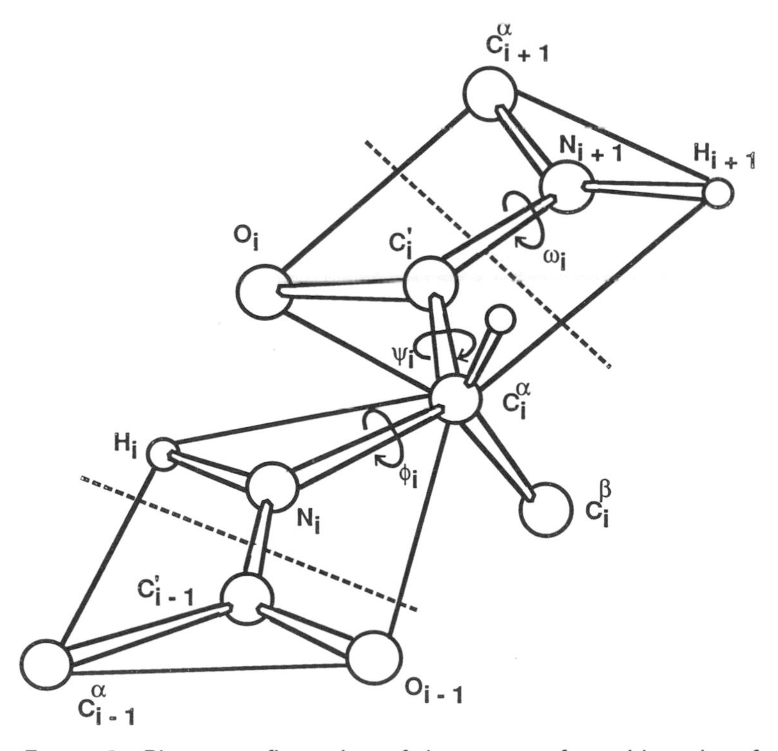

FIGURE 2 Planar configuration of the atoms of peptide units of a polypeptide backbone. ϕ and ψ are the dihedral (torsional) angles of C_α-N and C_α-C bonds. The side chains are located above or below the planes.

involving proline residues. Since the free energy change for *trans* → *cis* transformation of peptide bonds involving proline residues is only about 7.8 kJ/mol, at high temperatures these peptide bonds sometimes do undergo *trans–cis* isomerization.

Although the N-C_α and C_α-C bonds are truly single bonds, and thus the ϕ and ψ dihedral angles can theoretically have 360° rotational freedom, in reality their rotational freedoms are restricted by steric hindrances from side chain atoms at the C_α atom. These restrictions further decrease flexibility of the polypeptide chain.

6.3.1.2 Secondary Structure

Secondary structure refers to the periodic spatial arrangement of amino acid residues at certain segments of the polypeptide chain. The periodic structures arise when consecutive amino acid residues in a segment assume the same set of ϕ and ψ torsion angles. The twist of the ϕ and ψ angles is driven by near-neighbor or short-range noncovalent interactions between amino acid side chains, which leads to a decrease in local free energy. The aperiodic or random structure refers to those regions of the polypeptide chain where successive amino acid residues have different sets of ϕ and ψ torsion angles.

In general, two forms of periodic (regular) secondary structures are found in proteins. These are helical structures and extended sheet-like structures. The geometric characteristics of various regular structures found in proteins are given in Table 7.

Helical Structures

Protein helical structures are formed when the ϕ and ψ angles of consecutive amino acid residues are twisted to a same set of values. By selecting different combinations of ϕ and ψ angles, it is theoretically possible to create several types of helical structures with different geometries. However, in proteins, only three types of helical structures, namely, α-helix, 3_{10}-helix, and π-helix, are found (Fig. 3).

Among the three helical structures, the α-helix is the major form found in proteins and it is the most stable. The pitch of this helix, that is, the axial length occupied per rotation, is 5.4 Å. Each helical rotation involves 3.6 amino acid residues, with each residue extending the axial length by 1.5 Å. The angle of rotation per residue is 100° (i.e., 360°/3.6). The amino acid side chains are oriented perpendicular to the axis of the helix.

TABLE 7 Geometric Characteristics of Regular Secondary Structures in Proteins

Structure	ϕ	ψ	n	r	h(Å)	t
α-Right-handed helix	−58°	−47°	3.6	13	1.5	100°
α-Left-handed helix	+58°	+47°	3.6	13	1.5	100°
π-Helix	−57°.06	−69°.6	4.4	16	1.15	81°
3_{10}-Helix	−75°.5	−4°.5	3	10	2	120°
Fully extended chain	180°	180°	2	—	3.63	180°
β-Parallel sheet	−119°	+113°	2	—	3.25	—
β-Antiparallel sheet	−139°	+135°	2	—	3.5	—
Polyproline I (*cis*)	−83°	+158°				
Polyproline II (*trans*)	−78°	+149°				

Note: ϕ and ψ represent dihedral angles of the N-C_α and C_α-C bonds, respectively; n is number of residues per turn; r, number of backbone atoms within a hydrogen bonded loop of helix; h, rise of helix per amino acid residue; $t = 360°/n$, twist of helix per residue.
Source: Ref. 39.

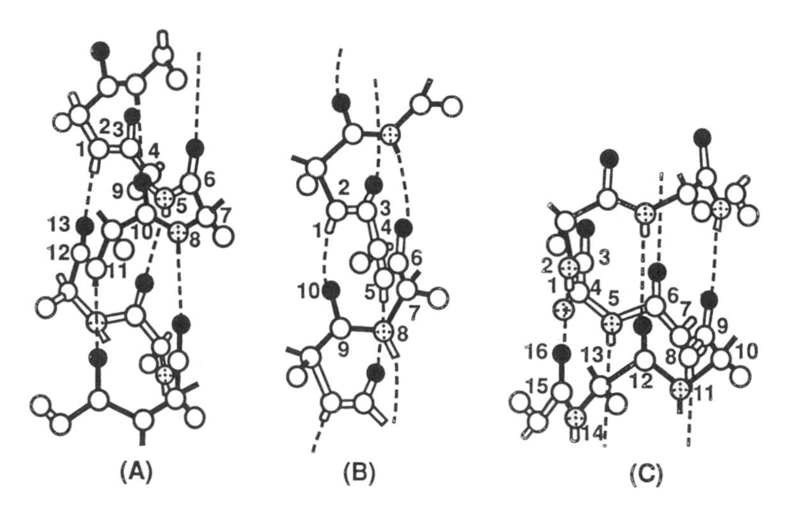

FIGURE 3 Spatial arrangement of polypeptides in (a) α-helix, (b) 3_{10}-helix, and (c) π-helix forms. (From Ref. 8; courtesy of Academic Press.)

α-Helices are stabilized by hydrogen bonding. In this structure, each backbone N-H group is hydrogen bonded to the C=O group of the fourth preceding residue. Thirteen backbone atoms are in this hydrogen-bonded loop; thus the α-helix is sometimes called the 3.6_{13} helix. The hydrogen bonds are oriented parallel to the helix axis, and the N, H, and O atoms of the hydrogen bond lie almost in a straight line; that is, the hydrogen bond angle is almost zero. The hydrogen bond length, that is, the N-H · · · O distance, is about 2.9 Å, and the strength of this bond is about 18.8 kJ/mol. The α-helix can exist in either a right- or left-handed orientation. However, the right-handed orientation is the more stable than the two.

The details for α-helix formation are embedded as a binary code in the amino acid sequence [52]. The binary code is related to the arrangement of polar and nonpolar residues in the sequence. Polypeptide segments with repeating heptet sequences of -P-N-P-P-N-N-P-, where P and N are polar and nonpolar residues, respectively, readily form α-helices in aqueous solutions. It is the binary code, and not the precise identities of the polar and nonpolar residues in the heptet sequence, that dictates α-helix formation. Slight variations in the binary code of the heptet are tolerated, provided other inter- or intramolecular interactions are favorable for α-helix formation. For example, tropomyosin, a muscle protein, exists entirely in a coiled-coil α-helical rod form. The repeating heptet sequence in this protein is -N-P-P-N-P-P-P-, which is slightly different from the preceding sequence. In spite of this variation, tropomyosin exists entirely in the α-helix form because of other stabilizing interactions in the coiled-coil rod [69].

Most of the α-helical structures found in proteins are amphiphilic in nature; that is, one side of the helical surface is occupied by hydrophobic side chains, and the other side by hydrophilic residues. This is schematically shown in the form of a helical wheel in Figure 4. In most proteins, the nonpolar surface of the helix faces the protein interior, and is generally engaged in hydrophobic interactions with other nonpolar surfaces.

Other types of helical structures found in proteins are the π-helix and the 3_{10}-helix. The π- and 3_{10}-helices are about 2.1 kJ/mol and 4.2 kJ/mol, respectively, less stable than the α-helix. These helices exist only as short segments involving a few amino acid residues, and they are not of major importance to the structures of most proteins.

In proline residues, because of the ring structure formed by covalent attachment of the

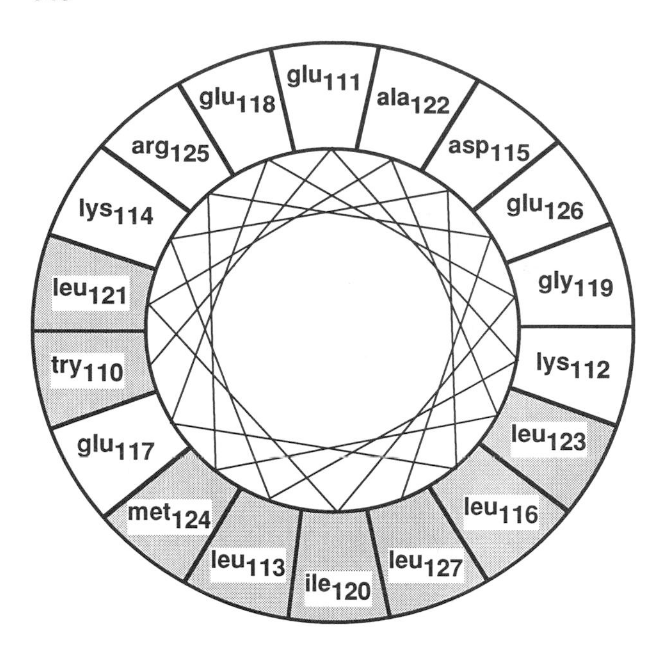

FIGURE 4 Cross-sectional view of the helical structure of residues 110–127 of bovine growth hormone. The top of the helical wheel (unfilled) represents the hydrophilic surface, and the bottom (filled) represents the hydrophobic surface of the amphiphilic helix. (From Ref. 10; courtesy of American Association for the Advancement of Science.)

propyl side chain to the amino group, rotation of the N-C_α bond is not possible, and therefore the ϕ angle has a fixed value of 70°. In addition, since there is no hydrogen at the nitrogen atom, it cannot form hydrogen bonds. Because of these two attributes, segments containing proline residues cannot form α-helices. In fact, proline is considered to be an α-helix breaker. Proteins containing high levels of proline residues tend to assume a random or aperiodic structure. For example, proline residues constitute about 17% of the total amino acid residues in β-casein, and 8.5% of the residues in α_{s1}-casein, and because of the uniform distribution of these residues in the primary structures of these proteins, α-helices are not present and the proteins have random structures. However, polyproline is able to form two types of helical structures, termed *polyproline I* and *polyproline II*. In polyproline I, the peptide bonds are in the *cis*-configuration, and in polyproline II they are in *trans*. Other geometric characteristics of these helices are given in Table 7. Collagen, which is the most abundant animal protein, exists as polyproline II-type helix. In collagen, every third residue is a glycine, which is preceded usually by a proline residue. Three polypeptide chains are entwined to form a triple helix, and the stability of the triple helix is maintained by interchain hydrogen bonds.

β-Sheet Structure

The β-sheet structure is an extended structure with specific geometries given in Table 7. In this extended form, the C=O and N-H groups are oriented perpendicular to the direction of the chain, and therefore hydrogen bonding is possible only between segments (i.e., intersegment), and not within a segment (i.e., intrasegment). The β-strands are usually about 5–15 amino acid residues long. In proteins, two β-strands of the same molecule interact via hydrogen bonds, forming a sheet-like structure known as β-pleated sheet. In the sheet-like structure, the side chains are oriented perpendicular (above and below) to the plane of the sheet. Depending on the N → C directional orientations of the strands, two types of β-pleated sheet structures, namely parallel β-sheet and antiparallel β-sheet, can form (Figure 5). In parallel β-sheet the directions of the β-strands run parallel to each other, whereas in the other they run opposite to each other. These differences in chain directions affect the geometry of hydrogen bonds. In antiparallel β-sheets the N-H · · · O atoms lie in a straight line (zero H-bond angle), which enhances the stability of the hydrogen bond, whereas in parallel β-sheets they lie at an angle, which reduces the stability of the hydrogen bonds. Antiparallel β-sheets are, therefore, more stable than parallel β-sheets.

The binary code that specifies formation of β-sheet structures in proteins is -N-P-N-P-N-P-N-P-. Clearly, polypeptide segments containing alternating polar and nonpolar residues have a high propensity to form β-sheet structures. Segments rich in bulky hydrophobic side chains, such as Val and Ile, also have a tendency to form a β-sheet structure. As expected, some variation in the code is tolerated.

The β-sheet structure is generally more stable than the α-helix. Proteins that contain large fractions of β-sheet structure usually exhibit high denaturation temperatures. Examples are β-lactoglobulin (51% β-sheet) and soy 11S globulin (64% β-sheet), which have thermal denaturation temperatures of 75.6 and 84.5°C, respectively. On the other hand, the denaturation temperature of bovine serum albumin, which has about 64% α-helix structure, is only about 64°C [19,20]. When solutions of α-helix-type proteins are heated and cooled, the α-helix is usually converted to β-sheet [19]. However, conversion from β-sheet to α-helix has not been observed in proteins.

Another common structural feature found in proteins is the β-bend or β-turn. This arises as a result of 180° reversal of the polypeptide chain involved in β-sheet formation (Fig. 6). The hairpin bend is the result of antiparellel β-sheet formation, and the crossover bend is the result of parallel β-sheet formation. Usually a β-bend involves a four residue segment folding

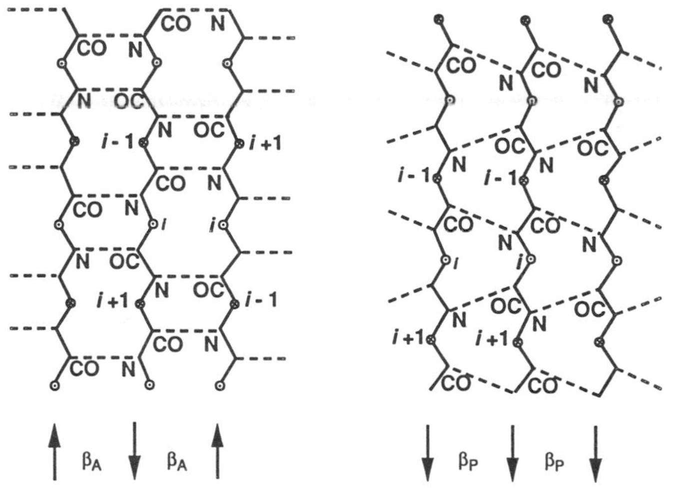

FIGURE 5 Anti-parallel (left) and parallel (right) β-sheets. The dotted lines represent hydrogen bonds between peptide groups. The arrows indicate N → C chain direction. The side chains at C_{α} atoms are oriented perpendicular (up or down) to the direction of the backbone. (From Ref. 99; courtesy of Springer-Verlag New York, Inc.)

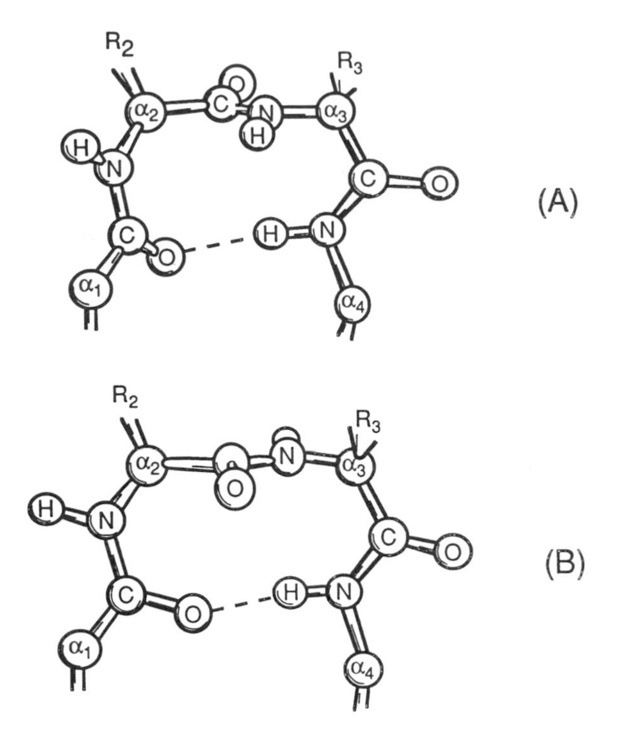

FIGURE 6 Conformations of type I (a) and type II (b) β-turns. (From Ref. 93; courtesy of Plenum Publishing Corp.)

back on itself, and the bend is stabilized by a hydrogen bond. The amino acid residues Asp, Cys, Asn, Gly, Tyr, and Pro are common in β-bends.

The secondary structure contents of several proteins are given in Table 8.

6.3.1.3 Tertiary Structure

Tertiary structure refers to the spatial arrangement attained when a linear protein chain with secondary structure segments folds further into a compact three-dimensional form. The tertiary structures of β-lactoglobulin and phaseolin (the storage protein in kidney beans) are shown in Figure 7. Transformation of a protein from a linear configuration into a folded tertiary structure is a complex process. At the molecular level, the details for formation of a protein structure are present in its amino acid sequence. From an energetics viewpoint, formation of tertiary structure involves optimization of various interactions (hydrophobic, electrostatic, and van der Waals), and hydrogen bonding between various groups in protein, so that the free energy of the molecule is reduced to the minimum value possible. The most important geometric rearrangement that accompanies the reduction in free energy during formation of tertiary structure is the relocation of most of the hydrophobic residues at the interior of the protein structure and relocation of most of the hydrophilic residues, especially charged residues, at the protein–water interface. Although there is a strong general tendency for hydrophobic residues to be buried in the protein interior, this often can be accomplished only partially. In fact, in most globular proteins, about 40–50% of the water accessible surface is occupied by nonpolar residues [63]. Also, some polar groups are inevitably buried in the interior of proteins; however, these buried polar groups are invariably hydrogen bonded to other polar groups, such that their free energies are minimized in the apolar environment of the protein interior.

TABLE 8 Secondary Structure Content of Selected Proteins

Protein	α-Helix (%)	β-Sheet (%)	β-Turns (%)	Aperiodic (%)
Deoxyhemoglobin	85.7	0	8.8	5.5
Bovine serum albumin	67.0	0	0	33.0
Chymotrypsinogen	11.0	49.4	21.2	18.4
Immunoglobulin G	2.5	67.2	17.8	12.5
Insulin (dimer)	60.8	14.7	10.8	15.7
Bovine trypsin inhibitor	25.9	44.8	8.8	20.5
Ribonuclease A	22.6	46.0	18.5	12.9
Lysozyme	45.7	19.4	22.5	12.4
Papain	27.8	29.2	24.5	18.5
α-Lactalbumin	26.0	14.0	—	60.0
β-Lactoglobulin	6.8	51.2	10.5	31.5
Soy 11S	8.5	64.5	0	27.0
Soy 7S	6.0	62.5	2.0	29.5
Phaseolin	10.5	50.5	11.5	27.5

Note: Compiled from various sources. The values represent percent of total number of amino acid residues.

The folding of a protein from a linear structure to a folded tertiary structure is accompanied by a reduction in interfacial area. The *accessible interfacial area* of a protein is defined as the total interfacial area of a three-dimensional space occupied by the protein, as determined by figuratively rolling a spherical water molecule of radius 1.4Å over the entire surface of the protein molecule. For native globular proteins, the accessible interfacial area (in Å^2) is a simple function of their molecular weight, as given by the equation [71]

$$A_s = 6.3M^{0.73} \tag{22}$$

The total accessible interfacial area of a nascent polypeptide in its extended linear state (i.e., fully stretched molecule with no secondary, tertiary, or quaternary structure) is also correlated to its molecular weight, M, by the equation [71]

$$A_t = 1.48M + 21 \tag{23}$$

The initial area (A_b) of a protein that has folded during formation of a globular tertiary structure (i.e., buried area) can be estimated from Equations 22 and 23.

The fraction and distribution of hydrophilic and hydrophobic residues in the primary structure affects several physicochemical properties of the protein. For instance, the shape of a protein molecule is dictated by its amino acid sequence. If a protein contains a large number of hydrophilic residues distributed uniformly in its sequence, it will assume an elongated or rod-like shape. This is because, for a given mass, an elongated shape has a large surface area to volume ratio so that more hydrophilic residues can be placed on the surface. On the other hand, if a protein contains a large number of hydrophobic residues, it will assume a globular (roughly spherical) shape. This minimizes the surface-area-to-volume ratio, enabling more hydrophobic residues to be buried in the protein interior. Among globular proteins, it is generally found that larger molecules contain larger fractions of nonpolar amino acids than do smaller molecules.

The tertiary structures of several single polypeptide proteins are made up of domains. Domains are defined as those regions of the polypeptide sequence that fold up into a tertiary form independently. These are, in essence, miniproteins within a single protein. The structural

(A)

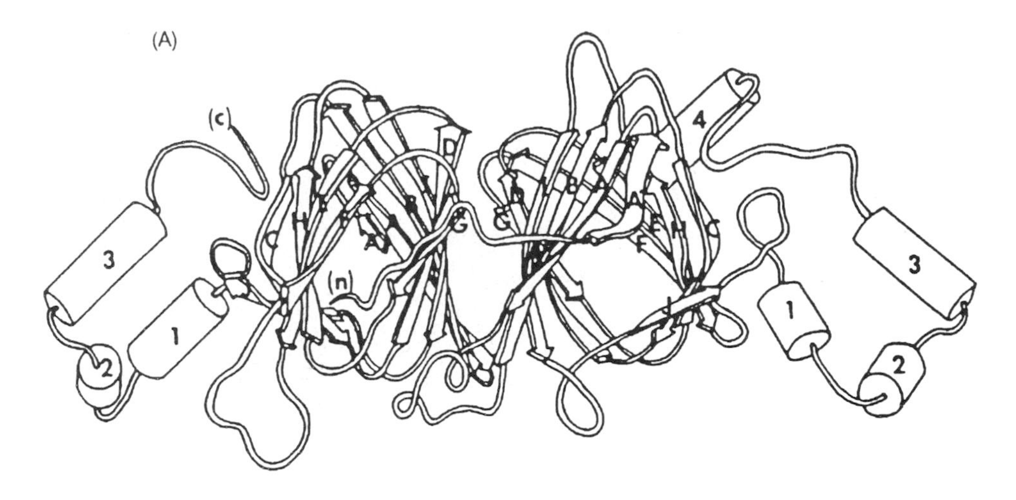

(B)

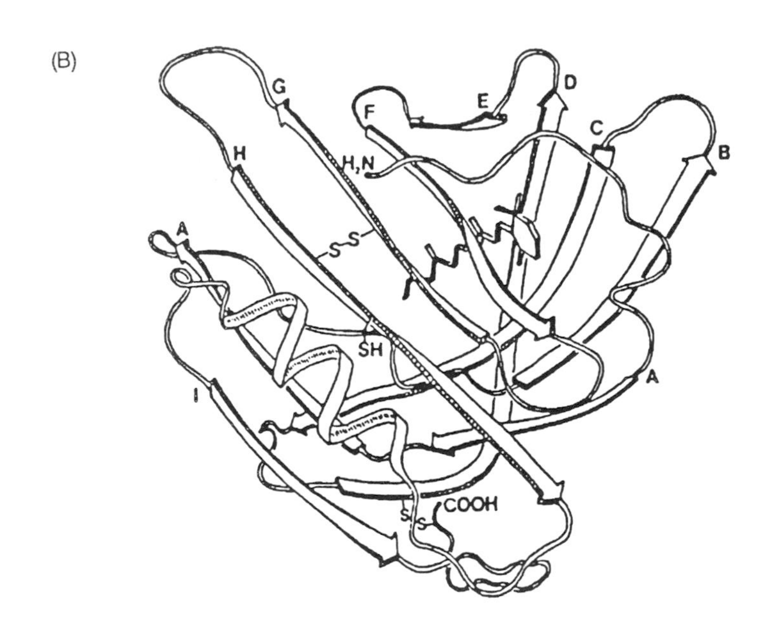

FIGURE 7 Tertiary structures of (A) phaseolin subunit and (B) β-lactoglobulin. The arrows indicate β-sheet strands, and the cylinders indicate α-helix (From Ref. 62a and Ref. 85, respectively).

stability of each domain is largely independent of the others. In most single-chain proteins, the domains fold independently and then interact with each other to form the unique tertiary structure of the protein. In some proteins, as in the case of phaseolin (Fig. 7), the tertiary structure may contain two or more distinct domains (structural entities) connected by a segment of the polypeptide chain. The number of domains in a protein usually depends on its molecular weight. Small proteins (e.g., lysozyme, β-lactoglobulin, and α-lactalbumin) with 100–150 amino acid residues usually form a single domain tertiary structure. Large proteins, such as immunoglobulin, contain multiple domains. The light chain of immunoglobulin G contains two domains, and the heavy chain contains four domains. The size of each of these domains is about 120 amino acid residues. Human serum albumin, which is made up of 585 amino acid residues, has three homologous domains, and each domain contains two subdomains [47].

6.3.1.4 Quaternary Structure

Quaternary structure refers to the spatial arrangement of a protein when it contains more than one polypeptide chain. Several biologically important proteins exist as dimers, trimers, tetramers, etc. Any of these quaternary complexes (also referred to as oligomers) can be made up of protein subunits (monomers) that are the same (homogeneous) or different (heterogeneous). For example, β-lactoglobulin exits as a dimer in the pH range 5–8, as an octomer in the pH range 3–5, and as a monomer above pH 8, and the monomeric units of these complexes are identical. On the other hand, hemoglobin is a tetramer made up of two different polypeptide chains that is, α and β chains.

Formation of oligomeric structures is the result of specific protein–protein interactions. These are primarily noncovalent interactions, such as hydrogen bonds, and both hydrophobic and electrostatic interactions. The fraction of hydrophobic amino acids appears to influence the tendency to form oligomeric proteins. Proteins that contain more than 30% hydrophobic amino acid residues exhibit a greater tendencey to form oligomeric structures than do those that contain fewer hydrophobic amino acid residues.

Formation of quaternary structure is primarily driven by the thermodynamic requirement to bury exposed hydrophobic surfaces of subunits. When the hydrophobic amino acid content of a protein is greater than 30%, it is physically impossible to form a structure that will bury all of the nonpolar residues. Consequently, there is a greater likelihood of hydrophobic patches to exist on the surface, and interaction of these patches between adjacent monomers can lead to the formation of dimers, trimers, etc. (Fig. 8).

Many food proteins, especially cereal proteins, exist as oligomers of different polypeptides. As would be expected, these proteins typically contain more than 35% hydrophobic amino acid residues (Ile, Leu, Trp, Tyr, Val, Phe, and Pro). In addition, they also contain 6–12% proline [12]. As a consequence, cereal proteins exist in complex oligomeric structures. The major storage proteins of soybean, namely, β-conglycinin and glycinin, contain about 41% and 39% hydrophobic amino acid residues, respectively. β-Conglycinin is a trimeric protein made up of three different subunits, and it exhibits complex association–dissociation phenomenon as a

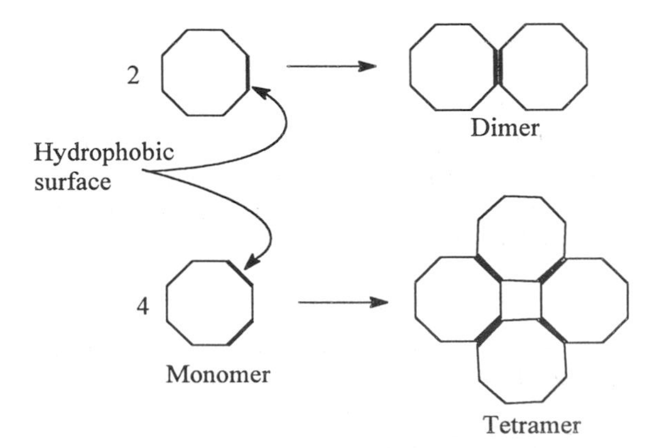

FIGURE 8 Schematic representation of formation of dimers and oligomers in proteins.

function of ionic strength and pH [76]. Glycinin is made up of 12 subunits, six of the subunits being acidic and the others basic. Each basic subunit is cross-linked to an acid subunit via a disulfide bond. The six acidic–basic pairs are held together in the oligomeric state by noncovalent interactions. Glycinin also exhibits complex association–dissociation behavior as a function of ionic strength [76].

In oligomeric proteins, the accessible surface area, A_S, is correlated to the molecular weight of the oligomer [71] by the equation

$$A_S = 5.3M^{0.76} \tag{24}$$

This relationship is different from that which applies to monomeric proteins. The surface area buried when the native oligomeric structure is formed from its constituent polypeptide subunits can be estimated from the equation

$$A_b = A_t - A_S = (1.48M + 21) - 5.3M^{0.76} \tag{25}$$

where A_t is the total accessible interfacial area of the nascent polypeptide subunits in their linear state.

6.3.2 Forces Involved in the Stability of Protein Structure

The process of folding of a random polypeptide chain into a unique three-dimensional structure is quite complex. As mentioned earlier, the basis for the biologically native conformation is encoded in the amino acid sequence of the protein. In the 1960s, Anfinsen and co-workers showed that when denatured ribonuclease was added to a physiological buffer solution, it refolded to its native conformation and regained almost 100% of its biological activity. A majority of enzymes have been subsequently shown to exhibit similar propensity. The slow but spontaneous transformation of an unfolded state to a folded state is facilitated by several intramolecular noncovalent interactions. The native conformation of a protein is a thermodynamic state in which various favorable interactions are maximized and the unfavorable ones are minimized such that the overall free energy of the protein molecule is at the lowest possible value.

The forces that contribute to protein folding may be grouped into two categories: (a) intramolecular interactions emanating from forces intrinsic to the protein molecule, and (b) intramolecular interactions affected by the surrounding solvent. van der Waals and steric interactions belong the former, and hydrogen bonding, electrostatic, and hydrophobic interactions belong to the latter.

Steric Strains

Although the ϕ and ψ angles theoretically have a 360° rotational freedom, their values are very much restricted because of steric hindrance from side-chain atoms. Because of this, segments of a polypeptide chain can assume only a limited number of configurations. Distortions in the planar geometry of the peptide unit, or stretching and bending of bonds, will cause an increase in the free energy of the molecule. Therefore, folding of the polypeptide chain can occur only in such a way that deformation of bond lengths and bond angles are avoided.

van der Waals Interactions

These are dipole-induced dipole and induced dipole-induced dipole interactions between neutral atoms in protein molecules. When two atoms approach each other, each atom induces a dipole in the other via polarization of the electron cloud. The interactions between these induced dipoles have an attractive as well as a repulsive component. The magnitudes of these forces are dependent on the interatomic distance. The attractive energy is inversely proportional to the sixth power of the interatomic distance, and the repulsive interaction is inversely proportional to the twelfth power of this distance. Therefore, at a distance r, the net interaction energy between two atoms is given by the potential energy function

$$E_{vdw} = E_a + E_r = \frac{A}{r^6} + \frac{B}{r^{12}} \tag{26}$$

where A and B are constants for a given pair of atoms, and E_a and E_r are the attractive and repulsive interaction energies, respectively. van der Waals interactions are very weak, decrease rapidly with distance, and become negligible beyond 6 Å. The van der Waals interaction energy for various pairs of atoms ranges from –0.17 to –0.8 kJ/mol. In proteins, however, since numerous pairs of atoms are involved in van der Waals interactions, the sum of its contribution to protein folding and stability is very significant.

Hydrogen Bonds

The hydrogen bond involves the interaction of a hydrogen atom that is covalently attached to an electronegative atom (such as N, O, or S) with another electronegative atom. Schematically, a hydrogen bond may be represented as D-H · · · A, where D and A are, respectively, the donor and acceptor electronegative atoms. The strength of a hydrogen bond ranges between 8.4 to 33 kJ/mol, depending on the pair of electronegative atoms involved and the bond angle.

Proteins contain several groups capable of forming hydrogen bonds. Some of the possible candidates are shown in Figure 9. Among these groups, the greatest number of hydrogen bonds are formed between the N-H and C=O groups of the peptide bonds in α-helix and β-sheet structures.

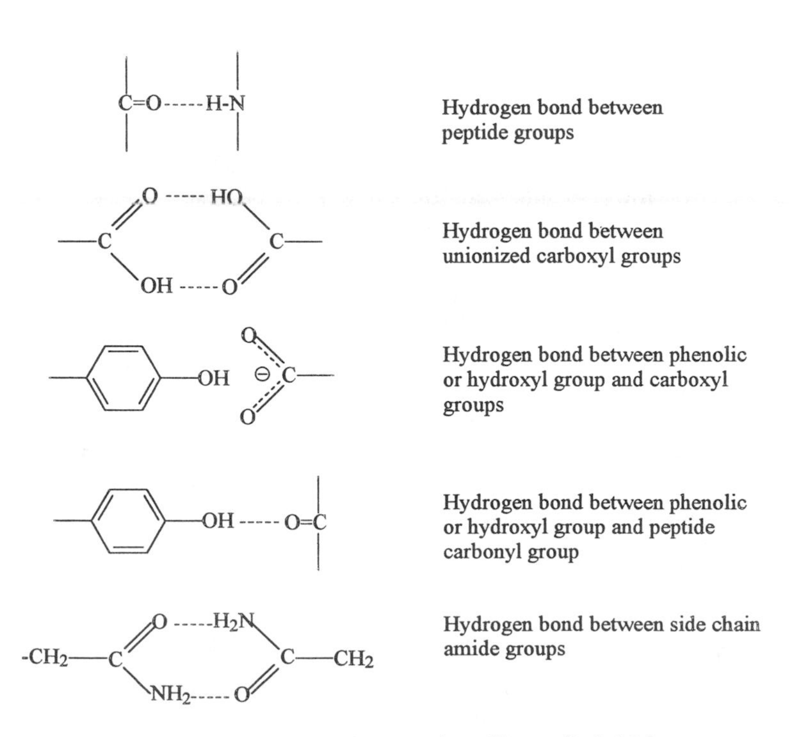

FIGURE 9 H-bonding groups in proteins. (From Ref. 98.)

The peptide hydrogen bond can be considered as a strong permanent dipole–dipole interaction between the $N^{\delta-}$-$H^{\delta+}$ and $C^{\delta+} = O^{\delta-}$ dipoles as shown:

$$\overset{\delta-}{N} \text{---} \theta \text{---} \overset{\delta-}{O} = \overset{\delta+}{C}, \quad \overset{\delta+}{H} \qquad (27)$$

The strength of the hydrogen bond is given by the potential energy function

$$E_{\text{H-bond}} = -\frac{\mu_1\mu_2}{\varepsilon r^3}\cos\theta \qquad (28)$$

where μ_1 and μ_2 are the dipole moments, ε is the dielectric constant of the medium, r is the distance between the electronegative atoms, and θ is the hydrogen bond angle. The hydrogen bond energy is directly proportional to the product of the dipole moments and to the cosine of the bond angle, and is inversely proportional to the third power of the N · · · O distance and to the dielectric constant of the medium. The strength of the hydrogen bond is maximum when θ is zero ($\cos 0 = 1$), and it is zero when θ is 90°. The hydrogen bonds in α-helix and antiparallel β-sheet structures have a θ value very close to zero, whereas those in parallel β-sheets have larger θ values. The optimum N · · · O distance for maximum hydrogen bond energy is 2.9 Å. At shorter distances the electrostatic repulsive interaction between the $N^{\delta-}$ and $O^{\delta-}$ atoms causes a significant decrease in the strength of the hydrogen bond. At longer distances weak dipole–dipole interaction between the N-H and C=O groups decreases the strength of the hydrogen bond. The strength of N-H · · · O=C hydrogen bonds in proteins is typically about 18.8 kJ/mol. The "strength" refers to the amount of energy needed to break the bond.

The existence of hydrogen bonds in proteins is well established. Since each hydrogen bond decreases the free energy of the protein by about −18.8 kJ/mol, it is commonly believed that they may act not only as the driving force for protein folding but also may contribute enormously to the stability of the native structure. However, this is not a valid assumption. Because water can compete for hydrogen bonding with N-H and C=O groups in proteins, hydrogen bonding between these groups cannot occur spontaneously, nor can formation of N-H · · · O=C hydrogen bonds be the driving force for formation of α-helix and β-pleated sheets in proteins. The hydrogen bonding interactions in α-helix and β-sheets are, therefore, a consequence of other favorable interactions that drive formation of these secondary hydrogen-bonded structures.

The hydrogen bond is primarily an ionic interaction. Like other ionic interctions, its stability also depends upon the dielectric constant of the environment. The stability of hydrogen bonds in secondary structures is mainly due to a local environment with a low permittivity (low dielectric constant) created by interaction between nonpolar residues. These bulky side chains prevent access of water to the N-H · · · O=C hydrogen bonds. They are only stable as long as they are protected from water.

Electrostatic Interactions

As noted earlier, proteins contain several amino acid residues with ionizable groups. At neutral pH, Asp and Glu residues are negatively charged, and Lys, Arg, and His are positively charged. At alkaline pH, Cys and Tyr residues assume a negative charge.

Depending upon the relative number of negatively and positively charged residues, proteins assume either a net negative or a net positive charge at neutral pH. The pH at which the net charge is zero is called the isoelectric pH (pI). The isoelectric pH is different from the isoionic point. The isoionic point is the pH of the protein solution in the absence of electrolytes. The isoelectric pH of a protein can be estimated from its amino acid composition and the pK_a values of the ionizable groups using the Hendersen–Hasselbach equation (Eq. 5).

With few exceptions, almost all charged groups in proteins are distributed on the surface of the protein molecule. Since at neutral pH proteins assume either a net positive or a net negative charge, one might expect that the net repulsive interaction between like charges would destabilize protein structure. It is also reasonable to assume that attractive interactions between oppositely charged groups at certain critical locations might contribute to the stability of the protein structure. In reality, however, the strength of these repulsive and attractive forces is minimized in aqueous solutions because of the high permittivity of water. The electrostatic interaction energy between two fixed charges q_1 and q_2 separated by distance r is given by

$$E_{ele} = \frac{q_1 q_2}{\varepsilon r} \tag{29}$$

where ε is the permittivity of the medium. In vacuum or air ($\varepsilon = 1$), the electrostatic interaction energy between two charges at a distance of 3 to 5 Å is about ±460 to ±277 kJ/mol. In water, however, this interaction energy is reduced to ±5.8 to ±3.5 kJ/mol, which is of the order of the thermal kinetic energy (*RT*) of the protein molecule at 37°C. Therefore, the attractive and repulsive electrostatic interactions between charges located on the protein surface do not contribute significantly to protein stability. However, charged groups partially buried in the protein interior, where the permittivity is lower than that of water, usually form salt bridges with strong interaction energy. The electrostatic interaction energy ranges between ±3.5 and ±460 kJ/mol depending on the distance and the local permittivity.

Although electrostatic interactions may not act as the primary force for protein folding, their penchant to remain exposed to the aqueous environment certainly would influence the folding pattern.

Hydrophobic Interactions

It is obvious from the foregoing discussions that in aqueous solutions the hydrogen bonding and electrostatic interactions between various polar groups in a polypeptide chain do not possess sufficient energy to act as driving forces for protein folding. These polar interactions in proteins are not very stable, and their stabilities depend on maintenance of an apolar environment. The major force driving protein folding comes from hydrophobic interactions among nonpolar groups.

In aqueous solutions, the hydrophobic interaction between nonpolar groups is the result of thermodynamically unfavorable interaction between water and nonpolar groups. When a hydrocarbon is dissolved in water, the free energy change (ΔG) is positive and the volume (ΔV) and enthalpy change (ΔH) are negative. Even though ΔH is negative, meaning that there is favorable interaction between water and the hydrocarbon, ΔG is positive. Since $\Delta G = \Delta H - T\,\Delta S$ (where T is the temperature and ΔS is the entropy change), the positive change in ΔG must result from a large negative change in entropy, which offsets the favorable change in ΔH. The decrease in entropy is caused by formation of a clathrate or cage-like water structure around the hydrocarbon. Because of the net positive change in ΔG, interaction between water and nonpolar groups is highly restricted. Consequently, in aqueous solutions, nonpolar groups tend to aggregate, so that the area of direct contact with water is minimized (see Chap. 2). This water structure-induced interaction between nonpolar groups in aqueous solutions is known as hydrophobic interaction. In proteins, hydrophobic interaction between nonpolar side chains of amino acid residues is the major reason that proteins fold into unique tertiary structures in which a majority of the nonpolar groups are removed from the aqueous environment.

Since the hydrophobic interaction is the antithesis of solution of nonpolar groups in water, ΔG for hydrophobic interaction is negative, and ΔV, ΔH, and ΔS are positive. Unlike other noncovalent interactions, hydrophobic interactions are endothermic; that is, hydrophobic interactions are stronger at high temperatures and weaker at low temperatures (opposite to that for hydrogen bonds). The variation of hydrophobic free energy with temperature usually follows a quadratic function, that is,

$$\Delta G_{H\phi} = a + bT + cT^2 \tag{30}$$

where a, b, and c are constants, and T is absolute temperature.

The hydrophobic interaction energy between two spherical nonpolar molecules can be estimated from the potential energy equation [50]

$$E_{H\phi} = 83.6\frac{R_1R_2}{(R_1 + R_2)}e^{-D/D_0}\ kJ/mol \tag{31}$$

where R_1 and R_2 are the radii of the nonpolar molecules, D is the distance (nm) between the molecules, and D_0 is the decay length (1 nm). Unlike electrostatic, hydrogen bonding, and van der Waals interactions, which follow a power-law relationship with distance between interacting groups, the hydrophobic interaction follows an exponential relationship with distance between interacting groups. Thus, it is effective over relatively long distances, such as 10 nm.

The hydrophobic free energy of proteins cannot be quantified using the preceding equation because of involvement of several nonpolar groups. It is possible, however, to estimate the hydrophobic free energy of a protein using other empirical correlations. The hydrophobic free energy of a molecule is directly proportional to the nonpolar surface area that is accessible to water (Fig. 10). The proportionality constant, that is, the slope, varies between 92 J mol^{-1} Å^{-2}

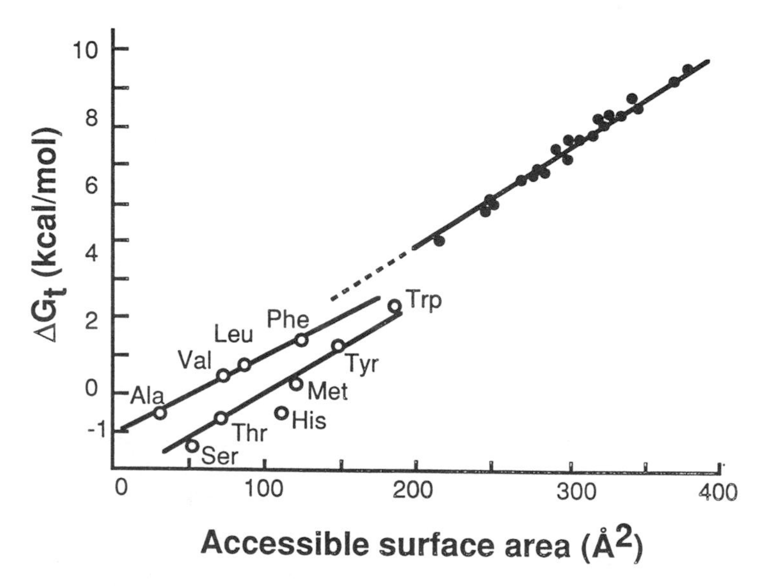

FIGURE 10 The relationship between hydrophobicity and accessible surface area of amino acid side chains (open circles) and hydrocarbons (filled circles); 1 kcal = 4.18 kJ. (From Ref. 92; courtesy of Annual Reviews, Inc.)

for Ala, Val, Leu, and Phe and 109 J mol^{-1} $Å^{-2}$ for Ser, Thr, Trp, and Met. On average, the hydrophobicity of apolar groups in amino acids or amino acid residues is about 100 J mol^{-1} $Å^{-2}$. This is close to the 104.5 J mol^{-1} $Å^{-2}$ value for alkanes. What this means is that for the removal of every 1 $Å^2$ area of nonpolar surface from the water environment, a protein will decrease its free energy by about 100 J/mol. Thus, the hydrophobic free energy of a protein can be estimated simply by multiplying the total buried surface area by 100 J mol^{-1} $Å^{-2}$.

The buried surface area in several globular proteins and the estimated hydrophobic free energies are shown in Table 9. It is evident that hydrophobic free energy contributes significantly to the stability of protein structure. The average hydrophobic free energy per amino acid residue in globular proteins amounts to about 10.45 kJ/mol.

Disulfide Bonds

Disulfide bonds are the only covalent side-chain cross-links found naturally in proteins. They can occur both intramolecularly and intermolecularly. In monomeric proteins, disulfide bonds are formed as a result of protein folding. When two Cys residues are brought into proximity with proper orientation, oxidation of the sulfhydryl groups by molecular oxygen results in disulfide bond formation. Once formed, disulfide bonds help stabilize the folded structure of proteins.

Protein mixtures containing cystine and Cys residues are able to undergo sulfhydryl–disulfide interchange reactions as shown:

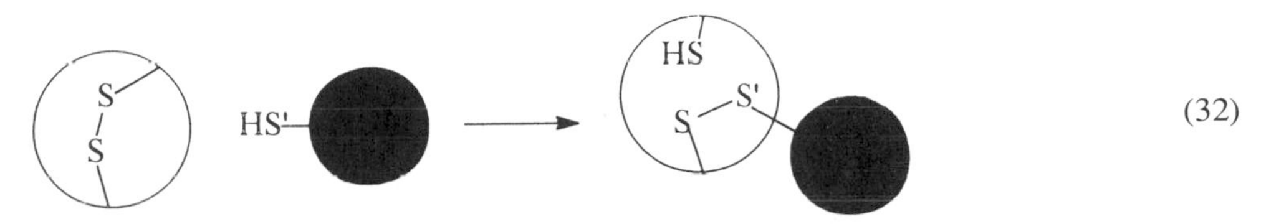

(32)

This interchange reaction also can occur within a single protein if it contains a free sulfhydryl group and a disulfide bond. The interchange reaction often leads to a decrease in stability of the protein molecule.

TABLE 9 Accessible Surface Area (A_s), Buried Surface Area (A_b), and Hydrophobic Free Energy of Proteins

Protein	Molecular mass (daltons)	A_s ($Å^2$)	A_b ($Å^2$)	$\Delta G_{H\phi}$ (kJ/mol)
Parvalbumin	11,450	5,930	11,037	1108
Cytochrome *c*	11,930	5,570	12,107	1212
Ribonuclease A	13,960	6,790	13,492	1354
Lysozyme	14,700	6,620	15,157	1521
Myoglobin	17,300	7,600	18,025	1810
Retinol binding protein	20,050	9,160	20,535	2061
Papain	23,270	9,140	25,535	2541
Chymotrypsin	25,030	10,440	26,625	2671
Subtilsin	27,540	10,390	30,390	3047
Carbonic anhydrase B	28,370	11,020	30,988	3110
Carboxypeptidase A	34,450	12,110	38,897	3900
Thermolysin	34,500	12,650	38,431	3854

Note: A_s values are from Ref. 71. A_b was calculated from Equations 22 and 23.

In summary, the formation of a unique three-dimensional protein structure is the net result of various repulsive and attractive noncovalent interactions and a few covalent disulfide bonds.

6.3.3 Conformational Stability and Adaptability of Proteins

The stability of the native protein structure is defined as the difference in free energy between the native and denatured (or unfolded) states of the protein molecule. This is usually denoted as ΔG_D.

All of the noncovalent interactions discussed already, except the repulsive electrostatic interactions, contribute to the stability of the native protein structure. The stabilizing influence on the native structure of the total free energy changes attributed to these interactions amounts to hundreds of kilojoules per mole. However, the ΔG_D of the majority of proteins is in the range of 20–85 kJ/mol. The major force tending to destabilize the native structure is the conformational entropy of the polypeptide chain. When a random-state polypeptide is folded into a compact state, the loss of translational, rotational, and vibrational motions of various groups of the protein molecule results in a decrease in conformational entropy. The entropy-derived increase in free energy as the protein is folded into its native state is more than offset by favorable noncovalent interactions, resulting in a net decrease in free energy. Thus, the difference in free energy between the native and denatured states can be expressed as

$$\Delta G_{D \rightarrow N} = \Delta G_{\text{H-bond}} + \Delta G_{\text{ele}} + \Delta G_{H\phi} + \Delta G_{\text{vdW}} - T\Delta S_{\text{conf}} \quad (33)$$

where $\Delta G_{\text{H-bond}}$, ΔG_{ele}, $\Delta G_{H\phi}$, and ΔG_{vdW}, respectively, are free energy changes for hydrogen bonding, electrostatic, hydrophobic, and van der Waals interactions, and ΔS_{conf} is the conforma-

TABLE 10 ΔG_D Values for Selected Proteins

Protein	pH	T (°C)	ΔG_D (kJ/mol)
α-Lactalbumin	7	25	18.0
Bovine β-lactoglobulin A + B	7.2	25	31.3
Bovine β-lactoglobulin A	3.15	25	42.2
Bovine β-lactoglobulin B	3.15	25	48.9
T4 Lysozyme	3.0	37	19.2
Hen egg-white lysozyme	7.0	37	50.2
Gactin	7.5	25	26.7
Lipase (from aspergillus)	7.0	—	46.0
Troponin	7.0	37	19.6
Ovalbumin	7.0	25	24.6
Cytochrome *c*	5.0	37	32.6
Ribonuclease	7.0	37	33.4
α-Chymotrypsin	4.0	37	33.4
Trypsin	—	37	54.3
Pepsin	6.5	25	45.1
Growth hormone	8.0	25	58.5
Insulin	3.0	20	26.7
Alkaline phosphatase	7.5	30	83.6

Note: ΔG_D represents $G_U - G_N$, where G_U and G_N are free energies of the denatured and native states, respectively, of a protein molecule. Compiled from several sources.

tional entropy of the polypeptide chain. The ΔS_{conf} of a protein in the unfolded state is about 8–42 J mol^{-1} K^{-1} per residue. Usually, an average value of 21.7 J mol^{-1} K^{-1} per residue is assumed. A protein with 100 amino acid residues at 310 K will have a conformational entropy of about $21.7 \times 100 \times 310 = 672.7$ kJ/mol. This destabilizing conformational energy will reduce the net stability of the native structure resulting from noncovalent interactions.

The ΔG_D values, that is, energy required to unfold, of various proteins are presented in Table 10. These values clearly indicate that in spite of numerous intramolecular interactions, proteins are only marginally stable. For example, ΔG_D values of most proteins correspond to an energy equivalent of one to three hydrogen bonds or about two to five hydrophobic interactions, suggesting that breakage of a few noncovalent interactions would destabilize the native structure of many proteins.

Conversely, it appears that proteins are not designed to be rigid molecules. They are highly flexible, their native state is in a metastable state, and breakage of one to three hydrogen bonds or a few hydrophobic interactions can easily cause a conformational change in proteins. Conformational adaptability to changing solution conditions is necessary to enable proteins to carry out several critical biological functions. For example, efficient binding of substrates or prosthetic ligands to enzymes invariably involves reorganization of polypeptide segments at the binding sites. Proteins that require high structural stability to function as catalysts usually are stabilized by intramolecular disulfide bonds, which effectively reduce conformational entropy (i.e., the tendency of the polypeptide chain to unfold).

6.4 PROTEIN DENATURATION

The native structure of a protein is the net result of various attractive and repulsive interactions emanating from various intramolecular forces as well as interaction of various protein groups with surrounding solvent water. However, native structure is largely the product of the protein's environment. The native state (of a single protein molecule) is thermodynamically the most stable with lowest feasible free energy at physiological conditions. Any change in its environment, such as pH, ionic strength, temperature, solvent composition, etc., will force the molecule to assume a new equilibrium structure. Subtle changes in structure, which do not drastically alter the molecular architecture of the protein, are usually regarded as "conformational adaptability," whereas major changes in the secondary, tertiary, and quaternary structures without cleavage of backbone peptide bonds are regarded as "denaturation." From a structural point of view, the denatured state of a protein molecule is an ill-defined state. "A major change" in structure may mean an increase in α-helix and β-sheet structure at the expense of random structure or vice versa. However, in most instances, denaturation involves a loss of ordered structure. Depending on the conditions of denaturation, proteins may assume several "denatured states," each differing only slightly in free energy. This is shown schematically in Figure 11. Some denatured states possess more residual folded structure than others. When fully denatured, globular proteins resemble a random coil. The intrinsic viscosity of a fully denatured protein is a function of the number of amino acid residues, and is expressed by the equation [105]

$$[\eta] = 0.716n^{0.66} \tag{34}$$

where n is the number of amino acid residues in the protein.

Often denaturation has a negative connotation, because it indicates loss of some properties. For example, many biologically active proteins lose their activity upon denaturation. In the case of food proteins, denaturation usually causes insolublization and loss of some functional properties. In some instances, however, protein denaturation is desirable. For example, thermal

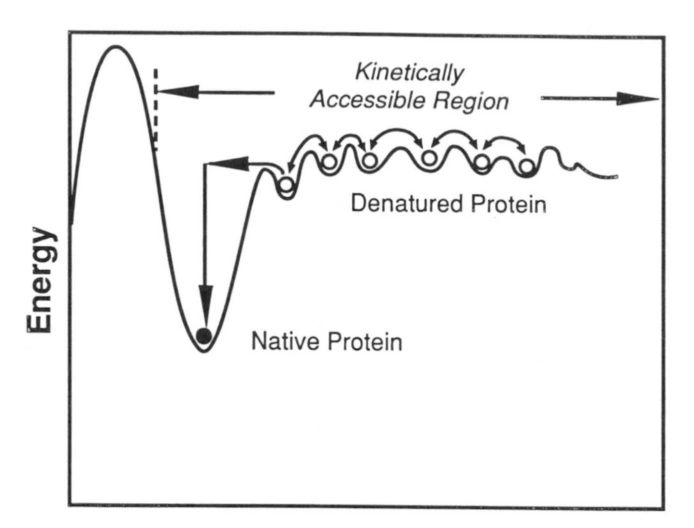

FIGURE 11 Schematic representation of the energy of a protein molecule as a function of its conformation. The conformation with the lowest energy is usually the native state. (Adapted from Ref. 107; courtesy of AVI Publishing Co.)

denaturation of trypsin inhibitors in legumes markedly improves digestibility and biological availability of legume proteins when consumed by some animal species. Partially denatured proteins are more digestible and have better foaming and emulsifying properties than do native proteins. Thermal denaturation is also a prerequisite for heat-induced gelation of food proteins.

6.4.1 Thermodynamics of Denaturation

Denaturation is a phenomenon that involves transformation of a well-defined, folded structure of a protein, formed under physiological conditions, to an unfolded state under nonphysiological conditions. Since structure is not an easily quantifiable parameter, direct measurement of the fractions of native and denatured protein in a solution is not possible. However, conformational changes in proteins invariably affect several of its chemical and physical properties, such as ultraviolet (UV) absorbance, fluorescence, viscosity, sedimentation coefficient, optical rotation, circular dichroism, reactivity of sulfhydryl groups, and enzyme activity. Thus, protein denaturation can be studied by monitoring changes in these physical and chemical properties.

When changes in a physical or chemical property, y, are monitored as a function of denaturant concentration or temperature, many monomeric globular proteins exhibit denaturation profiles as shown in Figure 12. The terms y_N and y_D are y values for the native and denatured states, respectively, of the protein.

For most proteins, as denaturant concentration (or temperature) is increased, the value of y remains unchanged initially, and above a critical point its value changes abruptly from y_N to y_D within a narrow range of denaturant concentration or temperature. The steepness of the transition curve observed for a majority of monomeric globular proteins indicates that protein denaturation is a cooperative process. That is, once a protein molecule begins to unfold, or once a few interactions in the protein are broken, the whole molecule completely unfolds with a further slight increase in denaturant concentration or temperature. This cooperative nature of unfolding suggests that globular proteins can exist only in the native and denatured states; that

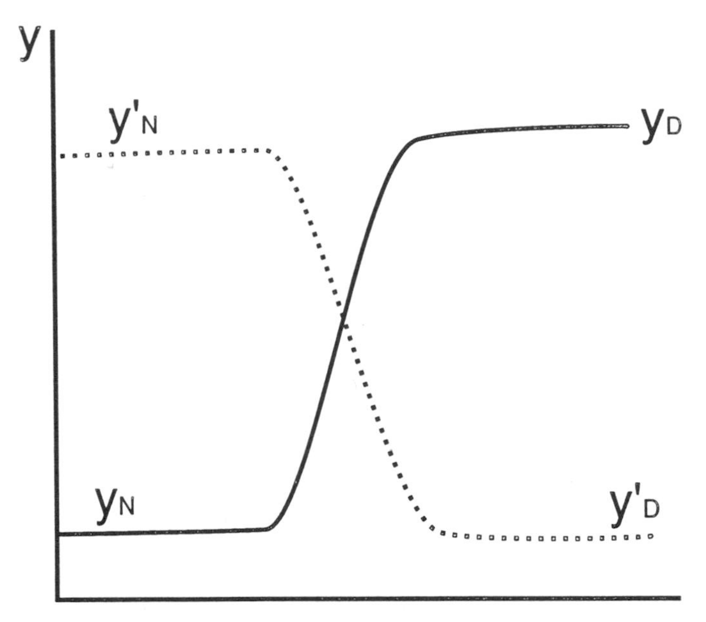

FIGURE 12 Typical protein denaturation curves; y represents any measurable physical or chemical property of the protein molecule that varies with protein conformation; y_N and y_D are the values of y for the native and denatured states, respectively.

is, intermediate states are not possible. This is known as a "two-state transition" model. For this two-state model, the equilibrium between the native and the denatured state in the cooperative transition region can be expressed as

$$N \underset{}{\overset{K_D}{\rightleftharpoons}} D \tag{35}$$

$$K_D = [D]/[N]$$

where K_D is the equilibrium constant. Since the concentration of denatured protein molecules in the absence of a denaturant (or critical input of heat) is extremely low (about 1 in 10^9), estimation of K_D is not possible. However, in the transition region, that is, at sufficiently high denaturant concentration (or sufficiently high temperature), an increase in the population of the denatured protein molecule permits determination of the apparent equilibrium constant, K_{app}. In the transition region, where both native and denatured protein molecules are present, the value of y is given by

$$y = f_N y_N + f_D y_D \tag{36}$$

where f_N and f_D are the fractions of the protein in the native and denatured states, and y_N and y_D are y values for the native and denatured states, respectively. From Figure 12,

$$f_N = (y_D - y)/(y_D - y_N) \tag{37}$$

$$f_D = (y - y_N)/(y_D - y_N) \tag{38}$$

The apparent equilibrium constant is given by

$$K_{app} = f_D/f_N = (y - y_N)/(y - y_D) \tag{39}$$

and the free energy of denaturation is given by

$$\Delta G_{app} = -RT \ln k_{app} \tag{40}$$

A plot of $-RT \ln K_{app}$ versus denaturant concentration in the transition region results in a straight line. The K_D and ΔG_D of the protein in pure water (or in buffer in the absence of denaturant) are obtained from the y-intercept. The enthalpy of denaturation, ΔH_D, is obtained from variation of the free energy change with temperature using the van't Hoff equation,

$$\Delta H_D = -R \frac{d \ln K_D}{d(1/T)} \tag{41}$$

Monomeric proteins that contain two or more domains with different structural stabilities usually exhibit multiple transition steps in the denaturation profile. If the transition steps are well separated, the stabilities of each domain can be obtained from the transition profile by using the preceding two-state model. Denaturation of oligomeric proteins proceeds via dissociation of subunits, followed by denaturation of the subunits.

Protein denaturation is reversible. When the denaturant is removed from the protein solution (or the sample is cooled), most monomeric proteins (in the absence of aggregation) refold to their native conformation under appropriate solution conditions, such as pH, ionic strength, redox potential, and protein concentration. Many proteins refold when the protein concentration is below 1 μM. Above 1 μM protein concentration, refolding is partially inhibited because of greater intermolecular interaction at the cost of intramolecular interactions. A redox potential comparable to that of biological fluid facilitates formation of the correct pairs of disulfide bonds during refolding.

6.4.2 Denaturing Agents

6.4.2.1 Physical Agents

Temperature and Denaturation

Heat is the most commonly used agent in food processing and preservation. Proteins undergo varying degrees of denaturation during processing. This can affect their functional properties in foods, and it is therefore important to understand the factors affecting protein denaturation.

When a protein solution is gradually heated above a critical temperature, it undergoes a sharp transition from the native state to the denatured state. The temperature at the transition midpoint, where the concentration ratio of native and denatured states is 1, is known either as the melting temperature T_m, or the denaturation temperature T_d. The mechanism of temperature-induced denaturation is highly complex and involves primarily destabilization of the major noncovalent interactions. Hydrogen bonding, electrostatic, and van der Waals interactions are exothermic (enthalpy driven) in nature. Therefore, they are destabilized at high temperatures and stabilized at low temperatures. However, since peptide hydrogen bonds in proteins are mostly buried in the interior, they remain stable over a wide range of temperature. On the other hand, hydrophobic interactions are endothermic (entropy driven). They are stabilized at high temperatures and destabilized at low temperatures. Therefore, as the temperature is increased, the changes in the stabilities of these two groups of noncovalent interactions oppose each other. However, the stability of hydrophobic interactions cannot increase infinitely with increasing temperature, because above a certain temperature, gradual breakdown of water structure will eventually destabilize hydrophobic interactions as well. The strength of hydrophobic interactions reaches a maximum at about 60–70°C[9].

Another major force that affects conformational stability of proteins is the conformational entropy, $-T\,\Delta S_{Conf}$, of the polypeptide chain. As temperature is increased, the increase in thermal kinetic energy of the polypeptide chain greatly facilitates unfolding of the polypeptide chain. The relative contributions of the major forces to stability of a protein molecule as a function of temperature are depicted in Figure 13. The temperature at which the sum of the free energies is zero (i.e., $K_D = 1$) is the denaturation temperature of the protein. The T_d values of some proteins are listed in Table 11.

It is often assumed that the lower the temperature, the greater will be the stability of a protein. This is not always true. For example (Fig. 14), the stability of lysozyme increases with lowering of temperature, whereas those of myoglobin and a mutant phage T4 lysozyme show maximum stability at about 30 and 12.5°C, respectively. Below and above these temperatures, myoglobin and T4 lysozyme are less stable. When stored below 0°C, these two proteins undergo cold-induced denaturation. The temperature of maximum stability (minimum free energy) depends on the relative magnitude of contributions from polar and nonpolar interactions. Those proteins in which polar interactions dominate over nonpolar interactions are more stable at or below refrigeration temperatures than they are at higher temperatures. On the other hand, proteins that are primarily stabilized by hydrophobic interactions are more stable at about ambient temperature than they are at refrigeration temperature.

Several food proteins undergo reversible dissociation and denaturation at low temperature.

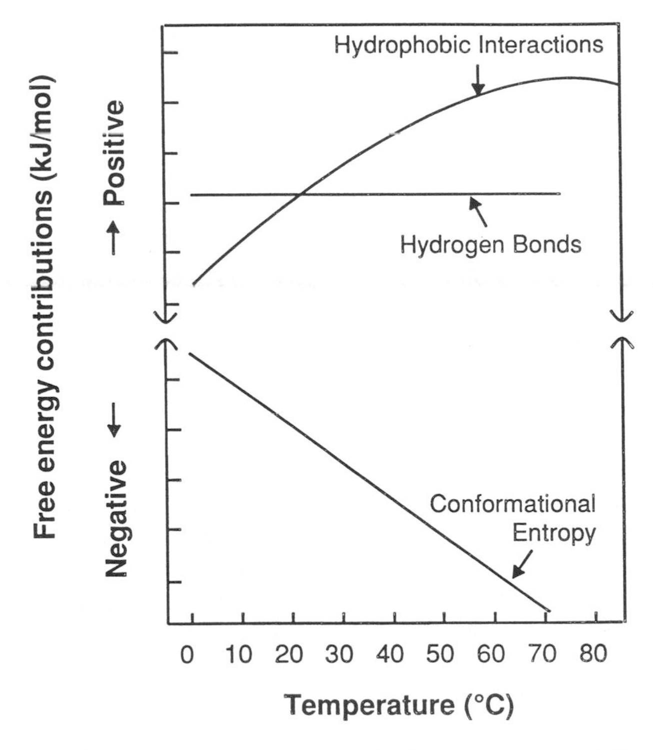

FIGURE 13 Relative changes in free energy contributions by hydrogen bonding, hydrophobic interactions, and conformational entropy to the stability of proteins as a function of temperature.

TABLE 11 Thermal Denaturation Temperatures (T_d) and Mean Hydrophobicities of Proteins

Protein	T_d	Mean hydrophobicity (kJ mol^{-1} residue^{-1})
Trypsinogen	55	3.68
Chymotrypsinogen	57	3.78
Elastase	57	
Pepsinogen	60	4.02
Ribonuclease	62	3.24
Carboxypeptidase	63	
Alcohol dehydrogenase	64	
Bovine serum albumin	65	4.22
Hemoglobin	67	3.98
Lysozyme	72	3.72
Insulin	76	4.16
Egg albumin	76	4.01
Trypsin inhibitor	77	
Myoglobin	79	4.33
α-Lactalbumin	83	4.26
Cytochrome *c*	83	4.37
β-Lactoglobulin	83	4.50
Avidin	85	3.81
Soy glycinin	92	
Broadbean 11S protein	94	
Sunflower 11S protein	95	
Oat globulin	108	

Source: Data were compiled from Ref. 11.

Glycinin, one of the storage proteins of soybean, aggregates and precipitates when stored at 2°C [58], then becomes soluble when returned to ambient temperature. When skim milk is stored at 4°C, β-casein dissociates from casein micelles, and this alters the physicochemical and rennetting properties of the micelles. Several oligomeric enzymes, such as lactate dehydrogenase and glyceraldehyde phosphate dehydrogenase, lose most of their enzyme activity when stored at 4°C, and this has been attributed to dissociation of the subunits. However, when warmed to and held at ambient temperature for a few hours, they reassociate and completely regain their activity [111].

The amino acid composition affects thermal stability of proteins. Proteins that contain a greater proportion of hydrophobic amino acid residues, especially Val, Ile, Leu, and Phe, tend to be more stable than the more hydrophilic proteins [118]. Proteins of thermophilic organisms usually contain large amounts of hydrophobic amino acid residues. However, this positive correlation between mean hydrophobicity and thermal denaturation temperature of proteins is only an approximate one (Table 11), suggesting that other factors, such as disulfide bonds and the presence of salt bridges buried in hydrophobic clefts, may also contribute to thermostability. A strong positive correlation also exists between thermostability and the number percent of certain amino acid residues. For example, statistical analysis of 15 different proteins has shown that thermal denaturation temperatures of these proteins are positively correlated ($r = 0.98$) to the number percent of Asp, Cys, Glu, Lys, Leu, Arg, Trp, and Tyr residues. On the other hand, thermal denaturation temperatures of the same set of proteins are negatively correlated

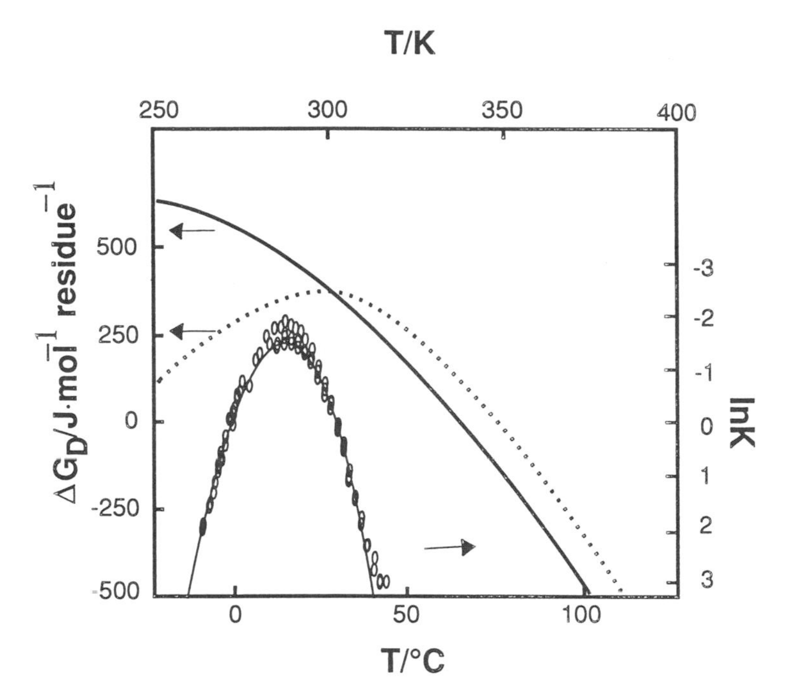

FIGURE 14 Variation of protein stability (ΔG_D) with temperature for myoglobin (· · ·), ribonuclease A (———), and a mutant ot T4 phage lysozyme (O). *K* is the equilibrium constant. (Compiled from Refs. 15 and 62.)

($r = -0.975$) to the number percent of Ala, Asp, Gly, Gln, Ser, Thr, Val, and Tyr (Fig. 15) [88]. Other amino acid residues have little influence on T_d. The underlying causes of these correlations are not clear. It seems, however, that thermostability of proteins is not simply dependent on either polar or nonpolar content, but on an optimum distribution of these two groups in the protein structure. An optimum distribution may maximize intramolecular interactions, decrease chain flexibility, and thus enhance thermostability. Thermostability is inversely correlated with protein flexibility [107].

Thermal denaturation of monomeric globular proteins is mostly reversible. For example, when many monomeric enzymes are heated above their denaturation temperatures, or even briefly held at 100°C, and then are immediately cooled to room temperature, they fully regain their activities [62]. However, thermal denaturation can become irreversible when the protein is heated at 90–100°C for a prolonged period even at neutral pH [2]. This irreversibility occurs because of several chemical changes in the protein, such as deamidation of Asn residues, cleavage of peptide bonds at Asp residues, destruction of Cys and cystine residues, and aggregation [2,109].

Water greatly facilitates thermal denaturation of proteins [37,95]. Dry protein powders are extremely stable to thermal denaturation. The value of T_d decreases rapidly as the water content is increased from 0 to 0.35 g water/g protein (Fig. 16). An increase in water content from 0.35 to 0.75 g water/g protein causes only a marginal decrease in T_d. Above 0.75 g water/g protein, the T_d of the protein is the same as in a dilute protein solution. The effect of hydration on thermostability is fundamentally related to protein dynamics. In the dry state, proteins have a static structure, that is, the mobility of polypeptide segments is restricted. As the water content is increased, hydration and partial penetration of water into surface cavities cause swelling of

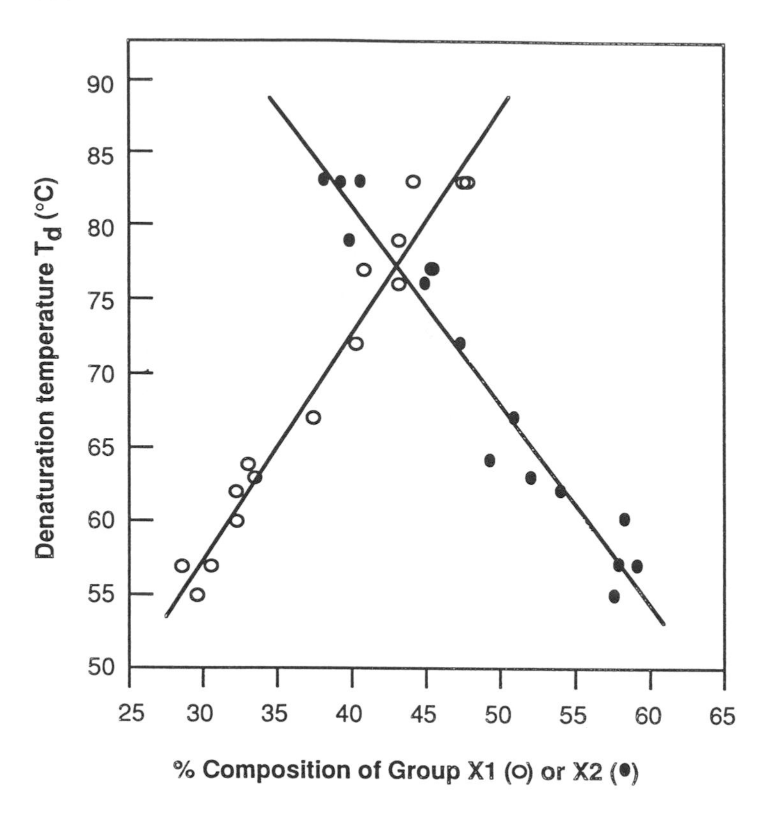

FIGURE 15 Group correlations of amino acid residues to thermal stability of globular proteins. Group X_1 represents Asp, Cys, Glu, Lys, Leu, Arg, Trp, and Tyr. Group X_2 represents Ala, Asp, Gly, Gln, Ser, Thr, Val, and Tyr. (Adapted from Ref. 88.)

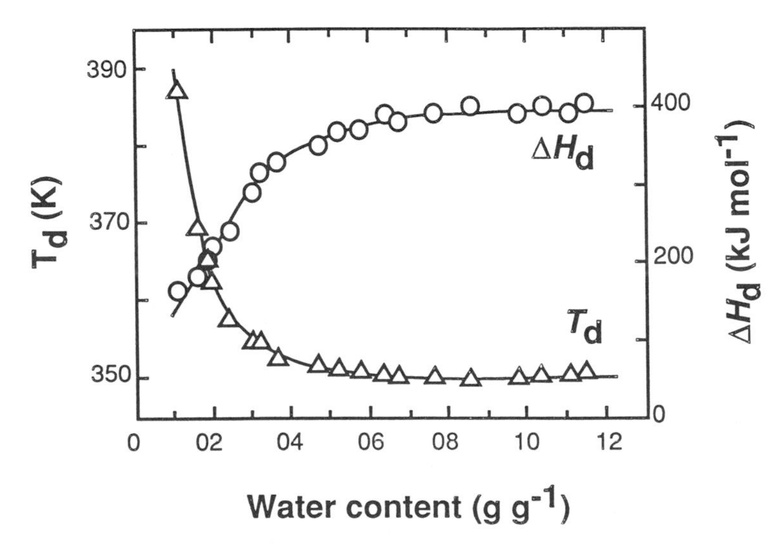

FIGURE 16 Influence of water content on the temperature (T_d) and enthalpy (ΔH_D) of denaturation of ovalbumin. (From Ref. 37.)

the protein. This swollen state presumably reaches a maximum value at a water content of 0.3–0.4 g/g protein at room temperature. The swelling of the protein increases chain mobility and flexibility, and the protein molecule assumes a more dynamic molten structure. When heated, this dynamic flexible structure provides greater access of water to salt bridges and peptide hydrogen bonds than is possible in the dry state, resulting in a lower T_d.

Additives such as salts and sugars affect thermostability of proteins in aqueous solutions. Sugars such as sucrose, lactose, glucose, and glycerol stabilize proteins against thermal denaturation [3,44]. Addition of 0.5 *M* NaCl to proteins such as β-lactoglobulin, soy proteins, serum albumin, and oat globulin significantly increases their T_d [19,20,44].

Hydrostatic Pressure and Denaturation

One of the thermodynamic variables that affects conformation of proteins is hydrostatic pressure. Unlike temperature-induced denaturation, which usually occurs in the range of 40–80°C at 1 atmospheric pressure (atm); pressure-induced denaturation can occur at 25°C if the pressure is sufficiently great. Most proteins undergo pressure-induced denaturation in the range of 1–12 kbar as evidenced from changes in their spectral properties. The midpoint of pressure-induced transition occurs at 4–8 kbar [48,112].

Pressure-induced denaturation of proteins occurs mainly because proteins are flexible and compressible. Although amino acid residues are densely packed in the interior of globular proteins, some void spaces invariably exist and this leads to compressibility. The average partial specific volume of globular proteins in the hydrated state, ν°, is about 0.74 ml/g. The partial specific volume can be considered as the sum of three components:

$$\nu^\circ = V_C + V_{Cav} + \Delta V_{Sol} \tag{42}$$

where V_C is the sum of the atomic volumes, V_{Cav} is the sum of the volumes of the void spaces in the interior of the protein, and ΔV_{Sol} is the volume change due to hydration [38]. The larger the ν° of a protein, the larger is the contribution of void spaces to partial specific volume, and the more unstable the protein will be when pressurized. Fibrous proteins are mostly devoid of void spaces, and hence they are more stable to hydrostatic pressure than globular proteins.

Pressure-induced denaturation of globular proteins is usually accompanied by a reduction in volume of about 30–100 ml/mol. This decrease in volume is caused by two factors: elimination of void spaces as the protein unfolds, and hydration of the nonpolar amino acid residues that become exposed during unfolding. The later event results in a decrease in volume (see Sec. 6.3.2). The volume change is related to the free energy change by the expression

$$\Delta V = d(\Delta G)/dp \tag{43}$$

where p is the hydrostatic pressure.

If a globular protein completely unfolds during pressurization, the volume change should be about 2%. However, the experimental value of 30–100 ml/mol volume decrease obtained with pressurized globular proteins corresponds to only about 0.5% volume decrease. This indicates that proteins only partially unfold even at hydrostatic pressure as high as 10 kbar.

Pressure-induced protein denaturation is highly reversible. Most enzymes, in dilute solutions, regain their activity once the pressure is decreased to atmospheric pressure. However, regeneration of near complete activity usually takes several hours. In the case of pressure-denatured oligomeric proteins and enzymes, subunits first dissociate at 0.001–2 kbar, and then subunits denature at higher pressures [111]; removal of pressure results in subunit reassociation and almost complete restoration of enzyme activity after several hours.

High hydrostatic pressures are being investigated as a food processing tool, for example, for microbial inactivation or gelation. Since high hydrostatic pressure (2–10 kbar) irreversibly damages cell membranes and causes dissociation of organelles in microorganisms, it will inactivate vegetative microorganisms [46]. Pressure gelation of egg white, 16% soy protein solution, or 3% actomyosin solution can be achieved by application of 1–7 kbar hydrostatic pressure for 30 min at 25°C. These pressure-induced gels are softer than thermally induced gels [82]. Also, exposure of beef muscle to 1–3 kbar hydrostatic pressure causes partial fragmentation of myofibrils, which may be useful as a means of tenderizing meat [102]. Pressure processing, unlike thermal processing, does not harm essential amino acids or natural color and flavor, nor does it cause toxic compounds to develop. Thus, processing of foods with high hydrostatic pressure may prove advantageous (except for cost) for certain food products.

Shear and Denaturation

High mechanical shear generated by shaking, kneading, whipping, etc. can cause denaturation of proteins. Many proteins denature and precipitate when they are vigorously agitated [81]. In this circumstance, denaturation occurs because of incorporation of air bubbles and adsorption of protein molecules to the air–liquid interface. Since the energy of the air–liquid interface is greater than that of the bulk phase, proteins undergo conformational changes at the interface. The extent of conformational change depends on the flexibility of the protein. Highly flexible proteins denature more readily at an air–liquid interface than do rigid proteins. The nonpolar residues of denatured protein orient toward the gas phase and the polar residues orient toward the aqueous phase.

Several food processing operations involve high pressure, shear, and high temperature, for example, extrusion, high-speed blending, and homogenization. When a high shear rate is produced by a rotating blade, subsonic pulses are created and cavitation also occurs at the trailing edges of the blade. Both of these events contribute to protein denaturation. The greater the shear rate, the greater is the degree of denaturation. The combination of high temperature and high shear force causes irreversible denaturation of proteins. For example, when a 10–20% whey protein solution at pH 3.5–4.5 and at 80–120°C is subjected to a shear rate of 7,500–10,000/sec, it forms insoluble spherical macrocolloidal particles of about 1 μm diameter. A hydrated material produced under these conditions, "Simplesse," has a smooth, emulsion-like organoleptic character [101].

6.4.2.2 Chemical Agents

pH and Denaturation

Proteins are more stable against denaturation at their isoelectric point than at any other pH. At neutral pH, most proteins are negatively charged, and a few are positively charged. Since the net electrostatic repulsive energy is small compared to other favorable interactions, most proteins are stable at around neutral pH. However, at extreme pH values, strong intramolecular electrostatic repulsion caused by high net charge results in swelling and unfolding of the protein molecule. The degree of unfolding is greater at extreme alkaline pH values than it is at extreme acid pH values. The former behavior is attributed to ionization of partially buried carboxyl, phenolic, and sulfhydryl groups, which cause unraveling of the polypeptide chain as they attempt to expose themselves to the aqueous environment. pH-induced denaturation is mostly reversible. However, in some cases, partial hydrolysis of peptide bonds, deamidation of Asn and Gln, destruction of sulfhydryl groups at alkaline pH, or aggregation can result in irreversible denaturation of proteins.

Organic Solvents and Denaturation

Organic solvents affect the stability of protein hydrophobic interactions, hydrogen bonding, and electrostatic interactions in different ways. Since nonpolar side chains are more soluble in organic solvents than in water, hydrophobic interactions are weakended by organic solvents. On the other hand, since the stability and formation of peptide hydrogen bonds are enhanced in a low-permittivity environment, certain organic solvents may actually strengthen or promote formation of peptide hydrogen bonds. For example, 2-chloroethanol causes an increase in α-helix content in globular proteins. The action of organic solvents on electrostatic interactions is twofold. By decreasing permittivity, they enhance electrostatic interactions between oppositely charged groups and also enhance repulsion between groups with like charge. The net effect of an organic solvent on protein structure, therefore, usually depends on the magnitude of its effect on various polar and nonpolar interactions. At low concentration, some organic solvents can stabilize several enzymes against denaturation [5]. At high concentrations, however, all organic solvents cause denaturation of proteins because of their solubilizing effect on nonpolar side chains.

Organic Solutes and Denaturation

Organic solutes, notably urea and guanidine hydrochloride (GuHC1), induce denaturation of proteins. For many globular proteins the midpoint of transition from the native to denatured state occurs at 4–6 *M* urea and 3–4 *M* GuHC1 at room temperature. Complete transition often occurs in 8 *M* urea and in about 6 *M* GuHC1. GuHC1 is a more powerful denaturant than urea because of its ionic character. Many globular proteins do not undergo complete denaturation even in 8 *M* urea, whereas in 8 *M* GuHC1 they usually exist in a random coil state (completely denatured).

Denaturation of proteins by urea and GuHC1 is thought to involve two mechanisms. The first mechanism involves preferential binding of urea and GuHC1 to the denatured protein. Removal of denatured protein as a protein–denaturant complex shifts the $N \leftrightarrow D$ equilibrium to the right. As the denaturant concentration is increased, continuous conversion of the protein to protein–denaturant complex eventually results in complete denaturation of the protein. Since binding of denaturant to denatured protein is very weak, a high concentration of denaturant is needed to cause complete denaturation. The second mechanism involves solubilization of hydrophobic amino acid residues in urea and GuHC1 solutions. Since urea and GuHC1 have the potential to form hydrogen bonds, at high concentration these solutes break down the hydrogen-bonded structure of water. This destructuring of solvent water makes it a better solvent for nonpolar residues. This results in unfolding and solubilization of apolar residues from the interior of the protein molecule.

Urea or GuHC1-induced denaturation can be reversed by removing the denaturant. However, complete reversibility of protein denaturation by urea is sometimes difficult. This is because some urea converts to cyanate and ammonia. Cyanate reacts with amino groups and alters the charge of the protein.

Detergents and Denaturation

Detergents, such as sodium dodecyl sulfate (SDS), are powerful protein denaturing agents. SDS at 3–8 m*M* concentration denatures most globular proteins. The mechanism involves preferential binding of detergent to the denatured protein molecule. This causes a shift in equilibrium between the native and denatured states. Unlike urea and GuHC1, detergents bind strongly to denatured proteins, which is the reason complete denaturation occurs at a relatively low detergent concentration of 3–8 m*M*. Because of this strong binding, detergent-induced denaturation is irreversible. Globular proteins denatured by SDS do not exist in a random coil state; instead, they assume an α-helical rod shape in SDS solutions. This rod shape is properly regarded as denatured.

Chaotropic Salts and Denaturation

Salts affect protein stability in two different ways. At low concentrations, ions interact with proteins via nonspecific electrostatic interactions. This electrostatic neutralization of protein charges usually stabilizes protein structure. Complete charge neutralization by ions occurs at or below 0.2 ionic strength, and it is independent of the nature of the salt. However, at higher concentrations (> 1 *M*), salts have ion specific effects that influence the structural stability of proteins. Salts such as Na_2SO_4 and NaF enhance, whereas NaSCN and $NaClO_4$ weaken it. Protein structure is influenced more by anions than by cations. For example, the effect of various sodium salts on the thermal denaturation temperature of β-lactoglobulin is shown in Figure 17. At equal ionic strength, Na_2SO_4 and NaCl increase T_d, whereas NaSCN and $NaClO_4$ decrease it. Regardless of their chemical makeup and conformational differences, the structural stability of macromolecules, including DNA, is adversely affected by high concentrations of salts [108]. NaSCN and $NaClO_4$ are strong denaturants. The relative ability of various anions at iso-ionic strength to influence the structural stability of protein (and DNA) in general follows the series $F^- < SO_4^{2-} < Cl^- < Br^- < I^- < ClO_4^- < SCN^- < Cl_3CCOO^-$. This ranking is known as the Hofmeister series or chaotropic series. Fluoride, chloride, and sulfate salts are structure stabilizers, whereas the salts of other anions are structure destabilizers.

The mechanism by which salts affect the structural stability of proteins is not well understood; however, their relative ability to bind to and alter hydration properties of proteins is probably involved. Salts that stabilize proteins enhance hydration of proteins and bind weakly, whereas salts that destabilize proteins decrease protein hydration and bind strongly [4]. These effects are primarily the consequence of energy perturbations at the protein–water

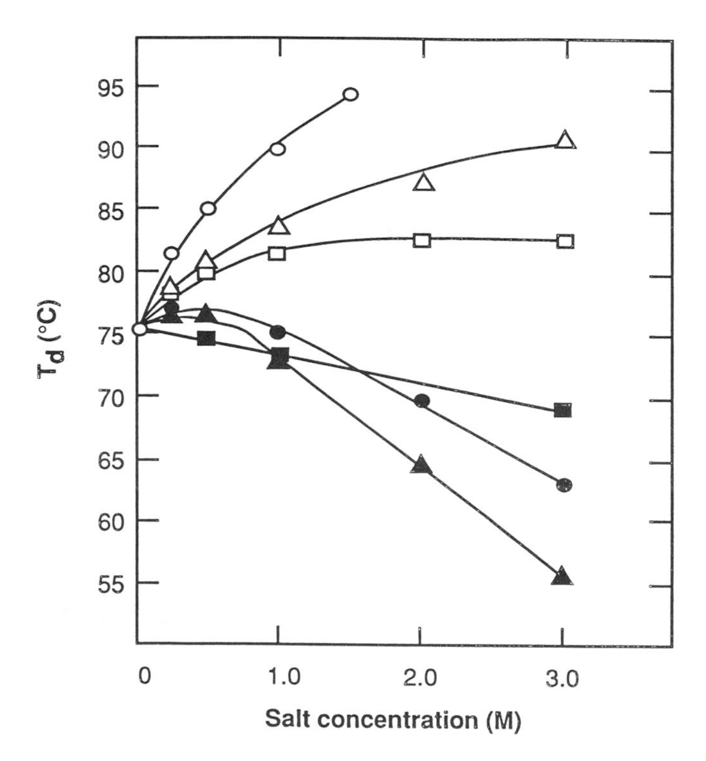

FIGURE 17 Effects of various sodium salts on the temperature of denaturation, T_d, of β-lactoglobulin at pH 7.0. △, NaCl; □, NaBr; ●, $NaClO_4$; ▲, NaSCN; ■, urea. (From Ref. 20.)

interface. On a more fundamental level, protein stabilization or denaturation by salts is related to their effect on bulk water structure. Salts that stabilize protein structure also enhance the hydrogen-bonded structure of water, and salts that denature proteins also break down bulk water structure and make it a better solvent for apolar molecules. In other words, the denaturing effect of chaotropic salts might be related to destabilization of hydrophobic interactions in proteins.

6.5 FUNCTIONAL PROPERTIES OF PROTEINS

Food preferences by human beings are based primarily on sensory attributes such as texture, flavor, color, and appearance. The sensory attributes of a food are the net effect of complex interactions among various minor and major components of the food. Proteins generally have a great influence on the sensory attributes of foods. For example, the sensory properties of bakery products are related to the viscoelastic and dough-forming properties of wheat gluten; the textural and succulence characteristics of meat products are largely dependent on muscle proteins (actin, myosin, actomyosin, and several water-soluble meat proteins); the textural and curd-forming properties of dairy products are due to the unique colloidal structure of casein micelles; and the structure of some cakes and the whipping properties of some dessert products depend on the properties of egg-white proteins. The functional roles of various proteins in different food products are listed in Table 12. "Functionality" of food proteins is defined as "those physical and chemical properties which affect the behavior of proteins in food systems during processing, storage, prepartion and consumption"[55].

The sensory attributes of foods are achieved by complex interactions among various functional ingredients. For instance, the sensory attributes of a cake emanate from gelling/heat-setting, foaming, and emulsifying properties of the ingredients used. Therefore, for a protein to be useful as an ingredient in cakes and other such products, it must possess multiple functionalities. Proteins of animal origin, such as milk (caseins), egg and meat proteins, are widely used in fabricated foods. These proteins are mixtures of several proteins with wide-ranging physico-chemical properties, and they are capable of performing multiple functions. For example, egg white possesses multiple functionalities such as gelation, emulsification, foaming, water binding, and heat coagulation, which makes it a highly desirable protein in many foods. The multiple functionalities of egg white arise from complex interactions among its protein constituents, namely, ovalbumin, conalbumin, lysozyme, ovomucin, and other albumin-type proteins. Plant proteins (e.g., soy and other legume and oilseed proteins), and other proteins, such as whey proteins, are used to a limited extent in conventional foods. Even though these proteins are also mixtures of several proteins, they do not perform as well as animal proteins in most food products. The exact molecular properties of proteins that are responsible for the various desirable functionalities in food are poorly understood.

The physical and chemical properties that govern protein functionality include size; shape; amino acid composition and sequence; net charge and distribution of charges; hydrophobicity/hydrophilicity ratio; secondary, tertiary, and quaternary structures; molecular flexibility/rigidity; and ability to interact/react with other components. Since proteins possess a multitude of physical and chemical properties, it is difficult to delineate the role of each of these properties with respect to a given functional property.

On an empirical level, the various functional properties of proteins can be viewed as manifestations of two molecular aspects of proteins: (a) hydrodynamic properties, and (b) protein surface-related properties [23]. The functional properties such as viscosity (thickening), gelation, and texturization are related to the hydrodynamic properties of proteins, which depend on size, shape, and molecular flexibility. Functional properties such as wettability, dispersibility, solubil-

TABLE 12 Functional Roles of Food Proteins in Food Systems

Function	Mechanism	Food	Protein type
Solubility	Hydrophilicity	Beverages	Whey proteins
Viscosity	Water binding, hydrodynamic size and shape	Soups, gravies, and salad dressings, desserts	Gelatin
Water binding	Hydrogen bonding, ionic hydration	Meat sausages, cakes, and breads	Muscle proteins, egg proteins
Gelation	Water entrapment and immobilization, network formation	Meats, gels, cakes, bakeries, cheese	Muscle proteins, egg and milk proteins
Cohesion–adhesion	Hydrophobic, ionic, and hydrogen bonding	Meats, sausages, pasta, baked goods	Muscle proteins, egg proteins, whey proteins
Elasticity	Hydrophobic bonding, disulfide cross-links	Meats, bakery	Muscle proteins, cereal proteins
Emulsification	Adsorption and film formation at interfaces	Sausages, bologna, soup, cakes, dressings	Muscle proteins, egg proteins, milk proteins
Foaming	Interfacial adsorption and film formation	Whipped toppings, ice cream, cakes, desserts	Egg proteins, milk proteins
Fat and flavor binding	Hydrophobic bonding, entrapment	Low-fat bakery products, doughnuts	Milk proteins, egg proteins, cereal proteins

Source: Ref. 56.

ity, foaming, emulsification, and fat and flavor binding are related to the chemical and topographical properties of the protein surface.

Although much is known about the physicochemical properties of several food proteins, prediction of functional properties from their molecular properties has not been successful. A few empirical correlations between molecular properties and certain functional properties in model protein systems have been established [74]. However, behavior in model systems often is not the same as behavior in real food products. This is attributable, in part, to denaturation of proteins during food fabrication. The extent of denaturation depends on pH, temperature, other processing conditions, and product characteristics. In addition, in real foods, proteins interact with other food components, such as lipids, sugars, polysaccharides, and minor components, and this modifies their functional behavior. Despite these inherent difficulties, considerable progress has been made toward understanding the relationship between various physicochemical properties of protein molecules and their functional properties.

6.5.1 Protein Hydration

Water is an essential constituent of foods. The rheological and textural properties of foods depend on the interaction of water with other food constituents, especially with macromolecules, such as proteins and polysaccharides. Water modifies the physicochemical properties of proteins. For example, the plasticizing effect of water on amorphous and semicrystalline food proteins changes their glass transition temperature (see Chap. 2) and T_D. The glass transition temperature refers to the conversion of a brittle amorphous solid (glass) to a flexible rubbery state, whereas the melting temperature refers to transition of a crystalline solid to a disordered structure.

Many functional properties of proteins, such as dispersibility, wettability, swelling, solubility, thickening/viscosity, water-holding capacity, gelation, coagulation, emulsification, and foaming, depend on water–protein interactions. In low and intermediate moisture foods, such as bakery and comminuted meat products, the ability of proteins to bind water is critical to the acceptability of these foods. The ability of a protein to exhibit a proper balance of protein–protein and protein–water interactions is critical to their thermal gelation.

Water molecules bind to several groups in proteins. These include charged groups (ion–dipole interactions); backbone peptide groups; the amide groups of Asn and Gln; hydroxyl groups of Ser, Thr, and Tyr residues (all dipole–dipole interactions); and nonpolar residues (dipole-induced dipole interaction, hydrophobic hydration).

The water binding capacity of proteins is defined as grams of water bound per gram of protein when a dry protein powder is equilibrated with water vapor at 90–95% relative humidity. The water binding capacities (also sometimes called hydration capacity) of various polar and nonpolar groups of proteins are given in Table 13. Amino acid residues with charged groups bind about 6 mol water/mol residue, the uncharged polar residues bind about 2 mol/mol residue, and the nonpolar groups bind about 1 mol/mol residue. The hydration capacity of a protein therefore is related, in part, to its amino acid composition—the greater the number of charged residues,

TABLE 13 Hydration Capacities[a] of Amino Acid Residues

Amino acid residue	Hydration (moles H_2O/mol residue)
Polar	
Asn	2
Gln	2
Pro	3
Ser, The	2
Trp	2
Asp (unionized)	2
Glu (unionized)	2
Tyr	3
Arg (unionized)	3
Lys (unionized)	4
Ionic	
Asp^-	6
Glu^-	7
Tyr^-	7
Arg^+	3
His^+	4
Lys^+	4
Nonpolar	
Ala	1
Gly	1
Phe	0
Val, Ile, Leu, Met	1

[a]Represents unfrozen water associated with amino acid residues based on nuclear magnetic resonance studies of polypeptide.
Source: Ref. 59.

the greater is the hydration capacity. The hydration capacity of a protein can be calculated from its amino acid composition using the empirical equation [59]

$$a = f_C + 0.4f_P + 0.2f_N \qquad (44)$$

where a is g water/g protein, and f_C, f_P, and f_N are the fractions of the charged, polar, and nonpolar residues, respectively, in the protein. The experimental hydration capacities of several monomeric globular proteins agree very well with those calculated from the preceding equation. This, however, is not true for oligomeric proteins. Since oligomeric structures involve partial burial of the protein surface at the subunit–subunit interface, calculated values are usually greater than experimental values. On the other hand, the experimental hydration capacity of casein micelles (~4 g water/g protein) is much larger than that predicted by the preceding equation. This is because of the enormous amount of void space within the casein micelle structure, which imbibes water through capillary action and physical entrapment.

On a macroscopic level, water binding to proteins occurs in a stepwise process. The high-affinity ionic groups are solvated first at low water activity, followed by polar and nonpolar groups. The sequence of steps involved at increasing water activity is presented in Figure 18 (see also Chap 2). The sorption isotherm of proteins, that is, the amount of water bound per gram of protein as a function of relative humidity, is invariably a sigmoidal curve (see Chap. 2).

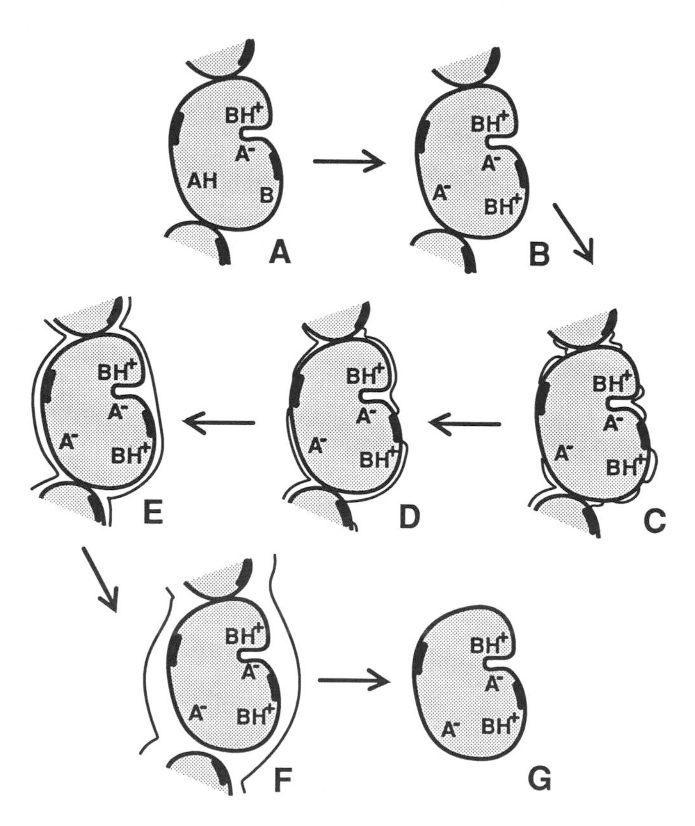

FIGURE 18 Sequence of steps involved in hydration of a protein. (A) Unhydrated protein. (B) Initial hydration of charged groups. (C) Water cluster formation near polar and charged sites. (D) Completion of hydration at the polar surfaces. (E) Hydrophobic hydration of nonpolar patches; completion of monolayer coverage. (F) Bridging between protein-associated water and bulk water. (G) Completion of hydrodynamic hydration. (From Ref. 96.)

For most proteins, so-called monolayer coverage occurs at a water activity (a_W) of 0.05–0.3, and multilayers of water are formed in the water activity range of 0.3–0.7. Water present in the monolayer associates primarily with ionic groups. This water is unfreezable, does not take part as a solvent in chemical reactions, and is often referred to as "bound" water, which should be understood to mean water with "hindered" mobility. The free energy change for desorption of water (i.e., for transfer from the protein surface to the bulk phase) in the monolayer hydration range of 0.07–0.27 g water/g protein is only about 0.75 kJ/mol at 25°C. Since the thermal kinetic energy of water at 25°C is about 2.5 kJ/mol (which is greater than the free energy change for desorption), water molecules in the monolayer are reasonably mobile.

At $a_W = 0.9$, proteins bind about 0.3–0.5 g water/g protein (Table 14). Much of this water is unfreezable at 0°C. At $a_W > 0.9$, liquid (bulk) water condenses into the clefts and crevices of protein molecules, or in the capillaries of insoluble protein systems, such as myofibrils. The properties of this water are similar to those of bulk water. This water is known as hydrodynamic water, and moves with the protein molecule.

Several environmental factors, such as pH, ionic strength, type of salts, temperature, and protein conformation, influence the water binding capacity of proteins. Proteins exhibit the least hydration at their isoelectric pH, where enhanced protein–protein interactions result in minimal interaction with water. Above and below the isoelectric pH, because of the increase in the net charge and repulsive forces, proteins swell and bind more water. The water binding capacity of most proteins is greater at pH 9–10 than at any other pH. This is due to ionization of sulfhydryl and tyrosine residues. Above pH 10, the loss of positively charged ε-amino groups of lysyl residues results in reduced water binding.

At low concentrations ($< 0.2\ M$), salts increase the water binding capacity of proteins. This is because hydrated salt ions bind (weakly) to charged groups on proteins. At this low concentration, binding of ions to proteins does not affect the hydration shell of the charged

TABLE 14 Hydration Capacities of Various Proteins

Protein	Hydration capacity (g water/g protein)
Pure proteins[a]	
Ribonuclease	0.53
Lysozyme	0.34
Myoglobin	0.44
β-Lactoglobulin	0.54
Chymotrypsinogen	0.23
Serum albumin	0.33
Hemoglobin	0.62
Collagen	0.45
Casein	0.40
Ovalbumin	0.30
Commercial protein preparations[b]	
Whey protein concentrates	0.45–0.52
Sodium caseinate	0.38–0.92
Soy protein	0.33

[a]At 90% relative humidity.
[b]At 95% relative humidity.
Source: Refs. 57 and 60.

groups on the protein, and the increase in water binding essentially comes from water associated with the bound ions. However, at high salt concentrations, much of the existing water is bound to salt ions, and this results in dehydration of the protein.

The water binding capacity of proteins generally decreases as the temperature is raised, because of decreased hydrogen bonding and decreased hydration of ionic groups. The water binding capacity of a denatured protein is generally about 10% greater than that of the native protein. This is due to an increase in surface area to mass ratio with exposure of some previously buried hydrophobic groups. However, if denaturation leads to aggregation of the protein, then its water binding capacity may actually decrease because of the protein–protein interactions. It should be pointed out that most denatured food proteins exhibit very low solubility in water. Their water binding capacities, however, are not drastically different from those in the native state. Thus, water binding capacity cannot be used to predict the solubility characteristics of proteins. In other words, the solubility of a protein is dependent not only on water binding capacity, but also on other thermodynamic factors.

In food applications, the water-holding capacity of a protein preparation is more important than the water binding capacity. Water-holding capacity refers to the ability of the protein to imbibe water and retain it against gravitational force within a protein matrix, such as protein gels or beef and fish muscle. This water refers to the sum of the bound water, hydrodynamic water, and the physically entrapped water. The contribution of the physically entrapped water to water holding capacity is much larger than those of the bound and hydrodynamic water. However, studies have shown that the water-holding capacity of proteins is positively correlated with water binding capacity. The ability of protein to entrap water is associated with juiciness and tenderness of comminuted meat products and desirable textural properties of bakery and other gel-type products.

6.5.2 Solubility

The functional properties of proteins are often affected by protein solubility, and those most affected are thickening, foaming, emulsifying, and gelling. Insoluble proteins have very limited uses in food.

The solubility of a protein is the thermodynamic manifestation of the equilibrium between protein–protein and protein–solvent interactions.

$$\text{Protein–Protein} + \text{Solvent–Solvent} \rightleftharpoons \text{Protein–Solvent} \tag{45}$$

The major interactions that influence the solubility characteristics of proteins are hydrophobic and ionic in nature. Hydrophobic interactions promote protein–protein interactions and result in decreased solubility, whereas ionic interactions promote protein–water interactions and result in increased solubility. Ionic residues introduce two kinds of repulsive forces between protein molecules in solution. The first involves electrostatic repulsion between protein molecules owing to a net positive or negative charge at any pH other than the isoelectric pH; the second involves repulsion between hydration shells around ionic groups.

Bigelow [6] proposed that the solubility of a protein is fundamentally related to the average hydrophobicity of the amino acid residues and the charge frequency. The average hydrophobicity is defined as

$$\Delta g = \sum \Delta g_{\text{residue}}/n \tag{46}$$

where $\Delta g_{\text{residue}}$ is the hydrophobicity of each amino acid side chain obtained from the free energy change for transfer from ethanol to water, and n is the total number of residues in the protein. The charge frequency is defined as

$$\sigma = (n^+ + n^-)/n \tag{47}$$

where n^+ and n^- are the total number of positively and negatively charged residues, respectively, and n is the total number of residues. According to Bigelow [6], the smaller the average hydrophobicity and the larger the charge frequency, the greater will be the solubility of the protein. Although this empirical correlation is true for most proteins, it is not an absolute one. The Bigelow approach is faulty because the solubility of a protein is dictated by the hydrophilicity and hydrophobicity of the protein surface that contacts with the surrounding water, rather than the average hydrophobicity and charge frequency of the molecule as a whole. The fewer the number of surface hydrophobic patches, the greater the solubility.

Based on solubility characteristics, proteins are classified into four categories. *Albumins* are those that are soluble in water at pH 6.6 (e.g., serum albumin, ovalbumin, and α-lactalbumin), *globulins* are those that are soluble in dilute salt solutions at pH 7.0 (e.g., glycinin, phaseolin, and β-lactoglobulin), *glutelins* are those that are soluble only in acid (pH 2) and alkaline (pH 12) solutions (e.g., wheat glutelins), and *prolamines* are those soluble in 70% ethanol (e.g., zein and gliadins). Both prolamines and glutelins are highly hydrophobic proteins.

In addition to these intrinsic physicochemical properties, solubility is influenced by several solution conditions, such as pH, ionic strength, temperature, and the presence of organic solvents.

pH and Solubility

At pH values below and above the isoelectric pH, proteins carry a net positive or a net negative charge, respectively. Electrostatic repulsion and hydration of charged residues promote solubilization of the protein. When solubility is plotted against pH, most food proteins exhibit a U-shaped curve. Minimum solubility occurs at about the isoelectric pH of proteins. A majority of food proteins are acidic proteins; that is, the sum of Asp and Glu residues is greater than the sum of Lys, Arg, and His residues. Therefore, they exhibit minimum solubility at pH 4–5 (isoelectric pH) and maximum solubility at alkaline pH. The occurrence of minimum solubility near the isoelectric pH is due primarily to the lack of electrostatic repulsion, which promotes aggregation and precipitation via hydrophobic interactions. Some food proteins, such as β-lactoglobulin (pI 5.2) and bovine serum albumin (pI 5.3), are highly soluble at their isoelectric pH. This is because these proteins contain a large ratio of surface hydrophilic residues to surface nonpolar groups. It should be remembered that even though a protein is electrically neutral at its pI, it is still charged, but the positive and negative charges on the surface are equal. If the hydrophilicity and the hydration repulsion forces arising from these charged residues are greater than the protein–protein hydrophobic interactions, then the protein will still be soluble at the pI.

Since most proteins are highly soluble at alkaline pH (8–9), protein extraction from plant sources, such as soybean flour, is carried out at this pH. The protein is then recovered from the extract by isoelectric precipitation at pH 4.5–4.8.

Heat denaturation changes the pH–solubility profile of proteins (Fig. 19). Native whey protein isolate is completely soluble in the pH range 2–9, but when heated at 70°C for 1 min to 10 min a typical U-shaped solubility profile develops with a solubility minimum at pH 4.5. The change in the solubility profile upon heat denaturation is due to an increase in the hydrophobicity of the protein surface as a consequence of unfolding. Unfolding alters the balance between protein–protein and protein–solvent interactions in favor of the former.

Ionic Strength and Solubility

The ionic strength of a salt solution is given by

$$\mu = 0.5 \sum C_i Z_i^2 \tag{48}$$

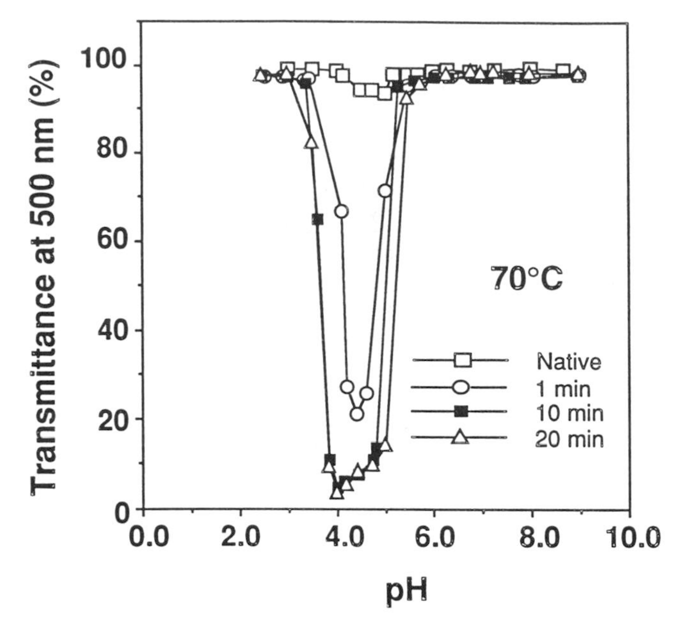

FIGURE 19 pH–solubility profile of whey protein isolate solutions heated at 70°C for various times. (from Ref. 119.)

where C_i is concentration of an ion, and Z_i is its valance. At low ionic strength (< 0.5), ions neutralize charges at the surface of proteins. This charge screening affects solubility in two different ways, depending on the characteristics of the protein surface. Solubility decreases for those proteins that contain a high incidence of nonpolar patches, and it increases for those that don't. The former behavior is typical for soy proteins, and the latter behavior is exhibited by β-lactoblobulin. While the decrease in solubility is caused by enhanced hydrophobic interactions, the increase in solubility is caused by a decrease in the ionic activity of the protein macroion. At ionic strength > 1.0, salts have ion specific effects on protein solubility. As salt concentration is increased up to $\mu = 1$; sulfate and fluoride salts progressively decrease solubility (salting out), whereas thiocyanate and perchlorate salts increase solubility (salting in). At constant μ, relative effectiveness of various ions on solubility follows the Hofmeister series, with anions promoting solubility in the order $SO_4^{2-} < F^- < Cl^- < Br^- < I^- < ClO_4^- < SCN^-$, and cations decreasing solubiltiy in the order $NH_4^+ < K^+ < Na^+ < Li^+ < Mg^{2+} < Ca^{2+}$. This behavior is analogous to the effects of salts on the thermal denaturation temperature of proteins (see Sec. 6.4).

Generally, solubility of proteins in salt solutions follows the relation

$$\log(S/S_0) = \beta - K_S C_S \tag{49}$$

where S and S_0 are solubilities of the protein in the salt solution and in water, respectively, K_S is the salting out constant, C_S is molar concentration of salt, and β is a constant. K_S is positive for salting-out type of salts and negative for salting-in type of salts.

Temperature and Solubility

At constant pH and ionic strength, the solubility of most proteins generally increases with temperature between 0 and 40°C. Exceptions occur with highly hydrophobic proteins, such as β-casein and some cereal proteins, which exhibit a negative relationship with temperature.

Above 40°C, the increase in thermal kinetic energy causes protein unfolding (denaturation), exposure of nonpolar groups, aggregation, and precipitation, that is, decreased solubility.

Organic Solvents and Solubility

Addition of organic solvents, such as ethanol or acetone, lowers the permittivity of an aqueous medium. This increases intra- and intermolecular electrostatic forces, both repulsive as well as attractive. The repulsive intramolecular electrostatic interactions cause unfolding of the protein molecule. In the unfolded state, the permittivity promotes intermolecular hydrogen bonding between the exposed peptide groups and attractive intermolecular electrostatic interactions between oppositely charged groups. These intermolecular polar interactions lead to precipitation of the protein in organic solvents or reduced solubility in an aqueous medium. The role of hydrophobic interactions in causing precipitation in organic solvents is minimal because of the solubilizing effect of organic solvents on nonpolar residues. However, even in aqueous media containing a low concentration of organic solvent, hydrophobic interactions between exposed residues may also contribute to insolublization.

Since solubility of proteins is intimately related to their structural states, it is often used as a measure of the extent of denaturation during extraction, isolation, and purification processes. It is also used as an index of the potential applications of proteins. Commercially prepared protein concentrates and isolates show a wide range of solubility. The solubility characteristics of these protein preparations are expressed as *protein solubility index* (PSI) or *protein dispersibility index* (PDI). Both of these terms express the percentage of soluble protein present in a protein sample. The PSI of commercial protein isolates varies from 25 to 80%.

6.5.3 Interfacial Properties of Proteins

Several natural and processed foods are either foam- or emulsion-type products. These types of dispersed systems are unstable unless a suitable amphiphilic substance is present at the interface between the two phases (see Chap. 3). Proteins are amphiphilic molecules, and they migrate spontaneously to an air–water interface or an oil–water interface. This spontaneous migration of proteins from a bulk liquid to an interface indicates that the free energy of proteins is lower at the interface than it is in the bulk aqueous phase. Thus, when equilibrium is established, the concentration of protein in the interfacial region is always much greater than it is in the bulk aqueous phase. Unlike low-molecular-weight surfactants, proteins form a highly viscoelastic film at an interface, which has the ability to withstand mechanical shocks during storage and handling. Thus, protein-stabilized foams and emulsions are more stable than those prepared with low-molecular-weight surfactants, and because of this, proteins are extensively used for these purposes.

Although all proteins are amphiphilic, they differ significantly in their surface-active properties. The differences in the surface-active properties among proteins cannot be attributed to differences in the ratio of hydrophobic to hydrophilic residues. If a large hydrophobicity/hydrophilicity ratio were the primary determinant of the surface activity of proteins, then plant proteins, which contain more than 40% hydrophobic amino acid residues, should be better surfactants than albumin-type proteins, such as ovalbumin and bovine serum albumin, which contain less than 30% hydrophobic amino acid residues. On the contrary, ovalbumin and serum albumin are better emulsifying and foaming agents than are soy proteins and other plant proteins. Furthermore, average hydrophobicities of most proteins fall within a narrow range [7], yet they exhibit remarkable differences in their surface activity. It must be concluded, therefore, that differences in surface activity are related primarily to differences in protein conformation. The conformational factors of importance include stability/flexibility of the polypeptide chain, ease of adaptability to changes in the environment, and distribution pattern of hydrophilic and hydrophobic groups on the protein surface. All of these conformational factors are interdependent, and they collectively have a large influence on the surface activity of proteins.

It has been shown that desirable surface-active proteins have three attributes: (a) ability to rapidly adsorb to an interface, (b) ability to rapidly unfold and reorient at an interface, and (c) an ability, once at the interface, to interact with the neighboring molecules and form a strong cohesive, viscoeleastic film that can withstand thermal and mechanical motions [26,87].

The first and most critical event in the creation of stable foams and emulsions during whipping or homogenization is the spontaneous and rapid adsorption of proteins at the newly created interface. The rapidity with which a protein can adsorb to air–water or oil–water interfaces depends on the distribution pattern of hydrophobic and hydrophilic patches on its surface. If the protein surface is extremely hydrophilic and contains no discernable hydrophobic patches, adsorption probably will not take place because its free energy will be lower in the aqueous phase than either at the interface or in the nonpolar phase. As the number of hydrophobic patches on the protein surface is increased, spontaneous adsorption to an interface becomes more probable (Fig. 20). Single hydrophobic residues randomly distributed on the protein surface do not constitute a hydrophobic patch, nor do they possess sufficient interaction energy to strongly anchor the protein at an interface. Even though more than 40% of a protein's overall accessible surface is covered with nonpolar residues, they will not enhance protein adsorption unless they exist as segregated regions or patches. In other words, the molecular characteristics of the protein surface have an enormous influence on whether a protein will spontaneously adsorb to an interface and how effective it will be as a stabilizer of dispersions.

The mode of adsorption of proteins at an interface is different from that of low-molecular-weight surfactants. In the case of low-molecular-weight surfactants, such as phospholipids and monoacylglycerols, conformational constraints for adsorption and orientation do not exist because hydrophilic and hydrophobic moieties are present at the ends of the molecule. In the case of proteins, however, the distribution pattern of hydrophobic and hydrophilic patches on the surface and the structural rigidity of the molecule cause constraints to adsorption and orientation. Because of the bulky, folded nature of proteins, once adsorbed, a large portion of the molecule remains in the bulk phase and only a small portion is anchored at the interface. The tenacity with

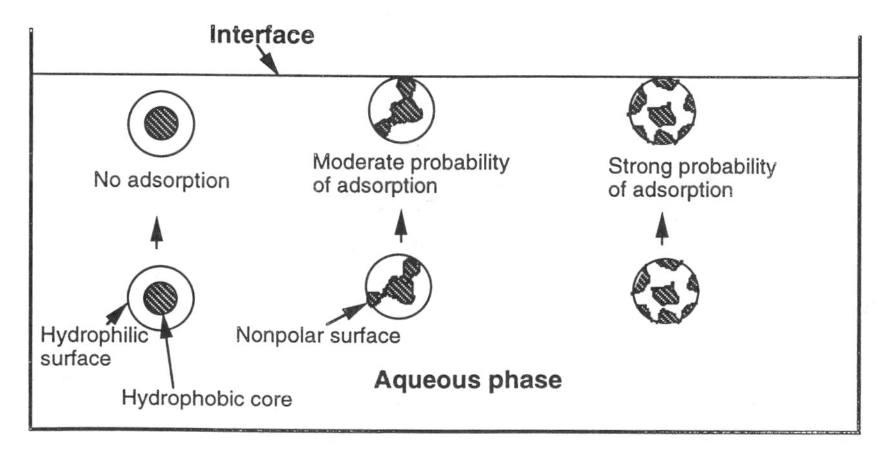

FIGURE 20 Schematic representation of the role of surface hydrophobic patches on the probability of adsorption of proteins at the air-water interface. (From Ref. 22.)

which this small portion of the protein molecule remains attached at the interface depends on the number of peptide segments anchored to the interface, and the energetics of interaction between these segments and the interface. The protein will be retained at the interface only when the sum of negative free energy changes of segment interactions is much greater than the thermal kinetic energy of the protein molecule. The number of peptide segments anchored at the interface depends, in part, on the conformational flexibility of the molecule. Highly flexible molecules, such as caseins, can undergo rapid conformational changes once they are adsorbed at the interface, enabling additional polypeptide segments to bind to the interface. On the other hand, rigid globular proteins, such as lysozyme and soy proteins, cannot undergo extensive conformational changes at the interface.

At interfaces, polypeptide chains assume one or more of three distinct configurations: trains, loops, and tails (Fig. 21). The trains are segments that are in direct contact with the interface, loops are segments of the polypeptide that are suspended in the aqueous phase, and tails are N- and C-terminal segments of the protein that are usually located in the aqueous phase. The relative distribution of these three configurations depends on the conformational characteristics of the protein. The greater the proportion of polypeptide segments in a train configuration, the stronger is the binding and the lower is the interfacial tension.

The mechanical strength of a protein film at an interface depends on cohesive intermolecular interactions. These include attractive electrostatic interactions, hydrogen bonding, and hydrophobic interactions. Interfacial polymerization of adsorbed proteins via disulfide–sulfhydryl interchange reactions also increase its viscoelastic properties. The concentration of protein in the interfacial film is about 20–25% (w/v), and the protein exists in almost a gel-like state. The balance of various noncovalent interactions is crucial to the stability and viscoelastic properties of this gel-like film. For example, if hydrophobic interactions are too strong, this can lead to interfacial aggregation, coagulation, and eventual precipitation of the protein, which will be detrimental to film integrity. If repulsive electrostatic forces are much stronger than attractive interactions, this may prevent formation of a thick, cohesive film. Therefore, a proper balance of attractive, repulsive, and hydration forces is required to form a stable viscoelastic film.

The basic principles involved in formation and stability of emulsions and foams are very similar. However, since the energetics of these interfaces are different, the molecular require-

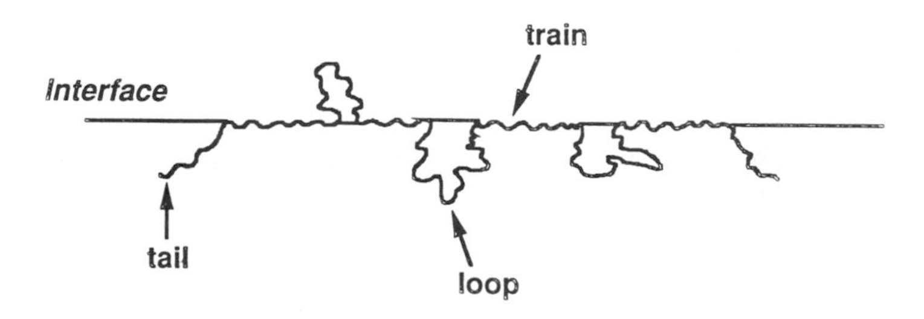

FIGURE 21 The various configurations of a flexible polypeptide at an interface. (From Ref. 22.)

ments for protein functionality at these interfaces are not the same. In other words, a protein that is a good emulsifier may not be a good foaming agent.

It should now be clear that the behavior of proteins at interfaces is very complex and not well understood. Therefore, the following discussion of the emulsifying and foaming properties of food proteins will be largely qualitative in nature.

6.5.3.1 Emulsifying Properties

The physical chemistry of emulsion formation and the factors affecting creaming, flocculation, coalescence, and stability were reviewed in Chapter 3.

Several natural and processed foods, such as milk, egg yolk, coconut milk, soy milk, butter, margarine, mayonnaise, spreads, salad dressings, frozen desserts, frankfurter, sausage, and cakes, are emulsion-type products where proteins play an important role as an emulsifier. In natural milk, the fat globules are stabilized by a membrane composed of lipoproteins. When milk is homogenized, the lipoprotein membrane is replaced by a protein film comprised of casein micelles and whey proteins. Homogenized milk is more stable against creaming than is natural milk because the casein micelle–whey protein film is stronger than the natural lipoprotein membrane.

6.5.3.2 Methods for Determining the Emulsifying Properties of Proteins

The emulsifying properties of food proteins are evaluated by several methods such as size distribution of oil droplets formed, emulsifying activity, emulsion capacity, and emulsion stability.

Emulsifying Activity Index

The physical and sensory properties of a protein-stabilized emulsion depend on the size of the droplets formed and the total interfacial area created.

The average droplet size of emulsions can be determined by several methods, such as light microscopy (not very reliable), electron microscopy, light scattering (photon correlation spectroscopy), or use of a Coulter counter. Knowing mean droplet size, total interfacial area can be obtained from the relation

$$A = \frac{3\phi}{R} \tag{50}$$

where ϕ is the volume fraction of the dispersed phase (oil) and R is the mean radius of the emulsion particles. If m is the mass of the protein, then the emulsifying activity index (EAI), that is, the interfacial area created per unit of mass protein, is

$$\text{EAI} = \frac{3\phi}{Rm} \tag{51}$$

Another simple and more practical method to determine EAI of proteins is the turbidimetric method [86]. The turbidity of an emulsion is given by

$$T = \frac{2.303A}{l} \tag{52}$$

where A is absorbance and l is path length. According to the Mie theory of light scattering, the interfacial area of an emulsion is two times its turbidity. If ϕ is the volume fraction of the oil and C is the weight of protein per unit volume of the aqueous phase, then the EAI of the protein is given by

$$\text{EAI} = \frac{2T}{(1 - \phi)C} \tag{53}$$

It should be mentioned that in the original article [86], ϕ instead of $(1 - \phi)$ was used in the denominator of this equation. The expression as given in Equation 53 is the correct one because ϕ is defined as the oil volume fraction, and thus $(1 - \phi)C$ is the total mass of protein in a unit volume of the emulsion [13]. Although this method is simple and practical, the main drawback is that it is based on measurement of turbidity at a single wavelength, 500 nm. Since the turbidity of food emulsions is wavelength dependent, the interfacial area obtained from turbidity at 500 nm is not very accurate. Therefore, use of this equation to estimate mean particle diameter or the number of emulsion particles present in the emulsion gives results that are not very reliable. However, the method can be used for qualitative comparison of emulsifying activities of different proteins, or changes in the emulsifying activity of a protein after various treatments.

Protein Load

The amount of protein adsorbed at the oil–water interface of an emulsion has a bearing on its stability. To determine the amount of protein adsorbed, the emulsion is centrifuged, the aqueous phase is separated, and the cream phase is repeatedly washed and centrifuged to remove any loosely adsorbed proteins. The amount of protein adsorbed to the emulsion particles is determined from the difference between the total protein initially present in the emulsion and the amount present in the wash fluid from the cream phase. Knowing the total interfacial area of the emulsion particles, the amount of protein adsorbed per square meter of the interfacial area can be calculated. Generally, the protein load is in the range of about 1–3 mg/m^2 of interfacial area. As the volume fraction of the oil phase is increased the protein load decreases at constant protein content in the total emulsion. For high-fat emulsions and small-sized droplets, more protein is obviously needed to adequately coat the interfacial area and stabilize the emulsion.

Emulsion Capacity

Emulsion capacity (EC) is the volume (ml) of oil that can be emulsified per gram of protein before phase inversion (a change from oil-in-water emulsion to water-in-oil) occurs. This method involves addition of oil or melted fat at a constant rate and temperature to an aqueous protein solution that is continuously agitated in a food blender. Phase inversion is detected by an abrupt change in viscosity or color (usually a dye is added to the oil), or by an increase in electrical resistance. For a protein-stabilized emulsion, phase inversion usually occurs when ϕ is about 0.65–0.85. Inversion is not instantaneous, but is preceded by formation of a water-in-oil-in-water double emulsion. Since emulsion capacity is expressed as volume of oil emulsified per gram protein at phase inversion, it decreases with increasing protein concentration once a point is reached where unadsorbed protein accumulates in the aqueous phase. Therefore, to compare emulsion capacities of different proteins, EC versus protein concentration profiles should be used instead of EC at a specific protein concentration.

Emulsion Stability

Protein-stabilized emulsions are often stable for days. Thus, a detectable amount of creaming or phase separation is usually not observed in a reasonable amount of time when samples are stored at atmospheric conditions. Therefore, drastic conditions, such as storage at elevated temperature or separation under centrifugal force, is often used to evaluate emulsion stability. If centrifugation is used, stability is then expressed as percent decrease in interfacial area (i.e., turbidity) of the emulsion, or percent volume of cream separated, or as the fat content of the cream layer. More often, however, emulsion stability is expressed as

$$ES = \frac{\text{volume of cream layer}}{\text{total volume of emulsion}} \times 100 \tag{54}$$

where the volume of the cream layer is measured after a standardized centrifugation treatment. A common centrifugation technique involves centrifugation of a known volume of emulsion in a graduated centrifuge tube at $1300 \times g$ for 5 min. The volume of the separated cream phase is then measured and expressed as a percentage of the total volume. Sometimes centrifugation at a relatively low gravitational force ($180 \times g$) for a longer time (15 min) is used to avoid coalescence of droplets.

The turbidimetric method (discussed earlier) can also be used to evaluate emulsion stability. In this case stability is expressed as emulsion stability index (ESI), which is defined as the time to achieve a turbidity of the emulsion that is one-half of the original value.

The methods used to determine emulsion stability are very empirical. The most fundamental quantity related to stability is the change in interfacial area with time, but this is difficult to measure directly.

6.5.3.3 Factors Influencing Emulsification

The properties of protein-stabilized emulsions are affected by several factors. These include intrinsic factors, such as pH, ionic strength, temperature, presence of low-molecular-weight surfactants, sugars, oil-phase volume, type of protein, and the melting point of the oil used; and extrinisic factors, such as type of equipment, rate of energy input, and rate of shear. Standardized methods for systematically evaluating the emulsifying properties of proteins have not been agreed upon. Therefore, results among laboratories cannot be accurately compared, and this has hampered the understanding of the molecular factors that affect emulsifying properties of proteins.

The general forces involved in the formation and stabilization of emulsion were discussed in Chapter 3. Therefore, only the molecular factors that affect protein-stabilized emulsions need be discussed here.

Solubility plays a role in emulsifying properties, but 100% solubility is not an absolute requirement. While highly insoluble proteins do not perform well as emulsifiers, no reliable relationship exists between solubility and emulsifying properties in the 25–80% solubility range [64]. However, since the stability of a protein film at the oil–water interface is dependent on favorable interactions with both the oil and aqueous phases, some degree of solubility is likely to be necessary. The minimum solubility requirement for good performance may vary among proteins. In meat emulsions, such as in sausage and frankfurter, solubilization of myofibrillar proteins in 0.5 *M* NaCl enhances their emulsifying properties. Commercial soy protein isolates, which are isolated by thermal processing, have poor emulsifying properties because of their very low solubility.

The formation and stability of protein-stabilized emulsions are affected by pH. Several mechanisms are involved. Generally, proteins that have high solubility at the isoelectric pH (e.g., serum albumin, gelatin, and egg-white proteins) show maximum emulsifying activity and emulsion capacity at that pH. The lack of net charge and electrostatic repulsive interactions at the isoelectric pH helps maximize protein load at the interface and promotes formation of a highly viscoelastic film, both of which contribute to emulsion stability. However, the lack of electrostatic repulsive interactions among emulsion particles can, in some instances, promote flocculation and coalescence, and thus decrease emulsion stability. On the other hand, if the protein is highly hydrated at the isoelectric pH (unusual), then hydration repulsion forces between emulsion particles may prevent flocculation and coalescence, and thus stabilize the emulsion. Because most food proteins (caseins, commercial whey proteins, meat proteins, soy proteins) at their isoelectric pH are sparingly soluble, poorly hydrated, and lack electrostatic repulsive forces, they are generally poor emulsifiers at this pH. These proteins may, however, be effective emulsifiers when moved away from their isoelectric pH.

The emulsifying properties of proteins show a weak positive correlation with surface hydrophobicity, but not with mean hydrophobicity (i.e., $Jmol^{-1}residue^{-1}$). The ability of various proteins to decrease interfacial tension at the oil–water interface and to increase the emulsifying activity index is related to their surface hydrophobicity values (Fig. 22). However, this relationship is by no means perfect. The emulsifying properties of several proteins, such as β-lactoglobulin, α-lactalbumin, and soy proteins, do not show a strong correlation with surface hydrophobicity.

The surface hydrophobicity of proteins is usually determined by measuring the amount of a hydrophobic fluorescent probe, such as *cis*-parinaric acid, that can bind to the protein [53]. Although this method provides some information on the nonpolarity of the protein surface, it is questionable whether the measured value truly reflects the "hydrophobicity" of the protein surface. The true definition of surface hydrophobicity is that portion of the nonpolar surface of the protein that makes contact with the surrounding bulk water. However, *cis*-parinaric acid is capable of binding only to hydrophobic cavities formed at the surface by association of nonpolar residues. These protein cavities are accessible to nonpolar ligands, but they are not accessible to water and may not be accessible to either phase in an oil–water emulsion, unless the protein is able to undergo rapid conformational rearrangement at the interface. The poor correlation of surface hydrophobicity (as measured by *cis*-parinaric acid binding) with the emulsifying properties of some proteins may be related to the fact that *cis*-parinaric acid provides no indication of molecular flexibility. Molecular flexibility at the oil–water interface may be the most important determinant of the emulsifying properties of proteins.

Partial denaturation (unfolding) of proteins prior to emulsification, which does not result

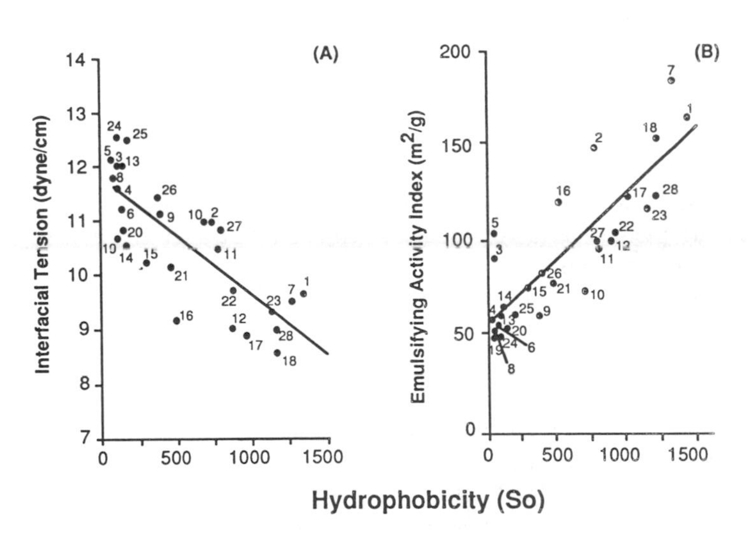

FIGURE 22 Correlations of surface hydrophobicity of various proteins with (A) oil/water interfacial tension, and (B) emulsifying activity index. Surface hydrophobicity was determined from the amount of hydrophobic fluorescent probe bound per unit weight of protein. The numbers in the plots represent (1) bovine serum albumin; (2) β-lactoblobulin; (3) trypsin; (4) ovalbumin; (5) conalbumin; (6) lysozyme; (7) κ-casein; (8–12) ovalbumin denatured by heating at 85°C for 1, 2, 3, 4, or 5 min, respectively; (13–18) lysozyme denatured by heating at 85°C for 1, 2, 3, 4, 5, or 6 min, respectively; (19–23) ovalbumin bound to 0.2, 0.3, 1.7, 5.7, or 7.9 mol dodecyl sulfate/mol protein, respectively; (24–28) ovalbumin bound to 0.3, 0.9, 3.1, 4.8 or 8.2 mol linoleate/mol protein, respectively. (From Ref. 53.)

in insolublization, usually improves their emulsifying properties. This is due to increased molecular flexibility and surface hydrophobicity. The rate of unfolding at an interface depends on the flexibility of the original molecule. In the unfolded state, proteins containing free sulfhydryl groups and disulfide bonds undergo slow polymerization via disulfide–sulfhydryl interchange reaction [27]. This leads to formation of a highly viscoelastic film at the oil–water interface. Heat denaturation that is sufficient to cause insolublization impairs emulsifying properties of proteins.

6.5.3.4 Foaming Properties

Foams consist of an aqueous continuous phase and a gaseous (air) dispersed phase. Many processed foods are foam-type products. These include whipped cream, ice cream, cakes, meringue, bread, souffles, mousses, and marshmallow. The unique textural properties and mouthfeel of these products stem from the dispersed tiny air bubbles. In most of these products, proteins are the main surface-active agents that help in the formation and stabilization of the dispersed gas phase.

Generally, protein-stabilized foams are formed by bubbling, whipping, or shaking a protein solution. The foaming property of a protein refers to its ability to form a thin tenacious film at gas–liquid interfaces so that large quantities of gas bubbles can be incorporated and stabilized. Foaming properties are evaluated by several means. The *foamability* or *foaming capacity* of a protein refers to the amount of interfacial area that can be created by the protein. It can be expressed in several ways, such as *overrun*(or steady-state foam value), or *foaming power* (or foam expansion). Overrun is defined as

$$\text{Overrun} = \frac{\text{volume of foam} - \text{volume of initial liquid}}{\text{volume of initial liquid}} \times 100 \qquad (55)$$

The foaming power, FP, is expressed as

$$\text{FP} = \frac{\text{volume of gas incorporated}}{\text{volume of liquid}} \times 100 \qquad (56)$$

Foaming power generally increases with protein concentration until a maximum value is attained. It is also affected by the method used for foam formation. FP at a given protein concentration is often used as a basis for comparing the foaming properties of various proteins. The foaming powers of various proteins at pH 8.0 are given in Table 15.

Foam stability refers to the ability of protein to stabilize foam against gravitational and mechanical stresses. Foam stability is often expressed as the time required for 50% of the liquid to drain from a foam or for a 50% reduction in foam volume. These are very empirical methods, and they do not provide fundamental information about the factors that affect foam stability. The most direct measure of foam stability is the reduction in foam interfacial area as a function of time. This can be done as follows. According to the Laplace principle, the internal pressure of a bubble is greater than the external (atmospheric) pressure, and under stable conditions the pressure difference, ΔP, is

$$\Delta P = P_{\mathrm{i}} - P_{\mathrm{o}} = \frac{4\gamma}{r} \qquad (57)$$

where P_{i} and P_{o} are the internal and external pressures, respectively, r is radius of the foam bubble, and γ is surface tension. According to this equation the pressure inside a closed vessel containing foam will increase when the foam collapses. The net change in the pressure is [79]:

TABLE 15 Comparative Foaming Power of Protein Solutions

Protein type	Foaming power[a] at 0.5% protein concentration (w/v)
Bovine serum albumin	280%
Whey protein isolate	600%
Egg albumen	240%
Ovalbumin	40%
Bovine plasma	260%
β-Lactoglobulin	480%
Fibrinogen	360%
Soy protein (enzyme hydrolyzed)	500%
Gelatin (acid-processed pigskin)	760%

[a]Calculated according to Equation 56.
Source: Ref. 89.

$$\Delta P = \frac{-2\gamma \Delta A}{3V} \tag{58}$$

where V is the total volume of the system, ΔP is the pressure change, and ΔA is the net change in interfacial area resulting from the fraction of collapsed foam. The initial interfacial area of the foam is given by

$$A_0 = \frac{3V \Delta P_\infty}{2\gamma} \tag{59}$$

where ΔP_∞ is the net pressure change when the entire foam is collapsed. The A_0 value is a measure of foamability, and the rate of decrease of A with time can be used as a measure of foam stability. This approach has been used to study the foaming properties of food proteins [117,119].

The *strength* or *stiffness* of the foam refers to the maximum weight a column of foam can withstand before it collapses. This property is also assessed by measuring foam viscosity.

6.5.3.5 Environmental Factors Influencing Foam Formation and Stability

pH

Several studies have shown that protein-stabilized foams are more stable at the isoelectric pH of the protein than at any other pH, provided there is no insolublization of the protein at pI. At or near the isoelectric pH region, the lack of repulsive interactions promotes favorable protein–protein interactions and formation of a viscous film at the interface. In addition, an increased amount of protein is adsorbed to the interface at the pI because of lack of repulsion between the interface and adsorbing molecules. These two factors improve both foamability and foam stability. If the protein is sparingly soluble at pI, as most food proteins are, then only the soluble protein fraction will be involved in foam formation. Since the concentration of this soluble fraction is very low, the amount of foam formed will be less, but the stability will be high. Although the insoluble fraction does not contribute to foamability, adsorption of these insoluble protein particles may stabilize the foam, probably by increasing cohesive forces in the protein film. Generally, adsorption of hydrophobic particles increases the stability of foams. At pH other than pI, foamability of proteins is often good, but foam stability is poor. Egg-white proteins exhibit good foaming properties at pH 8–9 and at their isoelectric pH 4–5.

Salts

The effects of salts on the foaming properties of proteins depends on the type of salt and the solubility characteristics of the protein in that salt solution. The foamability and foam stability of most globular proteins, such as bovine serum albumin, egg albumin, gluten, and soy proteins, increase with increasing concentration of NaCl. This behavior is usually attributed to neutralization of charges by salt ions. However, some proteins, such as whey proteins, exhibit the opposite effect: Both foamability and foam stability decrease with increasing concentration of NaCl (Table 16) [120]. This is attributed to salting-in of whey proteins, especially β-lactoglobulin, by NaCl. Proteins that are salted out in a given salt solution generally exhibit improved foaming properties, whereas those that are salted in generally exhibit poor foaming properties. Divalent cations, such as Ca^{2+} and Mg^{2+}, dramatically improve both foamability and foam stability at 0.02–0.4 *M* concentration. This is primarily due to cross-linking of protein molecules and creation of films with better viscoelastic properties [121].

Sugars

Addition of sucrose, lactose and other sugars to protein solutions often impairs foamability, but improves foam stability. The positive effect of sugars on foam stability is due to increased bulk-phase viscosity, which reduces the rate of drainage of the lamella fluid. The depression in foam overrun is mainly due to enhanced stability of protein structure in sugar solutions. Because of this, the protein molecule is less able to unfold upon adsorption at the interface. This decreases the ability of the protein to produce large interfacial areas and large foam volume during whipping. In sugar-containing, foam-type dessert products, such as meringues, soufflés, and cakes, it is preferable to add sugar after whipping when possible. This will enable the protein to adsorb, unfold, and form a stable film, and then the added sugar will increase foam stability by increasing the viscosity of the lamella fluid.

Lipids

Lipids, especially phospholipids, when present at concentrations greater than 0.5%, markedly impair the foaming properties of proteins. Because lipids are more surface-active than proteins, they readily adsorb at the air–water interface and inhibit adsorption of proteins during foam formation. Since lipid films lack the cohesive and viscoelastic properties necessary to withstand

TABLE 16 Effect of NaCl on Foamability and Stability of Whey Protein Isolate

NaCl concentration (*M*)	Total interfacial area (cm^2/ml of foam)	Time for 50% collapse of initial area (sec)
0.00	333	510
0.02	317	324
0.04	308	288
0.06	307	180
0.08	305	165
0.10	287	120
0.15	281	120

Source: Compiled from Ref. 120.

the internal pressure of the foam bubbles, the bubbles rapidly expand, then collapse during whipping. Thus, lipid-free whey protein concentrates and isolates, soy proteins, and egg proteins without egg yolk display better foaming properties than do lipid-contaminated preparations.

Protein Concentration

Several properties of foams are influenced by protein concentration. The greater the protein concentration, the stiffer the foam. Foam stiffness results from small bubble size and high viscosity. The stability of the foam is enhanced by greater protein concentration because this increases viscosity and facilitates formation of a multilayer, cohesive protein film at the interface. Foamability generally reaches a maximum value at some point during an increase in protein concentration. Some proteins, such as serum albumin, are able to form relatively stable foams at 1% protein concentration, whereas proteins such as whey protein isolate and soy conglycinin require a minimum of 2–5% to form a relatively stable foam. Generally, most proteins display maximum foamability at 2–8% concentration. The interfacial concentration of proteins in foams is about 2–3 mg/m^2.

Partial heat denaturation improves the foaming properties of proteins. For instance, heating of whey protein isolate (WPI) at 70°C for 1 min improves, whereas heating at 90°C for 5 min decreases foaming properties even though the heated proteins remain soluble in both instances [119]. The decrease in foaming properties of WPI heated at 90°C was attributed to extensive polymerization of the protein via disulfide–sulfhydryl interchange reactions. These very high-molecular-weight polymers are unable to adsorb to the air–water interface during foaming.

The method of foam generation influences the foaming properties of proteins. Air introduction by bubbling or sparging usually results in a "wet" foam with a relatively large bubble size. Whipping at moderate speed generally results in foam with small-sized bubbles because the shearing action results in partial denaturation of the protein before adsoprtion occurs. However, whipping at high shear rate or "overbeating" can decrease foaming power because of aggregation and precipitation of proteins.

Some foam-type food products, such as marshmallow, cakes, and bread, are heated after the foam is formed. During heating, expansion of air and decreased viscosity can cause bubble rupture and collapse of the foam. In these instances, the integrity of the foam depends on gelation of the protein at the interface so sufficient mechanical strength is developed to stabilize the foam. Gelatin, gluten, and egg white, which display good foaming and gelling properties, are highly suitable for this purpose.

6.5.3.6 Molecular Properties Influencing Foam Formation and Stability

For a protein to perform effectively as a foaming agent or an emulsifier it must meet the following basic requirements: (a) it must be able to rapidly adsorb to the air–water interface, (b) it must readily unfold and rearrange at the interface, and (c) it should be able to form a viscous cohesive film through intermolecular interactions. The molecular properties that affect foaming properties are molecular flexibility, charge density and distribution, and hydrophobicity.

The free energy of an air–water interface is significantly greater than that of an oil–water interface. Therefore, to stabilize an air–water interface, the protein must have the ability to rapidly adsorb to the freshly created interface, and instantaneously decrease the interfacial tension to a low level. The lowering of interfacial tension is dependent on the ability of the protein to rapidly unfold, rearrange, and expose hydrophobic groups at the interface. Studies have shown that β-casein, which is a random-coil-type protein, performs in this manner. On the other hand, lysozyme, which is a tightly folded globular protein with four intramolecular disulfide bonds, adsorbs very slowly, only partially unfolds, and reduces the surface tension only slightly (Fig. 23) [115]. Lysozyme is, therefore, a poor foaming agent. Thus, molecular flexibility at the interface is quintessential for good performance as a foaming agent.

Apart from molecular flexibility, hydrophobicity plays an important role in foamability of proteins. The foaming power of proteins is positively correlated with mean hydrophobicity

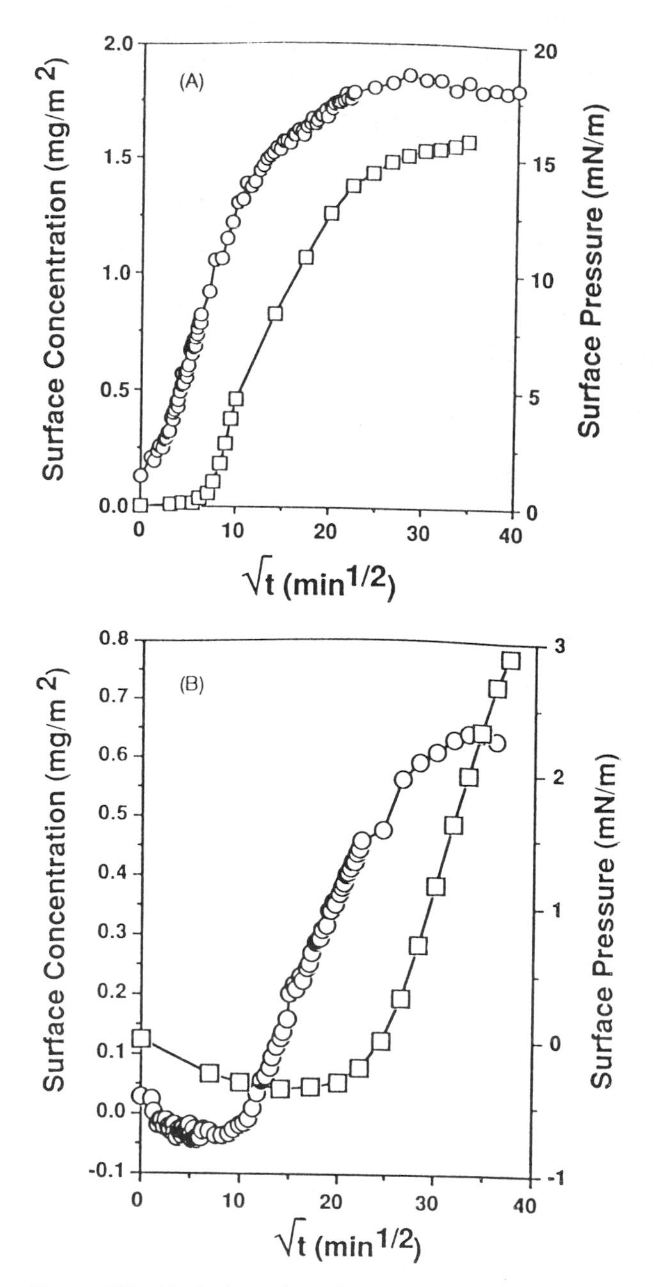

FIGURE 23 Variation of surface concentration (○) and surface pressure (□) with time during adsorption of β-casein (A) and egg-white lysozyme (B) at the air–water interface from a 1.5-μg/ml protein solution. (From Ref. 115.)

(Fig. 24A). However, the foaming power of proteins varies curvilinearly with surface hydrophobicity, and a significant correlation does not exist between these two properties at hydrophobicity values of greater than 1000 [54]. This indicates that a surface hydrophobicity of at least 1000 is needed for initial adsorption of proteins at the air–water interface, whereas, once adsorbed, the ability of the protein to create more interfacial area during foam formation depends on the mean hydrophobicity of the protein.

A protein that displays good foamability need not be a good foam stabilizer. For example, although β-casein exhibits excellent foamability, the stability of the foam is poor. On the other hand, lysozyme exhibits poor foamability, but its foams are very stable. Generally, proteins that possess good foaming power do not have the ability to stabilize a foam, and proteins that produce stable foams often exhibit poor foaming power. It appears that foamability and stability are influenced by two different sets of molecular properties of proteins that are often antagonistic. Whereas foamability is affected by rate of adsorption, flexibility, and hydrophobicity, stability depends on the rheological properties of the protein film. The rheological properties of films depend on hydration, thickness, protein concentration, and favorable intermolecular interactions. Proteins that only partially unfold and retain some degree of folded structure usually form thicker, denser films, and more stable foams (e.g., lysozyme and serum albumin) than do those that completely unfold (e.g., β-casein) at the air–water interface. In the former case, the folded structure extends into the subsurface in the form of loops. Noncovalent interactions, and possibly disulfide cross-linking, between these loops promote formation of a gel network, which has excellent viscoelastic and mechanical properties. For a protein to possess good foamability and foam stability it should have an appropriate balance between flexibility and rigidity, should easily undergo unfolding, and should engage in abundant cohesive interactions at the interface. However, what extent of unfolding is desirable for a given protein is difficult, if not impossible, to predict. In addition to these factors, foam stability usually exhibits an inverse relationship with the charge density of proteins (Fig. 24B). High charge density apparently interferes with formation of cohesive film.

Most food proteins are mixtures of various proteins, and therefore their foaming properties are influenced by interaction between the protein components at the interface. The excellent whipping properties of egg white are attributed to interactions between its protein components, such as ovalbumin, conalbumin, and lysozyme. Several studies have indicated that the foaming properties of acidic proteins can be improvd by mixing them with basic proteins, such as lysozyme and clupeine [89]. This enhancing effect seems to be related to the formation of an electrostatic complex between the acidic and basic proteins.

Limited enzymatic hydrolysis of proteins generally improves their foaming properties. This is because of increased molecular flexibility and greater exposure of hydrophobic groups. However, extensive hydrolysis impairs foamability because low-molecular-weight peptides cannot form a cohesive film at the interface.

6.5.4 Flavor Binding

Proteins themselves are odorless. However, they can bind flavor compounds, and thus affect the sensory properties of foods. Several proteins, especially oilseed proteins and whey protein concentrates, carry undesirable flavors, which limits their usefulness in food applications. These off flavors are due mainly to aldehydes, ketones, and alcohols generated by oxidation of unsaturated fatty acids. Upon formation, these carbonyl compounds bind to proteins and impart characteristic off flavors. For example, the beany and grassy flavor of soy protein preparations is attributed to the presence of hexanal. The binding affinity of some of these carbonyls is so strong that they resist even solvent extraction. A basic understanding of the mechanism of

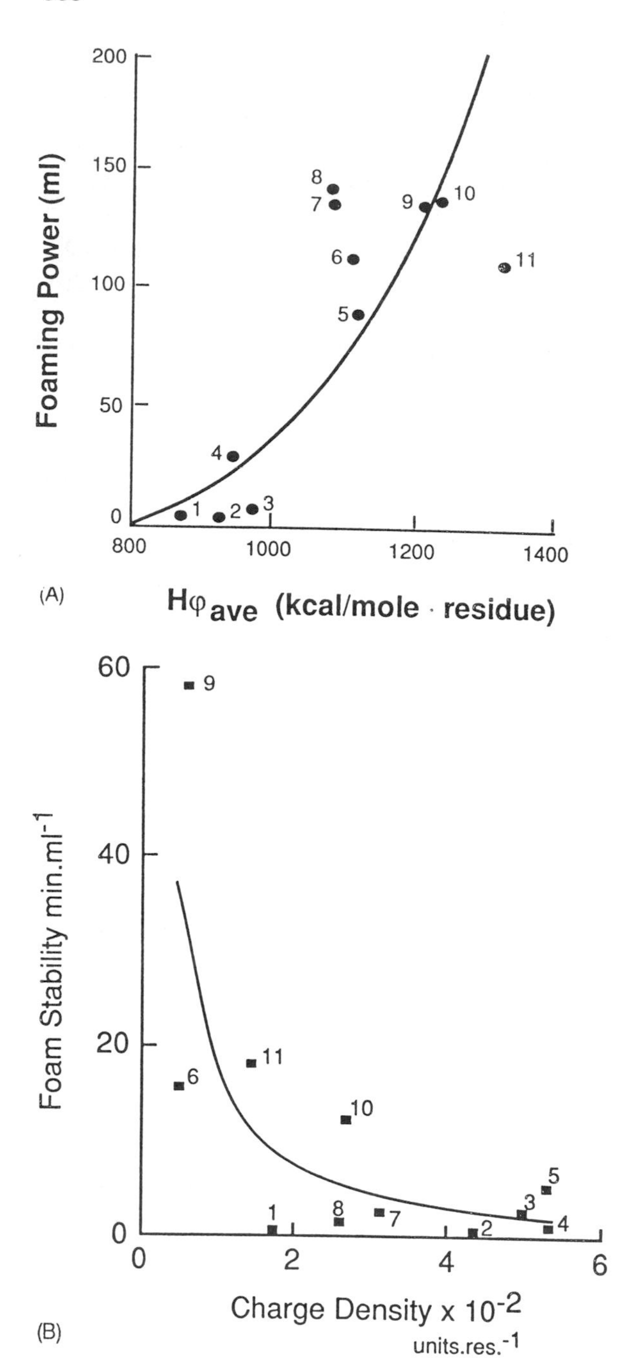

FIGURE 24 Correlations between foaming power and average hydrophobicity (A), and foam stability and charge density (B) of proteins. See original article for identities of the proteins; 1 kcal = 4.18 kJ. (From Ref. 106.)

binding of off flavors to proteins is needed so that appropriate methods can be developed for their removal.

The flavor-binding property of proteins also has desirable aspects, because they can be used as flavor carriers or flavor modifiers in fabricated foods. This is particularly useful in meat analogues containing plant proteins, where successful simulation of a meat-like flavor is essential for consumer acceptance. In order for a protein to function as a good flavor carrier, it should bind flavors tightly, retain them during processing, and release them during mastication of food in the mouth. However, proteins do not bind all flavor compounds with equal affinity. This leads to uneven and disproportionate retention of some flavors and undesirable losses during processing. Because protein-bound flavorants do not contribute to taste and aroma unless they are released readily in the mouth, knowledge of the mechanisms of interaction and binding affinity of various flavorants is essential if effective strategies for producing flavor–protein products or for removing off flavors from protein isolates are to be devised.

6.5.4.1 Thermodynamics of Protein–Flavor Interactions

In water–flavor model systems, addition of proteins has been shown to reduce the headspace concentration of flavor compounds [35]. This is attributed to binding of flavors to proteins. The mechanism of flavor binding to proteins depends upon the moisture content of the protein sample, but interactions are normally noncovalent. Dry protein powders bind flavors mainly via van der Waals, hydrogen bonding, and electrostatic interactions. Physical entrapment within capillaries and crevices of dry protein powders may also contribute to flavor properties of dry protein powders. In liquid or high-moisture foods, the mechanism of flavor binding by proteins primarily involves interaction of the nonpolar ligand with hydrophobic patches or cavities on the protein surface. In addition to hydrophobic interactions, flavor compounds with polar head groups, such as hydroxyl and carboxyl groups, may also interact with proteins via hydrogen bonding and electrostatic interactions. After binding to the surface hydrophobic regions, aldehydes and ketones may be able to diffuse into the hydrophobic interior of the protein molecule.

Flavor interactions with proteins are usually completely reversible. However, aldehydes can covalently bind to the amino group of lysyl side chains, and this interaction is nonreversible. Only the noncovalently bound fraction can contribute to aroma and taste of the protein product.

The extent of flavor binding by hydrated proteins depends on the number of hydrophobic binding regions available on the protein surface. The binding sites are usually made up of groups of hydrophobic residues segregated in the form of a well-defined cavity. Single nonpolar residues on the protein surface are less likely to act as binding sites. Under equilibrium conditions, the reversible noncovalent binding of a flavor compound with proteins follows the Scatchard equation,

$$\frac{\nu}{[\mathrm{L}]} = n\mathrm{K} - \nu K \tag{60}$$

where ν is moles of ligand bound per mole of protein, n is the total number of binding sites per mole of protein, [L] is the free ligand concentration at equilibrium, and K is the equilibrium binding constant (M^{-1}). According to this equation, a plot of ν/[L] versus ν will be a straight line; the values of K and n can be obtained from the slope and the intercept, respectively. The free energy change for binding of ligand to protein is obtained from the equation $\Delta G = -RT \ln K$, where R is the gas constant and T is absolute temperature. The thermodynamic constants for the binding of carbonyl compounds to various proteins are presented in Table 17. The binding constant increases by about threefold for each methylene group increment in chain length, with

TABLE 17 Thermodynamic Constants for Binding of Carbonyl Compounds to Proteins

Protein	Carbonyl compound	n (mol/mol)	K (M^{-1})	ΔG (kJ/mol)
Serum albumin	2-Nonanone	6	1800	−18.4
	2-Heptanone	6	270	−13.8
β-Lactoglobulin	2-Heptanone	2	150	−12.4
	2-Octanone	2	480	−15.3
	2-Nonanone	2	2440	−19.3
Soy protein				
Native	2-Heptanone	4	110	−11.6
	2-Octanone	4	310	−14.2
	2-Nonanone	4	930	−16.9
	5-Nonanone	4	541	−15.5
	Nonanal	4	1094	−17.3
Partially denatured	2-Nonanone	4	1240	−17.6
Succinylated	2-Nonanone	2	850	−16.7

Note: n, number of binding sites in native state; K, equilibrium binding constant.
Source: Refs. 24, 25, 83.

a corresponding free energy change of −2.3 kJ/mol per CH_2 group. This indicates that the binding is hydrophobic in nature.

It is assumed in the Scatchard relationship that all ligand binding sites in a protein have the same affinity, and that no conformational changes occur upon binding of the ligand to these sites. Contrary to the latter assumption, proteins generally do undergo a conformational change upon binding of flavor compounds. Diffusion of flavor compounds into the interior of the protein may disrupt hydrophobic interactions between protein segments, and thus destabilize the protein structure. Flavor ligands with reactive groups, such as aldehydes, can covalently bind the ε-amino groups of lysyl residues, change the net charge of the protein, and thus cause protein unfolding. Unfolding generally results in exposure of new hydrophobic sites for ligand binding. Because of these structural changes, Scatchard plots for protein are generally curvilinear. In the case of oligomeric proteins, such as soy proteins, conformational changes may involve both dissociation and unfolding of subunits. Denatured proteins generally exhibit a large number of binding sites with weak association constants. Methods for measuring flavor binding can be found in References 24 and 25.

6.5.4.2 Factors Influencing Flavor Binding

Since volatile flavors interact with hydrated proteins mainly via hydrophobic interactions, any factor that affects hydrophobic interactions or surface hydrophobicity of proteins will influence flavor binding. Temperature has very little effect on flavor binding, unless significant thermal unfolding of the protein occurs. This is because the association process is primarily entropy driven, not enthalpy driven. Thermally denatured proteins exhibit increased ability to bind flavors; however, the binding constant is usually low compared to that of native protein. The effects of salts on flavor binding are related to their salting-in and salting-out properties. Salting-in-type salts, which destabilize hydrophobic interactions, decrease flavor binding, whereas salting-out-type salts increase flavor binding. The effect of pH on flavor binding is generally related to pH-induced conformational changes in proteins. Flavor binding is usually enhanced more at alkaline pH that at acid pH; this is because proteins tend to denature more extensively at alkaline pH than at acid pH.

Breakage of protein disulfide bonds, which causes unfolding of proteins, usually increases flavor binding. Extensive proteolysis, which disrupts and decreases the number of hydrophobic regions, decreases flavor binding. This can be used as a way of removing off flavors from oilseed proteins. However, protein hydrolysis sometimes liberates bitter peptides. Bitterness of peptides is often related to hydrophobicity. Peptides with a mean hydrophobicity of less than 5.3 kJ/mol do not have a bitter taste. On the other hand, peptides with a mean hydrophobicity of greater than 5.85 kJ/mol often are bitter. Formation of bitter peptides in protein hydrolysates depends on the amino acid composition and sequence, and the type of enzyme used. Caseins and soy proteins hydrolyzed with several commercial proteases result in bitter peptides. The bitterness can be reduced or eliminated by using *endo*- and *exo*-peptidases, which further breakdown bitter peptides into fragments that have less than 5.3 kJ/mol mean hydrophobicity.

6.5.5 Viscosity

The consumer acceptability of several liquid and semisolid-type foods (e.g., gravies, soups, beverages, etc.) depends on the viscosity or consistency of the product. The viscosity of a solution relates to its resistance to flow under an applied force (or shear stress). For an ideal solution, the shear stress (i.e, force per unit area, F/A) is directly proportional to the shear rate (i.e., the velocity gradient between the layers of the liquid, dv/dr). This is expressed as

$$\frac{F}{A} = \eta \frac{dv}{dr} \tag{61}$$

The proportionality constant η is known as the viscosity coefficient. Fluids that obey this expression are called Newtonian fluids.

The flow behavior of solutions is greatly influenced by solute type. High-molecular-weight soluble polymers greatly increase viscosity even at very low concentrations. This again depends on several molecular properties such as size, shape, flexibility, and hydration. Solutions of randomly coiled macromolecules display greater viscosity than do solutions of compact folded macromolecules of the same molecular weight.

Most macromolecular solutions, including protein solutions, do not display Newtonian behavior, especially at high protein concentrations. For these systems, the viscosity coefficient decreases when the shear rate increases. This behavior is known as pseudoplastic or shear thinning, and follows the relationship

$$\frac{F}{A} = m\left(\frac{dv}{dr}\right)^n \tag{62}$$

where m is the consistency coefficient and n is an exponent known as the "flow behavior index." The pseudoplastic behavior of protein solutions arises because of the tendency of protein molecules to orient their major axes in the direction of flow. Dissociation of weakly held dimers and oligomers into monomers also contributes to shear thinning. When shearing or flow is stopped, the viscosity may or may not return to the original value, depending on the rate of relaxation of the protein molecules to random orientation. Solutions of fibrous proteins, such as gelatin and actomyosin, usually remain oriented, and thus do not quickly regain their original viscosity. On the other hand, solutions of globular proteins, such as soy proteins and whey proteins, rapidly regain their viscosity when flow is stopped. Such solutions are called *thixotropic*.

The viscosity (or consistency) coefficient of most protein solutions follows an exponential relationship with protein concentration because of both protein–protein interactions and interactions between the hydration spheres of protein molecules. An example involving soy protein fractions is shown in Figure 25 [91]. At high protein concentrations or in protein gels, where protein–protein interactions are numerous and strong, proteins display plastic viscoelastic behavior. In these cases, a specific amount of force, known as "yield stress," is required to initiate flow.

The viscosity behavior of proteins is a manifestation of complex interactions among several variables, including size, shape, protein–solvent interactions, hydrodynamic volume, and molecular flexibility in the hydrated state. When dissolved in water, proteins absorb water and swell. The volume of the hydrated molecules, that is, their hydrodynamic size or volume, is much larger than their unhydrated size or volume. The protein-associated water induces long-range effects on the flow behavior of the solvent. The dependence of viscosity on shape and size of protein molecules follows the relationship

$$\eta_{sp} = \beta C(v_2 + \delta_1 v_1) \tag{63}$$

where η_{sp} is specific viscosity, β is the shape factor, C is concentration, v_2 and v_1 are specific volumes of unhydrated protein and solvent, respectively, and δ_1 is grams of water bound per

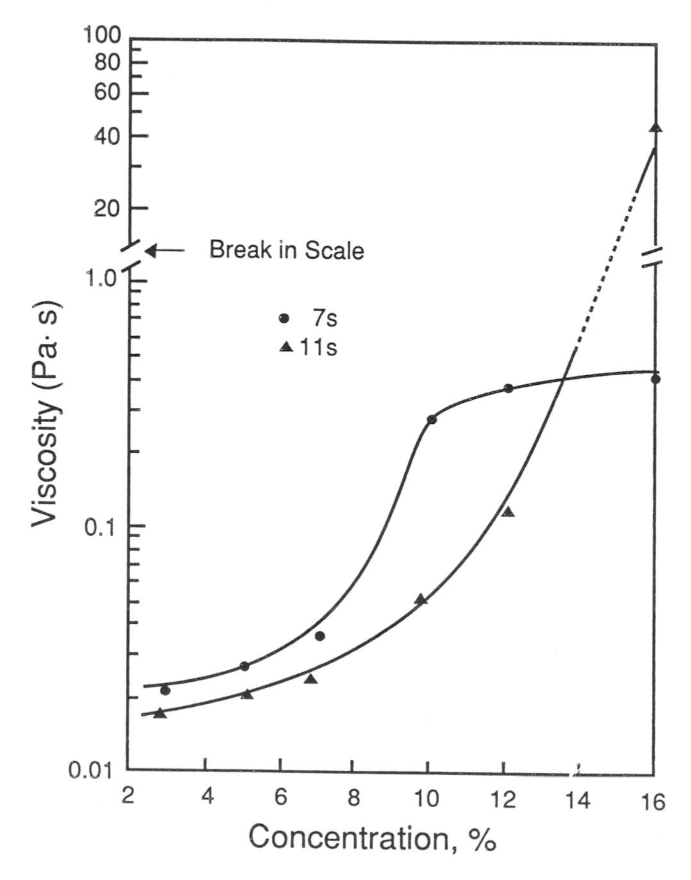

FIGURE 25 Effect of concentration on viscosity (or consistency index) of 7S and 11S soy protein solutions at 20°C (From Ref. 91.)

gram of protein. Here, v_2 is also related to molecular flexibility; the greater the specific volume of the protein, the greater its flexibility.

The viscosity of dilute protein solutions is expressed in several ways. *Relative viscosity*, η_{rel} refers to the ratio of viscosity of the protein solution to that of the solvent. It is measured in an Ostwald–Fenske type capillary viscometer, and is expressed as

$$\eta_{rel} = \frac{\eta}{\eta_o} = \frac{\rho t}{\rho_o t_o} \tag{64}$$

where ρ and ρ_o are densities of protein solution and solvent, respectively, and t and t_o are times of flow for a given volume of protein solution and solvent, respectively, through the capillary. Other forms of expressing viscosity can be obtained from the relative viscosity. *Specific viscosity* is defined as

$$\eta_{sp} = \eta_{rel} - 1 \tag{65}$$

Reduced viscosity is

$$\eta_{red} = \frac{\eta_{sp}}{C} \tag{66}$$

where C is the protein concentration, and *intrinsic viscosity* is

$$[\eta] - \text{Lim} \frac{\eta_{sp}}{C} \tag{67}$$

The intrinsic viscosity, $[\eta]$, is obtained by extrapolating a plot of reduced viscosity versus protein concentration to zero protein concentraton (Lim). Since protein–protein interactions are nonexistent at infinite dilution, intrinsic viscosity accurately depicts the effects of shape and size on the flow behavior of individual protein molecules. Changes in the hydrodynamic shape of proteins that result from heat and pH treatments can be studied by measuring their intrinsic viscosities.

6.5.6 Gelation

A gel is an intermediate phase between a solid and a liquid. Technically, it is defined as "a substantially diluted system which exhibits no steady state flow" [33]. It is made up of polymers cross-linked via either covalent or noncovalent bonds to form a network that is capable of entrapping water and other low-molecular-weight substances (see Chap. 3).

Protein gelation refers to transformation of a protein from the "sol" state to a "gel-like" state. This transformation is facilitated by heat, enzymes, or divalent cations under appropriate conditions. All these agents induce formation of a network structure; however, the types of covalent and noncovalent interactions involved, and the mechanism of network formation can differ considerably.

Most food protein gels are prepared by heating a protein solution. In this mode of gelation, the protein in a sol state is first transformed into a "progel" state by denaturation. The progel state is usually a viscous liquid state in which some degree of protein polymerization has already occurred. This step causes unfolding of the protein and exposure of a critical number of functional groups, such as hydrogen bonding and hydrophobic groups, so that the second stage, formation of a protein network, can occur. Creation of the progel is irreversible because many protein–protein interactions occur between the unfolded molecules. When the progel is cooled to ambient or refrigeration temperature, the decrease in the thermal kinetic energy facilitates formation of stable noncovalent bonds among exposed functional groups of the various molecules and this constitutes gelation.

The interactions involved in network formation are primarily hydrogen bonds, and hydrophobic and electrostatic interactions. The relative contributions of these forces vary with the type of protein, heating conditions, the extent of denaturation, and environmental conditions. Hydrogen bonding and hydrophobic interactions contribute more than electrostatic interactions to network formation except when multivalent ions are involved in cross-linking. Some proteins generally carry a net charge, electrostatic repulsion occurs among protein molecules and this is not usually conducive to network formation. However, charged groups are essential for maintaining protein–water interactions and water holding capacity of gels.

Gel networks that are sustained primarily by noncovalent interactions are thermally reversible; that is, upon reheating they will melt to a progel state, as is commonly observed with gelatin gels. This is especially true when hydrogen bonds are the major contributor to the network. Since hydrophobic interactions are strong at elevated temperatures, gel networks formed by hydrophobic interactions are irreversible, such as egg-white gels. Proteins that contain both cysteine and cystine groups can undergo polymerization via sulfhydryl–disulfide interchange reactions during heating, and form a continuous covalent network upon cooling. Such gels are usually thermally irreversible. Examples of gels of this type are ovalbumin, β-lactoglobulin, and whey protein gels.

Proteins form two types of gels, namely, coagulum (opaque) gels and translucent gels. The type of gel formed by a protein is dictated by its molecular properties and solution conditions. Proteins containing large amounts of nonpolar amino acid residues undergo

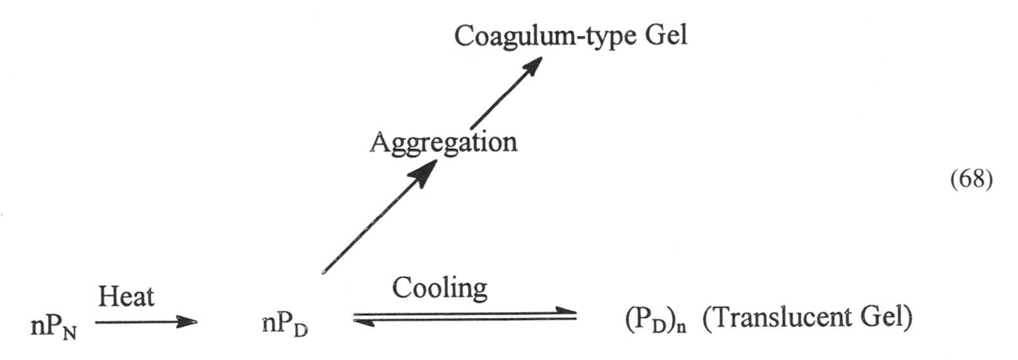

(68)

(P_N is native state, P_D is unfolded state, and *n* is the number of protein molecules taking part in cross-linking)

hydrophobic aggregation upon denaturation. These insoluble aggregates then randomly associate and set into an irreversible coagulum-type gel. Since the rate of aggregation and network formation is faster than the rate of denaturation, proteins of this type readily set into a gel network even while being heated. The opaqueness of these gels is due to light scattering caused by the unordered network of insoluble protein aggregates.

Proteins that contain small amounts of nonpolar amino acid residues form soluble complexes upon denaturation. Since the rate of association of these soluble complexes is slower than the rate of denaturation, and the gel network is predominantly formed by hydrogen bonding interactions, they often do not set into a gel until heating followed cooling has occurred (8–12% protein concentration assumed). Upon cooling, the slow rate of association of the soluble complexes facilitates formation of an ordered translucent gel network.

At the molecular level, coagulum-type gels tend to form when the sum of Val, Pro, Leu, Ile, Phe, and Trp residues of the protein exceeds 31.5 mole percent [100]. Those that contain less than 31.5 mole percent of the above hydrophobic residues usually form translucent gels when water is the solvent. There are several exceptions to this empirical rule. For example, the hydrophobic amino acid content of β-lactoblobulin is 32 mole percent, yet it forms a translucent gel in water. However, when NaCl is included, it forms a coagulum-type gel even when the salt concentration is as low as 50 m*M*. This occurs because of charge neutralization by NaCl, which promotes hydrophobic aggregation upon heating. Thus, gelation mechanism and gel appearance are fundamentally controlled by the balance between attractive hydrophobic interactions and repulsive electrostatic interactions. These two forces in effect control the balance between protein–protein and protein–solvent interactions in a gelling system. If the former is much greater than the latter, a precipitate or a coagulum is likely to form. If protein–solvent interactions predominate, the system may not gel. A coagulum gel or a translucent gel results when the hydrophobic and hydrophilic forces are somewhere in-between these two extremes.

Protein gels are highly hydrated systems, containing up to 98% water. The water entrapped in these gels has chemical potential (activity) similar to that in dilute aqueous solutions, but lacks fluidity and cannot be easily squeezed out. The mechanisms by which liquid water can be held in a nonflowable semisolid state in gels are not well understood. The fact that translucent gels, formed primarily by hydrogen-bonding interactions, hold more water than coagulum-type gels and are less prone to syneresis suggests that much of the water is hydrogen bonded to C=O and N-H groups of the peptide bonds, is associated with charged groups in the form of hydration shells, and/or exists in extensively hydrogen-bonded water–water networks. It is also possible that within the restricted environment of each cell of the gel structure, water may act as an hydrogen-bonding cross-linker between C=O and N-H groups of peptide segments (see Chap. 2). This may restrict the flowability of water within each cell, the more so as the cell size decreases. It is also likely that some water may be held as capillary water in the pores of the gel structure, especially in coagulum gels.

The stability of a gel network against thermal and mechanical forces is dependent on the number of types of cross-links formed per monomer chain. Thermodynamically, a gel network would be stable only when the sum of the interaction energies of a monomer in the gel network is greater than its thermal kinetic energy. This is dependent on several intrinsic (such as the size, net charge, etc.) and extrinsic factors (such as pH, temperature, ionic strength, etc.). The square root of the hardness of protein gels exhibits a linear relationship with molecular weight [110]. Globular proteins with molecular weight < 23,000 cannot form a heat-induced gel at any reasonable protein concentration, unless they contain at least one free sulfhydryl group or a disulfide bond. The sulfhydryl groups and disulfide bonds facilitate polymerization, and thus increase the effective molecular weight of polypeptides to > 23,000. Gelatin preparations with effective molecular weights of less than 20,000 cannot form a gel.

Another critical factor is protein concentration. To form a self-standing gel network, a minimum protein concentration, known as least concentration endpoint (LCE), is required. The LCE is 8% for soy proteins, 3% for egg albumin, and about 0.6% for gelatin. Above this mimimum concentration, the relationship between gel strength, *G*, and protein concentration, *C*, usually follows a power law,

$$G \propto (C - C_0)^n \tag{69}$$

where C_0 is the LCE. For proteins, the value of n varies from 1 to 2.

Several environmental factors, such as pH, salts, and other additives, also affect gelation of proteins. At or near isoelectric pH, proteins usually form coagulum-type gels. At extremes of pH, weak gels are formed because of strong electrostatic repulsion. The optimum pH for gel formation is about 7–8 for most proteins.

Formation of protein gels can sometimes be facilitated by limited proteolysis. A well-known example is cheese. Addition of chymosin (rennin) to casein micelles in milk results in the formation of a coagulum-type gel. This is achieved by cleavage of κ-casein, a micelle component, causing release of a hydrophilic portion, known as the glycomacropeptide. The remaining so-called *para*-casein micelles possess a highly hydrophobic surface that facilitates formation of a weak gel network.

Enzymic cross-linking of proteins at room temperature can also result in formation of a gel network. Transglutaminase is the enzyme often used to prepare these gels. This enzyme catalyzes formation of ε-(γ-glutamyl)lysyl cross-links between the glutamine and lysyl groups of protein molecules [77]. Using this enzymic cross-linking method, highly elastic and irreversible gels can be formed even at low protein concentration.

Divalent cations, such as Ca^{2+} and Mg^{2+}, can also be used to form protein gels. These ions form cross-links between negatively charged groups of protein molecules. A good example of this type of gel is tofu from soy proteins. Alginate gels also can be formed by this means.

6.5.7 Dough Formation [67,68]

Among food proteins, wheat protein is very unique because of its exceptional ability to form a viscoelastic dough. When a mixture of wheat flour and water (about 3:1 ratio) is kneaded, a dough with viscoelastic properties forms that is suitable for making bread and other bakery products. These unusual dough characteristics are mainly attributable to the proteins in wheat flour.

Wheat flour contains soluble and insoluble protein fractions. The soluble proteins, comprising about 20% of the total proteins, are primarily albumin- and globulin-type enzymes and certain minor glycoproteins. These proteins do not contribute to the dough-forming properties of wheat flour. The major storage protein of wheat is gluten. Gluten is a heterogeneous mixture of proteins, mainly gliadins and glutenins, with limited solubility in water. The formation of a viscoelastic dough capable of entrapping gas during fermentation is attributed entirely to the gluten proteins.

Gluten has a unique amino acid composition, with Glu/Gln and Pro accounting for more than 50% of the amino acid residues. The low water solubility of gluten is attributable to its low content of Lys, Arg, Glu, and Asp residues, which together amount to less than 10% of the total amino acid residues. About 30% of gluten's amino acid residues are hydrophobic, and the residues contribute greatly to its ability to form protein aggregates by hydrophobic interactions and to bind lipids and other nonpolar substances. The high glutamine and hydroxyl amino acid (~10%) contents of gluten is responsible for its water binding properties. In addition, hydrogen bonding between glutamine and hydroxyl residues of gluten polypeptides contributes to its cohesion–adhesion properties. Cysteine and cystine residues account for 2–3% of gluten's total amino acid residues. During formation of the dough, these residues undergo sulfhydryl–disulfide interchange reactions resulting in extensive polymerization of gluten proteins.

Several physical and chemical transformations occur during mixing and kneading of a mixture of wheat flour and water. Under the applied shear and tensile forces, gluten proteins absorb water and partially unfold. The partial unfolding of protein molecules facilitates hydrophobic interactions and sulfhydryl–disulfide interchange reactions, which result in formation of thread-like polymers. These linear polymers in turn are believed to interact with each other, presumably via hydrogen bonding, hydrophobic associations, and disulfide cross-linking, to form a sheet-like film capable of entrapping gas. Because of these transformations in gluten, the resistance of the dough increases with time until a maximum is reached, and this is followed by a decrease in resistance, indicative of a breakdown in the network structure. The breakdown involves alignment of polymers in the direction of shear and some scission of disulfide cross-links, which reduces the polymer size. The time it takes to reach maximum dough strength during kneading is used as a measure of wheat quality for bread making—a longer time indicating better quality.

The development of a viscoelastic dough is thought to be related to the extent of sulfhydryl–disulfide interchange reactions. This is supported by the fact that when reductants, such as cysteine, or sulfhydryl blocking agents, such as *N*-ethylmaleimide, are added to dough, viscoelasticity decreases greatly. On the other hand, addition of oxidizing agents, such as iodates and bromates, increase the elasticity of the dough. This implies that wheat varieties that are rich in SH and S-S groups might possess superior bread-making qualities, but this relationship is unreliable. Thus, the role of sulfhydryl–disulfide interchange reactions in the development of viscoelastic doughs is poorly understood.

Differences in bread-making qualities of different wheat cultivars may be related to the differences in the structure of gluten itself. As mentioned earlier, gluten is made up of gliadins and glutenins. Gliadins are comprised of four groups, namely α, β, γ, and ω-gliadins. In gluten these exist as single polypeptides with molecular weight ranging from 30,000 to 80,000. Although gliadins contain about 2–3% half-cystine residues, they apparently do not undergo extensive polymerization via sulfhydryl–disulfide interchange reactions. The disulfide bonds appear to remain as intramolecular disulfides during dough making. Dough made from isolated gliadins and starch is viscous but not viscoelastic.

Glutenins, on the other hand, are heterogeneous polypeptides with molecular weight ranging from 12,000 to 130,000. These are further classified into high-molecular-weight (M.W. $> 90{,}000$, HMW) and low-molecular-weight (M.W. $< 90{,}000$, LMW) glutenins. In gluten, these glutenin polypeptides are present as polymers joined by disulfide cross-links, with molecular weights ranging into the millions. Because of their ability to polymerize extensively via sulfhydryl–disulfide interchange reactions, glutenins contribute greatly to the elasticity of dough. Therefore, an optimum ratio of gliadins and glutenins seems to be necessary to form a viscoelastic dough. Some studies have shown a significant positive correlation between HMW glutenin content and bread-making quality in some wheat varieties, but not in others. Available information indicates that a specific pattern of disulfide cross-linked association among LMW and HMW glutenins in gluten structure may be far more important to bread quality than the amount of this protein. For example, association/polymerization among LMW glutenins gives rise to a structure similar to that formed by HMW gliadin. This type of structure contributes to viscosity of the dough, but not to elasticity. In contrast, if LMW glutenins link to HMW glutenins via disulfide cross-links (in gluten), then this is believed to contribute to dough elasticity. It is possible that in good-quality wheat varieties, more of the LMW glutenins may polymerize with HMW, whereas in poor-quality wheat varieties, most of the LMW glutenins may polymerize among themselves. These differences in associated states of glutenins in gluten of various wheat varieties may be related to differences in their conformational properties, such as surface hydrophobicity, and reactivity of sulfhydryl and disulfide groups.

In summary, hydrogen bonding among amide and hydroxyl groups, hydrophobic interactions, and sulfhydryl–disulfide interchange reactions all contribute to the development of the unique viscoelastic properties of wheat dough. However, culmination of these interactions into good dough-making properties may depend on the structural properties of each protein and the proteins with which it associates in the overall gluten structure.

Because polypeptides of gluten, especially the glutenins, are rich in proline, they have very little folded structure. Whatever folded structure initially exists in gliadins and glutenins is lost during mixing and kneading. Therefore, no additional unfolding occurs during baking.

Supplementation of wheat flour with albumin- and globulin-type proteins, such as whey proteins and soy proteins, adversely affects the viscoelastic properties and baking quality of the dough. These proteins decrease bread volume by interfering with formation of the gluten network. Addition of phospholipids or other surfactants to dough counters the adverse effects of foreign proteins on loaf volume. In this case, the surfactant/protein film compensates for the impaired gluten film. Although this approach results in acceptable loaf volume, the textural and sensory qualities of the bread are less desirable than normal.

Isolated gluten is sometimes used as a protein ingredient in nonbakery products. Its cohesion–adhesion properties make it an effective binder in comminuted meat and surimi-type products.

6.6 NUTRITIONAL PROPERTIES OF PROTEINS

Proteins differ in their nutritive value. Several factors, such as content of essential amino acids and digestibility, contribute to these differences. The daily protein requirement therefore depends on the type and composition of proteins in a diet.

6.6.1 Protein Quality

The "quality" of a protein is related mainly to its essential amino acid composition and digestibility. High-quality proteins are those that contain all the essential amino acids at levels greater than the FAO/WHO/UNU [30] reference levels, and a digestibility comparable to or better than those of egg-white or milk proteins. Animal proteins are better "quality" than plant proteins.

Proteins of major cereals and legumes are often deficient in at least one of the essential amino acids. While proteins of cereals, such as rice, wheat, barley, and maize, are very low in lysine and rich in methionine, those of legumes and oilseeds are deficient in methionine and rich or adequate in lysine. Some oilseed proteins, such as peanut protein, are deficient in both methionine and lysine contents. The essential amino acids whose concentrations in a protein are below the levels of a reference protein are termed *limiting amino acids*. Adults consuming only cereal proteins or legume proteins have difficulty maintaining their health; children below 12 years of age on diets containing only one of these protein sources cannot maintain a normal rate of growth. The essential amino acid contents of various food proteins are listed in Table 18.

Both animal and plant proteins generally contain adequate or more than adequate amounts of His, Ile, Leu, Phe + Tyr, and Val. These amino acids usually are not limiting in staple foods. More often, either Lys, Thr, Trp, or the sulfur-containing amino acids are the limiting amino acids.

The nutritional quality of a protein that is deficient in an essential amino acid can be improved by mixing it with another protein that is rich in that essential amino acid. For example, mixing of cereal proteins with legume proteins provides a complete and balanced level of essential amino acids. Thus, diets containing appropriate amounts of cereals and legumes (pulses) and otherwise nutritionally complete are often adequate to support growth and maintenance. The poor-quality protein also can be nutritionally improved by supplementing it with essential free amino acids that are under-represented. Supplementation of legumes with Met and of cereals with Lys usually improves their quality.

The nutritional quality of a protein or protein mixture is ideal when it contains all of the essential amino acids in proportions that produce optimum rates of growth and/or optimum maintenance capability. The ideal essential amino acid patterns for children and adults are given in Table 19. However, because actual essential amino acid requirements of individuals in a given population vary depending on their nutritional and physiological status, the essential amino acid requirements of preschool children (age 2–5) are generally recommended as a safe level for all age groups [31].

Overconsumption of any particular amino acid can lead to "amino acid antagonism" or

TABLE 18 Essential Amino Acid Contents and Nutritional Value of Proteins from Various Sources (mg/g Protein)

	Protein source												
Property (mg/g protein)	Egg	Cow's milk	Beef	Fish	Wheat	Rice	Maize	Barley	Soybean	Field bean (boiled)	Pea	Peanut	French bean
Amino acid concentration (mg/g protein)													
His	22	27	34	35	21	21	27	20	30	26	26	27	30
Ile	54	47	48	48	34	40	34	35	51	41	41	40	45
Leu	86	95	81	77	69	77	127	67	82	71	70	74	78
Lys	70	78	89	91	23[a]	34[a]	25[a]	32[a]	68	63	71	39[a]	65
Met + Cys	57	33	40	40	36	49	41	37	33	22[b]	24[b]	32	26
Phe + Tyr	93	102	80	76	77	94	85	79	95	69	76	100	83
Thr	47	44	46	46	28	34	32[b]	29[b]	41	33	36	29[b]	40
Trp	17	14	12	11	10	11	6[b]	11	14	8[a]	9[a]	11	11
Val	66	64	50	61	38	54	45	46	52	46	41	48	52
Total essential amino acids	512	504	480	485	336	414	422	356	466	379	394	400	430
Protein content, %	12	3.5	18	19	12	7.5	—	—	40	32	28	30	30
Chemical score (%) (based on FAO/WHO [30] pattern)	100	100	100	100	40	59	43	55	100	73	82	67	
PER	3.9	3.1	3.0	3.5	1.5	2.0	—	—	2.3	—	2.65	—	—
BV (on rats)	94	84	74	76	65	73	—	—	73	—	—	—	—
NPU	94	82	67	79	40	70	—	—	61	—	—	—	—

Note: Chemical score is defined as the ratio of the amount of a limiting essential amino acid in 1 g of a test protein to the amount of the same amino acid in 1 g of a reference protein. PER, protein efficiency ratio; BV, biological value; NPU, net protein utilization.

[a]Primary limiting acid.

[b]Second limiting acid.

Source: Refs. 28 and 30.

TABLE 19 Recommended Essential Amino Acid Pattern for Food Proteins

	Recommended pattern (mg/g protein)			
Amino acid	Infant (2–5 years)	Preschool child (10–12 years)	Preschool child	Adult
Histidine	26	19	19	16
Isoleucine	46	28	28	13
Leucine	93	66	44	19
Lysine	66	58	44	16
Met + Cys	42	25	22	17
Phe + Tyr	72	63	22	19
Threonine	43	34	28	9
Tryptophan	17	11	9	5
Valine	55	35	25	13
Total	434	320	222	111

Source: Ref. 30.

toxicity. Excessive intake of one amino acid often results in an increased requirement for other essential amino acids. This is due to competition among amino acids for absorption sites on the intestinal mucosa. For example, when the Leu level is relatively high, this decreases absorption of Ile, Val, and Tyr even if the dietary levels of these amino acids are adequate. This leads to an increased dietary requirement for the latter three amino acids. Overconsumption of other essential amino acids also can inhibit growth and induce pathological conditions.

6.6.2 Digestibility

Digestibility is defined as the proportion of food nitrogen that is absorbed after ingestion. Although the content of essential amino acids is the primary indicator of protein quality, true quality also depends on the extent to which these amino acids are utilized in the body. Thus, digestibility of amino acids can affect the quality of proteins. Digestibilities of various proteins in humans are listed in Table 20. Food proteins of animal origin are more completely digested than those of plant origin. Several factors affect digestibility of proteins.

TABLE 20 Digestibility of Various Food Proteins in Humans

Protein source	Digestibility (%)	Protein source	Digestibility (%)
Egg	97	Millet	79
Milk, cheese	95	Peas	88
Meat, fish	94	Peanut	94
Maize	85	Soy flour	86
Rice (polished)	88	Soy protein isolate	95
Wheat, whole	86	Beans	78
Wheat flour, white	96	Corn, cereal	70
Wheat gluten	99	Wheat, cereal	77
Oatmeal	86	Rice cereal	75

Source: Ref. 30.

1. Protein conformation: The structural state of a protein influences its hydrolysis by proteases. Native proteins are generally less completely hydrolyzed than partially denatured ones. For example, treatment of phaseolin (a protein from kidney beans) with a mixture of proteases results only in limited cleavage of the protein, resulting in liberation of a 22,000 MW polypeptide as the main product. When heat-denatured phaseolin is treated under similar conditions, it is completely hydrolyzed to amino acids an dipeptides. Generally, insoluble fibrous proteins and extensively denatured globular proteins are difficult to hydrolyze.
2. Antinutritional factors: Most plant protein isolates and concentrates contain trypsin and chymotrypsin inhibitors (Kunitz type and Bowman–Birk type) and lectins. These inhibitors impair complete hydrolysis of legume and oilseed proteins by pancreatic proteases. Lectins, which are glycoproteins, bind to intestinal mucosa cells and interfere with absorption of amino acids. Lectins and Kunitz-type protease inhibitors are thermolabile, whereas the Bowman–Birk type inhibitor is stable under normal thermal processing conditions. Thus, heat-treated legume and oilseed proteins are generally more digestible than native protein isolates (despite some residual Bowman–Birk type inhibitor). Plant proteins also contain other antinutritional factors, such as tannins and phytate. Tannins, which are condensed products of polyphenols, covalently react with ε-amino groups of lysyl residues. This inhibits trypsin-catalyzed cleavage of the lysyl peptide bond.
3. Binding: Interaction of proteins with polysaccharides and dietary fiber also reduces the rate and completeness of hydrolysis.
4. Processing: Proteins undergo several chemical alterations involving lysyl residues when exposed to high temperatures and alkaline pH. Such alterations reduce their digestibility. Reaction of reducing sugars with ε-amino groups also decreases digestibility of lysine.

6.6.3 Evaluation of Protein Nutritive Value

Since the nutritional quality of proteins can vary greatly and is affected by many factors, it is important to have procedures for evaluating quality. Quality estimates are useful for (a) determining the amount required to provide a safe level of essential amino acids for growth and maintenance, and (b) monitoring changes in the nutritive value of proteins during food processing, so that processing conditions that minimize quality loss can be devised.

The nutritive quality of proteins can be evaluated by several biological, chemical, and enzymatic methods.

Biological Methods

Biological methods are based on weight gain or nitrogen retention in test animals when fed a protein-containing diet. A protein-free diet is used as the control. The protocol recommended by FAO/WHO [31] is generally used for evaluating protein quality. Rats are the usual test animal, although humans are sometimes used. A diet containing about 10% protein on a dry weight basis is used to ensure that the protein intake is below daily requirements. Adequate energy is supplied in the diet. Under these conditions, protein in the diet is used to the maximum possible extent for growth. The number of test animals used must be sufficient to assure results that are statistically reliable. A test period of 9 days is common. During each day of test period, the amount (g) of diet consumed is tabulated for each animal, and the feces and urine are collected for nitrogen analysis.

The data from animal feeding studies are used in several ways to evaluate protein quality.

The *protein efficiency ratio* (PER) is the weight (in grams) gained per gram protein consumed. This is a simple and commonly used expression. Another useful expression is *net protein ratio* (NPR). This is calculated as follows:

$$\text{NPR} = \frac{(\text{weight gain}) - (\text{weight loss of protein–free group})}{\text{protein ingested}} \tag{70}$$

NPR values provide information on the ability of proteins to support both maintenance and growth. Rats are generally used as test animals in these methods. Since rats grow much faster than humans, and a larger percentage of protein is used for maintenance in growing children than in rats, it is often questioned whether PER and NPR values derived from rat studies are useful for estimating human needs [94]. Although this argument is a valid one, appropriate correction procedures are available.

Another approach to evaluating protein quality involves measuring nitrogen uptake and nitrogen loss. This allows two useful protein quality parameters to be calculated. *Apparent protein digestibility* or *coefficient of protein digestibility* is obtained from the difference between the amount of nitrogen ingested and the amount of nitrogen excreted in the feces. However, since total fecal nitrogen also includes metabolic or endogenous nitrogen, correction should be made to obtain *true protein digestibility.* True digestibility (TD) can be calculated in the following manner:

$$\text{TD} = \frac{I - (F_{\text{N}} - F_{\text{k,N}})}{I} \times 100 \tag{71}$$

where I is nitrogen ingested, F_{N} is total fecal nitrogen, and $F_{\text{k,N}}$ is endogenous fecal nitrogen. $F_{\text{k,N}}$ is obtained by feeding a protein-free diet.

TD gives information on the percentage of nitrogen intake absorbed by the body. However, it does not provide information on how much of the absorbed nitrogen is actually retained or utilized by the body.

Biological value, BV, is calculated as follows:

$$\text{BV} = \frac{I - (F_{\text{N}} - F_{\text{k,N}}) - (U_{\text{N}} - U_{\text{k,N}})}{I - (F_{\text{N}} - F_{\text{k,N}})} \times 100 \tag{72}$$

where U_{N} and $U_{\text{k,N}}$ are the total and endogenous nitrogen losses, respectively, in the urine.

Net protein utilization (NPU), the percentage of nitrogen intake retained as body nitrogen, is obtained from the product of TD and BV. Thus,

$$\text{NPU} = \text{TD} \times \text{BV} = \frac{I - (F_{\text{N}} - F_{\text{k,N}}) - (U_{\text{N}} - U_{\text{k,N}})}{I} \times 100 \tag{73}$$

The PER, BVs, and NPUs of several food proteins are presented in Table 18.

Other bioassays that are occasionally used to evaluate protein quality include assays for enzyme activity, changes in the essential amino acid content of plasma, levels of urea in the plasma and urine, and rate of repletion of plasma proteins or gain in body weight of animals previously fed a protein-free diet.

Chemical Methods

Biological methods are expensive and time-consuming. Quick assessment of a protein's nutritive value can be obtained by determining its content of amino acids, and comparing this with the essential amino acid pattern of an ideal reference protein. The ideal pattern of essential amino acids in proteins (reference protein) for preschool children (2–5 years) is given in Table 19 [30], and this pattern is used as the standard for all age groups except infants. Each essential amino acid in a test protein is given a *chemical score*, which is defined as

$$\frac{\text{mg amino acid/g test protein}}{\text{mg same amino acid/g reference protein}} \times 100 \tag{74}$$

The essential amino acid that shows the lowest score is the most limiting amino acid in the test protein. The chemical score of this limiting amino acid provides the chemical score for the test protein. As mentioned earlier, Lys, Thr, Trp, and sulfur amino acids are often the limiting amino acids in food proteins. Therefore, the chemical scores of these amino acids are often sufficient to evaluate the nutritive value of proteins. The chemical score enables estimation of the amount of a test protein or protein mix needed to meet the daily requirement of the limiting amino acid. This can be calculated as follows.

$$\text{Required intake of protein} = \frac{\text{recommended intake of egg or milk protein}}{\text{chemical score of protein}} \times 100 \tag{75}$$

One of the advantages of the chemical score method is that it is simple and allows one to determine the complementary effects of proteins in a diet. This also allows one to develop high-quality protein diets by mixing various proteins suitable for various feeding programs. There are, however, several drawbacks to use of the chemical score method. An assumption underlying chemical score is that all test proteins are fully or equally digestible, and that all essential amino acids are fully absorbed. Because this assumption is often violated, correlation between results from bioassays and chemical scores is often not good. However, the correlation improves when chemical scores are corrected for protein digestibility. The apparent digestibility of proteins can be rapidly determined in vitro using a combination of three or four enzymes, such as trypsin, chymotrypsin, peptidase, and bacterial protease.

Another shortcoming of chemical score is that it does not distinguish between D- and L-amino acids. Since only L-amino acids are usable in animals, the chemical score overestimates the nutritive value of a protein, especially in proteins exposed to high pH levels, which causes racemization. The chemical score method is also incapable of predicting the negative effects of high concentrations of one essential amino acid on the bioavailability of other essential amino acids, and it does not account for the effect of antinutritional factors that might be present in the diet. Despite these major drawbacks, recent findings indicate that chemical scores, when corrected for protein digestibility, correlate well with biological assays for those proteins having biological values above 40%; when the BV is below 40%, the correlation is poor [31].

Enzymic and Microbial Methods

In vitro enzymic methods are sometimes used to measure the digestibility and release of essential amino acids. In one method, test proteins are first digested with pepsin and then with pancreatin (freeze-dried powder of pancreatic extract) [70]. In another method, proteins are digested with three enzymes, namely, pancreatic trypsin, chymotrypsin, and porcine intestinal peptidase, under standard assay conditions [31]. These methods, in addition to providing information on innate digestibility of proteins, are useful for detecting processing-induced changes in protein quality.

Growth of several microorganisms, such as *Streptococcus zymogenes, Streptococcus faecalis, Leuconostoc mesenteroides, Clostridium perfringens*, and *Tetrahymena pyriformis* (a protozoan) also have been used to determine the nutritional value of proteins [34]. Of these microorganisms, *Tetrahymena pyriformis* is particularly useful, because its amino acid requirements are similar to those of rats and humans.

6.7 PROCESSING-INDUCED PHYSICAL, CHEMICAL, AND NUTRITIONAL CHANGES IN PROTEINS

Commercial processing of foods can involve heating, cooling, drying, applicaton of chemicals, fermentation, irradiation, or various other treatments. Of these, heating is most common. This is commonly done to inactivate microorganisms, to inactivate endogenous enzymes that cause oxidative and hydrolytic changes in foods during storage, and to transform an unappealing blend of raw food ingredients into a wholesome and organoleptically appealing food. In addition, proteins such as bovine β-lactoglobulin, α-lactalbumin, and soy protein, which sometimes cause allergenic or hypersensitive responses, can sometimes be rendered innocuous in this regard. Unfortunately, the beneficial effects achieved by heating proteinaceous foods are generally accompanied by changes that can adversely affect the nutritive value and functional properties of proteins. In this section, both desirable and undesirable effects of food processing on proteins will be discussed.

6.7.1 Changes in Nutritional Quality and Formation of Toxic Compounds

6.7.1.1 Effect of Moderate Heat Treatments

Most food proteins are denatured when exposed to moderate heat treatments (60–90°C, 1 h or less). Extensive denaturation of proteins often results in insolublization, which can impair those functional properties that are dependent on solubility. From a nutritional standpoint, partial denaturation of proteins often improves the digestibility and biological availability of essential amino acids. Several purified plant proteins and egg protein preparations, even though free of protease inhibitors, exhibit poor in vitro and in vivo digestibility. Moderate heating improves their digestibility without developing toxic derivatives.

In addition to improving digestibility, moderate heat treatment also inactivates several enzymes, such as proteases, lipases, lipoxygenases, amylases, polyphenoloxidase, and other oxidative and hydrolytic enzymes. Failure to inactivate these enzymes can result in development of off-flavors, rancidity, textural changes, and discoloration of foods during storage. For instance, oilseeds and legumes are rich in lipoxygenase. During crushing or cracking of these beans for extraction of oil or protein isolates, this enzyme, in the presence of molecular oxygen, catalyzes oxidation of polyunsaturated fatty acids to initially yield hydroperoxides. The hydroperoxides subsequently decompose and liberate aldehydes and ketones, which impart off-flavor to soy flour and soy protein isolates and concentrates. To avoid off-flavor formation, it is necessary to thermally inactivate lipoxygenase prior to crushing.

Moderate heat treatment is particularly beneficial for plant proteins, because they usually contain proteinaceous antinutritional factors. Legume and oilseed proteins contain several trypsin and chymotrypsin inhibitors. These inhibitors impair efficient digestion of proteins, and thus reduce their biological availability. Furthermore, inactivation and complexation of trypsin and chymotrypsin by these inhibitors induce overproduction and secretion of these enzymes by the pancreas, which can lead to pancreatic hypertropy (enlargement of the pancreas) and pancreatic adenoma. Legume and oilseed proteins also contain lectins, which are glycoproteins. These are also known as phytohemagglutinins because they cause agglutination of red blood cells. Lectins exhibit a high binding affinity for carbohydrates. When consumed by humans, lectins impair protein digestion [90] and cause intestinal malabsorption of other nutrients. The latter consequence results from binding of lectins to membrane glycoproteins of intestinal mucosa cells, which alters their morphology and transport properties [84]. Both protease inhibitors and lectins found in plant proteins are thermolabile. Toasting of legumes and oilseeds or moist heat treatment of soy flour inactivates both lectins and protease inhibitors, improves the digestibility and PER of these proteins (Fig. 26), and prevents pancreatic hypertropy [43]. These antinutritional factors do not pose problems in home-cooked or industrially processed legumes and flour-based products when heating conditions are adequate to inactivate them.

Milk and egg proteins also contain several protease inhibitors. Ovomucoid, which possesses antitryptic activity, constitutes about 11% of egg albumen. Ovoinhibitor, which inhibits trypsin, chymotrypsin, and some fungal proteases, is present at a 0.1% level in egg albumen. Milk contains several protease inhibitors, such as plasminogen activator inhibitor (PAI) and plasmin inhibitor (PI), derived from blood. All of these inhibitors lose their activity when subjected to moderate heat treatment in the presence of water.

The beneficial effects of heat treatment also include inactivation of proteinaceous toxins, such as botulinum toxin from *Clostridium botulinum* (inactivated by heating at 100°C) and enterotoxin from *Staphylococcus aureaus*.

6.7.1.2 Compositional Changes During Extraction and Fractionation

Preparation of protein isolates from biological sources involves several unit operations, such as extraction, isoelectric precipitation, salt precipitation, thermocoagulation, and ultrafiltration/diafiltration. It is very likely that some of the proteins in the crude extract might be lost during these operations. For example, during isoelectric precipitation, some sulfur-rich albumin-type

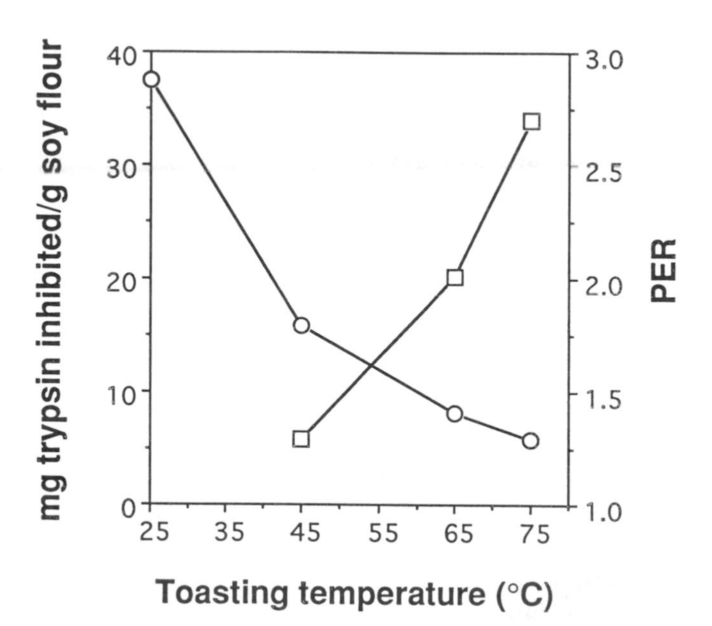

FIGURE 26 Effect of toasting on trypsin inhibitory activity (○) and PER (□) of soy flour. Adapted from Ref. 36.

proteins, which are usually soluble at isoelectric pH, might be lost in the supernatant fluid. Such losses can alter the amino acid composition and nutritional value of protein isolates compared to those of crude extracts. For example, the chemical scores of Met and Trp in crude coconut meal are about 100 and 89, respectively, whereas they are almost zero in coconut protein isolate obtained by isoelectric precipitation [42,61]. Similarly, whey protein concentrate (WPC) prepared by ultrafiltration/diafiltration and ion-exchange methods undergoes marked changes in their proteose–peptone contents. This markedly affects their foaming properties.

6.7.1.3 Chemical Alteration of Amino Acids

Proteins undergo several chemical changes when processed at high temperatures. These changes include racemization, hydrolysis, desulfuration, and deamidation. Most of these chemical changes are irreversible, and some of these reactions result in formation of amino acid types that are potentially toxic.

Racemization

Thermal processing of proteins at alkaline pH, as is done to prepare texturized foods, invariably leads to partial racemization of L-amino acid residues to D-amino acids [66]. Acid hydrolysis of proteins also causes some racemization of amino acids [32], as does roasting of proteins or protein containing foods above 200°C[45]. The mechanism at alkaline pH involves initial abstraction of the proton from the α-carbon atom by a hydroxyl ion. The resulting carbanion loses its tetrahedral asymmetry. Subsequent addition of a proton from solution can occur either from the top or bottom of the carbanion. This equal probability results in racemization of the amino acid residue [66]. The rate of racemization of a residue is affected by the electron-withdrawing power of the side chain. Thus, residues such as Asp, Ser, Cys, Glu, Phe, Asn, and Thr are racemized at a faster rate than are other amino acid residues [65]. The rate of racemization is also dependent on hydroxyl ion concentration, but is independent of protein concentration. Interestingly, the rate of racemization is about 10 times faster in proteins than in free amino acids [65], suggesting that intramolecular forces in a protein reduce the activation energy of racemization. In addition to racemization, the carbanion formed under alkaline pH also can undergo β-elimination reaction to yeild dehydroalanine. Cysteine and

$$\text{L-amino acid residue: } -NH-C(H)(CH_2R)-C(=O)- \xrightarrow{OH^{\ominus}} \text{Carbanion: } -NH-\overset{\ominus}{\ddot{C}}(CH_2R)-C(=O)- \tag{76}$$

$$\text{Carbanion} \xrightarrow{H^+} \text{D-amino acid residue: } -NH-C(H)(CH_2R)-C(=O)- \; + \text{ L-amino acid residue}$$

$$\text{Carbanion} \longrightarrow -NH-\overset{\ominus}{\ddot{C}}(CH_2-R)-C(=O)- \longrightarrow \text{Dehydroalanine: } -NH-C(=CH_2)-C(=O)-$$

phosphoserine residues display greater propensity for this route than do other amino acid residues. This is one of the reasons why a significant amount of D-cysteine is not found in alkali-treated proteins.

Racemization of amino acid residues causes a reduction in protein digestibility because the peptide bonds involving D-amino acid residues are less efficiently hydrolyzed by gastric and pancreatic proteases. This leads to loss of essential amino acids that have racemized, and impairs the nutritional value of the protein. D-Amino acids are also less efficiently absorbed through intestinal mucosa cells, and even if absorbed, they cannot be utilized in in vivo protein synthesis. Moreover, some D-amino acids, such as D-proline, have been found to be neurotoxic in chickens [16].

In addition to racemization and β-elimination reactions, heating of proteins at alkaline pH destroys several amino acid residues, such as Arg, Ser, Thr, and Lys. Arg decomposes to ornithine.

When proteins are heated above 200°C, as is commonly encountered on food surfaces during broiling, baking, and grilling, amino acid residues undergo decomposition and pyrolysis. Several of the pyrolysis products have been isolated and identified from broiled and grilled meat, and they are highly mutagenic as determined by the Ames test. The most carcinogenic/mutagenic products are formed from pyrolysis of Trp and Glu residues [14]. Pyrolysis of Trp residues gives rise to formation of carbolines and their derivatives. Mutagenic compounds are also produced in meats at moderate temperatures (190–200°C). These are known as amino-imidazoazaarenes (AIAs). One of the classes of compounds is imidazo quinolines (IQ compounds), which are condensation products of creatinine, sugars, and certain amino acids, such as Gly, Thr, Ala, and Lys [51]. The three most potent mutagens formed in broiled fish are shown here.

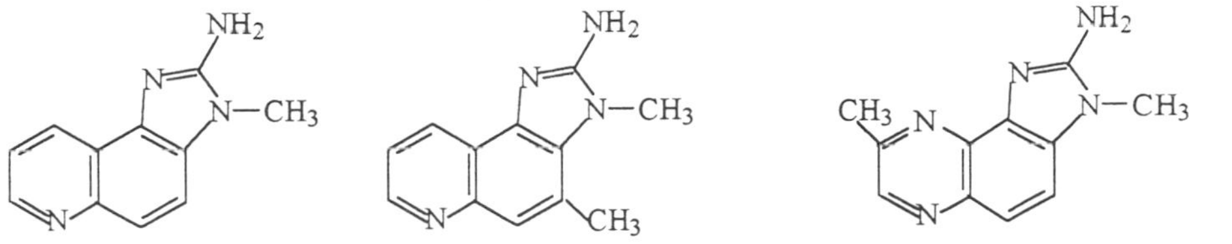

NH_2, $N-CH_3$ | NH_2, $N-CH_3$, CH_3 | CH_3, NH_2, $N-CH_3$ (77)

-Amino-3-methylimidazo-4,5-*f*]quinoline	2-Amino-3,4-dimethylimidazo-[4,5-*f*]quinoline	2-Amino-3,8-dimethylimidaz [4,5-*f*]quinoline
(IQ)	(MeIQ)	(MeIQx)

Following heating of foods according to recommended procedures, IQ compounds are generally present only at very low concentrations (microgram amounts).

6.7.1.4 Protein Cross-Linking

Several food proteins contain both intra- and intermolecular cross-links, such as disulfide bonds in globular proteins, desmosine, and isodesmosine; and di- and trityrosine-type cross-links in fibrous proteins such as keratin, elastin, resilin, and collagen. Collagen also contains ε-*N*-(γ-glutamyl)lysyl and/or ε-*N*-(γ-aspartyl)lysyl cross-links. One of the functions of these cross-links in native proteins is to minimize metabolic proteolysis. Processing of food proteins, especially at alkaline pH, also incudes cross-link formation. Such unnatural covalent bonds between polypeptide chains reduce digestibility and biological availability of essential amino acids that are involved in, or near, the cross-link.

As discussed in the previous section, heating of protein at alkaline pH, or heating above 200°C at neutral pH, results in abstraction of the proton from the α-carbon atom, resulting in formation of a carbanion. The carbanion derivative of Cys, cystine, and phosphoserine undergoes β-elimination reaction, leading to formation of a highly reactive dehydroalanine residue (DHA). DHA formation can also occur via a one-step mechanism without formation of the carbanion.

Cystine → Dehydroalanine + S + $S^{=}$ ← Cysteine (78)

Once formed, the highly reactive DHA residues react with nucleophilic groups, such as the ε-amino group of a lysyl residue, the thiol group of Cys residue, the δ-amino group of ornithine (formed by decomposition of arginine), or a histidyl residue, resulting in formation of lysinoalanine, lanthionine, ornithoalanine, and histidinylalanine cross-links, respectively, in proteins. Lysinoalanine is the major cross-link commonly found in alkali treated proteins because of the abundance of readily accessible lysyl residues.

Ornithine + DHA → Ornithoalanine; Lysine + DHA → Lysinoalanine; Cysteine + DHA → Lanthionine (79)

The formation of protein–protein cross-links in alkali-treated proteins decreases their digestibility and biological value. Both digestibility (and PER) and net protein utilization (NPU) decrease with an increase in lysinoalanine content. The decrease in digestibility is related to the inability of trypsin to cleave the peptide bond in the lysinoalanine cross-link. Moreover, the steric constraints imposed by the cross-links also prevent hydrolysis of other peptide bonds in the neighborhood of the lysinoalanine and similar cross-links. Evidence suggests that free lysinoalanine is absorbed in the intestine, but it is not utilized by the body and most of it is excreted in the urine. Some lysinoalanine is metabolized in the kidney.

Rats fed 100 ppm pure lysinoalanine or 3000 ppm protein-bound lysinoalanine exhibit nephrocytomegaly (i.e., kidney disorder). However, such nephrotoxic effects have not been observed in other animal species, such as quails, mice, hamsters, and monkeys. This has been attributed to differences in the types of metabolites formed in rats versus other animals. At levels encountered in foods, protein-bound lysinoalanine apparently does not cause nephrotoxicity in humans. Nevertheless, minimization of lysinoalanine formation during alkali processing of proteins is a desirable goal.

The lysinoalanine contents of several commercial foods are listed in Table 21. The extent of formation of lysinoalanine is dependent on pH and temperature. The higher the pH, the greater is the extent of lysinoalanine formation. High-temperature heat treatment of foods, such as milk, causes a significant amount of lysinoalanine to form even at neutral pH. Lysinoalanine formation in proteins can be minimized or inhibited by adding low-molecular-weight nucleophilic compounds, such as cysteine, ammonia, or sulfites. The effectiveness of cysteine results because the

TABLE 21 Lysinoalanine (LAL) Content of Processed Foods

Food	LAL (μg/g protein)
Corn chips	390
Pretzels	500
Hominy	560
Tortillas	200
Taco shells	170
Milk, infant formula	150–640
Milk, evaporated	590–860
Milk, UHT	160–370
Milk, HTST	260–1030
Milk, spray-dried powder	0
Skim milk, evaporated	520
Simulated cheese	1070
Egg-white solids, dried	160–1820
Calcium caseinate	370–1000
Sodium caseinate	430–6900
Acid casein	70–190
Hydrolyzed vegetable protein	40–500
Whipping agent	6500–50,000
Soy protein isolate	0–370
Yeast extract	120

Source: Ref. 103.

nucleophilic SH group reacts more than 1000 times faster than the ε-amino group of lysine. Sodium sulfite and ammonia exert their inhibitory effect by competing with the ε-amino group of lysine for DHA. Blocking of ε-amino groups of lysine residues by reaction with acid anhydrides prior to alkali treatment also decreases the formation of lysinoalanine. However, this approach results in loss of lysine activity and may be unsuitable for food applications.

Under normal conditions used for processing of several foods, only small amounts of lysinoalanine are formed. Thus, toxicity of lysinoalanine in alkali-treated foods is not believed to be a major concern. However, reduction in digestibility, loss of bioavailability of lysine, and racemization of amino acids (some of which are toxic) all are undesirable.

Excessive heating of pure protein solutions or proteinaceous foods low in carbohydrate content also results in formation of ε-*N*-(γ-glutamyl)lysyl and ε-*N*-(γ-aspartyl)lysyl cross-links. These involve a transamidation reaction between Lys and Gln or Asn residues. The resulting cross-links are termed isopeptide bonds because they are foreign to native proteins. Isopeptides resist enzymatic hydrolysis in the gut, and these cross-linkages therefore impair digestibility of proteins and bioavailability of lysine.

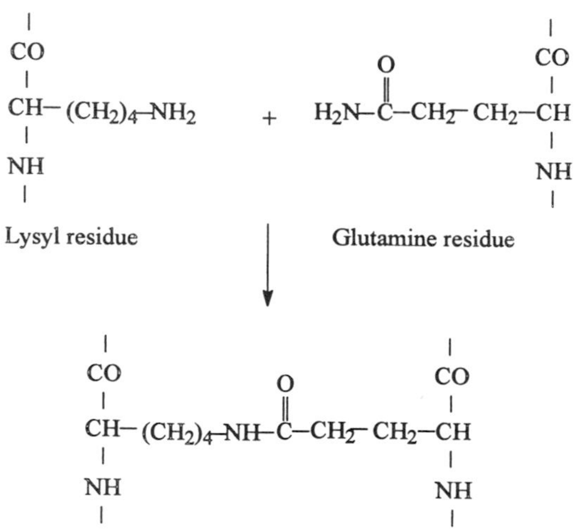

$$\underset{\text{Lysyl residue}}{\text{—NH—CH(CO—)—(CH}_2)_4\text{—NH}_2} + \underset{\text{Glutamine residue}}{\text{H}_2\text{N—C(=O)—CH}_2\text{—CH}_2\text{—CH(CO—)—NH—}} \longrightarrow \text{—NH—CH(CO—)—(CH}_2)_4\text{—NH—C(=O)—CH}_2\text{—CH}_2\text{—CH(CO—)—NH—} \quad (80)$$

Ionizing radiation of foods results in the formation of hydrogen peroxide through radiolysis of water in the presence of oxygen, and this, in turn, causes oxidative changes in, and polymerization of, proteins. Ionizing radiation also may directly produce free radicals via ionization of water.

$$H_2O \rightarrow H_2O^+ + e^- \quad (81)$$

$$H_2O^+ + H_2O \rightarrow H_3O^+ + \cdot OH \quad (82)$$

The hydroxyl free radical can induce formation of protein free radicals, which in turn may cause polymerization of proteins.

$$P + \cdot OH \rightarrow P^{\cdot} + H_2O \quad (83)$$

$$P^{\cdot} + P^{\cdot} \rightarrow P\text{–}P \quad (84)$$

Heating of protein solutions at 70–90°C and at neutral pH generally leads to sulfhydryl–disulfide interchange reactions (if these groups are present), resulting in polymerization of proteins. However, this type of heat-induced cross-link generally does not have an adverse effect on the digestibility and bioavailability of proteins and essential amino acids because these bonds can be broken in vivo.

6.7.1.5 Effects of Oxidizing Agents

Oxidizing agents such as hydrogen peroxide and benzoyl peroxide are used as bactericidal agents in milk, as bleaching agents in cereal flours, protein isolates, and fish protein concentrate, and for detoxification of oilseed meals. Sodium hypochlorite is also commonly used as a bactericidal and detoxifying agent in flours and meals. In addition to oxidizing agents that are sometimes added to foods, several oxidative compounds are endogenously produced in foods during processing. These include free radicals formed during irradiation of foods, during peroxidation of lipids, during photooxidation of compounds such as riboflavin and chlorophyll, and during nonenzymatic browning of foods. In addition, polyphenols present in several plant proteins can be oxidized by molecular oxygen to quinones at neutral to alkaline pH, and this will lead ultimately to peroxides. These highly reactive oxidizing agents cause oxidation of several amino acid residues and polymerization of proteins. The amino acid residues most susceptible to oxidation are Met, Cys/cystine, Trp, and His, and to a lesser extent Tyr.

Oxidation of Methionine

Methionine is easily oxidized to methionine sulfoxide by various peroxides. Incubation of protein bound mehtionine or free methionine with hydrogen peroxide (0.1 *M*) at elevated temperature for 30 min results in complete conversion of methionine to methionine sulfoxide [18]. Under strong oxidizing conditions, methionine sulfoxide is further oxidized to methionine sulfone, and in some cases to homocysteic acid.

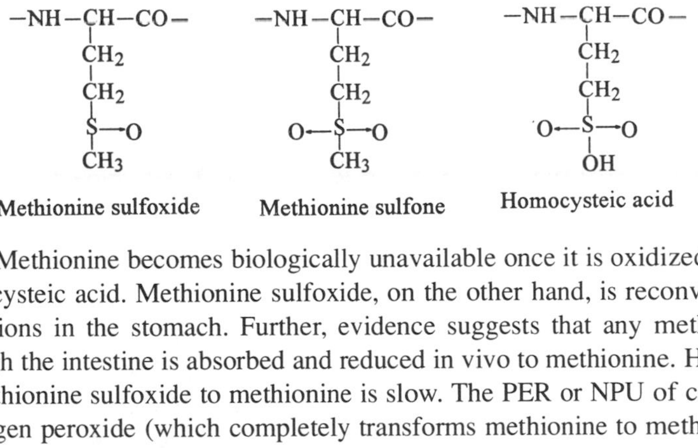

$$\underset{\text{Methionine sulfoxide}}{-NH-CH(CH_2-CH_2-S(\rightarrow O)-CH_3)-CO-} \quad \underset{\text{Methionine sulfone}}{-NH-CH(CH_2-CH_2-S(\rightarrow O)_2-CH_3)-CO-} \quad \underset{\text{Homocysteic acid}}{-NH-CH(CH_2-CH_2-S(\rightarrow O)_2-OH)-CO-} \tag{85}$$

Methionine becomes biologically unavailable once it is oxidized to methionine sulfone or homocysteic acid. Methionine sulfoxide, on the other hand, is reconverted to Met under acidic conditions in the stomach. Further, evidence suggests that any methionine sulfoxide passing through the intestine is absorbed and reduced in vivo to methionine. However, in vivo reduction of methionine sulfoxide to methionine is slow. The PER or NPU of casein oxidized with 0.1 *M* hydrogen peroxide (which completely transforms methionine to methionine sulfoxide) is about 10% less than that of control casein.

Oxidation of Cysteine and Cystine

Under alkaline conditions, cysteine and cystine follow the β-elimination reaction pathway to produce dehydroalanine residues. However, at acidic pH, oxidation of cysteine and cystine in simple systems results in formation of several intermediate oxidation products. Some of these derivatives are unstable.

Cystine mono- or disulfoxide

Cystine mono- or disulfone

Cystine

NH–CH–CO (cysteine) –H+→ Cystine: $CH\text{-}CH_2\text{-}S\text{-}S\text{-}CH_2\text{-}CH$ → $CH\text{-}CH_2\text{-}S(O)\text{-}S(O)\text{-}CH_2\text{-}CH$ → $CH\text{-}CH_2\text{-}S(O_2)\text{-}S(O_2)\text{-}CH_2\text{-}CH$ (86)

$CH\text{-}CH_2\text{-}SH$ –H+→ $CH\text{-}CH_2\text{-}SOH$ → $CH\text{-}CH_2\text{-}SO_2H$ → $CH\text{-}CH_2\text{-}SO_3H$

Cysteine sulfenic acid

Cysteine sulfinic acid

Cysteine sulfonic acid (cysteic acid)

Mono- and disulfoxides of L-cystine are biologically available, presumably because they are reduced back to L-cystine in the body. However, mono- and disulfone derivatives of L-cystine are biologically unavailable. Similarly, while cysteine sulfenic acid is biologically available, cysteine sulfinic acid and cysteic acid are not. The rate and extent of formation of these oxidation products in acidic foods are not well documented.

Oxidation of Tryptophan

Among the essential amino acids, Trp is exceptional because of its role in several biological functions. Therefore, its stability in processed foods is of major concern. Under acidic, mild, oxidizing conditions, such as in the presence of performic acid, dimethylsulfoxide, or *N*-bromosuccinimide (NBS), Trp is oxidized mainly to β-oxyindolylalanine. Under acidic, severe, oxidizing conditions, such as in the presence of ozone, hydrogen peroxide, or peroxidizing lipids, Trp is oxidized to *N*-formylkynurenine, kynurenine, and other unidentified products.

$HOOC\text{-}CH\text{-}NH_2$ / CH_2 — β-Oxyindolylalanine

Performic acid, dimethyl sulfoxide, or NBS

$HOOC\text{-}CH\text{-}NH_2$ / CH_2 — Tryptophan (87)

Ozone or oxygen + hν

$HOOC\text{-}CH\text{-}NH_2$ / $CO\text{-}CH_2$ / N(H)–CHO — N-formylkynurenine → $HOOC\text{-}CH\text{-}NH_2$ / $CO\text{-}CH_2$ / N(H)–H — Kynurenine

Exposure of Trp to light in the presence of oxygen and a photosensitizer, such as riboflavin, leads to formation of *N*-formylkynurenine and kynurenine as major products and several other minor ones. Depending upon the pH of the solution, other derivatives, such as 5-hydroxyformylkynurenine (pH > 7.0) and a tricyclic hydroperoxide (pH 3.6–7.1), are also formed [73]. In addition to the photoxidative products, Trp forms a photoadduct with riboflavin.

Both protein-bound and free tryptophan are capable of forming this adduct. The extent of formation of this photoadduct is dependent on availability of oxygen, being greater under anaerobic conditions [97].

The oxidation products of Trp are biologically active. In addition, kynurenines are carcinogenic in animals, and all other Trp photooxidative products as well as the carbolines formed during broiling/grilling of meat products exhibit mutagenic activities and inhibit growth of mammalian cells in tissue cultures. The tryptophan–riboflavin photoadduct shows cytotoxic effects on mammalian cells, and exerts hepatic dysfunctions during parenteral nutrition.

Riboflavin

hν

Tryptophan

(88)

Tryptophan free radical

Riboflavin free radical

Photoadduct

These undesirable products are normally present in extremely low concentration in foods unless an oxidation environment is purposely created.

Among the amino acid side chains, only those of Cys, His, Met, Trp, and Tyr are susceptible to sensitized photooxidation. In the case of Cys, cysteic acid is the end product. Met is photooxidized first to methionine sulfoxide, and finally to methionine sulfone and homocysteic acid. Photooxidation of histidine leads to the formation of aspartate and urea. The photooxidation products of tyrosine are not known. Since foods contain endogenous as well as supplemented riboflavin (vitamin B2), and usually are exposed to light and air, some degree of sensitized photooxidation of the preceding amino acid residues would be expected to occur. At equimolar concentrations, the rates of oxidation of the sulfur amino acids and Trp are likely to follow the order Met > Cys > Trp.

Oxidation of Tyrosine

Exposure of tyrosine solutions to peroxidase and hydrogen peroxide results in oxidation of tyrosine to dityrosine. Occurrence of this type of cross-link has been found in natural proteins, such as resilin, elastin, keratin, and collagen.

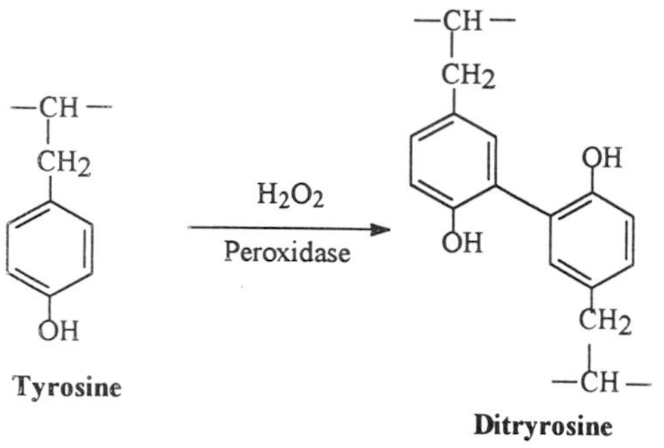

$$\text{Tyrosine} \xrightarrow[\text{Peroxidase}]{H_2O_2} \text{Ditryrosine} \qquad (89)$$

Tyrosine

Ditryrosine

6.7.1.6 Carbonyl-Amine Reactions

Among the various processing-induced chemical changes in proteins, the Maillard reaction (nonenzymatic browning) has the greatest impact on its sensory and nutritional properties. The Maillard reaction refers to a complex set of reactions initiated by reaction between amines and carbonyl compounds, which, at elevated temperature, decompose and eventually condense into insoluble brown products known as melanoidins (see Chap. 4). This reaction occurs not only in foods during processing, but also in biological systems. In both instances, proteins and amino acids typically provide the amino component, and reducing sugars (aldoses and ketoses), ascorbic acid, and carbonyl compounds generated from lipid oxidation provide the carbonyl component.

Some of the carbonyl derivatives from the nonenzymatic browning sequence react readily with free amino acids. This results in degradation of the amino acids to aldehydes, ammonia, and carbon dioxide, and the reaction is known as *Strecker degradation.* The aldehydes contribute to aroma development during the browning reaction. Strecker degradation of each amino acid produces a specific aldehyde with a distinctive aroma (see Chap. 11).

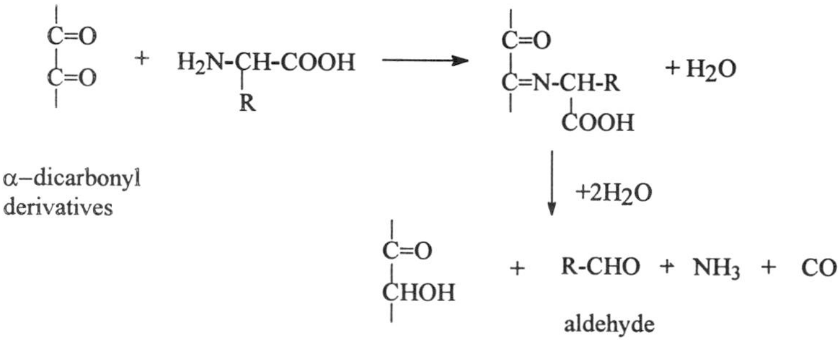

$$\begin{array}{c} | \\ C{=}O \\ | \\ C{=}O \\ | \end{array} + H_2N\text{-}\underset{R}{CH}\text{-}COOH \longrightarrow \begin{array}{l} | \\ C{=}O \\ | \\ C{=}N\text{-}CH\text{-}R \\ |\quad\;\; | \\ \quad\;\; COOH \end{array} + H_2O \qquad (90)$$

α–dicarbonyl derivatives

$$\downarrow +2H_2O$$

$$\begin{array}{c} | \\ C{=}O \\ | \\ CHOH \\ | \end{array} + \underset{\text{aldehyde}}{R\text{-}CHO} + NH_3 + CO$$

The Maillard reaction impairs protein nutritional value, and some of the products may be toxic, but probably are not hazardous at concentrations encountered in foods. Because the ε-amino group of lysine is the major source of primary amines in proteins, it is frequently involved in the carbonyl-amine reaction, and it typically suffers a major loss in bioavailability when this reaction occurs. The extent of Lys loss depends on the stage of the browning reaction. Lysine involved in the early stages of browning, including the Schiff's base, is biologically available. These early derivatives are hydrolyzed to lysine and sugar in the acidic conditions of the stomach. However, beyond the stage of ketosamine (Amadori product) or aldosamine (Heyns product), lysine is no longer biologically available. This is primarily because of poor absorption of these derivatives in the intestine [29]. It is important to note that no color has developed at this stage. Although sulfite inhibits formation of brown pigments [113], it cannot prevent loss of lysine availability, because it cannot prevent formation of Amadori or Heyns products.

Biological activity of lysine at various stages of the Maillard reaction can be determined chemically by addition of 1-fluoro-2,4-dinitrobenzene (FDNB), followed by acid hydrolysis of the derivatized protein. FDNB reacts with available ε-amino groups of lysyl residues. The hydrolysate is then extracted with ethyl ether to remove unreacted FDNB, and the concentration of ε-dinitrophenyl-lysyl (ε-DNP-lysine) in the aqueous phase is determined by measuring absorbance at 435 nm. Available lysine also can be determined by reacting 2,4,6-trinitrobenzene sufonic acid (TNBS) with the ε-amino group. In this case, the concentration of ε-trinitrophenyl-lysine (ε-TNP-lysine) derivative is determined from absorbance at 346 nm.

Nonenzymatic browning not only causes major losses of lysine, but reactive unsaturated carbonyls and free radicals formed during the browning reaction cause oxidation of several other essential amino acids, especially Met, Tyr, His, and Trp. Cross-linking of proteins by dicarbonyl compounds produced during browning decreases protein solubility, and impairs digestibility of proteins.

Some Maillard browning products are suspected mutagens. Although mutagenic compounds are not necessarily carcinogenic, all known carcinogens are mutants. Therefore, the formation of mutagenic Maillard compounds in foods is of concern. Studies with mixtures of glucose and amino acids have shown that the Maillard products of Lys and Cys are mutagenic, whereas those of Trp, Tyr, Asp, Asn, and Glu are not, as determined by the Ames test. It should be pointed out that pyrolysis products of Trp and Glu (in grilled and broiled meat) also are mutagenic (Ames test). As discussed earlier, heating of sugar and amino acids in the presence of creatine produces the most potent IQ-type mutagens (see Eq. 77). Although results based on model systems cannot be reliably applied to foods, it is possible that interaction of Maillard products with other low-molecular-weight constituents in foods may produce mutagenic and/or carcinogenic substances.

On a positive note, some Maillard reaction products, especially the reductones, do have antioxidative activity [75]. This is due to their reducing power, and their ability to chelate metals, such as Cu and Fe, which are pro-oxidants. The amino reductones formed from the reaction of triose reductones with amino acids such as Gly, Met, and Val show excellent antioxidative activity.

Besides reducing sugars, other aldehydes and ketones present in foods can also take part in the carbonyl-amine reaction. Notably, gossypol (in cotton seed), glutaraldehyde (added to protein meals to control deamination in the rumen of ruminants), and aldehydes (especially malonaldehyde) generated from the oxidation of lipids may react with amino groups of proteins. Bifunctional aldehydes, such as malonaldehyde, can cross-link and polymerize proteins.

$$\underset{\text{Protein}}{\text{P—NH}_2} + \underset{\text{malonaldehyde}}{\text{OHC-CH}_2\text{-CHO}} \longrightarrow \underset{\text{cross-linkage}}{\text{P—N=HC-CH}_2\text{-CH=N—P}} \qquad (91)$$

This may result in insolublization, loss of digestibility and bioavailability of lysine, and loss of functional properties of proteins. Formaldehyde also reacts with the ε-amino group of lysyl residues; the toughening of fish muscle during frozen storage is believed to be due to reactions of formaldehyde with fish proteins.

6.7.1.7 Other Reactions of Proteins in Foods

Reactions with Lipids

Oxidation of unsaturated lipids leads to formation of alkoxy and peroxy free radicals. These free radicals in turn react with proteins, forming lipid–protein free radicals. These lipid–protein conjugated free radicals can undergo polymerization cross-linking of proteins.

$$LH + O_2 \rightarrow LOOH \quad (92)$$

$$LOOH \rightarrow LO^* + {}^*OH \quad (93)$$

$$LOOH \rightarrow LOO^* + H^* \quad (94)$$

$$LO^* + PH \rightarrow LOP + H^* \quad (95)$$

$$LOP + LO^* \rightarrow {}^*LOP + LOH \quad (96)$$

$${}^*LOP + O_2 \rightarrow {}^*OOLOP \quad (97)$$

$${}^*OOLOP + PH \rightarrow POOLOP + H^* \quad (98)$$

$$LOO^* + PH \rightarrow LOOP + H^* \quad (99)$$

$$LOOP + LOO^* \rightarrow {}^*LOOP + LOOH \quad (100)$$

$${}^*LOOP + O_2 \rightarrow {}^*OOLOOP \quad (101)$$

$${}^*OOLOOP + PH \rightarrow POOLOOP + H^* \quad (102)$$

In addition, the lipid free radicals can also induce formation of protein free radicals at cysteine and histidine side chains, which may then undergo cross-linking and polymerization reactions.

$$LOO^* + PH \rightarrow LOOH + P^* \quad (103)$$

$$LO^* + PH \rightarrow LOH + P^* \quad (104)$$

$$P^* + PH \rightarrow P\text{-}P^* \quad (105)$$

$$P\text{-}P^* + PH \rightarrow P\text{-}P\text{-}P^* \quad (106)$$

$$P\text{-}P\text{-}P^* + P^* \rightarrow P\text{-}P\text{-}P\text{-}P \quad (107)$$

Lipid peroxides (LOOH) in foods can decompose, resulting in liberation of aldehydes and ketones, notably malonaldehyde. These carbonyl compounds react with amino groups of proteins via carbonyl-amine reaction and Schiff's base formation. As discussed earlier, reaction of malonaldehye with lysyl side chains leads to cross-linking and polymerization of proteins. The reaction of peroxidizing lipids with proteins generally has deleterious effects on nutritional value of proteins. Noncovalent binding of carbonyl compounds to proteins also imparts off flavors.

Reactions with Polyphenols

Phenolic compounds, such as *p*-hydroxybenzoic acid, catechol, caffeic acid, gossypol, and quercein, are found in all plant tissues. During maceration of plant tissues, these phenolic compounds can be oxidized by molecular oxygen at alkaline pH to quinones. This can also occur by the action of polyphenoloxidase, which is commonly present in plant tissues. These highly reactive quinones can irreversibly react with the sulfhydryl and amino groups of proteins.

Reaction of quinones with SH and α-amino groups (N-terminal) is much faster than it is with ε-amino groups. In addition, quinones the oxidation products of phenolic compound, can also undergo condensation reactions, resulting in formation of high molecular weight brown color pigments sometimes referred to as tannins. Tannins remain highly reactive and readily combine with SH and amino groups of proteins. Quinone-amino group reactions decrease the digestibility and bioavailability of protein-bound lysine and cysteine.

Reactions with Halogenated Solvents

Halogenated organic solvents are often used to extract oil and some antinutritive factors from oilseed products, such as soybean and cottonseed meals. Extraction with trichloroethylene results in formation of a small amount of *S*-dichlorovinyl-L-cysteine, which is toxic. On the other hand, the solvents dichloromethane and tetrachloroethylene do not seem to react with proteins. 1,2-Dichloroethane reacts with Cys, His, and Met residues in proteins. Certain fumigants, such as methyl bromide, can alkylate Lys, His, Cys, and Met residues. All of these reactions decrease the nutritional value of proteins, and some are of concern from a safety standpoint.

Reactions with Nitrites

Reaction of nitrites with secondary amines, and to some extent with primary and tertiary amines, results in formation of *N*-nitrosoamines, which are among the most carcinogenic compounds formed in foods. Nitrites are usually added to meat products to improve color and to prevent bacterial growth. The amino acids (or residues) primarily involved in this reaction are Pro, His, and Trp. Arg, Tyr, and Cys also can react with nitrites. The reaction occurs mainly under acidic conditions and at elevated temperatures.

Tryptophan
NH_2-CH-COOH
CH_2
$NaNO_2$
Acid
HONO
NH_2-CH-COOH
CH_2
SH
Cysteine
NH_2-CH-COOH
CH_2
S NO
S-nitrosocysteine
NH_2-CH-COOH
CH_2
NO
N-nitrosotryptophan
HN—CH-COOH
Proline
ON-N—CH-COOH
N-nitrosopyrrolidone

(108)

The secondary amines produced during the Maillard reaction, such as Amadori and Heyns products, also can react with nitrites. Formation of *N*-nitrosamines during cooking, grilling, and broiling of meat has been a major concern, but additives, such as ascorbic acid and erythorbate, are effective in curtailing this reaction.

Reaction with Sulfites

Sulfites reduce disulfide bonds in proteins to yield *S*-sulfonate derivatives. They do not react with cysteine residues.

$$\text{P-S-S-P} + SO_3^{-2} \rightleftharpoons \text{P-S-}SO_3^- + \text{P-S}^- \qquad (109)$$

In the presence of reducing agents, such as cysteine or mecaptoethanol, the *S*-sulfonate derivatives are converted back to cysteine residues. *S*-Sulfonates decompose under acidic (as in stomach) and alkaline pH to disulfides. The *S*-sulfonation does not decrease the bioavailability of cysteine. The increase in electronegativity and the breakage of disulfide bonds in proteins upon *S*-sulfonation causes unfolding of protein molecules, which affects their functional properties.

6.7.2 Changes in the Functional Properties of Proteins

The methods or processes used to isolate proteins can affect their functional properties. Minimum denaturation during various isolation steps is generally desired because this helps retain acceptable protein solubility, which is often a prerequisite to functionality of these proteins in food products. In some instances, controlled or partial denaturation of proteins can improve certain functional properties.

Proteins are often isolated using isoelectric precipitation. The secondary, tertiary, and quaternary structures of most globular proteins are stable at their isoelectric pH, and the proteins readily become soluble again when dispersed at neutral pH. On the other hand, protein entities such as casein micelles are irreversibly destabilized by isoelectric precipitation. The collapse of micellar structure in isoelectrically precipitated casein is due to several factors, including solubilization of colloidal calcium phosphate and the change in the balance of hydrophobic and electrostatic interactions among the various casein types. The compositions of isoelectrically precipitated proteins are usually altered from those of the raw materials. This is because some minor protein fractions are reasonably soluble at the isoelectric pH of the major component and are therefore do not precipitate. This change in composition affects the functional properties of the protein isolate.

Ultrafiltration (UF) is widely used to prepare whey protein concentrates (WPC). Both protein and nonprotein composition of WPC are affected by removal of small solutes during UF. Partial removal of lactose and ash strongly influences the functional properties of WPC. Furthermore, increased protein–protein interactions occur in the UF concentrate during exposure to moderate temperatures (50–55°C), and this decreases solubility and stability of the ultrafiltered protein, which in turn changes its water binding capacity and alters its properties with respect to gelation, foaming, and emulsification. Among the ash constituents, variations in calcium and phosphate content significantly affect the gelling properties of WPC. Whey protein isolates prepared by ion exchange contain little ash, and because of this they have functional properties that are superior to those of isolates obtained by ultrafiltration/diafiltration.

Calcium ions often induce aggregation of proteins. This is attributable to formation of ionic bridges involving Ca^{2+} ions and the carboxyl groups. The extent of aggregation depends on calcium ion concentration. Most proteins show maximum aggregation at 40–50 m*M* Ca^{2+} ion concentration. With some proteins, such as caseins and soy proteins, calcium aggregation leads to precipitation, whereas in the case of whey protein isolate a stable colloidal aggregate forms (Fig. 27).

Exposure of proteins to alkaline pH, particularly at elevated temperatures, causes irreversible conformational changes. This is partly because of deamidation of Asn and Gln residues, and β-elimination of cystine residues. The resulting increase in the electronegativity and breakage of disulfide bonds causes gross structural changes in proteins exposed to alkali. Generally, alkali-treated proteins are more soluble and possess improved emulsification and foaming properties.

Hexane is often used to extract oil from oilseeds, such as soybean and cottonseed. This treatment invariably causes denaturation of proteins in the meal, and thus impairs their solubility and other functional properties.

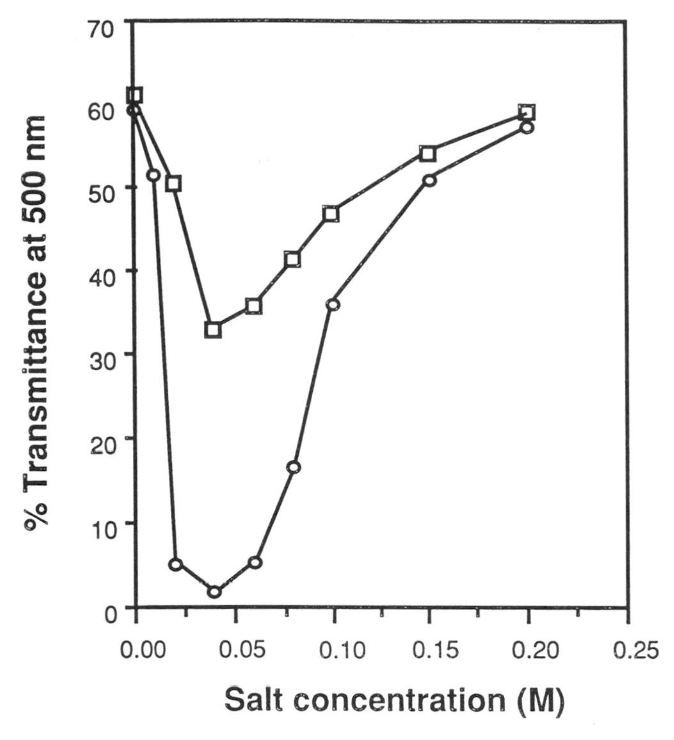

FIGURE 27 Salt concentration versus turbidity of whey protein isolate (5%) in $CaCl_2$ (○) and $MgCl_2$ (□) solutions after incubating for 24 h at ambient temperature. (From Ref. 121.)

The effects of heat treatments on chemical changes in, and functional properties of, proteins are described in Section 6.6. Scission of peptide bonds involving aspartyl residues during severe heating of protein solutions liberates low-molecular-weight peptides. Severe heating under alkaline and acid pH conditions also causes partial hydrolysis of proteins. The amount of low-molecular-weight peptides in protein isolates can affect their functional properties.

6.8 CHEMICAL AND ENZYMATIC MODIFICATION OF PROTEINS

6.8.1 Chemical Modifications

The primary structure of proteins contains several reactive side chains. The physicochemical properties of proteins can be altered, and their functional properties can be improved by chemically modifying the side chains. However, it should be cautioned that although chemical derivatization of amino acid side chains can improve functional properties of proteins, it can also impair nutritional value, create some amino acid derivatives that are toxic, and pose regulatory problems.

Since proteins contain several reactive side chains, numerous chemical modifications can be achieved. Some of these reactions are listed in Table 6. However, only a few of these reactions may be suitable for modification of food proteins. The ε-amino groups of lysyl residues and the

SH group of cyteine are the most reactive nucleophilic groups in proteins. The majority of chemical modification procedures involve these groups.

Alkylation

The SH and amino groups can be alkylated by reacting them with iodoacetate or iodoacetamide. Reaction with iodoacetate results in elimination of the positive charge of the lysyl residue, and introduction of negative charges at both lysyl and cyteine residues.

$$\text{P(SH)(NH}_2\text{)} \xrightarrow[\text{pH 8-9}]{\text{Iodoacetate, I-CH}_2\text{-COOH}} \text{P(S-CH}_2\text{-COOH)(NH-CH}_2\text{-COOH)}$$
$$\text{P(SH)(NH}_2\text{)} \xrightarrow[\text{pH 8-9}]{\text{Iodoacetamide, I-CH}_2\text{-CO-NH}_2} \text{P(S-CH}_2\text{-CO-NH}_2\text{)(NH-CH}_2\text{-CO-NH}_2\text{)} \quad (110)$$

The increase in the electronegativity of the iodoacetate-treated protein may alter its pH–solubility profile, and may also cause unfolding. On the other hand, reaction with iodoacetamide results only in the elimination of positive charges. This will also cause a local increase in electronegativity, but the number of negatively charged groups in proteins will remain unchanged. Reaction with iodoacetamide effectively blocks sulfhydryl groups so disulfide-induced protein polymerization cannot occur. Sulfhydryl groups also can be blocked by reaction with *N*-ethylmaleimide (NEM).

$$\text{P—SH} + \text{N-ethyl maleimide (N-C}_2\text{H}_5\text{)} \longrightarrow \text{P—S—(succinimide ring, N-C}_2\text{H}_5\text{)} \quad (111)$$

N-ethyl maleimide

Amino groups can also be reductively alkylated with aldehydes and ketones in the presence of reductants, such as sodium borohydride ($NaBH_4$) or sodium cyanoborohydride ($NaCNBH_3$). In this case, the Schiff base formed by reaction of the carbonyl group with the amino group is subsequently reduced by the reductant. Aliphatic aldehydes and ketones or reducing sugars can be used in this reaction. Reduction of the Schiff base prevents progression of the Maillard reaction, resulting in a glycoprotein as the end product (reductive glycosylation).

$$\text{R—CHO} + \text{NH}_2\text{—}\bullet \xrightarrow{\text{pH9}} \text{R—CH=N—}\bullet \xrightarrow{\text{NaBH}_4} \text{R—CH}_2\text{-HN—}\bullet \quad (112)$$

R = alkyl or polyol (sugar) group

The physicochemical properties of the modified protein will be affected by the reactant used. Hydrophobicity of the protein can be increased if an aliphatic aldehyde or ketone is selected for the reaction, and the degree of hydrophobicity can be varied by changing the chain length of the aliphatic group. On the other hand, if a reducing sugar is selected as the reactant, then the protein will become more hydrophilic. Since glycoproteins exhibit superior foaming and emulsifying properties (as in the case of ovalbumin), reductive glycosylation of proteins should improve solubility and interfacial properties of proteins.

Acylation

Amino groups can by acylated by reacting them with several acid anhydrides. The most common acylating agents are acetic anhydride and succinic anhydride. Reaction of protein with acetic anhydride results in elimination of the positive charges of lysyl residues, and a corresponding increase in electronegativity. Acylation with succinic or other dicarboxylic anhydrides results in replacement of positive charge with a negative charge at lysyl residues. This causes an enormous increase in the the electronegativity of proteins, and causes unfolding of the protein if extensive reaction is allowed to occur.

$$\bullet\!-\!\ddot{N}H_2 + \underset{\text{Acetic anhydride}}{(CH_3-CO)_2O} \xrightarrow{\text{pH 9}} \bullet\!-\!NH-\overset{O}{\overset{\|}{C}}-CH_3 + CH_3\text{-}COOH \qquad (113)$$

$$\bullet\!-\!\ddot{N}H_2 + \underset{\text{Succinic anhydride}}{(-CH_2-CO-)_2O} \xrightarrow{\text{pH 9}} \bullet\!-\!NH-C-CH_2-CH_2-\overset{O}{\overset{\|}{C}}-OH \qquad (114)$$

Acylated proteins are generally more soluble than native proteins. In fact, the solubility of caseins and other less soluble proteins can be increased by acylation with succinic anhydride. However, succinylation, depending on the extent of modification, usually impairs other functional properties. For example, succinylated proteins exhibit poor heat-gelling properties, because of the strong electrostatic repulsive forces. The high affinity of succinylated proteins for water also lessens their adsorptivity at oil–water and air–water interfaces, thus impairing their foaming and emulsifying properties. Also, because several carboxyl groups are introduced, succinylated proteins are more sensitive to calcium induced precipitation than is the parent protein.

Acetylation and succinylation reactions are irreversible. The succinyl–lysine isopeptide bond is resistant to cleavage catalyzed by pancreatic digestive enzymes. Furthermore, succinyl–lysine is poorly absorbed by the intestinal mucosa cells. Thus, succinylation and acetylation greatly reduce the nutritional value of proteins.

The amphiphilicity of proteins can be increased by attaching long-chain fatty acids to the ε-amino group of lysyl residues. This can be accomplished by reacting a fatty acylchloride or *N*-hydroxysuccinimide ester of a fatty acid with a protein. This type of modification can enhance lipophilicity and fat binding capacity of proteins, and can also facilitate formation of novel micellar structures and other types of protein aggregates.

$$\text{P}-\ddot{N}H_2 + R-\overset{O}{\overset{\|}{C}}-N(\text{succinimide}) \xrightarrow{pH\,9} \text{P}-NH-\overset{O}{\overset{\|}{C}}-R + HN(\text{succinimide}) \quad (115)$$

N-hydroxy succinimide ester

$$\text{P}-\ddot{N}H_2 + Cl-\overset{O}{\overset{\|}{C}}-R \xrightarrow{pH\,9} \text{P}-NH-\overset{O}{\overset{\|}{C}}-R + HCl \quad (116)$$

Fatty acyl chloride

Phosphorylation

Several natural food proteins, such as caseins, are phosphoproteins. Phosphorylated proteins are highly sensitive to calcium-ion-induced coagulation, which may be desirable in simulated cheese-type products. Proteins can be phosphorylated by reacting them with phosphorus oxychloride $POCl_3$. Phosphorylation occurs mainly at the hydroxyl group of seryl and threonine residues and at the amino group of lysyl residues. Phosphorylation greatly increases protein electronegativity. Phosphorylation of amino groups results in addition of two negative charges for each positive charge eliminated by the modification. Under certain reaction conditions, especially at high protein concentration, phosphorylation with $POCl_3$ can lead to polymerization of proteins, as shown here. Such polymerization reactions tend to minimize the increases in electronegativity and calcium sensitivity of the modified protein. The N-P bond is acid labile. Thus, under the conditions prevailing in the stomach, the *N*-phosphorylated proteins would be expected to undergo dephosphorylation and regeneration of lysyl residues. Thus, digestibility of lysine is probably not significantly impaired by chemical phosphorylation.

$$\text{P}(OH)(NH_2) + 2\,(POCl_3 + 2H_2O) \xrightarrow{pH\,8\text{-}9} \text{P}(O-PO_3^{2-})(NH-PO_3^{2-}) + 3\,HCl \quad (117)$$

O-phosphoserine

N-phospholysine

$$\text{P}(OH)(NH_2) \xrightarrow{POCl_3} \text{P}(O-POCl_2)(NH-POCl_2) \quad (118)$$

$$+\ \text{P}(NH_2)(OH) \rightarrow \text{P}(O-POCl)(NH-POCl-O-\text{P}(NH_2))$$

$$+\ \text{P}(HO)(NH_2) \rightarrow \text{P}(NH-POCl_2)(O-POCl-NH-\text{P}(OH))$$

Sulfitolysis

Sulfitolysis refers to conversion of disulfide bonds in proteins to the *S*-sulfonate derivative using a reduction–oxidation system involving sulfite and copper (Cu^{II}) or other oxidants. The mechanism is shown here. Addition of sulfite to protein initially cleaves the disulfide bond, resulting in the formation of one S-SO_3^- and one free thiol group. This is a reversible reaction, and the equilibrium constant is very small. In the presence of an oxidizing agent, such as copper (II), the newly liberated SH groups are oxidized back to either intra- or intermolecular disulfide bonds, and these, in turn, are again cleaved by sulfite ions present in the reaction mixture. The reduction–oxidation cycle repeats itself until all of the disulfide bonds and sulfhydryl groups are converted to the *S*-sulfonate derivative [41].

Disulfide ○–S–S–○ + SO_3^{2-} ⇌ Sulfocysteine ○–S–$SO_3^{\ominus}$ + Free thiol ○–SH

Oxidation (copper) (119)

Both cleavage of disulfide bonds and incorporation of SO_3^- groups cause conformational changes in proteins, which affect their functional properties. For example, sulfitolysis of proteins in cheese whey dramatically changes their pH-solubility profiles (Fig. 28) [41].

Amino acids

Several plant proteins are deficient in lysine and methionine. The nutritional value of such proteins can be improved by covalent attachment of methionine and lysine at the ε-amino group of lysyl residues. This can be accomplished by using the carbodiimide method or by reaction of *N*-carboxy anhydride of methionine or lysine with protein. Of these two approaches, the *N*-carboxy anhydride coupling method is preferable, because use of the carbodiimide is regarded as too toxic. *N*-Carboxy anhydrides of amino acids are extremely unstable in aqueous solutions, and they readily convert to the corresponding amino acid form even at low moisture levels. However, covalent coupling of amino acids with proteins can be achieved by mixing the protein directly with the anhydride. The isopeptide bond formed with the lysyl residues is susceptible to hydrolysis by pancreatic peptidases; thus, lysyl residues modified in this manner are biologically available.

N-carboxy anhydride of amino acid + H_2N–● (Protein) ⟶ ●–NH-CO-CH(R)-NH_2 + CO_2 (Isopeptide bond between lysyl side chain and amino acid) (120)

Esterification

Carboxyl groups of Asp and Glu residues in proteins are not highly reactive. However, under acidic conditions, these residues can be esterified with alcohols. These esters are stable at acid pH, but are readily hydrolyzed at alkaline pH.

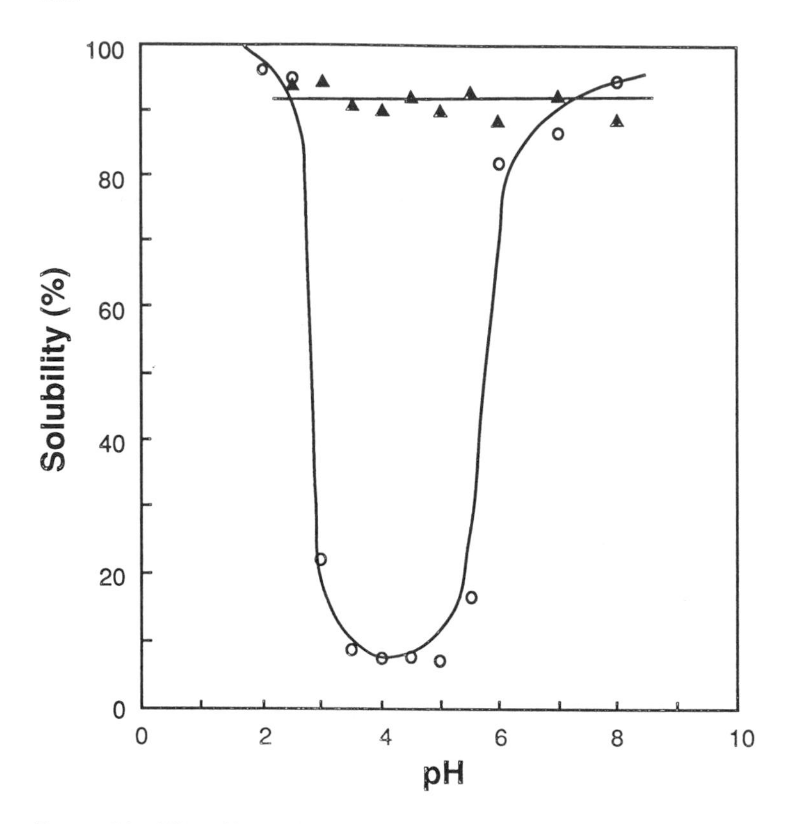

Figure 28 The pH versus protein solubility profile of (○) raw sweet whey and (▲) sulfonated sweet whey. (From Ref. 40.)

6.8.2 Enzymatic Modifications

Several enzymatic modifications of proteins/enzymes are known to occur in biological systems. These modifications can be grouped into six general categories: glycosylation, hydroxylation, phosphorylation, methylation, acylation, and cross-linking. Such enzymatic modifications of proteins in vitro can be used to improve their functional properties. Although numerous enzymatic modifications of proteins are possible, only a few of them are practical for modifying proteins intended for food use.

Enzymatic Hydrolysis

Hydrolysis of food proteins using proteases, such as pepsin, trypsin, chymotrypsin, papain, and thermolysin, alters their functional properties. Extensive hydrolysis by nonspecific proteases, such as papain, causes solubilizaiton of even poorly soluble proteins. Such hydrolysates usually contain low-molecular-weight peptides of the order of two to four amino acid residues. Extensive hydrolysis damages several functional properties, such as gelation, foaming, and emulsifying properties. These modified proteins are useful in liquid-type products, such as soups and sauces, where solubility is a primary criterion, and also in feeding persons who might not be able to digest solid foods. Partial hydrolysis of proteins either by using site-specific enzymes (such as trypsin or chymotrypsin) or by control of hydrolysis time often improves foaming and emulsification properties, but not gelling properties. With some proteins, partial hydrolysis may cause a transient decrease in solubility because of exposure of buried hydrophobic regions.

Certain oligopeptides released during protein hydrolysis have been shown to possess physiological activities, such as opioid activity, immunostimulating activity, and inhibition of angiotension-converting enzyme. The amino acid sequences of bioactive peptides found in peptic digests of human and bovine caseins are shown in Table 22. These peptides are not bioactive in the intact protein, but become active once they are released from the parent. Some of the physiological effects of these peptides include analgesia, catalepsy, sedation, respiratory depression, hypotension, regulation of body temperature and food intake, suppression of gastric secretion, and modification of sexual behavior [17].

When hydrolyzed, most food proteins liberate bitter-tasting peptides, which affect their acceptability in certain applications. The bitterness of peptides is associated with their mean hydrophobicity. Peptides that have a mean hydrophobicity value above 5.85 kJ/mol are bitter, whereas those with less than 5.43 kJ/mol are not. The intensity of bitterness depends on the amino acid composition, and sequence and the type of protease used. Hydrolysates of hydrophilic proteins, such as gelatin, are less bitter than the hydrolysates of hydrophobic proteins, such as caseins and soy protein. Proteases that show specificity for cleavage at hydrophobic residues produce hydrolysates that are less bitter than those enzymes that have broader specificity. Thus, thermolysin, which specifically attacks the amino side of hydrophobic residues, produces hydrolysates that are less bitter than those produced by low specificity trypsin, pepsin, and chymotrypsin [1].

Plastein Reaction

The plastein reaction refers to a set of reactions involving initial proteolysis, followed by resynthesis of peptide bonds by a protease (usually papain or chymotrypsin). The protein substrate, at low concentration, is first partially hydrolysed by papain. When the hydrolysate containing the enzyme is concentrated to 30–35% solids and incubated, the enzyme randomly recombines the peptides, generating new polypeptides. The plastein reaction also can be performed in a one-step process, in which a 30-35% protein solution (or a paste) is incubated with papain in the presence of L-cysteine [116]. Since the structure and amino acid sequence of plastein products are different from those of the original protein, they often display altered functional properties. When L-methionine is included in the reaction mixture, it is covalently incorporated into the newly formed polypeptides. Thus, the plastein reaction can be exploited to improve the nutritional quality of methionine- or lysine-deficient food proteins.

TABLE 22 Opioid Peptides from Caseins

Peptides	Name	Origin and position in the amino acid sequence
Tyr-Pro-Phe-Pro-Gly-Pro-Ile	β-Casomorphin 7	Bovine β-casein (60–66)
Tyr-Pro-Phe-Pro-Gly	β-Casomorphin 5	Bovine β-casein (60–64)
Arg-Tyr-Leu-Gly-Tyr-Leu-Glu	α-Casein exorphin	Bovine $\alpha_{\sigma1}$-casein (90–96)
Tyr-Pro-Phe-Val-Glu-Pro-Ile-Pro		Human β-casein (51–58)
Tyr-Pro-Phe-Val-Glu-Pro		Human β-casein (51–56)
Tyr-Pro-Phe-Val-Glu		Human β-casein (51–55)
Tyr-Pro-Phe-Val		Human β-casein (51–54)
Tyr-Gly-Phe-Leu-Pro		Human β-casein (59–63)

Source: Ref. 17.

TABLE 23 Covalent Attachment of Lysine and Methionine into Food Proteins by the Transglutaminase-Catalyzed Reaction

Protein	Amino acid content (g/100 g protein)	
	Control	Tranglutaminase-treated
Incorporation of methionine		
α_{s1}-Casein	2.7	5.4
β-Casein	2.9	4.4
Soybean 7S protein	1.1	2.6
Soybean 11S protein	1.0	3.5
Incorporation of lysine		
Wheat gluten	1.5	7.6

Source: Ref. 49.

Protein Cross-Linking

Transglutaminase catalyzes an acyl-transfer reaction that leads to covalent cross-linking of lysyl residues (acyl acceptor) with glutamine residues (acyl donor) via an isopeptide bond. This reaction can be used to cross-link different proteins, and to produce new forms of food proteins that might have improved functional properties. At high protein concentration, transglutaminase-catalyzed cross-linking leads to formation of protein gels and protein films at room temperature [72,77,78]. This reaction also can be used to improve nutritional quality of proteins by cross-linking lysine and/or methionine at the glutamine residues (Table 23) [49].

$$\text{P1}-(CH_2)_2-\overset{O}{\overset{\|}{C}}-NH_2 + NH_2-(CH_2)_2-\text{P2} \xrightarrow{\text{Transglutaminase}} \text{P1}-(CH_2)_2-\overset{O}{\overset{\|}{C}}-NH-(CH_2)_2-\text{P2} + NH_3 \quad (121)$$

Glutamine residue   Lysyl residue

BIBLIOGRAPHY

Bodwell, C. E., J. S., Adkins, and D. T. Hopkins (eds.)(1981). *Protein Quality in Humans: Assessment and In Vitro Estimation*,, AVI Publishing Co., Westport, CT.

Ghelis, C., and J. Yon (1982). *Protein Folding*, Academic Press, New York.

Hettiarachchy, N. S., and G. R. Ziegler (eds.)(1994). *Protein Functionality in Food systems*, Marcel Dekker, New York.

Kinsella, J. E., and W. G. Soucie (eds.)(1989). *Food Proteins*, American Oil Chemists' Society, Champaign, IL.

Mitchell, J. R., and D. A. Ledward (eds.)(1986). *Functional Properties of Food Macromolecules*, Elsevier Applied Science, New York.

Parris, N., and R. Barford (1991). *Interactions of Food Proteins*,, ACS Symposium Series 454, American Chemical Society, Washington, DC.

Phillips, R D., and J. W. Finley (eds.) (1989). *Protein Quality and the Effects of Processing*, Marcel Dekker, New York.

Schulz, G. E., and R. H. Schirmer (1980). *Principles of Protein Structure*, Springer-Verlag, New York.

Whitaker J. R., and S. R. Tannenbaum (1977). *Food Proteins*, AVI Publishing Co., Westport, CT.

Whitaker, J. R., and M. Fujimaki (eds.) (1980). *Chemical Deterioration of Proteins*, ACS Symposium Series, American Chemical Society, Washington, DC.

REFERENCES

1. Adler-Nissen, J. (1986). Relationship of structure to taste of peptides and peptide mixtures. In *Protein Tailoring for Food and Medical Uses* R. E. Feeney, and J. R. Whitaker, (eds.), Marcel Dekker, New york, pp. 97–122.
2. Ahren, T. J. and A. M. Klibanov (1985). The mechanism of irreversible enzyme inactivation at 100°C. *Science 228*:1280–1284.
3. Arakawa, T., and S. N. Timasheff (1982). Stabilization of protein structure by sugars. *Biochemistry 21*:6536–6544.
4. Arakawa, T., and S. N. Timasheff (1984). Mechanism of protein salting in and salting out by divalent cation salts: Balance between hydration and salt binding. *Biochemistry 23*:5912–5923.
5. Asakura, T., K. Adachi, and E. Schwartz (1978). Stabilizing effect of various organic solvents on protein. *J. Biol. Chem 253*:6423–6425.
6. Bigelow, C. C. (1967). On the average hydrophobicity of proteins and the relation between it and protein structure. *J. Theor. Biol. 16*:187–211.
7. Bigelow, C. C., and M. Channon (1976). Hydrophobicities of amino acids and proteins, in *Handbood of Biochemistry and Molecular Biology* (G. D. Fasman, ed.), 3rd ed. CRC Press, Boca Raton, FL pp. 209–243.
8. Blundell, T. L., and L. N. Johnson (1976). *Protein Crystallography*, Academic Press, London, p. 32.
9. Brandts, J. F. (1967). In *Thermobiology* (A. H. Rose, ed.), Academic Press, New York, pp. 25–72.
10. Brems, D. N. (1990). Folding of bovine growth hormone, in *Protein Folding* (L. M. Gierasch and J. King, eds.), American Association for the Advancement of Science, Washington, DC, p. 133.
11. Bull, H. B., and K. Breese (1973). Thermal stability of proteins. *Arch. Biochem. Biophys. 158*:681–686.
12. Bushuk, W., and F. MacRitchie (1989. Wheat Proteins: Aspects of structure that determine bread-making quality, in *Protein Quality and the Effects of Processing* (R. Dixon Phillips and J. W. Finley, eds.) Marcel Dekker, New York, pp. 345–369.
13. Cameron, D. R., M. E., Weber, E. S. Idziak, R. J. Neufeld, and D. G. Cooper (1991). Determination of interfacial areas in emulsions using turbidimetric and droplet size data: Correction of the formula for emulsifying activity index. *J. Agric. Food Chem. 39*:655–659.
14. Chen, C., A. M. Pearson, and J. I. Gray (1990). Meat mutagens. *Adv. Food Nutr Res. 34:* 387–449.
15. Chen, B., and J. A. Schellman (1989). Low Temperature unfolding of a mutant of phage T4 lysozyme. 1. Equilibrium studies. *Biochemistry 28*:685–691.
16. Cherkin, A. D., J. L. Davis, and M. W. Garman (1978). D-Proline: Sterospecific-sodium chloride dependent lethal convulsant activity in the chick. *Pharmacol. Biochem. Behav. 8*:623–625.
17. Chiba, H., and M. Yoshikawa (1986). Biologically functional peptides from food proteins: New opioid peptides from milk proteins, in *Protein Tailoring for Food and Medical Uses* (R. E. Feeney and J. R. Whitaker, eds.), Marcel Dekker, New York, pp. 123–153.
18. Cuq, J. L., M. Provansal, F. Uilleuz, and C. Cheftel (1973). Oxidation of methionine residues of casein by hydrogen peroxide. Effects on in vitro digestibility. *J. Food Sci. 38*:11–13.
19. Damodaran, S. (1988). Refolding of thermally unfolded soy proteins during the cooling regime of the gelation process: Effect on gelation. *J. Agric. Food Chem. 36*:262–269.
20. Damodaran, S. (1989). Influence of protein conformation on its adaptability under chaotropic conditions. *Int. J. Biol. Macromol. 11*:2–8.
21. Damodaran, S. (1989). Interrelationship of molecular and functional properties of food proteins, in *Food Proteins* (J. E. Kinsella and W. G. Soucie, eds.), American Oil Chemists' Society, Champaign, IL, pp. 21–51.
22. Damodaran, S. (1990). Interfaces, protein films, and foams. *Adv. Food Nutr. Res. 34*:1–79.
23. Damodaran, S. (1994). Structure-function relationship of food proteins, in *Protein Functionality in Food Systems* (N. S. Hettiarachchy and G. R. Ziegler, eds.), Marcel Dekker, New York, pp. 1–38.
24. Damodaran, S., and J. E. Kinsella (1980). Flavor-protein interactions: Binding of carbonyls to bovine serum albumin: Thermodynamic and conformational effects. *J. Agric. Food Chem. 28*:567–571.
25. Damodaran, S., and J. E. Kinsella (1981). Interaction of carbonyls with soy protein: Thermodynamic effects. *J. Agric. Food Chem. 29*:1249–1253.

26. Damodaran, S., and K. B. Song (1988). Kinetics of adsorption of proteins at interfaces: Role of protein conformation in diffusional adsorption. *Biochim. Biophys. Acta 954*:253–264.
27. Dickinson, E., and Y. Matsummura (1991). Time-dependent polymerization of b-lactoglobulin through disulphide bonds at the oil-water interface in emulsions. *Int. J. Biol. Macromol. 13*:26–30.
28. Eggum, B. O., and R. M. Beames (1983). The nutritive value of seed proteins, in *Seed Proteins* (W. Gottschalk and H. P. Muller, eds.), Nijhoff/Junk, The Hague, pp. 499–531.
29. Erbersdobler, H. F., M. Lohmann, and K. Buhl (1991). Utilizaiton of early Maillard reaction products by humans, in *Nutritional and Toxicological Consequences of Food Processing* (M. Friedman, ed.), Advances in Experimental Medicine and Biology, vol. 289, Plenum Press, New York, pp. 363–370.
30. FAO/WHO/UNU (1985). Energy and Protein Requirements, Report of a Joint FAO/WHO/UNU Expert Consultation. World Health Organization Technical Rep. Ser. 724, WHO, Geneva.
31. FAO/WHO (1991). Protein Quality Evaluation, Report of a Joint FAO/WHO Expert Consultation. FAO Food Nutr. Paper 51, FAO, Geneva, pp. 23–24.
32. Fay, L., U. Richli, and R. Liardon (1991). Evidence for the absence of amino acid isomerization in microwave-heated milk and infant formulas. *J. Agric. Food Chem. 39*:1857–1859.
33. Ferry, J. D. (1961). *Viscoelastic Properties of Polymers*, Wiley, New York, p. 391.
34. Ford, J. E. (1981). Microbiological methods for protein quality assessment, in *Protein Quality in Humans: Assessment and in vitro estimation* (C. E. Bodwell, J. S. Adkins, and D. T. Hopkins, eds.), AVI Publishing Co., Westport, CT. pp. 278–305.
35. Fransen, K., and J. E. Kinsella (1974). Parameters affecting the binding of volatile flavor compounds in model food system. I. Proteins. *J. Agric. Food Chem. 22*:675–678.
36. Friedman, M., and M. R. Gumbmann (1986). Nutritional improvement of legume proteins through disulfide interchange. *Adv. Exp. Med. Biol. 199*:357–390.
37. Fujita, Y., and Y. Noda (1981). The effect of hydration on the thermal stability of ovalbumin as measured by means of differential scanning calorimetry. *Bull. Chem. Soc. Jpn. 54*:3233–3234.
38. Gekko, K., and Y. Hasegawa (1986). Compressibility-structure relationship of globular proteins. *Biochemistry 25*:6563–6571.
39. Ghelis, C., and J. Yon (1982). *Protein Folding*, Academic Press, New York, p. 51.
40. Gonzalez, J. M., and S. Damodaran (1990). Recovery of proteins from raw sweet whey using a solid state sulfitolysis. *J Food Sci. 55*:1559–1563.
41. Gonzalez, J. M., and S. Damodaran (1990). Sulfitolysis of disulfide bonds in proteins using a solid state copper carbonate catalyst. *J. Agric. Food Chem. 38*:149–153.
42. Gonzalez, O. N., and R. H. Tanchuco (1977). Chemical composition and functional properties of coconut protein isolate. *Phil. J. Coco. Studies. 11*:21–29.
43. Gumbmann, M. R., W. L. Spangler, G. M. Dugan, and J. J. Rackis (1986). Safety of trypsin inhibitors in the diet: Effects on the rat pancreas of long-term feeding of soy flour and soy protein isolate. *Adv. Exp. Med. Biol. 199*:33–79.
44. Harwalkar, V. R. and C.-Y. Ma (1989). Effects of medium composition, preheating, and chemical modification upon thermal behavior of oat globulin and b-lactoglobulin, in *Food Proteins* (J. E. Kinsdella, and W. G. Soucie, eds.), American Oil Chemists' Society, Champaign, IL, pp. 210–231.
45. Hayase, F., H. Kato, and M. Fujimaki (1973). Racemization of amino acid residues in proteins during roasting. *Agric. Biol. Chem. 37*:191–192.
46. Hayashi, R. (1989). In *Engineering and Food* (W. E. L. Spiess and Schubert, H, eds.), Elsevier Applied Science, London, pp. 815–826.
47. He, X. M., and D. C. Carter (1992). Atomic structure and chemistry of human serum albumin. *Nature 358*:209–214.
48. Heremans, K. (1982). High pressure effects on proteins and other biomolecules. *Annu. Rev. Biophys. Bioeng. 11*:1–21
49. Ikura, K., M. Yoshikawa, R. Sasaki, and H. Chiba (1981). Incorporation of amino acids into food proteins by transglutaminase. *Agric. Biol. Chem. 45*:2587–2592.
50. Israelachvili, J., and R. Pashley (1982). The hydrophobic interaction is long range, decaying exponentially with distance. *Nature 300*:341–342.

51. Jagerstad, M., and K. Skog (1991). Formation of meat mutagens, in *Nutritional and Toxicological Consequences of Food Processing,* M. Friedman (ed.), Advances in Experimental Medicine and Biology, vol. 289, Plenum Press, New York, pp. 83–105.
52. Kamtekar, S., J. Schiffer, H. Xiong, J. M. Babik, and M. H. Hecht (1993). Protein design by binary patterning of polar and nonpolar amino acids. *Science 262*:1680–1685.
53. Kato, A., and S. Nakai (1980). Hydrophobicity determined by a fluorescence probe method and its correlation with surface properties of proteins. *Biochim. Biophys. Acta 624*:13–20.
54. Kato, S., Y. Osako, N. Matsudomi, and K. Kobayashi (1983). Changes in emulsifying and foaming properties of proteins during heat denaturation. *Agric. Biol. Chem. 47*:33–38.
55. Kinsella, J. E. (1976). Functional properties of food proteins: A review. *CRC Crit. Rev. Food Sci. Nutr. 7*:219–280.
56. Kinsella, J. E., S. Damodaran, and J. B. German (1985). Physicochemical and functional properties of oilseed proteins with emphasis on soy proteins, in *New Protein Foods: Seed Storage Proteins* (Altshul, A. M., and H. L. Wilcke, eds.), Academic Press, London, pp. 107–179.
57. Kinsella, J. E., and P. F. Fox (1986). Water sorption by proteins: Milk and whey proteins. *CRC Crit. Rev. Food Sci. Nutr. 24*:91–139.
58. Koshiyama, I. (1972). Purification and physicochemical properties of 11S globulin in soybean seeds. *Int. J. Peptide Protein Res. 4*:167–171.
59. Kuntz, I. D. (1971). Hydration of macromolecules. III. Hydration of polypeptides. *J. Am. Chem. Soc. 93*:514–516.
60. Kuntz, I. D., and W. Kauzmann (1974). Hydration of proteins and polypeptides. *Adv. Protein Chem. 28*:239–345.
61. Lachance, P. A., and M. R. Molina (1974). Nutritive value of a fiber-free coconut protein extract obtained by an enzymic-chemical method. *J. Food Sci. 39*:581–585.
62. Lapanje, S. (1978). *Physicochemical Aspects of Protein Denaturation,* Wiley-Intercience, New York.
62a. Lawrence, M. C., E. Suzuki, J. N. Varghese, P. C. Davis, A. Van Donkelaar, P. A. Tulloch, and P. M. Colman (1990). The three-dimensional structure of the seed storage protein phaseolin at 3 Å resolution. *EMBO J. 9*:9–15.
63. Lee, B., and F. M. Richards (1971). The interpretation of protein structure: Estimation of static accessibility. *J. Mol. Biol. 55*:379–400.
64. Liao, S. Y., and M. E. Mangino (1987). Characterization of the composition, physicochemical and functional properties of acid whey protein concentrates. *J. Food Sci. 52*:1033–1037.
65. Liardon, R. and D. Ledermann (1986). Racemization kinetics of free and protein-bound amino acids under moderate alkaline treatment. *J. Agric. Food Chem. 34*:557–565.
66. Liardon, R., and M. Friedman (1987). Effect of peptide bond cleavage on the racemization of amino acid residues in proteins. *J. Agric. Food Chem. 35*:661–667.
67. MacRitchie, F. (1992). Physicochemical properties of wheat proteins in relation to functionality. *Adv. Protein Chem. 36*:2–89.
69. Mak, A., L. B. Smillie, and G. Stewart (1980). A comparison of the amino acid sequences of rabbit skeletal muscle α- and β-tropomyosins. *J. Biol. Chem. 255*:3647–3655.
70. Marable, N. L., and G. Sanzone (1981). In vitro assays of protein quality assays utilizing enzymatic hydrolyses, in *Protein Quality in Humans: Assessment and In Vitro Estimation* (C. E. Bodwell, J. S. Adkins, and D. T. Hopkins, eds.), AVI Publishing Co., Westport, CT pp. 261–277.
71. Miller, S., J. Janin, A. M. Lesk, and C. Chothia (1987). Interior and surface of monomeric proteins. *J. Mol. Biol. 196*:641–656.
72. Motoki, M., H. Aso, K. Seguro, and N. Nio (1987). a_{s1}-Casein film prepared using transglutaminase. *Agric. Biol. Chem. 51*:993–996.
73. Nakagawa, M., Y. Yokoyama, S. Kato, and T. Hino (1985). Dye-sensitized photooxygenation of tryptophan. *Tetrahedron 41*:2125–2132.
74. Nakai, S., E. Li-Chan, M. Hirotsuka, M. C. Vazquez, and G. Arteaga (1991). Quantitation of hydrophobicity for elucidating the structure-activity relationships of food proteins, in *Interactions of Food Proteins* (N. Parris, and R. Barford, eds.), ACS Symposium Series 454, American Chemical Society, Washington, DC, pp. 42–58.

75. Namiki, M. (1988). Chemistry of Maillar reactions: Recent studies on the browning reaction mechanism and the development of antioxidants and mutagens. *Adv. Food Res. 32*:115–184.
76. Nielsen, N. C. (1985). Structure of soy proteins, in *New Protein Foods: Seed Storage Proteins,* vol. 5 (A. M. Altshul and H. L. Wilcke, eds.), Academic Press, New York, pp. 27–64.
77. Nio, N., Motoki, M., and K. Takinami (1985). Gelation of casein and soybean globulins by transglutaminase. *Agric. Biol. Chem. 49*:2283–2286.
78. Nio, N., Motoki, M., and K. Takinami (1986). Gelation mechanims of protein solution by transglutaminase. *Agric. Biol. Chem. 50*:851–855.
79. Nishioka, G. M., and S. Ross (1981). A new method and apparatus for measuring foam stability. *J. Colloid Interface Sci. 81*:1–7.
80. Nozaki, Y., and C. Tanford (1972). The solubility of amino acids and two glycine peptides in aqueous ethanol and dioxane solutions. *J. Biol. Chem. 246*:2211–2217.
81. Ohnishi, T., and T. Asakura (1976). Denaturation of oxyhemoglobin S by mechanical shaking. *Biochim. Boiophys. Acta 453*:93–100.
82. Okomoto, M., Y. Kawamura, and R. Hayashi (1990). Application of high pressure to food processing: Textural comparison of pressure- and heat-induced gels of food proteins. *Agric. Biol. Chem. 54*:183–189.
83. O'Neill, T. E., and J. E. Kinsella (1987). Binding of alkanone flavors to b-lactoglobulin: Effects of conformational and chemical modification. *J. Agric. Food Chem. 35*:770–774.
84. Ortiz, R., R. Sanchez, A. Paez, L. F. Montano, and E. Zenteno (1992). Induction of intestinal malabsorption syndrome in rats fed with *Agaricus bisporus* mushroom lectin. *J. Agric. Food Chem. 40*:1375–1378.
85. Papiz, M. Z., L. Sawyer, E. E. Eliopoulos, A. C. T. North, J. B. C. Findlay, R. Sivaprasadarao, T. A. Jones, M. E. Newcomer, and P. J. Kraulis (1986). The structure of β-lactoglobulin and its similarity to plasma retinol-binding protein. *Nature 324*:383–385.
86. Pearce, K. N., and J. E. Kinsella (1978). Emulsifying properties of proteins: Evaluation of a turbidimetric technique. *J. Agric. Food Chem. 26*:716–722.
87. Phillips, M. C. (1981). Protein conformation at liquid interfaces and its role in stabilizing emulsions and foams. *Food Technol. (Chicago) 35*:50–57.
88. Ponnuswamy, P. K., R. Muthusamy, and P. Manavalan (1982). Amino acid composition and thermal stability of proteins. *Int. J. Biol. Macromol. 4*:186–190.
89. Poole, S., S. I. West, and C. L. Walters (1984). Protein-protein interactions: Their importance in the foaming of heterogeneous protein systems. *J. Sci. Food Agric. 35*:701–711.
90. Pusztai, A., E. M. W. Clarke, G. Grant, and T. P. King (1981). The toxicity of *Phaseolus vulgaris* lectins. Nitrogen balance and immunochemicals studies. *J. Sci. Food Agric. 32*:1037–1046.
91. Rao, M. A., S. Damodaran, J. E. Kinsella, and H. J. Cooley (1986). Flow properties of 7S and 11S soy protein fractions. In *Food Engineering and Process Applications* (M. Le Maguer and P. Jelen, eds.), Elsevier Applied Science, New York, pp. 39–48.
92. Richards, F. M. (1977). Areas, volumes, packing, and protein structure. *Annu. Rev. Biophys. Bioeng. 6*:151–176.
93. Richardson, J. S., and D. C. Richardson (1989). In *Prediction of Protein Structure and the Principles of Protein Conformation* (G. D. Fasman, ed.), Plenum Press, New York, p. 24.
94. Ritchey, S. J. and L. J. Taper (1981). Estimating protein digestibility for humans from rat assays, in *Protein Quality in Humans: Assessment and In Vitro Estimation* (C. E. Bodwell, J. S. Adkins, and D. T. Hopkins, eds.), AVI Publishing Co., Westport, CT. pp. 306–315.
95. Ruegg, M., U. Moor, and B. Blanc (1975). Hydration and thermal denaturation of β-lactoglobulin. A calorimetric study. *Biochim. Biophys. Acta 400*:334–342.
96. Rupley, J. A., P.-H. Yang, and G. Tollin (1980). Thermodynamic and related studies of water interacting with proteins, in *Water in Polymers* (S. P. Rowland, ed.), ACS Symp. Ser. 127, American Chemical Society, Washington, DC, p. 91–139.
97. Salim-Hanna, M., A. M. Edwards, and E. Silva (1987). Obtention of a photo-incuded adduct between a vitamin and an essential amino acid: Binding of riboflavin to tryptophan. *Int. J. Vit. Nutr. Res. 57*:155–159.

98. Scheraga, H. A. (1963). In *The Proteins*, 2nd ed., Vol. 1 (H. Neurath, ed.), Academic Press, New York, pp. 478–594.
99. Schulz, G. E., and R. H. Schirmer (1979). *Principles of Protein Structure*, Springer-Verlag, New York, p. 76.
100. Shimada, K., and S. Matsushita (1980). Relationship between thermo-coagulation of proteins and amino acid compositions. *J. Agric. Food Chem. 28*:413–417.
101. Singer, N. S., J. Latella, and Y. Shoji (1990). Fat emulating protein products and processes. U.S. Patent No. 4,961,953.
102. Suzuki, A., M. Watanabe, K. Iwamura, Y. Ikeuchi, and M. Saito (1990). Effect of high pressure treatment on the ultrastructure and myofibrillar protein of beef skeletal muscle. *Agric. Biol. Chem. 54*:3085–3091.
103. Swaisgood, H. E., and G. L. Catignani (1991). Protein digestibility: In vitro methods of assessment. *Adv. Food Nutr. Res. 35*:185–236.
104. Tanford, C. (1962). Contribution of hydrophobic interactions to the stability of the globular conformation of proteins. *J. Am. Chem. Soc. 84*:4240–4247.
105. Tanford, C. (1957). Theory of protein titration curves. I. General equations for impenetrable spheres. *J. Am. Chem. Soc. 79*:5333–5339.
106. Townsend, A., and S. Nakai (1983). Relationship between hydrophobicity and foaming characteristics of food proteins. *J. Food Sci. 48*:588–594.
107. Vihinen, M. (1987). Relationship of protein flexibility to thermostability. *Protein Eng. 1*:477–480.
108. von Hippel, P. H. and T. Schleich (1969). The effects of neutral salats on the structure and conformational stability of macromolecules in solution, in *Structure and Stability of Biological Macromolecules* (S. N. Timasheff and G. D. Fasman, eds.), Marcel Dekker, New York, pp. 417–574.
109. Wang, C.-H., and S. Damodaran (1990). Thermal destruction of cysteine and cystine residues of soy protein under conditions of gelation. *J. Food Sci 55*:1077–1080.
110. Wang, C.-H. and S. Damodaran (1990). Thermal gelation of globular proteins: Weight average molecular weight dependence of gel strength. *J. Agric. Food Chem. 38*:1154–1164.
111. Weber, G. (1992). *Protein Interactions*, Chapman and Hall, New York, pp. 235–270.
112. Weber, G., and H. G. Drickamer (1983). The effect of high pressure upon proteins and other biomolecules. *Q. Rev. Biophys. 16*:89–112.
113. Wedzicha, B. L., I. Bellion, and S. J. Goddard (1991). Inhibition of browning by sulfites, in *Nutritional and Toxicological Consequences of Food Processing* (M. Friedman, ed.), Advances in Experimental Medicine and Biology, vol. 289, Plenum Press, New York, pp. 217–236.
114. Van Holde, K. E. (1977). Effect of amino acid composition and environment on protein structure, in *Food Proteins* (J. R. Whitaker, and S. R. Tannenbaum, eds.), AVI Publishing Co., Westport, CT. pp. 1–13.
115. Xu, S., and S. Damodaran (1993). Comparative adsorption of native and denatured egg-white, human and T4 phage lysozymes at the air-water interface. *J. Colloid Interface Sci. 159*:124–133.
116. Yamashita, M., S. Arai, Y. Imaizumi, Y. Amano, and M. Fujimaki (1979). A one-step process for incorporation of L-methionine into soy protein by treatment with papain. *J. Agric. Food Chem. 27*:52–56.
117. Yu, M.-A., and S. Damodaran (1991). Kinetics of protein foam destabilization: Evaluation of a method using bovine serum albumin. *J. Agric. Food Chem. 39*:1555–1562.
118. Zuber, H. (1988). Temperature adaptation of lactate dehydrogenase. Structural, functional and genetic aspects. *Biophys. Chem. 29*:171–179.
119. Zhu, H., and S. Damodaran (1994). Heat-induced conformational changes in whey protein isolate and its relation to foaming properties. *J. Agric. Food Chem. 42*:846–855.
120. Zhu, H., and S. Damodaran (1994). Proteose peptones and physical factors affect foaming properties of whey protein isolate. *J. Food Sci. 59*:554–560.
121. Zhu, H., and S. Damodaran (1994). Effects of calcium and magnesium ions on aggregation of whey protein isolate and its effect on foaming properties. *J. Agric. Food Chem. 42*:856–862.

7

Enzymes

JOHN R. WHITAKER
University of California, Davis, California

7.1 INTRODUCTION

7.1.1 Chemical Nature of Enzymes

Enzymes are proteins with catalytic activity due to their power of specific activation and conversion of substrates to products:

$$\text{Substrate(s)} \xrightarrow{\text{enzyme}} \text{product(s)} \tag{1}$$

Some of the enzymes are composed only of amino acids covalently linked via peptide bonds to give proteins that range in size from about 12,000 MW to those that are near 1,000,000 MW (Table 1). Other enzymes contain additional components, such as carbohydrate, phosphate, and cofactor groups. Enzymes have all the chemical and physical characteristics of other proteins (see Chap. 6). Composition-wise, enzymes are not different from all other proteins found in nature and they comprise a small part of our daily protein intake in our foods. However, unlike other groups of proteins, they are highly specific catalysts for the thousands of chemical reactions required by living organisms.

Enzymes are found in all living systems and make life possible, whether the organisms are adapted to growing near 0°C, at 37°C (humans), or near 100°C (in microorganisms found in some hot springs). Enzymes accelerate reactions by factors of 10^3 to 10^{11} times that of non-enzyme-catalyzed reactions (10^8 to 10^{20} over uncatalyzed reactions; Table 2). In addition, they are highly selective for a limited number of substrates, since the substrate(s) must bind stereospecifically and correctly into the active site before any catalysis occurs. Enzymes also control the direction of reactions, leading to stereospecific product(s) that can be very valuable by-products for foods, nutrition, and health or the essential compounds of life.

Enzymes are synthesized in vivo by living organisms, based on expression (translation) of specific genes. They are *wild type* ("nature" dictated sequences) enzymes that fit the definition already given. A few enzymes, such as ribonuclease, have been synthesized in the laboratory from the amino acids. However, the time, cost, and inefficiency of the process are prohibitive. *Isozymes* are enzymes from a single organism that have qualitatively the same enzymatic activity but differ quantitatively in activity and structure because of differences in amino acid sequences (different gene products) or differ quantitatively due to posttranslational modification (glycosylation, proteolytic activation, etc). They are found very frequently in organisms. There are also a few naturally occurring *ribozymes*, composed of ribonucleic acids, that slowly catalyze (one or two bonds per minute) the specific hydrolysis of phosphodiester bonds within the same

TABLE 1 Molecular Weights of a Few Selected Enzymes

Enzyme	Molecular weight	Enzyme	Molecular weight
Ribonuclease	13,683	Fumarase	194,000
Lysozyme	14,100	Catalase	232,000
Chymotrypsinogen	23,200	Aspartate transcarbamylase	310,000
β-Lactoglobulin	35,000	Urease	483,000
Alkaline phosphatase	80,000	β-Galactosidase	520,000
Polyphenol oxidase (mushrooom)	128,000	Glutamate dehydrogenase	2,000,000

Source: Adapted from Ref. 111, p. 41.

Table 2 Nature of Effect of Catalyst on E_a and on Relative Rates of Some Reactions

Substrate	Catalyst	E_a (kcal/mol)	n'/n[a] (25°C)	Relative rates (25°C)
H_2O_2	None	18.0	5.62×10^{-14}	1.00
	I^-	13.5	1.16×10^{-10}	2.07×10^{3}
	Catalase	6.4	1.95×10^{-5}	3.47×10^{8}
Sucrose	H^+	25.6	1.44×10^{-19}	1.00
	Invertase	11.0	8.04×10^{-9}	5.58×10^{10}
Carbonic acid	None	20.5	8.32×10^{-16}	1.00
	Carbonic anhydrase	11.7	2.46×10^{-9}	2.98×10^{6}
Urea	H^+	24.5	9.33×10^{-19}	1.00
	Urease	8.7	3.96×10^{-7}	4.25×10^{11}

[a]Fraction of molecules with E_a or greater.
Source: Ref. 111, p. 324.

ribonucleic acid or in different ribonucleic and deoxyribonucleic acids. These are important in vivo catalysts, but they will not be discussed further in this chapter.

Some enzymes result from in vivo or in vitro chemical modification of one or more nucleotide bases in genes of the wild-type enzymes or by chemical synthesis. These *mutated* enzymes may result from environmentally caused random in vivo modification (by free radicals, for example) of one or more nucleotide bases of a gene or in vitro by selective and deliberate chemical changes in the DNA nucleotide sequence of a gene. The latter is accomplished using specific restriction enzymes and ligases or by the polymerase chain reaction (PCR). When deliberately modified or synthesized in the laboratory, enzyme mimics are called *synzymes*. Other types of synzymes include the *abzymes* and the *pepzymes*. The abzymes are laboratory-made catalytic antibodies. The binding sites of catalytic antibodies are tailored in vivo by use of a specific hapten, while the catalytic groups are added in vitro by specific chemical modification of one or more of the amino acid side chains. The pepzymes are synthesized in the laboratory to mimic the sequence and stereochemistry of the active site of an enzyme. Enzymes may also be chemically modified by deliberate or unintentional changes (Maillard reaction, for example) in one or more side chain groups of a specific amino acid. This can lead to qualitatively different multiple forms of enzymes or total loss of activity.

7.1.2 Catalysis

Enzymes are positive catalysts; that is, they increase the rates of reactions by 10^3 to 10^{11} that of non-enzyme-catalyzed reactions (Table 2). A minimum of two events must occur for enzyme activity to be observed. First, the enzyme must bind (noncovalently) a compound stereospecifically into the active site. Second, there must be chemical conversion of the initial compound into a new compound. This is shown in Equation 2, where E is free enzyme, S is free substrate, E·S is a noncovalent complex of enzyme and substrate, and P is the new compound (product) formed.

$$\mathrm{E} + \mathrm{S} \underset{k_{-1}}{\overset{k_1}{\rightleftharpoons}} \mathrm{E{\cdot}S} \xrightarrow{k_2} \mathrm{E} + \mathrm{P} \tag{2}$$

The E is generated and participates repeatedly in the reaction. Some compounds only bind to the active site of the enzyme; they are not converted to new products. They are called competitive inhibitors, and no chemical changes occur in I.

$$E + I \underset{k_{-1}}{\overset{k_1}{\rightleftharpoons}} E{\cdot}I \tag{3}$$

7.1.3 Regulation of Enzyme Reactions

Enzyme activity can be controlled in a number of ways that are very important to food scientists. The velocity of an enzyme-catalyzed reaction is usually directly proportional to the active enzyme concentration, and dependent (in a complex way) on substrate, inhibitor, and cofactor concentrations, and on temperature and pH [111]. As an example, it is well known that enzyme-catalyzed reactions occur more slowly when a food is placed in a refrigerator (~ 4°C). But the reactions do not stop at 4°C (or at 0°C). Most enzyme-catalyzed reactions decrease 1.4–2 times/10°C decrease in temperature. Therefore, at 5°C, the velocities would be 0.5 to 0.25 times the velocity at 25°C.

Changing the pH of the system by 1 or 2 pH units from the pH-activity optimum can decrease enzymatic velocity to 0.5 or 0.1, respectively, of that at the pH optimum. Decreasing (by heating, breeding, or genetic engineering) the concentration of enzyme to 0.1 that normally present would decrease the enzyme activity to 0.1 that originally present.

7.1.4 Historical Aspects

Enzymes are essential for all living organisms; therefore, they have existed since the beginning of life. They are important in ripening of fruits, fermentation of fruit juices, souring of milk, tenderization of meat, etc., and humans were aware of these changes early on.

7.1.4.1 Nature of Enzymes

During the 1870–1890 period, there was a major debate between Pasteur [85], who believed that enzymes functioned only when associated with living organisms, and Liebig [72], who believed that enzymes continued to function in the absence of cells. This argument was settled in 1897 when Büchner [19] separated enzymes, including invertase, from yeast cells and showed that they were still active. The debate between Pasteur and Liebig had resulted in the terms *organized ferments* and *unorganized ferments* to differentiate between enzymes in yeasts and enzymes in the stomach. In 1878 Kühne proposed use of *en zyme* (Greek, "in yeast") to describe both types of enzymes [68].

7.1.4.2 Enzyme Specificity

During the period of 1890–1940 much work was done on the specificity of enzymes. Emil Fischer and his students, Bergmann, Fruton, and others, synthesized many carbohydrates, peptides, and other compounds of known structures and determined whether they were hydrolyzed, and at what rate, by carbohydrases and proteases. They found that small changes, such as removal or change in configuration of one group, could lead to marked effects on rates of hydrolyses. This led Fischer [38] to hypothesize the lock-and-key analogy of enzyme–substrate interactions in 1894 in which the active site of the enzyme was considered to be rigid. This concept remained essentially unchallenged until 1959, when Koshland [65] suggested the induced fit hypothesis to be better supported by currently available data, since there appeared to be some flexibility in accommodation between enzyme and substrate when they complexed. The induced fit hypothesis retained the concept of a stereospecific complex formation between enzyme and substrate, but it rejected the idea that the binding locus of the active site is a rigid structure that retains its exact shape even in the absence of substrate. Much data subsequently proved the induced fit concept to be more appropriate than the lock-and-key concept [78].

7.1.4.3 Enzyme Kinetics

Study of enzyme activities requires kinetic approaches, not equilibrium approaches. Therefore, the development of the kinetics to deal with these reactions was very important. In 1902 Henri [46;46a] and Brown [18] independently suggested that the hyperbolic relationship between velocity ($-dS/dt$ or dP/dt) and substrate concentration (Fig. 1) was due to an obligatory enzyme–substrate complex intermediate prior to conversion of the substrate to product (Eq. 2). When all the enzyme is saturated with substrate, the enzyme is operating at its maximum capacity and further increase in substrate concentration cannot increase the velocity further (Fig. 1). In 1913, Michaelis and Menten [79] showed that the hyperbolic relationship between velocity and substrate concentration (Fig. 1) can be expressed by

$$v = \frac{V_{\text{max}}[\text{S}]_t}{K_{\text{m}} + [\text{S}]_t} = -dS/dt = dP/dt \tag{4}$$

where v is the measured velocity, V_{max} is the maximum velocity when enzyme is saturated with substrate, K_{m} [$= (k_{-1} + k_2)/k_1$ for Eq. 2] is the substrate concentration at which $v = 0.5V_{\text{max}}$, and $[S]_t$ is the substrate concentration at any time t in the reaction. For reasons that will be detailed later, the initial velocity, v_{o}, determined early in the reaction ($< 5\%$ substrate conversion to product) is used, so that $[\text{S}]_t \cong [\text{S}]_{\text{o}}$, the starting substrate concentration.

Lineweaver and Burk [76] in 1934 developed a reciprocal transform ($1/v$ vs. $1/[\text{S}]$) of Equation 4 that gives a linear rather than a hyperbolic relationship, making it easier to determine K_{m} and V_{max}. Beginning in 1958, kinetic equations dealing with multisubstrate–multiproduct

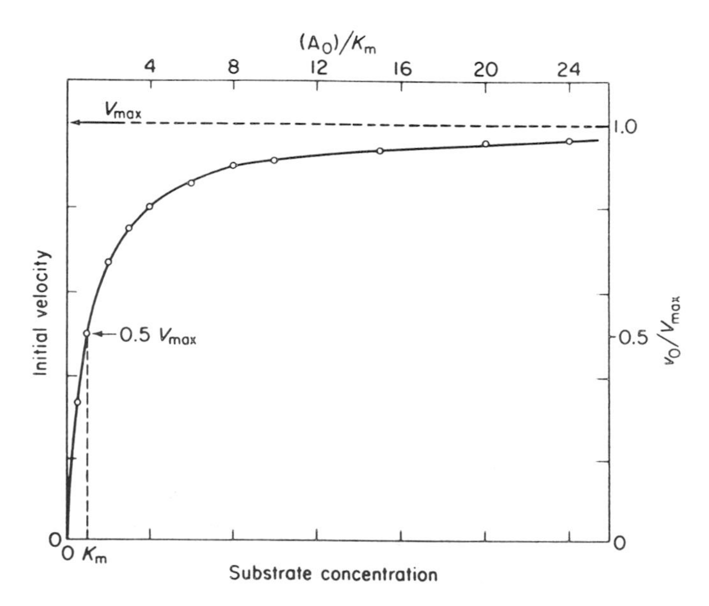

FIGURE 1 Variation of observed velocity with substrate concentration $[\text{A}_{\text{o}}]$ for an enzyme-catalyzed reaction that follows Michaelis–Menten kinetics. (From Ref. 111, p. 168.)

enzyme-catalyzed reactions were developed for more complex cases than that of Equation 2 [22, 27, 115]. However, Equation 4 still fits all Michaelis–Menten type reactions, where steady-state concepts are valid. The numerical definitions of V_{max} and K_m change as the complexity of the reaction changes. In 1965, Monod et al. [80] and Koshland et al. [66] developed equations for allosteric-behaving enzymes. Other workers [44] developed equations for pre-steady-state treatment of enzyme-catalyzed reactions.

7.1.4.4 Enzyme Purification

Purification of enzymes began only after 1920. Before 1920, it was generally thought that enzymes were proteins. However, during the period of 1922–1928, Willstätter and his colleagues purified horseradish peroxidase to the point where it had appreciable activity even when no protein could be detected by existing methods. Therefore, they concluded, incorrectly, that enzymes could not be proteins. In 1926, Sumner [98], at Cornell University, crystallized urease from jack beans and showed it to be a protein. This led to much debate between Willstätter, a noted German chemist, and Sumner, with many European biochemists and chemists siding with Willstätter. Sumner was later awarded the Noble prize for crystallization of the first enzyme and for showing that enzymes are proteins. Beginning in the 1930s, and continuing to the present, much work has been devoted to enzyme purification using techniques such as chromatography (ion exchange, gel filtration, chromatofocusing, affinity, hydrophobic), several types of electrophoresis [regular polyacrylamide gel electrophoresis (PAGE), SDS-PAGE, isoelectric focusing], and techniques based on solubility and stability differences. Now nearly 3000 enzymes have been purified. All are proteins.

7.1.4.5 Enzyme Structure

The first primary sequence of a protein, insulin (of 6000 MW), was determined by Sanger and co-workers in 1955 [90] after almost 10 years of methods development and application. In 1960 [49, 95], the primary sequence of the first enzyme, ribonuclease (13,683 MW), was determined. Many primary sequences of enzymes are now known, many via gene sequencing. The secondary and tertiary structures of ribonuclease were determined in 1967 [58]. Now several hundred secondary and tertiary structures of enzymes are known, including structures in solutions, by nuclear magnetic resonance (NMR). The first enzyme, ribonuclease, was completely synthesized chemically from amino acids in 1969 by two groups. Now, wild-type and mutant enzymes can be synthesized almost overnight by the PCR (polymerase chain reaction) method, once the gene has been isolated. Those of us who have observed advances in protein and enzyme chemistry since the 1940s are astounded and awed by the remarkable ease and speed at which enzyme sequences and structures can now be determined.

7.1.5 Literature on Enzymology

Perhaps no other single topic has received so much attention from researchers as enzymes. Papers on enzymes can be found in almost any journal in the physical, chemical, or biological sciences, since they are of interest to chemists, physicists, mathematicians, chemical engineers, and the life scientists, including food scientists and nutritionists. The bibliography at the end of this chapter includes selected journals, books, and monographs on enzymes. Obviously, no scientist can keep up completely in this field. Food scientists should, by all means, be familiar with current progress in fundamental aspects of enzymes, as well as with applications.

7.2 ENZYME NOMENCLATURE

The first enzyme to be named was catalase, which converts hydrogen peroxide to water and O_2 (Eq. 5). Its name was derived by using the stem of catalyst and adding ase. This generic type name is unfortunate considering the many enzymes that have since been discovered.

$$2H_2O_2 \xrightarrow{\text{catalase}} 2H_2O + O_2 \tag{5}$$

The name of the second enzyme, diastase, was derived from the Greek word diastasis, meaning to separate. The name of the third enzyme, peroxidase, was based on one of the substrates being peroxide. Polyphenol oxidase was so named because it oxidizes numerous phenols, while invertase inverts the optical rotation of a solution of sucrose ($[\alpha] = +66.5°$) to $[\alpha] = -19.7°$ due to formation of equimolar concentrations of glucose ($[\alpha]) = +52.7°$) and fructose ($[\alpha] = -92°$). Note that two general principles began to be used in naming some enzymes: *-ase* to designate an enzyme, while in two cases the stem was derived from the name of one of the substrates (polyphenol oxidase from phenols and peroxidase from peroxide). In the case of polyphenol oxidase, it is clear the enzyme catalyzes oxidation of phenols.

7.2.1 The Enzyme Commission

As more enzymes were discovered nomenclature became an increasingly severe problem. Therefore, an International Commission on Enzymes of the International Union of Biochemistry was established and its charge was "to consider the classification and nomenclature of enzymes and coenzymes, their units of activity and standard methods of assays, together with the symbols used in the description of enzyme kinetics." The Report of the Commission (1961) contained 712 enzymes. Subsequent updates were issued, the most recent being Enzyme Nomenclature (1992), which contains 3,196 enzymes [35].

7.2.1.1 Rules for Naming

The basis for the classification adopted by the Enzyme Commission was to divide the enzymes into groups on the basis of the type of reaction catalyzed, and this, together with the name(s) of the substrate(s), provided a basis for naming individual enzymes.

There are six types of reactions catalyzed by enzymes: (a) oxidoreduction, (b) transfer, (c) hydrolysis, (d) formation of double bonds without hydrolysis, (e) isomerization, and (f) ligation. The names of the type of enzymes are formed by adding -ase to the stem of the type of reaction catalyzed. Thus the corresponding six groups of enzymes are (a)oxidoreductases, (b) transferases, (c) hydrolases, (d) lyases, (e) isomerases, and (f) ligases.

7.2.1.2 Overall Reaction as Basis of Nomenclature

The overall reaction, as expressed by the formal chemical equation (Eq. 6), is the basis for the nomenclature. Intermediate steps in the reaction reflecting the mechanism are not taken into account. Therefore, an enzyme cannot be named properly by this system until the substrate(s) and chemical nature of the reaction catalyzed are known.

$$\underset{\text{Ethanol}}{CH_3CH_2OH} + NAD^+ \rightleftharpoons \underset{\text{Acetaldehyde}}{CH_3CHO} + NADH + H^+ \tag{6}$$

7.2.1.3 Three-Tier Classification

The Enzyme Commission assigned each enzyme three designations: a systematic name, a trivial name, and an Enzyme Commission (EC) number. Generally, the systematic name is composed of two principal parts. The first consists of the name(s) of the substrate(s); in the case of two or more substrates (or reactants), the name of the substrates are separated by a colon. The second part of the systematic name, ending in -ase, is based on one of the six types of chemical reactions catalyzed (see Sec. 7.2.1.1). When the overall reaction involves two different chemical reactions, such as oxidative demethylation, the second type of reaction is listed in parenthesis following the first chemical reaction—for example, "sarcosine:oxygen oxidoreductase (demethylating)."

The trivial name is one that is generally recognized and in common use such as α-amylase, cellulase, trypsin, chymotrypsin, peroxidase, or catalase. It is usually a shorter name than the systematic name.

The number system derives directly from the classification scheme, and each number contains four digits, separated by periods and preceded by EC. The numbers are permanent. Newly discovered enzymes are placed at the end of the list under appropriate headings. If the classification of an enzyme is changed, the number remains in the listing, but the user is directed to the new listing, including the new number of the enzyme.

7.2.1.4 Six Main Types of Enzymes

The six main types of enzymes, based on the chemical reaction catalyzed, are further explained in this section.

Oxidoreductases

Oxidoreductases are enzymes that oxidize or reduce substrates by transfer of hydrogens or electrons or by use of oxygen. The systematic name is formed as "donor:acceptor oxidoreductase." An example, including systematic name followed by trivial name and EC number in parenthesis, is [35]

$$H_2O_2 + H_2O_2 = O_2 + 2H_2O$$

which gives hydrogen peroxide:hydrogen peroxide oxidoreductase (catalase, EC 1.11.1.6).

Transferases

Transferases are enzymes that remove groups (not including H) from substrates and transfer them to acceptor molecules (not including water). The systematic name is formed as "donor:acceptor group-transferred-transferase." An example is:

$$\text{ATP} + \text{D-glucose} = \text{ADP} + \text{D-glucose 6-phosphate}$$

which gives ATP:D-glucose 6-phosphotransferase (glucokinase, EC 2.7.1.2). Note that the position to which the group is transferred is given in the systematic name when more than one possibility exists.

Hydrolases

Hydrolases are enzymes in which water participates in the breakage of covalent bonds of the substrate, with concurrent addition of the elements of water to the principles of those bonds. The systematic name is formed as "substrate hydrolase." Water is not listed as a substrate, even though it is, because it is 55.6 *M* and the concentration does not change significantly during the reaction. When the enzyme specificity is limited to removal of a single group, the group is named as a prefix, for example, "adenosine aminohydrolase." Another example is

Triacylglycerol + H_2O = diacylglycerol + a fatty acid anion

catalyzed by triacylglycerol acylhydrolase (triacylglycerol lipase, EC 3.1.1.3).

Lyases

Lyases are enzymes that remove groups from their substrates (not by hydrolysis) to leave a double bond, or which conversely add groups to double bonds. The systematic name is formed as "substrate prefix-lyase." Prefixes such as "hydro-" and "ammonia-" are used to indicate the type of reaction—for example, "L-malate hydro-lyase" (EC 4.2.1.2). Decarboxylases are named as carboxy-lyases. A hyphen is always written before "lyase" to avoid confusion with hydrolases, carboxylases, etc. An example is:

(*S*)-Malate = fumarate + H_2O

using the enzyme (*S*)-malate hydro-lyase (fumarate hydratase, EC 4.2.1.2; formerly known as fumarase).

Isomerases

Isomerases are enzymes that bring about an isomerization of substrate. The systematic name is formed as "substrate prefix-isomerase." The prefix indicates the type of isomerization involved, for example, "maleate *cis-trans*-isomerase" (EC 5.2.1.1) or "phenylpyruvate keto-enol-isomerase" (EC 5.3.2.1). Enzymes that catalyze an aldose–ketose interconversion are known as "ketol-isomerases," for example, "L-arabinose ketol-isomerase" (EC 5.3.1.4). When the isomerization consists of an intramolecular transfer of a group, such as 2-phospho-D-glycerate = 3-phospho-D-glycerate, the enzyme is named a "mutase," for example, "D-phosphoglycerate 2,3-phosphomutase" (EC 5.4.2.1). Isomerases that catalyze inversions of asymmetric groups are termed "racemases" or "epimerases," depending on whether the substrate contains one or more than one center of asymmetry, respectively. A numerical prefix is attached to the word "epimerase" to show the position of inversion. An example is

L-Alanine = D-alanine

with alanine racemase (alanine racemase, EC 5.1.1.1).

Ligases

Ligases are enzymes that catalyze the covalent linking together of two molecules, coupled with the breaking of a pyrophosphate bond as in ATP. This group of enzymes has previously been referred to as the "synthetases." The systematic name is formed as "X:Y ligase (Z)," where X and Y are the two molecules to be joined together. The compound Z is the product formed from the triphosphate during the reaction.

An example is:

ATP + L-aspartate + NH_3 = AMP + pyrophosphate + L-asparagine

with L-aspartate:ammonia ligase (AMP-forming) (aspartate-ammonia ligase, EC 6.3.1.1).

7.2.2 Numbering and Classification of Enzymes

A key to the numbering and classification of enzymes, in the extensive listings in *Enzyme Nomenclature* (1992) [35], is given in Table 3. The subclassification of the enzymes is best determined from this type of key. For example, it is easy to see how the number system is derived. As noted previously, the first EC number indicates one of the six types of chemical reactions catalyzed by enzymes of that type. Using the EC 1. category, the second number indicates the nature of the donor substrate, the third number the nature of the acceptor molecule (not shown in Table 3) and the fourth number (not shown in Table 3) is the number of the specific enzyme. For example, EC 1.1.1.1 is alcohol dehydrogenase (alcohol:NAD^+ oxidoreductase), where the second number indicates the donor substrate is a primary alcohol, the third number that the acceptor molecule is NAD^+ or $NADP^+$, and the fourth number is the specific enzyme alcohol dehydrogenase. EC 1.6.2.x would be NADH (or NADPH):cytochrome oxidoreductases.

7.3 ENZYMES IN ORGANISMS

All organisms contain many enzymes. The raw materials that are converted to foods contain hundreds of different enzymes. These enzymes are important in the growth and maturation of the raw materials, and the enzymes continue to be active after harvest until all the substrate is exhausted, or the pH changes to one where the enzyme is no longer active, or the enzyme is denatured by an imposed treatment (pH, heating, chemicals, etc.). Often, enzyme activity increases during storage because of disintegration of cellular structure that separates enzymes and their substrates. As an example, softening of tomatoes accelerates following optimum maturity due to increased enzyme activity. Plant tissues contaminated with microorganisms, such as brown rot fungi containing polyphenol oxidase, can literally be lost overnight due to the rapid growth and multiplication of enzyme-containing microorganisms. Bruising of a fruit (i.e. apple) or vegetable (i.e. potato) leads to rapid browning due to polyphenol oxidase. This is because O_2 from the atmosphere, one of the required substrates, can more easily penetrate the damaged skin of the fruit or vegetable. An intact skin is one of the best protections against browning. Unfortunately, up to 50% of tropical fruits are lost due to improper handling, shipping, and storage, and this is caused by uncontrolled activity of endogenous enzymes or exogenous enzymes present in attached microorganisms.

7.3.1 Location of Enzymes

Enzymes are not uniformly distributed in organisms. A particular enzyme sometimes is found only in one type of organelle within a cell, as shown in Table 4. The nucleus (Fig. 2 and Table 4) contains primarily enzymes involved with nucleic acid biosynthesis and hydrolytic degradation. The mitochondria (Fig. 2) contain oxidoreductases associated with oxidative phosphorylation and formation of ATP, and the lysosomes and the pancreatic zymogen granules contain primarily hydrolases (Table 4). Each type of organelle in a cell is specialized in carrying out limited types of enzyme-catalyzed reactions.

Organs may also be specialized in the types of enzymes found. In animals, the gastrointestinal tract contains primarily hydrolases designed to hydrolyze complex α-1,4-glucose-type carbohydrates, lipids, proteins, and nucleic acids to glucose, glycerol and fatty acids, amino acids, and purine and pyrimidines, respectively. Hydrolysis (digestion) of starch and glycogen by α-amylase begins in the mouth and is completed in the small intestine. Hydrolysis of proteins begins in the stomach (with pepsin) and is completed in the small intestine (with trypsin, chymotrypsin, carboxypeptidases A and B, aminopeptidases, and di- and tripeptidases). The first four proteolytic enzymes are synthesized in and secreted by the pancreas into the small intestine. Lipids are hydrolyzed by gastric (stomach) lipases and small-intestinal lipases. Nucleic acids are hydrolyzed by nucleases in the small intestine. Plant organs also contain specialized enzymes. For example, the seed contains substantial amounts of hydrolytic enzymes to hydrolyze starch and proteins, especially to feed the new seedling following seed germination.

TABLE 3 Key to Numbering and Classification of Enzymes[a]

1.	**Oxidoreductases**
1.1	Acting on the CH—OH group of donors
1.2	Acting on the aldehyde or oxo group of donors
1.3	Acting on the CH—CH group of donor
1.4	Acting on the CH—NH_2 group of donors
1.5	Acting on the CH—NH group of donors
1.6	Acting on NADH or NADPH
1.7	Acting on other nitrogenous compounds as donors
1.8	Acting on a sulfur group of donors
1.9	Acting on a heme group of donors
1.10	Acting on diphenols and related substances as donors
1.11	Acting on hydrogen peroxide as acceptor
1.12	Acting on hydrogen as donor
1.13	Acting on single donors with incorporation of molecular oxygen (oxygenases)
1.14	Acting on paired donors with incorporation of molecular oxygen
1.15	Acting on superoxide radicals as acceptor
1.16	Oxidizing metal ions
1.17	Acting on CH_2 groups
1.18	Acting on reduced ferredoxin as donor
1.19	Acting on reduced flavodoxin as donor
1.97	Other oxidoreductases
2.	**Transferases**
2.1	Transferring one-carbon groups
2.2	Transferring aldehyde or ketone residues
2.3	Acyltransferases
2.4	Glycosyltransferases
2.5	Transferring alkyl or aryl groups, other than methyl groups
2.6	Transferring nitrogenous groups
2.7	Transferring phosphorus-containing groups
2.8	Transferring sulfur-containing groups
3.	**Hydrolases**
3.1	Acting on ester bonds
3.2	Glycosidases
3.3	Acting on ether bonds
3.4	Acting on peptide bonds (peptidases)
3.5	Acting on carbon-nitrogen bonds, other than peptide bonds
3.6	Acting on acid anhydrides
3.7	Acting on carbon-carbon bonds
3.8	Acting on halide bonds
3.9	Acting on phosphorus-nitrogen bonds
3.10	Acting on sulfur-nitrogen bonds
3.11	Acting on carbon-phosphorus bonds
4.	**Lyases**
4.1	Carbon-carbon lyases
4.2	Carbon-oxygen lyases
4.3	Carbon-nitrogen lyases
4.4	Carbon-sulfur lyases
4.5	Carbon-halide lyases
4.6	Phosphorus-oxygen lyases
4.99	Other lyases

5. Isomerases
   - 5.1 Racemases and epimerases
   - 5.2 *cis–trans*-Isomerases
   - 5.3 Intramolecular oxidoreductases
   - 5.4 Intramolecular transferases (mutases)
   - 5.5 Intramolecular lyases
   - 5.99 Other isomerases
6. Ligases
   - 6.1 Forming carbon-oxygen bonds
   - 6.2 Forming carbon-sulfur bonds
   - 6.3 Forming carbon-nitrogen bonds
   - 6.4 Forming carbon-carbon bonds
   - 6.5 Forming phosphoric ester bonds

[a]The third and fourth levels of classification are given in Ref. 35.
Source: Ref. 35, pp. v–xi, by courtesy of Academic Press.

If we want to find a particular enzyme in plants, animals, and microorganisms, it is important to know in which organelle or organ an enzyme is located. Two techniques are important for this. The organs containing a particular enzyme can be determined by histochemical techniques (Fig. 3). As shown in Figure 3, alkaline phosphatase and polyphenol oxidase are not distributed uniformly in cells of porcine longissimus muscle and the beet, respectively. These sensitive enzyme histochemical techniques are used routinely by medical and veterinary histologists and pathologists [100]. Unfortunately, they are rarely used by food scientists.

TABLE 4 Subcellular Location of Several Enzymes in Animal Cells

Location	Enzymes
Nucleus	*DNA-dependent RNA polymerase,*[a] polyadenylate synthetase
Mitochondria	*Succinate dehydrogenase, cytochrome oxidase,* glutamate dehydrogenase, malate dehydrogenase, α-ketoglutarate dehydrogenase, α-glyerol phosphate dehydrogenase, pyruvate decarboxylase
Lysosomes	Cathepsins A, B, C, D, and E, collagenase, *acid ribonuclease, acid phosphatase,* β-galactosidase, sialidase, lysozyme, triglyceride lipase
Peroxisomes (microbodies)	Catalase, *urate oxidase,* D-amino acid oxidase
Reticulum, golgi, etc.	*Glucose 6-phosphatase, nucleoside diphosphatase, TPNH-linked lipid peroxidase,* nucleoside phosphatases
Soluble	*Lactate dehydrogenase, phosphofructokinase, glucose 6-phosphate dehydrogenase,* transketolase, transaldolase
Pancreatic zymogen granules	*Trypsinogen,*[b] *chymotrypsinogen,*[b] lipase, amylase, ribonuclease

[a]Enzymes in italics are most frequently used as indicator enzymes for type of cellular organelles present in a homogenate.
[b]Zymogens activated to trypsin and chymotrypsin, respectively.
Source: Ref. 111, p. 68.

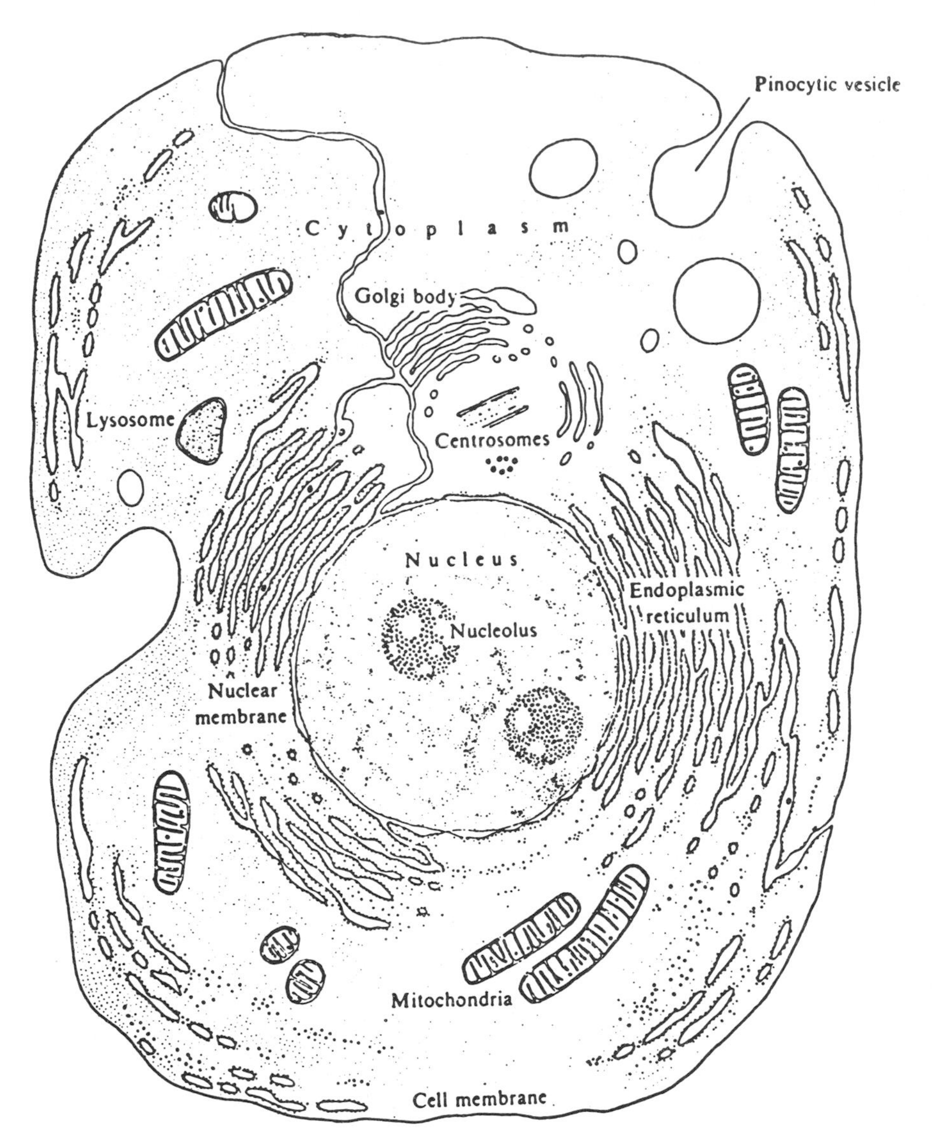

FIGURE 2 "Typical" cell. Diagram of a typical cell based on data from electron micrographs. (From Ref. 14, p. 55, by courtesy of Scientific American, Inc.)

The second method used to determine which organelle contains a particular enzyme is to use differential centrifugation, following careful disintegration of the tissue in a buffer to control pH and osmotic pressure so that the organelles remain intact. The separated organelles are then disintegrated and the enzymes present are determined by use of specific substrates to detect their presence and concentration. The results of this type of study are shown in Table 4.

7.3.2 Compartmentation and Access to Substrate

The activities of enzymes in intact cells are controlled by compartmentation in subcellular membranes, by organelles, by membrane- or cell-wall-bound enzymes and/or substrates, and by

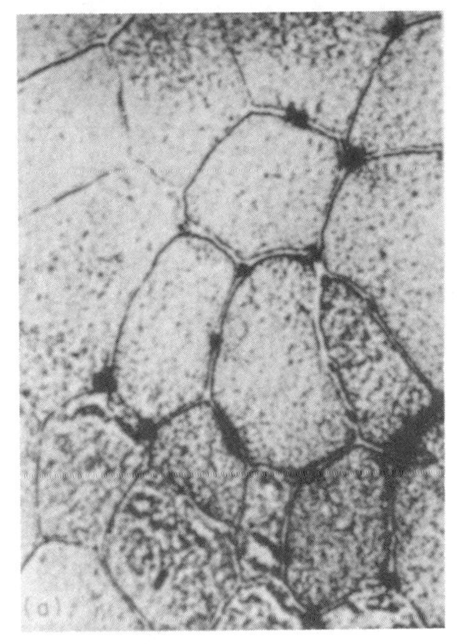
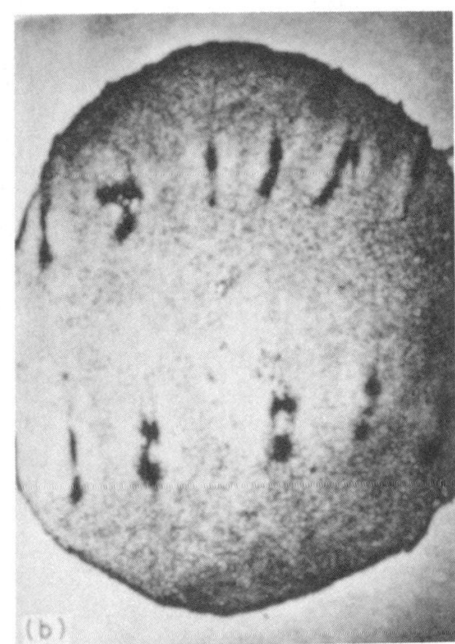

FIGURE 3 Histochemical localization of enzyme activity. (a) Alkaline phosphatase activity in frozen section of porcine longissimus muscle detected with α-naphthyl phosphate and diazotized 4′-amino-2′, 5′-diethoxybenzanilide. (From Ref. 24, p. 300, by courtesy of Institute of Food Technologists.) (b) Transverse section of fresh beef tissue after treatment with dihydroxyphenylalanine for detection of polyphenol oxidase. (From Ref. 13, p. 579, by courtesy of the Institute of Food Technologists.)

separation of the substrate from the enzyme, including exclusion of free O_2 from the tissues. Other controls of activity include proenzyme (inactive until activated) biosynthesis (such as found in the blood and pancreas of animals) and by physiologically important endogenous enzyme inhibitors. Partial disintegration of the tissue, by aging or bruising, by insects or microorganisms, by intentional peeling, cutting, slicing, or blending, or by freezing and thawing, brings the enzymes and substrates together, allowing the enzymes to act very rapidly on their substrates. This can cause rapid changes in the color, texture, flavor, aroma, and nutritional qualities of food. Heat treatment, storage at low temperature, and/or use of enzyme inhibitors is necessary to stabilize the product.

7.3.3 Typical Concentrations of Enzymes in Some Foods

Relative amounts of three enzymes, polygalacturonase, lipoxygenase, and peroxidase, in plants are shown in Table 5. Polygalacturonase, responsible for softening, varies widely in content in the plant sources listed, being very high in concentration in tomato and zero in cranberry, carrot, and grape. Lipoxygenase is very high in concentration in soybeans, accounting for the beany taste of cooked soybeans, but occurs at barely detectable levels in wheat and peanuts. Peroxidase is found in essentially all fruits, but varies some sevenfold from English green peas to lima beans.

Polyphenol oxidase is one of the most noticeable enzymes in plants. It is present at high concentrations in some grapes (dark raisins), prune plums, Black Mission figs, dates, tea leaves, and coffee beans, where its action is desired. It is present at moderate concentrations in peaches, apples, bananas, potatoes, and lettuce, where its activity is undesirable, and is not present in peppers.

TABLE 5 Relative Amounts of Some Enzymes in Various Sources

Enzyme	Source	Relative amounts
Polygalacturonase[a]	Tomato	1.00
	Avocado	0.065
	Medlar	0.027
	Pear	0.016
	Pineapple	0.024
	Cranberry	0
	Carrot	0
	Grape	0
Lipoxygenase[b]	Soybeans	1.00
	Urd beans	0.60
	Mung beans	0.47
	Peas	0.35
	Wheat	0.02
	Peanuts	0.01
Peroxidase[c]	Green peas	1.00
	Pea pods	0.72
	String beans	0.62
	Spinach	0.32
	Lima beans	0.15

[a]Adapted from Ref. 51.
[b]Adapted from Ref. 93.
[c]Adapted from Ref. 54.

The levels of a given enzyme activity can be highly variable in raw foods, since a cultivar of the same fruit, for example, can be bred to have more or less enzyme. The age (maturity) of an organism (Fig. 4), and the environmental conditions of growing (especially plants), including temperature, water supply, soil, and fertilization, all affect the level of enzyme activity. This makes it very difficult for the food processor to produce uniform-quality food products. Fortunately, the rate of denaturation of enzymes is generally first order; therefore, the time required to inactivate a fixed percentage of the enzyme is independent of its concentration. However, at different initial concentrations of active enzyme, the absolute concentrations of active enzyme left will be different. For example, at 5 half-lives, there will be 3.13% of the original active enzyme left regardless of initial active enzyme concentration. But at initial concentrations of $1 \times 10^{-3} M$ and $1 \times 10^{-5} M$ active enzyme, after 5 half-lives there will be $3.13 \times 10^{-5} M$ and $3.13 \times 10^{-7} M$ active enzyme left, respectively. So both the two initial and two final concentrations of active enzyme differ by 100. It takes longer to inactivate all the enzyme with higher initial concentration. The conditions just listed also affect the relative levels of isozymes (for example, those of peroxidase or lipoxygenase). These isozymes often have different temperature stabilities, as shown for the isoenzymes of peroxidase and lipoxygenase (Fig. 5). The presence of two or more isozymes is indicated by the inability to inactivate completely each enzyme at 60°C; at 70°C, lipoxygenase is completely inactivated but peroxidase still retains about 40% activity.

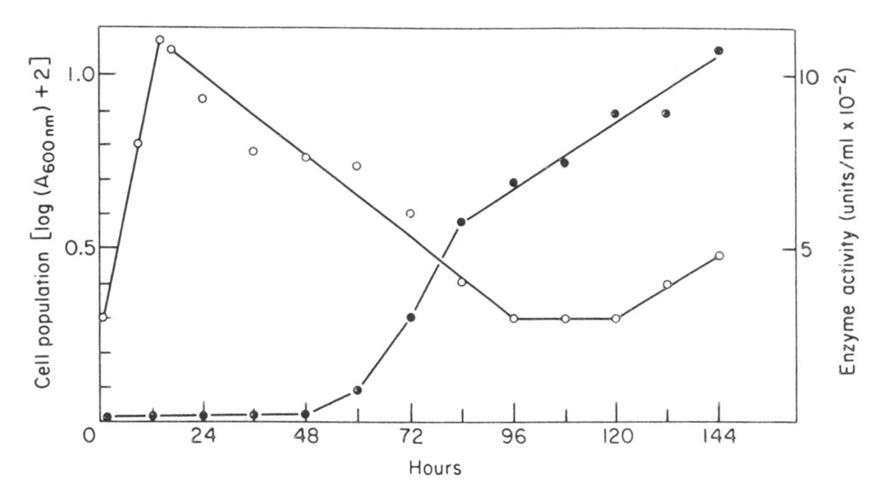

Figure 4 Changes in the polygalacturonic acid lyase activity of *Bacillus pumilis* as a function of age of culture: ○, absorbance at 600 nm of a 1:20 dilution as a measure of cell population; ●, activity of enzyme (units/ml). (From Ref. 28, p. 41, by courtesy of B. A. Dave.)

7.4 RATES OF ENZYME-CATALYZED REACTIONS

7.4.1 Rate Determination

The distinguishing feature of enzymes from other proteins is that enzymes bind their substrates stereospecifically into the active site *and* convert the substrates to products. Therefore, the presence of an enzyme is determined by whether it converts a compound to a product faster than that which occurs in its absence. Assays for enzyme activity are based on physical or chemical

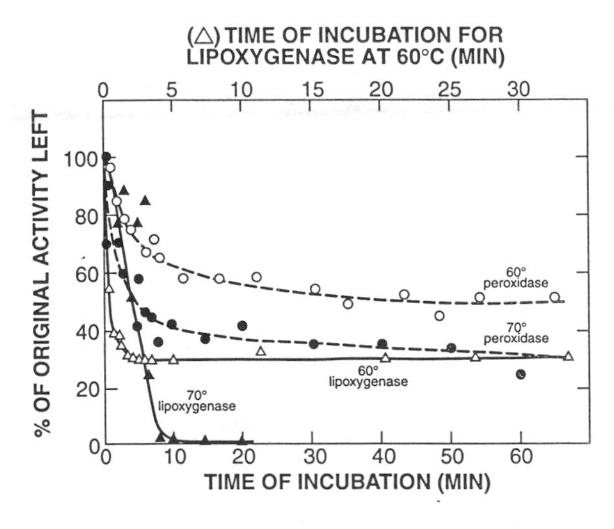

Figure 5 Rates of inactivation of peroxidase and lipoxygenase in English green pea homogenates incubated at 60 and 70°C. (From Ref. 114, p. 132, with permission of the Institute of Food Technologists.)

changes in substrate(s) and product(s) in the reaction mixture. Continuous methods (changes in absorbance, fluorescence, or pH) are very much preferred over methods that require taking aliquots at selected times, stopping the reaction by inactivating the enzyme, and adding one or more reagents to form a measurable derivative of either the substrate or product (preferable).

Since an enzyme is highly specific for only one or a few substrates, it follows that it is relatively easy to measure that enzyme selectively even among hundreds of other enzymes of different specificities. The enzyme activity can be used to determine not only the presence of an enzyme (qualitative), but also how much enzyme is present (quantitative), based on its rate under controlled conditions.

Some substrates and products do not differ greatly in physical and/or chemical properties; therefore it is difficult to determine the first-order rate constant, k_1 (Eq. 7). In these cases, coupled enzyme assays should be considered, where the product, P_1, of the first enzyme is a substrate of an added second enzyme. The product, P_2, of the second enzyme is then chosen for measurement (Eq. 7). In order to measure k_1 correctly, the relation $k_2[E_2] \cong 100\ k_1[E_1]$ should be used.

$$\mathrm{A} \xrightarrow{\mathrm{E_1},\, k_1} \mathrm{P_1} \xrightarrow{\mathrm{E_2},\, k_2} \mathrm{P_2} \tag{7}$$

7.4.2 Steady-State Rates

The velocity of a typical enzyme-catalyzed reaction is shown in Fig. 6, where concentration of product formed is plotted versus time. In the first few milliseconds, the velocity (dP/dt) of product formation accelerates depending on how fast the enzyme and substrate combine to give the enzyme · substrate complex (Eq. 2). This is the *pre-steady-state* period (0.1–2 msec) and can only be measured by very fast mixing and measuring systems (i.e., stopped-flow spectrophotometers). After ~2 msec, the concentration of enzyme–substrate complex [E·S] reaches a steady

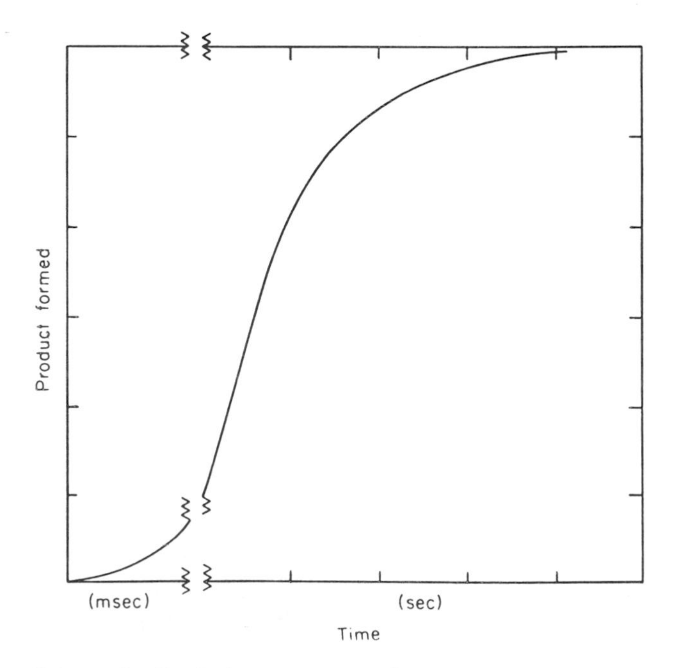

FIGURE 6 Typical enzyme-catalyzed reaction showing the pre-steady-state (msec), the constant rate, and the declining rate parts of the progress curve. (From Ref. 111, p. 161.)

state where $d[\text{E·S}]/dt = -d[\text{E·S}]/dt$ over the time required to measure a change in [P] or [S]. The steady-state concept applies from the end of the pre-steady-state period to the end of conversion of all substrate. Enzyme activity is measured under steady-state conditions in most laboratories (see later discussion). Why then does the velocity of the reaction become slower and approach zero as shown in Figure 6? The decrease in velocity with time can be a result of these: (a) the [S] becomes the rate-limiting factor (i.e., $[S] < 100K_m$); (b) as [P] increases it may partially inhibit the enzyme; (c) the reaction is reversible and the reverse reaction becomes more noticeable as [P] increases; and/or (d) instability of the enzyme. These possible causes can be distinguished by appropriate methods [111].

Since steady-state velocities of enzyme-catalyzed reactions decrease with reaction time, often in a complex manner for the reasons already given, enzymologists generally determine the initial velocity, v_o, of enzyme-catalyzed reactions. This is done as shown in Figure 7. If done with proper control(s), all reactions should begin at zero time with zero [P]. The v_o is determined from a tangent drawn (dashed lines) to the initial part of the reaction. No more than 5% of the substrate should be converted to product during the time required to obtain the tangent. Note that accurate tangents are easier to obtain when the velocity is relatively slow. Experimentally, the velocity can be controlled by controlling the enzyme concentration, and this must be done if accurate, reproducible v_o values are expected. Please note that the slope of two point assays (zero at zero time and another point later on) does not yield accurate values of v_o!

7.4.2.1 Kinetics and Reaction Order

Reaction order is determined from the dependence of velocity (dP/dt, or $-dS/dt$) on the concentration(s) of the reactant(s). Therefore, we begin with an equation. For an enzyme,

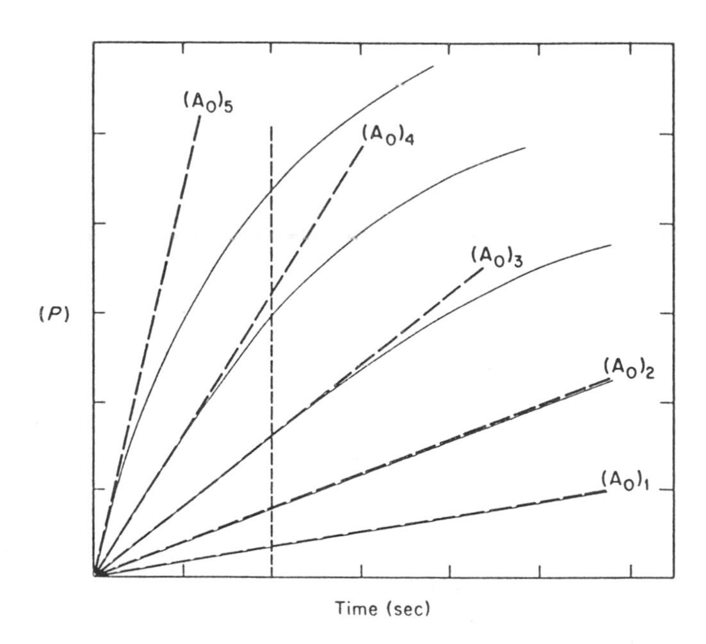

FIGURE 7 Method of determining initial velocities of reactions. The solid lines are the experimentally determined data, and the dashed lines are tangents drawn to the initial slope of the experimental data. Note the marked difference between the actual concentration of product formed at the time indicated by vertical dashed line for reactions $(A_o)_4$ and $(A_o)_5$ compared with that predicted from initial rates. (From Ref. 111, p. 163.)

appropriate kinetic description is given by Equation 8 (same as Eq. 1), where E is enzyme, S is substrate, E·S is the enzyme–substrate complex, and P is product. The velocity of

$$\mathrm{E} + \mathrm{S} \underset{k_{-1}}{\overset{k_1}{\rightleftharpoons}} \mathrm{E{\cdot}S} \underset{k_{-1}}{\overset{k_1}{\rightleftharpoons}} \mathrm{E} + \mathrm{P} \tag{8}$$

formation of E·S (d[E·S]/dt) is k_1[E][S], while the velocity of disappearance of E·S (–d[E·S]/dt) is k_{-1}(E·S) + k_2[E·S]. Steady-state conditions exist whenever d(E·S)/dt = –d(E·S)/dt for ~ 5 msec.

The Michaelis–Menten equation (Eq. 9), derived from Equation 8, is based on the

$$v_o = \frac{V_{max}[\mathrm{S}]_o}{K_m + [\mathrm{S}]_o} = \frac{k_2[\mathrm{E}]_o[\mathrm{S}]_o}{K_m + [\mathrm{S}]_o} \tag{9}$$

following assumptions [111]:

1. Initial velocity, v_o, is used so that $[\mathrm{S}]_o \approx [\mathrm{S}]$. As discussed earlier, this is common practice.
2. $[\mathrm{S}]_o >> [\mathrm{E}]_o$, so that there is little change in $[\mathrm{S}]_o$. Practically, this is the case with most enzymes where $[\mathrm{S}]_o$ is of the order of 10^{-4} to 10^{-2} *M* since K_m is generally within this range and $[\mathrm{E}]_o$ is on the order of 10^{-8}–10^{-6} *M*.
3. The step controlled by k_2 (Eq. 8) is irreversible in reality or because v_o is used in ([P] is essentially zero). Therefore, k_{-2} is ~0.
4. *d*(E·S/*dt* = –*d*(E·S)/*dt*, so steady-state conditions prevail.
5. k_2 controls the velocity of formation of product (*dP*/*dt*). If $k_2 > k_1$ then k_1 controls velocity of product formation. Note that V_{max} is not really a constant in Equation 9, since it is dependent on $[\mathrm{E}]_o$ ($V_{max} = k_2[\mathrm{E}]_o$); k_2 is a constant, but V_{max} will change when $[\mathrm{E}]_o$ is changed.
6. If any one of these assumptions is not true then the form of the Michaelis–Menten equation will be more complex, even though a plot of v_o versus $[\mathrm{S}]_o$ will be hyperbolic (Fig. 1). For example, if there is an additional intermediate in the reaction, such as an acylenzyme (Eq. 10); [111], then Equation 10 applies:

$$\mathrm{E} + \mathrm{S} \underset{k_{-1}}{\overset{k_1}{\rightleftharpoons}} \mathrm{E{\cdot}S} \xrightarrow{k_2} \mathrm{E} - \mathrm{X} \xrightarrow{k_3} \mathrm{E} + \mathrm{P} \tag{10}$$

and the Michaelis–Menten equation is

$$v_o = \frac{V_{max}[\mathrm{S}]_o}{K_m + [\mathrm{S}]_o} = \frac{[k_2k_3/(k_2 + k_3)][\mathrm{E}]_o[\mathrm{S}]_o}{(k_{-1} + k_2)k_3/k_1(k_2 + k_3) + [\mathrm{S}]_o} \tag{11}$$

In the preceding equations, $[\mathrm{E}]_o$ is total enzyme concentration, [E] is free enzyme concentration, [E·S] is enzyme–substrate concentration, $[\mathrm{S}]_o$ is initial substrate concentration, [S] is substrate concentration at any time *t*, and $K_m = (k_2 + k_{-1})/k_1$ for Equation 9 and $(k_{-1} + k_2)k_3/k_1(k_2 + k_3)$ for Equation 11.

What is the order of an enzyme-catalyzed reaction and how does one determine the order? Since $[\mathrm{E}]_o$ is constant throughout the reaction, assuming it is stable, the order of an enzyme-catalyzed reaction is determined by the relationship between $[\mathrm{S}]_o$ and K_m (as long as $[\mathrm{S}]_o >> [\mathrm{E}]_o$; see assumption 2). When $[\mathrm{S}]_o \le 0.01\ K_m$, the order of the reaction is first order with respect to $[\mathrm{S}]_o$. This can be readily seen from Equation 9. When $[\mathrm{S}]_o < 0.01\ K_m$, $[\mathrm{S}]_o$ in the denominator can be ignored and $v_o = V_{max}[\mathrm{S}]_o/K_m$ ($v_o = k'[\mathrm{S}]_o$), where $k' = Vmax/K_m$, showing direct dependence on $[\mathrm{S}]_o$. A plot of $\ln([\mathrm{S}]_o/[\mathrm{S}])$ versus time gives a straight-line plot over several half-lives of substrate concentration disappearance, with the slope equal to V_{max}/K_m, and the half-life time independent of $[S]_o$.

When $[S]_o \geq 100K_m$, the order of the reaction with respect to $[S]_o$ is zero. This also can be seen from Equation 9. When $[S]_o >> K_m$, Equation 9 reduces to $v_o = V_{max}$, indicating that v_o is independent of $[S]_o$ and all enzyme is saturated with substrate. A plot of [P] versus time gives a straight-line plot, with a slope V_{max} $(= k_2[E]_o)$.

When $[S]_o > 0.01$ K_m but < 100 K_m, the order of the reaction with respect to $[S]_o$ is a mixture of first and zero order and the integrated form of the Michaelis–Menten equation is:

$$k_2[E]_o t = K_m \ln([S]_o/[S]) + ([S]_o - [S]) \tag{12}$$

The relative contributions of the terms $\ln([S]_o/[S])$ and $([S]_o - [S])$ will depend on the relation of $[S]_o$ to K_m. Section 7.5 on factors influencing enzyme reactions will provide experimental procedures for evaluating the effect of $[S]_o$ on v_o.

Some typical rate constants of enzyme-catalyzed reactions are given in Table 6. The value of k_1, the rate constant for formation of the enzyme–substrate complex, ranges from 10^9 to 10^4 M^{-1} sec^{-1} (where 10^9 is the limiting rate of diffusion-controlled reactions). The value of k_{-1}, the rate constant for dissociation of the enzyme–substrate complex, ranges from about 4.5×10^4 to about 1.4 scc^{-1}. The value of k_o, the observed rate constant for conversion of enzyme–substrate complex to product, ranges from 10^7 to 10^1 sec^{-1}. In reactions that follow Equation 8 $k_o = k_2$, but not for those that follow Equation 10.

7.4.3 Pre-Steady-State Reactions [17, 44]

Pre-steady-state kinetic experiments are rarely done in food science, yet they provide details regarding additional intermediate steps in enzyme reactions that are difficult to obtain by other

TABLE 6 Rate Constants for Some Selected Enzymes

Enzyme	Substrate	k_1[a] (M^{-1} sec^{-1})	k_1[b] (sec^{-1})	k_o[c] (sec^{-1})
Fumarase	Fumarate	$>10^9$	$>4.5 \times 10^4$	10^3
Acetylcholinesterase	Acetylcholine	10^9	—	10^3
Liver alcohol	NAD^+	5.3×10^5	74	10^3
dehydrogenase	NADH	1.1×10^7	3.1	
	Ethanol	$>1.2 \times 10^4$	>74	
Catalase	H_2O_2	5×10^6	—	10^7
Peroxidase	H_2O_2	9×10^6	<1.4	10^6
Hexokinase	Glucose	3.7×10^6	1.5×10^3	10^3
Urease	Urea	$>5 \times 10^6$	—	10^4
Chymotrypsin				10^2 to 10^3
Trypsin				10^2 to 10^3
Ribonuclease				10^2
Papain				10^1

[a]Rate constant for formation of enzyme–substrate complex.
[b]Rate constant for dissociation of enzyme–substrate complex.
[c]Order of magnitude of the turnover number in moles of substrate converted to product per second per mole of enzyme: k_o ($\equiv k_{cat}$) is the observed rate constant and may or may not involve a single rate-limiting step.
Source: Ref. 33, p. 14, courtesy of Academic Press.

methods. Why are they not used in food science? Primarily because of requirements of relatively large amounts of pure enzyme, specialized instrumentation, and lack of appreciation of the nature and simplicity of interpreting the results.

The instrumentation for following pre-steady-state reactions requires mixing times of <1 msec for substrate and enzyme and recording times of milliseconds. This is because concentrations of enzyme and substrate must be in the range of 10^{-4} M and both must be present at about equal concentration in order to detect intermediate steps.

Bray et al. [17] used low-temperature pre-steady-state kinetic methods to identify and place in order the intermediate steps in the reaction of milk xanthine oxidase with xanthine (Eq. 13), a very important reaction in milk and in gout.

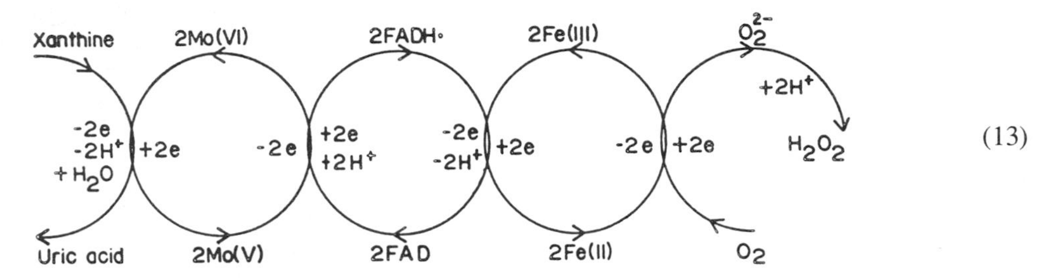

(13)

Another way in which intermediate steps in enzyme-catalyzed reactions can be studied is by doing steady-state kinetics using several substrates that have different rate-determining steps.

7.4.4 Immobilized Enzyme Reactions [52, 88, 117]

The preceding discussions apply to systems where enzyme and substrate are soluble and both are free to diffuse independently until the two collide with enough energy and proper orientation to form the enzyme–substrate complex. However, enzymes are frequently attached to cell walls and membranes in vivo or are bound to insoluble supports for commercial conversion of substrate to product. When the enzyme is immobilized, additional factors affect formation of the enzyme–substrate complex and catalysis. These include: (a) only the substrate is free of diffuse; (b) the enzyme support is surrounded by the Nerst (diffusion) layer, which acts as a boundary, decreasing the substrate concentration in proximity to the enzyme, compared to the bulk phase; (c) there are electrostatic factors due to charges on the substrate, enzyme, and support that may increase binding (opposite charges) or decrease binding (same charges); and (d) initial velocity, v_o, conditions do not apply as maximum conversion of substrate to product is desired.

Hornby et al. [52] developed equations to take factors (a)–(c) into account. The velocity for a flow column reactor is given in Equation 14:

$$v = \frac{V_{\max}[S]}{K_m^* + [S]} \tag{14}$$

where

$$K_m^* = \left(K_m + \frac{xV_{\max}}{D}\right)\frac{RT}{RT - xzFV} \tag{15}$$

and K_m^* is the apparent Michaelis constant, x is the thickness of the Nerst boundary layer, D is the diffusion coefficient for substrate, T is the temperature (Kelvin), z is the valence of substrate, F is Faraday's constant, R is the universal gas constant, and V is the potential gradient around the enzyme support.

Since v is not v_o and as much substrate conversion as possible is desired, zero-, mixed-, and first-order rate kinetics are observed in the reactor at different times of reaction. Therefore,

$$k_2[E]_o t = K_m^* \ln([S]_o/[S]_t) + ([S]_o - [S]_t) \tag{16}$$

describes the kinetic process in the column reactor.

The fraction p of S reacted at any time is

$$p = \frac{[S]_o - [S]_t}{[S]_o} \tag{17}$$

Also required is the void volume, V_o, of the column and the flow rate, Q, of the substrate through the column. The ratio of V_o/Q is equal to the residence time t in the reactor. Therefore, the reactor output can be expressed as

$$[S]_o p = K_m^* \ln(1 - p) + \frac{k_2[E]_o V_o}{Q} \tag{18}$$

Equation 18 is the equation of a straight line, where $y = [S]_o p$ and $x = \ln(1 - p)$. The slope of the plot is K_m^*.

Lilly and Sharp [73] developed an analogous kinetic equation applicable to a continuous-feed stirred tank reactor:

$$[S]_o p = -\frac{p}{1 - p} K_m^* + \frac{k_2[E]_o V_o}{Q} \tag{19}$$

In some cases, such as the hydrolysis of insoluble cellulose, biomass, lipid micelles, and cell walls, the substrate is insoluble and the soluble enzyme must diffuse to and bind to the insoluble substrate as shown in Equation 20 where E is the soluble enzyme and E_S is bound to the insoluble substrate:

$$E \underset{k_d}{\overset{k_p}{\rightleftharpoons}} E_S \tag{20}$$

The size of the insoluble substrate (micelle droplets or insoluble particles) also affects $V_{max}^{\#}$ and $K_m^{\#}$. Verger et al. [102] and others have shown that the kinetics of soluble enzyme action on insoluble substrate fits Michaelis–Menten kinetics where

$$v = \frac{V_{max}^{\#}[S]}{K_m^{\#} + [S]} \tag{21}$$

The value of [S] is dependent on the surface of the insoluble substrate and

$$K_m^{\#} = \frac{k_d}{k_p} \frac{K_m[S]}{K_m + [S]_o} \tag{22}$$

where k_d and k_p are from Equation 20. The values of $V_{max}^{\#}$ and $K_m^{\#}$ can be determined from Lineweaver–Burk plots. The $V_{max}^{\#}$ and $K_m^{\#}$ will be greatly dependent on experimental conditions.

7.5 FACTORS INFLUENCING ENZYME REACTIONS

In the previous sections, we developed the basic concepts and kinetic equations required to quantify the velocity (v_o and v) of enzyme-catalyzed equations. In developing Section 7.4, only the substrate concentration was considered as a variable. All other conditions were assumed to be held constant. In Section 7.5, we will expand further on the effect of substrate concentration on the velocity of enzyme-catalyzed reactions. In addition, the effect of enzyme concentrations, pH, temperature, water activity, and organic solvents on velocity of the reactions will be discussed. The effect of activators and inhibitors on velocity of enzyme-catalyzed reactions will be discussed in Sections 7.6 and 7.7, respectively.

7.5.1 Substrate Concentration

The effect of substrate concentration on velocity of product formation is shown in Figure 1 for one-substrate reactions when the reactions follow Michaelis–Menten kinetics. Determination of V_{max} and K_m from data plotted as in Figure 1 is difficult at best because V_{max} is achieved only when $[S]_o > 100K_m$ (zero order with respect to substrate concentration). The substrate may be insoluble and/or expensive at the concentrations needed, the substrate may inhibit the reaction at high substrate concentrations (Fig. 8), or it may activate the reaction at high substrate concentrations (Fig. 9).

For these and other reasons, Lineweaver and Burk [76] in 1934 showed that the Michaelis–Menten equation (Eq. 9) can be transformed from a right hyperbola to a straight line by taking the reciprocal to give

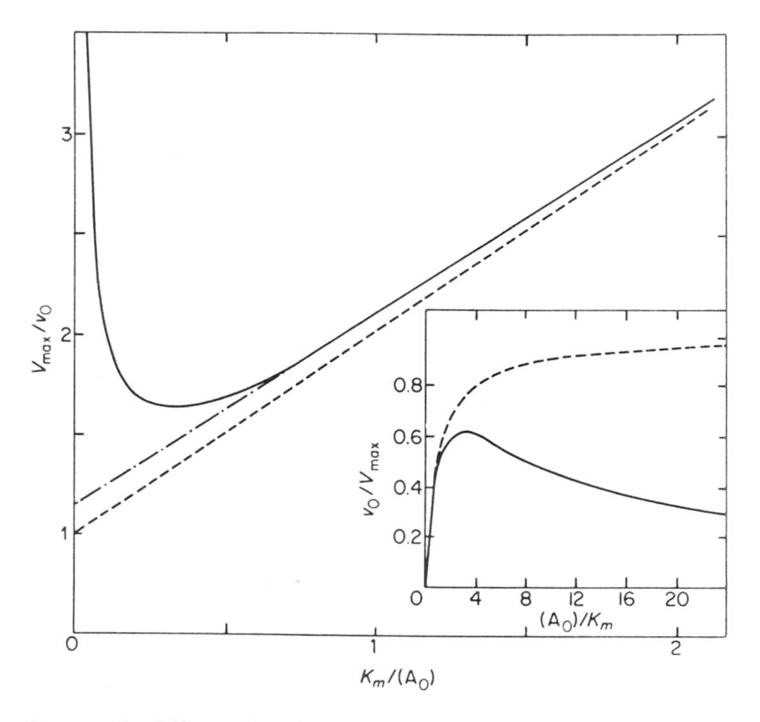

FIGURE 8 Effect of substrate inhibition on the initial velocity of an enzyme-catalyzed reaction. The dashed line shows the normal curve in absence of inhibition, and the solid line shows inhibition by a second substrate molecule for which the dissociation constant, K_s', is $10K_m$. The solid line is calculated according to the equation $v_o = V_{max}/\{[1 + K_m/(A_o)] + [A_o/K_s']\}$. $E{\cdot}A_2$ does not form product. Large graph plotted by Lineweaver–Burk method; the insert is a Michaelis–Menten plot. (From Ref. 111, p. 193).

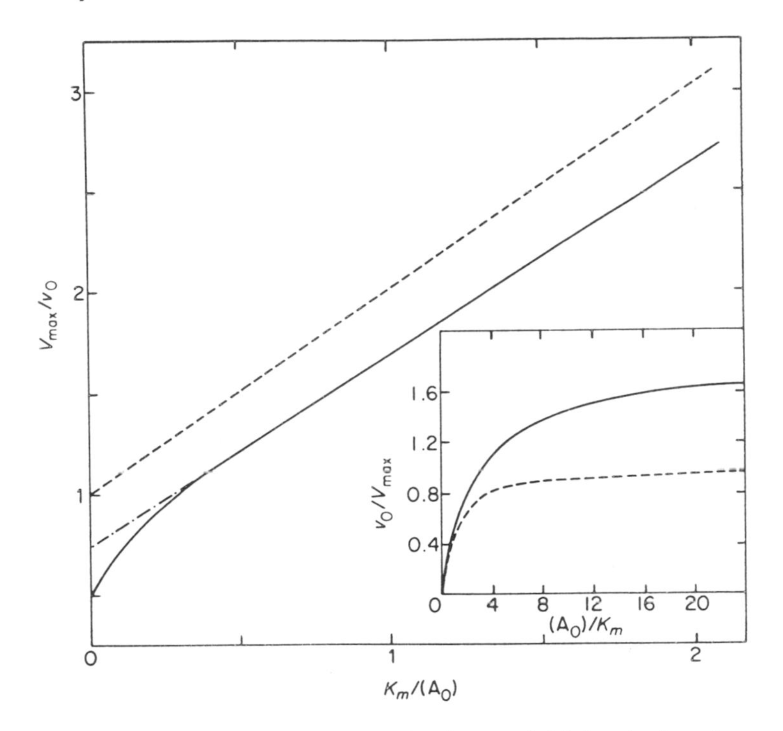

FIGURE 9 Effect of substrate activation on initial velocity of an enzyme-catalyzed reaction. The dashed line shows a normal reaction in absence of activation, and the solid line shows a reaction in the presence of activation. The solid line is calculated on the assumptions that K_s' for the second substrate molecule is $10K_m$, that V_{max} is doubled when all the enzyme is saturated with a second substrate molecule (i.e., $E{\cdot}S_2$ goes to product twice as fast as E·S), and that the second substrate molecule does not form product. The large graph is plotted by the Lineweaver–Burk method; the insert is a Michaelis–Menten plot. (From Ref. 111, p. 194.)

$$\frac{1}{v_o} = \frac{K_m}{V_{max}[S]_o} + \frac{1}{V_{max}} \tag{21}$$

where $1/v_o = y$, $1/[S]_o = x$, K_m/V_{max} is the slope and $1/V_{max}$ is the y-intercept of the straight-line plot, and $-1/K_m$ is the x-intercept (Fig. 10). Therefore, V_{max} and K_m can be readily determined, using all the experimental data. For best results the $[S]_o$ used to obtain data for Figure 10 should range from $0.2K_m$ to $5K_m$, if possible. The Lineweaver–Burk method for calculating V_{max} and K_m is, by far, the most used. Other linear transforms include the methods of Augustinsson (Eq. 22) [111] and of Eadie-Hofstee (Eq. 23) [111].

$$\frac{[S]_o}{v_o} = \frac{K_m}{V_{max}} + \frac{[S]_o}{V_{max}} \tag{22}$$

$$v_o = V_{max} - \frac{v_o}{[S]_o}K_m \tag{23}$$

It is often useful to plot the experimental data by all three methods; easily done with available computer software programs for enzyme kinetics. Linear plots by all three methods are

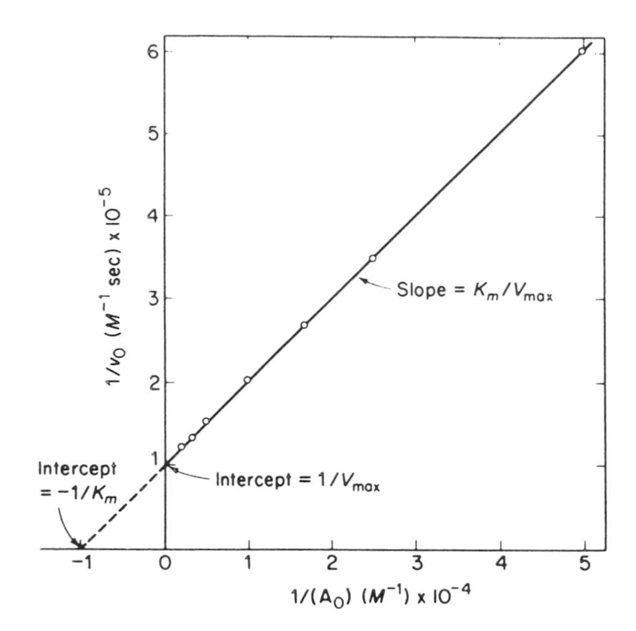

FIGURE 10 Plot of substrate-velocity data according to the Lineweaver–Burk method as shown in Equation 21. (From Ref. 111, p. 176.)

not always found, indicating complexities (multiple intermediate steps, etc.) not readily apparent from only one of the plots.

7.5.1.1 Multiple Substrate Reactions

Most enzyme-catalyzed reactions involve more than one substrate. Hydrolase-catalyzed reactions require water as the second substrate, but under most conditions where the water concentration is 55.6 *M*, its effect is ignored in developing the kinetic equations. However, in all other cases, the effect of concentration of the second substrate cannot be ignored. Several investigators have developed kinetic equations for handling these multi-substrate reactions [22, 27, 115]. Whitaker [111] has summarized the experimental procedures and kinetic approaches to be used. The experimental procedures are no more complex than those described for the one substrate reactions, and most of the reactions fit Michaelis–Menten type kinetics. Experiments must be done so that the effects of all substrates can be determined, varying the concentration of one of them at a time. The interpretation of V_{max} and K_m is more complex but the individual rate constants can be determined readily by the King–Altman method [62], among other methods.

The multiple substrate reactions that fit Michaelis–Menten kinetics belong to three main types [22]. Using examples for two-substrate/two-product reactions, two of these types are (a) ordered sequential bi bi mechanism, in which the two substrates must bind with the active site of the enzyme in an ordered manner and the two products are released from the enzyme in an ordered manner (Eq. 24), and (b) random sequential bi bi mechanism, in which the two substrates can bind into the active site of an enzyme in either order and the two products come off the enzyme in either order (Eq. 25). In the sequential mechanisms, all substrates must bind into the active site of the enzyme before any products are formed. In the third type, (c) the ping-pong bi bi mechanism, the first substrate must bind in the active site and one product is released before the second substrate can bind into the active site and be catalyzed to product (Eq. 26). These different types of mechanisms are best determined by plotting $1/v_o$ versus $1/[S]_o$ according to the

Lineweaver–Burk method. The two sequential mechanisms [(a) and (b)] are distinguished by use of equilibrium dialysis to determine the order of substrate binding into the active site.

Ordered sequential bi bi mechanism

```
     A        B                          P         Q
     ↓        ↓                          ↑         ↑
E    E·A      E·A·B ⇌ E·P·Q              E·Q       E
```
(24)

Random sequential bi bi mechanism

```
        A    B          P    Q
        ↓    ↓          ↑    ↑
          EA              EQ
E                                  E
          EB   (EAB)      EP
               (EPQ)
        ↑    ↑          ↓    ↓
        B    A          Q    P
```
(25)

Ping-pong bi bi mechanism

```
   A                 P      B                 Q
   ↓                 ↑      ↓                 ↑
E    E·A ⇌ F·P       F      F·B ⇌ E·Q         E
```
(26)

The reader is referred to Whitaker [111] or Cleland [22] for further details on experimental procedures and analyses of data.

7.5.1.2 Substrate Inhibition

One substrate molecule binds into the active site of the enzyme and is catalytically converted to product according to normal kinetics. At higher substrate concentrations, a second molecule binds at a site different than the active site. It is not catalyzed to product, but its binding results in a decrease in catalytic efficiency of the enzyme (Fig. 8). Note that when plotted by the Michaelis–Menten method (insert), v_o reaches a maximum but is lower than the true V_{max}, and v_o decreases as $[S]_0$ is increased beyond the maximum v_o. The Lineweaver–Burk plot shows an upward deviation from linearity at high substrate concentrations. Isozymes sometimes show differences in substrate inhibition that appear to be a regulatory control of activity.

7.5.1.3 Substrate Activation

Binding of one substrate molecule into the enzyme active site results in product formation kinetics in the normal manner. The binding of a second substrate molecule at another site does not result in conversion of the substrate molecule to product, but it does enhance the rate of conversion of the substrate molecule in the active site to product (Fig. 9). Note that substrate activation results in downward deviation of the line from linearity at higher substrate concentrations when plotted by the Lineweaver–Burk method.

Substrate activation can be confused when two enzymes act on the same substrate, if the K_m values for the two enzymes are somewhat different. This possibility can be verified by showing that two isozymes are present by electrophoretic separation (for example) of the two.

7.5.1.4 Allosteric Behavior

Allosteric behavior is defined by a plot of v_o versus $[S]_o$ giving a sigmoidal shaped curve (Fig. 11; solid line), in contrast to a right hyperbola (Fig. 11; dashed line) found for Michaelis–Menten

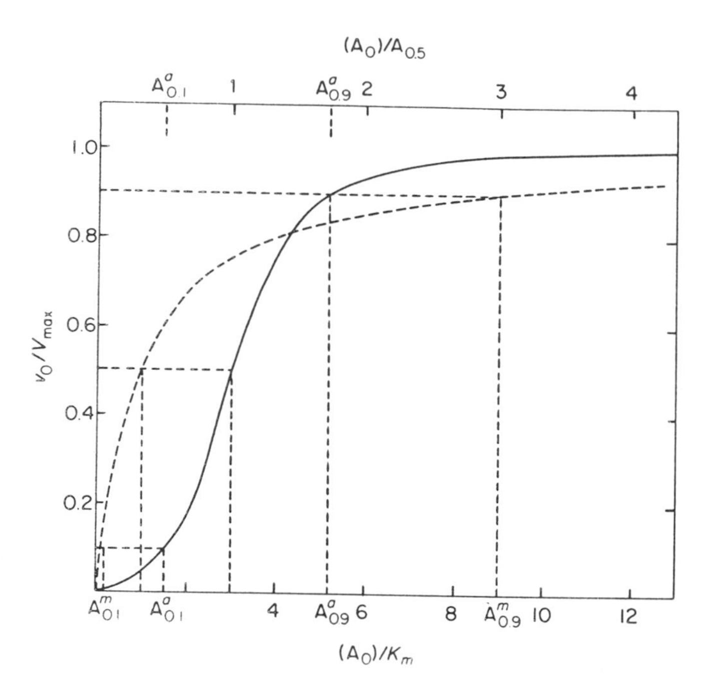

FIGURE 11 Comparison of effect of substrate concentration on the intial velocities of two enzyme-catalyzed reactions, one that obeys Michaelis–Menten kinetics (dashed line) and one that shows allosteric behavior (solid line). The designations $A^{m}_{0.1}$ and $A^{m}_{0.9}$ indicate the substrate concentrations at which the initial velocity is 0.1 and 0.9 V_{max}, respectively, for Michaelis–Menten kinetics, while $A^{a}_{0.1}$ and $A^{a}_{0.9}$ have the same meaning for the allosteric behaving system. The Hill value was 4 in the calculation of the allosteric curve (solid line). (From Ref. 111, p. 195.)

behavior. There is a smaller than expected effect of increasing $[S]_o$ on v_o at $[S]_o << K_m$, followed by larger than expected effect of $[S]_o$ on v_o at $[S]_o$ around K_m. The example shown in Figure 11 is for positive allosteric kinetic behavior. In some cases, the effect of $[S]_o$ on v_o is greater than expected at lower $[S]_o$, and then v_o does not increase as much as $[S]_o$ increases. This is negative allosteric kinetic behavior.

Positive allosteric kinetic behavior was discovered by Monod et al. [80], who subsequently received the Nobel Prize. Koshland et al. [66] discovered negative allosteric kinetic behavior, and developed methods for distinguishing allosteric behavior from Michaelis–Menten behavior, including a mathematical verification for occurrence of allosteric behavior.

To distinguish allosteric behavior from Michaelis–Menten behavior, Koshland et al. [66] proposed an equation for measuring cooperativity:

$$R_S = [S]_{0.9V_{max}}/[S]_{0.1V_{max}} = \text{cooperativity factor} \tag{27}$$

where $[S]_{0.9}$ is the substrate concentration required to give $v_o = 0.9V_{max}$ and $[S]_{0.1}$ is the substrate concentration required to give $v_o = 0.1V_{max}$. These points are marked on Figure 11 for $A^{m}_{0.1}$ and $A^{m}_{0.9}$ for Michaelis–Menten behaving enzymes and $A^{a}_{0.1}$ and $A^{a}_{0.9}$ for positive allosteric behavior. For all Michaelis–Menten behaving enzymes, R_S is 81, while for positive allosteric behaving enzymes $R_S < 81$ (often as low as 30–40), and for negative allosteric behaving enzymes $R_S > 81$.

Allosteric behavior is most often observed with multi-subunit enzymes. Koshland et al. [66] proposed that for allosteric behavior two or more subunits of an enzyme must show cooperativity:

$$\square\square \underset{}{\overset{K_1,\text{S}}{\rightleftharpoons}} \text{Ⓢ}\square \underset{}{\overset{K_2,\text{S}}{\rightleftharpoons}} \text{ⓈⓈ} \qquad (28)$$

Consider K_1 and K_2 as binding constants. If $K_2 > K_1$ then the second substrate molecule binds more easily (tightly) than the first substrate molecule and positive allosteric behavior is seen. If $K_1 > K_2$ the first substrate molecule binds more tightly than the second one, and negative allosteric behavior is seen. If $K_1 = K_2$, the two substrate molecules bind equally well with no effect on each other's binding, and Michael–Menten behavior is observed.

Allosteric behavior is one method of regulating enzymatic activity by small changes in substrate concentration, especially around K_m. The metabolic enzymes in the glycolytic and TCA cycles often show allosteric behavior. Allosteric behavior can also be elicited in some enzymes by non-substrate compounds that act as allosteric inhibitors or activators.

7.5.2 Enzyme Concentration

The relationship between v_o and $[E]_o$ is usually linear when all other factors, such as $[S]_o$, pH, and temperature, are kept constant (Fig. 12). Doubling the $[E]_o$ doubles v_o; tripling the $[E]_o$ triples v_o. This is expected since $v_o = k_2[E{\cdot}S] = dP/dt = -dS/dt$ (see Eq. 2). This linear relationship between $[E]_o$ and v_o at all $[S]_o$ is very valuable, since we can determine how much enzyme is present by measuring its activity under standard conditions without purifying the enzyme. Of course, the enzyme should be extracted from the tissues and should be free of any insoluble materials and/or other compounds that might interfere with determination of activity (such as those having intense absorbance at the same wavelength, in vivo activators and inhibitors that occur in variable amounts, or other enzymes that act on the same substrate). Enzyme activity determinations are important in clinical medicine, nutritional abnormalities, quality control in food processing, and in analytical uses of enzymes to determine concentration of compounds that are substrates, activators, or inhibitors (see Sec. 7.12).

There are at least five exceptions to the expected linear relationship between $[E]_o$ and v_o [111]. Two of these are (a) limitations on solubility of a substrate, such as O_2, and (b) conversion of substrates to products that either are less good substrates or are competitive inhibitors. Both lead to a decrease in v_o as [P] increases. The third (c) is coupled enzyme

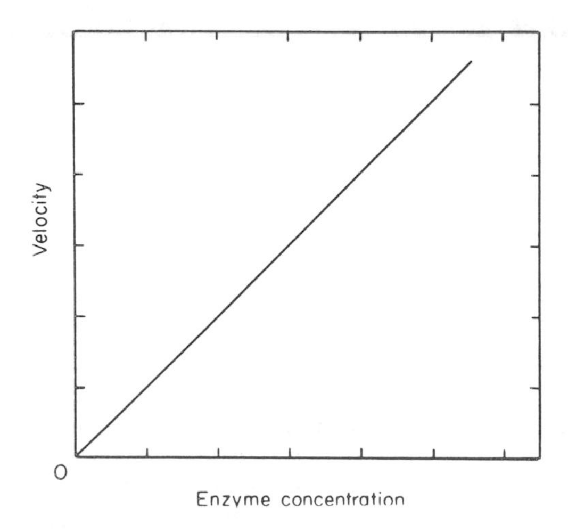

FIGURE 12 Expected relationship between enzyme concentration and observed velocity of reaction. Substrate concentration, pH, temperature, and buffer are kept constant. (From Ref. 111, p. 202.)

reactions where the product of enzyme-1 is the substrate of enzyme-2. In this instance, product-2 is measured to determine v_o, yet the concentration of enzyme-1 is desired:

$$A \xrightarrow{E_1, k_1} P_1 \xrightarrow{E_2, k_2} P_2 \tag{29}$$

To obtain linearity between v_o (dP_2/dt) and $[E_1]_o$, $k_2[E_2]_o \geq 100k_1[E_1]_o$ should be used. There may also be (d) irreversible inhibitor, such as Hg^{2+}, Ag^+, or Pb^{2+} in the substrate, buffer, or water, which inactivates a fixed amount of the enzyme; and (e) a dissociable essential cofactor in the enzyme solution.

7.5.3 Effect of pH

The pH has a marked effect on activity of most enzymes. As shown in Figure 13, pepsin, peroxidase, trypsin, and alkaline phosphatase have optimal activity at about pH 2, 6, 8, and 10, respectively. All the activity–pH curves are bell-shaped, with activity decreasing to near zero at 2 pH units below or above the pH optimum. Note that the left-hand side of the pepsin curve and the right-hand side of the alkaline phosphatase curve (Fig. 13) decrease abruptly as a function of pH (the bell-shaped curves are skewed). The pH optima of several enzymes found in raw food products are listed in Table 7. The pH optima range from 2 for pepsin to 10 for alkaline phosphatase. Catalase (Table 7) has maximum activity from pH 3 to 10.

The apparent complexity of the effect of pH on enzyme-catalyzed reactions is shown in Figure 14, with milk alkaline phosphatase as an example. Figures 14a and 14d illustrate the effect of different concentrations of the substrate phenyl phosphate on the shape and height, respectively, of the pH versus v_o curves. The pH optimum shifts from pH 8.4 at 2.5×10^{-5} M phenyl phosphate to pH 10 at 2.5×10^{-2} and 7.5×10^{-2} M phenyl phosphate. The pH optimum also is different for different substrates (Fig. 14c). The pH optima are 9.3, 9.5, and 9.8 for 0.02 M

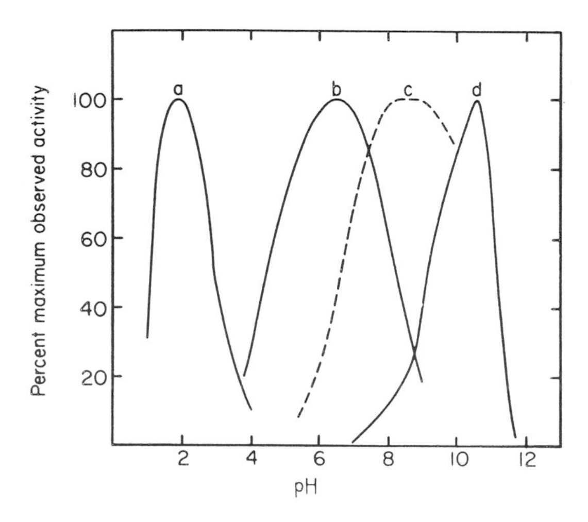

FIGURE 13 pH effect on activity of several enzymes. (a) Pepsin acting on *N*-acetyl-L-phenylalanyl-L-diiodotyrosine at 36.7°C and 5×10^{-4} M pepsin; reaction time 15 min. (b), *Ficus glabrata* peroxidase acting on 0.03 M guaiacol and 0.005 M H_2O_2 at 30.0°C. (c) Trypsin acting on 0.1% casein. (d) Hydrolysis of 5×10^{-4} M *p*-nitrophenyl phosphate by crude milk alkaline phosphatase at 35°C in 0.1 M sodium glycinate buffer; reaction time 15 min. Times of reaction are given for one-point assays, not when v_o was determined. (From Ref. 111, p. 273.)

TABLE 7 pH–Activity Optimum of Several Enzymes[a]

Enzyme	pH optimum
Acid phosphatase (prostate gland)	5
Alkaline phosphatase (milk)	10
α-Amylase (human salivary)	7
β-Amylase (sweet potato)	5
Carboxypeptidase A (bovine)	7.5
Catalase (bovine liver)	3–10
Cathepsins (liver)	3.5–5
Cellulase (snail)	5
α-Chymotrypsin (bovine)	8
Dextransucrase (*Leuconostoc mesenteroides*)	6.5
Ficin (fig)	6.5
Glucose oxidase (*Penicillium notatum*)	5.6
Lactate dehydrogenase (bovine heart)	7 (forward reaction) 9 (backward reaction)
Lipase (pancreatic)	7
Lipoxygenase-1 (soybean)	9
Lipoxygenase-2 (soybean)	7
Pectin esterase (higher plants)	7
Pepsin (bovine)	2
Peroxidase (fig)	6
Polygalacturonase (tomato)	4
Polyphenol oxidase (peach)	6
Rennin (calf)	3.5
Ribonuclease (pancreatic)	7.7
Trypsin (bovine)	8

[a]The pH optimum will vary with source and experimental conditions. These pH values should be taken as approximate values.
Source: Ref. 111, p. 274.

β-glycerophosphate, 0.015 *M* phosphocreatine, and 0.025 *M* phenyl phosphate, respectively. The pH optimum of alkaline phosphatase is dependent on cofactor type, with pH optima of 8.0 for Mn^{2+} and 9.4 for Mg^{2+}. The in vivo cofactor is Zn^{2+}.

One can conclude from the data of Figures 13 and 14 that the pH optimum of an enzyme is dependent on the nature of the enzyme and the conditions used to measure the activity as a function of pH. But is it possible to get valuable information from pH versus v_o curves? The answer is yes, provided the experiments are properly done. The four factors that primarily affect the nature of the pH versus v_o curves are (a) whether v_o data are used, (b) pH stability of the enzyme, (c) equilibria (important ionizations and dissociations) of the system, and (d) the relationship of $[S]_o$ to K_m. These factors deserve discussion.

7.5.3.1 How Velocity Data Are Obtained

The importance of obtaining initial velocities, v_o, in the kinetic investigations of enzymes cannot be overstressed. The data of Figure 15 illustrate that the difference in determined pH optimum is about 0.5 pH unit when different times are used to measure v, for an enzyme that is unstable either below or above the pH optimum. Since the enzyme is more stable on the acid side of the

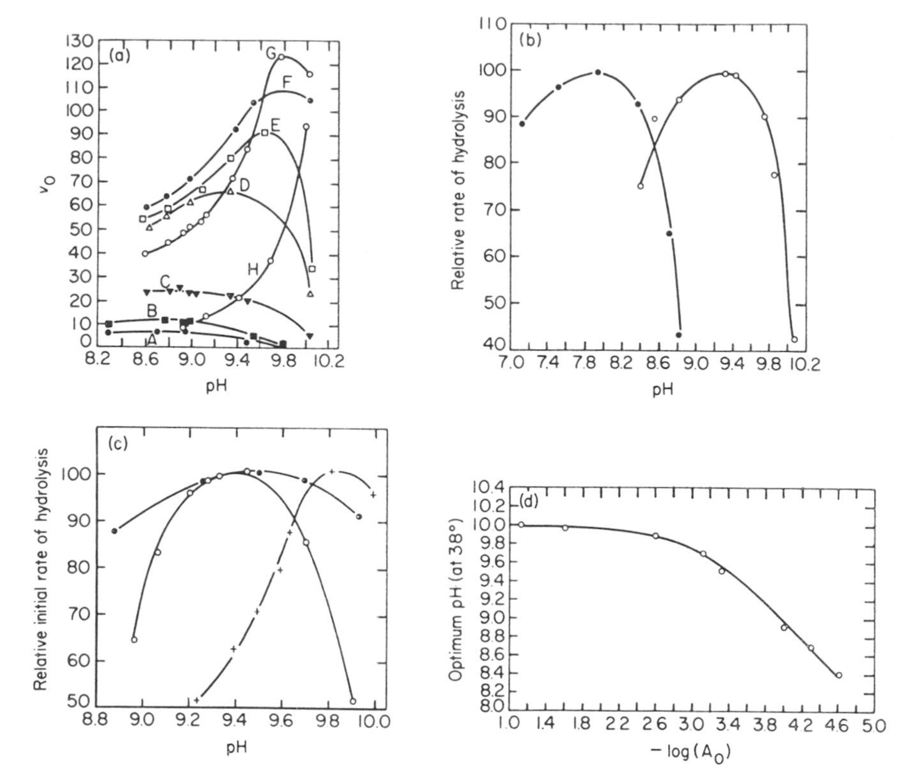

FIGURE 14 Effect of experimental conditions on the pH optimum of calf intestinal mucosa alkaline phosphatase. (a) Effect of initial substrate concentration on pH optimum. The curves are for the following substrate (phenyl phosphate) concentrations (*M*): A, 2.5×10^{-5}; B, 5×10^{-5}; C, 1×10^{-4}; D, 5×10^{-4}; E, 7.5×10^{-4}; F, 2.5×10^{-3}; G, 2.5×10^{-2}; H, 7.5×10^{-2}. (From Ref. 82, p. 675, by courtesy of the Biochemical Society.) (b) Effect of nature of activating cation on pH optimum. The cations were: ○, 5×10^{-3} *M* $MgCl_2$ and ●, 1×10^{-3} *M* $MnCl_2$ with 2.5×10^{-4} *M* phenyl phosphate as substrate. (From Ref. 82, p. 677, by courtesy of the Biochemical Society.) (c) Effect of nature of substrate on pH optimum. The substrates were: ●, 0.015 *M* phosphocreatine: ○, 0.02 *M* β-glycerophosphate; +, 0.025 *M* phenyl phosphate. (From Ref. 81, p. 235, by courtesy of the Biochemical Society.) (d) pH optima from data of part (a) plotted against $-\log[A]_o$. The lowest substrate concentration is at the right of the figure. (From Ref. 82, p. 675, by courtesy of the Biochemical Society.) The temperature was 38.0°C in all cases.

optimum than on the alkaline side, the observed pH optimum shifts to the left as a function of time between v_o and v_t.

7.5.3.2 Stability of the Enzyme

Enzymes generally are not stable at all pH levels. It is important to determine the pH range over which an enzyme is stable before determining the pH optimum. This can be done by at least three methods.

The best way is to incubate the enzyme under the same conditions (pH, temperature, buffer, enzyme concentration, time) to be used to determine v_o at different pH, but in the absence of substrate. At various times, aliquots are removed, added to tubes containing substrate buffered at a pH at or near the pH optimum, and v_o is determined. The control (100% activity) is a sample

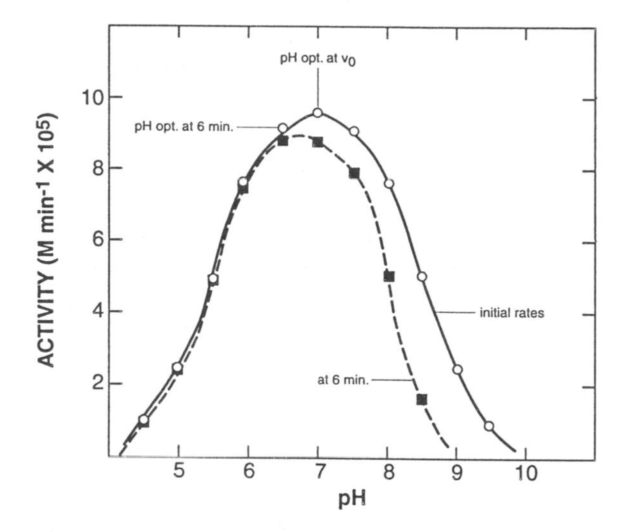

FIGURE 15 pH–activity curves for alkaline phosphatase. The solid line is rate data from initial rates; the dashed line is rate data determined at 6-min reaction times. (From Ref. 112, with permission of authors.)

of the enzyme solution incubated at 0°C, rather than at the temperature used to measure stability. The v_0 values of the samples incubated at higher temperatures are compared with v_0 for the sample held at 0°C.

Evidence for pH instability can also be obtained from plots of v_0 versus pH. The pH–v_0 curve will be skewed below or above the pH optimum, if there is instability (Fig. 15). This is the explanation for the skewed pH versus v_0 curves for pepsin and alkaline phosphatase in Figure 13 and the pH versus v_0 curve E in Figure 14a for alkaline phosphatase.

A third way to determine enzyme instability is to determine product formation (at different pH) as a function of time when $[S]_0 >> K_m$. If the enzyme is stable, v will be constant, while if it is unstable v will decrease with time.

7.5.3.3 Effect of Equilibria

Several of the constituents of an enzyme reaction system ionize or dissociate, depending on pH. Constituents that may ionize include the buffer, the substrate, the cofactor (if required), and the essential ionizable groups in the active site of the enzyme. Association/dissociation occurs in the binding of substrate and cofactor into the active site. Nothing can be done with respect to ionization of the buffer. Certainly, the ionic strength of the solution should be kept constant at all levels by use of NaCl or KCl. When different buffers must be used, it is essential to overlap the two buffers, for one or more pH levels. Alternatively, the buffers can be combined at all pH levels.

For fundamental studies, the substrate should be neutral, so that ionization does not affect combination of substrate with enzyme. Practically, this may be impossible. The control of ionic strength is again very important. If a cofactor is required, the enzyme should be saturated with the cofactor at all pH levels ($[CoF] >> K_{CoF}$). For further details on this subject see Whitaker [111].

7.5.3.4 The Relationship of $[S]_0$ to K_m

This is a major problem in most experiments reported in the literature. Consider the minimum steps involved in enzyme-catalyzed reactions (Eq. 7) and the possible effect of pH on these steps. First, there is binding of substrate and enzyme to form the enzyme–substrate complex (E·S) as controlled by k_1. Second, there is the dissociation of the E·S complex to E + S (controlled by k_{-1}) and there is the catalytic conversion of E·S complex to E + P, controlled by k_2 (or more than one rate constant if there are additional intermediate steps). The ionization of each of the reactants (E, S, E·S, and P) will affect the pH versus v_o curve. The substrate should be selected (if possible) so as not to undergo ionization in the pH range used.

By proper design of the experiments, the effect of pH on stability of enzyme, the binding of substrate into the active site of the enzyme, and the pK_a values of the ionizable groups in the active site of the free enzyme, the enzyme–substrate complex, and of other intermediate species (such as acylenzyme) can be determined [111]. Improperly designed experiments give data, such as in Figure 14 for alkaline phosphatase, but the results are more confusing than enlightening as to what is going on between the enzyme and the substrate.

7.5.4 Effect of Temperature

Experimental data on the effect of temperature on velocities of enzyme-catalyzed reactions can be just as confusing and uninterpretable as the effect of pH, unless the experiments are designed properly. Temperature affects not only the velocity of catalysis of $E{\cdot}S \xrightarrow{k_2} E + P$, but also stability of the enzyme; the equilibria of all association/dissociation reactions (ionization of buffer, substrate, product and cofactors (if any); association/disassociation of enzyme–substrate complex; reversible enzyme reactions ($S \rightleftharpoons P$); solubility of substrates, especially gases; and ionization of prototropic groups in the active site of the enzyme and enzyme–substrate complex. To some extent, proper design of the experiments for temperature effects is easier than for pH effects.

There are usually three reasons temperature effects on enzymes are studied: (a) to determine stability of the enzyme; (b) to determine the activation energy, E_a, of the enzyme-catalyzed reaction; and (c) to determine the chemical nature of the essential prototropic groups in the active site of the enzyme. The design of the experiments is not different, except the buffer must always be made up to have the same pH at all temperatures used. This requires the buffer to be made up at the temperature to be used. Otherwise, the pH of the buffer is an uncontrolled variable.

7.5.4.1 Stability of Enzyme

This can be determined in the following manner. Tubes with a fixed concentration of enzyme (similar to those to be used in the kinetic experiments), in the buffer adjusted to pH desired when made up at each temperature to be used, are incubated at the selected temperatures (usually starting at 25°C). Aliquots are removed at various times, substrate in a buffer near the pH optimum is added, and the activity left is determined at a constant temperature and pH. The control enzyme activity (100%) is determined on enzyme maintained at 0°C. The data are plotted as shown in Figure 16, since the rate of loss of activity is usually first order for a pure enzyme (no isozymes present). The slope of the line is k, the rate constant for denaturation of the enzyme.

As shown (Fig. 16), the enzyme is stable at 20–35°C, but the rates of loss of activity increase as temperature increases. Within the temperature range where there is loss of activity, a plot of ln k versus $1/T(K)$ should give a linear relationship with a slope E_a/R, where E_a is the activation energy for denaturation and R is the universal gas constant in cal mol^{-1} deg^{-1} (or J mol^{-1} deg^{-1}).

Typical transition state denaturation constants for some enzymes are shown in Table 8 (where $\Delta H^{\ddagger} = E_a - RT$).

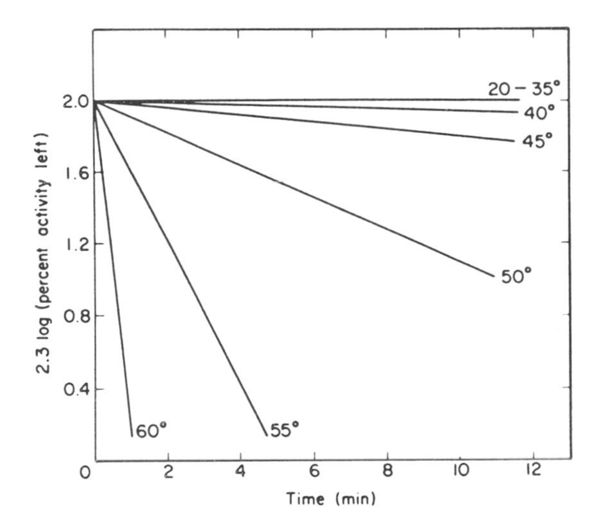

FIGURE 16 Rate of denaturation of an enzyme at various temperatures. The data are calculated for E_a of 60,000 cal/mol, where the first-order rate constant, *k*, of denaturation is 0.005, 0.020, 0.090, 0.395, and 1.80 min^{-1} at 40, 45, 50, 55, and 60°C, respectively. Calories × 4.186 = Joules. (From Ref. 111, p. 305.)

7.5.4.2 Activation Energy of the Enzyme-Catalyzed Reaction

To determine activation energies, two plots are required, the first is a plot of experimentally determined product concentration versus time at various temperatures (Fig. 17), and the second a plot of log *k*, the zero-order reaction rate constant versus $1/T$ (in K; Fig. 18). For the first plot,

TABLE 8 Transition-State Denaturation Constants for Various Enzymes

Substance	$\Delta H^{\ddagger}$ (cal/mol)	$\Delta S^{\ddagger}$ (eu)[a]	Number of bonds broken[b]	$\Delta G^{\ddagger}$ (25°C) (cal/mol)
Lipase, pancreatic	45,400	68.2	9	25,100
Amylase, malt	41,600	52.3	8	26,000
Pepsin	55,600	113.3	11	21,800
Peroxidase, milk	185,300	466.0	37	46,400
Rennin	89,300	208.0	18	27,300
Trypsin	40,200	44.7	8	26,900
Invertase, yeast				
pH 5.7	52,400	84.7	10	27,200
pH 5.2	86,400	185.0	17	31,300
pH 4.0	110,400	262.5	22	32,200
pH 3.0	74,400	152.4	15	29,000

[a]$\Delta S^{\ddagger}$ in cal/mol degree.
[b]Number of noncovalent bonds broken on denaturation = $\Delta H^{\ddagger}/5000$, where the average $\Delta H^{\ddagger}$ per bond is assumed to be 5000 cal/mol. Calories × 4.186 = Joules.
Source: Modified from Ref. 96.

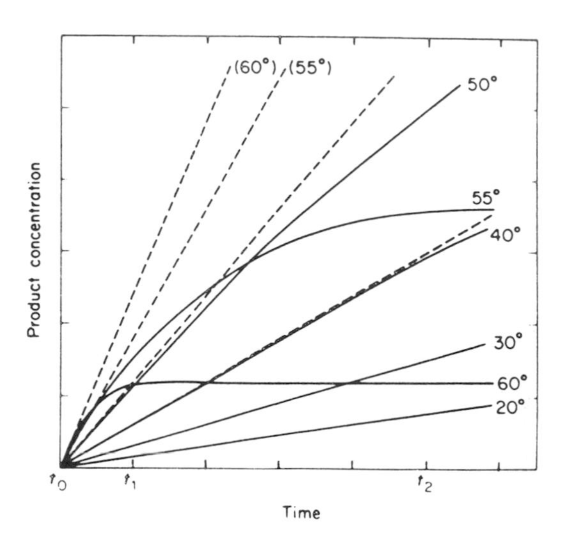

FIGURE 17 Effect of temperature on rate of product formation. The solid lines are for experimental data; the dashed lines are based on initial rates (i.e., lines drawn tangent to experimental data at time close to zero). The data shown were calculated for the following conditions: $[S]_o >> K_m$; E_a for transformation of reactant to product, 12,000 cal/mol; E_a for denaturation of enzyme, 60,000 cal/mol; first-order rate constant, *k*, for denaturation of enzyme, 9.0×10^{-4} min^{-1} at 40°C. Calories × 4.186 = Joules. (From Ref. 111, p. 303.)

temperatures must be low enough to obtain good v_o values before rate of denaturation of the enzyme becomes a problem, $[S]_o$ must be $>> K_m$ so the enzyme is saturated with substrate at all temperatures, and the pH (same value at all temperatures) must be at the pH optimum. If all enzyme is not in the form E·S, then the temperature dependence will also include K_s (association/dissociation of $E{\cdot}S \rightleftharpoons E + S$). If the pH is not at the pH optimum, then the effect of temperature on ionization of groups in the active site of the E·S complex will also be measured. Each slope of a dashed line in Figure 17 is a *k* value.

Using the *k* values and temperatures from Figure 17, the second plot can be prepared (Fig. 18). E_a, the activation energy, is obtained from the slope of this plot (slope = $-E_a/2.3R$).

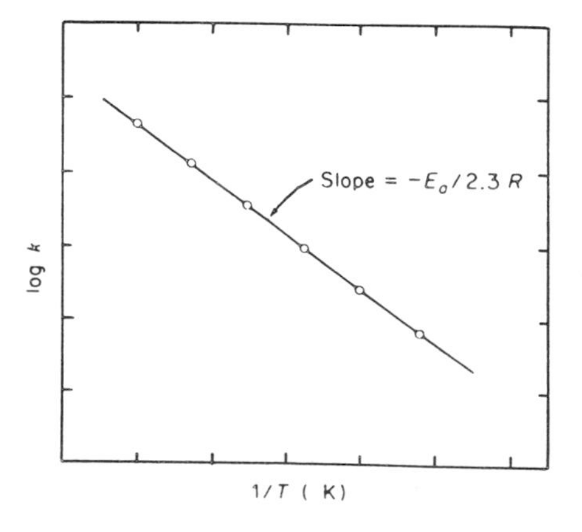

FIGURE 18 Effect of temperatures on rate constant of a reaction. Plotted as log *k* versus 1/*T*(K) to permit determination of E_a as shown. (From Ref. 111, p. 315.)

Typical E_a values for enzyme-catalyzed reactions range from about 25,000 to 50,000 J/mol. A comparison of E_a values for noncatalyzed, non-enzyme-catalyzed, and enzyme-catalyzed reactions are shown in Table 2.

The value of $\Delta H^{\ddagger}$ at a fixed temperature can be determined from the relationship $E_a = \Delta H^{\ddagger} + RT$; $\Delta G^{\ddagger}$ is calculated from

$$\Delta G^{\ddagger} = -RT \ln(kh/k_B T) \tag{30}$$

where k_B is the Boltzman constant (1.380×10^{-16} erg deg^{-1}) and h is the Planck constant (6.624×10^{-27} erg sec). The $\Delta S^{\ddagger}$, the entropy of activation, is calculated for a fixed temperature according to

$$\Delta G^{\ddagger} = \Delta H^{\ddagger} - T\Delta S^{\ddagger} \tag{31}$$

7.5.4.3 Chemical Nature of Prototropic Group(s) in Active Site of Enzyme

The effect of temperature on ionization of prototropic groups in the active site of the free enzyme or E·S complex can be determined from the change in pK_a values for the groups. This is important in determining the chemical nature of ionizable groups. Shown in Figure 19 are the effects of temperature and pH on relative V_{max}/K_m for hydrolysis of α-*N*-benzoyl-L-argininamide by papain [97]. There is no effect of temperature on ionization of the prototropic group of p$K_{a1} = 4$ in the active site of the free enzyme, consistent with the group being a carboxyl group, active in the carboxylate form. The second prototropic group has pK_{a2} values of 9.0, 8.2, and 7.4 at 5, 38, and 66°C, respectively. The ΔH_{ion} calculated from a plot of $-\log K_{ion}$ versus $1/T$(K) is 33,000 J/mol (Fig. 20 gives an example of this type of plot). Both pK_{a2} values and ΔH_{ion} (Table 9) indicate that the group is probably an –SH group of Cys_{25}, known to be involved in activity of papain.

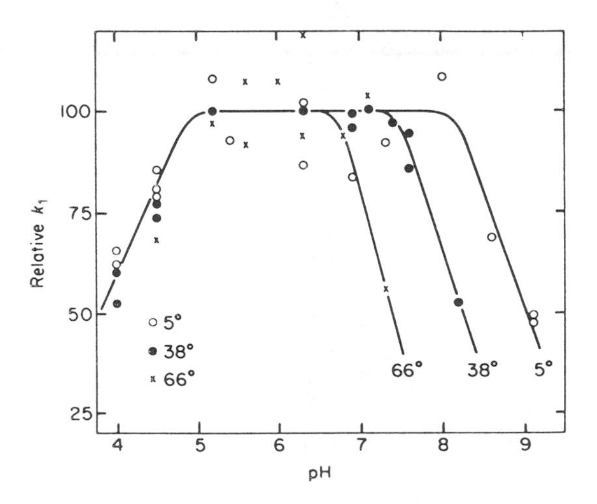

FIGURE 19 Effect of temperature (and pH) on relative k_1 for papain-catalyzed hydrolysis of α-N-benzoyl-L-argininamide. The k_1 values were calculated from $V_{max}/K_m^{H^+}$, where the maximum value at each temperature (5, 38, and 66°C) was set to 100. (From Ref. 97, p. 20, by courtesy of the American Society of Biological Chemists.)

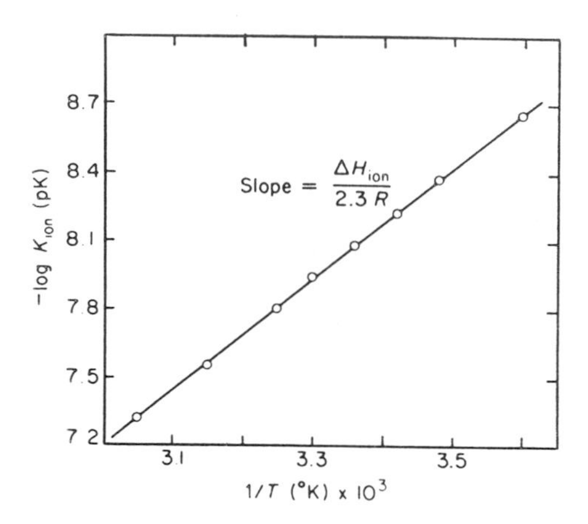

FIGURE 20 Effect of temperature on ionization constant, K_{ion}, of Tris. pK (–log K_{ion}) values are plotted as a function of $1/T$(K) to permit calculation of ΔH_{ion}. (From Ref. 111, p. 310.)

7.5.4.4 Enzyme Activity at Low Temperatures

(See also Chap. 2.) Intuitively, we might suppose that enzyme activity ceases at temperatures below 0°C, especially after the solution appears to be frozen. If so, this would be an important way of preserving our food indefinitely. Also, perhaps enzymes are denatured by freezing. Figure 21 shows the effect of temperature on invertase-catalyzed hydrolysis of sucrose from 49.6 to –19.4°C, and β-galactosidase-catalyzed hydrolysis of *o*- and *p*-nitrophenyl-β-galactosides from 25 to –60.2°C. Note that these two enzymes have activity over the temperature ranges studied,

TABLE 9 Prototropic Groups That May Be Involved in Enzyme Catalysis

Group	Ionization		pK_a	ΔH_{ion} (kcal/mol)
Carboxyl	$-COOH \rightleftharpoons -COO^{\ominus} + H^{\oplus}$	α,[a] β,γ,	3.0–3.2 3.0–4.7	±1.5
Imidazolium	HN–C(H)=NH⁺ (imidazolium ring) ⇌ HN–C(H)=N (imidazole ring) $+ H^{\oplus}$		5.6–7.0	6.9–7.5
Sulfhydryl	$-SH \rightleftharpoons -S^{\ominus} + H^{\oplus}$		8.0–8.5	6.5–7.0
Ammonium	$-NH_3^{\oplus} \rightleftharpoons -NH_2 + H^{\oplus}$	α[a], ε,	7.6–8.4 9.4–10.6	10–13
Phenolic hydroxyl	$-C_6H_4-OH \rightleftharpoons -C_6H_4-O^{\ominus} + H^{\oplus}$		9.8–10.4	6
Guanidinium	$-NHC(=NH_2^{\oplus})NH_2 \rightleftharpoons -NHC(=NH)NH_2 + H^{\oplus}$		11.6–12.6	12–13

[a]Located at the end of polypeptide chain. Calories × 4.186 = Joules.

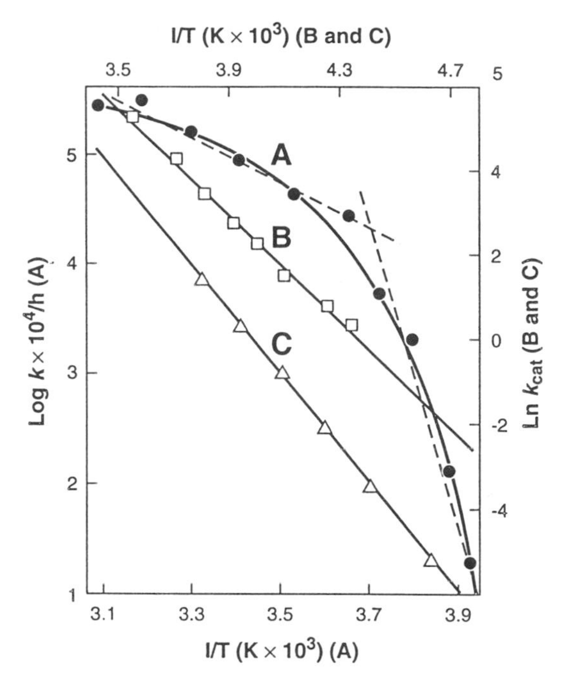

FIGURE 21 Effect of temperature (49.6 to –60.2°C) on velocities of enzyme-catalyzed reactions. (A) ●, Invertase-catalyzed hydrolysis of sucrose in aqueous buffer system. (Adapted from Refs. 61 and 94.) The intersection of the two dashed lines indicate the point of change in rates at the freezing point of the solution. (B) □, β-Galactosidase-catalyzed hydrolysis of *o*-nitrophenyl-β-galactoside, pH 6.1. (C) △, β-Galactosidase-catalyzed hydrolysis of *p*-nitrophenyl-β-galactoside, pH 7.6. Solvent in (B) and (C) was 50% dimethyl sulfoxide–water. (Adapted from Ref. 36.)

even though the activities decreased by 10^5 for invertase and $10^{4.3}$ for β-galactosidase. There is a change in slope of plot A at –3°C, where the solution began to freeze. One interpretation for the change in slope is the phase change. Others [29,30] have suggested that the change in slope is due to a change in the rate-determining step, formation or stabilization of intracellular hydrogen bonds in the enzyme, association of the enzyme into polymers, or increased hydrogen bonding between the substrate and water.

The β-galactosidase-catalyzed hydrolysis of *o*- and *p*-nitrophenyl-β-galactosides was performed in 50% dimethyl sulfoxide to prevent freezing. There were no changes in the slopes, and k_{cat} and K_m changed in a linear fashion. Therefore β-galactosidase acted in a predictable fashion over the range of 25 to –60.2°C, provided ice was absent.

Storage of foods at or just below the freezing point of water should be avoided. As water freezes the enzyme and substrate become more concentrated (solute is rejected from the ice phase), which may lead to enhanced activity. In addition, freezing and thawing disrupt the tissues, permitting greater access of enzyme to substrate. As shown in Figure 22, phospholipase activity in cod muscle is about five times greater at –4°C, below the freezing point, than at –2.5°C.

7.5.5 Water Activity/Concentration

Enzyme activities usually occur in aqueous media in vitro although in vivo enzyme reactions can occur not only in the cytoplasm but in cell membranes, in lipid depots, and in the electron

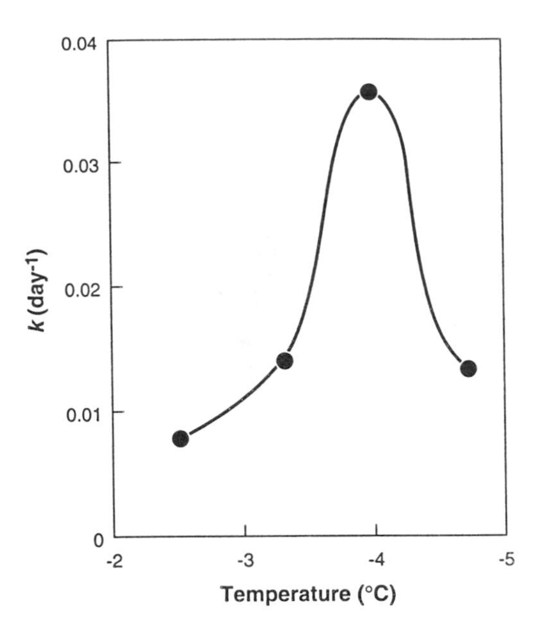

FIGURE 22 Rate constants (k) of phospholipase-catalyzed hydrolysis of phospholipids in cod muscle at subfreezing temperatures. (From Ref. 77, with permission of Institute of Food Technologists.)

transport system, where transfer of electrons is known to occur in a lipid matrix. There are three major ways of studying the effect of water activity on enzyme activity. The first method is to carefully dry an unheated biological sample (or model system) containing active enzymes, then to equilibrate it to various water activities and measure the velocity of enzyme activity. An example of this approach is shown in Figure 23. Below $0.35a_W$ (< 1% total water) there is no phospholipase activity on lecithin. Above $0.35a_W$, there is a nonlinear increase in activity. Maximum activity was still not reached at $0.9a_W$ (about 12% total water content). β-Amylase had no activity on starch until about $0.8a_W$ (~ 2% total water); activity then increased 15 times

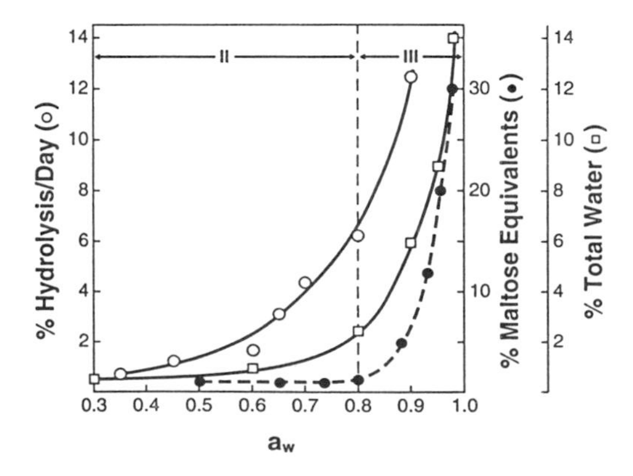

FIGURE 23 Enzyme activity as a function of a_W. (A) ○, Phospholipase-catalyzed hydrolysis of lecithin (adapted from data of Ref. 1). (B) □, Percent total water content of soluble solids (50:50 (w/w) protein and carbohydrate). (Adapted from data of Ref. 32.) (C) •, β-Amylase-catalyzed hydrolysis (as maltose equivalents) of starch. (Adapted from data of Ref. 31.)

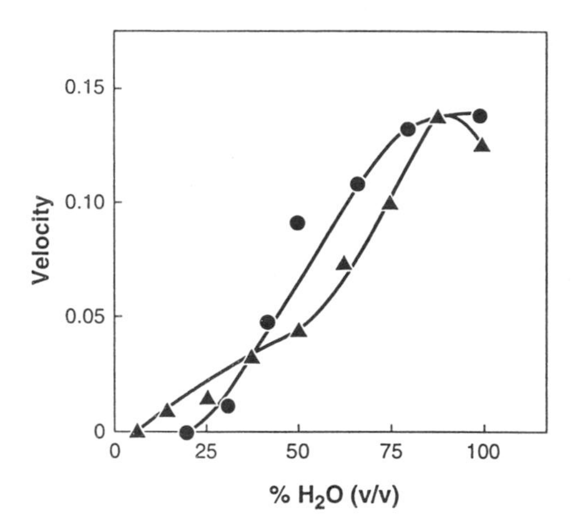

FIGURE 24 Effect of glycerol concentration in water on peroxidase (▲) and lipoxygenase (●) reaction velocities. (Adapted from Ref. 11.)

by 0.95 a_W (~ 12% total water). From these examples, it can be concluded that the total water content must be < 1–2% to prevent enzyme activity.

The second method of determining the concentration of water needed for enzymatic activity is to replace some of the water with organic solvents. Replacement of water with water-miscible glycerol reduces the activities of peroxidase and lipoxygenase when water content is reduced below 75% (Fig. 24). At 20% and 10% water, lipoxygenase and peroxidase have zero activity. Viscosity and specific effects of glycerol may have a bearing on these results.

In the third method, most of the water can be replaced by immiscible organic solvents in lipase-catalyzed transesterification of tributyrin with various alcohols [118]. The "dry" lipase particles (0.48% water), suspended in dry *n*-butanol at 0.3, 0.6, 0.9, and 1.1% water (w/w) overall, gave initial velocities of 0.8, 3.5, 5, and 4 μmol transesterification/h·100 mg lipase. Therefore, porcine pancreatic lipase had a maximum v_o for transesterification at 0.9% water concentration.

Organic solvents can have two major effects on enzyme-catalyzed reactions: an effect on stability, and an effect on direction of reversible reactions. These effects are different in water-immiscible and water-miscible solvents. In immiscible organic solvents there is a shift in specificity from hydrolysis to synthesis. Rates of lipase-catalyzed lipid transesterification reactions are increased more than sixfold while there is up to 16 times decrease in rate of hydrolysis when "dry" (~1% water) enzyme particles are suspended in the immiscible solvent [4, 91, 118]. Alkylation of lipases and trypsin to make them more hydrophobic has similar effects in shifting from hydrolysis to synthesis as does the use of immiscible solvents [4, 91]. The rate of trypsin-catalyzed esterification of sucrose by oleic acid was increased six times by alkylation of some of the amino groups of trypsin [4]. There is also a change in the stereospecificity of the products formed in organic solvents [40]. The ratio of v^R/v^S chiral isomers was 75 and 6 when lipase-catalyzed transesterification of vinyl butyrate with *sec*-phenethyl alcohol was performed in nitromethane versus decane, respectively [40].

Enzymes can be more stable in organic solvents than in aqueous buffers. Ribonuclease and lysozyme become much more stable as the water content is reduced, whether this is done by drying (ribonclease) or by adding water-immiscible organic solvents (Table 10). At 6% water content, ribonuclease has a thermal transition temperature (T_m) of 124°C and a half-life of 2.0 h at 145°C. The stability decreases as the water content increases; a dilute solution of ribonuclease has a T_m of 61°C and a half-life too short to measure [103]. Lysozyme is nearly as stable in

Table 10 Stability of Lysozyme and Ribonuclease as a Function of Water Content

Enzyme	Water content (%)	Thermal transition temperature, T_m	Half-life of activity (h)
Ribonuclease[a]	6	124	2.0
	11	111	0.83
	13	106	0.5
	16	99	0.17
	20	92	0.07
	Dilute solution	61	Too fast to measure
Lysozyme[b]	Dry powder		200
	Cyclohexane		140
	Hexadecane		120
	1-heptanol		100
	Buffer,[c] pH 4		1.42
	Buffer, pH 6		0.17
	Buffer, pH 8		0.01

[a]Adapted from Ref. 103. Half-life determined at 145°C.
[b]Adapted from Ref. 64. The dry powder contained 0.5% moisture. The dry powder was dispersed in anhydrous organic solvent.
[c]The lysozyme powder was dissolved in aqueous buffer to give 1% solution.

water-immiscible organic solvents as in the dry powder, but the half-life of dilute solutions is very short [64]. Subtilisin crystals placed in acetonitrile have the same crystal structure as does subtilisin crystallized from aqueous solutions [41]. Enzymes do not turn inside-out when placed in immiscible organic solvents.

The stability and catalytic activity of enzymes in water/water-miscible organic solvent systems are different from those in water-immiscible organic solvent systems. Kang et al. [55–57] showed that protease-catalyzed hydrolysis of casein, in either 5% ethanol/95% aqueous buffer or 5% acetonitrile/95% aqueous buffer systems, caused an increase in K_m, a decrease in V_{max}, and a decrease in stability (determined by circular dichroism and differential scanning calorimetry methods) compared to controls in aqueous buffer only. It is well known that protic solvents, such as alcohols and amines, compete with water in hydrolytic enzyme reactions [5, 9].

7.5.6 Why Enzymes Are Effective Catalysts

Enzymes are very effective catalysts as indicated by their ability to lower the activation energy, E_a, of reactions (Table 2). This results in larger values of v_0. But these facts provide no indication of how the enzymes achieve such "magic." Be assured that enzymes act by the same principles of all chemical reactions. They just do it better [67].

Listed in Table 11 are the factors that account for the catalytic efficiency of enzymes. Not all enzymes use all possibilities listed, but all enzymes do bind substrates stereospecifically into the active site where essential groups (side chain of specific amino acids or required cofactors) perform the reaction chemistry either by general acid–general base or nucleophilic–electrophilic processes. It is apparent that the most impressive enhancement of the rate is due to formation of the enzyme–substrate complex, where the rate enhancement is on the order of 10^4 for a two-substrate reaction, 10^9 for a three-substrate reaction, and 10^{15} (or more) for a four-substrate reaction.

The seven factors listed in Table 11 can account for 10^{18} to 10^{36} rate enhancements, if all

TABLE 11 Factors Accounting for Catalytic Effectiveness of Enzymes

Factor	Rate enhancement[a]
1. Formation of stereospecific enzyme–substrate complex (conversion from inter- to intramolecular reaction)	10^4; 10^9; 10^{15}[b]
2. Decreased entropy of reaction	10^3
3. Concentration of reactive catalytic groups	10^3–10^4
4. Distortion of substrate	10^2–10^4
5. General acid/general base catalysis	10^2–10^3
6. Nucleophilic/electrophilic catalysis	10^2–10^3
7. Using several steps	10^2–10^4
Overall rate enhancement	10^{18}–10^{36}

[a]In some cases, these values can be approximated by model experiments. In others, they are the best estimates available.
Source: Adapted from Ref. 111, pp. 129–142.

factors are operative. An important factor not included is the role of the increased hydrophobicity of the active site when the substrates are bound. The role of hydrophobicity in the active site during catalysis is one of the least known and appreciated factors in enzyme catalysis. A good example is synthesis of glutathione from γ-glutamylcysteine and glycine, using ATP as the energy source (a three-substrate reaction). This reaction is catalyzed by glutathione synthetase. After the three substrates bind stereospecifically into the active site, the active site is closed by a "lid" consisting of a 17-amino-acid loop in the enzyme. This protects the transition state intermediates from competition with water [60]. The rate constant k_o is 151 sec^{-1} for the synthesis. Using recombinant DNA technology, Kato et al. [60] replaced the 17-amino-acid loop with a sequence of three glycine residues, so that the active site could not close. The k_o for the mutant enzyme is 0.163 sec^{-1}, which is 1×10^{-3} that of the wild-type enzyme. Loop replacement did not appear to cause any other change in the physical structure of the enzyme.

7.6 ENZYME COFACTORS

Some enzymes, especially the hydrolases, are composed only of amino acids. The catalytic activity of the active site is the result of specific binding and catalysis due to specific side chains of amino acid residues (Table 9). These prototropic groups alone, such as the imidazole group of His_{57}, carboxyl group of Asp_{102}, and the hydroxyl group of Ser_{195} of chymotrypsin, are sufficient for the catalytic performance of some enzymes (Fig. 25).

The pathway for chymotrypsin-catalyzed reactions has been described [10]. A stereospecific adsorptive complex is formed between the substrate and enzyme (Fig. 25, structure A). The imidazole group of His_{57} acts as a general base to stretch the H-O bond of the hydroxyl group of Ser_{195}, thus facilitating the nucleophilic attack of the O of Ser-0H on the carbonyl carbon of the peptide bond of the substrate. This leads initially to formation of a tetrahedral intermediate (structure B) followed by formation of the acylenzyme intermediate by expulsion of the R′ group of the substrate (structure C). In the deacylation process, the imidazole group acts as a general base to extract a proton from water to facilitate the attack of the active site-generated hydroxide ion at the carbonyl group of the acylenzyme (structure D). In formation of the second transition-state intermediate (structure E), the imidazole group acts as a general acid in deacylation of the acylenzyme (structure F). The action of His_{57} as a general base (acylation step) and

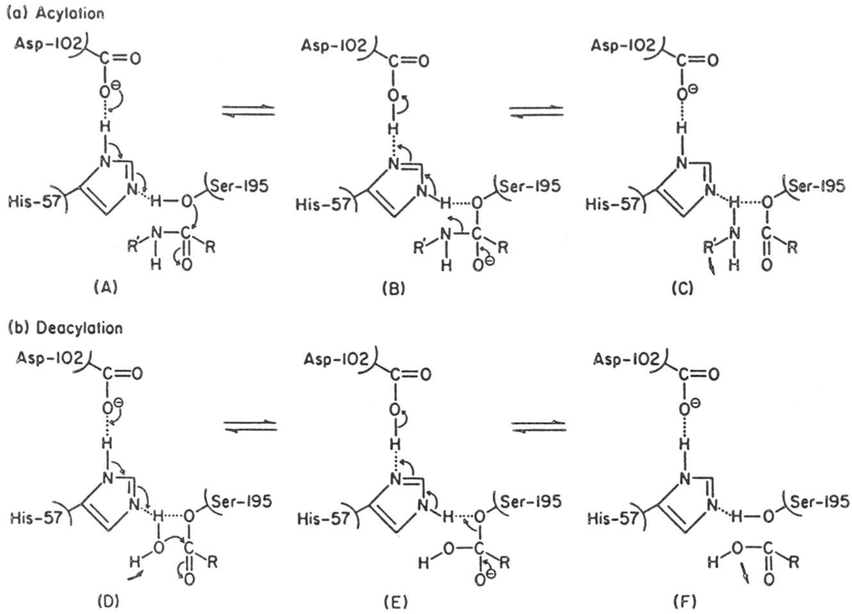

FIGURE 25 Proposed mechanism for α-chymotrypsin-catalyzed reactions. The top reactions show acylation of the Ser-OH of the enzyme by the substrate; the bottom reactions show removal of the acyl group by hydrolysis. (Modified from Ref. 12.)

general acid (deacylation step) is facilitated by hydrogen bonding with the Asp_{102} carboxyl group. The acyl part of the substrate diffuses away and the enzyme is ready to accept another substrate molecule (structure A). Each cycle requires about 10 msec to complete, with either the acylation (for amide bonds) or deacylation (for ester bonds) being the rate-determining step.

Many enzymes require cofactors (nonprotein organic compounds or inorganic ions) for activity. Cofactors include coenzymes, prosthetic groups, and the inorganic ions. Many of the coenzymes and prosthetic groups require a vitamin and often phosphate, ribose, and a nucleotide as part of the cofactor (Table 12). The nucleotide binds into the active site, specifically placing the cofactor so it can participate in the binding and/or catalytic step. Enzymes associated with these cofactors are also listed in Table 12.

The essential vitamins, cations, and anions must come from our foods, since we cannot synthesize them. Zn^{2+}, one of the essential cations, is part of the active site of at least 154 different enzymes in our bodies.

7.6.1 Distinguishing Features of Organic Cofactors

The coenzymes and prosthetic groups can be distinguished in two important ways. The coenzymes are loosely bound to the active site and dissociate from the enzyme at the end of each catalytic cycle, as shown for the reaction catalyzed by alcohol dehydrogenase (Eq. 32). They also are lost during purification of coenzyme-requiring enzymes and must be added back to the enzymes in the in vitro systems.

$$\text{E} \xrightarrow{\text{A}\downarrow} \text{E·A} \xrightarrow{\text{B}\downarrow} \text{E·A·B} \rightleftharpoons \text{E·P·Q} \xrightarrow{\text{P}\uparrow} \text{E·Q} \xrightarrow{\text{Q}\uparrow} \text{E} \tag{32}$$

TABLE 12 Importance of Phosphate, Ribose, and Purine and Pyrimidine Bases in Cofactors

Enzyme	Cofactor	Vitamin	Phosphate	Ribose	Base
Oxidoreductases	NAD^+	Niacin	+	+	Adenine
Oxidoreductases	$NADP^+$	Niacin	+	+	Adenine
Oxidoreductases	FMN	Riboflavin	+	+	—
Oxidoreductases	FAD	Riboflavin	+	+	Adenine
Ligases	ATP	—[a]	+	+	Adenine
Ligases	UTP	—	+	+	Uridine
Ligases	CTP	—	+	+	Cytidine
Transferases	CoA	Pantothenic acid	+	+	Adenine
Transferases	Acetyl phosphate	—	+	—	—
Transferases	Carbamyl phosphate	—	+	—	—
Transferases	*S*-Adenosyl methionine	—	—	+	Adenine
	Adenosine-3′-phosphate-5′-phosphosulfate	—	+	+	Adenine
Transferases and ligases	Pyridoxal phosphate	Pyridoxine	+	—	—
	Thiamine pyrophosphate	Thiamine	+	—	—

[a]—, Not present.
Source: Adapted from Ref. 111, p. 331.

where $CH_3CH_2OH + NAD^+ \rightleftharpoons CH_3CHO + NADH + H^+$
(A) (B) (P) (Q)

Kinetically, the NAD^+ behaves as a second substrate in the reaction (Eq. 32). It forms the product NADH, which can be recycled back to NAD^+ only by a second enzyme, not alcohol dehydrogenase. The velocity of the reaction is best followed spectrophotometrically based on absorbance of NADH at 340 nm ($\varepsilon_M = 6.2 \times 10^3\ M^{-1}\ cm^{-1}$).

In contrast to coenzymes, prosthetic groups are tightly bound (often covalently) to the enzyme, and they end up at the end of the cycle in the same oxidation state as they started. A typical example is flavin adenine dinucleotide (FAD), the prosthetic group of glucose oxidase (Eq. 33).

$$\begin{array}{ccccc} & \overset{A}{\downarrow} & \overset{P}{\uparrow} & \overset{B}{\downarrow} & \overset{Q}{\uparrow} \\ \text{E-FAD} & \text{E-FAD·A} \rightleftharpoons \text{E-FADH}_2\text{·P} & \text{E-FADH}_2 & \text{E-FADH}_2\text{·B} \rightleftharpoons \text{E-FAD·BH}_2 & \text{E-FAD} \end{array} \quad (33)$$

7.6.2 Coenzymes

We shall continue with the role of NAD^+ as coenzyme for alcohol dehydrogenase. Figure 26 is a schematic diagram of NAD^+ bound in the active site of yeast alcohol dehydrogenase, with the adenine ribose phosphate (ADPR) moiety bound at the ADPR binding site. The nicotinamide moiety, the functional part of NAD^+ in accepting H from the ethanol (R–C(H)(H)–O^+H), is bound into the lipophobic binding site. Zn^{2+}, a cation cofactor, is required also; it is shown liganded to three amino acid side chains on the protein and coordinately bound to the O of the OH group of ethanol. A tetrahedral intermediate is formed in step 2 followed by transfer of the H from ethanol to form NADH. The turnover number is about 10^3 moles of substrate converted to product per mole enzyme per second. Therefore, the recycle time is about 1 msec.

The primary role of the alcohol dehydrogenase is as a specific template, binding ethanol, NAD^+, and Zn^{2+} stereospecifically at the active site. As discussed in Section 7.5.7, that is a very significant role, accounting for ~10^{15} rate enhancement. It must also be remembered that the enzyme provides stereospecific treatment of the substrate. Even though ethanol does not have an asymmetric carbon atom, the enzyme always recognizes ethanol as being asymmetric because of the three-point attachment in the transition state (Fig. 26). The protein also provides the environment, often hydrophobic, in which the reaction takes place.

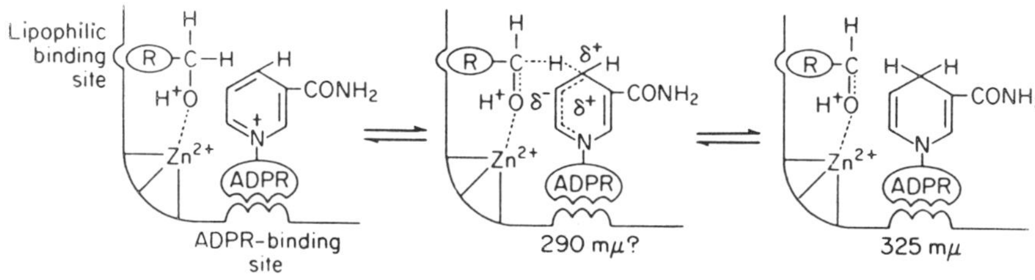

FIGURE 26 Schematic representation of binding of NAD^+ and ethanol in the active site of yeast alcohol dehydrogenase followed by oxidation of ethanol to acetaldehyde. (From Ref. 99, p. 552, by courtesy of Springer-Verlag.)

7.6.3 Prosthetic Group

Pyridoxal phosphate is selected as the prosthetic group to discuss because of its involvement in five types of reactions, one of them being the development of off aroma in broccoli and cauliflower and another the development of the desired aroma of onions and garlic.

The reactions catalyzed by pyridoxal phosphate (PALP) are shown in Figure 27. The first step, common to all five reactions, is reaction of the aldehyde group of PALP with the α-amino group of serine (used as an example) to give a Schiff base intermediate. This is stabilized by an acidic group B+ in the enzyme active site. Then, depending on the specific nature of the protein to which PALP is bound, decarboxylation (decarboxylase), deamination (transaminase), α,β-elimination (lyase) or racemization (isomerase) of serine occurs. The intermediate of serine

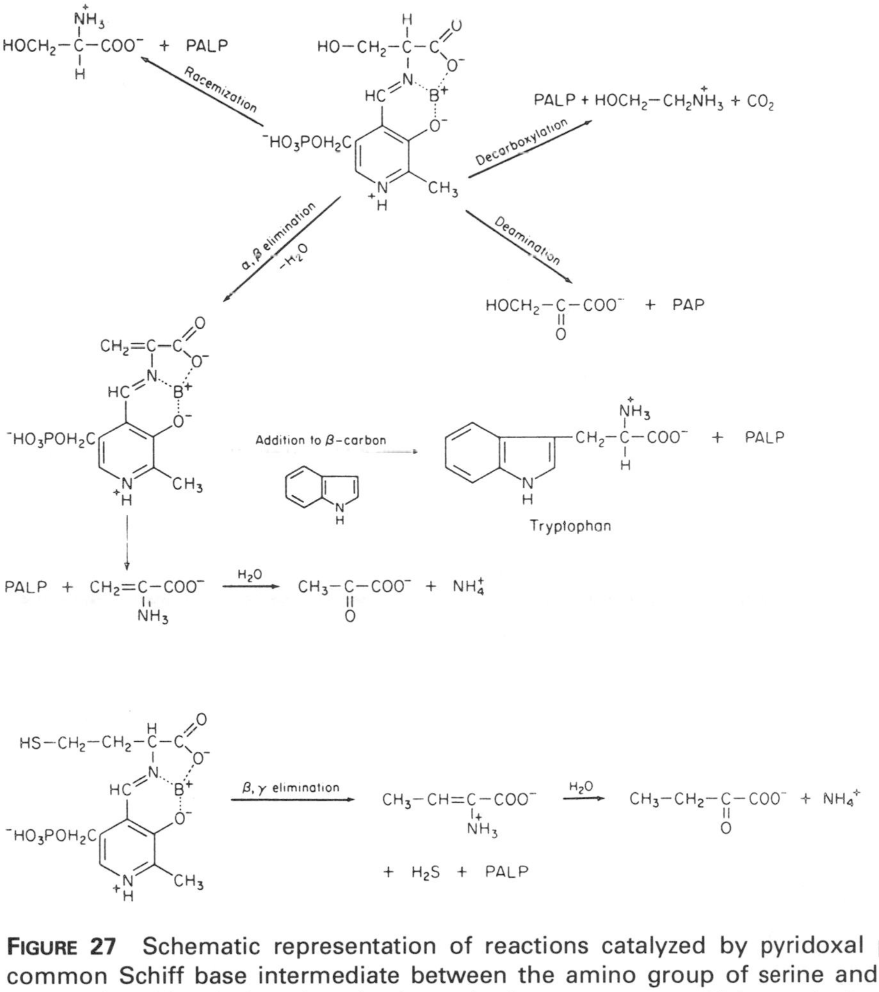

FIGURE 27 Schematic representation of reactions catalyzed by pyridoxal phosphate. A common Schiff base intermediate between the amino group of serine and the carbonyl group of pyridoxal phosphate is shown in the upper center. This results in conversion of substrate to one of five types of products, depending on the enzyme involved. The lower scheme shows β,γ-elimination involving the –SH group of cysteine. The intermediate is also a Schiff base. B^+ is an acidic group on the enzyme. PALP and PAP are pyridoxal phosphate and pyridoxamine phosphate, respectively. (From Ref. 111, p. 351.)

formed in the α,β-elimination reaction can react with indole to give tryptophan, or pyruvate and ammonia (one off-aroma compound in broccoli and cauliflower). If serine is replaced with cystine the products are H_2S, NH_3, pyruvate and thiolcysteine. H_2S and NH_3 are the two major off-aroma compounds of broccoli.

Alliinase is the key enzyme in aroma and flavor development in onions and garlic when they are cut or mashed. Alliinase is a PALP-requiring enzyme. The substrates *S*-1-propenyl-L-cysteine sulfoxide and *S*-allyl-L-cysteine sulfoxide undergo α,β-elimination by alliinase to produce 1-propenylsulfenic acid (a lachrymator) or allylsulfenic acid which then nonenzymatically forms the aroma and flavor components of onions and garlic, respectively [113].

7.6.4 Inorganic Ions

Both cations and anions can serve as cofactors. The cations include Ca^{2+}, Mg^{2+}, Zn^{2+}, Fe^{2+}, Cu^{2+}, Co^{2+}, Ni^{2+}, Na^+, K^+, and others. The anions include Cl^-, Br^-, F^-, and I^-, among others.

The cations may be involved directly in catalysis, as shown earlier for yeast alcohol dehydrogenase (Fig. 26), in binding of substrate to the active site, in maintaining the conformation and stability of the substrate, such as Ca^{2+} in α-amylase, or as part of the substrate, such as $MgATP^{2-}$ as required by kinases.

Zn^{2+} in *Escherichia coli* milk alkaline phosphatase (APase) will be used as an example because Zn^+ is a major essential cation in enzyme systems. Alkaline phosphatase requires four Zn^{2+} ions per molecule of enzyme (80,000 MW). Two of the Zn^{2+} serve a structural function, helping to hold the two subunits of the enzyme together (maintaining quaternary structure). The other two Zn^{2+} are located in the two active sites of the enzyme, where the role of the Zn^{2+} is as a "super acid" group to activate the Ser-OH group involved in the hydrolysis of the *p*-nitrophenylphosphate substrate:

$$O_2N{-}C_6H_4{-}O{-}P(=O)(O^-){-}O^- + H_2O \xrightarrow{\text{APase}} O_2N{-}C_6H_4{-}OH + HPO_4^{=} \quad (34)$$

The in vivo role of alkaline phosphatase is to help recycle phosphate, an essential compound. Alkaline phosphatase is ideally designed for this, as it is specific for the orthophosphate group only and it does not really matter what the other group (*p*-nitrophenyl here) is.

Chloride ion is an important anionic factor. Cl^- is essential for the activity of salivary and pancreatic α-amylases of humans and other animals and for some microbial α-amylases. Its role in α-amylase appears to be twofold. The binding constant for Ca^{2+} (another cation cofactor) in pancreatic α-amylase is increased ~100 times in the presence of Cl^-, thereby increasing the effectiveness of Ca^{2+} as a stabilizer of the tertiary structure of the protein [71]. Cl^- also causes a shift in the pH optimum of human salivary α-amylase [83] from pH 6 in the absence of Cl^- (much lower levels of activity) to pH 6.8 in the presence of Cl^- (0.005–0.04 *M* Cl^- gives maximum activity). It is postulated that the pH shift results from Cl^- masking an unwanted positively charged group in or near the active site of the enzyme [83]. Perhaps this represents a primitive control mechanism in some α-amylases.

7.7 ENZYME INACTIVATION AND CONTROL

Enzymes are responsible for the myriad reactions associated with reproduction, growth, and maturation of all organisms. In most cases, these are desired activities. In some cases, too much enzyme activity, such as polyphenol oxidase-caused browning, can lead to major losses in fresh fruits and vegetables. In many humans, absence of or too little of an enzyme is responsible for many genetically related diseases [104].

Microbially caused diseases present another problem. The best way to treat these types of diseases is through inhibition of one or more key enzymes of the microorganisms, resulting in their death. The inhibitors might compete reversibly with substrates or cofactors for binding to the active site, or the inhibitor might form a covalent bond with active site groups (affinity labeling inhibitor), or the compound might be treated as a substrate and be catalyzed to a product that, while still in the active site, forms a covalent bond with a group in the active site (k_{cat} inhibitors) [8]. The last type is the most specific and desirable in medicine and in food because the inhibitor can be targeted specifically for the enzyme.

Enzymes continue to catalyze reactions in raw food materials after they reach maturity. These reactions can lead to loss of color, texture, flavor and aroma, and nutritional quality. Therefore, there is need for control of these enzymes to stabilize the product as food. Enzyme inhibitors are also important in the control of insects and microorganisms that attack raw food. They also are used as herbicides in the control of unwanted weeds, grasses, and shrubs.

Enzyme inhibitors are an important means of controlling enzyme activity. An enzyme inhibitor is any compound that decreases v_o when added to the enzyme–substrate reaction. There are many enzyme inhibitors, both naturally occurring and synthetic [106]. Some inhibitors bind reversibly to enzymes and others form irreversible, covalent bonds with the enzymes. Some inhibitors are large proteins or carbohydrates, and others are as small as HCN. Products of enzyme-catalyzed reactions can be inhibitory.

Change in pH can alter activity by making conditions less optimum for enzyme activity. Elevated temperatures can decrease enzyme activity by denaturing some of the enzyme, but at the same time increasing the velocity of conversion of substrate to product by the active enzyme. Most enzymologists do not consider either of these variables to be enzyme inhibitors.

Denaturation of the enzyme eliminates its activity, and this can be accomplished by shear forces, very high pressures, irradiation, or miscible organic solvents. Enzyme activity can also be decreased by chemical modification of essential active site groups of the enzyme. Enzymes are also inactive when their substrate(s) are removed. All of these inhibitory approaches are valid ways of controlling enzyme activities in foods.

7.7.1 Reversible Inhibitors

Reversible inhibitors are distinguished from irreversible inhibitors by the following criteria: (a) reversible inhibitors rapidly (within milliseconds) form noncovalent diffusion-controlled equilibrium complexes with enzymes; the complex can be dissociated and enzyme activity restored by displacing the equilibrium by dialysis or by gel filtration; (b) irreversible inhibitors slowly form covalent derivatives of the enzyme that cannot be dissociated by dialysis or by gel filtration.

The reversible inhibitors can be treated kinetically by the methods described in Section 7.4 provided the equilibrium dissociation constant (K_i) is not less then 10^{-8} M. Four types of reversible inhibitors can be identified based on their effects on the slopes and intercepts of Lineweaver–Burk plots, V_{max} and K_m, or allosteric effects: (a) competitive inhibitors, (b) noncompetitive inhibitors, (c) uncompetitive inhibitors, and (d) allosteric inhibitors. Allosteric inhibitors not only decrease the velocity of the enzyme-catalyzed reactions, they cause allosteric (sigmoidal) plots of v_o versus [I], similar to the effect of $[S]_o$ (Fig. 11).

The type of reversible inhibitor must be determined by kinetic methods. Fortunately, all the mathematical equations differ in slope and/or y-intercept only by $1 + [I]_o/K_i$ from the Michaelis–Menten and Lineweaver–Burk equations in the absence of inhibitor.

7.7.1.1 Competitive Inhibition

In addition to the assumptions made in derivation of the Michaelis–Menten equation (Sec. 7.4.2.2), it is assumed for competitive inhibition that $[I]_o >> [E]_o$, such that $[I] \cong [I]_o$. In competitive inhibition the substrate and the competitive inhibitor compete for the binding to the enzyme, as shown in Figure 28 and Equations 35 and 36:

$$E + S \underset{k_{-1}}{\overset{k_1}{\rightleftharpoons}} E{\cdot}S \xrightarrow{k_2} E + P \tag{35}$$

$$E + I \underset{k_{-3}}{\overset{k_3}{\rightleftharpoons}} E{\cdot}I \tag{36}$$

where [I] is the free inhibitor concentration, and E·I is the enzyme–inhibitor complex, and $K_i = k_{-3}/k_3$. E·I does not bind S, and it does not form a product. The conservation equation with respect to enzyme concentration is

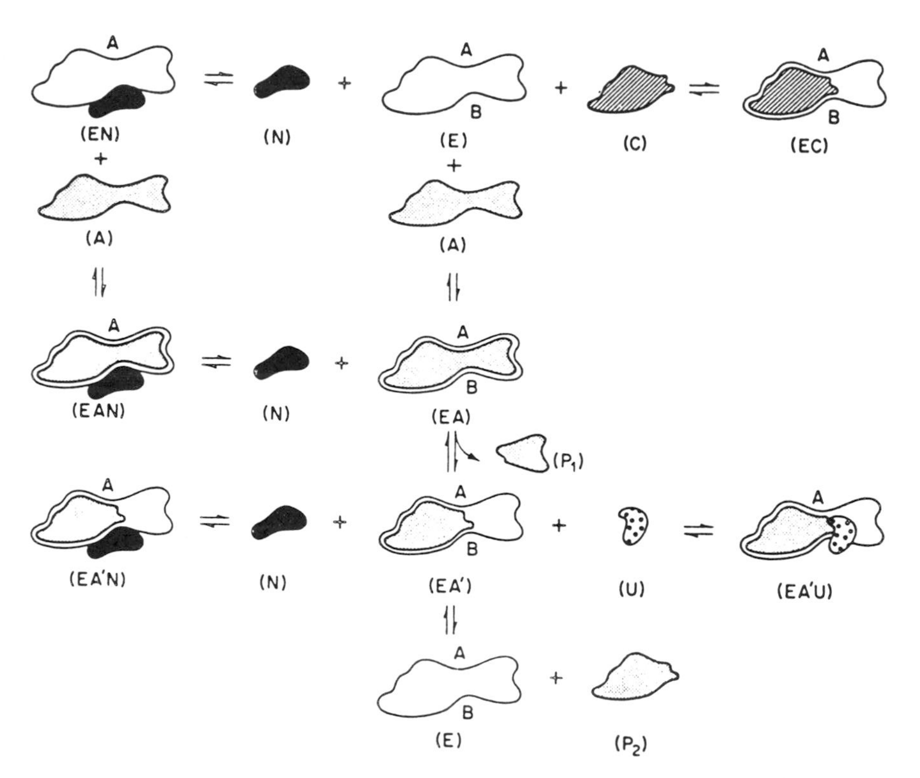

FIGURE 28 Schematic representation of inhibition of enzymes by various types of inhibitors. The model of E shows only the active site with the binding locus (inner void) and the transforming locus, with catalytic groups A and B. The symbols for the species involved (in parentheses) are: E, free enzyme; E·A, enzyme–substrate complex; E-A′, acylenzyme intermediate; C, competitive inhibitor; N, noncompetitive inhibitor; U, uncompetitive inhibitor; P_1 and P_2 are products formed from the substrate, A, and E·N, E·A·N, E·A′·N, E·C, and E·A′·U are complexes with respective enzyme species and inhibitors. All complexes formed involve noncovalent bonds except E-A′, where a covalent bond is formed with catalytic group A. (From Ref. 111, p. 235.)

$$[E]_o = [E] + [E{\cdot}S] + [E{\cdot}I] \tag{37}$$

Inclusion of Equations 36 and 37 in derivation of the Michaelis–Menten equation gives

$$v_o' = \frac{V_{max}[S]_o}{(1 + \frac{[I]_o}{K_i})K_m + [S]_o} \tag{38}$$

where v_o' is the initial velocity in the presence of a fixed concentration of inhibitor. The Lineweaver–Burk equation derived from Equation 38 is

$$\frac{1}{v_o'} = \left(1 + \frac{[I]_o}{K_i}\right)\frac{K_m}{V_{max}[S]_o} + \frac{1}{V_{max}} \tag{39}$$

A plot of experimental data in the presence and absence of a competitive inhibitor is shown in Figure 29. Reactions with variable $[S]_o$ must be run in the absence and presence of a fixed concentration of $[I]_o$, and v_o and v_o' must be determined. Linear competitive inhibition is indicated experimentally when the y-intercept is the same in the presence and absence of inhibitor, but the slope in the presence of inhibitor is greater than that in the absence of inhibitor by $1 + [I]_o/K_i$.

7.7.1.2 Noncompetitive Inhibition

In this type of inhibition, the inhibitor does not compete with the substrate for binding with enzyme, so the inhibitor and substrate can bind to the enzyme simultaneously (Fig. 28; Eqs. 40–43).

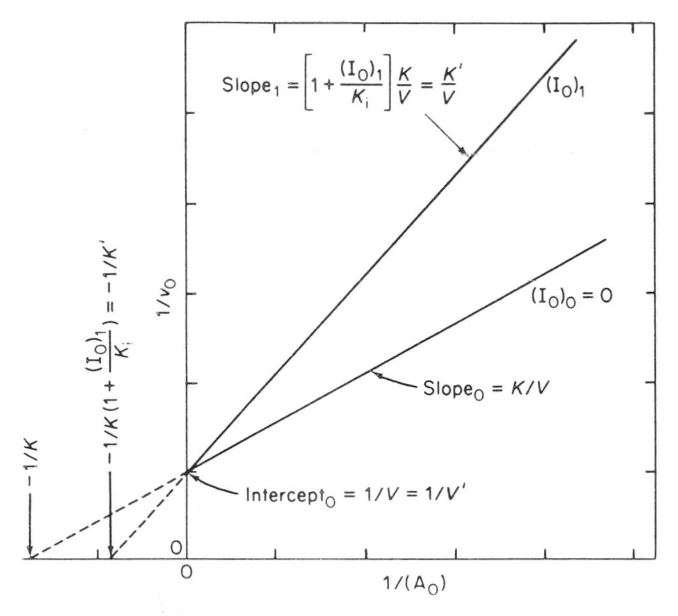

FIGURE 29 Linear competitive inhibition. The $1/v_o$ and $1/v_o'$ values are plotted versus $1/[A]_o$. Data are plotted in reciprocal form according to Equation 39. $[I]_{o1} = K_i$. $[A]_o$ is initial substrate concentration. (From Ref. 111, p. 228.)

$$E + S \underset{k_{-1}}{\overset{k_1}{\rightleftharpoons}} E{\cdot}S \xrightarrow{k_2} E + P \tag{40}$$

$$E + I \underset{k_{-3}}{\overset{k_3}{\rightleftharpoons}} E{\cdot}I \tag{41}$$

$$E{\cdot}S + I \underset{k_{-3}}{\overset{k_3}{\rightleftharpoons}} E{\cdot}S{\cdot}I \tag{42}$$

$$E{\cdot}I + S \underset{k_{-3}}{\overset{k_3}{\rightleftharpoons}} E{\cdot}S{\cdot}I \tag{43}$$

The assumptions are that k_{-1}/k_1 is the dissociation constant, K_S, for both E·S and E·S·I, and k_{-3}/k_3 is the dissociation constant, K_i, for both E·I and E·S·I. Further, E·S·I does not form products from the substrate. These assumptions lead to simple linear noncompetitive inhibition where

$$v_o' = \frac{V_{max}[S]_o}{\left(1 + \frac{[I]_o}{K_i}\right)(K_m + [S]_o)} \tag{44}$$

and the corresponding Lineweaver–Burk equation is

$$\frac{1}{v_o'} = \left(1 + \frac{[I]_o}{K_i}\right)\left(\frac{K_m}{V_{max}[S]_o} + \frac{1}{V_{max}}\right) \tag{45}$$

The enzyme conservation equation is

$$[E]_o = [E] + [E{\cdot}S] + [E{\cdot}I] + [E{\cdot}S{\cdot}I] \tag{46}$$

In the case of simple linear noncompetitive inhibition, both the y-intercept and the slope are increased by $1 + [I_o]/K_i$ (Fig. 30).

7.7.1.3 Uncompetitive Inhibition

Unlike the cases of competitive and noncompetitive inhibition, in uncompetitive inhibition, the inhibitor cannot bind to the free enzyme but only with one or more of the intermediate complexes as shown in Fig. 28. Equations 47 and 48 apply:

$$E + S \underset{k_{-1}}{\overset{k_1}{\rightleftharpoons}} E{\cdot}S \xrightarrow{k_2} E + P \tag{47}$$

$$E{\cdot}S + I \underset{k_{-3}}{\overset{k_3}{\rightleftharpoons}} E{\cdot}S{\cdot}I \tag{48}$$

where E·S·I does not form product from the substrate. The appropriate form of the Michaelis–Menten equation is

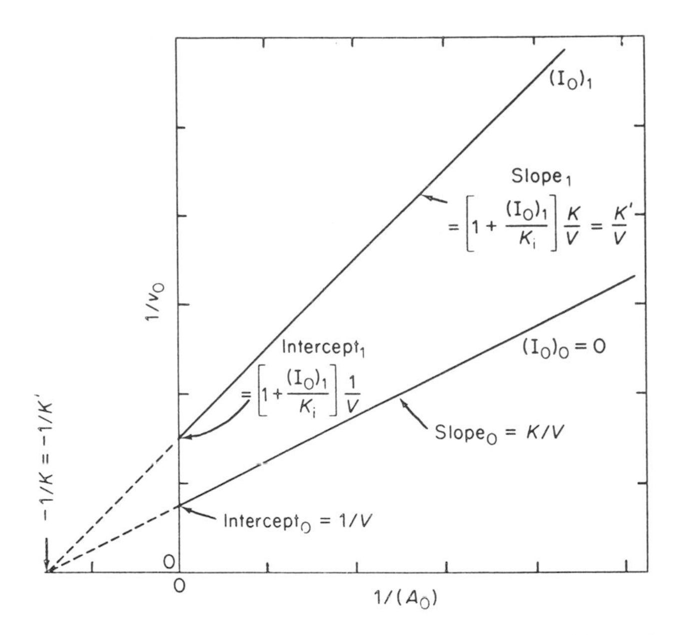

FIGURE 30 Simple linear noncompetitive inhibition. The $1/v_o$ and $1/v_o'$ values are plotted versus $1/[A]_o$ according to Equation 45. $[I_o]_1 = K_i$, where $K_{intercept}' = K_{i,slope}'$. $[A]_o$ is initial substrate concentration. (From Ref. 111, p. 229.)

$$v_o' = \frac{V_{max}[S]_o}{K_m + [S]_o\left(1 + \frac{[I]_o}{K_i}\right)} \tag{49}$$

and the corresponding Lineweaver–Burk and conservation equations are Equations 50 and 51:

$$\frac{1}{v_o'} = \frac{K_m}{V_{max}[S]_o} + \frac{1 + [I]_o/K_i}{V_{max}} \tag{50}$$

$$[E]_o = [E] + [E{\cdot}S] + [E{\cdot}S{\cdot}I] \tag{51}$$

The associated plot is shown in Figure 31. The y-intercept changes by $1 + [I]_o/K_i$ from the reaction with no inhibitor, while the slopes in the presence and absence of an uncompetitive inhibitor are the same.

7.7.1.4 Allosteric Inhibition

Allosteric inhibition usually results from the binding of inhibitor to multi-subunit enzymes, in the same way as described for allosteric behavior on substrate binding (Eq. 28). Whenever K_{i2} and subsequent K_i values are smaller (tighter binding) than K_{i1}, positive allosteric inhibition results. Negative allosteric inhibition results when K_{i1} is smaller than K_{i2} and subsequent K_i values.

Allosteric inhibition can be quantified by the methods of Monod et al. [80] and Koshland et al. [66] or by the modified Hill equation [47]. The Hill equation, developed for O_2 or CO_2 binding to hemoglobin, is applicable to substrate or inhibitor binding to enzymes,

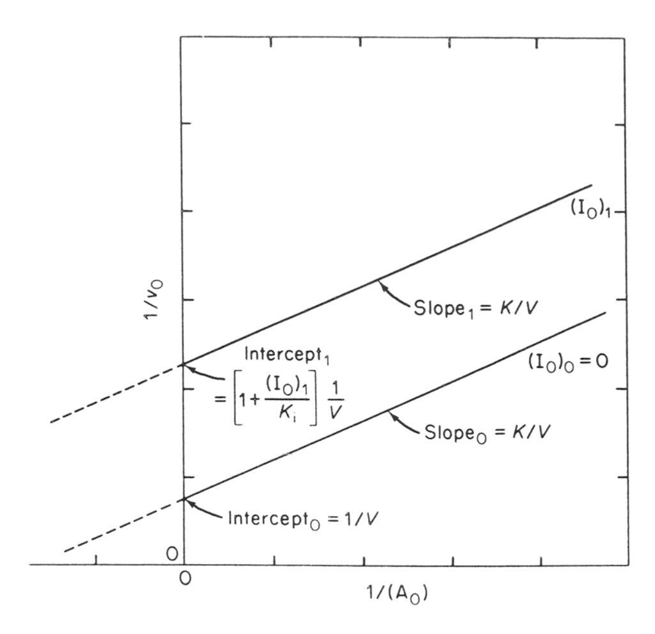

FIGURE 31 Linear uncompetitive inhibition. Data plotted in reciprocal form according to Equation 49. $[I_o]_1 = 2K_i$. $[A]_o$ is initial substrate concentration. (From Ref. 111, p. 232.)

to give allosteric kinetics. The Hill equation for effect of $[S]_o$ on an enzyme-catalyzed reaction is

$$v_o = \frac{V_{max}[S]_o^n}{K_m + [S]_o^n} \tag{52}$$

where n is the apparent number of interacting binding sites. Equation 52 can be written in a linear transform as

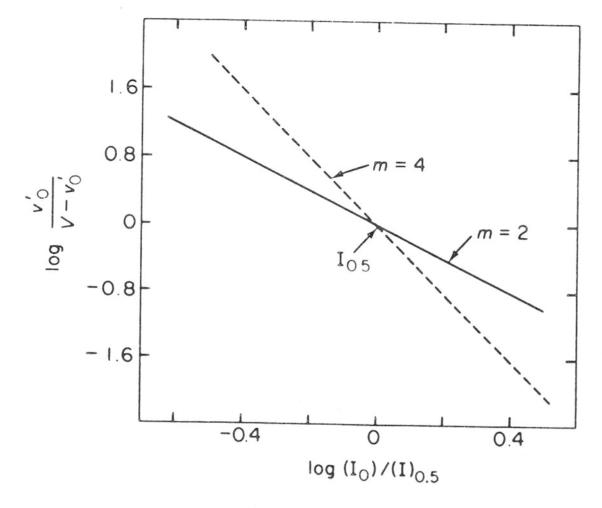

FIGURE 32 Effect of inhibitor concentration on activity of an allosteric-behaving enzyme system plotted according to Equation 54. (From Ref. 111, p. 233.)

$$\ln\left(\frac{v_o}{V_{max} - v_o}\right) = n \ln[S]_o - \ln K_m \quad (53)$$

with the analogous equation for an inhibitor being

$$\ln\left(\frac{v_o'}{V'_{max} - v_o'}\right) = n \ln[I]_o - \ln K_i \quad (54)$$

A plot of experimentally obtained v'_o values as a function of $[S]_o$ concentration is shown in Figure 32. The $[I]_{0.5}$ for 50% inhibition of the enzyme is shown at the common cross point of two sets of data obtained at two different, not saturating, inhibitor concentrations. The slope values, m indicate the number of apparent interacting binding sites with inhibitor, modified by the strength of interaction, n.

7.7.2 Irreversible Inhibitors

Many inhibitors form covalent derivatives with enzymes; therefore, they should be treated by rate equations, not by K_i. Kinetically, irreversible inhibition also occurs when $K_i \leq 1 \times 10^{-9}\ M$, since the $t_{0.5}$ of the dissociation rate (from k_{-1} of Eq. 55) of the E·I is 1.2 min. So very tight-binding inhibitors cannot be treated by the methods described in Section 7.7.1. Sometimes there is confusion about this because some inhibitors bind very rapidly to enzymes, but form very tight noncovalent complexes or covalent derivatives at slower rates (Eq. 55). These are called slow, tight-binding inhibitors.

$$E + I \underset{k_{-1}}{\overset{k_1}{\rightleftharpoons}} E{\cdot}I \xrightarrow{k_2} E - I \quad (55)$$

One needs to know which step(s) leads directly to loss of enzyme activity (rate = $k_1[E]_o[I]_o$ or k_2 [E·I]) in order to interpret the data and identify the mechanism of the reaction. The reaction shown by Equation 55 is typical of affinity labeling inhibitors.

Sometimes when compounds bind into the active site of an enzyme they are catalyzed to products, which then react covalently with the enzyme to irreversibly inactivate it:

$$\begin{array}{c} E + S \underset{k_{-1}}{\overset{k_1}{\rightleftharpoons}} E{\cdot}S \xrightarrow{k_2} E{\cdot}P \xrightarrow{k_3} E - P \\ \quad\quad\quad k_{-4} \Big\Uparrow\Big\Downarrow k_4 \\ \quad\quad\quad E + P \end{array} \quad (56)$$

When $k_3 > k_4$ the enzyme is rapidly inactivated. Should the substrate be called an inhibitor or a substrate? These types of inhibitors are called k_{cat} inhibitors and are among the most specific inhibitors known. They are highly valued in medicine, and could be very useful in control of enzymes of foods.

Other irreversible enzyme inhibitors are specific for certain amino acid side-chain groups of the enzyme, such as the sulfhydryl group, the amino group, the carboxyl group, and the imidazole group. Generally, these types of reactions are called protein modification. A lot of different compounds have been studied in this connection, in an attempt to determine the nature of one or more groups in the active site of enzyme or to label a specific group so that its location in the primary amino acid sequence can be determined. Some examples are given in Table 13. They are often not specific for a single type of group and may react also with the same type of specific group outside the active site, unless the one in the active site is more reactive than the

TABLE 13 Some Reagents Used in Chemical Modification of Enzymes

	Group modified							
Reagent	Amino	Carboxyl	Sulfhydryl	Ser Thr	His	Trp	Tyr	Met
Acetic anhydride	+		+	+			+	
Acetyl imidazole	±		+				+	
Aryl sulfonyl chlorides	+		+	±	+		+	
N-Bromosuccinimide			+		+	+	+	
Carbodiimides		+						
Diazomethane	+	+	+				+	
2,4-Dinitrofluoro-benzene	+		+		+		+	
N-Ethyl maleimide	±		+		±			
Haloacetate, haloamide	+	+	+		+			
Hydrogen peroxide			±			±		+
Hydroxylamine		+						
Hydroxynitrobenzyl bromide						+		
Mercuribenzoate	±		+		±			
Photooxidation			+		+	+	+	±
Tetranitromethane			+				+	±
Trinitrobenzene sulfonate	+							

Source: Adapted from Ref. 48.

same groups outside the active site. The reader is referred to Whitaker [111] for more detailed discussion of these types of inhibitors.

7.7.3 Naturally Occurring Inhibitors

Some organisms have evolved enzyme inhibitors as metabolic and physiological regulatory systems. They also produce inhibitors to prevent premature activation of proenzymes and to protect the organism against insect and microbial attacks. Ryan [89] has shown that insect larvae feeding on leaves or stems of tomatoes and potatoes stimulates rapid protease inhibitor production, which causes the larvae to quit feeding. The elicitors appear to be pectin fragments produced by action of pectic enzymes, in response to tissue damage. Most of the elicited protease inhibitors are proteins.

7.7.3.1 Protease Inhibitors

Trypsin inhibitors appear to be ubiquitous in all tissues, and they have been most intensively studied in the legumes and cereals. Soybean seeds contain two types of trypsin inhibitors, the Kunitz inhibitor of 21,000 MW that is specific for trypsin only (1:1 complex) and the Bowman–Birk inhibitor of 8300 MW that binds independently and simultaneously to trypsin and chymotrypsin (1:1:1 complex). Most legumes contain the Bowman–Birk type inhibitors, with considerable homology among them, while most legumes do not produce the Kunitz-type inhibitor. The Kunitz inhibitor, with two disulfide bonds, is much more heat labile than the Bowman–Birk inhibitors, with seven disulfide bonds. About 1 h of cooking is required to completely inactivate the Bowman–Birk inhibitor, unless a reducing compound, such as cysteine, is added [39]. Several isoinhibitors of trypsin are often found in higher plants [116].

The legume inhibitors are known to be significant, nutritionally, at least in some animals. As shown in Fig. 33, the PER (protein efficiency ratio) increases as the amount of inhibitor decreases. As stated earlier, the inhibitor can be inactivated by prolonged cooking. The four red kidney bean protease inhibitors have K_i values for bovine trypsin ranging from 3.4×10^{-10} to 8.4×10^{-10} M, while K_i for bovine chymotrypsin is 5.5×10^{-10} to 4.0×10^{-9} M [116]. These are slow, very tight-binding inhibitors. The K_i values can differ significantly with trypsin from different sources.

Chicken egg white contains three types of protease inhibitors, the ovomucoids specific for trypsin, the ovoinhibitor with specificity for trypsin, chymotrypsin, subtilisin, and *Aspergillus oryzae* protease, and the papain inhibitor (cystatin) with specificity for papain and ficin [107]. Protease inhibitors are found in the pancreas and in the blood, where they protect against premature activation of the pro-forms of the digestive proteases and the blood clotting proteases, respectively.

Microorganisms contain a variety of low-molecular-weight peptide inhibitors of proteases. These inhibitors include the leupeptins active against trypsin, plasmin, papain, and cathepsin B, the chymostatins active against chymotrypsin and papain, elastatinal active against elastase, pepstatins active against pepsin and several other carboxyl proteases and phosphoramidon active against thermolysin. The reader is referred to Whitaker [107] for more detailed information. These small peptide inhibitors are very stable and can be produced economically by fermentation. Therefore, they could be used to control some enzyme activities in foods.

7.7.3.2 α-Amylase Inhibitors

There are three types of α-amylase inhibitors: (a) proteins produced by higher plants, (b) small polypeptides produced by several species of *Streptomyces*, and (c) small N-containing carbohydrates produced by *Streptomyces* [50].

Several higher plants contain inhibitors of α-amylases of mammals and insects, with a few

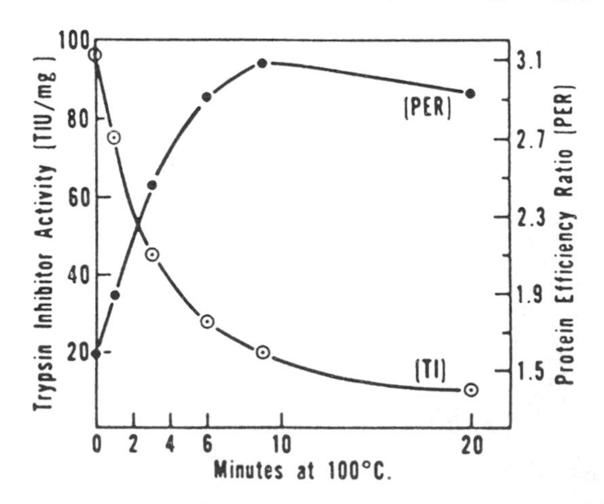

FIGURE 33 Effect of heat treatment on the trypsin inhibitory activity and nutritive value, as measured by the protein efficiency ratio (PER), of soybean meal. (From Ref. 87, p. 164A, with permission of American Oil Chemists Society.)

of the inhibitors active on microbial α-amylases and host-specific α-amylases. Sources include wheat, corn, barley, millet, sorghum, peanut, and beans. The proteins range from 9000 to 63,000 MW. Most slowly form 1:1 stoichiometric complexes with the α-amylases. The K_i values for the bean α-amylase inhibitors are 10^{-10} to 10^{-11} *M*; therefore, these inhibitors are slow, tight-binding inhibitors. They are quite heat stable, so that some of the inhibitor can survive breadmaking (for example) and up to 1 h cooking of the seeds.

Whether these α-amylase inhibitors are nutritionally important is controversial. They certainly slow down the rate of digestion of starch in saliva and small intestine of humans and rats, and some undigested starch is present in the feces of rats fed the inhibitors. Glucose is not released as rapidly from starch digestion into the blood of humans and rats in the presence of the inhibitors. However, there is little effect of the inhibitor on growth of rats and chickens.

Relatively small polypeptide inhibitors of α-amylase ranging from 3936 to 8500 MW are produced by several species of *Streptomyces*. There is much homology among the five polypeptide inhibitors that have received the most attention. The microbially derived polypeptide α-amylase inhibitors do not appear to have been tested with plant, insect, and most animal α-amylases. Their reaction with human α-amylases has been studied extensively at a clinical level.

The *Streptomyces* also produce three types of N-containing carbohydrates that are effective α-amylase and α-glucosidase inhibitors. These are the oligostatins, the amylostatins, and the trestatins. All have in common a pseudodisaccharide unit, oligobioamine or dehydro-oligobioamine, a variable number of α-D-glucose units linked α-1,4, and in one type α-1,1. Substantial clinical research has been done on their effects in modulating the effects of diabetes and hyperglycemia.

7.7.3.3 Invertase Inhibitors

Irish potatoes and sweet potatoes (yams) contain invertase inhibitors [107]. Invertase inhibitor in the Irish potato is a 17,000-MW protein. It inhibits potato invertase and several other plant invertases [86]. The level of invertase inhibitor in stored potatoes is temperature dependent, increasing at higher temperatures and decreasing at lower temperatures. The in vivo function of the inhibitor in potatoes is thought to be regulation of invertase activity. Potatoes stored at lower, nonfreezing temperatures are sweeter (more sucrose) than those stored at higher temperatures (where the potato is more starchy). These enzymatically caused changes are important both for quality and cost.

7.7.3.4 Other Enzyme Inhibitors

Inhibitors against several other enzymes have been reported [107]. Probably the best understood, physiologically, are the protein phosphatase inhibitors and the protein kinase inhibitors. Phosphorylase and glycogen synthase are the key enzymes involved in the metabolic breakdown of glycogen to glucose, and biosynthesis of glucose glycogen in animal tissues, respectively. The role of the protein phosphatase inhibitors and the protein kinase inhibitors appears to be regulation of the amount of phosphorylase *a* present by the mechanism shown in Equation 57.

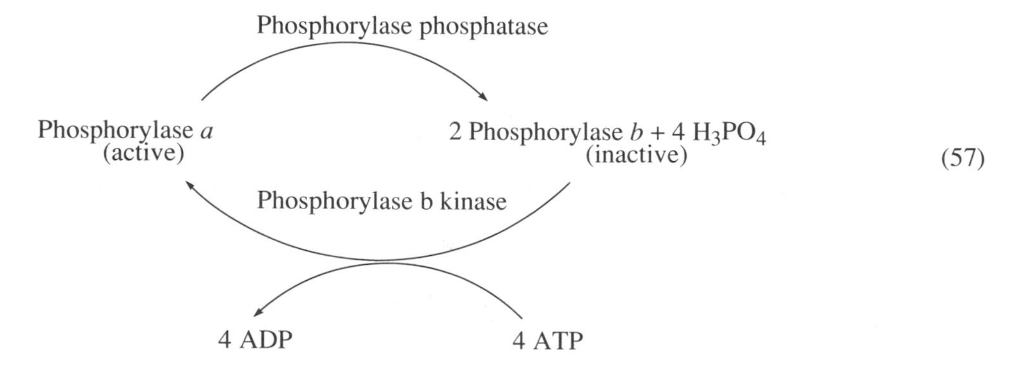

7.7.3.5 Chemically Tailored Inhibitors

Chemical tailoring of inhibitors is a major activity in pharmaceutical research laboratories. In the 1950–1970 era the affinity labeling inhibitors (Eq. 55) were the big thing because of the increased specificity in covalent binding to the enzyme that needed to be inhibited. Then came the k_{cat}-type inhibitors (Eq. 56) with their double specificity of binding selectively to the active site, and catalytic conversion to a product that irreversibly inactivates the enzyme while still in the active site. Now two new thrusts are to target malignant cells, as in cancer, specifically with appropriate antibodies, and to destroy the cells either by a radio label or by tailoring the target molecule to enter the cell and inhibit a key enzyme(s) in the cell division process. The other major area is the use of antisense *m*RNA designed to "turn off" a specific gene [25], as is done in the Flavr Savr tomato in suppressing the translation of the polygalacturonase gene. This is a general method that, in theory, can be used to suppress the production of any detrimental enzyme in our raw food products or those responsible for diseases of plants and animals. This method is much more specific than decreasing the level of an enzyme by breeding, as has been done, for example, with polyphenol oxidase in peaches.

7.7.4 Inactivation and/or Control by Physical Methods

Several handling methods used in food processing inactivate enzymes, intentionally or not. These include exposure to interfaces formed along the walls of transport tubes or the effects of stirring, whipping, shearing, high temperatures, pressure, and heat. However, not all of these lead to "instantaneous" or complete inactivation, and there can be an initial period when the activity of the enzyme, in contact with its substrate, can be enhanced. Chemical changes, enzymatically catalyzed or not, intended or not, are of major importance to the quality and safety of foods.

7.7.4.1 Interfacial Inactivation of Enzymes

Air/water and lipid/water interfaces are regions of high energy in which proteins tend to align themselves, with hydrophilic segments oriented into the water phase and hydrophobic segments oriented into the air or lipid phase. Unfolding of the protein at the interface is thermodynamically favored. Proteins are ideal surface-active molecules and very important in foaming and emulsifying in foods (see Chaps. 3 and 6). Mild shaking of a solution of enzyme can cause rapid denaturation. This effect can be a disadvantage in orange and tomato juice products and in the cold break process in tomato paste production. Sometimes, foaming is used in enzyme or protein purification, where one protein is more stable to foaming than the other. For example, the purification of invertase inhibitor from Irish potatoes is a problem because the inhibitor complexes with the invertase present. By controlled foaming the invertase can be denatured, enhancing recovery of undenatured inhibitor [15].

7.7.4.2 Pressure Effect on Enzymes

Normal food processing practices do not create high enough pressures alone to inactivate enzymes. In combination processes, such as pressure combined with high-temperature treatment and/or high shear rates, the latter two aspects are primarily responsible for any enzyme inactivation. Extruders operate at pressures greater than those normally encountered during food processing, and this results in greater fluidity, texturization of the food through shear forces, chemical reactions, and high temperatures. It is doubtful that most enzymes would survive extruder conditions.

Research has been done on the use of high pressures to inactivate microorganisms.

Hydrostatic pressures that are compatible with intact food tissues almost certainly will not completely inactivate enzymes of interest.

It is anticipated that pressure treatment of food will have greater effect on multi-subunit enzymes than on single polypeptide enzymes, since pressure, if high enough, will favor the monomer form of a protein, that is, will cause more dissociation to the monomer. Gutfreund and his colleagues [44] studied the effect of a pressure jump on the dissociation of hemoglobin. They reported that 10^7 Pa pressure caused 4% dissociation into two α,β-dimers, a relatively easy dissociation (i.e., 0.1 *M* NaCl will do this).

7.7.4.3 Inactivation of Enzymes by Shearing

Considerable shearing occurs in food processing during mixing, transfer through tubes, and extrusion. Some attention has been given to the shear conditions required for enzyme inactivation. Charm and Matteo [20] determined that when the shear value [shear rate (sec^{-1}) times the exposure time (sec)] is greater than 10^4 some inactivation of rennet is detectable (Fig. 34), and the rate of inactivation is very much faster at 7×10^5.

About 50% inactivation of catalase, rennet and carboxypeptidase was found at shear values of about 10^7 [20]. Rennet regained some activity after shear inactivation, similar to regaining of activity following heat inactivation.

7.7.4.4 Ionizing Radiation and Enzyme Inactivation

Much attention has been given to the effect of ionizing radiation of food on the inactivation of enzymes [34]. It soon became clear that about 10 times greater dose of ionizing radiation is required to inactivate enzymes than is needed to destroy microbial spores. For example, irradiated meat in which the microbial population was reduced to quite low numbers still developed textural defects due to protease activity during storage.

Some of the factors that influence the rate of trypsin inactivation by irradiation are shown in Figure 35. Trypsin is more stable in the dry state than in the wet state, because the indirect effect of free radicals (HO· and H·) produced from water are very important in enzyme

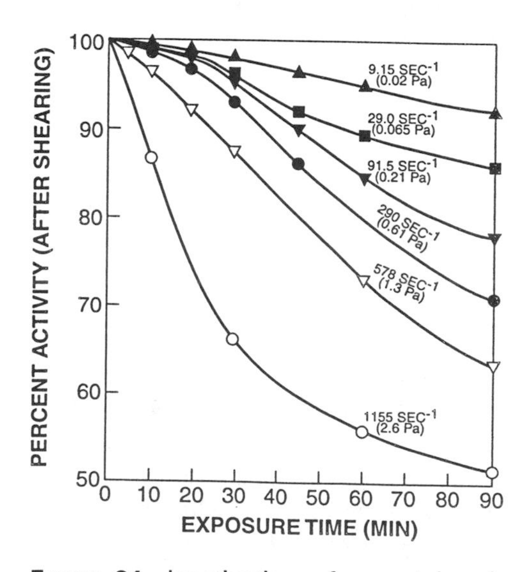

FIGURE 34 Inactivation of rennet by shearing at 4°C. (Reprinted from Ref. 20, p. 505, couurtesy of Academic Press.)

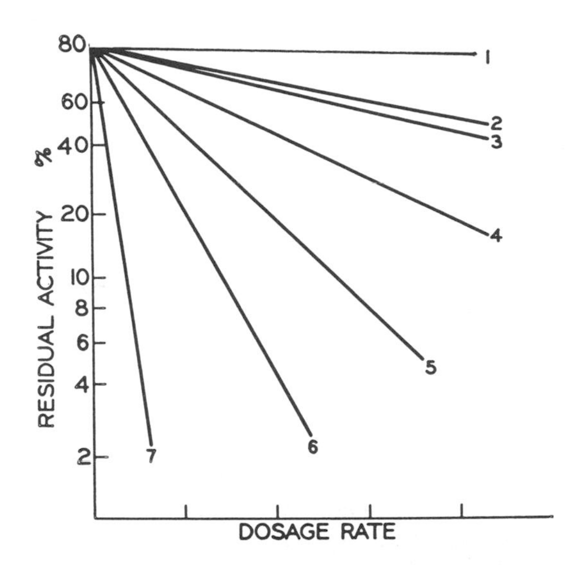

FIGURE 35 Inactivation of trypsin by irradiation under various conditions: (1) dry trypsin, (2) 10 mg/ml of trypsin solution at pH 8 and –78°C, (3) 80 mg/ml of trypsin solution at pH 8 and ambient temperature, (4) 10 mg/ml of trypsin at pH 2.6 and –78°C, (5) 80 mg/ml of trypsin at pH 2.7 and ambient temperature, (6) 10 mg/ml of trypsin at pH 2.6 and ambient temperature, (7) 1 mg/ml of trypsin at pH 2.6 and ambient temperature. (Redrawn from Ref. 69.)

inactivation. In general, the more dilute the trypsin solution, the more effective the radiation treatment. A given dose of ionizing radiation is also more effective at ambient temperature than at –78°C. At least part of this difference in effect results from immobilization of water free radicals at –78°C (frozen state). Differences in pH do not seem to have a pronounced effect on irradiation inactivation of trypsin, but these differences have been shown to be important in activating enzymes by irradiation in meat. Other factors that enhance inactivation by ionizing radiation are the presence of O_2, metal ions such as Cu^{2+}, unsaturated lipids, fatty acids, and purity of the enzyme, since other proteins protect against inactivation. Individual enzymes differ substantially in their resistance to inactivation by ionizing radiation, with catalase being about 60 times more resistant than carboxypeptidase (both metal-containing enzymes; Fe^{3+} is the cofactor in catalase, Zn^{2+} is the cofactor in carboxypeptidase).

Ionizing radiation causes most damage to tyrosyl, histidyl, tryptophanyl, and cysteinyl residues in proteins. Sulfhydryl compounds in food tend to protect enzymes against inactivation by ionizing radiation.

In order to avoid using higher doses of ionizing radiation than required to destroy microorganisms, heat treatment (to inactivate enzymes) is usually combined with ionizing radiation in those foods where this preservation method is permitted.

7.7.4.5 Solvent Inactivation of Enzymes

The effect of solvents on enzyme activity was discussed for model systems in Section 7.5.6. In general, water immiscible solvents, by displacing water, stabilize enzymes, just as does removal of water by drying. However, water-miscible solvents, when present at concentrations exceeding about 5–10%, generally inactivate enzymes. This effect is, of course, temperature dependent (more stable at lower temperatures).

Solvent treatment, for example with ethanol, can be quite effective in inactivating microorganisms on the surface of grains and legume seeds. However, the efficacy of this treatment probably does not depend on inactivation of enzymes.

7.7.5 Removal of Substrate and/or Cofactor

It is obvious that removal of one or more of the required substrates or a required cofactor will eliminate enzyme activity. It is well known that polyphenol oxidase browning can be prevented by excluding O_2. This is effectively done by natural selective permeability of the surface of most fruits and vegetables. Browning catalyzed by polyphenol oxidase can also be prevented by removing or altering the phenols (substrates). Binding of phenols to polyethylene glycol, polyvinylpyrrolidone, or Sephadex is effective. Polyphenol oxidase activity can be prevented by enzymatic methylation of one or more of the hydroxyl groups of the phenolic substrate [37]. This might be effective in juice, for example.

Another approach to preventing browning caused by polyphenol oxidase is to reduce the initial product, *o*-benzoquinone, back to the substrate before the *o*-benzoquinone can undergo further (non-enzyme-catalyzed) oxidative polymerization to melanin. Compounds that reduce *o*-benzoquinone include ascorbic acid, sodium bisulfite, and thiol compounds. It has been shown that these compounds also directly inactivate polyphenol oxidase by free radical degradation of histidine residues at the active site (ascorbic acid and Cu^{2+}), and by reducing Cu^{2+} to Cu^{+} in the active site of the enzyme, thereby causing Cu^{+} to dissociate more readily from the enzyme [84].

Enzymes are generally more stable in the presence of substrates, cofactors and competitive inhibitors. The general conclusion is that binding of these compounds into the active site of the enzyme has a stabilizing effect.

7.8 FOOD MODIFICATION BY ENZYMES

Enzymes have a very important inpact on the quality of our foods. In fact, without enzymes there would be no food. But then there would be no need for food, since no organism could live without enzymes. They are the catalysts that make life possible, as we know it.

For any organism, life begins with enzyme action in the gestation and fertilization processes. The growth and maturation of our foods depend on enzyme actions. While we have known for some time that environmental conditions during growing affect the composition, including enzymes, of our plant foods, a recent review [16] has detailed just how much effect moisture deficiency has on the expression of genes during growth and maturation of plants.

Following maturation, the harvesting, storage, and processing conditions can markedly affect the rate of food deterioration. Enzymes can also be added to foods during processing to change their characteristics, and some of these changes will be discussed. Microbial enzymes, left after destruction of the microorganisms, continue to affect the quality of processed and reformulated foods. For example, starch-based sauces can undergo undesirable changes in consistency because of heat-stable microbial α-amylases that survive a heat treatment sufficient to destroy the microorganisms. Because of their high specificity, enzymes are also the ideal catalysts for the biosynthesis of highly complex chemicals.

7.8.1 Role of Endogenous Enzymes in Food Quality

7.8.1.1 Color

(See also Chap. 10.) Color is probably the first attribute the consumer associates with quality and acceptability of foods. A steak must be red, not purple or brown. Redness is due only to oxymyoglobin, the main pigment in meat. Deoxymyoglobin is responsible for the purple color of meat. Oxidation of the Fe(II) present in oxymyoglobin and deoxymyoglobin, to Fe(III) producing metmyoglobin, is responsible for the brown color of meat. Enzyme-catalyzed reactions in meat can compete for oxygen, can produce compounds that alter the oxidation–reduction state and water content, and can thereby influence the color of meat.

The quality of many fresh vegetables and fruits is judged on the basis of their "greenness." On ripening, the green color of many of our fruits decreases and is replaced with red, orange, yellow, and black colors. In green beans and English green peas, maturity leads to a decrease in chlorophyll level. All of these changes are a result of enzyme action. Three key enzymes responsible for chemical alterations of pigments in fruits and vegetables are lipoxygenase, chlorophyllase, and polyphenol oxidase.

Lipoxygenase

Lipoxygenase (lineoleate:oxygen oxidoreductase; EC 1.13.11.12) has six important effects on foods, some desirable and others undesirable. The two desirable functions are (a) bleaching of wheat and soybean flours and (b) formation of disulfide bonds in gluten during dough formation (eliminates the need to add chemical oxidizers, such as potassium bromate). The four undesirable actions of lipoxygenase in food are (a) destruction of chlorophyll and carotenes, (b) development of oxidative off flavors and aromas, often characterized as haylike, (c) oxidative damage to compounds such as vitamins and proteins, and (d) oxidation of the essential fatty acids, lineoleic, linolenic, and arachidonic acids.

All six of these reactions result from the direct action of lipoxygenase in oxidation of polyunsaturated fatty acids (free and lipid-bound) to form free radical intermediates (Fig. 36). In steps 2, 3, and 4 (Fig. 36) free radicals are formed, while in step 5 a hydroperoxide is formed. Further nonenzymatic reactions lead to formation of aldehydes (including malondialdehyde) and other components that contribute to off flavors and off aromas. The free radicals and hydroperoxide are responsible for loss of color (chlorophyll; the orange and red colors of the carotenoids), disulfide bond formation in gluten of doughs, and damage to vitamins and proteins. The most oxidation-sensitive amino acid residues in proteins are cysteine, tyrosine, histidine, and tryptophan. Antioxidants, such as vitamin E, propyl gallate, benzoylated hydroxytoluene, and nordihydroguaiacetic acid, protect foods from damage from free radicals and hydroperoxides.

Chlorophyllase

Chlorophyllase (chlorophyll chlorophyllido-hydrolase, EC 3.1.1.14) is found in plants and chlorophyll-containing microorganisms. It hydrolyzes the phytyl group from chlorophyll to give phytol and chlorophyllide (see Chap. 10). Although this reaction has been attributed to a loss of green color, there is no evidence to support this as chlorophyllide is green. Furthermore, there is no evidence that the chlorophyllide is any less stable to color loss (loss of Mg^{2+}) than is chlorophyll. The role of chlorophyllase in vivo in plants is not known. Very few studies have been done on chlorophyllase-catalyzed hydrolysis of chlorophyll during storage of raw plant foods (Chap. 10).

Polyphenol Oxidase

Polyphenol oxidase (1,2-benzenediol:oxygen oxidoreductase; EC 1.10.3.1) is frequently called tyrosinase, polyphenolase, phenolase, catechol oxidase, cresolase, or catecholase, depending on the substrate used in its assay or found in the greatest concentration in the plant that serves as a source of the enzyme. Polyphenol oxidase is found in plants, animals and some microorganisms, especially the fungi. It catalyzes two quite different reactions with a large number of phenols, as shown in Equations 58 and 59.

$$p\text{-Cresol} + O_2 + BH_2 \longrightarrow \text{4-Methylcatechol} + B + H_2O \tag{58}$$

$$2\ \text{Catechol} + O_2 \longrightarrow 2\ o\text{-Benzoquinone} + 2H_2O \tag{59}$$

The 4-methyl-*o*-benzoquinone (Eq. 59) is unstable and undergoes further non-enzyme-catalyzed oxidation by O_2, and polymerization, to give melanins. The latter is responsible for the undesirable brown discoloration of bananas, apples, peaches, potatoes, mushrooms, shrimp, and humans (freckles), and the desirable brown and black colors of tea, coffee, raisins, prunes, and human skin pigmentation. The *o*-benzoquinone reacts with the ε-amino group of lysyl residues of proteins, leading to loss of nutritional quality and insolubilization of proteins. Changes in texture and taste also result from the browning reactions.

It is estimated that up to 50% of some tropical fruits are lost due to enzymatic browning. This reaction is also responsible for deterioration of color in juices and fresh vegetables such as lettuce, and in taste and nutritional quality. Therefore, much effort has gone into developing methods for control of polyphenol oxidase activity. As discussed earlier in Section 7.7.5, elimination of O_2 and the phenols will prevent browning. Ascorbic acid, sodium bisulfite, and thiol compounds prevent browning due to reduction of the initial product, *o*-benzoquinone, back to the substrate, thereby preventing melanin formation. When all of the reducing compound is consumed, browning still may occur, since the enzyme may still be active. Ascorbic acid, sodium sulfite, and thiol compounds also have a direct effect in inactivating polyphenol oxidase due to destruction of the active-site histidines (ascorbic acid) or in removal by (sodium bisulfite and thiols) of the essential Cu^{2+} in the active site [84].

4-Hexylresorcinol, benzoic acid, and some other nonsubstrate phenols are effective inhibitors of some polyphenol oxidases, with K_i values around 1–10 μ*M*. Most likely, inhibition results from competitive binding of this type of inhibitor into the reducing compound (BH_2) binding site (Eq. 58).

7.8.1.2 Texture

Texture is a very important quality attribute in foods. In fruits and vegetables, texture is due primarily to the complex carbohydrates: pectic substances, cellulose, hemicelluloses, starch, and lignin. There are one or more enzymes that act on each of the complex carbohydrates that are important in food texture. Proteases are important in the softening of animal tissues and high-protein plant foods.

Pectic Enzymes

Three types of pectic enzymes that act on pectic substances are well described. Two (pectin methylesterase and polygalacturonase) are found in higher plants and microorganisms and one type (the pectate lyases) is found in microorganisms, especially certain pathogenic microorganisms that infect plants [109].

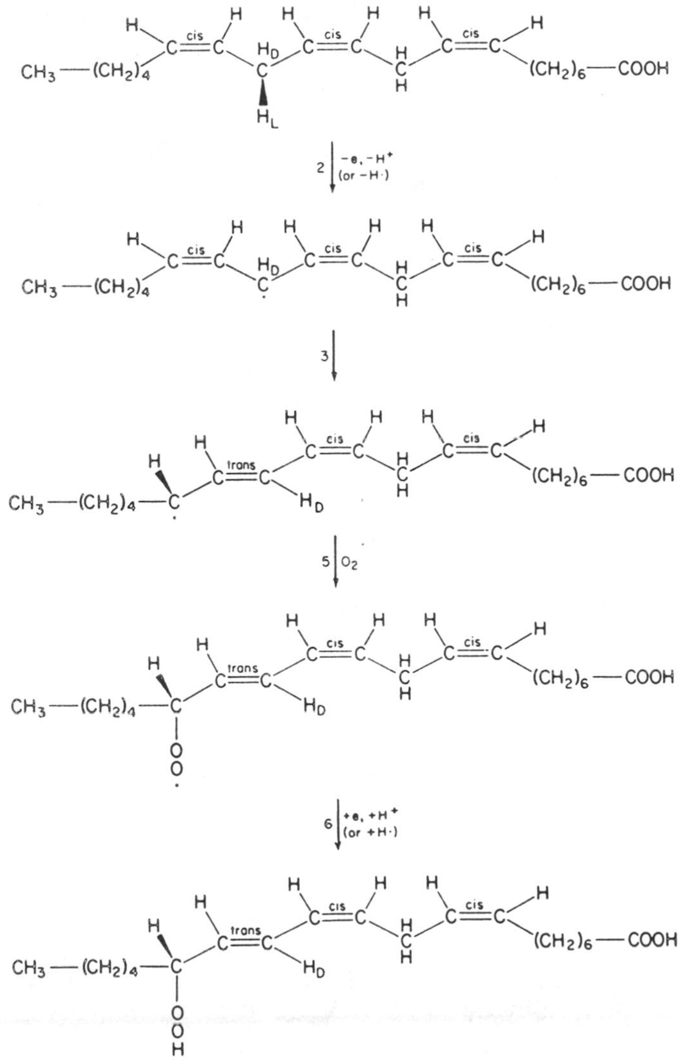

H H H H H
H cis cis cis H
CH_3—$(CH_2)_4$ C=C H_D C C=C H C C=C $(CH_2)_6$—COOH
H
H_L

2 ↓ −e, −H^+ (or −H·)

H H H H H
H cis cis cis H
CH_3—$(CH_2)_4$ C=C H_D C· C=C H C C=C $(CH_2)_6$—COOH
H

3 ↓

H H H
H cis cis H
H trans H
CH_3—$(CH_2)_4$—C· C=C C=C C C=C $(CH_2)_6$—COOH
H_D H

5 ↓ O_2

H H H
H cis cis H
H trans H
CH_3—$(CH_2)_4$—C C=C C=C C C=C $(CH_2)_6$—COOH
H_D H
O
O·

6 ↓ +e, +H^+ (or +H·)

H H H
H cis cis H
H trans H
CH_3—$(CH_2)_4$—C C=C C=C C C=C $(CH_2)_6$—COOH
H_D H
O
O
H

FIGURE 36 Reaction catalyzed by lipoxygenase. See the test for a description of the sequence of reactions. (From Ref. 111, p. 588.)

Pectin methylesterase (pectin pectylhydrolase, EC 3.1.1.11) hydrolyzes the methyl ester bond of pectin to give pectic acid and methanol (Eq. 60).

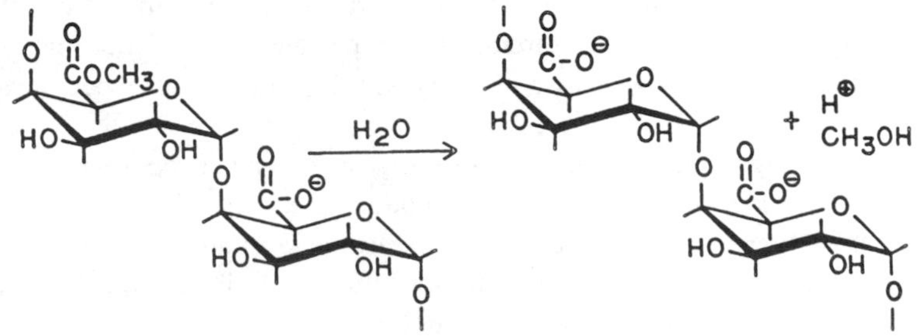

O COCH3 O HO OH O C-O⊖ O HO OH O —H_2O→ O C-O⊖ O HO OH O C-O⊖ O HO OH O + $H^{\oplus}$ CH_3OH (60)

The enzyme is sometimes referred to as pectinesterase, pectase, pectin methoxylase, pectin demethoxylase, and pectolipase. Hydrolysis of pectin to pectic acid in the presence of divalent ions, such as Ca^{2+}, leads to an increase in textural strength, due to the cross bridges formed between Ca^{2+} and the carboxyl groups of pectic acid.

Polygalacturonase (poly-α-1,4-galacturonide glycano-hydrolase, EC 3.2.1.15) hydrolyzes the α-1,4-glycosidic bond between the anhydrogalacturonic acid units (Eq. 61).

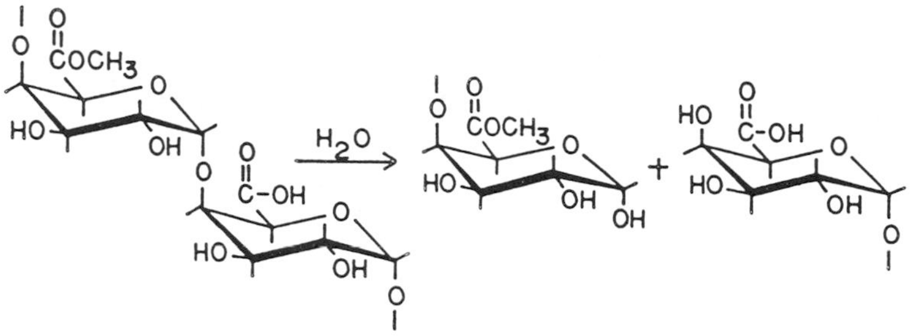

(61)

Both *endo*- and *exo*-polygalacturonases exist; the *exo* type hydrolyzes bonds at the ends of the polymer and the *endo* type acts in the interior. There are differences in opinion as to whether plants contain both polymethylgalacturonases (act on pectins) and polygalacturonases (act on pectic acid), since pectin methylesterase in the plant rapidly converts pectin to pectic acid. Action of polygalacturonase results in hydrolysis of pectic acid, leading to important decreases in texture of some raw food materials, such as tomatoes.

The pectate lyases [poly(1,4-α-D-galacturonide) lyase, EC 4.2.2.2] split the glycosidic bond of both pectin and pectic acid, not with water, but by β-elimination (Eq. 62). They are found in microorganisms, but not in higher plants.

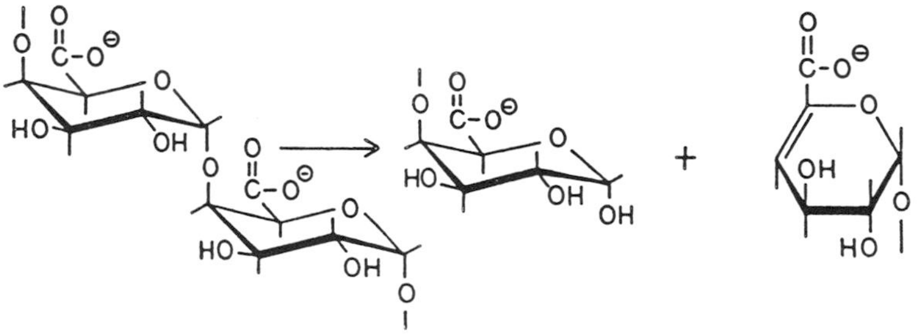

(62)

Splitting of the glycosidic bond gives a product with a reducing group and another product with a double bond. Polygalacturonases and pectate lyases both produce reducing groups when the glycosidic bond is split, resulting in a decrease in texture, so they cannot be distinguished by this method. However, the double bond in the second product (Eq. 62) has an extinction coefficient of 4.80×10^3 M^{-1} cm^{-1} at 235 nm. Therefore, a change in absorbance at 235 nm is the method of choice for distinguishing between polygalacturonases and pectate lyases. There are both *endo*- and *exo*-splitting pectate lyases, as well as pectate lyases that act on either pectin or on pectic acid.

A fourth type of pectin-degrading enzyme, protopectinase, has been reported in a few microorganisms. Protopectinase hydrolyzes protopectin, producing pectin. However, it is not clear yet whether the protopectinase activity in plants is due to the combined action of pectin methylesterase and polygalacturonase or to a true protopectinase. It is unlikely that protopectinase is an "H-bondase" as suggested by some researchers.

Cellulases

Cellulose is abundant in trees and cotton. Fruits and vegetables contain small amounts of cellulose, which has a role in the structure of cells. Whether cellulases are important in the softening of green beans and English green pea pods is still a matter of controversy. Abundant information is available on the microbial cellulases because of their potential importance in converting insoluble cellulosic waste to glucose [70].

Pentosanases

Hemicelluloses, which are polymers of xylose (xylans), arabinose (arabans), or xylose and arabinose (arabinoxylans), with small amounts of other pentoses or hexoses, are found in higher plants. Pentosanases in microorganisms [109], and in some higher plants, hydrolyze the xylans, arabans, and arabinoxylans to smaller compounds. The microbial pentosanases are better characterized than those in higher plants.

Several *exo*- and *endo*-hydrolyzing pentosanases also exist in wheat at very low concentrations, but little is known about their properties. It is important that these pentosanases receive more attention from food scientists.

Amylases

Amylases, the enzymes that hydrolyze starches, are found not only in animals, but also in higher plants and microorganisms. Therefore, it is not surprising that some starch degradation occurs during maturation, storage, and processing of our foods. Since starch contributes in a major way to viscosity and texture of foods, its hydrolysis during storage and processing is a matter of importance. There are three major types of amylases: α-amylases, β-amylases, and glucoamylases. They act primarily on both starch and glycogen. Other starch-splitting enzymes also exist (Table 14).

The α-amylases, found in all organisms, hydrolyze the interior α-1,4-glucosidic bonds of starch (both amylose and amylopectin), glycogen, and cyclodextrins with retention of the α-configuration of the anomeric carbon. Since the enzyme is *endo*-splitting, its action has a major effect on the viscosity of starch-based foods, such as puddings, cream sauces, etc. The salivary and pancreatic α-amylases are very important in digestion of starch in our foods. Some microorganisms contain high levels of α-amylases. Some of the microbial α-amylases have high inactivation temperatures, and, if not activated, they can have a drastic undesirable effect on the stability of starch-based foods.

β-Amylases, found in higher plants, hydrolyze the α-1,4-glucosidic bonds of starch at the nonreducing end to give β-maltose. Since they are *exo*-splitting enzymes, many bonds must be hydrolyzed before an appreciable effect on viscosity of starch paste is observed. Amylose can be hydrolyzed to 100% maltose by β-amylase, while β-amylase cannot continue beyond the first α-1,6-glycosidic bond encountered in amylopectin. Therefore, amylopectin is hydrolyzed only to a limited extent by β-amylase alone. "Maltose" syrups, of about DP 10, are very important in the food industry. β-Amylase, along with α-amylase, is very important in brewing, since the maltose can be rapidly converted to glucose by yeast maltase. β-Amylase is a sulfhydryl enzyme and can be inhibited by a number of sulfhydryl group reagents, unlike α-amylase and glucoamylase. In malt, β-amylase is often covalently linked, via disulfide bonds, to other sulfhydryl groups; therefore, malt should be treated with a sulfhydryl compound, such as cysteine, to increase its activity in malt.

Proteases

Texture of food products is changed by hydrolysis of proteins by endogenous and exogenous proteases. Gelatin will not gel when raw pineapples is added, because the pineapple contains

TABLE 14 Some Starch- and Glycogen-Degrading Enzymes

Type	Configuration of glucosidic bond	Comments
Endo-splitting (configuration retained)		
α-Amylase (EC 3.2.1.1)	α-1,4	Initial major products are dextrins; final major products are maltose and maltotriose
Isoamylase (EC 3.2.1.68)	α-1,6	Products are linear dextrins
Isomaltase (EC 3.2.1.10)	α-1,6	Acts on products of α-amylase hydrolysis of amylopectin
Cyclomaltodextrinase (EC 3.2.1.54)	α-1,4	Acts on cyclodextrins and linear dextrins to give maltose and maltotriose
Pullulanase (EC 3.2.1.41)	α-1,6	Acts on pullulan to give maltotriose and on starch to give linear dextrins
Isopullulanase (EC 3.2.1.57)	α-1,4	Acts on pullulan to give isopanose and on starch to give unknown products
Neopullulanase	α-1,4	Acts on pullulan to give panose and on starch to give maltose
Amylopullulanase	α-1,6	Acts on pullulan to give maltotriose
	α-1,4	Acts on starch to give DP 2–4 products
Amylopectin 6-glucano hydrolase (EC 3.2.1.41)	α-1,6	Acts only on amylopectin to hydrolyze α-1,6-glucosidic linkages
Exo-splitting (nonreducing end)		
β-Amylase (EC 3.2.1.2)	α-1,4	Product is β-maltose
α-Amylase	α-1,4	Product is α-maltose; there are specific *exo*-α-amylases that produce maltotriose, maltotetraose, maltopentaose, and maltohexaose, with retention of configuration
Glucoamylase (EC 3.2.1.3)	α-1,6	β-Glucose is produced
α-Glucosidase (EC 3.2.1.20)	α-1,4	α-Glucose is produced; there are a number of α-glucosidases
Transferase		
Cyclomaltodextrin glucanotransferase (EC 2.4.1.19)	α-1,4	α- and β-Cyclodextrins formed from starch with 6–12 glucose units

bromelain, a protease. Chymosin causes milk to gel, as a result of its hydrolysis of a single peptide bond between Phe_{105}-Met_{106} in κ-casein. This specific hydrolysis of κ-casein destabilizes the casein micelle, causing it to aggregate to form a curd (cottage cheese). Action of intentionally added microbial proteases during aging of brick cheeses assists in development of flavors (flavors in Cheddar cheese vs. blue cheese, for example). Protease activity on the gluten proteins of wheat bread doughs during rising is important not only in the mixing characteristics and energy requirements but also in the quality of the baked breads.

The effect of proteases in the tenderization of meat is perhaps best known and is economically most important. After death, muscle becomes rigid due to rigor mortis (caused by extensive interaction of myosin and actin). Through action of endogenous proteases (Ca^{2+}-activated proteases, and perhaps cathepsins) on the myosin–actin complex during storage (7–21 days) the muscle becomes more tender and juicy. Exogenous enzymes, such as papain and ficin, are added to some less choice meats to tenderize them, primarily due to partial hydrolysis of elastin and collagen.

7.8.1.3 Flavor and Aroma Changes in Foods

Chemical compounds contributing to the flavor and aroma of foods are numerous, and the critical combinations of compounds are not easy to determine. It is equally difficult to identify the enzymes instrumental in the biosynthesis of flavors typical of food flavors and in the development of undesirable flavors.

Enzymes cause off flavors and off aromas in foods, particularly during storage. Improperly blanched foods, such as green beans, English green peas, corn, broccoli, and cauliflower, develop very noticeable off flavors and off aromas during frozen storage. Peroxidase, a relatively heat-resistant enzyme not usually associated with development of defects in food, is generally used as the indicator for adequate heat treatment of these foods. It is clear now that a higher quality product can be produced by using the primary enzyme involved in off flavor and off aroma development as the indicator enzyme.

With this in mind, Whitaker and Pangborn [74,75,101] and their students determined that lipoxygenase is responsible for off flavor and off aroma development in English green peas, green beans, and corn, and that cystine lyase is the primary enzyme responsible for off flavor and off aroma development in broccoli and cauliflower. Evidence to support the important role of lipoxygenase as a catalyst of off-flavor development in green beans is shown in Figure 37. Additional work indicated that the flavor stability of frozen food (of types mentioned above), blanched to the endpoint of the responsible enzyme, is better than that of comparable samples blanched to the peroxidase endpoint.

Naringin is responsible for the bitter taste of grapefruit and grapefruit juice. Naringin can be destroyed by treating the juice with naraginase. Some research is underway to eliminate naringin biosynthesis by recombinant DNA techniques.

7.8.1.4 Nutritional Quality

There is relatively little data available with respect to the effects of enzymes on nutritional quality of foods. Lipoxygenase oxidation of linoleic, linolenic, and arachidonic acids certainly decreases the amounts of these essential fatty acids in foods. The free radicals produced by lipoxygenase-catalyzed oxidation of polyunsaturated fatty acids decrease the carotenoid (vitamin A precursors), tocopherols (vitamin E), vitamin C, and folate content of foods. The free radicals also are damaging to cysteine, tyrosine, tryptophan, and histidine residues of proteins. Ascorbic acid is destroyed by ascorbic acid oxidase found in some vegetables such as squash. Thiaminase destroys thiamine, an essential cofactor involved in amino acid metabolism. Riboflavin hydrolase, found in some microorganisms, can degrade riboflavin. Polyphenol oxidase-caused browning decreases the available lysine content of proteins.

7.8.2 Enzymes Used as Processing Aids and Ingredients

Enzymes are ideal for producing key changes in the functional properties of food, for removal of toxic constituents, and for producing new ingredients. This is because they are highly specific, act at low temperatures (25–45°C), and do not produce side reactions. Full utilization of the high

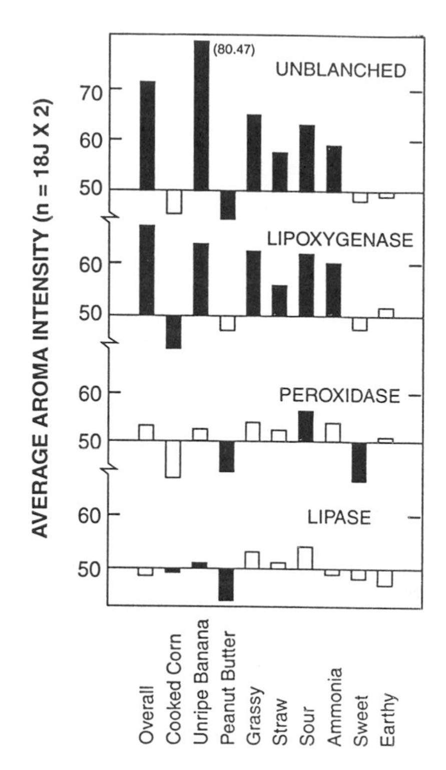

FIGURE 37 Effect of added enzymes on development of undesirable aroma descriptors in green beans. Average aroma differences are shown between sample and a blanched reference sample to which no enzymes were added. An average aroma intensity of 50 is the same as the reference. The shaded bars indicate aroma descriptor intensity differences statistically different from the reference. (From Ref. 114, p. 134, with permission of Institute of Food Technologists.)

specificity of enzymes is compromised currently because relatively crude enzyme systems are used, because of cost. Crude preparations contain other enzymes that may produce unwanted products. Of major concern is that the crude enzyme preparations often differ substantially from batch to batch. With recombinant DNA methods, the time is closer when pure enzymes can be used in food processing.

There are some major successes in the use of food-related enzymes. Production of high-fructose corn syrup is one example. This involves a relatively heat-stable α-amylase, glucoamylase, and glucose isomerase:

$$\text{starch} \xrightarrow{\alpha\text{-amylase}} \text{dextrins} \xrightarrow{\text{glucoamylase}} \text{glucose} \underset{\text{isomerase}}{\overset{\text{glucose}}{\rightleftharpoons}} \text{fructose} \tag{63}$$

Starch is heated to 105°C, *Bacillus licheniformis* α-amylase is added, and dextrins of DP 10-12 are produced by the *endo*-splitting enzyme. The soluble digest is passed through giant columns (6–10 ft in diameter and 20 ft high) of immobilized glucoamylase where glucose is produced. The glucose-containing stream is then run through giant columns of immobilized glucose isomerase where approximately equimolar concentrations of glucose and fructose are produced. The fructose is separated from glucose by differential crystallization and is used as a major sweetener in the food industry (~100 billion tons/year). The glucose or a mixture of

TABLE 15 Biocatalytic Production of Sweeteners

Feedstock	Product	Enzyme(s)
Starch	Corn syrups	α-Amylases, pullulanases
	Glucose	α-Amylases, glucoamylase
	Fructose	α-Amylase, glucoamylase, glucose isomerase
Starch + sucrose	Sucrose derivatives	Cyclodextringlucosyltransferase and pullulanase (or isoamylase)
Sucrose	Glucose + fructose	Invertase
Sucrose	Isomaltulose	β-Glucosyltransferase and isomaltulose synthetase
Sucrose + fructose	Leucrose	α-1,6-Glycosyltransferase
Lactose	Glucose + galactose	β-Galactosidase
Galactose	Galacturonic acid	Galactose oxidase
	Glucose	Galactose epimerase
Several	Aspartame	Thermolysin, penicillinacylase
Stevioside	α-Glycosylated stevioside	α-Glucosidase
	Ribaudioside-A	β-Glycosyltransferase

Source: Ref. 110, p. 676.

glucose and fructose is used as sweetener, or the glucose is recycled to produce more fructose. Other sweeteners can also be produced enzymatically (Table 15).

A second example is the use of aminoacylases to separate racemic mixtures of DL-amino acids, in multi-ton lots. A third example is the use of specific lipases to tailor-make lipids with respect to melting point, unsaturation, or specific location of a fatty acid in a triacylglycerol. This requires a beginning lipid, the required concentrations of free fatty acids, appropriate enzymes, and appropriate conditions (Table 16). For example, a 1,3-specific lipase catalyzes transesterification of fatty acids at the 1- and 3-positions of glycerol, a 2-specific lipase catalyzes transesterification only at the 2-position of glycerol, and a nonspecific lipase catalyzes transesterification at the 1-, 2-, and 3-positions of glycerol. An example using 1,3-specific lipase is:

$$2\,\mathrm{O}\!-\!\left[\begin{matrix}\mathrm{X}\\ \\ \mathrm{X}\end{matrix}\right. + 3\mathrm{L} \underset{\text{1,3-specific lipase}}{\rightleftharpoons} \mathrm{O}\!-\!\left[\begin{matrix}\mathrm{L}\\ \\ \mathrm{X}\end{matrix}\right. + \mathrm{O}\!-\!\left[\begin{matrix}\mathrm{L}\\ \\ \mathrm{L}\end{matrix}\right. + 3\mathrm{X} \tag{64}$$

TABLE 16 Enzymes for Modifying Lipids

Enzyme	Products
Lipases	Rearranged triglycerides
	Waxes via esterification
	Monoglycerides
Phospholipase A	Lysophospholipid
Phospholipase D	Phosphotidylglycerol

Source: Ref. 110, p. 685.

7.8.2.1 Specialty Products and Ingredients via Enzymology

The following text and tables illustrate some of the current or potential uses of enzymes on an industrial scale.

Enzymatic Production of Valuable Compounds

Listed in Table 17 are 14 enzymes used to make specialty compounds as additives in foods.

Enzymatic Removal of Undesirable Compounds

Raw food materials often contain toxic or anti-nutrient compounds that are sometimes removed by proper heat treatment, extraction or by enzymatic reactions (Table 18). There are more than 12,000 plants in the world that may have potential as food sources. Many are not used because of undesirable properties, some of which could be overcome by the proper use of enzymes.

Enzymes in Milk and Dairy Products

Bovine milk contains many enzymes (see Chap. 14), and other enzymes are added during processing (Table 19). Of most importance economically is the use of chymosin (rennet) in production of several kinds of cheese. β-Galactosidase is potentially of great importance in the commercial hydrolysis of lactose in milk and dairy products, so that these products can be consumed by individuals who are deficient in β-galactosidase. β-Galactosidase is also used to convert whey lactose to glucose and galactose, sweeteners with greater commercial demand.

Enzymes in Baking

Several enzymes are used in breadmaking (Table 20). Additions of amylases and proteases has been common for years. Several enzyme preparations are available for the stated purpose of reducing the rate of staling in bread. These include (a) a debranching enzyme (1,6-splitting;

TABLE 17 Enzymatic Production of Wanted Compounds or Creation of Desirable Effects

Enzymes	Purpose
Aminoacylases	Resolve DL-amino acids
Aspartase	Produce aspartate
Proteases	Surfactants
Peroxidase	Phenol resins
5′-Phosphodiesterases	5′-Nucleotides for flavor enhancers
5′-Adenylic deaminase	Produce 5′-inosinic acid for flavor
Lipases (pregastric)	Cheese/butter flavors
Proteases	Decrease ripening time of cheeses Tenderization of meat
Lipases/esterases	Flavor esters
Proteases, nucleases	Meaty flavors from yeast hydrolysis
Fumarase	Fumaric acid as acidulant
Cyclomaltodextrin glucanotransferase	Cyclodextrins for inclusion complexation
Tannase	Antioxidants, such as propylgallate
α-Galactosidase	Modified food gums

Source: Ref. 110, p. 674.

TABLE 18 Enzymatic Removal of Unwanted Constituents

Enzymes	Unwanted constituent
α-Galactosidase	Raffinose
β-Galactosidase	Lactose
Glucose oxidase	Glucose
	O_2
Phytase	Phytic acid
Thioglycosidases	Thioglycosides
Oxalate oxidase	O_2
Alcohol oxidase	O_2
Oxyrase	O_2
Catalase	H_2O_2
Sulfhydryl oxidase	Oxidized flavors
Urease	Carbamates
Cyanidase	Cyanide
Pepsin, chymotrypsin, carboxypeptidase A	Bitter peptides
Naringinase	Bitter compounds in citrus
Proteases	Phenylalanine
α-Amylases	Amylase inhibitors
Proteases	Protease inhibitors

Source: Ref. 110, p. 678.

pullulanase-type) from *Bacillus acidopullulyticus* along with α-amylase; (b) a genetically engineered α-amylase from *Bacillus megaterium* that rearranges the branched chains of amylopectin, via chain transfer, to give a linear polymer; (c) a mixture of cellulase, β-glucanase, and pentosanase produced by *Trichoderma reesei*; and (d) a mixture of cellulase, hemicellulases, and pentosanases used in conjunction with fungal α-amylase. Preparations containing pentosanases are reported to increase the moisture content of rye bread products by decreasing the high pentosan content of rye flour.

TABLE 19 Enzymes in Milk and Diary Products

Enzymes	Function
Chymosin	Milk coagulation
	For rennet puddings
Chymosin, fungal proteases	For cottage cheese
	For brick cheeses
Proteases	Flavor improvement, decrease ripening time of cheeses
Lipases	Flavor improvement; decrease ripening time of cheeses
Sulfhydryl oxidase	Remove cooked flavor
β-Galactosidase	Lactose removal
Microbial proteases	Soybean milk coagulation

Source: Ref. 110, p. 687.

TABLE 20 Enzymes in Baking

Enzymes	Purpose
Amylase	To maximize fermentation process; prevent staling
Proteases	Improve handling and rheological properties
a-Glutamyl transferase, glutathione oxidase, cysteinyl glycine dipeptidase	Improve dough elasticity, loaf volume, crumb structure, and retention of bread softness on storage
Pentosanases	In rye bread products to decrease dough development time and power requirements, increase moistness
Sulfhydryl oxidase	Strengthen weak doughs by -S-S- formation

Source: Ref. 110, p. 689.

Enzymes in Brewing

Several recent advances have been made with regard to the use of enzymes in brewing (Table 21). There is increasing interest in the possible use of blended α-amylase and protease preparations as replacements for malt in brewing. This is attributable to the expense and limited supplies of malt and the possibility that better quality control might be achieved. The industry has also used amyloglucosidases recently to make "light" beer. Amyloglucosidases hydrolyze the α-1,6-glucosidic bonds of the amylopectin fraction, permitting the complete fermentation of starch. Use of β-glucanases may solve the high viscosity/slow filtration rate problems caused by mannans from cell walls. The most exciting advance, however, is the use of acetolactate decarboxylase, now cloned into brewers yeast, to shorten the fermentation time by avoiding diacetyl formation (Eq. 65) [83a].

$$\underset{\text{Acetolactate}}{CH_3-\overset{O}{\overset{\|}{C}}-\underset{OH}{\underset{|}{\overset{COOH}{\overset{|}{C}}}}-CH_3} \xrightarrow[\searrow CO_2]{(A)} \underset{\text{Acetoin}}{CH_3-\overset{O}{\overset{\|}{C}}-\underset{OH}{\underset{|}{\overset{H}{\overset{|}{C}}}}-CH_3} \xrightarrow{(B)} \underset{\text{2,3-butylene glycol}}{CH_3-\underset{OH}{\underset{|}{\overset{H}{\overset{|}{C}}}}-\underset{OH}{\underset{|}{\overset{H}{\overset{|}{C}}}}-CH_3} \quad (65)$$

(A = acetolactate decarboxylase; B = dehydrogenase)

Enzymes for Control of Microorganisms

Enzymes have potential for destroying microorganisms by several means (Table 22). The means range from hydrolysis of cell-wall compounds, such as β-glucans, chitin, and peptidoglycans, to production of H_2O_2 and $O_2^{\dot{-}}$, which oxidize the essential –SH group of key sulfhydryl enzymes or polyunsaturated fatty acids in cell walls. These are interesting possibilities, worthy of being tested at the commercial level.

7.9 IMMOBILIZED ENZYMES IN FOOD PROCESSING

7.9.1 Advantages and Disadvantages

Why immobilized enzymes? The objectives of immobilizing enzymes are to permit their repeated use to make products, and at the end, to have a product free of enzyme. In Section 7.8.2,

TABLE 21 Enzymes in Brewing

Enzymes	Purpose
Amylases (α and β)	Convert nonmalt starch to maltose and dextrins; fermented by yeast to alcohol and CO_2
Proteases (*endo* and *exo*)	Hydrolyze proteins to amino acids; used by yeast to grow
Papain	Chillproofing beer
Amyloglucosidase	Hydrolyze 1,6-linkages in amylopectin permitting complete fermentation of starch (light beer)
β-Glucanases	Hydrolyze glucans to reduce viscosity and aid filitration
Acetolactate decarboxylase	Decrease fermentation time, by avoiding diacetyl formation

Source: Ref. 110, p. 690.

the discussion focused on the many applications and potential applications of enzymes in food processing, their use in developing food ingredients and in the removing of unwanted compounds. It was mentioned that crude enzyme preparations produce numerous side reactions because of contaminating enzymes. An example of this problem can occur during fruit juice production. Pectic enzyme preparations from *Aspergillus niger* are often used to increase yield and clarity. However, there are several polymer-hydrolyzing enzymes in the *A. niger* enzyme preparation (Table 23). The net effect observed after enzyme treatment of the juice is a composite of the activities of many of the enzymes present. For example, in apple juice an arabinosidase removes the side chain arabinose units from the linear araban chain, which then aggregates to

TABLE 22 Enzymes for Control of Microorganisms

Enzyme	Function
Oxidases	Removal of O_2, NADH or NADPH; produce H_2O_2 and $O_2^{\bullet-}$, which oxidize -SH groups and polyunsaturated lipids
Xylitol phosphorylase	Conversion of xylitol to xylitol 5-phosphate, which kills microorganism
Lipases	Liberation of free fatty acids, which are toxic to protozoa *Giardia lamblia*
Lactoperoxidase	Uses $O_2^{\bullet-}$ & H_2O_2, produced by oxidase, to convert SCN^- to $SCNO^-$ and -SH groups to -S-S-, -S-SCN, or -S-OH
Myeloperoxidase	With added H_2O_2 and Cl^-, produces HOCl and chloroamines
Lysozyme	Effective against a number of gram-positive organisms via hydrolysis of cell-wall peptidoglycans; hydrolysis of protein-mannan outer layer of yeast when added with an endo-β-1,3-glucanase
Mannanase	Lysis of β-glucans-protein cell wall of yeast
Chitinase	Effective against chitin in cell wall of several fungi
Antienzymes	Proteases, sulfhydryl oxidase, dehydrogenases, lipoxygenases

Source: Adapted from Ref. 92.

TABLE 23 Enzymes Produced by *Aspergillus niger* That Hydrolyze Polymers

Substrate	Enzymes
Arabinans	α-L-Arabinofuranosidase
Cellulose	Cellulases[a]
Dextran	Dextranase
DNA, RNA	Deoxyribonuclease, ribonuclease
β-Glucans	β-Glucanase
Hemicellulose	Hemicellulases (pentosanases)
Inulin	Inulinase
Mannans	β-Mannanase
Pectic substances	Pectin methylesterase, pectate lyase, polygalacturonase
Proteins	Proteases
Starch	α-Amylase, glucoamylase
Xylans	Xylanase

[a]All types of cellulases (see Section 7.11.2).
Source: Ref. 110, p. 684.

give turbidity. Also, the next batch of *A. niger* pectic enzymes may perform differently unless the growth conditions are controlled exactly.

At the present time, the cost of a more purified enzyme preparation would be prohibitive if used only once. The advantage of immobilized enzymes is that they can be used repeatedly, greatly decreasing the overall cost. The cost of fructose production from corn starch would be prohibitive if the glucoamylase and glucose isomerase were used for only one batch. Contrast that with the current system using immobilized glucoamylase and glucose isomerase where the cost of producing fructose is about 65% of that of producing sucrose from sugar cane and sugar beets.

Why are immobilized enzyme systems not used more in food processing? The answer is that most food systems are too complex, physically. Chemical complexity is not a problem. Because of the high specificity of enzymes, a single compound such as glucose, for example, can be selected from a complex mixture, bound by glucose oxidase, and converted to product. The thousands of other compounds are not changed. In an immobilized enzyme system, special arrangements are necessary to bring the enzyme and substrate into contact. This can be done by adding an immobilized enzyme preparation and providing thorough circulation, or by flowing the food past a stationary immobilized enzyme (column, coated tube wall, etc). The immobilized enzyme can be used for several months.

Immobilized enzymes are especially valuable for analytical processes. This topic will be discussed in Section 7.12.4.

7.9.2 Methods of Immobilizing Enzymes

Enzymes can be immobilized by several methods: (a) covalent attachment to insoluble support materials such as metals, glass, ceramic, nylon, cellulose, Sepharose, or Sephadex; (b) entrapment in a gel matrix made from, for example, polyacrylamide, Sephadex or agar; (c) adsorption on an insoluble matrix by hydrophobic, electrostatic or other noncovalent affinity methods; (d) adsorption on an insoluble matrix and then covalent cross-linking to the matrix; (e) intermolecular cross-linking of enzymes to form an insoluble matrix that is used in granular form; (f) attachment to semipermeable membranes; and (g) suspension of insoluble enzyme particles in immiscible organic solvents.

Some of the criteria used in deciding on the method of immobilization include: (a) stability of the enzyme during repeated use; (b) retention of activity of the enzyme following immobilization; (c) stability of the matrix to other enzymes, flow, pressure, temperature and other conditions encountered; (d) accessibility of the substrate to the enzyme (diffusion of substrate to the enzyme and products away); (e) susceptibility of the enzyme system to fouling by the substrate feed; (f) efficiency of converting substrate to product; (g) compatibility of the system with subsequent steps in conversion and recovery of product; (h) need for cofactors; (i) compatibility of pure enzyme, crude enzyme, or cells to be immobilized; (j) susceptibility to microbial contamination; (k) renewal efficiency; and (l) cost.

Ideally the immobilized enzyme system should be reusable hundreds of times (3–6 months) before there is 50% loss of enzyme activity. Do such systems exist for food applications? Yes, they do. Three will be described.

One of the best types of systems might involve countercurrent flow of two immiscible liquids, one containing an enzyme able to function at an interface and the other containing a substrate that can concentrate at the interface. This type of system could potentially be used to transesterify triacylglycerols. The substrates (triacylglycerol and fatty acids) are soluble in organic phase, and the lipase is soluble in the aqueous phase and functions at the interface of micellular substrates. Such systems are now in use.

Another system potentially could involve enzymes immobilized to an insoluble magnetic-sensitive matrix. After reaction with the desired food components in a stirred reactor, the insoluble magnetic matrix could easily be removed and reused. A third method might involve use of a semipermeable membrane through which the enzyme is circulated while immersed in the food. The substrate must diffuse into the semipermeable membrane and the product must diffuse out.

7.9.3 Applications

The starch to fructose conversion described above is almost an ideal system. The starch granules are ruptured by heating at 105°C to achieve solubility; however, the resulting solution is too viscous to manage. Hydrolysis of starch with the relatively heat-stable *Bacillus lichenformis* α-amylase to DP ≈ 10 solves that problem. Also, heating at 105°C destroys any microorganisms in the starch and solvents. Therefore, the remainder of the hydrolysis and isomerization can be carried out in giant columns of glucoamylase and glucose isomerase, which are also relatively stable. Fouling and regeneration of the columns are not major problems. Other major commercial applications of immobilized enzymes also have similar features (Table 24).

How many other applications of enzymes are awaiting commercial application? Immobilized β-galactosidase for hydrolysis of lactose in milk is being considered in the United States. Column-based systems would be hampered by fouling, but it is surprising that some of the other ways in which immobilized enzymes can be used (discussed earlier) are not yet commercially acceptable. Only Italy commercially processes milk to hydrolyze lactose enzymatically.

7.9.4 Immobilized Cells

In some cases, immobilized microbial cells can be employed to produce useful compounds (Table 25) or to remove toxic compounds from waste products (Table 26). Japanese scientists have contributed especially to developments in this area. In theory, this should be a very important method, since the desired enzyme is already immobilized inside the cell, cofactors can be regenerated by appropriate metabolic enzymes, the cell is often living and can replenish enzyme losses, and complex reactions requiring several steps can be carried out. It has also been suggested that other enzymes required for hydrolysis of feedstock, cell metabolism, or cofactor

TABLE 24 Commercial Applications of Immobilized Enzymes in the Food and Related Industries

Immobilized enzymes or cells	Substrates	Products
Aminoacylase	Synthetic acyl-DL-amino acid + H_2O	L-Amino acid + acyl-D-amino acid
Aspartase, *Escherichia coli*	Fumarate + NH_3	L-Aspartate
Fumarase, *Brevibacterium ammoniagenes*	Fumarate + H_2O	L-Malate
Glucose isomerase *Streptomyces* sp. *Bacillus coagulans* *Actinoplanes missouriensis* *Arthrobacter* sp.	D-Glucose	D-Fructose
α-Galactosidase (used to hydrolyze raffinose in sugar beet juice, which hinders crystallization of sucrose)	Raffinose	D-Galactose and sucrose

Source: Ref. 88, p. 436.

production might be immobilized on the surface of the microbial cell [45]. For example, if endocellulases were immobilized on microbial cells they would convert insoluble cellulose to soluble oligosaccharides; these would diffuse into the cell, be further hydrolyzed by exo-cellulases and cellobiase to glucose, and glucose would then be metabolized to ethanol and CO_2, both useful products.

However, there are formidable conditions that must be met before immobilized microbial

TABLE 25 Production of Useful Compounds by Immobilized Living Microbial Cells

Useful compounds	Microbial cells	Carriers for immobilization
Ethanol	*Saccharomyces carlsbergensis*	Carrageenan
	Saccharomyces cerevisiae	Polyacrylamide
	Saccharomyces carlsbergensis	Polyvinyl chloride
	Saccharomyces cerevisiae	Calcium alginate
L-Isoleucine	*Serratia marcescens*	Carrageenan
Acetic acid	*Acetobacter* sp.	Hydrous titanium oxide
Lactic acid	Mixed culture of lactobacilli and yeasts	Gelatin
	Arthrobacter oxidans	Polyacrylamide
α-Ketogluconic acid	*Serratia marcescens*	Collagen
Bacitracin	*Bacillus* sp.	Polyacrylamide
Amylase	*Bacillus subtilis*	Polyacrylamide
Hydrogen gas	*Clostridium butyricum*	Polyacrylamide
	Rhodospirillium rubrum	Agar

Source: Ref. 21, p. 88, courtesy of International Union of Biochemistry and Elsevier/North-Holland.

TABLE 26 Decomposition of Poisonous Compounds Using Immobilized Living Microbial Cells

Applications	Poisonous compounds	Microbial cells	Carriers for immobilization
Waste treatment	Phenol	*Candida tropicalis*	Polyacrylamide
	Benzene	*Pseudomonas putida*	Polyacrylamide
	Waste water	Mixed culture of various microorganisms	Bentonite and Mg^{2+} Polyurethane sponge
Denitrification	Nitrate, nitrite	*Micrococcus denitrificans*	Liquid membrane
		Pseudomonas sp.	Active carbon

Source: Ref. 21, p. 89, courtesy of International Union of Biochemistry and Elsevier/North Holland.

cells can be used commercially: (a) the microorganisms must be approved for use in foods and/or in food ingredient manufacture; (b) the substrate must diffuse readily into the cell and the product out (while cell walls can be made more permeable by chemical treatments, there is a limit to how much can be done before the cell is killed and/or becomes too fragile); (c) the "living" cell must not duplicate, but must remain viable in terms of enzyme and cofactor generation; (d) the cell must be rugged enough to permit repeated use; (e) the cell must not make toxic or significant amounts of unwanted products; and (f) the desired product must not be a substrate for other enzymes.

It has been proposed repeatedly that plant cells might be immobilized and packed into columns to make flavor constituents, colorants, vitamins, and other valuable products, but this has not been done except in the laboratory. The advantage would be a relatively pure compound that needs little further purification to be of commercial use. Other advantages would be energy efficiency, small space requirement, and independence from climate. Problems with cell fragility, getting feedstock in and product out of the cells, long-term utilization, and efficient heat removal have not been solved. But one day, this approach will be successful and, hopefully, food scientists and the food industry will contribute greatly to this occurrence.

7.10 SOLVENT-PARTITIONED ENZYME SYSTEMS

Continuous-flow systems for mass manufacturing of all kinds of products have important advantages over batch systems. Thus, immobilized enzyme systems used in continuous systems have advantages over batch-wise use of enzymes. In solvent-partitioned enzyme systems, there are two phases, but unlike immobilized enzyme columns, both phases can be in motion simultaneously, by counterflow or in stirred reactors. For success, substrates, products, and enzyme(s) must be easily separated when appropriate. Two systems appear to meet these expectations: counterflow systems of two immiscible solutions, and systems of "dry" enzyme suspended in an inert organic solvent.

7.10.1 Activity and Stability of Enzymes in Organic Solvents

Can hydrolases synthesize bonds? Do hydrolases have activity in water immiscible organic solvents? Are enzymes stable in water-immiscible organic solvents? Some years ago, it became clear to researchers that hydrolases can catalyze the biosynthesis of esters, amides, peptides, and carbohydrates under appropriate experimental conditions. Papain, for example, can form peptide bonds during hydrolysis of proteins, provided the pH is about 6 and the peptide concentration is 30–35%. Later, it was shown that this biosynthesis is consistent with the intermediate acylenzyme mechanism where an acyl group is transferred to a nucleophilic acceptor molecule (usually water). However, if the concentration of another nucleophile, such as an amine group (amino acid), is high enough, the acyl group of the acylenzyme is partitioned between water and the amino group in accord with their nucleophilicity and relative concentrations. If some of the water is replaced by immiscible organic solvents, biosynthesis occurs even more rapidly (Eqs. 66 and 67), since the concentration of water is decreased, relative to that of H_2NR'.

$$\mathrm{E{-}O{-}\overset{\overset{\displaystyle O}{\|}}{C}{-}R} + \mathrm{H_2NR'} \underset{}{\overset{K_1}{\rightleftharpoons}} \mathrm{E{-}OH} + \mathrm{R{-}\overset{\overset{\displaystyle O}{\|}}{C}{-}NH{-}R'} \tag{66}$$

$$\mathrm{E{-}O{-}\overset{\overset{\displaystyle O}{\|}}{C}{-}R} + \mathrm{H_2O} \overset{K_2}{\rightleftharpoons} \mathrm{E{-}OH} + \mathrm{R{-}\overset{\overset{\displaystyle O}{\|}}{C}{-}OH} \tag{67}$$

It also helps to have the acid protonated. These competitive hydrolysis/biosynthesis reactions have been used successfully in the plastein reaction [6] to alter characteristics of soy protein. This is accomplished by adding methionine methyl ester to the reaction mixture and causing its incorporation into the soy protein. The result is a product that is firmer, chewier, and has improved nutritional quality. Similarly, superior surfactant products have been created by using a protease to form lauryl derivatives of gelatin. Again, replacement of some of the water with immiscible organic solvents can facilitate the biosynthetic reaction.

The same considerations apply to reformation of triacylglycerols to yield better melting, greater solubility and unsaturation, and better nutritional properties than those of the original triacylglycerol. An example of reformation of a triacylglycerol by addition of a fatty acid (L) and a 1,3-specific lipase is shown in Eq. 64. If a non-specific lipase is used, the triacylglycerol can be reformulated to several products (Eq. 68), where E is a nonspecific lipase.

$$5\ \mathrm{O}{-}\left[\begin{matrix}\mathrm{X}\\ \\ \mathrm{X}\end{matrix}\right. + 9\,\mathrm{L} \overset{\mathrm{E}}{\rightleftharpoons} \mathrm{O}{-}\left[\begin{matrix}\mathrm{L}\\ \\ \mathrm{X}\end{matrix}\right. + \mathrm{O}{-}\left[\begin{matrix}\mathrm{L}\\ \\ \mathrm{L}\end{matrix}\right. + \mathrm{L}{-}\left[\begin{matrix}\mathrm{X}\\ \\ \mathrm{X}\end{matrix}\right. + \mathrm{L}{-}\left[\begin{matrix}\mathrm{L}\\ \\ \mathrm{X}\end{matrix}\right. + \mathrm{L}{-}\left[\begin{matrix}\mathrm{L}\\ \\ \mathrm{L}\end{matrix}\right. + 6\mathrm{X} + 3\mathrm{O} \tag{68}$$

By using two or more fatty acids in controlled amounts, many types of intermediate products with interesting functionalities can be obtained.

Therefore, hydrolases can perform biosynthetic reactions under appropriate conditions (Section 7.5.5) [4, 91, 118], with biosynthesis favored by a low water activity in conjunction with an organic solvent. Rates of lipase-catalyzed lipid transesterification reactions can be increased more than sixfold with a simultaneous 16-fold decrease in the rate of hydrolysis.

If it is important not to use organic solvents in the biosynthesis, similar biosynthesis-favored reactions can be accomplished with alkylated lipases [91, 118]. Apparently, the increased hydrophobicity of the lipase surface is responsible for this shift in the ratio of biosynthesis to hydrolysis. Alkylated trypsin is a more effective catalyst for esterification of sucrose by oleic acid than is trypsin [4].

Are the enzymes stable enough to permit biosysthesis in organic solvents? This question was addressed in Section 7.5.6. Ribonuclease and lysozyme are more stable when suspended in immiscible organic solvents than when dissolved in water (Table 10). Therefore, performing transesterification reactions with enzymes suspended in immiscible organic solvents is entirely feasible. An additional advantage is the avoidance of microbial problems.

7.10.2 Advantages of Solvent-Partitioned Enzyme Systems

By now, several of the advantages of performing hydrolase-catalyzed reactions in solvent-partitioned systems rather than aqueous systems should be obvious: (a) equilibrium is shifted toward biosynthesis; (b) the enzyme is more stable; (c) enzyme is readily removed from the system; (d) product of high purity is more easily obtainable, particularly if it partitions into the organic phase; (e) product is not likely to be contaminated by the enzyme; and (f) microbial problems are minimized.

In some situations there is another advantage to enzyme-catalyzed reactions in immiscible organic solvents; that is, the system can favor one stereoisomeric form over another. Table 27 illustrates the enantioselectivity of transesterification between vinyl butyrate and *sec*-phenethyl alcohol as catalyzed by porcine pancreatic lipase in different immiscible solvents. The enantioselectivity, υ^R/υ^S, (where υ^R and υ^S are chiralities) decreases from 75 in nitromethane to 6 in decane. Therefore, the enantioselectivity is greater with greater polarity (dielectric constant) of the solvent.

7.10.3 Applications

Currently, the most important application of solvent-partitioned enzyme systems is in the restructuring of triacylglycerols for food use. Restructuring of triacylglycerols is important not only for changing physical properties but also for improving nutritional attributes. This area is important enough that genetic engineering companies have invested millions of dollars into tailoring the genes in higher plants to make specialty triacylglycerols.

Another application is in the manufacture of surfactants and detergents. Long-chain fatty acids can be covalently linked to amides, such as amino acids, or to glucose to give highly functional surfactants for use in soaps and shampoos, as well as in foods. This can result in improved foaming and emulsifying properties, and in easier incorporation of water, insoluble

TABLE 27 Enantioselectivity of Porcine Pancreatic Lipase in Transesterification Between Vinyl Butyrate and *sec*-Phenethyl Alcohol[a]

Solvent	v^{Rb} (m*M*/h)	v^{Sb} (m*M*/h)	Enantioselect, (v^R/v^S)
Nitromethane	9.7	0.13	75
Dimethylformamide	2.3	0.038	61
Triethylamine	4.2	0.099	42
tert-Amyl alcohol	6.9	0.20	35
Butanone	8.6	0.32	27
Acetonitrile	14	0.64	22
Benzene	10	0.67	15
Cyclohexane	43	3.3	13
Decane	5.5	0.85	6

[a]100 mM sec-phenethyl alcohol and 100 mg/ml lipase.
[b]Chirality of product.
Source: Reprinted with permission from Ref. 40, p. 3170, copyright 1991 American Chemical Society.

flavors, colors and antioxidants in systems containing incompatible components. Similar advantages can accrue from using enzymes to make totally biodegradable soaps, shampoos, and detergents from naturally occurring compounds, such as glucose.

The discovery that enzymes can efficiently catalyze reactions in immiscible solvents, including two-phase solvent systems, surely is one of the major advances in application of enzymology in the last twenty years.

7.11 ENZYMES IN WASTE MANAGEMENT

It is estimated that the annual production of major carbohydrate feedstock for use as potential fuels or for manufacturing chemicals in the United States is about 1160 million tons, versus 50 million tons of organic chemical feedstocks [42]. This number includes 160 million tons of municipal solid waste, 400 million tons of agricultural residue, 400 million tons of forest residue, and 200 million tons of corn and grains. These residues pose major environmental problems, since a large amount of the agricultural residues are burned (where permitted), the forest residues result in major fires each year, and the massive municipal landfills are often sources of groundwater pollution. Therefore, if some of these materials could be converted to other forms of fuel (ethanol and methanol) or to fermentation feedstock (glucose) to produce proteins, ethanol, and CO_2, disposal problems would be alleviated and nonrenewable fuel supplies would be conserved. The U.S. government is spending considerable money on research on how to efficiently tap these potential sources of fuel and chemicals.

What are the compounds that need to be converted? They are largely starch, cellulose, lignin, lipids, nucleic acids and proteins. What enzymes are needed? Primarily the amylases, cellulases, lignin peroxidases, lipases, nucleases, and proteases. These are enzymes we know quite a lot about. So, what is the problem? The major problems are the insolubility of the potential substrates just listed, and the poor stability and catalytic inefficiency of the enzymes required. Some methods exist for increasing the solubility of the substrates and thereby the rate of hydrolysis. But all methods require considerable energy expenditure that makes the process uneconomical at the moment, when petroleum and coal are relatively inexpensive.

Do commercially feasible solutions exist for converting biomass to valuable chemicals? Nature does this efficiently, given the huge 1160 million tons to be converted annually. The natural conversion of these compounds occurs by action of the usual classes of enzymes, but the process is slow. Improvements in rate are possible through redesign of enzymes with greater efficiency and synergistic properties.

Rapid, economical biomass conversion is a difficult task so the approach taken must be multifaceted. More efficient enzymes are needed, and this requires knowledge of why these enzymes are so inefficient in attacking insoluble substrates. Once this is better understood, better enzymes can be searched for or developed by recombinant DNA techniques. One needed property is greater stability, especially at temperatures sufficiently high microbial growth will not be a problem. The pentose-containing polymers are especially difficult to hydrolyze and even more difficult to ferment to ethanol and CO_2. Lignin severely limits hydrolysis and fermentation of cellulose, because it is so inert. Enzymes that efficiently attack these substrates are needed.

Solution of the biomass conversion problem is a must. Conversion processes must be environmentally acceptable, and all components, even minor ones, must be economically converted to usable or innocuous products to meet waste minimization goals pursued by the U.S. Environmental Protection Agency.

7.11.1 Starch and Amylases

Conversion of starch from wet-milled corn to usable products is a huge success story. Starch conversion from food-processing waters should also be considered. The amylases, glucoamylases, and glucose isomerases (see Sections 7.8.1.4 and 7.8.2), each of which represents a group of enzymes with similar catalytic properties yet quite diverse chemical and physical properties, are the tools that allow starch to be commercially converted to other useful products. The products are food-grade syrups of various DPs (degrees of polymerization), glucose, and fructose. This industry has over 1 billion dollars of sales per year in the United States alone, and all of this growth has occurred in the last 25 years.

There is also a fledging alcohol fuels industry based on corn starch, with the annual output being about 800 million gallons in 1988. Why has starch conversion to useful products developed so well? In part this is because starch is a reasonably tractable substrate, becoming soluble, but viscous, on heating at 105°C. But also of major importance was the intense research that led to a fundamental understanding of the three major groups of enzymes needed and how these enzymes could be immobilized for use in continuous systems.

7.11.2 Cellulose and Cellulases

Belatedly, a major worldwide effort is underway to understand the structure and function of cellulases, to identify microorganisms that produce large amounts of cellulases with high efficiency, to determine their primary structures through gene identification and cloning, and to genetically engineer cellulases that have highly desirable properties. Another major problem with respect to cellulose conversion is its insolubility. It is also unknown whether crystalline cellulose, oligosaccharides, or synthetic substrates, such as the *o*- and *p*-nitrophenyl oligosaccharides, are the best substrates to use as a standard for interlaboratory comparisons of cellulose conversion methods.

The enzymes that act on cellulose and degradation intermediates can be divided into four groups [26].

1. The endoglucanases [1,4(1,3;1,4)-β-D-glucan 4-glucanohydrolases, EC 3.2.1.4] are relatively inactive against crystalline regions of cotton and Avicel, but they hydrolyze amorphous regions of these substrates, including filter paper, and soluble substrates such as carboxymethylcellulose and hydroxymethylcellulose. Endoglucanase activity is characterized by random hydrolysis of β-glucosidic bonds, resulting in a rapid decrease in viscosity relative to the rate of increase in reducing groups. The products, especially late in the reaction sequence, include glucose, cellobiose, and cellodextrins of various sizes.
2. The second group of cellulases is the cellobiohydrolases (1,4-β-D-glucan cellobiohydrolases, EC 3.2.1.91), which are *exo*-splitting enzymes. They degrade amorphous cellulose by consecutive removal of cellobiose from the nonreducing ends of the cellulose. When pure, they usually have little activity on cotton, but can hydrolyze about 40% of the hydrolyzable bonds in Avicel, a microcrystalline cellulose. The rate of decrease in viscosity is much lower in relation to the increase in reducing groups than it is for the endoglucanases. Endoglucanases and cellobiohydrolases show synergism in the hydrolysis of crystalline cellulose, for reasons not clearly understood.
3. The third group of cellulases is the exoglucohydrolases (1,4-β-D-glucan glucobiohydrolase, EC 3.2.1.74), which hydrolyze consecutively the removal of glucose units from the nonreducing end of cellodextrins. The rate of hydrolysis decreases as the chain length of the substrate decreases.

4. The fourth group of enzymes is the β-glucosidases (β-D-glucoside glucohydrolase, EC 3.2.1.21), which cleave cellobiose to glucose and remove glucose from the nonreducing end of small cellodextrins. Unlike the exoglucohydrolases, the rate of β-glucosidase hydrolysis increases as the size of the substrate decreases, with cellobiose being the most rapidly hydrolyzed substrate.

7.11.3 Hemicelluloses and Pentosanases

Because of the complexity of the hemicelluloses, several endo- and exo-splitting (by hydrolysis) pentosanases exist. The endo-splitting enzymes are of two types: the arabanases that hydrolyze the 1,4-α-D-arabinopyranosyl glycosidic bonds in the linear backbone of the arabans, and the xylanases that hydrolyze the 1,4-β–D-xylanopyranosyl glycosidic bonds in xylans and arabinoxylans. *Exo*-splitting enzymes of one type act at the non-reducing ends of arabans (the arabinosidases), xylans (the xylosidases), and the arabinoxylans (xylanosidases). *Exo*-splitting enzymes of a second type (arabinosidases) remove arabinofuranosyl side-chain units from arabinoxylans.

Some fungi produce substantial amounts of pentosanases. Several *endo*- and *exo*-pentosanases also exist in wheat at very low levels. Some pentosanases increase 6–10 times in concentration during germination of seeds. The pentosanases are thought to be located primarily in the bran, but it is not known where the enzymes newly synthesized during germination are located.

7.11.4 Proteins and Proteolytic Enzymes

All biomass contains proteins. The level of proteins is relatively low in most waste materials from higher plants, whereas waste from meat, beans, nuts, and cereal seed products contains about 20–35% protein. In general, proteins are at least partially soluble in water. Since they provide a source of essential amino acids for microorganisms, they are degraded rapidly by a host of proteases found in plants and microorganisms.

All biological materials probably contain both *endo*- and *exo*-proteases. The *endo*-proteases hydrolyze specific peptide bonds in the interior of the protein chain for which they have specificity. For example, trypsin-like enzymes hydrolyze peptide bonds where Arg and Lys residues furnish the carbonyl group of the peptide bond. No other peptide bonds are hydrolyzed. The resulting peptides are acted on by carboxypeptidases from the carboxyl-terminal end and by aminopeptidases from the amino-terminal end. Di- and tripeptidases are found in most organisms. Therefore, proteins are rapidly and specifically hydrolyzed to amino acids, in an efficient manner. There are no major problems in degradation of the protein in waste materials.

7.11.5 Lipids and Lipases

Lipases hydrolyze triacylglycerols to fatty acids, mono- and/or diacylglycerols and glycerol. As discussed in Section 7.8.2, there are 1,3-type, 2-type, and nonspecific lipases that can efficiently hydrolyze lipids in waste materials. Lipases act on the micellular form of the lipid, preferably in a liquid form. Solid lipids are hydrolyzed slowly. There appears to be no restraint on the effective hydrolysis of lipids in wastes, such as in household garbage. In fact, problems with hydrolysis of lipids are rarely mentioned in discussions of waste product conversions.

7.12 ENZYMES IN FOOD ANALYSIS [108]

Enzyme assays are often the method of choice for chemical analyses, because of high sensitivity and great specificity. The cost of reagents, especially for the pure enzyme, is more than compensated for by savings in time and expense since there is no need to do partial or complete separation of a compound prior to analysis. For example, glucose can be determined in blood, urine, and plant extracts by glucose oxidase without any preparation other than removal of insoluble materials that interfere with measuring absorbance. Enzymatic reactions are usually performed near room temperature and neutral pH and only a few minutes are required, thereby minimizing non-enzyme-caused changes in the compounds. Undesirable enzyme-catalyzed side reactions are not a problem provided other contaminating enzymes are removed or inactivated.

Enzyme assays of biological materials are used to assess adequacy of blanching or processing, maturity, storage time, stability, cultural practices, and breeding or genetic modifications. They are used to make quantitative determinations of food composition, including compounds that function as substrates, activators, or inhibitors of specific enzymes. Many genetically related or environmentally induced diseases result in changes in levels of key enzyme systems in the human body, and this in turn influences food choice. Two examples of human enzyme deficiency are the inability to tolerate phenylalanine (phenylketonuria) and lactose (β-galactosidase deficiency).

In using enzyme assays, it is important to be familiar with the factors that affect enzyme activity (Section 7.5), since all variables must be appropriately controlled. Standard curves and internal standards must be used, just as in any analytical analyses.

7.12.1 Enzyme Concentration

This is the easiest of all analytical determinations involving enzymes. There is usually a linear relationship between v_0 and $[E]_0$ (Fig. 12), regardless of $[S]_0$, pH, and temperature (all three must be kept constant), provided there is sufficient activity to measure. For the majority of enzymes, loss of activity due to denaturation or reaction with product is not a problem if v_0 is obtained with less than 5% conversion of substrate to product.

The relationship between v_0 and $[E]_0$ (Fig. 12) becomes the basis of the analytical determination of enzyme concentration in biological materials. Doubling the $[E]_0$ should double v_0. It is important, and easy, to verify this relationship, just by adding different size aliquots of an enzyme extract using constant buffer and substrate concentrations. Don't assume a linear relationship, as several factors can lead to nonlinearity [111].

Absence of contaminating enzymes that compete for the same substrate is important. So, use as specific a substrate as possible. For example, in determining trypsin concentration use *N*-tosyl-*L*-arginine ethyl ester, rather than casein, which is a substrate for all proteases.

Table 28 lists a few examples of assays for enzyme concentration. For a complete list see Reference 108.

TABLE 28 Analytical Determination of Some Enzyme Concentrations

Enzyme	Substrate	Detection[a]
Alcohol dehydrogenase	Ethanol	S; $NAD^+ \rightarrow NADH$
Catalase	Hydrogen Peroxide	E; O_2
β-Galactosidase	*o*-Nitrophenyl-β-D-galactoside	S; *o*-nitrophenol
Peroxidase	H_2O_2, *o*-diansidine	S; oxidized *o*-dianisidine
Polyphenol oxidase	Pyrocatechol	S; benzoquinone

[a]E, electrometric method; S, spectrophotometric method.
Source: Adapted from Ref. 108, pp. 331–335.

7.12.2 Compounds That Are Substrates

The relationship between v_o and $[S]_o$ is not linear, but rather is a right hyperbolic one (Fig. 1). As discussed earlier (Section 7.4.2), when $[S]_o > 100\ K_m$ the rate of an enzyme-catalyzed reaction is independent of $[S]_o$, so this condition cannot be used to determine the concentration of a compound that serves as a substrate. Therefore, the concentration of the compound must be less than $100K_m$, preferably less than $5K_m$. When $[S]_o < 100K_m$ then Equation 69,

$$v_o = \frac{V_{max}[S]_o}{K_m + [S]_o} \tag{69}$$

can be used to calculate $[S]_o$. Better yet, a standard curve of v_o versus $[S]_o$ should be prepared, and plotted in reciprocal form, according to Figure 13. The standard curve can be used to determine $1/[S]_o$ required to give $1/v_o$. The Lineweaver–Burk equation (Eq. 21) can also be used.

The concentration of compounds that are substrates can also be determined by the total change method (Fig. 38). The reaction is required to go to completion; then the concentration of the product is determined by the total change in absorbance or fluorescence. Such assays are less demanding in terms of precise control of pH and temperature; however, they require more

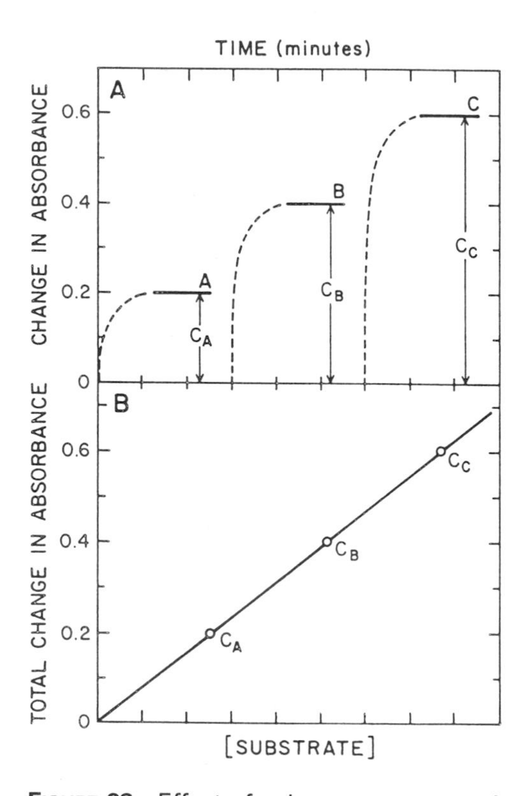

FIGURE 38 Effect of substrate concentration on change in absorbance by the total change method. (A) Effect of three concentrations, C_A, C_B, and C_C, of substrate on change in absorbance. Note that the initial change in absorbance is very fast so the reaction is completed within a few minutes. It is only necessary to know the initial and final absorbance in the total change method. (B) Relationship between substrate concentration and total change in absorbance, as replotted from A. (From Ref. 108, p. 318.)

enzyme in order to complete the reaction within a reasonable time (2–10 min). In some cases, the reaction does not go to completion because an equilibrium is reached. In this instance two options are available. One is to prepare the standard curve under exactly the same conditions used for assay of the compound. The second method is to trap one or more of the products as they are formed so that the reaction goes to completion (such as trapping pyruvate with oxamate in the lactate dehyrogenase reaction, using lactate and NAD^+ as substrates).

A few examples of compounds than can be quantitatively determined by enzymatic assay are shown in Table 29. For a more complete list, the reader is referred to Reference 108.

7.12.3 Other Assays

Some enzymes, but not all, require cofactors for activity (see Section 7.6). The cofactors may be organic compounds, such as pyridoxal phosphate or NAD^+, or inorganic, such as Zn^{2+} or Mg^{2+}. Therefore, whenever v_0 of an enzyme catalyzed reaction is increased, in a reproducible manner, by addition of a compound, the extent of this increase can be used to determine the concentration of that compound. The principle of this determination is shown by Equation 70 where Act is activator:

$$\text{E} + \text{Act} \overset{K_d}{\rightleftharpoons} \text{E·Act} \underset{k_{-1}}{\overset{k_1,\,\text{S}}{\rightleftharpoons}} \text{E·Act·S} \overset{k_2}{\rightleftharpoons} \text{E·Act} + \text{P} \tag{70}$$

Two types of relationships between activator concentration and v_0 are found. If the activator binds tightly to the enzyme (prosthetic group, for example; $K_d < 10^{-9}$ *M*), then a plot of v_0 versus [activator] is linear. If the activator binds loosely to the enzyme ($K_d > 10^{-8}$ *M*), then the relationship between v_0 and [activator] is hyperbolic. Determination of compounds that are loose-binding activators is quite similar to the determination of substrate concentration. For more examples, the reader is referred to Reference 108.

Concentrations of compounds in foods that inhibit enzymes (inhibitors) can also be determined analytically by use of enzymes. An inhibitor is any compound that, when added to an enzyme-catalyzed reaction, decreases the v_0 of that reaction. Inhibitors may be inorganic compounds such as Pb^{2+}, Hg^{2+}, or Ag^{2+}, or organic compounds that bind reversibly to the enzyme (competitive, noncompetitive, uncompetitive inhibitors) or irreversibly to the enzyme (see Sec. 7.7). Because of the diversity of types of enzyme inhibitors, the design of experiments and interpretation of the results can be more complex than for determining enzyme, substrate, or activator concentrations. Analytical methods are described in Reference 108.

TABLE 29 Analytical Determination of Some Compounds That Serve as Enzyme Substrates

Compound	Enzyme	Detection[a]
Ammonia	Glutamate dehydrogenase	S; $NADH \rightarrow NAD^+$
Ethanol	Alcohol dehydrogenase	S; $NAD^+ \rightarrow NADH$
D-Glucose	Glucose oxidase	S; H_2O_2 with peroxidase
Nitrate	Nitrate reductase	E; NH_3 electrode
Urea	Urease	E; NH_3 electrode

[a]E, electrometric method; S, spectrophotometric method.
Source: Adapted from Ref. 108, pp. 340–344.

7.12.4 Immobilized Enzymes in Food Analysis

Immobilized enzymes permit repeated use of the same enzyme, thus reducing the cost per analysis. The immobilized enzyme can make the assays faster and easier. There are at least four possible arrangements that work well. These are immobilized enzyme columns, enzyme electrodes, enzyme-containing chips, and enzyme-linked immunosorbent assay (ELISA) type assay systems [43, 105, 108].

7.12.5 Indicators of Adequate Blanching or Pasteurization

Blanching is a mild heat treatment of raw fruits and vegetables conducted primarily to stabilize them against enzymatic deterioration and/or microbial growth during storage. The products may be minimally processed foods that will maintain safety and quality for 1 or 2 weeks of refrigerated storage, or after frozen or dry storage for 1 to 2 years. Vegetative microorganisms are destroyed at lower temperatures than are most enzymes. An analogous heat treatment of milk and some other food products is called pasteurization.

Indicators of blanching adequacy are needed. The most used indicators are peroxidase in fruits and vegetables, and alkaline phosphatase in milk, dairy products and ham. β-Acetylglucosaminidase has been used for eggs. Why are peroxidase and alkaline phosphatase used? Because they are the most heat-stable enzymes found in fruits/vegetables and milk, respectively. Although they are not the cause of quality loss, their activities do correlate with the stability of the stored foods. Better indicators of storage stability are sometimes available, and these were discussed in Section 7.8. Table 30 is a summary of various uses of enzymes as indices of food quality.

7.13 RECOMBINANT DNA TECHNOLOGY AND GENETIC ENGINEERING

7.13.1 Introduction

Recombinant DNA technology and genetic engineering are of major importance to food scientists because they are powerful tools for altering the properties and processing conditions for foods. The Flavr Savr tomato, in which polygalacturonase expression is repressed, has received final approval and is on the market. There are more than 80 other changes in the genetic composition of our food crops poised to be on the market in less than 5 years.

Genetic engineering involves the alteration of one or more genes (removal or addition, suppression or enhancement of expression, modification of functionality) in an organism to achieve improvements in quality and/or to lengthen the storage life of raw food crops. Breeders have done this for many decades, by crosses among plants combined with field trials. Breeding changes several genes simultaneously, and not all changes are desirable. Genetic engineering permits change of one gene at a time, as desired. This gene, for example, may control properties of an enzyme such as concentration or stability, as well as expression level.

Recombinant DNA technology involves a change in one or more of the thousands of nucleotides of the gene in order to change properties of the expressed protein. For example, the change may result in more or less stability, increased catalytic efficiency or more or less strict substrate specificity (see also Sec. 7.13.4).

The process of inserting genetic material (for chymosin, for example) from a donor (calf) into a host plasmid (from a microorganism, for example) is shown in Figure 39. The donor gene is replicated in the host, and the product of that gene is secreted and can be obtained in high yield (for example, chymosin, which is already on the market).

Two groups of enzymes that make genetic engineering and recombinant DNA technology

TABLE 30 Use of Enzymes to Determine Indices of Food Quality

Purpose	Enzyme	Food material
Proper heat treatment	Peroxidase	Fruits and vegetables
	Alkaline phosphatase	Milk, dairy products, ham
	β-Acetylglucosaminidase	Eggs
Freezing and thawing	Malic enzyme	Oysters
	Glutamate oxaloacetate transaminase	Meat
Bacterial contamination	Acid phosphatase	Meat, eggs
	Catalase	Milk
	Reductase	Milk
	Glutamate decarboxylase	Milk
	Catalase	Green beans
Insect infestation	Uricase	Stored cereals
	Uricase	Fruit products
Freshness	Lysolecithinase	Fish
	Xanthine oxidase	Hypoxanthine in fish
Maturity	Sucrose synthetase	Potatoes
	Pectinase	Pears
Sprouting	Amylase	Flour
	Peroxidase	Wheat
Color	Polyphenol oxidase	Coffee, wheat
	Polyphenol oxidase	Peaches, avocado
	Succinic dehydrogenase	Meat
Flavor	Alliinase	Onions, garlic
	Glutaminyl transpeptidase	Onions
Nutritional quality	Proteases	Digestibility
	Proteases	Protein inhibitors
	L-Amino acid decarboxylases	Essential amino acids
	Lysine decarboxylase	Lysine

Source: Ref. 108, p. 337.

possible are the Type II DNA restriction endonucleases (EC 3.1.21.4) and the DNA ligases (polymerases; *E. coli* enzyme EC 6.5.1.2 and T4 enzyme EC 6.5.1.1).

7.13.1.1 Restriction Enzymes

As shown in Figure 39, foreign DNA is obtained from the donor by cutting the double-stranded DNA in two places by two highly specific DNA restriction endonucleases. A segment is cut out of the host donor DNA of the plasmid, using the same two restriction enzymes, and is replaced by the donor DNA containing the desired gene. The segment is covalently attached to the host DNA by one of the DNA ligases, to reform the phosphodiester linkage. The reactions involved are shown in Figures 40 and 41.

Shown in Figure 40 are details of how the restriction endonuclease catalyzes hydrolysis of the 3′-O-P bond between two nucleotides in the donor and host DNAs (Fig. 39). More than 600 restriction endonucleases are known, and more than 100 of them are commercially available in pure form. The restriction endonucleases are highly specific, recognizing a sequence of several nucleotides, both along the DNA chain and on the complementary DNA chain. They act only on double-stranded DNA.

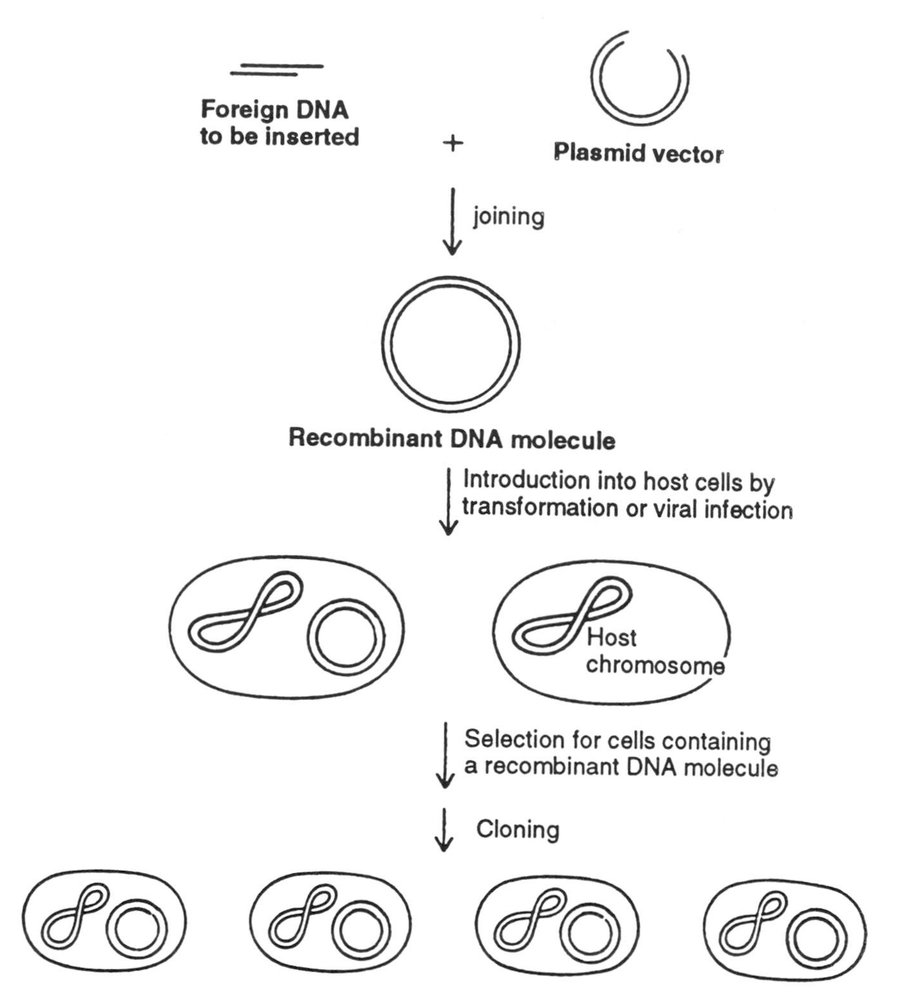

FIGURE 39 Schematic diagram of insertion of a foreign DNA fragment (from donor) into a plasmid vector (from host) and cloning of DNA molecules in host cells. (From Ref. 111, p. 462.)

7.13.1.2 DNA Polymerases

DNA polymerases reform the phosphodiester bonds when all the appropriate reactants (Fig. 41) are present. The reactants are the primer strand, one or more of the four nucleotide triphosphates, a template (pattern giving sequence of nucleotides to be copied; Eq. 71), the DNA polymerase, and Mg^{2+} (required cofactor). If all the nucleotide triphosphates are supplied, the polymerase can copy the entire DNA template chain (the polymerase chain reaction, known as PCR).

$$\begin{array}{c} \text{5'}-\text{C}-\text{C}-\text{G}^{\text{OH}} \\ \vdots\;\;\vdots\;\;\vdots \\ \text{3'}-\text{G}-\text{G}-\text{C}-\text{T}-\text{A}-\text{T}-\text{C}-\text{G}-\text{A}\ldots \end{array} + \begin{array}{c} \text{2dATP} \\ \text{dCTP} \\ \text{dGTP} \\ \text{2dTTP} \end{array} \xrightarrow[\searrow 6\text{PP}_i]{Mg^{2+}} \begin{array}{c} \text{5'}-\text{C}-\text{C}-\text{G}-\text{A}-\text{T}-\text{A}-\text{G}-\text{C}-\text{T}\ldots \\ \vdots\;\;\vdots\;\;\vdots\;\;\vdots\;\;\vdots\;\;\vdots\;\;\vdots\;\;\vdots\;\;\vdots \\ \text{3'}-\text{G}-\text{G}-\text{C}-\text{T}-\text{A}-\text{T}-\text{C}-\text{G}-\text{A}\ldots \end{array} \quad (71)$$

(Deoxyribonucleotide)$_n$ (top) DNA template (bottom) — Nucleotides — (Deoxyribonucleotide)$_{n+6}$ (top) DNA template (bottom)

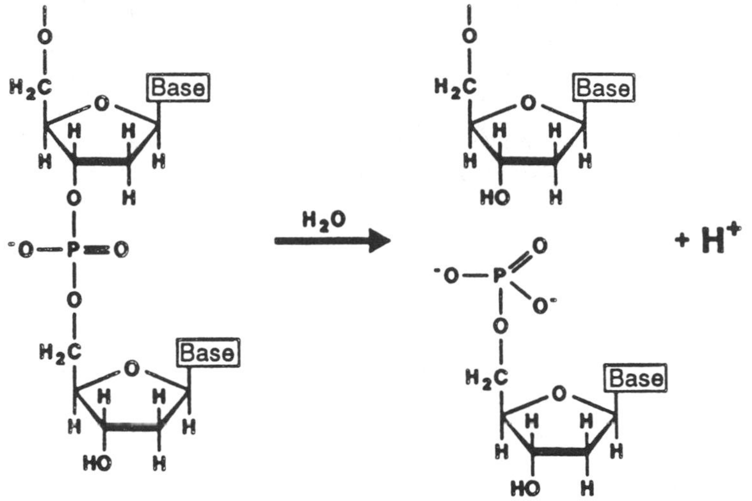

FIGURE 40 Restriction endonuclease-catalyzed hydrolysis of the 3′-O-P bond between two nucleotides. (From Ref. 111, p. 462.)

7.13.2 Applications: Raw Food Materials

Many possibilities exist for improving our raw food materials by genetic engineering. Proteins from legumes are deficient in the essential amino acid methionine, while the cereal proteins are deficient in lysine. As an example, the gene for a high-methionine-containing protein could be incorporated into the genome for the legume (common beans, for example) to improve its nutritional quality. Researchers are working on incorporating a high-methionine-containing protein from Brazil nuts or from soybeans into common beans. In practice, the nucleotide sequence of a gene of an existing protein in common beans might be modified to contain more methionine. A potential problem is that the functional properties might be changed such that the

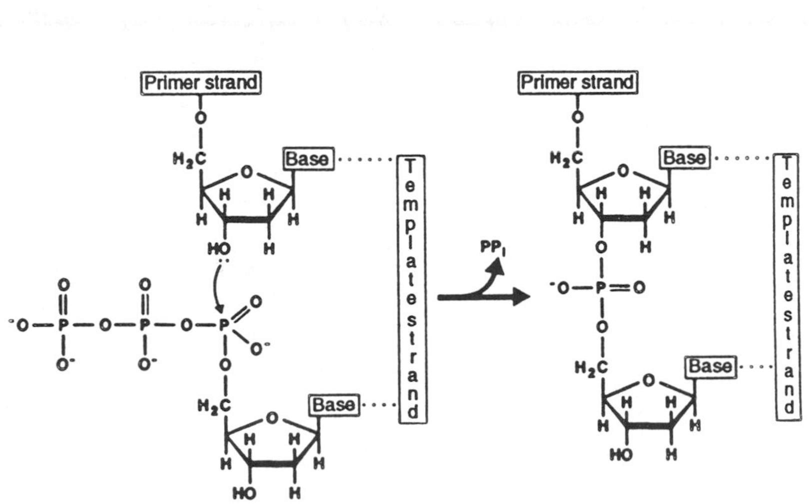

FIGURE 41 Proposed mechanism for DNA polymerase I-catalyzed reaction. (From Ref. 111, p. 467.)

protein could not perform the expected physiological functions and/or would be hydrolyzed rapidly in vivo by proteases.

Some plants contain naturally occurring protease and α-amylase inhibitors, which protect them against insects and microorganisms. It has already been shown that genes for these inhibitors can be expressed in other host plants, protecting them against specific types of insects. In other instances, it may be desirable to reduce the level of protease inhibitors (~2% of total proteins in soybeans, for example), since protease inhibitors can have a negative effect on protein digestibility in animals (Fig. 33).

The functional properties of proteins can be modified by changing as few as one amino acid residue [59]. Some proteins are too stable and do not unfold readily at air/water (foaming) or lipid/water (emulsifying) interfaces. Potentially, this shortcoming can be overcome by genetic engineering. Some companies now supply partial protein hydrolysates as functional ingredients. In the near future, functional ingredients, including proteins, carbohydrates, and gums, will be genetically engineered.

The triacylglycerol compounds of plants can be genetically engineered to alter the type and location of the fatty acids on the glycerol, to provide better nutrition, or to alter the freezing and melting points. The opportunity also exists, through genetic engineering, to remove unwanted flavors (bitterness of naringin) and colors, and to alter other quality factors of plant and animal products. Resistance of plants to pesticides and herbicides, tolerance to salt, and perhaps nitrogen-fixation capability also are properties potentially achievable by genetic engineering. This is by no means a complete listing of possibilities.

7.13.3 Applications: Control of Endogenous Enzymes

Unwanted endogenous enzyme activities that affect the color, flavor and aroma, texture, and nutritional qualities of our foods were mentioned in Section 7.8.2. Of particular importance are polyphenol oxidase, lipoxygenase, and polygalacturonase. Calgene has developed the Flav Savr tomato in which less polygalacturonase is expressed, using the RNA-antisense approach to prevent expression of one or more of the genes for polygalacturonase. It is important to not turn off all of the polygalacturonase expression, since this is an important enzyme for plant growth and maturation.

Researchers are also working on other techniques that might control the rate of maturation of fruits and vegetables. In many plants, ethylene production is essential. By modifying (decreasing) one or more of the enzymes involved in ethylene production, the rate of maturation and ripening can be delayed and the fruits and vegetables could then be stored longer.

Control of gene expression by antisense RNA is a general tool that can be used to modify the expression of any enzyme, in principle. Most likely, methods of controlling polyphenol oxidase and lipoxygenase expression will appear soon. Already, soybeans have been bred to contain much less lipoxygenase [63].

The level of production of enzymes (and any other compound) is controlled by the number of copies of the gene in cells, and regulation of expression of that gene. By decreasing the number of copies, the level of expression can be lowered; by increasing the number of copies, the level of expression can be increased. This ability is very important in the production of ingredients, including enzymes, for food use.

7.13.4 Applications: Tailoring Enzymes for Food Uses

It is important to have adequate supplies of reasonably pure enzymes for food uses. The cost of preparing reasonably pure enzymes for industrial use is prohibitive, unless very large amounts (low cost per unit) are used efficiently. This occurs, for example, in sweetener production from starch, production of amino acids, and resolution of racemic amino acid mixtures. The cost of reasonably pure enzymes is dependent on the enzyme's expression level by the organism. When the organism is genetically engineered to produce a 100-fold greater concentration of a desired enzyme than it normally does, a great reduction in cost is achieved.

There is another advantage to having reasonably pure enzymes—that is, fewer unwanted side reactions occur. There is no reason why some of these side reactions that are catalyzed by other enzymes cannot be reduced by genetic engineering.

Other enzyme properties that are potentially altered by genetic engineering include (a) stability, (b) pH for optimum activity, (c) cofactor binding and regeneration characteristics, (d) substrate specificity, (e) efficiency (increased rate of conversion of substrate to product), and (f) ease of immobilization (for example, extra amino acid residues at the N- or C-terminal end to permit easier covalent coupling to insoluble supports).

7.14 ENZYMES AND HEALTH/NUTRITION/FOOD ISSUES

In the recent past most of us ate foods primarily for their sensory properties, as dictated by individual experience and preferences. Now much greater attention is given to the nutritional content of the diet and the number of calories consumed. Enzymes play an important role in the development of foods that are nutritionally better and safer.

7.14.1 Enzyme Deficiencies

The levels of enzymes produced by our bodies are controlled by our genetic make up, by age and by diet. A surprisingly large amount of enzymes is produced each day by our bodies. Our digestive system alone produces 15–20 g of proteases, carbohydrases, lipases, and nucleases to digest our foods. Myosin, 55% of the protein of our muscle, is an enzyme, and a significant percentage of it is turned over each day. The liver is the major enzyme-requiring "factory" of the body. It is fortunate that most of the enzymes produced each day are hydrolyzed by proteases, and returned as amino acids for biosynthesis into new enzymes, and other proteins. This system recycles 50–100 g protein/day, as compared to about 70 g/day provided by food intake.

There are more than 100 known genetic defects that result in less than normal enzyme production (Table 31) [104] or the production of defective enzymes that undergo proteolysis. Some of these enzyme deficiencies, such as alcaptonuria, phenylketonuria, galactosemia, Gaucher's disease, etc., are life-threatening. Others, such as lactase deficiency and gout, require modification of the diet. Food allergies/intolerances are poorly understood, but they affect many individuals and involve food products including milk, wheat, and eggs. The food industry and food scientists do not give appropriate attention to understanding these genetically caused food intolerances and developing specialty foods to help alleviate them.

Several of these diseases can be prevented by increasing the level of the deficit enzyme in the affected individual. In some cases the means for doing this is known. However, ethical issues must be resolved before this approach becomes common. In the meantime, we should, as food scientists, help meet the special dietary needs of these individuals.

7.14.2 Special Foods for Infants and the Elderly

The enzyme composition of our bodies changes with age. Infants produce chymosin in the stomach to coagulate milk, so that it is retained in the stomach long enough to begin the digestion of proteins and lipids. Later the chymosin is replaced with pepsin, also a protease for digestion of proteins. Most infants also have relatively high levels of β-galactosidase in the mucosa of the small intestine. This is present to hydrolyze lactose to glucose and galactose, both important for

TABLE 31 Some Genetic Diseases Caused by Enzymatic Defects

Disease	Defective enzymes
Alcaptonuria	Homogentisic acid oxidase
Phenylketonuria	Phenylalanine hydroxylase
Hyperammonemia	Ornithine transcarbamylase
Hemolytic anemia	Erythrocyte glucose-6-phosphate dehydrogenase
Acatalasia	Erythrocyte catalase
Congenital lysine intolerance	L-Lysine-NAD-oxidoreductase
Gaucher's disease	Glucocerebrosidase
Refsum's disease	Phytanic acid oxidase
McArdle's syndrome	Muscle phosphorylase
Hypophosphatasia	Alkaline phosphatase
Congenital lactase deficiency	Lactase
Gout	Hypoxanthine-guanine phosphoribosyl transferase
Xanthinuria	Xanthine oxidase
Hereditary fructose intolerance	Fructose 1-phosphate aldolase

Source: Adapted from Ref. 104, p. 22.

energy, but also important for biosynthesis of cerebrosides. With advancing age the level of β-galactosidase decreases. It is estimated that 80% of all adults cannot consume milk and several dairy foods without some discomfort. Some ethnic groups are much less tolerant to milk and dairy products than are other ethnic groups. These are well-known examples of how levels of some enzymes change with age and ethnicity. Other examples undoubtedly exist, but the number is not known accurately.

Elderly people often develop food intolerances. Some of these intolerances may relate to changes in biosynthesis of enzymes. Food scientists need to work closely with nutritionists and medical scientists to better understand these changes in metabolism during aging and to provide appropriate diets that fully meet the nutritional requirements of this segment of the population.

The adequacy of enzyme biosynthesis is also related to diet. More than 150 enzymes require Zn^{2+} as an essential factor. Many other enzymes require other cofactors. Biosynthesis of enzymes requires that essential amino acids be present. These constituents must be available in appropriate amounts in the diet.

7.14.3 Toxicants and Antinutrients [7, 23, 53]

Raw plant materials contain substances necessary to survival of the plant but not desirable in food. These substances include inhibitors against enzymes, especially proteases and amylases, a variety of alkaloids, such as solanine in potatoes, cyanogenic glycosides, hemagglutinins (lectins), and avidin of egg white. Ames [2, 3] determined that these naturally occurring substances are more toxic and should be of more concern than the pesticide residues at concentrations found in foods.

In some instances, the concentrations of these undesirable, naturally occurring substances must be reduced before the food can be safely consumed. Some processing treatments stimulate enzyme action to degrade the unwanted constituent, or inactivate enzymes that catalyze formation of unwanted substances (e.g., HCN in cassava).

7.14.4 Allergenic Substances

We know very little about what makes individuals respond adversely to food constituents, especially proteins. Wheat gluten, egg-white protein, and milk proteins all cause significant allergenic reactions in susceptible individuals. Do these individuals become sensitive to these proteins because they "leak" across the intestinal mucosa, building up sensitivity by immune reactions to these compounds? Do peptide fragments that accumulate from action of proteases have allergenic properties? Can prehydrolyzed proteins be consumed safely by these individuals? We need much more research on these and other questions, before we can determine the role, if any, that enzymes, especially proteases, play in causing allergenic reactions and food intolerances.

BIBLIOGRAPHY

Food Enzymology

Bergmeyer, H. U., ed. (1983–1986). *Methods of Enzymatic Analysis*, Academic Press, New York.
Fogarty, W. M., and C. T. Kelly, eds.(1990). *Microbial Enzymes and Biotechnology*, 2nd. ed., Elsevier Applied Science, New York.
Fox, P. F., ed. (1991). *Food Enzymology*, vols. 1 and 2, Elsevier Applied Science, New York.
Guilbault, G. G. (1984). *Analytical Uses of Immobilized Enzymes*, Marcel Dekker, New York.
Leatham, G. F., and M. E. Himmel, eds. (1991). *Enzymes in Biomass Conversion*, ACS Symposium Series 460, American Chemical Society, Washington, DC.
Nagodawithana, T., and G. Reed (1993). *Enzymes in Food Processing*, Academic Press, San Diego.
Robinson, D. S., and N. A. M. Eskin, eds. (1991). *Oxidative Enzymes in Foods*, Elsevier Applied Science, New York.
Schwimmer, S. (1981). *Source Book of Food Enzymology*, Avi Publishing Co., Westport, CT.
Wagner, G., and G. G. Guilbault (1994). *Food Biosensor Analysis*, Marcel Dekker, New York.
Whitaker, J. R. (1994). *Principles of Enzymology for the Food Sciences*, 2nd ed., Marcel Dekker, New York.

Enzyme Chemistry, Kinetics, and Mechanisms

Kuby, S. A. (1991). *A Study of Enzymes*, vols. 1 and 2, CRC Press, Boca Raton, FL.
Page, M. I., and A. Williams, eds. (1987). *Enzyme Mechanisms*, Royal Society of Chemistry, London.
Silverman, R. B. (1988). *Mechanism-Based Enzyme Inactivation: Chemistry and Enzymology*, vols. 1 and 2, CRC Press, Boca Raton, FL.

Problem Solving, Enzyme Kinetics

Cleland, W. W. (1963). The kinetics of enzyme-catalyzed reactions with two or more substrates or products. I. Nomenclature and rate equations. II. Inhibition: Nomenclature and theory. III. Prediction of initial velocity and inhibition by inspection. *Biochim. Biophys. Acta 67*: 104–137, 173–187, 188–196.
Segel, I. H. (1976). *Biochemical Calculations*, 2nd ed., John Wiley and Sons, New York.

Journals

Archives of Biochemistry and Biophysics
Biochemical and Biophysical Research Communications
Biochemical Journal
Biochemistry
Biochimica et Biophysica Acta
Bioenzymology
European Journal of Biochemistry

FEBS Letters
Journal of Biochemistry
Journal of Biological Chemistry
Journal of Food Biochemistry

REFERENCES

1. Acker, L., and H. Kaiser (1961). Uber den Einfluss der Feuchtigkeit auf den Ablauf Enzymatischer Reaktioner in Wasserarmen Lebensmiteln. III. Mitt: Uber das Verhalten der Phosphatase in Modellgemischen. *Zietschrift Leberns.-Unter.-Forsch. 115*:201–210.
2. Ames, B. N. (1983). Dietary carcinogens and anticarcinogens. *Science 221*: 1256–1264.
3. Ames, B. N., R. Magaw, and L. S. Gold (1987). Ranking possible carcinogenic hazards. *Science 236*:271–280.
4. Ampon, K., A. B. Salleh, A. Teoh, W. M. Z. Wan Yunus, C. N. A. Razak, and M. Basri (1991). Sugar esterification catalyzed by alkylated trysin in dimethyl formamide. *Biotechnol. Lett. 13*:25–30.
5. Arai, S., M. Yamashita, and M. Fujimaki (1981). Protease action on proteins at low water concentration, in *Water Activity: Influence on Food Quality* (L. B. Rockland and C. F. Stewart, eds.), Academic Press, New York, pp. 489–510.
6. Arai, S., M. Watanabe, and N. Hirao (1986). Modification to change physical and functional properties of food proteins, in *Protein Tailoring for Food and Medical Uses* (R. E. Feeney and J. R. Whitaker, eds.), Marcel Dekker, New York, pp. 75–95.
7. Ayres, J. C., and J. C. Kirschman (1981). *Impact on Toxicology on Food Processing*, Avi, Westport, CT.
8. Baker, B. R. (1967). *Design of Active-Site-Directed Irreversible Enzyme Inhibitors*, Wiley, New York.
9. Bender, M. L., and W. A. Glasson (1960). The kinetics of the α-chymotrypsin-catalyzed hydrolysis and methanolysis of acetyl-L-phenylalanine methyl ester. Evidence for the specific binding of water on the enzyme surface. *J. Am. Chem. Soc. 82*:3336–3342.
10. Bender, M. L., G. E. Clement, F. J. Kezdy, and H. D'A. Heck (1964). The mechanism of action of proteolytic enzymes. XXIX. The correlation of the pH (PD) dependence and the stepwise mechanism of α-chymotrypsin-catalyzed reactions. *J. Am. Chem. Soc. 86*:3680–3690.
11. Blain, J. A. (1962). Moisture levels and enzyme activity. *Recent Adv. Food Sci., Papers, Glasgow 1960, 2*:41–45.
12. Blow, D. M., J. J. Birktoft, and B. S. Hartley (1969). Role of a buried group in the mechanism of action of chymotrypsin. *Nature 221*: 337–340.
13. Boscan. L., W. D. Powrie, and O. Fennema (1962). Darkening of red beets, *beta vulgaris. J. Food Sci. 27*:574–582.
14. Brachet, J. (1961). The living cell. *Sci. Am. 205* (3):50–62.
15. Bracho, G., and J. R. Whitaker (1990). Purification and partial characterization of potato (*Solanum tuberosum*) invertase and its endogenous proteinaceous inhibitor. *Plant Physiol. 92*:386–394.
16. Bray, E. A. (1993). Molecular responses to water deficit. *Plant Physiol. 103*:1035–1040.
17. Bray, R. C., G. Palmer, and H. Beinert (1964). Direct studies on the electron transfer sequence in xanthine oxidase by electron paramagnetic resonance spectroscopy. *J. Biol. Chem. 239*:2667–2676.
18. Brown, A. J. (1902). Enzyme action. *J. Chem. Soc. 81*:373–388.
19. Büchner, E. (1897). Alkoholische Gahrung ohne Hefezellen. *Ber. 30*:117–124, 335, 1110–1113.
20. Charm, S. E., and S. E. Matteo (1971). Scale up of protein isolation, in *Methods in Enzymology*, Vol. 22 (W. B. Jakoby, ed.), Academic Press, New York, pp. 476–556.
21. Chibata, I., and T. Tosa (1980). Immobilized microbial cells and their applications. *Trends Biochem. Sci. 5*:88–90.
22. Cleland, W. W. (1963). The kinetics of enzyme-catalyzed reactions with two or more substrates or products. I. Nomenclature and rate questions. II. Inhibition: Nomenclature and theory. III. Prediction of initial velocity and inhibition by inspection. *Biochim. Biophys. Acta 67*:104–137; 173–187; 188–196.

23. Committee on Food Protection (1973). *Toxicants Occurring Naturally in Foods*, 2nd. ed., National Academy of Sciences, Washington, DC.
24. Cooper, C. C., R. G. Cassens, and E. J. Briskey (1969). Capillary distribution and fiber characterisitics in skeletal muscle of stress-susceptible animals. *J. Food Sci. 34*:299–302.
25. Crooke, S. T., and B. Lebleu, eds. (1993). *Antisense Research and Applications*, CRC Press, Boca Raton, FL.
26. Coughlan, M. D. (1990). Cellulose degradation by fungi, in *Microbial Enzymes and Biotechnology*, 2nd ed. (W. M. Fogarity and C. T. Kelley, eds.), Elsevier Applied Science, New York, pp. 1–36.
27. Dalziel, K. (1957). Initial steady state velocities in the evaluation of enzyme-coenzyme-substrate reaction mechanisms. *Acta Chem. Scand. 11*:1706–1723.
28. Dave, B. A. (1968). Studies on polygalacturonic acid *trans*-eliminase produced by *Bacillus pumulis*, Ph. D. Dissertation, University of California, Davis, CA.
29. Douzou, P. (1971). Aqueous-organic solutions of enzymes at subzero temperatures. *Biochemie 53*: 1135–1145.
30. Douzou, P. (1973). Enzymology at sub-zero temperatures. *Mol. Cell. Biochem. 1*:15–27.
31. Drapon, R. (1972). Enzymic reactions in systems with a low amount of water. *Ann. Technol. Agric. 21*:487–499.
32. Duckworth, R. B., J. Y. Allison, and J. A. Clapperton (1976). The aqueous environment for chemical changes in intermediate moisture foods, in *Intermediate Moisture Foods* (R. Davies, G. G. Birch, and K. J. Parker, eds.), Applied Science Publishers, London, pp. 89–99.
33. Eigen, M., and G. G. Hammes (1963). Elementary steps in enzyme reactions. *Adv. Enzymol. 25*:1–38.
34. Elias, P. S., and A. J. Cohen (1977). *Radiation Chemistry of Major Food Components*, Elsevier Scientific, New York.
35. *Enzyme Nomenclature* (1992). Recommendations of the Nomenclature Committee of the International Union of Biochemistry, published for the International Union of Biochemistry, Academic Press, San Diego.
36. Fink, A. L., and K. Magnusdottir (1984). Cryoenzymology of β-galactosidase: Effect of cryosolvents. *J. Protein Chem. 3*:229–241.
37. Finkle, B. J., and R. F. Nelson (1963). Enzyme reactions with phenolic compounds: Effect of *O*-methyltransferase on a natural substrate of fruit polyphenoloxidase. *Nature 197*:902–903.
38. Fischer, E. (1984). Einfluss der Configuration auf die Wirkung der Enzyme. *Berichte der Chemische Gesselschaft 27*:2985–2993.
39. Friedman, M., O. K. K. Grosjean, and J. C. Zahnley (1982). Inactivation of soybean trypsin inhibitors by thiols. *J. Sci. Food Agric. 33*:165–172.
40. Fitzpatrick, P. A., and A. M. Klibanov (1991). How can solvent affect enzyme enantioselectivity? *J. Am. Chem. Soc. 113*:3166–3171.
41. Fitzpatrick, P. A., A. C. U. Steinmetz, D. Ringe, and A. M. Klibanov (1993). Enzyme crystal structure in a neat organic solvent. *Proc. Natl. Acad. Sci. USA 90*:8653–8657.
42. Grohmann, K., and M. E. Himmel (1991). Enzymes for fuels and chemical feedstocks, in *Enzymes in Biomass Conversion* (G. R. Leatham and M. E. Himmel, eds.), *ACS Symposium Series 460*, American Chemical Society, Washington, DC.
43. Guilbault, G. G. (1984). *Analytical Uses of Immobilized Enzymes*, Marcel Dekker, New York.
44. Gutfreund, H. (1972). *Enzymes: Physical Principles*, Wiley-Interscience, London.
45. Hagerdal, B. (1980). Enzyme co-immobilized with microorganisms for the microbial conversion on non-metabolizable substrates. *Acta Chem. Scand. B34*:611–613.
46. Henri, V. (1902). Action de quelques sels neutres sur l'inversion du saccharose par la sucrase. *C. R. Soc. Biol. 54*, 353–354.
46a. Henri, V. (1903). *Lois generales de l'action des diastases*, Hermann, Paris.
47. Hill, A. V. (1913). XLVII. The combinations of haemoglobin with oxygen and with carbon dioxide. *Biochem. J. 7*:471–480.
48. Hirs, C. H. W. and S. N. Timasheff, eds. (1972). *Enzyme Structure: Part B. Methods Enzymol. 25*.
49. Hirs, C. H. W., S. Moore, and W. H. Stein (1960). The sequence of the amino acid residues in performic acid-oxidized ribonuclease. *J. Biol. Chem. 235*:633–647.

50. Ho, M. F., X. Yin, F. F. Filho, F. Lajolo, and J. R. Whitaker (1994). Naturally occurring α-amylase inhibitors: Structure/function relationships, in *Protein Structure-Function Relationships in Foods* (R. Y. Yada, R. L. Jackman, and J. L. Smith, eds.), Blackie Academic and Professional, Glasgow, pp. 89–119.
51. Hobson, G. E. (1962). Determination of polygalacturonase in fruits. *Nature 195*:804–805.
52. Hornby, W. E., M. D. Lilley, and E. M. Crook (1968). Some changes in the reactivity of enzymes resulting from their chemical attachment of water-insoluble derivatives of cellulose. *Biochem. J. 107*:669–674.
53. Jelliffe, E. F. P., and D. B. Jelliffe, eds. (1982). *Adverse Effects of Foods*, Plenum, New York.
54. Joslyn, M. A., and E. Zuegg (1956). Report on peroxidase in frozen vegetables. Ascorbic acid oxidation method for determining peroxidase activity. *J. Assoc. Offic. Agric. Chem. 39*:267–283.
55. Kang, Y., A. G. Marangoni, and R. Y. Yada (1994). Effect of two polar organic-aqueous solvent systems on the structure-function relationships of proteases. I. Pepsin. *J. Food Biochem. 17*:353–369.
56. Kang, Y., A. G. Marangoni, and R. Y. Yada (1994). Effect of two polar organic-aqueous solvent systems on the structure-function relationships of proteases. II. Chymosin and *Mucor miehei* proteinase. *J. Food Biochem. 17*:371–387.
57. Kang, Y., A. G. Marangoni, and R. Y. Yada (1994). Effect of two polar organic-aqueous solvent systems on the structure-function relationship of proteases. III. Papain and trypsin. *J. Food Biochem. 17*:389–405.
58. Kartha, G., J. Bello, and D. Harker (1967). Tertiary structure of ribonuclease. *Nature 213*:862–865.
59. Kato, A., and K. Yutani (1988). Correlation of surface properties with conformational stabilities of wild-type and six mutant tryptophan synthase α-subunits substituted at the same position. *Protein Eng. 2*:153–156.
60. Kato, H., T. Tanaka, H. Yamaguchi, T. Hara, T. Nishioka, Y. Katsube, and J. Oda (1994). Flexible loop that is novel catalytic machinery in a ligase. Atomic structure and function of the loopless glutathione synthetase. *Biochemistry 33*:4995–4999.
61. Kavanau, J. L. (1950). Enzyme kinetics and rate of biological processes. *J. Gen. Physiol. 34*:193–209.
62. King, E. L., and C. Altman (1956). A schematic method of deriving the rate laws for enzyme-catalyzed reactions. *J. Phys. Chem. 60*:1375–1378.
63. Kitamura, K. (1984). Biochemical characterization of lipoxygenase lacking mutants, L-1-less, L-2-less, and L-3-less soybeans. *Agric. Biol. Chem. 48*:2339–2346.
64. Klibanov, A. M., and T. J. Ahern (1987). Thermal stability of proteins, in *Protein Engineering* (D. L. Oxender and C. F. Fox, eds.), Alan R. Liss, New York, pp. 213–218.
65. Koshland, D. E., Jr. (1959). Mechanisms of transfer enzymes, in *The Enzymes*, 1st ed. (P. D. Boyer, H. Lardy and K. Myrback, eds.), vol. 1, Academic Press, New York, pp. 305–346.
66. Koshland, D. E., Jr., G. Nemethy, and D. Filmer (1966). Comparison of experimental data and theoretical models in proteins containing subunits. *Biochemistry 5*:365–385.
67. Knowles, J. R. (1991). Enzyme catalysis: Not different, just better. *Nature 350*:121–124.
68. Kuhne, W. (1877). Ueber das Verhalten verschiedener organisirter und sog. ungeformter fermente. *Nat. Med. Verhandl. 1*:190–193.
69. Kuzin, A. M. (1964). *Radiation Biochemistry*, D. Davey, New York.
70. Leatham, G. F., and M. E. Himmel, eds. (1991). *Enzymes in Biomass Conversion*, ACS Symposium Series 460, American Chemical Society, Washington, DC.
71. Levitzki, A., and M. L. Steer (1974). Allosteric activation of mammalian α-amylase by chloride. *Eur. J. Biochem. 41*:171–180.
72. von Liebig, J. (1872). Ueber den Kochsalzgehalt des Extractum carnis. *Anal. Chem. Pharm. 157*:369–373.
73. Lilly, M. D., and A. K. Sharp (1968). The kinetics of enzymes attached to water-insoluble polymers. *Chem. Eng. (Lond.) 215*:CE12–CE18.
74. Lim, M. H., A. O. Chen, R. M. Pangborn, and J. R. Whitaker (1989). Effect of enzymes on aroma changes in vegetables during frozen storage, in *Trends in Food Biotechnology* (A. H. Ghee, N.B. Hen, and L. K. Kong, eds.), Singapore Institute of Food Science and Technology, Singapore, pp. 169–176.

75. Lim, M. H., P. J. Velasco, R. M. Pangborn, and J. R. Whitaker (1989). Enzymes involved in off-aroma formation in broccoli, in *Quality Factors of Fruits and Vegetables* (J. J. Jen, ed.), *ACS Symp. Ser. 405*:72–83.
76. Lineweaver, H., and D. Burk (1934). Determination of enzyme dissociation constants. *J. Am. Chem. Soc. 56*:658–666.
77. Lovern, J. A., and J. Olley (1962). Inhibition and promotion of post-mortem lipid hydrolysis in the flesh of fish. *J. Food Sci. 27*:551–559.
78. Maehler, R., and J. R. Whitaker (1982). Dynamics of ligand binding to α-chymotrypsin and to *N*-methyl α-chymotrypsin. *Biochemistry 21*:4621–4633.
79. Michaelis, L., and M. L. Menten (1913). Kinetics of invertase action. *Biochem. Z. 49*:333–369.
80. Monod, J., J. Wyman, and J. P. Changeux (1965). On the nature of allosteric transitions: a plausible model. *J. Mol. Biol. 12*:88–118.
81. Morton, R. K. (1955). The substrate specificity and inhibition of alkaline phosphatases of cow's milk and calf intestinal mucosa. *Biochem. J. 61*:232–240.
82. Morton, R. K. (1957). The kinetics of hydrolysis of phenyl phosphate by alkaline phosphatases. *Biochem. J. 65*:674–682.
83. Myrback, K. (1926). Uber verbindungen einiger enzyme mit inaktivierenden stoffen. II. *Z. Physiol. Chem. 159*:1–84.
83a. Olsen, F., and K. Aunstrup (1986). Alpha-acetolactate decarboxylase enzyme and preparation thereof. U. S. Patent 4,617,273.
84. Osuga, D., and J. R. Whitaker (1994). Mechanisms of some reducing compounds that inactivate polyphenol oxidase, in *Enzymatic Browning and Its Prevention* (C. Y. Lee and J. R. Whitaker, eds.), *ACS Symp. Ser. 600:* 210–222.
85. Pasteur, L. (1875). Nouvelles observations sur la nature de la fermentation alcoolique. *C. R. Acad. Sci. 80*:452–457.
86. Pressey, R. (1967). Invertase inhibitor from potatoes: Purification, characterization and reactivity with plant invertases. *Plant Physiol. 42*:1780–1786.
87. Rackis, J. J. (1974). Biological and physiological factors in soybeans. J. Am. Oil Chem. Soc. 51:161A–174A.
88. Richardson, T., and D. B. Hyslop (1985). Enzymes, in *Food Chemistry*, 2nd ed. (O. R. Fennema, ed.), Marcel Dekker, New York, pp. 390–393.
89. Ryan, C. A. (1973). Proteolytic enzymes and their inhibitors in plants. *Annu. Rev. Plant Physiol. 24*:173–196.
90. Ryle, A. P., F. Sanger, L. F. Smith, and R. Kitai (1955). The disulfide bonds of insulin. *Biochem. J. 60*:541–556.
91. Salleh, A. B., K. Ampon, F. Salam, W. M. Z. Wan Yunus, C. N. A. Rozak, and M. Basri (1990). Modification of lipase for reactions in organic solvents. *Ann. NY Acad. Sci. 613*:521–522.
92. Scott, D. (1989). Specialty enzymes and products for the food industry, in *Biocatalysis in Agricultural Biotechnology* (J. R. Whitaker and P. E. Sonnet, eds.), *ACS Symposium Ser. 389*, American Chemical Society, Washington, DC, pp. 176–192.
93. Siddiqi, A. M. and A. L. Tappel (1957). Comparison of some lipoxidases and their mechanism of action. *J. Am. Oil Chem. Soc. 34*:529–533.
94. Sizer, I. W., and E. S. Josephson (1942). Kinetics as a function of temperature of lipase, trypsin and invertase activity from –70 to 50°C (–94 to 122°F). *Food Res. 7*:201–209.
95. Spackman, D. H., W. H. Stein, and S. Moore (1960). The disulfide bonds of ribonuclease. *J. Biol. Chem. 235*:648–659.
96. Stearn, A. E. (1949). Kinetics of biological reactions with specific reference to enzyme processes. *Adv. Enzymol. 9*:25–74.
97. Stockwell, A., and E. L. Smith (1957). Kinetics of papain action. I. Hydrolysis of benzoyl-L-argininamide. *J. Biol. Chem. 227*:1–26.
98. Sumner, J. B. (1926). The isolation and crystallization of the enzyme urease. *J. Biol. Chem. 69*:435–441.
99. Theorell, H., and T. Yonetani (1963). Liver alcohol dehydrogenase-diphosphopyridine nucleotide

(DPN)-pyrazole complex; A model of a ternary intermediate in the enzyme reaction. *Biochem. Z. 338*:537–553.

100. Van Noorden, C. J. F., and W. M. Fredericks (1992). *Enzyme Histochemistry. A Laboratory Manual of Current Methods*, Oxford University Press, New York.
101. Velasco, P. J., M. H. Lim, R. M. Pangborn, and J. R. Whitaker (1989). Enzymes responsible for off-flavor and off-aroma in blanched and frozen-stored vegetables. *Biotechnol. Appl. Biochem. 11*:118–127.
102. Verger, R., J. Rietsch, F. Pattus, F. Ferrato, G. Peroni, G. H. deHaase, and P. Desneulle (1978). Studies of lipase and phospholipase A_2 acting on lipid monolayers. *Adv. Exp. Med. Biol. 101*:79–94.
103. Volkin, D. B., A. Staubli, R. Langer, and A. M. Klibanov (1991). Enzyme thermoinactivation in anhydrous organic solvents. *Biotech. Bioeng. 37*:843–853.
104. Wacher, W. E. C., and T. L. Coombs (1969). Clinical biochemistry: Enzymatic methods: Automation and atomic absorption spectroscopy. *Annu. Rev. Biochem. 38*:539–568.
105. Wagner, G., and G. G. Guilbault, eds. (1994). *Food Biosensor Analysis*, Marcel Dekker, New York.
106. Webb, J. L. (1967). *Enzyme and Metabolic Inhibitors*, vol. I–III, Academic Press, New York.
107. Whitaker, J. R. (1981). Naturally occurring peptide and protein inhibitors of enzymes, in *Impact of Toxicology on Food Processing* (J. C. Ayres and J. C. Kirschman, eds.), Avi, Westport, CT, pp.57–104.
108. Whitaker, J. R. (1984). Analytical uses of enzymes, in *Food Analysis: Principles and Techniques*, vol. 3, *Biological Techniques* (D. W. Gruenwedel and J. R. Whitaker, eds.), Marcel Dekker, New York, pp. 297–377.
109. Whitaker, J. R. (1990). Microbial pectolytic enzymes, in *Microbial Enzymes and Biotechnology*, 2nd ed. (W. M. Fogarty and C. T. Kelly, eds.), Elsevier Applied Science, London, pp. 133–176.
110. Whitaker, J. R. (1990). New and future uses of enzymes in food processing. *Food Biotechnol. 4*:669–697.
111. Whitaker, J. R. (1994). *Principles of Enzymology for the Food Sciences*, 2nd ed., Marcel Dekker, New York.
112. Whitaker, J. R., and R. A. Bernhard (1991). *Experiments for an Introduction to Enzymology*, University of California, Davis, p. 83.
113. Whitaker, J. R., and M. Mazelis (1991). Enzymes important in flavor development in the alliums, in *Food Enzymology*, vol. 1 (P. F. Fox, ed.), Elsevier, London, pp. 479–497.
114. Williams, D. C., M. H. Lim, A. O. Chen, R. M. Pangborn, and J. W. Whitaker (1986). Blanching of vegetables for freezing. Which indicator enzyme to choose. *Food Technol. 40*:130–140.
115. Wong, J. T. F., and C. S. Hanes (1962). Kinetic formulations for enzymic reactions involving two substrates. *Can. J. Biochem. Physiol. 40*:763–804.
116. Wu, C., and J. R. Whitaker (1990). Purification and partial characterization of four trypsin/chymotrypsin inhibitors from red kidney beans (*Phaseolus vulgaris*, var. Linden), *J. Agric. Food Chem. 38*:1523–1529.
117. Zaborsky, O. R. (1973). *Immobilized Enzymes*, CRC Press, Boca Raton, FL.
118. Zaks, A., and A. M. Klibanov (1984). Enzymatic catalysis in organic media at 100°C. *Science 224*:1249–1251.

8

Vitamins

JESSE F. GREGORY III
University of Florida, Gainesville, Florida

8.1 INTRODUCTION

8.1.1 Objectives

Since the discovery of the basic vitamins and their many forms, a wealth of information has been generated and published on their retention in foods during postharvest handling, commercial processing, distribution, storage, and preparation, and many reviews have been written on this topic. A good summary of this information is *Nutritional Evaluation of Food Processing* [61, 62, 72], to which the reader is encouraged to refer.

The major objective of this chapter is to discuss and critically review the chemistry of the individual vitamins and our understanding of the chemical and physical factors that influence vitamin retention and bioavailability in foods. A secondary objective is to indicate gaps in our understanding and to point out factors that affect the quality of data with respect of our understanding of vitamin stability. It should be noted that there is an unfortunate state of inconsistency of nomenclature in the vitamin literature, with many obsolete terms still being used. Throughout this chapter, terminology recommended by the International Union of Pure and Applied Chemistry (IUPAC) and the American Institute of Nutrition [1] will be used.

8.1.2 Summary of Vitamin Stability

The vitamins comprise a diverse group of organic compounds that are nutritionally essential micronutrients. Vitamins function in vivo in several ways, including (a) as coenzymes or their precursors (niacin, thiamin, riboflavin, biotin, pantothenic acid, vitamin B6, vitamin B12, and folate); (b) as components of the antioxidative defense system (ascorbic acid, certain carotenoids, and vitamin E); (c) as factors involved in genetic regulation (vitamins A, D, and potentially several others); and (d) in specialized functions such as vitamin A in vision, ascorbate in various hydroxylation reactions, and vitamin K in specific carboxylation reactions.

Vitamins are quantitatively minor constituents of foods. From the viewpoint of food chemistry, we are mainly interested in maximizing vitamin retention by minimizing aqueous extraction (leaching) and chemical changes such as oxidation and reaction with other food constituents. In addition, several of the vitamins influence the chemical nature of food, by functioning as reducing agents, radical scavengers, reactants in browning reactions, and flavor precursors. Although much is known about the stability and properties of vitamins, our knowledge of their behavior in the complex milieu of food is limited. Many published studies have, sometimes by necessity, involved the use of chemically defined model systems (or even just buffer solutions) to simplify the investigation of vitamin stability. Results of such studies should be interpreted with caution because, in many cases, the degree to which these model systems simulate complex food systems is not known. While these studies have provided important insight into chemical variables affecting retention, they are sometimes of limited value for predicting the behavior of vitamins in complex food systems. This is so because complex foods often differ markedly from model systems in terms of physical and compositional variables including water activity, ionic strength, pH, enzymatic and trace metallic catalysts, and other reactants (protein, reducing sugars, free radicals, active oxygen species, etc.). Throughout this chapter, emphasis will be placed on the behavior of vitamins under conditions relevant to actual foods.

Most of the vitamins exist as groups of structurally related compounds exhibiting similar nutritional function. Many attempts have been made to summarize the stability of the vitamins, such as that shown in Table 1 [60]. The major limitation of such exercises is the marked variation in stability that can exist among the various forms of each vitamin. Various forms of each vitamin can exhibit vastly different stability (e.g., pH of optimum stability and susceptibility to oxidation) and reactivity. For example, tetrahydrofolic acid and folic acid are two folates that exhibit nearly identical nutritional properties. As described later, tetrahydrofolic acid (a naturally occurring form) is extremely susceptible to oxidative degradation, while folic acid (a synthetic form used in food fortification) is very stable. Thus, attempts to generalize or summarize the properties of the vitamins are at best imprecise and at worst highly misleading.

TABLE 1 Summary of Vitamin Stability

Nutrient	Neutral	Acid	Alkaline	Air or oxygen	Light	Heat	Maximum cooking loss
Vitamin A	S	U	S	U	U	U	40
Ascorbic acid	U	S	U	U	U	U	100
Biotin	S	S	S	S	S	U	60
Carotenes	S	U	S	U	U	U	30
Choline	S	S	S	U	S	S	5
Vitamin B12	S	S	S	U	U	S	10
Vitamin D	S	S	U	U	U	U	40
Folate	U	U	U	U	U	U	100
Vitamine K	S	U	U	S	U	S	5
Niacin	S	S	S	S	S	S	75
Pantothenic acid	S	U	U	S	S	U	50
Vitamin B6	S	S	S	S	U	U	40
Riboflavin	S	S	U	S	U	U	75
Thiamin	U	S	U	U	S	U	80
Tocopherols	S	S	S	U	U	U	55

Note: S, stable (no important destruction); U, unstable (significant destruction). Caution: These conclusions are oversimplifications and may not accurately represent stability under all circumstances.
Source: Adapted from Ref. 60.

8.1.3 Toxicity of Vitamins

In addition to the nutritional role of vitamins, it is important to recognize their potential toxicity. Vitamins A, D, and B6 are of particular concern in this respect. Episodes of vitamin toxicity are nearly always associated with overzealous consumption of nutritional supplements. Toxic potential also exists from inadvertent excessive fortification, as has occurred in an incident with vitamin D-fortified milk. Instances of intoxication from vitamins occurring endogenously in food are exceedingly rare.

8.1.4 Sources of Vitamins

Although vitamins are consumed in the form of supplements by a growing fraction of the population, the food supply generally represents the major and most critically important source of vitamin intake. Foods, in their widely disparate forms, provide vitamins that occur naturally in plant, animal, and microbial sources as well as those added in fortification. In addition, certain dietetic and medical foods, enteric formulas, and intravenous solutions are formulated so that the entire vitamin requirements of the individual are supplied from these sources.

Regardless of whether the vitamins are naturally occurring or added, the potential exist for losses by chemical or physical (leaching or other separations) means. Losses of vitamins are, to some degree, inevitable in the manufacturing, distribution, marketing, home storage, and preparation of processed foods, and losses of vitamins, can also occur during the post-harvest handling and distribution of fruits and vegetables and during the post-slaughter handling and distribution of meat products. Since the modern food supply is increasingly dependent on processed and industrially formulated foods, the nutritional adequacy of the food supply depends, in large measure, on our understanding of how vitamins are lost and on our ability to control these losses.

Although considerable information is available concerning the stability of vitamins in foods, our ability to use such information is frequently limited by a poor understanding of reaction mechanisms, kinetics, and thermodynamics under various circumstances. Thus, it is frequently difficult on the basis of our present knowledge to predict the extent to which given processing, storage, or handling conditions will influence the retention of many vitamins. Without accurate information regarding reaction kinetics and thermodynamics, it is also difficult to select conditions and methods of food processing, storage, and handling to optimize vitamin retention. Thus, there is a great need for more thorough characterization of the basic chemistry of vitamin degradation as it occurs in complex food systems.

8.2 ADDITION OF NUTRIENTS TO FOODS

Throughout the early 20th century, nutrient deficiency represented a major public health problem in the United States. Pellagra was endemic in much of the rural South, while deficiencies of riboflavin, niacin, iron, and calcium were widespread. The development of legally defined standards of identity under the 1938 Food, Drug, and Cosmetic Act provided for the direct addition of several nutrients to foods, especially certain dairy and cereal-grain products. Although technological and historical aspects of fortification are beyond the scope of this chapter, the reader is referred to *Nutrient Additions to Food: Nutritional, Technological and Regulatory Aspects* [7] for a comprehensive discussion of this topic. The nearly complete eradication of overt vitamin-deficiency disease provides evidence of the exceptional effectiveness of fortification programs and the general improvement in the nutritional quality of the U.S. food supply.

Definitions of terms associated with the addition of nutrients of foods include:

1. Restoration: Addition to restore the original concentration of key nutrients.
2. Fortification: Addition of nutrients in amounts significant enough to render the food a good to superior source of the added nutrients. This may include addition of nutrients not normally associated with the food or addition to levels above that present in the unprocessed food.
3. Enrichment: Addition of specific amounts of selected nutrients in accordance with a standard of identity as defined by the U. S. Food and Drug Administration (FDA).
4. Nutrification: This is a generic term intended to encompass any addition of nutrients to food.

The addition of vitamins and other nutrients to food, while clearly beneficial in current practice, also carries with it the potential for abuse and, thus, risk to consumers. For these reasons, important guidelines have been developed that convey a reasonable and prudent approach. These U. S. Food and Drug Administration guidelines [21 CFR Sec. 104.20(g)] state that the nutrient added to a food should be:

1. Stable under customary conditions of storage, distribution, and use.
2. Physiologically available from the food.
3. Present at a level where there is assurance that there will not be excessive intake.
4. Suitable for its intended purpose and in compliance with provisions (i.e., regulations) governing safety.

Further, it is stated in these guidelines that "the FDA does not encourage the indiscriminate addition of nutrients to foods." Similar recommendations have been developed and endorsed jointly by the Council on Foods and Nutrition of the American Medical Association (AMA), the Institute of Food Technologists (IFT) and the Food and Nutrition Board (FNB) of the National Academy of Sciences–National Research Council [4].

Additionally, AMA–IFT–FNB guidelines recommend that the following prerequisites be met to justify fortification: (a) the intake of the particular nutrient is inadequate for a substantial portion of the population; (b) the food (or category) is consumed by most individuals in the target population; (c) there is reasonable assurance that excessive intake will not occur; and (d) the cost is reasonable for the intended population. The joint statement also included the following endorsement of enrichment programs:

> Specifically the following practices in the United States continue to be endorsed: The enrichment of flour, bread, degerminated and white rice (with thiamin, riboflavin, niacin, and iron); the retention or restoration of thiamin, riboflavin, niacin, and iron in processed food cereals; the addition of vitamin D to milk, fluid skimmed milk, and nonfat dry milk, the addition of vitamin A to margarine, fluid skim milk, and nonfat dry milk, and the addition of iodine to table salt. The protective action of fluoride against dental caries is recognized and the standardized addition of fluoride is endorsed in areas in which the water supply has a low fluoride content.

At the time of this writing, the merits of the proposed addition of folic acid to foods are being evaluated by the FDA with respect to inclusion of this vitamin in enriched cereal-grain products (i.e., wheat flours, rice, corn meals, breads, and pastas). Enrichment appears to be a viable approach to providing supplemental folic acid for the purpose of reducing the risk of certain birth defects (spina bifida and anencephaly); however, excessive intake of total folate must be avoided in order to avoid masking the diagnosis of vitamin B12 deficiency.

The stability of vitamins in fortified and enriched foods has been thoroughly evaluated. As

TABLE 2 Stability of Vitamins Added to Cereal Grain Products

			Storage time (months at 23°C)		
Vitamin	Claim	Found	2	4	6
In 1 lb of white flour					
Vitamin A (IU)	7500	8200	8200	8020	7950
Vitamin E (IU)[a]	15.0	15.9	15.9	15.9	15.9
Pyridoxine (mg)	2.0	2.3	2.2	2.3	2.2
Folate (mg)	0.30	0.37	0.30	0.35	0.3
Thiamin (mg)	2.9	3.4	—	—	3.4
In 1 lb of yellow corn meal					
Vitamin A (IU)	—	7500	7500	—	6800
Vitamin E (IU)[a]	—	15.8	15.8	—	15.9
Pyridoxine (mg)	—	2.8	2.8	—	2.8
Folate (mg)	—	0.30	0.30	—	0.29
Thiamin (mg)	—	3.5	—	—	3.6
	After baking		5 Days of storage (23°C)		
In 740 g of bread					
Vitamin A (IU)	7500	8280		8300	
Vitamin E (IU)[a]	15	16.4		16.7	
Pyridoxine (mg)	2	2.4		2.5	
Folate (mg)	0.3	0.34		0.36	

[a]Vitamin E is expressed as *dl*-α-tocopherol acetate.
Source: Ref. 25.

shown in Table 2, the stability of added vitamins in enriched cereal grain products under conditions of accelerated shelf-life testing is excellent [3, 25]. Similar results have been reported with fortified breakfast cereals (Table 3). This excellent retention is due, in part, to the stability of the chemical forms of these vitamins used, as well as the favorable environment with respect to water activity and temperature. The stability of vitamins A and D in fortified milk products also has been shown to be satisfactory.

8.3 DIETARY RECOMMENDATIONS

To assess the impact of food composition and intake patterns on the nutritional status of individuals and populations, and to determine the nutritional effects of particular food processing and handling practices, a nutritional reference standard is essential. In the United States, the Recommended Dietary Allowances (RDA) have been developed for these purposes. The RDA values have been defined as "the level of intake of essential nutrients considered, in the judgement of the Committee on Dietary Allowances of the Food and Nutrition Board on the basis of available scientific knowledge, to be adequate to meet the known nutrition needs of practically all healthy persons" [44]. To the extent possible, the RDA values are formulated to include allowances for variability within the population with respect to nutrient requirements as well as the potential for incomplete bioavailability of nutrients. However, limitations in our current knowledge of the bioavailability of vitamins in foods render such allowances highly uncertain. Many other countries and several international organizations such as the FAO/WHO have developed reference values similar to the RDAs, and these differ quantitatively because of differences in scientific judgement or philosophy.

Table 3 Stability of Vitamins Added to Breakfast Cereal Products

Vitamin content (per gram of product)	Initial value	Storage time: 3 months, 40°C	Storage time: 6 months, 23°C
Vitamin A (IU)	193	168	195
Ascorbic acid (mg)	2.6	2.4	2.5
Thiamin (mg)	0.060	0.060	0.064
Riboflavin (mg)	0.071	0.074	0.67
Niacin (mg)	0.92	0.85	0.88
Vitamin D	17.0	15.5	16.6
Vitamin E (IU)	0.49	0.49	0.46
Pyridoxine (mg)	0.085	0.088	0.081
Folate (mg)	0.018	0.014	0.018
Vitamin B12 (μg)	0.22	0.21	0.21
Pantothenic acid (mg)	0.42	0.39	0.39

Source: Ref. 3.

For food labeling to be meaningful, the concentration of micronutrients is best expressed relative to reference values. In the United States, nutrition labeling data for micronutrients has been traditionally expressed as a percentage of a "U.S. RDA" value, a practice that was originated at the onset of nutrition labeling. These were derived from the 1968 RDA values and differ somewhat from the current RDA values reported by the Food and Nutrition Board (Table 4). These differences, although not readily evident to the consumer, should be recognized and understood. Federal regulations permit modification of U.S. RDAs by the FDA "from time to time as more information on human nutrition becomes available" [21 CFR Sec.101.0(c)(7)(b)(ii)], although no changes were implemented. Under the revised labeling regulation implemented by the Food and Drug Administration in 1994, the U.S. RDA term has been replaced by the "Reference Daily Intake (RDI)," which is currently equivalent to the previous U.S. RDAs. In the current nutrition labeling format, vitamin content is expressed as a percentage of the RDI and is listed as "% Daily Value."

8.4 ANALYTICAL METHODS AND SOURCES OF DATA

The major sources of information regarding the content of vitamins in U.S. foods are the U.S. Department of Agriculture's National Nutrient Data Bank and the Agricultural Handbook No. 8 series. An important limitation of these data and most other data bases is the uncertain adequacy of the analytical methods used. It is frequently unclear how the data were obtained, what methods were used, and whether the results were based on truly representative samples. Issues regarding the merits of information in nutrient databases have been discussed in several reviews [8, 9].

The adequacy of analytical methods is a serious problem with respect to many vitamins. While current analytical methods are generally acceptable for some vitamins (e.g., ascorbic acid, thiamin, riboflavin, niacin, vitamin B6, vitamin A, and vitamin E), they are less adequate for others (e.g., folate, pantothenic acid, biotin, carotenoids, vitamin B12, vitamin D, and vitamin K). Factors that limit the suitability of analytical methods may involve a lack of specificity of traditional chemical methods, interferences in microbiological assays, incomplete extraction of the analyte(s) from the food matrix, and incomplete measurement of complexed forms of a vitamin. Improvement of analytical data for vitamins will require additional support for methods

TABLE 4 Comparison of Recommended Dietary Allowances for Vitamins and "Reference Daily Intake" (RDI) Currently Used in Nutritional Labeling in the United States

Category	Age (years)	Vitamin A (μg RE)[a]	Vitamin D (μg)	Vitamin E (mg α-TE)	Vitamin K (μg)	Vitamin C (mg)	Thiamin (mg)	Riboflavin (mg)	Niacin (mg NE)	Vitamin B6 (mg)	Folate (μg)	Vitamin B12 (μg)
Infants	0.0–0.5	375	7.5	3	5	30	0.3	0.4	5	0.3	25	0.3
	0.5–1.0	375	10	4	10	35	0.4	0.5	6	0.6	35	0.5
Children	1–3	400	10	6	15	40	0.7	0.8	9	1.0	50	0.7
	4–6	500	10	7	20	45	0.9	1.1	12	1.1	75	1.0
	7–10	700	10	7	30	45	1.0	1.2	13	1.4	100	1.4
Males	11–14	1000	10	10	45	50	1.3	1.5	17	1.7	150	2.0
	15–18	1000	10	10	65	60	1.5	1.8	20	2.0	200	2.0
	19–24	1000	10	10	70	60	1.5	1.7	19	2.0	200	2.0
	25–50	1000	5	10	80	60	1.5	1.7	19	2.0	200	2.0
	51+	1000	5	10	80	60	1.2	1.4	15	2.0	200	2.0
Females	11–14	800	10	8	45	50	1.1	1.3	15	1.4	150	2.0
	15–18	800	10	8	55	60	1.1	1.3	15	1.5	180	2.0
	19–24	800	10	8	60	60	1.1	1.3	15	1.6	180	2.0
	25–50	800	5	8	65	60	1.1	1.3	15	1.6	180	2.0
	51+	800	5	8	65	60	1.0	1.2	13	1.6	180	2.0
Pregnant		800	10	10	65	70	1.5	1.6	17	2.2	400	2.2
Lactating	0–0.5[c]	1300	10	12	65	95	1.6	1.8	20	2.1	280	2.6
	0.5–1.0[c]	1200	10	11	65	90	1.6	1.7	20	2.1	260	2.6
RDI[b] (used in food labeling)		1000 (5000 IU)	10 (400 IU)	20 (30 IU)	no RDI	60	1.5	1.7	20	2.0	400	6.0

[a]Units: RE, retinol equivalent (1 RE = 1 μg retinol or 6 μg β-carotene); α-TE, α-tocopherol equivalent (1 mg α-TE = 1 mg d-α–tocopherol); NE, niacin equivalent (1 mg NE = 1 mg niacin or 60 mg tryptophan).
[b]Reference Daily Intake (RDI) is the reference unit used in U.S. nutrition labeling of foods, formerly termed the U.S. RDA.
[c]Duration of lactation.
Source: Ref. 44.

development research, improved training of analysts, development of quality control protocols (i.e., validation and standardization of procedures), and development of standard reference materials for vitamin analysis. The strengths and limitations of analytical methods for each vitamin will be briefly addressed in this chapter.

8.5 BIOAVAILABILITY OF VITAMINS

The term bioavailability refers to the degree to which an ingested nutrient undergoes intestinal absorption and metabolic function or utilization within the body. Bioavailability involves both absorption and utilization of the nutrient *as consumed*; this concept does not refer to losses that may occur prior to consumption. For a complete description of the nutritional adequacy of a food, three factors must be known: (a) the concentration of the vitamin at the time of consumption; (b) the identity of various chemical species of the vitamin present; (c) and the bioavailability of these forms of the vitamin as they exist in the meal consumed.

Factors that influence the bioavailability of vitamins include (a) composition of the diet, which could influence intestinal transit time, viscosity, emulsion characteristics, and pH; (b) form of the vitamin [forms may differ in rate or extent of absorption, ease of conversion to metabolically (e.g., coenzymic) active form, or metabolic functionality]; and (c) interactions between a vitamin and components of the diet (e.g., proteins, starches, dietary fiber, lipids) that interfere with intestinal absorption of the vitamin. Although our understanding of the relative bioavailability of the various species of each vitamin is rapidly improving, the complex influences of food consumption on vitamin bioavailability remain poorly understood. In addition, the effects of processing and storage on vitamin bioavailability have been only partially determined.

The application of information regarding bioavailibility of vitamins is limited at this time. Bioavailibility is generally considered in the development of dietary recommendations (e.g., RDA values), but this involves only the use of estimated mean bioavailability values. At the present time, our knowledge is too fragmentary to permit vitamin bioavailability data to be included in food composition tables. However, even if our understanding of the bioavailability of vitamins in individual foods were much more complete, such data regarding individual foods may be of little use. A far greater need is for a better understanding of vitamin bioavailability in the diet as a whole (including interactive effects of individual foods) and the sources of variation in this respect among individual people.

8.6 GENERAL CAUSES OF VARIATION/LOSSES OF VITAMINS IN FOOD

Beginning at the time of harvesting, all foods inevitably undergo some loss of vitamins. The nutritional significance of partial loss of vitamins depends on the nutritional status of the individual (or population) for the vitamin of interest, the importance of the particular food as a source of that vitamin, and the bioavailability of the vitamin. Most processing, storage, and handling methods are intended to minimize vitamin losses. The following is a summary of the various factors responsible for variation in the vitamin content of foods.

8.6.1 Inherent Variation in Vitamin Content

The concentration of vitamins in fruits and vegetables often varies with stage of maturity, site of growth, and climate. During maturation of fruits and vegetables, vitamin concentration is determined by the rates of synthesis and degradation. Information on the time course of vitamin concentration in most fruits and vegetables is not available except for ascorbic acid and β-carotene in a few products. In the example shown in Table 5, the maximum concentration of ascorbic acid in tomatoes occurred prior to full maturity. A study of carrots showed that carotenoid concentration varied markedly with variety but was not influenced significantly by stage of maturity.

Little is known about developmental changes in vitamin content of cereal grains and legumes. In contrast to fruits and vegetables, cereal grains and legumes are harvested at a fairly uniform stage of maturity.

Agricultural practices and environmental conditions undoubtedly influence the content of vitamins in plant-derived foods, but few data are available on this subject. Klein and Perry [78] determined the content of ascorbic acid and vitamin A activity (from carotenoids) in selected fruits and vegetables sampled from six different locations across the United States. In their study, wide variation was found among sampling sites, possibly as a result of geographic/climatic effects, varietal differences, and effects of local agricultural practices. Agricultural practices including type and amount of fertilizer and irrigation regimen can influence vitamin content of plant-derived foods.

The vitamin content of animal products is governed both by biological control mechanisms and the diet of the animal. In the case of many B vitamins, the concentration of the vitamin in tissues is limited by the capacity of the tissues to take up the vitamin from the blood and to convert it to the conezymic form(s). A nutritionally inadequate animal diet can yield reduced tissue concentrations of both water-soluble and fat-soluble vitamin(s) involved. In contrast to the situation with water-soluble vitamins, dietary supplementation with fat-soluble vitamins can more readily increase tissue concentrations. This has been examined as a means of increasing the vitamin E concentration of certain animal products to improve oxidative stability and color retention.

8.6.2 Postharvest Changes in Vitamin Content of Foods

Fruits, vegetables, and animal tissues retain enzymes that contribute to postharvest changes in the vitamin content of foods. The release of oxidative and hydrolytic enzymes, as a result of deterioration of cellular integrity and enzymatic compartmentation, can cause changes in the distribution of chemical forms and activity of vitamins. For example, dephosphorylation of vitamin B6, thiamin, or flavin coenzymes, deglycosylation of vitamin B6 glucosides, and the deconjugation of polyglutamyl folates can cause differences between postharvest distributions and those occurring naturally in the plant or animal prior to harvest or slaughter. The extent of such changes will depend on physical damage encountered during handling, possible temperature abuse, and length of time between harvest and processing. Such changes will have little influence on the net concentration of a vitamin but may influence its bioavailability. In contrast,

TABLE 5 Influence of Degree of Maturity on Ascorbic Acid Content of Tomatoes

Weeks from anthesis	Mean weight (g)	Color	Ascorbic acid (mg/100 g)
2	33.4	Green	10.7
3	57.2	Green	7.6
4	102	Green-yellow	10.9
5	146	Yellow-red	20.7
6	160	Red	14.6
7	168	Red	10.1

Source: Ref. 89.

oxidative changes caused by the action of lipoxygenases can reduce the concentration of many vitamins, while ascorbic acid oxidase can specifically reduce the concentration of ascorbic acid.

Changes in vitamin concentration appear to be small but significant when proper procedures are followed during postharvest handling fruit and vegetables. The mishandling of plant products through prolonged holding or shipment at ambient temperatures can contribute to major losses of labile vitamins. Continued metabolism of plant tissues postharvest can be responsible for changes in total concentration as well as distribution of chemical forms of certain vitamins depending on the storage conditions. Postharvest losses of vitamins in meat products are usually minimal under typical conditions of refrigerated storage.

8.6.3 Preliminary Treatments: Trimming, Washing, Milling

The peeling and trimming of fruits and vegetables can cause losses of vitamins to the extent that they are concentrated in the discarded stem, skin, or peel fractions. Although this can be a source of significant loss relative to the intact fruit or vegetable, in most cases this must be considered to be an inevitable loss regardless of whether it occurs in industrial processing or home preparation.

Alkaline treatments to enhance peeling can cause increased losses of labile vitamins such as folate, ascorbic acid, and thiamin at the surface of the product. However, losses of this kind tend to be small compared to the total vitamin content of the product.

Any exposure of cut or otherwise damaged tissues of plant or animal products to water or aqueous solutions causes the loss of water-soluble vitamins by extraction (leaching). This can occur during washing, transportation via flumes, and exposures to brines during cooking. The extent of such losses depends on factors that influence the diffusion and solubility of the vitamin, including pH (can affect solubility and dissociation of vitamins from binding sites within the tissue), ionic strength of the extractant, temperature, the volume ratio of food to aqueous solution, and the surface-to-volume ratio of the food particles. Extractant properties that affect destruction of the vitamin once extracted include concentration of dissolved oxygen, ionic strength, concentration and type of catalytic trace metals, and the presence of solutes that are destructive (e.g., chlorine) or protective (e.g., certain reducing agents).

The milling of cereal grains involves grinding and fractionation to remove the bran (seed coat) and embryo. Because many vitamins are concentrated in the embryo and bran, major losses of vitamins can occur during their removal (Fig. 1). Such losses, as well as the prevalence of vitamin deficiency diseases, provided the rationale for the legalization of enriched cereal grain products containing several added nutrients (riboflavin, niacin, thiamin, iron and calcium). The beneficial impact of this enrichment program on public health has been enormous.

8.6.4 Effects of Blanching and Thermal Processing

Blanching, a mild heat treatment, is an essential step in the processing of fruits and vegetables. The primary purposes are to inactivate potentially deleterious enzymes, reduce microbial loads, and decrease interstitial gasses prior to further processing. Inactivation of enzymes often has a beneficial effect on the stability of many vitamins during subsequent food storage.

Blanching can be accomplished in hot water, flowing steam, hot air, or with microwaves. Losses of vitamins occur primarily by oxidation and aqueous extraction (leaching), with heat being a factor of secondary importance. Blanching in hot water can cause large losses of water-soluble vitamins by leaching. An example is shown in Figure 2.

It has been well documented that high-temperature, short-time treatments improve retention of labile nutrients during blanching and other thermal processes. Specific effects of blanching have been thoroughly reviewed by Selman [118].

Change in the vitamin content of foods during processing has been the topic of extensive

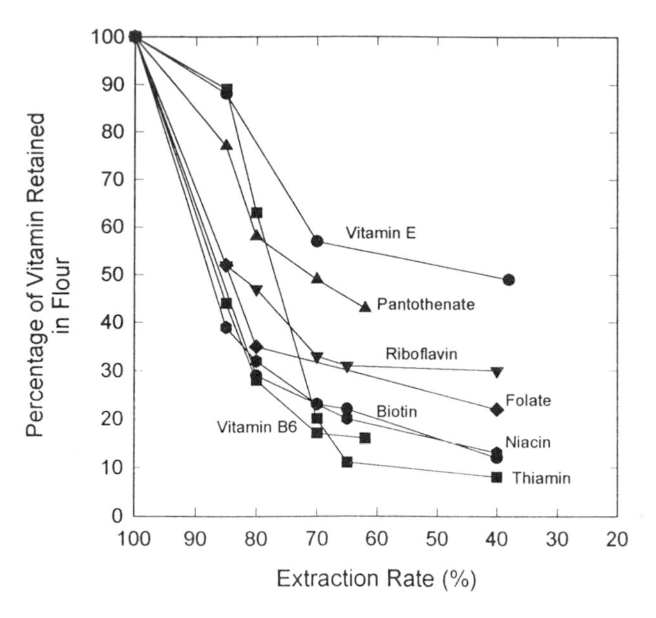

FIGURE 1 Retention of selected nutrients as a function of degree of refining in production of wheat flour. Extraction rate refers to the percentage recovery of flour from whole grain during milling (redrawn by Ref. 98).

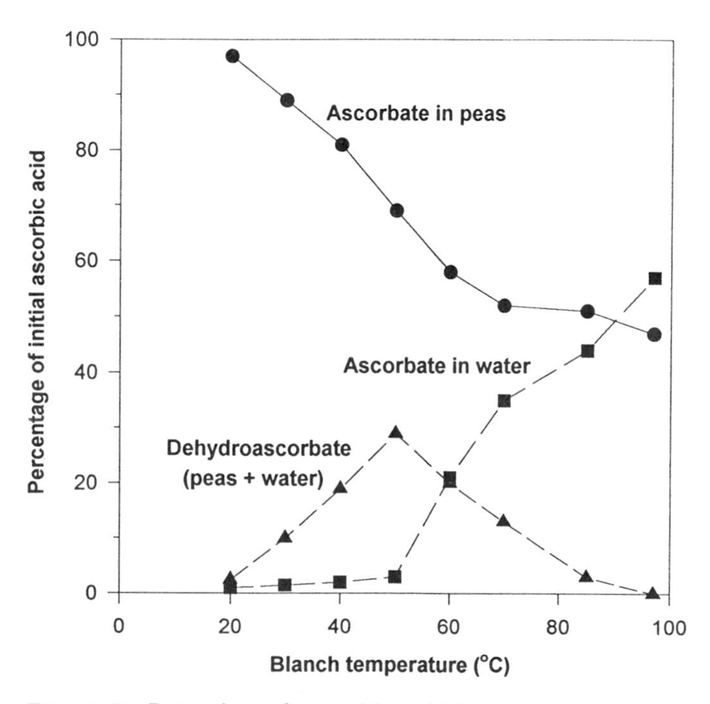

FIGURE 2 Retention of ascorbic acid in peas during experimental water blanching for 10 min at various temperatures (redrawn from Ref. 118.)

study and thorough review [61, 62, 72, 113]. The elevated temperature of thermal processing accelerates reactions that would otherwise occur more slowly at ambient temperature. Thermally induced losses of vitamins depend on the chemical nature of the food, its chemical environment (pH, relative humidity, transition metals, other reactive compounds, concentration of dissolved oxygen, etc.), the stabilities of the individual forms of vitamins present, and the opportunity for leaching. The nutritional significance of such losses depends on the degree of loss and the importance of the food as a source of the vitamin in typical diets. Although subject to considerable variation, representative data for losses of vitamins during the canning of vegetables are shown in Table 6.

8.6.5 Losses of Vitamins Following Processing

Compared to loss of vitamins during thermal processing, subsequent storage often has a small but significant effect on vitamin content. Several factors contribute to small postprocessing losses: (a) reaction rates are relatively slow at ambient or reduced temperature, (b) dissolved oxygen may be depleted, and (c) pH may change during processing (pH usually declines) because of thermal effects or concentrative effects (drying or freezing), and this can have a favorable effect on the stability of vitamins such as thiamin and ascorbic acid. For example, Figure 3 illustrates how vitamin C retention in potatoes can be affected by thermal processing. The relative importance of leaching, chemical degradation, and the type of container (cans or pouches) is apparent from these data.

In reduced-moisture foods, vitamin stability is strongly influenced by water activity in addition to the other factors to be discussed. In the absence of oxidizing lipids, water-soluble vitamins generally exhibit little degradation at water activity less than or equal to monolayer hydration (~0.2–0.3 a_W). Degradation rates increase in proportion to water activity in regions of multilayer hydration, which reflects greater solubility of the vitamin, potential reactants and catalysts. In contrast, the influence of water activity on the stability of fat-soluble vitamins and carotenoids parallels the pattern for unsaturated fats, that is, a minimum rate at monolayer hydration and increased rates above or below this value (see Chap. 2). Substantial losses of oxidation-sensitive vitamins can occur if foods are overdried.

TABLE 6 Typical Losses of Vitamins during canning[a]

Product	Biotin	Folate	B6	Pantothenic acid	A	Thiamin	Riboflavin	Niacin	C
Asparagus	0	75	64	—	43	67	55	47	54
Lima beans	—	62	47	72	55	83	67	64	76
Green beans	—	57	50	60	52	62	64	40	79
Beets	—	80	9	33	50	67	60	75	70
Carrots	40	59	80	54	9	67	60	33	75
Corn	63	72	0	59	32	80	58	47	58
Mushrooms	54	84	—	54	—	80	46	52	33
Green peas	78	59	69	80	30	74	64	69	67
Spinach	67	35	75	78	32	80	50	50	72
Tomatoes	55	54	—	30	0	17	25	0	26

[a]Includes blanching.
Source: From various sources, compiled by Lund [87].

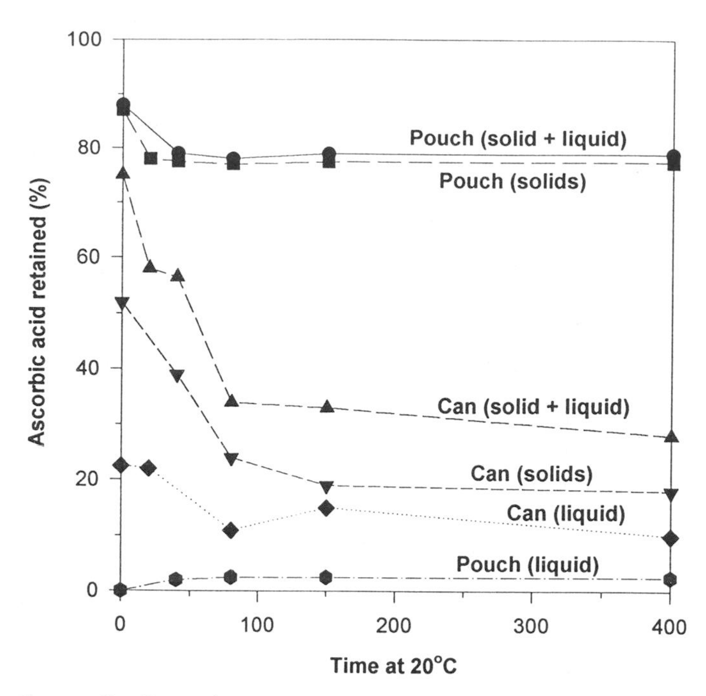

FIGURE 3 Retention and distribution of ascorbic acid in potatoes thermally processed in cans or flexible pouches. Values show the content of ascorbic acid, relative to that present prior processing, in the potatoes and liquid of the containers. Lethality values (F_o) were not provided. (Redrawn from Ref. 113. Original data from Ref. 32a.)

8.6.6 Influence of Processing Chemicals and Other Food Components

The chemical composition of food can strongly influence the stability of vitamins. Oxidizing agents directly degrade ascorbic acid, vitamin A, carotenoids, and vitamin E and may indirectly affect other vitamins. The extent of their impact is dictated by concentration of the oxidant and its oxidation potential. In contrast, reducing agents such as ascorbic and isoascorbic acids and various thiols add to the stability of oxidizable vitamins, like tetrahydrofolates, by their reducing action and as scavengers of oxygen and free radicals. The following is a brief discussion of the influence of several other processing chemicals on vitamins. See later sections for vitamin-specific details.

Chlorine is applied to foods as hypochlorous acid (HOCl), hypochlorite anion (OCl^-), molecular chlorine (Cl_2), or chlorine dioxide (ClO_2). These compounds can interact with vitamins by electrophilic substitution, by oxidation or by chlorination of double bonds. The extent of vitamin loss caused by treatments of food with chlorinated water has not been thoroughly studied; however, one would predict relatively minor effects if the application is confined to the product surface. Chlorination of cake flour presumably has little influence on vitamins in other ingredients used in baking because residual chlorine would be negligible. Reaction products of various forms of chlorine with vitamins are, for the most part, unknown.

Sulfite and other sulfiting agents (SO_2, bisulfite, metabisulfite), as used in wines for antimicrobial effects and in dried food to inhibit enzymatic browning, have a protective effect on ascorbic acid and a deleterious effect on several other vitamins. Sulfite ions directly react with thiamin, causing its inactivation. Sulfite also reacts with carbonyl groups and is known to convert vitamin B6 aldehydes (pyridoxal and pyridoxal phosphate) to their presumably inactive sulfonated derivatives. The extent to which sulfiting agents affect other vitamins has not been extensively studied.

Nitrite is used in the preservation and curing of meats and may develop through microbial reduction of naturally occurring nitrate. Ascorbic acid or isoascorbic acid is added to nitrite-containing meats to prevent formation of *N*-nitrosamines. This is accomplished by forming NO and preventing formation of undesirable nitrous anhydride (N_2O_3, the primary nitrosating agent). The proposed reactions [84] are:

$$\text{Ascorbic acid} + HNO_2 \rightarrow \text{2-nitrite ester of ascorbic acid} \rightarrow \text{semidehydroascorbate radical} + NO$$

Formation of NO is desirable because it is the desired ligand for binding to myoglobin to form the cured meat color. The residual semidehydroascorbate radical retains partial vitamin C acitivity.

Chemical sterilants are used in highly specific applications such as treating spices with ethylene and propylene oxides for deinfestation. The biocidal function of these compounds occurs by alkylation of proteins and nucleic acids. Similar effects have been observed to occur with some vitamins, although loss of vitamin activity by this means is insignificant in the overall food supply.

Chemicals and food ingredients that influence pH will directly affect the stability of vitamins such as thiamin and ascorbic acid, particularly in the neutral to mildly acidic pH range. Acidulation increases the stability of ascorbic acid and thiamin. In contrast, alkalizing compounds reduce the stability of ascorbic acid, thiamin, pantothenic acid, and certain folates.

8.7 FAT-SOLUBLE VITAMINS

8.7.1 Vitamin A

8.7.1.1 Structure and General Properties

Vitamin A refers to a group of nutritionally active unsaturated hydrocarbons, including retinol and related compounds (Fig. 4) and certain carotenoids (Fig. 5). Vitamin A activity in animal tissues is predominantly in the form of retinol or its esters, retinal, and to a lesser extent, retinoic acid. The concentration of vitamin A is greatest in liver, the major body pool, in which retinol and retinol esters are the primary forms present. The term retinoids refers to the class of compounds including retinol and its chemical derivatives having four isoprenoid units. Several retinoids that are analogues of the nutritionally active forms of vitamin A exhibit useful pharmocological properties. In addition, synthetic retinyl acetate and retinyl palmitate are used widely in synthetic form for food fortification.

Carotenoids may contribute significant vitamin A activity to foods of both plant and animal origin. Of approximately 600 known carotenoids, ~50 exhibit some provitamin A activity (i.e., are partially converted to vitamin A in vivo). Preformed vitamin A does not exist in plants and fungi; their vitamin A activity is associated with certain carotenoids. The structure of selected carotenoids, along with their vitamin A activities determined by rat bioassay, are presented in Figure 5. The reader is referred to Chapter 10 for further discussion of the properties of the carotenoids in the context of their role as food pigments.

For a compound to have vitamin A or provitamin A activity, it must exhibit certain structural similarities to retinol, including (a) at least one intact nonoxygenated β-ionone ring, and (b) an isoprenoid side chain terminating in an alcohol, aldehyde, or carboxyl function (Fig. 4). The vitamin A-active carotenoids such as β-carotene (Fig. 5) are considered to have pro-vitamin A activity until they undergo oxidative enzymatic cleavage of the central $C^{15}–C^{15'}$

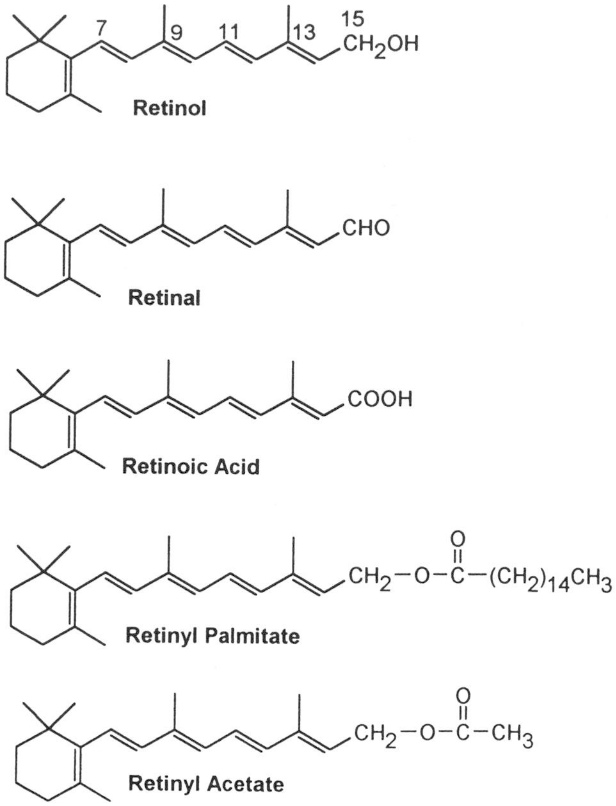

Retinol

Retinal

Retinoic Acid

Retinyl Palmitate

Retinyl Acetate

FIGURE 4 Structures of common retinoids.

bond in the intestinal mucosa to release two molecules of active retinal. Among the carotenoids, β-carotene exhibits the greatest pro-vitamin A activity. Carotenoids with ring hydroxylation or the presence of carbonyl group exhibit less pro-vitamin A activity than β-carotene if only one ring is affected, and have no activity if both rings are oxygenated. Although two molecules of vitamin A are potentially produced from each molecule of dietary β-carotene, the inefficiency of the process accounts for the fact that β-carotene exhibits only ~50% of the vitamin A activity exhibited by retinol, on a mass basis. Considerable variation exists among various animal species and humans with respect to the efficiency of utilization of carotenoids and the extent of absorption of carotenoid molecules in intact form. The in vivo antioxidative function attributed to dietary carotenoids requires absorption of the intact molecule [18].

The retinoids and pro-vitamin A carotenoids are very lipophilic compounds because of their nonpolar structures. Consequently, they associate with lipid components, specific organelles, or carrier proteins in foods and living cells. In many food systems, retinoids and carotenoids are found associated with lipid droplets or micelles dispersed in an aqueous environment. For example, both retinoids and carotenoids are present in the fat globules of milk, while in orange juice the carotenoids associate with dispersed oils. The conjugated double bond systems of retinoids gives strong and characteristic ultraviolet absorption spectra, while the additional conjugated double bond system of carotenoids causes absorption in the visible spectrum and the yellow-orange color of these compounds. All-*trans* isomers exhibit the greatest vitamin A activity and are the predominant naturally occurring forms of retinoids and carotenoids

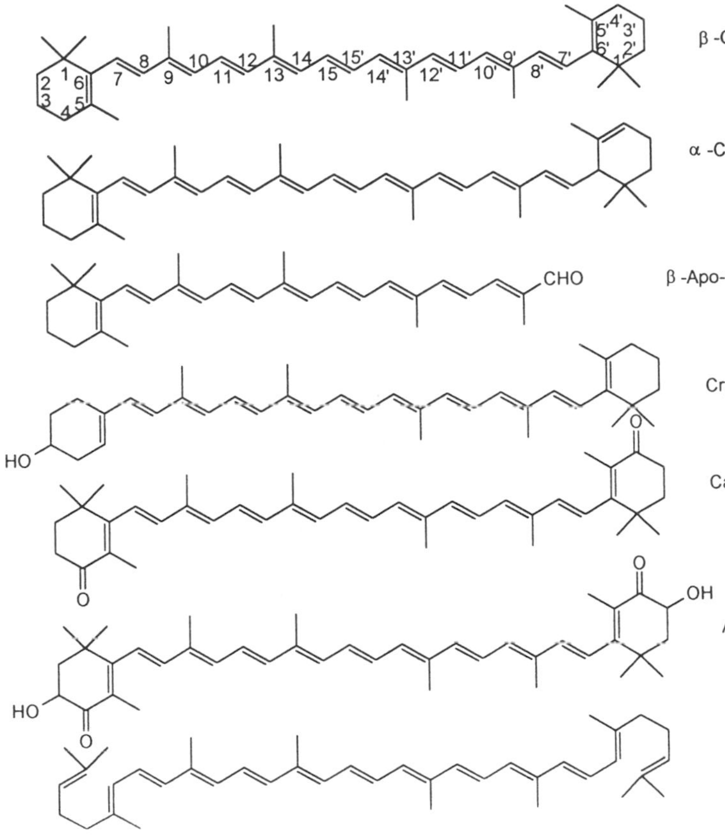

Compound	Relative Activity
β-Carotene	50
α-Carotene	25
β-Apo-8'-Carotenal	25-30
Cryptoxanthin	0
Canthaxanthin	0
Astacene	0
Lycopene	0

FIGURE 5 Structures and provitamin A activities of selected carotenoids.

in foods (Table 7 and 8). Conversion to *cis* isomers, which can occur during thermal processing, causes a loss of vitamin A activity.

8.7.1.2 Stability and Modes of Degradation

The degradation of vitamin A (retinoids and vitamin A-active cartenoids) generally parallels the oxidative degradation of unsaturated lipids. Factors that promote oxidation of unsaturated lipids enhance degradation of vitamin A, either by direct oxidation or by indirect effects of free radicals. Changes in the β-carotene content of cooked dehydrated carrots illustrate typical extents of degradation during processing and typical exposure to oxygen during associated handling (Table 9). It should be noted, however, that extended storage of vitamin A in foods such as fortified breakfast cereal products, infant formulas, fluid milk, fortified sucrose, and condiments is usually not highly detrimental to the retention of added vitamin A.

Losses of vitamin A activity of retinoids and carotenoids in foods occur mainly through reactions involving the unsaturated isoprenoid side chain, by either autoxidation or geometric isomerization. Retinoid and carotenoid molecules remain chemically intact during thermal processing, although they do undergo some isomerization. High-performance liquid chromatography (HPLC) analysis has revealed that many foods contain a mixture of all-*trans* and *cis* isomers of retinoids and carotenoids. As summarized in Table 10, conventional cannning of fruits and vegetables is sufficient to induce isomerization and ensuing losses of vitamin A activity. In addition

TABLE 7 Relative Vitamin A Activity of Stereoisomeric Forms of Retinol Derivatives

	Relative vitamin A activity[a]	
Isomer	Retinyl acetate	Retinal
All-*trans*	100	91
13-*cis*	75	93
11-*cis*	23	47
9-*cis*	24	19
9,13-Di-*cis*	24	17
11,13-Di-*cis*	15	31

[a]Molar vitamin A activity relative to all-*trans*-retinyl acetate in rat bioassays.
Source: Ref. 2.

to thermal isomerization, the conversion of all-*trans* forms of retinoids and carotenoids to various *cis* isomers can be induced by exposure to light, acid, chlorinated solvents (e.g., chloroform), or dilute iodine. Chlorinated solvents often used in lipid analysis enhance the photochemical isomerization of retinyl palmitate and, presumably, other retinoids and carotenoids.

The occurrence of *cis* isomers of carotenoids has been known for many years (Fig. 6). Previous nomenclature for β-carotene isomers was derived from chromatographic separations and included neo-β-carotene U (9-*cis*-β-carotene) and neo-β-carotene B (13-*cis*-β-carotene). Confusion exists in the literature because neo-β-carotene B was originally identified incorrectly as 9,13′-di-*cis*-β-carotene [133]. Analogous isomerization occurs with other carotenoids. The maximum extent of thermal isomerization generally observed in canned fruits and vegetables is about 40% 13-*cis*-β-carotene and 30% 9-*cis*-β-carotene (Table 10). The values observed for *cis* isomers of β-carotene in processed foods are similar to the equilibrium values observed in the iodine catalyzed isomerization of β-carotene, which suggests that the extent and specificity of isomerization is similar regardless of the mechanism.

Photochemical isomerization of vitamin A compounds occurs both directly and indirectly

TABLE 8 Relative Vitamin A Activity of Stereoisomeric Forms of Carotenes

Compound and isomer	Relative vitamin A activity[a]
B-Carotene	
All-*trans*	100
9-*cis* (neo-U)	38
13-*cis* (neo-B)	53
α-Carotene	
All-*trans*	53
9-*cis* (neo-U)	13
13-*cis* (neo-B)	16

[a]Activity relative to that of all-*trans*-β-carotene in rat bioassays.
Source: Ref. 145.

TABLE 9 Concentration of β-Carotene in Cooked Dehydrated Carrots

Sample	β-Carotene concentration (μg/g solids)
Fresh	980–1860
Explosive puff-dried	805–1060
Vacuum freeze-dried	870–1125
Conventional air-dried	636–987

Source: Ref. 30.

via a photosensitizer. The ratios and quantities of *cis* isomers produced differ with the means of photoisomerization. Photoisomerization of all-*trans*-β-carotene involves a series of reversible reactions, and each isomerization is accompanied by photochemical degradation (Fig. 7). Similar rates of photoisomerization and photodegradation have been observed in aqueous dispersions of β-carotene and in carrot juice. These photochemical reactions also have been observed when retinoids in foods are exposed to light (e.g., milk). The type of packaging material can have a substantial effect on net retention of vitamin A activity in food exposed to light during storage.

Oxidative degradation of vitamin A and carotenoids in foods can occur by direct peroxida-

TABLE 10 Distribution of β-Carotene Isomers in Selected Fresh and Processed Fruits and Vegetables

		Percentage of total β-carotene		
Product	Status	13-*cis*	All-*trans*	9-*cis*
Sweet potato	Fresh	0.0	100.0	0.0
Sweet potato	Canned	15.7	75.4	8.9
Carrot	Fresh	0.0	100.0	0.0
Carrot	Canned	19.1	72.8	8.1
Squash	Fresh	15.3	75.0	9.7
Squash	Canned	22.0	66.6	11.4
Spinach	Fresh	8.8	80.4	10.8
Spinach	Canned	15.3	58.4	26.3
Collard	Fresh	16.3	71.8	11.7
Collard	Canned	26.6	46.0	27.4
Cucumber	Fresh	10.5	74.9	14.5
Pickle	Pasteurized	7.3	72.9	19.8
Tomato	Fresh	0.0	100.0	0.0
Tomato	Canned	38.8	53.0	8.2
Peach	Fresh	9.4	83.7	6.9
Peach	Canned	6.8	79.9	13.3
Apricot	Dehydrated	9.9	75.9	14.2
Apricot	Canned	17.7	65.1	17.2
Nectarine	Fresh	13.5	76.6	10.0
Plum	Fresh	15.4	76.7	8.0

Source: Ref. 23.

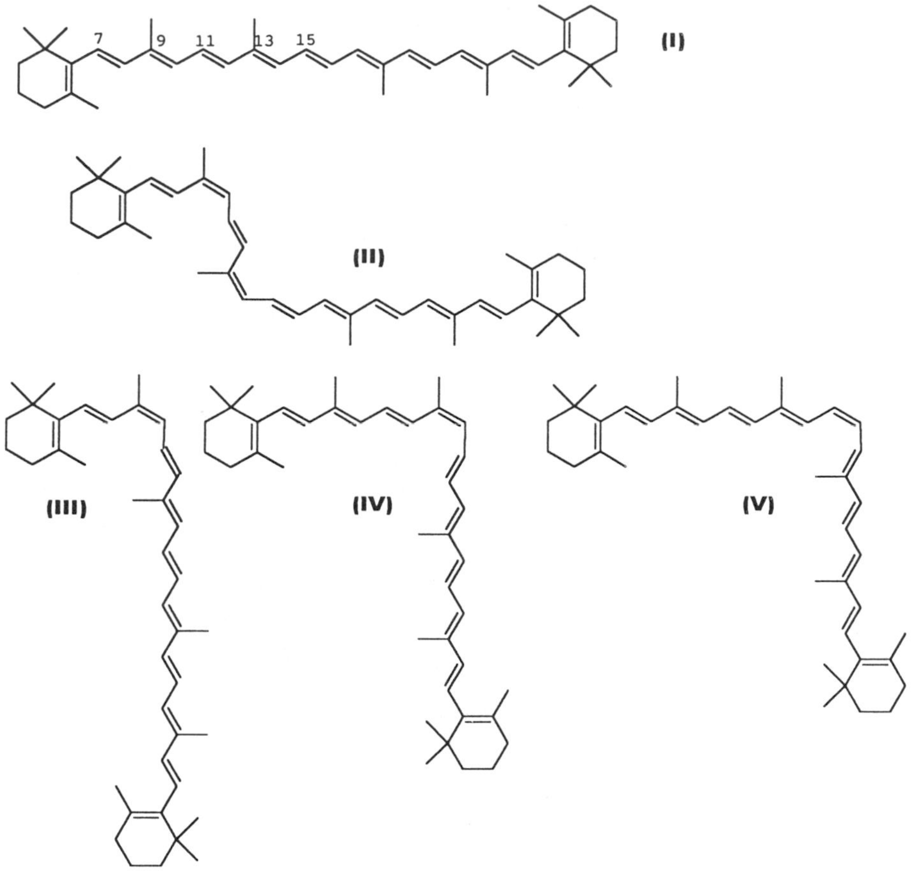

Figure 6 Structures of selected *cis* isomers of β-carotene: (I) all-*trans*; (II) 11, 15-di-*cis*; (III) 9-*cis*; (IV) 13-*cis*; (V) 15-*cis*.

tion or by indirect action of free radicals produced during oxidation of fatty acids. β-Carotene and probably other carotenoids have the ability to act as antioxidants under conditions of reduced oxygen concentration (< 150 torr O_2), although they may act as pro-oxidants at higher oxygen concentrations [19]. β-Carotene can act as an antioxidant by scavenging singlet oxygen, hydroxyl, and superoxide radicals, and by reacting with peroxyl radicals ($ROO^{\cdot}$). Peroxyl radicals attack β-carotene to form an adduct postulated to be ROO-β-carotene, in which the peroxyl radical bonds to the C^7 position of β-carotene, while the unpaired electron is delocalized across the conjugated double bond system. β-Carotene apparently does not act as a chain-breaking radical (donating $H^{\cdot}$) as do phenolic antioxidants. This antioxidant behavior of β-carotene, and

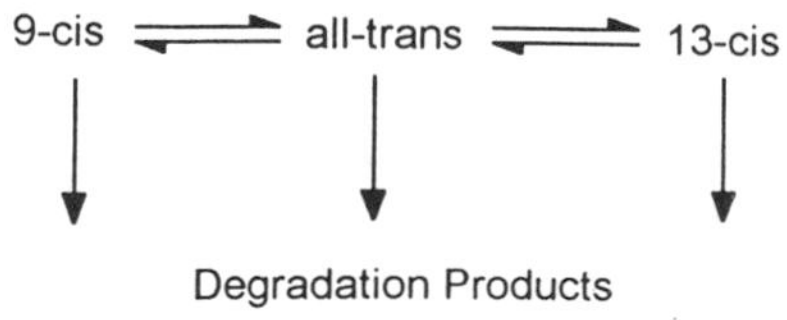

9-cis ⇌ all-trans ⇌ 13-cis

↓ ↓ ↓

Degradation Products

Figure 7 Model of photochemically induced reactions of β-carotene. (From Ref. 104.)

presumably other carotenoids, causes a reduction or total loss of vitamin A activity regardless of the mechanisms by which free radical initiation occurs. For retinol and retinyl esters, however, the attack of free radicals occurs at the C^{14} and C^{15} positions.

Oxidation of β-carotene involves the formation of the 5,6-epoxide, which may isomerize to the 5,8-epoxide (mutachrome). Photochemically induced oxidation yields mutachrome as the primary degradation product. Fragmentation of β-carotene to many lower molecular weight compounds can occur especially during high-temperature treatments. Resulting volatiles can have a significant effect on flavor. Such fragmentation also occurs during oxidation of retinoids. An overview of these reactions and other aspects of the chemical behavior of vitamin A is shown in Figure 8.

8.7.1.3 Bioavailability

Retinoids are absorbed effectively except under conditions in which malabsorption of fat occurs. Retinyl acetate and palmitate are as effectively utilized as nonesterified retinol. Diets containing nonabsorbable hydrophobic materials such as certain fat substitutes may contribute to malab-

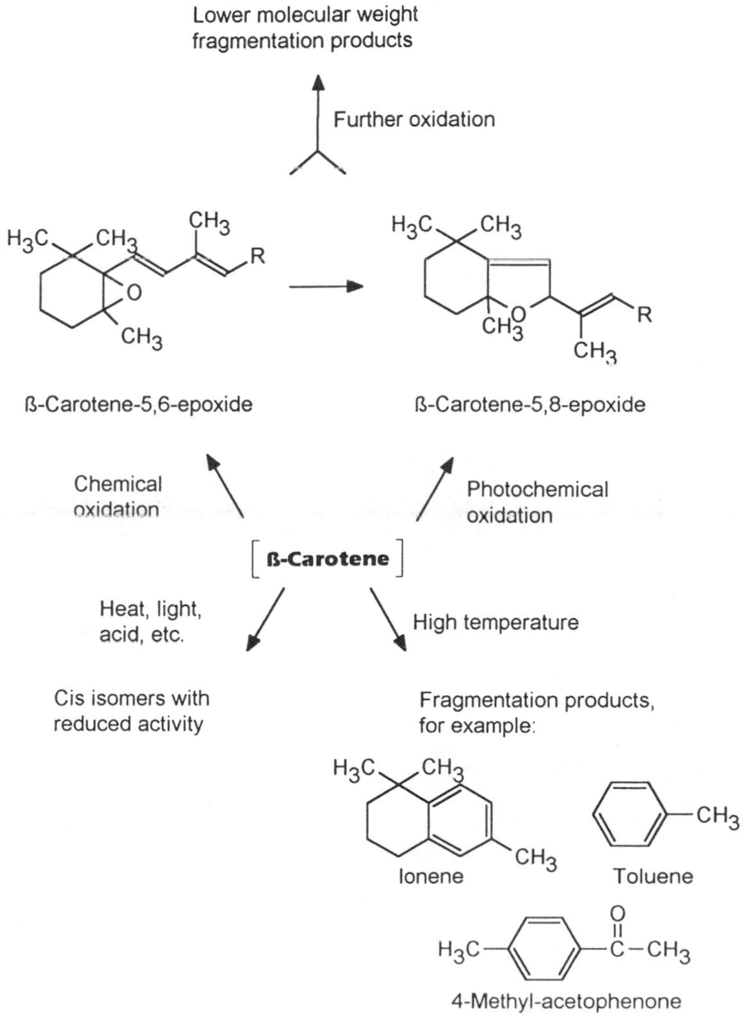

FIGURE 8 Overview of vitamin A degradation.

sorption of vitamin A. The bioavailability of vitamin A added to rice has been demonstrated in human subjects.

Aside from the inherent difference in utilization between retinol and the provitamin A carotenoids, carotenoids in many foods undergo markedly less intestinal absorption. Absorption may be impaired by the specific binding of carotenoids as carotenoproteins or by entrapment in poorly digestible vegetable matrices. In studies with human subjects, β-carotene from carrots yielded only ~21% of the plasma β-carotene response obtained from an equivalent dose of pure β-carotene, while β-carotene in broccoli exhibited a similarly low bioavailability [15].

8.7.1.4 Analytical Methods

Early methods of vitamin A analysis centered on the reactions of retinoids with Lewis acids such as antimony trichloride and trifluoroacetic acid to yield a blue color. In addition, fluorometric methods have been used to measure vitamin A [131]. Interferences often occur when these methods are applied to foods. Furthermore, these methods do not detect *trans–cis* isomerization that may occur during processing or storage of foods. Because the *cis* isomers exhibit less nutritional activity than the all-*trans* compound, it is inaccurate to regard total vitamin A or provitamin A activity simply as the sum of all isomeric forms. HPLC is the method of choice because it enables individual retinoids to be determined with considerable accuracy [131]. Accurate measurement of carotenoids is a very complex task in view of the many naturally occurring chemical forms present in foods [73].

8.7.2 Vitamin D

8.7.2.1 Structure and General Properties

Vitamin D activity in foods is associated with several lipid-soluble sterol analogues including cholecalciferol (vitamin D3) from animal sources and ergocalciferol (vitamin D2) produced synthetically (Fig. 9). Both of these compound are used in synthetic form for food fortification. Cholecalciferol forms in human skin upon exposure to sunlight, and this is a multistep process involving photochemical modification of 7-dehydrocholesterol followed by nonenzymatic isomerization. Because of this in vivo synthesis, the requirement for dietary vitamin D will depend on the extent of exposure to sunlight. Ergocalciferol is an exclusively synthetic form of vitamin D that is formed by commercial irradiation of phytosterol (a plant sterol) with UV light. Several hydroxylated metabolites of vitamin D2 and D3 form in vivo. The 1,25-dihydroxy derivative of cholecalciferol is the main physiologically active form, and it is involved in the regulation of calcium absorption and metabolism. 25-Hydroxycholecalciferol, in addition to cholecalciferol,

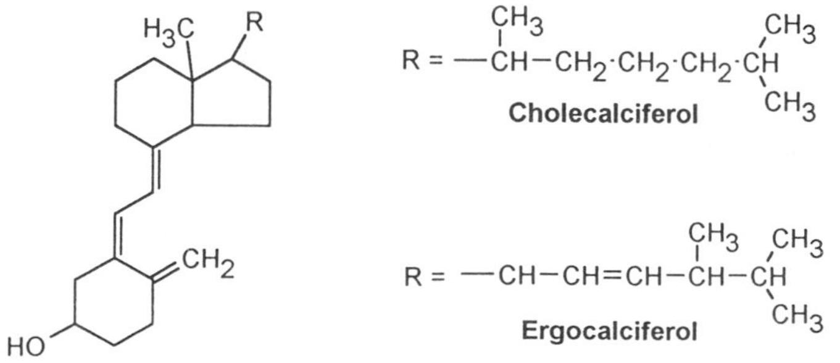

FIGURE 9 Structure of ergocalciferol (vitamin D2) and cholecalciferol (vitamin D3).

comprises a significant amount of the naturally occurring vitamin D activity in meat and milk products.

Fortification of most fluid milk products with either ergocalciferol or cholecalciferol D makes a significant contribution to dietary needs. Vitamin D is susceptible to degradation by light, and this may occur in glass-packaged milk during retail storage. For example, approximately 50% of cholecalciferol added to skim milk is lost during 12 days of continual exposure to fluorescent light at 4°C. It is not known whether this degradation involves direct photochemical degradation or occurs as an indirect effect of light-induced lipid oxidation. Like other unsaturated fat-soluble components of foods, vitamin D compounds are susceptible to oxidative degradation. Overall, however, the stability of vitamin D in foods, especially under anaerobic conditions, is not a major concern.

8.7.2.2 Analytical Methods

Measurement of vitamin D is performed primarily by HPLC methods [135]. Alkaline conditions yield rapid degradation of vitamin D; thus, saponification as used widely in the analysis of lipid-soluble materials cannot be employed. Various preparative chromatographic methods have been developed for the purification of food extracts prior to HPLC analysis.

8.7.3 Vitamin E

8.7.3.1 Structure and General Properties

Vitamin E is the generic term for tocols and tocotrienols that exhibit vitamin activity similar to that of α-tocopherol. Tocols are 2-methyl-2(4′,8′,12′-trimethyltridecyl)chroman-6-ols, while tocotrienols are identical except for the presence of double bonds at positions 3′, 7′, and 11′ of the side chain (Fig. 10). Tocopherols, which are typically the main compounds having vitamin E activity in foods, are derivatives of the parent compound tocol, and have one or more methyl groups at position 5, 7, or 8 of the ring structure (chromanol ring) (Fig. 10). The α, β, γ forms of tocopherol and tocotrienol differ according to the number and position of the methyl groups and thus differ significantly in vitamin E activity (Table 10).

There are three asymmetric carbons (2, 4′, and 8′) in the tocopherol molecule, and the stereochemical configuration at these positions influences the vitamin E activity of the compound. Early nomenclature for vitamin E compounds is confusing with regard to the vitamin activity of the stereoisomers. The naturally occurring configuration of α-tocopherol exhibits the greatest vitamin E activity and is now designated RRR-α-tocopherol; other terminology, such as the term *d*-α-tocopherol, should be discontinued. Synthetic forms of α-tocopheryl acetate are used widely in food fortification. The presence of the acetate ester greatly improves the stability of the compound. Synthetic forms that are racemic mixtures consisting of eight possible combinations of geometric isomers involving positions 2, 4′, and 8′ should be designated all-*rac*-α-tocopheryl acetate rather than the previously used term *dl*-α-tocopheryl-acetate. Vitamin E activity of tocopherols and tocotrienols varies according to the particular form present (α, β, γ, or δ) (Table 11), in addition to stereochemical nature of the tocopherol side chain (Table 12). The lower vitamin E activity of all-*rac*-α-tocopheryl acetate, relative to naturally occurring RRR isomers of the vitamin, should be recognized and compensated for when using these compounds for food fortification. While α-tocopherol is the major form of vitamin E in most animal products, other tocopherols and tocotrienols occur in varying proportions in plant products (Table 13).

The tocopherols and tocotrienols are very nonpolar and exist mainly in the lipid phase of foods. All tocopherols and tocotrienols, when not esterified, have the ability to act as antioxi-

α-Tocopherol

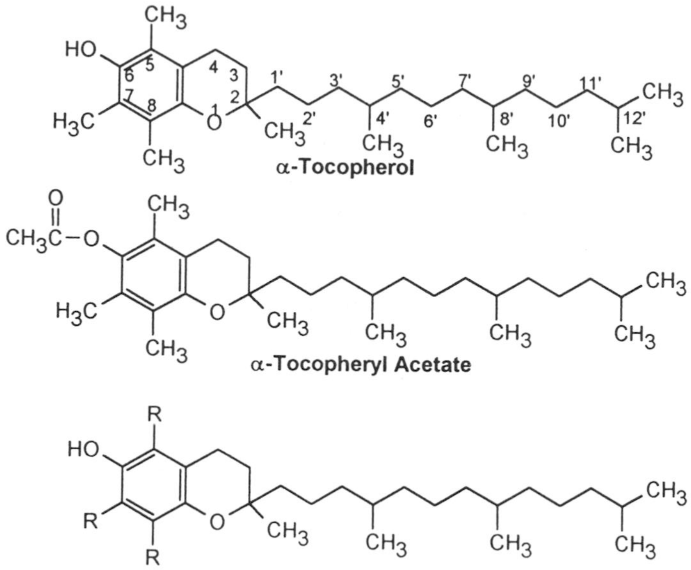

α-Tocopheryl Acetate

	R_1	R_2	R_3
α	CH_3	CH_3	CH_3
β	CH_3	H	CH_3
γ	H	CH_3	CH_3
δ	H	H	CH_3

FIGURE 10 Structures of tocopherols. The structures of tocotrienols are identical to the corresponding tocopherols, except for the presence of double bonds at positions 3′, 7′, and 11′.

dants; they quench free radicals by donating the phenolic H and an electron. Tocopherols are a natural constituent of all biological membranes and are thought to contribute to membrane stability through their antioxidant activity. Naturally occurring tocopherols and tocotrienols also contribute to the stability of highly unsaturated vegetable oils through this antioxidant action. In contrast, α-tocopheryl acetate added in food fortification has no antioxidant activity because the acetate ester has replaced the phenolic hydrogen atom. α-Tocopheryl acetate does exhibit vitamin E activity and in vivo antioxidant effects as a result of enzymatic cleavage of the ester. The concentration of dietary vitamin E in animals has been shown to influence the oxidative stability of meats after slaughter. For example, it has been shown that the susceptibility of pork muscle products to oxidation of cholesterol and other lipids is inversely related to α-tocopheryl acetate intake of the pigs.

8.7.3.2 Stability and Mechanism of Degradation

Vitamin E compounds exhibit reasonably good stability in the absence of oxygen and oxidizing lipids. Anaerobic treatments in food processing, such as retorting of canned foods, have little effect on vitamin E activity. In contrast, the rate of vitamin E degradation increases in the presence of molecular oxygen and can be especially rapid when free radicals are also present. Oxidative degradation of vitamin E is strongly influenced by the same factors that influence oxidation of unsaturated lipids. The a_w dependence of α-tocopherol degradation is similar to that

TABLE 11 Relative Vitamin E Activity of Tocopherols and Tocotrienols

	Bioassay method			
Compound	Rat fetal resorption	Rat erythrocyte hemolysis	Muscular dystrophy (chicken)	Muscular dystrophy (rat)
α-Tocopherol	100	100	100	100
β-Tocopherol	25–40	15–27	12	
γ-Tocopherol	1–11	3–20	5	11
δ-Tocopherol	1	0.3–2		
α-Tocotrienol	27–29	17–25		28
β-Tocotrienol	5	1–5		

Source: Adapted from Ref. 88.

of unsaturated lipids, with a rate minimum occurring at the monolayer moisture value and greater rates at either higher or lower a_w (see Chap. 2). The use of intentional oxidative treatments, such as the bleaching of flour, can lead to large losses of vitamin E.

An interesting nonnutritional use of α-tocopherol in foods is in the curing of bacon to reduce formation of nitrosamines. It is thought that α-tocopherol serves as a lipid-soluble phenolic compound to quench nitrogen free radicals ($NO^{\cdot}$, $NO_2^{\cdot}$) in a radical-mediated nitrosation process.

Reactions of vitamin E compounds, especially α-tocopherol, in foods have been studied extensively. As summarized in Figure 11, α-tocopherol can react with a peroxyl radical (or other free radicals) to form a hydroperoxide and an α-tocopheryl radical. As with other phenolic radicals, this is relatively unreactive because the unpaired electron resonates across the phenolic ring system. Radical termination reactions can occur to form covalently linked tocopheryl dimers and trimers, while additional oxidation and rearrangement can yield tocopheroxide, tocopheryl hydroquinone, and tocopheryl quinone (Fig. 11). Rearrangement and further oxidation can yield many other products. Although α-tocopheryl acetate and other vitamin E esters do not participate

TABLE 12 Vitamin E Activity of Isomeric Forms of α-Tocopheryl Acetate

Form of α-tocopheryl acetate[a]	Relative vitamin E activity (%)
RRR	100
All-*rac*	77
RRS	90
RSS	73
SSS	60
RSR	57
SRS	37
SRR	31
SSR	21

[a]R and S refer to the chiral configuration of the 2, 4′, and 8′ positions, respectively. R is the naturally occurring chiral form. All-*rac* signifies fully racemic.
Source: Adapted from Ref. 139.

TABLE 13 Concentration of Tocopherols and Tocotrienols in Selected Vegetable Oils and Foods

Food	α-T	α-T3	β-T	β-T3	γ-T	γ-T3	δ-T	δ-T3
Vegetable oils (mg/100 g)								
Sunflower	56.4	.013	2.45	0.207	0.43	0.023	0.087	
Peanut	14.1	0.007	0.396	0.394	13.1	0.03	0.922	
Soybean	17.9	0.021	2.80	0.437	60.4	0.078	37.1	
Cottonseed	40.3	0.002	0.196	0.87	38.3	0.089	0.457	
Corn	27.2	5.37	0.214	1.1	56.6	6.17	2.52	
Olive	9.0	0.008	0.16	0.417	0.471	0.026	0.043	
Palm	9.1	5.19	0.153	0.4	0.84	13.2	0.002	
Other foods (μg/ml or g)								
Infant formula (saponified)	12.4		0.24		14.6		7.41	
Spinach	26.05	9.14						
Beef	2.24							
Wheat flour	8.2	1.7	4.0	16.4				
Barley	0.02	7.0		6.9		2.8		

Note: T, tocopherol; T3, tocotrienol.
Source: Adapted from Refs. 132 and 135.

in radical quenching, they are subject to oxidative degradation but at a slower rate than that of nonesterified compounds. The degradation products of vitamin E exhibit little or no vitamin activity. Through their ability to act as phenolic antioxidants, nonesterified vitamin E compounds contribute to the oxidative stability of food lipids.

Vitamin E compounds also can contribute indirectly to oxidative stability of other compounds by scavenging singlet oxygen while being concurrently degraded. As shown in Figure 12, singlet oxygen directly attacks the tocopherol molecule ring system to form a transient hydroperoxydieneone derivative. This can rearrange to form both the tocopheryl quinone and the tocopheryl quinone 2,3-oxide, which have little vitamin E activity. The order of reactivity toward singlet oxygen is $\alpha > \beta > \gamma > \delta$, and antioxidative potency is in the reverse order. Tocopherols also can physically quench singlet oxygen, which involves deactivation of the singlet-state oxygen without oxidation of the tocopherol. These attributes of tocopherols are consistent with the fact that the tocopherols are potent inhibitors of photosensitized, singlet-oxygen-mediated oxidation of soybean oil.

8.7.3.3 Bioavailability

The bioavailability of vitamin E compounds is usually quite high in individuals who digest and absorb fat normally. On a molar basis, bioavailability of α-tocopheryl acetate is nearly equivalent to that of α-tocopherol [19] except at high doses, where the enzymatic deesterification of α-tocopheryl acetate can be limiting. Previous studies indicating that α-tocopheryl acetate was more potent than α-tocopherol on a molar basis may have been biased by the susceptibility of α-tocopherol to undergo oxidation prior to testing.

8.7.3.4 Analytical Methods

HPLC methods for determination of vitamin E have largely superseded previous spectrophotometric and direct fluorometric procedures. The use of HPLC permits the measurement of

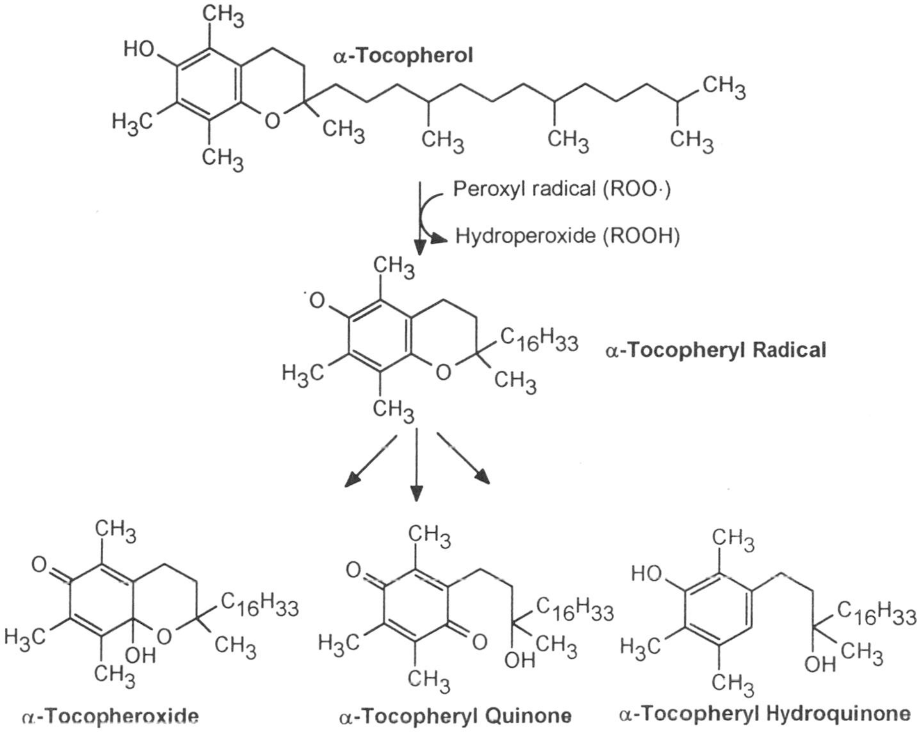

FIGURE 11 Overview of the oxidative degradation of vitamin E. In addition to the initial oxidation products shown, many other compounds are formed as a result of further oxidation and rearrangement.

specific forms of vitamin E (e.g., α-, β-, γ-, δ-tocopherols and tocotrienols) and, thus, estimation of total vitamin E activity in a product based on relative potencies of the specific compounds [26]. Detection can be accomplished using either ultraviolet (UV) absorbance or fluorescence. When saponification is used to aid in the separation of lipids from vitamin E, any vitamin E ester will be hydrolyzed to free α-tocopherol. Care must be taken to prevent oxidation during extraction, saponification, and other preliminary treatments.

8.7.4 Vitamin K

8.7.4.1 Structure and General Properties

Vitamin K consists of a group of naphthoquinones that exist with or without a terpenoid side chain in the 3-position (Fig. 13). The unsubstituted form of vitamin K is menadione, and it is of primary significance as a synthetic form of the vitamin that is used in vitamin supplements and food fortification. Phylloquinone (vitamin K_1) is a product of plant origin, while menaquinones (vitmain K_2) of varying chain length are products of bacterial synthesis, mainly by intestinal microflora. Phylloquinones occur in relatively large quantities in leafy vegetables including spinach, kale, cauliflower, and cabbage, and they are present, but less abundant, in tomatoes and certain vegetable oils. Vitamin K deficiency is rare in healthy individuals because of the widespread presence of phylloquinones in the diet and because microbial menoquinones are absorbed from the lower intestine. Vitamin K deficiency is ordinarily associated with malabsorp-

α-Tocopherol

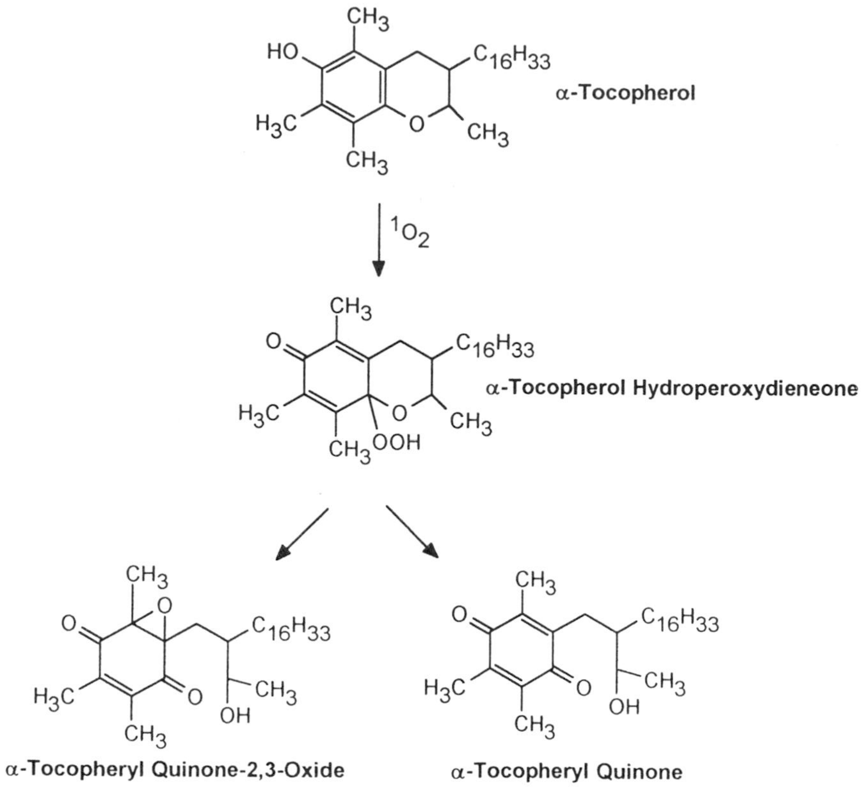

$^1O_2$

α-Tocopherol Hydroperoxydieneone

α-Tocopheryl Quinone-2,3-Oxide

α-Tocopheryl Quinone

FIGURE 12 Reaction of singlet oxygen and α-tocopherol.

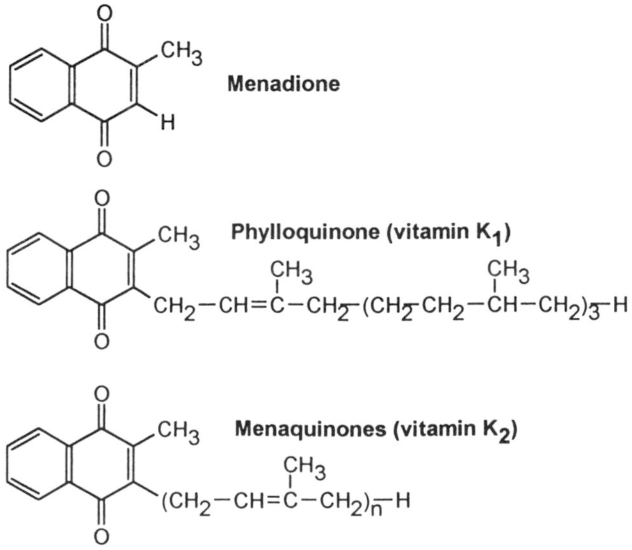

Menadione

Phylloquinone (vitamin K_1)

Menaquinones (vitamin K_2)

FIGURE 13 Structure of various forms of vitamin K.

tion syndromes or the use of pharmacological anticoagulants. Although the use of certain fat substitutes has been reported to impair vitamin K absorption, moderate intakes of these substitutes have no significant effect on vitamin K utilization.

The quinone structure of vitamin K compounds can be reduced to the hydroquinone form by certain reducing agents, but vitamin K activity is retained. Photochemical degradation can occur, but the vitamin is quite stable to heat.

8.7.4.2 Analytical Methods

Spectrophotometric and chemical assays based on measurement of oxidation–reduction properties of vitamin K lack the specificity required for food analysis. Various HPLC methods exist that provide satisfactory specificity and permit individual forms of vitamin K to be measured [135].

8.8 WATER-SOLUBLE VITAMINS

8.8.1 Ascorbic Acid

8.8.1.1 Structure and General Properties

L-Ascorbic acid (AA) (Fig. 14) is a carbohydrate-like compound whose acidic and reducing properties are contributed by the 2,3-enediol moiety. This compound is highly polar; thus, it is readily soluble in aqueous solution and insoluble in less nonpolar solvents. AA is acidic in character as a result of ionization of the C-3 hydroxyl group ($pK_{a1} = 4.04$ at 25°C). A second ionization, dissociation of the C-2 hydroxyl, is much less favorable ($pK_{a2} = 11.4$). Two-electron oxidation and hydrogen dissociation convert L-ascorbic acid to L-dehydroascorbic acid (DHAA).

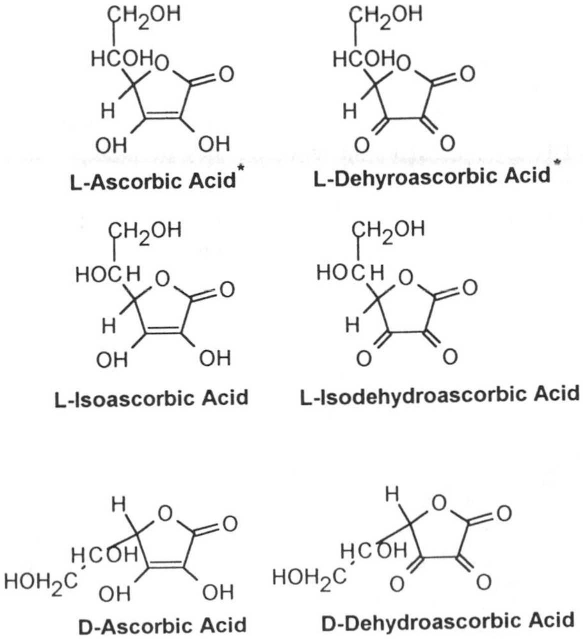

L-Ascorbic Acid* **L-Dehyroascorbic Acid***

L-Isoascorbic Acid **L-Isodehydroascorbic Acid**

D-Ascorbic Acid **D-Dehydroascorbic Acid**

FIGURE 14 Structures of L-ascorbic acid and L-dehydroascorbic acid and their isomeric forms. (Asterisk indicates vitamin C activity.)

DHAA exhibits approximately the same vitamin activity as AA because it is almost completely reduced to AA in the body.

L-Isoascorbic acid, the C-5 optical isomer, and D-ascorbic acid, the C-4 optical isomer (Fig. 14), behave in a chemically similar manner to AA, but these compounds have essentially no vitamin C activity. L-Isoascorbic acid and AA are widely used as food ingredients for their reducing and antioxidative activity (e.g., in the curing of meats and for inhibiting enzymatic browning in fruits and vegetables), but isoascorbic acid (or D-ascorbic acid) has no nutritional value.

AA occurs naturally in fruits and vegetables and, to a lesser extent, in animal tissues and animal-derived products. It occurs naturally almost exclusively in the reduced L-ascorbic acid (i.e., AA) form. The concentration of DHAA found in foods is almost always substantially lower than AA and is a function of the rates of ascorbate oxidation and DHAA hydrolysis to 2,3-diketogulonic acid. Dehydroascorbate reductase and ascorbate free radical reductase activity exists in certain animal tissues. These enzymes are believed to conserve the vitamin through recycling and contribute to low DHAA concentrations. A significant but currently unknown fraction of the DHAA in foods and biological materials appears to be an analytical artifact that arises from oxidation of AA to DHAA during sample preparation and analysis. The instability of DHAA further complicates this analysis.

AA may be added to foods as the undissociated acid or as the neutralized sodium salt (sodium ascorbate). Conjugation of AA with hydrophobic compounds confers lipid solubility to the ascorbic acid moiety. Fatty acid esters such as ascorbyl palmitate and ascorbic acid acetals (Fig. 15) are lipid soluble and can provide a direct antioxidative effect in lipid environments.

Oxidation of AA takes place as either two one-electron transfer processes or as a single two-electron reaction without detection of the semihydroascorbate intermediate (Fig. 16). In one-electron oxidations, the first step involves transfer of an electron to form the free radical semidehydroascorbic acid. Loss of an additional electron yields dehydroascorbic acid, which is highly unstable because of the susceptibility to hydrolysis of the lactone bridge. Such hydrolysis, which irreversibly forms 2,3-diketogulonic acid (Fig. 16), is responsible for loss of vitamin C activity.

AA is highly susceptible to oxidation, especially when catalyzed by metal ions such as Cu^{2+} and Fe^{3+}. Heat and light also accelerate the process, while factors such as pH, oxygen concentration, and water activity strongly influence the rate of reaction. Since hydrolysis of DHAA occurs very readily, oxidation to DHAA represents an essential and frequently rate-limiting aspect of the oxidative degradation of vitamin C.

A frequently overlooked property of AA is its ability, at low concentrations, to act as a pro-oxidant with high oxygen tension. Presumably this occurs by ascorbate-mediated generation of hydroxyl radicals ($OH^{\cdot}$) or other reactive species. This appears to be of minor importance in most aspects of food chemistry.

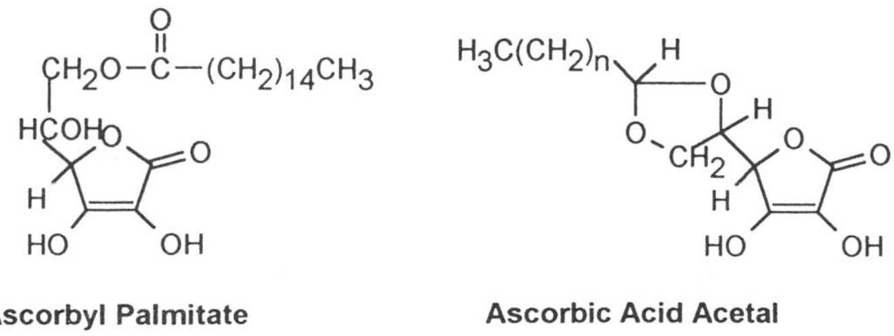

Ascorbyl Palmitate

Ascorbic Acid Acetal

FIGURE 15 Structures of ascorbyl palmitate and acetals.

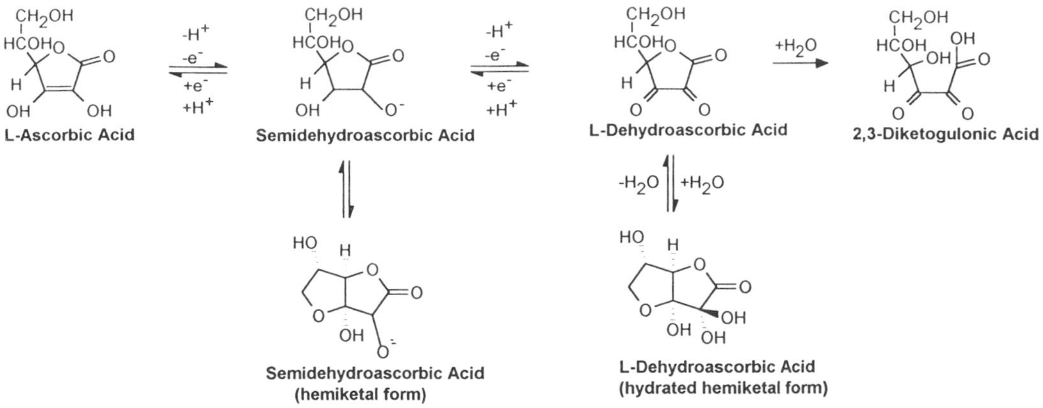

FIGURE 16 Sequential one-electron oxidations of L-ascorbic acid. All have vitamin C activity except 2,3-diketogulonic acid.

8.8.1.2 Stability and Modes of Degradation

Overview

Because of the high solubility of AA aqueous solutions, the potential exists for significant losses by leaching from freshly cut or bruised surfaces of fruits and vegetables. Chemical degradation primarily involves oxidations to DHAA, followed by hydrolysis to 2,3-diketogulonic acid and further oxidation, dehydration, and polymerization to form a wide array of other nutritionally inactive products. The oxidation and dehydration processes closely parallel dehydration reactions of sugars that lead to many unsaturated products and polymers. The primary factor affecting the rate, mechanism, and qualitative nature of the AA generation products include pH, oxygen concentration, and the presence of trace metal catalysts.

The rate of oxidative degradation of the vitamin is a nonlinear function of pH because of the various ionic forms of the AA differ in their susceptibility to oxidation: fully protonated (AH_2) < ascorbate monoanion (AH^-) < ascorbate dianion (A^{2-}) [16]. Under conditions relevant to most foods, pH dependence of oxidation is governed mainly by the relative concentration of AH_2 and AH^- species, and this, in turn, is governed by pH (pK_{a1} 4.04). The presence of significant concentrations of the A^{2-} form, as controlled by pK_{a2} of 11.4, yields an increase in rate at pH ≥ 8.

Catalytic Effects of Metal Ions

The overall scheme of AA degradation depicted in Figure 17 is an integrated view of the effects of metal ions and the presence or absence of oxygen on the mechanism of ascorbic acid degradation. The rate of oxidative degradation of AA is generally observed to be first order with respect to the concentration of the ascorbate monoanion (HA^-), molecular oxygen, and the metal ion. It was once believed that oxidative degradation of AA at neutral pH and in the absence of metal ions (i.e., the "uncatalyzed" reaction) occurred at a rate that was slow but significant. For example, a first-order rate constant of 5.87×10^{-4} sec^{-1} has been reported for the assumed spontaneous uncatalyzed oxidation of ascorbate at neutral pH. However, more recent evidence indicates a much smaller rate constant of 6×10^{-7} sec^{-1} for AA oxidation in an air-saturated solution at pH 7.0 [16]. This difference suggests that uncatalyzed oxidation is essentially negligible and that trace metals in foods or experimental solutions are responsible for much of the oxidative degradation. Rate constants obtained in the presence of metal ions at concentrations

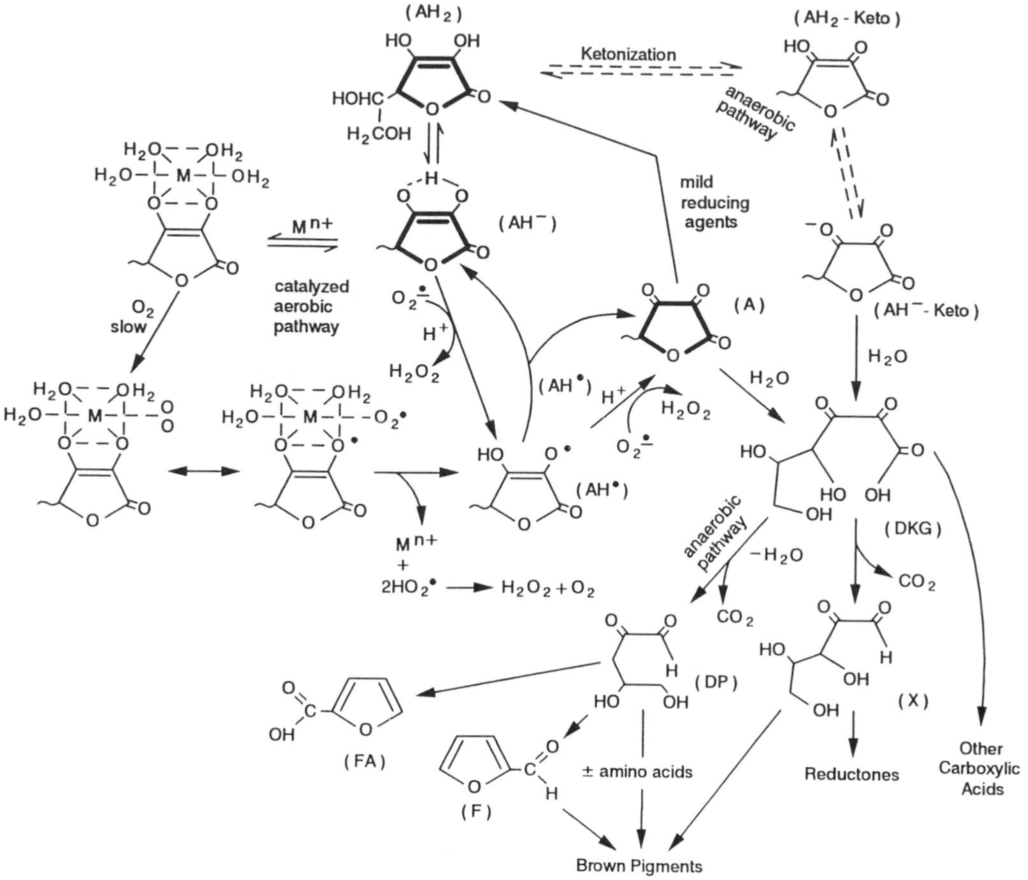

FIGURE 17 Overview of mechanisms for the oxidative and anaerobic degradation of ascorbic acid. Structures with bold lines are primary sources of vitamin C activity. Abbreviations: AH_2, fully protonated ascorbic acid; AH^-, ascorbate monoanion; $AH^{\bullet}$; semidehydroascorbate radical; A, dehydroascorbic acid; FA, 2-furoic acid; F, 2-furaldehyde; DKG, diketogulonic acid; DP, 3-deoxypentosone, X, xylosone; Mn^+, metal catalyst; $HO_2^{\bullet}$, perhydroxyl radical. (Based on Refs. 16,17,74,75,84, and 128.)

of several parts per million are several orders of magnitude greater than those obtained in solutions nearly devoid of metal ions.

The rate of metal-catalyzed oxidation of AA is proportional to the partial pressure of dissolved oxygen over the range of 1.0–0.4 atm, and is independent of oxygen concentration at partial pressures < 0.20 atm [74]. In contrast, the oxidation of AA catalyzed by metal chelates is independent of oxygen concentration [75].

The potency of metal ions in catalyzing ascorbate degradation depends on the metal involved, its oxidation state, and the presence of chelators. Catalytic potency is as follows: Cu(II) is about 80 times more potent than Fe(III), and the chelate of Fe(III) and ethylenediaminetetraacetic acid (EDTA) complex is ~4 times more catalytic than free Fe(III) [16]. When the rate expression of ascorbate oxidation is presented as

$$-d[\mathrm{TA}]/dt = k_{\mathrm{cat}} \times [\mathrm{AH}^-] \times [\mathrm{Cu(II)\ or\ Fe(III)}]$$

the metal ion concentration and the k_{cat} for metal ions can be used to estimate the rate of AA degradation (where [TA] = concentration of total ascorbic acid). In pH 7.0 phosphate buffers (20°C), k_{cat} values for Cu(II), Fe(III), and Fe(III)-EDTA are 880, 42, and 10 (M^{-1} sec^{-1}), respectively. It should be noted that the relative and absolute values of these catalytic rate constants in simple solutions may differ from those of actual food systems. This is likely because trace metals may associate with other constituents (e.g., amino acids) or may participate in other reactions, some of which may generate reactive free radicals or active oxygen species that may hasten oxidation of ascorbic acid.

In contrast to the enhanced catalytic potency of Fe(III) when chelated by EDTA, Cu(II)-catalyzed oxidation of ascorbate is largely inhibited in the presence of EDTA [16]. Thus, the influence of EDTA or other chelators (e.g., citric acid and polyphosphates) on the oxidation of ascorbic acid in foods is not fully predictable.

Mechanisms of AA Degradation

Oxidation of AA can be initiated by formation of a ternary complex (ascorbate monoanion, metal ion, and O_2), as described already, or by a variety of one-electron oxidations. As reviewed by Buettner [17], there are many ways in which the one-electron oxidation of AH^- to $A^{-\cdot}$, and $A^{-\cdot}$ to DHAA can occur. A ranking of the reduction potential, that is, reactivity, of relevant oxidants is summarized in Table 14. This illustrates the interrelationships in antioxidative function of several vitamins including ascorbic acids, α-tocopherol, and riboflavin.

The mechanism of AA degradation may differ depending on the nature of the food system or reaction medium. Metal-catalyzed degradation of AA has been proposed to occur through formation of a ternary complex of ascorbate monoanion, O_2, and a metal ion (Fig. 17). The ternary complex of ascorbate, oxygen, and metal catalyst appears to yield directly DHAA as the product, without detectable formation of the product of one-electron oxidation, semidehydro-ascorbate radical.

The loss of vitamin C activity during oxidative degradation of AA occurs with the hydrolysis of the DHAA lactone to yield 2,3-diketogulonic acid (DKG). This hydrolysis is favored by alkaline conditions, and DHAA is most stable at pH 2.5–5.5. The stability of DHAA at pH > 5.5 is very poor, and becomes more so as pH increases. For example, half-time values for DHAA hydrolysis at 23°C are 100 and 230 min at pH 7.2 and 6.6, respectively [11]. The rate of DHAA hydrolysis markedly increases with increasing temperature but is unaffected by the presence or absence of oxygen. In view of the labile nature of DHAA at neutral pH, analytical data showing significant quantities of DHAA in foods should be viewed with caution because elevated DHAA concentrations may also reflect uncontrolled oxidation during the analysis.

Although the ternary complex, as proposed by Khan and Martell [74], is apparently an accurate model of AA oxidation, recent findings have expanded our knowledge of the mechanism. Scarpa et al. [117] observed that metal-catalyzed oxidation of the ascorbate monoanion forms superoxide ($O_2^{-\cdot}$) in the rate-determining step:

$$AH^- + O_2 \xrightarrow{\text{catalyst}} AH^{\cdot} + O_2^{-\cdot}$$

Subsequent steps of the reaction involve superoxide as a rate enhancer, effectively doubling the overall rate ascorbate oxidation:

$$AH^- + O_2^{-\cdot} \xrightarrow{H^+} AH^{\cdot} + H_2O$$

$$AH^{\cdot} + O_2^{-\cdot} \xrightarrow{H^+} AH^{\cdot} + H_2O_2$$

TABLE 14 Reduction Potential of Selected Free Radicals and Antioxidants Arranged from the Most Highly Oxidizing (Top) to the Most Highly Reducing; each Oxidized Species in an Oxidation–Reduction Couple Is Capable of Abstracting an Electron or H Atom from Any Reduced Species Below It

Couple[a]		
Oxidized	Reduced	$\Delta E^{o\cdot}$ (Mv)
$HO^{\cdot}$, H^+	H_2O	2310
$RO^{\cdot}$, H^+	ROH	1600
$HO_2^{\cdot}$, H^+	H_2O_2	1060
$O_2^{\cdot-}$, $2H^+$	H_2O_2	940
$RS^{\cdot}$	RS^-	920
O_2 ($^1\Delta_g$)	$O_2^{\cdot-}$	650
$PUFA^{\cdot}$, H^+	PUFA-H	600
α–Tocopheroxyl$^{\cdot}$, H+	α-Tocopherol	500
H_2O_2, H^+	H_2O, $OH^{\cdot}$	320
Ascorbate$^{\cdot}$, H+	Ascorbate monoanion	282
Fe(III)EDTA	Fe(II)EDTA	120
Fe(III)aq	Fe(II)aq	110
Fe(III)citrate	Fe(II)citrate	~100
Dehydroascorbate	Ascorbate$^{\cdot-}$	~100
Riboflavin	Riboflavin$^{\cdot-}$	–317
O_2	$O_2^{\cdot-}$	–330
O_2, H+	$HO_2^{\cdot}$	–460

[a]*Note:* ascorbate$^{\cdot-}$, semidehydroascorbate radical; PUFA, polyunsaturated fatty acid radical; PUFA-H, polyunsaturated fatty acid, bis-allylic H; $RO^{\cdot}$, aliphatic alkoxy radical. $\Delta E^{o\cdot}$ is the standard one-electron reduction potential (mV).
Source: Adapted from Ref. 17.

As shown in Figure 17, termination also can occur through the reaction of two ascorbate radicals as

$$2AH^{\cdot} \xrightarrow{-H^+} DHAA + AH^-$$

Anaerobic degradation of AA (Fig. 17) is relatively insignificant as a means of loss of the vitamin in most foods. The anaerobic pathway becomes most significant in canned products, such as vegetables, tomatoes, and fruit juices, after depletion of residual oxygen, but even in these products loss of AA through anaerobic means typically occurs very slowly. Surprisingly, the anaerobic pathway has been identified as the predominant mechanism for loss of AA during storage of dehydrated tomato juice in the presence or absence of oxygen. Trace-metal catalysis of anaerobic degradation has been demonstrated, with the rate increasing in proportion to copper concentration.

The mechanism of anaerobic degradation of AA has not been fully established. Direct cleavage of the 1,4-lactone bridge without prior oxidation to DHAA appears to be involved, perhaps following an enol-keto tautomerization as shown in Figure 17. Unlike degradation of AA under oxidative conditions, anaerobic degradation exhibits a maximum rate at pH ~3–4. This maximum rate in the mildly acidic range may reflect the effects of pH on the opening of the lactone ring and on the concentration of the monoanionic ascorbate species.

The complexity of the anaerobic degradation mechanism, and the influence of food composition, is suggested by the significant change in activation energy at 28°C for the loss of total vitamin C from single strength orange juice during storage. In contrast, the Arrhenius plot for the degradation of total vitamin C during the storage of canned grapefruit juice is linear over the same range (~4–50°C), which suggests that a single mechanism predominates [101]. The reasons for this kinetic and/or mechanistic difference in such similar products are not known.

In view of the residual oxygen present in many food packages, degradation of ascorbic acid in sealed containers, especially cans and bottles, would typically occur by both oxidative and anaerobic pathways. In most cases, rate constants for anaerobic degradation of ascorbic acid will be two to three orders of magnitude less than those for the oxidative reaction.

Products of AA Degradation

Regardless of the mechanism of degradation, opening of the lactone ring irreversibly destroys vitamin C activity. Although devoid of nutritional relevance, the many reactions involved in the terminal phases of ascorbate degradation are important because of their involvement in producing flavor compounds or precursors and through their participation in nonenzymatic browning.

Over 50 low-molecular-weight products of ascorbic acid degradation have been identified. The kinds and concentrations of such compounds, and the mechanisms involved are strongly influenced by factors such as the temperature, pH, water activity, concentrations of oxygen and metal catalysts, and presence of active oxygen species. Three general types of decomposition products have been identified: (a) polymerized intermediates, (b) unsaturated carboxylic acids of five- and six-carbon chain length, and (c) fragmentation products having five or fewer carbons. Generation of formaldehyde during thermal degradation of ascorbate at neutral pH also has been reported. Some of these compounds are likely contributors to the changes in flavor and odor that occur in citrus juices during storage or excessive processing.

The degradation of sugars and ascorbic acid is strikingly similar, and in some cases mechanistically identical. Qualitative differences between aerobic and anaerobic conditions occur in the pattern of AA degradation, and pH exerts an influence in all circumstances. The major AA breakdown products in neutral and acidic solution include L-xylosone, oxalic acid, L-threonic acid, tartaric acid, 2-furaldehyde (furfural), and furoic acid, as well as a wide variety of carbonyls and other unsaturated compounds. As with sugar degradation, the extent of fragmentation increases under alkaline conditions.

AA degradation is associated with discoloration reactions in both the presence and absence of amines. DHAA as well as the dicarbonyls formed during its degradation can participate in Strecker degradation with amino acids. Following Strecker degradation of DHAA with an amino acid, the sorbamic acid product (Fig. 18) can form dimers, trimers, and tetramers, several of which are reddish or yellowish in color. In addition, 3,4-dihydroxy-5-methyl-2(5*H*)-furanone, an intermediate product of dehydration following decarboxylation during anaerobic degradation of AA, has a brownish color. Further polymerization of these or other unsaturated products forms either melanoidins (nitrogenous polymers) or nonnitrogenous caramel-like pigments. Although the nonenzymatic browning of citrus juices and related beverages is a complex process, the contribution of AA to browning has been clearly demonstrated [69].

Other Environmental Variables

Aside from the factors affecting ascorbate stability as discussed previously (e.g., oxygen, catalysts, pH, etc.), many other variables influence retention of this vitamin in foods. As with many other water-soluble compounds, the rate of oxidation of AA in low-moisture food systems simulating breakfast cereals has been found to increase progressively over the range of ~0.10–0.65 water activity [31, 71, 80] (Fig. 19). This apparently is associated with increased availability of

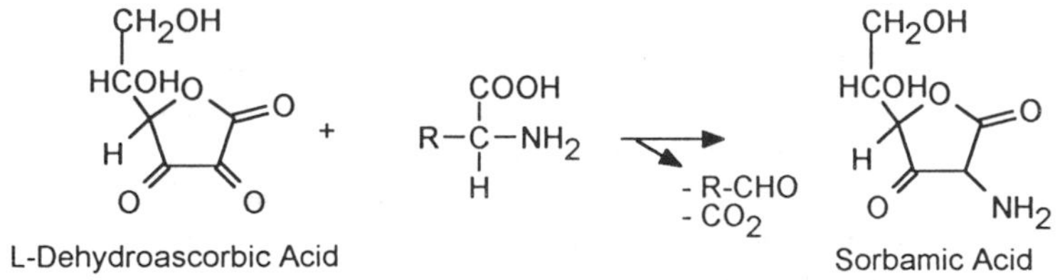

L-Dehydroascorbic Acid

Sorbamic Acid

FIGURE 18 Participation of dehydroascorbic acid in the Strecker degradation reaction.

water to act as a solvent for reactants and catalysts. The presence of certain sugars (ketoses) can increase the rate of anaerobic degradation. Sucrose has a similar effect at low pH, consistent with its pH-dependent generation of fructose. In contrast, some sugars and sugar alcohols exert a protective effect against the oxidative degradation of AA, possibly by binding metal ions and reducing their catalytic potency. The significance of these observations to actual foods remains to be determined.

8.8.1.3 Functions of Ascorbic Acid in Foods

In addition to its function as an essential nutrient, AA is widely used as a food ingredient/additive because of its reducing and antioxidative properties. As discussed elsewhere in this book, AA effectively inhibits enzymatic browning by reducing *ortho*-quinone products. Other functions include (a) reductive action in dough conditioners, (b) protection of certain oxidizable com-

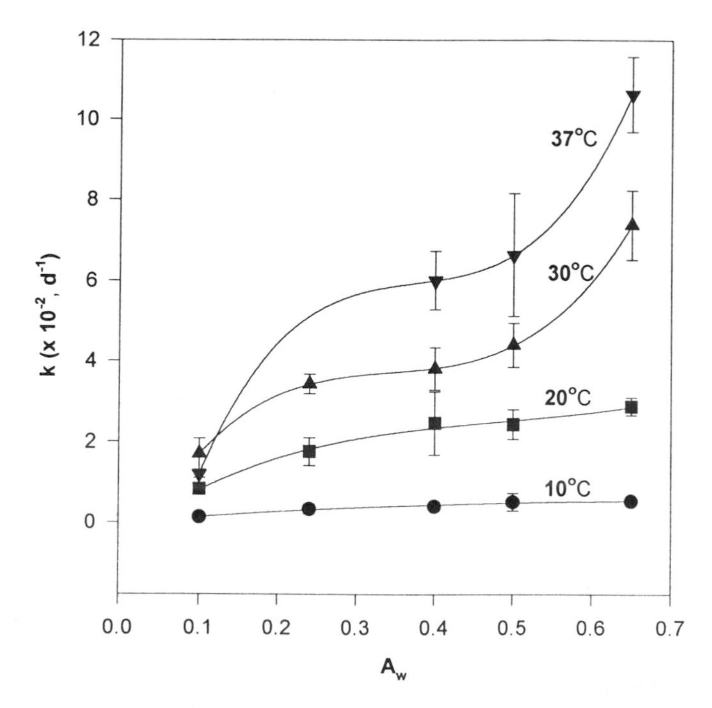

FIGURE 19 Degradation of ascorbic acid as a function of storage temperature and water activity in dehydrated model food systems simulating breakfast cereal products. Data (means ±SD) are expressed as apparent first-order rate constants for the loss of total ascorbic acid (AA + DHAA). (Data from Ref. 77).

pounds (e.g., folates) by reductive effects, free radical scavenging, and oxygen scavenging, (c) inhibition of nitrosamine formation in cured meats, and (d) reduction of metal ions.

The antioxidative role of ascorbic acid is multifunctional, with ascorbate inhibiting lipid autoxidation by several mechanisms [17, 84, 122]. These include (a)scavenging singlet oxygen, (b) reduction of oxygen- and carbon-centered radicals, with formation of a less reactive semidehydroascorbate radical or DHAA, (c) preferential oxidation of ascorbate, with concurrent depletion of oxygen, and (d) regeneration of other antioxidants, such as through reduction of the tocopherol radical.

AA is a very polar compound and is, therefore, essentially insoluble in oils. However, AA is surprisingly effective as an antioxidant when dispersed in oils as well as in emulsions [46]. Combinations of ascorbic acid and tocopherol are especially effective in oil-based systems, while the combination of α-tocopherol and the lipophilic ascorbyl palmitate is more effective in oil-in-water emulsions. Similarly, ascorbyl palmitate has been shown to act synergistically with α-tocopherol and other phenolic antioxidants.

8.8.1.4 Bioavailability of Ascorbic Acid in Foods

The principal dietary sources of AA are fruits, vegetables, juices, and fortified foods (e.g., breakfast cereals). The bioavailability of AA in cooked broccoli, orange sections, and orange juice has been shown to be equivalent to that of vitamin-mineral tablets for human subjects [90]. Bioavailability of AA in raw broccoli is 20% lower than that in cooked broccoli. This may be caused by incomplete chewing and/or digestion. The relatively small difference in AA bioavailability in raw broccoli and potentially other raw vegetables, relative to their cooked forms, may have little nutritional significance. Based on a recent review of the literature, it is clear that AA in most fruits and vegetables is highly available to humans [50].

8.8.1.5 Analytical Methods

Many procedures exist for the measurement of AA in foods, and selection of a suitable analytical method is essential to obtain accurate results [103]. Ascorbic acid absorbs UV light strongly ($\lambda_{max} \approx 245$ nm), although direct spectrophotometric analysis is precluded by the many other chromophores found in most foods. DHAA absorbs only weakly at its $\lambda_{max} \approx 300$ nm. Traditional analytical procedures involve redox titration of the sample with a dye such as 2,6-dichlorophenolindophenol, during which oxidation of AA accompanies reduction of the redox dye to its colorless form. A limitation of this approach is the interference by other reducing agents and the lack of a response to DHAA. Sequential analysis of the sample before and after saturation with H_2S gas or treatment with a thiol reagent to reduce DHAA to L-ascorbic acid permits the measurement of total ascorbic acid. Measurement of DHAA by difference lacks the precision of direct analysis, however.

An alternative approach involves condensation of DHAA (formed by controlled oxidation of L-ascorbic acid in the sample) with various carbonyl reagents. Direct treatment with phenylhydrazine to form the spectrophotometrically detectable ascorbyl-bis-phenylhydrazine derivative permits the simple measurement of L-AA in pure solution. Many carbonyl compounds in foods will interfere with this procedure. A similar method involves the reaction of DHAA with *o*-phenylenediamine, which forms a tricyclic, highly fluorescent condensation product. Although more specific and sensitive than the phenylhydrazine method, the *o*-phenylenediamine procedure is also subject to interference by certain dicarbonyls in foods. Foods containing isoascorbic acid cannot be analyzed for vitamin C by redox titration or condensation with carbonyl reagents because these methods respond to this nutritionally inactive compound.

Many HPLC methods permit accurate and sensitive measurement of total ascorbic acid (before and after treatment with a reducing agent), and certain methods permit direct measurement of L-ascorbic acid and DHAA. The coupling of chromatographic separation with spectrophotometric, fluorometric, or electrochemical detection makes HPLC analysis far more specific than traditional redox methods. HPLC methods have been reported that permit the simultaneous determination of ascorbic and isoascorbic acids as well as their dehydro forms [134]. A method based on gas chromatography–mass spectrometry has been reported, but extensive sample preparation is a disadvantage of the procedure [33].

8.8.2 Thiamin

8.8.2.1 Structure and General Properties

Thiamin is a substituted pyrimidine linked through a methylene bridge ($-CH_2-$) to a substituted thiazole (Fig. 20). Thiamin is widely distributed in plant and animal tissues. Most naturally occurring thiamin exists as thiamin pyrophosphate (Fig. 20), with lesser amounts of nonphosphorylated thiamin, thiamin monophosphate, and thiamin triphosphate. Thiamin pyrophosphate functions as a coenzyme of various α-keto acid dehydrogenases, α-keto acid decarboxylases, phosphoketolases, and transketolases. Thiamin is commercially available as the hydrochloride and mononitrate salts, and these forms are widely used for food fortification and as nutritional supplements (Fig. 20).

The thiamin molecule exhibits unusual acid–base behavior. The first pK_a (~4.8) involves dissociation of the protonated pyrimidine N^1 to yield the uncharged pyrimidyl moiety of thiamin free base (Fig. 21). In the alkaline pH range another transition is observed (apparent pK_a 9.2) that corresponds to the uptake of two equivalents of base to yield the thiamin pseudobase. The pseudobase can undergo opening of the thiazol ring to yield the thiol form of thaimin, accompanied by dissociation of a single proton. Another characteristic of thiamin is the quaternary N of the thiazole ring, which remains cationic at all pH values. The marked pH dependence of thiamin degradation corresponds to the pH-dependent changes in ionic form. Protonated thiamin is far more stable than free base, pseudobase, and thiol forms, which accounts for the greater stability observed in acidic media (Table 15). Although thiamin is relatively stable to oxidation and light, it is among the least stable of the vitamins when in solution at neutral or alkaline pH.

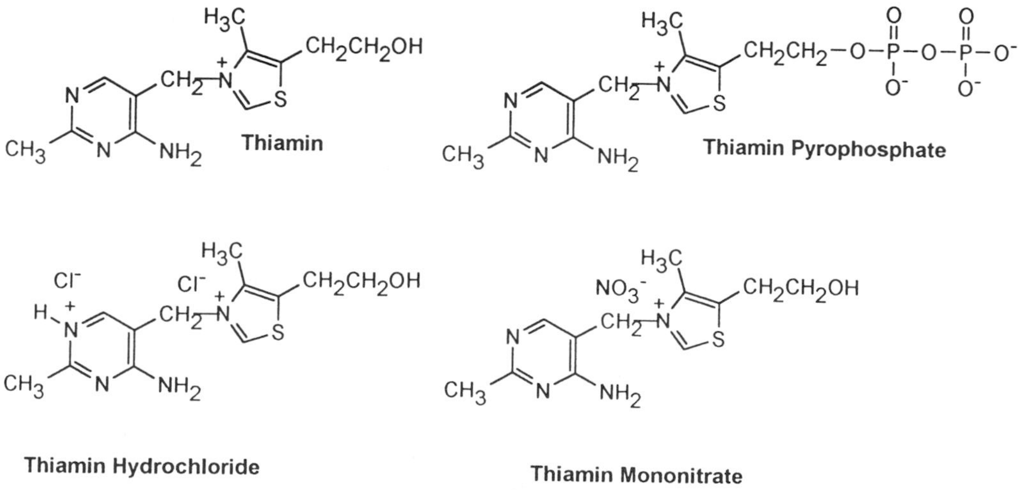

Thiamin

Thiamin Pyrophosphate

Thiamin Hydrochloride

Thiamin Mononitrate

FIGURE 20 Structures of various forms of thiamin. All have thiamin (vitamin B1) activity.

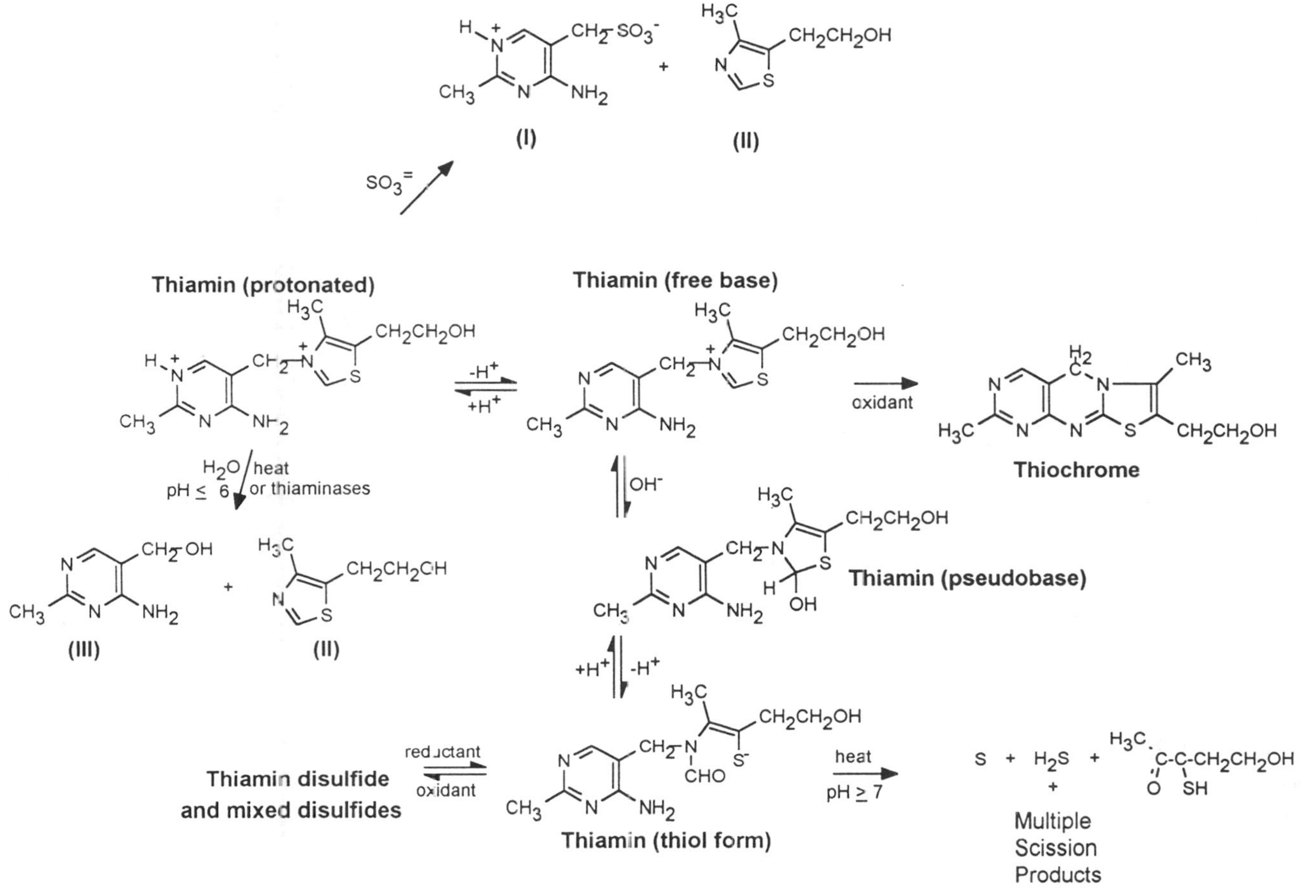

FIGURE 21 Summary of major pathways for ionization and degradation of thiamin (Adapted in modified form from Refs. 128 and 38.)

TABLE 15 Comparison of Thermal Stability of Thiamin and Thiamin Pyrophosphate in 0.1 *M* Phosphate Buffer at 265°C

	Thiamin		Thiamin pyrophosphate	
Solution pH	k^a (min^{-1})	$t_{1/2}$ (min)	k^a (min^{-1})	$t_{1/2}$ (min)
4.5	0.0230	30.1	0.0260	26.6
5.0	0.0215	32.2	0.0236	29.4
5.5	0.0214	32.4	0.0358	19.4
6.0	0.0303	22.9	0.0831	8.33
6.5	0.0640	10.8	0.1985	3.49

Note: k is the first-order rate constant and $t_{1/2}$ is the time for 50% thermal degradation.
Source: Adapted from Ref. 100.

8.8.2.2 Stability and Modes of Degradation

Stability Properties

A wealth of published data exists concerning the stability of thiamin in foods [42, 92]. Representative studies, as summarized previously by Tannenbaum et al. [128], illustrate the potential for large losses under certain conditions (Table 16). Losses of thiamin from foods are favored when (a) conditions favor leaching of the vitamin into surrounding aqueous media,

TABLE 16 Representative Rates of Degradation (Half-Life at Reference Temperature of 100°C) and Energy of Activation for Losses of Thiamin from Foods During Thermal Processing

Food system	pH	Temperature range studied (°C)	Half-life (h)	Energy of activation (kJ/mol)
Beef heart purée	6.10	109–149	4	120
Beef liver purée	6.18	109–149	4	120
Lamb purée	6.18	109–149	4	120
Pork purée	6.18	109–149	5	110
Ground meat product	Not reported	109–149	4	110
Beef purée	Not reported	70–98	9	110
Whole milk	Not reported	121–138	5	110
Carrot purée	6.13	120–150	6	120
Green bean purée	5.83	109–149	6	120
Pea purée	6.75	109–149	6	120
Spinach purée	6.70	109–149	4	120
Pea purée	Not reported	121–138	9	110
Peas in brine purée	Not reported	121–138	8	110
Peas in brine	Not reported	104–133	6	84

Note: Water activity estimated to be 0.98–0.99. Half-life and energy of activation values rounded to one and two significant figures, respecitvely.
Source: Ref. 92 (data compiled from multiple sources).

(b) the pH is approximately neutral or greater, and/or (c) exposure to a sulfiting agent occurs. Losses of thiamin also can occur in fully hydrated foods during storage at moderate temperatures, although at predictably lower rates than those observed during thermal processing (Table 17). Thiamin degradation in foods almost always follows first-order kinetics. Because degradation of thiamin can occur by several possible mechanisms, multiple mechanisms sometimes occur simultaneously. The occurrence of nonlinear Arrhenius plots for thermal losses of thiamin in certain foods is evidence of multiple degradation mechanisms that have different temperature dependence.

Thiamin exhibits excellent stability under conditions of low water activity at ambient temperature. Thiamin in dehydrated model systems simulating breakfast cereals underwent little or no loss at temperatures less than 37°C at a_W 0.1–0.65 (Fig. 22). In contrast, thiamin degradation was markedly accelerated at 45°C, especially at a_W 0.4 or greater (i.e., above the monomolecular moisture value of $a_W \approx 0.24$). In these model systems, the maximum rate of thiamin degradation occurred at water activities of 0.5–0.65 (Fig. 23). In similar model systems, the rate of thiamin degradation declined as the a_W was increased from 0.65 to 0.85 [5].

Thiamin is somewhat unstable in many fish and crustaceans postharvest, and this has been attributed to the presence of thiaminases. However, at least part of this thiamin degradation activity is caused by heme proteins (myoglobin and hemoglobin) that are thermostable nonenzymatic catalysts of thiamin degradation [106]. The presence of thiamin-degrading heme proteins in tuna, pork, and beef muscle suggests that denatured myoglobin may be involved in the degradation of thiamin during food processing and storage. This nonenzymatic, thiamin-modifying activity apparently does not cause cleavage of the thiamin molecule, as is common in thiamin degradation. The antithiamin component of fish viscera, previously reported to be a thaiminase, is now believed to be a thermostable and probably nonenzymatic catalyst.

Other components of food can influence degradation of thiamin in foods. Tannins can inactivate thiamin apparently by the formation of several biologically inactive adducts. Various flavonoids may alter the thiamin molecule, but the apparent product of flavonoid oxidation in the presence of thiamin is thiamin disulfide, a compound that has thiamin activity. Proteins and carbohydrates can reduce the rate of thiamin degradation during heating or in the presence of bisulfite, although the extent of this effect is difficult to predict in complex food systems. A part of the stabilizing effect of protein may occur through the formation of mixed disulfides with the thiol form of thiamin, a reaction that appears to retard further modes of degradation. Chlorine

TABLE 17 Typical Losses of Thiamin from Canned Foods During Storage

	Retention after 12 months of storage (%)	
Food	38°C	1.5°C
Apricots	35	72
Green beans	8	76
Lima beans	48	92
Tomato juice	60	100
Peas	68	100
Orange juice	78	100

Source: Ref. 47.

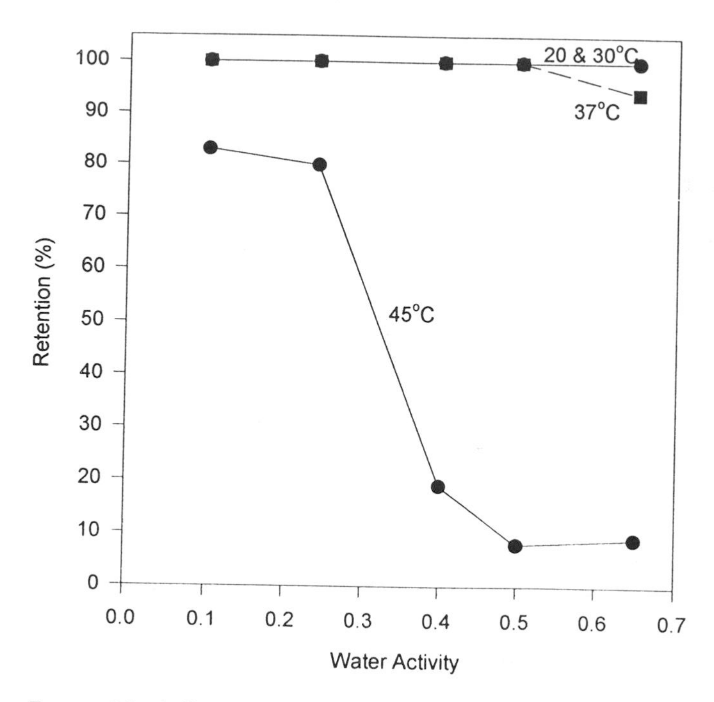

FIGURE 22 Influence of water activity and temperature on the retention of thiamin in a dehydrated model food system simulating a breakfast cereal product. Percentage retention values apply to an 8-month storage period. (From Ref. 32.)

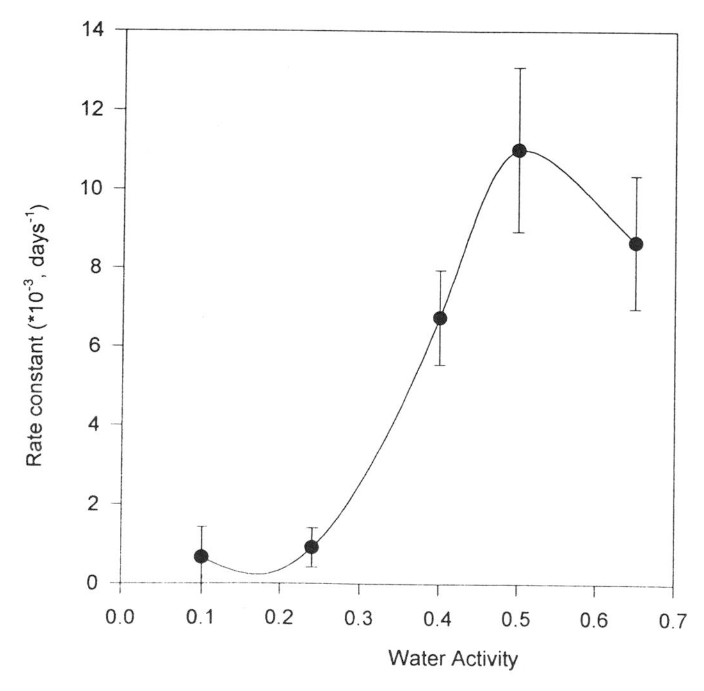

FIGURE 23 Influence of water activity on the first-order rate constant of thiamin degradation in a dehydrated model food system simulating a breakfast cereal product stored at 45°C. (From Ref. 32.)

(as hypochlorite ion), at levels present in water used in food formulation and processing, can cause rapid degradation of thiamin by a cleavage process that is apparently identical to the thermal cleavage of thiamin under acidic conditions.

Another complicating factor in the assessment and prediction of thiamin stability is the inherent difference in stability and pH dependence between free thiamin and the major naturally occurring form, thiamin pyrophosphate. Although thiamin and thiamin pyrophosphate exhibit nearly equivalent rates of thermal degradation at pH 4.5, thiamin pyrophosphate degrades almost three times faster at pH 6.5 (Table 15).

Significant differences exist in the stability of hydrochloride and mononitrate forms of synthetic thiamin. Thiamin HCl is more soluble than the mononitrate, and this is advantageous for fortification of liquid products. Because of differing energies of activation, thiamin mononitrate is more stable at temperatures less than 95°C, while the hydrochloride exhibits greater stability at temperatures > 95–110°C (Table 18).

Mechanisms of Degradation

The rate and mechanism of thermal degradation of thiamin are strongly influenced by pH of the reaction medium, but degradation usually involves cleavage of the molecule at the central methylene bridge.

In acidic conditions (i.e., $pH \leq 6$) thermal degradation of thiamin occurs slowly and involves cleavage of the methylene bridge to release the pyrimidine and thiazole moieties largely in unchanged form. Between pH 6 and 7, thiamin degradation accelerates along with a large increase in the extent of fragmentation of the thiazole ring, and at pH 8 intact thiazole rings are not found among the products. Thiamin degradation is known to yield a large number of sulfur-containing compounds that presumably arise from fragmentation and rearrangement of the thiazole ring. These compounds have been shown to contribute to meat flavor. Products from

TABLE 18 Kinetic Values for Thiamin Loss in Semolina Dough Subjected to High Temperatures

a_w	Temperature (°C)	k ($\times 10^4$ min^{-1}) ± 95% CI[a]	Half life (min)	Energy of activation (kcal/mol)[b]
Hydrochloride				
0.58	75	3.72 ± 0.01	1863	95.4
	85	11.41 ± 3.64	607	
	95	22.45 ± 2.57	309	
0.86	75	5.35 ± 2.57	1295	92.1
	85	12.20 ± 4.45	568	
	95	30.45 ± 8.91	228	
Mononitrate				
0.58	75	2.88 ± 0.01	2406	109
	85	7.91 ± 0.01	876	
	96	22.69 ± 2.57	305	
0.86	75	2.94 ± 0.01	2357	111
	85	8.31 ± 0.01	834	
	95	23.89 ± 0.01	290	

[a]First-order rate constant ± 95% confidence interval.
[b]kcal × 4.186 = kJ.
Source: Ref. 81.

thiazole fragmentation are thought to arise from the small amounts of thiamin that exist in the thiol or pseudobase forms at $pH > 6$.

Thiamin degrades rapidly in the presence of bisulfite ions, a phenomenon that stimulated federal regulations prohibiting the use of sulfiting agents in foods that are significant sources of dietary thiamin. The cleavage of thiamin by bisulfite is similar to that occurring at $pH \le 6$, although the pyrimidine product is sulfonated (Fig. 21). This reaction is described as a base exchange or nucleophilic displacement at the methylene carbon, by which the bisulfite ion displaces the thiazole moiety. It is unclear whether other nucleophiles relevant to foods can have a similar effect. Cleavage of thiamin by bisulfite occurs over a broad pH range, with the maximum rate occurring at $pH \approx 6$ [146]. A bell-shaped pH profile of this reaction occurs because the sulfite ion primarily reacts with the protonated form of thiamin.

Several researchers have noted a correspondence of the conditions (e.g., pH and water activity) favoring degradation of thiamin and progress of the Maillard reaction. Specifically, thiamin has a primary amino group on its pyrimidyl moiety, shows a maximum rate of degradation at an intermediate water activity, and exhibits greatly increased reaction rates at neutral and alkaline pH values. Early studies demonstrated the ability of thiamin to react with sugars under certain conditions; however, sugars often tend to increase the stability of thiamin. Despite the similarity of conditions favoring thiamin degradation and Maillard browning, there appears to be little or no direct interaction of thiamin with the reactants or intermediates of the Maillard reaction in foods.

8.8.2.3 Bioavailability

Although the bioavailability of thiamin has not been fully evaluated, its utilization appears to be nearly complete in most foods examined. As mentioned previously, formation of thiamin disulfide and mixed thiamin disulfides during food processing apparently has little effect on thiamin bioavailability. Thiamin disulfide exhibits 90% of the activity of thiamin in animal bioassays.

8.8.2.4 Analytical Methods

Although microbiological growth methods exist for measurement of thiamin in foods, they are rarely used because of the availability of fluorometric and HPLC procedures [43]. Thiamin is generally extracted from the food by heating (e.g., autoclaving) a homogenate in dilute acid. For analysis of total thiamin, treatment of the buffered extract with a phosphatase hydrolyzes phosphorylated forms of the vitamin. Following chromatographic removal of nonthiamin fluorophores, treatment with an oxidizing agent converts thiamin to the highly fluorescent thiochrome that is easily measured (Fig. 21).

Total thiamin can be determined by HPLC following phosphatase treatment. Fluorometric HPLC analysis can be used following conversion of thiamin to thiochrome or, alternatively, postcolumn oxidation to thiochrome can permit fluorometric detection. Individual phosphate esters of thiamin can be determined simultaneously by HPLC.

8.8.3 Riboflavin

8.8.3.1 Structure and General Properties

Riboflavin, formerly known as vitamin B2, is the generic term for the group of compounds that exhibit the biological activity of riboflavin (Fig. 24). The parent compound of the riboflavin family is 7,8-dimethyl-10(1′-ribityl)isoalloxazine, and all derivatives of riboflavin are given the generic name flavins. Phosphorylation of the 5′-position of the ribityl side chain yields flavin

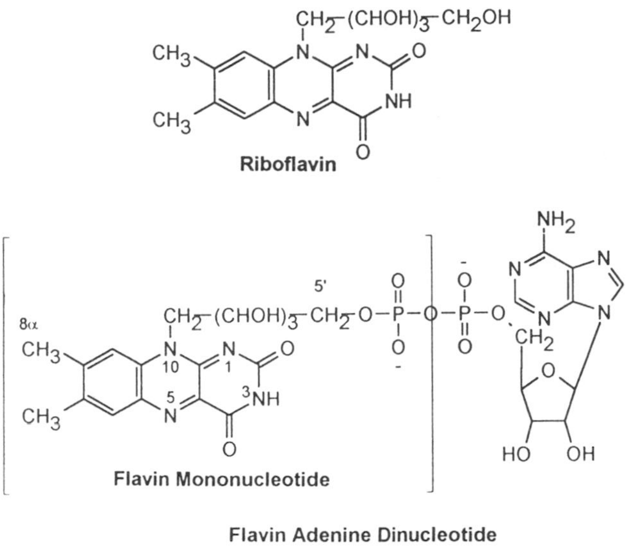

Riboflavin

Flavin Mononucleotide

Flavin Adenine Dinucleotide

FIGURE 24 Structures of riboflavin, flavin mononucleotide, and flavin adenine dinucleotide.

mononucleotide (FMN), whereas flavin adenine dinucleotide (FAD) has an additional 5′-adenosyl monophosphate moiety (Fig. 24). FMN and FAD function as conezymes in a large number of flavin-dependent enzymes that catalyze various oxidation reduction processes. Both forms are readily convertible to riboflavin by action of phosphatases that are present in foods and those of the digestive system. A relatively minor (< 10%) of the FAD in biological materials exists in a covalently bound enzyme form in which position 8α is covalently linked to an amino acid residue of the enzyme protein.

The chemical behavior of riboflavin and other flavins is complex, with each form able to exist in several oxidation states as well as multiple ionic forms [93]. Riboflavin, as the free vitamin and during coenzymic function, undergoes redox cycling among three chemical species. These include the native (fully oxidized) yellow flavoquinone (Fig. 25), the flavosemiquinone (red or blue depending on pH), and the colorless flavohydroquinone. Each conversion in this sequence involves a one-electron reduction and H^+ uptake. The flavosemiquinone N^5 has a pK_a of ~8.4, while the flavohydroquinone N^1 has a pK_a of ~6.2.

Several minor forms of riboflavin also exist in foods, although their chemical origin and quantitative significance in human nutrition have not been fully determined. As shown in Table 19, FAD and free riboflavin account for over 80% of the total flavins in cow's and human milk [111, 112]. Of the minor forms present, most interesting is 10-hydroxyethylflavin, a product of bacterial flavin metabolism. 10-Hydroxyethylflavin is a known inhibitor of mammalian flavokinase and may inhibit the uptake of riboflavin into tissues. Other minor derivatives (such as lumiflavin) may also act as antagonists. Thus, foods contain flavins such as riboflavin, FAD, and FMN that exhibit vitamin activity, but in addition they may contain compounds that act as antagonists of riboflavin transport and metabolism. This illustrates the need for a thorough analysis of the forms of riboflavin and other vitamins in order to assess accurately the nutritional properties of foods.

8.8.3.2 Stability and Modes of Degradation

Riboflavin exhibits its greatest stability in acidic medium, is somewhat less stable at neutral pH, and rapidly degrades in alkaline environments. Retention of riboflavin in most foods is moderate

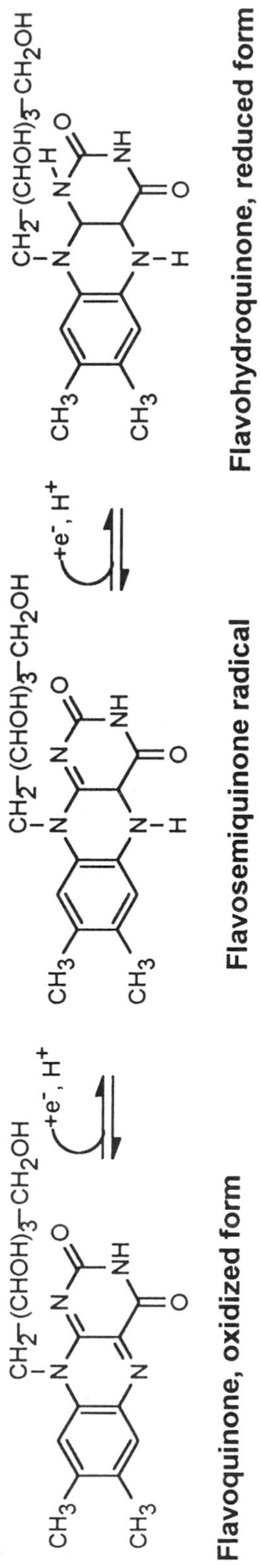

$CH_2-(CHOH)_3-CH_2OH$

Flavoquinone, oxidized form

$+e^-, H^+$

$CH_2-(CHOH)_3-CH_2OH$

Flavosemiquinone radical

$+e^-, H^+$

$CH_2-(CHOH)_3-CH_2OH$

Flavohydroquinone, reduced form

FIGURE 25 Oxidation–reduction behavior of flavins.

TABLE 19 Distribution of Riboflavin Compounds in Fresh Human and Cow's Milk

Compound	Human milk (%)	Cow's milk (%)
FAD	38–62	23–46[a]
Riboflavin	31–51	35–59
10-Hydroxyethylflavin	2–10	11–19
10-Formylmethyflavin	Trace	Trace
7α-Hydroxyriboflavin	Trace–0.4	0.1–0.7
8α-Hydroxyriboflavin	Trace	Trace–0.4

[a]Following pasteurization, FAD in bulk raw milk decreases from 26 to 13%, with a corresponding increase in the percentage of riboflavin.
Source: Adapted from Refs. 111 and 112.

to very good during conventional thermal processing, handling, and preparation. Losses during storage of riboflavin in various dehydrated food systems (breakfast cereals and model systems) are usually negligible. Rates of degradation increase measurably at a_w above the monolayer value when temperatures are above ambient [32].

The typical mechanism of degradation of riboflavin is photochemical, which yields two biologically inactive products, lumiflavin and lumichrome (Fig. 26), and an array of free radicals [142]. Exposure of solutions of riboflavin to visible light has been used for many years as an experimental technique to generate free radicals. Photolysis of riboflavin yields superoxide and riboflavin radicals ($R^{\cdot}$), and the reaction of O_2 with $R^{\cdot}$ provides peroxy radicals and a wide range of other products. The extent to which photochemical degradation of riboflavin is responsible for photosensitized oxidation reactions in food has not been quantitatively determined, although this process assuredly contributes significantly. Sunlight-induced off flavor in milk, which is no longer common, is a riboflavin-mediated photochemical process. Although the mechanism of off-flavor formation has not been fully determined, light-induced (probably radical-mediated) decarboxylation and deamination of methionine to form methional (CH_3-S-CH_2-CH_2=O) is at least partially responsible. Concurrent mild oxidation of milk lipids also occurs. Changes in packaging and commercial distribution have minimized this problem.

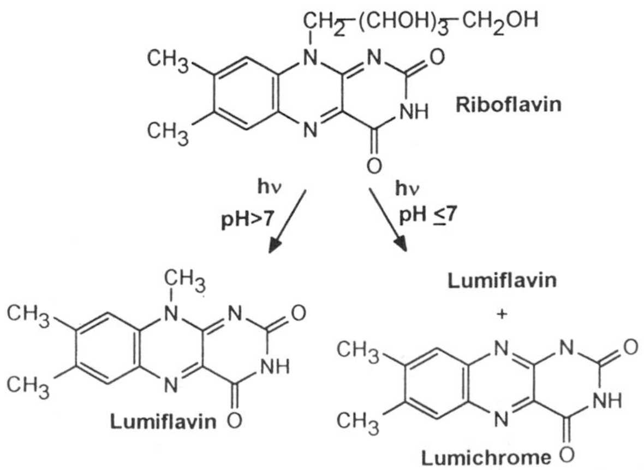

FIGURE 26 Photochemical conversion of riboflavin to lumichrome and lumiflavin.

8.8.3.3 Bioavailability

Relatively little is known regarding the bioavailability of naturally occurring forms of riboflavin; however, there is little evidence of problems associated with incomplete bioavailability. The covalently bound forms of FAD coenzymes have been shown to exhibit very low availability when administered to rats, although these are minor forms of the vitamin. The nutritional significance of dietary riboflavin derivatives that have potential antivitamin activity has not yet been determined in animals or humans.

8.8.3.4 Analytical Methods

Flavins are highly fluorescent compounds in their fully oxidized flavoquinone form (Fig. 25), and this property serves as the basis for most analytical methods. The traditional assay procedure for the measurement of total riboflavin in foods involves measurement of fluorescence before and after chemical reduction to the nonfluorescent flavohydroquinone [120]. Fluorescence is a linear function of concentration in dilute solution, although certain food components can interfere with accurate measurement. Several HPLC methods are also suitable for measurement of total riboflavin in food extracts [43, 135]. These HPLC procedures and the fluorometric method require extraction by autoclaving in dilute acid followed by a phosphatase treatment to release riboflavin from FMN and FAD. HPLC can also be used to measure the individual riboflavin compounds in foods [111].

8.8.4 Niacin

8.8.4.1 Structure and General Properties

Niacin is the generic term for pyridine 3-carboxylic acid (nicotinic acid) and derivatives that exhibit similar vitamin activity (Fig. 27). Nicotinic acid and the corresponding amide (nicotinamide; pyridine 3-carboxamide) are probably the most stable of the vitamins. The coenzyme forms of niacin are nicotinamide adenine dinucleotide (NAD) and nicotinamide adenine dinucleotide phosphate (NADP), either of which can exist in oxidized or reduced form. NAD and NADP function as coenzymes (in the transfer of reducing equivalents) in many dehydrogenase reactions. Heat, especially under acid or alkaline conditions, converts nicotinamide to nicotinic acid without loss of vitamin activity. Niacin is not affected by light, and no thermal losses occur under conditions relevant to food processing. As with other water-soluble nutrients, losses can occur by leaching during washing, blanching, and processing/preparation and by exudation of fluids from tissues (i.e., drip). Niacin is widely distributed in vegetables and foods of animal origin. Niacin deficiency is rare in the United States, partially as a result of programs to enrich cereal grain products with this nutrient. Diets high in protein reduce the requirement for dietary niacin because of the metabolic conversion of tryptophan to nicotinamide.

In certain cereal grain products, niacin exists in several chemical forms that, unless hydrolyzed, exhibit no niacin activity. These inactive niacin forms include poorly characterized complexes involving carbohydrates, peptides, and phenols. Analysis of these nutritionally unavailable, chemically bound forms of niacin has revealed chromatographic heterogeneity and variation in chemical composition, indicating that many bound forms of niacin exist naturally. Alkaline treatments release niacin from these complex derivatives, which permits measurement of total niacin. Several esterified forms of nicotinic acid exist naturally in cereal grains, and these compounds contribute little to niacin activity in foods.

Trigonelline, or *N*-methyl-nicotinic acid, is a naturally occurring alkaloid found at relatively high concentrations in coffee and at lower concentrations in cereal grains and legumes.

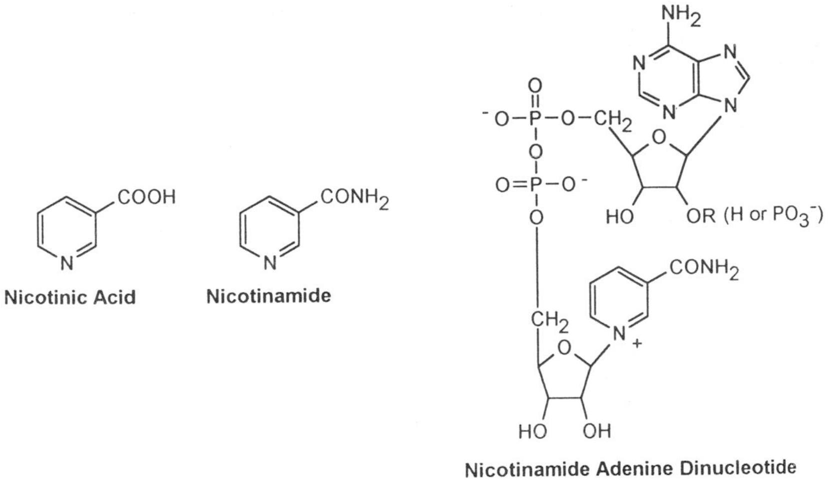

Nicotinic Acid

Nicotinamide

Nicotinamide Adenine Dinucleotide

FIGURE 27 Structures of nicotinic acid, nicotinamide, and nicotinamide adenine dinucleotide (phosphate).

Under the mildly acidic conditions that prevail during roasting of coffee, trigonelline is demethylated to form nicotinic acid, yielding a 30-fold increase in the niacin concentration and activity of coffee.

Cooking also changes the relative concentration of certain niacin compounds through interconversion reactions [137, 138]. For example, heating releases free nicotinamide from NAD and NADP during the boiling of corn. In addition, the distribution of niacin compounds within a product varies as a function of variety (e.g., sweet corn vs. field corn) and stage of maturity.

8.8.4.2 Bioavailability

The existence of nutritionally unavailable forms of niacin in many foods of plant origin has been known for many years, although the chemical identities of the unavailable forms of the vitamin are poorly characterized. In addition to the chemically bound forms discussed earlier, several other forms of niacin contribute to its incomplete availability in foods of plant origin [137]. NADH, the reduced form of NAD, and presumably NADPH exhibit very low bioavailability because of their instability in the gastric acid environment. This may be of little nutritional significance because of the low concentration of these reduced forms in many foods. The primary factor affecting niacin bioavailability is the proportion of the total niacin that is chemically bound. As shown in Table 20, there is often much more niacin measurable following alkaline extraction than there is by rat bioassays (biological available niacin) or by direct analysis (free niacin).

8.8.4.3 Analytical Methods

Niacin can be measured by microbiological assay. The principal chemical assay involves a reaction of niacin with cyanogen bromide to yield an N-substituted pyridine that is then coupled to an aromatic amine to form a chromophore [39]. Several HPLC methods are available for measurement of nicotinic acid and nicotinamide in foods [43, 135], and HPLC has been used to determine individual free and bound forms of niacin in cereal grains [137, 138].

TABLE 20 Concentration of Niacin in Selected Foods as Determined by Chemical Assay (Acidic or Alkaline Extraction Methods) or Rat Bioassay

Food	Type of chemical assay		
	Free niacin (μg/g)[a]	Total niacin (alkaline extraction, μg/g)[a]	Rat bioassay (μg/g)[a]
Corn	0.4	25.7	0.4
Boiled corn	3.8	23.8	6.8
Corn after alkaline heating (liquid retained)	24.6	24.6	22.3
Tortillas	11.7	12.6	14
Sweet corn (raw)	—	54.5	40
Steamed sweet corn	45	56.4	48
Boiled sorghum grain	1.1	45.5	16
Boiled rice	17	70.7	29
Boiled wheat	—	57.3	18
Baked potatoes	12	51	32
Baked liver	297	306	321
Baked beans	19	24	28

Note: Analysis of HCl extract yields a measure of "free niacin," assay of the alkaline extract provides a measure of total niacin, and the rat bioassay is a measure of biologically available niacin.
[a]Wet weight basis.
Source: Adapted from Ref. 137.

8.8.5 Vitamin B6

8.8.5.1 Structure and General Properties

Vitamin B6 is a generic term for the group of 2-methyl-3-hydroxy-5-hydroxymethylpyridines having the vitamin activity of pyridoxine. The various forms of vitamin B6 differ according to the nature of the one-carbon substituent at the 4-position, as shown in Figure 28. For pyridoxine (PN) the substituent is an alcohol, for pyridoxal (PL) it is an aldehyde, and for pyridoxamine (PM) it is an amine. These three basic forms can also be phosphorylated at the 5′-hydroxymethyl group, yielding pyridoxine 5′-phosphate (PNP), pyridoxal 5′-phosphate (PLP), or pyridoxamine 5′-phosphate (PMP). Vitamin B6, in the form of PLP, and to a lesser extent PMP, functions as a coenzyme in over 100 enzymatic reactions involved in the metabolism of amino acids, carbohydrates, neurotransmitters and lipids. All of the mentioned forms of vitamin B6 possess vitamin activity because they can be converted in vivo to these conenzymes. The use of "pyridoxine" as a generic term for vitamin B6 has been discontinued. Similarly, the term "pyridoxol" has been discontinued in favor of pyridoxine.

Glycosylated forms of vitamin B6 are present in most fruits, vegetables, and cereal grains, generally as pyridoxine-5′-β-D-glucoside (Fig. 28) [55, 144]. These comprise 5–75% of the total vitamin B6 and account for 15–20% of the vitamin B6 in typical mixed diets. Pyridoxine glucoside becomes nutritionally active only after hydrolysis of the glucoside by the action of β-glucosidases in the intestine or other organs. Several other glycosylated forms of vitamin B6 are also found in certain plant products.

Vitamin B6 compounds exhibit complex ionization that involves several ionic sites (Table 21). Because of the basic character of the pyridinium N ($pK_a \approx 8$) and the acidic nature

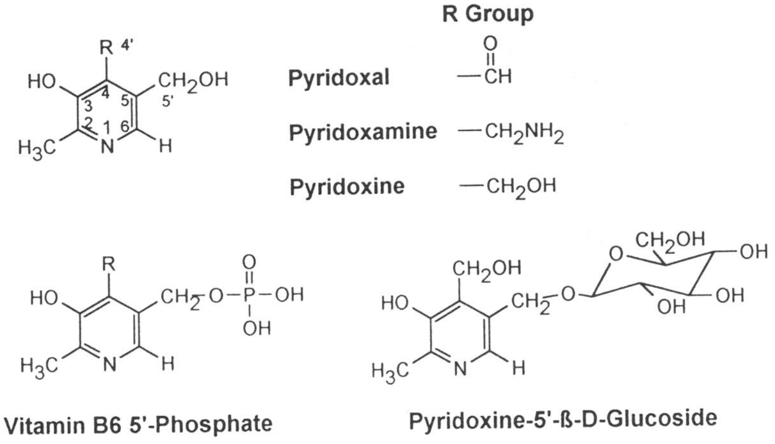

FIGURE 28 Structures of vitamin B6 compounds.

of the 3-OH (p$K_a \approx 3.5$–5.0), the pyridine system of vitamin B6 molecules mainly exists in Zwitterionic form at neutral pH. The net charge on vitamin B6 compounds varies markedly as a function of pH. The 4′-amino group of PM and PMP (p$K_a \approx 10.5$) and the 5′-phosphate ester of PLP and PMP (p$K_a < 2.5$, ~6, and ~12) also contribute to the charge of these forms of the vitamin.

All chemical forms of vitamin B6 exist in foods, although the distribution varies markedly. PN glucoside exists only in the plant products, although most plant products also contain all other forms of the vitamin. Vitamin B6 in muscle and organ meats is predominantly (> 80%) PLP and PMP, with minor amounts of the nonphosphorylated species. Disruption of raw plant tissues by freeze–thaw cycling or homogenization releases phosphatases and β-glucosidases that can alter the forms of vitamin B6 compounds by catalyzing dephosphorylation and deglycosylation reactions. Similarly, disruption of animal tissues prior to cooking can cause extensive dephosphorylation of PLP and PMP. PNP is a transient intermediate in vitamin B6 metabolism and is usually a negligible component of the total vitamin B6 content. Pyridoxine (as the HCl salt) is the form of vitamin B6 used for food fortification and in nutritional supplements because of its good stability.

TABLE 21 pK_a Values of Vitamin B6 Compounds

	pK_a				
Ionization	PN	PL	PM	PLP	PMP
3-OH	5.00	4.20–4.23	3.31–3.54	4.14	3.25–3.69
Pyridinium N	8.96–8.97	8.66–8.70	7.90–8.21	8.69	8.61
4′-Amino group			10.4–10.63		ND
5′-Phosphate ester					
pK_{a1}				<2.5	<2.5
pK_{a2}				6.20	5.76
pK_{a3}				ND	ND

Note: PN, pyridixine; PL, pyridoxal; PM, pyridoxamine; PLP, pyridoxal 5′-phosphate; PMP, pyridoxamine 5′-phosphate. ND, not determined.
Source: Ref. 123.

The aldehyde and amine forms of vitamin B6 readily participate in carbonyl-amine reactions: PLP or PL with amines, and PMP or PM with aldehydes or ketones (Fig. 29). The coenzymic action of PLP in most B6-dependent enzymes involves a carbonyl-amine condensation. PLP and PL readily form Schiff bases with the neutral amino groups of amino acids, peptides, or proteins. Coordinate covalent bonding to a metal ion increases the stability of the Schiff base in nonenzymatic systems, although Schiff bases can exist in solutions devoid of metal ions. PLP forms Schiff bases much more readily than PL because the phosphate group of PLP blocks the formation of an internal hemiacetal and maintains the carbonyl in reactive form (Fig. 30) [141]. As with other carbonyl-amine reactions, the nonenzymatic formation of a vitamin B6 Schiff base is strongly pH dependent and exhibits an alkaline pH optimum. The stability of the Schiff base complexes is also strongly pH dependent, with dissociation occurring in acidic environments. Thus, Schiff-base forms of vitamin B6 would be expected to dissociate fully in acidic media such as the postprandial gastric contents. In addition to the Schiff base in Figure 29, several other tautomeric and ionic forms exist in an equilibrium situation.

Depending on the chemical nature of the amino compound condensing with PLP or PL in the Schiff base, further rearrangement to various cyclic structures can occur. For example, cysteine condenses with PL or PLP to form a Schiff-base; then the SH group attacks the Schiff base 4′-C to form the cyclic thiazolidine derivative (Fig. 31). Histidine, tryptophan, and several related compounds (e.g., histamine and tryptamine) can form similar cyclic complexes with PL or PLP bases through reactions of the imidazolium and indolyl side chains, respectively.

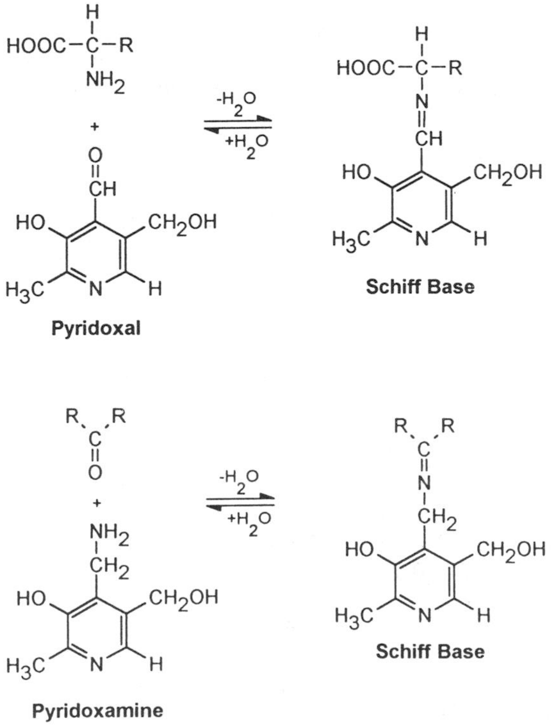

HOOC–CH(R)–NH_2 + Pyridoxal $\underset{+H_2O}{\overset{-H_2O}{\rightleftharpoons}}$ Schiff Base

Pyridoxamine + $R_2C{=}O$ $\underset{+H_2O}{\overset{-H_2O}{\rightleftharpoons}}$ Schiff Base

FIGURE 29 Formation of Schiff base structures from pyridoxal (PL) and pyridoxamine (PM). Analogous reactions occur with PLP and PMP.

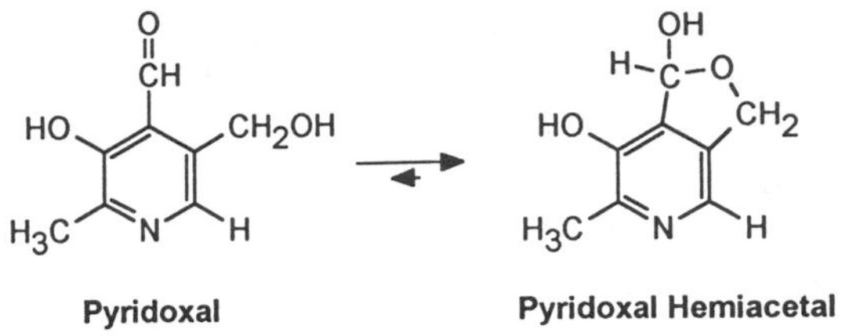

FIGURE 30 Formation of pyridoxal hemiacetal.

8.8.5.2 Stability and Modes of Degradation

Thermal processing and storage of foods can influence the vitamin B6 content in several ways. As with other water-soluble vitamins, exposure to water can cause leaching and consequent losses. Chemical changes can involve interconversion of chemical forms of vitamin B6, thermal or photochemical degradation, or irreversible complexation with proteins, peptides, or amino acids.

The interconversion of vitamin B6 compounds occurs mainly by nonenzymatic transamination, which involves formation of a Schiff base and migration of the Schiff base double bond, followed by hydrolysis and dissociation. Nonenzymatic transamination has been studied extensively as a model of PLP-mediated enzymatic transaminations. This process occurs extensively during the thermal processing of foods that contain the aldehyde or amine forms of vitamin B6. For example, increases in the proportion of PM and PMP are frequently observed in the cooking or thermal processing of meat and dairy products [14, 56] and in studies of protein-based liquid model systems [54]. The occurrence of such transamination has no adverse nutritional effect. Similar transamination has been shown to occur during the storage of intermediate moisture, model-food systems ($a_w \approx 0.6$). PL-mediated nonenzymatic elimination of H_2S and methyl mercaptan from sulfur-containing amino acids can also occur during food processing. This can be a significant source of flavor and can cause discoloration in canned foods through formation of black FeS [57].

All vitamin B6 compounds are susceptible to light-induced degradation, which can cause losses during food processing, preparation and storage and during analysis. The mechanism of vitamin B6 degradation by light is not well understood and the relationship between reaction rate and wavelength is not known. Light-mediated oxidation appears to be involved, presumably with a free radical intermediate. Exposure of vitamin B6 to light causes formation of the nutritionally inactive derivatives 4-pyridoxic acid (from PL and PM) and 4-pyridoxic acid

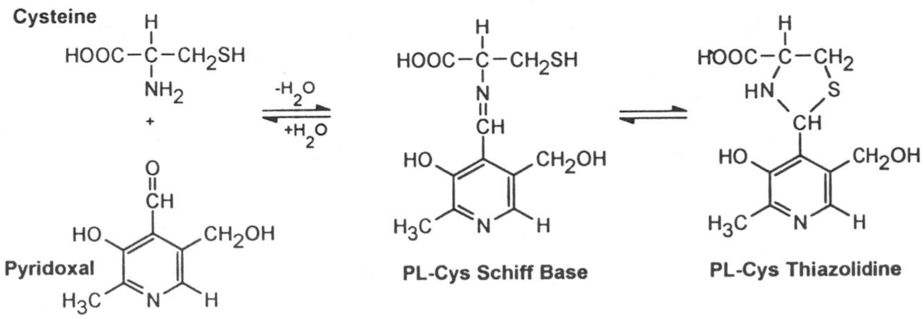

FIGURE 31 Formation of the Schiff base and thiazolidine complexes of pyridoxal (PL) and cysteine.

5′-phosphate (from PLP and PMP), providing evidence of susceptibility to photochemical oxidation [108, 114]. However, the rates of photochemical degradation of PLP, PMP, and PM and the amounts of degradation products obtained differ only slightly in the presence or absence of air, suggesting that initiation of oxidation does not require direct attack by O_2. Photochemical degradation of PL in low-moisture model food systems occurs at a greater rate than that of PL and PN. The reactions are first-order in PL concentration, are strongly influenced by temperature, but are affected only slightly by water activity (Table 22).

The rate of nonphotochemical degradation of vitamin B6 is strongly dependent on the form of the vitamin, temperature, pH of the solution, and the presence of other reactive compounds (e.g., proteins, amino acids, and reducing sugars). All forms of vitamin B6 exhibit excellent stability at very low pH (e.g., 0.1 *M* HCl), a condition used during extraction for vitamin B6 analysis. During incubation of vitamin B6 compounds at 40 or 60°C in aqueous solutions buffered at pH 4–7 for up to 140 days, PN exhibited no loss, PM exhibited the greatest loss at pH 7, and PL showned its greatest loss at pH 5 (Table 23) [114]. Degradation of PL and PM followed first-order kinetics in these studies. In contrast, in similar studies of the degradation of PL, PN, and PM at 110–145°C in aqueous solution buffered at pH 7.2, these compounds exhibited kinetics best described as second order, 1.5 order, and pseudo-first order, respectively [102]. In parallel studies, thermal degradation of vitamin B6 in cauliflower purée also did not conform to first-order kinetics. The reasons responsible for the kinetic differences of these studies are not clear, although the initial concentration of the vitamin B6 compounds may be a factor of importance. Under conditions of dry heat simulating toasting processes, degradation of PN in dehydrated model systems exhibited consistent first-order kinetics [41].

Studies of the thermal stability of vitamin B6 in foods are complicated because the multiple forms of the vitamin can undergo various degradation reactions, and interconversions can also occur among the various forms of the vitamin. Losses of total vitamin B6 in food processing or storage are similar to those observed with other water-soluble vitamins. For example, garbanzo and lima beans exhibit approximately 20–25% loss of total vitamin B6 during the blanching and canning process.

The development of HPLC methods has facilitated studies of the chemical behavior of vitamin B6 compounds during processing and storage. The simultaneous interconversion of PL and PM, along with first-order loss of total vitamin B6, has been observed in studies of thermal processing and storage of intermediate-moisture model food systems and in liquid model systems simulating infant formula (Fig. 32 and Table 24) [54]. PN exhibited greater stability than PL or

TABLE 22 Influence of Temperature, Water Activity, and Light Intensity on Degradation of Pyridoxal in a Dehydrated Model Food System

Light intensity (lumens/m^2)	a_w	Temperature, (°C)	k (day^{-1})	$t_{1/2}$ (days)
4300	0.32	5	0.092	7.4
		28	0.1085	6.4
		37	0.2144	3.2
		55	0.3284	2.1
4300	0.44	5	0.0880	7.9
		28	0.1044	6.6
		55	0.3453	2.0
2150	0.32	27	0.0675	10.3

Note: k, First-order rate constant; $t_{1/2}$, time for 50% degradation.
Source: Adapted from Ref. 114.

TABLE 23 Influence of pH and Temperature on the Degradation of Pyridoxal and Pyridoxamine in Aqueous Solution

Compound	Temperature (°C)	pH	k (day^{-1})	$t_{1/2}$ (days)
Pyridoxal	40	4	0.0002	3466
		5	0.0017	407
		6	0.0011	630
		7	0.0009	770
Pyridoxal	60	4	0.0011	630
		5	0.0225	31
		6	0.0047	147
		7	0.0044	157
Pyridoxamine	40	4	0.0017	467
		5	0.0024	289
		6	0.0063	110
		7	0.0042	165
Pyridoxamine	60	4	0.0021	330
		5	0.0044	157
		6	0.0110	63
		7	0.0108	64

Note: No significant degradation of pyridoxine was found at pH 4–7 at 40 or 60°C for up to 140 days. k, First-order rate constant; $t_{1/2}$ time for 50% degradation.
Source: Adapted from Ref. 114.

PM, although the magnitude of the difference varied with temperature (Table 24). Although differences in energy of activation suggest a difference in degradation mechanism for these three forms of the vitamin, calculation of the activation entropy, enthalpy, and free energy provides evidence of a common rate-limiting step for loss of PL, PM, and PN [51]. There is a need for further kinetic and thermodynamic studies of this type to assess more fully the behavior of naturally occurring vitamin B6 in various foods.

Extensive examination of the thermal stability of vitamin B6 in milk products was prompted by an unfortunate incident involving infant formulas. In the early 1950s over 50 cases of convulsive seizures occurred in infants that had consumed a commercially available milk-based infant formula [27], whereas thousands of infants consumed the same formula without ill effects. These convulsive disorders were corrected by administration of PN to the infants. The problem of inadequate vitamin B6 content in the processed formulas was corrected by fortification with PN, which is much more stable than PL, the major naturally occurring form of the vitamin in milk [63]. Commercial sterilization of evaporated milk or unfortified infant formula causes 40–60% loss of the naturally occurring vitamin B6. Little or no loss of added PN was found during a comparable thermal process. This incident highlights the need for complete and thorough assessment of the nutritional quality of foods, especially when new formulations and processing methods are employed.

The occurrence of vitamin B6 deficiency in the case of these unfortified infant formulas has been attributed, at least in part, to the interaction of PL with milk proteins during processing to form a sulfur-containing derivative, bis-4-pyridoxyl-disulfide (Fig. 33). Bis-4-pyridoxyl-disulfide has been observed to form slowly after heating a concentrated solution of PL and cysteine [140], and it exhibits only partial (~20%) vitamin activity in rat bioassays. Evidence of involvement of sulfhydryl groups in the interaction of PL with milk proteins also has been

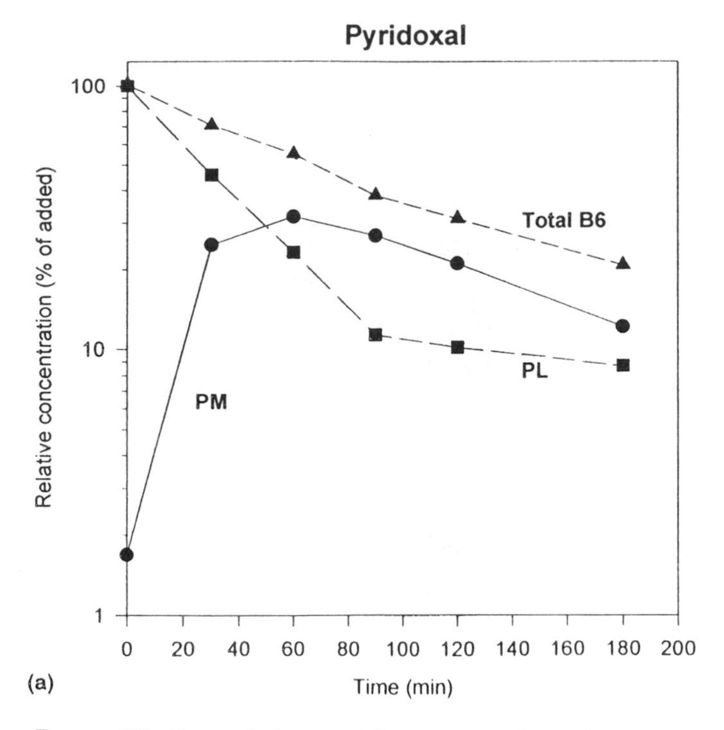

FIGURE 32 Degradation and interconversion of vitamin B6 compounds during thermal processing at 118°C of a liquid model food system simulating infant formula. Original vitamin B6 content 100% pyridoxal (PL) (a), 100% pyridoxine (PN) (b), and 100% pyridoxamine (PM) (c). (From Ref. 54.)

observed [124]. However, HPLC analysis of thermal processing of milk containing radiolabeled PL and PLP revealed no evidence of the formation of bis-4-pyridoxyl-disulfide [56]. In this same study, PL and PLP were found to undergo extensive binding to lysyl ε-amino groups of milk proteins by reduction of the Schiff base -C=N-linkage (Fig. 34). The formation of such pyridoxyllysyl residues also has been detected in thermally processed muscle and liver and during storage of intermediate-moisture model food systems. The mechanism by which the reduction of the Schiff base linkage occurs has not been determined.

Pyridoxyllysyl residues associated with food proteins have been shown to exhibit approximately 50% of the vitamin B6 activity of PN [48]. When administered to vitamin B6-deficient rats, this compound exacerbates the deficiency. This effect may have been involved in the deficiencies associated with consumption of an unfortified thermally processed infant formula mentioned earlier.

The role of protein sulfhydryl groups in the interaction of PL with protein has not been fully resolved. Sulfhydryls, as is true of ε-amino or imidazolium groups of amino acid side chains, can reversibly interact with the Schiff base linkage of protein-bound PL to form a substituted aldamine in a manner analogous to that shown in Figure 31.

Vitamin B6 also can be converted to biologically inactive compounds by reactions with free radicals. Hydroxyl radicals generated during degradation of ascorbic acid can directly attack the C^6 position of PN to form the 6-hydroxy derivative [126]. Presumably this reaction could occur with all other forms of vitamin B6. 6-Hydroxypyridoxine totally lacks vitamin B6 activity.

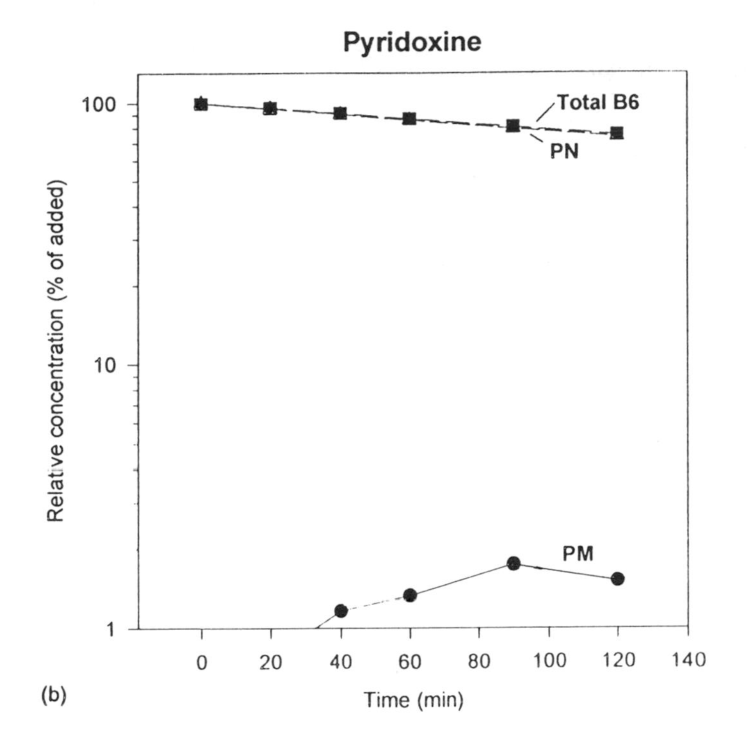
Pyridoxine
Total B6
PN
PM
Relative concentration (% of added)
100
10
1
0
20
40
60
80
100
120
140
Time (min)

(b)

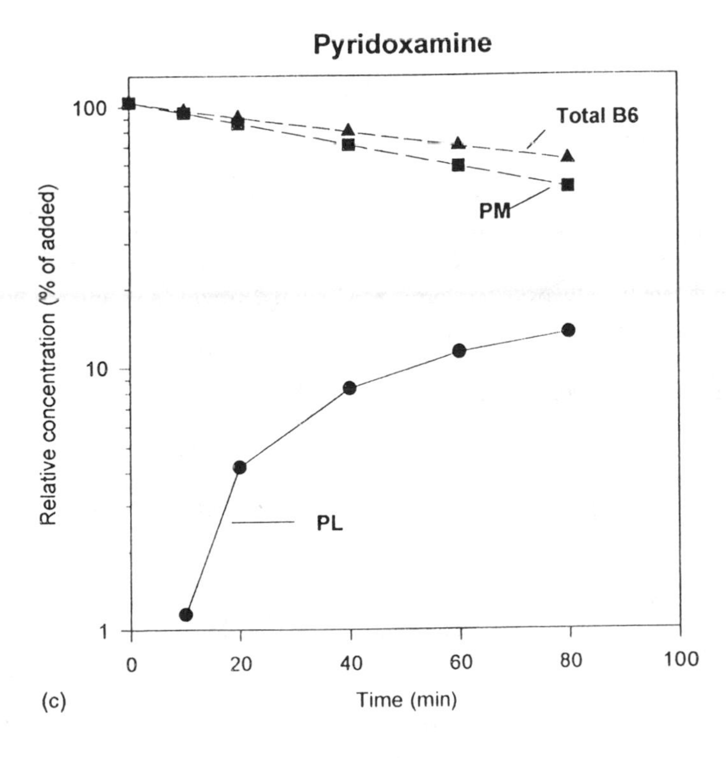
Pyridoxamine
Total B6
PM
PL
Relative concentration (% of added)
100
10
1
0
20
40
60
80
100
Time (min)

(c)

TABLE 24 Rate Constants and Energies of Activation for the Thermal Loss of Total Vitamin B6 in Liquid Model System Simulating Infant Formula

Form of vitamin B6 added	Temperature (°C)	k (min^{-1})	$t_{1/2}$ (min)	Energy of activation (kJ/mol)
Pyridoxine	105	0.0006	1120	114
	118	0.0025	289	
	133	0.0083	62	
Pyridoxamine	105	0.0021	340	99.2
	118	0.0064	113	
	133	0.0187	35	
Pyridoxal	105	0.0040	179	87.0
	118	0.0092	75	
	133	0.0266	24	

Note: k, First-order rate constant; $t_{1/2}$,time for 50% degradation.
Source: Ref. 54.

8.8.5.3 Bioavailability of Vitamin B6

The bioavailability of the total vitamin B6 content of a typical mixed diet has been estimated to be approximately 75% for adult humans [129]. Dietary PL, PN, PM, PLP, PMP, and PNP, if the latter is present, appear to be efficiently absorbed and effectively function in vitamin B6 metabolism. Schiff base forms a PL, PLP, PM, and PMP dissociate in the acidic environment of the stomach and exhibit high bioavailability.

PN glucoside and other glycosylated forms of vitamin B6 are inefficiently utilized in humans. The bioavailability of PN glucoside is approximately 60% relative to PN, although wide variation is observed among individuals. The importance of the incomplete bioavailability of PN glucoside in human diets depends largely on the total quantity of vitamin B6 consumed and the selection of foods. However, even foods with high percentages of PN glucoside can be quite effective sources of dietary vitamin B6 because of this reasonably good bioavailability. PN glucoside bioavailability varies markedly among animal species. It is essential that analytical methods for vitamin B6 are able to detect the glycosylated forms and, ideally, provide a specific measurement of their relative quantities.

Determining the bioavailability of PL or PLP in the form of a pyridoxylamino compound (e.g., pyridoxyllysine, Fig. 34) is not simple. In contrast to the Schiff base forms, pyridoxylamino

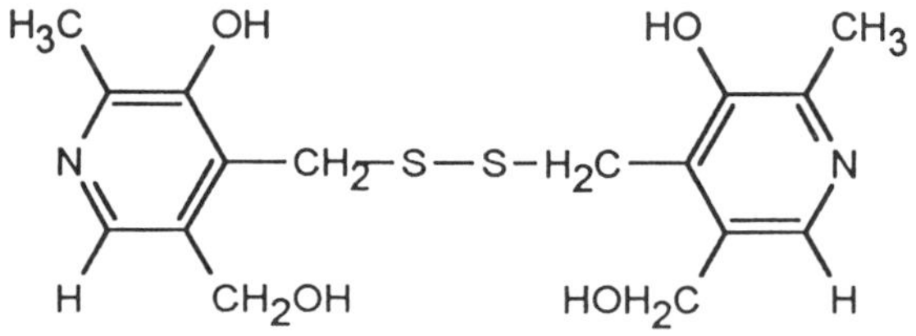

FIGURE 33 Structure of bis-4-pyridoxyl-disulfide.

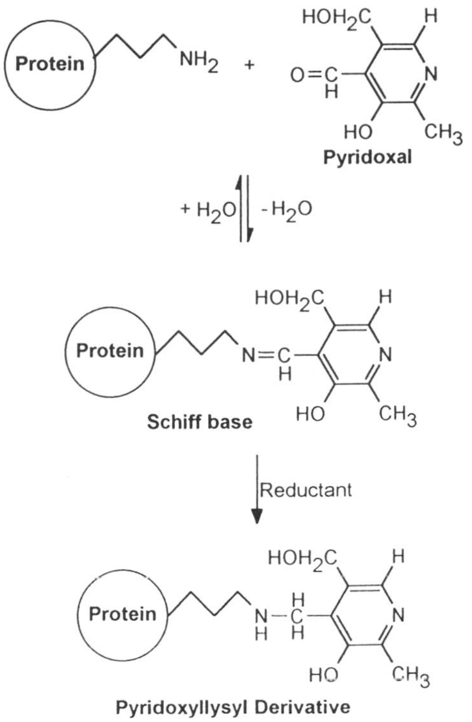

FIGURE 34 Interaction of PL with the ε-amino group of a lysyl residue of a food protein to form a Schiff base, followed by reduction to a pyridoxylamino complex.

forms have a very stable reduced linkage between PL or PLP and the lysine ε-amino group. Because of this stable covalent linkage, there is little or no dissociation of the B6 moiety of pyridoxylamino compounds during extraction conditions typically used in vitamin B6 analysis. Thus, the reductive binding of PL or PLP to protein appears in most food analyses to be a mode of vitamin B6 degradation and is seen as a loss of measurable vitamin B6. However, studies of the bioavailability of PLP bound reductively (pyridoxyllysyl residues) to dietary protein have indicated a bioavailability of approximately 50% [48]. Mammals can partially utilize enzymatic phosphorylation and oxidative cleavage, which frees the PLP moiety. Pyridoxyllsine can exert a weak antivitamin B6 activity, although this contributes to vitamin B6 deficiency only when the diet is marginal with respect to total vitamin B6 content [48].

8.8.5.4 Measurement of Vitamin B6

Vitamin B6 can be measured by microbiological assay methods or by HPLC [52]. Microbiological assays for total vitamin B6 can be performed using the yeasts *Saccharomyces uvarum* or *Kloeckera brevis*. Yeast growth assays involve a prior acid hydrolysis to extract vitamin B6 from food, and to hydrolyze phosphate esters and β-glucosides. Care must be taken when using microbiological assays because the organisms used may underestimate PM. HPLC methods are based mainly on reverse-phase or ion-exchange separation with fluorometric detection, and direct measurement of each form of the vitamin can be achieved.

8.8.6 Folate

8.8.6.1 Structure and General Properties

The generic term "folate" refers to the class of compounds having chemical structure and nutritional activity similar to that of folic acid (pteroyl-L-glutamic acid). The various components of this class are designated as "folates." The use of "folacin" and "folic acid" as generic terms is no longer recommended. Folic acid consists of L-glutamic acid that is coupled through its α-amino group to the carboxyl group of *para*-aminobenzoic acid, which, in turn, is linked to a 2-amino-4-hydroxypteridine (Fig. 35). In folic acid, the pteridine moiety is fully oxidized; that is, the pteridine exists as a fully double-bonded conjugated system. All folates contain an amide-like structure involving N^3 and C^4 that resonates between the two forms shown in Figure 36.

Folic acid (pteroyl-L-glutamic acid) exists naturally only in trace quantities. The major naturally occurring forms of folate in materials of plant, animal, and microbial sources are polyglutamyl species of 5,6,7,8-tetrahydrofolates (H_4folates) (Fig. 35), in which two double bonds of the pteridine ring system are reduced. Small amounts of 7,8-dihydrofolates (H_2folates) also exist naturally (Fig. 35). Metabolically H_4folates are mediators of one-carbon transformations, that is, the transfer, oxidation, and reduction of one-carbon units, which account for the presence of folates having various one-carbon-substituted forms in living cells. One-carbon substituents can exist on either the N^5 or N^{10} positions (predominantly as methyl or formyl groups), or as methylene ($-CH_2-$) or methenyl ($-CH=$) units bridging between N^5 and N^{10} (Fig. 35).

Most of the naturally occurring folates in plant and animal tissues, and in foods derived from plant and animal sources, have a side chain of five to seven glutamate residues with γ-peptide linkages. It is generally assumed that approximately 80% of food folate exists in polyglutamyl form. Limited data exist concerning the distribution of the various folates in foods. All folates, regardless of oxidation state of the pteridine ring system, N^5 or N^{10} one-carbon substituent, or polyglutamyl chain length, exhibit vitamin activity in mammals, including

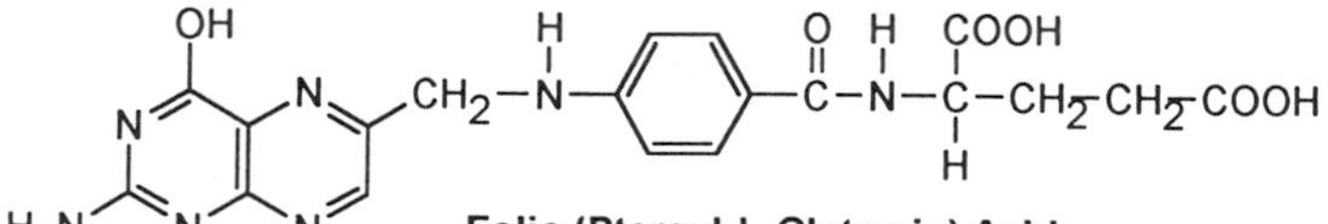

Folic (Pteroyl-L-Glutamic) Acid

Polyglutamyl Tetrahydrofolates

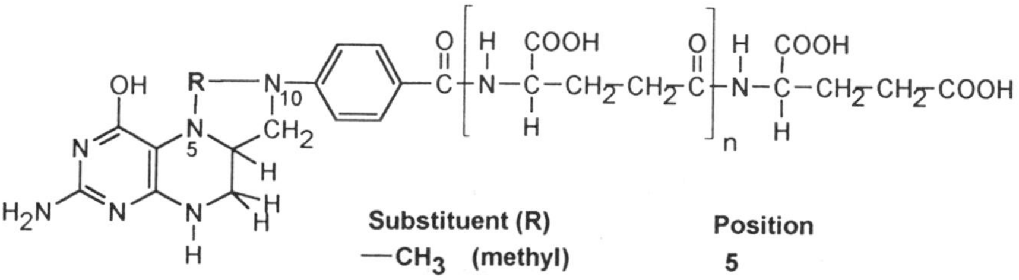

Substituent (R)	Position
—CH_3 (methyl)	5
—CHO (formyl)	5 or 10
—CH=NH (formimino)	5
—CH_2— (methylene)	5 and 10
—CH= (methenyl)	5 and 10

FIGURE 35 Structures of folates.

FIGURE 36 Resonance of the 3,4-amide site of folate pteridine ring system. The fully oxidized pteridine system of folic acid is shown; H_4folates and H_2folates exhibit identical behavior.

humans. Folate analogues with a 4-amino group or other modifications frequently are potent antagonists used in chemotherapy for cancer and autoimmune disease.

Folates undergo changes in ionic form as a function of pH (Table 25). Changes in charge of the pteridine ring system account, in part, for the pH dependence of folate stability, and for the pH-dependent behavior of folates during chromatographic separation.

Asymmetric carbons (glutamyl α-carbon of all folates and pteridine C^6 of H_4folates) can each exist in either of two configurations. The glutamic acid moiety must be in the L isomeric form for vitamin activity, while C^6 must be in the 6S isomeric form for tetrahydrofolates to exhibit vitamin activity. Tetrahydrofolates synthesized by chemical reduction of folic acid contain a racemic carbon at C^6, while those reduced enzymatically are fully in the 6S configuration. Since the 6S configuration is required for vitamin activity in animals (including humans) and in microbiological assays for folate, racemic (6R + 6S) preparations of H_4folates only exhibit 50% nutritional activity.

The 5-formyl or 10-formyl folates have an aldehyde group as the one-carbon substituent. Formyl forms of H_4folate are interconvertible through the 5,10-methenyl intermediate. Formation of the methenyl species from either 5-formyl or 10-formyl-H_4folate is favored only at pH < 2; thus this form is but a minor constituent of folates in most foods. The transient existence of 5,10-methenyl H_4folate at pH > 2 accounts for the conversion of 10-formyl-H_4folate to the more stable 5-formyl-H_4folate when heated in weak acid, and for the pH-dependent formation of 10-formyl-H_4folate from 5-formyl-H_4folate [110].

Large differences in stability exist among the various H_4folates as a result of the influence of the one-carbon substituent on susceptibility to oxidative degradation. In most cases, folic acid (with the fully oxidized pteridine ring system) exhibits substantially greater stability than the H_4folates or H_2folates. The order of stability of the H_4folates is 5-formyl-H_4folate > 5-methyl-H_4folate > 10-formyl-H_4folate ≥ H_4folate. Stability of each folate is dictated only by the chemical nature of the pteridine ring system, with no influence of polyglutamyl chain length. The inherent differences in stability among folates, as well as chemical and environmental variables influencing folate stability, will be discussed further in the next subsection.

TABLE 25 pK_a Values for Ionizable Groups of Folates

Folate compound	Amide	N^1	N^5	N^{10}	α-COOH	γ-COOH
5,6,7,8-H_4folate	10.5	1.24	4.82	−1.25	3.5	4.8
7,8-H_2folate	9.54	1.38	3.84	0.28	ND	ND
Folic acid	8.38	2.35	<−1.5	0.20	ND	ND

Note: ND, not determined due to insufficient solubility. It is assumed that the pK_a values of these carboxyl groups are similar for all folates. Amide refers to dissociation of the N^3–C^4 amide-like site.
Source: Adapted from Ref. 105.

All folates are subject to oxidative degradation, although the mechanism and the nature of the products varies among the various chemical species of the vitamin. Reducing agents such as ascorbic acid and thiols exert multiple protective effects on folates through their actions as oxygen scavengers, reducing agents, and free radical scavengers.

Aside from molecular oxygen, other oxidizing agents found in foods can have deleterious effects on folate stability. For example, at concentrations similar to those used for antimicrobial treatments, hypochlorite causes oxidative cleavage of folic acid, H_2folate, and H_4folate to nutritionally inactive products. Under the same oxidizing conditions, certain other folates (e.g., 5-methyl-H_4folate) are converted to forms that may retain at least partial nutritional activity. Light is also known to promote the cleavage of folates, although the mechanism has not been determined.

Relative to the nutritional needs of humans, folate is frequenly among the most limiting of the vitamins. This is probably due to (a) poor diet selection, especially with respect to foods rich in folate (e.g., fruits, especially citrus, green leafy vegetables, and organ meats), (b) losses of folate during food processing and/or home preparation by oxidation, leaching, or both, and (c) incomplete bioavailability of folate in many human diets.

Folic acid, because of its excellent stability, is the sole form of folate added to foods, albeit in relatively few types of products (e.g., many breakfast cereals, infant formulas, nutritional supplements, and meal replacements), and it is also used in vitamin pills. In clinical situations requiring use of a reduced folate, 5-formyl-H_4folate is employed because of its stability (similar to folic acid).

8.8.6.2 Stability and Modes of Degradation

Folate Stability

Folic acid exhibits excellent retention during the processing and storage of fortified foods and premixes [49]. As shown in Tables 2 and 3, little degradation of this form of the vitamin occurs during extended low-moisture storage. Similar good retention of added folic acid has been observed during the retorting of fortified infant formulas and medical formulas.

Many studies have shown the potential for extensive losses of folate during processing and home preparation of foods. In addition to susceptibility to oxidative degradation, folates are readily extracted from foods by aqueous media (Table 26). By either means, large losses of naturally occurring folate can occur during food processing and preparation. The overall loss of folate from a food depends on the extent of extraction, forms of folate present, and the nature of

TABLE 26 Effect of Cooking on Folate Content of Selected Vegetables

	Total folate[a] (μg/100 g fresh weight)		
Vegetable (boiled 10 min in water)	Raw	Cooked	Folate in cooking water
Asparagus	175 ± 25	146 ± 16	39 ± 10
Broccoli	169 ± 24	65 ± 7	116 ± 35
Brussels sprouts	88 ± 15	16 ± 4	17 ± 4
Cabbage	30 ± 12	16 ± 8	17 ± 4
Cauliflower	56 ± 18	42 ± 7	47 ± 20
Spinach	143 ± 50	31 ± 10	92 ± 12

[a]Mean ± SD, n = 4.
Source: Adapted from Ref. 82.

the chemical environment (catalysts, oxidants, pH, buffer ions, etc.). Thus, folate retention is difficult to predict for a given food.

Degradation Mechanisms

The mechanism of folate degradation depends on the form of the vitamin and the chemical environment. As mentioned previously, folate degradation generally involves changes at the C^9-N^{10} bond, the pteridine ring system, or both. Folic acid, H_4folate, and H_2folate can undergo C^9-N^{10} cleavage and resulting inactivation in the presence of either oxidants or reductants [91]. Dissolved SO_2 has been found to cause cleavage of certain folates, although few other reducing agents relevant to foods can induce such cleavage. There is little direct oxidative conversion of H_4folate or folic acid.

It is well known that oxidative cleavage of H_4folates, H_2folate, and, to a lesser extent, folic acid yields nutritionally inactive products (*p*-aminobenzoylglutamate and a pterin). The mechanism of oxidation and the exact nature of the pterin produced during oxidative cleavage of H_4folate vary with pH, as shown in Figure 37.

The major naturally occurring form of folate in many foods is 5-methyl-H_4folate. Oxidative degradation of 5-methyl-H_4folate occurs by conversion to at least two products (Fig. 38). The first has been identified tentatively as 5-methyl-5,6-dihydrofolate (5-methyl-H_2folate), which retains vitamin activity since it can be readily reduced back to 5-methyl-H_4folate by weak reductants such as thiols or ascorbate. 5-Methyl-H_2folate undergoes cleavage of the C^9-N^{10} bond in acidic medium, which causes losses of vitamin activity. The alternate product of 5-methyl-H_4folate degradation was originally identified as 4a-hydroxy-5-methyl-H_4folate, although spectral data are more consistent with a pyrazino-*s*-triazine structure formed by rearrangement of the pteridine ring (Fig. 38) [68]. Whether 5-methyl-H_2folate is an intermediate in the formation of the pyrazino-*s*-triazine has not been determined.

Blair et al. [10] reported that the pH dependence of 5-methyl-H_4folate oxidation is pronounced. Stability (as monitored by oxygen uptake) increases as pH is reduced from 6 to 4, this range corresponding to the range of protonation of the N^5 position. Contrary results have been reported [95], and factors responsible for this contradiction have not been determined.

In certain animal products, such as liver, 10-formyl-H_4folate may account for as much as one-third of the total folate. Oxidative degradation of 10-formyl-H_4folate can occur either by oxidation of the pteridine moiety to yield 10-formyl-folic acid or by oxidative cleavage to form a pterin and *N*-formyl-*p*-aminobenzoylglutamate (Fig. 39). 10-Formyl-folic acid has nutritional activity while the cleavage products do not. Factors that influence the relative importance of these oxidative pathways in foods have not been determined. In contrast to 10-formyl-H_4folate, 5-formyl-H_4folate exhibits excellent thermal and oxidative stability.

Factors Affecting Folate Stability

Many studies have been conducted to compare the relative stability of folates in buffered solution as a function of pH, oxygen concentration, and temperature. Stability of folates in complex foods is less well understood.

Folic acid is generally the most stable form. It is resistant to oxidation, although reduced stability occurs in acidic media. H_4folate is the least stable form of the vitamin. Maximal stability of H_4folate is observed between pH 8 and 12, and 1 and 2, while the stability is minimal between pH 4 and 6. However, even in the favorable pH zones, H_4folate is extremely unstable. H_4folates having a substituent at the N^5 position exhibit much greater stability than does unsubstituted H_4folate. This suggests that the stabilizing effect of the N^5 methyl group is due, at least in part, to steric hindrance in restricting access of oxygen or other oxidants to the pteridine ring. The stabilizing effect of the N^5-substituent is more pronounced with 5-formyl-H_4folate than with

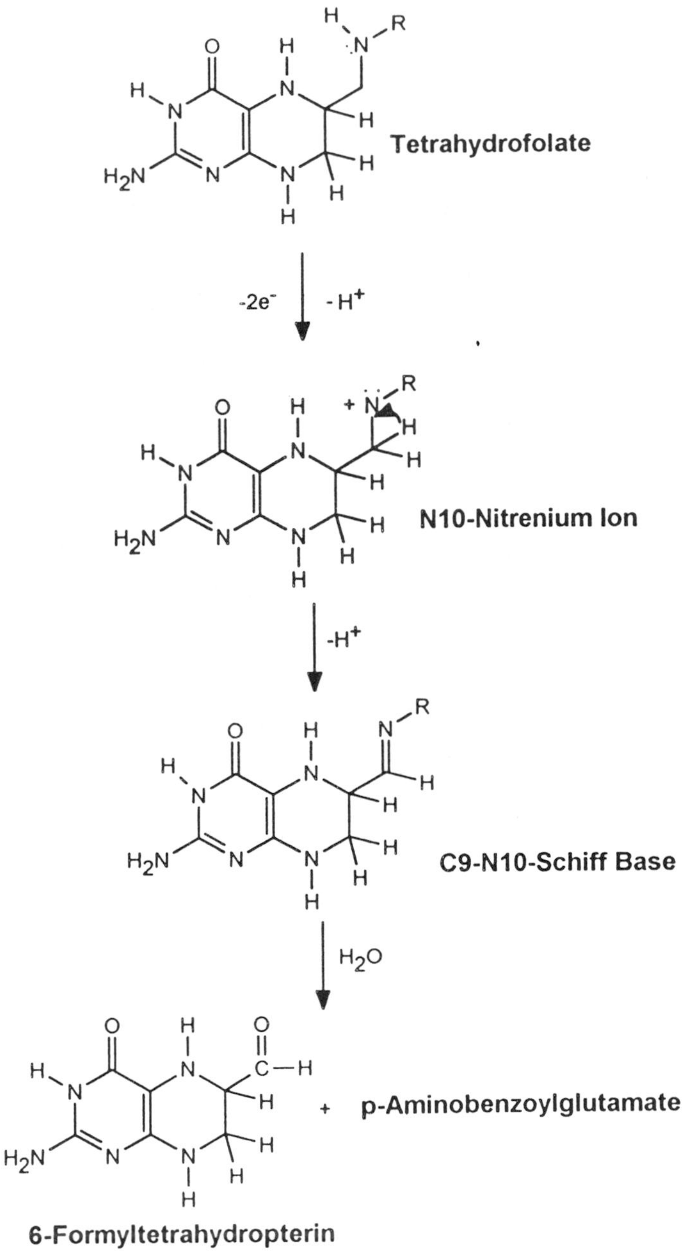

(a)

FIGURE 37 Oxidative degradation of H_4folate. (Adapted from Ref. 107)

5-methyl-H_4folate, and both exhibit much greater stability than H_4folate or 10-formyl-H_4folate. Under conditions of low oxygen concentration, 5-methyl-H_4folate and folic acid exhibit similar stability during thermal processing.

The influence of oxygen concentration on the stability of folates in foods, buffer solutions, and model food systems has been widely studied. As mentioned previously, the rate of oxidation

Tetrahydrofolate

$-2e^-$ $-2H^+$

Quinonoid Dihydrofolate

7,8-Dihydropterin + p-Aminobenzoylglutamate + HCHO

(b)

of 5-methyl-H_4folate is dependent on the concentration of dissolved oxygen in accord with a pseudo-first-order relationship. In relatively anaerobic conditions, the presence of added components such as ascorbate, ferrous iron, and reducing sugar tends to improve the oxidative stability of folic acid and 5-methyl-H_4folate. These components apparently function by reducing the concentration of dissolved oxygen through their own oxidation reactions (Fig. 40). These findings indicate that complex foods can contain components that influence folate stability by consuming oxygen, acting as reducing agents, or both.

Barrett and Lund [6] studied thermal degradation of 5-methyl-H_4folate in neutral buffer solution and observed both aerobic and anaerobic degradation. Surprisingly, rate constants for aerobic and anaerobic degradation reactions are of similar magnitude (Table 27). The extent to which other folates conform to this behavior has not been determined.

The ionic composition of the medium also significantly influences the stability of most folates. Phosphate buffers have been reported to accelerate oxidative degradation of folates, while this effect can be overcome by addition of citrate ions. The frequent presence of Cu(II) as a contaminant in phosphate buffer salts may explain this effect because metal catalysts are known

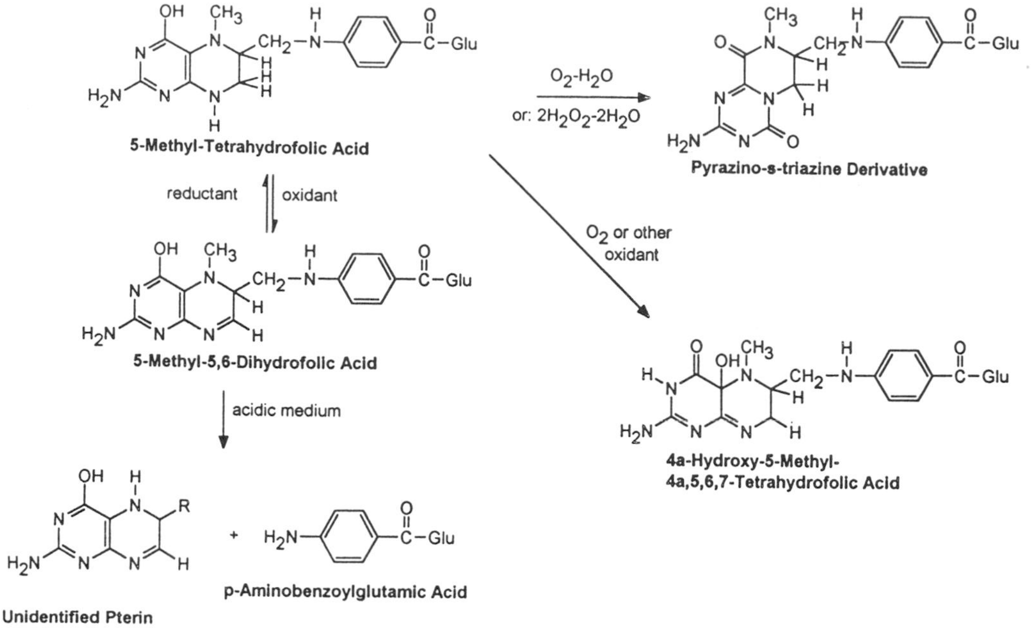

FIGURE 38 Oxidative degradation of 5-methyl-H_4folate.

to accelerate folate oxidations. For example, in aerobic solutions of 5-methyl-H_4folate in water, addition of 0.1 m*M* Cu(II) causes nearly a 20-fold acceleration in oxidation rate, although Fe(III) causes only a 2-fold increase [10]. Under anaerobic conditions, Fe(III) catalyzes oxidation of H_4pteridines (e.g., H_4folate) → H_2pteridines (e.g., H_2folate) → fully oxidized pteridines (e.g., folic acid). The reason for the differences in the catalytic efficiency of these metals is not known. Folates undergo degradation by superoxide ions [121, 127], although the extent of such radical-mediated losses of folates in foods has not been determined.

Several reactive components of foods may accelerate degradation of folates. Dissolved SO_2 can cause reductive cleavage of folates, as stated previously. Exposure to nitrite ions

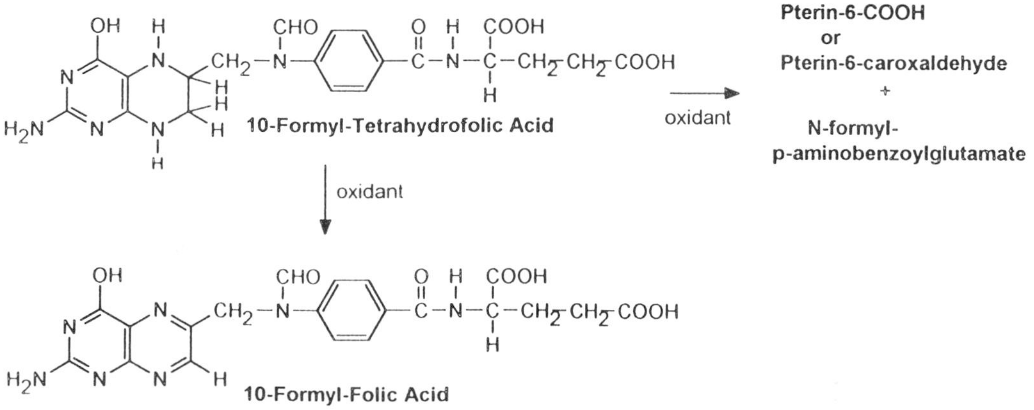

FIGURE 39 Oxidative degradation of 10-formyl-H_4folate.

contributes to the oxidative cleavage of 5-methyl-H_4folate and H_4folate. In contrast, nitrite reacts with folic acid to yield 10-nitroso-folic acid, a weak carcinogen. The significance of the latter reaction in foods is minimal because folic acid does not occur significantly in foods containing nitrite. Oxidative degradation of folates by exposure to hypochlorite may yield significant losses of folates in certain foods.

8.8.6.3 Bioavailability of Folate in Foods

The absorption of folates takes place mainly in the jejunum and requires hydrolysis of the polyglutamyl chain by a specific peptidase (pteroylpolyglutamate hydrolase), followed by absorption via a carrier-mediated transport process [49, 53]. Bioavailability of naturally occurring folate in foods is incomplete, often averaging 50% or less [115]. Moreover, the bioavailability of naturally occurring folates in most foods has not been fully determined under conditions of actual consumption, including the consequences of interactions among various foods. The mean bioavailability of polyglutamyl folates is typically 70% relative to the monoglutamyl species, which indicates the rate-limiting nature of intestinal deconjugation. In addition, the bioavailability of folic acid added to cereal grain products has been reported to be only 30–60% [24].

Factors responsible for incomplete bioavailability include (a) effects of the food matrix, presumably through noncovalent binding of folates; (b) possible degradation of labile H_4folates in the acidic gastric environment; and (c) incomplete intestinal enzymatic conversion of polyglutamyl folates to the absorbable monoglutamyl forms. Many foods contain compounds that inhibit intestinal pterolypolyglutamate hydrolase when studied in vivo; however, the significance of these effects with respect to in vivo folate bioavailability is unclear. Many raw fruits, vegetables, and meats also contain active hydrolases capable of deconjugating polyglutamyl folates. Homogenization, freezing and thawing, and other procedures that disrupt cells may release these enzymes and promote the deconjugation process. The extent to which this would improve bioavailability of dietary folates has not been determined. Little or no deconjugation of polyglutamyl folates occurs during food preparation and processing, unless endogenous conjugases are present.

8.8.6.4 Analytical Methods

Techniques potentially suitable for the measurement of folate in foods include microbiological growth methods, HPLC methods, and competitive-binding radioassay procedures [49]. Measurement of folate is complicated by the need to account for all forms of the vitamin, which could easily include several dozen compounds if each form of the folate nucleus existed in all possible combinations with several different polyglutamate chain lengths. Prior to the early 1960s, folate assays yielded grossly inaccurate results because a necessary reducing agent in the extraction buffer and in the microbiological assay medium was not included. Either ascorbate, a thiol reagent such as mercaptoethanol, or a combination of ascorbate and thiol is needed to stabilize folates during extraction and analysis.

Extraction of folate from food samples involves (a) disruption of the food matrix and cellular structure by homogenization in a buffer solution; (b) heat (typically 100°C) to release folate from folate-binding proteins, to inactivate enzymes capable of catalyzing interconversion of folates, and to deproteinate the sample; (c) centrifugation to yield a clarified extract; and (d) treatment with a pteroylglutamate hydrolase ("conjugase"), if the assay responds only to monoglutamyl or other short-chain folates. Other enzymes such as a protease or amylase may be useful in improving extraction of folate from certain foods. A need exists to standardize extraction and enzymatic pretreatment methods so interlaboratory precision and accuracy of folate assays can be improved.

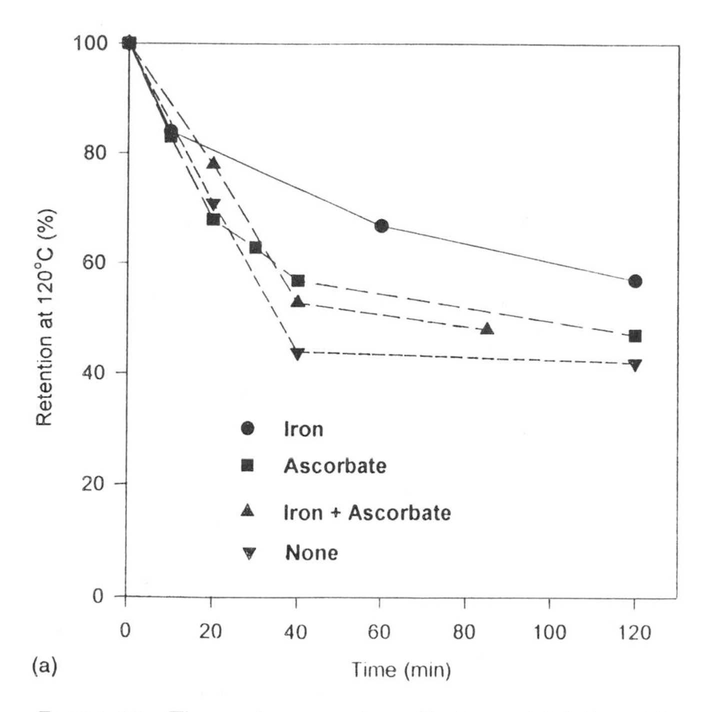

FIGURE 40 Thermal processing effects on (a) folic acid and (b) 5-methyl-H_4folate in liquid model food systems simulating infant formula. The model system consisted of 1.5% (w/v) potassium caseinate and 7% (w/v) lactose in 0.1 *M* phosphate buffer, pH 7.0. When present, iron was added at 6.65 mg/100 ml ferrous sulfate heptahydrate and ascorbate was added at 6.38 mg/100 ml sodium ascorbate. Initial concentration of folates was 10 mg/ml. (From Ref. 29.)

Microbiological growth assays serve as the traditional method of folate analysis and are based on the nutritional requirements of microorganisms (*Lactobacillus casei*, *Pediococcus cerevisiae*, and *Streptococcus faecium*). *Pediococcus cerevisiae* and *S. faceium* (used in the Association of Official Analytical Chemists official method) have little use in food analysis because they do not respond to all forms of the vitamin. In contrast, *L. casei* responds to all forms of folate and serves as the most appropriate test organism for microbiological assays of total folate in food. With appropriate control of pH in the growth medium, *L. casei* yields equivalent response to all forms of folate. Since foods typically contain several folates, verification of equivalent response in microbiological assays is essential.

Competitive-binding assays involve competition between folate in the sample or standard, with radiolabeled folate for the binding site of a folate-binding protein, typically from milk. In spite of the speed and convenience of these assays, their application to food analysis is limited because of varying affinity for different forms of folate. Comparisons of competitive-binding assays with the *L. casei* methods have yielded poor agreement, presumably for this reason.

Several HPLC methods have been developed for measurement of folates in foods and other biological materials. Several approaches to HPLC analysis are potentially suitable, although extractions, enzymatic pretreatments, and sample cleanup must be optimized to permit application to various types of foods. An important advance in HPLC is the use of affinity chromatography with a folate-binding protein column to yield a highly purified extract that is amenable to HPLC analysis [119]. HPLC analysis can be performed on extracted folates in intact polyglutamyl form or following enzymatic deconjugation to the monoglutamyl form [49, 119].

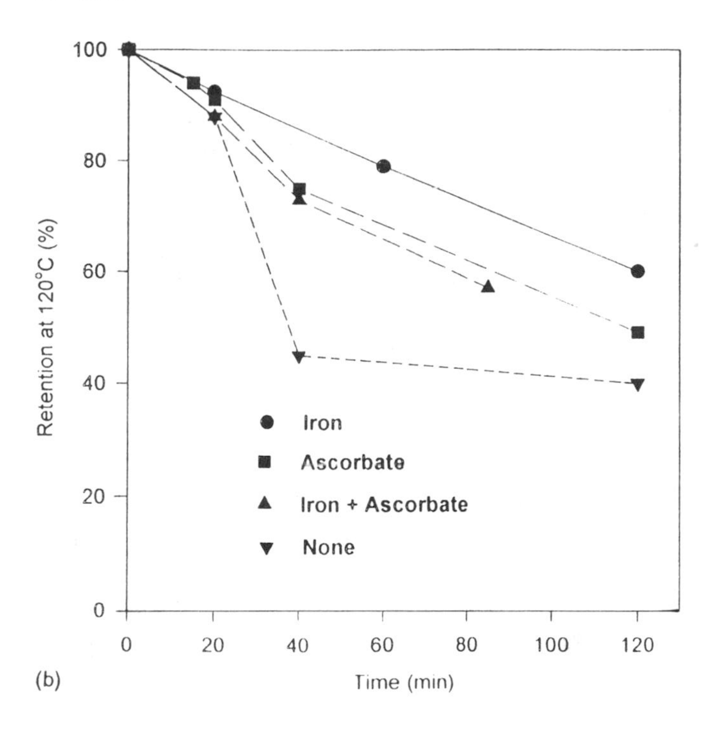

(b)

The folate content of foods is highly variable, presumably as a function of climate, agricultural practices, and postharvest handling, and of susceptibility to losses by leaching and oxidation. Thus, mean values for folate content in food composition data bases must be considered only as rough approximations. Furthermore, because of analytical limitations, especially with respect to extraction methods, much of the data used in such databases is of questionable accuracy.

TABLE 27 Reaction Rate Constants for the Degradation of 5-Methyl-H_4folate by Oxidative and Nonoxidative Processes in 0.1 *M* Phosphate Buffer, pH 7.0

Temperature (°C)	$k_{(O_2+N_2)}$ (combined oxidative + nonoxidative, min^{-1})	k_{N_2} (nonoxidative, min^{-1})	k_{O_2} (oxidative, min^{-1})
40	0.004 ± 0.0002	0.0004 ± 0.00001	0.004 ± 0.00005
60	0.020 ± 0.0005	0.009 ± 0.00004	0.011 ± 0.0001
80	0.081 ± 0.010	0.046 ± 0.003	0.035 ± 0.009
92	0.249 ± 0.050	0.094 ± 0.009	0.155 ± 0.044

Note: Values are means ±95% confidence intervals. Values for apparent first-order rate constants: k_{N_2}, degradation by nonoxidative process (in N_2-saturated environment); k_{O_2}, degradation by oxidative process (in O_2-saturated environment); $k_{(O_2+N_2)}$, degradation by both oxidative and nonoxidative processes.
Source: Ref. 6.

8.8.7 Biotin

8.8.7.1 Structure and General Properties

Biotin is a bicyclic, water-soluble vitamin that functions coenzymatically in carboxylation and transcarboxylation reactions. The two naturally occurring forms are free D-biotin and biocytin (ε-*N*-biotinyl-L-lysine) (Fig. 41). Biocytin functions as the coenzyme form and actually consists of a biotinylated lysyl residue covalently incorporated in a protein chain of various carboxylases. The ring system of biotin can exist in eight possible stereoisomers, only one of which (D-biotin) is the natural, biologically active form. Both free biotin and protein-bound biocytin exhibit activity when consumed in the diet. Biotin is widely distributed in plant and animal products, and biotin deficiency is rare in normal humans.

8.8.7.2 Stability of Biotin

Biotin is very stable to heat, light, and oxygen. Extremes of high or low pH can cause degradation, possibly because they promote hydrolysis of an -N-C=O (amide) bond of the biotin ring system. Oxidizing conditions such as exposure to hydrogen peroxide can oxidize the sulfur to form biologically inactive biotin sulfoxide or sulfone. Reaction of the biotin ring carbonyl with amines also may occur, although this has not been examined. Losses of biotin during food processing and subsequent storage have been documented and summarized [12, 67, 86]. Such losses may occur by chemical degradation processes as mentioned earlier and by leaching of free biotin. Little degradation of biotin occurs during low-moisture storage of fortified cereal products. Overall, biotin is quite well retained in foods.

The stability of biotin during storage of human milk also has been examined [174]. The biotin concentration of the milk samples did not change over 1 week at ambient temperature, 1 month at 5°C, or at –20°C or lower for 1.5 years.

8.8.7.4 Analytical Methods

Measurement of biotin in foods is performed by microbiological assay (usually with *Lactobacillus plantarum*) or by various ligand binding procedures involving avidin as the binding

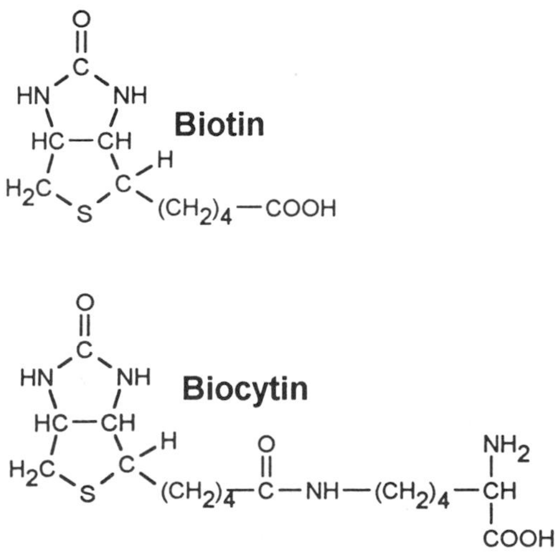

FIGURE 41 Structures of biotin and biocytin.

protein. Several HPLC methods also have been developed but their applicability to foods has not been fully established. The microbiological and ligand binding assays respond to free biotin and biocytin, but biocytin cannot be determined unless it is first released from the protein by cleavage of the peptide bond by enzymatic or acid hydrolysis [96]. Care should be taken because acid hydrolysis can degrade a substantial proportion of the biotin [97]. The existence of nutritionally inactive biotin analogs, such as bis-norbiotin and biotin sulfoxide detected in human urine, may complicate analyses. Such analogs may respond in avidin binding procedures and certain microbiological assays.

8.8.7.3 Bioavailability

Relatively little is known about the bioavailability of biotin in foods. There appears to be sufficient biotin in normal diets so that incomplete bioavailability usually has little adverse nutritional impact. Bacterial synthesis of biotin in the lower intestine provides an additional source of partially available biotin for humans. The majority of biotin in many foods exists as protein-bound biocytin. This is released by biotinidase in pancreatic juice and in the intestinal mucosa to convert the bound biotin to the functionally active free form; however, some absorption of biotinyl peptides also may occur [96].

Absorption of biotin is almost totally prevented by the consumption of raw egg albumen which contains the biotin-binding protein avidin. Avidin is a tetrameric glycoprotein in egg albumen that is capable of binding one biotin per subunit. This protein binds very tightly (dissociation constant $\approx 10^{-15}$ *M*) and resists digestion. Little or no avidin-bound biotin is absorbed. Chronic consumption of raw eggs or egg albumen will thus impair biotin absorption and can lead to deficiency. Small amounts of avidin in the diet have no nutritional consequence. The use of dietary avidin (or egg albumen) permits the experimental development of biotin deficiency in laboratory animals. Cooking denatures avidin and eliminates its biotin-binding properties.

While little information exist regarding the bioavailability of biotin in humans, much more is known about its bioavailability in animal feedstuffs. As shown in Table 28, bioavailability of biotin is low in some materials.

TABLE 28 Bioavailabiliy of Biotin in Feedstuffs for Pigs and Turkeys

	Biotin bioavailability (%)	
Material	Pigs [116]	Turkeys [94]
Soybean meal	55.4	76.8
Meat and bone meal	2.7	ND
Canola meal	3.9	65.4
Barley	4.8	19.2
Corn	4.0	95.2
Wheat	21.6	17.0
Supplemental biotin	93.5	ND
Sorghum	ND	29.5

Note: ND, not determined.

8.8 Pantothenic Acid

8.8.8.1 Structure and General Properties

Pantothenic acid, or D-*N*-(2,4-dihydroxy-3,3-dimethyl-butyryl)-β-alanine, is a water soluble vitamin comprised of β-alanine in amide linkage to 2,4-dihydroxy-3,3-dimethyl-butyric (pantoic) acid (Fig. 42).Pantothenic acid functions metabolically as a component of a coenzyme A (Fig. 42) and as a covalently bound prosthetic group (without the adenosyl moiety of coenzyme A) of acyl carrier proteins in fatty acid synthesis. Formation of a thioester derivative of coenzyme A (CoA) with organic acids facilitates a wide variety of metabolic processes, mainly involving addition or removal of acyl groups, in an array of biosynthetic and catabolic reactions. Pantothenic acid is essential for all living things and is distributed widely among meats, cereal grains, eggs, milk, and many fresh vegetables.

Pantothenic acid in many foods and most biological materials is mainly in the form of coenzyme A, the majority of which exists as thioester derivatives of a wide variety of organic acids. Although analytical data are quite limited with respect to the free and coenzyme A forms of pantothenic acid in foods, free pantothenic acid has been found to account for only half of the total of this vitamin in beef muscle and peas [58]. Coenzyme A is fully available as a source of pantothenic acid because it is converted to free pantothenic acid in the small intestine by the action of alkaline phosphatase and an amidase. Intestinal absorption occurs through a carrier-mediated absorption process.

Synthetic pantothenic acid is used in food fortification and in vitamin supplements in the

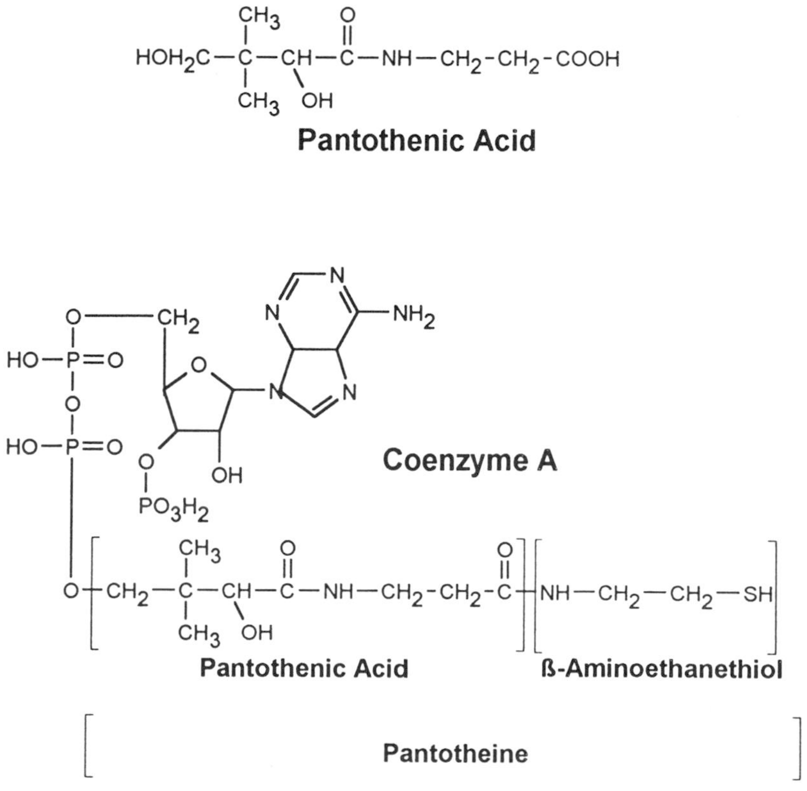

Pantothenic Acid

Coenzyme A

Pantothenic Acid

ß-Aminoethanethiol

Pantotheine

FIGURE 42 Structure of various forms of pantothenic acid.

form of calcium pantothenate. This compound is a white crystalline material that exhibits greater stability and is less hygroscopic than the free acid. Panthenol, the corresponding alcohol, also has been used as a feed supplement for animals.

8.8.8.2 Stability and Modes of Degradation

In solution, pantothenic acid is most stable at pH 5–7. Pantothenic acid exhibits relatively good stability during food storage, especially at reduced water activity. Losses occur in cooking and thermal processing in proportion to the severity of the treatment and extent of leaching and range from 30 to 80%. Leaching of pantothenic acid or its loss in tissue fluids can be very significant. Although the mechanism of thermal loss of pantothenic acid has not been fully determined, an acid-catalyzed hydrolysis of the linkage between β-alanine and the 1,4-dihydroxy,3,3-butyryl-carboxylic acid group appears likely. The pantothenic acid molecule is otherwise quite unreactive and interacts little with other food components. Coenzyme A is susceptible to the formation of mixed disulfides with other thiols in foods; however, this exerts little effect on the net quantity of available pantothenic acid.

Degradation of pantothenic acid during thermal processing conforms to first-order kinetics [58]. Rate constants for degradation of free pantothenic acid in buffered solutions increase with decreasing pH over the range of pH 6.0–4.0, while the energy of activation decreases over this range. The rates of degradation reported for pantothenic acid are substantially less than those for other labile nutrients (e.g., thiamin). These findings suggest that losses of pantothenic acid reported in other studies of food processing may be predominantly due to leaching rather than actual destruction. The net result would be the same, however.

8.8.8.3 Bioavailability

The mean bioavailability of pantothenate for humans consuming a mixed diet has been reported to be 51% [129]. There is little concern regarding any adverse consequences of this incomplete bioavailability. No evidence of nutritionally significant problems of incomplete bioavailability has been reported, and the complexed coenzymic forms of the vitamin are readily digested and absorbed.

8.8.8.4 Analytical Methods

Pantothenic acid in foods may be measured either by microbiological assay using *Lactobacillus plantarum* or by radioimmunoassay [143]. A key factor that affects the validity of pantothenic acid analysis is the pretreatment needed to release bound forms of the vitamin. Various combinations of proteases and phosphatases have been used to release pantothenic acid from the many coenzyme A derivatives and protein-bound forms.

8.8.9 Vitamin B12

8.8.9.1 Structure and General Properties

Vitamin B12 is the generic term for the group of compounds having vitamin activity similar to that of cyanocobalamin. These compounds are corrinoids, tetrapyrrole structures in which a cobalt ion is chelated by the four pyrrole nitrogens. The fifth coordinate covalent bond to Co is with a nitrogen of the dimethylbenzimidazole moiety, while the sixth position may be occupied by cyanide, a 5′-deoxyadenosyl group, a methyl group, water, a hydroxyl ion, or other ligands such as nitrite, ammonia or sulfite (Fig. 43). All forms of vitamin B12 shown in Figure 43 exhibit vitamin B12 activity. Cyanocobalamin, a synthetic form of vitamin B12 used in food fortification and nutrient supplements, exhibits superior stability and is readily available commercially. The coenzyme forms of vitamin B12 are methylcobalamin and 5′-deoxyadenosylcobalamin. Methylcobalamin

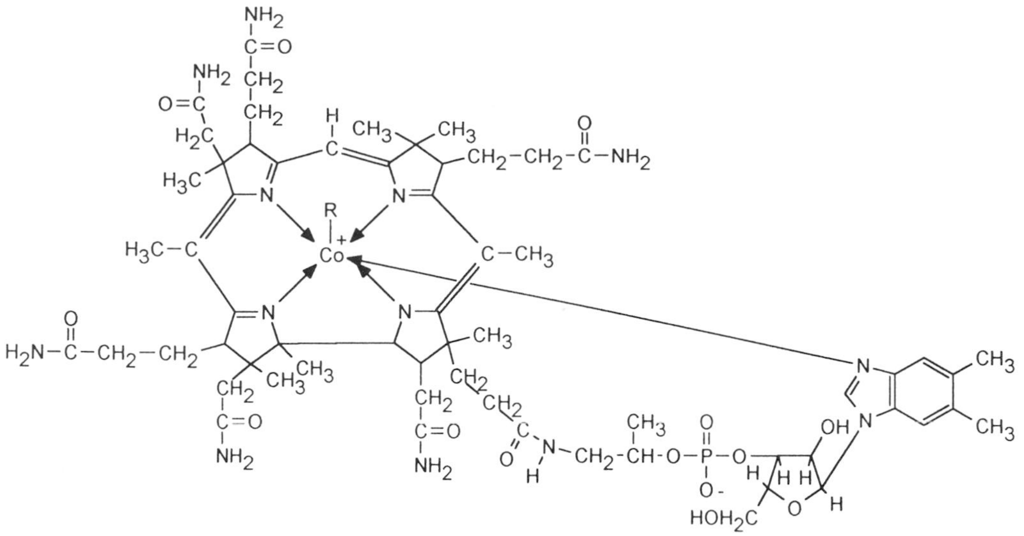

FIGURE 43 Structure of various forms of vitamin B12.

functions coenzymatically in the transfer of a methyl group (from 5-methyltetrahydrofolate) in methionine synthetase, while 5′-deoxyadenosylcobalamin serves as the coenzyme in an enzymatic rearrangement reaction catalyzed by methylmalonyl-CoA mutase. Little or no naturally occurring cyanocobalamin exists in foods; in fact, the original identification of vitamin B12 as cyanocobalamin involved its formation as an artifact of the isolation procedure. Cyanocobalamin has a reddish color in the crystalline state and in solution. This coloration may pose a limitation in the possible addition of cyanocobalamin to certain foods, especially light colored products (e.g., white bread).

Unlike other vitamins that are synthesized primarily in plants, cobalamins are produced only by microbial biosynthesis. Certain legumes have been reported to absorb small amounts of vitamin B12 produced by bacteria associated with root nodules, but little enters the seeds [99] Most plant-derived foods are devoid of vitamin B12 unless contaminated by fecal material, such as from fertilizer [64, 65]. The vitamin B12 in most animal tissues consists mainly of the coenzyme forms, methylcobalamin and 5′-deoxyadenosylcobalamin, in addition to aquocobalamin. Herbert [65] has classified foods according to their vitamin B12 content, as shown in Table 29.

Approximately 20 naturally occurring analogs of vitamin B12 have been identified. Some of these have no biological activity in mammals, while others exhibit at least partial vitamin activity but are often poorly absorbed.

8.8.9.2 Stability and Modes of Degradation

Under most conditions of food processing, preservation, and storage, there is little nutritionally significant loss of vitamin B12. Cyanocobalamin added to breakfast cereal products has been reported to undergo an average loss of 17% during processing, with an additional 17% loss during storage for 12 months at ambient temperature [125]. In studies of the processing of fluid milk, 96% mean retention has been observed during high-temperature, short-time (HTST) pasteurization, and similar retention ($> 90\%$) was found in milk processed using various modes of ultra-high-temperature (UHT) processing [45]. Although refrigerated storage of milk has little impact on vitamin B12 retention, storage of UHT-processed milk at ambient temperature for up to 90 days causes progressive losses that can approach 50% of the initial vitamin B12

Table 29 Classification of Foods According to Their Vitamin B12 Concentration

Food	Vitamin B12 (μg/100 g wet weight)
Rich sources: organ meats (liver, kidney, heart), bivalves (clams and oysters)	>10
Moderately rich sources: nonfat dry milk, some fish and crabs, egg yolks	3–10
Moderate sources: muscle meats, some fish, fermenting cheeses	1–3
Other: fluid milk, cheddar cheese, cottage cheese	<1

Source: Adapted from Ref. 65.

concentration [20]. Sterilization of milk for 13 min at 120°C has been reported to cause 77% loss of vitamin B12 [71], and prior concentration (as in production of evaporated milk) contributes to more severe losses. This indicates the potential for substantial losses of vitamin B12 during prolonged heating of foods at or near neutral pH. Typical oven heating of commercially prepared convenience dinners has been shown to yield 79–100% retention of vitamin B12.

Ascorbic acid has long been known to accelerate the degradation of vitamin B12, although this may be of little practical significance because foods containing vitamin B12 usually do not contain significant amounts of ascorbic acid. Douglass et al. [37] examined the influence of the use of ascorbate or erthyorbate in curing solutions for ham and found that these compounds have no influence on vitamin B12 retention. Thiamin and nicotamide in solution can accelerate degradation of vitamin B12, but the relevance of this phenomenon to foods is unclear.

The mechanism of vitamin B12 degradation has not been fully determined, in part because of the complexity of the molecule and the very low concentration in foods. Photochemical degradation of vitamin B12 coenzymes yields aquocobalamin. Such a reaction interferes with experimental studies of B12 metabolism and function, but this conversion has no influence on the total vitamin B12 activity of foods because aquocobalamin retains vitamin B12 activity. The overall stability of vitamin B12 is greatest at pH 4–7. Exposure to acid causes the hydrolytic removal of the nucleotide moiety, and additional fragmentation occurs as the severity of the acidic conditions increases. Exposure to acid or alkaline conditions causes hydrolysis of amides, yielding biologically inactive carboxylic acid derivatives of vitamin B12.

Interconversions among cobalamins can occur through exchange of the ligand bonded to the Co atom. For example, bisulfite ions cause conversion of aquocobalamin to sulfitocobalamin, while similar reactions can occur to form cobalamins substituted with ammonia, nitrite, or hydroxyl ions. These reactions have little influence on the net vitamin B12 activity of foods.

8.8.9.3 Bioavailability

The bioavailability of vitamin B12 has been examined mainly in the context of the diagnosis of vitamin B12 deficiency associated with malabsorption. Little is known about the influence of food composition on the bioavailability of vitamin B12. Several studies have shown that pectin and, presumably, similar gums reduce vitamin B12 bioavailability in rats. The significance of this effect in humans is unclear, however. Although little vitamin B12 is present in most plants, certain forms of algae do contain significant quantities of the vitamin. Algae is not recommended as a source of vitamin B12 because of its very low bioavailability [28].

In normal human beings, absorption of vitamin B12 from eggs has been shown to be less than half that of cyanocobalamin adminstered in the absence of food [35]. Similar results have been obtained regarding vitamin B12 bioavailability in studies with fish and various meats [34, 36]. Certain individuals are marginally deficient in vitamin B12 because of poor digestion and incomplete release of cobalamins from the food matrix even though they absorb the pure compound normally [21].

8.8.9.4 Analytical Methods

The concentration of vitamin B12 in foods is determined primarily by microbiological growth assays using *Lactobacillus leichmannii* or by radioligand binding procedures [40]. Although the various forms of vitamin B12 can be separated chromatographically, HPLC methods are not readily suitable for food analysis because of the very low concentrations typically found. Early types of radioligand binding assays for vitamin B12 in clinical specimens and foods were often inaccurate because the binding protein employed could bind to active forms of vitamin B12 as well as biologically inactive analogues. The specificity of such assays has been greatly improved through the use of a vitamin B12-binding protein (generally porcine intrinsic factor) that is specific for the biologically active forms of the vitamin. Microbiological assays with *L. leichmannii* may be subject to interference if samples contain high concentrations of deoxyribonucleosides.

Food samples are generally prepared by homogenization in a buffered solution, followed by incubation at elevated temperature (~60°C) in the presence of papain and sodium cyanide. This treatment releases protein-bound forms of vitamin B12 and converts all cobalamins to the more stable cyanocobalamin form. Conversion to cyanocobalamin also improves the performance of assays that may differ in response to the various forms of the vitamin.

8.9 CONDITIONALLY ESSENTIAL VITAMIN-LIKE COMPOUNDS

8.9.1 Choline

Choline (Fig. 44) exists in all living things as the free compound and as a constituent of a number of cellular components including phosphatidylcholine (the most prevalent dietary source of choline), sphingomyelin, and acetylcholine. Although choline synthesis occurs in humans and other mammals, there is a growing body of evidence that dietary choline is also required [22]. Choline is used as chloride and bitartrate salts in fortification of infant formulas. Choline is not ordinarily added to other foods except as an ingredient, for example, phosphatidylcholine as an emulsifier. Choline is a very stable compound. No significant loss of choline during food storage, handling, processing, or preparation.

8.9.2 Carnitine

Carinitine (Fig. 45) can be synthesized by the human body; however, certain individuals appear to benefit from additional dietary carnitine [13]. No nutritional requirements have been estab-

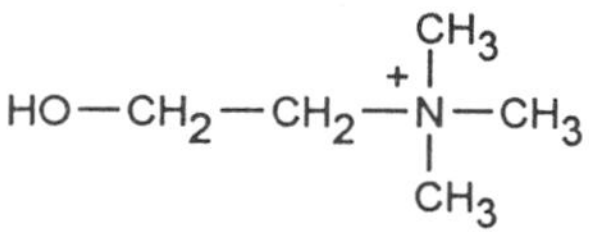

$$HO-CH_2-CH_2-\overset{+}{N}(CH_3)_3$$

FIGURE 44 Structure of choline.

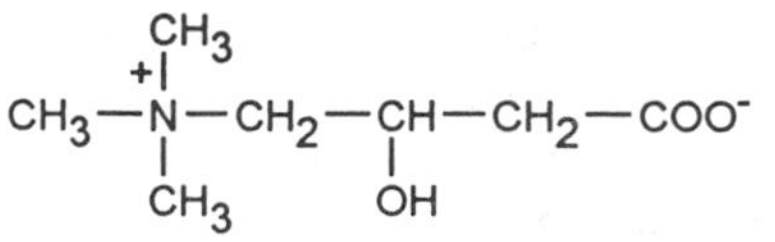

FIGURE 45 Structure of carnitine.

lished for carnitine. Although little or no carnitine is found in plants and plant products, it is widely distributed in foods of animal origin. Carnitine functions metabolically in the transport of organic acids across biological membranes and thus facilitates their metabolic utilization and/or disposal. Carnitine also facilitates transport of certain organic acids to lessen the potential for toxicity in certain cells. In animal-derived foods, carnitine exists in free and acylated form. The acyl carnitines occur with various organic acids esterified to the C-3 hydroxyl group. Carnitine is highly stable and undergoes little or not degradation in foods.

Synthetic carnitine is used in certain clinical applications as the biologically active L-isomer. D-Carnitine has no biological activity. L-Carnitine is added to infant formulas to raise their carnitine concentration to that of human milk [13].

8.9.3 Pyrroloquinoline Quinone

Pyrroloquinoline quinone (PQQ) is a tricylic quinone (Fig. 46) that functions as a coenzyme in several bacterial oxidoreductases and has been reported to be a coenzyme in mammalian lysyl oxidase and amine oxidases. However, recent findings indicate that the coenzyme originally identified as PQQ in these mammalian enzymes was misidentified and is probably 6-hydroxy-dihydroxyphenylalanine quinone [59]. Although no function of PQQ is currently known in mammals, several studies have shown a very small nutritional requirement for rats and mice that appears to be associated with the formation of connective tissue and normal reproduction [76]. Thus, the function of PQQ in mammalian species is an enigma. Because of the ubiquitous nature of PQQ and its synthesis by intestinal bacteria, the development of spontaneous deficiency of PQQ in rodents or humans is unlikely.

8.10 OPTIMIZATION OF VITAMIN RETENTION

To varying degrees, inevitable losses of nutritional value occur during the course of the postharvest handling, cooking, processing, and storage of foods. Such losses occur in the food processing industry, in food service establishments, and in the home. Optimization of nutrient retention is a responsibility of food manufacturers and processors and is in the mutual interest

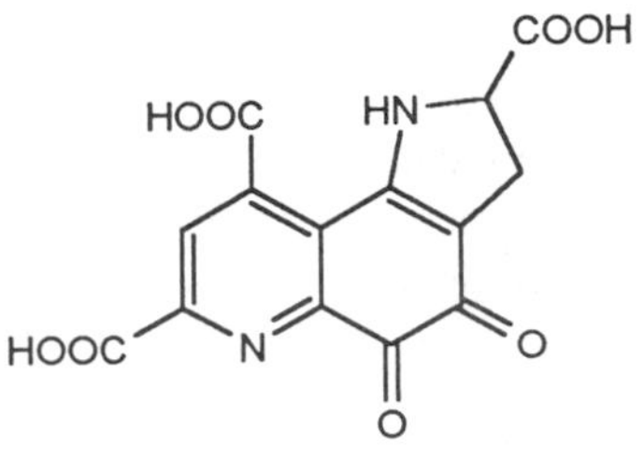

FIGURE 46 Structure of pyrroloquinoline quinone.

of the industry and the public. Likewise, maximization of nutrient retention in the home, and in institutional and retail food service, is an opportunity that should not be overlooked.

Many approaches to optimization of vitamin retention are based on the chemical and physical properties of the particular nutrients involved. For example, the use of acidulants, if compatible with the product, would promote the stability of thiamin and ascorbic acid. Reduction in pH would decrease the stability of certain folates, however, which illustrates the complexity of this approach. Cooking or commercial processing under conditions that minimize exposure to oxygen and excess liquid lessens the oxidation of many vitamins and the extraction (i.e., leaching) of vitamins and minerals. High-temperature, short-time (HTST) conditions will, in many instances, cause less vitamin degradation than will conventional thermal processes of equal thermal severity (based on microbial inactivation). In addition, certain combinations of ingredients can enhance retention of several nutrients (e.g., the presence of natural antioxidants would favor retention of many vitamins).

Several examples of nutrient optimization follow. The reader is referred to additional discussions of this topic [70, 85].

8.10.1 Optimization of Thermal Processing Conditions

Losses of nutrients frequently occur during thermal processing procedures intended to provide a shelf-stable product. Such losses often involve both chemical degradation and leaching. The kinetics and thermodynamics of chemical changes involving the destruction of microorganisms and vitamins differ markedly. Thermal inactivation of microorganisms occurs largely by denaturation of essential macromolecules and involves large energies of activation (typically 200–600 kJ/mol). In contrast, reactions associated with the degradation of vitamins generally exhibit activation energies of 20–100 kJ/mol. Thus, rates of microbial inactivation and rates of vitamin degradation have temperature dependencies that differ significantly. Consequently, the rate of microbial inactivation increases as a function of temperature much more rapidly than does the rate of vitamin degradation. These principles of reaction kinetics and thermodynamics form the basis of enhancement of nutrient retention when HTST conditions are used. Classical studies of Teixeira et al. [130] involved a variety of thermal processing conditions, all of which provided equivalent microbial lethality. These authors showed that thiamin retention during thermal processing of pea purée could be enhanced at least 1.5-fold through selection of the proper time–temperature combination. Although many other vitamins are less labile than thiamin during the processing of low-acid foods, a similar enhancement of their retention would be predicted.

8.10.2 Prediction of Losses

Predicting the magnitude of losses of vitamins requires accurate knowledge of degradation kinetics and temperature dependence of the particular form(s) of the vitamin(s) considered in the chemical milieu of the food(s) of interest. Different chemical forms of vitamins react differently to various food compositions and to specific processing conditions. One must first determine whether kinetic studies of total content (i.e., sum of all forms) of the vitamin of interest yield useful information, or whether more specific information on the various forms of the vitamin is needed [54]. Processing studies must be conducted under conditions identical to those prevailing during the actual commercial processing or storage condition being modeled because of the sensitivity of many nutrients to their chemical and physical environments. As summarized previously [66, 83], reaction kinetics should be obtained at several temperatures to permit calculation of rate constants and an energy of activation. In addition, the experimental conditions should be selected to provide sufficient loss of the vitamin being studied so that the rate constant can be determined with appropriate accuracy [66]. Accelerated storage studies may be performed if the kinetics and mechanisms at elevated temperature are consistent with those occurring under the actual storage conditions. Because temperature fluctuates during actual storage and transportation of foods, models of vitamin stability should include temperature fluctuation [79].

8.10.3 Effects of Packaging

Packaging influences vitamin stability in several ways. In canning, foods that transmit heat energy primarily by conduction (solids or semisolids) will undergo greater overall loss of nutrients than will foods that transmit heat by convection, especially if large containers are used. This difference is caused by the requirement that the thermal process must be based on the "slowest to heat" portion of the product, which, for conduction-heating foods, is the geometric center of the container. Such losses are minimized by using containers with a large surface-to-mass ratio, that is, small cans and noncylindrical containers such as pouches [109]. Pouches also offer the advantage of requiring less liquid for filling; thus, leaching of nutrients during the processing of particulate foods can be minimized.

The permeability of packaging material also can have a substantial effect on the retention of vitamins during food storage. Ascorbic acid in juices and fruit beverages has been shown to exhibit much greater stability when packages with low permeability to oxygen are used [69]. In addition, use of opaque packaging materials prevents the photochemical degradation of photolabile vitamins such as vitamin A and riboflavin.

8.11 SUMMARY

As discussed in this chapter, vitamins are organic chemicals that exhibit a wide range of properties with respect to stability, reactivity, susceptibility to environmental variables, and influence on other constituents of foods. Prediction of net vitamin retention or mechanisms of degradation under a given set of circumstances is often fraught with difficulty because of the multiple forms of most vitamins. With that caveat, the reader is referred to Table 1 as an overview of the general characteristics of each vitamin.

BIBLIOGRAPHY

Augustin, J., B. P. Klein, D. A. Becker, and P. B. Venugopal (eds.) (1985). *Methods of Vitamin Assay*, 4th ed., John Wiley & Sons, New York.

Bauerfeind, J. C., and P. A. Lachance (1992). *Nutrient Additions to Food. Nutritional, Technological and Regulatory Aspects*, Food and Nutrition Press, Trumbull, CT.

Brody, T. (1994). *Nutritional Biochemistry*, Academic Press, Orlando, FL.

Brown, M. L. (ed.) (1990). *Present Knowledge in Nutrition*, International Life Sciences Institute–Nutrition Foundation, Washington, DC.

Chytyl, F., and D. B. McCormick (eds.) (1986). *Methods in Enzymology*, vols. 122 and 123, Parts G and H (respectively), *Vitamins and Coenzymes*, Academic Press, San Diego.

Davidek, J., J. Velisek, and J. Pokorny (eds.) (1990). Vitamins, in *Chemical Changes During Food Processing*, Elsevier, Amsterdam, pp. 230–301.

Harris, R. S., and E. Karmas (1975). *Nutritional Evaluation of Food Processing*, 2nd. ed., AVI, Westport, CT.

Harris, R. S., and H. von Loesecke (1971). *Nutritional Evaluation of Food Processing*, AVI, Westport, CT.

Institute of Medicine (1990). *Nutrition Labeling. Issues and Directions for the 1990s* (D. V. Porter and R. O. Earl, eds.), National Academy Press, Washington, DC.

Karmas, E., and R. S. Harris (1988). *Nutritional Evaluation of Food Processing*, 3rd ed., Van Nostrand Reinhold, New York.

Machlin, L. J. (ed.) (1991). *Handbook of Vitamins*, 2nd ed., Marcel Dekker, New York.

McCormick, D. B. (1991). Coenzymes, biochemistry, in *Encyclopedia of Human Biology*, vol. 2, Acad. Press, San Diego, pp. 527–545.

Zapsalis, C., and R. A. Beck (1985). Vitamins and vitamin-like substances, in *Food Chemistry and Nutritional Biochemistry*, John Wiley and Sons, New York, pp. 189–313.

REFERENCES

1. American Institute of Nutrition (1987). Nomenclature policy: generic descriptions and trivial names for vitamins and related compounds. *J. Nutr. 120*:12–19.
2. Ames, S. R. (1965). Bioassay of vitamin A compounds. *Fed. Proc. Fed. Am. Soc. Exp. Biol. 24*:917–923.
3. Anderson, R. H., D. L. Maxwell, A. E. Mulley, and C. W. Fritsch (1976). Effects of processing and storage on micronutrients in breakfast cereals. *Food Technol. 30*:110–114.
4. Anonymous (1982). The nutritive quality of processed food. General policies for nutrient addition. *Nutr. Rev. 40*:93–96.
5. Arabshahi, A., and D. B. Lund (1988). Thiamin stability in simulated intermediate moisture food. *J. Food Sci. 53*:199–203.
6. Barrett, D. M., and D. B. Lund (1989). Effect of oxygen on thermal degradation of 5-methyl-5,6,7,8-tetrahydrofolic acid. *J. Food Sci. 54*:146–149.
7. Bauernfeind, J. C., and P. A. Lachance (1992). *Nutrient Additions to Food Nutritional Technological and Regulatory Aspects*, Food and Nutrition Press, Trumbull, CT.
8. Beecher, G. R., and R. H. Matthews. 1990. Nutrient composition of foods, in *Present Knowledge in Nutrition*, 6th ed. (M. L. Brown, ed.), International Life Sciences Institute–Nutrition Foundation, Washington, DC, pp. 430–439.
9. Bergström, L. (1988). Nutrient data banks for nutrient evaluation in foods, in *Nutritional Evaluation of Food Processing*, 3rd ed. (E. Karmas and R. S. Harris, eds.), Van Nostrand Reinhold, New York, pp. 745–764.
10. Blair, J. A., A. J. Pearson, and A. J. Robb (1975). Autoxidation of 5-methyl-5,6,7,8-tetrahydrofolate. *J. Chem. Soc. Perkin Trans. II*:18–21.
11. Bode, A. M., L. Cunningham, and R. C. Rose. (1990). Spontaneous decay of oxidized ascorbic acid (dehydro-L-ascorbic acid) evaluated by high-pressure liquid chromatography. Clin. Chem. 36:1807–1809.
12. Bonjour, J. -P. (1991). Biotin, in *Handbook of Vitamins*, 2nd ed. (L. J. Machlin, ed.), Marcel Dekker, New York, pp. 393–427.
13. Borum, P. R. (1991). Carnitine, in *Handbook of Vitamins*, 2nd ed. (L. J. Machlin, ed.), Marcel Dekker, New York, pp. 557–563.
14. Bowers, J. A., and J. Craig (1978). Components of vitamin B6 in turkey breast muscle. *J. Food Sci. 43*:1619–1621.
15. Brown, E. D., M. S. Micozzi, N. E. Craft, J. G. Bieri, G. Beecher, B. K. Edwards, A. Rose, P. R. Taylor, and J. C. Smith (1989). Plasma carotenoids in normal men after a single ingestion of vegetables or purified β-carotene. *Am. J. Clin. Nutr. 49*:1258–1265.
16. Buettner, G. R. (1988). In the absence of catalytic metals ascorbate does not autoxidize at pH 7: Ascorbate as a test for catalytic metals. *J. Biochem. Biophys. Methods 16*:27–40.
17. Buettner, G. R. (1993). The pecking order of free radicals and antioxidants: Lipid peroxidation, α-tocopherol, and ascorbate. *Arch. Biochem. Biophys. 300*:535–543.
18. Burton, G. W., and K. U. Ingold (1984). β-Carotene: An unusual type of lipid antioxidant. *Science 224*:569–573.
19. Burton, G. W., and M. G. Traber (1990). Vitamin E: Antioxidant activity, biokinetics, and bioavailability. *Annu. Rev. Nutr. 10*:357–382.
20. Burton, H., J. E. Ford, J. G. Franklin, and J. W. G. Porter (1967). Effects of repeated heat treatments on the levels of some vitamins of the B-complex in milk. *J. Dairy Res. 34*:193–197.
21. Carmel, R., R. M. Sinow, M. E. Siegel, and I. M. Samloff (1988). Food cobalamin malabsorption

occurs frequently in patients with unexplained low serum cobalamin levels. *Arch. Intern. Med. 148*:1715–1719.

22. Chan, M. M. (1991). Choline, in *Handbook of Vitamins*, 2nd. ed. (L. J. Machlin, ed.), Marcel Dekker, New York, pp. 537–556.
23. Chandler, L. A., and S. J. Schwartz (1987). HPLC separation of cis-trans carotene isomers in fresh and processed fruits and vegetables. *J. Food Sci. 52*:669–672.
24. Colman, N., R. Green, and J. Metz (1975). Prevention of folate deficiency by food fortification. II. Absorption of folic acid from fortified staple foods. *Am. J. Clin. Nutr. 28*:459–464.
25. Cort, W. M., B. Borenstein, J. H. Harley, M. Osadca, and J. Scheiner (1976). Nutrient stability of fortified cereal products. *Food Technol 30*:52–62.
26. Cort, W. M., T. S. Vicente, E. H. Waysek, and B. D. Williams (1983). Vitamin E content of feedstuffs determined by high-performance liquid chromatography. *J. Agric. Food Chem. 31*:1330–1333.
27. Coursin, D. B. (1954). Convulsive seizures in infants with pyridoxine deficient diet. *J. Am. Med. Assoc. 154*:406–408.
28. Dagnelie, P. C., W. A. van Staveren, and H. van den Berg (1991). Vitamin B12 from algae appears not to be bioavailable. *Am. J. Clin. Nutr. 53*:695–697.
29. Day, B. P. F., and J. F. Gregory 1983. Thermal stability of folic acid and 5-methyltetrahydrofolic acid in liquid model food systems. *J. Food Sci. 48*:581–587, 599.
30. Dellamonica, E. S., and P. E. McDowell (1965). Comparison of beta-carotene content of dried carrots prepared by three dehydrated processes. *Food Technol. 19*:1597–1599.
31. Dennison, D. B., and J. R. Kirk (1978). Oxygen effect on the degradation of ascorbic acid in a dehydrated food system. *J. Food Sci. 43*:609–618.
32. Dennison, D., J. Kirk, J. Bach, P. Kokoczka, and D. Heldman (1977). Storage stability of thiamin and riboflavin in a dehydrated food system. *J. Food Proc. Preserv. 1*:43–54.

32a. De Saucedo, S. M. U. (1982). Losses of thiamin, ascorbic acid and lysine in thermally sterilized foods. PhD thesis, University of Leeds (UK).

33. Deutsch, J. C., and J. F. Kolhouse (1993). Ascorbate and dehydroascorbate measurements in aqueous solutions and plasma determined by gas chromatography–mass spectrometry. Anal. Chem. 65:321–326.
34. Doscherholmen, A., J. McMahon, and P. Economon (1981). Vitamin B12 absorption from fish. *Proc. Soc. Exp. Biol. Med. 167*:480–484.
35. Doscherholmen, A., J. McMahon, and D. Ripley (1975). Vitamin B12 absorption from eggs. *Proc. Soc. Exp. Biol. Med. 149*:987–990.
36. Doscherholmen, A., J. McMahon, and D. Ripley (1978). Vitamin B12 assimilation from chicken meat. *Am. J. Clin. Nutr. 31*:825–830.
37. Douglass, J. S., F. D. Morrow, K. Ono, J. T. Keeton, J. T. Vanderslice, R. C. Post, and B. W. Willis (1989). Impact of sodium ascorbate and sodium erthorbate use in meat processing on the vitamin B12 content of cured ham. *J. Food Sci 54*:1473–1474.
38. Dwivedi, B. K., and R. G. Arnold (1973). Chemistry of thiamine degradation in food products and model systems: A review. *J. Agric. Food Chem. 21*:54–60.
39. Eitenmiller, R. R., and S. de Souza (1985). Niacin, in *Methods of Vitamin Assay*, 4th ed. (J. Augustin, B. P. Klein, D. A. Becker, and P. B. Venugopal, eds.), John Wiley & Sons, New York, pp. 385–398.
40. Ellenbogen, L., and B. A. Cooper (1992). Vitamin B12, in *Handbook of Vitamins, Second Edition* (L. J. Machlin, ed.), Marcel Dekker, New York, pp. 491–536.
41. Evans, S. R., J. F. Gregory, and J. R. Kirk (1981). Thermal degradation kinetics of pyridoxine hydrochloride in dehydrated model food systems. *J. Food Sci. 46*:555–558, 563.
42. Farrer, K. T. H. (1955). The thermal destruction of vitamin B1 in foods. *Adv. Food Res. 6*:257–311.
43. Finglas, P. M., and R. M. Faulks (1987). Critical review of HPLC methods for the determination of thiamin, riboflavin and niacin in food. *J. Micronutr. Anal. 3*:251–283.
44. Food and Nutrition Board (1989). *Recommended Dietary Allowances*, 10th ed., National Research Council, National Academy of Sciences, Washington, DC.
45. Ford, J. E., J. W. G. Porter, S. Y. Thompson, J. Toothill, and J. Edwards-Webb (1969). Effects of

ultra-high-temperature (UHT) processing and of sequence storage on the vitamin content of milk. *J. Dairy Res. 36*:447–454.

46. Frankel, E. N., S.-W. Huang, J. Kanner, and J. B. German (1994). Interfacial phenomena in the evaluation of antioxidants: Bulk oils vs. emulsions. *J. Agric. Food Chem. 42*:1054–1059.
47. Freed, M., S. Brenner, and V. O. Wodicka. (1948). Prediction of thiamine and ascorbic acid stability in canned stored foods. *Food Technol. 3*:148–151.
48. Gregory, J. F. (1980). Effects of ε-pyridoxyllysine bound to dietary protein on the vitamin B-6 status of rats. *J. Nutr. 110*:995–1005.
49. Gregory, J. F. (1989). Chemical and nutritional aspects of folate research: analytical procedures, methods of folate synthesis, stability, and bioavailability of dietary folates. *Adv. Food Nutr. Res. 33*:1–101.
50. Gregroy, J. F. (1993). Ascorbic acid bioavailability in foods and supplements. *Nutr. Rev. 51*:301–309.
51. Gregory, J. F. (1985). Chemical changes of vitamins during processing, in *Chemical Changes in Food During Processing* (T. Richardson and J. W. Finley, eds.), AVI, Westport, CT., pp. 373–408.
52. Gregory, J. F. (1993). Methods for determination of vitamin B6 in foods and other biological materials: A critical review. *J. Food Comp. Anal. 1*:105–123.
53. Gregory, J. F. (1995). The bioavailability of folate, in *Folate in Health and Disease* (L. B. Bailey, ed.), Marcel Dekker, New York, pp. 195–235.
54. Gregory, J. F., and M. E. Hiner (1983). Thermal stability of vitamin B6 compounds in liquid method food systems. *J. Food Sci. 48*:1323–137, 1339.
55. Gregroy, J. F., and S. L. Ink (1987). Identification and quantification of pyridoxine-β-glucoside as a major form of vitamin B6 in plant-derived foods. *J. Agric. Food Chem. 35*:76–82.
56. Gregory, J. F., S. L. Ink, and D. B. Sartain (1986). Degradation and binding to food proteins of vitamin B-6 compounds during thermal processing. *J. Food Sci. 51*:1345–1351.
57. Greundwedel, D. W., and R. K. Patnaik (1971). Release of hydrogen sulfide and methyl mercaptan from sulfur-containing amino acids. *J. Agric. Food Chem. 19*:775–779.
58. Hamm, D. J., and D. B. Lund (1978). Kinetics parameters for thermal inactivation of pantothenic acid. *J. Food Sci.* 43:631–633.
59. Harris, E. D. (1992). The pyrroloquinoline quinone (PQQ) coenzymes: a case of mistaken identity. *Nutr. Rev. 50*:263–274.
60. Harris, R. S. (1971). General discussion on the stability of nutrients, in *Nutritional Evaluation of Food Processing* (R. S. Harris and H. von Loesecke, eds.), AVI, Westport, CT, pp. 1–4.
61. Harris, R. S., and E. Karmas (1975). *Nutritional Evaluation of Food Processing*, 2nd. ed., AVI, Westport, CT.
62. Harris, R. S., and H. von Loesecke (1971). *Nutritional Evaluation of Food Processing*, AVI, Westport, CT.
63. Hassinen, J. B., G. T. Durbin, and F. W. Bernhart (1954). The vitamin B6 content of milk products. *J. Nutr. 53*:249–257.
64. Herbert, V. (1988). Vitamin B12: Plant sources, requirements, and assay. *Am. J. Clin. Nutr. 48*:852–858.
65. Herbert, V. (1990). Vitamin B-12, in *Present Knowledge in Nutrition*, 6th ed. (M. L. Brown, ed.), International Life Sciences Institute, Nutrition Foundation, Washington, DC, pp. 170–178.
66. Hill, M. K., and R. A. Grieger-Block (1980). Kinetic data: Generation, interpretation, and use. *Food Technol. 34*:56–66.
67. Hoppner, K., and B. Lampi (1993). Pantothenic acid and biotin retention in cooked legumes. *J. Food Sci. 58*:1085, 1089.
68. Jongehan, J. A., H. I. X. Mager, and W. Berends (1979). Autoxidation of 5-alkyl-tetrahydropteridines. The oxidation product of 5-methyl-THF, in *The Chemistry and Biology of Pteridines* (R. Kisliuk, ed.), Elsevier North Holland, pp. 241–246.
69. Kacem, B., J. A. Cornell, M. R. Marshall, R. B. Shireman, and R. F. Matthews (1987). Nonenzymatic browning in aseptically packaged orange juice and orange drinks. Effect of ascorbic acid, amino acids and oxygen. *J. Food Sci. 52*:1668–1672.

70. Karel, M. (1979). Prediction of nutrient losses and optimization of processing conditions, in *Nutritional and Safety Aspects of Food Processing* (S. R. Tannenbaum, ed.), Marcel Dekker, New York, pp. 233–263.
71. Karlin, R., C. Hours, C. Vallier, R. Bertoye, N. Berry, and H. Morand (1969). Sur la teneur en folates des laits de grand mélange. Effects de divers traitment thermiques sur les taux de folates, B12 et B6 de ces laits. *Int. Z. Vitaminforsch.* 39:359–371.
72. Karmas, E., and R. S. Harris. (1988). *Nutritional Evaluation of Food Processing*, 3rd ed., Van Nostrand Reinhold, New York.
73. Khachik, F., G. R. Beecher, and W. R. Lusby (1989). Separation, identification, and quantification of the major carotenoids in extracts of apricots, peaches, cantaloupe, and pink grapefruit by liquid chromatography. *J. Agric. Food Chem. 37*:1465–1473.
74. Khan, M. M. T., and A. E. Martell (1967). Metal ion and metal chelate catalyzed oxidation of ascorbic acid by molecular oxygen. I. Cupric and ferric ion oxidation. *J. Am. Chem. Soc. 89*:4176–4185.
75. Khan, M. M. T., and A. E. Martell (1967). Metal ion and metal chelate catalyzed oxidation of ascorbic acid by molecular oxygen. II. Cupric and ferric chelate catalyzed oxidation. *J. Am. Chem. Soc. 89*:7104–7111.
76. Kilgore, J., C. Smidt, L. Duich, N. Romero-Chapman, D. Tinker, M. Melko, D. Hyde, and R. Rucker (1989). Nutritional importance of pyrroloquinoline quinone. *Science 245*:850–852.
77. Kirk, J., D. Dennison, P. Kokoczka, and D. Heldman (1977). Degradation of ascorbic acid in a dehydrated food system. *J. Food Sci. 42*:1274–1279.
78. Klein, B. P., and A. K. Perry (1982). Ascorbic acid and vitamin A activity in selected vegetables from different geographical areas of the United States. *J. Food Sci. 47*:941–945.
79. Labuza, T. P. (1979). A theoretical comparison of losses in foods under fluctuating temperature sequences. *J. Food Sci. 44*:1162–1168.
80. Labuza, T. P. (1980). The effect of water activity on reaction of kinetics of food deterioration. *Food Technol. 34*:36–41, 59.
81. Labuza, T. P., and J. F. Kamman (1982). A research note. Comparison of stability of thiamin salts at high temperature and water activity. *J. Food Sci. 47*:664–665.
82. Leichter, J., V. P. Switzer, and A. F. Landymore (1978). Effect of cooking on folate content of vegetables. *Nutr. Rep. Int. 18*:475–479.
83. Lenz, M. K., and D. B. Lund (1980). Experimental procedures for determining destruction of kinetics of food components. *Food Technol. 34*:51–55.
84. Liao, M.-L, and P. A. Seib (1987). Selected reactons of L-ascorbic acid related to foods. *Food Technol. 41*:104–107, 111.
85. Lund, D. B. (1977). Designing thermal processes for maximizing nutrient retention. *Food Technol. 31*:71–78.
86. Lund, D. (1979). Effect of commercial processing on nutrients. *Food Technol 33*:28–34.
87. Lund, D. (1988). Effects of heat processing on nutrients, in *Nutritional Evaluation of Food Processing*, 3rd. ed. (E. Karmas and R. S. Harris, eds.), Van Nostrand Reinhold, New York, pp. 319–354.
88. Machlin, L. J. (1991). Vitamin E, in *Handbook of Vitamins*, 2nd ed. (L. J. Machlin, ed.), Marcel Dekker, New York, pp. 99–144.
89. Malewski, W., and P. Markakis (1971). A research note. Ascorbic acid content of developing tomato fruit. *J. Food Sci. 36*:537–539.
90. Mangels, A. R., G. Block, C. M. Frey, et al. (1993). The bioavailability to humans of ascorbic acid from oranges, orange juice and cooked broccoli is similar to that of synthetic ascorbic, *J. Nutr. 123*:1054–1061.
91. Maruyama, T., T. Shiota, and C. L. Krumdieck (1978). The oxidative cleavage of folates. A critical study. *Anal. Biochem. 84*:277–295.
92. Mauri, L. M., S. M. Alzamora, J. Chirife, and M. J. Tomio (1989). Review: Kinetic parameters for thiamine degradation in foods and model solutions of high water activity. *Int. J. Food Sci. Technol. 24*:1–9.

93. McCormick, D. B. (1990). Riboflavin, in *Present Knowledge in Nutrition* (M. L. Brown, ed.), International Life Sciences Institute–Nutrition Foundation, Washington, DC, pp. 146–154.
94. Misir, R., and R. Blair (1988). Biotin, bioavailability of protein supplements and cereal grains for starting turkey poults. *Pout. Sci. 67*:1274–1280.
95. Mnkeni, A. P., and T. Beveridge 1983. Thermal destruction of 5-methyltetrahydrofolic acid in buffer and model food systems. *J. Food Sci. 48*:595–599.
96. Mock, D. M. (1990). Biotin, in *Present Knowledge in Nutrition*, 6th ed. (M. L. Brown, ed.), International Life Sciences Institute, Nutrition Foundation, Washington, DC, pp. 189–207.
97. Mock, D. M., N. I. Mock, and S. E. Langbehn (1992). Biotin in human milk: methods, location, and chemical form. *J. Nutr. 122*:535–545.
98. Moran, R. (1959). Nutritional significance of recent work on wheat flour and bread. *Nutr. Abstr. Rev. 29*:1–10.
99. Mozafar, A., and J. J. Oertli (1992). Uptake of a microbially-produced vitamin (B12) by soybean roots. *Plant Soil 139*:23–30.
100. Mulley, E. A., C. R. Strumbo, and W. M. Hunting (1975). Kinetics of thiamine degradation by heat. Effect of pH and form of the vitamin on its rate of destruction. *J. Food Sci. 40*:989–992.
101. Nagy, S. (1980). Vitamin C contents of citrus fruit and their products: A review. *J. Agric. Food Chem. 28*:8–18.
102. Navankasattusas, S., and D. B. Lund (1982). Thermal destruction of vitamin B6 vitamers in buffer solution and cauliflower puree. *J. Food Sci. 47*:1512–1518.
103. Pelletier, O. (1985). Vitamin C, in *Methods of Vitamin Assay*, 4th ed. (J. Augustin, B. P. Klein, D. A. Becker, and P. B. Venugopal, eds.), John Wiley & Sons, New York, pp. 303–347.
104. Pesek, C. A., and J. J. Warthesen (1990). Kinetic model for photoisomerization and concomitant photodegradation of β-carotenes. *J. Agric. Food Chem. 38*:1313–1315.
105. Poe, M. (1977). Acidic dissociation constants of folic acid, dihydrofolic acid, and methotrexate. *J. Biol. Chem. 252*:3724–3728.
106. Porzio, M. A., N. Tang, and D. M. Hilker (1973). Thiamine modifying properties of heme proteins from skipjack tuna, pork, and beef. *J. Agric. Food Chem. 21*:308–310.
107. Reed, L. S., and M. C. Archer (1980). Oxidation of tetrahydrofolic acid by air. *J. Agric. Food Chem. 28*:801–805.
108. Reiber, H. (1972). Photochemical reactions of vitamin B6 compounds, isolation and properties of products. *Biochim. Biophys. Acta 279*:310–315.
109. Rizvi, S. S. H., and J. C. Acton (1982). Nutrient enhancement of thermostabilized foods in retort pouches. *Food Technol 36*:105–109.
110. Robinson, D. R. (1971). The nonenzymatic hydrolysis of N^5,N^{10}-methenyltetrahydrofolic acid and related reactions, in *Methods in Enzymology* (F. Chytyl, ed.), vol. 18B, Academic Press, San Diego, pp. 716–725.
111. Roughead, Z. K., and D. B. McCormick (1991a). Qualitative and quantitative assessment of flavins in cow's milk. *J. Nutr. 120*:382–388.
112. Roughead, Z. K., and D. B. McCormick (1990b). Flavin composition of human milk. *Am. J. Clin. Nutr. 52*:854–857.
113. Ryley, J., and P. Kajda (1993). Vitamins in thermal processing. *Food Chem. 49*:119–129.
114. Saidi, B., and J. J. Warthesen (1983). Influence of pH and light on the kinetics of vitamin B6 degradation. *J. Agric. Food Chem. 31*:876–880.
115. Sauberlich, H. E., M. J. Kretsch, J. H. Skala, H. L. Johnson, and P. C. Taylor (1987). Folate requirements and metabolism in nonpregnant women. *Am. J. Clin. Nutr. 46*:1016–1028.
116. Sauer, W. C., R. Mosenthin, and L. Ozimek (1988). The digestibility of biotin in protein supplements and cereal grains for growing pigs. *J. Anim. Sci. 66*:2583–2589.
117. Scarpa, M., R. Stevanato, P. Viglino, and A. Rigo (1983). Superoxide ion as active intermediate in the autoxidation of ascorbate by molecular oxygen. *J. Biol. Chem. 258*:6695–6697.
118. Selman, J. D. (1993). Vitamin retention during blanching of vegetables. *Food Chem. 49*:137–147.

119. Seyoum, E., and J. Selhub 1993. Combined affinity and ion pair column chromatographies for the analysis of food folate. *J. Nutr. Biochem. 4*:488–494.
120. Shah, J. J. (1985). Riboflavin, in *Methods of Vitamin Assay*, 4th ed. (J. Augustin, B. P. Klein, D. A. Becker, and P. B. Venugopal, eds.), John Wiley & Sons, New York, pp. 365–383.
121. Shaw, S., E. Jayatilleke, V. Herbert, and N. Colman (1989). Cleavage of folates during ethanol metabolism. *Biochem. J. 257*:277–280.
122. Sies, H., W. Stahl, and A. R. Sundquist (1992). Antioxidant functions of vitamins. Vitamins E and C, beta-carotene, and other carotenoids, in *Beyond Deficiency. New Views on the Functions and Health Effects of Vitamins* (H. E. Sauberlich and L. J. Machlin, eds.), *Ann. N. Y. Acad. Sci. 669*:7–20.
123. Snell, E. E. (1963). Vitamin B6. *Comp. Biochem. 2*:48–58.
124. Srncova, V., and J. Davidek (1972). Reaction of pyridoxal and pyridoxal-5-phosphate with proteins. Reaction of pyridoxal with milk serum proteins. *J. Food Sci. 37*:310–312.
125. Steele, C. J. (1976). Cereal fortification—Technological problems. *Cereal Foods World 21*:538–540.
126. Tadera, K., M. Arima, and F. Yagi (1988). Participation of hydroxyl radical in hydroxylation of pyridoxine by ascorbic acid. *Agric. Biol. Chem. 52*:2359–2360.
127. Taher, M. M., and N. Lakshmaiah (1987). Hydroperoxide-dependent folic acid degradation by cytochrome c. *J. Inorg. Biochem. 28*:133–141.
128. Tannenbaum, S. R., V. R. Young, and M. C. Archer (1985). Vitamins and minerals, in *Food Chemistry, Second Edition, Revised and Expanded* (O. R. Fennema, ed.), Marcel Dekker, New York, pp. 477–544.
129. Tarr, J. B., T. Tamura, and E. L. R. Stokstad (1981). Availability of vitamin B6 and pantothenate in an average American diet in man. *Am. J. Clin. Nutr. 34*:1328–1337.
130. Teixeira, A. A., J. R. Dixon, J. W. Zahradnik, and G. E. Zinsmeister (1969). Computer optimization of nutrient retention in the thermal processing of conduction-heating foods. *Food Technol 23*:845.
131. Thompson, J. N. (1986). Problems of official methods and new techniques for analysis of foods and foods for vitamin A. *J. Assoc. Offic. Anal. Chem. 69*:727–738.
132. Thompson, J. N., and G. Hatina (1979). Determination of tocopherols and tocotrienols in foods and tissues by high performance liquid chromatography. *J. Liq. Chromatogr. 2*:327–344.
133. Tsukida, K., K. Saiki, and M. Suguira (1981). Structural elucidation of the main cis-β-carotenes. *J. Nutr. Sci. Vitaminol. 27*:551–556.
134. Vanderslice, J. T., and D. J. Higgs (1988). Chromatographic separation of ascorbic acid, isoascorbic acid, and dehydroascorbic acid and dehydro-isoascorbic acid and their quantitation in food products. *J. Micronutr. Anal. 4*:109–118.
135. Van Niekerk, P. J. (1988). Determination of vitamins, in *HPLC in Food Analysis* 2nd ed. (R. Macrae, ed.), Academic Press, San Diego, pp. 133–184.
136. Van Niekerk, P. J., and A. E. C. Burger (1985). The estimation of the composition of edible oil mixtures. *J. Am. Oil Chem. Soc. 62*:531–538.
137. Wall, J. S., and K. J. Carpenter (1988). Variation in availability of niacin in grain products. Changes in chemical composition during grain development and processing affect the nutritional availability of niacin. *Food Technol. 42*:198–204.
138. Wall, J. S., M. R. Young, and K. S. Carpenter (1987). Transformation of niacin-containing compounds in corn during grain development. Relationship of niacin nutritional availability. *J. Agric. Food Chem. 35*:752–758.
139. Weiser, H., and M. Vecchi (1982). Stereoisomers of α-tocopheryl acetate. II. Biopotencies of all eight stereoisomers, individually or in mixtures, as determined by rat resorption-gestation tests. *Int. J. Vit. Nutr. Res. 32*:351–370.
140. Wendt, G., and F. W. Bernhart (1960). The structure of a sulfur-containing compound with vitamin B6 activity. *Arch. Biochem. Biophys. 88*:270–272.
141. Weisinger, H., and H.-J. Hinz (1984). Kinetic and thermodynamic parameters for Schiff base formation of vitamin B6 derivatives with amino acids. *Arch. Biochem. Biophys. 235*:34–40.
142. Woodcock, E. A., J. J. Warthesen, and T. P. Labuza (1982). Riboflavin photochemical degradation in pasta measured by high performance liquid chromatography. *J. Food Sci. 47*:545–555.

143. Wyse, B. W., W. O. Song, J. H. Walsh, and R. G. Hansen (1991). Pantothenic acid in *Methods of Vitamin Assay*, 4th ed. (J. Augustin, B. P. Klein, D. A. Becker, and P. B. Venugopal, eds.), John Wiley & Sons, New York, pp. 399–416.
144. Yasumoto, K., H. Tsuji, K. Iwami, and H. Mitsuda (1977). Isolation from rice bran of a bound form of vitamin B6 and its identification of 5′-*O*-(β-D-glucopyranosyl)pyridoxine. *Agri. Biol. Chem. 41*:1061–1067.
145. Zechmeister, L. (1949). Stereoisomeric provitamins A. *Vitam. Hormones (NY) 7*:57–81.
146. Zoltevicz, J. A., and G. M. Kauffman (1977). Kinetics and mechanism of the cleavage of thiamin, 2-(1-hydroxyethyl)thiamin, and a derivative by bisulfite ion in aqueous solution. Evidence for an intermediate. *J. Am. Chem. Soc. 99*:3134–3142.

9

Minerals

DENNIS D. MILLER
Cornell University, Ithaca, New York

9.1 INTRODUCTION

Ninety chemical elements occur naturally in the earth's crust. About 25 of them are known to be essential to life and thus are present in living cells (Fig. 1). Since our food is ultimately derived from living plants or animals, we can expect to find these 25 elements in foods as well. Foods also contain other elements, since living systems can accumulate nonessential elements from their environment. Moreover, elements may enter foods as contaminants during harvest, processing, and storage, or they may be present in intentional food additives.

While there is no universally accepted definition of "mineral" as it applies to food and nutrition, the term usually refers to elements other than C, H, O, and N that are present in foods. These four nonmineral elements are present primarily in organic molecules and water, and constitute about 99% of the total number of atoms in living systems [10]. Thus mineral elements are present in relatively low concentrations in foods. Nevertheless, they play key functional roles in both living systems and foods.

Historically, minerals have been classified as either major or trace, depending on their concentrations in plants and animals. This classification arose at a time when analytical methods were not capable of measuring small concentrations of elements with much precision. Thus the term "trace" was used to indicate the presence of an element that could not be measured accurately. Today, modern methods and instruments allow for extremely precise and accurate measurement of virtually all of the elements in the periodic table (Fig. 1) [28]. Nevertheless, the terms major and trace continue to be used to describe mineral elements in biological systems. Major minerals include calcium, phosphorus, magnesium, sodium, potassium, and chloride. Trace elements include iron, iodine, zinc, selenium, chromium, copper, fluorine, lead, and tin.

9.2 PRINCIPLES OF MINERAL CHEMISTRY

Many different chemical forms of mineral elements are present in foods. These forms are commonly referred to as *species* and include compounds, complexes, and free ions [44]. Given the diversity of chemical properties among the mineral elements, the number and diversity of nonmineral compounds in foods that can bind mineral elements, and the chemical changes that occur in foods during processing and storage, it is not surprising that the number of different mineral species in foods is large indeed. Moreover, since foods are so complex and since many mineral species are unstable, it is very difficult to isolate and characterize mineral species in

I-A	II-A	III-B	IV-B	V-B	VI-B	VII-B	VIII	VIII	VIII	I-B	II-B	III-A	IV-A	V-A	VI-A	VIIA	O
H																	He
Li	Be											B	C	N	O	F	Ne
Na	Mg											Al	Si	P	S	Cl	Ar
K	Ca	Sc	Ti	V	Cr	Mn	Fe	Co	Ni	Cu	Zn	Ga	Ge	As	Se	Br	Kr
Rb	Sr	Y	Zr	Nb	Mo	Tc	Ru	Rh	Pd	Ag	Cd	In	Sn	Sb	Te	I	Xe
Cs	Ba	Ln	Hf	Ta	W	Re	Os	Ir	Pt	Au	Hg	Tl	Pb	Bi	Po	At	Rn
Fr	Ra	Ac	Th	Pa	U												

FIGURE 1 Periodic table of the naturally occurring elements. Shaded elements are believed to be essential nutrients for animals and humans.

foods. Thus, our understanding of the exact chemical forms of minerals in foods remains limited. Fortunately, principles and concepts from the vast literature in inorganic, organic, and biochemistry can be very useful in guiding predictions about the behavior of mineral elements in foods.

9.2.1 Solubility of Minerals in Aqueous Systems

All biological systems contain water, and most nutrients are delivered to and metabolized by organisms in an aqueous environment. Thus the availabilities and reactivities of minerals depend, in large part, on solubility in water. This excludes the elemental form of nearly all elements (dioxygen and nitrogen are exceptions) from physiological activity in living systems since these forms, such as elemental iron, are insoluble in water and therefore are unavailable for incorporation into organisms or biological molecules.

The species (forms) of elements present in food vary considerably depending on the chemical properties of the element. Elements in groups IA and VIIA (Fig.1) exist in foods predominantly as free ionic species (Na^+, K^+, Cl^-, and F^-). These ions are highly water soluble and have low affinities for most ligands; thus they exist primarily as free ions in aqueous systems. Most other minerals are present as complexes, chelates, or oxygen-containing anions (see later discussion of complexes and chelates).

The solubilities of mineral complexes and chelates may be very different from that of inorganic salts. For example, if ferric chloride is dissolved in water, the iron will soon precipitate as ferric hydroxide. On the other hand, ferric iron chelated with citrate is quite soluble. Conversely, calcium as calcium chloride is quite soluble, while calcium chelated with oxalate ion is insoluble.

9.2.2 Minerals and Acid/Base Chemistry

Much of the chemistry of the mineral elements can be understood by applying the concepts of acid/base chemistry. Moreover, acids and bases may profoundly influence functional properties and stabilities of other foods components by altering the pH of the food. Thus acid/base chemistry is of critical importance to food scientists. A brief review of acid/base chemistry follows. For a more complete treatment of this topic, see Shriver et al. [38] or other textbooks on inorganic chemistry.

9.2.2.1 Bronsted Theory of Acids and Bases

A Bronsted acid is any substance capable of donating a proton.
A Bronsted base is any substance capable of accepting a proton.

Many acids and bases occur naturally in foods, and they may be used as food additives or processing aids. Common organic acids include acetic, lactic, and citric acids. Phosphoric acid is an example of a mineral acid present in foods. It is used as an acidulant and flavoring agent in some carbonated soft drinks. It is a tribasic acid (contains three available protons).

$$H_3PO_4 \rightleftarrows H_2PO_4^- + H^+ \qquad pK_1 = 2.12$$

$$H_2PO_4^- \rightleftarrows HPO_4^{2-} + H^+ \qquad pK_2 = 7.1$$

$$HPO_4^{2-} \rightleftarrows PO_4^{3-} + H^+ \qquad pK_3 = 12.4$$

Other common mineral acids include HCl and H_2SO_4. They are not added to foods directly, although they may be generated in foods during processing or cooking. For example,

H_2SO_4 is produced when the common leavening acid sodium aluminum sulfate is heated in the presence of water:

$$Na_2SO_4 \cdot Al_2(SO_4)_3 + 6H_2O \rightarrow Na_2SO_4 + 2Al(OH)_3 + 3H_2SO_4$$

Some of the intermediate steps in this reaction are:

$$Al_2(SO_4)_3 \rightleftarrows 2Al^{3+} + 2SO_4^{2-}$$

Aluminum ions may then form complexes with water to produce hydrated aluminum ions:

$$Al^{3+} + 6H_2O \rightleftarrows Al(H_2O)_6^{3+}$$

The Bronsted acidity of the coordinated water molecules is increased as a result of the electron-withdrawing capabilities of the metal ion:

$$Al(H_2O)_6^{3+} + H_2O \rightarrow Al(H_2O)_5(OH^-)^{2+} + H_3O^+$$

Acid strength varies as a function of the electronegativity of the metal ion. The acid strength of a hydrated aluminum ion is about the same as that of acetic acid; pK_a values of $Al(H_2O)_6^{3+}$ and acetic acid are 5.0 and 4.8, respectively.

9.2.2.2 Lewis Theory of Acids and Bases

An alternative, and more general, definition of an acid and a base was developed by G. N. Lewis in the 1930s [38]:

A Lewis acid is an electron pair acceptor.
A Lewis base is an electron pair donor.

By convention, Lewis acids are often represented as "A" and Lewis bases as "B". The reaction between a Lewis acid and a Lewis base then becomes:

$$A + :B \rightarrow A\text{-}B$$

It is important to remember that this reaction does not involve a change in the oxidation state of either A or B; that is, it is not a redox reaction. Thus, A must possess a vacant low-energy orbital and B must possess an unshared pair of electrons. The bonding results from the interaction of orbitals from the acid and the base to form new molecular orbitals. The stability of the complex depends in large part on the reduction of electronic energy that occurs when orbitals from A and :B interact to form bonding molecular orbitals. The electronic structures of these complexes are very intricate since multiple atomic orbitals may be involved. The *d*-metals, for example, can contribute up to nine atomic orbitals (1*s*, 3*p*, and 5*d* orbitals) to the formation of molecular orbitals. See Shriver et al. [38] for an excellent discussion of molecular orbital theory. The product of the reaction between a Lewis acid and a Lewis base is commonly referred to as a complex where A and :B are bonded together through the sharing of the electron pair donated by :B.

The Lewis acid/base concept is key to understanding the chemistry of minerals in foods because metal cations present are Lewis acids and they are bound to Lewis bases. The complexes resulting from reactions between metal cations and food molecules range from metal hydrates, to metal-containing pigments such as hemoglobin and chlorophyll, to metalloenzymes.

The number of Lewis base molecules that may bind to a metal ion is more or less independent of the charge on the metal ion. This number, usually referred to as the coordination number, may range from 1 to 12 but is most commonly 6. For example, Fe^{3+} binds six water molecules to form hexaaquo iron, which takes on an octahedral geometry (Fig.2).

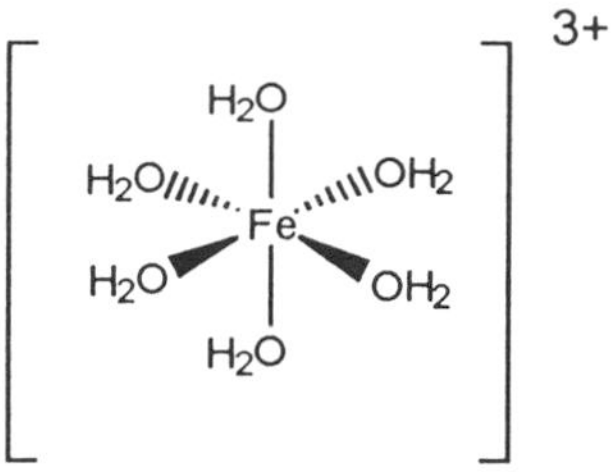

FIGURE 2 Ferric iron with six coordinated water molecules. This is the predominant form of Fe^{3+} in acidic (pH < 1) aqueous solutions.

The electron-donating species in these complexes are commonly referred to as *ligands*. The principal electron donating atoms in ligands are oxygen, nitrogen, and sulfur. Thus many food molecules including proteins, carbohydrates, phospholipids, and organic acids are ligands for mineral ions. Ligands may be classified according to the number of bonds they can form with a metal ion. Those that form one bond are monodentate ligands; those that form two bonds are bidentate; and so on. Ligands that form two or more bonds are referred to collectively as multidentate ligands. Some common examples of ligands are shown in Figure 3.

Stabilities of metal complexes may be expressed as the equilibrium constant for the reaction representing the formation of the complex. The terms "stability constant," k, and "formation constant" are often used interchangeably. The generalized reaction for formation of a complex of a metal ion (M) and a ligand (L) is [38]:

$$\mathrm{M} + \mathrm{L} \rightleftarrows \mathrm{ML} \qquad K_1 = \frac{[\mathrm{ML}]}{[\mathrm{M}][\mathrm{L}]}$$

$$\mathrm{ML} + \mathrm{L} \rightleftarrows \mathrm{ML_2} \qquad K_2 = \frac{[\mathrm{ML_2}]}{[\mathrm{ML}][\mathrm{L}]}$$

$$\downarrow \quad \downarrow \quad \downarrow$$

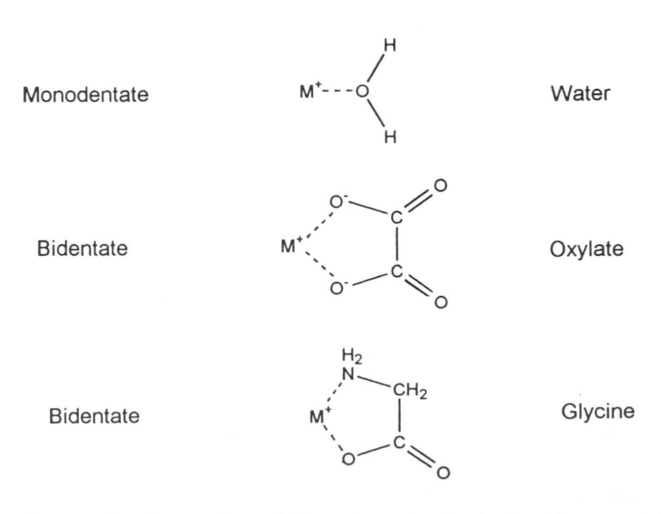

FIGURE 3 Examples of ligands coordinated with a metal ion (M^+)

TABLE 1 Stability Constants (log *K*) for Selected Metal Complexes and Chelates

Ligand	Cu^{2+}	Fe^{3+}
OH^-	6.3	11.8
Oxalate	4.8	7.8
Histidine	10.3	10.0
EDTA	18.7	25.1

Note: Values are corrected to a constant ionic strength.
Source: Adapted from Ref. 10.

$$ML_{n-1} + L \rightleftarrows ML_n \qquad K_n = \frac{[ML_n]}{[ML_{n-1}][L]}$$

When more than one ligand is bound to one metal ion, the overall formation constant may be expressed as:

$$\beta_n = \frac{[ML_n]}{[M][L]^n}$$

where $\beta_n = K_1K_2 \cdots K_n$.

The brackets indicate the equilibrium activities of the bracketed species. Thus K (or β) is an expression of the affinity of ligands for metal ions. Ligands forming complexes with large stability constants will displace metals from complexes with smaller stability constants. Most stability constants are very large, so values for K are often expressed as the logarithms of the actual values. Table 1 shows some stability constants for Cu^{2+} and Fe^{3+}.

9.2.2.3 Activity Versus Concentration

The behavior of ions in solution is influenced by ionic strength. At high ionic strength the "activity" of an ion with respect to its participation in complex formation or other chemical reactions will be different than it would be at the same concentration in a similar solution of lower ionic strength. Recall that the ionic strength of a solution is given by

$$\text{Ionic strength} = \mu = \tfrac{1}{2}(m_1Z_1^2 + m_2Z_2^2 + m_3Z_3^2 + \cdots)$$

where m_1, m_2, and m_3 are the molar concentrations of various ions in solution and Z is the ionic charge.

In order to account for the effect of ionic strength on stability constants, activities rather than actual concentrations should be used in making equilibrium calculations. Activity in this regard is defined as:

$$a_A = f_A[A]$$

where a_A is the activity of species A, f_A is the activity coefficient, and [A] is the concentration of species A. In general, f_A decreases as ionic strength increases. At very low ionic strengths, f_A approaches 1 and activity equals concentration. Data in Figure 4 indicate that ionic strength can have a marked effect on activity coefficients, especially when the charge on the ion is large.

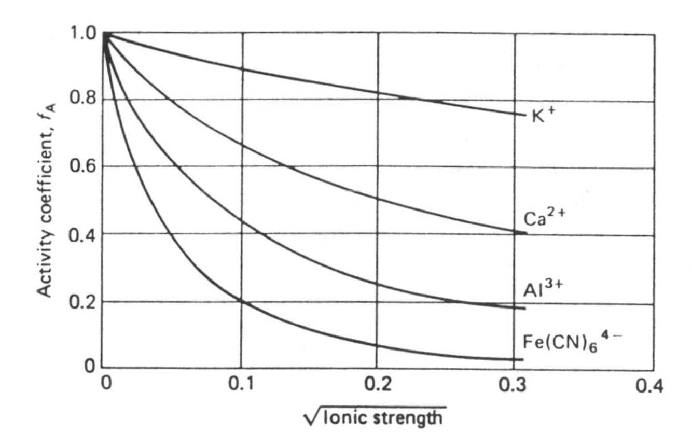

FIGURE 4 The effect of ionic strength on activity coefficients of selected ions. (From Ref. 39.)

Unfortunately, it is difficult to determine activity coefficients, especially in complex systems like foods, so activity coefficients generally are assumed, improperly, to be 1. This being the case, one may wonder about the utility of stability constants for predicting the forms of metal ions in foods. The problem can be partly overcome by expressing stability constants at a given ionic strength and assuming that effects of ionic strength will be similar for all species. Thus, it is reasonable to conclude from the large stability constant for the Fe(EDTA) complex (Table 1) that EDTA (ethylenediamine tetraacetic acid) added to a food will complex much of the iron in that food. On the other hand, it may not be reasonable to conclude that copper will displace iron from a histidine complex since the stability constants, while different, are similar in magnitude.

9.2.2.4 The Chelate Effect

The stabilities of complexes resulting from reactions between Lewis acids and bases are proportional to the driving force of the reaction. This can be described quantitatively by the Gibbs equation for free energy:

$$\Delta G = \Delta H - T\,\Delta S$$

where ΔG is the free energy change in a system undergoing transformation, ΔH is the enthalpy change, T is absolute temperature, and ΔS is the entropy change.

Reactions will occur spontaneously (but at unspecified rates) when ΔG is negative, and as ΔG becomes more negative, the reaction becomes more favorable. Note that for a spontaneous reaction to occur, enthalpy of the system must decrease or entropy must increase. A decrease in enthalpy means that the total electronic energy of the system decreases. An increase in entropy means the randomness of the system increases.

A chelate is a complex resulting from the combination of a metal ion and a multidentate ligand such that the ligand forms two or more bonds with the metal, resulting in a ring structure that includes the metal ion. The term chelate is derived from *chele*, the Greek word for claw. Thus a chelating ligand (also called a chelating agent) must contain at least two functional groups capable of donating electrons. In addition, these functional groups must be spatially arranged so that a ring containing the metal ion can form. Chelates have greater thermodynamic stabilities than similar complexes that are not chelates, a phenomenon referred to as the "chelate effect." Several factors interact to affect the stability of a chelate. Kratzer and Vohra [23] summarized these factors as follows:

1. Ring size. Five-membered unsaturated rings and six-membered saturated rings tend to be more stable than larger or smaller rings.
2. Number of rings. The greater the number of rings in the chelate, the greater the stability.
3. Lewis base strength. Stronger Lewis bases tend to form stronger chelates.
4. Charge of Ligand. Charged ligands form more stable chelates than uncharged ligands. For example, citrate forms more stable chelates than citric acid.
5. Chemical environment of the donating atom. Relative strengths of metal–ligand bonds are shown below in decreasing order.

   Oxygen as donor: $H_2O > ROH > R_2O$

   Nitrogen as donor: $H_3N > RNH_2 > R_3N$

   Sulfur as donor: $R_2S > RSH > H_2S$

6. Resonance in chelate ring. Enhanced resonance tends to increase stability.
7. Stearic hindrance. Large bulky ligands tend to form less stable chelates.

Thus, chelate stabilities are affected by many factors and are difficult to predict. However, the concept of Gibbs free energy is useful for explaining the chelate effect. Consider the example of Cu^{2+} complexing with either ammonia or ethylenediamine [38]:

$$Cu(H_2O)_6^{2+} + 2NH_3 \rightarrow [Cu(H_2O)_4(NH_3)_2]^{2+} + 2H_2O$$

($\Delta H = -46$ kJ mol^{-1}; $\Delta S = -8.4$ J K^{-1} mol^{-1}; and log $\beta = 7.7$.)

$$Cu(H_2O)_6^{2+} + NH_2CH_2CH_2NH_2 \rightarrow [Cu(H_2O)_4(NH_2CH_2CH_2NH_2)]^{2+} + 2H_2O$$

($\Delta H = -54$ kJ mol^{-1}; $\Delta S = +23$ J K^{-1} mol^{-1}; and log $K = 10.1$)

Both complexes have two nitrogens bound to a single copper ion (Fig. 5), and yet the stability of the ethylenediamine complex is much greater than that of the ammonia complex (log of formation constant is 10.1 and 7.7, respectively). Both enthalpy and entropy contribute to the difference in stabilities, but the entropy change is a major factor in the chelate effect. Ammonia, a monodentate ligand, forms one bond to copper while ethylenediamine, a bidentate ligand, forms two. The difference in entropy change is due to the change in the number of independent molecules in solution. In the first reaction, the number of molecules is equal on both sides of the equation so the entropy change is small. The chelation reaction, on the other hand, results in a net increase in the number of independent molecules in solution, and, thus, an increase in entropy.

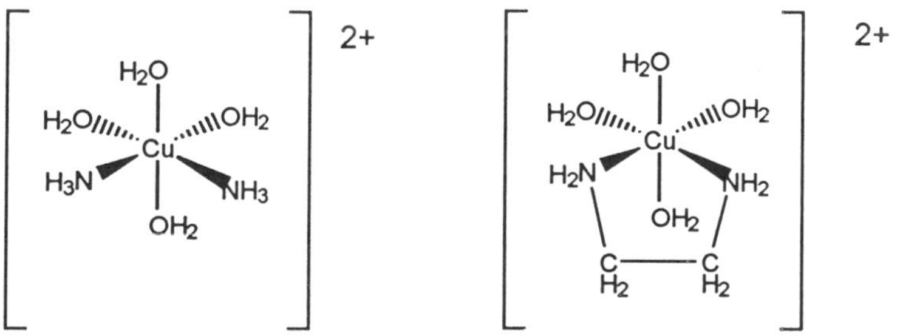

FIGURE 5 Cupric ion (Cu^{2+}) complexed with ammonia (left) and chelated with ethylene (right) in an aqueous system.

Ethylenediamine tetraacetate ion (EDTA) provides an even more dramatic illustration of the chelate effect [34]. EDTA is an hexadentate ligand. When it forms a chelate with a metal ion in solution, it displaces six water molecules from the metal, and this has a large effect on the entropy of the system (Fig. 6):

$$Ca(H_2O)_6^{2+} + EDTA^{4-} \rightarrow Ca(EDTA)^{2-} + 6H_2O \qquad \Delta S = +118\ J\ K^{-1}\ mol^{-1}$$

Moreover, EDTA chelates contain five-membered rings, which also enhance stability. EDTA forms stable chelates with many metal ions and is widely used as a food additive.

Chelates are very important in foods and in all biological systems. Chelating agents may be added to foods to sequester mineral ions, such as iron or copper, to prevent them from acting as prooxidants. Furthermore, most complexes resulting from metal ions and food molecules are chelates.

9.3 NUTRITIONAL ASPECTS OF MINERALS

9.3.1 Essential Mineral Elements

What do we mean when we say an element is "essential for life"? Several definitions have been proposed. A widely accepted definition is: An element is essential for life if its removal from the diet or other route of exposure to an organism "results in a consistent and reproducible impairment of a physiological function" [42]. Thus, essentiality can be demonstrated by feeding diets low in a particular element to humans or experimental animals and watching for signs of impaired function.

Human requirements for essential minerals vary from a few micrograms per day up to about 1 g/day. If intakes are low for some period of time, deficiency signs will develop. Conversely, excessively high intakes can result in toxicity. Fortunately, for most minerals the range of safe and adequate intake is fairly wide, so deficiency or toxicity is relatively rare provided a varied diet is consumed. This concept is illustrated in Figure 7.

This broad range of safe and adequate intakes is possible because higher organisms have homeostatic mechanisms for dealing with both low and high exposures to essential nutrients.

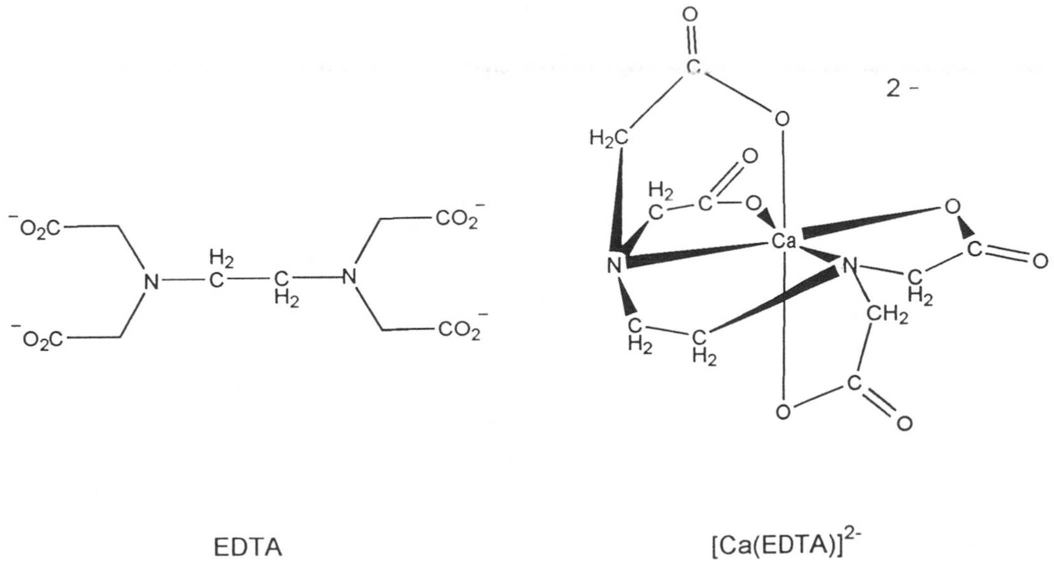

EDTA

$[Ca(EDTA)]^{2-}$

FIGURE 6 Ethylenediamine tetraacetic acid (EDTA) and calcium (Ca^{2+}) chelated with EDTA.

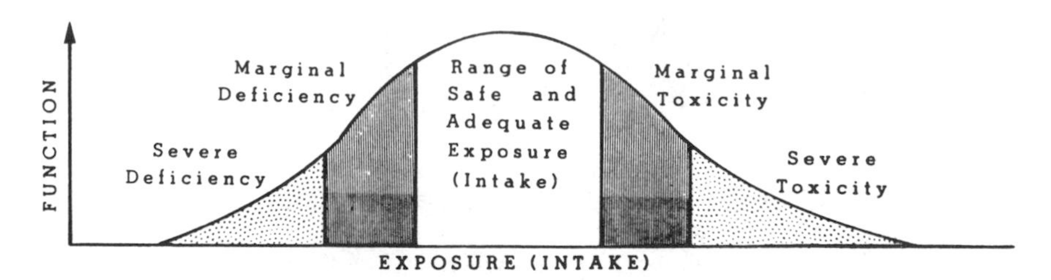

FIGURE 7 Dose-response relationship between mineral intake (dose) and an associated physiological function. Function is impaired at both excessively low and excessively high intakes of most mineral nutrients. (From Ref. 42.)

Homeostasis may be defined as the processes whereby an organism maintains tissue levels of nutrients within a narrow and constant range. In higher organisms, homeostasis is a highly complex set of processes involving regulation of absorption, excretion, metabolism, and storage of nutrients. Without homeostatic mechanisms, intakes of nutrients would have to be very tightly controlled to prevent deficiency or toxicity. Homeostasis can be overridden when dietary levels are excessively low or high for extended periods of time. Persistently low intakes of mineral nutrients is not uncommon, especially in poor populations where access to a variety of foods is often limited. Toxicities caused by high dietary intakes of minerals are less common, although much has been made recently of possible links between chronic diseases and high dietary levels of iron.

9.3.2 Recommended Dietary Allowances for Mineral Nutrients (U.S.)

Recommended dietary allowances (RDAs) are defined as "the levels of intake of essential nutrients that, on the basis of scientific knowledge, are judged by the Food and Nutrition Board to be adequate to meet the known nutrient needs of practically all healthy persons" [14]. RDAs are established by estimating the requirement for the *absorbed* nutrient, adjusting for incomplete utilization of the ingested nutrient, and incorporating a safety factor to account for variability among individuals. Thus RDA values are greater than the requirement, usually by two standard deviations above the mean. This means that individuals with nutrient intakes below the RDA do not necessarily have an inadequate intake. However, when a significant fraction of a population has an intake that is significantly below the RDA for an extended period of time, the probability that deficiency will occur in some individuals increases.

The RDA committee has published RDAs for only 7 of the 20 known essential minerals (Table 2). An RDA is established only when data are sufficient to determine a reliable value. The RDA Committee has, however, published "estimated minimum requirements" for several minerals where data to support an RDA are not adequate. These are sodium, chloride, and potassium (Table 3). "Estimated safe and adequate intakes" have been published for copper, manganese, fluoride, chromium, and molybdenum (Table 4).

Deficiencies of the minerals listed in Table 3 and 4 are rare in the United States. However, for many people, intakes of sodium tend to be higher than desirable for optimal health. Epidemiological data show that blood pressure is positively associated with level of salt intake. Populations with salt intakes greater than 6 g/day have progressively increasing blood pressure with age, while blood pressure rise with age is not seen in populations with salt intakes below 4.5 g/day [8]. Thus, dietary guidelines for the prevention on chronic disease usually include the recommendation that intakes of salt be less than 6 g/day. Sodium chloride is 40% sodium by weight, so 6 g salt provides 2400 mg sodium.

TABLE 2 Recommended Dietary Allowances (RDAs) for the Mineral Nutrients

		Minerals						
Category	Age (years) or condition	Calcium (mg)	Phosphorus (mg)	Magnesium (mg)	Iron (mg)	Zinc (mg)	Iodine (μg)	Selenium (mg)
Infants	0.0–0.5	400	300	40	6	5	40	10
	0.5–1.0	600	500	60	10	5	50	15
Children	1–3	800	800	80	10	10	70	20
	4–6	800	800	120	10	10	90	20
	7–10	800	800	170	10	10	120	30
Males	11–14	1200	1200	270	12	15	150	40
	15–18	1200	1200	400	12	15	150	50
	19–24	1200	1200	350	10	15	150	70
	25–50	800	800	350	10	15	150	70
	51+	800	800	350	10	15	150	70
Females	11–14	1200	1200	280	15	12	150	45
	15–18	1200	1200	300	15	12	150	50
	19–24	1200	1200	280	15	12	150	55
	25–50	800	800	280	15	12	150	55
	51+	800	800	280	10	12	150	55
Pregnant		1200	1200	320	30	15	175	65
Lactating	1st 6 months	1200	1200	355	15	19	200	75
	2nd 6 months	1200	1200	340	15	16	200	75

Source: Adapted from Ref. 14.

9.3.3 Bioavailability

It has been recognized for at least a century that the concentration of a nutrient in a food is not necessarily a reliable indicator of the value of that food as a source of the nutrient in question. This led nutritionists to develop the concept of nutrient bioavailability. Bioavailability may be defined as the proportion of a nutrient in ingested food that is available for utilization in metabolic processes. In the case of mineral nutrients, bioavailability is determined primarily by the efficiency of absorption from the intestinal lumen into the blood. In some

TABLE 3 Estimated Minimum Requirements for Electrolytes

Age	Weight (kg)	Sodium (mg)	Chloride (mg)	Potassium (mg)
Months				
0.–5	4.5	120	180	500
6–11	8.9	200	300	700
Years				
1	11.0	225	350	1000
2–5	16.0	300	500	1400
6–9	25.0	400	600	1600
10–18	50.0	500	750	2000
> 18	70.0	500	50	2000

Source: Adapted from Ref. 14.

TABLE 4 Estimated Safe and Adequate Daily Dietary Intakes of Selected Minerals

Category	Age (years)	Cu (mg)	Mn (mg)	F (mg)	Cr (μg)	Mo (μg)
Infants	0.–0.5	0.4–0.6	0.3–0.6	0.1–0.5	10–40	15–30
	0.5–1	0.6–0.7	0.6–1.0	0.2–1.0	20–60	20–40
Children	1–3	0.7–1.0	1.0–1.5	0.5–1.5	20–80	25–50
and adolescents	4–6	1.0–1.5	1.5–2.0	1.0–2.5	30–120	30–75
	7–10	1.0–2.0	2.0–3.0	1.5–2.5	50–200	50–150
	11+	1.5–2.5	2.0–5.0	1.5–2.5	50–200	5–250
Adults		1.5–3.0	2.0–5.0	1.5–4.0	50–200	75–250

Source: Adapted from Ref. 14.

cases, however, absorbed nutrients may be in a form that cannot be utilized. For example, iron is bound so tightly in some chelates that even if the iron chelate is absorbed, the iron will not be released to cells for incorporation into iron proteins; rather, the intact chelate will be excreted in the urine.

Bioavailabilities of mineral nutrients vary from less than 1% for some forms of iron to greater than 90% for sodium and potassium. The reasons for this wide range are varied and complex, since many factors interact to determine the ultimate bioavailability of a nutrient.

9.3.3.1 Nutritional Utilization of Minerals

The process of mineral nutrient utilization may be described as follows [29]. In the mouth, the food is masticated while salivary amylase begins the process of starch digestion. At this stage, only limited changes in mineral species occur. Next, the food is swallowed and enters the stomach where the pH is gradually lowered to about 2 by gastric acid. At this stage, dramatic changes occur in mineral species. Stabilities of complexes are changed by the altered pH and by protein denaturation and hydrolysis. Minerals may be released into solution and may reform complexes with different ligands. In addition, transition metals such as iron may undergo a valance change when the pH is reduced. The redox behavior of iron is strongly pH dependent. At neutral pH, even in the presence of excess reducing agents like ascorbic acid, ferric iron will not be reduced. However, when the pH is lowered, ascorbic acid rapidly reduces Fe^{3+} to Fe^{2+}. Since Fe^{2+} has lower affinity than Fe^{3+} for most ligands, this reduction will promote the release of iron from complexes in food.

In the next stage of digestion, the partially digested food in the stomach is emptied into the proximal small intestine, where pancreatic secretions containing sodium bicarbonate and digestive enzymes raise the pH and continue the process of protein and starch digestion. In addition, lipases begin digesting triacylglycerols. As digestion proceeds, more new ligands are formed and existing ligands are altered in ways that undoubtedly affect their affinities for metal ions. Thus a further reshuffling of mineral species occurs in the lumen of the small intestine, resulting in a complex mixture of soluble and insoluble and high and low molecular weight species. Soluble species may diffuse to the brush border surface of the intestinal mucosa where they may be taken up by the mucosal cell or pass between cells (the paracellular route). Uptake can be facilitated by a membrane carrier or ion channel; may be an active, energy requiring process; may be saturable; and may be regulated by physiological processes.

Clearly, the process of mineral absorption and the factors that affect it are extremely complex. Moreover, speciation of minerals in the gastrointestinal tract, although known to occur, is poorly understood. Nevertheless, results from hundreds of studies allow us to identify factors that may influence mineral bioavailability. Some of these are summarized in Table 5.

9.3.3.2 Calcium Bioavailability

The absorption of calcium from foods is determined by the concentration of calcium in the food and the presence of inhibitors or enhancers of calcium absorption [27, 45]. Calcium absorption is inversely related to the log of ingested calcium concentration over a wide range of intakes [18]. The main dietary inhibitors of calcium absorption are oxalate and phytate, with oxalate being the most potent of the two. Fiber does not appear to have a major impact on calcium absorption [27, 45].

Listed in Table 6 for selected foods are the calcium content, absorption adjusted for calcium load, and the number of servings equivalent to the absorbable calcium in one serving of milk. Only fortified fruit juices supply more absorbable calcium per serving than milk. These data show that it is difficult to achieve recommended intakes of calcium without consuming milk or other calcium-rich dairy products.

It is apparent from Table 6 that both calcium content of foods and absorbability vary greatly. The percent absorption of calcium from milk is lower than that for some other foods not because it is bound in an unavailable form but because it is present at a high concentration. The

TABLE 5 Factors That May Influence Mineral Bioavailability from Foods

1. Chemical form of the mineral in food
   - Highly insoluble forms are poorly absorbed.
   - Soluble chelated forms may be poorly absorbed if chelate has high stability.
   - Heme iron is absorbed more efficiently than nonheme iron.
2. Food ligands
   - Ligands that form soluble chelates with metals may enhance absorption from some foods (e.g., EDTA enhances iron absorption).
   - High-molecular-weight ligands that are poorly digestible may reduce absorption (e.g., dietary fiber, some proteins).
   - Ligands that form insoluble chelates with minerals may reduce absorption (e.g., oxalate inhibits calcium absorption, phytic acid inhibits iron, zinc, and calcium absorption).
3. Redox activity of food components
   - Reductants (e.g., ascorbic acid) enhance absorption of iron but have little effect on other minerals.
   - Oxidants inhibit the absorption of iron.
4. Mineral–mineral interactions
   - High concentrations of one mineral in the diet may inhibit the absorption of another (e.g., calcium inhibits iron absorption, iron inhibits zinc absorption, lead inhibits iron absorption).
5. Physiological state of consumer
   - Homeostatic regulation of minerals in the body may operate at the site of absorption, resulting in enhanced absorption in deficiency and reduced absorption in adequacy. This is the case for iron, calcium, and zinc.
   - Malabsorption disorders may reduce absorption of minerals.
   - Iron and calcium absorption are reduced in achlorhydria (reduced gastric acid secretion).
   - Age (absorption efficiencies may decline in the elderly)

TABLE 6 Calcium Content and Bioavailability in Selected Foods

Food	Serving size (g)	Calcium content (mg)	Fractional absorption[a] (%)	Estimated absorbable Ca/serving (mg)	Servings to equal 240 ml milk (*n*)
Milk	240	300	32.1	96.3	1.0
Almonds	28	80	21.2	17.0	5.7
Pinto beans	86	44.7	17.0	7.6	12.7
Broccoli	71	35	52.6	18.4	5.2
Cabbage, green	75	25	64.9	16.2	5.9
Cauliflower	62	17	68.6	11.7	8.2
Citrus punch, with CCM[b]	240	300	50.0	150	0.064
Kale	65	47	58.8	27.6	3.5
Soy milk	120	5	31.0	1.6	60.4
Spinach	90	122	5.1	6.2	15.5
Tofu, Ca set	126	258	31.0	80.0	1.2
Turnup greens	72	99	51.6	31.1	1.9
Water cress	17	20	67.0	13.4	7.2

[a]Percent absorption adjusted for calcium load.
[b]Calcium-citrate-maleate.
Source: Ref. 45.

poor bioavailability of calcium from spinach and pinto beans is probably due to high concentrations of oxalate and phytate, respectively.

9.3.3.3 Iron Bioavailability

Iron bioavailability is determined almost totally by the efficiency of iron absorption in the intestine. Total iron intake, composition of the diet, and iron status of the individual consuming the diet all play a role in determining the amount of iron absorbed.

Diets in industrialized countries like the United States consistently provide about 6 mg iron per 1000 kcal (4186 kJ) [6]. Iron species in foods may be broadly grouped as either heme or nonheme. Heme iron is firmly bound in the center of a porphyrin ring and does not dissociate from this ligand until after it is taken up by intestinal mucosal cells. It occurs primarily as hemoglobin or myoglobin and thus is found exclusively in meat, poultry, and fish. Virtually all of the iron in plant foods and approximately 40–60% of the iron in animal tissues is nonheme iron. It is bound primarily to proteins but may also be complexed with citrate, phytate, oxalate, polyphenolics, or other ligands.

The bioavailability of heme iron is relatively unaffected by composition of the diet and is generally significantly greater than that of nonheme iron. The bioavailability of nonheme iron varies markedly depending on composition of the diet. It is widely assumed that nonheme iron from all sources in a meal (foods as well as fortification iron) enters a common pool during digestion and that absorption of iron from this pool is determined by the totality of ligands present in the small intestine at the site of absorption.

Several enhancers and inhibitors of nonheme iron absorption have been identified. Enhancers include meat, poultry, fish, ascorbic acid, and EDTA (in diets where bioavailabilities are low). Inhibitors include polyphenolics (tannins in tea, legumes, and sorghum), phytates (present in legumes and whole-grain cereals), some plant proteins (especially legume proteins), calcium, and phosphates.

The overall bioavailability of iron in a diet is determined by complex interactions of the enhancers and inhibitors present. Iron absorption from diets composed primarily of roots, tubers, legumes, and cereals, with limited meat and ascorbic acid, may be only about 5% even in people with poor iron status. Such a diet would provide only about 0.7 mg absorbable iron per day, a quantity too small to meet the needs of many individuals. Iron absorption from diets based on roots, cereals, and legumes that contain some meat, poultry, or fish and some foods high in ascorbic acid may be about 10%. These diets provide about 1.4 mg of absorbable iron per day, an amount that is adequate for most men and postmenopausal women but inadequate for up to 50% of women of child-bearing age. Diets composed of generous quantities of meat, poultry, fish, and foods high in ascorbic acid provide over 2 mg absorbable iron per day, an amount sufficient to meet the needs of nearly all healthy persons [6].

9.3.4 Minerals of Particular Nutritional Concern

For various reasons, deficiencies are common for some mineral elements and rare or nonexistent for others. Moreover, there are large variations in prevalences of specific deficiencies across geographical and socioeconomic divisions. Human dietary deficiencies have been reported for calcium, cobalt (as vitamin B_{12}), chromium, iodine, iron, selenium, and zinc [17]. Calcium, chromium, iron, and zinc occur in bound form in foods, and bioavailabilities may be low depending on the composition of the food or meal. Thus, deficiencies of these minerals result from a combination of poor bioavailability and low intakes.

Iodine is present in foods and water predominantly as the ionic, unbound form and has high bioavailability. Iodine deficiency is caused primarily by low intakes. Selenium is present in foods as selenomethionine, a chelate, but it is efficiently utilized so deficiency is caused by low intakes. Vitamin B_{12} deficiency is a problem only with persons on strict vegetarian diets, which are low in this vitamin. These observations further illustrate the complexities of the mineral bioavailability story. Some bound forms have low bioavailability while other bound forms have high bioavailability. Unbound forms generally have high bioavailability. Current thinking on bioavailability and mineral deficiencies is summarized in Figure 8.

In the United States, deficiencies of calcium and iron have received the most attention. In

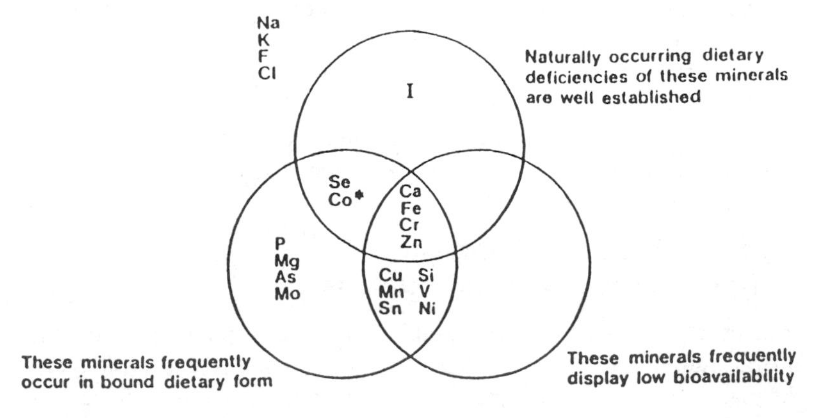

FIGURE 8 Essential minerals grouped by chemical form (free ions in solution or bound to food ligands), bioavailability, and occurrence of nutritional deficiency. *As vitamin B_{12}. (From Ref. 17.)

developing countries, iron and iodine have been targeted because of high prevalences of deficiencies among these populations.

9.4 MINERAL COMPOSITION OF FOODS

9.4.1 Ash

"Ash" is included in nutrient data bases as one of the proximate components of foods. It provides an estimate of the total mineral content of foods [16]. Methods for determination of ash in specific foods and food groups are described in official publications [3]. Minerals in the ash are in the form of metal oxides, sulfates, phosphates, nitrates, chlorides, and other halides. Thus, ash content overestimates total mineral content by a considerable extent since oxygen is present in many of the anions. It does, however, provide a crude idea of mineral content and it is required for calculation of total carbohydrate in the proximate analysis scheme.

9.4.2 Individual Minerals

Individual minerals in foods are determined by ashing the food, dissolving the ash (usually in acid), and measuring mineral concentrations in the resulting solution [16, 19]. Both chemical and instrumental methods are used to measure mineral concentrations, but instrumental methods are generally more rapid, precise, and accurate. Atomic absorption spectroscopy has been available since the 1960s and is still widely used. It is a reliable technique but can measure only one mineral at a time. Inductively coupled plasma spectrometers have gained popularity in recent years primarily because they are capable of quantifying several mineral elements simultaneously from a single sample [28].

Listed in Table 7 are concentrations of some minerals in selected foods. Sources include the U.S. Department of Agriculture database, journal articles, and manufacturers' data. Values are means; data from individual foods may vary substantially from the data reported.

9.4.3 Factors Affecting the Mineral Composition of Foods

Many factors interact to affect the mineral composition of foods, so compositions can vary greatly.

9.4.3.1 Factors Affecting the Mineral Composition of Plant Foods

In order for plants to grow, they must take up water and essential mineral nutrients from the soil. Once taken up by plant roots, nutrients are transported to other parts of the plant. The ultimate composition of the edible parts of plants is influenced and controlled by fertility of the soil, genetics of the plant, and the environment in which it grows (Fig. 9). The degree to which mineral content can vary even within a plant species is illustrated by wheat grain. For grain grown in Australia, North America, an the United Kingdom, zinc concentrations range from 4.5 to 37.2 mg/kg and iron from 23.6 to 74.7 mg/kg [5].

9.4.3.2 Adequacy of Plant Foods for Supplying the Nutrient Needs of Animals and Humans

Several questions are pertinent. Do plants and humans require the same mineral nutrients? Are the concentrations of mineral nutrients in plants sufficient to meet human requirements? Can mineral concentrations in plants be altered by agricultural or genetic means to enhance the nutritional quality of plants? Are plants grown on depleted soils nutritionally inferior to plants grown on more fertile soils?

The list of essential minerals for plants is similar but not identical to the list for humans.

F, Se, and I are essential for humans but not for most plants. Thus, we might expect to see human deficiencies of these elements in populations that depend on plants grown locally where soil concentrations of these elements are low. In fact, serious human deficiencies of selenium and iodine do exist in several areas of the world [41].

For nutrients required by both plants and animals, we might expect human deficiencies to be less of a problem because the elements will necessarily be present in plant foods. Unfortunately, concentrations of minerals in plants are sometimes too low to meet human needs, or the minerals may be present in forms that cannot be efficiently utilized by humans (see earlier section on bioavailability). These situations apply, respectively, to calcium an iron. The calcium content of some plants is extremely low. Rice, for example, contains only about 10 mg calcium per 100 kcal. Thus persons consuming rice-based diets must depend on other foods to meet calcium requirements. Iron is more uniformly distributed in plant foods than calcium but its bioavailability can be extremely poor, so diets based on cereals and legumes are often inadequate in iron.

While it is possible in some cases to enhance the nutritional quality of crops through agronomic practices and plant breeding, the movement of mineral nutrients from the soil to the plant and from the plant to the animal or human is an extemely complicated process. Soils differ considerably in their mineral composition. Moreover, the concentration of an element in the soil may not be a good indicator of the amount that can be taken up by plant roots, since the chemical form of the element and soil pH have marked effects on mineral bioavailability to plants. For example, increasing soil pH by ading lime will lower availability of iron, zinc, manganese, and nickel to plants and will increase availability of molybdenum and selenium [47]. Also, plants generally possess physiological mechanisms for regulating amounts of nutrients taken up from the soil. Therefore, we might expect that attempts to alter the mineral composition of food crops would meet with mixed results. For example, application of fertilizer does not significantly increase iron, manganese, or calcium content of food crops [47]. On the other hand, fertilization with zinc at levels in excess of the zinc requirement of the plant has been shown to increase the level of zinc in pea seeds [48].

9.4.3.3 Factors Affecting the Mineral Composition of Animal Foods

Mineral concentrations in animal foods vary less than mineral concentrations in plant foods. In general, changes in dietary intake of the animal have only a small effect on mineral concentrations in meat, milk, and eggs. This is because homeostatic mechanisms operating in the animal regulate tissue concentrations of essential nutrients.

9.4.3.4 Adequacy of Animal Foods for Supplying the Nutrient Needs of Humans

The composition of animal tissues is similar to that of humans; thus we might expect animal foods to be good sources of nutrients. Meat, poultry, and fish are good sources of iron, zinc, phosphate, and cobalt (as vitamin B_{12}). These products are not good sources of calcium unless bones are consumed, which is usually not the case. Also, the iodine content of animal foods may be low. Dairy products are excellent sources of calcium. Thus, consumption of a variety of animal foods along with a variety of plant foods is the best way to ensure adequate intakes of all essential minerals.

9.4.4 Fortification

Fortification of the U.S. food supply began in 1924 with the addition of iodine to salt to prevent goiter, a prevalent public health problem in the United States at the time. In 1933, the American

TABLE 7 Mineral Composition of Selected Foods

Quantity	Food	Weight (g)	kcal[2]	Ca	Mg	P	Na	K	Fe	Zn	Cu	Se
1 egg	Scrambled	100	157	57	13	269	290	138	2.1	2.0	0.06	8
1 slice	White bread	28	75	35	6	30	144	31	0.8	0.2	0.04	8
1 slice	Whole wheat bread	28	70	20	26	74	180	50	1.5	1.0	0.10	16
0.5 cup	Spaghetti, cooked without salt	70	99	5	13	38	1	22	1.0	0.4	0.07	19.0
0.5 cup	Brown rice, cooked	98	108	10	42	81	5	42	0.4	0.6	0.01	13.0
0.5 cup	White rice, parboiled, cooked	88	100	17	11	37	3	32	1.0	0.3	0.08	8.30
0.5 cup	Black beans, cooked	86	113	24	61	120	1	305	2.0	1.0	0.18	6.9
0.5 cup	Red kidney beans	89	112	25	40	126	2	356	3.0	0.9	0.21	1.9
1 cup	Whole milk	244	150	291	33	228	120	370	0.1	0.9	0.05	3.0
1 cup	Skim milk/nonfat milk	245	86	302	28	247	126	406	0.1	0.9	0.05	6.6
1.5 oz.	American cheese, processed	43	159	261	10	316	608	69	0.2	1.3	0.01	3.8
1.5 cup	Cheddar cheese	43	171	305	12	219	264	42	0.3	1.3	0.01	6.0
0.5 cup	Cottage cheese, creamed, small curd	105	108	63	6	139	425	89	0.1	0.4	0.03	6.3
1 cup	Yogurt, low-fat, plain	227	144	415	10	326	159	531	0.2	2.0	0.10	5.5
0.5 cup	Ice cream, regular vanilla	67	134	88	9	67	58	128	0.1	0.7	0.01	4.7
1 each	Baked potato with skin	202	220	20	55	115	16	844	2.8	0.7	0.62	1.8

1 each	Peeled potato, boiled	135	116	10	26	54	7	443	0.4	0.4	0.23	1.2
3 each	Broccoli, raw spears	453	126	216	114	297	123	1470	4.0	2.0	0.40	0.9
3 each	Broccoli spear, cooked from fresh	540	151	249	130	318	141	1575	4.5	2.1	0.23	1.1
0.5 cup	Raw carrot, grated	55	24	15	8	24	19	178	0.3	0.1	0.03	0.8
0.5 cup	Cooked carrots, from frozen	73	26	21	7	19	43	115	0.4	0.2	0.05	0.9
1 each	Tomato, fresh, whole, average	123	26	6	14	30	11	273	0.6	0.1	0.09	0.6
0.75 cup	Tomato juice, canned	183	31	17	20	35	661	403	1.0	0.3	0.18	0.4
0.75 cup	Orange juice prepared from frozen	187	83	17	18	30	2	356	0.2	0.1	0.08	0.4
1 each	Orange, average, 2⅝ in diameter	131	60	52	13	18	0	237	0.1	0.1	0.06	1.2
1 each	Apple with peel, 2.75 in diameter	138	80	10	6	10	1	159	0.3	0.1	0.06	0.6
1 each	Banana (peeled weight)	114	85	7	32	22	1	451	0.4	0.2	0.12	1.1
3 oz.	Beef round, roasted	85	205	5	21	176	50	305	1.6	3.7	0.08	—
3 oz.	Veal, round, roasted	85	160	6	28	234	68	389	0.9	3.0	0.13	—
3 oz.	Chicken, white meat, roasted	85	140	13	25	194	63	218	0.9	0.8	0.04	—
3 oz.	Chicken, leg meat, roasted	85	162	10	20	156	77	206	1.1	2.4	0.07	—
3 oz.	Salmon, cooked	85	183	6	26	234	56	319	0.5	0.4	0.06	—
3 oz.	Salmon, canned, with bones	85	130	203	25	277	458	231	0.9	0.9	0.07	—

[2]kcal × 4.186 = kJ.

Note: Values are mg per serving, except Se is μg per serving.

Source: Refs. 12 and 43.

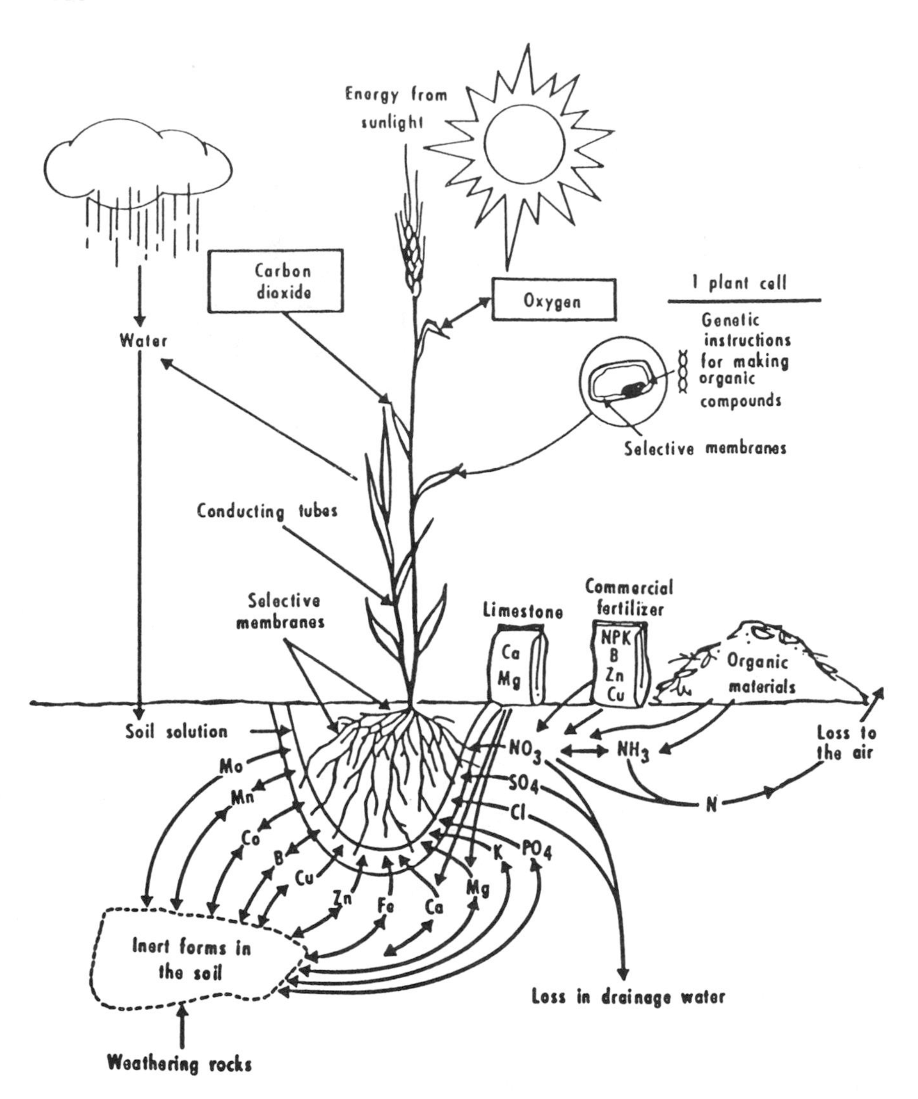

FIGURE 9 Plants obtain mineral nutrients from the soil solution surrounding the roots. Sources of these minerals include fertilizer, decaying organic materials, and weathering rocks. The minerals are taken up in the roots by a selective process and transported upward to all parts of the plant. The whole process is regulated according to instructions encoded in the plants genetic material. (From Ref. 1.)

Medical Association recommended that vitamin D be added to milk to prevent rickets. In the early 1940s, food fortification was expanded further when it became apparent that many young adults were failing Army physical exams due to poor nutritional status. In 1943, the government issued an order making enrichment of flour with iron, riboflavin, thiamin, and niacin mandatory.

Since the introduction of fortification back in the 1920s there has been a dramatic reduction in the prevalences of many nutrient deficiency diseases in the United States, including iron deficiency and goiter. While general improvements in diets were major factors in this improvement in nutritional status, fortification undoubtedly deserves much of the credit for the low prevalences of nutrient deficiency diseases in the United States today.

In the United States, fortification of foods with iron and iodine remains widespread. In addition, calcium, zinc, and other trace minerals are sometimes added to breakfast cereals and other products. Infant formulas contain the largest number of added minerals since they must be nutritionally complete.

Iron has received far more attention than the other minerals because of the high prevalences of iron deficiency anemia and because of technological and stability problems associated with the addition of iron to foods. Thus, iron fortification deserves detailed attention here.

The first recorded recommendation for iron fortification was made in 4,000 BC by a Persian physician named Melampus [35]. He recommended that sailors consume sweet wine laced with iron filings to strengthen their resistance to spears and arrows and to enhance sexual potency. Widespread iron fortification began in the United States in 1943 when War Food Order No. 1 made enrichment of white flour sold in interstate commerce mandatory. Federal regulations no longer require flour enrichment, but many state regulations do. If flour and other cereal products are enriched, permissible levels of addition are specified in FDA regulations (Table 8).

Addition of iron to foods is a difficult balancing act because some forms of iron catalyze oxidation of unsaturated fatty acids and vitamins A, C, and E. In many cases, forms that are highly bioavailable are also the most active catalytically, and forms that are relatively chemically inert tend to have poor bioavailability. Numerous iron sources have been studied for their effect on oxidative stability and nutritional efficacy. These sources are listed in Table 9.

Ferrous sulfate is the cheapest, most bioavailable, and most widely used iron source for food fortification. It is routinely used as the reference standard in iron bioavailability studies because of its high bioavailability (Table 9). Results of several studies have indicated that off odors and off flavors occur in bakery products made from flour that was heavily fortified with ferrous sulfate and stored for extended periods of time [4]. Barrett and Ranum [4] made the following recommendations for minimizing oxidation problems in bakery products that have been fortified with ferrous sulfate:

1. Ferrous sulfate is the preferred iron source for addition at the bakery.
2. Ferrous sulfate may be used to fortify wheat flour provided iron levels are kept below 40 ppm and the flour is stored at moderate temperatures and humidities for periods not to exceed 3 months.
3. Ferrous sulfate should not be used to fortify flour that may be stored for extended

TABLE 8 FDA Standards for the Enrichment of Cereal Products with Iron and Calcium

Food	Iron (mg/lb) (shall contain)	Calcium (mg/lb) (may contain)
Enriched flour	20	960
Enriched rice	Not less than 16 Not more than 32	Not less than 500 Not more than 750
Enriched corn grits	Not less than 13 Not more than 26	Not less than 500 Not more than 1000
Enriched macaroni products	Not less than 13 Not more than 16.5	Not less than 500 Not more than 625
Enriched bread, rolls, and buns	12.5	600

Note: Forms of iron and calcium used must be harmless and assimilable.
Source: Ref. 13.

TABLE 9 Iron Sources Used in Food Fortification and Their Bioavailabilities

Chemical name	Formula	Fe content (g/kg fortificant)	Relative biological value[a] Human	Rat
Ferrous sulfate	$FeSO_4 \cdot 7H_2O$	200	100	100
Ferrous lactate	$Fe(C_3H_5O_3)_2 \cdot 3H_2O$	190	106	—
Ferric phosphate	$FePO_4 \cdot xH_2O$	280	31	3–46
Ferric pyrophosphate	$Fe_4(P_2O_7)_3 \cdot 9H_2O$	250	—	45
Ferric sodium pyrophosphate	$FeNaP_2O_3 \cdot 2H_2O$	150	15	14
Ferric ammonium citrate	$Fe_xNH_4(C_6H_8O_7)_x$	165–185	—	107
Elemental Fe	Fe	960–980	13–90	8–76

[a]Relative biological value is the bioavailability relative to ferrous sulfate which is set at 100.
Source: Adapted from Refs. 35 and 21.

periods of time (as is the case with all purpose flour intended for home use) or for flour that is to be used in mixes containing added fats, oils, or other easily oxidized ingredients.

4. Concentrated premixes containing ferrous sulfate and wheat flour for later addition to flour should not be used because rancidity may develop in the premix.

When fortification with ferrous sulfate is likely to cause problems in a food, other sources are commonly used. In recent years, elemental iron powders have been the sources of choice for fortification of flour for home use, breakfast cereals, and infant cereals. These are all products with long shelf lives.

As the name implies, elemental iron powders consist of elemental iron in a finely divided form. These forms are nearly pure iron with some contamination with other trace minerals and iron oxides. Elemental iron is insoluble in water and thus it is likely that it must be oxidized to a higher oxidation state before it can be absorbed from the intestine. Presumably, this oxidation occurs in the stomach when the iron is exposed to stomach acid:

$$Fe^{\circ} + 2H^+ \rightarrow Fe^{2+} + H_2 \uparrow$$

Three different types of elemental iron powders are available [32].

Reduced iron: This form is produced by reducing iron oxide with hydrogen or carbon monoxide gas and then milling to a fine powder. It is the least pure of the three types, and purity depends largely on the purity of the iron oxide used [32].

Electrolytic iron: This form is produced by the electrolytic deposition of iron onto a cathode made of flexible sheets of stainless steel. The deposited iron is removed by flexing the sheets and it is then milled to a fine powder. The purity of electrolytic iron is greater than that of reduced iron. The main impurity is the iron oxide that forms on the surface during grinding and storage [32].

Carbonyl iron: This form is produced by heating scrap or reduced iron in the presence of CO under high pressure to form iron pentacarbonyl, $Fe(CO)_5$. The pentacarbonyl is then decomposed by heating to yield a very fine powder of high purity [32].

Elemental iron powders are relatively stable and do not appear to cause serious problems with oxidation in foods. However, the bioavailability of the powders is variable, probably due to differences in particle size. Iron powders are dark gray in color and do cause a slight darkening of white flour, but this is not considered to be a problem [4].

Sodium iron EDTA has not yet been approved by the U.S. Food and Drug Administration for use as an iron fortificant in the United States. However, disodium and calcium disodium EDTA have been approved and are widely used. Sodium iron EDTA is attractive as an iron fortificant because it is relatively stable in foods, it does not catalyze the formation of undesirable flavors and odors in most foods, it is not markedly affected by dietary inhibitors of iron absorption, and it may actually enhance the bioavailability of intrinsic food iron in some low-bioavailability foods [49].

There has been reluctance to used iron EDTA in foods because of concern over possible adverse effects of excessive levels of EDTA in the diet. However, recent surveys have shown that dietary levels of EDTA in the United States are substantially lower than previously thought. This finding may cause the FDA to approve iron EDTA for use in the United States [49].

9.4.5 Effects of Processing

Mineral elements, unlike vitamins and amino acids, cannot be destroyed by exposure to heat, light, oxidizing agents, extremes in pH, or other factors that affect organic nutrients. In essence, minerals are indestructible. Minerals can, however, be removed from foods by leaching or physical separation. Also, the bioavailabilities of minerals may be altered by the factors mentioned earlier.

The most important factor causing mineral loss in foods is milling of cereals. Mineral elements in grain kernels tend to be concentrated in the bran layers and the germ. Thus, removal of bran and germ leaves pure endosperm, which is mineral poor. Mineral concentrations in whole wheat, white flour, wheat bran, and wheat germ are shown in Table 10. Similar losses occur during milling of rice and other cereals. These are substantial losses. During fortification of milled products in the United States, iron is the only mineral commonly added.

Retention of calcium in cheese can be dramatically affected by manufacturing conditions. In cheeses where the pH is low, substantial losses of calcium occur when the whey is drained. Calcium contents of various cheeses are shown in Table 11. Compositions are expressed both as mg/100 g cheese and as a Ca:protein ratio. The latter expression gives a better comparison of Ca losses because the water content of cheeses varies from one variety to another. Cottage cheese has the smallest calcium concentration because the pH at time of whey removal is typically less than 5 [15]. In

TABLE 10 Concentrations of Selected Trace Minerals in Wheat and Milled Wheat Products

Mineral	Whole heat	White flour	Wheat germ	Millfeeds (bran)	Loss from wheat to flour (%)
Iron	43	10.5	67	47–78	76
Zinc	35	8	101	54–130	78
Manganese	46	6.5	137	64–119	86
Copper	5	2	7	7–17	68
Selenium	0.6	0.5	1.1	0.5–0.8	16

Note: Values are mg mineral/kg product.
Source: Adapted from Ref. 36.

TABLE 11 Protein, Calcium, and Phosphate Contents of Selected Cheeses

Cheese variety	Protein (%)	Ca (mg/100 g)	Ca:protein (mg:g)	PO_4 (mg/100 g)	PO_4:protein (mg:g)
Cottage	15.2	80	5.4	90	16.7
Cheddar	25.4	800	31.5	860	27.3
Emmenthal	27.9	920	33.1	980	29.6

Source: Adapted from Refs. 15 and 25.

cheddar and Emmenthal cheeses, the whey is normally drained at pH 6.1 and 6.5, respectively, and this causes the calcium content of Emmenthal cheese to be greater than that of cheddar [25].

Since many minerals do have significant solubility in water, it is reasonable to expect that cooking in water would result in some losses of minerals. Unfortunately, few controlled studies have been done. In general, boiling in water causes greater loss of minerals from vegetables than steaming [24]. Mineral losses during cooking of pasta are minimal for iron but more than 50% for potassium [24]. This is predictable because potassium is present in foods as the free ion while iron is bound to proteins and other high-molecular-weight ligands in the food.

9.5 CHEMICAL AND FUNCTIONAL PROPERTIES OF MINERALS IN FOODS

Even though minerals are present in foods at relatively low concentrations, they often have profound effects on physical and chemical properties of foods because of interactions with other food components. Details of mineral–food interactions for the broad array of minerals found in foods are given mainly in other chapters, and these interactions as well as their roles are summarized in Table 12. A more detailed treatment of selected minerals follows.

9.5.1 Calcium

Besides its structural role in plants and animals, calcium plays a major regulatory role in numerous biochemical and physiological processes. For example, calcium is involved in photosyntheses, oxidative phosphorylation, blood clotting, muscle contraction, cell division, transmission of nerve impulses, enzyme activity, cell membrane function, and hormone secretion.

Calcium is a divalent cation with a radius of 0.95 Å. Its multiple roles in living cells are related to its ability to form complexes with proteins, carbohydrates, and lipids. Calcium binding is selective. Its ability to bind to neutral oxygens, including those of alcohols and carbonyl groups, and to bind to two centers simultaneously allow it to function as a cross-linker of proteins and polysaccharides [10]. This property has numerous consequences in foods.

The functional role of calcium in milk and milk products has been studied extensively and serves as an example of mineral interactions in a food system (see Chap. 14). Milk contains a complex mixture of minerals including calcium, magnesium, sodium, potassium, chloride, sulfate, and phosphate. Calcium in milk is distributed between the milk serum and the casein micelles. The calcium in serum is in solution and comprises about 30% of the total milk calcium. The remainder of the calcium is associated with casein micelles and is present primiarily as colloidal calcium phosphate. It is likely that association of submicelles involves calcium bridges between phosphate groups esterified to serine residues in casein and inorganic phosphate ions.

Calcium and phosphate play an important functional role in the manufacture of cheese.

Addition of calcium prior to renneting shortens coagulation time [25]. Curds with lower Ca contents tend to be crumbly, while cheeses higher in Ca are more elastic.

9.5.2 Phosphates

Phosphates occur in foods in many different forms, both as naturally occurring components of biological molecules and as food additives with specific functions.

In living systems, phosphates serve a variety of functions. For example adenosine triphosphate (ATP) is the principle source of energy in cells. Phosphoproteins (ferritins) are involved in iron storage. Phospholipids are major components of membranes. Hydroxyapatite, $Ca_{10}(PO_4)_6(OH)_2$, comprises the mineral phase of bone. Sugar phosphates, such as glucose 6-phosphate, are key intermediates in carbohydrate metabolism.

A voluminous literature exists on the use of phosphates in foods. See Ellinger [11] and Molins [30] for in-depth treatments of this topic. Several phosphates are approved food additives. These include phosphoric acid, the orthophosphates, pyrophosphates, tripolyphosphates, and higher polyphosphates. Structures are shown in Figure 10.

Phosphate food additives serve many functions including acidification (soft drinks), buffering (various beverages), anticaking, leavening, stabilizing, emulsifying, water binding, and protection against oxidation. The chemistry responsible for the wide array of functional properties of phosphates is not fully understood but undoubtedly is related to the acidity of protons associated with phosphates an the charge on phosphate ions. At pH levels common in foods, phosphates carry negative charges and polyphosphates behave as polyelectrolytes. These negative charges give phosphates strong Lewis-base character and thus strong tendencies to bind metal cations. An ability to bind metal ions may underlie several of the functional properties noted earlier. It should be mentioned, however, that there is considerable controversy about mechanisms of phosphate functionality, particularly as it relates to enhanced water-holding capacity in meats and fish.

9.5.3 Iron

Iron is the fourth most abundant element in the earth's crust and is an essential nutrient for nearly all living species. In biological systems, iron is present almost exclusively as chelates with proteins. Iron plays many key roles in biological systems, including oxygen transport and storage in higher animals (hemoglobin and myoglobin), ATP generation (iron-sulfur proteins and cytochromes), DNA synthesis (ribonucleotide reductase), and chlorophyll synthesis. Unfortunately, free iron can be toxic to living cells. Presumably, this toxicity results from the generation of activated species of oxygen, which in turn can promote lipid oxidation or attack DNA molecules (see later discussion).

In order to avoid the toxic consequences of free iron, virtually all living cells have a mechanism for storing extra iron intracellularly in a nontoxic form. The iron is sequestered in the interior of a hollow protein shell called apoferritin. This protein shell is composed of 24 polypeptide subunits arranged as a sphere. Iron is deposited in the cavity of the shell as polymeric ferric oxyhydroxide. Up to 4500 atoms of iron can be stored in one ferritin shell [46]. Ferritin iron is essentially a cellular reserve that can be mobilized when iron is needed for the synthesis of hemoglobin, myoglobin, or other iron proteins.

In spite of its abundance in the environment, iron deficiency in humans, some farm animals, and crops grown on some soils is a problem of major proportions. For example, Schrimshaw [37] estimated that two-thirds of children and women of childbearing age in most developing countries suffer from iron deficiency. The presence of iron deficiency in the United

TABLE 12 Nutritional and Functional Roles of Minerals and Mineral Salts/Complexes in Foods

Mineral	Food sources	Function
Aluminum	Low and variable in foods, component of some antacids and leavening agents	Essential nutrient: Possibly essential, evidence not conclusive. Deficiency unknown. Leavening agent: As sodium aluminum sulfate ($Na_2SO_4 \cdot Al_2(SO_4)_3$) Texture modifier
Bromine	Brominated flour	Essential nutrient: Not known to be essential to humans. Dough improver: $KBrO_3$ improves baking quality of wheat flour. It is the most used dough improver.
Calcium	Dairy products, green leafy vegetables, tofu, fish bones	Essential nutrient: Deficiency leads to osteoporosis in later life. Texture modifier: Forms gels with negatively charged macromolecules such as alginates, low-methoxy pectins, soy proteins, caseins, etc. Firms canned vegetables when added to canning brine.
Copper	Organ meats, seafoods, nuts, seeds	Essential nutrient: Deficiency rare. Catalyst: Lipid peroxidation, ascorbic acid oxidation, nonenzymatic oxidative browning. Color modifier: May cause black discoloration in canned, cured meats. Enzyme cofactor: Polyphenoloxidase. Texture stabilizer: Stabilizes egg-white foams.
Iodine	Iodized salt, seafood, plants and animals grown in areas where soil iodine is not depleted	Essential nutrient: Deficiency produces goiter and cretinism. Dough improver: KIO_3 improves baking quality of wheat flour.
Iron	Cereals, legumes, meat, contamination from iron utensils and soil, enriched products	Essential nutrient: Deficiency leads to anemia, impaired immune response, reduced worker productivity, impaired cognitive development in children. Excessive iron stores may increase risk of cancer and heart disease. Catalyst: Fe^{2+} and Fe^{3+} catalyze lipid peroxidation in foods. Color modifier: Color of fresh meat depends on valence of Fe in myoglobin and hemoglobin: Fe^{2+} is red, Fe^{3+} is brown. Forms green, blue, or black complexes with polyphenolic compounds. Reacts with S^{2-} to form black FeS in canned foods. Enzyme cofactor: Lipoxygenase, cytochromes, ribonucleotide reductase, etc.
Magnesium	Whole grains, nuts, legumes, green leafy, vegetables	Essential nutrient: Deficiency rare. Color modifier: Removal of Mg from chlorophyll changes color from green to olive-brown

Table 12 Continued

Mineral	Food sources	Function
Manganese	Whole grains, fruits, vegetables	Essential nutrient: Deficiency extremely rare. Enzyme cofactor: pyruvate carboxylase, superoxide dismutase.
Nickel	Plant foods	Essential nutrient: Deficiency in humans unknown. Catalyst: Hydrogenation of vegetable oils—finely divided, elemental Ni is the most widely used catalyst for this process.
Phosphates	Ubiquitous, animal products tend to be good sources	Essential nutrient: Deficiency rare due to presence in virtually all foods. Acidulent: H_3PO_4 in soft drinks. Leavening acid: $Ca(HPO_4)_2$ is a fast-acting leavening acid. Moisture retention in meats: Sodium tripolyphosphate improves moisture retention in cured meats. Emulsification aid: Phosphates are used to aid emulsification in comminuted meats and in process cheeses.
Potassium	Fruits, vegetables, meats	Essential nutrient: Deficiency rare. Salt substitute: KCl may be used as a salt substitute. May cause bitter flavor. Leavening agent: Potassium acid tartrate.
Selenium	Seafood, organ meats, cereals (levels vary depending on soil levels)	Essential nutrient: Keshan disease (endemic cardiomyopathy in China) was associated with selenium deficiency. Low selenium status may be associated with increased risk for cancer and heart disease. Enzyme cofactor: Glutathione peroxidase.
Sodium	NaCl, MSG, other food additives, milk; low in most raw foods	Essential nutrient: Deficiency is rare; excessive intakes may lead to hypertension. Flavor modifier: NaCl elicits the classic salty taste in foods. Preservative: NaCl may be used to lower water activity in foods. Leavening agents: Many leaving agents are sodium salts, e.g., sodium bicarbonate, sodium aluminum sulfate, sodium acid pyrophosphate.
Sulfur	Widely distributed	Essential nutrient: A constituent of the essential amino acids methionine and cystine. Sulfur amino acids may be limiting in some diets. Browning inhibitor: Sulfur dioxide and sulfites inhibit both enzymatic and nonenzymatic browning. Widely used in dried fruits. Antimicrobial: Prevents, controls microbial growth. Widely used in wine making.
Zinc	Meats, cereals	Essential nutrient: Deficiency produces loss of appetite, growth retardation, skin changes. Marginal deficiency exists in United States but extent is unknown. Pronounced deficiency was documented in populations in the Middle East. ZnO is used in the lining of cans for proteinaceous foods to lessen formation of black FeS during heating. Zn can be added to green beans to help stabilize the color during canning.

phosphoric acid

orthophosphate

pyrophosphate

tripolyphosphate

higher polyphosphates

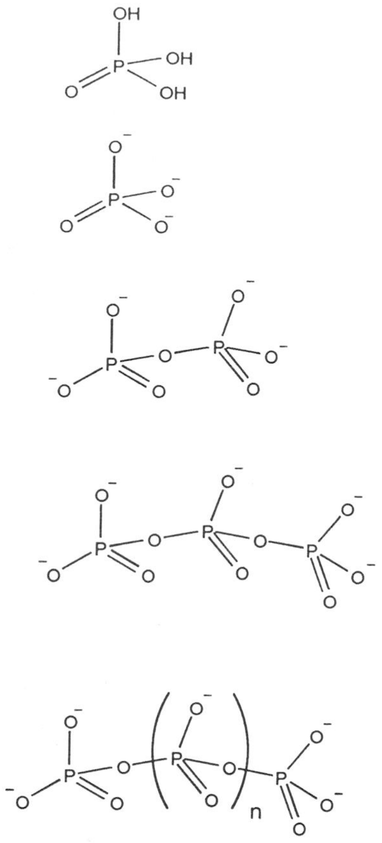

FIGURE 10 Structures of phosphoric acid and phosphate ions important in foods.

States and other industrialized countries is lower than that in many developing countries but remains a persistent problem.

The paradox of high prevalences of nutritional deficiency of a nutrient present in such abundance in the environment may be explained by the behavior of iron in aqueous solutions. Iron is a transition element, which means that it has unfilled d orbitals. Its oxidation state in most natural forms is either +2 (ferrous) or +3 (ferric). Ferrous iron has six *d* electrons while ferric iron has five. In aqueous solutions under reducing conditions, the ferrous form preominates. Ferrous iron is quite soluble in water at physiological pH levels. In the presence of molecular oxygen, however, aqueous Fe^{2+} may be oxidized to Fe^{3+}.

$$Fe^{2+}_{aq} + O_2 \rightarrow Fe^{3+}_{aq} + O_2^-$$

The hydrated Fe^{3+} will then undergo progressive hydrolysis to yield increasingly insoluble ferric hydroxide species [7]:

$$Fe(H_2O)_6^{3+} + H_2O \rightarrow Fe(H_2O)_5(OH^-)^{2+} + H_3O^+ \rightarrow \rightarrow \rightarrow Fe(OH)_3$$

Because this hydrolysis reaction occurs readily except at very low pH, the concentration of free ferric ions in aqueous systems is vanishingly small. The predominance of low-solubility forms of iron explains why it is so poorly available.

It is well established that iron can promote lipid peroxidation in foods. Iron appears to catalyze both the initiation and propagation stages of lipid peroxidation. The chemistry is exceedingly complex, but several probable mechanisms hae been suggested. In the presence of thiol groups, ferric iron promotes the formation of the superoxide anion [50]:

$$Fe^{3+} RSH \rightarrow Fe^{2+} + RS^{\cdot} + H^{+}$$

$$RSH + RS^{\cdot} + O_2 \rightarrow RSSR + H^{+} + {}^{\cdot}O_2^{-}$$

The superoxide anion may then react with protons to form hydrogen peroxide or reduce ferric iron to the ferrous form:

$$2H^{+} + 2{}^{\cdot}O_2^{-} \rightarrow H_2O_2 + O_2$$

$$Fe^{3+} + {}^{\cdot}O_2^{-} \rightarrow Fe^{2+} + O_2$$

Ferrous ion promotes decomposition of hydrogen peroxide to hydroxyl radicals by the Fenton reaction:

$$Fe^{2+} + H_2O_2 \rightarrow Fe^{3+} + OH^{-} + {}^{\cdot}OH$$

The hydroxyl radical is highly reactive and may rapidly generate lipid free radicals by abstracting hydrogen atoms from unsaturated fatty acids. This initiates the lipid peroxidation chain reaction.

Iron can also catalyze lipid peroxidation by accelerating decomposition of lipid hydroperoxides present in foods:

$$Fe^{2+} + LOOH \rightarrow Fe^{3+} + LO^{\cdot} + OH^{-}$$

or

$$Fe^{3+} + LOOH \rightarrow Fe^{2+} + LOO^{\cdot} + H^{+}$$

The rate of the first reaction is greater than the second by an order of magnitude. This explains why ascorbic acid may function as a prooxidant in some food systems since it can reduce ferric iron to the ferrous form.

9.5.4 Nickel

While nickel deficiency has never been documented in humans, there is substantial evidence of its essentiality in several animal species [14]. The primary significance of nickel from a food processing perspective is its use as a catalyst for the hydrogenation of edible oils [31] (see Chap. 5). The hydrogenation reaction takes place on the surface of the catalyst. Thus catalytic activity is a function of the surface area of the catalyst. For this reason, catalysts are prepared as a finely divided power of elemental Ni dispersed in a solid support. When the catalyst is mixed with the oil, and hydrogen gas is introduced, double bonds in the fatty acid residues in the triacylglycerol molecules attach to the surface of the Ni through a π-bonding interaction and the bond opens. H_2 also binds to the Ni and the H-H bond is cleaved. Thereafter, several possibilities exist (Fig. 11): (a) transfer of two hydrogens from the Ni to carbons on the fatty acid and subsequent dissociation of the fatty acid from the Ni leaving a saturated C-C bond; (b) dissociation of the fatty acid from the Ni before transfer of hydrogens, leaving the original CIS-unsaturated species; (c) rotation around the C-C bond on the carbon attached to the catalyst followed by dissociation from the catalyst before transfer of hydrogens, leaving a *trans* double bond; or (d) transfer of hydrogens to the double bond followed by removal of different hydrogens

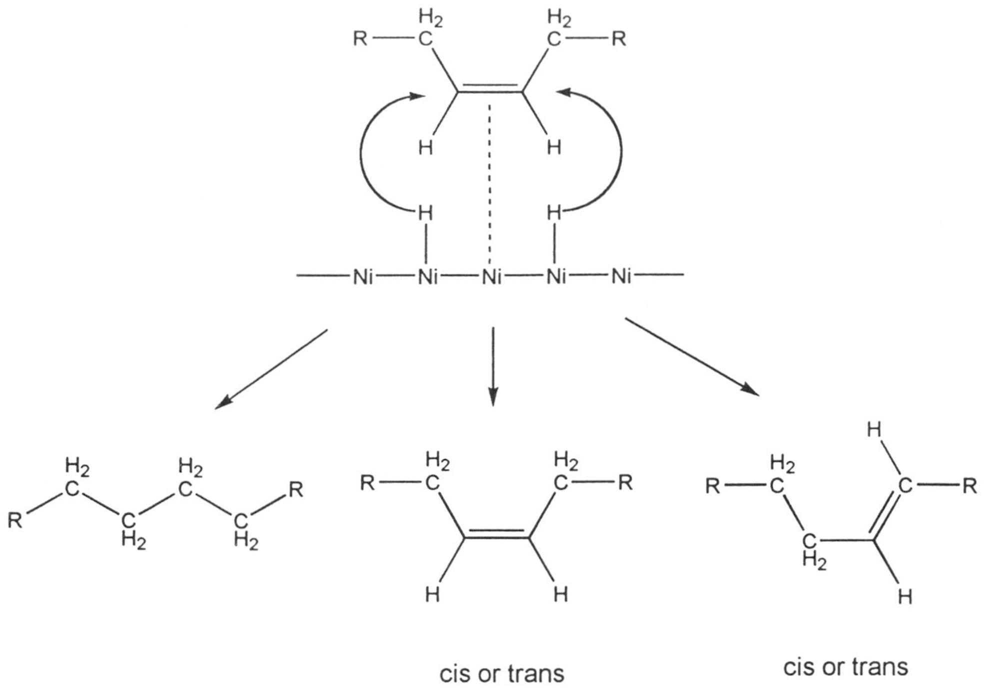

FIGURE 11 Possible products of hydrogenation of a monounsaturated fatty acid using a nickel catalyst. R denotes hydrocarbon chain with methyl or carboxyl end group.

before the fatty acid disassociates, leaving a double bond (either *cis* or *trans*) shifted one position. Thus hyrogenation results in a mixture of products.

The situation is even more complicated when the fatty acid contains two or more double bonds. This is because methylene hydrogens are labile and may be easily removed by the catalyst. Rearrangement of electrons then results in conjugation of double bonds, and subsequent hydrogenation of one of the double bonds yields a mixture of *cis* and *trans* isomers in different positions in the molecule.

The physical and nutritional properties of the hydrogenated oil are determined by the degree of saturation, the location of the double bonds, and the fraction of *trans* double bonds. Thus, it is important to adjust operating conditions (mainly hydrogen pressure and temperature) to achieve a desirable final product [2].

9.5.5 Copper

Copper, like iron, is a transition element and exists in foods in two oxidation states Cu^{1+} and Cu^{2+}. It is a cofactor in many enzymes including phenolase and is at the active center of hemocyanin, an oxygen carrying protein in some arthropods. Both Cu^{1+} and Cu^{2+} bind tightly to organic molecules and thus exist primarily as complexes and chelates in foods. On the negative side, copper is a potent catalyst of lipid oxidation in foods.

An intriguing functional role of copper has been exploited in Western cuisine for at least 300 years [26]. Many recipes for meringues specify copper bowls as the preferred vessel for whipping egg whites. A common problem with egg white foams is collapse resulting from overwhipping. Presumably, foam stability is reduced when the proteins at the air–liquid interface are excessively denatured by whipping. Egg white contains conalbumin, a protein analogous to the plasma iron-binding protein transferrin. Conalbumin binds Cu^{2+} as well as Fe^{3+}, and the presence of bound copper or iron stabilizes conalbumin against excessive denaturation [33].

9.6 SUMMARY

Minerals are present in foods at low but variable concentrations and in multiple chemical forms. These species undergo complex changes during processing, storage, and digestion of foods. With the exception of group IA and VIIA elements, minerals exist in foods as complexes, chelates, or oxyanions. While understanding of the chemical forms and properties of many of these mineral species remains limited, their behavior in foods often can be predicted by applying principles of inorganic, organic, physical, and biological chemistry.

The primary role of minerals in foods is to provide a reliable source of essential nutrients in a balanced and bioavailable form. In cases where concentrations and/or bioavailabilities in the food supply are low, fortification has been used to help assure adequate intakes by all members of the population. Fortification with iron and iodine has dramatically reduced deficiency diseases associated with these nutrients in the United States and other industrialized countries. Unfortunately, it has not been possible to fortify appropriate staple foods in many developing countries, leaving hundreds of millions of people in these countries to suffer the tragic consequences of iron and/or iodine deficiency.

Minerals also play key functional roles in foods. For example, minerals may dramatically alter the color, texture, flavor, and stability of foods. Thus minerals may be added to or removed from foods to achieve a particular functional effect. When manipulation of concentrations of minerals in foods is not practical, chelating agents such as EDTA (when allowed) can be used to alter their behavior.

BIBLIOGRAPHY

da Silva, J. J. R. F., and R. J. P. Williams (1991). *The Biological Chemistry of the Elements: The Inorganic Chemistry of Life*, Clarenon Press, Oxford.

Hazell, T. (1985). Minerals in foods: Dietary sources, chemical forms, interactions, bioavailability. *World Rev. Nutr. Diet 46*:1–123.

Mertz, W. (1987). *Trace Elements in Human and Animal Nutrition,* 5th ed., vols. 1 and 2. Academic Press, San Diego.

Shriver, D. F., P. W. Atkins, and C. H. Langford (1994). *Inorganic Chemistry,* 2nd ed., W. H. Freeman, New York.

Smith, K. T., ed. (1988) *Trace Minerals in Foods,* Marcel Dekker, New York.

REFERENCES

1. Allaway, W. H. (1975). *The Effect of Soils and Fertilizers on Human and Animal Nutrition.* Agriculture Information Bulletin No. 378. U.S. Department of Agriculture. Washington, DC.
2. Allen, R. R. (1987). Theory of hydrogenation and isomerization, in *Hydrogenation: Proceedings of an AOCS Colloquium* (R. Hastert, ed.), American Oil Chemists' Society, Champaign, IL, pp. 1–10.
3. AOAC (1990). *Official Methods of Analysis,* 15th ed., Association of Official Analytical Chemists, Washington, DC.
4. Barrett, F., and P. Ranum (1985). Wheat and blended cereal foods, in *Iron Fortification of Foods* (F. M. Clydesdale and K. L. Wiemer, eds.), Academic Press, Orlando, FL, pp. 75–109.
5. Batten, G. D. (1994). Concentrations of elements in wheat grains grown in Australia, North America, and the United Kingdom. *Austr. J. Exp. Agric. 34*:51–56.

6. Baynes, R. D., and T. H. Bothwell (1990). Iron deficiency. *Annu. Rev. Nutr. 10*:133–148.
7. Chrichton, R. R. (1991). *Inorganic Biochemistry of Iron Metabolism,* Ellis Horwood Series in Inorganic Chemistry (J. Burgess, series ed.), Ellis Horwood, New York.
8. Committee on Diet and Health (1989). *Diet and Health: Implications for Reducing Chronic Disease Risk,* National Academy Press, Washington, DC.
9. Dallman, P. R. (1990). Iron, in *Present Knowledge in Nutrition* (M. L. Brown, ed.), International Life Sciences Institute, Washington, DC, pp. 241–250.
10. da Silva, J. J. R. F., an R. J. P. Williams (1991). *The Biological Chemistry of the Elements: The Inorganic Chemistry of Life,* Clarendon Press, Oxford.
11. Ellinger, R. H. (1972). Phosphates in food processing, in *Handbook of Food Additives,* 2nd ed. (T. E. Furia, ed.), CRC Press, Cleveland, pp. 617–780.
12. ESHA Research (1992). *The Food Processor Plus,* ESHA Research, Salem, OR.
13. Food and Drug Administration (1994). *Code of Federal Regulations.* U.S. Government Printing Office, Washington, DC.
14. Food and Nutrition Board. National Research Council (1989). *Recommended Dietary Allowances,* 10th ed., National Academy Press, Washington, DC.
15. Guinee, T. P., P. D. Pudja, and N.Y. Farkye. (1993). Fresh acid-curd cheese varieties, in *Cheese: Chemistry, Physics and Microbiology,* vol. 2, 2nd ed. (P. F. Fox, ed.), Chapman & Hall, London, pp. 369–371.
16. Harbers, L. H. (1994). Ash analysis, in *Introduction to the Chemical Analysis of Foods* (S. S. Nielson, ed.), Jones and Bartlett, Boston, pp. 113–121.
17. Hazell, T. (1985). Minerals in foods: Dietary Sources, Chemical Forms, Interactions, Bioavailability. *World Rev. Nutr. Diet 46*:1–123.
18. Heaney, R. P., C. M. Weaver, and M. L. Fitzsimmons (1990). Influence of calcium load on absorption fraction. *J. Bone Miner. Res. 5*:1135–1138.
19. Hendricks, D. G. (1994). Mineral analysis by traditional methods, in *Introduction to the Chemical Analysis of Foods* (S. S. Nielson, ed.), Jones and Bartlett, Boston, pp. 123–135.
20. Hetzel, B. S. (1993). The control of iodine deficiency. *Am. J. Public Health 83*:494–495.
21. Hurrell, R. F. (1985). Nonelemental sources, in *Iron Fortification of Foods* (F. M. Clydesdale and K. L. Wiemer, eds.), Academic Press, Orlando, FL, pp. 39–53.
22. Hurrell, R. F., and J. D. Cook (1990). Strategies for iron fortification of foods. *Trends in Food Sci. Technol. 1*(3):56–61.
23. Kratzer, F. H. and P. Vohra (1986). *Chelates in Nutrition,* CRC Press, Boca Raton, FL.
24. Lachance, P. A., and M. C. Fisher (1988). Effects of food preparation procedures in nutrient retention with emphasis on food service practices, in *Nutritional Evaluation of Food Processing* (E. Karmas and R. S. Harris, eds.), Van Nostrand Reinhold, New York, pp. 505–556.
25. Lucey, J. A., and P. F. Fox (1993). Importance of calcium and phosphate in cheese manufacture: a review. *J. Dairy Sci. 76*(6):1714–1724.
26. McGee, H. J., S. R. Long, an W. R. Briggs (1984). Why whip egg whites in copper bowls? *Nature 308*:667–668.
27. Miller, D. D. (1989) Calcium in the diet: Food sources, recommended intakes, and nutritional bioavailability. *Adv. Food Nutr. Res. 33*:103–156.
28. Miller, D. D. (1994). Atomic absorption an emission spectroscopy, in *Introduction to the Chemical Analysis of Foods* (S. S. Nielson, ed.), Jones and Bartlett, Boston, pp. 353–370.
29. Miller, D. D., and L. A. Berner (1989). Is solubility in vitro a reliable predictor of iron bioavailability? *Biological Trace Elem. Res. 19*:11–24.
30. Molins, R. A. (1910). *Phosphates in Food,* CRC Press, Boca Raton, FL.
31. Mounts, T. L. (1987). Alternative catalysts for hydrogenation of edible oils, in *Hydrogenation: Proceedings of an AOCS Colloquium* (R. Hastert, ed.), American Oil Chemists' Society, Champaign, IL.
32. Patrick, J. (1985). Elemental sources, in *Iron Fortification of Foods* (F. M. Clyesdale and K. L. Wiemer, eds.), Academic Press, Orlando, FL, pp. 31–38.
33. Phillips, L. G., Z. Haque, and J. E. Kinsella (1987). A method for measurement of foam formation and stability. *J. Food Sci. 52*:1047–1049.

34. Porterfield, W. (1993). *Inorganic Chemistry: A Unified Approach,* 2nd ed., Academic Press, San Diego.
35. Richardson, D. P. (1990). Food fortification. *Proc. Nutr. Soc. 49*:39–50.
36. Rotruck, J. T. (1982). Effect of processing on nutritive value of food: Trace elements, in *Handbook of Nutritive Value of Processed Food* (M. Rechcigl, Jr., ed.) CRC Press, Boca Raton, Fl, vol. I, pp. 521–528.
37. Scrimshaw, N. S. (1991). Iron deficiency. *Sci. Am. 265*(1):46–52.
38. Shriver, D. F., P. W. Atkins, and C. H. Langford (1994). *Inorganic Chemistry,* 2nd ed., W. H. Freeman, New York.
39. Skoog, D. A., and D. M. West. (1976). *Fundamentals of Analytical Chemistry,* 3rd ed., Holt, Rinehart, an Winston, New York.
40. Smallwood, N.J. (1987). Edible oil hydrogenation process design and operation, in *Hydrogenation: Proccedings of an AOCS Colloquium* (R. Hastert, ed.), American Oil Chemists' Society, Champaign, IL.
41. Trowbridge, F. L., S. S. Harris, J. Cook, J. T. Dunn, R. F. Florentino, B. A. Kodyat, B. A. Underwood, and R. Yip (1993). Coordinated strategies for controlling micronutrient malnutrition: a technical workship. *J. Nutr. 123*:775–787.
42. Underwood, E. J., and W. Mertz (1987). Introduction in *Trace Elements in Human and Animal Nutrition,* 5th ed. (W. Mertz, ed.), Academic Press, San Diego, pp. 1–19.
43. U.S. Department of Agriculture (1976–1986). *Composition of Foods. Agriculture Handbook* Nos. 8-1 to 8-16. Human Nutrition Information Service, USDA, Hyattsville, MD.
44. van Dokkum, W. (1989). The significance of speciation for predicting mineral bioavailability, in *Nutrient Availability: Chemical and Biological Aspects* (D. Southgate, I. Johnson, and G. R. Fenwick, eds.), Royal Society of Chemistry, Cambridge, pp. 89–96.
45. Weaver, C. M., an K. L. Plawecki (1994). Dietary calcium: Adequacy of a vegetarian diet. *Am. J. Clin. Nutr. 59*(suppl.):1238S–1241S.
46. Weinberg, E. D. (1991). Cellular iron metabolism in health and disease. Drug Metab. Rev. 22(5):531–579.
47. Welch, R. M., and W. A. House (1984). Factors affecting the bioavailability of mineral nutrients in plant foods, in *Crops as Sources of Nutrients for Humans* (R. M. Welch, ed.), Soil Science Society of America, Madison, WI, pp. 37–54.
48. Welch, R. M., W. A. House, and W. H. Allaway (1974). Availaility of zinc from pea seeds to rats. *J. Nutr. 104*:733–740.
49. Whittaker, P., J. E. Vanderveen, M. J. Dinovi, P. M. Kuznesof, and V. C. Dunkel (1993). Toxicological profile, current use, and regulatory issues on EDTA compounds for assessing use of sodium iron EDTA for food fortification. *Regul. Toxicol. Pharmacol. 18*:419–427.
50. Wong, D. W. S. (1989). *Food Chemistry,* Van Nostrand Reinhold, New York.

10

Colorants

J. H. VON ELBE
University of Wisconsin—Madison, Madison, Wisconsin

STEVEN J. SCHWARTZ*
North Carolina State University, Raleigh, North Carolina

10.1 INTRODUCTION

To understand colorants in foods some terms need to be defined. Color refers to human perception of colored materials—red, green, blue, etc. A colorant is any chemical, either natural or synthetic, that imparts color. Foods have color because of their ability to reflect or emit different quantities of energy at wavelengths able to stimulate the retina in the eye. The energy

---

Present affiliation: The Ohio State University, Columbus, Ohio.

range to which the eye is sensitive is referred to as visible light. Visible light, depending on an individual's sensitivity, encompasses wavelengths of approximately 380–770 nm. This range makes up a very small portion of the electromagnetic spectrum (Fig. 1). In addition to obvious colors (hues), black, white, and intermediate grays are also regarded as colors.

Pigments are natural substances in cells and tissues of plants and animals that impart color. Dyes are any substances that lend color to materials. The term dye is commonly used in the textile industry. In the U.S. food industry, a dye is a food-grade water-soluble colorant certified by the U.S. Food and Drug Administration (FDA). These specific dyes are referred to as "certified colors," and each one is assigned an FD&C number. The FD&C designation means that the dye may be used in foods, drugs, and cosmetics. Added to the approved list of certified colors are the FD&C lakes. Lakes are dyes extended on a substratum and they are oil dispersible. The dye/substratum combination is achieved by adsorption, coprecipitation, or chemical reaction. The complex involves a salt of a water-soluble primary dye and an approved insoluble base stratum. Alumina is the only approved substratum for preparing FD&C lakes. In addition, there are other dyes or lakes approved for use in other countries, where specifications are established by the European Economic Community (EEC) or the World Health Organization (WHO). Colorants exempt from certification may also be used. These are natural pigments or substances synthesized, but identical to the natural pigment. A classification of colorants and an example within each class are given in Table 1.

Color and appearance are major, if not the most important, quality attributes of foods. It is because of our ability to easily perceive these factors that they are the first to be evaluated by the consumer when purchasing foods. One can provide consumers the most nutritious, safest, and most economical foods, but if they are not attractive, purchase will not occur. The consumer also relates specific colors of foods to quality. Specific colors of fruits are often associated with maturity—while redness of raw meat is associated with freshness, a green apple may be judged immature (although some are green when ripe), and brownish-red meat as not fresh.

Color also influences flavor perception. The consumer expects red drinks to be strawberry, raspberry, or cherry flavored, yellow to be lemon, and green to be lime flavored. The impact of color on sweetness perception has also been demonstrated. It should also be noted that some substances such as β-carotene or riboflavin are not only colorants but nutrients as well. It is clear

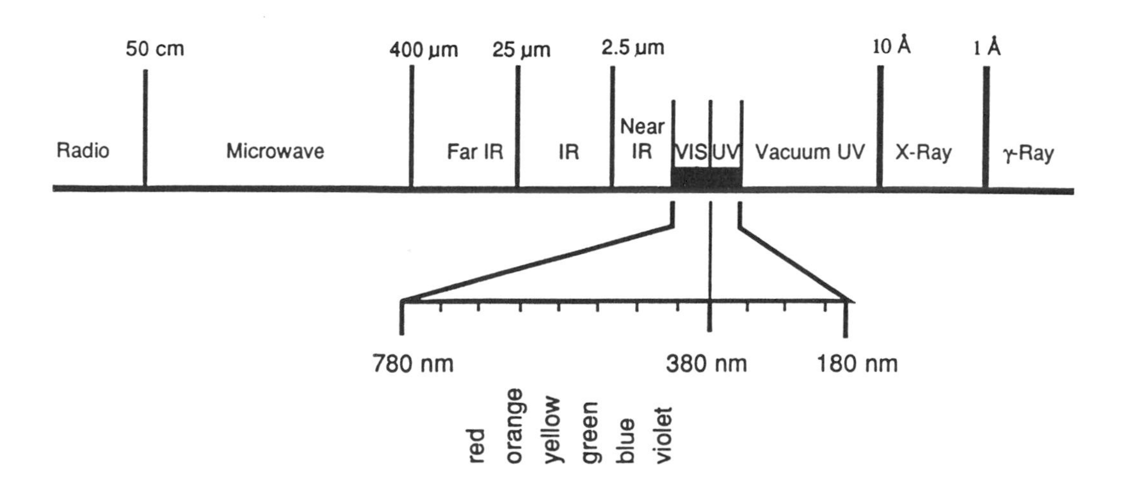

FIGURE 1 Electromagnetic spectrum.

TABLE 1 Classification of Colorants

Colorant	Example
A Certified	
1. Dye	FD&C Red No. 40
2. Lake	Lake of FD&C Red No. 40
B Exempt from certification	
1. Natural pigments	Anthocyanin, juice concentrate, annatto extract
2. Synthetic (nature identical)	β-Carotene

therefore that color of foods has multiple effects on consumers, and it is wrong to regard color as being purely cosmetic.

Many food pigments are, unfortunately, unstable during processing and storage. Prevention of undesirable changes is usually difficult or impossible. Depending on the pigment, stability is impacted by factors such as the presence or absence of light, oxygen, heavy metals, and oxidizing or reducing agents; temperature and water activity; and pH. Because of the instability of pigments, colorants are sometimes added to foods.

The purpose of this chapter is to provide an understanding of colorant chemistry—an essential prerequisite for controlling the color and color stability of foods.

10.2 PIGMENTS IN ANIMAL AND PLANT TISSUE

10.2.1 Heme Compounds

Heme pigments are responsible for the color of meat. Myoglobin is the primary pigment and hemoglobin, the pigment of blood, is of secondary importance. Most of the hemoglobin is removed when animals are slaughtered and bled. Thus, in properly bled muscle tissue myoglobin is responsible for 90% or more of the pigmentation. The myoglobin quantity varies considerably among muscle tissues and is influenced by species, age, sex, and physical activity. For example, pale-colored veal has a lower myoglobin content than red-colored beef. Muscle-to-muscle differences within an animal also are apparent, and these differences are caused by varying quantities of myoglobin present within the muscle fibers. Such is the case with poultry, where light-colored breast muscle is easily distinguished from the dark muscle color of leg and thigh muscles. Listed in Table 2 are the major pigments found in fresh, cured, and cooked meat. Other minor pigments present in muscle tissue include the cytochrome enzymes, flavins, and vitamin B_{12}.

10.2.1.1 Myoglobin/hemoglobin

Structure of Heme Compounds

Myoglobin is a globular protein consisting of a single polypeptide chain. Its molecular mass is 16.8 kD and it is comprised of 153 amino acids. This protein portion of the molecule is known as globin. The chromophore component responsible for light absorption and color is a porphyrin known as heme. Within the porphyrin ring, a centrally located iron atom is complexed with four tetrapyrrole nitrogen atoms. Thus, myoglobin is a complex of globin and heme. The heme porphyrin is present within a hydrophobic pocket of the globin protein and bound to a histidine residue (Fig. 2). The centrally located iron atom shown possesses six coordination sites, four of which are occupied by the nitrogen atoms within the tetrapyrrole ring. The fifth coordination site

TABLE 2 Major Pigments Found in Fresh, Cured, and Cooked Meat

Pigment	Mode of formation	State of iron	State of hematin nucleus	State of globin	Color
1. Myoglobin	Reduction of metmyoglobin; deoxygenation of oxymyoglobin	Fe^{2+}	Intact	Native	Purplish-red
2. Oxymyoglobin	Oxygenation of myoglobin	Fe^{2+}	Intact	Native	Bright red
3. Metmyoglobin	Oxidation of myoglobin, oxymyoglobin	Fe^{3+}	Intact	Native	Brown
4. Nitric oxide myoglobin (nitrosomyoglobin)	Combination of myoglobin with nitric oxide	Fe^{2+}	Intact	Native	Bright red (pink)
5. Nitric oxide metmyoglobin (nitrosometmyoglobin)	Combination of metmyoglobin with nitric oxide	Fe^{3+}	Intact	Native	Crimson
6. Metmyoglobin nitrite	Combination of metmyoglobin with excess nitrite	Fe^{3+}	Intact	Native	Reddish-brown
7. Globin myohemochromogen	Effect of heat, denaturing agents on myoglobin, oxymoglobin; irradiation of globin hemichromogen	Fe^{2+}	Intact (usually bound to denatured protein other than globin)	Denatured (usually detached)	Dull red

8.	Globin myohemichromogen	Effect of heat, denaturing agents on myoglobin, oxymoglobin, metmyoglobin, hemochromogen	Fe^{3+}	Intact (usually bound to denatured protein other than globin)	Denatured (usually detached)	Brown (sometimes greyish)
9.	Nitric oxide myohemochromogen	Effect of heat, denaturing agents on nitric oxide myoglobin	Fe^{2+}	Intact	Denatured	Bright red (pink)
10.	Sulfmyoglobin	Effect of H_2S and oxygen on myoglobin	Fe^{3+}	Intact but one double bond saturated	Native	Green
11.	Metsulfmyoglobin	Oxidation of sulfmyoglobin	Fe^{3+}	Intact but one double bond saturated	Native	Red
12.	Choleglobin	Effect of hydrogen peroxide on myoglobin or oxymyoglobin; effect of ascorbine or other reducing agent on oxymyoglobin	Fe^{2+} or Fe^{3+}	Intact but one double bond saturated	Native	Green
13.	Nitrihemin	Effect of large excess nitrite and heat on 5	Fe^{3+}	Intact but reduced	Absent	Green
14.	Verdohaem	Effect of reagents as in 7–9 in excess	Fe^{3+}	Porphyrin ring opened	Absent	Green
15.	Bile pigments	Effect of reagents as in 7–9 in large excess	Fe absent	Porphyrin ring destroyed Chain of porphyrins	Absent	Yellow or colorless

From Lawrie (61) courtesy of Pergammon Press, New York

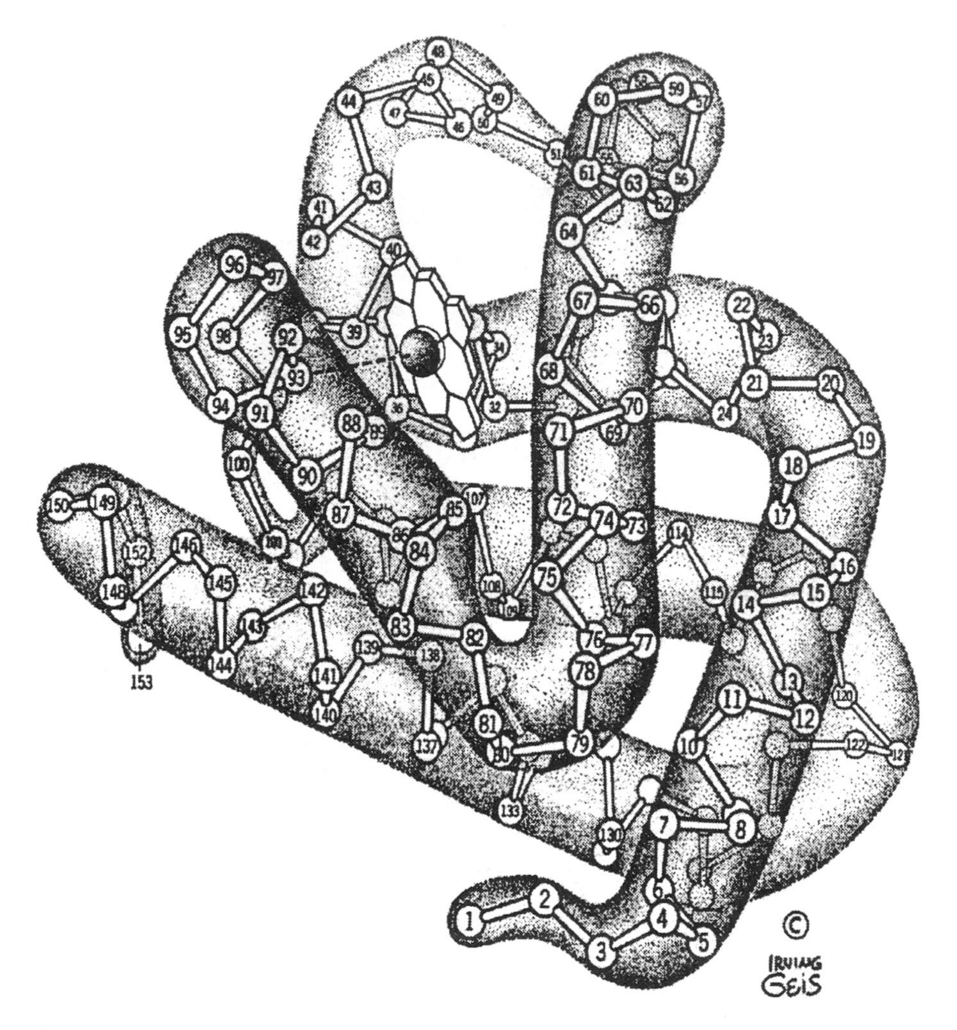

FIGURE 2 Tertiary structure of myoglobin. (Ref. 35.)

is bound by the histidine residue of globin, leaving the sixth site available to complex with electronegative atoms donated by various ligands.

Hemoglobin consists of four myoglobins linked together as a tetramer. Hemoglobin, a component of red blood cells, forms reversible complexes with oxygen in the lung. This complex is distributed via the blood to various tissues throughout the animal where oxygen is absorbed. It is the heme group that binds molecular oxygen. Myoglobin within the cellular tissue acts in a similar fashion, accepting the oxygen carried by hemoglobin. Myoglobin thus stores oxygen within the tissues, making it available for metabolism.

Chemistry and Color—Oxidation

Meat color is determined by the chemistry of myoglobin, its state of oxidation, type of ligands bounds to heme, and state of the globin protein. The heme iron within the porphyrin ring may exist in two forms: either reduced ferrous (+2) or oxidized ferric (+3). This state of oxidation for the iron atom within heme should be distinguished form oxygenation of myoglobin. When molecular oxygen binds to myoglobin, oxymyoglobin (MbO_2) is formed and this is referred to as oxygenation. When oxidation of myoglobin occurs, the iron atom is converted to the ferric (+3) state, forming metmyoglobin (MMb). Heme iron in the +2 (ferrous) state, which lacks a bound ligand in the sixth position, is called myoglobin.

Meat tissue that contains primarily myoglobin (also referred to as deoxymyoglobin) is purplish-red in color. Binding of molecular oxygen at the sixth ligand yields oxymyoglobin (MbO_2) and the color of the tissue changes to the customary bright red. Both the purple myoglobin and the red oxymyoglobin can oxidize, changing the state of the iron from ferrous to ferric. If this change in state occurs through autooxidation, these pigments acquire the undesirable brownish-red color of metmyoglobin (MMb). In this state, metmyoglobin is not capable of binding oxygen and the sixth coordination position is occupied by water [25]. Shown in Fig. 3 are the various reactions of the heme pigment. Color reactions in fresh meat are dynamic and determined by conditions in the muscle and the resulting ratios of myoglobin (Mb), metmyoglobin (MMb), and oxymyoglobin (MbO_2). Interconversion among these forms can occur readily.

Shown in Fig. 4 is the relationship between oxygen partial pressure and the percentage of each type of heme pigment. A high partial pressure of oxygen favors oxygenation, forming bright-red oxymyoglobin. Conversely, at low oxygen partial pressures myoglobin and metmyoglobin are favored. In order to enhance oxymyoglobin formation, saturation levels of oxygen in the environment are useful. The rate of metmyoglobin formation, caused by heme oxidation ($Fe^{2+} \rightarrow Fe^{3+}$), can be minimized if oxygen is totally excluded. Muscles have varying oxygen partial pressures, causing the ratios of each of the pigment forms to vary.

Presence of the globin protein is known to decrease the rate of heme oxidation ($Fe^{2+} \rightarrow Fe^{3+}$). In addition, oxidation occurs more rapidly at lower pH values. Furthermore, the rate of autoxidation of oxymyoglobin occurs more slowly than that of myoglobin. The presence of trace metals, especially copper ions, is known to promote autoxidation.

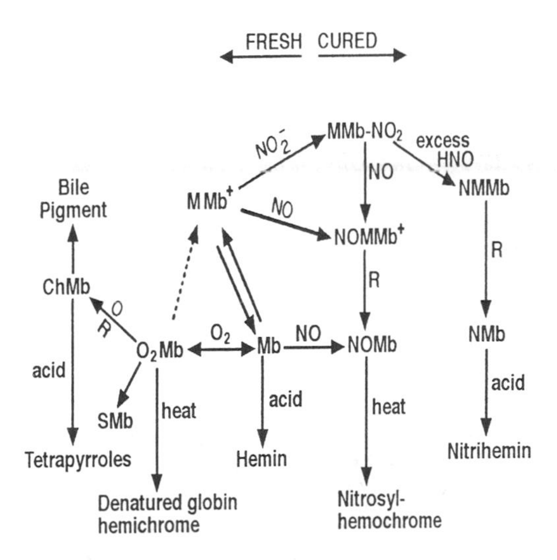

FIGURE 3 Myoglobin reactions in fresh and cured meats. ChMb = cholemyglobin (oxidized porphyrin ring); O_2Mb = oxymyoglobin (Fe^{2+}); MMb = metmyoglobin (Fe^{3+}); Mb = myoglobin (Fe^{2+}); MMb-NO_2 = metmyoglobin nitrite; NOMMb = nitrosylmetmyoglobin; NOMb = nitrosylmyoglobin; NMMb = nitrometmyoglobin; NMb = nitromyoglobin, the latter two being reaction products of nitrous acid and the heme portion of the molecule; R = reductant; O = strong oxidizing conditions. (From Ref. 31.)

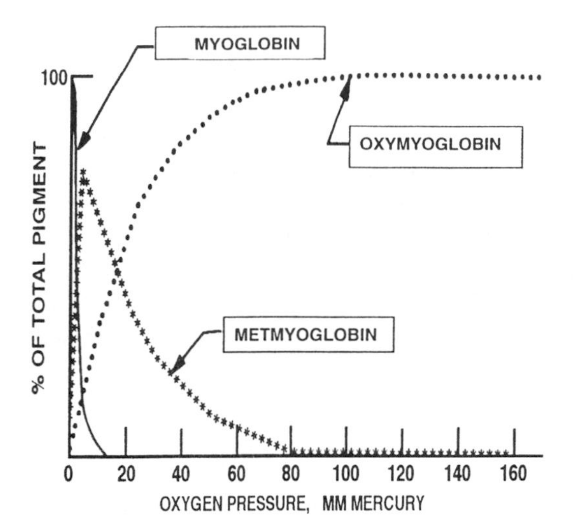

FIGURE 4 Influence of oxygen partial pressure on the three chemical states of myoglobin. (From Ref. 29.)

Chemistry and Color—Discoloration

Two different reactions can cause green discoloration of myoglobin [61]. Hydrogen peroxide can react with either the ferrous or ferric site of heme, resulting in choleglobin, a green-colored pigment. Also, in the presence of hydrogen sulfide and oxygen, green sulfomyoglobin can form. It is thought that hydrogen peroxide and/or hydrogen sulfide arise from bacterial growth. A third mechanisms for green pigmentation occurs in cured meats and is mentioned later.

10.2.1.2 Cured Meat Pigments

During the curing process, specific reactions occur that are responsible for the stable pink color of cured meat products. These reactions are outlined in Fig. 3 and the compounds responsible for the reactions are listed in Table 2.

The first reaction occurs between nitric oxide (NO) and myoglobin to produce nitric oxide myoglobin ($MbNO_2$), also know as nitrosylmyoglobin. $MbNO_2$ is bright red and unstable. Upon heating, the more stable nitric oxide myohemochromogen (nitrosylhemochrome) forms. This product yields the desirable pink color of cured meats. Heating of this pigment denatures globin, but the pink color persists. If metmyoglobin is present, it has been postulated that reducing agents are required to convert metmyoglobin to myoglobin before the reaction with nitric oxide can take place. Alternatively, nitrite can interact directly with metmyoglobin. In the presence of excess nitrous acid, nitrimyoglobin (NMb) will form. Upon heating in a reducing environment, NMb is converted to nitrihemin, a green pigment. This series of reactions causes a defect known as "nitrite burn."

In the absence of oxygen, nitric oxide complexes of myoglobin are relatively stable. However, especially under aerobic conditions, these pigments are sensitive to light. If reductants are added, such as ascorbate or sulfhydryl compounds, the reductive conversion of nitrite to nitric oxide is favored. Thus, under these conditions, formation of nitric oxide myoglobin occurs more readily. Detailed reviews on the chemistry of cured meat pigments are available [31, 32].

10.2.1.3 Stability of Meat Pigments

Many factors operative in a complex food system can influence the stability of meat pigments. In addition, interactions between these factors are critical and make it difficult to determine cause and effect relationships. Some environmental conditions that have important effects on meat color and pigment stability include exposure to light, temperature, relative humidity, pH, and the presence of specific bacteria. Review papers on this subject are available [25, 57].

Specific reactions, such as lipid oxidation, are known to increase the rate of pigment oxidation [26]. Similarly, color stability can be improved by the addition of antioxidants such as ascorbic acid, vitamin E, butylated hydroxyanisole (BHA), or propyl gallate [38]. These compounds have been shown to delay lipid oxidation and improve retention of color in tissues. Other biochemical factors, such as the rate of oxygen consumption prior to slaughter and activity of metmyoglobin reductase, can influence the color stability of fresh meat [67].

10.2.1.4 Packaging Considerations

An important means of stabilizing meat color is to store it under appropriate atmospheric conditions. The problem of discoloration caused by heme oxidation ($Fe^{2+} \rightarrow Fe^{3+}$) can be resolved by using modified atmosphere (MA) packaging. This technique requires the use of packaging films with low gas permeabilities. After packaging, air is evacuated from the package and the storage gas is injected. By employing oxygen-enriched or -devoid atmospheres, color stability can be enhanced [75]. Muscle tissue stored under conditions devoid of O_2 (100% CO_2) and in the presence of an oxygen scavenger also exhibits good color stability [86]. Use of modified atmosphere packaging techniques may, however, result in other chemical and biochemical alterations that can influence the acceptability of meat products. Part of the influence of modified atmospheres on pigment stability no doubt relates to its influence on microbial growth. Further information on use of MA for fresh meat storage can be found in a review article by Seideman and Durland [96].

10.2.2 Chlorophyll

10.2.2.1 Structure of Chlorophylls and Derivatives

Chlorophylls are the major light-harvesting pigments in green plants, algae, and photosynthetic bacteria. They are magnesium complexes derived from porphin. Porphin is a fully unsaturated macrocyclic structure that contains four pyrrole rings linked by single bridging carbons. The rings are numbered I through IV or A through D according to the Fisher numbering system (Fig. 5). Pyrrole carbon atoms on the periphery of the porphin structure are numbered 1 through 8. Carbon atoms of the bridging carbons are designated alpha, beta, gamma, and delta. The IUPAC numbering system for porphin is shown in Figure 5b. The more common numbering system is the Fisher system.

Substituted porphins are name porphyrins. Phorbin (Fig. 5c) is considered to be the nucleus of all chlorophylls and is formed by the addition of a fifth isocyclic ring (V) to porphin. A porphyrin is any macrocyclic tetrapyrrole pigment in which the pyrrole rings are joined by methyne bridges and the system of double bonds forms a closed, conjugated loop. Chlorophylls, therefore, are classified as porphyrins.

Several chlorophylls are found in nature. Their structures differ in the substituents around the phorbin nucleus. Chlorophyll *a* and *b* are found in green plants in an approximate ratio of 3:1. They differ in the carbon C-3 substituent. Chlorophyll *a* contains a methyl group while chlorophyll *b* contains a formyl groups (Fig. 5d). Both chlorophylls have a vinyl and an ethyl group at the C-2 and C-4 position, respectively; a carbomethoxy group at the C-10 position of the isocylic rings, and a phytol group esterfied to propionate at the C-7 position. Phytol is a 20-carbon monounsaturated isoprenoid alcohol. Chlorophyll *c* is found in association with chlorophyll *a* in marine algae, dinoflagellates, and diatoms. Chlorophyll *d* is a minor constituent accompanying chlorophyll *a* in red algae. Bacteriochlorophylls and chlorobium chlorophylls are principal chlorophylls found in purple photosynthetic bacteria and green sulfur bacteria, respec-

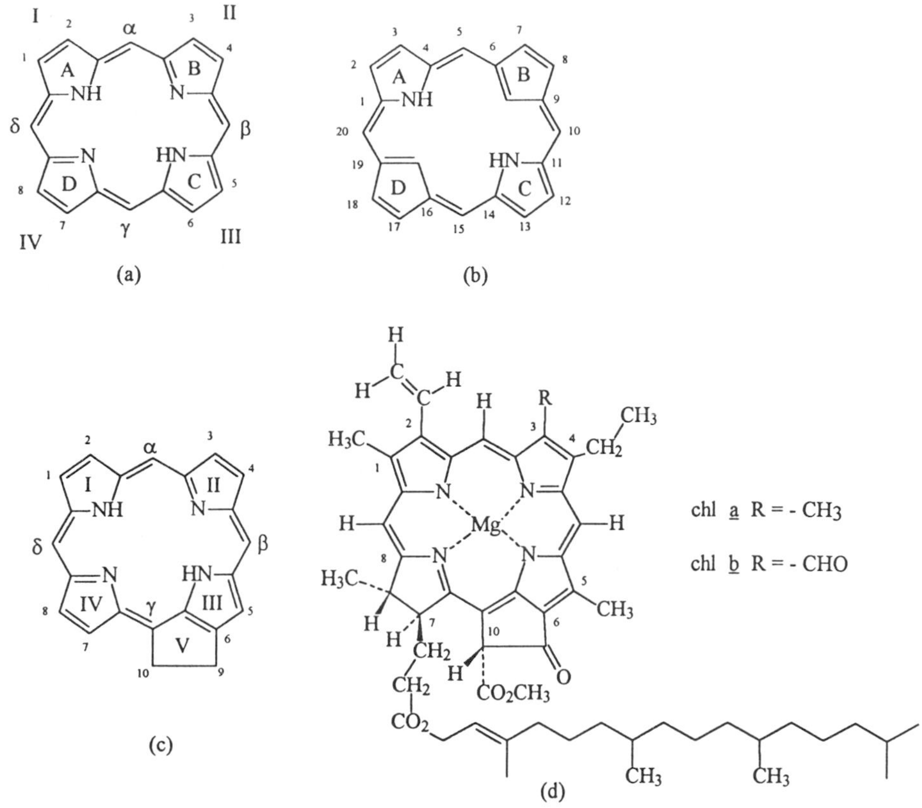

FIGURE 5 Structure of porphin (a,b), phorbin (c), and chlorophyll [chl] (d).

TABLE 3 Nomenclature of Chlorophyll Derivatives

Phyllins:	Chlorophyll derivatives containing magnesium
Pheophytins:	The magnesium-free derivatives of the chlorophylls
Chlorophyllides:	The products containing a C-7 propionic acid resulting from enzymic or chemical hydrolysis of the phytyl ester
Pheophorbides:	The products containing a C-7 propionic acid resulting from removal of magnesium and hydrolysis of the phytyl ester
Methyl or ethyl pheophorbides:	The corresponding 7-propionate methyl of ethyl ester
Pyro compounds:	Derivatives in which the C-10 carbomethoxy group has been replaced by hydrogen
Meso compounds:	Derivatives in which the C-2 vinyl group has been reduced to ethyl
Chlorins e:	Derivatives of pheophorbide *a* resulting from cleavage of the isocyclic ring
Rhodins g:	The corresponding derivatives from pheophorbide *b*

tively. Trivial names are widely used for chlorophylls and their derivatives [49]. Listed in Table 3 are the names used most often. Figure 6 is a schematic representation of the structural relationships of chlorophyll and some of its derivatives.

10.2.2.2 Physical Characteristics

Chlorophylls are located in the lamellae of intercellular organelles of green plants known as chloroplasts. They are associated with carotenoids, lipids, and lipoproteins. Weak linkages (non-covalent bonds) exist between these molecules. The bonds are easily broken; hence chlorophylls can be extracted by macerating plant tissue in organic solvents. Various solvents have been used. Polar solvents such as acetone, methanol, ethanol, ethyl acetate, pyridine, and dimethylformamide are most effective for complete extraction of chlorophylls. Nonpolar solvents such as hexane or petroleum ether are less effective. High-performance liquid chromatography (HPLC) today is the method of choice for separating individual chlorophylls and their derivatives. In the case of chlorophyll *a* and *b*, for example, the increase in polarity contributed by the C-3 formyl substituent of chlorophyll *b* causes it to be more strongly adsorbed on a normal-phase column and more weakly absorbed on a reverse-phase column than chlorophyll *a* [14].

Identification of chlorophyll and its derivatives is, to a large extent, based on absorption characteristics of visible light. Visible spectra of chlorophyll *a* and *b* and their derivatives are characterized by sharp light absorption bands between 600 and 700 nm (red regions) and between 400 and 500 nm (blue regions) (Table 4). The wavelength of maximum absorption for chlorophylls *a* and *b* dissolved in ethyl ether are, respectively, 660.5 and 642 nm in the red region, and 428.5 and 452.5 nm in the blue region [100].

10.2.2.3 Alterations of Chlorophyll

Enzymatic

Chlorophyllase is the only enzyme known to catalyze the degradation of chlorophyll. Chlorophyllase is an esterase, and in vitro, it catalyzes cleavage of phytol from chlorophylls and its Mg-free derivatives (pheophytins), forming chlorophyllides and pheophorbides, respectively (Fig. 6). Its activity is limited to porphyrins with a carbomethoxy groups at C-10 and hydrogens at positions C-7 and C-8 [73]. The enzyme is active in solutions containing water, alcohols, or acetone. In the presence of large amounts of alcohols such as methanol or ethanol, the phytol group is removed and the chlorophyllide is esterified to form either methyl or ethyl chlorophyllide. Formation of chlorophyllides in fresh leaves does not occur until the enzyme has been heat activated postharvest. The optimum temperature for chlorophyllase activity in vegetables ranges

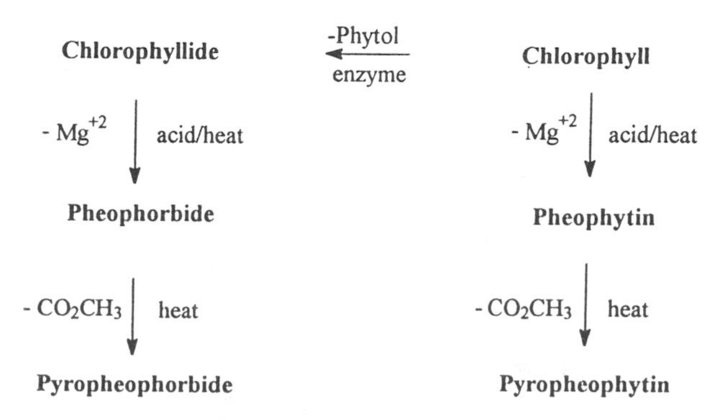

FIGURE 6 Chlorophyll and its derivatives.

TABLE 4 Spectral Properties in Ethyl Ether of Chlorophyll *a* and *b* and Their Derivatives

Compound	Absorption maxima (nm) "Red" region	"Blue" region	Ratio of absorption ("blue"/"red")	Molar absorptivity ("red" region)
Chlorophyll *a*	660.5	428.5	1.30	86,300[a]
Methyl chlorophyllide *a*	660.5	427.5	1.30	83,000[b]
Chlorophyll *b*	642.0	452.5	2.84	56,100[a]
Methyl chlorophyllide *b*	641.5	451.0	2.84	—[b]
Pheophytin *a*	667.0	409.0	2.09	61,000[b]
Methyl pheophorbide *a*	667.0	408.5	2.07	59,000[b]
Pheophytin *b*	655	434	—	37,000[c]
Pyropheophytin *a*	667.0	409.0	2.09	49,000[b]
Zinc pheophytin *a*	653	423	1.38	90,300[d]
Zinc pheophytin *b*	634	446	2.94	60,200[d]
Copper pheophytin *a*	648	421	1.36	67,900[d]
Copper pheopytin *b*	627	438	2.53	49,800[d]

[a]Strain et al. [99].
[b]Pennington et al. [80].
[c]Davidson [20].
[d]Jones et al. [55].

between 60 and 82.2°C [64]. Enzyme activity decreases when plant tissue is heated above 80°C, and chlorophyllase loses its activity if heated to 100°C. Chlorophyllase activity in spinach during growth and fresh storage is shown in Figure 7. Maximum activity is observed at the time the plant begins flowering (solid line). Postharvest storage of fresh spinach at 5°C decreases enzyme activity compared to activities measured at the time of harvest (broken line) [122].

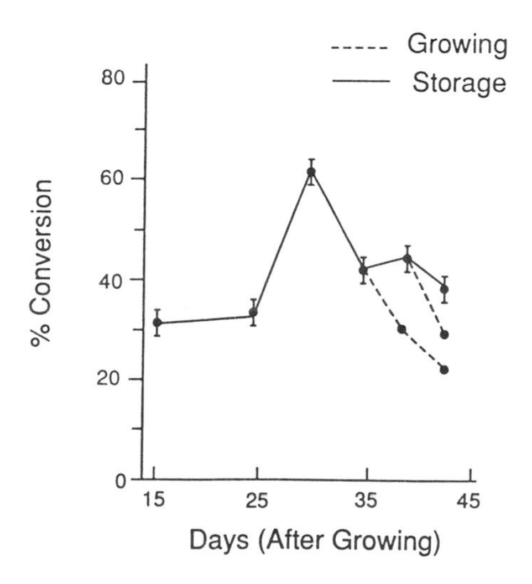

FIGURE 7 Chlorophyllase activity, expressed as percent conversion of chlorophyll to chloro phyllide, in spinach during growth (solid line) and after storage at 5°C (broken line). (From Ref. 112.)

Heat and Acid

Chlorophyll derivatives formed during heating or thermal processing can be classified into two groups based on the presence or absence of the magnesium atom in the tetrapyrrole center. Mg-containing derivatives are green in color, while Mg-free derivatives are olive-brown in color. The latter are chelators and when, for example, sufficient zinc or copper ions are available they will form green zinc or copper complexes.

The first change observed when the chlorophyll molecule is exposed to heat is isomerization. Chlorophyll isomers are formed by inversion of the C-10 carbomethoxy group. The isomers are designated as *a′* and *b′*. They are more strongly absorbed on a C-18 reverse-phase HPLC column than are their parent compounds, and clear separation can be achieved. Isomerization occurs rapidly in heated plant tissue or in organic solvents. Establishment of equilibrium in leaves results in conversion of 5–10% of chlorophyll *a* and *b* to *a′* and *b′* after heating for 10 min at 100°C [3, 92, 113}. Chromatograms of a spinach extract of fresh versus blanched spinach in Figure 8 show isomer formation during heating.

The magnesium atom in chlorophyll is easily displaced by two hydrogen ions, resulting in the formation of olive-brown pheophytin (Fig. 9). The reaction is irreversible in aqueous solution. Compared to their parent compounds, pheophytin *a* and *b* are less polar and more strongly absorbed on a reverse-phase HPLC column. Formation of the respective pheophytins occur more rapidly from chlorophyll *a* than from chlorophyll *b*. Chlorophyll *b* is more heat stable

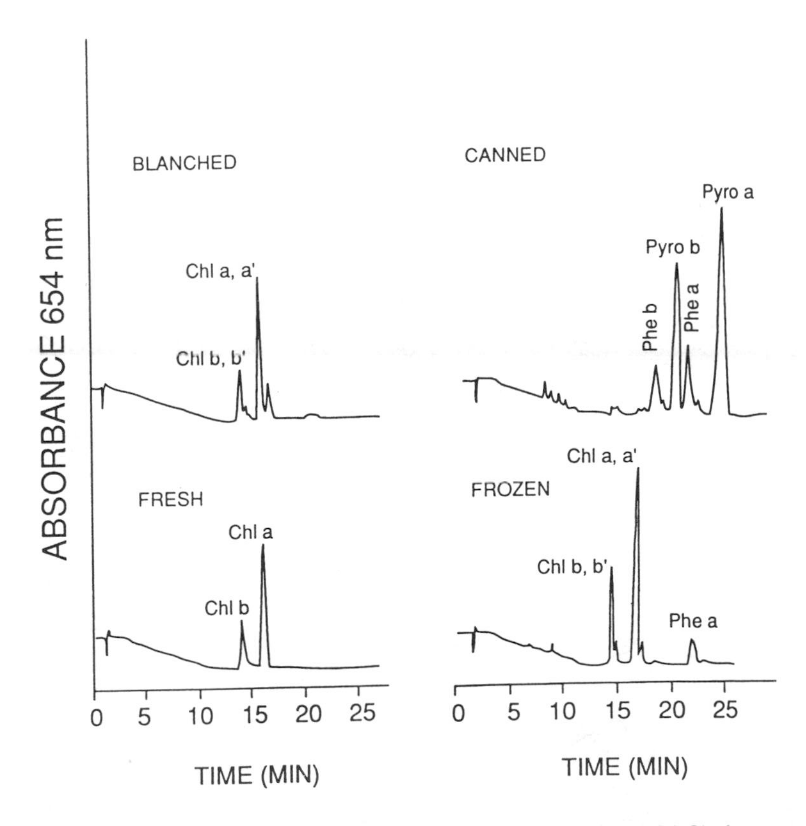

FIGURE 8 High-performance liquid chromatography (HPLC) chromatograms of chlorophylls (chl) and derivatives in fresh, blanched, frozen, and canned spinach. Phe = pheophytin, Pyro = pyropheophytin. (From Ref. 92.)

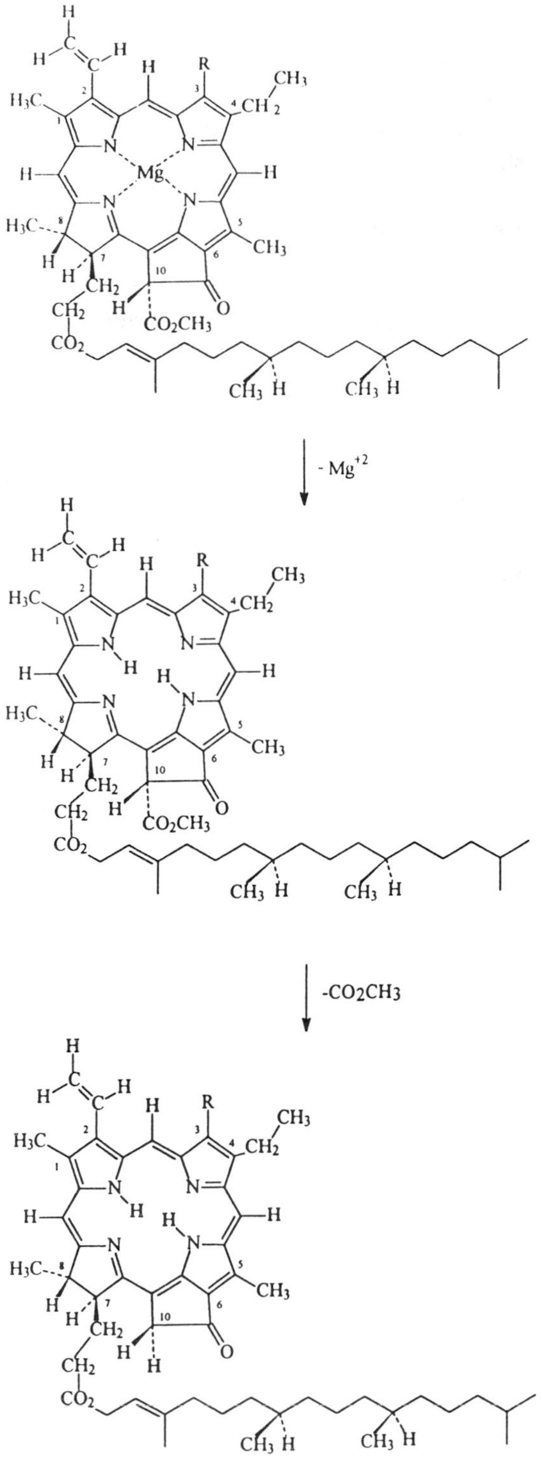

CHLOROPHYLL (chl)

chl a, R = $-CH_3$

chl b, R = $-CHO$

$- Mg^{+2}$

PHEOPHYTIN (phe)

phe a, R = $-CH_3$

phe b, R = $-CHO$

$-CO_2CH_3$

PYROPHEOPHYTIN (pyro)

pyro a, R = $-CH_3$

pyro b, R = $-CHO$

FIGURE 9 Formation of pheophytin and pyropheophytin from chlorophyll.

than is chlorophyll *a*. The greater stability of *b* is attributed to the electron-withdrawing effect of its C-3 formyl group. Transfer of electrons away from the center of the molecule occurs because of the conjugated structure of chlorophyll. The resulting increase in positive charge on the four pyrrole nitrogens reduces the equilibrium constant for the formation of the reaction intermediate. Results of rate studies of chlorophyll degradation to pheophytin in heated plant tissue also have shown greater heat stability of chlorophyll *b*. Reported activation energies for the reaction range from 12.6 to 35.2 kcal/mol. This variance has been attributed to differences in media composition, pH, temperature range, and method of analysis. With respect to the latter, using spectrophotometric methodology, for example, would not allow individual compounds to be distinguished, whereas HPLC will.

Chlorophyll degradation in heated vegetable tissue is affected by tissue pH. In a basic media (pH 9.0) chlorophyll is very stable toward heat, whereas in an acidic media (pH 3.0) it is unstable. A decrease of 1 pH unit can occur during heating of plant tissue through the release of acids, and this has an important detrimental effect on the rate of chlorophyll degradation. Pheophytin formation in intact plant tissue, postharvest, appears to be mediated by cell membranes. In a study by Haisman and Clarke [41], chlorophyll degradation in sugar beet leaves held in heated buffer was not initiated until the temperature reached 60°C or higher. Conversion of chlorophyll to pheophytin after holding for 60 min at 60 or 100°C was 32 and 97%, respectively. It was proposed that pheophytin formation in plant cells is initiated by a heat-induced increase in permeability of hydrogen ions across cell membranes. The critical temperature for initiation of pheophytin formation coincided with gross changes in membrane organization as observed with an electron microscope.

The addition of chloride salts of sodium, magnesium, or calcium decreases pheophytinization in tobacco leaves heated at 90°C by about 47, 70, and 77%, respectively. The decrease in chlorophyll degradation was attributed to the electrostatic shielding effect of the salts. It was proposed that the addition of cations neutralizes the negative surface charge of the fatty acids and protein in the chloroplast membrane and thereby reduces the attraction of hydrogen ions to the membrane surface [74].

The permeability of hydrogen across the membrane can also be affected by the addition of detergents that absorb on the surface of the membrane. Cationic detergents repel hydrogen ions at the membrane surface, limiting their diffusion into the cell and thereby decreasing chlorophyll degradation. Anionic detergents attract hydrogen ions, increasing their concentration at the membrane surface, increasing the rate of hydrogen diffusion, and increasing the degradation of chlorophyll. In the case of neutral detergents, the negative surface charge on the membrane is diluted and therefore the attraction for hydrogen ions and consequent degradation of chlorophyll is decreased [17, 41].

Replacement of the C-10 carbomethoxy group of pheophytin with a hydrogen atom results in the formation of olive-colored pyropheophytin. The wavelengths of maximum light absorption by pyropheophytin are identical to those for pheophytin in both the red and blue regions (Table 4). Retention times for pyropheophytin *a* and *b* using reverse-phase HPLC are greater than those for the respective pheophytins.

Chlorophyll alteration during heating is sequential, and proceeds according to the following kinetic sequence:

Chlorophyll $\rightarrow$ pheophytin $\rightarrow$ pyropheophytin

The data in Table 5 show that during heating for the first 15 min, chlorophyll decreases rapidly and pheophytin increases rapidly. With further heating pheophytin decreases and phyropheophytin rapidly increases, although a small amount of pyropheophytin is evident after 4 min of heating. Accumulation does not become appreciable until after 15 min, thus supporting a sequen-

TABLE 5 Concentration (mg/g Dry Weight)[a] of Chlorophylls *a* and *b* and Pyropheophytins *a* and *b* in Fresh, Blanched, and Heated Spinach Processed at 121°C for Various Times

	Chlorophyll		Pheophytin		Pyropheophytin		
	a	*b*	*a*	*b*	*a*	*b*	pH[b]
Fresh	6.98	2.49					
Blanched	6.78	2.47					7.06
Processed (min)[c]							
2	5.72	2.46	1.36	0.13			6.90
4	4.59	2.21	2.20	0.29	0.12		6.77
7	2.81	1.75	3.12	0.57	0.35		6.60
15	0.59	0.89	3.32	0.78	1.09	0.27	6.32
30		0.24	2.45	0.66	1.74	0.57	6.00
60			1.01	0.32	3.62	1.24	5.65

[a]Estimated error ±2%; each value represents a mean of 3 determinations.
[b]The pH was measured after processing and before pigment extraction.
[c]Times listed were measured after the internal product temperature reached 121°C.
Source: Ref. 90.

tial mechanism. First-order rate constant for conversion of pheophytin *b* to pyropheophytin *b* are 25–40% greater than those for conversion of pheophytin *a* to pyropheophytin *a* [91]. Activation energies for removal of the C-10 carbomethoxy group from either pheophytin *a* or *b* are smaller than those for formation of pheophytin *a* and *b* from chlorophyll *a* and *b*, indicating a slightly lower temperature dependency for formation of the pyropheophytins.

Listed in Table 6 are the concentrations of pheophytins *a* and *b* and pyropheophytins *a* and *b* in commercially canned vegetable products. These data indicate that pyropheophytins *a* and *b* are the major chlorophyll derivatives responsible for the olive-green color in many canned vegetables. It is also significant that the amount of pyropheophytin derivatives formed is an indication of the severity of the heat treatment. Comparative heat treatments for commercial sterility of spinach, green beans, cut asparagus, and green peas in 303 cans processed at 121°C are approximately 51, 11, 13, and 17 min, respectively. From Table 6, the percents of pyropheo-

TABLE 6 Pheophytins *a* and *b* in Commercially Canned Vegetables

	Pheophytin[a] (μg/g dry weight)		Pyropheophytin (μg/g dry weight)	
Product	*a*	*b*	*a*	*b*
Spinach	830	200	4000	1400
Beans	340	120	260	95
Asparagus	180	51	110	30
Peas	34	13	33	12

[a]Estimated error = ±2%.
Source: Ref. 90.

phytin compared to the total pheo compounds are 84, 44, 38, and 49%, respectively, corresponding fairly well to the heating times.

Replacement of the magnesium atom in chlorophyllide (green) with hydrogen ions results in the formation of olive-brown pheophorbide. Pheophorbide *a* and *b* are more water-soluble than the respective pheophytins and have the same spectral characteristics. Removal of the C-10 phytol chain appears to affect the rate of loss of magnesium from the tetrapyrrole center. Degradation of chlorophyllides *a* and *b* and their respective methyl and ethyl esters in acidic acetone increases as the length of the chain is decreased, suggesting that steric hindrance from the C-10 chain affects the rate of hydrogen ion attack [87].

Past rate studies using green plant tissue have generally supported the view that chlorophyllides are slightly more heat stable than chlorophylls. More recently [15], however, it was shown that the reverse is true. Chlorophyllide *a* degraded 3.7 times faster than chlorophyll *a* at 145°C. The more water-soluble chlorophyllide apparently is more likely to contact hydrogen ions in solution and is likely to react more rapidly. The latter result is believed to be the correct one because, in this study, chlorophyllase was activated more completely, and therefore greater amounts of chlorophyllides were available for kinetic studies, and a very sensitive HPLC technique was used.

Metallo Complex Formation

The two hydrogen atoms within the tetrapyrrole nucleus of the magnesium-free chlorophyll derivative are easily displaced by zinc or copper ions to form green metallo complexes. Formation of metallo complexes from pheophytins *a* and *b* causes the red maximum to shift to a shorter wavelength and the blue maximum to a longer wavelength (Table 4) [53]. Spectral characteristics of the phytol-free metal complexes are identical to their parent compounds.

The Zn and Cu complexes are more stable in acid solutions than in alkaline. Magnesium, as has been pointed out, is easily removed by the addition of acid at room temperature, while zinc pheophytin *a* is stable in solution at pH 2. Removal of copper is achieved only at pH values sufficiently low to begin degradation of the porphyrin ring.

Incorporation of metal ions into the neutral porphyrin is a bimolecular reaction. It is believed that the reaction begins with attachment of the metal ion to a pyrrole nitrogen atom, followed by the immediate and simultaneous removal of two hydrogen atoms. Formation of metallo complexes is affected by substituent groups because of the highly resonant structure of the tetrapyrrole nucleus. [24].

Metallo complexes are known to form in plant tissue, and the *a* complexes form faster than do *b* complexes. The slower formation of the *b* complexes has been attributed to the electron-withdrawing effect of the C-3 formyl group. Migration of electrons through the conjugated porphyrin ring system causes pyrrole nitrogen atoms to become more positively charged and therefore less reactive with metal cations. Steric hindrance of the phytol chain also decreases the rate of complex formation. Pheophorbide *a* in ethanol reacts four times faster with copper ions than does pheophytin *a* [52].

The kinetics of zinc complex formation have been studied. Formation of Zn^{2+} pyropheoporbide *a* occurs most rapidly in acetone/water (80/20), followed by pheophorbide *a*, methyl pheophorbide *a*, ethyl pheophorbide *a*, pyropheophytin *a*, and pheophytin *a*. Reaction rates decrease as the length of the alkyl chain esterified at the C-7 carbon increases, suggesting that steric hindrance is important. Similarly, the greater rate of formation of zinc pyropheophytin *a* as compared to pheophytin *a* is attributed to interference from the C-10 carbomethoxy group of pheophytin *a* [81,108].

Comparative studies on the formation of metallo complexes in vegetable purées indicate copper is chelated more rapidly than zinc. Copper complexes are detectable in pea purée when the concentration of copper is as low as 1–2 ppm. In contrast, zinc complex formation under similar conditions does not occur in purée containing less than 25 ppm Zn^{2+}. When both Zn^{2+} and Cu^{2+} are present, formation of copper complexes dominates [88].

The pH is also a factor in the rate of complex formation. Increasing the pH of spinach purée from 4.0 to 8.5 results in an 11-fold increase in the amount of zinc pyropheophytin *a* formed during heating for 60 min at 121°C. A decrease in complex formation occurs when the pH is raised to 10, presumably because of precipitation of Zn^{2+} [58].

These metallo complexes are of interest because the copper complexes, based on their stability under most conditions experienced in food processing, are used as colorants in the European Economic Community. This technology is not approved in the United States. A process that improves the green color of canned vegetables is based on the formation of zinc metallo complexes. Green vegetables canned with this process were introduced into the United States in 1990.

Allomerization

Chlorophylls oxidize when dissolved in alcohol or other solvents and exposed to air. This process is referred to as allomerization. It is associated with uptake of oxygen equimolar to the chlorophylls present [94]. The products are blue-green in color and do not give a yellow-brown ring in the Molisch phase test [42]. The lack of a color response indicates that the cyclopentanone ring (ring V, figure 5c) has been oxidized or that the carbomethoxy group at C-10 has been removed. The products of allomerization have been identified as 10-hydroxychlorophylls and 10-methoxylactones (Fig. 10). The principal allomerization product of chlorophyll *b* is the 10-methoxylactone derivative.

Photodegradation

Chlorophyll is protected from destruction by light during photosynthesis in healthy plant cells by surrounding carotenoids and other lipids. Once this protection is lost by plant senescence, by pigment extraction from the tissue, or by cell damage caused during processing, chlorophylls are susceptible to photodegradation [62, 63]. When these conditions prevail and light and oxygen are present, chlorophylls are irreversibly bleached.

Many researchers have tried to identify colorless photodegradation products of chlorophylls. Methyl ethyl maleimide has been identified by Jen and Mackinney [50, 51]. In a study by Llewllyn et al. [62, 63] glycerol was found to be the major breakdown product, with lactic, citric, succininc and malonic acids, and alanine occurring in lesser amounts. The reacted pigments were completely bleached.

It is believed that photodegradation of chlorophylls results in opening of the tetrapyrrole ring and fragmentation into lower molecular weight compounds. It has been suggested that photodegradation begins with ring opening at one of the methine bridges to form oxidized linear tetrapyrroles [102]. Singlet oxygen and hydroxyl radicals are known to be produced during exposure of chlorophylls or similar porphyrins to light in the presence of oxygen [28]. Once formed, singlet oxygen or hydroxyl radicals will react further with tetrapyrroles to form peroxides and more free radicals, eventually leading to destruction of the porphyrins and total loss of color.

10.2.2.4 Color Loss During Thermal Processing

Loss of green color in thermally processed vegetables results from formation of pheophytin and pyropheophytin. Blanching and commercial heat sterilization can reduce chlorophyll content by as much as 80–100% [92]. Evidence that a small amount of pheophytin is formed during blanching before commercial sterilization is provided in Figure 8. The greater amount of pheophytin detected in frozen spinach as compared to spinach blanched for canning is most

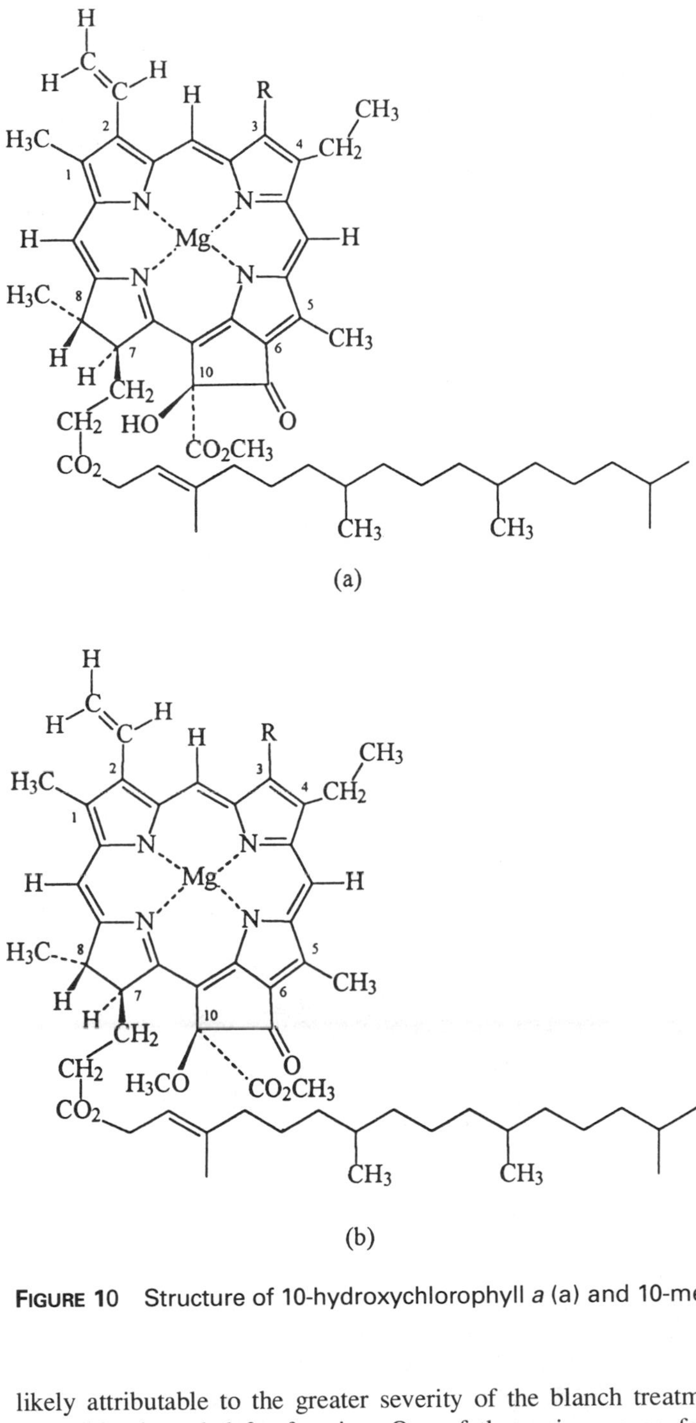

(a)

(b)

FIGURE 10 Structure of 10-hydroxychlorophyll *a* (a) and 10-methoxylactone of chlorophyll (b).

likely attributable to the greater severity of the blanch treatment that is generally applied to vegetables intended for freezing. One of the major reason for blanching of spinach prior to canning is to wilt the tissue and facilitate packaging, whereas blanching prior to freezing must be sufficient not only to wilt the tissue but also to inactivate enzymes. The pigment composition shown for the canned sample indicated that total conversion of chlorophylls to pheophytins and pyropheophytins has occurred.

Degradation of chlorophyll within plant tissues postharvest is initiated by heat-induced decompartmentalization of cellular acids as well as the synthesis of new acids [41]. In vegetables several acids have been identified, including oxalic, malic, citric, acetic, succinic, and pyrrolidone carboxylic acid (PCA). Thermal degradation of glutamine to form PCA is believed to be the major cause of the increase in acidity of vegetables during heating [18]. Other contributors to increased acidity may be fatty acids formed by lipid hydrolysis, hydrogen sulfide liberated from proteins or amino acids, and carbon dioxide from browning reactions. The pH decrease occurring during thermal processing of spinach purée is shown in Table 5.

10.2.2.5 Technology of Color Preservation

Efforts to preserve green color in canned vegetables have concentrated on retaining chlorophyll, forming or retaining green derivatives of chlorophyll, that is, chlorophyllides, or creating a more acceptable green color through the formation of metallo complexes.

Acid Neutralization to Retain Chlorophyll

The addition of alkalizing agents to canned green vegetables can result in improved retention of chlorophylls during processing. Techniques have involved the addition of calcium oxide and sodium dihydrogen phosphate in blanch water to maintain product pH or to raise the pH to 7.0. Magnesium carbonate or sodium carbonate in combination with sodium phosphate has been tested for this purpose. However, all of these treatments result in softening of the tissue and an alkaline flavor.

Blair in 1940 recognized the toughening effect of calcium and magnesium when added to vegetables. This observation led to the use of calcium or magnesium hydroxide for the purpose of raising pH and maintaining texture. This combination of treatments became known as the "Blair process" [7]. Commercial application of these processes has not been successful because of the inability of the alkalizing agents to effectively neutralize interior tissue acids over a long period of time, resulting in substantial color loss after less than 2 months of storage.

Another technique involved coating the can interior with ethycellulose and 5% magnesium hydroxide. It was claimed that slow leaching of magnesium oxide from the lining would maintain the pH at or near 8.0 for a longer time and would therefore help stabilize the green color [68, 69]. These efforts were only partially successful, because increasing the pH of canned vegetables can also cause hydrolysis of amides such as glutamine or asparagine with formation of undesirable ammonia-like odors. In addition, fatty acids formed by lipid hydrolysis during high-pH blanching may oxidize to form rancid flavors. In peas an elevated pH (8.0 or above) can cause formation of struvite, a glass-like crystals consisting of a magnesium and ammonium phosphate complex. Struvite is believed to result from the reaction of magnesium with ammonium generated from the protein in peas during heating.

High-Temperature Short-Time Processing

Commercially sterilized foods processed at a higher than normal temperature for a relatively short time (HTST) often exhibit better retention of vitamins, flavor and color than do conventionally processed foods. The greater retention of these constituents in HTST foods results because their destruction is more temperature dependent than that for inactivation of *Clostridium botulinum* spores. Temperature dependence can be expressed in terms of z-value or activation energy. The z-value is the change in °C required to effect a tenfold change in the destruction rate. The z-values for the formation of pheophytins a and b in heated spinach purée have been determined to be 51 and 98°C, respectively [40]. The large values for both as compared to that for *Clostridium botulinum* spores ($z = 10$°C), result in greater color retention when HTST processing is used [90]. However, this advantage of HTST processing is lost after about 2 months of storage, apparently because of a decrease in product pH during storage [66, 106].

Other studies of vegetable tissue have combined HTST processing with pH adjustment. Samples treated in this manner were initially greener and contained more chlorophyll than control samples (typical processing and pH). However, the improvement in color, as previously mentioned, was generally lost during storage [40, 11].

Enzymatic Conversion of Chlorophyll to Chlorophyllides to Retain Green Color

Blanching at lower temperatures than conventionally used to inactivate enzymes has been suggested as a means of achieving better retention of color in green vegetables, in the belief that the chlorophyllides produced have greater thermal stability than their parent compounds. Early studies showed that when spinach was blanched for canning at 71°C (160°F) for a total of 20 min better color retention resulted. This occurred as long as the blanch temperature was maintained between 54°C (130°F) and 76°C (168°F). It was concluded that the better color of processed spinach blanched under low temperature conditions (65°C for up to 45 min) was caused by heat-induced conversion of chlorophyll to chlorophyllides by the enzyme chlorophyllase [64]. However, the improvement in color retention achieved by this approach was insufficient to warrant commercialization of the process [17].

The progress of conversion of chlorophyll to chlorophyllides in heated spinach leaves is shown in Figure 11. The extract of unblanched spinach contained only chlorophylls *a* and *b*. Activation of chlorophyllase in spinach blanched at 71°C is illustrated by the formation of chlorophyllides (Figure 11b), while the absence of almost all chlorophyllides in spinach blanched at 88°C results from inactivation of the enzyme. These data also illustrate the previously mentioned point that isomerization of chlorophyll occurs during heating.

Commercial Application of Metallo Complex

Current efforts to improve the color of green processed vegetables and to prepare chlorophylls that might be used as food colorants have involved the use of either zinc or copper complexes of chlorophyll derivatives. Copper complexes of pheophytin and pheophorbide are available commercially under the names copper chlorophyll and copper chlorophyllin, respectively. The chlorophyll derivatives cannot be used in foods in the United States. Their use in canned foods, soups, candy, and dairy products is permitted in most European countries under regulatory control of the European Economic Community. The Food and Agriculture Organization (FAO) of the United Nations has certified their use as safe in foods, provided no more than 200 ppm of free ionizable copper is present.

Commercial production of the Cu pigments was described by Humphry [45]. Chlorophyll is extracted from dried grass or alfalfa with acetone or chlorinated hydrocarbons. Sufficient water is added, depending on the moisture content of the plant material, to aid penetration of the solvent while avoiding activation of chlorophyllase. Some pheophytin forms spontaneously during extraction. Copper acetate is added to form oil-soluble copper chlorophyll. Alternatively, pheophytin can be acid hydrolyzed before copper ion is added, resulting in formation of water-soluble copper chlorophyllin. The copper complexes have greater stability than comparable Mg complexes; for example, after 25 hr at 25°C, 97% of the chlorophyll degrades while only 44% of the copper chlorophyll degrades.

Regreening of Thermal Processed Vegetables

It has been observed that when vegetables purées are commercially sterilized, small bright-green areas occasionally appear. It was determined that pigments in the bright-green areas contained

a)

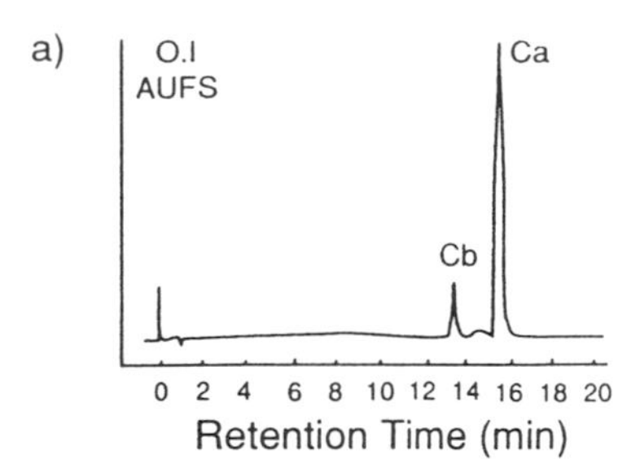

b)

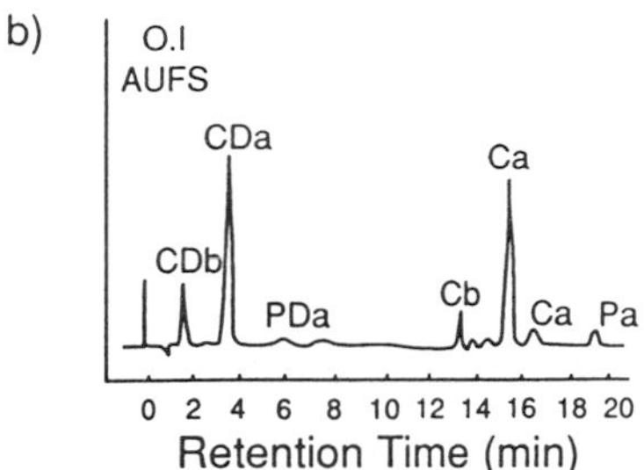

c)

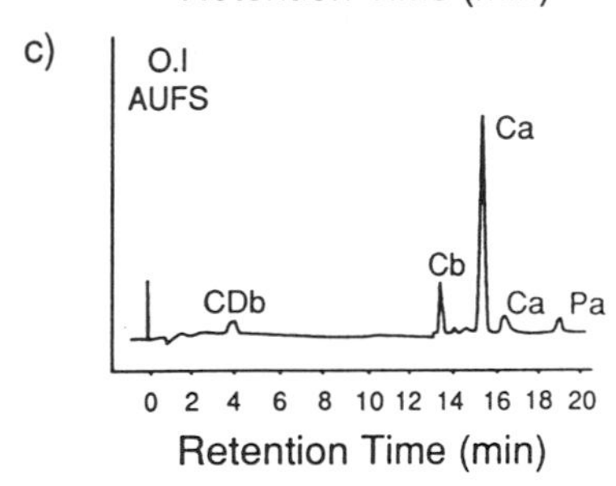

FIGURE 11 Chlorophyll and chlorophyll derivatives of spinach: (a) unblanched, (b) blanched 3 min at 71°C, (c) blanched 3 min at 88°C. C_a = chlorophyll *a*, C_b = chlorophyll *b*, P_a = pheophytin *a*, CD_a = chlorophyllide *a*, CD_b = chlorophyllide *b*, PD_a = pheophorbide *a*. (From Ref. 112.)

zinc and copper. This formation of bright-green areas in vegetable purées was termed "regreening." Regreening of commercially processed vegetables has been observed when zinc and/or copper ions are present in process solutions. Okra when processed in brine solution containing zinc chloride retains its bright green color, and this is attributed to the formation of zinc complexes of chlorophyll derivatives [27, 103, 105].

A patent was issued to Continental Can Company (now Crown Cork & Seal Company) for the commercial canning of vegetables with metal salts in blanch or brine solution. The process involved blanching vegetables in water containing sufficient amounts of Zn^{2+} or Cu^{2+} salts to raise the tissue concentration of the metal ions to between 100 and 200 ppm. The direct addition of zinc chloride to the canning brine has no significant effect on the color of vegetables (green beans and peas). Green vegetables processed in modified blanch water were claimed to be greener than conventionally processed vegetables. Other bi- or trivalent metal ions were either less effective or ineffective [95]. This approach is known as the Veri-Green process. Pigments present in canned green beans processed by the Veri-Green process were identified as zinc pheophytin and zinc pyropheophytin [111].

Commercial production of zinc-processed green beans and spinach is presently conducted by several processors, but results have been mixed. Shown in Figure 12 is the sequence of pigment changes occurring when pea purée is heated in the presence of 300 ppm Zn^{2+}. Chlorophyll *a* decreases to trace levels after only 20 min of heating. Accompanying this rapid decrease in chlorophyll is the formation of zinc complexes of pheophytin *a* and pyropheophytin *a*. Further heating increases the zinc pyropheophytin concentration at the expense of a decrease in zinc pheophytin. In addition, zinc pyropheophytin may form through decarboxymethylation of zinc pheophytin or by reaction of pyropheophytin with Zn^{2+}. These results suggest that the green color in vegetables processed in the presence of zinc is largely due to the presence of zinc pyropheophytin.

Shown in Figure 13 is a proposed sequence of reactions for conversion of chlorophyll *a* to zinc pyropheophytin *a*. Formation of Zn-complexes occurs most rapidly between pH 4.0 and 6.0, and the rate decreases markedly at pH 8.0. The reason for the decrease is that chlorophyll is retained at the high pH, thereby limiting the amount of chlorophyll derivatives available for complex formation (Figure 14) [59]. It has further been shown that zinc complex formation can be influenced by the presence of surface-active anionic compounds. Such compounds adsorb onto the chloroplast membranes, increasing the negative surface charge and thereby increasing complex formation.

Currently, the best process for attaining a desirable green color in canned vegetables involves adding zinc to the blanch solution, increasing membrane permeability by heating the tissue prior to blanching at or slightly above 60°C, choosing a pH that favors formation of metallo complexes, and using anions to alter surface charge on the tissue.

10.2.3 Carotenoids

Carotenoids are nature's most widespread pigments, with the earth's annual biomass production estimated at 100 million tons. A large majority of these pigments are biosynthesized by the ocean algae population. In higher plants, carotenoids in chloroplasts are often masked by the more dominant chlorophyll pigments. In the autumn season when chloroplasts decompose during plant senescene, the yellow-orange color of carotenoids becomes evident [4].

For several decades, it has been known that carotenoids play important functions in

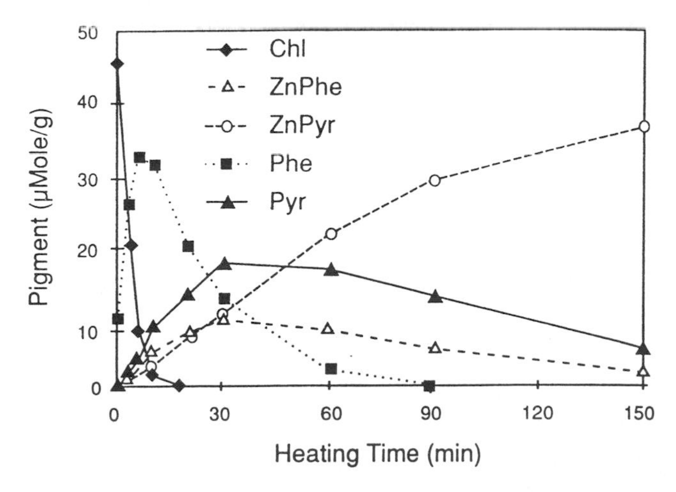

FIGURE 12 Pigments in pea purée containing 300 ppm of Zn^{2+} after heating at 121°C for up to 150 min. Chl = chlorophyll, ZnPhe = zinc pheophytin, ZnPyr = zinc pyropheophytin, Phe = pheophytin, Pyr = pyropheotphyin. (From Ref. 112.)

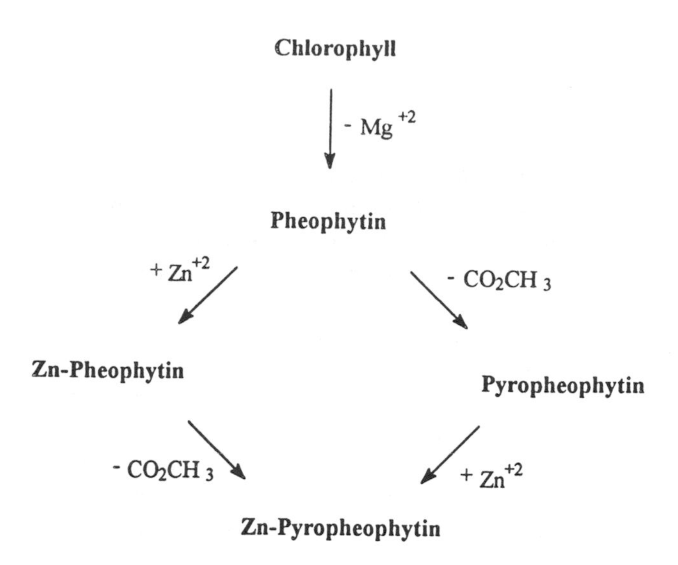

FIGURE 13 Chemical reactions occurring in heated green vegetables containing zinc.

photosynthesis and photoprotection in plant tissues [37]. In all chlorophyll-containing tissues, carotenoids function as secondary pigments in harvesting light energy. The photoprotection role of carotenoids stems from their ability to quench and inactivate reactive oxygen species formed by exposure to light and air. In addition, specific carotenoids present in roots and leaves serve as precursors to abscisic acid, a compound that functions as a chemical messenger and growth regulator [22, 78].

The most prominent role of carotenoid pigments in the diet of humans and other animals is their ability to serve as precursors of vitamin A. Although the carotenoid beta-carotene possesses the greatest provitamin A activity because of its two β-ionone rings (Fig. 15), other commonly consumed carotenoids, such as α-carotene and β-cryptoxanthin, also possess provi-

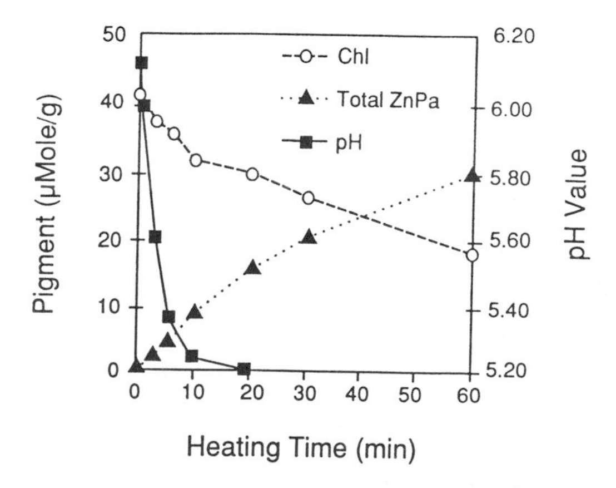

FIGURE 14 Conversion of chlorophyll *a* (chl) to total zinc complexes (ZnPa) and change in pH in pea puree heated at 121°C for up to 60 min (From Ref. 58.)

tamin A activity. Provitamin A carotenoids present in fruits and vegetables are estimated to provide 30–100% of the vitamin A requirement in human populations. A prerequisite to vitamin A activity is the existence of the retinoid structure (with the β-ionone ring) in the carotenoid. Thus, only a few carotenoids possess vitamin activity. This topic is covered thoroughly in Chapter 8.

In 1981, Peto et al. [83] drew attention to these pigments because of their epidemiological findings that consumption of fruits and vegetables high in carotenoid content is associated with a decreased incidence of specific cancers in humans. More recently, interest has focused on the presence of processing-induced *cis* isomeric carotenoids in the diet and on their physiological significance. These findings have stimulated a substantial increase in carotenoid research.

10.2.3.1 Structures of Carotenoids

Carotenoids are comprised of two structural groups: the hydrocarbon carotenes and the oxygenated xanthophylls. Oxygenated carotenoids (xanthophylls) consist of a variety of derivatives frequently containing hydroxyl, epoxy, aldehyde, and keto groups. In addition, fatty acid esters of hydroxylated carotenoids are also widely found in nature. Thus, over 560 carotenoid structures have been identified and compiled [101]. Furthermore, when geometric isomers of *cis* (Z) or trans (E) forms are considered, a great many more configurations are possible.

The basic carotenoid structural backbone consists of isoprene units linked covalently in either a head-to-tail or a tail-to-tail fashion to create a symmetrical molecule (Fig. 16). Other carotenoids are derived from this primary structure of forty carbons. Some structures contain cyclic end groups (β-carotene, Fig. 15) while others possess either one or no cyclization (lycopene, the prominent red pigment in tomatoes) (Fig. 16). Other carotenoids may have shorter carbon skeletons and are known as apocarotenals. Although rules exist for naming and numbering all carotenoids [46, 47] the trivial names are commonly used and presented in this chapter.

The most common carotenoid found in plant tissues is β-carotene. This carotenoid is also used as a colorant in foods. Both the naturally derived and synthetic forms can be added to food products. Some carotenoids found in plants are shown in Figure 15, and they include α-carotene (carrots), capsanthin (red peppers, paprika), lutein, a diol of α-carotene, its esters (marigold petals), and bixin (annatto seed). Other common carotenoids found in foods include zeaxanthin (a diol of β-carotene), violaxanthin (an epoxide carotenoid), neoxanthin (an allenic triol), and β-cryptoxanthin (a hydroxylated derivative of β-carotene).

Animals derive carotenoids pigments by consumption of carotenoid-containing plant materials. For example, the pink color of salmon flesh is due mainly to the presence of astaxanthin, which is obtained by ingestion of carotenoid-containing marine plants. It is also well known that some carotenoids in both plants and animals are bound to or associated with proteins. The red astaxanthin pigment of shrimp and lobster exoskeletons is blue in color when complexed with proteins. Heating denatures the complex and alters the spectroscopic and visual properties of the pigment, thus changing the color from blue to red. Other examples of carotenoid–protein complexes are ovoverdin, the green pigment found in lobster eggs, and the carotenoid–chlorophyll–protein complexes in plant chloroplasts. Other unique structures include carotenoid glycosides, some of which are found in bacteria and other microorganisms. One example of a carotenoid glycoside present in plants is the carotenoid crocein found in saffron.

10.2.3.2 Occurrence and Distribution

Edible plant tissues contain a wide variety of carotenoids [39]. Red, yellow, and orange fruits, root crops, and vegetables are rich in carotenoids. Prominent examples include tomatoes

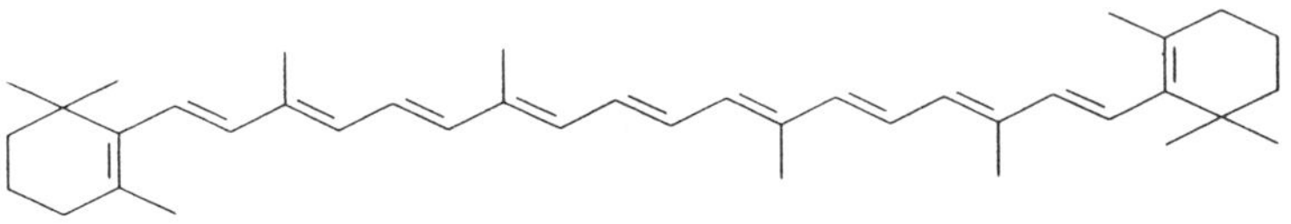

(ß-ionone ring) (ß-ionone ring)

β-CAROTENE

$(C_{40}H_{56})$

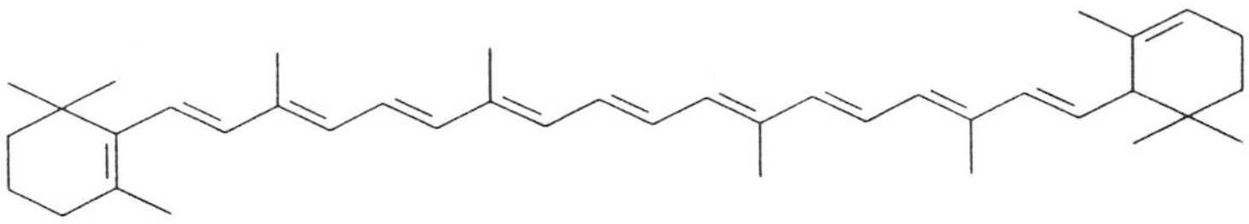

α-CAROTENE

$(C_{40}H_{56})$

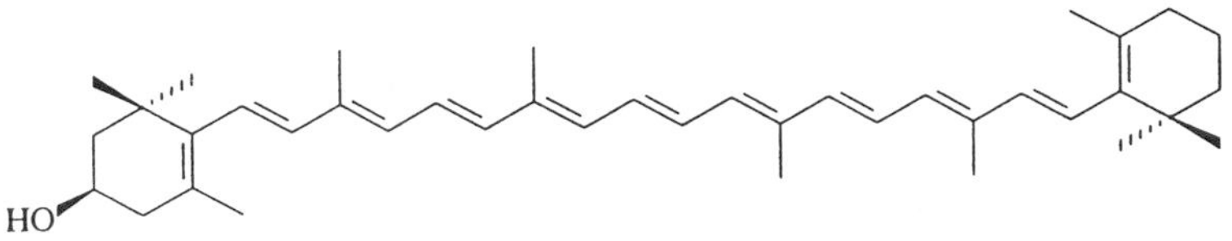

HO

β-CRYPTOXANTHIN

$(C_{40}H_{56}O)$

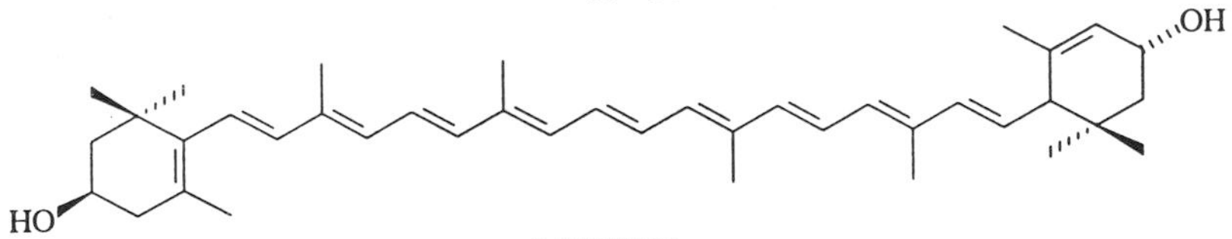

OH

HO

LUTEIN

$(C_{40}H_{56}O_2)$

OH

HO

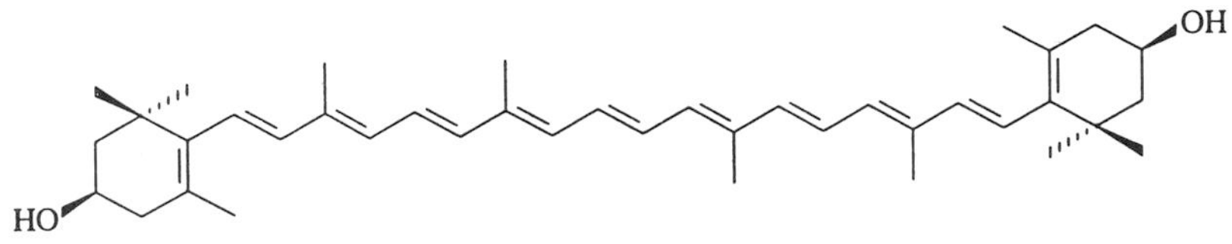

ZEAXANTHIN

$(C_{40}H_{56}O_2)$

FIGURE 15 Structures of commonly occurring carotenoids. Lutein (yellow) from green leaves, corn, and marigold; zeaxanthin (yellow) from corn and saffron; β-cryptoxanthin (yellow) from corn; β-carotene (yellow) from carrots; β-carotene (yellow) from carrots and sweet potatoes; neoxanthin (yellow) from green leaves; capsanthin (red) from red peppers; violaxanthin (yellow) from green leaves; and bixin (yellow) from annatto seeds.

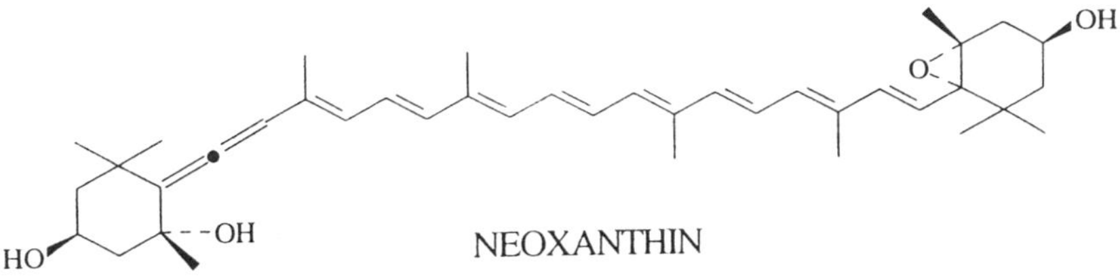

NEOXANTHIN

$(C_{40}H_{56}O_4)$

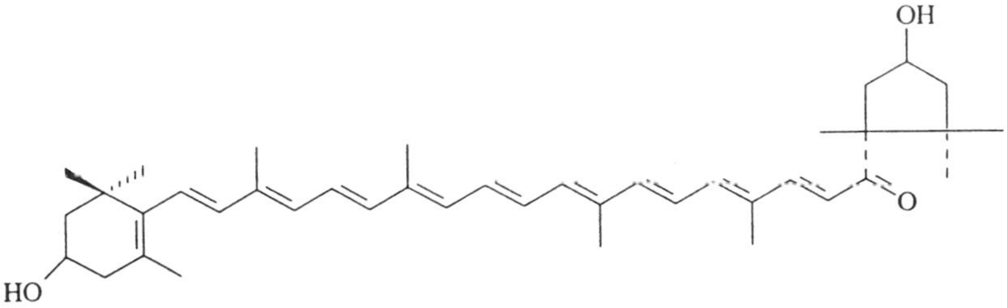

CAPSANTHIN

$(C_{40}H_{56}O_3)$

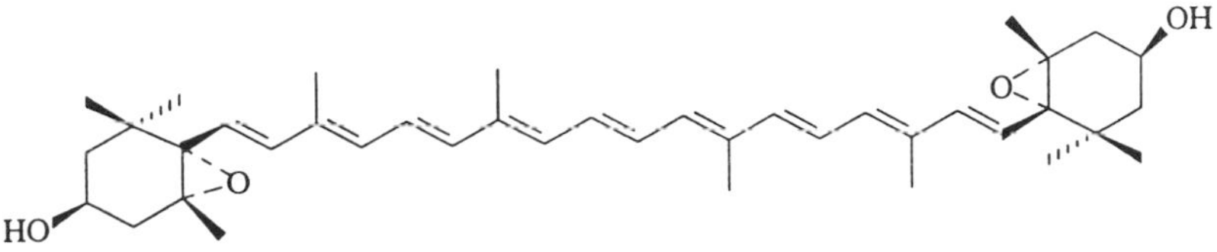

VIOLAXANTHIN

$(C_{40}H_{56}O_4)$

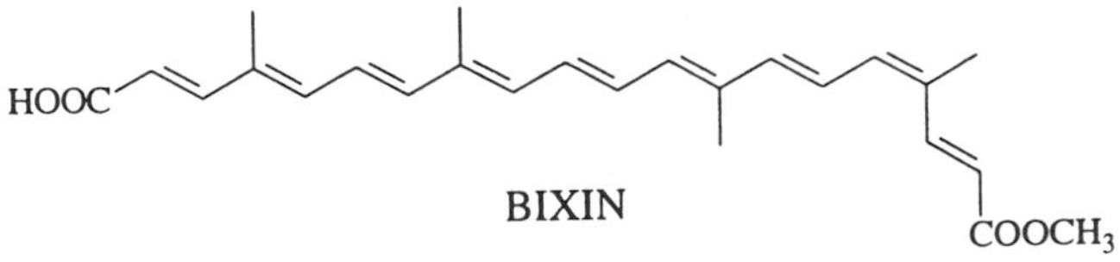

BIXIN

$(C_{25}H_{30}O_4)$

(lycopene), carrots (α- and β-carotenes), red peppers (capsanthin), pumpkins (β-carotene), squashes (β-carotene), corn (lutein and zeaxanthin), and sweet potatoes (β-carotene). All green leafy vegetables contain carotenoids but their color is masked by the green chlorophylls. Generally, the largest concentrations of carotenoids exist in those tissues with the largest amount of chlorophyll pigments. For example, spinach and kale are rich in carotenoids, and peas, green beans, and asparagus contain significant concentrations.

Many factors influence the carotenoid content of plants. In some fruits, ripening may bring about dramatic changes in carotenoids. For example, in tomatoes, the carotenoid content,

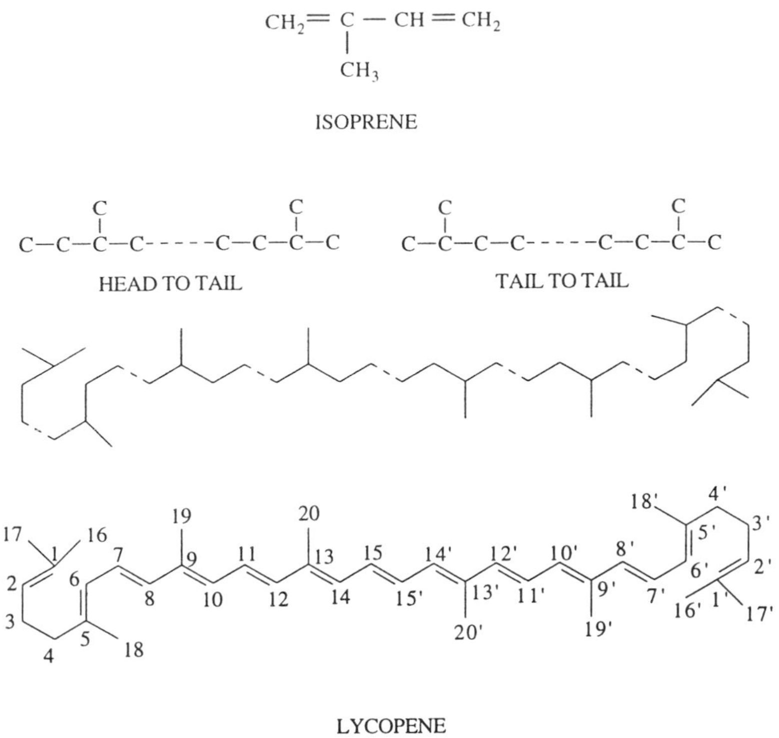

$CH_2{=}C(CH_3){-}CH{=}CH_2$

ISOPRENE

C–C–C(C)–C- - - - -C–C–C(C)–C

HEAD TO TAIL

C–C(C)–C–C- - - - -C–C–C(C)–C

TAIL TO TAIL

LYCOPENE

FIGURE 16 Joining of isoprenoid units to form lycopene (primary red pigment of tomatoes) (From Ref. 39.)

especially lycopene, increases significantly during the ripening process. Thus, concentrations differ depending on the stage of plant maturity. Even after harvest, tomato carotenoids continue to be synthesized. Since light stimulates biosynthesis of carotenoids, the extent of light exposure is known to affect their concentration. Other factors that alter carotenoid occurrence or amount include growing climate, pesticide and fertilizer use, and soil type [39].

10.2.3.3 Physical Properties, Extraction, and Analysis

All classes of carotenoids (hydrocarbons: carotenes and lycopene, and oxygenated xanthophylls) are lipophilic compounds and thus are soluble in oils and organic solvents. They are moderately heat stable and are subject to loss of color by oxidation. Carotenoids can be easily isomerized by heat, acid, or light. Since they range in color from yellow to red, detection wavelengths for monitoring carotenoids typically range from approximately 430 to 480 nm. The higher wavelengths are usually used for some xanthophylls to prevent interference from chlorophylls. Many carotenoids exhibit spectral shifts after reaction with various reagents, and these spectral changes are useful to assist in identification.

The complex nature and diversity of carotenoid compounds present in plant foods necessitate chromatographic separation [89]. Extraction procedures for quantitative removal of carotenoids from tissue utilize organic solvents that must penetrate a hydrophilic matrix. Hexane–acetone mixtures are commonly employed for this purpose, but special solvents and treatments are sometimes needed to achieve satisfactory separation [54].

Many chromatographic procedures, including HPLC, have been developed for separating carotenoids [107, 93]. Special analytical challenges occur when carotenoid esters, *cis*/*trans* isomers, and optical isomers need to be separated and identified.

10.2.3.4 Chemical Properties

Oxidation

Carotenoids are easily oxidized because of the large number of conjugated double bonds. Such reactions cause color loss of carotenoids in foods and are the major degradation mechanisms of concern. The stability of a particular pigment to oxidation is highly dependent on its environment. Within tissues, the pigments are often compartmentalized and protected from oxidation. However, physical damage to the tissue or extraction of the carotenoids increases their susceptibility to oxidation. Furthermore, storage of carotenoid pigments in organic solvents will often accelerate decomposition. Because of the highly conjugated, unsaturated structure of carotenoids, the products of their degradation are very complex. These products are largely uncharacterized, except for β-carotene. Shown in Figure 17 are various degradation products of β-carotene during oxidation and thermal treatments. During oxidation, epoxides and carbonyl compounds are initially formed. Further oxidation results in formation of short-chain mono- and dioxygenated compounds including epoxy-β-ionone. Generally, epoxides form mainly within the end rings, while oxidative scission can occur at a variety of sites along the chain. For provitamin A carotenoids, epoxide formation in the ring results in loss of the provitamin activity. Extensive autoxidation will result in bleaching of the carotenoid pigments and loss in color. Oxidative destruction of β-carotene is intensified in the presence of sulfite and metal ions [82].

Enzymatic activity, particularly lipoxygenase, hastens oxidative degradation of carotenoid pigments. This occurs by an indirect mechanisms. Lipoxygenase first catalyzes oxidation of unsaturated or polyunsaturated fatty acids to produce peroxides, and these in turn react readily with carotenoid pigments. In fact, this coupled reaction scheme is quite efficient, and the loss of

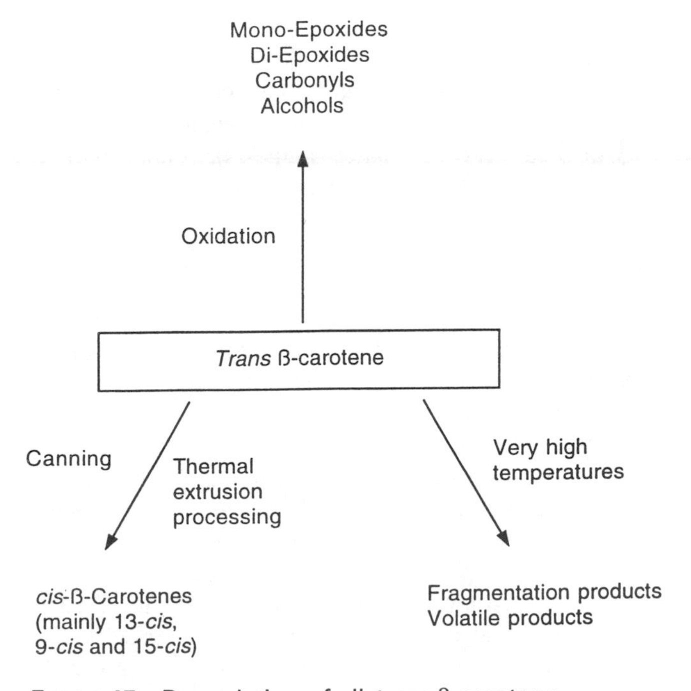

FIGURE 17 Degradation of all-*trans*-β-carotene.

carotene color and decreased absorbance in solution are often used as an assay for lipoxygenase activity [5].

Antioxidative Activity

Because carotenoids can be readily oxidized, it is not surprising that they have antioxidant properties. In addition to cellular and in vitro protection against singlet oxygen, carotenoids, at low oxygen partial pressures, inhibit lipid peroxidation [12]. At high oxygen partial pressures, β-carotene has pro-oxidant properties [13]. In the presence of molecular oxygen, photosensitizers (i.e., chlorophyll), and light, singlet oxygen may be produced, which is a highly reactive active oxygen species. Carotenoids are known to quench singlet oxygen and thereby protect against cellular oxidative damage. Not all carotenoids are equally effective as photochemical protectors. For example, lycopene is known to be especially efficient in quenching singlet oxygen relative to other carotenoid pigments [23, 98].

It has been proposed that the antioxidant functions of carotenoids play a role in limiting cancer, cataracts, atherosclerosis, and the processes of aging [16]. A detailed overview on the antioxidant role of carotenoid compounds is beyond the scope of this discussion, and the reader is referred to several excellent reviews [13, 34, 56, 77].

Cis/Trans Isomerization

In general, the conjugated double bonds of carotenoid compounds exist in an all-*trans* configuration. The *cis* isomers of a few carotenoids can be found naturally in plant tissues, especially in algae sources, which are currently being harvested as a source of carotenoid pigments. Isomerization reactions are readily induced by thermal treatments, exposure to organic solvents, contact for prolonged periods with certain active surfaces, treatment with acids, and illumination of solutions (particularly with iodine present). Iodine-catalyzed isomerization is a useful means in the study of photoisomerization because an equilibrium mixture of isomeric configurations is formed. Theoretically, large numbers of possible geometrical configurations could result from isomerization because of the extensive number of double bonds present in carotenoids. For example, β-carotene has potentially 272 different *cis* forms. Because of the complexity of various *cis/trans* isomers within a single carotenoid, only recently have accurate methods been developed to study these compounds in foods [76]. *Cis/trans* isomerization affects the provitamin A activity of carotenoids. The provitamin A activity of β-carotene *cis* isomers ranges, depending on the isomeric form, from 13 to 53% as compared to that of all-*trans*-β-carotene [114]. The reason for this reduction in provitamin A activity is not known [4, 114].

10.2.3.6 Stability During Processing

Carotenoids are relatively stable during typical storage and handling of most fruits and vegetables. Freezing causes little change in carotene content. However, blanching is known to influence the level of carotenoids. Often blanched plant products exhibit an apparent increase in carotenoid content relative to raw tissues. This is caused by inactivation of lipoxygenase, which is known to catalyze oxidative decomposition of carotenoids, the loss of soluble constituents into the blanch water, or the mild heat treatments traditionally used during blanching may enhance the efficiency of extraction of the pigments relative to fresh tissue. Lye peeling, which is commonly used for sweet potatoes, causes little destruction or isomerization of carotenoids.

Although carotene historically has been regarded as fairly stable during heating, it is now known that heat sterilization induces *cis/trans* isomerization reactions as shown in Figure 17. To lessen excessive isomerization, the severity of thermal treatments should be minimized when possible. In the case of extrusion cooking and high-temperature heating in oils, not only will carotenoids isomerize but thermal degradation will also occur. Very high temperatures can yield fragmentation products that are volatile. Products arising from severe heating of β-carotene in the presence of air are similar to those arising from severe heating of β-carotene oxidation (Fig. 17). In contrast, air dehydration exposes carotenoids to oxygen, which can cause extensive degradation of carotenoids. Dehydrated products that have large surface-to-mass ratios, such as carrot or sweet potato flakes, are especially susceptible to oxidative decomposition during drying and storage in air.

When *cis* isomers are created, only slight spectral shifts occur and thus color of the product is mostly unaffected; however, a decrease in provitamin A activity occurs. These reactions have important nutritional effects that should be considered when selecting analytical measurements for provitamin A. The older methods for vitamin A determination in foods did not account for the differences in the provitamin A activity of individual carotenoids or their isomeric forms. Therefore, older nutritional data for foods are in error, especially for foods that contain high proportions of provitamin A carotenoids other than β-carotene and those that contain a significant amount of *cis* isomers.

10.2.4 Flavonoids and Other Phenols

10.2.4.1 Anthocyanins

Phenolic compounds comprise a large group of organic substances, and flavonoids are an important subgroup. The flavonoid subgroup contains the anthocyanins, one of the most broadly distributed pigment groups in the plant world. Anthocyanins are responsible for a wide range of colors in plants, including blue, purple, violet, magenta, red, and orange. The word anthocyanin is derived from two Greek words: *anthos,* flower, and *kyanos,* blue. These compounds have attracted the attention of chemists for years, two of the most notable investigators being Sir Robert Robinson (1886–1975) and Professor Richard Willstätter (1872–1942). Both were awarded Nobel prizes in chemistry, in part for their work with plant pigments.

Structure

Anthocyanins are considered flavonoids because of the characteristic $C_6C_3C_6$ carbon skeleton. The basic chemical structure of the flavonoid group and the relationship to anthocyanin are shown in Figure 18. Within each group there are many different compounds with their color depending on the substituents on rings A and B.

The base structure of anthocyanin is the 2-phenylbenzopyrylium of flavylium salt (Figure 19). Anthocyanins exist as glycosides of polyhydroxy and/or polymethoxy derivatives of the salt. Anthocyanins differ in the number of hydroxyl and/or methoxy groups present, the types, numbers, and sites of attachment of sugars to the molecule, and the types and numbers of aliphatic or aromatic acids that are attached to the sugars in the molecule. The most common sugars are glucose, galactose, arabinose, xylose, and homogeneous or heterogeneous di- and trisaccharides formed by combining these sugars. Acids most commonly involved in acylation of sugars are caffeic, *p*-coumaric, sinapic, *p*-hydroxybenzoic, ferulic, malonic, malic, succinic, and acetic.

When the sugar moiety of an anthocyanin is hydrolyzed, the aglycone (the nonsugar hydrolysis product) is referred to as an anthocyanidin. The color of anthocyanins and anthocyanidins results from excitation of a molecule by visible light. The ease with which a molecule is excited depends on the relative electron mobility in the structure. Double bonds, which are abundant in anthocyanins and anthocyanidins, are excited very easily, and their presence is essential for color. Anthocyanins occurring in nature contain several anthocyanidins, but only six occur commonly in foods (Figure 20). It should be noted that increasing substitution on the molecule results in a deeper hue. The deepening of hue is the result of a bathochromic change (longer wavelength), which means that the light absorption band in the visible spectrum

8 9 7 A 4 C—C—C B 3 2 1 6 5

Basic $C_6C_3C_6$ Structure

FLAVANONES

O

Chalcone

O

O

Flavanones

O

O

Flavones

O

OH

O

Flavanonols

O

OH

O

Flavonols

FLAVANS

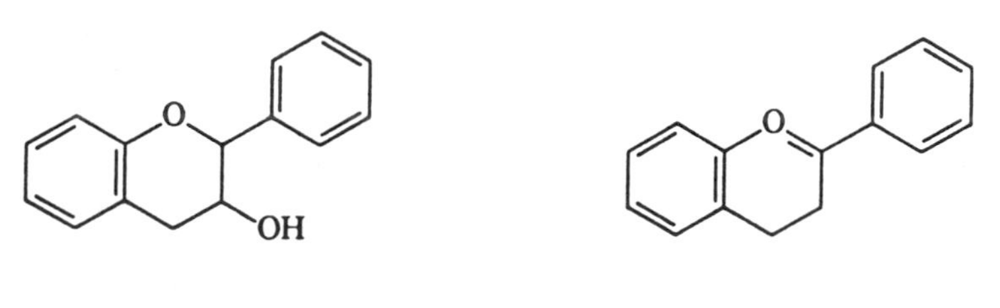

Catechins

Anthocyanidins

FIGURE 18 Carbon skeleton of the flavonoid group.

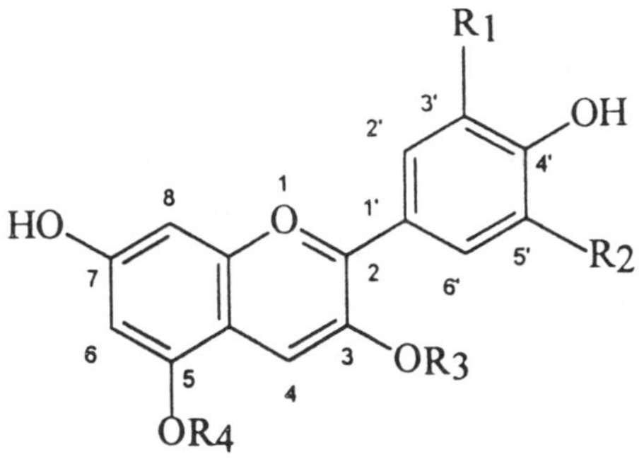

FIGURE 19 The flavylium cation. R_1 and R_2 = -H, -OH, or OCH_3, R_3 = -glycosyl, R_4 = -H or -glycosyl.

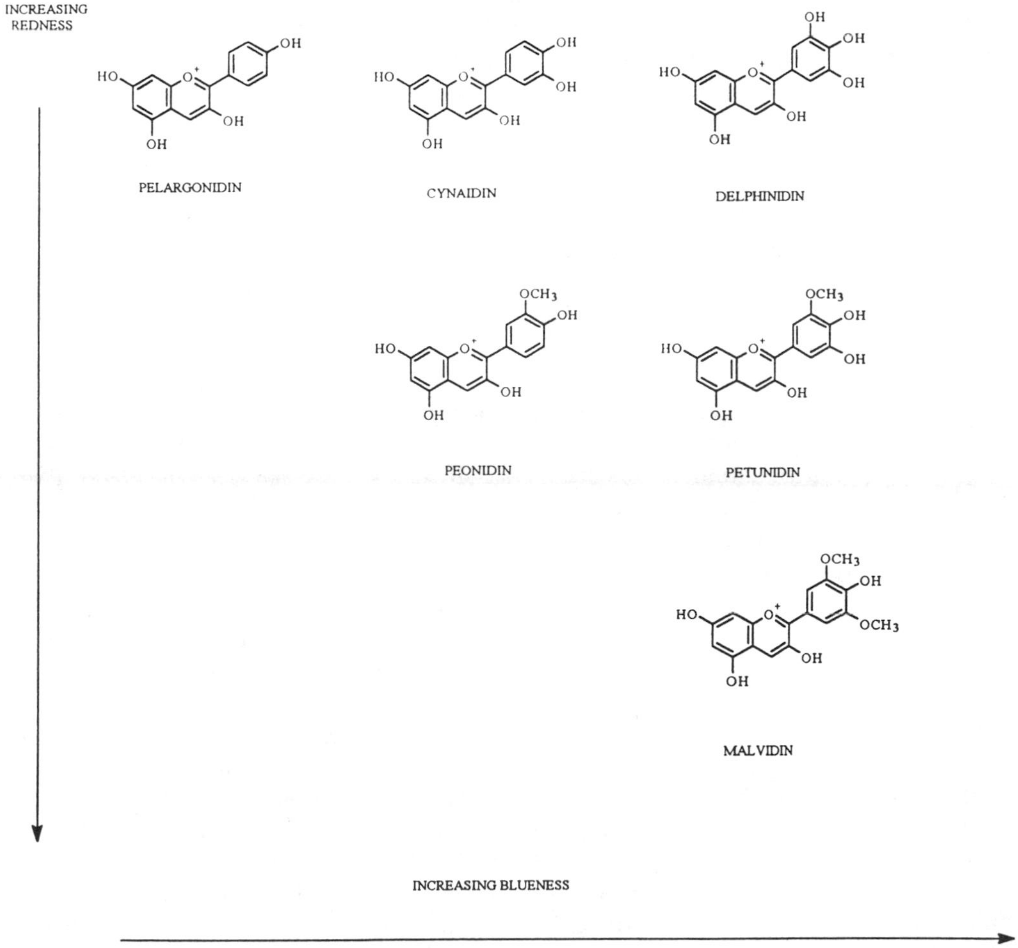

FIGURE 20 Most common anthocyanidins in foods, arranged in increasing redness and blueness.

shifts from violet through red to blue. An opposite change is referred to as a hypsochromic shift. Bathochromic effects are caused by auxochrome groups, groups that by themselves have no chromophoric properties but cause deepening in the hue when attached to the molecule. Auxochrome groups are electron-donating groups, and in the case of anthocyanidins they are the hydroxyl and methoxy groups. The methoxy groups, because their electron-donating capacity is greater than that of hydroxyl groups, cause a greater bathochromic shift than do hydroxyl groups. The effect of the number of methoxy groups on redness is illustrated in Figure 20. In anthocyanin, several other factors, such as change in pH, metal complex formation, and copigmentation, cause changes in hue. These will be discussed later.

Anthocyanidins are less water soluble than their corresponding glycosides (anthocyanins) and they are not, therefore, found free in nature. Those reported, other than the 3-deoxy form, which is yellow, are most likely hydrolysis products formed during isolation procedures. The free 3-hydroxyl group in the anthocyanidin molecule destabilizes the chromophore; therefore, the 3-hydroxyl group is always glycosylated. Additional glycosylation is most likely to occur at C-5 and can also occur at C-7, -3′, -4′, and/or -5′ hydroxyl group (Fig. 19). Steric hindrance precludes glycosylation at both C-3′ and C-4′ [10].

With this structural diversity, it is not surprising that more than 250 different anthocyanins have been identified in the plant world. Plants not only contain mixtures of anthocyanins but the relative concentrations vary among cultivars and with maturity. Total anthocyanin content varies among plants and ranges from about 20 mg/100 g fresh weight to as high as 600. The reader is referred to a book by Mazza and Miniali [72] for a more detailed account of anthocyanins in fruits, vegetables, and grains.

Color and Stability of Anthocyanins

Anthocyanin pigments are relatively unstable, with greatest stability occurring under acidic conditions. Both the hue of the pigment and its stability are greatly impacted by substituents on the aglycone. Degradation of anthocyanins occurs not only during extraction from plant tissue but also during processing and storage of foods tissues.

Knowledge of the chemistry of anthocyanins can be used to minimize degradation by proper selection of processes and by selection of anthocyanin pigments that are most suitable for the intended application. Major factors governing degradation of anthocyanins are pH, temperature, and oxygen concentration. Factors that are usually of less importance are the presence of degradative enzymes, ascorbic acid, sulfur dioxide, metal ions, and sugars. In addition, copigmentation may affect or appear to affect the degradation rate.

Structural Transformation and pH Degradation rates vary greatly among anthocyanins because of their diverse structures. Generally, increased hydroxylation decreases stability, while increased methylation increases stability. The color of foods containing anthocyanins that are rich in pelargonidin, cyanidin, or delphinidin aglycones is less stable than that of foods containing anthocyanins that are rich in petunidin or malvidin aglycones. The increased stability of the latter group occurs because reactive hydroxyl groups are blocked. It follows that increased glycosylation, as in monoglucosides and diglucosides, increases stability. It has also been shown, although it is not fully understood why, that the type of sugar moiety influences stability. Starr and Francis [99] found that cranberry anthocyanins that contained galactose were more stable during storage than those containing arabinose. Cyanidin 3-2 glucosylrutinoside at pH 3.5, 50°C, has a half-life of 26 hr compared to 16 hr for cyanidin-3-rutinoside [110]. These examples illustrate that substituents have a marked effect on anthocyanin stability although they themselves do not react.

In an aqueous medium, including foods, anthocyanins can exist in four possible structural forms depending on pH (Fig. 21, I): the blue quinonoidal base (A), the red flavylium cation

(AH$^+$), the colorless carbinol pseudobase (B), and the colorless chalcone (C). Shown in Figure 21 are the equilibrium distributions of these four forms in the pH range 0–6 for malvidin-3-glucoside (Fig. 21, II) dihydroxyflavylium chloride (Fig. 21, III) and 4 methoxy-4-methyl-7-hydroxyflavylium chloride (Fig. 21, IV). For each pigment only two of the four species are important over this pH range. In a solution of malvidin-3-glucoside at low pH the flavylium

I

II

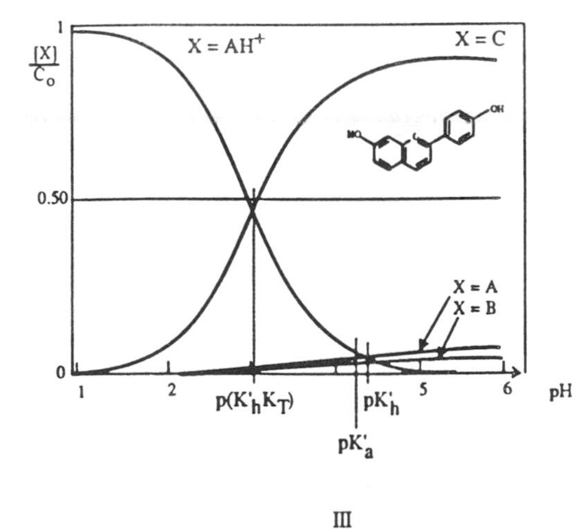

III

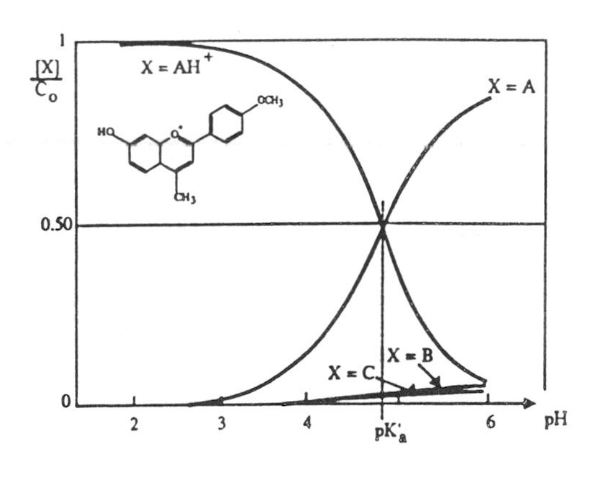

IV

FIGURE 21 (I) The four anthocyanin structures present in aqueous acidic solution at room temperatures: A, quinonoidal base (blue); (AH$^+$) flavylium salt (red); B, pseudobase or carbinol (colorless); C, chalcone (colorless). (II–IV) Equilibrium distribution at 25°C of AH$^+$, A, B, and C as a function of pH: (II) for malvidin 3- glucoside; (III) for 4′,7-hydroxyflavylium chloride; (IV) for 4′-methoxyl-4-methyl-7-hydroxy flavylium chloride. (From Ref. 10.)

structure dominates, while at pH 4–6 the colorless carbinol dominates. A similar situation exists with 4′,7-hydroxyflavylium except the equilibrium mixture consists mainly of the flavylium and the chalcone structure. Thus, as the pH approaches 6 the solution becomes colorless. In a solution of 4′-methoxy-4-methyl-7-hydroxyflavylium chloride an equilibrium exists between the flavylium cation and the quinonoidal base. This solution therefore is colored throughout the 0–6 pH range, turning from red to blue as pH is increased in this range.

To further demonstrate the effect of pH on the color of anthocyanins, the spectra for cyanidin-3-rhamnoglucoside in buffer solutions at pH levels between 0.71 to 4.02 are shown in Figure 22. Although the absorption maximum remains the same over this pH range, the intensity of absorption decreases with increasing pH. Color changes in a mixture of cranberry anthocyanins as a function of pH are shown in Figure 23. In aqueous medium, as in cranberry cocktail, changes in pH can cause major changes in color. Anthocyanins show their greatest tinctorial strength at approximately pH 1.0, when the pigment molecules are mostly in the unionized form. At pH 4.5 anthocyanins in fruit juices are nearly colorless (slightly bluish) if yellow flavonoids are not present. If yellow pigments are present, as is common in fruits, the juice will be green.

Temperature Anthocyanin stability in foods is greatly affected by temperature. Rates of degradation are also influenced by the presence or absence of oxygen and, as already pointed out, by pH and structural conformation. In general, structural features that lead to increased pH stability also lead to increased thermal stability. Highly hydroxylated anthocyanidins are less stable than methylated, glycosylated, or acylated anthocyanidins. For example, the half-life of 3,4′,5,5′,7-pentahydroxyflavylium at pH 2.8 is 0.5 days compared to 6 days for the 3,4′,5,5′,7-pentamethoxyflavylium [72]. Under similar conditions the half-life for cyanidin-3-rutinoside is 65 days compared to 12 hr for cyanidin [70].

It should be noted that comparison of published data for pigment stability is difficult because of differing experimental conditions used. One of the errors in published data involves

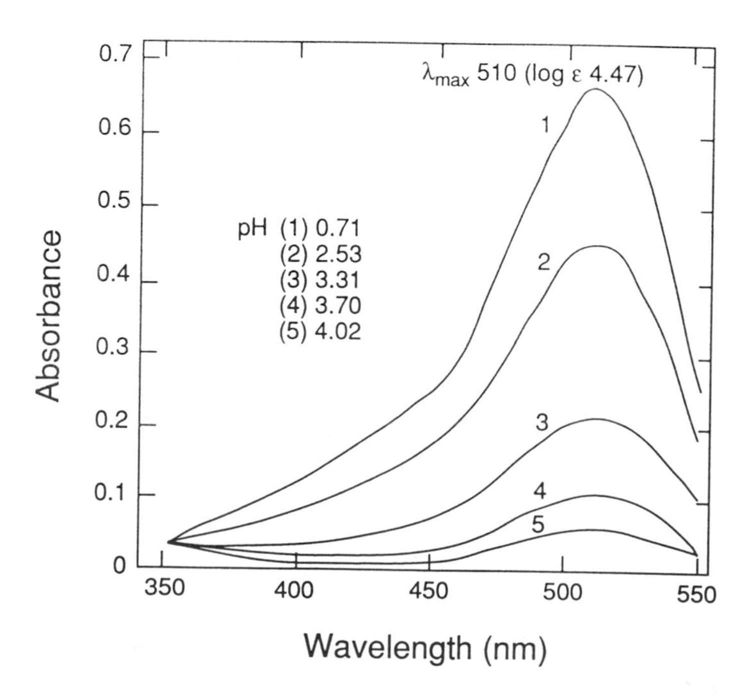

FIGURE 22 Absorption spectra of cyanidin-3-rhamnoglucosde in buffer solutions at pH 0.71–4.02. Pigment concentration 1.6×10^{-2} g/L. (From Ref. 33).

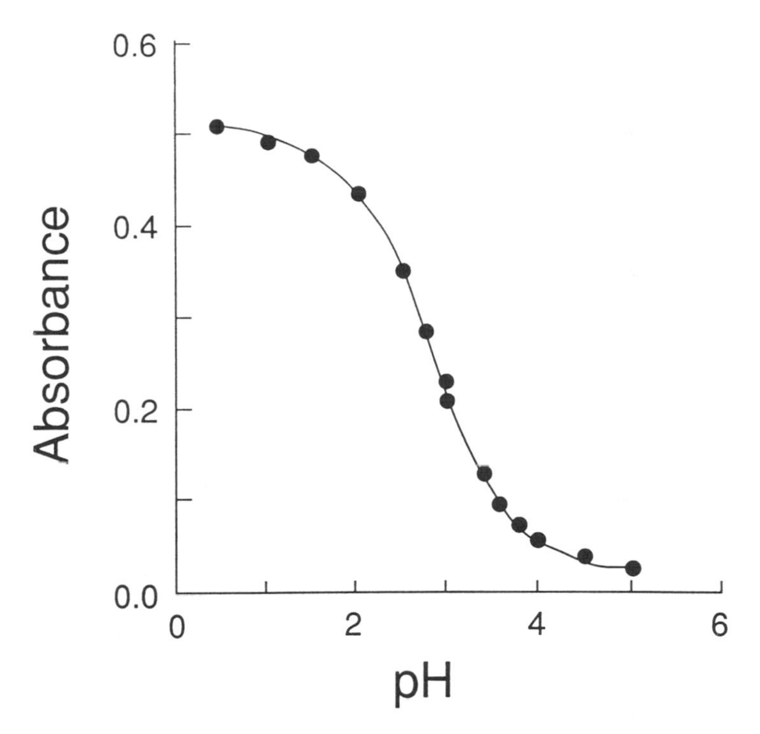

FIGURE 23 Absorbance of cranberry anthocyanins versus pH. (Adapted from Ref. 33.)

a failure to consider the equilibrium reactions among the four known anthocyanin structures (Figure 21, I).

(A)	⇌	(AH+)	⇌	(B)	⇌	(C)
quinonoid		flavylium		carbinol base		chalcone
(blue)		(red)		(colorless)		(colorless)

Heating shifts the equilibria toward the chalcone and the reverse reaction is slower than the forward reaction. It takes, for example, 12 hr for the chalcone of 1, 3, 5-diglycoside to reach equilibrium. Since determination of the amount of pigment remaining is generally based on measurement of the flavylium salt, an error is introduced if insufficient time is allowed for equilibrium to be attained [70].

The exact mechanism of thermal degradation of anthocyanin has not been fully elucidated. Three pathways have been suggested. Coumarin 3, 5-diglycoside is a common degradation product for anthocyanidin (cyanidin, peonidin, delphinidin, petunidin, and malvidin) 3,5-diglycoside (Fig. 24). In path (a) the flavylium cation is first transformed to the quinonoidal base, then to several intermediates, and finally to the coumarin derivative and a compound corresponding to the B-ring.

In path (b) (Fig. 24) the flavylium cation is first transformed to the colorless carbinol base, than to the chalcone and finally to brown degradation products. Path (c) (Fig. 24) is similar except degradation products of chalcone are first inserted. These three proposed mechanisms suggest that thermal degradation of anthocyanins depends on the type of anthocyanin involved and the degradation temperature.

Oxygen and Ascorbic Acid The unsaturated nature of the anthocyanidin structure makes

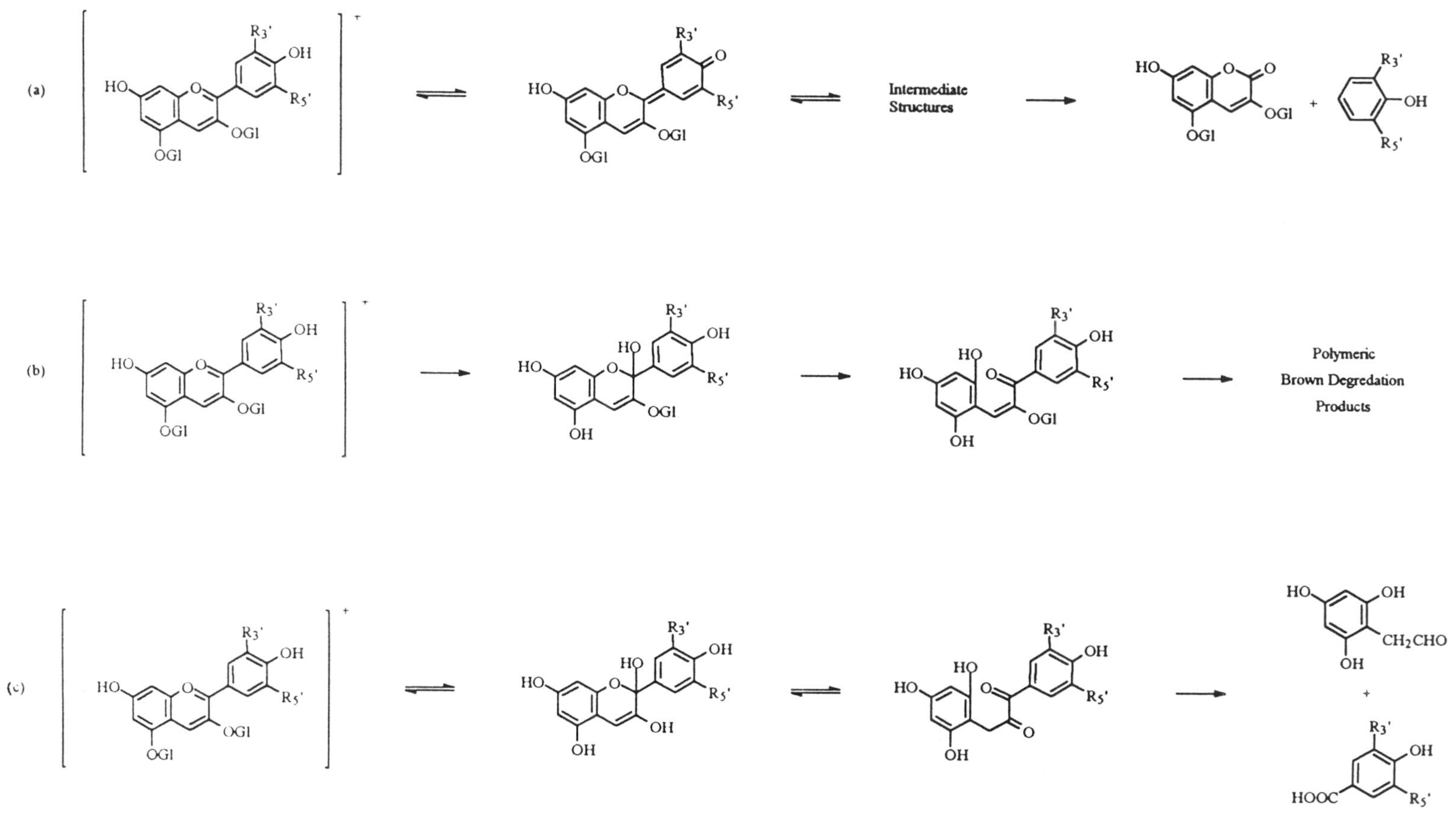

Figure 24 Mechanisms of degradation for anthocyanidin 3,5-diglycoside and for anthocyanidin 3-diglucosides. $R_{3'}$, $R_{5'}$ = -OH, -H -OCH_3 or -OGL; GL = glucosyl group. (From Ref. 48.)

it susceptible to molecular oxygen. It has been known for many years that when grape juice is hot-filled into bottles, complete filling of the bottles will delay degradation of the color from purple to dull brown. Similar observations have been made with other anthocyanin-containing juices. The positive effect of oxygen removal on retention of anthocyanin color has been further demonstrated by processing anthocyanin-pigmented fruit juices under nitrogen or vacuum [20, 99]. Also, the stability of pigments from Concord grape juice in a dry beverage is greatly enhanced when the product is packaged in a nitrogen atmosphere.

Though little information exists relating a_w to anthocyanin stability, stability was found to be greatest at a_w values in the range of 0.63 to 0.79 (Table 7).

It is known that ascorbic acid and anthocyanins disappear simultaneously in fruit juices, suggesting some direct interaction between the two molecules. This, however, has been discounted, and it is believed instead that ascorbic acid-induced degradation of anthocyanin results indirectly from hydrogen peroxide that forms during oxidation of ascorbic acid [48]. The latter reaction is accelerated by the presence of copper and inhibited by the presence of flavonols such as quercetin and quercitrin [97]. Conditions that do not favor formation of H_2O_2 during oxidation of ascorbic acid therefore account for anthocyanin stability in some fruit juices. H_2O_2 cleavage of the pyrylium ring by a nucleophilic attack at the C-2 position of the anthocyanin produces colorless esters and coumarin derivatives. These breakdown products may further degrade or polymerize and ultimately lead to a brown precipitate that is often observed in fruit juices.

Light It is generally recognized that light accelerates degradation of anthocyanins. This adverse effect has been demonstrated in several fruit juices and red wine. In wine it has been determined that acylated, methylated diglycosides are more stable than nonacylated diglycosides, which are more stable than monoglycosides [109]. Copigmentation (anthocyanin condensation with themselves or other organic compound) can either accelerate or retard the degradation, depending on the circumstances. Polyhydroxylated flavone, isoflavone, and aurone sulfonates exert a protective effect against photodegradation [104]. The protective effect is attributable to the formation of intermolecular ring interactions between the negatively charged sulfonate and the positively charged flavylium ion (Fig. 25). Anthocyanins substituted at the C-5 hydroxyl groups are more susceptible to photodegradation than those unsubstituted at this position. Unsubstituted or monosubstituted anthocyanins are susceptible to nucleophilic attack at the C-2 and/or C-4 positions.

Other forms of radiant energy such as ionizing radiation can also result in anthocyanin degradation [71].

TABLE 7 Effect of a_w on Color Stability of Anthocyanins[a] During Heating as Measured by Absorbance

Holding time at 43°C (min)	Absorbance at water activities of:						
	1.00	0.95	0.87	0.74	0.63	0.47	0.37
0	0.84	0.85	0.86	0.91	0.92	0.96	1.03
60	0.78	0.82	0.82	0.88	0.88	0.89	0.90
90	0.76	0.81	0.81	0.85	0.86	0.87	0.89
160	0.74	0.76	0.78	0.84	0.85	0.86	0.87
Percent change in absorbance (0–160 min)	11.9	10.5	9.3	7.6	7.6	10.4	15.5

[a]Concentration 700 mg/100 ml (1g commercially dried pigment powder).
Source: Adapted from Ref. 54.

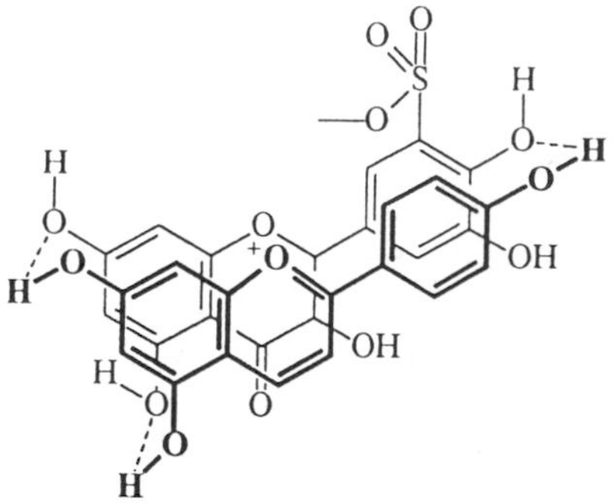

FIGURE 25 Molecular complex between anthocyanin and a polyhydroxyflavone sulfonate. (From Ref. 104.)

Sugars and Their Degradation Products Sugars at high concentrations, as in fruit preserves, stabilize anthocyanins. This effect is believed to result from a lowering of water activity (see Table 7). Nucleophilic attack of the flavylium cation by water occurs at the C-2 position, forming the colorless carbinol base. When sugars are present at concentrations sufficiently low to have little effect on a_w, they or their degradation products sometimes can accelerate anthocyanin degradation. At low concentrations, fructose, arabinose, lactose, and sorbose have a greater degradative effect on anthocyanins than do glucose, sucrose, and maltose. The rate of anthocyanin degradation follows the rate of sugar degradation to furfural. Furfural, which is derived form aldo-pentoses, and hydroxylmethylfurfural, which is derived from keto-hexoses, result from the Maillard reaction or from oxidation of ascorbic acid. These compounds readily condense with anthocyanin, forming brown compounds. The mechanism of this reaction is unknown. The reaction is very temperature dependent, is hastened by the presence of oxygen, and is very noticeable in fruit juices.

Metals Metal complexes of anthocyanin are common in the plant world and they extend the color spectrum of flowers. Coated metal cans have long been found to be essential for retaining typical colors of anthocyanins of fruits and vegetables during sterilization in metal cans. Anthocyanins with vicinal, phenolic hydroxyl groups can sequester several multivalent metals. Complexation produces a bathochromic shift toward the blue. Addition of $AlCl_3$ to anthocyanin solutions has been used as an analytical tool to differentiate cyanidin, petunidin, and delphinidin from pelargonidin, peonidin, and malvidin. The latter group of anthocyanidins does not possess vicinal phenolic hydroxyls and will not react with Al^{3+} (Fig. 20). Some studies have shown that metal complexation stabilizes the color of anthocyanin-containing foods. Ca, Fe, Al, and Sn have been shown to offer some protection to anthocyanins in cranberry juice; however, the blue and brown discoloration produced by tannin–metal complexes negates any beneficial effect [33].

A fruit discoloration problem referred to as "pinking" has been attributed to formation of metal anthocynin complexes. This type of discoloration has been reported in pears, peaches, and lychees. It is generally believed that pinking is caused by heat-induced conversion of colorless proanthocyanidins to anthocyanins under acid conditions, followed by complex formation with metals [65].

Sulfur Dioxide One step in the production of maraschino, candied and glacé cherries involves bleaching of anthocyanins by SO_2 at high concentrations (0.8–1.5%). The bleaching effect can be reversible or irreversible. An example of reversible bleaching occurs when fruits containing anthocyanins are preserved against microbial spoilage by holding them in a solution containing 500–2000 ppm SO_2. During storage the fruit loses its color, but the color can be restored by "desulfuring" (thorough washing) before further processing. In the reversible reaction a colorless complex is first formed. This reaction has been extensively studied, and it is believed that the reaction involves attachment of SO_2 at position C-4 (Fig. 26). The reason for suggesting involvement of the 4 position is that SO_2 in this position results in loss of color. The rate constant (k) for the discoloration reaction of cyanidin 3-glucoside has been calculated as 25,700 $^1/_{\mu Amps}$.

$$AH^+ + SO_2 \rightleftarrows AHSO_2^+ \quad k = \frac{[AH^+SO_2]}{[AH^+][SO_2]} = 25{,}700\ ^1/_{\mu Amps}$$

The large rate constant means that a small amount of SO_2 can quickly decolorize a significant amount of anthocyanin. Anthocyanins that are resistant to SO_2 bleaching either have the C-4 position blocked or exist as dimers linked through their 4 position [9]. The bleaching that occurs during production of maraschino or candied cherries is irreversible. The reason for this is not known.

Copigmentation Anthocyanins are known to condense with themselves (self-association) and other organic compounds (copigmentation). Weak complexes form with proteins, tannins, other flavonoids, and polysaccharides. Although most of these compounds themselves are not colored, they augment the color of anthocyanins by causing bathochromic shifts and increased light absorption at the wavelength of maximum light absorption. These complexes also tend to be more stable during processing and storage. The stable color of wine is believed to result from self-association of anthocyanin. Such polymers are less pH sensitive and, because the association occurs through the 4 position, are resistant to discoloration by SO_2.

Absorption of the flavylium cation and/or the quinonoidal base on a suitable substrate, such as pectins or starches, can stabilize anthocyanins. This stabilization should enhance their utility as potential food color additives. Other condensation reactions can lead to color loss. Certain nucleophiles, such as amino acids, phloroglucinol, and catechin, can condense with flavylium cations to yield colorless 4-substituted flav-2-enes [70]. Proposed structures are shown in Figure 27.

Enzyme Reactions

Enzymes have been implicated in the decolorization of anthocyanins. Two groups have been identified: glycosidases and polyphenol oxidases. Together they are generally referred to as anthocyanases. Glycosidases, as the name implies, hydrolyze glycosidic linkages, yielding sugars(s) and the agylcone. Loss of color intensity results from the decreased solubility of the anthocyanidins and their transformation to colorless products. Polyphenol oxidases act in the presence of *o*-diphenols and oxygen to oxidize anthocyanins. The enzyme first oxidizes the *o*-diphenol to *o*-benzoquinone, which in turn reacts with the anthocyanins by a nonenzymatic mechanism to form oxidized anthocyanins and degradation products (Fig. 28) [70].

Although blanching of fruits is not a general practice, anthocyanin destroying enzymes can be inactivated by a short blanch treatment (45–60 sec at 90–100°C). This has been suggested for sour cherries before freezing.

Very low concentrations of SO_2 (30 ppm) have been reported to inhibit enzymatic

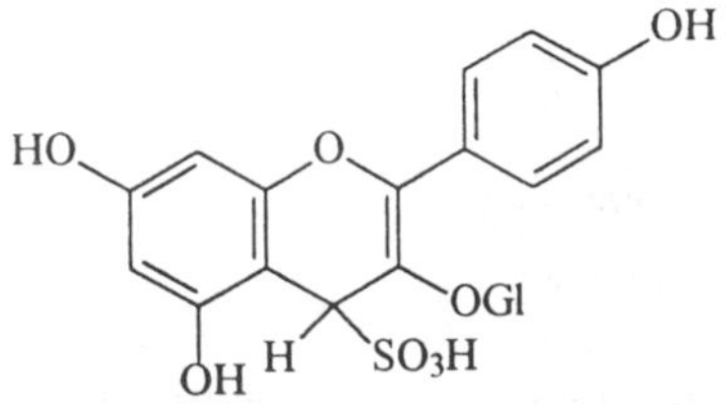

HO OH O OGl OH H SO_3H

FIGURE 26 Colorless anthocyanin–sulfate (-SO_2) complex.

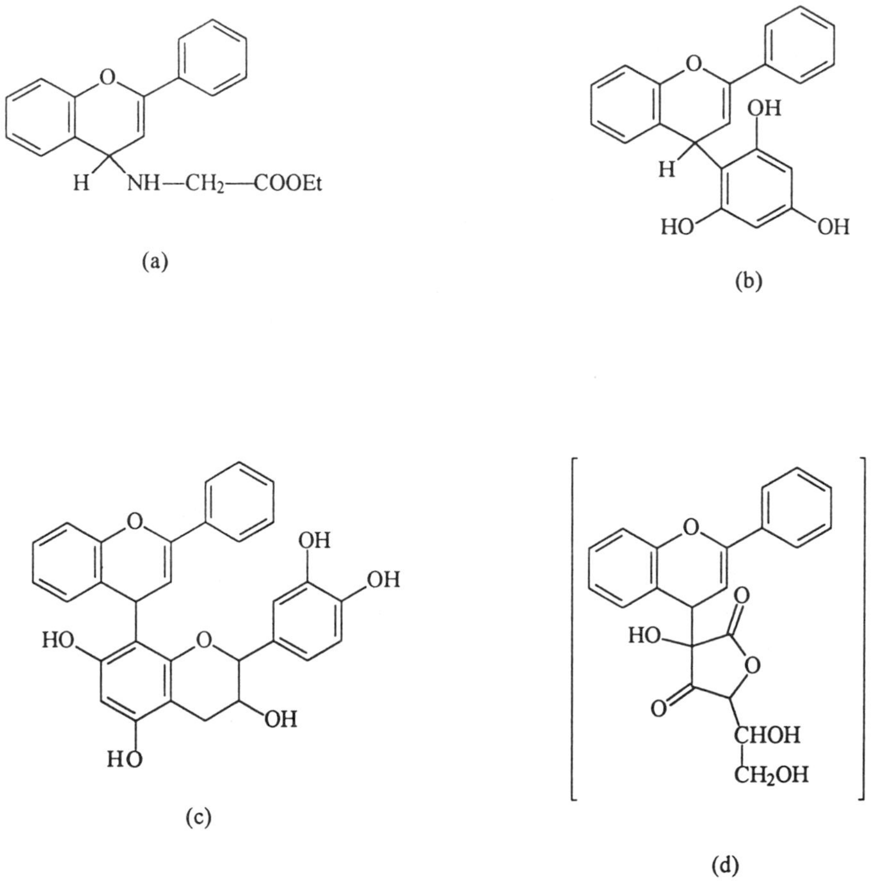

H NH—CH₂—COOEt

(a)

(b)

(c)

(d)

FIGURE 27 Colorless 4-substituted flav-2-enes resulting from the condensation of flavylium with (a) ethylglycine, (b) phloroglucinol, (c) catechin, and (d) ascorbic acid. (From Ref. 70.)

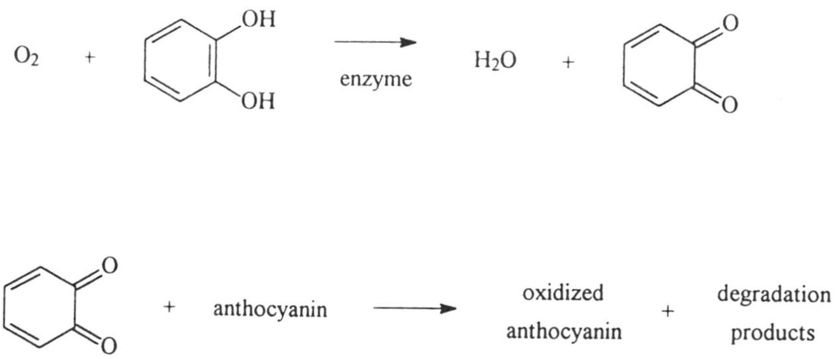

O_2 + [catechol] $\xrightarrow{\text{enzyme}}$ H_2O + [o-quinone]

[o-quinone] + anthocyanin ⟶ oxidized anthocyanin + degradation products

FIGURE 28 Proposed mechanisms of anthocyanin degradation by polyphenol oxidase. (From Ref. 80.)

degradation of anthocyanin in cherries [36]. Similarly, a heat-stabilization effect on anthocyanin has been noted when Na_2SO_3 is present [1].

10.2.4.2 Other Flavonoids

Anthocyanins, as previously mentioned, are the most prevalent flavonoids. Although most yellow colors in food are attributable to the presence of carotenoids, some are attributable to the presence of nonanthocyanin-type flavonoids. In addition, flavonoids also account for some of the whiteness of plant materials, and the oxidation products of those containing phenolic groups contribute to the browns and blacks found in nature. The term anthoxanthin (Gr. *anthos*, flower; *xanthos*, yellow) is also sometimes used to designate certain groups of yellow flavonoids. Differences among classes of flavonoids relate to the state of oxidation of the 3-carbon link (Fig. 18). Structures commonly found in nature vary from flavan-3-ols (catechin) to flavonols (3-hydroxyflavones) and anthocyanins. The flavonoids also include flavanone, flavononols or dihydroflavonol, and flavan-3, 4-diols (proanthocyanidin). In addition, there are five classes of compounds that do not possess the basic flavonoid skeleton, but are chemically related, and therefore are generally included in the flavonoid group. These are the dihydrochalcones, chalcones, isoflavones, neoflavones, and aurones. Individual compounds within this group are distinguished, as with anthocyanins, by the number of hydroxyl, methoxyl, and other substituents on the two benzene rings. Many flavonoid compounds carry a name related to the first source from which they were isolated, rather than being named according to the substituents of the respective aglycone. This inconsistent nomenclature has brought about confusion in assigning compounds to various classes.

Physical Properties

The light absorption characteristics of flavonoid classes clearly demonstrates the relationship of color and unsaturation within a molecule and the impact of auxochromes (groups present in a molecule which deepens the color). In the hydroxy-substituted flavans catechin and proanthocyanin, the unsaturation is interrupted between the two benzene rings, and therefore, the light absorption is similar to that of phenols, which exhibit maximum light absorption between 275 and 280 nm (Fig. 29a). In the flavanone naringenin, the hydroxyl groups only occur in conjunction with the carbonyl group at C-4 and therefore do not exert their auxochromic characteristics Fig. 29b. Therefore its light absorption is similar to that of flavans. In the case of the flavone luteolin (Fig. 29c), the hydroxyl groups associated with both benzene rings exert their auxochormic characteristics through the conjugation of C-4. Light absorption of longer wavelength (350 nm) is associated with the B-ring, while that of shorter wavelength is associated with the A-ring. The hydroxyl group at C-3 in the flavonol quercetin causes a further shift to a still longer wavelength (380 nm) for maximum light absorption, compared to that of the flavones (Fig. 29c). The flavonols therefore appear yellow if present at high enough concentration. Acylation and/or glycosylation results in further shifts in light absorption characteristics.

As previously mentioned, flavonoids of these types can become involved in copigmentation, and this occurrence has a major impact on many hues in nature. In addition, flavonoids, like anthocyanins, are chelators of metals. Chelation with iron or aluminum increases the yellow saturation. Luteolin when chelated with aluminum is an attractive yellow (390 nm).

Importance in Foods

Non-anthocyanin (NA) flavonoids make some contribution to color in foods; however, the paleness of most NA- flavonoids generally restricts their overall contribution. The whiteness of vegetables such as cauliflower, onion, and potato is attributable largely to NA- flavonoids, but their contribution to color through copigmentation is more important. The chelation characteris-

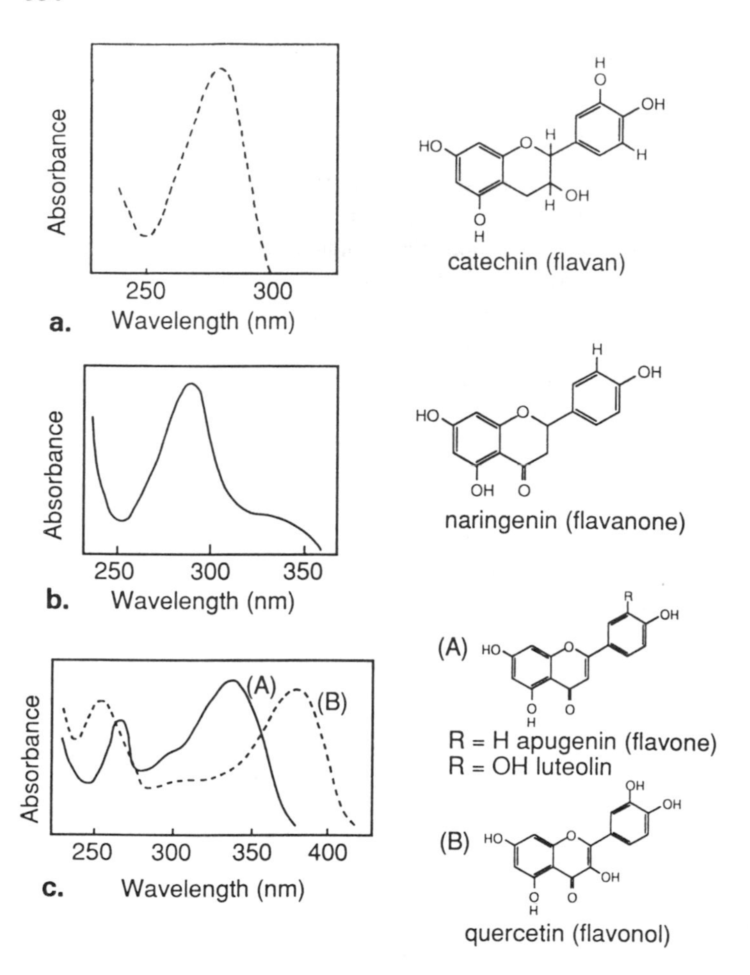

FIGURE 29 Absorption spectra of specific flavonoids.

tics of these compounds can contribute both positively or negatively to the color of foods. For example, rutin (3-rutiniside of quercetin) causes a greenish-black discoloration in canned asparagus when it complexes with ferric-state iron. The addition of a chelator such as ethylenediamine tetraacetic acid (EDTA) will inhibit this undesirable color. The tin complex of rutin has a very attractive yellow color, which contributed greatly to the acceptance of yellow wax beans until the practice of canning wax beans in plain tin cans was eliminated. The tin–rutin complex is more stable than the iron complex; thus the addition or availability of only very small amounts of tin would favor formation of the tin complex.

The color of black ripe olives is due in part to the oxidative products of flavonoids. One of the flavonoids involved is luteolin 7-glucoside. Oxidation of this compound and formation of the black color occur during fermentation and subsequent storage [8]. Other very important functions of flavonoids in foods are their antioxidant properties and their contribution to flavors, particularly bitterness.

Proanthocyanadins

Consideration of proanthocyanidins under the general topic of anthocyanins is appropriate. Although these compounds are colorless they have structural similarities with anthocyanidins. They

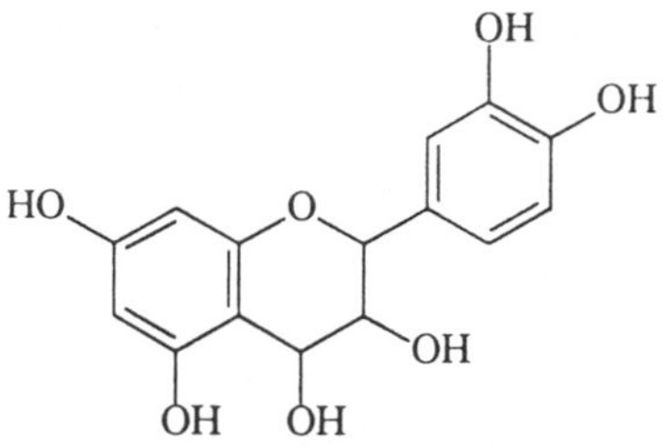

flavan-3,4-diol (leucoanthocyanidin)

FIGURE 30 Basic structure of proanthocyanidin.

can be converted to colored products during food processing. Proanthocyanidins are also referred to as leucoanthocyanidins or leucoanthocyanins. Other terms used to describe these colorless compounds are anthoxanthin, anthocyanogens, flavolans, flavylans, and flaylogens. The term leucoanthocyanidin is appropriate if it is used to designate the monomeric flaven-3, 4-diol (Fig. 30), which is the basic building block proanthocyanidins. The latter can occur as dimers, trimers, or higher polymers. The intermonomer linkage is generally through carbons 4 → 8 or 4 → 6.

Proanthocyanidins were first found in cocoa beans, where upon heating under acidic conditions they hydrolzye into cyanidin and (–)-epicatechin (Fig. 31) [30]. Dimeric proanthocyanidins have been found in apples, pears, kola nuts, and other fruits. These compounds are known to degrade in air or under light to red-brown stable derivatives. They contribute significantly to the color of apple juice and other fruit juices, and to astringency in some foods. To produce astringency, proanthocyanidins of two to eight units interact with proteins. Other proanthocyanidins found in nature will yield on hydrolysis the common anthocyanidins—pelargonidin, petunidin, or delphinidin.

Tannins

A rigorous definition of tannins does not exist, and many substances varying in structure are included under this name. Tannins are special phenolic compounds and are given this name

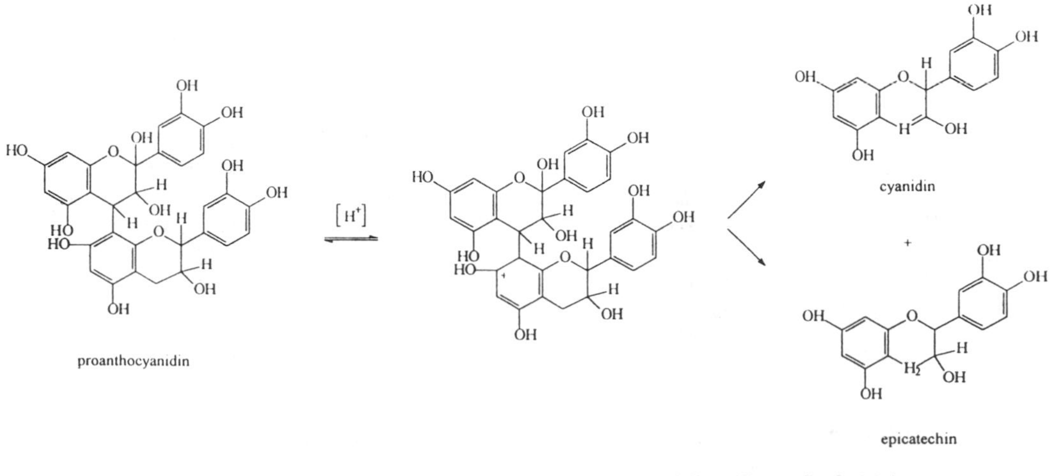

proanthocyanidin

cyanidin

epicatechin

FIGURE 31 Mechanism of acid hydrolysis of proanthocyanidin. (From Ref. 30.)

simply by virtue of their ability to combine with proteins and other polymers such as polysaccharides, rather than their exact chemical nature. They are functionally defined, therefore, as water-soluble polyphenolic compounds with molecular weights between 500 and 3000 that have the ability to precipitate alkaloids, gelatin, and other proteins. They occur in the bark of oak trees and in fruits. The chemistry of tannins is complex. They are generally considered as two groups: (a) proanthocyanidins, also referred to as "condensed tannins" (previously discussed), and (b) glucose polyesters of gallic acid of hexahydroxydiphenic acids (Fig. 32). The latter group is also known as hydrolyzable tannins, because they consist of a glucosidic molecule bonded to different phenolic moities. The most important example is glucose bonded to gallic acid and the lactone of its dimer, ellagic acid.

Tannins range in color from yellowish-white to light brown and contribute to astringency in foods. They contribute to the color of black teas when catechins are converted to aflavins and arubigins during fermentation. Their ability to precipitate proteins makes them valuable as clarifying agents.

10.2.4.4 Quinoids and Xanthones

Quinones are phenolic compounds varying in molecular weight from a monomer, such as 1,4-benzoquinone, to a dimer, 1,4-naphthaquinone, to a trimer, 9,10-anthraquinone, and finally to a polymer represented by hypericin (Fig. 33). They are widely distributed in plants, specifically trees, where they contribute to the color of wood. Most quinones are bitter in taste. Their contribution to color of plants is minimal. They contribute to some of the darker colors, to the yellows, oranges, and browns of certain fungi and lichens, and to the reds, blues, and purples of sea lilies and coccid insects. Compounds with complex substitutents such as naphthaquinone and anthraquinones occur in plants, and these have deep purple to black hues. Further color changes can occur in vitro under alkaline conditions by the addition of hydroxyl groups. Xanthone pigments are yellow, phenolic pigments and they are often confused with quinones and flavones because of their structural characteristics. The xanthone mangiferin (Fig. 34) occurs as a glucoside in mangoes. They are easily distinguishable from quinones by their spectral characteristics.

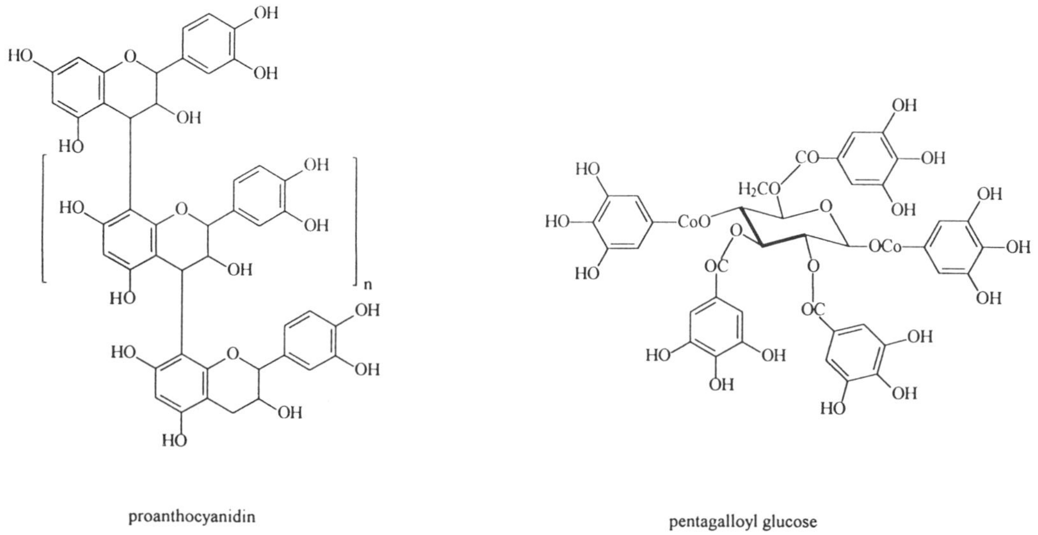

proanthocyanidin

pentagalloyl glucose

FIGURE 32 Structure of tannins.

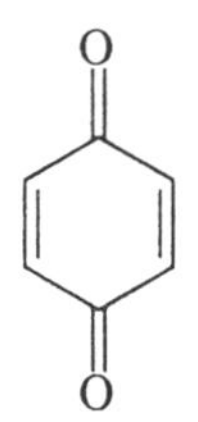

1,4-benzoquinone

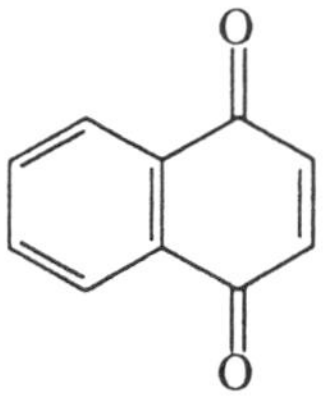

1,4-napthoquinone

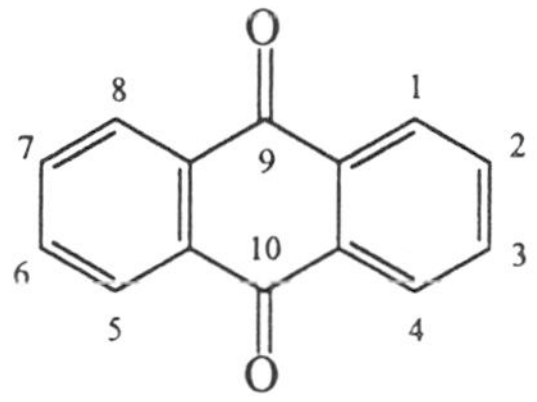

9,10-anthraquinone

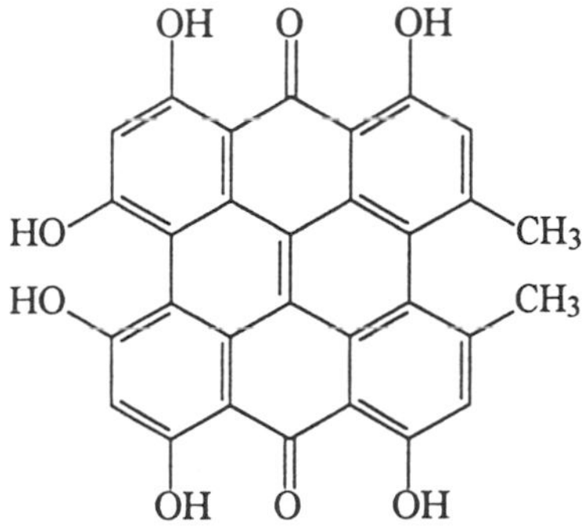

hypericin

FIGURE 33 Structure of quinones.

10.2.5 Betalaines

10.2.5.1 Structure

Plants containing betalaines have colors similar to plants containing anthocyanins. Betalaines are a group of pigments containing betacyanins (red) and betaxanthins (yellow) and their color is not affected by pH, contrary to the behavior of anthocyanins. They are water soluble and exist as internal salts (zwitterions) in the vacuoles of plant cells. Plants containing these pigments are restricted to 10 families of the order *Centrospermae*. The presence of betalaines in plants is mutually exclusive of the occurrence of anthocyanins. The general formula for betalaines (Fig. 10.35a) represents condensation of a primary or secondary amine with betalamic acid (Fig. 35b). All betalaine pigments can be described as a 1,2,4,7,7-pentasubstituted 1,7-diazaheptamethin system (Fig. 35c). When R′ does not extend conjugation of the 1,7-diazaheptamethine system, the compound exhibits maximum light absorption at about 480 nm, characteristic of yellow

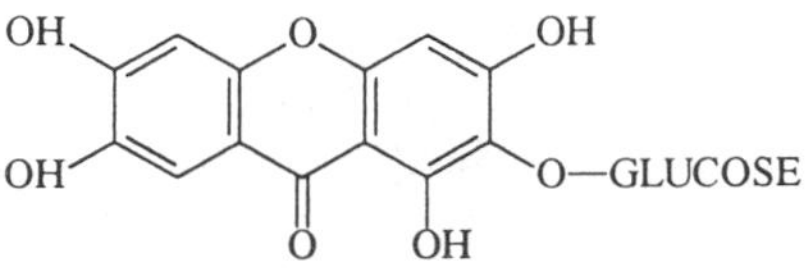

FIGURE 34 Structure of mangiferin.

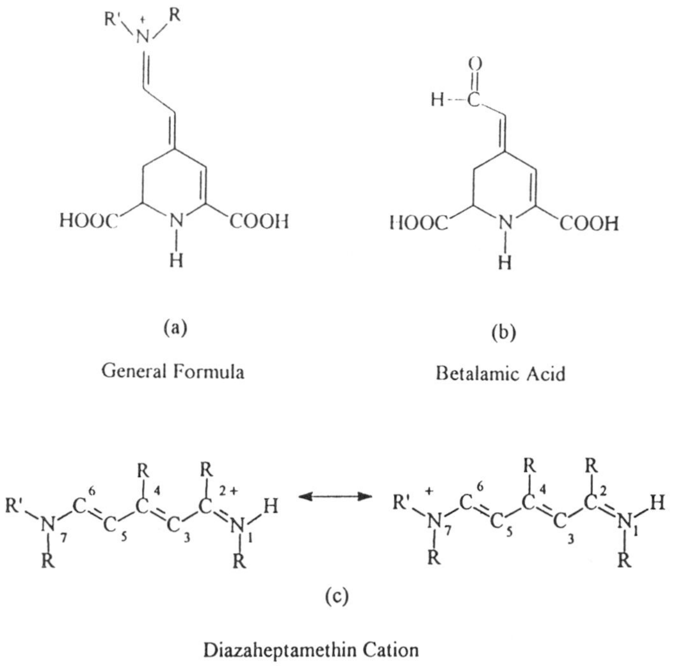

(a) General Formula

(b) Betalamic Acid

(c) Diazaheptamethin Cation

FIGURE 35 General formulas of betalaines.

betaxanthins. If the conjugation is extended at R′ the maximum light absorption shifts to approximately 540 nm, characteristic of red betacyanines.

Betacyanins are optically active because of the two chiral carbons C-2 and C-15 (Fig. 36). Hydrolysis of betacyanin leads to either betanidin (Fig. 36), or the C-15 epimer isobetanidine (Fig. 36d), or a mixture of the two isomeric aglycones. These aglycones are shared by all betacyanines. Differences between betacyanins are found in their glucoside residue. Common vegetables containing betalaines are red beet and amaranth. The latter is either consumer fresh as "greens" or at the mature state as grain. The most studied betalaines are those of the red beet. The major betacyanins in the red beet are betanin and isobetanin (Fig. 36b,e), while in amaranth they are amarathin and isoamaranthin (Fig. 36c,f).

The first betaxanthin isolated and characterized was indicaxanthin (Fig. 37a). Structurally these pigments are very similar to betacyanins. Betaxathins differ from betacyanins in that the indole nucleus is replaced with an amino acid. In the case of indicaxanthin the amino acid is proline. Two betaxanthines have been isolated from beet, vulgaxanthin I and II (Fig. 37b). They differ from indicaxanthine in that the proline has been replaced by glutamine or glutamic acid, respectively. Although only a few betaxanthines have been characterized to date, considering the number of amino acids available, it is likely a large number of different betaxanthins exist.

10.2.5.2 Physical Properties

Betalaines absorb light strongly. The absorptivity value ($A^{1\%}_{1cm}$) is 1120 for betanin and 750 for vulgaxanthine, suggesting high tinctorial strength in the pure state. The spectra of betanine solutions at pH values between 4.0 and 7.0 do not change, and they exhibit maximum light absorbance at 537–538 nm. No change in hue occurs between these pH values. Below pH 4.0, the absorption maximum shifts toward a slightly shorter wavelength (535 nm at pH 2.0). Above pH 7.0, the absorption maximum shifts toward a longer wavelength (544 nm at pH 9.0).

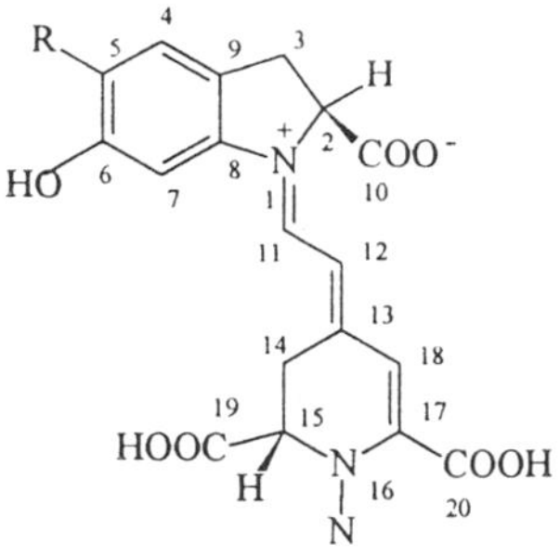

(a) Betanidin, R = -OH
(b) Betanin, R = -Glucose
(c) Amaranthin, R = 2'-Glucuronic acid-Glucose

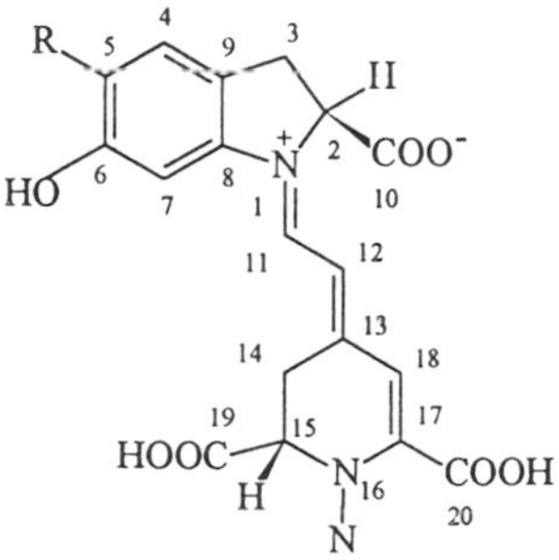

(d) Isobetanidin, R = -OH
(e) Isobetanin, R = -Glucose
(f) Isoamaranthin, R = 2'-Glucuronic acid-Glucose

FIGURE 36 Structure of betacyanins.

10.2.5.3 Chemical properties

Like other natural pigments, betalaines are affected by several environmental factors.

Heat and/or Acidity

Under mild alkaline conditions betanin degrades to betalamic acid (BA) and cyclodopa-5-*O*-glucoside (CDG) (Fig. 38). These two degradation products also form during heating of acidic betanin solutions or during thermal processing of products containing beet root, but more slowly.

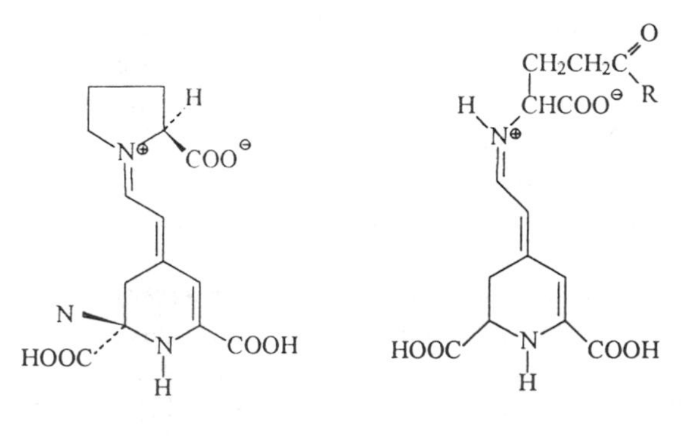

(a) Indicaxanthin

(b) Vulgaxanthin-I, R = $-NH_2$
Vulgaxanthin-II, R = -OH

FIGURE 37 Structure of betaxanthins.

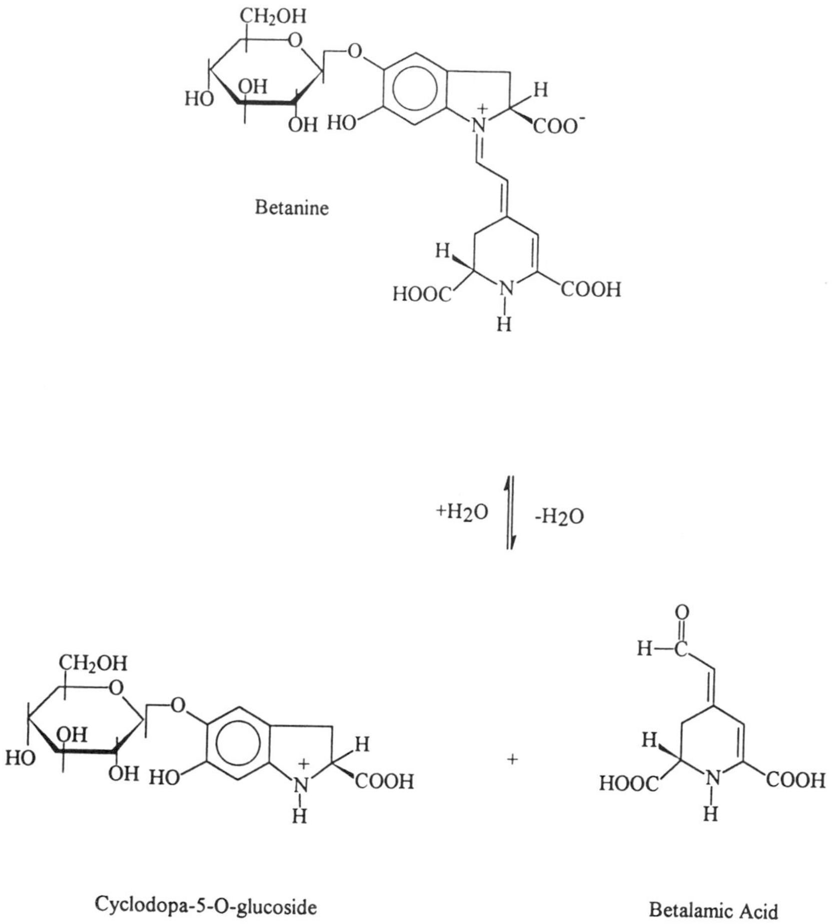

Betanine

$+H_2O$ ⇌ $-H_2O$

Cyclodopa-5-O-glucoside + Betalamic Acid

FIGURE 38 Degradation reaction of betanin.

The reaction is pH dependent (Table 8), and the greatest stability is in the pH range of 4.0–5.0. It should be noted that the reaction requires water; thus when water is unavailable or limited, betanine is very stable. It follows that a decrease in water activity will cause a decrease in the degradation rate of betanin [79]. An a_w of 0.12 (moisture content of about 2%; dry weight basis), is recommended for optimal storage stability of the pigments in beet powder [19].

No studies have been done concerning the degradation mechanism of betaxanthines. Since both betacyanines and betaxanthines possess the same general structure, the mechanisms for degradation of betanin is likely to apply.

Degradation of betanin to BA and CDG is reversible, and therefore partial regeneration of the pigment occurs following heating. The mechanism proposed for regeneration involves a Schiff-base condensation of the aldehyde group of BA and the nucleophilic amine of CDG (Fig. 38). Regeneration of betanin is maximized at an intermediate pH range (4.0–5.0) [43]. It is of interest that canners traditionally, for reasons not necessarily known to them, examine canned beets several hours after processing to evaluate color, thus taking advantage of regeneration of the pigment(s).

Betacyanins, because of the C-15 chiral center (Fig. 36), exist in two epimeric forms.

TABLE 8 Effect of Oxygen and pH on the Half-Life Values of Betanine in Aqueous Solution at 90°C

	Half-life values of betanine (min)	
pH	Nitrogen	Oxygen
3.0	56 ± 6	11.3 ± 0.7
4.0	115 ± 10	23.3 ± 1.5
5.0	106 ± 8	22.6 ± 1.0
6.0	41 ± 4	12.6 ± 0.8
7.0	4.8 ± 0.8	3.6 ± 0.3

Source: Adapted from Ref. 43.

Epimerization is brought about by either acid or heat. It would therefore be expected that during heating of a food containing betanine the ratio of isobetanine to betanine would increase.

It has also been shown that when betanine in aqueous solution is heated, decarboxylation can occur. Evidence for this evolution of carbon dioxide is loss of the chiral center. The rate of decarboxylation increases with increasing acidity [44]. Degradation reactions of betanin under acid and/or heat are summarized in Figure 39.

Oxygen and Light

Another major factor that contributes to degradation of betalaines is the presence of oxygen. Oxygen in the head space of canned beets has long been known to accelerate pigment loss. In solutions containing a molar excess of oxygen over betanin, betanin loss follows apparent first-order kinetics. Betanin degradation deviates from first-order kinetics when the molar oxygen concentration is reduced to near that of betanin. In the absence of oxygen, stability is

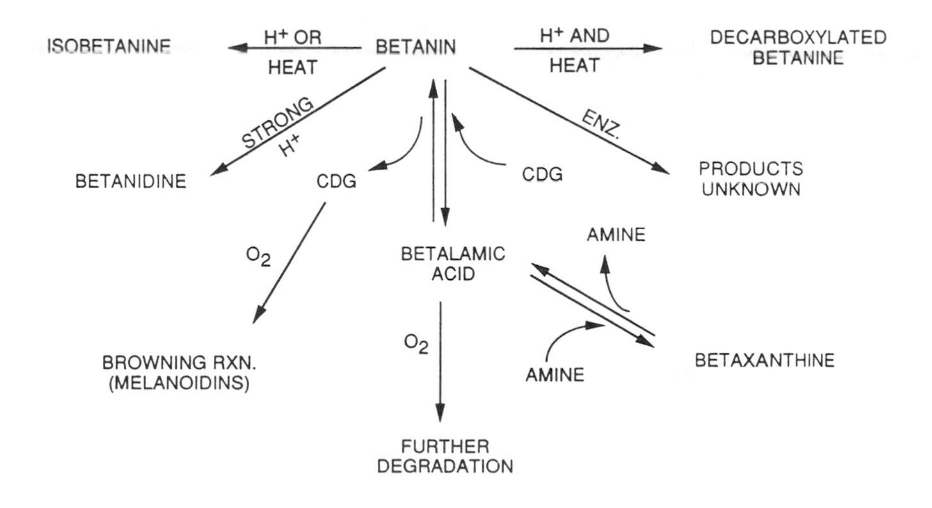

FIGURE 39 Degradation of betanin under acid and/or heat.

increased. Molecular oxygen has been implicated as the active agent in oxidative degradation of betanine. Active oxygen species such as singlet oxygen or superoxide anion are not involved. Degradation of betanine in the presence of oxygen is influenced by pH (Table 8).

Oxidation of betalaines accelerates in the presence of light. The presence of antioxidants, such as ascorbic acid and isoascorbic acid, improves their stability. Because copper and iron cations catalyze oxidation of ascorbic acid by molecular oxygen, they detract from the effectiveness of ascorbic acid as a protector of betalaines. The presence of metal chelators (EDTA or citric acid) greatly improves the effectiveness of ascorbic acid as a stabilizer of betalaines [2, 6].

Several phenolic antioxidants, including butylated hydroxyanisole, butylated hydroxytoluene, catechin, quercetin, nordihydroguaiartic acid, chlorogenic acid, and alpha-tocopherol, inhibit free-radical chain autoxidation. Since free-radical oxidation does not seem to be involved in betalaine oxidation, these antioxidants are, not surprisingly, ineffective stabilizers of betanine. Similarly, sulfur-containing antioxidants such as sodium sulfite and sodium metabisulfite are not only ineffective stabilizers, they hasten loss of color. Sodium thiosulfite, a poor oxygen scavenger, has no effect on betanine stability. Thioproprionic acid and cysteine also are ineffective as stabilizers of betanine. These observations confirm that betanine does not degrade by a free-radical mechanism. The susceptibility of betalaines to oxygen has limited their use as food colorants.

Conversion of Betacyanin to Betaxanthin

In 1965 it was shown that the betaxanthin indicaxanthin could be formed from betacyanin, betanine, and an excess of proline in the presence of 0.6 *N* ammonium hydroxide under vacuum. This was the first conclusive evidence of a structural relationship between betacyanin and betaxanthin. It was further demonstrated that formation of betaxanthin from betanin involved condensation of the betanin hydrolysis product betalamic acid, and an amino acid (Fig. 40) [84, 85].

Information on the stability of betaxanthins is limited. Similar to betacyanins, their stability is pH dependent. Shown in Figure 41 are the heat stability differences between the betacyanin, betanin, and the betaxanthin vulgaxanthin under similar experimental conditions. The mechanism in Figure 40 suggests that an excess of the appropriate amino acid will shift the equilibrium toward the corresponding betaxanthin and will reduce the quantity of betalamic acid in solution. An excess of an amino acid increases the stability of the betaxanthin formed by reducing the amount

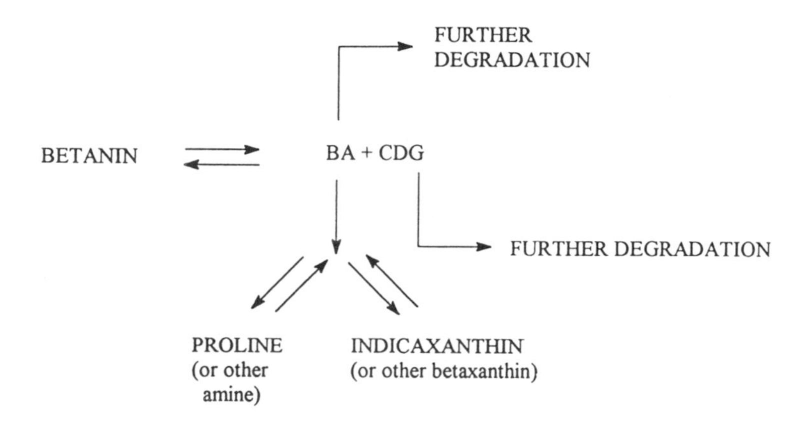

FIGURE 40 Formation of indicaxanthin from betanin in excess of proline.

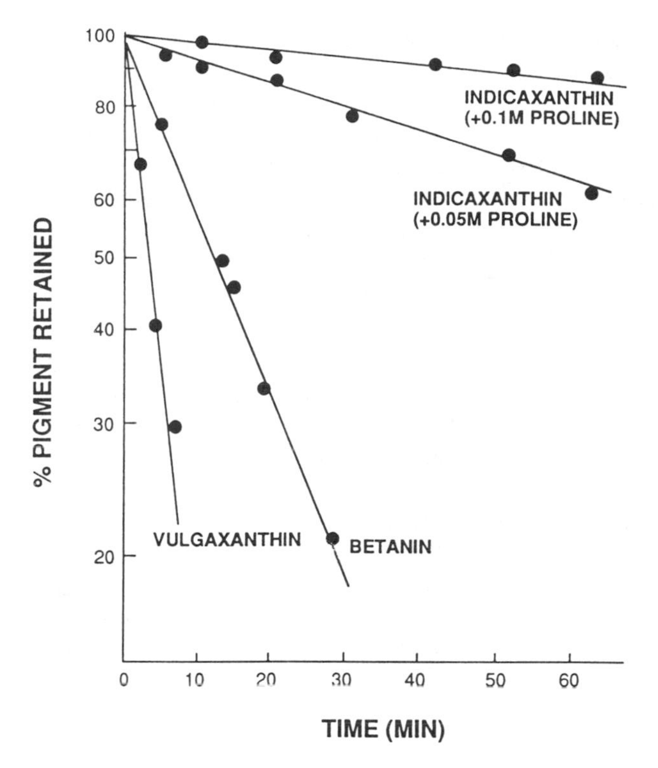

FIGURE 41 Stability comparison of betanine vulgaxanthin I and of indicaxanthin in presence of proline in solution at pH 5.0, 90°C, under atmospheric conditions.

of betalamic acid available for degradation. This effect is illustrated in the two upper curves of Figure 41. Conversion of betacyanin to betaxanthin can occur in protein-rich foods, and this accounts for the loss of red color in these types of foods that are colored with betalaines.

10.3 FOOD COLORANTS

10.3.1 Regulatory Aspects

10.3.1.1 United States

In the United States, colorant use is controlled by the 1960 Color Additive Amendment to the U.S. Food, Drug and Cosmetic Act of 1938. The amendment deals with two categories of colorants, certified colors and colors exempt from certification. Certified colors are synthetic dyes, and with the exception of one dye, they are not found in nature. Certification means that the dye meets specific government quality standards. Samples of each production batch must be submitted to an FDA laboratory for determination of compliance. If the batch is in compliance it is assigned an official lot number. Certified dyes are further classified as either permanently or provisionally listed. A "provisionally" approved certified dye can be legally used pending completion of all scientific investigation needed for determination for or against permanent approval. The same considerations apply to lakes.

Colorants exempt from certification are natural pigments or specific synthetic dyes that are nature identical. An example of the latter is β-carotene, which is widely distributed in nature but also can be synthesized to achieve a "nature identical" substance.

The Color Additive Amendment includes a simplified nomenclature for certified dyes. Rather than the use of long and difficult common names, certified dyes are referred to by a number and the abbreviation FD&C, D&C, or Ext. (external) D&C. FD&C stands for food, drugs, and cosmetics, and these colorants may be used in either foods, drugs, or cosmetics. D&C and Ext. (external) D&C dyes can be used only in drugs or cosmetics. Thus, the certified dye sunset yellow FC has the designation FD&C Yellow No. 6. The current list of permitted certified dyes contains seven colorants for general use (Table 9). Two additional dyes, FD&C Orange B and FD&C Citrus No. 2, may be used; however, their use is restricted to specific applications. FD&C Orange B may be used only for coloring the casings or surfaces of frankfurters and sausages, and its use in these applications is restricted to no more than 150 ppm by weight of the finished product. FD&C Citrus Red No. 2 may be used only for coloring the skins of oranges not intended or used for processing, and its use in this application is restricted to no more than 2 ppm based on the whole fruit.

The amendment further requires that when colorants are used, the foods to which they are added must be labeled with "color added." The Nutritional Labeling Regulations of 1973 require that if a food contains FD&C Yellow No. 5, the dye must appear on the label by name; for all other dyes the collective term "color added" is acceptable. The reason yellow no. 5 received special treatment is because a small percentage of the population is allergic to the dye. The regulation also allows use of abbreviated FD&C color designation as listed in Table 9. Adoption of the Nutritional Labeling and Education Act of 1990, which became effective in 1994, makes mandatory the individual listing of certified colors, by their abbreviated names, as well as the listing of colors exempt from certification. Color additives currently exempt form certification are listed in Table 10.

10.3.1.2 International

Colors are added to foods in many countries of the world, but the type of colorants permitted for use varies greatly among countries. Since international trade is becoming increasingly important, color legislation is now of international concern. Unfortunately, a worldwide list of permitted color additives does not exist; therefore, color additives have, in some instance, become trade barriers for foods. In the United States, for example, FD&C Red No. 40 is permitted for food use, whereas FD&C Red No. 2, since 1976, is no longer permitted. The United States and

TABLE 9 Certified Color Additives Currently Permitted for General Use

Federal name	Status		Abbreviated name
	Dye	Lake	
FD&C Blue No. 1	Permanent	Provisional	Blue 1
FD&C Blue No. 2	Permanent	Provisional	Blue 2
FD&C Green No. 3	Permanent	Provisional	Green 3
FD&C Red No. 3	Permanent	[a]	Red 3
FD&C Red No. 40	Permanent	Provisional	Red 40
FD&C Yellow No. 5	Permanent	Provisional	Yellow 5
FD&C Yellow No. 6	Permanent	Provisional	Yellow 6

[a]Use of lake of FD&C Red No. 3 was terminated effective January 29, 1990.

TABLE 10 U.S. Color Additives Currently Exempt from Certification

Color	Use limitation[a]
Annatto extract	—
Dehydrated beet (beet powder)	—
Canthaxanthin	Not to exceed 66 mg/kg of solid or pint of liquid food
β-Apo-8′-carotenal	Not to exceed 15 mg/lb or liter of food
β-Carotene	—
Caramel	33 mg/kg
Cochineal extract; carmine	—
Toasted, partially defatted, cooked cottonseed flour	—
Ferrous gluconate	Colorant for ripe olives
Grape-skin extract (enocianina)	Colorant for beverages only
Synthetic iron oxide	Not to exceed 0.25% by weight of pet food
Fruit juice	—
Vegetable juice	—
Dried algae metal	Chicken feed only; to enhance yellow color of skin and eggs
Tagetes (Aztec marigold) and extract	Chicken feed only; to enhance yellow color of skin and eggs
Carrot oil	—
Corn endosperm oil	Chicken feed only; to enhance yellow color of skin and eggs
Paprika	—
Paprika oleoresin	—
Riboflavin	—
Saffron	—
Titanium dioxide	Not to exceed 1% by weight of food
Turmeric	—
Turmeric oleoresin	—

[a]Unless otherwise stated the colorant may be used in an amount consistent with good manufacturing practices, except for foods where use is specified by a standard of identity.

Canada are currently the only nations in which FD&C red no. 40 is a legal color additive. At the other extreme, Norway prohibits the use of any synthetic dye in the manufacturing of foods.

Legislative authorities of the European Economic Community (EEC) have attempted to achieve uniform color additive legislation for Common Market Countries. Each permitted color additive has been assigned an E-number (E = Europe). Listed in Table 11 are the currently permitted synthetic dyes, their E-number, equivalent FD&C number, if applicable, and the current regulatory status in various countries. Similar information for EEC natural colorants is given in Table 12. In reviewing these tables, it must be remembered that a colorant may be restricted to one or more specific products. An EEC general-use colorant also may not be approved by every country in the EEC. It is apparent that greater latitude of use of both synthetic and natural colorants is currently allowed among EEC countries than in the United States, Canada, and Japan. The student is referred to the Food Additives Tables (Elsevier, Amsterdam) for further details. These tables are periodically updated, and Appendix I of these tables is a list of all permitted colorants for countries having specific color additive legislation.

TABLE 11 Synthetic Color Additives Currently Permitted by the European Economic Community (EEC)

Name	EEC number	FDA number	Regulatory legal status[a,b]			
			EEC	United States	Canada	Japan
Erythrosine	E123	FD&C Red No. 3	+	+	+	+
Brilliant blue FCF	—	FD&C Blue No. 1	–[c]	+	+	+
Indigotine	E132	FD&C Blue No. 2	+	+	+	+
Tartrazine	E102	FD&C Yellow No. 5	+[d]	+	+	+
Quinoline yellow	E104	FD&C Yellow No. 6	+[e]	–	–	–
Allura red	—	FD&C Red No. 40	–	+	+	–
Yellow 2G	E107	—	+[f]	–	–	–
Ponceau 4R	E124	—	+[d]	–	–	+
Carmoisine	E122	—	+[d,e,g]	–	–	–
Amaranth	E123	FD&C Red No. 2	+[d]	–	+	–
Red 2G	E128	—	+[f]	–	–	–
Patent blue	E131	—	+[e]	–	–	–
Green S	E142	—	+[d,e,g]	–	–	–
Brown FK	E154	—	+[f]	–	–	–
Chocolate brown HT	E155	—	+[h]	–	–	–
Black PN	E151	—	+[d,e]	–	–	–

[a]+, Permitted for food use, (in some countries limited to specific foods); –, prohibited for food use.
[b]No synthetic colorants permitted in Norway.
[c]Permitted in Denmark, Ireland, Netherlands.
[d]Not permitted in Finland.
[e]Not permitted in Portugal.
[f]Permitted in Ireland only.
[g]Not permitted in Sweden.
[h]Permitted in Ireland and Netherlands.

TABLE 12 Some Natural Color Additives Currently Permitted by the European Economic Community (EEC)

Name	EEC number	Registered status[a]			
		EEC	United States	Canada	Japan
Anthocyanin, juice concentrate	E163	+[b]	+	+	+
Beet pigment (beet red)	E162	+	+	+	+
Carbon black	E153	+	−	+	+
β-Apo-8′-Carotenal	E160e	+[b]	+	+	−
Ethyl ester of β-apo-8′-carotenic acid	E160f	+[b]	−	+	−
Annatto	E160b	+	+	+	+
β-Carotene	E160a	+	I	+	+
Canthaxanthin	E161	+	−	+	+
Chlorophyll	E140	+[d]	−	+	+
Copper complexes of chlorophyll	E141	+	−	−	+
Caramel	E150	+	+	+	+
Cochineal	E120	+[d,e]	+	+	+
Iron oxide	E172	+[b,d]	−	+	+
Titanium dioxide	E171	+[b]	+	+	+
Riboflavin	E101	+[b]	+	+	+
Cucumin	E100	+	+	+	+

[a]+, Permitted for food use in some countries limited to specific foods; −, prohibited for food use.
[b]Not permitted in Portugal.
[c]Not permitted in Austria, Norway.
[d]Not permitted in Finland.
[e]Not permitted in Sweden.

The Food and Agriculture Organization (FAO) and the World Health Organization (WHO) have attempted to harmonize food regulations among countries through their Codex Alimentarius. FAO and WHO have also devised "acceptable daily intakes" (ADI) for food additives, including colorants (Table 13). Additives have been placed in one of three categories or "lists," A, B, or C.

TABLE 13 Acceptable Daily Intake of Some Synthetic and Natural Colorants

Synthetic color	Intake[a] (mg/kg)	Natural color	Amount (mg/kg)
Tartrazine	7.5	β-Apo-8′-carotenal	2.5
Sunset yellow FCF	5.0	β-Carotene	5.0
Amaranth	1.5	Canathaxanthin	25.0
Erythrosine	1.25	Riboflavin	0.5
Brilliant blue FCF	12.5	Chlorophyll	GMP[b]
Indigotine	2.5	Caramel	GMP

[a]Daily intake over lifetime without risk.
[b]Amount consistent with "good manufacturing practice."

List A has two subparts: A-1 and A-2. Listed in subpart A-1 are those additives for which an ADI value has been established and approved. This list contains 15 colorants, including six artificial dyes. Additives in this list are considered not to present a hazard to consumers. Listed in subpart A-2 are those additives for which the safety evaluation is not complete, and therefore these additives have a provisional use status. Included in this category are colorants derived from beet, annatto, and turmeric, all of which are approved for use in the United States.

List B includes additives for which evaluation is pending. List C, like list A, has two subparts. Listed in subpart C-1 are those additives that, in the opinion of the FAO/WHO, are unsafe for use in foods. Listed in subpart C-2 are additives for which use is restricted and specific limitations have been established.

Worldwide efforts toward establishing safety of colorants hopefully will lead to internationally accepted regulations for colorant use in foods.

10.3.2 Properties of Certified Dyes

The safety of certified colors has received much public attention in recent years. The root of the concern can in part be attributed to the unfortunate association of synthetic colors to the original term "coal-tar" dyes. The public's concept of coal-tar is a thick black substance unsuitable for use in foods. The fact is that raw materials for synthesis of colors are highly purified before use. The final product is a specific chemical that bears little relationship to the term coal tar.

Certified dyes fall into four basic chemical classes: azo, triphenylmethane, xanthine, or indigo type dyes. Listed in Table 14 are the FD&C dyes, their chemical class, and some of their properties. The structures are shown in Figure 42. Listed in Table 15 are solubility and stability data for EEC dyes.

A simplified sequence for chemical synthesis of FD&C Green No. 3, a triphenylmethane dye, is given in Figure 43. In manufacturing any dye the major difficulty is to meet the specifications of purity given for certification in the United states (Code of Federal Regulations, Title 21, Part 70-83). The color manufacturing industry not only meets these purity specifications, but most manufacturers exceed them.

The pure dye content of a typical certified colorant is 86–96%. Variation of 2–3% in total dye content of a basic color is of little practical significance since such variation has no significant effect on the ultimate color of a product. The moisture content of dye powders is between 4 and 5%. The salt (ash) content of the dye powder is approximately 5%. The high ash content comes from the salt used to crystallize (salt out) the colorant. Although it would be technically possible to remove the sodium chloride used, such steps would be costly and would have minimal benefit.

All water-soluble FD&C azo dyes are acidic, and their physical properties are quite similar. Chemically they are reduced easily by strong reducing agents and therefore are susceptible to oxidizing agents. FD&C triphenylmethane dyes (FD&C Green No. 3 and FD&C Blue No. 1) are similar in structure, differing only in one –OH group. Differences in solubility and stability are therefore minor. Substitution of a sulfonic acid group for a hydroxyl group in either of these dyes improves stability to light and resistance to alkali. Alkali decolorization of a triphenylmethane dye involves formation of a colorless carbinol base (Figure 44). The *ortho*-substituted sulfonic acid group sterically hinders access of the hydroxyl ion to the central carbon atom, thus preventing formation of the carbinol base.

FD&C Red No. 3 is the only xanthine type-dye. The structure of red 3 suggests that the dye is insoluble in acids, quite stable to alkali, and exhibits strong fluorescence. The lake of red 3 is no longer permitted for use in foods because of toxicologic concerns. Although the dye is permanently listed, its long term future is questionable.

TABLE 14 Certified Colorants and Their Chemical and Physical Properties

Common name and FD&C number	Dye type	Solubility (g/100 ml)[a]								Stability[b]									
		Water		Propolyne glycol		Alcohol		Glycerine		At pH:									
		25°C	60°C	25°C	60°C	25°C	60°C	25°C	60°C	3.0	5.0	7.0	8.0	Light	10% AcOH	10% NaOH	250 ppm SO_2	1% Ascorbic acid	1% Sodium benzoate
Brilliant blue, blue no. 1	Triphenyl-methane	20.0	20.0	20.0	20.0	0.35	0.20	20.0	20.0	4	5	5	5	3	5	4	5	4	6
Indigotine, blue no. 2	Indigo	1.6	2.2	0.1	0.1	In	0.007	20.0	20.0	3	3	2	1	1	1	2^{b2}	1	2	4
Fast green, green no. 3	Triphenyl-methane	20.0	20.0	20.0	20.0	0.01	0.03	1.0	1.3	4	4	4	4^{b1}	3	5	2^{b1}	5	4	6
Erythrosine, red no. 3	Xanthine	9.0	17.0	20.0	20.0	In	0.01	20.0	20.0	In	In	6	6	2	In	2	In	In	5
Allura red, red no. 40	Azo	22.0	26.0	1.5	1.7	0.001	0.113	3.0	8.0	6	6	6	6	5	5	3^{b1}	6	6	6
Tartrazine, yellow no. 5	Azo	20.0	20.0	7.0	7.0	In	0.201	18.0	18.0	6	6	6	6	5	5	4	3	3	6
Sunset yellow, yellow no. 6	Azo	19.0	20.0	2.2	2.2	In	0.001	20.0	20.0	6	6	6	6	3	5	5	3	2	6

[a]In, Insoluble.

[b]1 = fades; 2 = considerable fade; 3 = appreciable fade; 4 = slight fade; 5 = very slight fade; 6 = no change; b1 = hue turns blue; b2 = hue turns yellow.

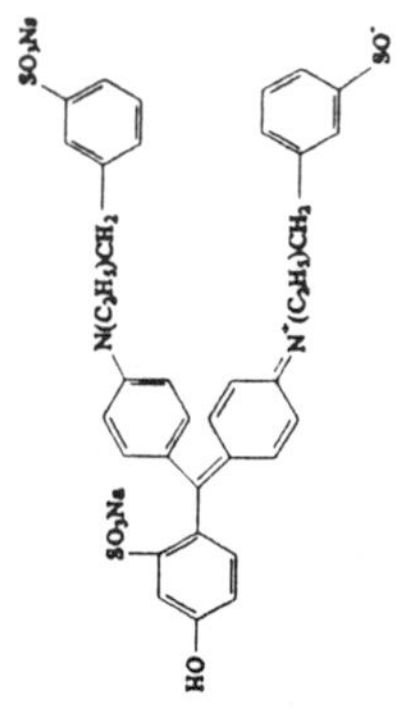

FD&C Blue No. 1

FD&C Blue No. 2

FD&C Green No. 3

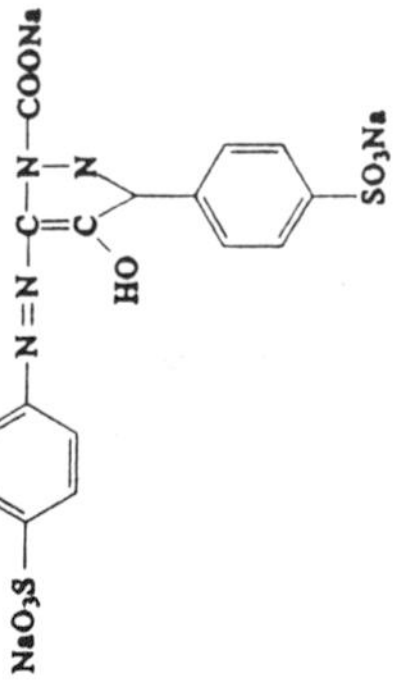

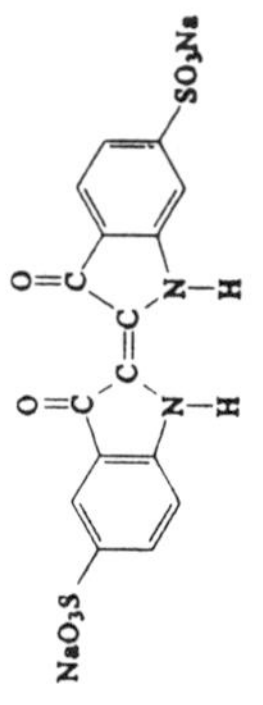

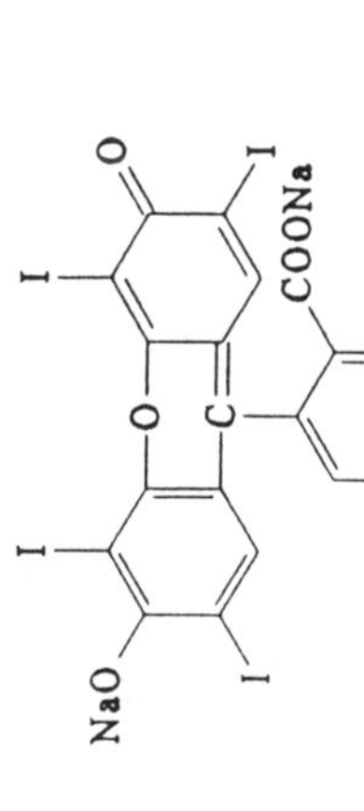

FD&C Red No. 3

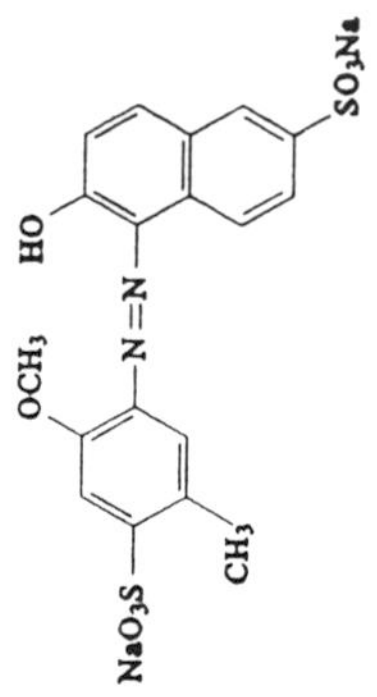

FD&C Red No. 40

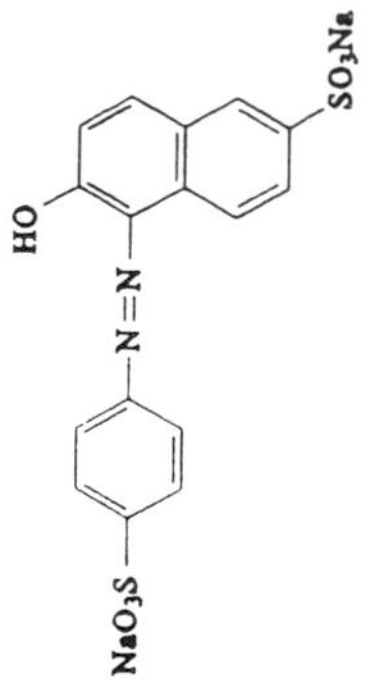

FD&C Yellow No. 5

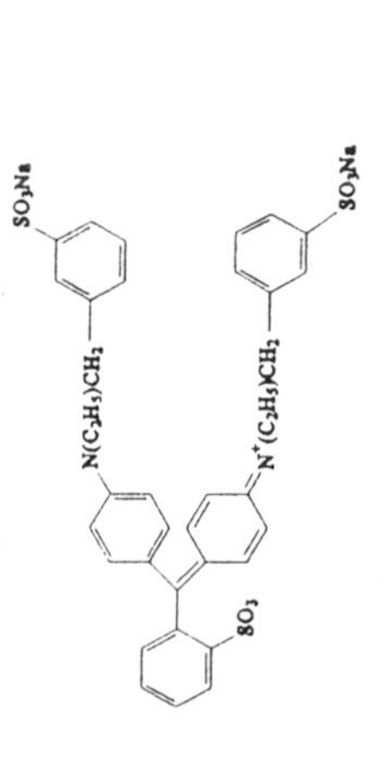

FD&C Yelllow No. 6

FD&C Blue No. 2 is the only indigoid-type dye. It is made from indigo, one of the oldest known and most extensively utilized natural pigments. The pigment is derived from various species of the indigo plant found in India. Until its synthesis, extracted dye was a major product of commerce. Blue 2 is made by sulfonating indigo, yielding 5,5′-indigotin disulfonate (Fig. 45). The color is a deep blue, compared to the greenish-blue of FD&C blue no. 1. The dye has the lowest water solubility and poorest light resistance of any of the FD&C dyes, but is relatively resistant to reducing agents.

In general, conditions most likely to cause discoloration or precipitation of certified dyes are the presence of reducing agents or heavy metals, exposure to light, excessive heat, or exposure to acid or alkali. Many of the conditions causing failure of the dyes can be prevented in foods. Reducing agents are most troublesome. Reduction of the chromophores of azo and triphenylmethane dyes is shown in Figure 46. Azo dyes are reduced to the colorless hydrazo form or sometimes to the primary amine. Triphenylmethane dyes are reduced to the colorless leuco base. Common reducing agents in foods are monosaccharides (glucose, fructose), aldehydes, ketones, and ascorbic acid.

Free metals can combine chemically with many dyes causing loss of color. Of most concern are iron and copper. The presence of calcium and/magnesium can result in the formation of insoluble salts and precipitates.

10.3.3 Use of Certified Dyes

Better uniformity in incorporating a water-soluble dye in foods is achieved if the dye is first dissolved in water. Distilled water should be used to prevent precipitation. Liquid colors of various strengths can be purchased from manufacturers. Dye concentration in these preparations usually does not exceed 3% to avoid overcoloring. Citric acid and sodium benzoate are commonly added to liquid preparations to prevent spoilage.

Many foods contain low levels of moisture making it impossible to completely dissolve and uniformly distribute a dye. The result is a weak color and/or a speckled effect. This is a potential problem in hard candy that has a moisture content of $< 1\%$. The problem is averted by using solvents other than water, such as glycerol or propylene glycol (Table 14). A second approach to overcoming problems of poor dispersion of dyes in foods with low moisture contents is the use of "lakes." Lakes exist as dispersions in food rather than in solution. They range in dye content from 1 to 40%. A large dye content does not necessarily lead to intense color. Particle size is of key importance—the smaller the particle size, the finer will be the dispersion and the more intense will be the color. Special grinding techniques used by color manufacturers have made it possible to prepare lakes with a mean particle size of less than 1 μm.

As with dyes, predispersion of lakes in glycerol, propylene glycol, or edible oils is often required. Predispersion helps prevent agglomeration of particles, and thereby helps develop full color intensity and reduces the incidence of speckled products. Lake dispersions vary in dye content from 15 to 35%. A typical lake dispersion may contain 20% FD&C lake A, 20% FD&C lake B, 30% glycerol, and 30% propylene glycol, resulting in a final dye content of 16%.

Color manufacturers also prepared color pastes or solid cubes of dyes or lakes. A paste is made with the addition, for example, of glycerol as a solvent and powdered sugar to increase viscosity. Colorants, in the form of cubes, are achieved by adding gums and emulsifiers to color dispersions in the manufacturing process.

FIGURE 42 Structures of certified color additives currently permitted for general use in the United States.

TABLE 15 Chemical and Physical Properties of Common EEC Dyes

	Solubility (g/100 ml) at 16°C				Stability[a]				
								pH	
Name and EEC number	Water	Propylene glycol	Alcohol	Glycerine	Light	Heat	SO_2	3.5/4.0	8.0/9.0
Quinoline yellow, E104	14	<0.1	<0.1	<0.1	6	5	4	5	2
Ponceau 4R, E124	30	4	<0.1	0.5	4	5	3	4	1
Carmoisine, E122	8	1	<0.1	2.5	5	5	4	4	3
Amaranth, E123	5	0.4	<0.1	1.5	5	5	3	4	3
Patent blue, E131	6	2	<0.1	3.5	7	5	3	1	2
Green S, E142	5	2	0.2	1.5	3	5	4	4	3
Chocolate brown HT, E156	20	15	insoluble	5	5	5	3	4	4
Brilliant black BN, E151	5	1	<0.1	<0.5	6	1	1	3	4

[a]1 = fades; 2 = considerable fade; 3 = appreciable fade; 4 = slight fade; 5 = very slight fade; 6 = no change.

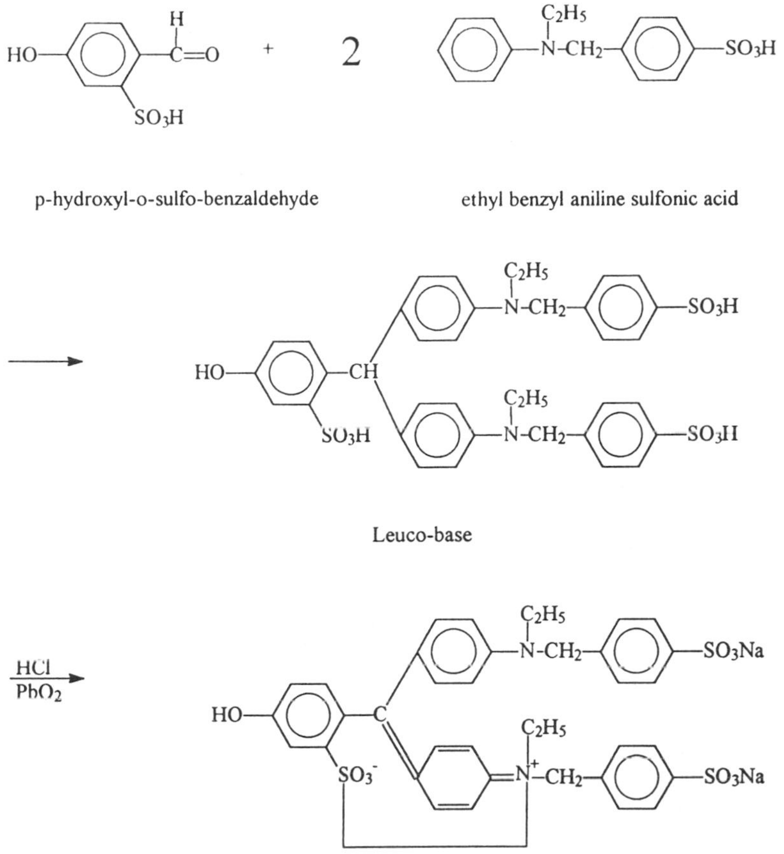

HO— —C=O (H) + 2 SO3H

—N-CH2— —SO3H (C2H5)

p-hydroxyl-o-sulfo-benzaldehyde

ethyl benzyl aniline sulfonic acid

→ HO— —CH SO3H

C2H5 —N-CH2— —SO3H

C2H5 —N-CH2— —SO3H

Leuco-base

HCl / PbO2 → HO— —C SO3⁻

C2H5 —N-CH2— —SO3Na

C2H5 =N⁺-CH2— —SO3Na

FD&C Green No. 3 (Fast Green)

FIGURE 43 Synthesis of FD&C Green No. 3 (fast green).

10.3.4 Colors Exempt from Certification

A brief description of each of the colorants listed in Table 10 follows.

Annatto extract is the extract prepared from Annatto seed, *Bixa orellana* L. Several food-grade solvents can be used for extraction. Supercritical carbon dioxide extraction has been tested as an alternative to using standard organic solvents. This technology, however, has not yet

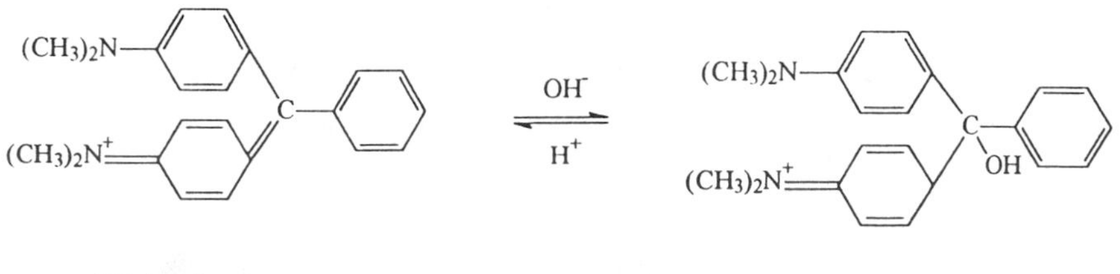

$(CH_3)_2N$—

$(CH_3)_2N^+$=

C—

OH^- / H^+

$(CH_3)_2N$—

$(CH_3)_2N^+$=

C— OH

Triphenylmethane dye

Carbinal base

FIGURE 44 Formation of a colorless carbinol base from a triphenylmethane dye.

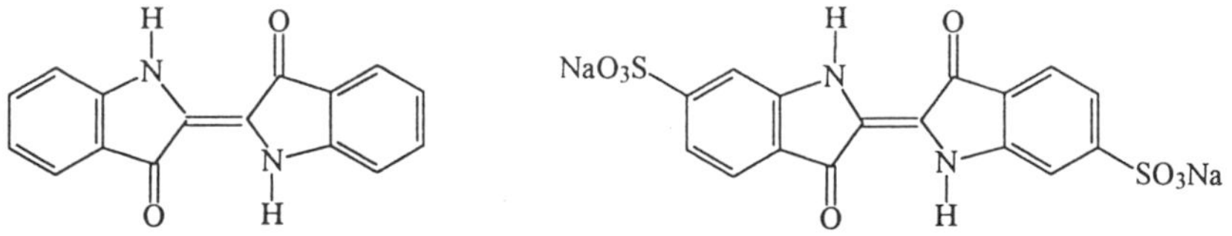

indigo blue (indigotin)

FD&C Blue No. 2
(Indigo Carmine)
5,5'-indigotin disulfonate

FIGURE 45 Structures of indigoid type dyes.

been commercialized. The main pigment in annatto extract is the carotenoid bixin. Upon saponification of bixin the methyl ester group is hydrolyzed and the resulting diacid is called norbixin (Fig. 47). Bixin and norbixin differ in solubility and form the basis for oil- and water-soluble annatto colors, respectively.

Dehydrated beet is obtained by dehydrating the juice of edible whole beets. The pigments in beet colorants are betalains [betacyanins (red) and betaxanthins (yellow)]. The ratio of betacyanin/betaxanthin will vary depending on the cultivar and maturity of beets. Beet colorant can also be produced under the category of "vegetable juice." This type of beet colorant is obtained by concentrating beet juice under vacuum to a solid content sufficient to prevent spoilage (about 60% solids).

Canthaxanthin (β-carotene-4,4′dione), β-apo-8′-carotenal, and β-carotene are synthesized carotenoids and are regarded as "nature identical." Structures of these compounds are shown in Figure 48.

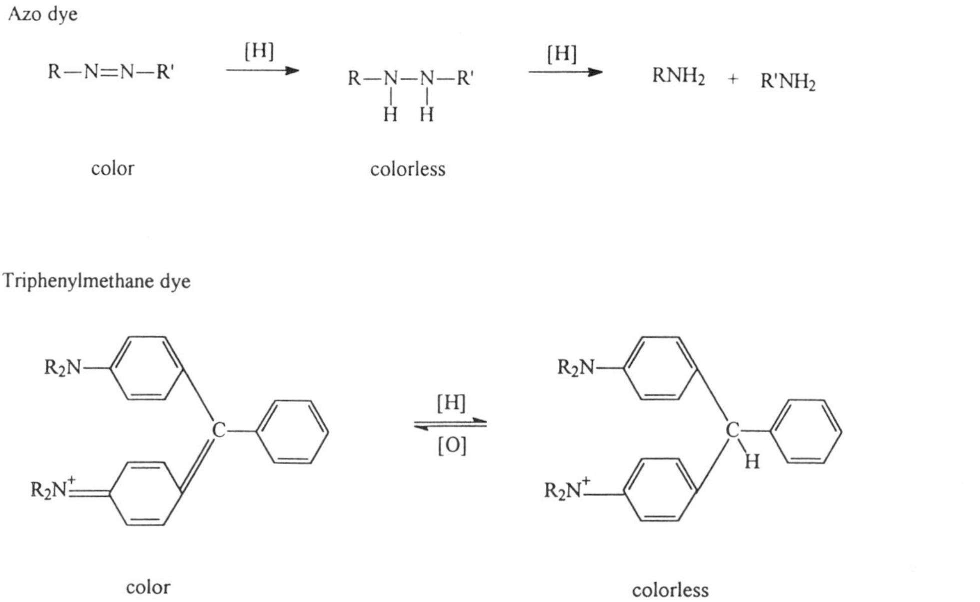

Azo dye

$$R{-}N{=}N{-}R' \xrightarrow{[H]} R{-}NH{-}NH{-}R' \xrightarrow{[H]} RNH_2 + R'NH_2$$

color — colorless

Triphenylmethane dye

color $\underset{[O]}{\overset{[H]}{\rightleftharpoons}}$ colorless

FIGURE 46 Reduction of azo or triphenylmethane dyes to colorless products.

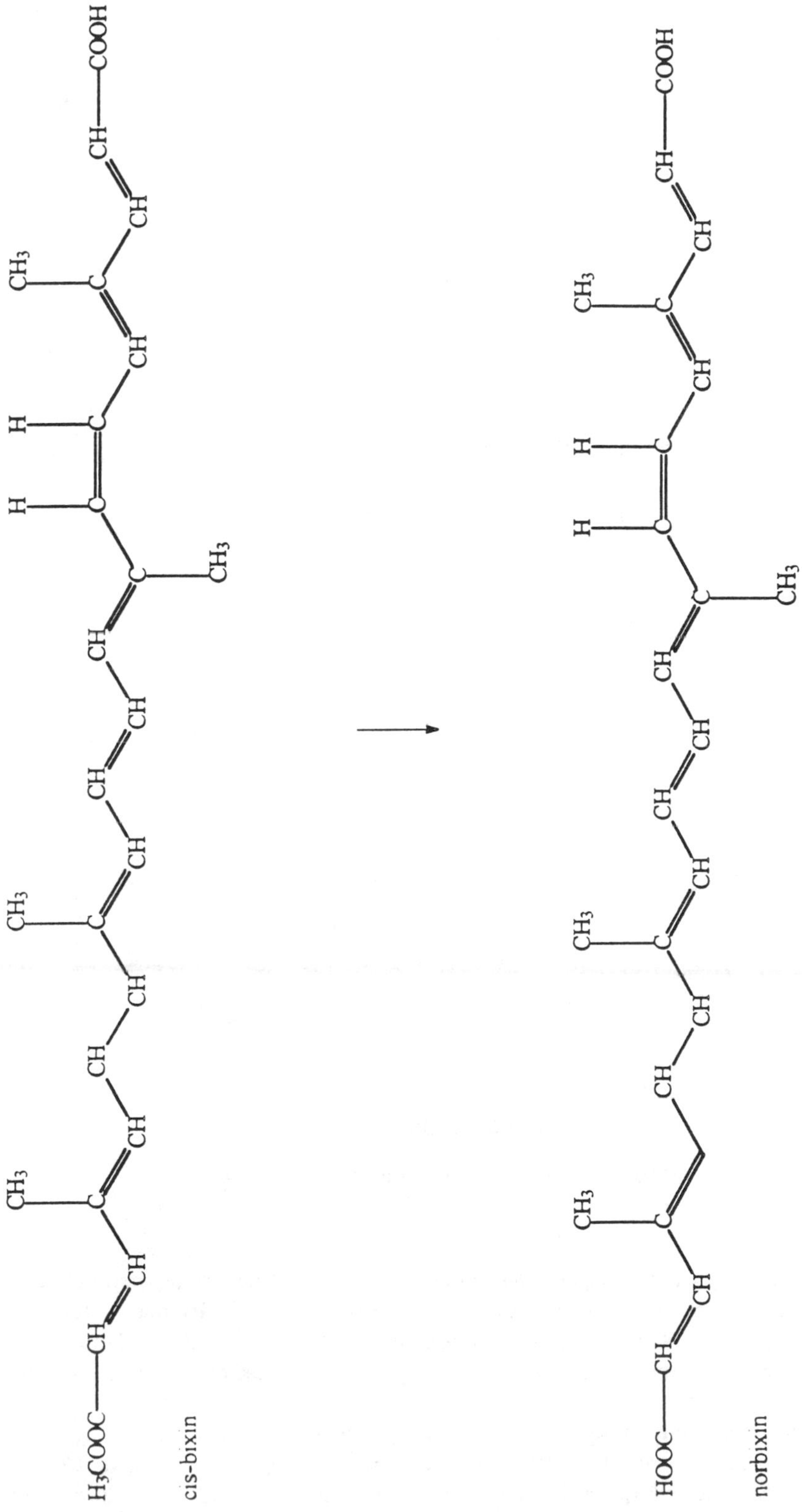

cis-bixin

H_3COOC—CH=CH—C(CH_3)=CH—CH=CH—C(CH_3)=CH—CH=CH—CH=C(CH_3)—CH=CH(cis)—CH=C(CH_3)—CH=CH—COOH

→

norbixin

HOOC—CH=CH—C(CH_3)=CH—CH=CH—C(CH_3)=CH—CH=CH—CH=C(CH_3)—CH=CH(cis)—CH=C(CH_3)—CH=CH—COOH

FIGURE 47 Formation of norbixin from bixin.

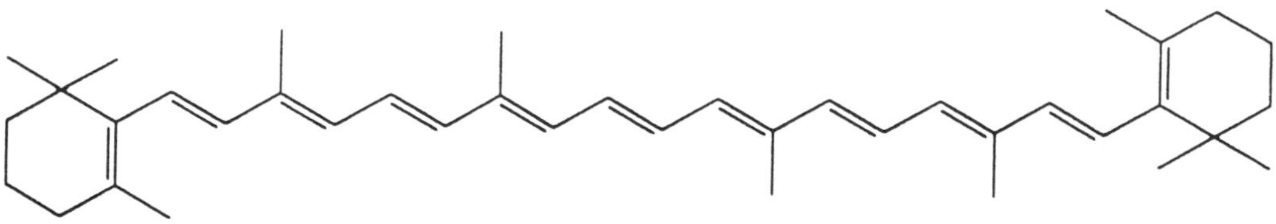

β-carotene (trans)

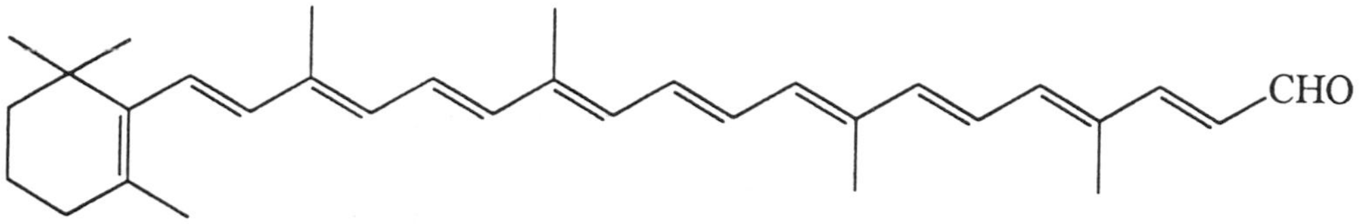

CHO

β-Apo-8'-carotenal (C_{30}) (trans)

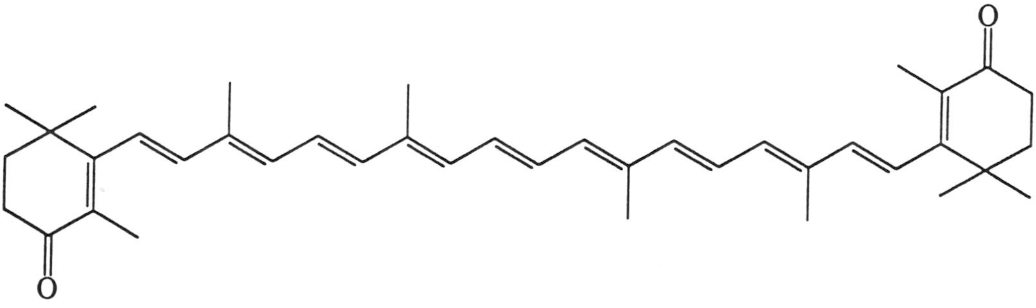

O

O

Canthaxanthin (trans)

Figure 48 Structure of "nature identical" carotenoids.

Cochineal extract is the concentrate produced from an aqueous–alcoholic extract of the cochineal insect, *Dactylopius coccus costa*. The coloring is principally due to carminic acid, a red pigment (Figure 49). The extract contains about 2–3% carminic acid. Colorants with carminic acid concentrations of up to 50% are also produced. These colorants are sold under the name carmine colors.

Caramel is a dark-brown liquid produced by heat-induced caramelization of carbohydrates.

Toasted, partially defatted, cooked cottonseed flour is prepared as follows: Cottonseed is delineated and decorticated; the meats are screened, aspirated, and rolled; moisture is adjusted;

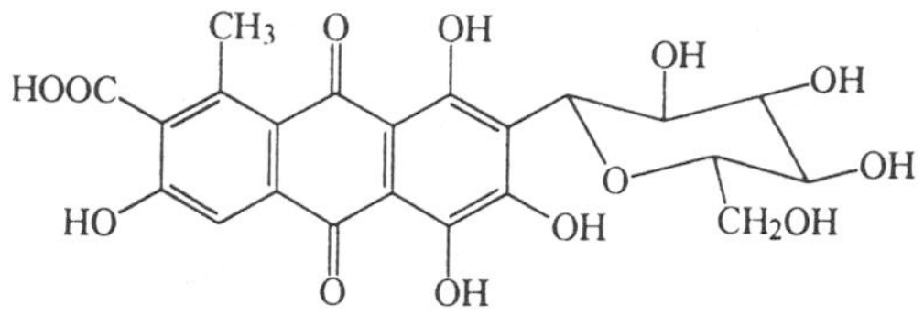

FIGURE 49 Structure of carminic acid.

the meats are heated and the oil is expressed; and the cooked meats are cooled, ground, and reheated to obtain a product varying in shade from light brown to dark brown.

Ferrous gluconate is a yellowish-gray powder with a slight odor resembling that of burnt sugar.

Grape skin extract is a purplish-red liquid prepared from an aqueous extract of pomace remaining after grapes have been pressed to remove the juice. The coloring matter of the extract consists mainly of anthocyanins. It is sold under the name "enocianina" and is restricted for coloring noncarbonated and carbonated drinks, beverage bases, and alcoholic beverages.

Fruit and vegetable juices are acceptable color additives, and they can be used at single strength or as concentrated liquids. Depending on the source of the juice, pigments from many of the previously described classes can be involved. Beet and grape juice concentrates have been produced and marketed as colorants in this category. Grape juice concentrate, in contrast to grape skin extract, may be used in nonbeverage foods.

Carrot oil is produced by extracting edible carrots with hexane. The hexane is subsequently removed by vacuum distillation. The colorant is mainly α- and β-carotene and other minor carotenoids found in carrots.

Paprika or paprika oleoresin is either the ground dried pods of paprika (*Capsicum annuum* L.) or an extract of this plant. In the production of oleoresin, several solvents may be used. The main colorant in paprika is capxanthin, a carotenoid.

Riboflavin or vitamin B_2 is an orange-yellow powder.

Saffron is the dried stigma of *Crocus sativul* L. Its yellow color is attributable to croxin, the digentiobioside of crocetin.

Titanium oxide is synthetically prepared. It often contains silicon dioxide and/or aluminum oxide to aid dispersion in foods. These diluents may not exceed 2% of the total.

Turmeric and turmeric oleoresin are the ground rhizomes or an extract of turmeric (*Curcuma longa* L.). The coloring matter in turmeric is curcumin. Several organic solvents may be used in the production of turmeric oleoresin.

Other Iron oxide, dried algae meal (dried algae cells of the genus *Spongiococcum*), tagetes meal (dried ground flower petals of the Aztic marigold, *Tagetes erecta* L.), and corn endosperm oil are of little interest here since these colorants are restricted to use in animal feeds. They can, however, indirectly affect the color of foods.

The labeling declaration of added colors that are exempt from certification is somewhat illogical. Although colors exempt from certification are natural or nature-identical, they must be listed as "artificial color added." This is required because, in the vast majority of uses, the colorant added is not natural to the food product. Similar to certified colors, colors exempt from certification must be declared by name when used in foods in the United States.

10.3.5 Use of Colors Exempt from Certification

With the exception of synthetic, nature-identical pigments, exempt-from-certification colorants are chemically crude preparations. They are either totally unpurified materials or crude plant or animal extracts. Because of their impurity, relatively large amounts are needed to achieve the desired color. This has caused some to suggest that these pigments lack tinctorial strength and contribute undesirable flavors to a product. Neither criticism is necessarily true. Many pure natural pigments have high tinctorial strengths. This can be illustrated by comparing the 1% absorptivity values ($A^{1\%}_{1cm}$) of a natural pigment with that of a synthetic dye. At wavelengths of maximum light absorption, the $A^{1\%}_{1cm}$ values for FD&C red no. 40 and yellow no. 6 are 586 and 569, respectively, whereas the $A^{1\%}_{1cm}$ for betanin, the main pigment component in beet powder, and β-carotene are 1120 and 2400, respectively.

Furthermore, most pure pigments do not contribute to product flavor. The lack of tinctorial strength and the possible contribution to flavor by the unpurified natural colorants can easily be overcome by applying available technologies of separating and purification. Unfortunately, these advances in technology have not been sanctioned.

BIBLIOGRAPHY

Bauernfeind, J. B., ed. (1981). *Carotenoids as Colorants and vitamin A Precursors*, Academic Press, New York.

Counsell, J. N. ed. (1981). *Natural Colours for Food and Other Uses*, Applied Science, Essex.

Hendry, G. A. F., and J. D. Haughton, eds. (1992). *Natural Food Colorants*, Van Nostrand Reinhold, New York.

Hutchings, J. B. (1994). *Food Colour and Appearance*, Blackie, London.

Marmion, D. M. (1979). *Handbook of U.S. Colorants for Foods, Drugs and Cosmetics*, John Wiley & Sons, New York.

Mazza, G., and E. Miniati (1993). *Anthocyanins in Fruits, Vegetables and Grains*, CRC Press, Boca Raton, FL.

Walford, J., ed. (1980). *Developments in Food Colours—1*, Applied Science, London.

Walford, J., ed. (1984). *Development in Food Colours—2*, Elsevier Applied Science, London.

Schwartz, S. J., and T. V. Lorenzo (1990). *Chlorophylls in Foods. Food Science and Nutrition*. CRC Press, Boca Raton, FL.

REFERENCES

1. Adams, J. B. (1973). Colour stability of red fruits. *Food Manufact. 48*(2):19–20, 41.
2. Attoe, E. L., and J. H. von Elbe (1984). Oxygen involvement in betanine degradation: Effect of antioxidants. *J. Food Sci. 50*:106–110.
3. Bacon, M. F., and M. Holden (1967). Changes in chlorophylls resulting from various chemical and physical treatments of leaf extracts. *Phytochemistry 6*:193–210.
4. Bauernfeind, J. C. (1972). Carotenoid vitamin A precursors and analogs in foods and feeds. *J. Agric. Food Chem. 20*:456–473.
5. Ben Aziz, A., S. Grossman, I. Ascarelli, and P. Budowski (1971). Carotene-bleaching activities of lipoxygenase and heme proteins as studied by a direct spectrophotometric method. *Phytochemistry. 10*:1445–1452.
6. Bilyk, A., M. A. Kolodijand, and G. M. Sapers (1981). Stabilization of red beet pigments with isoascorbic acid. *J. Food Sci., 46*:1616–1617.
7. Blair, J. S. (1940). Color stabilization of green vegetables. U. S. Patent 2, 186, 003, March 3, 1937.
8. Brenes, P., M. C. Garcia Duran, and A. Garrido (1993). Concentration of phenolic compounds in storage brines or ripe olives. *J. Food Sci. 58*:347–350.
9. Bridle, P., K. G. Scott, and C. F. Timberlake (1973). Anthocyanins in *Salix* species. A new anthocyanin in *Salix purpurea* bark. *Phytochemistry 12*:1103–1106.
10. Brouillard, R. (1982). Chemical structures of anthocyanins, in *Anthocyanins as Food Colors* (P. Markakis, ed.), Academic Press, New York, pp. 1–40.

11. Buckle, K. A., and R. A. Edwards (1979). Chlorophyll: Color and pH changes in HTST processed green pea puree. *J. Food Technol. 5*:173–186.
12. Burton, G. W. (1989). Antioxidant action of carotenoids. *J. Nutr.* 119:109–111.
13. Burton, G. W., and K. U. Ingold (1984). Beta-carotene: An unusual type of lipid antioxidant. *Science 224*:569–573.
14. Canjura, F. L., and S. J. Schwartz (1991). Separation of chlorophyll compounds and their polar derivatives by high-performance liquid chromatography. *J. Agric. Food Chem. 39*:1102–1105.
15. Canjura, F. L., S. J. Schwartz, and R. V. Nunes (1991). Degradation kinetics of chlorophylls and chlorophyllides. *J. Food Sci. 56*:1639–1643.
16. Canfield, L. M., N. I. Krinsky, and J. A. Olson (1993). *Carotenoids in Human health*, (vol. 691), New York Academy of Sciences, New York.
17. Clydesdale, F. M., D. L. Fleischmann, and F. J. Francis (1970). Maintenance of color in processed green vegetables. *Food Prod. Dev. 4*:127–138.
18. Clydesdale, F. M., Y. D. Lin, and F. J. Francis (1972). Formation of 2-pyrroldone-*S*-carboxylic acid from glutamine during processing and storage of spinach puree. *J. Food Sci. 37*:45–47.
19. Cohen, E., and I. Saguy (1983). Effect of water activity and moisture content on the stability of beet powder pigments. *J. Food Sci. 48*:703–707.
20. Daravingas, G., and R. F. Cain (1965). Changes in anthocyanin pigments of raspberries during processing and storage. *J. Food Sci. 33*:400–405.
21. Davidson, J. (1954). Procedure for the extraction, separation, and estimation of major fat-soluble pigments of hay. *J. Sci. Food Agric. 5*:1–7.
22. Davies, W. J., and J. Zhang (1991). Root signals and the regulation of growth and development of plants in drying soil. *Annu. Rev. Plant Mol. Biol. 42*:55–76.
23. Di Mascio, P., S. Kaiser, and H. Sies (1989). Lycopene as the most efficient biological carotenoid singlet oxygen quencher. *Arch. Biochem. Biophys. 274*:532–538.
24. Falk, J. E., and J. N. Phillips (1964). Physical and coordination chemistry of the tetrapyrrole pigments, in *Chelating Agents and Metal Chelates* (F. P. Dwyer and D. P. Mellor, eds.), Academic Press, New York, pp. 441–490.
25. Faustman, C., and R. G. Cassens (1990). The biochemical basis of discoloration in fresh meat: A review. *J. Muscle Foods 1*:217–243.
26. Faustman, C., R. G. Cassens, D. M. Schaeffer, D. R. Buege, and K. K. Scheller (1989). Improvement of pigment and lipid stability in Holstein steer beef by dietary suuplementaion with vitamin E. *J. Food Sci. 54*:858–862.
27. Fischbach, H. (1943). Microdeterminations for organically combined metal in pigment of okra. *J. Assoc. Off. Agric. Chem. 26*:139–143.
28. Foote, C. S. (1968). Mechanisms of photosensitized oxidation. *Science 162*:963–970.
29. Forrest, J. C., E. D. Aberle, H. B. Hedrick, M. D. Judge, and R. A. Merkel (1975). *Principles of Meat Science*, W. H. Freeman, San Francisco.
30. Forsyth, W. G. C., and J. B. Roberts (1958). Cocoa "leucocyanidin." *Chem. Ind. (Lond.)* 755.
31. Fox, J. B., Jr. (1966). The chemistry of meat pigments. *J. Agric. Food Chem. 14*:207–210.
32. Fox, J. B., Jr., and S. A. Ackerman (1968). Formation of nitric oxide myoglobin: Mechanisms of the reaction with various reductants. *J. Food Sci. 33*:364–370.
33. Franics, F. J. (1977). Anthocyanins, *Current Aspects of Food Colorants* (T. E. Furia, ed.), CRC Press, Cleveland, OH, pp. 19–27.
34. Frankel, E. N. (1989). The antioxidant and nutritional effects of tocopherols, ascorbic acid and β-carotene in relation to processing of edible oils, in *Nutritional Impact of Food Processing* (J. C. Somogyi and H. R. Muller, eds.), Bibl. Nutr. Dieta., Karger, Basel, pp. 297–312.
35. Geis, I. (1983). *Biochemistry*, 2nd ed. (F. B. Armstrong), Oxford University Press, New York, p. 108.
36. Goodman, L. P., and P. Markakis (1965). Sulfur dioxide inhibition of anthocyanin degradation by phenolase. *J. Food Sci. 130*:135–137.
37. Goodwin, T. W. (1980). Functions of carotenoids, in *The Biochemistry of the Carotenoids*, vol. 1, 2nd ed. (T. W. Goodwin, ed.), Chapman and Hall, New York, pp. 77–95.

38. Govindarajan, S., H. O. Hultin, and A. W. Kotula (1977). Myoglobin oxidation in ground beef: Mechanistic studies. *J. Food Sci.* 42:571–577, 582.
39. Gross, J. (1991). *Pigments in Vegetables: Chlorophylls and Carotenoids*, Van Nostrand Reinhold, New York.
40. Gupte, S. M., H. M. El-Bisi, and F. J. Francis (1964). Kinetics of thermal degradation of chlorophyll in spinach puree. *J. Food Sci. 29*:379–382.
41. Haisman, D. R., and M. W. Clarke (1975). The interfacial factor in heat induced conversion of chlorophyll to pheophytin in green leaves. *J. Sci. Food Agric. 26:*1111–1126.
42. Holt, A. S. (1958). The phase test intermediate and the allomerization of chlorophyll *a. Can J. Biochem. Physiol. 36*:439–456.
43. Huang, A. S., and J. H. von Elbe (1985). Kinetics of the degradation and regeneration of betanine. *J. Food Sci. 50*:1115–1120, 1129.
44. Huang, A. S., and J. H. von Elbe (1987). Effect of pH on the degradation and regeneration of betanine. *J. Food Sci. 52*:1689–1693.
45. Humphry, A. M. (1980). Chlorophyll. *Food. Chem. 5*:57–67.
46. Isler, O. (1971). *Carotenoids*, Birkhauser Verlag, Basel.
47. IUPAC-IUB (International Union of Pure and Applied Chemistry-International Union of Biochemistry) (1975). Nomenclature of carotenoids. *Pure Appl. Chem. 41*:407–419.
48. Jackman, R. L., and J. L. Smith (1992). Anthocyanins and betalains, in *Natural Food Colorants.* (G. A. F. Hendry and J. D. Houghton, eds.), Van Norstrand Reinhold, New York, pp. 183–231.
49. Jackson, A. H. (1976). Structure, properties and distribution of chlorophyll, in *Chemistry and Biochemistry of Plant Pigments*, vol. 1, 2nd ed. (T. W. Goodwin, ed.), Academic Press, New York, pp. 1–63.
50. Jen, J. J., and G. Mackinney (1970). On the photodecomposition of chlorophyll in vitro. I. Reaction rates. *Photochem. Photobiol. 11*:297–302.
51. Jen, J. J., and G. Mackinney (1970). On the photodecomposition of chlorophyll in vitro. II. Intermediates and breakdown products. *Photochem. Photobiol. 11*:303–308.
52. Jones, I. D., R. C. White, E. Gibbs, L. S. Butler, and L. A. Nelson (1977). Experimental formation of zinc and copper complexes of chlorophyll derivatives in vegetable tissue by thermal processing. *J. Agric. Food Chem. 25*:149–153.
53. Jones, I. D., R. C. White, E. Gibbs, and C. D. Denard (1968). Absorption spectra of copper and zinc complexes of pheophytins and pheophorbides. *J. Agric. Food Chem. 16*:80–83.
54. Khachik, F., G. R. Beecher, and N. F. Whittaker (1986). Separation, identification, and quanitifcation of the major carotenoid and chlorophyll constituents in extracts of several green vegetables by liquid chromatography. *J. Agric. Food Chem. 34*:603–616.
55. Kearsley, M. W., and N. Rodriguez (1981). The stability and use of natural colours in foods: anthocyanin, β-carotene, and riboflavin. *J. Food Technol. 16*:421–431.
56. Krinsky, N. I. (1989). Antioxidant functions of carotenoids. *Free Radical Biol Med. 7*:617–635.
57. Kropf, D. H. (1980). Effects of retail display conditions on meat color. *Proc. Recip. Meat Conf. 33*:15–32.
58. La Borde, L. F., and J. H. von Elbe (1990). Zinc complex formation in heated vegetable purees. *J. Agric. Food Chem. 38*:484–487.
59. La Borde, L. F., and J. H. von Elbe (1994). Chlorophyll degradation and zinc complex formation with chlorophyll derivatives in heated green vegetables. *J. Agric. Food Chem. 42*:1100–1102.
60. La Borde, L. F., and J. H. von Elbe (1994). Effect of solutes on zinc complex formation in heated green vegetables. *J. Agric. Food Chem. 42*:1096–1099.
61. Lawrie, R. A. (1985). *Meat Science*, 4th ed., Pergamon Press, New York.
62. Llewellyn, C. A., R. F. C. Mantoura, and R. G. Brereton (1990). Products of chlorophyll photodegradation. 1. Detection and separation. *Photochem. Photobiol. 52*:1037–1041.
63. Llewellyn, C. A., R. F. C. Mantoura, and R. G. Brereton (1990). Products of chlorophyll photodegradation. 2. Structural identification. *Photochem. Photobiol. 52*:1043–1047.
64. Loef, H. W. and S. B. Thung (1965). Ueber den Einfluss von Chlorophyllase auf die Farbe von Spinat waehrend und nach der Verwertung. *Z. Lebensm. Unters. Forsch. 126*:401–406.

65. Luh, B. S., S. J. Leonard, and D. S. Patel (1960). Pink discoloration in canned Barlett pears. *Food Technol. 14*:53–56.
66. Luh, B. S., S. Leonard, M. Simone, and F. Villarreal (1964). Aseptic canning of foods. VII. Strained lima beans. *Food Technol. 18*:363–366.
67. Madhavi, D. L., and C. E. Carpenter 91993). Aging and processing affect color, metmyoglobin reductase and oxygen consumption of beef muscles. *J. Food. Sci., 58*:939–942, 947.
68. Malecki, G. J. (1965). Blanching and canning process for green vegetables. U. S. Patent 3, 183, 102, June 19, 1961.
69. Malecki, G. J. 1978. Processing of green vegetables for color retention in canning, U. S. Patent 4, 104, 410, September 9, 1976.
70. Markakis, P. (1982). Stability of anthocyanins in foods, in *Anthocyanins as Food Colors* (P. Markakis, ed.), Academic Press, New York, pp. 163–180.
71. Markakis, P., G. E. Livingstone, and R. C. Fellers (1957). Quantitative aspects of strawberry-pigment degradation. *Food Res. 22*:117–130.
72. Mazza, G., and E. Miniati (1993). *Anthocyanins in Fruits, Vegetables, and Grains*, CRC Press, Boca Raton, FL.
73. McFeeters, R. F. (1975). Substrate specificity of chlorophyllase. *Plant Physiol. 158*:377–381.
74. Nakatani, H. Y., J. Barber, and J. A. Forrester (1979). Surface charges on chloroplast membranes as studied by particle electrophoresis. *Biochim. Biophys. Acta 504*:215–225.
75. Okayama, T. (1987). Effect of modified gas atmosphere packaging after dip treatment on myoglobin and lipid oxidation of beef steaks. *Meat Sci. 19*:179–185.
76. O'Neil, C. A., and S. J. Schwartz (1992). Chromatographic analysis of cis/trans carotenoid isomers. *J. Chromatogr. 624*:235–252.
77. Palozza, P., and N. I. Krinsky (1992). Antioxidant effects of carotenoids *in vivo* and *in vitro:* An overview. *Methods Enzymol. 213*:403–421.
78. Parry, A. D., and R. Horgan (1992). Abscisic acid biosythesis in roots: The identification of potential abscisic acid precursors, and other carotenoids. *Planta 187*:185–191.
79. Pasch, J. H., and J. H. von Elbe (1975). Betanine degradation as inlfuenced by water activity. *J. Food Sci. 44*:72–73.
80. Peng, C. Y., and P. Markakis (1963). Effect of prenolase on anthocyanins. *Nature (Lond.) 199*:597–598.
81. Pennington, F. C., H. H. Strain, W. A. Svec, and J. J. Katz (1964). Preparation and properties of pyrochlorophyll *a*, methyl pyrochlorophyllide *a*, pyropheophytin *a*, and methyl pyropheophorbide *a* derived from chlorophyll by decarbomethoxylation. *J. Am. Chem. Soc. 86*:1418–1426.
82. Peiser, G. D., and S. F. Yang (1979). Sulfite-mediated destruction of β-carotene. *J. Agric. Food Chem. 27*:446–449.
83. Peto, R., R. Doll, J. D. Buckley, and M. B. Sporn (1981). Can dietary betacarotene materially reduce human cancer rates? *Nature 290*:201–208.
84. Piatelli, M., L. Minale, and G. Prota (1965). Pigments of Centospermae—III. Betaxanthines from *Beta vulgaris* L. *Phytochemistry 4*:121–125.
85. Piatelli, M., L. Minale, and R. A. Nicolaus (1965). Pigments of Centrospermae—V. Betaxanthines from *Mirabilis jalapa* L. *Phytochemistry 4*:817–823.
86. Sante, V., M. Renerre, and A. Lacourt (1994). Effect of modified gas atmosphere packaging on color stability and on microbiology of turkey breast meat. *J. Food Qual. 17*:177–195.
87. Schanderl. S. H., C. O. Chichester, and G. L. Marsh (1962). Degradation of chlorophyll and several derivatives in acid solution. *J. Org. Chem. 27*:3865–3868.
88. Schanderl, S. H., G. L. Marsh, and C. O. Chichester (1965). Color reversion in processed vegetables. I. Studies on regreened pea puree. *J. Food Sci. 30*:312–316.
89. Schwartz, S. J. (1994). Pigment analysis, in *Introduction to the Chemical Analysis of Foods*, (S. Nielson, ed.), Jones and Bartlett, Boston, pp. 261–271.
90. Schwartz, S. J., and T. V. Lorenzo (1991). Chlorophyll stability during aseptic processing and storage. *J.Food Sci. 56*:1056–1062.
91. Schwartz, S. J., and J. H. von Elbe (1983). Kinetics of chlorophyll degradation to pyropheophytin in vegetables. *J. Food Sci. 48*:1303–1306.

92. Schwartz, S. J., S. L. Woo, and J. H. von Elbe (1981). High performance liquid chromatography of chlorophylls and their derivatives in fresh and processed spinach. *J. Agric. Food Chem. 29*:533–535.
93. Scott, K. J. (1992). Observations on some of the problems associated with the analysis of carotenoids in foods by HPLC. *Food Chem. 45*:357–364.
94. Seely, G. R. (1966). The structure and chemistry of functional groups, in *The Chlorophylls* (L. P. Vernon and G. R. Seely, eds.), Academic Press, New York, pp. 67–109.
95. Segner, W. P., T. J. Ragusa, W. K. Nank, and W. C. Hayle (1984). Process for the preservation of green color in canned vegetables. U.S. Patent 4, 473, 591, December 15, 1982.
96. Seidman, S. C., and P. R. Durland (1984). The utilization of modified gas atmosphere packaging in fresh meat: A review. *J. Food Qual. 6*:239–252.
97. Shrikhande, A. J., and F. J. Francis (1974). Effect of flavonols on ascorbic acid and anthocyanin stability in model systems. *J. Food Sci. 39*:904–906.
98. Sies, H. (1992). Antioxidant functions of vitamins: Vitamins E and C, beta-carotene and other carotenoids. *Ann. NY Acad. Sci., 669*:7–20.
99. Starr, M. S., and F. J. Francis (1968). Oxygen and ascorbic acid effect on the relative stability of four anthocyanins pigments in cranberry juice. *Food Technol. 22*:1293–1295.
100. Strain, H. H., M. F. Thomas, and J. J. Katz (1963). Spectral absorption properties of ordinary and fully deuteriated chlorophylls *a* and *b*. *Biochim. Biophys. Acta 75*:306–311.
101. Straub, O. (1987). *Key to Carotenoids*, 2nd ed. (H. Pfander, M. Gerspacher, M. Rychener, and R. Schwabe, eds.), Birkhauser Verlag, Boston.
102. Struck, A., E. Cmiel, S. Schneider, and H. Scheer (1990). Photochemical ring-opening in meso-chlorinated chlorophylls. *Photochem. Photobiol. 51*:217–222.
103. Sweeny, J. G., and M. E. Martin (1958). Determination of chlorophyll and pheophytin in broccoli heated by various procedures. *Food Res. 23*:635–647.
104. Sweeny, J. G., M. M. Wilkinson, and G. A. Iacobucci (1981). Effect of flavonoid sulfonates on the photobleaching of anthocyanins in acid solution. *J. Agric. Food Chem. 29*:563–567.
105. Swiski, M. A., R. Alloug, A. Fuimard, and H. Cheftel (1969). A water-soluble green pigment, originating during processing of canned brussels sprouts picked before the first autumn frost. *J. Agric. Food Chem. 17*:799–801.
106. Tan, C. T., and F. J. Francis (1962). Effect of processing temperature on pigments and color of spinach. *J. Food Sci. 27*:232–240.
107. Tee, E. -S., and C. -L. Lim (1991). Carotenoid composition and content of Malaysian vegetables and fruits by the AOAC and HPLC methods. *Food Chem. 41*:309–339.
108. Tonucci, L. H., and J. H. von Elbe (1991). Kinetics of the formation of zinc complexes of chlorophyll derivatives. *J. Agric. Food Chem. 40*:2341–2344.
109. Van Buren, J. P., J. J. Bertino, and W. B. Robinson (1968). The stability of wine anthocyanins on exposure to heat and light. *Am. J. Enol. Vitic. 19*:147–154.
110. von Elbe, J. H. (1963). Factors affecting the color stability of cherry pigments and cherry juice. PhD thesis, University of Wisconsin, Madison.
111. von Elbe, J. H., A. S. Huang, E. L. Attoe, and W. K. Nank (1986). Pigment composition and color of conventional and Veri-Green canned peas. *J.Agric. Food Chem. 34*:52–54.
112. von Elbe, J. H., and L. F. LaBorde (1989). Chemistry of color improvement in thermally processed green vegetables, *Quality Factors of Fruit and Vegetables Chemistry and Technology* (J. J. Jen, ed.), Am. Chem. Soc. Symp. Ser. 405, Washington, DC, pp. 12–28.
113. Watanaba, T., M. Nakazato, H. Mazaki, A. Hongu, M. Kono, S. Saitoh, and H. Kenichi (1985). Chlorophyll *a* epimer and pheophytin *a* in green leaves. *Biochim. Biophys. Acta 807*:110–117.
114. Zechmeister, L. (1962). *Cis-Trans Isomeric Carotenoids, Vitamins A and Arylpolyenes*, Academic Press, New York.

11

Flavors

Robert C. Lindsay
University of Wisconsin—Madison, Madison, Wisconsin

11.1 INTRODUCTION

11.1.1 General Philosophy

Knowledge of the chemistry of flavors is commonly perceived as a relatively recent development in food chemistry that evolved from the use of gas chromatography and fast-scan mass spectrometry. Although the availability of these instrumental tools has provided the means to definitively investigate the entire range of flavor substances, classic chemical techniques were elegantly applied in early studies, especially for essential oils and spice extractives [25]. This extensive and somewhat separate focus of attention to perfumery, combined with a rapid, seemingly disorganized development about the chemistry of food flavors beginning in the early 1960s, has contributed to the slow evolution of a discipline-oriented identity for the field of flavors. Even though flavor substances represent an extremely wide range of chemical structures derived from every major constituent of foods, their common feature of "stimulating taste or aroma receptors to produce an integrated psychological response known as flavor" remains adequate as a definition of this applied discipline.

Generally, the term "flavor" has evolved to a usage that implies an overall integrated perception of all of the contributing senses (smell, taste, sight, feeling, and sound) at the time of food consumption. The ability of specialized cells of the olfactory epithelium of the nasal cavity to detect trace amounts of volatile odorants accounts for the nearly unlimited variations in intensity and quality of odors and flavors. Taste buds located on the tongue and back of the oral cavity enable humans to sense sweetness, sourness, saltiness, and bitterness, and these sensations contribute to the taste component of flavor. Nonspecific or trigeminal neural responses also provide important contributions to flavor perception through detection of pungency, cooling, umami, or delicious attributes, as well as other chemically induced sensations that are incompletely understood. The nonchemical or indirect senses (sight, sound, and feeling) influence the perception of tastes and smells, and hence food acceptances, but a discussion of these effects is beyond the scope of this chapter. Thus, materials presented in this chapter will be confined to discussions about substances that yield taste and/or odor responses, but a clear distinction between the meaning of these terms and that of flavor will not always be attempted.

11.1.2 Methods for Flavor Analysis

As noted at the beginning of this chapter, flavor chemistry often has been equated with the analysis of volatile compounds using gas chromatography combined with fast-scan mass spectrometry, but this impression is much too restrictive. Here, only limited attention will be directed to this topic although extensive discussions can be found elsewhere [45, 57, 66].

Several factors make the analysis of flavors somewhat demanding, and these include their presence at low concentrations (ppm, 1×10^6; ppb, 1×10^9; ppt, 1×10^{12}), complexity of mixtures (e.g., over 450 volatiles identified in coffee), extremely high volatility (high vapor pressures), and instability of some flavor compounds in dynamic equilibria with other constituents in foods. Identification of flavor compounds requires initial isolation from the bulky constituents of foods combined with substantial concentration, and this should occur with minimal distortion of the native composition if flavor quality is being studied. The adsorption of flavor compounds on porous polymers followed by either thermal desorption or solvent elution has provided a means to minimize destruction of sensitive compounds during isolation. However, higher boiling compounds and some compounds present in very low concentrations still require distillation techniques to assure adequate recovery for identification. Successful identifications of compounds also require separation of flavor isolates into individual components, and advances in the technology of preparing fused-silica capillary columns for gas chromatography have proven highly successful in achieving this goal. Similarly, advances in high-performance liquid chromatography have provided powerful means for separating many higher boiling compounds and precursors.

11.1.3 Sensory Assessment of Flavors

Sensory assessments of flavor compounds and foods are essential for achieving objectives of flavor investigations regardless of ultimate goals. Some situations call for sensory characterization of samples by skilled individuals (experienced flavorists or researchers). In other instances, it is necessary to use formal panels for sensory analysis followed by statistical analysis of data. Excellent reviews and books are available on this subject [1, 2, 14, 53, 62], and these should be consulted for detailed information on this extremely important aspect of flavor assessment.

Sensory assessments are essential for the determination of detection thresholds, which provide a measure of the potency of flavor provided by individual compounds. Threshold values are usually determined using individuals representative of the general population. A range of concentrations of a selected flavor compound in a defined medium (water, milk, air, etc.) is presented to sensory panelists, and each panelist indicates whether or not the compound can be detected. The concentration range where at least half (sometimes greater) of the panelists can detect the compound is designated as the flavor threshold [18]. Compounds vary greatly in their flavor or odor potency, and thus, a minute amount of a compound having a very low threshold will have substantially greater influence on the flavor of a food than one that is quite abundant but that possesses a high flavor threshold. The calculation of the odor units (OU) involves dividing the concentration of the flavor compound by its flavor threshold (OU = concentration present/threshold concentration), and provides an estimate of the contribution by a flavor compound. More recently, aroma extract dilution analysis (AEDA) has been used extensively to identify the most potent odorants in foods [24], and this involves sensory detection of individual compounds (flavor dilution factor) in gas-chromatographic effluents resulting from serial dilutions of aroma extracts from foods. Such methods provide information about the most potent flavor compounds present in foods and beverages, but provide this information in the absence of influences of food matrices and interactive psychophysical effects of the perception of mixtures of flavor compounds. Therefore, extrapolation of such data to actual food systems is severely limited.

Measurement of chemical flavor parameters to provide definitive information about flavor intensity and quality of foods has long been an idealized goal of flavor investigations [55]. Much progress has been made in the application of methods for correlation of subjective sensory information with objective flavor chemical data, but routine assessment of flavor quality by purely analytical means remains limited.

Attention is usually directed to important flavor compounds that provide the "characterizing" or "character-impact" features of a particular flavor. These substances are reasonably limited in number, but are frequently present in extremely low concentrations, and they may be extremely unstable. The seemingly impossible task of defining character impact compounds led, early on, to the so-called component balance theory for explaining the chemical basis of certain flavors. However, as methods of analysis have become more sensitive, more and more foods with different flavors but similar low-resolution flavor-compound profiles have been found to contain one or more character-impact compounds responsible for the principal flavor. Identification of character-impact compounds can be expected for many flavors that remain a mystery for the present, but for some transient, unstable flavor compounds, identification by the methods currently available may not be possible. For example, the key flavor compounds responsible for cheddar cheese aroma, the dihydropyrazines of freshly roasted nuts, and furfuryl thioaldehyde believed present in freshly brewed coffee may not yield themselves to isolation and identification techniques used now, and their definition must await the development of new techniques.

This chapter deals primarily with the chemistry of important character-impact compounds that have been selected to illustrate the chemistry of food systems and the chemical basis for the existence of flavor compounds in foods. Where appropriate, and when information is available, structure–activity relationships for flavor compounds are noted. Limited attention is given to listings of profiles of flavor compounds present in various foods. Comprehensive lists of flavor compounds for foods are available elsewhere [42, 69], as are tabulations of threshold concentrations for individual compounds [18]. These sources of information should prove useful in searches for further details about specific flavor compounds. Finally, a choice exists on whether information pertaining to the flavor chemistry of major food constituents is dealt with here or in the chapters devoted to those major food constituents. It has been deemed appropriate to conduct these discussions in the major-constituent chapters. For example, flavors deriving from the Maillard reaction are discussed in Chapter 4 and those deriving from free radical oxidation of lipids are discussed in Chapter 5. Information on low-calorie sweeteners and on binding of flavors by macromolecules is, of necessity, partly covered here and partly covered in Chapters 4 and 6 (binding by macromolecules) and in Chapter 12 (low-calorie sweeteners).

11.2 TASTE AND NONSPECIFIC SAPOROUS SENSATIONS

Frequently, substances responsible for these components of flavor perception are water soluble and relatively nonvolatile. As a general rule, they are also present at higher concentrations in foods than those responsible for aromas, and have been often treated lightly in coverages of flavors. Because of their extremely influential role in the acceptance of food flavors, it is appropriate to examine the chemistry of substances responsible for taste sensations as well as those responsible for some of the less defined flavor-taste sensations.

11.2.1 Taste Substances: Sweet, Bitter, Sour, and Salty

Sweet substances have been the focus of much attention because of interest in sugar alternatives and the desire to find suitable replacements for the low-calorie sweeteners saccharin and cyclamate (see Chaps. 4 and 12). The bitterness sensation appears to be closely related to sweetness from a molecular structure–receptor relationship, and as a result much has been learned about bitterness in studies directed primarily toward sweetness. Development of bitterness in protein hydrolysates and aged cheeses is a troublesome problem, and this problem has stimulated research on the causes of bitterness in peptides. With regard to saltiness, national policies that encourage a reduction of sodium in diets have stimulated renewed interest in the mechanisms of the salty taste.

11.2.1.1 Structural basis of the sweet modality

Before modern sweetness theories were advanced, it was popular to deduce that sweetness was associated with hydroxyl (-OH) groups because sugar molecules are dominated by this feature.

However, this view was soon subject to criticism because polyhydroxy compounds vary greatly in sweetness, and many amino acids, some metallic salts, and unrelated compounds, such as chloroform ($CHCl_3$) and saccharin (Chap. 12), also are sweet. Still, it was apparent that some common characteristics existed among sweet substances, and over the past 75 years a theory relating molecular structure and sweet taste has evolved that satisfactorily explains why certain compounds exhibit sweetness.

Shallenberger and Acree [59] first proposed the AH/B theory for the saporous (taste eliciting) unit common to all compounds that cause a sweet sensation (Fig. 1). The saporous unit was initially viewed as a combination of a covalently bound H-bonding proton and an electronegative orbital positioned at a distance of about 3 Å from the proton. Thus, vicinal electronegative atoms on a molecule are essential for sweetness. Further, one of the atoms must possess a hydrogen bonding proton. Oxygen, nitrogen, and chlorine atoms frequently fulfill these roles in sweet molecules, and hydroxyl-group oxygen atoms can serve either the AH or B function in a molecule. Simple AH/B relationships are shown for chloroform (I), saccharin (II), and glucose (III).

(I) Chloroform

(II) Saccharin

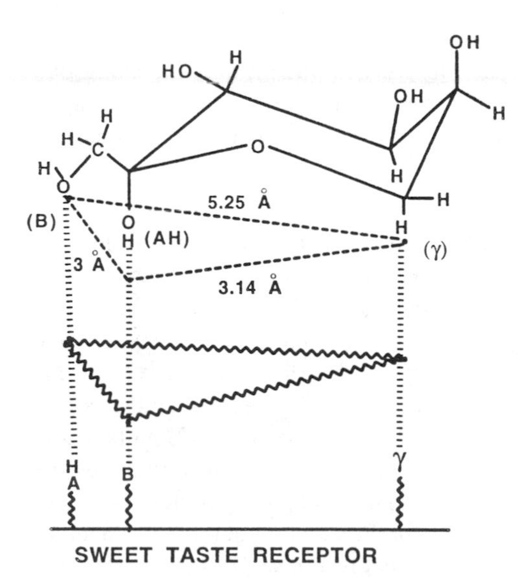

FIGURE 1 Schematic showing the relationship between AH/B and γ sites in the saporous sweet unit for β-D-fructopyranose.

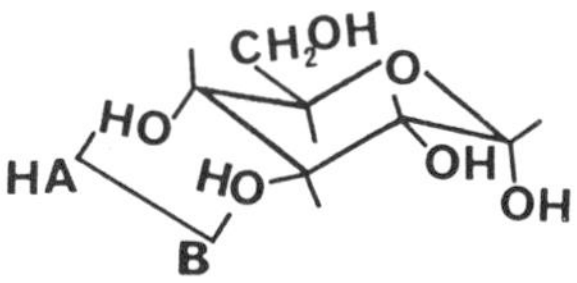

(III) Glucose

As indicated in Figure 1, however, stereochemical requirements are also imposed on the AH/B components of the saporous unit so that they will align suitably with the receptor site. The interaction between the active groups of the sweet molecule and the taste receptor is currently envisioned to occur through H-bonding of the AH/B components to similar structures in the taste receptor. A third feature also has been added to the theory to extend its validity to intensely sweet substances. This addition incorporates appropriate stereochemically-arranged lipophilic regions of sweet molecules, usually designated as γ, which are attracted to similar lipophilic regions of the taste receptor. Lipophilic portions of sweet molecules are frequently methylene ($-CH_2-$), methyl ($-CH_3$), or phenyl ($-C_6H_5$) groups. The complete sweet saporous structure is geometrically situated so that triangular contact of all active units (AH, B, and γ) with the receptor molecule occurs for intensely sweet substances, and this arrangement forms the rationale for the tripartite structure theory of sweetness.

The γ-site is an extremely important feature of intensely sweet substances, but plays a lesser role in sugar sweetness [7]. It appears to function through facilitating the accession of certain molecules to the taste receptor site, and as such, effects the perceived intensity of sweetness. Since sugars are largely hydrophilic, this feature comes into play in a limited sense only for some sugars, such as fructose. This component of the saporous sweet unit probably accounts for a substantial portion of the variation in sweetness quality that is observed between different sweet substances. Not only is it important in the time–intensity or temporal aspects of sweetness perception, but it also appears to relate to some of the interactions between sweet and bitter tastes observed for some compounds.

Sweet–bitter sugar structures possess features that apparently allow them to interact with either or both types of receptors, thus producing the combined taste sensation. Bitterness properties in structures depress sweetness even if the concentration in a test solution is below that for the bitter sensation. Bitterness in sugars appears to be imparted by a combination of effects involving the configuration of the anomeric center, the ring oxygen, the primary alcohol group of hexoses, and the nature of any substituents. Often, changes in the structure and stereography of a sweet molecule lead to the loss or suspension of sweetness or the induction of bitterness.

11.2.1.2 The Bitter Taste Modality

Bitterness resembles sweetness because of its dependence on the stereochemistry of stimulus molecules, and the two sensations are triggered by similar features in molecules, causing some molecules to yield both bitter and sweet sensations. Although sweet molecules must contain two polar groups that may be supplemented with a nonpolar group, bitter molecules appear to have a requirement for only one polar group and a hydrophobic group [8]. However, some [4, 6, 11] believe that most bitter substances possess an AH/B entity identical to that found in sweet molecules as well as the hydrophobic group. In this concept, the orientation of AH/B units within specific receptor sites, which are located on the flat bottom of receptor cavities, provides the discrimination between sweetness and bitterness for molecules possessing the required molecular features. Molecules that fit into sites that were oriented for bitter compounds give a bitter response; those fitting the orientation for sweetness elicit a sweet response. If the geometry of a molecule were such that it could orient in either direction, it would give bitter–sweet responses. Such a model appears especially attractive for amino acids where D-isomers are sweet and

L-isomers are bitter [37]. Since the hydrophobic or γ-site of the sweet receptor is nondirectional lipophilicity, it could participate in either sweet or bitter responses. Molecular bulkiness factors serve to provide stereochemical selectivity to the receptor sites located in each receptor cavity. It can be concluded that there is a very broad structural basis for the bitter taste modality, but most empirical observations about bitterness and molecular structure can be explained by current theories.

11.2.1.3 Bitter Compounds That Are Important in Foods

Bitterness may be desirable or undesirable in food flavors, and because of genetic differences in humans, individuals vary in their ability to perceive certain bitter substances. A compound may be bitter or bitter–sweet depending on the individual. Saccharin is perceived as purely sweet by some individuals, but others find it to range from only slightly bitter and sweet to quite bitter and sweet. Many other compounds also show marked variations in the manner which individuals perceive them, and frequently either taste bitter or are not perceived at all. Phenylthiocarbamide or PTC (IV) is one of the most notable compounds in this category [1]:

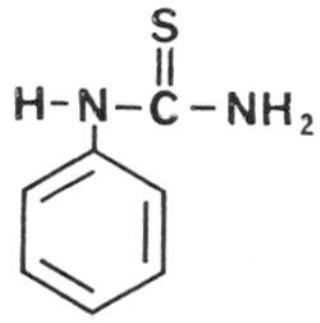

S
||
H-N-C-NH$_2$

(IV) Phenylthiocarbamide (PTC)

For PTC, about 40% of the Caucasian American population is taste-blind to the bitterness attribute that is perceived by the other 60% of the Caucasian American population. Although PTC is a novel compound that does not occur in foods, creatine (V), shown here,

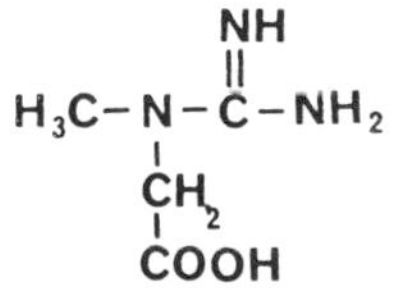

NH
||
H$_3$C-N-C-NH$_2$
|
CH$_2$
|
COOH

(V) Creatine

is a constituent of muscle foods that exhibits similar properties of varied taste sensitivity in the population. Creatine may occur at levels up to about milligrams per gram in lean meats [1], and this is adequate to make soups taste bitter to sensitive individuals. As with other bitter substances, these molecules contain AH/B sites suitable for inducing the bitter sensation. Genetic variation in the perception of specific molecules appears to relate to the size of receptor site cavities and to the arrangement and nature of atoms on cavity walls, with these atoms governing which molecules are allowed in the site and which are rejected.

Quinine is an alkaloid that is generally accepted as the standard for the bitter taste sensation. The detection threshold for quinine hydrochloride (VI) is about 10 ppm. In general, bitter substances have lower taste thresholds than other taste substances, and they also tend to be less soluble in water than taste-active materials. Quinine is permitted as an additive in beverages,

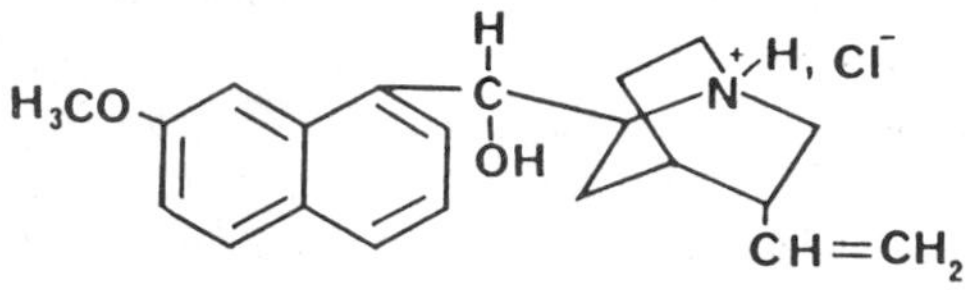

H$_3$CO — H-C-OH — N$^+$-H, Cl$^-$ — CH=CH$_2$

(VI) Quinine Cl$^-$

such as soft drinks, that also have tart–sweet attributes. The bitterness blends well with the other tastes and provides a refreshing gustatory stimulation in these beverages. The practice of mixing quinine into soft-drink beverages apparently stems from efforts to suppress or mask the bitterness of quinine when it was prescribed as a drug for malaria.

In addition to soft drinks, bitterness is an important flavor attribute of several other beverages consumed in large quantities, including coffee, cocoa, and tea. Caffeine (VII) is moderately bitter at 150–200 ppm in water, and occurs in coffee, tea, and cola nuts. Theobromine (VIII) is very

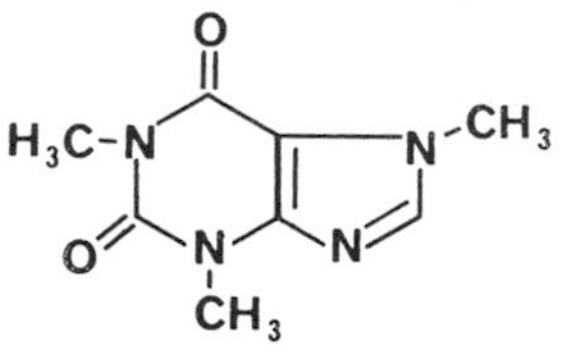

(VII) Caffeine

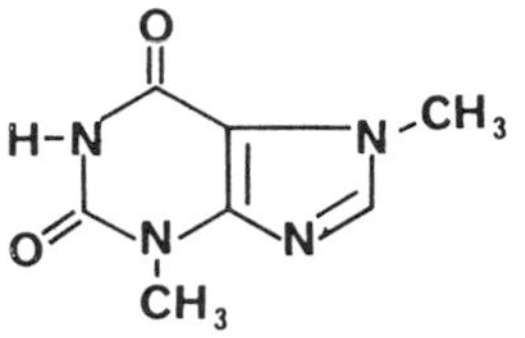

(VIII) Theobromine

similar to caffeine and is present most notably in cocoa, where it contributes to bitterness. Caffeine is added in concentrations up to 200 ppm to soft cola beverages, and much of the caffeine employed for this purpose is obtained from extractions of green coffee beans, which is carried out in the preparation of decaffeinated coffee.

Large amounts of hops are employed in the brewing industry to provide unique flavors to beer. Bitterness, contributed by some unusual isoprenoid-derived compounds, is a very important aspect of hop flavor. These substances can generally be categorized as derivatives of humulone or lupulone, that is, α-acids or β-acids, respectively, as they are known in the brewing industry. Humulone is the most abundant substance, and it is converted during wort boiling to isohumulone by an isomerization reaction (Eq. 1) [13].

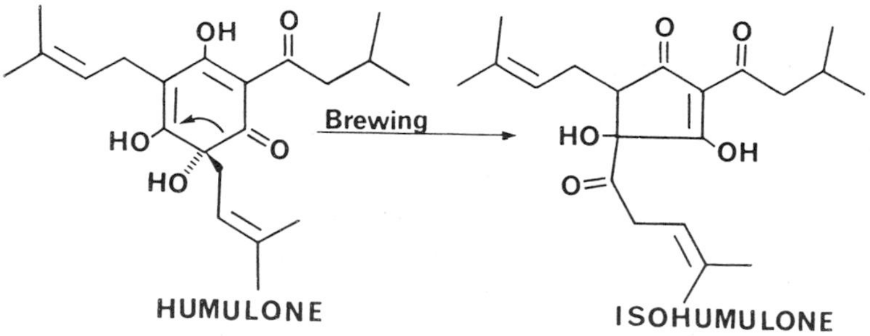

HUMULONE —Brewing→ ISOHUMULONE (1)

Isohumulone is the precursor for the compound that causes the sunstruck or skunky flavor in beer exposed to light. In the presence of hydrogen sulfide from yeast fermentation, a photocatalyzed reaction occurs at the carbon adjacent to the keto group in the isohexenyl chain. This gives rise to 3-methyl-2-butene-1-thiol (prenylmercaptan), which has a skunky aroma [38]. Selective reduction of the ketone in preisomerized hop extracts prevents this reaction and allows packaging of beer in clear glass without development of the skunky or sunstruck flavor. Whether volatile hop aroma compounds survive the wort-boiling process was a controversial topic for a number of years. However, it is now well documented that influential compounds do survive the wort-boiling process and others are formed from hop bitter substances; together they contribute to the kettle-hop aroma of beer.

The development of excessive bitterness is a major problem of the citrus industry, especially in processed products. In the case of grapefruit, some bitterness is desirable and expected, but frequently the intensity of bitterness in both fresh and processed fruit exceeds that preferred by many consumers. The principal bitter component in navel and Valencia oranges is a triterpenoid dilactone (A- and D-rings) called limonin, and it is also found as a bittering agent in grapefruit. Limonin is not present to any extent in intact fruits, but rather a flavorless limonin derivative produced by enzymic hydrolysis of limonin's D-ring lactone is the predominant form (Fig. 2). After juice extraction, acidic conditions favor the closing of the D-ring to form limonin, and the phenomenon of delayed bitterness occurs, yielding serious economic consequences.

Methods for debittering orange juice have been developed using immobilized enzymes from *Arthrobacter* sp. and *Acinetobacter* sp. [30]. Enzymes that simply open the D-ring lactone provide only temporary solutions to the problem because the ring closes again under acidic conditions. However, the use of limonoate dehydrogenase to convert the open D-ring compound to nonbitter 17-dehydrolimonoate A-ring lactone (Fig. 2) provides an irreversible means to debitter orange juice, although the process has not yet been commercialized. Methods to debitter citrus juices also include the use of polymeric adsorbants, which are the preferred methods for commercial processors [34].

Citrus fruits also contain several flavonone glycosides, and naringen is the predominant flavonone found in grapefruit and bitter orange (*Citrus auranticum*). Juices that contain high levels of naringen are extremely bitter, and are of little economic value except in instances where they can be extensively diluted with juices containing low bitterness levels. The bitterness of naringen is associated with the configuration of the molecule that develops from the 1 to 2 linkage between rhamnose and glucose. Naringenase is an enzyme that has been isolated from commercial citrus pectin preparations and from *Aspergillus* sp., and this enzyme hydrolyzes the 1 to 2 linkage (Fig. 3) to yield nonbitter products. Immobilized enzyme systems have also been developed to debitter grapefruit juices containing excessive levels of naringen. Naringen has also been commercially recovered from grapefruit peels and is used instead of caffeine for bitterness in some food applications.

Pronounced, undesirable bitterness is frequently encountered in protein hydrolysates and cured cheeses, and this effect is caused by the overall hydrophobicity of amino acid side chains in peptides. All peptides contain suitable numbers of AH-type polar groups that can fit the polar receptor site, but individual peptides vary greatly in the size and nature of their hydrophobic groupings and thus in the ability of these hydrophobic groups to interact with the essential hydrophobic sites of bitterness receptors. Ney [50] has shown that the bitter taste of peptides can

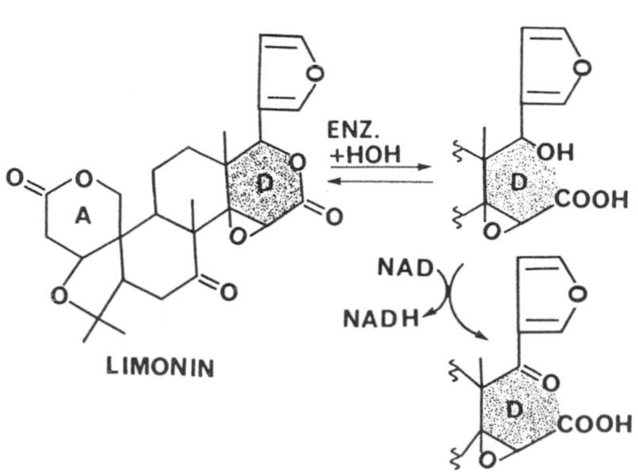

FIGURE 2 Structure of limonin and reactions leading to enzymatically induced nonbitter derivatives (the remainder of the molecule, including the A ring, remains unchanged).

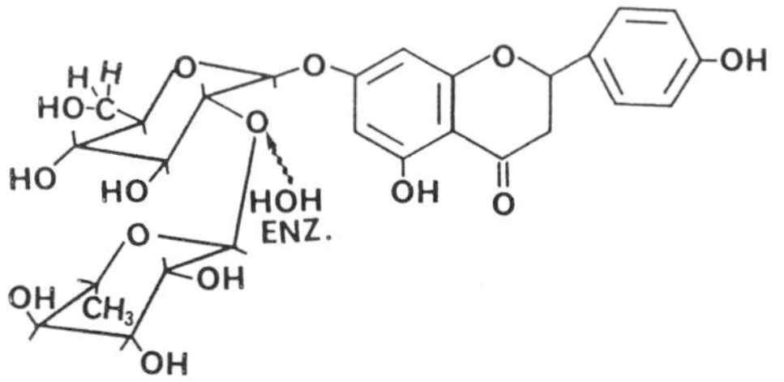

FIGURE 3 Structure of naringen showing site of enzymatic hydrolysis leading to nonbitter derivatives.

be predicted by calculation of a mean hydrophobicity value, termed Q. The ability of a protein to engage in hydrophobic associations is related to the sum of the individual hydrophobic contributions of the nonpolar, amino acid side chains, and these interactions contribute mainly to the free energy (ΔG) associated with protein unfolding. Thus, by summation of ΔG values for the individual amino acid side chains in a peptide, it is possible to calculate the mean hydrophobicity Q using Equation 2:

$$Q = \frac{\Sigma \Delta G}{n} \quad (2)$$

where n is the number of amino acid residues. Individual ΔG values for amino acids have been determined from solubility data [65], and these are summarized in Table 1. The Q values above

TABLE 1 Calculated ΔG Values for Individual Amino Acids

Amino acids	ΔG value[a] (kJ mol^{-1})	
Glycine	0	(0)
Serine	167.3	(40)
Threonine	1839.9	(440)
Histidine	2090.8	(500)
Aspartic acid	2258.1	(540)
Glutamic acid	2299.9	(550)
Arginine	3052.6	(730)
Alanine	3052.6	(730)
Methionine	5436.1	(1300)
Lysine	6272.4	(1500)
Valine	7066.9	(1690)
Leucine	10,119.5	(2420)
Proline	10,955.8	(2620)
Phenylalanine	11,081.2	(2650)
Tyrosine	12,001.2	(2870)
Isoleucine	12,419.4	(2970)
Tryptophan	12,544.8	(3000)

[a]ΔG values in kcal mol^{-1} are shown in parentheses; 1 kcal = 4.1814 kJ.
Source: From Ref. 50.

5855 based on kJ (1400 based on kcal) indicate that the peptide will be bitter; values below 5436 based on kJ (1300 based on kcal) assure that the peptide will not be bitter. The molecular weight of a peptide also influences its ability to produce bitterness, and only those with molecular weights below 6000 have a potential for bitterness. Peptides larger than this apparently are denied access to receptor sites because of their bulkiness (also see Chap. 6).

The peptide shown in Figure 4 is derived from the cleavage of α_S1-casein between residue 144–145 and residue 10–151 [50], and has a calculated Q value of 9576 based on kJ (2290 based on kcal). This peptide is very bitter, and is illustrative of the strongly hydrophobic peptides that can be derived readily from α_S1-casein. Such peptides are responsible for bitterness that develops in aged cheeses.

Ney [51] has also extended this general approach to predict the bitterness of lipid derivatives and sugars. Hydroxylated fatty acids, particularly some trihydroxy derivatives, are frequently bitter. The ratio of the number of C atoms to the numberof OH groups, or the R value (n_C/n_{OH}), of a molecule gives an indication of the bitterness of these substances. Sweet compounds yield R values of 1.00–1.99, bitter compounds 2.00–6.99, and nonbitter compounds have values above 7.00.

Bitterness in salts appears to involve a different mechanism of reception that seems to be related to the sum of the ionic diameters of the anion and cation components [4, 5]. Salts with ionic diameters below 6.5 Å are purely salty in taste (LiCl = 4.98 Å; NaCl = 5.56 Å; KCl = 6.28 Å), although some individuals find KCl to be somewhat bitter. As the ionic diameter increases (CsCl = 6.96 Å; CsI = 7.74 Å), salts become increasingly bitter. Magnesium chloride (8.50 Å) is therefore quite bitter.

11.2.1.4 The Salty and Sour Taste Modalities

Classic salty taste is represented by sodium chloride (NaCl), and is also given by lithium chloride (LiCl). National policies encouraging reduction in sodium consumption have stimulated interest in foods in which sodium salts have been replaced by alternative substances, particularly those containing potassium and ammonium ions. Since these foods have different, usually less desirable, tastes than those flavored with NaCl, renewed efforts are being expended to better understand mechanisms of the salty taste, in the hope that low-sodium products with near normal salty taste can be devised.

Salts have complex tastes, consisting of psychological mixtures of sweet, bitter, sour, and salty perceptual components. Furthermore, it has been shown recently that the tastes of salts often fall outside the traditional taste sensation [56] and are difficult to describe in classic terms. Nonspecific terms, such as chemical and soapy, often seem to more accurately describe the sensations produced by salts than do the classic terms.

Chemically, it appears that cations cause salty tastes and anions modify salty tastes [3].

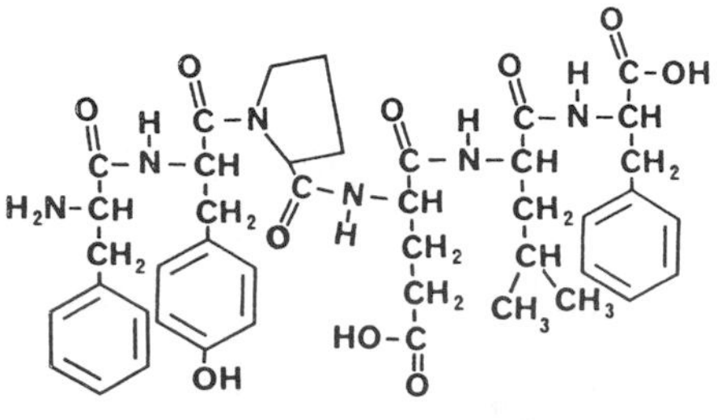

FIGURE 4 Structure of bitter peptide (phe-tyr-pro-glu-leu-phe) derived from α_S1-casein, showing strong nonpolar features.

Sodium and lithium produce only tastes, while potassium and other alkaline earth cations produce both salty and bitter tastes. Among the anions commonly found in foods, the chloride ion is least inhibitory to the salty taste, and the citrate anion is more inhibitory than orthophosphate anions. Anions not only inhibit tastes of cations, but also contribute tastes of their own. The chloride ion does not contribute a taste, and the citrate anion contributes less than the orthophosphate anion. Anion taste effects impact the flavor of foods, such as processed cheese, where citrate and phosphate anions contained in emulsifying salts (Chap. 12) suppress perceived saltiness contributed by sodium ions. Similarly, soapy tastes caused by sodium salts of long-chain fatty acids (IX) and detergents or long-chain sulfates (X) result from specific tastes elicited by the anions, and these tastes can completely mask the taste of the cation.

$$H_3C-(CH_2)_{10}-C(=O)-O^-, Na^+$$

(IX) Sodium laurate

$$H_3C-(CH_2)_n-S(=O)_2-O^-, Na^+$$

(X) Sodium lauryl sulfate

The most acceptable model for describing the mechanism of salty taste perception involves the interaction of hydrated cation–anion complexes [4] with AH/B-type receptor sites (discussed earlier). The individual structures of such complexes vary substantially, so that both water OH groups and salt anions or cations associate with receptor sites.

Similarly, the perception of sour compounds is believed to involve an AH/B-type receptor site [4]. However, data are not sufficient to determine whether hydronium ions (H_3O^+), dissociated inorganic or organic anions, or nondissociated molecular species are most influential in the sour response. Contrary to popular belief, the acid strength in a solution does not appear to be the major determinant of the sour sensation; rather, other poorly understood molecular features appear to be of primary importance (e.g., molecular weight, size, and overall polarity).

11.2.2 Flavor Enhancers

Compounds eliciting this unique effect have been utilized by humans since the inception of food cooking and preparation, but the actual mechanism of flavor enhancement remains largely a mystery. These substances contribute a delicious or umami taste to foods when used at levels in excess of their independent detection threshold, and they simply enhance flavors at levels below their independent detection thresholds. Their effects are prominent and desirable in the flavors of vegetables, dairy products, meats, poultry, fish, and other seafoods. The best known members of this group of substances are the 5′-ribonucleotides, of which 5′-inosine monophosphate or 5′-IMP (XI) serves as a suitable example, and monosodium L-glutamate (MSG) (XII).

(XI) 5′-Inosine monophosphate

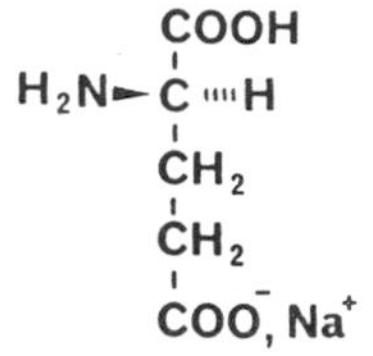

$$\text{H}_2\text{N}-\overset{\overset{\text{COOH}}{|}}{\text{C}}\cdots\text{H}-\text{CH}_2-\text{CH}_2-\text{COO}^-,\ \text{Na}^+$$

(XII) L-Glutamate, Na^+ (MSG)

D-Glutamate and the 2′- or the 3′-ribonucleotides do not exhibit flavor enhancing activity. Although MSG and 5′-IMP and 5′-guanosine monophosphate are the only flavor enhancers used commercially, 5′-xanthine monophosphate and a few natural amino acids, including L-ibotenic acid and L-tricholomic acid, are potential candidates for commercial use [74]. Much of the flavor contributed to foods by yeast hydrolysates results from the 5′-ribonucleotides present. Large amounts of purified flavor enhancers employed in the food industry are derived from microbial sources, including phosphorylated (in vitro) nucleosides derived from RNA [37].

Several synthetic derivatives of the 5′-ribonucleotides have strong flavor-enhancing properties [37]. Generally, these derivatives have substitutions on the purine moiety in the 2-position. The strong dependence of flavor-enhancement activity on specific structural characteristics suggests the involvement of a receptor site for these substances, or possible a joint occupancy of receptor sites usually involved in perception of sweet, sour, salty, and bitter sensations. It is well documented that a synergistic interaction occurs between MSG and the 5′-ribonucleotides in providing both the umami taste and in enhancing flavors. This suggests that some common structural features exist among active compounds. Undoubtedly, efforts will remain concentrated on gaining an understanding of the fundamental mechanisms involved in this extremely important component of taste.

Although most attention has been directed toward the 5′-ribonucleotides and MSG, other flavor-enhancing compounds have been claimed to exist. Maltol and ethyl maltol are worthy of mention because they are used commercially as flavor enhancers for sweet goods and fruits. At high concentrations maltol possesses a pleasant, burnt caramel aroma. It provides a smooth, velvety sensation to fruit juices when it is employed in concentrations of about 50 ppm. Maltol belongs to a group of unique compounds that exist in a planar enolone form [52]. The planar enolone is completely favored over the cyclic diketone form because the enolone form allows strong intramolecular H-bonding to occur (Eq. 3).

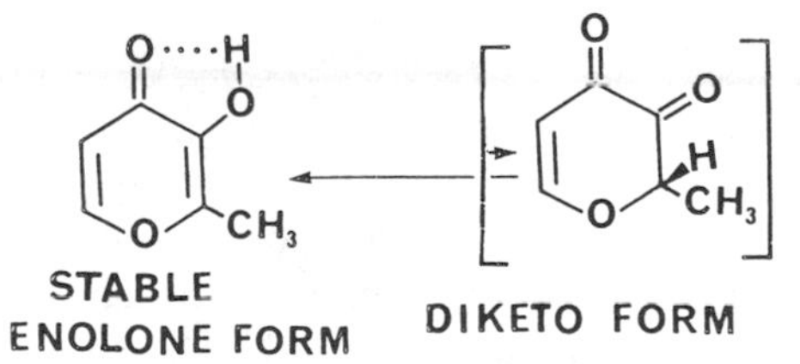

STABLE ENOLONE FORM ← DIKETO FORM (3)

Maltol and ethyl maltol (-C_2H_5 instead of -CH_3 on the ring) both could fit the AH/B portion of the sweet receptor (Fig. 1), but ethyl maltol is more effective as a sweetness enhancer than maltol. Maltol lowers the detection threshold concentration for sucrose by a factor of two. The actual mechanism for the enhancing effects of these compounds is unknown. Similar compounds of this type are derived naturally from browning reactions and are noted later in the chapter in the section on the development of "reaction" flavors.

Excellent discussions on flavor enhancement can be found in several reviews [37, 74].

11.2.3 Astringency

Astrigency is a taste-related phenomenon, perceived as a dry feeling in the mouth along with a coarse puckering of the oral tissue [39]. Astringency usually involves the association of tannins

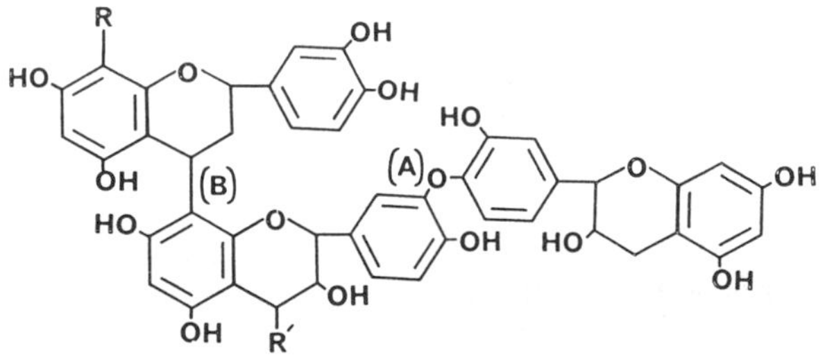

FIGURE 5 Structure of a procyanidin-type tannin showing condensed tannin linkage (B) and hydrolyzable tannin linkage (A) and also showing large hydrophobic areas capable of associating with proteins to cause astringency.

or polyphenols with proteins in the saliva to form precipitates or aggregates. Additionally, sparingly soluble proteins such as those found in certain dry milk powders also combine with proteins and mucopolysaccharides of saliva and cause astringency. Astringency is often confused with bitterness because many individuals do not clearly understand its nature, and many polyphenols or tannins cause both astringent and bitter sensations.

Tannins (Fig. 5) have broad cross-sectional areas suitable for hydrophobic associations with proteins. They also contain many phenolic groups that can convert to quinoid structures; these in turn can cross-link chemically with proteins [49]. Such cross-links have been suggested as a possible contributor to astringency activity.

Astringency may be a desirable flavor property, such as in tea. However, the practice of adding milk or cream to tea removes astringency through binding of polyphenols with milk proteins. Red wine is a good example of a beverage that exhibits both astringency and bitterness cause by polyphenols. However, too much astringency is considered sensorily undesirable in wines, and means are often taken to reduce polyphenol tannins, which are related to the anthocyanin pigments. Astringency derived from polyphenols in unripe bananas can also lead to an undesirable taste in product to which the bananas have been added [20].

11.2.4 Pungency

Certain compounds found in several spices and vegetables cause characteristic hot, sharp, and stinging sensations that are known collectively as pungency [12]. Although these sensations are difficult to separate from those of general chemical irritation and lachrymatory effects, they are usually considered separate flavor-related sensations. Some pungent principles, such as those found in chili peppers, black pepper, and ginger, are not volatile and exert their effects on oral tissues. Other spices and vegetables contain pungent principles that are somewhat volatile, and produce both pungency and characteristic aromas. These include mustard, horseradish, vegetable radishes, onions, garlic, watercress, and the aromatic spice, clove, which contains eugenol as the active component. All these spices and vegetables are used in foods to provide characteristic flavors or to generally enhance the palatability. Usage at low concentrations in processed foods frequently provides a liveliness to flavors through subtle contributions that fill out the perceived flavors. Only the three major pungent spices, chili peppers, ginger, and pepper, are discussed in this chapter in discussions about plant-derived flavors (i.e., eugenol, isothiocyanates, and thiopropanal *S*-oxide). Comprehensive reviews on pungent compounds [e.g., 21] should be consulted for in-depth information.

Chili peppers (*Capsicum* sp.) contain a group of substances known as capsaicinoids, which are vanillylamides of monocarboxylic acids with varying chain length (C_8–C_{11}) and unsaturation. Capsaicin (XIII) is representative of these pungent principles. Several capsaicinoids containing

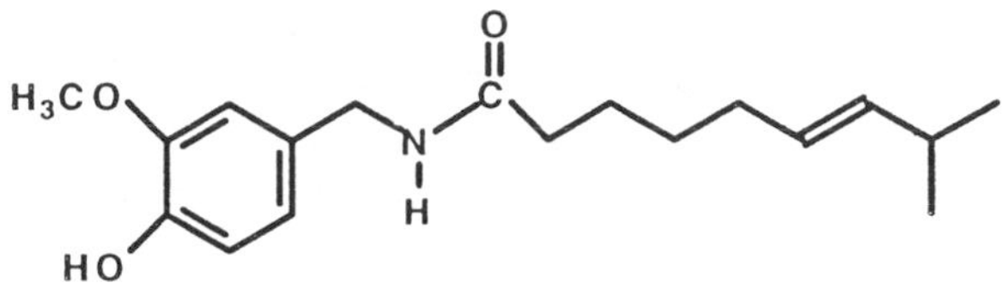

(XIII) Capsaicin

saturated straight-chain acid components are synthesized as substitutes for natural chili extractives or oleorisins. The total capsaicinoid content of world *Capsicum* species vary widely [21]; for example, red pepper contains 0.06%, cayenne red pepper, 0.2%, Sannam (India), 0.3%, Uganda (Africa), 0.85%. Sweet paprika contains a very low concentration of pungent compounds and is used mainly for its coloring effects and subtle flavor. Chili peppers also contain volatile aroma compounds that become part of the overall flavor of foods seasoned with them.

Black and white pepper are made from the berries of *Piper nigrum*, and differ only in that black pepper is prepared from immature, green berries and white pepper is made from more mature berries usually harvested at the time they are changing from green to yellow in color, but before they become red. The principal pungent compound in pepper is piperine (XIV), an amide. The trans geometry of the alkyl unsaturation is necessary for strong pungency, and loss of

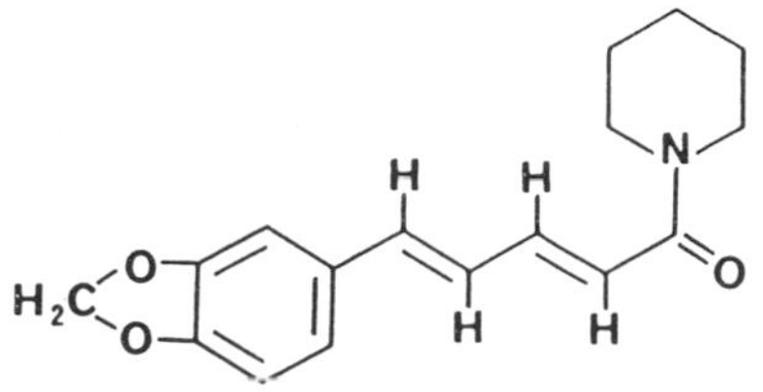

(XIV) Piperine

pungency during exposure to light and storage is attributed mainly to isomerizations of these double bonds [21]. Pepper also contains volatile compounds, including 1-formylpiperdine and piperonal (heliotropin), which contribute to flavors of foods seasoned with pepper spice or oleoresins. Piperine is also synthesized for use in flavoring foods.

Ginger is a spice derived from the rhizome of a tuberous perennial, *Zingiber officinale* Roscoe, and it possesses pungent principles as well as some volatile aroma constituents. The pungency of fresh ginger is caused by a group of phenylalkyl ketones called gingerols, and [6]-gingerol (XV) is most active of these compounds. Gingerols vary in chain length (C_5–C_9)

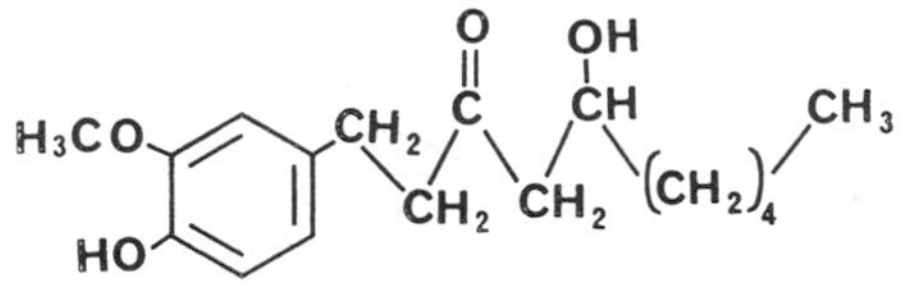

(XV) [6]-Gingerol

external to the hydroxyl substituted C atom. During drying and storage gingerols tend to dehydrate to form an external double bond that is in conjugation with the keto group. The reaction results in a group of compounds known as shogoals, which are even more potent than the gingerols. Exposure of [6]-gingerol to an elevated temperature leads to cleavage of the alkyl chain external to the keto group, yielding a methyl ketone, zingerone, which exhibits only mild pungency.

11.2.5 Cooling

Cooling sensations occur when certain chemicals contact the nasal or oral tissues and stimulate a specific saporous receptor [70]. These effects are most commonly associated with mint-like flavors, including peppermint, spearmint, and wintergreen. Several compounds cause the sensation but (–)-menthol (XVI), in the natural form (l-isomer), is most commonly used in flavors. The fundamental mechanism of the cooling sensation is not known, but compounds yielding this sensation also produce accompanying aromas. Camphor (XVII) is

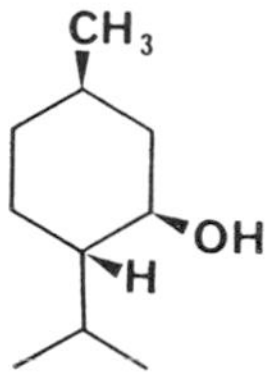

(XVI) (–)-Menthol

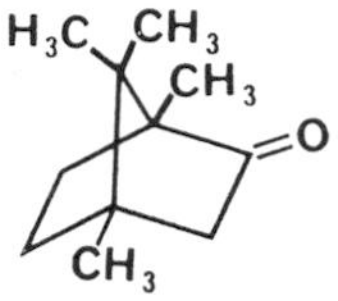

(XVII) d-Camphor

cited often as the model for this group of compounds, and it produces a distinctive camphoraceous odor in addition to a cooling sensation. The cooling effect produced by the mint-related compounds is mechanistically different from the slight cooling sensation produced when polyol sweeteners (Chap. 4 and 12), such as xylitol, are tasted as crystalline materials. In the latter case, it is generally believed that an endothermic dissolution of the materials gives rise to the effect.

11.3 VEGETABLE, FRUIT, AND SPICE FLAVORS

Categorization of vegetable and fruit flavors in a reasonably small number of distinctive groups is not easy, since logical groupings are not necessarily available for vegetables and fruits. For example, some information on plant-derived flavors was presented in the section on pungency, and some are covered in the section dealing with the development of "reaction" flavors. Emphasis in this section is on the biogenesis and development of flavors in important vegetables and fruits. For information on the other fruit and vegetable flavors, the reader is directed to the general references.

11.3.1 Sulfur-Containing Volatiles in *Allium* sp.

Plants in the genus *Allium* are characterized by strong, penetrating aromas, and important members are onions, garlic, leek, chives, and shallots. These plants lack the strong characterizing aroma unless the tissue is damaged and enzymes are decompartmentalized so that flavor precursors can be converted to odorous volatiles. In the case of onions (*A. cepa* L.), the precursor of the sulfur compounds that are responsible for the flavor and aroma of this vegetable is *S*-(l-propenyl)-L-cysteine sulfoxide [60, 71]. This precursor is also found in leeks. Rapid hydrolysis of the precursor by allinase yields a hypothetical sulfenic acid intermediate along with ammonia and pyruvate (Fig. 6). The sulfenic acid undergoes further rearrangements to yield the lachrymator, thiopropanal S-oxide, which is associated with the overall aroma of fresh onions. The pyruvic acid produced by the enzymatic conversion of the precursor compound is a stable product of the reaction and serves as a good index of the flavor intensity of onion products. Part

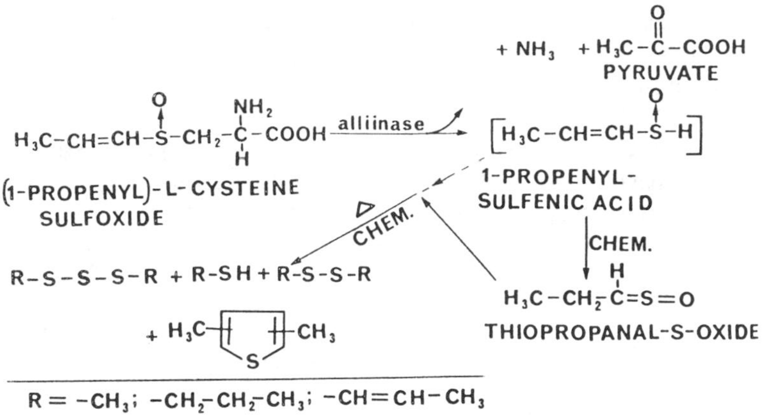

$H_3C{-}CH{=}CH{-}\overset{O}{\overset{\uparrow}{S}}{-}CH_2{-}\underset{H}{\overset{NH_2}{C}}{-}COOH \xrightarrow{\text{alliinase}} [H_3C{-}CH{=}CH{-}\overset{O}{\overset{\uparrow}{S}}{-}H]$ + NH_3 + $H_3C{-}\overset{O}{\overset{\|}{C}}{-}COOH$ (PYRUVATE)

S-(1-PROPENYL)-L-CYSTEINE SULFOXIDE

1-PROPENYL-SULFENIC ACID

CHEM. → $H_3C{-}CH_2{-}\overset{H}{C}{=}S{=}O$ THIOPROPANAL-S-OXIDE

Δ CHEM. → R-S-S-S-R + R-SH + R-S-S-R + dimethylthiophene

R = $-CH_3$; $-CH_2{-}CH_2{-}CH_3$; $-CH{=}CH{-}CH_3$

FIGURE 6 Reactions involved in the formation of onion flavor (chem. = nonenzymic).

of the unstable sulfenic acid also rearranges and decomposes to a rather large number of compounds in the classes of mercaptans, disulfides, trisulfides, and thiophenes. These compounds also comprise the flavor substances which provide cooked onion flavors.

The flavor of garlic (*Allium sativum* L.) is formed by the same general type of mechanism that functions in onion, except that the precursor is *S*-(2-propenyl)-L cysteine sulfoxide [60]. Diallyl thiosulfinate (allicin) (Fig. 7) contributes to flavor of garlic, and an *S*-oxide lachyramator similar to that formed in onion is not formed. The thiosulfinate flavor compound of garlic decomposes and rearranges in much the same manner as indicated for the sulfenic acid of onion (Fig. 6). This results in methyl allyl and diallyl disulfides and other principles in garlic oil and cooked garlic flavors.

11.3.2 Sulfur-Containing Volatiles in the *Cruciferae*

The Cruciferae family contains *Brassica* plants such as cabbage (*Brassica oleracea capitata* L.), brussel sprouts (*Brassica oleracea* var. *gemmifera* L.), turnips (*Brassica rapa* var. *rapa* L.), and brown mustard (*Brassica juncea* Coss.), as well as watercress (*Nasturtium officinale* R. Br.), radishes (*Raphanus sativus* L.), and horseradish (*Armoracia lapathifolia* Gilib). As noted in the discussion about pungent compounds, the active pungent principles in the Cruciferae are also volatile and therefore contribute to characteristic aromas. Further, the pungency sensation frequenly involves irritation sensations, particularly in the nasal cavity, and lachrymatory effects. The flavor compounds in these plants are formed through enzymic processes in disrupted tissues and through cooking. The fresh flavors of the disrupted tissue are caused mainly by isothio-

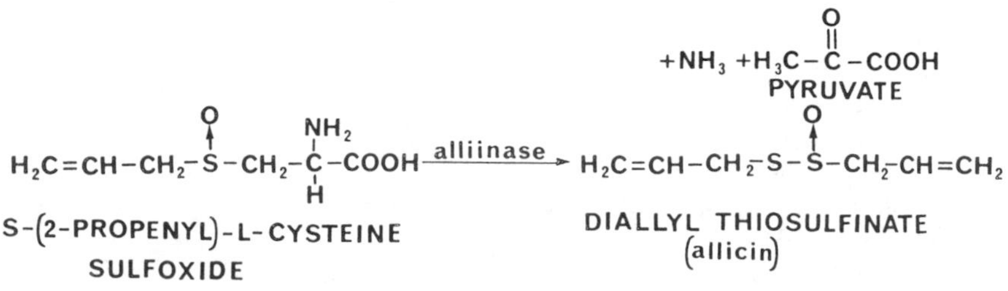

$H_2C{=}CH{-}CH_2{-}\overset{O}{\overset{\uparrow}{S}}{-}CH_2{-}\underset{H}{\overset{NH_2}{C}}{-}COOH \xrightarrow{\text{alliinase}} H_2C{=}CH{-}CH_2{-}S{-}\overset{O}{\overset{\uparrow}{S}}{-}CH_2{-}CH{=}CH_2$ + NH_3 + $H_3C{-}\overset{O}{\overset{\|}{C}}{-}COOH$ (PYRUVATE)

S-(2-PROPENYL)-L-CYSTEINE SULFOXIDE

DIALLYL THIOSULFINATE (allicin)

FIGURE 7 Formation of the principal fresh garlic flavor compounds.

cyanates resulting from the action of glucosinolases on thioglycoside precursors. The reaction shown in Figure 8, yielding allyl isothiocyanate, is illustrative of the flavor-forming mechanism in Cruciferae, and the resulting compound is the main source of pungency and aroma in horseradish and black mustard [21].

Several glucosinolates (*S*-glycosides, Chaps. 4 and 13) occur in the cruciferae [60], and each gives rise to characteristic flavors. The mild pungency of radishes is caused by the aroma compound 4-methylthio-3-*t*-butenylisothiocyanate (XVIII). In addition to the isothiocyanates, glucosinolates also yield thiocyanates (R-S=C=N) and nitriles (Fig. 8).

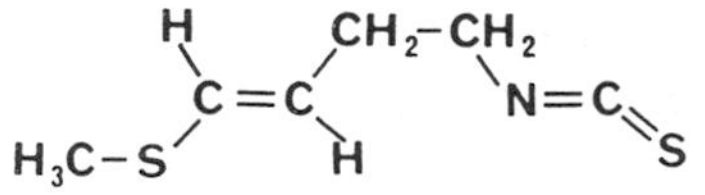

(XVIII) 4-Methylthio-3-*t*-butenylisothiocyanate

Cabbage and brussels sprouts contain both allyl isothiocyanate and allyl nitrile, and the concentration of each varies with the stage of growth, location in the edible part, and the severity of processing encountered. Processing at temperatures well above ambient (cooking and dehydrating) tends to destroy the isothiocyanates and enhance the amount of nitriles and other sulfur-containing degradation and rearrangement compounds. Several aromatic isothiocyanates occur in Cruciferae; for example, 2-phenylethyl isothiocyanate is one of the main aroma compounds of watercress. This compound also contributes a tingling pungent sensation.

11.3.3 Unique Sulfur Compound in Shiitake Mushrooms

A novel C-S lyase enzyme system has been discovered in Shiitake mushrooms (*lentinus edodes*), which are prized in Japan for their delicious flavor. The precursor for the major flavor contributor, lentinic acid, is an *S*-substituted L-cysteine sulfoxide bound as a γ-glutamyl peptide [75]. The initial enzyme reaction in flavor development involves a γ-glutamyl transpeptidase, which releases a cysteine sulfoxide precursor (lentinic acid). Lentinic acid is then attacked by *S*-alkyl-L-cysteine sulfoxide lyase (Fig. 9) to yield to lenthionine, the active flavor compound. These reactions are initiated only after the tissue is disrupted, and the flavor develops only after drying and rehydration or after holding freshly mascerated tissue for a short period of time. Other polythiepanes in addition to lenthionine are formed [60], but the flavor is ascribed

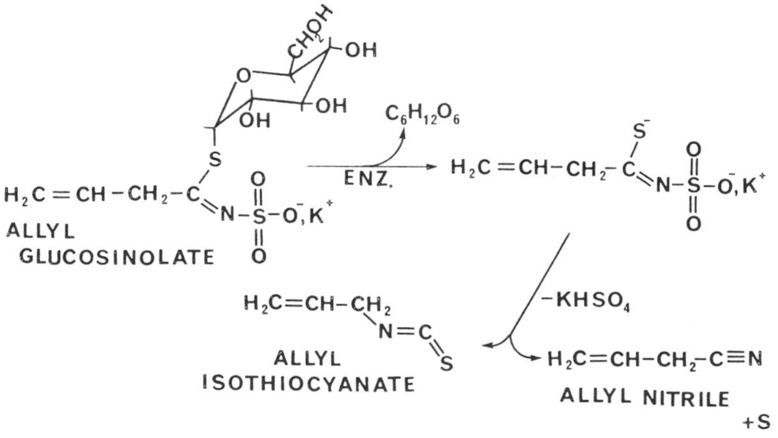

ALLYL GLUCOSINOLATE
ENZ.
$C_6H_{12}O_6$
$-KHSO_4$
ALLYL ISOTHIOCYANATE
ALLYL NITRILE
+S

FIGURE 8 Reactions involved in the formation of *Cruciferae* flavors.

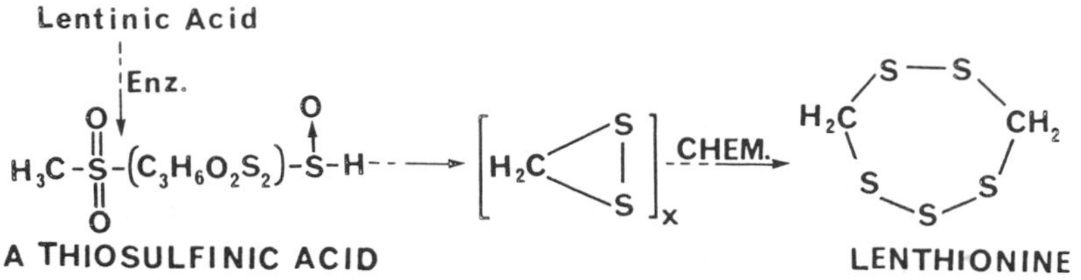

FIGURE 9 Formation of lenthionine in Shiitake mushrooms (chem. = nonenzymic).

to lenthionine. Flavor roles for other mixed-atom polysulfides with related structural features, remain largely unexplored.

11.3.4 Methoxy Alkyl Pyrazine Volatiles in Vegetables

Many fresh vegetables exhibit green-earthy aromas that contribute strongly to their recognition, and it has been found that the methoxy alkyl pyrazines are frequently responsible for this property [71]. These compounds have unusually potent and penetrating odors, and they provide vegetables with strong identifying aromas. 2-Methoxy-3-isobutylpyrazine was the first of this class discovered, and it exhibits a powerful bell pepper aroma detectable at a threshold level of 0.002 ppb. Much of the aroma of raw potatoes and green pea pods is contributed by 2-methoxy-3-isopropyl pyrazine, and 2-methoxy-3-*s*-butylpyrazine contributes to the aroma of raw red beet roots. These compounds arise biosynthetically in plants, and some strains of microorganisms (*Pseudomonas perolens* and *Pseudomonas tetrolens*) also actively produce these unique substances [44]. Branched-chain amino acids serve as precursors for methoxy alkyl pyrazine volatiles, and the mechanistic scheme shown in Figure 10 has been proposed.

11.3.5 Enzymically Derived Volatiles from Fatty Acids

Enzymically generated compounds derived from long-chain fatty acids play an extremely important role in the characteristic flavors of fruits and vegetables. In addition, these types of reactions can lead to important off flavors, such as those associated with processed soybean proteins. Further information about these reactions can be found in the discussions about lipids (Chap. 5) and enzymes (Chap. 7).

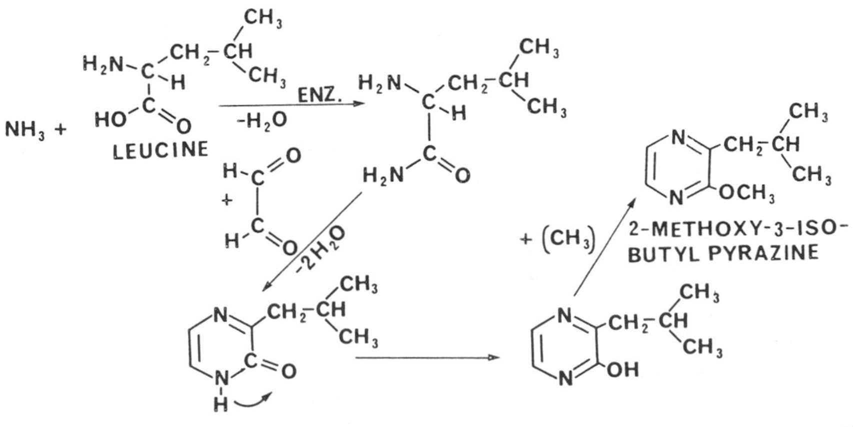

FIGURE 10 Proposed enzymatic scheme for the formation of methoxy alkyl pyrazines.

11.3.5.1 Lipoxygenase-Derived Flavors in Plants

In plant tissues, enzyme-induced oxidative breakdown of unsaturated fatty acids occurs extensively, and this yields characteristic aromas associated with some ripening fruits and disrupted tissues [16]. In contrast to the random production of lipid-derived flavor compounds by purely autoxidizing systems, very distinctive flavors occur when the compounds produced are enzyme determined. The specificity for flavor compounds is illustrated in Figure 11, where the production of 2-*t*-hexenal and 2-*t*,6-*c*-nonadienal by site-specific hydroperoxidation of a fatty acid is dictated by a lipoxygenase and a subsequent lyase cleavage reaction. Upon cleavage of the fatty acid molecule, oxoacids are also formed, but they do not appear to influence flavors. Decompartmentalization of enzymes is required to initiate this and other reactions, and since successive reactions occur, overall aromas change with time. For example, the lipoxygenase-derived aldehydes and ketones are converted to corresponding alcohols (Fig. 12), which usually have higher detection thresholds and heavier aromas than the parent carbonyl compounds. Although not shown, *cis-trans* isomerases are also present that convert *cis*-3 bonds to *trans*-2 isomers. Generally, C_6 compounds yield green plant-like aromas like fresh-cut grass, C_9 compounds smell like cucumbers and melons, and C_8 compounds smell like mushrooms or violet and geranium leaves [67]. The C_6 and C_9 compounds are primary alcohols and aldehydes; the C_8 compounds are secondary alcohols and ketones.

11.3.5.2 Volatiles from β-Oxidation of Long-Chain Fatty Acids

The development of pleasant, fruity aromas is associated with the ripening of pears, peaches, apricots, and other fruits, and these aromas are frequently dominated by medium chain-length (C6–C12) volatiles derived from long-chain fatty acids by β-oxidation [68]. The formation of ethyl deca-2-*t*,4-*c*-dienoate by this means is illustrated in Figure 13. This ester is the impact or characterizing aroma compound in the Bartlett pear. Although not included in the figure, hydroxy acids (C8–C12) are also formed through this process, and they cyclize to yield γ and δ-lactones. Similar reactions occur during degradation of milk fat, and these reactions are discussed in more detail in Section 11.5. The C8–C12 lactones possess distinct coconut-like and peach-like aromas characteristic of these respective fruits.

11.3.6 Volatiles from Branched-Chain Amino Acids

Branched-chain amino acids serve as important flavor precursors for the biosynthesis of compounds associated with some ripening fruits. Bananas and apples are particularly good

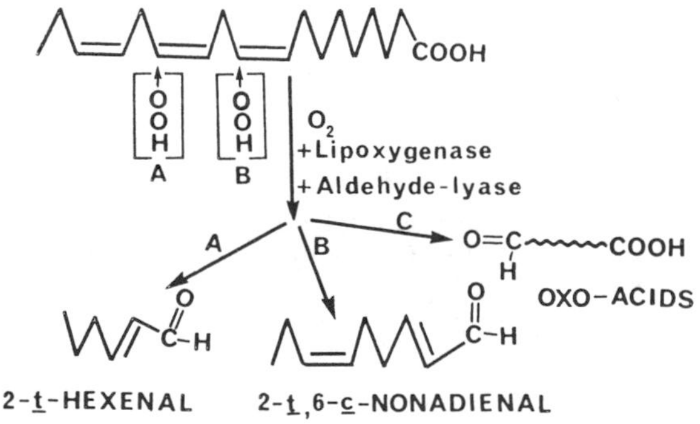

COOH
OOH A
OOH B
O_2 +Lipoxygenase +Aldehyde-lyase
A B C
O=C~~~COOH H OXO-ACIDS
2-t-HEXENAL 2-t,6-c-NONADIENAL

FIGURE 11 Formation of lipoxygenase-directed aldehydes from linolenic acid: (A) important in fresh tomatoes; (B) important in cucumbers.

2-t-6-c-NONADIENAL —Alcohol Dehydrogenase→ 2-t-6-c-NONADIENOL

FIGURE 12 Conversion of aldehyde to alcohol resulting in subtle flavor modification.

examples for this process because much of the ripe flavor of each is caused by volatiles from amino acids [68]. The initial reaction involved in flavor formation (Fig. 14) is sometimes referred to as enzymic Strecker degradation because transamination and decarboxylation occur that are parallel to those occuring during non-enzymatic browning. Several microorganisms, including yeast and malty flavor-producing strains of *Streptoccus lactis*, can also modify most of the amino acids in a fashion similar to that shown in Figure 14. Plants can also produce similar derivatives from amino acids other than leucine, and the occurrence of 2-phenethanol with a rose- or lilac-like aroma in blossoms is attributed to these reactions.

Although the aldehydes, alcohols, and acids from these reactions contribute directly to the flavors of ripening fruits, the esters are the dominant character-impact compounds. It has long been known that isoamyl acetate is important in banana flavor, but other compounds are also required to give full banana flavor. Ethyl 2-methylbutyrate is even more apple-like than ethyl 3-methylbutyrate, and is the dominant note in the aroma of ripe Delicious apples.

11.3.7 Flavors Derived from the Shikimic Acid Pathway

In biosynthetic systems, the shikimic acid pathway provides the aromatic portion of compounds related to shikimic acid, and the pathway is best known in its role in the production of phenylalanine and the aromatic amino acids. In addition to flavor compounds derived from aromatic amino acids, the shikimic acid pathway provides other volatile compounds that are frequently associated with essential oils (Fig. 15). It also provides the phenyl propanoid skeleton to lignin polymers that are the main structural elements of plants. As indicated in Figure 15, lignin yields many phenols during pyrolysis [72], and the characteristic aroma of smokes used in foods is largely caused by compounds developed from precursors in the shikimic acid pathway.

Also apparent from Figure 15 is that vanillin, the most important characterizing compound in vanilla extracts, can be obtained via this pathway or as a lignin by-product during processing of wood pulp and paper. Vanillin is also biochemically synthesized in the vanilla bean, where it initially is present largely as vanillin glucoside until the glycoside is hydrolyzed during fermentation. The methoxylated aromatic rings of the pungent principles in ginger, pepper, and

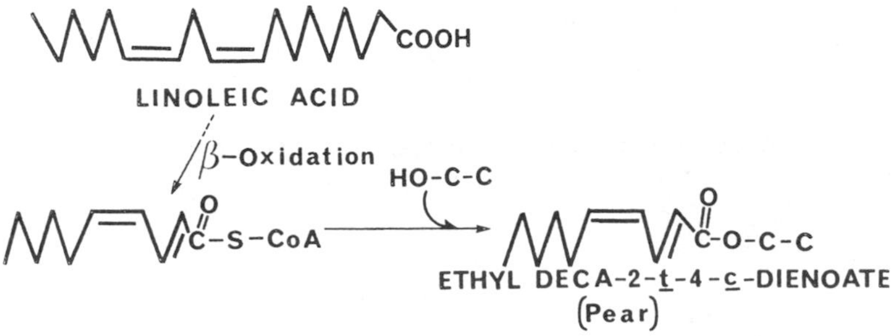

LINOLEIC ACID

β-Oxidation

HO-C-C

ETHYL DECA-2-t-4-c-DIENOATE (Pear)

FIGURE 13 Formation of a key aroma substance in pears through β-oxidation of linoleic acid followed by esterification.

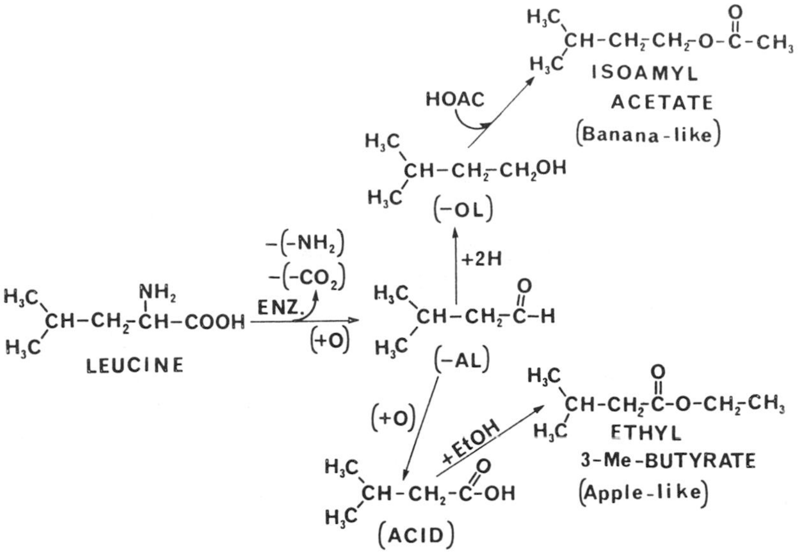

FIGURE 14 Enzymatic conversion of leucine to volatiles illustrating the aroma compounds formed from amino acids in ripening fruits.

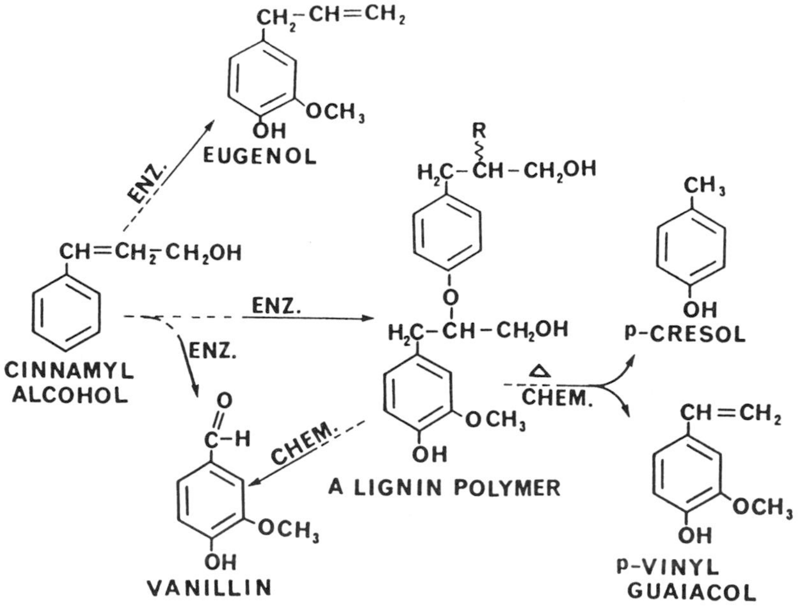

FIGURE 15 Some important flavor compounds derived from shikimic acid pathway precursors (chem. = nonenzymic).

chili peppers, discussed earlier in this chapter, also contain the essential features of those compounds in Figure 15. Cinnamyl alcohol is an aroma constituent of cinnamon spice, and eugenol is the principal aroma and pungency element in cloves.

11.3.8 Volatile Terpenoids in Flavors

Because of the abundance of terpenes in plant materials used in the essential oil and perfumery industries, their importance in other plant-associated flavors is sometimes underestimated. They are largely responsible, however, for the flavors of citrus fruits and many seasonings and herbs. Terpenes are present in low concentrations in several fruits, and are responsible for much of the flavor of raw carrot roots. Terpenes are biosynthesized through the isoprenoid (C_5) paths (Fig. 16), and monoterpenes contain 10 C atoms; the sesquiterpenes contain 15 C atoms.

Sesquiterpenes are also important characterizing aroma compounds, and β-sinensal (XIX) and nootkatone (XX) serve as good examples because they provide characterizing flavors to oranges and grapefruit, respectively. The diterpenes (C_{20}) are too large and nonvolatile to contribute directly to aromas.

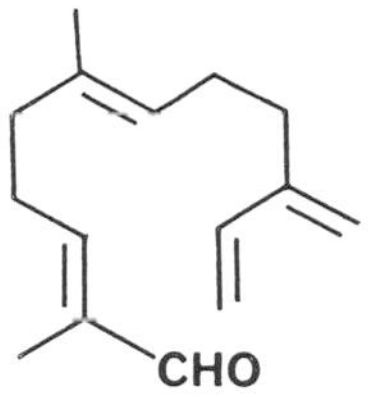

CHO

(XIX) β-Sinesal

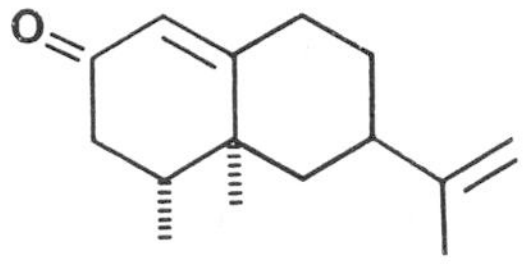

O

(XX) Nootkatone

Terpenes frequently possess extremely strong character-impact properties, and many can be easily identified by one experienced with natural product aromas. Optical isomers of terpenes,

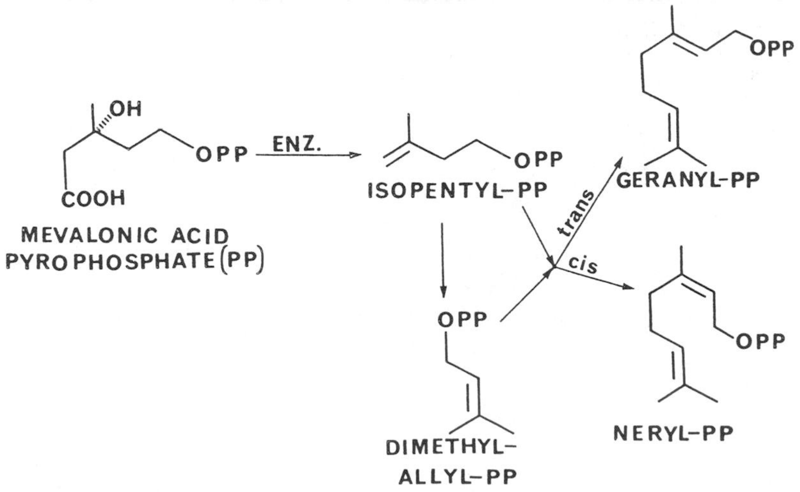

OH

OPP

ENZ.

COOH

MEVALONIC ACID
PYROPHOSPHATE (PP)

OPP

ISOPENTYL–PP

OPP

GERANYL–PP

trans

cis

OPP

OPP

DIMETHYL–
ALLYL–PP

NERYL–PP

FIGURE 16 Generalized isoprenoid scheme for the biosynthesis of monoterpenes.

as well as optical isomers of other nonterpenoid compounds, can exhibit extremely different odor qualities [10, 36]. In the case of terpenes, the carvones have been studied extensively from this perspective, and the aroma of *d*-carvone [4*S*-(+)carvone] (XXI) has the characteristic aroma of caraway spice; *l*-carvone [4*R*-(N)carvone] (XXII) possesses a strong, characteristic spearmint aroma. Studies on such pairs of compounds are of interest since they provide information on the fundamental process of olfaction and structure–activity relationships for molecules.

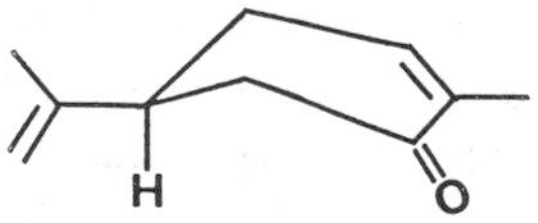

(XXI) 4*S*-(+)carvone

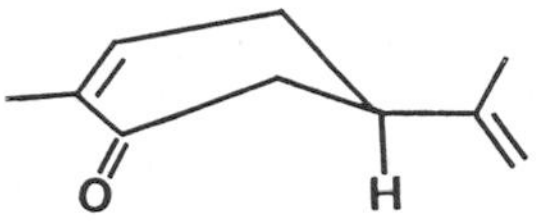

(XXII) 4*R*-(N)carvone

11.3.9 Citrus Flavors

Citrus flavors are among the most popular fresh fruits as well as flavors for beverages, and most information about the flavor chemistry of natural citrus flavors stems from research on processed juices, peel essential oils, essential oils, and aqueous essences used to flavor juice products. Several classes of flavor components serve as major contributors to citrus flavors, including terpenes, aldehydes, esters, and alcohols, and large numbers of volatile compounds have been identified in the various extracts from each citrus fruit [61]. However, the important flavor compounds, including character-impact compounds, for citrus fruits generally are limited to relatively few, and those considered important to some major citrus fruits are shown in Table 2.

The flavors of orange and mandarin (tangerine is used interchangeably with mandarin in the United States) are delicate, and quite easily changed. As can be noted from Table 2, relatively few aldehydes and terpenes are considered essential for these flavors, although a large number of other compounds are present. Both α- and β-sinensal (XIX) are present in orange and mandarin flavors, and α-sinensal is considered especially important to providing a ripe orange-

TABLE 2 Some Volatile Compounds Considered Important in Citrus Flavors

Orange	Mandarin	Grapefruit	Lemon
Ethanal	Ethanal	Ethanal	Neral
Octanal	Octanal	Decanal	Geranial
Nonanal	Decanal	Ethyl acetate	β-Pinene
Citral	α-Sinensal	Methyl butanoate	Geraniol
Ethyl butanoate	γ-Terpinene	Ethyl butanoate	Geranyl acetate
d-Limonene	β-Pinene	*d*-Limonene	Neryl acetate
α-Pinene	Thymol	Nootkatone	Bergamotene
	Methyl-*N*-methylanthranilate	1-*p*-Menthene-8-thiol	Caryophyllene
			Carvyl ethyl ether
			Linalyl ethyl ether
			Fenchyl ethyl ether
			Methyl epijasmonate

Source: From Ref. 61.

citrus flavor to mandarin flavors. Grapefruit contains two character-impact compounds, nootkatone (XX) and 1-*p*-methene-8-thiol, that provide much of the easily recognized flavor to this fruit. Nootkatone is used extensively to provide artificial grapefruit flavor, and *p*-menthene-8-thiol is one of a few sulfur compounds that are influential in citrus flavors.

Lemon flavor requires contributions from a large number of important compounds, and lemon flavors benefit from the present of several terpene ethers. Similarly, the number of essential compounds in lime flavor is large, and two types of lime oil are commonly used. The major lime oil in commerce is distilled Mexican lime oil, which possesses a harsh, strong lime flavor that is popular in lemon-lime and cola soft drinks. Cold-pressed Persian lime oil and centrifuged Mexican lime oils are becoming more popular because they possess a more natural flavor. The less rigorous processing involved in cold-pressing and centrifugation, compared to distillation, results in the survival of some of the more sensitive but important fresh lime flavor compounds. For example, citral, which has a desirable more fresh aroma, is degraded to *p*-cymene and α-*p*-dimethylstyrene, which have harsh flavors, under acidic distillation conditions, and this leads to the more harsh flavors of distilled lime oil [61].

Citrus essential oils, or flavor extracts containing terpenes, can be separated into non oxygenated (hydrocarbon) and oxygenated fractions using silicic acid chromatography and nonpolar and polar solvent elutions, respectively. Terpeneless orange oil, for example, contains principally the oxygenated terpenes, aldehydes, and alcohols, that have been recovered from orange oils. Oxygenated fractions are frequently more desired in flavors than nonoxygenated fractions.

11.3.10 Flavors of Herbs and Spices

Although some variations occur between industrial and the various domestic and international regulatory agencies regarding definitions, spices and herbs are known and regulated as spices and condiments, which are natural vegetable products used for flavoring, seasoning, and imparting aroma to foods. The U. S. Food and Drug Administration excludes alliaceous products, such as onions and garlic, from classification as spices, but international and industrial spice classifications commonly include these materials. The term condiments has been retained in regulatory definitions, and is defined as something to enhance the flavor of food, or a pungent seasoning. It has been argued that the term should be dropped from usage, but some favor retention of the term for certain situations. However, the term generally does not provide a basis for inclusion in classification schemes for aromatic plant materials.

In some botanically based classification schemes, culinary herbs are separated from spices, and include such aromatic, soft-stemmed plants as basil, marjoram, mints, oregano, rosemary, and thyme, as well as aromatic shrub (sage) and tree (laurel) leaves. In such classifications, spices comprise all other aromatic plant materials used in the flavoring or seasoning of foods. Such spices generally lack chlorophyll, and include rhizomes or roots (ginger), barks (cinnamon), flower buds (cloves), fruits (dill, pepper) and seeds (nutmeg, mustard).

Spices and herbs have been used since antiquity for adding savoriness, tanginess, and zestiness, as well as characterizing flavors to foods and beverages. Some also have been used extensively in perfumery and for medicinal purposes, and many exhibit antioxidant or microbial inhibitory effects. While many herbs and spices exist throughout the world, some of which are used for perfumery and herbal medicines, only about 70 are officially recognized as useful ingredients for foods. However, the flavor characteristics often vary for a given spice depending on location of growth and genetic variations; thus this group provides a wide range of flavors for foods. Here, only those spices and herbs that are commonly used in the food industry for flavoring are considered.

Spices generally are derived from tropical plants, while herbs are generally derived from subtropical or nontropical plants. Spices also generally contain high concentrations of phenylpropanoids from the shikimic acid pathway (e.g., eugenol in cloves; XXIII), herbs generally contain higher concentrations of *p*-menthanoids from terpene biosynthesis (e.g, menthol in peppermint; XXIV).

(XXIII) Eugenol

(XXIV) Menthol

Typically, spices and herbs contain a large number of volatile compounds, but in most instances certain compounds, either abundant or minor volatile constituents, provide character-impact aromas and flavor to the material. Important flavor compounds found in the principal herbs and spices used by the food industry are summarized in Tables 3 and 4, respectively. Successful applications of herbs and spices in foods and beverages require either a personal knowledge of the materials or knowledge of the dominant and subtle flavor notes provided by flavor chemicals. Evaluation of the important flavor compounds listed in Table 3 and 4 for spices and herbs shows which spices and herbs provide related flavors, and also indicates the type of flavors

TABLE 3 Important Flavor Compounds Found in Some Culinary Herbs Commonly Used for Flavoring in the Food Industry

Herbs	Plant part	Important flavor compounds
Basil, sweet	Leaves	Methylchavicol, linalool, methyl eugenol
Bay laurel	Leaves	1,8-Cineole
Marjoram	Leaves, flowers	*c*- and *t*-Sabinene hydrates, terpinen-4-ol
Oregano	Leaves, flowers	Carvacrol, thymol
Origanum	Leaves	Thymol, carvacrol
Rosemary	Leaves	Verbenone, 1,8-cineole, camphor, linalool
Sage, clary	Leaves	Salvial-4(14)-en-1-one, linalool
Sage, Dalmation	Leaves	Thujone, 1,8-cineole, camphor
Sage, Spanish	Leaves	*c*- and *t*-Sabinyl acetate, 1,8-cineole, camphor
Savory	Leaves	Carvacrol
Tarragon	Leaves	Methyl chavicol, anethole
Thyme	Leaves	Thymol, carvacrol
Peppermint	Leaves	*l*-Menthol, menthone, menthofuran
Spearmint	Leaves	*l*-Carvone, carvone derivatives

Source: From Refs. 10 and 54.

TABLE 4 Important Flavor Compounds Found in Some Spices Commonly Used for Flavoring in the Food Industry

Spice	Plant part	Important flavor compounds
Allspice (pimento)	Berry, leaves	Eugenol, β-caryophyllene
Anise	Fruit	(*E*)-Anethole, methyl chavicol
Capsicum peppers	Fruit	Capsaicin, dihydrocapsaicin
Caraway	Fruit	*d*-Carvone, carvone derivatives
Cardamom	Fruit	α-Terpinylacetate, 1,8-cineole, linalool
Cinnamon, cassia	Bark, leaves	Cinnamaldehyde, eugenol
Clove	Flower bud	Eugenol, eugenylacetate
Coriander	Fruit	*d*-Linalool, C_{10}–C_{14} 2-alkenals
Cumin	Fruit	Cuminaldehyde, *p*-1,3-menthadienal
Dill	Fruit, leaves	*d*-Carvone
Fennel	Fennel	(*E*)-Anethole, fenchone
Ginger	Rhizome	Gingerol, shogaol, neral, geranial
Mace	Aril	α-Pinene, sabinene, 1-terpenin-4-ol
Mustard	Seed	Allyl isothiocyanate
Nutmeg	Seed	Sabinine, α-pinene, myristicin
Parsley	Leaves, seed	Apiol
Pepper	Fruit	Piperine, δ-3-carene, β-carophyllene
Saffron	Stigma	Safranal
Turmeric	Rhizome	Turmerone, zingeriberene, 1,8-cineole
Vanilla	Fruit, seed	Vanillin, *p*-OH-benzyl methyl ether

Source: From Refs. 10 and 54.

to be expected from their usage. Still, it must be remembered that a large number of compounds of lesser influence are present and also contribute to the unique flavors of herbs and spices.

11.4 FLAVORS FROM LACTIC ACID–ETHANOL FERMENTATIONS

Involvement of microorganisms in flavor production is extensive, but often their specific or definitive role in the flavor chemistry of fermentations is not well known or the flavor compounds do not have great character impact. Much attention has been given to cheese flavor, but apart from the distinctive flavor properties given by methyl ketones and secondary alcohols to blue-mold cheeses and moderate flavor properties given by certain sulfur compounds to surface-ripened cheeses, microbially derived flavor compounds cannot be classified in the character-impact category. Similarly, yeast fermentations, carried out extensively for beer, wines, spirits, and yeast-leavened breads, do not appear to yield strong, distinctive character-impact flavor compounds. Ethanol in alcoholic beverages, however, should be considered as having character impact.

The primary fermentation products of heterofermentative lactic acid bacteria (e.g., *Leuconostoc citrovorum*) are summarized in Figure 17, and the combination of acetic acid, diacetyl, and acetaldehyde provides much of the characteristic aroma of cultured butter and buttermilk. Homofermentative lactic acid bacteria (e.g., *Lactococcus lactis, Streptococcus thermophilus*) produce only lactic acid, acetaldehyde, and ethanol in milk cultures. Acetaldehyde is the character-impact compound found in yogurt, a product prepared by a homofermentative process.

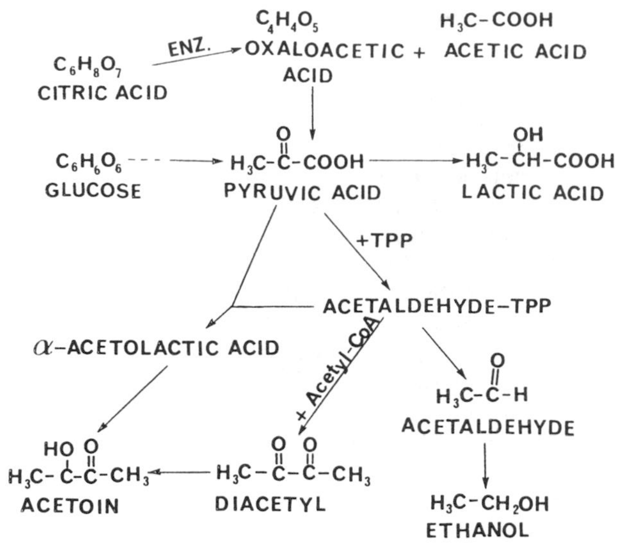

$C_6H_8O_7$ CITRIC ACID —ENZ.→ $C_4H_4O_5$ OXALOACETIC ACID + H_3C-COOH ACETIC ACID

$C_6H_6O_6$ GLUCOSE → H_3C-CO-COOH PYRUVIC ACID → H_3C-CH(OH)-COOH LACTIC ACID

PYRUVIC ACID —+TPP→ ACETALDEHYDE–TPP

ACETALDEHYDE–TPP + PYRUVIC ACID → α-ACETOLACTIC ACID → H_3C-CH(OH)-CO-CH_3 ACETOIN

ACETALDEHYDE–TPP —+ Acetyl-CoA→ H_3C-CO-CO-CH_3 DIACETYL → ACETOIN

ACETALDEHYDE–TPP → H_3C-CO-H ACETALDEHYDE → H_3C-CH_2OH ETHANOL

FIGURE 17 Principal volatile fermentation products from heterofermentative lactic acid bacterial metabolism.

Diacetyl is the character-impact compound in most mixed-strain lactic fermentations, and has become universally known as a dairy or butter-type flavorant. Lactic acid contributes sourness to cultured or fermented dairy products. Acetoin, although essentially odorless, can undergo oxidation to diacetyl.

Viewed in general terms, lactic acid bacteria produce very little ethanol (parts per million levels), and they use pyruvate as the principal final H receptor in metabolism. On the other hand, yeast produce ethanol as a major end product of metabolism. Malty strains of *Streptococcus lactis* and all brewer's yeast (*Saccharomyces cerevisiae*, *Saccharomyces carlsbergensis*) also actively convert amino acids to volatile compounds through transaminations and decarboxylations (Figure 18). These reactions are analogous to those discussed for branched-chain amino acids in Section 11.3.6. However, these organisms tend to produce mainly the reduced forms of the derivative (alcohols), although some oxidized compounds (aldehydes and acids) also appear. Wine and beer flavors, which can be ascribed directly to fermentations, involve complex mixtures of these volatiles and interaction products of these compounds with ethanol, such as mixed esters and acetals. These mixtures give rise to familiar yeasty and fruity flavors associated with fermented beverages.

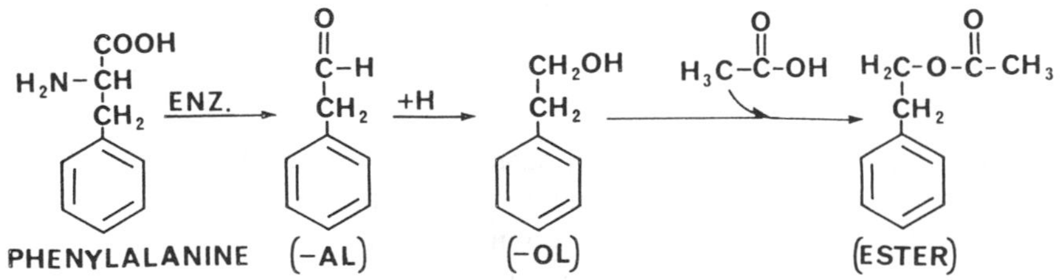

PHENYLALANINE (H_2N-CH(COOH)-CH_2-C_6H_5) —ENZ.→ (-AL) (O=CH-CH_2-C_6H_5) —+H→ (-OL) (CH_2OH-CH_2-C_6H_5) —H_3C-CO-OH→ (ESTER) (H_2C-O-CO-CH_3, CH_2-C_6H_5)

FIGURE 18 Enzymatic formation of volatiles from amino acids by microorganisms using phenylalanine as a model precursor compound.

11.5 FLAVOR VOLATILES FROM FATS AND OILS

Fats and oils are notorious for their role in the development of off flavors through autoxidation and the chemistry of lipid-derived flavors has been well summarized [23, 33]. Details of lipid autoxidation and other degradations are also presented in Chapter 5. Aldehydes and ketones are the main volatiles from autoxidation, and these compounds can cause painty, fatty, metallic, papery, and candle-like flavors in foods when their concentrations are sufficiently high. However, many of the desirable flavors of cooked and processed foods derive from modest concentrations of these compounds.

Hydrolysis of plant acyglycerols and animal depot fats leads mainly to the formation of potentially soapy-tasting fatty acids. Milkfat, on the other hand, serves as a rich source of volatile flavor compounds that are influential in the flavor of dairy products and foods prepared with milkfat or butter. The classes of volatiles obtained by hydrolysis of milkfat are shown in Figure 19, with specific compounds selected to illustrate each class. The even C-numbered, short-chain fatty acids (C_4–C_{12}) are extremely important in the flavor of cheese and other dairy products, with butyric acid being the most potent and influential of the group. Hydrolysis of hydroxy fatty acids leads to the formation of lactones, which provide desirable coconut-like or peach-like flavor notes to baked goods, but cause staling in stored, sterile concentrated milks. Methyl ketones are produced from β-ketoacids by heating after hydrolysis, and they contribute to the flavor of dairy products in much the same manner as the lactones. In blue-mold cheeses, however, methyl ketones are much more abundantly produced by the metabolic activities of *Penicillium roquefort* on fatty acids than by the conversion of those bound in acylglycerols.

Although hydrolysis of fats other than milkfat does not yield distinct flavors as noted earlier, animal fats are believed to be inextricably involved in the species-related flavors of meats. The role of lipids in determining species-related aspects of meat flavors is discussed in the next section.

11.6 FLAVOR VOLATILES IN MUSCLE FOODS AND MILK

The flavor of meats have attracted much attention, but in spite of considerable research, knowledge about the flavor compounds causing strong character impacts for meats of various species is limited [58]. Nevertheless, the concentrated research efforts on meat flavors have produced a wealth of information about compounds that contribute to cooked meat flavors. The

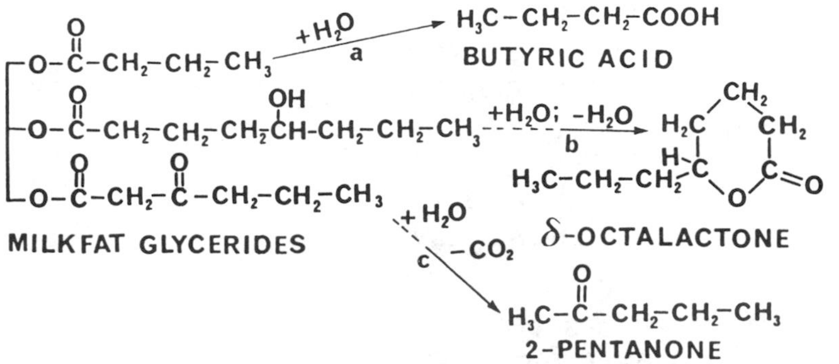

MILKFAT GLYCERIDES
+H₂O a → $H_3C-CH_2-CH_2-COOH$ BUTYRIC ACID
+H₂O; −H₂O b → δ-OCTALACTONE
+H₂O −CO₂ c → $H_3C-CO-CH_2-CH_2-CH_3$ 2-PENTANONE

FIGURE 19 Formation of influential volatile flavor compounds from milk fat obtained through hydrolytic cleavage of acylglycerols.

somewhat distinctive flavor qualities of meat flavor compounds that are not species-related are very valuable to the food and flavor industry, but chemical functions of lightly cooked and species-related flavors are still eagerly sought. Some details of the chemistry relating to the flavors of well-cooked meat are discussed in Section 11.7.

11.6.1 Species-Related Flavors of Meats and Milk from Ruminants

The characterizing flavors of at least some meats are inextricably associated with the lipid fraction. Progress on defining species-related flavors was initiated by Wong and coworkers [73] in relation to lamb and mutton flavors. These workers showed that a characteristic sweat-like flavor of mutton was closely associated with some volatile, medium chain-length fatty acids of which some methyl-branched members are highly significant. The formation of one of the most important branched-chain fatty acids in lamb and mutton, 4-methyloctanoic acid, is shown in Figure 20. Ruminal fermentations yield acetate, propionate, and butyrate, but most fatty acids are biosynthesized from acetate, which yields nonbranched chains. Some methyl branching occurs routinely because of the presence of propionate, but when dietary and other factors enhance the propionate concentrations in the rumen, greater methyl branching occurs [63]. Ha and Lindsay [26, 27] have shown that several medium-chain, methyl-branched fatty acids are important contributors to species-related flavors, and 4-ethyloctanoic acid (threshold = 1.8 ppb in water) is particularly important for conveying goat-like flavors to both meat and milk products. Additionally, several alkylphenols (methylphenol isomers, ethylphenol isomers, isopropylphenol isomers, and methyl-isopropylphenol isomers) contribute very characteristic cow-like and sheep-like species-related flavors to meats and milks [28, 41]. Alkylphenols are present as free and conjugate-bound substances in meat and milk, and are derived from shikimic acid pathway biochemical intermediates found in forages. Sulfate (XXV), phosphate (XXVI), and glucuronide (XXVII) conjugates of alkylphenols are formed in vivo, and are distributed via the circulatory system. Both enzymic and thermal hydrolysis of conjugates release phenols that enhance flavor development during fermentation and cooking of meat and dairy products.

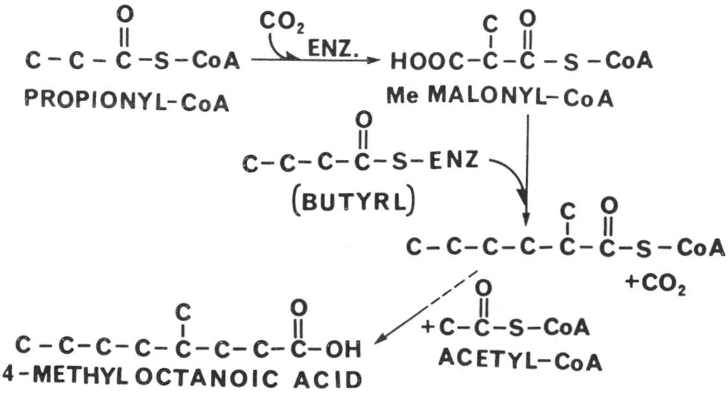

C–C–C(=O)–S–CoA
PROPIONYL-CoA
CO_2 ENZ.
HOOC–C(C)–C(=O)–S–CoA
Me MALONYL-CoA
C–C–C–C(=O)–S–ENZ
(BUTYRL)
C–C–C–C–C(C)–C(=O)–S–CoA
$+CO_2$
+C–C(=O)–S–CoA
ACETYL-CoA
C–C–C–C–C(C)–C–C–C(=O)–OH
4-METHYL OCTANOIC ACID

FIGURE 20 Ruminant biosynthesis of methyl-branched, medium-chain fatty acids.

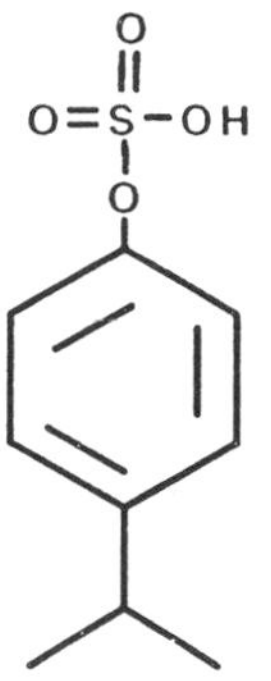

(XXV) *p*-Isopropylphenyl sulfate

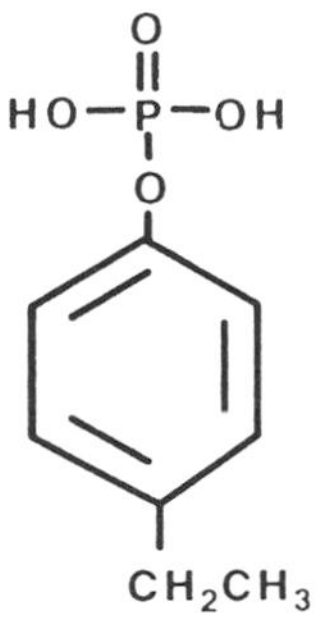

(XXVI) *p*-Ethylphenyl phosphate

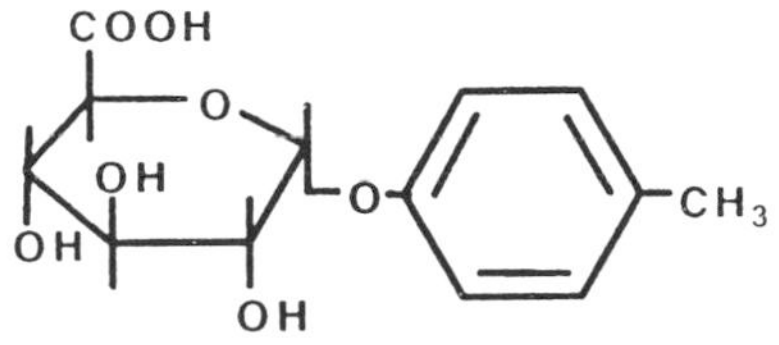

(XXVII) *p*-Cresyl glucuronide

11.6.2 Species-Related Flavors of Meats from Nonruminants

Species-related aspects of the flavor chemistry of nonruminant meats remain somewhat incomplete. However, studies have shown that the γ-C_5, C_9, and C_{12} lactones are reasonably abundant in pork [14], and these compounds likely contribute to some of the sweet-like flavor of pork. The distinct pork-like or piggy flavor, noticeable in lard or cracklings and in some pork, is caused by *p*-cresol and isovaleric acid that are produced from microbial conversions of corresponding amino acids in the lower gut of swine [26, 28]. Similar formation of indole and skatole from tryptophan may also intensify unpleasant piggy flavors in pork.

Much interest has centered on the aroma compounds responsible for the swine sex odor that causes serious off flavors in pork. Two compounds are responsible for the flavor, 5α-androst-16-en-3-one (Fig. 21), which has a urinous aroma, and 5α-androst-16-en-3α-ol, which has a musklike aroma [22]. The swine sex odor compounds are mainly associated with males, but may occur in castrated males and in females. The steroid compounds are particularly offensive to some individuals, especially women, and yet others are genetically odor-blind to them. Since the compounds responsible for the swine sex odor have only been found to cause off flavors in pork, they can be regarded as species-related flavor compounds for swine.

The distinctive flavors of poultry have also been the subject of many studies, and lipid oxidation appears to yield the character impact compounds for chicken [29]. The carbonyls *c*-4-decenal,*t*-2-*c*-5-undecadienal, and *t*-2-*c*-4-*t*-7-tridecatrienal reportedly may contribute the

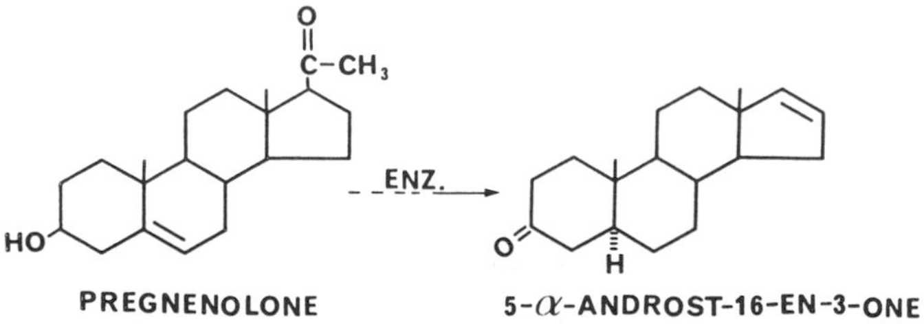

PREGNENOLONE ENZ. 5-α-ANDROST-16-EN-3-ONE

FIGURE 21 Formation of the steroid compound responsible for the urinous aroma associated with the swine sex odor defect of pork.

characteristic flavor of stewed chicken, and they are derived from lineolic and arachidonic acids. Chickens accumulate α-tocopherol (an antioxidant), but turkeys do not, and during cooking, such as roasting, carbonyls are formed to a much greater extent in turkey than in chicken. Additionally, certain chemical environments greatly affect the outcome of some lipid autoxidations. For example, Swoboda and Peers [64] have shown that the presence of copper ions and α-tocopherol results in selective oxidations of milkfat, producing octa-l,*c*-5-dien-3-one, the cause of metallic faints found in butter. Directed lipid oxidations may also occur in the development of species-related poultry flavors, leading to species-related flavors.

11.6.3 Volatiles in Fish and Seafood Flavors

Characterizing flavors in seafoods cover a somewhat broader range of flavor qualities than those occurring in other muscle foods. The broad range of animals involved (finfish, shellfish, and crustaceans) and the variable flavor and aroma qualities related to freshness each account for the different flavors encountered. Historically, trimethylamine has been associated with fish and crab-like aromas, and alone it exhibits an ammoniacal, fishy aroma. Trimethylamine and dimethylamine are produced through enzymic degradation of trimethylamine oxide (Fig. 22), which is found in significant quantities only in saltwater species of seafoods. Since very fresh fish contain essentially no trimethylamine, this compound modifies and contributes to the aroma of staling fish, in which it enhances fishhouse-type aromas. Trimethylamine oxide serves as part of the buffer system in marine fish species [32]. The formaldehyde produced concurrently with dimethylamine is believed to facilitate protein cross-linking, and thereby contributes to the toughening of fish muscle during frozen storage.

Fishy aromas characterized by such terms as "oxidized fish oil" and "cod liver oil-like" are largely caused by carbonyl compounds produced from the autoxidation of long-chain

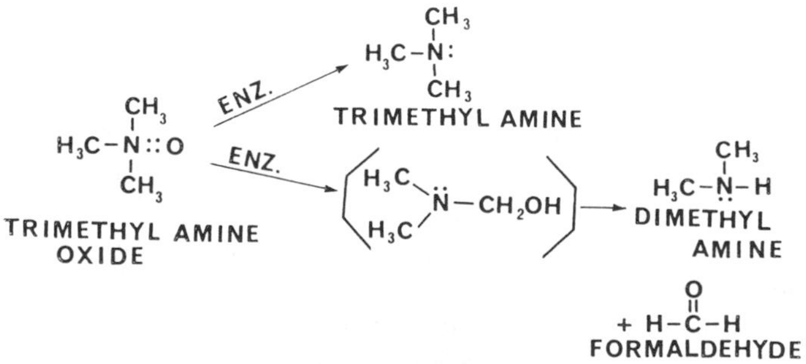

TRIMETHYL AMINE OXIDE ENZ. TRIMETHYL AMINE ENZ. DIMETHYL AMINE + FORMALDEHYDE

FIGURE 22 Microbial formation of principal volatile amines in fresh saltwater fish species.

ω-3-unsaturated fatty acids. These characteristic aromas result from 2,4,7-decatrienal isomers, and *c*-4-heptenal potentiates the fishy character of the decatrienals [40].

Because the fresh flavors and aromas of seafoods frequently have been greatly diminished or lost from fresh, frozen, and processed products available through commercial channels, many consumers associate the fishy flavors just described with all fish and seafoods. However, very fresh seafoods exhibit delicate aromas and flavors quite different from those usually evident in "commercially fresh" seafoods. It has been discovered that a group of enzymatically derived aldehydes, ketones, and alcohols provides the characterizing aroma of fresh fish [40], and these are very similar to the C_6, C_8, and C_9 compounds produced by plant lipoxygenases (see Sec. 11.3.5.1). Collectively, these compounds provide melony, heavy plant-like, and fresh fish aromas, and they are derived through a lipoxygenase system. The lipoxygenase systems found in fish and seafoods perform enzymic oxidations related to leukotriene synthesis, and flavor compound production is a by-product of those reactions. Hydroperoxidation followed by disproportionation reactions apparently leads first to the alcohol (Fig. 23), and then to the corresponding carbonyl. Some of these contribute to the distinctive flavors of very fresh cooked fish, either directly or as reactants that lead to new flavors during cooking.

The flavors of crustaceans and mollusks rely heavily on nonvolatile taste substances in addition to contributions from volatiles. For example, the taste of cooked snow crab meat has been largely duplicated with a mixture of 12 amino acids, nucleotides, and salt ions [35]. Good imitation crab flavors can be prepared from the taste substances just mentioned, along with some contributions from carbonyls and trimethylamine. Dimethyl sulfide provides a characterizing top-note aroma to cooked clams and oysters, and it arises principally from the thermal degradation of dimethyl-β-propiothetin present in ingested marine microflora [47].

11.7 DEVELOPMENT OF PROCESS OR REACTION FLAVOR VOLATILES

Many flavor compounds found in cooked or processed foods occur as the result of reactions common to all types of foods, regardless of whether they are of animal, plant, or microbial derivation. These reactions take place when suitable reactants are present and appropriate conditions (heat, pH, light) exist. Process or reaction flavors are discussed separately in this section because of their broad importance to all foods, and because they comprise a large volume of natural flavor concentrates that are used widely in foods, especially when meat or savory

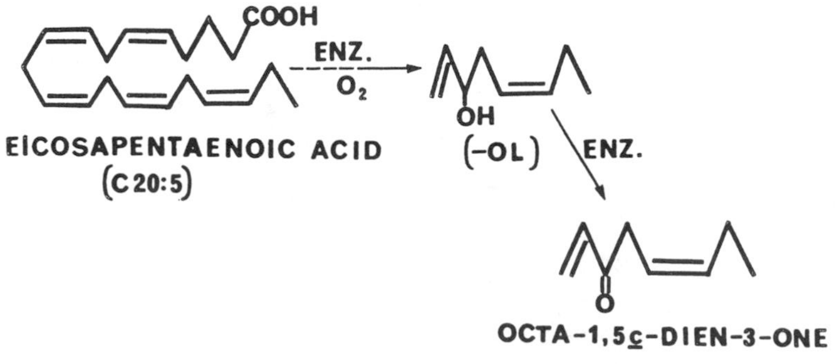

COOH
ENZ.
O_2
OH
EICOSAPENTAENOIC ACID
(C 20:5)
(-OL)
ENZ.
O
OCTA-1,5c-DIEN-3-ONE

FIGURE 23 Enzymatic formation of influential volatiles in fresh fish aroma from long-chain ω-3-unsaturated fatty acid.

flavors are desired. Related information can be found in discussions dealing with carbohydrates (Chap. 4), lipids (Chap. 5), and vitamins (Chap. 8).

11.7.1 Thermally Induced Process Flavors

Traditionally, these flavors have been broadly viewed as products from browning reactions because of early discoveries showing the role of reducing sugars and amino acid compounds in the induction of a process that ultimately leads to the formation of brown pigments (see Chap. 4 for details of Maillard browning reactions). Although browning reactions are almost always involved in the development of process flavors in foods, the interactions between (a) degradation products of the browning reaction and (b) other food constituents are also important and extensive. By taking a broad approach to discussions of thermally induced flavors, the aforementioned interactions, as well as reactions that occur following heat treatment, can be appropriately considered.

Although many of the compounds associated with process flavors possess potent and pleasant aromas, relatively few of these compounds seem to provide truly distinguishing character-impact flavor effects. Instead they often contribute general nutty, meaty, roasted, toasted, burnt, floral, plant, or caramel odors. Some process flavor compounds are acyclic, but many are heterocyclic, with nitrogen, sulfur, or oxygen substituents common (Fig. 24). These process flavor compounds occur in many foods and beverages, such as roasted meats, boiled meats, coffee, roasted nuts, beer, bread, crackers, snack foods, cocoa, and most other processed foods. The distribution of individual compounds does, however, depend on factors such as the availability of precursors, temperature, time, and water activity.

Production of process flavor concentrates is accomplished by selecting reaction mixtures and conditions so that those reactions occurring in normal food processing are duplicated. Selected ingredients (Table 5), usually including a reducing sugar, amino acids, and compounds with sulfur atoms, are processed under elevated temperatures to produce a distinctive profile of flavor compounds [31]. Thiamin is a popular ingredient because it provides both nitrogen and sulfur atoms already in ring structures (see Chap. 8).

Because of the large number of process flavor compounds produced during normal food processing or process simulation, it is unrealistic to cover the chemistry of their formation

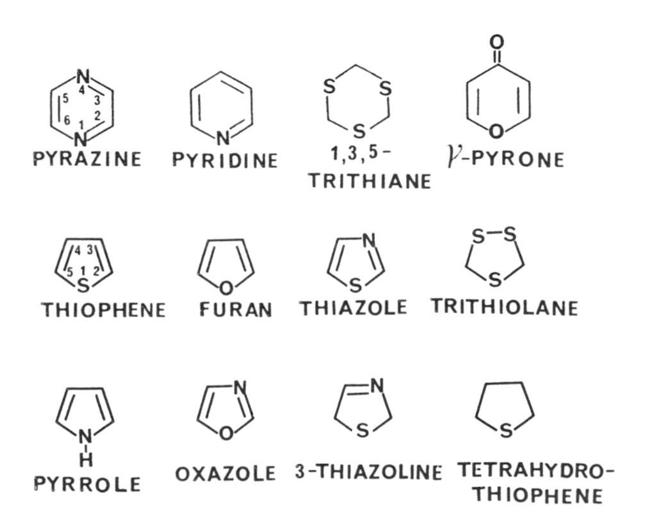

FIGURE 24 Some heterocyclic skeletons found commonly in flavor componds associated with thermally-induced or browning flavors.

TABLE 5 Some Common Ingredients Used in Process Flavor Reaction Systems of the Development of Meat-Like Flavors

Hydrolyzed vegetable protein	Thiamin
Yeast autolysate	Cysteine
Beef extract	Glutathione
Specific animal fats	Glucose
Chicken egg solids	Arabinose
Glycerol	5′-Ribonucleotides
Monosodium glutamate	Methionine

in depth. Rather, examples are given to illustrate some of the more important flavor volatiles formed and the mechanisms of their formation. Alkyl pyrazines were among the first compounds to be recognized as important contributors to the flavors of all roasted, toasted, or similarly thermally processed foods. The most direct route of their formation results from the interaction of α-dicarbonyl compounds (intermediate products in the Maillard reaction) with amino acids through the Stecker degradation reaction (Fig. 25). Transfer of the amino group to the dicarbonyl provides a means for integrating amino acid nitrogen into small compounds destined for any of the condensation reaction mechanisms envisioned in these reactions. Methionine has been selected as the amino acid involved in the Strecker degradation reaction because it contains a sulfur atom and it leads to formation of methional, which is an important characterizing compound in boiling potatoes and cheese-cracker flavors. Methional also readily decomposes further to yield methanethiol (methyl mercaptan), which oxidizes to dimethyl disulfide, thus providing a source of reactive, low-molecular-weight sulfur compounds that contribute to the overall system of flavor development.

Hydrogen sulfide and ammonia are very reactive ingredients in mixtures intended for the development of process flavors, and they are often included in model systems and assist in determining reaction mechanisms. Thermal degradation of cysteine (Fig. 26) yields both ammonia and hydrogen sulfide as well as acetaldehyde. Subsequent reaction of acetaldehyde with a

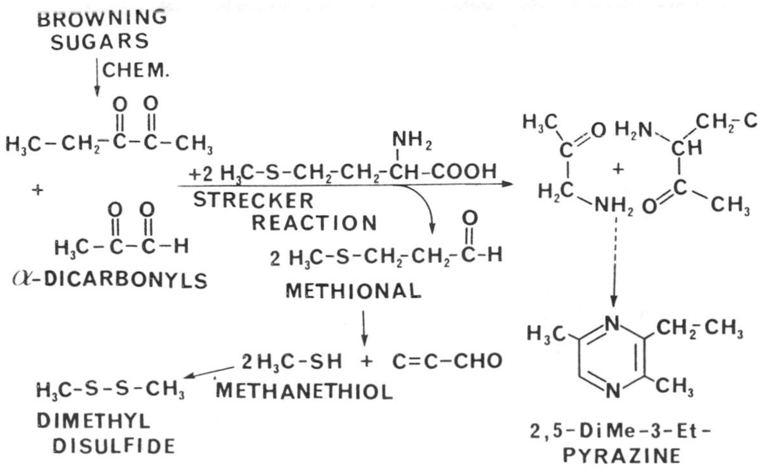

FIGURE 25 Formation of an alkyl pyrazine and small sulfur compounds through reactions occurring in the development of process flavors.

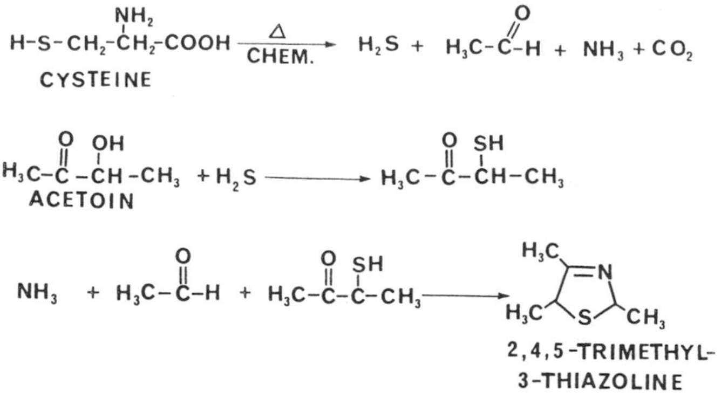

$$\underset{\text{CYSTEINE}}{H\text{-}S\text{-}CH_2\text{-}\overset{NH_2}{\overset{|}{C}}H\text{-}COOH} \xrightarrow[\text{CHEM.}]{\Delta} H_2S + H_3C\text{-}\overset{O}{\overset{\|}{C}}\text{-}H + NH_3 + CO_2$$

$$\underset{\text{ACETOIN}}{H_3C\text{-}\overset{O}{\overset{\|}{C}}\text{-}\overset{OH}{\overset{|}{C}}H\text{-}CH_3} + H_2S \longrightarrow H_3C\text{-}\overset{O}{\overset{\|}{C}}\text{-}\overset{SH}{\overset{|}{C}}H\text{-}CH_3$$

$$NH_3 + H_3C\text{-}\overset{O}{\overset{\|}{C}}\text{-}H + H_3C\text{-}\overset{O}{\overset{\|}{C}}\text{-}\overset{SH}{\overset{|}{C}}\text{-}CH_3 \longrightarrow \text{2,4,5-TRIMETHYL-3-THIAZOLINE}$$

FIGURE 26 Formation of a thiazoline found in cooked beef through the reaction of fragments from cysteine and sugar-amino browning (chem. = nonenzymic).

mercapto derivative of acetoin (from the Maillard reaction) gives rise to thiazoline, which contributes to the flavor of boiled beef [48].

Some heterocyclic flavor compounds are quite reactive and tend to degrade or interact further with components of foods or reaction mixtures. An interesting example of flavor stability and carry-through in foods is provided by the compounds shown in Equation 4, both of which provide distinct but different meat-like aromas [17]. A roasting meat aroma is exhibited by

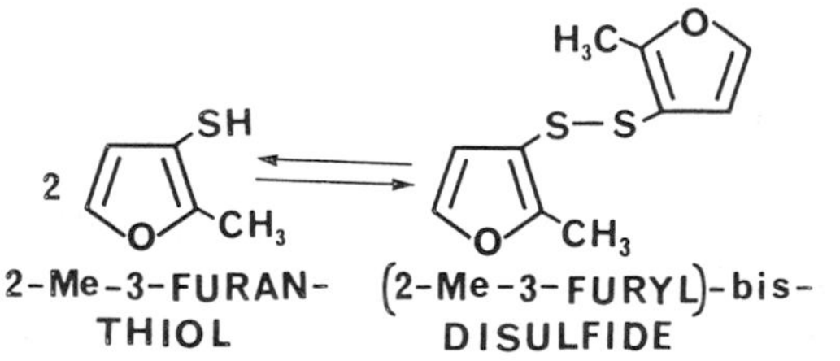

$$2\ \underset{\text{2-Me-3-FURANTHIOL}}{\text{(furan ring with SH, } CH_3)} \rightleftharpoons \underset{\text{(2-Me-3-FURYL)-bis-DISULFIDE}}{\text{(furyl)-S-S-(furyl)}} \tag{4}$$

2-methyl-3-furanthiol (reduced form), but upon oxidation to the disulfide form, the flavor becomes more characteristic of fully cooked meat that has been held for some time. Chemical reactions, such as the one just mentioned, are responsible for the subtle changes in meat flavor that occur because of the degree of cooking and the time interval after cooking.

During processing of complex systems, sulfur as such, or as thiols or polysulfides, can be incorporated in various compounds, resulting in the generation of new flavors. However, even though dimethyl sulfide is often found in processed foods, it does not react readily. In plants, dimethyl sulfide originates from biologically synthesized molecules, especially *S*-methylmethionine sulfonium salts (Fig. 27). *S*-Methylmethionine is quite labile to heat, and dimethyl sulfide is readily released. Dimethyl sulfide provides characterizing top-note aromas to fresh and canned sweet corn, tomato juice, and stewing oysters and clams.

Some of the most pleasant aromas derived from process reactions are provided by the compounds shown in Figure 28. These compounds exhibit caramel-like aromas and have been found in many processed foods. The planar enol-cyclic-ketone structure (see also Eq. 3) is usually derived from the sugar precursors, and this structural component appears responsible for the caramel-like aroma quality [52]. Cyclotene is used widely as a synthesized maple syrup flavor substance, and maltol is used widely as a flavor enhancer for sweet foods and beverages (see Sec. 11.2.2). Both furanones have been found in boiled beef where they appear to enhance meatiness. 4-Hydroxy-2,5-dimethyl-3(2*H*)-furanone is sometimes known as the "pineapple

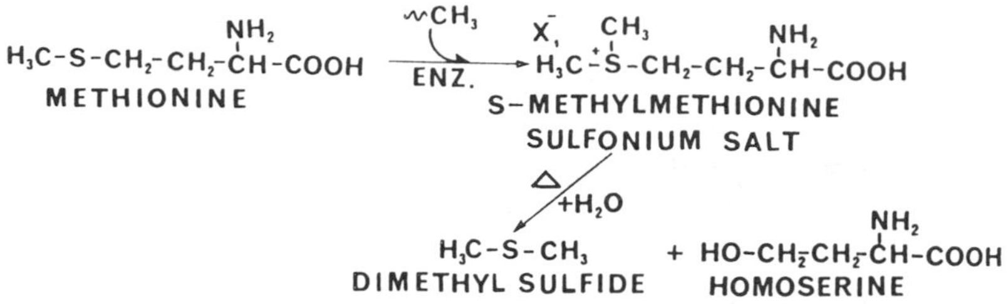

$H_3C-S-CH_2-CH_2-CH(NH_2)-COOH$ (METHIONINE) $\xrightarrow[\text{ENZ.}]{\sim CH_3}$ X^-, $H_3C-S^+(CH_3)-CH_2-CH_2-CH(NH_2)-COOH$ (S-METHYLMETHIONINE SULFONIUM SALT)

$\xrightarrow{\Delta,\ +H_2O}$ $H_3C-S-CH_3$ (DIMETHYL SULFIDE) + $HO-CH_2-CH_2-CH(NH_2)-COOH$ (HOMOSERINE)

FIGURE 27 Formation of dimethyl sulfide from thermal degradation of *S*-methylmethionine sulfonium salts.

compound" because it was first isolated from processed pineapple, where it contributes strongly to its characteristic flavor.

The flavor of chocolate and cocoa has received much attention because of the high demand for these flavors. After harvesting, cocoa beans are fermented under somewhat poorly controlled conditions. The beans are then roasted, sometimes with an intervening alkali treatment that darkens the color and yields a less harsh flavor. The fermentation hydrolyzes sucrose to reducing sugars, frees amino acids, and oxidizes some polyphenols. During roasting, many pyrazines and other heterocyclics are formed, but the unique flavor of cocoa is derived from an interaction between aldehydes from the Strecker degradation reaction. The reaction shown in Figure 29 between phenylacetaldehye (from phenylalanine) and 3-methylbutanal (from leucine) constitutes an important flavor-forming reaction in cocoa. The product of this aldol condensation, 5-methyl-2-phenyl-2-hexenal, exhibits a characterizing persistent chocolate aroma. This example also serves to show that reactions in the development of process flavors do not always yield heterocyclic aroma compounds.

11.7.2 Volatiles Derived from Oxidative Cleavage of Carotenoids

Oxidations focusing on triacylglycerols and fatty acids are discussed in another section, but some extremely important flavor compounds that are oxidatively derived from carotenoid precursors

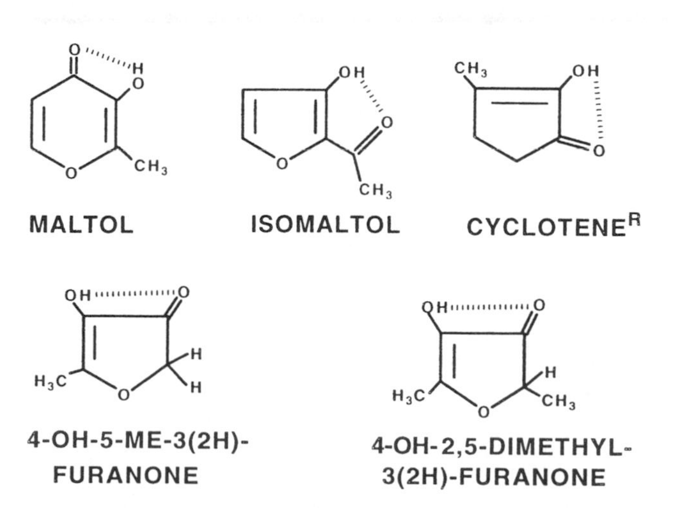

FIGURE 28 Structures of some important caramel-like flavor compounds derived from reactions occurring during processing.

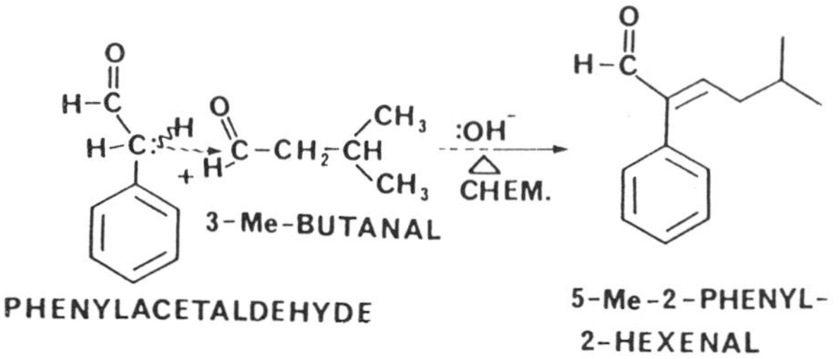

FIGURE 29 Formation of an important cocoa aroma volatile through an aldol condensation of two Strecker reaction-derived aldehydes (chem. = nonenzymic).

have not been covered and deserve attention here. Some of these reactions require singlet oxygen through chlorophyll sensitization; others are photooxidation processes. A large number of flavor compounds, derived from oxidizing carotenoids (or isoprenoids), have been identified in curing tobacco [15], and many of these are considered important for characterizing tobacco flavors. However, relatively few compounds in this category (three representative compounds are shown in Fig. 30) are currently considered highly important as food flavors. Each of these compounds exhibits unique sweet, floral, and fruit-like characteristics that vary greatly with concentration. They also blend nicely with aromas of foods to produce subtle effects that may be highly desirable or very undesirable. β-Damascenone exerts very positive effects on the flavors of wines, but in beer this compound at only a few parts per billion results in a stale, raisin-like note. β-Ionone also exhibits a pleasant violet, floral aroma compatible with fruit-type flavors, but it is also the principal off-flavor compound present in oxidized, freeze-dried carrots. Furthermore, these compounds have been found in black tea, where they make positive contributions to the flavor. Theaspirane and related derivatives contribute importantly to the sweet, fruity, and earthy notes of tea aroma. Although usually present in low concentrations, these compounds and related ones appear to be widely distributed, and it is likely that they contribute to the full, well-blended flavors of many foods.

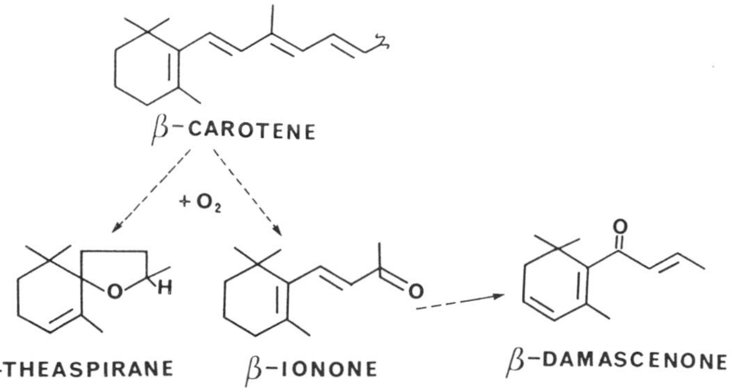

FIGURE 30 Formation of some important tea flavor compounds through oxidative cleavages of carotenoids.

11.8 FUTURE DIRECTIONS OF FLAVOR CHEMISTRY AND TECHNOLOGY

Knowledge about the chemistry and technology of flavors has expanded greatly in the past 35 years, and information has accumulated at this point where control and manipulation of many flavors in foods is possible. However, some of the areas of research likely to be fruitful in the foreseeable future are binding of flavors to macromolecules (see Chap. 4 and 6), structure–activity relationships in taste and olfaction as determined using computer techniques, control of glycoside precursors of plant-derived flavors, control of reaction flavor chemistry, and flavor development as related to genetics of plant cultivars and cultured cells (microbial and tissue cultures).

The issue of authenticity of natural flavors continues to attract considerable attention, and research on the various aspects of analysis and structural relationships of closely related molecules can be expected to enhance knowledge about subtle molecular influences on the flavor quality of such optical isomers. Isotopic mass spectrometry (carbon-13 and deuterium) techniques have made it possible in many cases to differentiate between natural and synthetic molecules, but carbon-13 enrichment alterations of synthetic molecules can render such approaches for detecting adulteration invalid. However, site-specific natural isotope fractionation measured by nuclear magnetic resonance is developing to the point where it provides an isotropic fingerprint that makes it possible to detect adulteration of natural origin flavors [43].

The chemical accuracy of nature-identical synthetic substances continues to be challenged by new information that is accumulating on the unique odor and flavor properties of enantiomers and other chiral compounds [46]. Such differences in odor quality between enantiomers not only have relevance to nature-identical issues, but also impact strongly on basic understandings of structure–function relationships in the olfaction of molecules, and new information should further this field.

Enzymic production of flavors within foods and ingredients will undoubtedly form a significant area of flavor technology in the future. Early in the modern era of analytical flavor chemistry, that is, 35 years ago, the flavorese concept was offered as a means for regenerating flavors in processed foods, and this type of approach is certain to become more important. However, because of the complex nature of natural flavors as they are now understood, it becomes apparent that the flavorese concept as initially developed would meet with great difficulties when applied to foods. Still, with recent developments in making encapsulated flavor enzyme systems, it is possible to maintain substrate–enzyme proximity and to control the amounts of flavor compounds produced so that unbalanced flavors can be avoided. With the increasing emphasis on high-quality formulated, complex foods, flavor development by enzymes should find a role of increasing importance. Finally, efforts to identify flavor compounds will continue, particularly in cases in which important characterizing flavor compounds appear to be present.

BIBLIOGRAPHY

Amerine, M. A., R. M. Pangborn, and E. B. Roessler (1965). *Principles of Sensory Evaluation of Food*, Academic Press, New York.

Beets, M. G. J. (1978). *Structure-Activity Relationships in Human Chemoreception*, Applied Science, London.

Fazzalari, F. A. (ed.) (1978). *Compilation of Odor and Taste Threshold Values Data*, American Society for Testing Materials, Philadelphia.

Furia, T. E., and N. Bellanca (eds.) (1975). *Fenaroli's Handbook of Flavor Ingredients*, Chemical Rubber Co., Cleveland.

Ho, C. T., and T. G. Hartman (eds.) (1994). *Lipids in Food Flavors*, ACS Symposium Series 558, American Chemical Society, Washington, DC.

Maarse, H. (1991). *Volatile Compounds in Foods and Beverages*, Marcel Dekker, New York.

Morton, I. D., and A. J. Macleod (eds.) (1982). *Food Flavours*, Part A, *Introduction, Elsevier,* New York.

O'Mahony, M. (1986). *Sensory Evaluation of Food: Statistical Methods and Procedures*, Marcel Dekker, New York.

Parliment, T. H., M. J. Morello, and R. J. McGorrin (1994). *Thermally Generated Flavors: Maillard, Microwave, and Extrusion Processes*, American Chemical Society, ACS Symposium Series 543, Washington, D. C.

Rouseff, R. L. (ed.) (1990). *Bitterness in Foods and Beverages*, Elsevier, Amsterdam.

Shahidi, F. (ed.) (1994). *Flavor of Meat and Meat Products*, Blackie, London.

Stone, H., and J. L. Sidel (1985). *Sensory Evaluation Practices,* Academic Press, New York.

Teranishi, R., G. R. Takeoka, and M. Guntert (eds.) (1992). *Flavor Precursors: Thermal and Enzymatic Conversions*, ACS Symposium Series 490, American Chemical Society, Washington, DC.

REFERENCES

1. Amerine, M. A., R. M. Pangborn, and E. B. Roessler (1965). *Principles of Sensory Evaluation of Food*, Academic Press, New York, p. 106.
2. *ASTM Manual of Sensory Testing Methods*, STP 434 (1968). American Society of Testing Materials, Philadelphia.
3. Bartoshuk, L. M. (1980). Sensory analysis in the taste of NaCl, in *Biological and Behaviorial Aspects of Salt Intake* (R. H. Cagen and M. R. Kare, eds.), Academic Press, New York, pp. 83–96.
4. Beets, M. G. J. (1978). The sweet and bitter modalities, in *Structure-Activity Relationships in Human Chemoreception*, Applied Science, London, pp. 259–303.
5. Beets, M. G. J. (1978). The sour and salty modalities, in *Structure-Activity Relationships in Human Chemoreception*, Applied Science, London, pp. 348–362.
6. Belitz, H. D., and H. Wieser (1985). Bitter compounds: Occurrence and structure-activity relationships. *Food Rev. Int. 1*(2):271–354.
7. Birch, G. G. (1981). Basic tastes of sugar molecules, in *Criteria of Food Acceptance* (J. Solms and R. L Hall, eds.), Forster Verlag, Zurich, pp. 282–291.
8. Birch, G. G., C. K. Lee, and A. Ray (1978). The chemical basis of bitterness in sugar derivatives, in *Sensory Properties of Foods* (G. G. Birch, J. G. Brennan, and K. J. Parker, eds.), Applied Science, London, pp. 101–111.
9. Boelens, M. H. (1991). Spices and condiments II, in *Volatile Compounds in Foods and Beverages* (H. Maarse, ed.), Marcel Dekker, New York, pp. 449–482.
10. Boelens, M. H., H. Boelens, and L. J. van Gemert (1993). Sensory properties of optical isomers. *Perfumer Flavorist 18*(6):1–16.
11. Brieskorn, C. H. (1990). Physiological and therapeutical aspects of bitter compounds in *Bitterness in Foods and Beverages* (R. L. Rouseff, ed.), Elsevier, Amsterdam, pp. 15–33.
12. Cliff, M., and H. Heymann (1992). Descriptive analysis of oral pungency. *J. Sensory Studies 7*:279–290.
13. DeTaeye, D. DeKeukeleire, E. Siaeno, and M. Verzele (1977). Recent developments in hop chemistry, in *European Brewery Convention Proceedings,* European Brewing Congress, Amsterdam, pp. 153–156.
14. Dwivedi, B. K. (1975). Meat flavor. *Crit. Rev. Food Technol. 5*:487–535.
15. Enzell, C. R. (1981). Influence of curing on the formation of tobacco flavour, in *Flavour '81* (P. Schreier, ed.), Walter de Gruyer, Berlin, pp. 449–478.
16. Eriksson, C. E. (1979) Review of biosynthesis of volatiles in fruits and vegetables since 1975, in *Progress in Flavour Research* (D. G. Land and H. E. Nursten, eds.), Applied Science, London, pp. 159–174.
17. Evers, W. J., H. H. Heinsohn, B. J. Mayers, and A. Sanderson (1976). Furans substituted at the three positions with sulfur, in *Phenolic, Sulfur, and Nitrogen Compounds in Food Flavors* (G. Charalambous and I. Katz, eds.), American Chemical Society, Washington, DC, pp. 184–193.

18. Fazzalari, F. A. (ed.) (1978). *Compilation of Odor and Taste Threshold Values Data*, American Society for Testing Materials, Philadelphia.
19. Forss, D. A. (1981). Sensory characterization, in *Flavor Research, Recent Advances* (R. Teranishi, R. A. Flath, and H. Sugisawa, eds.), Marcel Dekker, New York, pp. 125–174.
20. Forsyth, W. G. C. (1981). Tannins in solid foods, in *The Quality of Foods and Beverages*, vol. 1, *Chemistry and Technology* (G. Charalambous and G. Inglett, eds.), Academic Press, New York, pp. 377–388.
21. Govindarajan, V. S. (1979). Pungency: The stimuli and their evaluation, in *Food Taste Chemistry* (J. C. Boudreau, ed.), American Chemical Society, Washington, DC, pp. 52–91.
22. Gower, D. B., M. R. Hancock, and L. H. Bannister (1981). Biochemical studies on the boar pheromones, 5α-androst-16-en-3-one and 5α-androst-16-en-3α-ol, and their metabolism by olfactory tissue, in *Biochemistry of Taste and Olfaction* (R. H. Cagan and M. R. Kare, eds.), Academic Press, New York, pp. 7–31.
23. Grosch, W. (1982). Lipid degradation products and flavour, in *Food Flavours*, Part A, *Introduction* (I. D. Morton and A. J. Macleod, eds.), Elsevier, Amsterdam, pp.325–398.
24. Grosch, W. (1994). Determination of potent odourants in foods by aroma extract dilution analysis (AEDA) and calculation of odour activity values (OAVs). *Flavour Fragrance J. 9*:147–158.
25. Guenther, E., (1948). *The Essential Oils*, vols. 1–6, Van Nostrand, Reinhold, New York.
26. Ha, J. K., and R. C. Lindsay (1990). Distribution of volatile branched-chain fatty acids in perinephric fats of various red meat species. *Lebensm. Wissen. Tech. 23*:433–439.
27. Ha, J. K., and R. C. Lindsay (1991). Contributions of cow, sheep, and goat milks to characterizing branched-chain fatty acid and phenolic flavors in varietal cheeses. *J. Dairy Sci. 74*:3267–3274.
28. Ha, J. K., and R. C. Lindsay (1991). Volatile alkylphenols and thiophenol in species-related characterizing flavors of red meats. *J. Food Sci. 56*:1197–1202.
29. Harkes, P. D., and W. J. Begemann (1974). Identification of some previously unknown aldehydes in cooked chicken. *J. Am. Oil Chem. Soc. 51*:356–359.
30. Hasegawa, S., M. N. Patel, and R. C. Snyder (1982). Reduction of limonin bitterness in navel orange juice serum with bacterial cells immobilized in acrylamide gel. *J. Agric. Food Chem. 30*:509–511.
31. Heath, H. B. (1981). *Source Book of Flavors*, AVI, Westport, CT, p. 110.
32. Hebard, C. E., G. J. Flick, and R. E. Martin (1982). Occurrence and significance of trimethylamine oxide and its derivatives in fish and shellfish, in *Chemistry and Biochemistry of Marine Products* (R. E. Martin, G. J. Flick, and D. R. Ward, eds.), AVI, Westport, CT, pp. 149–304.
33. Ho, C. T., and T. G. Hartmann (eds.) (1994). *Lipids in Food Flavors*, American Chemical Society, Washington, DC, pp. 1–333.
34. Kimball, D. A., and S. I. Norman (1990). Processing effects during commercial debittering of California navel orange juice. *J. Agric. Food Chem. 38*:1396–1400.
35. Konosu, S. (1979). The taste of fish and shellfish, in *Food Taste Chemistry* (J. C. Boudreau, ed.), American Chemical Society, Washington, DC, pp. 203–213.
36. Koppenhoefer, B., R. Behnisch, U. Epperlein, H. Holzschuh, and R. Bernreuther (1994). Enantiomeric odor differences and gas chromatographic properties of flavors and fragrances. *Perfumer Flavorist 19*(5):1–14.
37. Kuninaka, A. (1981). Taste and flavor enhancers, in *Flavor Research Recent Advances* (R. Teranishi, R. A. Flath, and H. Sugisawa, eds.), Marcel Dekker, New York, pp. 305–353.
38. Kuroiwa, Y., and N. Hashimoto (1961). Composition of sunstruck flavor substance and mechanism of its evolution, *ASBC Proc. 19*:28–36.
39. Lee, C. B., and H. T. Lawless (1991). Time-course of astringent sensations. *Chem. Senses 16*:225–238.
40. Lindsay, R. C. (1990). Fish flavors. *Food Rev. Int. 6*(4): 437–455.
41. Lopez, V., and R. C. Lindsay (1993). Metabolic conjugates as precursors for characterizing flavor compounds in ruminant milks. *J. Agric. Food Chem. 41*:446–454.
42. Maarse, H. (1991). *Volatile Compounds in Foods and Beverages*, Marcel Dekker, New York.
43. Martin, G., G. Remaud, and G. J. Martin (1993). Isotopic methods for control of natural flavours authenticity. *Flavor Fragrance J. 8*:97–107.

44. Morgan, M. E., L. M. Libbey, and R. A. Scanlan (1972). Identity of the musty-potato aroma compound in milk cultures of *Pseudomonas taetrolens*. *J. Dairy Sci. 55*. 666.
45. Morton, I. D., and A. J. Macleod (eds.), (1982). *Food Flavours*, Part A, *Introduction*. Elsevier, Amsterdam.
46. Mosandl, A. (1988). Chirality in flavor chemistry—Recent developments in synthesis and analysis. *Food Rev. Int. 4*(1):1–43.
47. Motohito, T. (1962). Studies on the petroleum odor in canned chum salmon. *Mem. Fac. Fisheries, Hokkaido Univ.10*:1–5.
48. Mussinan, C. J., R. A. Wilson, I. Katz, A. Hruza, and M. H. Vock (1976). Identification and some flavor properties of some 3-oxazolines and 3-thiazolines isolated from cooked beef, in *Phenolic, Sulfur, and Nitrogen Compounds in Food Flavors* (G. Charalambous and I. Katz, eds.), American Chemical Society, Washington, DC, pp. 133–145.
49. Neucere, N. J., T. J. Jacks, and G. Sumrell (1978). Interactions of globular proteins with simple polyphenols. *J. Agric. Food Chem. 26*:214–216.
50. Ney, K. H. (1979). Bitterness of peptides: Amino acid composition and chain length, in *Food Taste Chemistry* (J. C. Boudreau, ed.), American Chemical Society, Washington, DC, pp. 149–173.
51. Ney, K. H. (1979). Bitterness of lipids. *Fette. Seifen. Anstrichm. 81*:467–469.
52. Ohloff, G. (1981). Bifunctional unit concept in flavor chemistry, in *Flavour '81* (P. Schreier, ed.), Walter de Gruyter, Berlin, pp. 757–770.
53. O'Mahony, M. (1986). *Sensory Evaluation of Food: Statistical Methods and Procedures*, Marcel Dekker, New York.
54. Richard, H. M. J. (1991). Spices and condiments I, in *Volatile Compounds in Foods and Beverages* (H. Maarse, ed.), Marcel Dekker, New York, pp. 411–447.
55. Scanlan, R. A. (ed.) (1977). *Flavor Quality: Objective Measurement*, American Chemical Society, Washington, DC.
56. Schiffman, S. S. (1980). Contribution of the anion to the taste quality of sodium salts, in *Biological and Behaviorial Aspects of Salt Intake* (R. H. Cagan and M. R. Kare, eds.), Academic Press, New York, pp. 99–114.
57. Schreier, P. (ed.) (1981). *Flavour '81*, Walter de Gruyter, Berlin, 1981.
58. Shahidi, F.(ed.) (1994). *Flavor of Meat and Meat Products*, Blackie, London, pp. 1–298.
59. Shallenberger, R. S., and T. E. Acree (1967). Molecular theory of sweet taste. *Nature* (*Lond.*) *216*:480–482.
60. Shankaranarayana, M. L., B. Raghaven, K. O. Abraham, and C. P. Natarajan (1982). Sulphur compounds in flavours, in *Food Flavours*, Part A, *Introduction* (I. D. Morton and A. J. Macleod, eds.), Elsevier, Amsterdam, pp. 169–281.
61. Shaw, P. E. (1991). Fruits II, in *Volatile Compounds in Foods and Beverages* (H. Maarse, ed.), Marcel Dekker, New York, pp. 305–327.
62. Stone, H., and J. L. Sidel (1985). *Sensory Evaluation Practices*, Academic Press, New York.
63. Smith, A., and W. R. H. Duncan (1979). Characterization of branched-chain lipids from fallow deer perinephric triacylglycerols by gas chromatography-mass spectrometry. *Lipids 14*:350–355.
64. Swoboda, P. A. T., and K. E. Peers (1979). The significance of octa-1, *cis 5-dien-3-one*, in *Progress in Flavour Research* (D. G. Land and H. E. Nursten, eds.), Applied Science Publishers, London, pp. 275–280.
65. Tanford, C. (1960). Contribution of hydrophobic interactions to the stability of globular conformations of proteins. *J. Amer. Chem. Soc. 84*:4240–4247.
66. Teranishi, R., R. A. Flath, and H. Sugisawa (eds.) (1981). *Flavor Research: Recent Advances*, Marcel Dekker, New York.
67. Tressl, R., D. Bahri, and K. H. Engel (1982). Formation of eight-carbon and ten-carbon components in mushrooms (*Agaricus campestris*). *J. Agric. Food Chem. 30*:89–93.
68. Tressl, R., M. Holzer, and M. Apetz (1975). Biogenesis of volatiles in fruits and vegetables, in *Aroma Research: Proceedings of the International Symposium on Aroma Research*, Zeist (H. Maarse and P. J. Groenen, eds.), Centre for Agricultural Publishing and Documentation, PUDOC, Wageningen, pp. 41–62.

69. van Straten, S., F. de Vrijer, and J. C. de Beauveser (eds.) (1977). *Volatile Compounds in Food*, 4th ed., Central Institute for Nutrition and Food Research, Ziest.
70. Watson, H. R. (1978). Flavor characteristics of synthetic cooling compounds, in *Flavor: Its Chemical, Behavioral, and Commercial Aspects* (C. M. Apt, ed.), Westview Press, Boulder, CO, pp. 31–50.
71. Whitfield, F. B., and J. H. Last (1991). Vegetables, in *Volatile Compounds in Foods and Beverages* (H. Maarse, ed.), Marcel Dekker, New York, pp. 203–269.
72. Wittkowski, R., J. Ruther, H. Drinda, and F. Rafiei-Taghanaki (1992). Formation of smoke flavor compounds by thermal lignin degradation, in *Flavor Precursors: Thermal and Enzymatic Conversions* (R. Teranishi, G. R. Takeoka, and M. Guntert, eds.), American Chemical Society, Washington, DC, pp. 232–243.
73. Wong, E., L. N. Nixon, and C. B. Johnson (1975). Volatile medium chain fatty acids and mutton flavor. *J. Agric. Food Chem. 23*:495–498.
74. Yamaguchi, S. (1979). The umami taste, in *Food Taste Chemistry* (J. C. Boudreau, ed.), American Chemical Society, Washington, DC. pp. 33–51.
75. K. Yashimoto, K., K. Iwami, and H. Mitsuda (1971). A new sulfur-peptide from *Lentinus edodes* acting as a precursor for lenthionine. *Agric. Biol. Chem. 35*:2059–2069.

12

Food Additives

Robert C. Lindsay
University of Wisconsin—Madison, Madison, Wisconsin

12.1 INTRODUCTION

Many substances are incorporated into foods for functional purposes, and in many cases these ingredients also can be found occurring naturally in some food. However, when they are used in processed foods, these chemicals have become known as "food additives." From a regulatory standpoint, each of the food additives must provide some useful and acceptable function or attribute to justify its usage. Generally, improved keeping quality, enhanced nutritional value, functional property provision and improvement, processing facilitation, and enhanced consumer acceptance are considered acceptable functions for food additives. The use of food additives to conceal damage or spoilage to foods or to deceive consumers is expressly forbidden by regulations governing the use of these substances in foods. Additionally, food additive usages are discouraged where similar effects can be obtained by economical, good manufacturing practices.

Since natural counterparts exist for many food additives, and frequently they are derived commercially from natural sources, further discussions about the chemistry of this group of substances can be found in the appropriate chapters of this book. For example, natural substances sometimes used as food additives are discussed in Chapter 5 (antioxidants), Chapter 8 (vitamins), Chapter 9 (minerals), Chapter 10 (colorants), and Chapter 11 (flavors). This chapter focuses on the role of both natural and synthetic substances that are added to foods and provides an integrating view of their functionalities. Emphasis is given to substances not covered elsewhere in this book.

12.2 ACIDS

12.2.1 General Attributes

Both organic and inorganic acids occur extensively in natural systems, where they function in a variety of roles ranging from intermediary metabolites to components of buffer systems. Acids are added for numerous purposes in foods and food processing, where they provide the benefits of many of their natural actions. One of the most important functions of acids in foods is participation in buffering systems, and this aspect is discussed in a following section. The use of acids and acid salts in chemical leavening systems, the role of specific acidic microbial inhibitors (e.g., sorbic acid, benzoic acid) in food preservation, and the function of acids as chelating agents are also discussed in subsequent sections of this chapter. Acids are important in the setting of pectin gels (Chap. 4), they serve as defoaming agents and emulsifiers, and they induce coagulation of milk proteins (Chaps. 6 and 14) in the production of cheese and cultured dairy products such as sour cream. In natural culturing processes, lactic acid (CH_3-CHOH-COOH) produced by streptococci and lactobacilli causes coagulation by lowering the pH to near the isoelectric point of casein. Cheeses can be produced by adding rennet and acidulants such as citric acid and hydrochloric acids to cold milk (4–8°C). Subsequent warming of the milk (to 35°C) produces a uniform gel structure. Addition of acid to warm milk results in a protein precipitate rather than a gel.

δ-Gluconolactone also can be used for slow acid production in cultured dairy products and chemical leavening systems because it slowly hydrolyzes in aqueous systems to form gluconic acid (Fig. 1). Dehydration of lactic acid yields lactide, a cylic dilactone (Fig. 2), which also can be used as a slow-release acid in aqueous systems. The dehydration reaction occurs under conditions of low water activity and elevated temperature. Introduction of lactide into foods with high water activity causes a reversal of the process with the production of two moles of lactic acid.

Acids such as citric are added to some moderately acid fruits and vegetables to lower the pH to a value below 4.5. In canned foods this permits sterilization to be achieved under less severe thermal conditions than is necessary for less acid products and has the added advantage of precluding the growth of hazardous microorganisms (i.e., *Clostridium botulinum*).

Acids, such as potassium acid tartrate, are employed in the manufacture of fondant and fudge to induce limited hydrolysis (inversion) of sucrose (Chap. 4). Inversion of sucrose yields fructose and glucose, which improve texture through inhibition of excessive growth of sucrose crystals. Monosaccharides inhibit crystallization by contributing to the complexity of the syrup and by lowering its equilibrium relative humidity.

One of the most important contributions of acids to foods is their ability to produce a sour or tart taste (Chap. 11). Acids also have the ability to modify and intensify the taste perception of other flavoring agents. The hydrogen ion or hydronium ion (H_3O^+) is in-

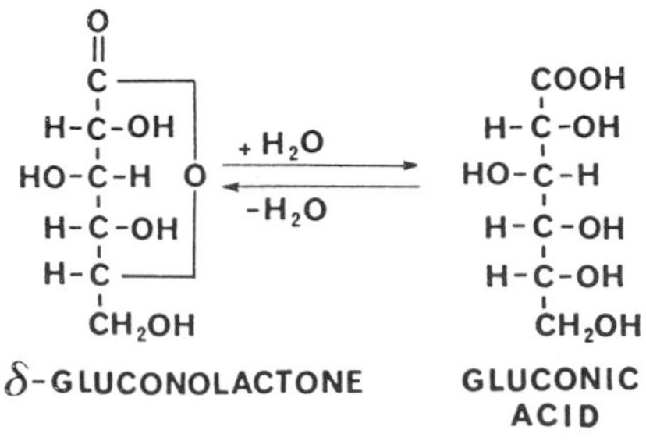

```
    O
    ||
    C ───┐                       COOH
    |    |                        |
  H-C-OH |      + H2O           H-C-OH
    |    |    ────────►           |
 HO-C-H  O    ◄────────        HO-C-H
    |    |      -H2O              |
  H-C-OH |                      H-C-OH
    |    |                        |
  H-C ───┘                      H-C-OH
    |                             |
   CH2OH                         CH2OH

δ-GLUCONOLACTONE              GLUCONIC
                                ACID
```

FIGURE 1 Formation of gluconic acid from the hydrolysis of δ-gluconolactone.

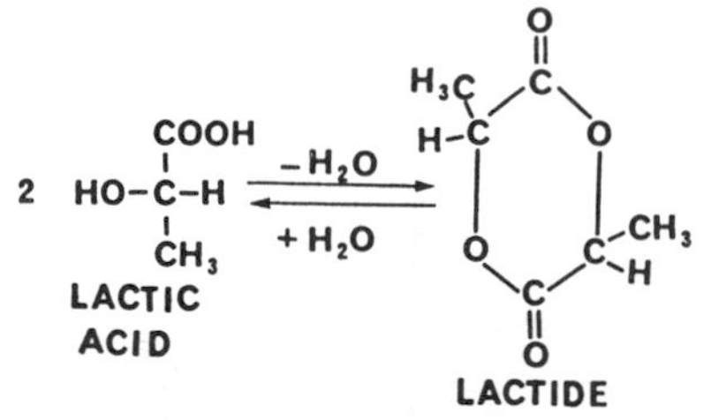

FIGURE 2 Equilibrium reaction showing formation of lactic acid from the hydrolysis of lactide.

volved in the generation of the sour taste response. Furthermore, short-chain free fatty acids (C_2–C_{12}) contribute significantly to the aroma of foods. For example, buytric acid at relatively high concentrations contributes strongly to the characteristic flavor of hydrolytic rancidity, but at lower concentrations contributes to the typical flavor of products such as cheese and butter.

Numerous organic acids are available for food applications [19]. Some of the more commonly used acids are acetic (CH_3COOH), lactic (CH_3-CHOH-COOH), citric [HOOC-CH_2-COH(COOH)-CH_2-COOH], malic (HOOC-CHOH-CH_2-COOH), fumaric (HOOC-CH=CH-COOH), succinic (HOOC-CH_2-CH_2-COOH), and tartaric (HOOC-CHOH-CHOH-COOH). Phosphoric acid (H_3PO_4) is the only inorganic acid extensively employed as a food acidulant. Phosphoric acid is an important acidulant in flavored carbonated beverages, particularly in colas and root beer. The other mineral acids (e.g., HCl, H_2SO_4) are usually too highly dissociated for food applications, and their use may lead to problems with quality attributes of foods. Dissociation constants for some food acids are shown in Table 1.

TABLE 1 Dissociation Constants at 25°C for Some Acids Used in Foods

Acid	Step	pK_a	Acid	Step	pK_a
Organic acids					
Acetic		4.75	Propionic		4.87
Adipic	1	4.43	Succinic	1	4.16
	2	5.41		2	5.61
Benzoic		4.19	Tartaric	1	3.22
n-Butyric		4.81		2	4.82
Citric	1	3.14	Inorganic acids		
	2	4.77	Carbonic	1	6.37
	3	6.39		2	10.25
Formic		3.75	o-Phosphoric	1	2.12
Fumaric	1	3.03		2	7.21
	2	4.44		3	12.67
Hexanoic		4.88	Sulfuric	2	1.92
Lactic		3.08			
Malic	1	3.40			
	2	5.10			

Source: From Ref. 56.

12.2.2 Chemical Leavening Systems

Chemical leavening systems are composed of compounds that react to release gas in a dough or batter under appropriate conditions of moisture and temperature. During baking, this gas release, along with expansion of entrapped air and moisture vapor, imparts a characteristic porous, cellular structure to finished goods. Chemical leavening systems are found in self-rising flours, prepared baking mixes, household and commercial baking powders, and refrigerated dough products [22].

Carbon dioxide is the only gas generated from currently used chemical leavening systems, and it is derived from a carbonate or bicarbonate salt. The most common leavening salt is sodium bicarbonate ($NaHCO_3$), although ammonium carbonate [$(NH_4)_2CO_3$] and bicarbonate (NH_4HCO_3) are sometimes used in cookies. Both of the ammonium salts decompose at baking temperatures and thus do not require, as does sodium bicarbonate, an added leavening acid for functionality. Potassium bicarbonate ($KHCO_3$) has been employed as a component of leavening systems in reduced-sodium diets, but its application is somewhat limited because of its hygroscopic nature and slightly bitter flavor.

Sodium bicarbonate is quite soluble in water (619 g/100 ml), and ionizes completely (Eqs. 1, 2, and 3).

$$NaHCO_3 \rightleftharpoons Na^+ + HCO_3^- \tag{1}$$

$$HCO_3^- + H_2O \rightleftharpoons H_2CO_3 + OH^- \tag{2}$$

$$HCO_3^- \rightleftharpoons Co_3^{2-} + H^+ \tag{3}$$

These reactions, of course, apply only to simple water solutions. In dough systems the ionic distribution becomes much more complex since proteins and other naturally occurring ionic species are available to participate in the reactions. In the presence of hydrogen ions provided mainly by leavening acids, and to some extent by the dough, sodium bicarbonate reacts to release carbon dioxide (Eq. 4).

$$R\text{-}O^-, H^+ + NaHCO_3 \rightarrow R\text{-}O^-, Na^+ + H_2O + CO_2 \tag{4}$$

The proper balance of acid and sodium bicarbonate is essential because excess sodium bicarbonate imparts a soapy taste to bakery products; an excess of acid leads to tartness and sometime bitterness. The neutralizing power of leavening acids is not uniform and the relative activity of an acid is given by its neutralizing value. The neutralizing value of an acid is determined by calculating the parts by weight of sodium bicarbonate that will neutralize 100 parts by weight of the leavening acid [50]. However, in the presence of natural flour ingredients, the amount of leavening acid required to give neutrality or any other desired pH in a baked product may be quite different from the theoretical amount determined for a simple system. Still, neutralizing values are useful in determining initial formulations for leavening systems. Residual salts from a properly balanced leavening process help stabilize the pH of finished products.

Leavening acids are often not easily recognized as acids in the usual sense, yet they must provide hydrogen ions to release carbon dioxide. The phosphates and potassium acid tartrate are metal salts of partially neutralized acids; sodium aluminum sulfate reacts with water to yield sulfuric acid (Eq. 5).

$$Na_2SO_4 \cdot Al_2(SO_4)_3 + 6\,H_2O \rightarrow Na_2SO_4 + 2\,Al(OH)_3 + 3\,H_2SO_4 \tag{5}$$

As mentioned earlier, δ-gluconolactone is an intramolecular ester (or lactone) that hydrolyzes slowly in aqueous systems to yield gluconic acid.

Leavening acids generally exhibit limited water solubility at room temperature, but some are less soluble than others. This difference in solubility or availability accounts for the initial rate of carbon dioxide release at room temperature and is the basis for classifying leavening acids according to speed. For example, if the compound is moderately soluble, carbon dioxide is rapidly evolved and the acid is referred to as fast-acting. Conversely, if the acid dissolves slowly, it is a slow-acting leavening acid. Leavening acids usually release a portion of the carbon dioxide prior to baking and the remainder under elevated temperatures of the baking process.

General patterns of carbon dioxide release at 27°C for fast-acting monocalcium phosphate monohydrate [$Ca(HPO_4)_2 \cdot H_2O$] and slow-acting 1-3-8 sodium aluminum phosphate [$NaH_{14}Al_3(PO_4)_8 \cdot 4H_2O$] are show in Figure 3. Over 60% of the carbon dioxide is released very quickly from the more soluble monocalcium phosphate monohydrate; only 20% of the potential carbon dioxide is released from the slow-acting 1-3-8 sodium aluminum phosphate during a 10-min reaction period. Because of a hydrated alumina coating, the latter leavening acid reacts to only a small extent until activated by heat. Also shown in Figure 3 is the low-temperature release pattern of carbon dioxide from coated anhydrous monocalcium phosphate [$Ca(HPO_4)_2$]. The crystals of this leavening acid were coated with compounds of slightly soluble alkali metal phosphates. The gradual release of carbon dioxide over the 10-min reaction period corresponds to the time required for water to penetrate the coating. This behavior is very desirable in some products that encounter a delay prior to baking.

The release of the remainder of the carbon dioxide from leavening systems during baking provides the final modifying action on texture. In most leavening systems the rate at which carbon dioxide is released greatly accelerates as the temperature is elevated. The effect of elevated temperatures on the release rate of carbon dioxide from slow-acting sodium acid pyrophosphate ($Na_2H_2P_2O_7$) is presented in Figure 4. Even a slight increase in temperature (from 27 to 30°C) noticeably accelerates gas production. Temperatures near 60°C cause a complete release of carbon dioxide within 1 min. Some leavening acids are less sensitive to high temperatures and do not exhibit vigorous activity until temperatures near the maximum baking

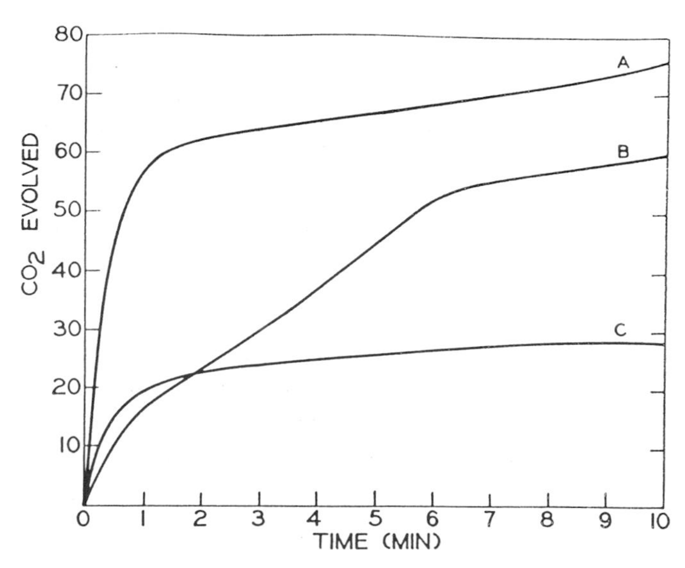

FIGURE 3 Carbon dioxide production at 27°C from the reaction of $NaHCO_3$ with (A) monocalcium phosphate $\cdot H_2O$, (B) coated anhydrous monocalcium phosphate, and (C) 1-3-8 sodium aluminum phosphate. Data from Ref. 50.

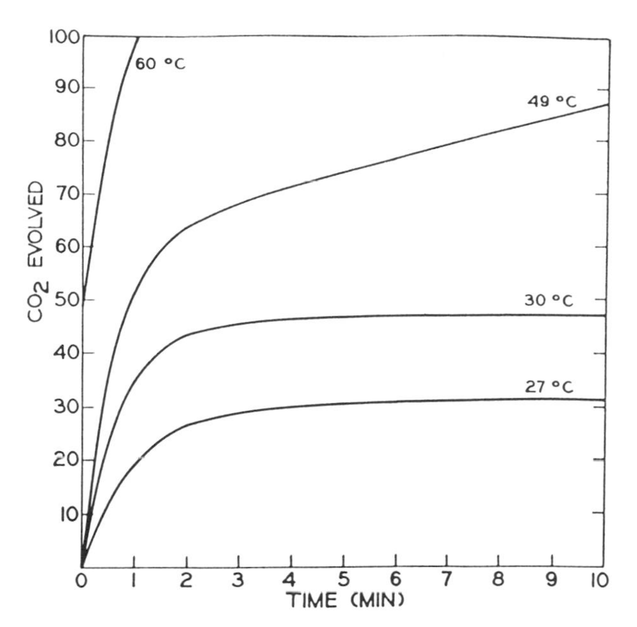

FIGURE 4 Effect of temperature on the rate of carbon dioxide evolution from the reaction of $NaHCO_3$ and slow-speed acid pyrophosphate. Reprinted from Ref. 50, p. 201.

temperature are obtained. Dicalcium phosphate ($CaHPO_4$) is unreactive at room temperature because it forms a slightly alkaline solution at this temperature. However, upon heating above approximately 60°C, hydrogen ions are released, thereby activating the leavening process. This slow action confines its use to products requiring long baking times, such as some types of cakes. Formulations of leavening acids employing one or more acidic components are common, and systems are often tailored for specific dough or batter applications.

Leavening acids currently employed include potassium acid tartrate, sodium aluminum sulfate, δ-gluconolatone, and ortho- and pyrophosphates. The phosphates include calcium phosphate, sodium aluminum phosphate, and sodium acid pyrophosphate. Some general properties of commonly used leavening acids are given in Table 2. It must be remembered that these are only examples, and that an extensive technology has developed for modification and control of the phosphate leavening acids [50].

Baking powders account for a large part of the chemical leaveners used both in the home and in bakeries. These preparations include sodium bicarbonate, suitable leavening acids, and starch and other extenders. Federal standards for baking powder require that the formula must yield at least 12% by weight of available carbon dioxide, and most contain 26–30% by weight of sodium bicarbonate. Traditional baking powders of the potassium acid tartrate type have been largely replaced by double-acting preparations. In addition to $NaHCO_3$ and starch, these baking powders usually contain monocalcium phosphate monohydrate [$Ca(HPO_4)_2 \cdot H_2O$], which provides rapid action during the mixing stage, and sodium aluminum sulfate [$Na_2SO_4 \cdot Al_2(SO_4)_3$], which does not react appreciably until the temperature increases during baking.

The increase in convenience foods has stimulated sales of prepared baking mixes and refrigerated dough products. In white and yellow cake mixes, the most widely used blend of leavening acids contains anhydrous monocalcium phosphate [$Ca(HPO_4)_2$] and sodium aluminum phosphate [$NaH_{14}Al_3(PO_4)_8 \cdot 4H_2O$]; chocolate cake mixes usually contain anhydrous monocalcium phosphate and sodium acid pyrophosphate ($Na_2H_2P_2O_7$) [50]. Typical blends of acids

Table 2 Some Properties of Common Leavening Acids

Acid	Formula	Neutralizing value[a]	Relative reaction rate at room temperature[b]
Sodium aluminum sulfate	$Na_2SO_4 \cdot Al_2(SO_4)_3$	100	Slow
Dicalcium phosphate dihydrate	$CaHPO_4 \cdot 2H_2O$	33	None
Monocalcium phosphate monohydrate	$Ca(HPO_4)_2 \cdot H_2O$	80	Fast
1-3-8 Sodium aluminum phosphate	$NaH_{14}Al_3(PO_4)_8 \cdot 4H_2O$	100	Slow
Sodium acid pyrophosphate (slow-type)	$Na_2H_2P_2O_7$	72	Slow
Potassium acid tartrate	$KHC_4H_4O_6$	50	Medium
δ-Gluconolactone	$C_6H_{10}O_6$	55	Slow

[a]In simple model systems; parts by weight of $NaHCO_3$ that will neutralize 100 parts by weight of the leavening acid.
[b]Rate of CO_2 evolution in the presence of $NaHCO_3$.
Source: Ref. 50.

contain 10–20% fast-acting anhydrous monophosphate compounds and 80–90% of the slower acting sodium aluminum phosphate or sodium acid pyrophosphate. The leavening acids in prepared biscuit mixes usually consist of 30–50% anhydrous monocalcium phosphate and 50–70% sodium aluminum phosphate or sodium acid pyrophosphate. The earliest self-rising flours and cornmeal mixes contained monocalcium phosphate monohydrate [$Ca(HPO_4)_2 \cdot H_2O$], but coated anhydrous monocalcium phosphate and sodium aluminum phosphate are in common use [50].

Refrigerated doughs for biscuit and roll products require limited initial carbon dioxide release during preparation and packaging, and considerable gas release during baking. Formulations for biscuits usually contain 1.0–1.5% sodium bicarbonate and 1.4–2.0% slow-acting leavening acids, such as coated monocalcium phosphate and sodium acid pyrophosphate, based on total dough weight. The pyrophosphates are useful in dough because they can be manufactured with a wide range of reactivities. For example, pyrophosphatase in flour is capable of hydrolyzing sodium acid pyrophosphate to orthophosphate (Fig. 5), and the reaction of sodium bicarbonate and pyrophosphate yields some trisodium monohydrogen pyrophosphate, which also can be hydrolyzed to orthophosphates. This enzymic action leads to gas production that assists in sealing

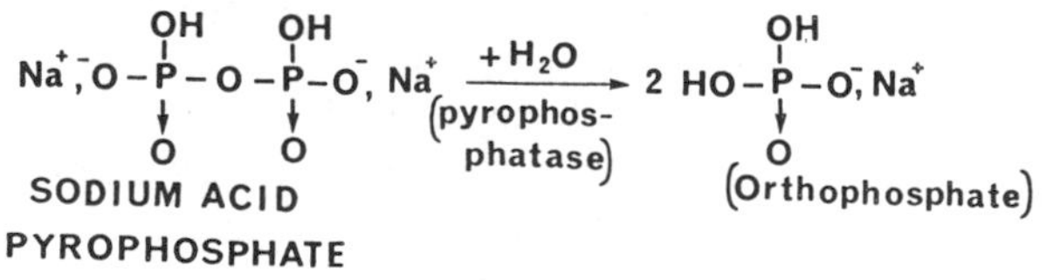

$$Na^+,{}^-O-\underset{\underset{O}{\downarrow}}{\overset{\overset{OH}{|}}{P}}-O-\underset{\underset{O}{\downarrow}}{\overset{\overset{OH}{|}}{P}}-O^-,Na^+ \xrightarrow[\text{(pyrophosphatase)}]{+H_2O} 2\ HO-\underset{\underset{O}{\downarrow}}{\overset{\overset{OH}{|}}{P}}-O^-,Na^+$$

SODIUM ACID PYROPHOSPHATE (Orthophosphate)

Figure 5 Enzymic hydrolysis of sodium acid pyrophosphate.

packages of refrigerated dough, but it can also lead to formation of large crystals of orthophosphates that may be mistaken for broken glass by the consumer.

12.3 BASES

Basic or alkaline substances are used in a variety of applications in foods and food processing. Although the majority of applications involve buffering and pH adjustments, other functions include carbon dioxide evolution, enhancement of color and flavor, solubilization of proteins, and chemical peeling. The role of carbonate and bicarbonate salts in carbon dioxide production during baking has been discussed previously.

Alkali treatments are imposed on several food products for the purpose of color and flavor improvement. Ripe olives are treated with solutions of sodium hydroxide (0.25–2.0%) to aid in the removal of the bitter principal and to develop a darker color. Pretzels are dipped in a solution of 1.25% sodium hydroxide at 87–88°C (186–190°F) prior to baking to alter proteins and starch so that the surface becomes smooth and develops a deep-brown color during baking. It is believed that the NaOH treatment used to prepare hominy and tortilla dough destroys disulfide bonds, which are base labile, and improves the flavor. Soy proteins are solubilized through alkali processing, and concern has been expressed about alkaline-induced racemization of amino acids (Chap. 6) and losses of other nutrients. Small amounts of sodium bicarbonate are used in the manufacture of peanut brittle candy to enhance caramelization and browning, and to provide, through release of carbon dioxide, a somewhat porous structure. Bases, usually potassium carbonate, are also used in cocoa processing for the production of dark (Dutched) chocolate [38]. The elevated pH enhances sugar-amino browning reactions and polymerization of flavonoids (Chap. 10), resulting in a smoother, less acid and less bitter chocolate flavor, a darker color, and a slightly improved solubility.

Food systems sometime require adjustment to higher pH values to achieve more stable or more desirable characteristics. For example, alkaline salts, such as disodium phosphate, trisodium phosphate, and trisodium citrate are used in the preparation of processed cheese (1.5–3%) to increase the pH (from 5.7 to 6.3) and to effect protein (casein) dispersion. This salt–protein interaction improves the emulsifying and water-binding capabilities of the cheese proteins [45] because the salts bind the calcium components of the casein micelles forming chelates.

Popular instant milk-gel puddings are prepared by combining dry mixes containing pregelatinized starch with cold milk, and allowing them to stand for a short time at refrigerator temperatures. Alkaline salts, such as tetrasodium pyrophosphate ($Na_4P_2O_7$) and disodium phosphate (Na_2HPO_4), in the presence of calcium ions in milk cause milk proteins to gel in combination with the pregelatinized starch. The optimum pH for acceptable puddings falls between 7.5 and 8.0. Although some of the necessary alkalinity is contributed by alkaline phosphate salts, other alkalizing agents are often added [10].

The addition of phosphates and citrates changes the salt balance in fluid milk products by forming complexes with calcium and magnesium ions from casein. The mechanism is incompletely understood, but depending on the type and concentration of salt added, the milk protein system can undergo stabilization, gelation, or destabilization.

Alkaline agents are used to neutralize excess acid in the production of foods such as cultured butter. Before churning, the cream is fermented by lactic acid bacteria so that it contains about 0.75% titratable acidity expressed as lactic acid. Alkalis are then added to achieve a titratable acidity of approximately 0.25%. The reduction in acidity improves churning efficiency and retards the development of oxidative off flavors. Several materials, including sodium bicarbonate ($NaHCO_3$), sodium carbonate (Na_2CO_3), magnesium carbonate ($MgCO_3$), magnesium oxide (MgO), calcium hydroxide [$Ca(OH)_2$], and sodium hydroxide ($NaOH$), are utilized alone or in combination as neutralizers for foods. Solubility, foaming as a result of carbon dioxide release, and strength of the base influence the selection of the alkaline agent. The use of alkaline agents or bases in excessive amounts leads to soapy or neutralizer flavors, especially when substantial quantities of fatty acids are present.

Strong bases are employed for peeling various fruits and vegetables. Exposure of the product to hot solutions at 60–82°C (140–180°F) of sodium hydroxide (about 3%), with subsequent mild abrasion, effects peel removal with substantial reductions in plant wastewater as compared to the conventional peeling techniques. Differential solubilization of cell and tissue constituents (pectic substances in the middle lamella are particularly soluble) provides the basis for caustic peeling processes.

12.4 BUFFER SYSTEMS AND SALTS

12.4.1 Buffers and pH Control in Foods

Since most foods are complex materials of biological origin, they contain many substances that can participate in pH control and buffering systems. Included are proteins, organic acids, and weak inorganic acid-phosphate salts. Lactic acid and phosphate salts, along with proteins, are important for pH control in animal tissue; polycarboxylic acids, phosphate salts, and proteins are important in plant tissues. The buffering effects of amino acids and proteins and the influence of pH and salts on their functionalities are discussed in Chapter 6. In plants, buffering systems containing citric acid (lemons, tomatoes, rhubarb), malic acid (apples, tomatoes, lettuce), oxalic acid (rhubarb, lettuce), and tartaric acid (grapes, pineapple) are common, and they usually function in conjunction with phosphate salts in maintaining pH control. Milk acts as a complex buffer because of this content of carbon dioxide, proteins, phosphate, citrate, and several other minor constituents.

In situations where the pH must be altered, it is usually desirable to stabilize the pH at the desired level through a buffer system. This is accomplished naturally when lactic acid is produced in cheese and pickle fermentations. Also, in some instances where substantial amounts of acids are used in foods and beverages, it is desirable to reduce the sharpness of acid tastes, and obtain smoother product flavors without inducing neutralization flavors. This usually can be accomplished by establishing a buffer system in which the salt of a weak organic acid is dominant. The common ion effect is the basis for obtaining pH control in these systems, and the system develops when the added salts contain an ion that is already present in an existing weak acid. The added salt immediately ionizes resulting in repressed ionization of the acid with reduced acidity and a more stable pH. The efficiency of a buffer depends on the concentration of the buffering substances. Since there is a pool of undissociated acid and dissociated salt, buffers resist changes in pH. For example, relatively large additions of a strong acid, such as hydrochloric acid, to an acetic–sodium acetate system causes hydrogen ions to react with the acetate ion pool to increase the concentration of slightly ionized acetic acid, and the pH remains relatively stable. In a similar manner, addition of sodium hydroxide causes hydroxyl ions to react with hydrogen ions to form undissociated water molecules.

Titration of buffered systems and resulting titration curves (i.e., pH vs. volume of base added) reveal their resistance to pH change. If a weak acid buffer is titrated with a base, there is a gradual but steady increase in the pH as the system approaches neutralization; that is, the change in pH per milliliter of added base is small. Weak acids are only slightly dissociated at the beginning of the titration. However, the addition of hydroxyl ions shifts the equilibrium to the dissociated species and eventually the buffering capacity is overcome.

In general, for an acid (HA) in equilibrium with ions (H^+) and (A^-), the equilibrium is expressed as in Equation 6.

$$HA \rightleftharpoons H^+ + A^- \tag{6}$$

$$K_a^1 = \frac{[H^+][A^-]}{[HA]} \tag{7}$$

The constant K_a^1 is the apparent dissociation constant, and is characteristic of the particular acid. The apparent dissociation constant K_a^1 becomes equal to the hydrogen ion concentration $[H^+]$ when the anion concentration $[A^-]$ becomes equal to the concentration of undissociated acid [HA]. This situation gives rise to an inflection point on a titration curve, and the pH corresponding to this point is referred to as the pK_a^1 of the acid. Therefore, for a weak acid, pH is equal to pK_a^1 when the concentration of the acid and conjugate base are equal:

$$pH = pK_a^1 = -\log_{10}[H^+] \tag{8}$$

A convenient method for calculation the approximate pH of a buffer mixture, given the pK_a^1 of the acid, is provided by Equation 11. This equation is arrived at by first solving Eq. 12.7 for $[H^+]$ to yield Equation 9.

$$[H^+] = K_a^1\frac{[HA]}{[A^-]} = K_a^1\frac{[acid]}{[salt]} \tag{9}$$

Since the salt in solution is almost completely dissociated, it is almost equal to the concentration of the conjugate base $[A^-]$. The negative logarithms of the terms yield Equation 10. By substituting pH for $-\log[H^+]$ and pK_a^1 for $-\log K_a^1$, Equation 11 is obtained. The pH of a buffer system derived from any weak acid that dissociates to H^+ and A^- can be calculated by using Equation 11.

$$-\log[H^+] = -\log K_a^1 - \log - \frac{[acid]}{[salt]} \tag{10}$$

$$pH = pK_a^1 + \log - \frac{[salt]}{[acid]} = pK_a^1 + \log\frac{[A^-]}{[HA]} \tag{11}$$

In calculating the pH values of buffer solutions, it is important to recognize that the apparent dissociation constant K_a^1 differs from K_a, the true dissociation constant. However, for any buffer, the value of K_a^1 remains constant as the pH is varied, provided the total ionic strength of the solution remains unchanged.

The sodium salts of gluconic, acetic, citric, and phosphoric acids are commonly used for pH control and tartness modification in the food industry. The citrates are usually preferred over phosphates for tartness modification since they yield smoother sour flavors. When low-sodium products are required, potassium buffer salts may be substituted for sodium salts. In general, calcium salts are not used because of their limited solubilities and incompatibilities with other components in the system. The effective buffering ranges for combinations of common acids and salts are pH 2.1–4.7 for citric acid-sodium citrate, pH 3.6–5.6 for acetic acid–sodium acetate, and pH 2.0–3.0, 5.5–7.5, and 10–12, respectively for the three ortho- and pyrophosphate anions.

12.4.2 Salts in Processed Dairy Foods

Salts are used extensively in processed cheeses and imitation cheeses to promote a uniform, smooth texture. These additives are sometimes referred to as emulsifying salts because of their ability to aid in dispersion of fat. Although the emulsifying mechanism remains somewhat less than fully defined, anions from the salts when added to processed cheese combine with and remove calcium from the *para*-casein complex, and this causes rearrangement and exposure of both polar and nonpolar regions of the cheese proteins. It is also believed that the anions of these salts participate in ionic bridges between protein molecules, and thereby provide a stabilized matrix that entraps the fat in processed cheese [45]. Salts used for cheese processing include mono-, di-, and trisodium phosphate, dipotassium phosphate, sodium hexametaphosphate, sodium acid pyrophosphate, tetrasodium pyrophosphate, sodium aluminum phosphate and other condensed phosphates, trisodium citrate, tripotassium citrate, sodium tartrate, and sodium potassium tartrate.

The addition of certain phosphates, such as trisodium phosphate, to evaporated milk prevents separation of the milkfat and aqueous phases. The amount required varies with the season of the year and the source of milk. Concentrated milk that is sterilized by a high-temperature, short-time method frequently gels upon storage. The addition of polyphosphates, such as sodium hexametaphosphate and sodium tripolyphosphate, prevents gel formation through a protein denaturation and solubilization mechanism that involves complexing of calcium and magnesium by phosphates.

12.4.3 Phosphates and Water-Binding in Animal Tissues

The addition of appropriate phosphates increases the water-holding capacity of raw and cooked meats [24], and these phosphates are used in the production of sausages, in the curing of ham, and to decrease drip losses in poultry and seafoods. Sodium tripolyphosphate ($Na_5P_3P_{10}$) is the most common phosphate added to processed meat, poultry, and seafoods. It is often used in blends with sodium hexametaphosphate [$(NaPO_3)_n$, $n = 10–15$] to increase tolerance to calcium ions that exist in brines used in meat curing. Ortho- and pyrophosphates often precipitate if used in brines containing substantial amounts of calcium.

The mechanism by which alkaline phosphates and polyphosphates enhance meat hydration is not clearly understood despite extensive studies. The action may involve the influence of pH changes (Chap. 6), effects of ionic strength, and specific interactions of phosphate anions with divalent cations and myofibrillar proteins. Many believe that calcium complexing and a resulting loosening of the tissue structure is a major function of polyphosphates. It is also believed that binding of polyphosphate anions to proteins and simultaneous cleavage of cross-linkages between actin and myosin results in increased electrostatic repulsion between peptide chains and a swelling of the muscle system. If exterior water is available, it then can be taken up in an immobilized state within the loosened protein network. Further, because the ionic strength has been increased, the interaction between proteins is perhaps reduced to a point where part of the myofibrillar proteins form a colloidal solution. In communited meat products, such as bologna and sausage, the addition of sodium chloride (2.5–4.0%) and polyphosphate (0.35–0.5%) contributes to a more stable emulsion, and after cooking to a cohesive network of coagulated proteins.

If the phosphate-induced solubilization occurs primarily on the surface of tissues, as is the case with polyphosphate-dipped (6–12% solution with 0.35–0.5% retention) fish fillets, shellfish, and poultry, a layer of coagulated protein is formed during cooking, and this improves moisture retention [35].

12.5 CHELATING AGENTS (SEQUESTRANTS)

Chelating agents or sequestrants play a significant role in food stabilization through reactions with metallic and alkaline earth ions to form complexes that alter the properties of the ions and their effects in foods. Many of the chelating agents employed in the food industry are natural substances, such as polycarboxylic acids (citric, malic, tartaric, oxalic, and succinic), polyphosphoric acids (adenosine triosphate and pyrophosphate), and macromolecules (porphyrins, proteins). Many metals exist in a naturally chelated state. Examples, include magnesium in chlorophyll; copper, zinc, and manganese in various enzymes; iron in proteins such as ferritin; and iron in the porphyrin ring of myoglobin and hemoglobin. When these ions are released by hydrolytic or other degradative reactions, they are free to participate in reactions that lead to discoloration, oxidative rancidity, turbidity, and flavor changes in foods. Chelating agents are sometimes added to form complexes with these metal ions, and thereby stabilize the foods.

Any molecule or ion with an unshared electron pair can coordinate or form complexes with metal ions. Therefore, compounds containing two or more functional groups, such as -OH, -SH, -COOH, $-PO_3H_2$, C=O, $-NR_2$, -S-, and -O-, in proper geometrical relation to each other, can chelate metals in a favorable physical environment. Citric acid and its derivatives, various phosphates, and salts of ethylenediamine tetraacetic acid (EDTA) are the most popular chelating agents used in foods. Usually the ability of a chelating agent (ligand) to form a five- or six-membered ring with a metal is necessary for stable chelation. For example, EDTA forms chelates of high stability with calcium because of an initial coordination involving the electron pairs of its nitrogen atoms and the free electron pairs of the anionic oxygen atoms of two of the four carboxyl groups (Fig. 6). The spatial configuration of the calcium–EDTA complex is such that it allows additional coordination of the calcium with the free electron pairs of the anionic oxygen atoms of the remaining two carboxyl groups, and this results in an extremely stable complex utilizing all six electron donor groups.

In addition to steric and electronic considerations, factors such as pH influence the formation of strong metal chelates. The nonionized carboxylic acid group is not an efficient donor group, but the carboxylate ion functions effectively. Judicious raising of the pH allows dissociation of the carboxyl group and enhances chelating efficiency. In some instances hydroxyl ions compete for metal ions and reduce the effectiveness of chelating agents. Metal ions exist in solution as hydrated complexes (metal $\cdot$ H_2O^{M+}), and the rate at which these complexes are disrupted influences the rate which they can be complexed with chelating agents. The relative attraction of chelating agents for different ions can be determined from stability or equilibrium constants ($K =$ [metal $\cdot$ chelating agent]/[metal][chelating agent]) (Chap. 9). For example, for calcium the stability constant (expressed as log K) is 10.7 with EDTA, 5.0 with pyrophosphate, and 3.5 with citric acid [17]. As the stability constant (K) increases, more of the metal is complexed, leaving less metal in cation form (i.e., the metal in the complex is more tightly bound).

Chelating agents are not antioxidants in the sense that they arrest oxidation by chain termination or serve as oxygen scavengers. They are, however, valuable antioxidant synergists since they remove metal ions which catalyze oxidation (Chap. 5). When selecting a chelating agent for an antioxidant synergist role, its solubility must be considered. Citric acid and citrate esters (20–200 ppm) in propylene glycol solution are solubilized by fats and oils and thus are effective synergists in all-lipid systems. On the other hand, Na_2EDTA and Na_2Ca-EDTA dissolve to only a limited extent, and are not effective in pure fat systems. The EDTA salts (to 500 ppm),

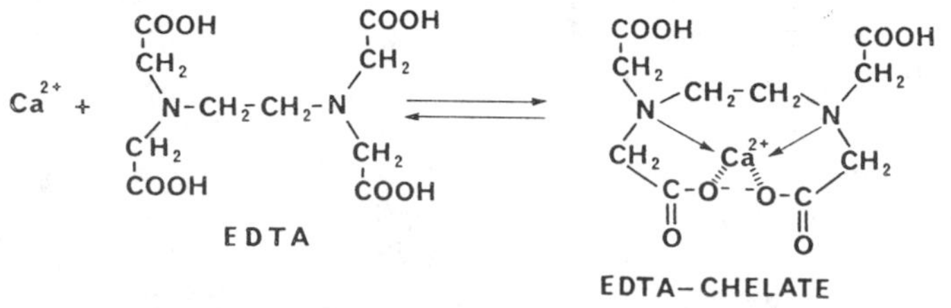

Ca^{2+} + EDTA ⇌ EDTA–CHELATE

FIGURE 6 Schematic representation of chelation of calcium by EDTA.

however, are very effective antioxidants in emulsion systems. such as salad dressings, mayonnaise, and margarine, because they can function in the aqueous phase.

Polyphosphates and EDTA are used in canned seafoods to prevent the formation of glassy crystals of struvite or magnesium ammonium phosphate ($MgNH_4PO_4 \cdot 6H_2O$). Seafoods contain substantial amounts of magnesium ions, which sometimes react with ammonium phosphate during storage to give crystals that may be mistaken as glass contamination. Chelating agents complex magnesium and minimize struvite formation. Chelating agents also can be used to complex iron, copper, and zinc in seafoods to prevent reactions, particularly with sulfides, that lead to product discoloration.

The addition of chelating agents to vegetables prior to blanching can inhibit metal-induced discolorations, and can remove calcium from pectic substances in cell walls and thereby promote tenderness.

Although citric and phosphoric acids are employed as acidulants in soft drink beverages, they also chelate metals that otherwise could promote oxidation to flavor compounds, such as terpenes, and catalyze discoloration reactions. Chelating agents also stabilize fermented malt beverages by complexing copper. Free copper catalyzes oxidation of polyphenolic compounds, which subsequently interact with proteins to form permanent hazes or turbidity.

The extremely efficient chelating abilities of some agents, notably EDTA, has caused speculation that excessive usage in foods could lead to the depletion of calcium and other minerals in the body. To deal with this concern, levels and applications are regulated, and in some instances calcium is added to food systems through the use of the Na_2Ca salt of EDTA rather than the all-sodium (Na, Na_2, Na_3, or Na_4EDTA) or acid forms. However, there appears to be little concern about using these chelators in the amounts permitted, considering the natural concentrations of calcium and other divalent cations that are present in foods.

12.6 ANTIOXIDANTS

Oxidation occurs when electrons are removed from an atom or group of atoms. Simultaneously, there is a corresponding reduction reaction that involves the addition of electrons to a different atom or group of atoms. Oxidation reactions may or may not involve the addition of oxygen atoms or the removal of hydrogen atoms from the substance being oxidized. Oxidation–reduction reactions are common in biological systems, and also are common in foods. Although some oxidation reactions are beneficial in foods, others can lead to detrimental effects including degradation of vitamins (Chap. 8), pigments (Chap. 10), and lipids (Chap. 5), with loss of nutritional value and development of off flavors. Control of undesirable oxidation reactions in foods is usually achieved by employing processing and packaging techniques that exclude oxygen or involve the addition of appropriate chemical agents.

Before the development of specific chemical technology for the control of free-radical-mediated lipid oxidation, the term antioxidant was applied to all substances that inhibited oxidation reactions regardless of the mechanism. For example, ascorbic acid was considered an antioxidant and was employed to prevent enzymic browning of the cut surfaces of fruits and vegetables (Chap. 7). In this application ascorbic acid functions as a reducing agent by transferring hydrogen atoms back to quinones that are formed by enzymic oxidation of phenolic compounds. In closed systems ascorbic acid reacts readily with oxygen and thereby serves as an oxygen scavenger. Likewise, sulfites are readily oxidized in food systems to sulfonates and sulfate, and thereby function as effective antioxidants in foods such as dried fruits (Sec. 12.7.1). The most commonly employed food antioxidants are phenolic substances. More recently the term "food antioxidants" often has been applied to those compounds that interrupt the free-radical chain reaction involved in lipid oxidation and those that scavenge singlet oxygen; however, the term should not be used in such a narrow sense.

Antioxidants often exhibit variable degrees of efficiency in protecting food systems, and combinations often provide greater overall protection than can be accounted for through the simple additive effect [52]. Thus, mixed antioxidants sometimes have a synergistic action, the mechanisms for which are not completely understood. It is believed, for example, that ascorbic acid can regenerate phenolic antioxidants by supplying hydrogen atoms to the phenoxy radicals that form when the phenolic antioxidants yield hydrogen atoms to the lipid oxidation chain reaction. In order to achieve this action in lipids ascobic acid must be made less polar so it will dissolve in fat. This is done by esterification to fatty acids to form compounds such as ascorbyl palmitate.

The presence of transition-state metal ions, particularly copper and iron, promotes lipid oxidation through catalytic actions (Chaps. 5 and 9). These metallic pro-oxidants are frequently inactivated by adding chelating agents, such as citric acids or EDTA (Sec. 12.5). In this role chelating agents are also referred to as synergists since they greatly enhance the action of the phenolic antioxidants. However, they are often ineffective as antioxidants when employed alone.

Many naturally occurring substances possess antioxidant capabilities, and the tocopherols are noted examples that are widely employed (Chap. 5). Recently, extractives of spices, particularly rosemary, also have been successfully commercially exploited as natural antioxidants. Gossypol, which occurs naturally in cottonseed, is an antioxidant, but it has toxic properties (Chap. 13). Other naturally occurring antioxidants are coniferyl alcohol (found in plants) and guaiaconic and guaiacic acid (from gum guaiac). All of these are structurally related to butylated hydroxyanisole (BHA), butylated hydroxytoluene (BHT), propyl gallate, and di-*t*-butylhydroquinone (TBHQ), which are synthetic phenolic antioxidants currently approved for use in foods (Chap. 5). Nordihydroguaiaretic acid, a compound related to somc of the constituents of gum guaiac, is an effective antioxidant, but its use directly in foods has been suspended because of toxic effects. All of these phenolic substances serve as oxidation terminators by participating in the reactions through resonance stabilized free-radical forms [52], but they also are believed to serve as singlet oxygen scavengers. However, β-carotene is considered to function more efficiently as a singlet oxygen scavenger than phenolic substances.

Thiodipropionic acid and dilauryl thiodiproprionate remain as approved food antioxidants, even though removal of these compounds from the approved list was recently proposed because they were not being used in foods. The presence of a sulfur atom in the thiodipropionates has led to speculation that they could cause off flavors, but this view is unfounded. A more compelling reason that the thiodipropionates have not been used in foods is their failure to inhibit lipid oxidation in foods, as measured by peroxide value, when used at permitted levels (to 200 ppm) [29]. The classic role of thiodipropionates is as secondary antioxidants, where at high concentrations (> 1000 ppm) they degrade hydroperoxides formed during olefin oxidation to relatively stable end products, and when used as such, they are useful in stabilizing synthetic polyolefins.

Although thiodipropionates, at levels allowed in foods, are ineffective in reducing peroxide values, they are highly effective in decomposing peracids (Fig. 7) found in oxidizing lipids [29]. Peracids are very efficient substances for mediating the oxidation of double bonds to epoxides, and in the presence of water, epoxides formed by this reaction readily hydrolyze to form diols. When these reactions occur with cholesterol, both cholesterol epoxide and the cholesterol-triol derivative are formed, and these cholesterol oxides are widely considered potentially mutagenic and atherogenic, respectively, to humans [39]. Because the thiodipropionates readily inhibit the accumulation of peracids, they have been retained as approved antioxidants by the US Food and Drug Administration.

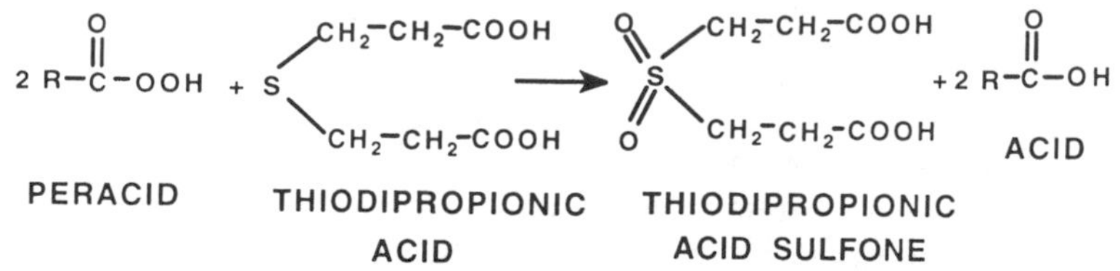

$$2\,R{-}\overset{O}{\overset{\|}{C}}{-}OOH + S(CH_2{-}CH_2{-}COOH)_2 \longrightarrow O_2S(CH_2{-}CH_2{-}COOH)_2 + 2\,R{-}\overset{O}{\overset{\|}{C}}{-}OH$$

PERACID THIODIPROPIONIC ACID THIODIPROPIONIC ACID SULFONE ACID

FIGURE 7 Mechanism of hydroperoxide decomposition by thiodipropionic acid.

A chemical structure similar to that in the thiodipropionates occurs in methionine (Chap. 6), and accounts, presumably by an analogous mechanism, for some of the antioxidant properties shown by proteins. Reaction of a sulfide with one peracid or hydroperoxide yields a sulfoxide, while reactions with two peracids or hydroperoxides yields a sulfone. Additional aspects of antioxidant chemistry are presented in Chapter 5.

12.7 ANTIMICROBIAL AGENTS

Chemical preservatives with antimicrobial properties play an important role in preventing spoilage and assuring safety of many foods. Some of these are discussed in this section.

12.7.1 Sulfites and Sulfur Dioxide

Sulfer dioxide (SO_2) and its derivatives long have been used in foods as general food preservatives. They are added to food to inhibit nonenzymic browning, to inhibit enzyme catalyzed reactions, to inhibit and control microorganisms, and to act as an antioxidant and a reducing agent. Generally, SO_2 and its derivatives are metabolized to sulfate and excreted in the urine without any obvious pathologic results [57]. However, because of somewhat recently recognized severe reactions to sulfur dioxide and its derivatives by some sensitive asthmatics, their use in foods is currently regulated and subject to rigorous labeling restrictions. Nonetheless, these preservatives serve key roles in contemporary foods.

The commonly used forms in foods include sulfur dioxide gas and the sodium, potassium, or calcium salts of sulfite (SO_3^{2-}), bisulfite (HSO_3^-), or metabisulfite ($S_2O_5^{2-}$). The most frequently used sulfiting agents are the sodium and potassium metabisulfites because they exhibit good stability toward autoxidation in the solid phase. However, gaseous sulfur dioxide is employed where leaching of solids causes problems or where the gas may also serve as an acid for the control of pH.

Although the traditional names for the anions of these salts are still widely used (sulfites, bisulfites, and metabisulfites), they have been designated by IUPAC as the sulfur(IV) oxoanions, sulfites (SO_3^{2-}), hydrogen sulfites (HSO_3^-), and disulfites ($S_2O_5^{2-}$), respectively. The oxoacids, H_2SO_3 and $H_2S_3O_5$, are designated as sulfurous and disulfurous acids, respectively [57].

Widely held views also have changed somewhat on the existence of sulfurous acid in aqueous solutions. Earlier, it was presumed that when sulfur dioxide was dissolved in water, it formed sulfurous acid, because the salts of simple oxoanions of sulfur (IV) (valence +4) are salts of this acid (H_2SO_3; sulfurous acid). However, evidence for the existence of free sulfurous acid has not been found, and it has been estimated that it accounts for less than 3% of nondissociated dissolved SO_2. Instead, solution of SO_2 yields only weak interactions with water, which results in a nondissociated complex that is particularly abundant below pH 2. This complex has been denoted $SO_2 \cdot H_2O$, and a distinction between this complex and sulfurous acid is not generally made [57] (see Eq. 12).

Since the acidity of solutions of sulfur dioxide is significant, and free sulfurous acid is not found, it is argued that the strong acid, $HSO_2(OH)$, exists in small amounts, and its dissociation leads predominantly to the HSO_3^- (bisulfite) ion rather than the $SO_2(OH)^-$ ion, which has been estimated to occur to an extent of only 2.5% of the hydrogen sulfite species (Eq. 13). The HSO_3^- ion predominates from pH 3 to 7, but above pH 7, the sulfite (SO_3^{2-}) ion is most abundant (Eq. 14). The pK_a for the first dissociation of "sulfurous acid" is 1.86, and the second dissociation has a pK_a of 7.18. In dilute solutions of HSO_3^- (10^{-2} *M*), little metabisulfite (disulfite) ion exists, but as the concentration of bisulfite increases, the proportion increases rapidly (Eq. 15). Thus, the relative proportion of each form depends on the pH of the solution, the ionic strength of the sulfur(IV) oxospecies, and the concentration of neutral salts [57].

$$SO_2 + H_2O \rightleftharpoons SO_2 \cdot H_2O \tag{12}$$

$$SO_2 \cdot H_2O \rightleftharpoons HSO_2(OH) \rightleftharpoons HSO_3^- + H^+ \tag{13}$$

$$HSO_3^- \rightleftharpoons H + SO_3^{2-} \tag{14}$$

$$2HSO_3^- \rightleftharpoons S_2O_5^{2-} + H_2O \tag{15}$$

Sulfur dioxide is most effective as an antimicrobial agent in acid media, and this effect may result from conditions that permit undissociated compounds to penetrate the cell wall. At high pH, it has been noted that the HSO_3^- ion is effective against bacteria but not against yeasts. Sulfur dioxide acts as both a biocidal and biostatic agent, and is more active against bacteria than molds and yeasts. Also, it is more effective against gram-negative bacteria than gram-positive bacteria.

The nucleophilicity of the sulfite ion is believed responsible for much of the effectiveness of sulfur dioxide as a food preservative in both microbial and chemical applications [58]. Some evidence has accumulated that the interaction of sulfur(IV) oxospecies with nucleic acids cause the biostatic and biocidal effects [57]. Other postulated mechanisms by which sulfur(IV) oxospecies inhibit microorganisms include the reaction of bisulfite with acetaldehyde in the cell, the reduction of essential disulfide linkages in enzymes, and the formation of bisulfite addition compounds that interfere with respiratory reactions involving nicotinamide dinucleotide.

Of the known inhibitors of nonenzymic browning in foods (Chap. 4), sulfur dioxide is probably the most effective. Multiple chemical mechanisms are involved in sulfur dioxide inhibition of nonenzymic browning (Fig. 8), but one of the most important involves the reaction of sulfur(IV) oxoanions (bisulfite) with carbonyl groups of reducing sugars and other compounds participating in browning. These reversible bisulfite addition compounds thus bind carbonyl groups to retard the browning process, but it also has been proposed that the reaction removes carbonyl chromophores in melanoidin structures, leading to a bleaching effect on the pigment. Sulfur(IV) oxoanions also irreversibly react with hydroxyl groups, especially those in the 4-position, on sugar and ascorbic acid intermediates in browning reactions to yield sulfonates ($R—CHSO_3^-—CH_2$-R′). The formation of relatively stable sulfonate derivatives retards the overall reaction and interferes with pathways that are particularly prone to producing colored pigments [57].

Sulfur dioxide also inhibits certain enzyme-catalyzed reactions, notably enzymic browning, that are important in food preservation. The production of brown pigments by enzyme-catalyzed oxidation of phenolic compounds can lead to a serious quality problem during the handling of some fresh fruits and vegetables (Chap. 7). However, the use of sulfite or metabisulfite sprays or dips with or without added citric acid provides effective control of enzymic browning in prepeeled and presliced potatoes, carrots, and apples.

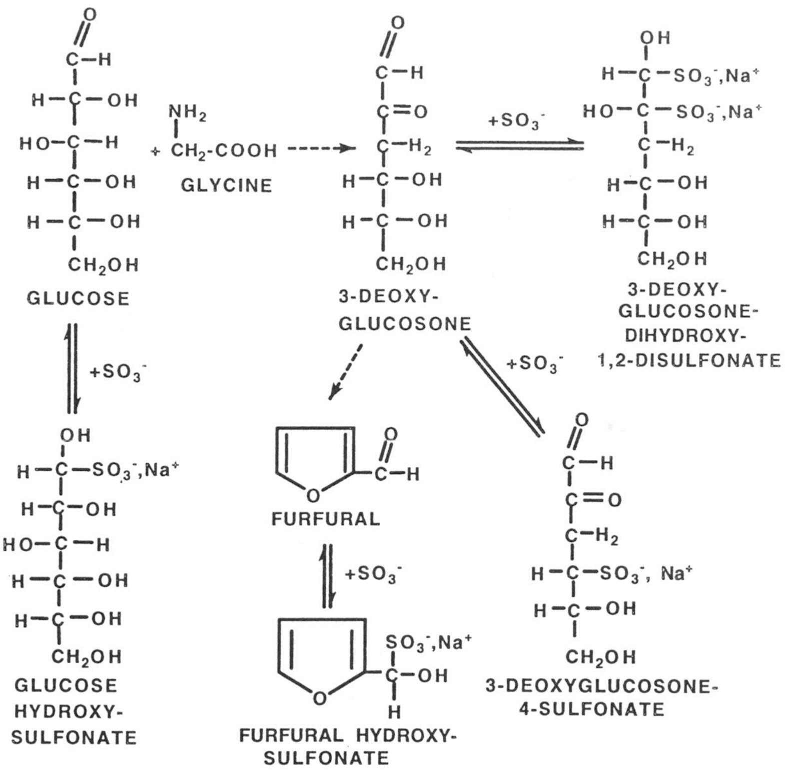

FIGURE 8 Mechanisms of inhibition of Maillard (carbonyl-amino) browning by some sulfur (IV) oxoanions (bisulfite, sulfite).

Sulfur dioxide also functions as an antioxidant in a variety of food systems, but it is not usually employed for this purpose. When sulfur dioxide is added to beer, the development of oxidized flavors is inhibited significantly during storage. The red color of fresh meat also can be effectively maintained by the presence of sulfur dioxide. However, this practice is not permitted because of the potential for masking deterioration in abused meat products.

When added during manufacture of wheat flour doughs, sulfur dioxide effects a reversible cleavage of protein disulfide bonds. In the instance of cookie manufacture, the addition of sodium bisulfite reduces mixing time and the elasticity of the dough that facilitates dough sheeting, and it also reduces variations caused by different lots of flour [57]. Prior to drying of fruits, gaseous sulfur dioxide is often applied, and this is sometime done in the presence of buffering agents (i.e., $NaHCO_3$). This treatment prevents browning and induces oxidative bleaching of anthocyanin pigments. The resulting properties are desired in products, such as those used to make white wines and maraschino cherries. Levels of sulfur dioxide encountered in dried fruits immediately following processing sometimes approach 2000 ppm. However, much lower amounts are found in most other foods because concentrations above 500 ppm give noticeably disagreeable flavors, and because sulfites tend to volatilize and/or react during storage and cooking.

12.7.2 Nitrite and Nitrate Salts

The potassium and sodium salts of nitrite and nitrate are commonly used in curing mixtures for meats to develop and fix the color, to inhibit microorganisms, and to develop characteristic flavors. Nitrite rather than nitrate is apparently the functional constituent. Nitrites in meat form nitric oxide, which reacts with heme compounds to form nitrosomyoglobin, the pigment responsible for the pink color of cured meats (Chap. 10). Sensory evaluations also indicate that nitrite contributes to cured meat flavor, apparently through an antioxidant role, but the details of this chemistry are poorly understood [43]. Furthermore, nitrites (150–200 ppm) inhibit clostridia in canned-comminuted and cured meats. In this regard, nitrite is more effective at pH 5.0–5.5 than it is at higher pH values. The antimicrobial mechanism of nitrite is unknown, but it has been suggested that nitrite reacts with sulfhydryl groups to create compounds that are not metabolized by microorganisms under anaerobic conditions.

Nitrites have been shown to be involved in the formation of low, but possibly toxic, levels of nitrosamines in certain cured meats. The chemistry and health implications of nitrosamines are discussed in Chapter 13. Nitrate salts also occur naturally in many foods, including vegetables such as spinach. The accumulation of large amounts of nitrate in plant tissues grown on heavily fertilized soils is of concern, particularly in infant foods prepared from these tissues. The reduction of nitrate to nitrite in the intestine, with subsequent absorption, could lead to cyanosis due to methemoglobin formation. For these reasons, the use of nitrites and nitrates in foods has been questioned. The antimicrobial capability of nitrite provides some justification for its use in cured meats, especially where growth of *Clostridium botulinum* is possible. However, in preserved products where botulism does not present a hazard, there appears to be little justification for adding nitrates and nitrites.

12.7.3 Sorbic Acid

Straight-chain, monocarboxylic, aliphatic fatty acids exhibit antimycotic activity, and α-unsaturated fatty acid analogs are especially effective. Sorbic acid (C-C=C-C=C-COOH) and its sodium and potassium salts are widely used to inhibit mold and yeasts in a wide variety of foods including cheese, baked products, fruit juices, wine, and pickles. Sorbic acid is particularly effective in preventing mold growth, and it contributes little flavor at the concentrations employed (up to 0.3% by weight). The method of application may involve direct incorporation, surface coatings, or incorporation in a wrapping material. The activity of sorbic acid increases as the pH decreases, indicating that the undissociated form is more inhibitory than the dissociated form. In general, sorbic acid is effective up to pH 6.5, which is considerably above the effective pH ranges for propionic and benzoic acids.

The antimycotic action of sorbic acid appears to arise because molds are unable to metabolize the α-unsaturated diene system of its aliphatic chain. It has been suggested that the diene structure of sorbic acid interferes with cellular dehyrogenases that normally dehydrogenate fatty acids as the first step in oxidation. Saturated short-chain (C_2–C_{12}) fatty acids are also moderately inhibitory to many molds, such as *Penicillium roqueforti*. However, some of these molds are capable of mediating β-oxidation of saturated fatty acids to corresponding β-keto acids, especially when the concentration of the acid is only marginally inhibitory. Decarboxylation of the resulting β-keto acid yields the corresponding methyl ketone (Fig. 9) which does not exhibit antimicrobial properties. A few molds also have been shown to metabolize sorbic acid, and it has been suggested that this metabolism proceeds through β-oxidation, similar to that in mammals. All evidence indicates that animals and humans metabolize sorbic acid in much the same ways as they do other naturally occurring fatty acids.

Although sorbic acid might at first appear quite stable and unreactive, it is quite often microbiologically or chemically altered in foods. Two other mechanisms for deactivating the antimicrobial properties of sorbic acid are shown in Figure 10. The reaction labeled "a" in Figure 10 has been demonstrated in molds, especially *P. roqueforti*. This involves direct decarboxylation

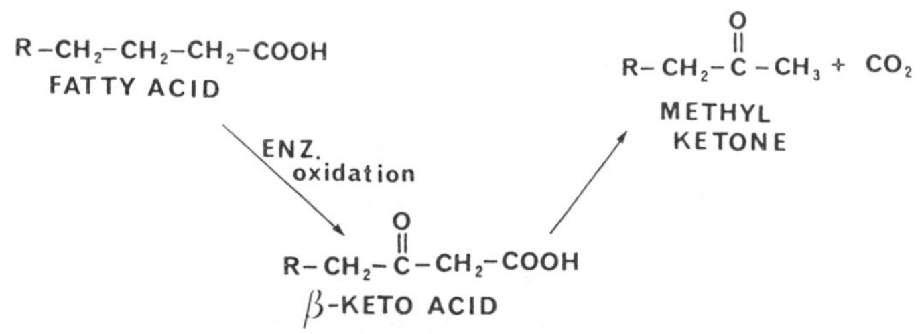

R–CH₂–CH₂–CH₂–COOH FATTY ACID → (ENZ. oxidation) → R–CH₂–CO–CH₂–COOH β-KETO ACID → R–CH₂–CO–CH₃ + CO₂ METHYL KETONE

FIGURE 9 Formation of a methyl ketone via mold-mediated enzymic oxidation of a fatty acid followed by a decarboxylation reaction.

of sorbic acid to yield the hydrocarbon 1,3-pentadiene. The intense aroma of this compound can cause gasoline or hydrocarbon-like off flavors when mold growth occurs in the presence of sorbic acid, especially on the surface of cheese treated with sorbate.

If wine containing sorbic acid undergoes spoilage in the bottle by lactic acid bacteria, an off flavor described as geranium-like develops [8]. Lactic acid bacteria reduce sorbic acid to sorbyl alcohol, and then, because of the acid conditions they have created, cause a rearrangement to a secondary alcohol (Fig. 10,b). The final reaction involves the formation of an ethoxylated hexadiene, which has a pronounced, easily recognized aroma of geranium leaves.

Sorbic acid is sometimes used in combination with sulfur dioxide, and this leads to reactions that deplete both sorbic acid and sulfur(IV) oxoanions (Fig. 11) [20]. Under aerobic conditions, especially in the presence of light, $\cdot SO_3^-$ radicals are formed, and these radicals sulfonate olefin bonds as well as promote oxidation of sorbic acid. This reaction, uniquely involving sorbic acid, is not noticeably affected by the presence of conventional antioxidants, and aerobically held foods containing sulfur dioxide and sorbic acid are very susceptible to autoxidation. Under anaerobic conditions, the combination of sorbic acid and sulfur dioxide in foods results in a much slower nucleophilic reaction of the sulfite ion (SO_3^-) with the diene (1,4-addition) in sorbic acid to yield 5-sulfo-3-hexenoic acid (Fig. 11).

Reactions between sorbic acid and proteins occur when sorbic acid is used in certain foods, such as wheat dough, whose proteins contain substantial amounts of oxidized or reduced thiol

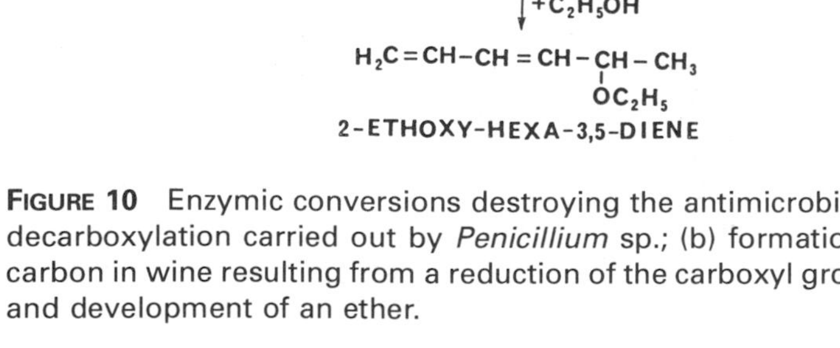

H₃C–CH=CH–CH=CH–COOH SORBIC ACID → (a, ENZ.) → H₃C–CH=CH–CH=CH₂ 1,3-PENTADIENE + CO₂

H₃C–CH=CH–CH=CH–COOH SORBIC ACID → (b, ENZ.) → H₃C–CH=CH–CH=CH–CH₂OH → H₂C=CH–CH=CH–CH(OH)–CH₃ → (+C₂H₅OH) → H₂C=CH–CH=CH–CH(OC₂H₅)–CH₃ 2-ETHOXY-HEXA-3,5-DIENE

FIGURE 10 Enzymic conversions destroying the antimicrobial properties of sorbic acid: (a) decarboxylation carried out by *Penicillium* sp.; (b) formation of ethoxylated diene hydrocarbon in wine resulting from a reduction of the carboxyl group followed by rearrangement and development of an ether.

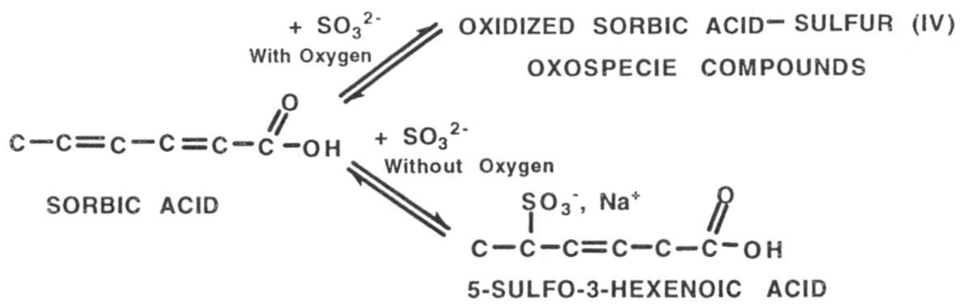

+ SO_3^{2-} With Oxygen → OXIDIZED SORBIC ACID– SULFUR (IV) OXOSPECIE COMPOUNDS

C–C=C–C=C–C(=O)–OH
SORBIC ACID

+ SO_3^{2-} Without Oxygen → C–C(SO_3^-, Na^+)–C=C–C–C(=O)–OH
5-SULFO-3-HEXENOIC ACID

FIGURE 11 Reactions of sorbic acid with sulfur dioxide [sulfur(IV) anions].

groups (R-S-S-R in cystine and R-SH in cysteine, respectively). The thiol groups (R-SH) dissociate to thiolate ions ($R\text{-}S^-$) that are reactive nucleophiles, and they react mainly by 1,6-addition to the conjugated diene of sorbic acid. This reaction binds the protein to the sorbic acid, and the reaction readily occurs at higher pH (> 5) and elevated temperatures, such as occur during bread baking. Although the reaction is reversible under very acidic conditions (pH < 1), the usual consequence at the higher pH values of foods is the loss of the preservative action of sorbic acid [30].

While sorbic acid and potassium sorbate have gained wide recognition as antimycotics, more recent research has established that sorbate has broad antimicrobial activity that extends to many bacterial species that are involved in spoilage of fresh poultry, fish and meats. It is especially effective in retarding toxigenesis of *Clostridium botulinum* in bacon and refrigerated fresh fish packaged in modified atmospheres.

12.7.4 Natamycin

Natamycin or pimaricin (CAS No. 768-93-8) is a polyene macrolide antimycotic (I) that has been approved in the United States for use against molds on cured cheeses. This mold inhibitor is

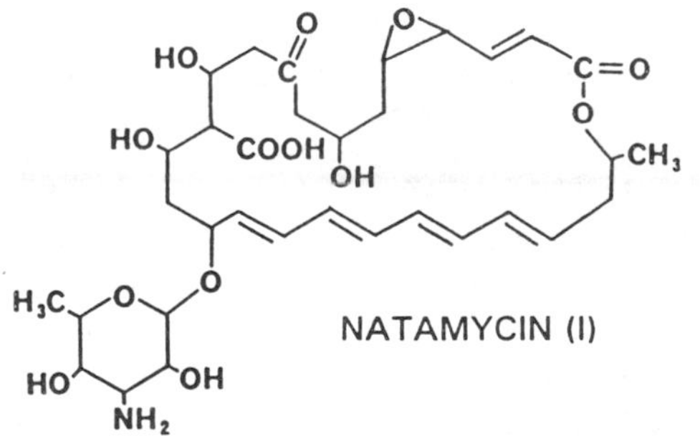

NATAMYCIN (I)

highly effective when applied to surfaces of foods exposed directly to air where mold has a tendency to proliferate. Natamycin is especially attractive for application on fermented foods, such as cured cheeses, because it selectively inhibits molds while allowing normal growth and metabolism of ripening bacteria.

12.7.5 Glyceryl Esters

Many fatty acids and monoglycerides show pronounced antimicrobial activity against gram-positive bacteria and some yeasts [28]. Unsaturated members, especially those with 18 carbon atoms, show strong activity as fatty acids; the medium chain-length members (12 carbon atoms) are most inhibitory when esterified to glycerol. Glyceryl monolaurate (II), also known under the tradename of Monolaurin, is inhibitory against several potentially pathogenic staphylococcus and streptococcus when present at concentrations of 15–250 ppm. It is commonly used in cosmetics, and because of its lipid nature can be used in some foods.

$$\begin{array}{l} \quad\quad\;\; O \\ \quad\quad\;\; \| \\ C-O-C-(CH_2)_{10}-CH_3 \\ | \\ CHOH \\ | \\ CH_2OH \end{array}$$

GLYCERYL MONOLAURATE (II)

Lipophilic agents of this kind also exhibit inhibitory activity against *C. botulinum*, and glyceryl monolaurate, serving this function, may find applications in cured meats and in refrigerated, packaged fresh fish. The inhibitory effect of lipophilic glyceride derivatives apparently relates to their ability to facilitate the conduction of protons through the cell membranes, which effectively destroys the proton motive force that is needed for substrate transport [15]. Cell-killing effects are observed only at high concentrations of the compounds, and death apparently results form the generation of holes in cell membranes.

12.7.6 Propionic Acid

Propionic acid (CH_3-CH_2-COOH) and its sodium and calcium salts exert antimicrobial activity against molds and a few bacteria. This compound occur naturally in Swiss cheese (up to 1% by weight), where it is produced by *Propionibacterium shermanii* [5]. Propionic acid has found extensive use in the bakery field where it not only inhibits molds effectively, but also is active against the ropy bread organism, *Bacillus mesentericus*. Levels of use generally range up to 0.3% by weight. As with other carboxylic acid antimicrobial agents, the undissociated form of propionic acid is active, and the range of effectiveness extends up to pH 5.0 in most applications. The toxicity of propionic acid to molds and certain bacteria is related to the inability of the affected organisms to metabolize the three carbon skeleton. In mammals, propionic acid is metabolized in a manner similar to that of other fatty acids, and it has not been shown to cause any toxic effects at the levels utilized.

12.7.7 Acetic Acid

The preservation of foods with acetic acid (CH_3COOH) in the form of vinegar dates to antiquity. In addition to vinegar (4% acetic acid) and acetic acid, also used in food are sodium acetate (CH_3COONa), potassium acetate (CH_3COOK), calcium acetate [$(CH_3COO)_2Ca$], and sodium diacetate ($CH_3COONa \cdot CH_3$-$COOH \cdot \frac{1}{2}H_2O$). The salts are used in bread and other baked goods (0.1–0.4%) to prevent ropiness and the growth of molds without interfering with yeasts [5]. Vinegar and acetic acid are used in pickled meats and fish products. If fermentable carbohydrates are present, at least 3.6% acid must be present to prevent growth of lactic acid bacilli and yeasts. Acetic acid is also used in foods such as catsup, mayonnaise, and pickles, where it serves a dual function of inhibiting microorganisms and contributing to flavor. The antimicrobial activity of acetic acid increases as the pH is decreased, a property analogous to that found for other aliphatic fatty acids.

12.7.8 Benzoic Acid

Benzoic acid (C_6H_5COOH) has been widely employed as an antimicrobial agent in food [5], and it occurs naturally in cranberries, prunes, cinnamon, and cloves. The undissociated acid is the form with antimicrobial activity, and it exhibits optimum activity in the pH range of 2.5–4.0,

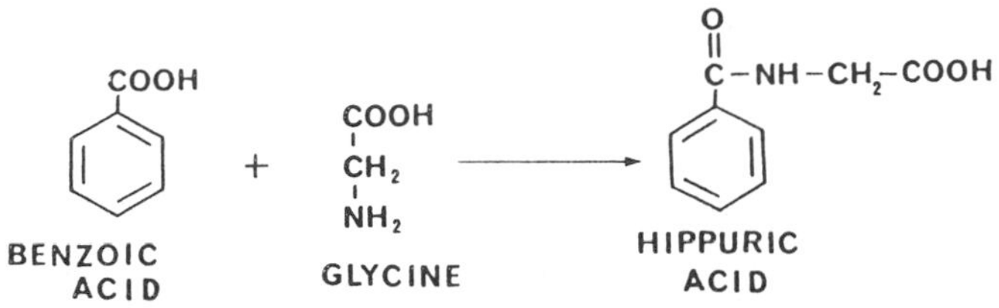

BENZOIC ACID + GLYCINE → HIPPURIC ACID

FIGURE 12 Conjugation of benzoic acid with glycine to facilitate excretion.

making it well suited for use in acid foods such as fruit juices, carbonated beverages, pickles, and sauerkraut. Since the sodium salt of benzoic acid is more soluble in water than the acid form, the former is generally used. Once in the product, some of the salt converts to the active acid form, which is most active against yeasts and bacteria and least active against molds. Often benzoic acid is used in combination with sorbic acid or parabens, and levels of use usually range from 0.05 to 0.1% by weight.

Benzoic acid has been found to cause no deleterious effects in humans when used in small amounts. It is readily eliminated from the body primarily after conjugation with glycine (Fig. 12) to form hippuric acid (benzoyl glycine). This detoxification step precludes accumulation of benzoic acid in the body.

12.7.9 *p*-Hydroxybenzoate Alkyl Esters

The parabens are a group of alkyl esters of *p*-hydroxybenzoic acid that have been used widely as antimicrobial agents in foods, pharmaceutical products, and cosmetics. The methyl (III) propyl, and heptyl (IV) esters are used domestically, and in some other countries the ethyl and butyl esters are used as well.

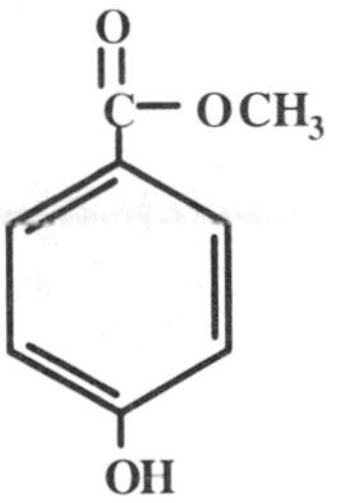

METHYL PARABEN (III)

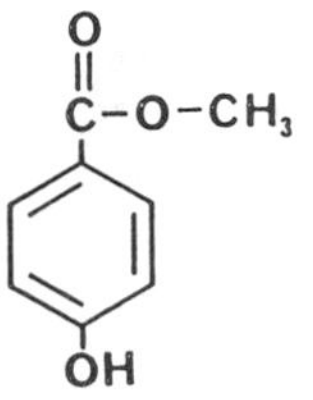

PROPYL PARABEN

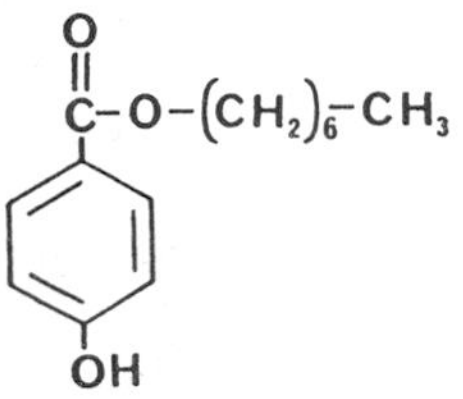

HEPTYL PARABEN (IV)

Parabens are used as microbial preservatives in baked goods, soft drinks, beer, olives, pickles, jams and jellies and syrups. They have little effect on flavor, are effective inhibitors of molds and yeasts (0.5–0.1% by weight), and are relatively ineffective against bacteria, especially gram-negative bacteria [5]. The antimicrobial activity of parabens increases and their solubility in water decreases with increases in the length of the alkyl chain. The shorter chain members often are used because of their solubility characteristics. In contrast to other antimycotic agents, the parabens are active at pH 7 and higher, apparently because of their ability to remain undissociated at these pH values. The phenolic group provides a weak acid character to the molecule. The ester linkage is stable to hydrolysis even at temperatures used for sterilization. The parabens have many properties in common with benzoic acid and they are often used together. Parabens exhibit a low order of toxicity to humans and are excreted in the urine after hydrolysis of the ester group and subsequent metabolic conjugation.

12.7.10 Epoxides

Most antimicrobial agents used in foods exhibit inhibitory rather than lethal effects at the concentrations employed. However, exceptions occur with ethylene (V) and propylene oxides (VI). These chemical sterilants are used to treat certain low-moisture foods and to sterilize

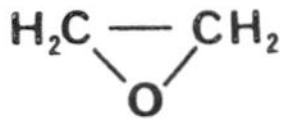

H_2C — CH_2 (ring closed through O)

ETHYLENE OXIDE (V)

H_3C — CH_2–CH_3 (ring closed through O)

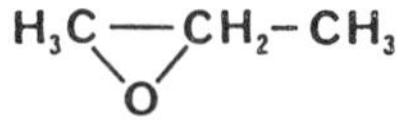

PROPYLENE OXIDE (VI)

aseptic packaging materials. In order to achieve intimate contact with microorganisms the epoxides are used in a vapor state, and after adequate exposure, most of the residual unreacted epoxide is removed by flushing and evacuation.

The epoxides are reactive cyclic ethers that destroy all forms of microorganisms, including spores and even viruses, but the mechanism of action of epoxides is poorly understood. In the case of ethylene oxide it has been proposed that alkylation of essential intermediary metabolites having a hydroethyl group ($-CH_2-CH_2-OH$) could account for the lethal results [5]. The site of attack would be any labile hydrogen in the metabolic system. The epoxides also react with water to form corresponding glycols (Fig. 13). However, the toxicity of the glycols is low, and therefore cannot account for the inhibitory effect.

Since the majority of the active epoxide is removed from the treated food, and the glycols formed are of low toxicity, it might appear that these gaseous sterilants would be used extensively. Their use, however, is limited to dry items, such as nutmeats and spices, because reaction with water rapidly depletes the concentration of epoxides in high-moisture foods. Spices often contain high microbial loads and are destined for incorporation into perishable foods. Thermal sterilization of spices is unsuitable because important flavor compounds are volatile and the product is generally unstable to heat. Thus, treatment with epoxides is a suitable method for reducing the microbial load.

The potential formation of relatively toxic chlorohydrins as a result of reactions between epoxides and inorganic chlorides (Fig. 13) is a point of some concern. However, there are reports that dietary chlorohydrin in low concentrations causes no ill effect [59]. another consideration in the use of epoxides is their possible adverse effects on vitamins, including riboflavin, niacin, and pyridoxine.

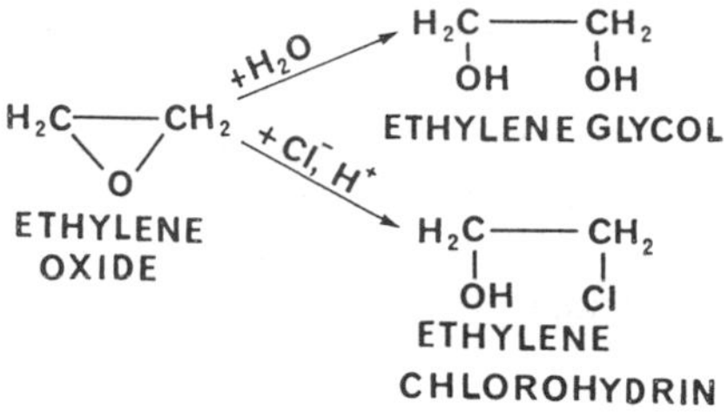

$H_2C—CH_2$ (O bridge) ETHYLENE OXIDE $\xrightarrow{+H_2O}$ $H_2C(OH)—CH_2(OH)$ ETHYLENE GLYCOL

ETHYLENE OXIDE $\xrightarrow{+Cl^-, H^+}$ $H_2C(OH)—CH_2(Cl)$ ETHYLENE CHLOROHYDRIN

FIGURE 13 Reactions of ethylene oxide with water and chloride ion, respectively.

Ethylene oxide (boiling point 13.2°C) is more reactive than propylene oxide and is also more volatile and flammable. For safety purposes, ethylene oxide is often supplied as a mixture consisting of 10% ethylene oxide and 90% carbon dioxide. The product to be sterilized is placed in a closed chamber, and the chamber is evacuated, then pressurized to 30 lb with the ethylene oxide-carbon dioxide mixture. This pressure is needed to provide a concentration of epoxide sufficient to kill microorganisms in a reasonable time. When propylene oxide (boiling point 34.3°C) is used, sufficient heat must be applied to maintain the epoxide in a gaseous state.

12.7.11 Antibiotics

Antibiotics comprise a large group of antimicrobial agents produced naturally by a variety of microorganisms. They exhibit selective antimicrobial activity, and their applications in medicine have contributed significantly to the field of chemotherapy. The successes of antibiotics in controlling pathogenic microorganisms in living animals have led to extensive investigations into their potential applications in food preservation. However, because of the fear that routine use of antibiotics will cause resistant organisms to evolve, their application to foods, with one exception (nisin), is not currently permitted in the United States. The development of resistant strains of organisms would be of particular concern if an antibiotic proposed for use in food is also used in a medical application.

Nisin, a polypeptide antibiotic, is produced by lactic streptococci, and in the United States, it is now permitted in high-moisture processed cheese products where it is used to prevent potential outgrowth of *C. botulinum.* Nisin has been explored extensively for applications in food preservation. It is active against gram-positive organisms, especially in preventing the outgrowth of spores [46], and it is not used in medical applications. Nisin is also used in other parts of the world for prevention of spoilage of dairy products, such as processed cheese and condensed milk. Nisin is not effective against gram-negative spoilage organisms, and some strains of clostridia are resistant. However, nisin is essentially nontoxic to humans, does not lead to cross-resistance with medical antibiotics, and is degraded harmlessly in the intestinal tract.

Some other countries allow limited use of a relatively few other antibiotics. These include chlortetracycline and oxytetracycline [5]. Most actual or proposed applications of antibiotics in foods involve their use as adjuncts to other methods of food preservation. Notably, this includes delaying spoilage of refrigerated, perishable foods, and reducing the severity of thermal processes. Fresh meats, fish, and poultry comprise a group of perishable products that could benefit from the action of broad-spectrum antibiotics. In fact, many years ago, the U.S. Food and Drug Administration permitted dipping whole poultry carcasses into solutions of chlortetracycline or oxytetracycline. This increased the shelf life of the poultry, and residual antibiotics were destroyed by usual cooking methods.

The biochemical modes of actions for antibiotics are just coming into focus, with research effort emphasizing molecular mechanisms. In addition, there is a continuing search for natural preservatives, which hopefully will be suitable for application to foods. However, the necessarily stringent requirements placed on substances for food applications indicate that acceptable substances will be difficult to find.

12.7.12 Diethyl Pyrocarbonate

Diethyl pyrocarbonate has been used as an antimicrobial food additive for beverages such as fruit juices, wine, and beer. The advantage of diethyl pyrocarbonate is that it can be used in a cold pasteurization process for aqueous solutions, following which it readily hydrolyzes to ethanol and carbon dioxide (Fig. 14). Usage levels between 120 and 300 ppm in acid beverages (below pH 4.0) cause complete destruction of yeasts in about 60 min. Other organisms, such as lactic acid bacteria, are more resistant and sterilization is achieved only when the microbial load is low (less than 500 ml^{-1}) and the pH is below 4.0. The low pH retards the rate of diethyl pyrocarbonate decomposition and intensifies its effectiveness.

Concentrated diethyl pyrocarbonate is an irritant. However, since hydrolysis is essentially complete within 24 hr in acid beverages, there is little concern for direct toxicity. Unfortunately, diethyl pyrocarbonate reacts with a variety of compounds to form carbethoxy derivatives and ethyl esters. Specifically, diethyl pyrocarbonate reacts readily with ammonia to yield urethane (ethyl carbamate; Fig. 14). Ostensibly this reaction was considered responsible for urethane found in foods treated with diethyl pyrocarbonate, and a ban on the use of diethyl pyrocarbonate was issued in 1972 because urethane is a known carcinogen. Since ammonia is ubiquitous in plant and animal tissues, it seemed reasonable that foods treated with diethyl pyrocarbonate will contain some urethane.

However, it was shown later that urethane occurs intrinsically in fermented foods and beverages, and it is usually present below 10 parts per billion in most fermented foods, including bread, wine, and beer [23]. It has been suggested that the major pathway for its production in these foods is the reaction of urea, from arginine metabolism, with ethanol. Alcoholic beverages contain much higher levels of urethane than nonalcoholic foods, and levels to 10 parts per million have been reported in stone fruit brandies. In spite of these natural occurrences of urethane, diethyl pyrocarbonate is no longer permitted in foods in the United States because of the potential for elevating levels of a carcinogen in foods.

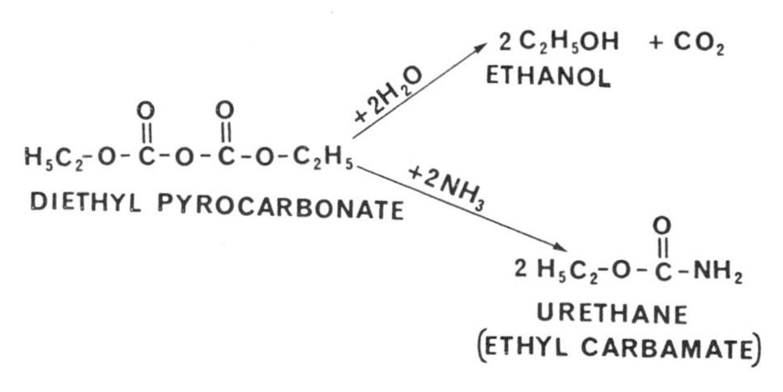

FIGURE 14 Reactions showing the hydrolysis and amidization of diethyl pyrocarbonate.

12.8 INTENSELY SWEET NONNUTRITIVE AND LOW-CALORIE SWEETENERS

Nonnutritive and low-calorie sweeteners encompass a broad group of substances that evoke a sweet taste or enhance the perception of sweet tastes (see Chap. 11). The ban on the use of cyclamates in the United States along with questions raised about the safety of saccharin initiated an intensive search for alternate low-calorie sweeteners to meet the demand for low-calorie foods and beverages. This has led to the discovery of many new sweet molecules, and the number of viable potentially commercially useful, low-calorie sweeteners is growing. The relative sweetness values for some of these substances are given in Table 3.

12.8.1 Cyclamate

Cyclamate (cyclohexyl sulfamate) was approved as a food additive in the United States in 1949, and before the substances were prohibited by the U.S. Food and Drug Administration in late 1969, the sodium and calcium salts and the acid form of cyclamic acid were widely employed as sweeteners. Cyclamates are about 30 times sweeter than sucrose, taste much like sucrose without significant interfering taste sensations, and are heat stable. Cyclamate sweetness has a slow onset, and persists for a period of time that is longer than that for sucrose.

Some early experimental evidence with rodents had suggested that cyclamate and its hydrolysis product, cyclohexylamine (Fig. 15), caused bladder cancer [3, 42]. However, extensive testing subsequently has not substantiated the early reports, and petitions have been filed in the United States for reinstatement of cylcamate as an approved sweetener [36]. Currently, cyclamate is permitted for use in low-calorie foods in 40 countries and Canada. Still, for various reasons, even though extensive data supporting the conclusion that neither cyclamate nor cyclohexylamine are carcinogen or genotoxic [2], the U.S. Food and Drug Administration has chosen not to reapprove cyclamates for use in foods.

TABLE 3 Relative Sweetness of Some Intensely Sweet Substances

Substance	Relative sweetness values[a] (sucrose = 1, weight basis)
Acesulfame K	200
Alitame	2000
Aspartame	180–200
Cyclamate	30
Glycyrrhizin	50–100
Monellin	3000
Neohesperitin dihydrochalcone	1600–2000
Saccharin	300–400
Stevioside	300
Sucralose	600–800
Thaumatin	1600–2000

[a]Commonly cited relative sweetness values are listed; however, the concentration and the food or beverage matrix may greatly influence actual relative sweetness values for sweeteners.

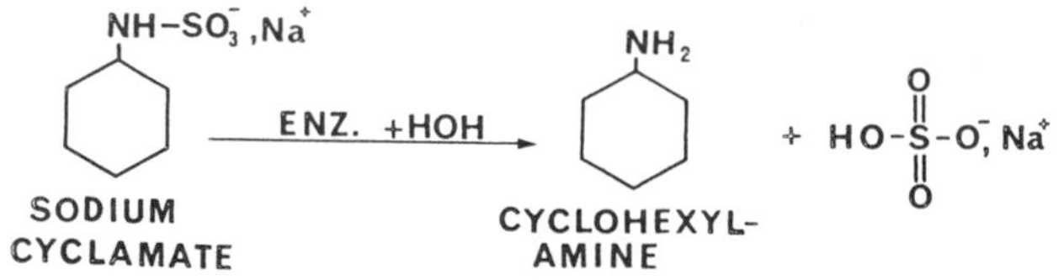

NH–SO_3^-,Na^+ — ENZ. +HOH → NH_2 + HO–S(=O)(=O)–O^-, Na^+

SODIUM CYCLAMATE — CYCLOHEXYL-AMINE

FIGURE 15 Formation of cyclohexylamine by the hydrolysis of cyclamate.

12.8.2 Saccharin

The calcium and sodium salts and free acid form of saccharin (3-oxo-2,3-dihydro-1,2-benzisothiazole-1, 1 dioxide) are available as nonnutritive sweeteners (VII).

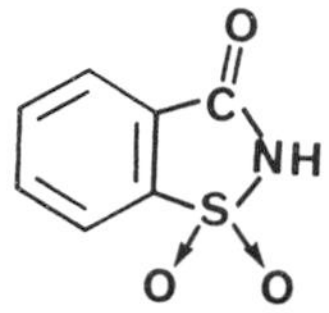

SACCHARIN (VII)

The commonly accepted rule of thumb is that saccharin is about 300 times as sweet as sucrose in concentrations up to the equivalent of a 10% sucrose solution, but the range is from 200 to 700 times the sweetness of sucrose depending on the concentration and the food matrix [44]. Saccharin exhibits a bitter, metallic aftertaste, especially to some individuals, and this effect becomes more evident with increasing concentration.

The safety of saccharin has been under investigation for over 50 years, and it has been found to cause a low incidence of carcinogenesis in laboratory animals. However, many scientists argue that the animal data are not relevant to humans. In humans saccharin is rapidly absorbed, and then is rapidly excreted in the urine. Although current regulations in the United States prohibit the use of food additives that cause cancer in any experimental animals, a ban on saccharin in the United States, proposed by the FDA in 1977, has been stayed by congressional legislation pending further research. However, a health warning statement is required on packages of saccharin-containing foods. Saccharin is approved for use in more than 90 countries around the world.

12.8.3 Aspartame

Aspartame or L-aspartyl-L-phenylalanine methyl ester (Fig. 16) is a caloric sweetener because it is a dipeptide that is completely digested after consumption. However, its intense sweetness (about 200 times as sweet as sucrose) allows functionality to be achieved at very low levels that provide insignificant calories. It is noted for a clean, sweet taste that is similar to that of sucrose. Aspartame was first approved in the United States in 1981, and now is approved for use in over 75 countries where it is used in over 1700 products.

Two disadvantages of aspartame are its instability under acid conditions, and its rapid degradation when exposed to elevated temperatures. Under acid conditions, such as carbonated soft drinks, the rate of loss of sweetness is gradual and depends on the temperature and pH. The peptide nature of aspartame makes it susceptible to hydrolysis, and this feature also permits other chemical interactions and microbial degradations. In addition to loss of sweetness resulting from hydrolysis of either the methyl ester on phenylalanine or the peptide bond between the two amino

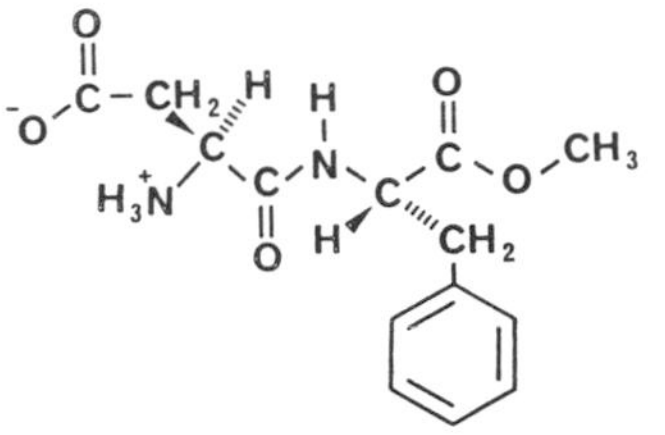

L-ASPARTYL-L-PHENYLALANINE
METHYL ESTER
(ASPARTAME)

FIGURE 16 Stereochemical configuration of aspartame.

acids, aspartame readily undergoes an intramolecular condensation, especially at elevated temperatures, to yield the diketopiperazine (5-benzyl-3,6-dioxo-2-piperazine acetic acid) shown in Figure 17. This reaction is especially favored at neutral and alkaline pH values because nonprotonated amine groups on the molecule are more available for reaction under these conditions. Similarly, alkaline pH values promote carbonyl-amino reactions, and aspartame has been shown to react readily with glucose and vanillin under such conditions. With the glucose reaction, loss of aspartame's sweetness during storage is the principal concern, while loss of vanilla flavor is the main concern in the latter case.

Even though aspartame is composed of naturally-occurring amino acids and its daily intake is projected to be very small (0.8 g/person), concern has been expressed about its potential safety as a food additive. Aspartame-sweetened products must be labelled prominently about their phenylalanine content to allow avoidance of consumption by phenylketonuric individuals who lack 4-monooxygenase that is involved in the metabolism of phenylalanine. However, aspartame consumption of aspartame by the normal population is not associated with adverse health effects. Extensive testing has similarly shown that the diketopiperazine poses no risk to humans at concentrations potentially encountered in foods [26].

12.8.4 Acesulfame K

Acesulfame K [6-methyl-1,2,3-oxathiazine-4(3*H*)-one-2,2-dioxide] was discovered in Germany, and was first approved for use as a nonnutritive sweetener in the United States in 1988. The complex chemical name of this substance led to the creation of the trademarked common name,

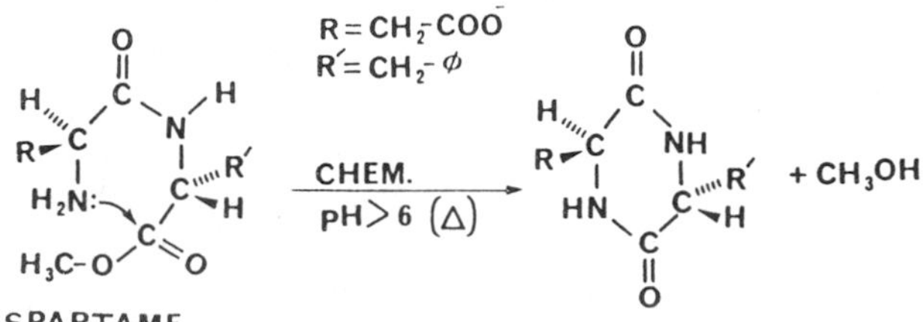

$R = CH_2{-}COO^-$
$R' = CH_2{-}\phi$

CHEM.
pH > 6 (Δ)

$+ CH_3OH$

ASPARTAME

A DIKETOPIPERAZINE

FIGURE 17 Intramolecular condensation of aspartame yielding a diketopiperazine degradation product.

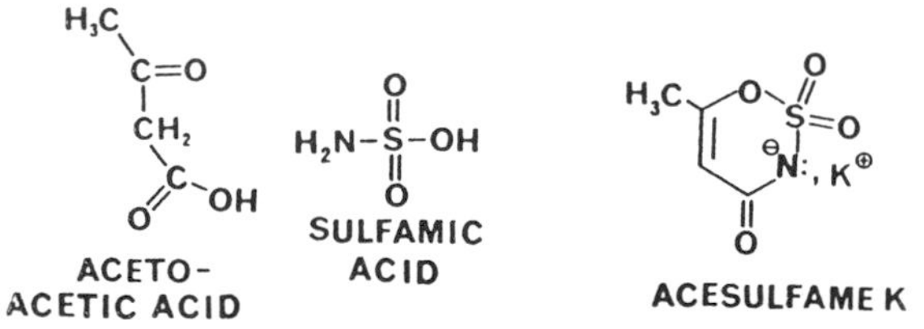

ACETO-ACETIC ACID

SULFAMIC ACID

ACESULFAME K

FIGURE 18 Structurally related compounds that form the basis for the derived name of the nonnutritive sweetener Acesulfame K.

Acesulfame K, which is based on its structural relationships to acetoacetic acid and sulfamic acid, and to its potassium salt nature (Fig. 18).

Acesulfame K is about 200 times as sweet as sucrose at a 3% concentration in solution, and it exhibits a sweetness quality between that of cyclamates and saccharin. Since acesulfame K possesses some metallic and bitter taste notes at higher concentrations, it is especially useful when blended with other low-calorie sweeteners, such as aspartame. Acesulfame K is exceptionally stable at elevated temperatures encountered in baking, and it is also stable in acidic products, such as carbonated soft drinks. Acesulfame K is not metabolized in the body, thus providing no calories, and is excreted by the kidneys unchanged. Extensive testing has shown no toxic effects in animals, and exceptional stability in food applications.

12.8.5 Sucralose

Sucralose (1,6-dichloro-1,6-dideoxy-β-fructofuranosyl-4-chloro-α-D-galactopyranoside) (VIII) is a noncaloric sweetener produced by the selective chlorination of the sucrose molecule, and it is

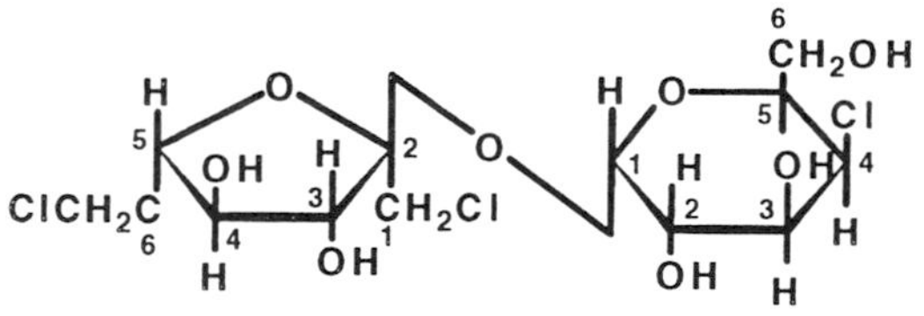

SUCRALOSE (VIII)

about 600 times sweeter than sucrose. Food additive petitions for use of sucralose were filed in the United States in 1987 and 1989, but sucralose has not yet been approved in this country. Sucralose, however, was approved for some applications in Canada in 1991.

Sucralose exhibits a high degree of crystallinity, high water solubility, and very good stability at high temperatures, thus making it an excellent ingredient for bakery applications. It is also quite stable at the pH of carbonated soft drinks, and only limited hydrolysis to monosaccharide units occurs during usual storage of these products. Sucralose possesses a sweetness time–intensity profile similar to sucrose, and exhibits no bitterness or other unpleasant aftertastes. Extensive studies have been conducted on the safety of sucralose, and these have generally demonstrated that the substance is safe at the expected usage levels.

12.8.6 Alitame

Alitame [L-α-aspartyl-*N*-(2,2,4,4-tetramethyl-3-thietanyl)-D-alaninamide] (IX) is an amino acid-based sweetener that possesses a sweetening power of about 2000 times that of

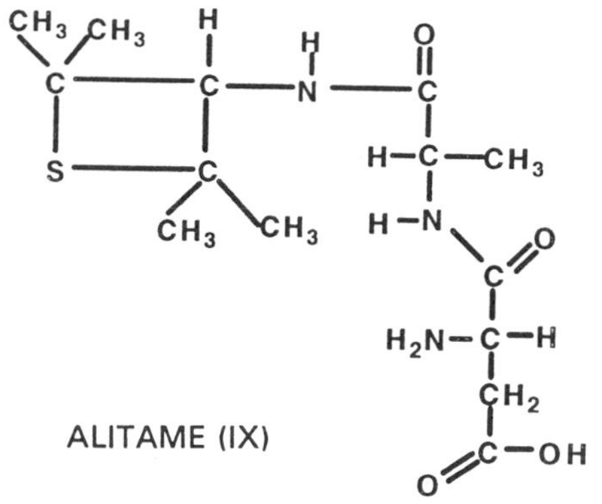

ALITAME (IX)

sucrose, and it exhibits a clean sweet taste similar to sucrose. It is highly soluble in water and has good thermal stability and shelf life, but prolonged storage in some acidic solutions may result in off flavors. Generally, alitame has the potential for use in most foods where sweeteners are employed, including baked goods.

Alitame is prepared from the amino acids L-aspartic acid and D-alanine and a novel amine. Although the aspartic acid component of alitame is metabolized, its caloric contribution is insignificant because alitame is such an intense sweetener. The alanine amide moiety of alitame passes through the body with minimal metabolic changes. Extensive testing indicates that alitame is safe for human consumption, and a petition for its use in foods was filed in 1986 with the U.S. Food and Drug Administration. Although not yet approved for use in foods in the United States, alitame has been approved in Australia, New Zealand, China, and Mexico.

12.8.7 Other Intensely Sweet Nonnutritive or Low-Calorie Sweeteners

The intensive search for alternative sweeteners over the past decade has led to the discovery of a large number of new sweet compounds, and many of these are undergoing further development and safety studies to determine if they are suitable for future commercialization. Included are β-substituted β-amino acids that are up to 20,000 times as sweet as sucrose, and trisubstituted guanidines with sweetness potencies up to 170,000 times that of sucrose. These compounds join a substantial list of less well known but emerging intensely sweet compounds, and some of the latter group are discussed here.

Glycyrrhizin (glycyrrhizic acid) is a triterpene saponin that is found in licorice root, and is 50–100 times sweeter than sucrose. Glycyrrhizin is a glycoside that on hydrolysis yields two moles of glucuronic acid and one mole of glycyrrhetinic acid, a triterpene related to aleanolic acid. Ammonium glycyrrhizin, the fully ammoniated salt of glycyrrhizic acid, is commercially available, and is approved for use only as a flavor and as a surfactant, but not as a sweetener. Glycyrrhizic acid is used primarily in tobacco products and to some extent in foods and beverages. Its licorice-like flavor influences its suitability for some applications.

A mixture of glycosides found in the leaves of the South American plant *Stevia rebaudiana* Bertoni is the source of stevioside and rebaudiosides. Pure stevioside is about 300 times as sweet as sucrose. Stevioside exhibits some bitterness and undesirable aftertastes at higher concentrations, and rebaudioside A exhibits the best taste profile of the mixture. However, extracts produced from *S. rebaudiana* are used as commercial forms of this sweetener, and they are employed extensively in Japan. Extensive safety and toxicological testing has indicated that the extracts are safe for human consumption, but they are not approved in the United States.

Neohesperidin dihydrochalcone is a nonnutritive sweetener that is 1500–2000 times as sweet as sucrose, and it is derived from the bitter flavonones of citrus fruits. Neohesperidin dihydrochalcone exhibits a slow onset in sweetness and a lingering sweet aftertaste, but it decreases the perception of concurrent bitterness. This intensely sweet substance and other similar compounds are produced by hydrogenation of (1) naringin to yield naringin dihydrochalcone, (2) neohesperidin to yield neohesperidin dihydrochalcone, or (3) hesperidin to yield hesperidin dihydrochalcone 4′-*O*-glucoside [41]. Neohesperidin dihydrochalcone has been extensively tested for safety, and the studies have generally confirmed its safety. It is approved for use in Belgium and Argentina, but the U.S. Food and Drug Administration has requested additional toxicology testing.

Several sweet proteins are now known, and thaumatins I and II obtained from the tropical African fruit katemfe (*Thaumatococcus daniellii*) have been well characterized. Thaumatins I and II are alkaline proteins, each with a molecular weight of about 20,000 [55], and on a mass basis they are about 1600–2000 times as sweet as sucrose. An extract of katemfe fruit is marketed under the tradename of Talin in the United Kingdom, and its use as a sweetener and flavor enhancer has been approved in Japan and Great Britain. It is also permitted as a flavor enhancer in chewing gum in the United States. Talin exhibits a long-lasting sweetness with a slight licorice-like taste, which limits it use along with its high cost.

Another sweet protein, monellin, is obtained from the serendipity berry, and it has a molecular weight of about 11,500. Monellin is about 3000 times as sweet as sucrose on a mass basis, and the sweetness of natural monellin is destroyed by boiling. The potential use of sweet proteins is limited because the compounds are expensive, unstable to heat, and lose sweetness when held in solution below pH 2 at room temperature.

Another basic protein, miraculin, has been isolated from miracle fruit (*Richadella dulcifica*), and this protein is tasteless. However, it has the peculiar property of changing sour taste into sweet taste; that is, it makes lemons taste sweet. Miraculin is a glycoprotein with a molecular weight of 42,000 [55], and, similar to other protein sweeteners, miraculin is heat labile and inactivated at low pH values. The sweetness induced by 0.1 *M* citric acid after tasting 1 μ*M* miraculin solution is equivalent to a 0.4 *M* sucrose solution; thus, the sweetness of a miraculin solution induced by 0.1 *M* citric acid solution has been calculated to be 400,000 times that of a sucrose solution. The taste effects of miraculin persist for over 24 hrs after placing it in the mouth, and this limits its potential use. In the 1970s, miraculin was introduced in the United States as a sweetening aid for diabetics, but it was banned by the U.S. Food and Drug Administration because of insufficient safety data.

12.9 POLYHYDRIC ALCOHOL TEXTURIZERS AND REDUCED-CALORIE SWEETENERS

Polyhydric alcohols are carbohydrate derivatives that contain only hydroxyl groups as functional groups (Chap. 4), and as a result, they are generally water-soluble, hygroscopic materials that exhibit moderate viscosities at high concentrations in water. While the number of available polyhydric alcohols is substantial, relatively few have been important in food application. However, the usage of some polyhydric alcohols is growing because of demands for their reduced-calorie sweetener properties.

This class of substances includes synthetic propylene glycol ($CH_2OH\text{-}CHOH\text{-}CH_3$) and naturally produced glycerol ($CH_2OH\text{-}CHOH\text{-}CH_2OH$). Additionally, xylitol ($CH_2OH\text{-}CHOH\text{-}CHOH\text{-}CHOH\text{-}CH_2OH$), sorbitol, and mannitol ($CH_2\text{-}CHOH\text{-}CHOH\text{-}CHOH\text{-}CHOH\text{-}CH_2OH$ isomers) are produced by hydrogenation of xylose, glucose, and mannose, respectively. More recently, hydrogenated starch hydrolysates have entered the market, and these contain sorbitol from glucose, maltitol from maltose, and various polymeric polyols from oligosaccharides. Many polyhydric alcohols occur naturally, but because of their limited concentrations, they usually do not exhibit functional roles in food. For example, free glycerol exists in wine and beer as a result of fermentation, and sorbitol occurs in fruits such as pears, apples, and prunes.

The polyhydroxy structures of these compounds provide water-binding properties that have been exploited in foods. Specific functions of polyhydric alcohols include control of viscosity and texture, addition of bulk, retention of moisture, reduction of water activity, control of crystallization, improvement or retention of softness, improvement of rehydration properties of dehydrated foods, and use as a solvent for flavor compounds [21].

Sugars and polyhydric alcohols are structurally similar, except that sugars also contain aldo or keto groups (free or bound) that adversely affect their chemical stability, especially at high temperatures. Many applications of polyhydric alcohols in foods rely on concurrent contributions of functional properties from sugars, proteins, starches and gums. Polyhydric alcohols generally are sweet, but less so than sucrose (Table 4). Short-chain members, such as glycerol, are slightly bitter at high concentrations. When used in the dry form, sugar alcohols (polyols) contribute a pleasant cooling sensation because of their negative heat of solution.

Historically, the energy value of simple polyols derived from sugars, like sugars, has been considered to be 16.7 kJ (4 kcal) g^{-1} (joules = calories × 4.1814) for labeling purposes in the United States. However, this view has changed very recently following a 1990 European Union lead of assigning an energy value of 10 kJ (2.4 kcal) g^{-1} to polyols as a group. The U.S. Food and Drug Administration has accepted caloric contents ranging from 6.7 to 12.5 kJ (1.6–3.0 kcal)

TABLE 4 Relative Sweetness and Energy Values of Some Polyols and Sugars

Substance	Relative sweetness[a] (Sucrose = 1, weight basis)	Energy value[b] ($kJ\ g^{-1}$)
Polyols		
Mannitol	0.6	6.69
Lactitol	0.3	8.36
Isomalt[c]	0.4–0.6	8.36
Xylitol	1.0	10.03
Sorbitol	0.5	10.87
Maltitol	0.8	12.54
Hydrogenated corn syrup	0.3–0.75	12.54
Sugars		
Xylose	0.7	16.72
Glucose	0.5–0.8	16.72
Fructose	1.2–1.5	16.72
Galactose	0.6	16.72
Mannose	0.4	16.72
Lactose	0.2	16.72
Maltose	0.5	16.72
Sucrose	1.0	16.72

[a]Commonly cited relative sweetness values are listed; however, the concentration and the food or beverage matrix may greatly influence actual relative sweetness values for sweeteners.
[b]Energy values accepted by the U.S. Food and Drug Administration; 1 kcal = 4.1814 kJ.
[c]Equimolar mixture of α-D-glucopyranosyl-1→6-mannitol and α-D-glucopyranosyl-1→6-sorbitol derived by enzymic conversion of sucrose and followed by hydrogenation.

g^{-1} for the various commercially available polyols (Table 4). This has markedly changed the positioning of polyols as food ingredients, and it can be anticipated that their presence in low-calorie, reduced-fat and sugar-free foods will markedly increase in the future. Although there is some controversy relating to the influence of polyols on diabetics, there also appears to be an emerging philosophy that they are suitable for diets of these individuals.

Attention also has been given to the development of polymeric forms of polyhydric alcohols for food applications. Whereas ethylene glycol (CH_2OH-CH_2OH) is toxic, polyethylene glycol 6000 is allowed in some food coating and plasticizing applications. Polyglycerol [CH_2OH-CHOH-CH_2-(O-CH_2CHOH-$CH_2)_n$-O-CH_2-CHOH-CH_2OH], formed from glycerol through an alkaline-catalyzed polymerization, also exhibits useful properties. It can be further modified by esterification with fatty acids to yield materials with lipid-like characteristics. These polyglycerol materials have been approved for food use because the hydrolysis products, glycerol and fatty acids, are metabolized normally.

Intermediate moisture (IM) foods deserve some discussion since polyhydric alcohols can make an important contribution to the stability of these products. IM foods contain substantial moisture (15–30%), yet are stable to microbiological deterioration without refrigeration. Several familiar foods, including dried fruits, jams, jellies, marshmallows, fruit cake, and jerky, owe their stability to IM characteristics. Some of these items may be rehydrated prior to consumption, but all possess a plastic texture and can be consumed directly. Although moist shelf-stable pet foods have found ready acceptance, new forms of intermediate moisture foods for human consumption have not as yet become popular. Nevertheless, meat, vegetable, fruit, and combination prepared dishes are under development and may eventually become important forms of preserved foods.

Most IM foods possess water activities of 0.70–0.85 and those containing humectants contain moisture contents of about 20 g of water per 100 g of solids (82% H_2O by weight). If IM foods with a water activity of about 0.85 are prepared by desorption, they are still susceptible to attack by molds and yeasts. To overcome this problem, the ingredients can be heated during preparation and an antimycotic agent, such as sorbic acid, can be added.

To obtain the desired water activity it is usually necessary to add a humectant that binds water and maintains a soft palatable texture. Relatively few substances, mainly glycerol, sucrose, glucose, propylene glycol, and sodium chloride, are sufficiently effective in lowering the water activity while being tolerable organoleptically to be of value in preparing IM foods. Figure 19 illustrates the effectiveness of the polyhydric alcohol glycerol on the water activity of a cellulose model system. Note that in a 10% glycerol system, a water activity of 0.9 corresponds to a moisture content of only 25 g H_2O/100 g solids, whereas the same water activity in a 40% glycerol systems corresponds to a moisture content of 75 g H_2O/100 g solids.

The principal flavor criticism of glycerol is its sweet–bitter sensation. Similarly, a problem of excessive sweetness exists for sucrose and glucose when used in IM foods. However, many IM foods show promise when combinations of glycerol, salt, propylene glycol, sucrose, and other bulking substances are employed.

12.10 STABILIZERS AND THICKENERS

Many hydrocolloid materials are widely used for their unique textural, structural, and functional characteristics in foods where they provide stabilization for emulsions, suspensions and foams, and general thickening properties. Most of these materials, sometimes classes as gums, are derived from natural sources although some are chemically modified to achieve desired characteristics. Many stabilizers and thickeners are polysaccharides, such as gum arabic, guar gum,

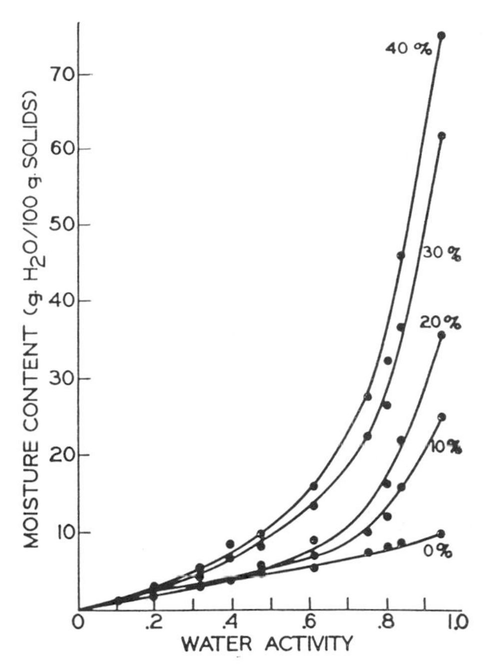

FIGURE 19 Moisture sorption isotherms of cellulose model systems containing various amounts of glycerol at 37°C. Reprinted from Ref. 32, p. 86.

carboxymethycellulose, carrageenan, agar, starch, and pectin. The chemical properties of these and related carbohydrates are discussed in Chapter 4. Gelatin, a protein derived from collagen, is one of the few noncarbohydrate stabilizers used extensively, and it is discussed in Chapter 6. All effective stabilizers and thickeners are hydrophilic and are dispersed in solution as colloids, which leads to the designation hydrocolloid. General properties of useful hydrocolloids include significant solubility in water, a capability to increase viscosity, and in some cases an ability to form gels. Some specific functions of hydrocolloids include improvement and stabilization of texture, inhibition of crystallization (sugar and ice), stabilization of emulsions and foams, improvement (reduced stickiness) of icings on baked goods, and encapsulation of flavors [31]. Hydrocolloids are generally used at a concentration of about 2% or less because many exhibit limited dispersibility, and the desired functionality is provided at these levels. The efficacy of hydrocolloids in many applications is directly dependent on their ability to increase viscosity. For example, this is the mechanism by which hydrocolloids stabilize oil-in-water emulsions. They cannot function as true emulsifiers since they lack the necessary combination of strong hydrophilic and lipophilic properties in single molecules.

12.11 FAT REPLACERS

Although fat is an essential dietary component, too much fat in the diet has been linked with a higher risk of coronary heart disease and certain types of cancer. Consumers are being advised to eat lean meats, especially fish and skinless poultry, low-fat dairy products, and to restrict their consumption of fried foods, high-fat baked goods, and sauces and dressings. However, consumers want substantially reduced-calorie foods that possess the sensory properties of traditional high-fat foods.

Although the increasing availability of complex prepared foods has contributed to the overabundance of fat in the diets of developed countries, it also has provided an opportunity to develop the complex technologies required for the manufacture and mass marketing of reduced-fat foods that simulate full-fat counterparts. Over the past two decades, a great deal of progress has been made in the adaptation and development of ingredients for use in reduced-fat foods. The types of ingredients suggested for various reduced-fat food applications vary widely, and are derived form several chemical groups, including carbohydrates, proteins, lipids, and purely synthetic compounds.

When fat is either partially or completely omitted from foods, the properties of the foods are altered, and it is necessary to replace it by some other ingredient or component. Hence, the term "fat replacers" has been spawned to broadly indicate the ingredients that are functionally used in this capacity. When the substances provide identical physical and sensory properties to fats, but without calories, they are designated "fat substitutes." These ingredients both convey fat-like sensory properties in foods and perform physically in various applications, such as frying foods.

Other ingredients, which do not possess full functional equivalency to fats, are termed "fat mimetics," because they can be made to mimic the effects of fat in certain applications. An example of this is the simulation of pseudomoistness provided by fat to certain high-fat bakery products. Certain substances, such as specially modified starches, can be used to provide the desired simulated fat properties by contributing to sensory properties arising from bulking and moisture retention.

12.11.1 Carbohydrate Fat Mimetics

Modestly processed starches, gums, hemicelluloses, and cellulose are used in many forms for providing partial fat functionality in reduced fat foods, and the chemistry of these substances is discussed in Chapter 4 and in Sections 12.9 and 12.10 (this chapter). Additional information about their applications in reduced fat foods also can be found in recent reviews [1, 34]. Generally, some carbohydrate fat mimetics provide essentially no calories (for example, gums, celluloses), while others provide up to 16.7 kJ (4 kcal) g^{-1} (for example, modified starches) rather than the 37.6 kJ (9kcal) g^{-1} of traditional fats. These substances mimic the smoothness or creaminess of fats in foods primarily by moisture retention and bulkiness of their solids, which assist in providing fat-like sensations, such as moistness in baked goods and the textural bite of ice cream. Tradenames for some of these products are Avicel, Oatrim, Kelcogel, Stellar, and Slendid.

12.11.2 Protein Fat Mimetics

Several proteins (Chap. 6) have been exploited as fat mimetics [36], and have GRAS (generally recognized as safe; United States) approval. However, the functionality of these proteins as fat mimetics is limited because they do not perform like fats at highly elevated temperatures, such as is required in frying applications. Nevertheless, these protein [16.7 kJ (4kcal) g^{-1}] ingredients are valuable for replacing fat in foods, especially in oil-in-water emulsions. For these applications, they can be variously prepared into microparticulates (< 3 μm) diameter), where they simulate the physical nature of fats through a means that has been described as being similar to flexible ball bearings. Proteins in solution also provide thickening, lubricity, and mouth-coating effects. Gelatin is quite functional in reduced-fat, solid products, such as margarine, where it provides thermally reversible gelation during manufacture, and subsequently it provides thickness to the margarine mass.

The manufacture of protein-based fat mimetics involves several strategies that each utilize soluble proteins as the starting materials. Particulate proteins are obtained from soluble proteins by inducing one of the following events: (a) hydrophobic interactions, (b) isoelectric precipitation, (c) heat denaturation and/or coagulation, (d) protein–protein complex formation, or (e) protein–polysaccharide complex formation [36]. These processes are often accompanied by physical shearing action, which assures the formation of microparticles. Some tradenames for protein fat mimetic are Simplesse, Trailblazer, and Lita.

12.11.3 Reduced-Calorie Synthetic Triacylglycerol Fat Substitutes

Recently, advantage has been taken of certain triglycerides (triacylglycerols; Chap. 5) that because of unique structural features, do not yield full caloric value when consumed by humans and other monogastrics. These triglycerides are variously synthesized utilizing hydrogenation and directed esterifications or interesterifications. One member of this group of lipids is the medium-chain triglycerides (MCTs), which have long been used in the treatment of certain lipid metabolism disorders. MCTs are composed of saturated fatty acids with chain lengths of C_6 to C_{12}, and they provide about 34.7 kJ (8.3 kcal) g^{-1}, compared to regular triglycerides, which contain 37.6 kJ (9 kcal) g^{-1} [34].

The incorporation of saturated short-chain fatty acids (C_2 to C_5) along with a long-chain saturated fatty acid (C_{14} to C_{24}) in a triglyceride molecule is another strategy, and this greatly reduces the caloric value. The caloric reduction results in part because short-chain fatty acids provide fewer calories per unit weight than long-chain fatty acids. In addition, the position of the long-chain fatty acid on the glycerol molecule greatly influences the absorption of the long-chain fatty acid. In some positional combinations of short- and long-chain saturated fatty acids, the absorption of the long-chain fatty acid may be reduced by over half.

A family of triglycerides based on the above principles, tradenamed Salatrim (Short and long acyltriglyceride molecule), has recently been described and petitioned for GRAS use in foods [49]. Salatrim is a mixture of triglycerides composed of mainly stearic acid (C18) as the long-chain fatty acid obtained from hydrogenated vegetable fats and various proportions of acetic, propionic, and butyric acids (C_2, C_3, and C_4, respectively) as the short-chain fatty acids (X). Humans realize between 19.6 and 21.3 kJ (4.7 and 5.1 kcal) g^{-1} from various Salatrim products, and the fatty acid composition can be controlled to provide the desired physical properties, such as melting points.

Caprenin is the tradename of a similarly synthesized, reduced-calorie triglyceride [about 20.9 kJ (5 kcal) g^{-1}] product that contains the medium-chain fatty acids, caprylic (C_6) and capric (C_{10}) acids, along with the long-chain fatty acid, behenic acid (C_{22}) (XI). Caprylic and capric acids are obtained from coconut and palm oils, and behenic acid can be obtained from hydrogenated marine

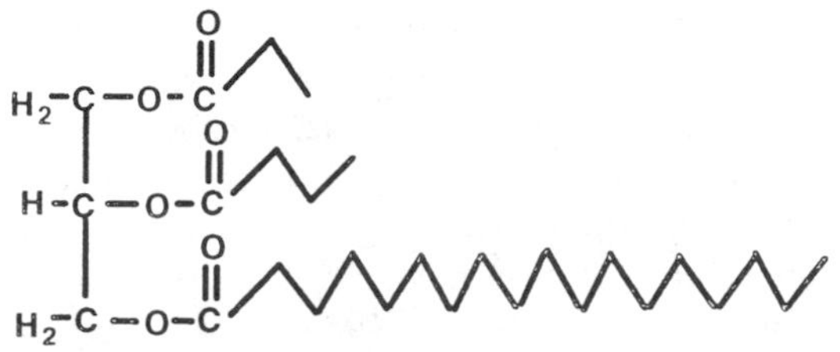

SALATRIM ISOMER (X)
(1-propionyl-2-butyryl-3-stearoyl-sn-glycerol)

CAPRENIN ISOMER (XI)
(1-caprylyl-2-capryl-3-behenyl-sn-glycerol)

oils, hydrogenated rapeseed oil, and peanut oil. Peanut oil contains about 3% behenic acid whereas rapeseed oil contains about 35% erucic acid ($C_{22:1}$), which is converted to behenic acid by hydrogenation. Marine oils often contain over 10% of docosahexaenoic acid (DHA), which also is converted to behenic acid by hydrogenation. Caprenin has been used in candy bars, and a petition has been made for its GRAS use in foods.

12.11.4 Synthetic Fat Replacers

A great number of synthetic compounds have been found to provide either fat mimetic or fat substitute properties [1]. Many of them contain triacylglycerol-like structural and functional groups, such as the trialkoxycarballates, which in effect have the ester groups reversed compared to conventional fats (i.e., a tricarboxylic acid is esterified to saturated alcohols rather than glycerol being esterified to fatty acids). Because of their synthetic nature, these compounds are resistant to enzymic hydrolysis, and are largely undigested in the gut. U.S. Food and Drug Administration approval for many of these substances is proving to be difficult to obtain, and it remains to be seen whether some of these compounds will find an ultimate role in the food supply.

12.11.4.1 Polydextrose

Although principally used as a reduced-calorie, carbohydrate bulking ingredient, polydextrose behaves as a fat mimetic in some applications. Since polydextrose yields only 4.18 kJ (1kcal) g^{-1}, it is especially attractive as a dual purpose ingredient which reduces calories from carbohydrates as well as fats. Contemporary polydextrose (tradename Litesse) is manufactured by randomly polymerizing glucose (minimum 90%), sorbitol (maximum 2%), and citric acid, and it contains minor amounts of glucose monomer and 1,6-anhydroglucose [1] (Fig. 20). To maintain suitable water solubility, the molecular weights of polydextrose polymers are controlled below 22,000.

12.11.4.2 Sucrose Polyester

Sucrose polyester (tradename Olestra) is a member of a family of carbohydrate fatty acid polyesters that are lipophilic, nondigestible, nonabsorbable fat-like molecules with physical and chemical properties of conventional fats. Sucrose polyester is manufactured by using various means of esterification of sucrose with long-chain fatty acids obtained from vegetable fats (XII). Sucrose polyesters intended for fat substitute applications have a high degree of esterification, while those intended for emulsifier applications have a lower degree of esterification. Sucrose polyester emulsifiers have been approved in the United States since 1983. Sucrose polyester has been extensively studied for safety and health aspects for over two decades and it has recently (1996) been approved for limited use in foods in the U.S.

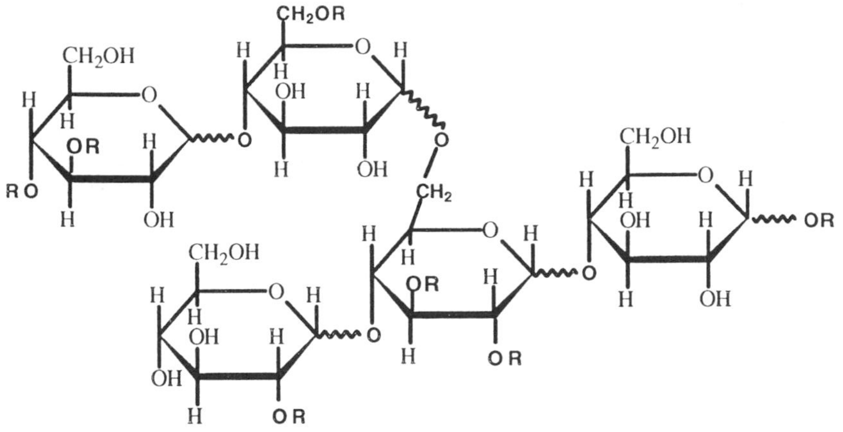

FIGURE 20 Structural features of polydextrose; R = glucose, remainder of polymer, sorbitol, or citric acid; note that linkages are random.

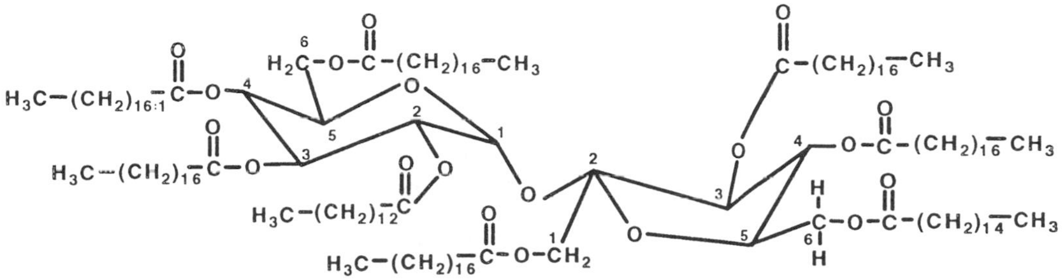

SUCROSE POLYESTER ISOMER (XII)

12.12 MASTICATORY SUBSTANCES

Masticatory substances are employed to provide the long-lasting, pliable properties of chewing gum. These substances are either natural products or the result of organic synthesis, and both kinds are quite resistant to degradation. Synthetic masticatory substances are prepared by the Fischer-Tropsch process involving carbon monoxide, hydrogen, and a catalyst, and after further processing to remove low-molecular-weight compounds, the product is hydrogenated to yield synthetic paraffin [7]. Chemically modified masticatory substances are prepared by partially hydrogenating wood rosin, which is largely composed of diterpenes, and then esterifying the products with pentaerythritol or glycerol. Other polymers similar to synthetic rubbers have also been prepared for use as masticatory substances, and these substances are prepared from ethylene, butadiene, or vinyl monomers.

Much of the masticatory base employed in chewing gum is derived directly from plant gums. These gums are purified by extensive treatments involving heating, centrifuging, and filtering. Chicle from plants in the *Sapotaceae* (Sapodilla) family, gums from Gutta Katiau from *Palaquium* sp., and latex solids (natural rubber) from *Henea brasiliensis* are widely used, naturally derived, masticatory substances.

12.13 FIRMING TEXTURIZERS

Thermal processing or freezing of plant tissues usually causes softening because the cellular structure is modified. Stability and integrity of these tissues are dependent on maintenance of intact cells and firm molecular bonding between constituents of cell walls. The pectic substances (Chaps. 4 and 16) are extensively involved in structure stabilization through cross-linking of their free carboxyl groups via polyvalent cations. Although considerable amounts of polyvalent cations are naturally present, calcium salts (0.1–0.25% as calcium) are frequently added. This increases firmness since the enhanced cross-linking results in increased amounts of relatively insoluble calcium pectinate and pectate. These stabilized structures support the tissue mass, and integrity is maintained even through heat processing. Fruits, including tomatoes, berries, and apple slices, are commonly firmed by adding one or more calcium salts prior to canning or freezing. The most commonly used salts include calcium chloride, calcium citrate, calcium sulfate, calcium lactate, and monocalcium phosphate. Most calcium salts are sparingly soluble, and some contribute a bitter flavor at higher concentrations.

Acidic alum salts, sodium aluminum sulfate [$NaAl(SO_4)_2 \cdot 12H_2O$], potassium aluminum sulfate, ammonium aluminum sulfate, aluminum sulfate [$Al_2(SO_4)_3 \cdot 18H_2O$], potassium aluminum sulfate, ammonium aluminum sulfate, and aluminum sulfate [$Al_2(SO_4)_3 \cdot 18H_2O$], are added to fermented, salt-brined pickles to make cucumber products that are crisper and firmer than those prepared without these salts. The trivalent aluminum ion is believed to be involved in the crisping process through the formation of complexes with pectin substances. However, some investigations have demonstrated that aluminum sulfate has a softening effect on fresh-pack or pasteurized pickles and should not be included in these products [11]. The reasons for the softening are not understood, but the presence of aluminum sulfate counteracts the firming effects normally provided by adjusting the pH to near 3.8 with acetic or lactic acids.

The firmness and texture of some vegetables and fruits can be manipulated during processing without the use of direct additives. For example, an enzyme, pectin methylesterase, is activated during low-temperature blanching (70–82°C for 3–15 min), rather than inactivated as is the case during usual blanching (88–100°C for 3 min). The degree of firmness produced following low-temperature blanching can be controlled by the holding time prior to retorting [54]. Pectin methylesterase hydrolyzes esterified methanol (sometimes referred to as methoxyl groups) from carboxyl groups on pectin to yield pectinic and pectic acids. Pectin, having relatively few free carboxyl groups, is not strongly bound, and because it is water soluble, it is free to migrate from the cell wall. On the other hand, pectinic acid and pectic acid possess large numbers of free carboxyl groups and they are relatively insoluble, especially in the presence of endogenous or added calcium ions. As a result they remain in the cell wall during processing and produce firm textures. Firming effects through activation of pectin methylesterase have been observed for snap beans, potatoes, cauliflower, and sour cherries. Addition of calcium ions in conjunction with enzyme activation leads to additional firming effects.

12.14 APPEARANCE CONTROL AND CLARIFYING AGENTS

In beer, wine, and many fruit juices the formation of hazes or sediments and oxidative deterioration are long-standing problems. Natural phenolic substances are involved in these phenomena. The chemistry of this important group, including anthocyanins, flavonoids, proanthocyanidins, and tannins, is discussed in Chapter 10. Proteins and pectic substances participate with polyphenols in the formation of haze-forming colloids. Specific enzymes have been utilized to partially hydrolyze high-molecular-weight proteins (Chap. 7), and thereby reduce the tendency toward haze formation. However, in some instances excess enzymic activity can adversely affect other desirable properties, such as foam formation in beer.

An important means of manipulating polyphenolic composition to control both its desirable and undesirable effects is to use various clarifying ("fining") agents and adsorbents. Preformed haze can be at least partially removed by filter aids, such as diatomaceous earth. Many of the clarifying agents that have been used are nonselective and they affect the polyphenolic content more or less incidentally. Adsorption is usually maximal when solubility of the adsorbate is minimal, and suspended or nearly insoluble materials such as tannin–protein complexes tend to collect at any interface. As the activity of the adsorbent increases, the less soluble substances still tend to be adsorbed preferentially, but more soluble compounds are also adsorbed.

Bentonite, a montmorillonite clay, is representative of many similar and moderately effective minerals that have been employed as clarifying agents. Montmorillonite is a complex hydrated aluminum silicate with exchangeable cations, frequently sodium ions. In aqueous suspension bentonite behaves as small platelets of insoluble silicate. The bentonite platelets have a negative charge and a very large surface area of about 750 m^2 g^{-1}. Bentonite is a rather selective adsorbent for protein, and evidently this adsorption results from an attraction between the positive charges of the protein and the negative charges of the silicate. A particle of bentonite covered with adsorbed protein will adsorb some phenolic tannins on or along with the protein [48]. Bentonite is used as a clarifying or fining agent for wines to preclude protein precipitation. Doses of the order of a few pounds per thousand gallons usually reduce the protein content of wine from 50–100 mg/L to a stable level of less than 10 mg/L. Bentonite rapidly forms a heavy compact sediment and is often employed in conjunction with final filtration to remove precipitated colloids.

The important clarifying agents that have a selective affinity for tannins, proanthocyanidins, and other polyphenols include proteins and certain synthetic resins, such as the polyamides and polyvinylpyrrolidone (PVP). Gelatin and isinglass (obtained form the swim bladder of fish) are the proteins most commonly used to clarify beverages. It appears that the most important type of linkage between tannins and proteins, although probably not the only type, involves hydrogen bonding between phenolic hydroxyl groups and amide bonds in proteins. The addition of a small amount of gelatin (40–170 g per 380 L) to apple juice causes aggregation and precipitation of a gelatin–tannin complex, which on settling enmeshes and removes other suspended solids. The exact amount of gelatin for each use must be determined at the time of processing. Juices containing low levels of polyphenolics are supplemented with added tannin or tannic acid (0.005–0.01%) to facilitate flocculation of the gelatin.

At low concentrations, gelatin and other soluble clarifying agents can act as protective colloids, at higher concentrations they can cause precipitation, and at still higher concentrations they can again fail to cause precipitation. Hydrogen bonding between the colloidal clarifying agents and water accounts for their solubilities. Molecules of the clarifying agent and polyphenol can combine in different proportions to either neutralize or enhance the hydration and solubility of a given colloidal particle. The most nearly complete disruption of H bonding between water and either the protein or the polyphenol gives the most complete precipitation. This would be expected to occur when the amount of dissolved clarifying agent roughly equals the weight of the tannin being removed.

The synthetic resins (polyamides and polyvinylpyrrolidone or PVP) have been used to prevent browning in white wines [4] and to remove haze for beers [9]. These polymers are available in both soluble and insoluble forms, but requirements for little or no residual polymer in beverages has stimulated use of the high-molecular-weight cross-linked forms that are insoluble. The synthetic resins have been particularly useful in the brewing industry where reversible refrigeration-induced haze (chill haze) and permanent haze (that which is associated with the development of oxidized flavors) are serious problems. These hazes are caused by formation of complexes between native proteins and proanthocyanidins from malted barley. Excessive removal of proteins leads to defective foam character, but the selective removal of polyphenols extends the stability of beer. Initial applications involved polyamides (nylon 66), but greater efficiency has been achieved with cross-linked polyvinylpyrrolidone (XIII). Treatment with 1.4–2.3 kg of insoluble PVP per 100 barrels of beer provides control of chill haze and improves storage stability

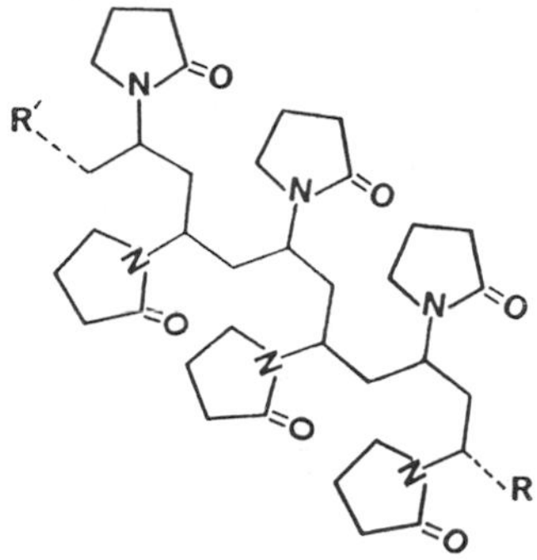

POLYVINYL PYRROLIDONE (XIII)

[9]. PVP is added after fermentation and prior to filtration, and it rapidly adsorbs polyphenols. Just as bentonite removes some tannins along with preferentially adsorbed protein, selective tannin adsorbents remove some proteins along with the phenolics.

In addition to the adsorbents already discussed, activated charcoal and some other materials have been employed. Activated charcoal is quite reactive but it adsorbs appreciable amounts of smaller molecules (flavors, pigments) along with the larger compounds that contribute to haze formation. Tannic acid (tannin) is used to precipitate proteins, but its addition can potentially lead to the undesirable effects described previously. Other proteins with low solubility (keratin, casein, and zein) and soluble proteins (sodium caseinate, egg albumen, and serum albumin) also have selective adsorptive capacities for polyphenols, but they have not been extensively employed.

12.15 FLOUR BLEACHING AGENTS AND BREAD IMPROVERS

Freshly milled wheat flour has a pale yellow tint, and yields a sticky dough that does not handle or bake well. When the flour is stored, it slowly becomes white and undergoes an aging or maturing process that improves its baking qualities. It is a usual practice to employ chemical treatments to accelerate these natural processes [51], and to use other additives to enhance yeast leavening activity and to retard the onset of staling.

Flour bleaching involves primarily the oxidation of carotenoid pigments. This results in disruption of the conjugated double bond system of carotenoids to a less conjugated colorless system. The dough-improving action of oxidizing agents is believed to involve the oxidation of sulfhydryl groups in gluten proteins. Oxidizing agents employed may participate in bleaching only, in both bleaching and dough improvement, or in dough improvement only. One commonly used flour bleaching agent, benzoyl peroxide [$(C_6H_5CO)_2O_2$], exhibits a bleaching or decolorizing action but does not influence baking properties. Materials that act as both bleaching and improving agents include chlorine gas (Cl_2), chlorine dioxide (ClO_2), nitrosyl chloride (NOCl), and oxides of nitrogen (nitrogen dioxide, NO_2, and nitrogen tetroxide, N_2O_4) [40]. These oxidizing agents are gaseous and exert their action immediately upon contact with flour. Oxidizing agents that serve primarily as dough improvers exert their action during the dough stages rather than in the flour. Included in this group are potassium bromate ($KBrO_3$), potassium iodate (KIO_3), calcium iodate [$Ca(IO_3)_2$], and calcium peroxide (CaO_2).

Benzoyl peroxide is usually added to flour (0.25–0.075%) at the mill. It is a powder and is usually added along with diluting or stabilizing agents such as calcium sulfate, magnesium carbonate, dicalcium phosphate, calcium carbonate, and sodium aluminum phosphate. Benzoyl peroxide is a free radical initiator (see Chap. 5), and it requires several hours after addition to decompose into available free radicals for initiation of carotenoid oxidation.

The gaseous agents for oxidizing flour show variable bleaching efficiencies, but effectively improve baking qualities of suitable flours. Treatment with chlorine dioxide improves flour color only slightly, but yields flour with improved dough handling properties. Chlorine gas, often containing a small amount of nitrosyl chloride, is used extensively as a bleach and improver for soft wheat cake flour. Hydrochloric acid is formed from oxidation reactions of chlorine, and the resulting slightly lowered pH values lead to improved cake baking properties. Nitrogen tetroxide (N_2O_4) and other oxides of nitrogen, produced by passing air through an intense electric arc, are only moderately effective bleaching agents, but they produce good baking qualities in treated flours.

Oxidizing agents that function primarily as dough improvers can be added to flour (10–40 ppm) at the mill. They are, however, often incorporated into a dough conditioner mix containing several inorganic salts, and then added at the bakery. Potassium bromate, an oxidizing agent used extensively as a dough improver, remains unreactive until yeast fermentation lowers the pH of the dough sufficiently to activate it. As a result it acts rather late in the process and causes increased loaf volume, improved loaf symmetry, and improved crumb and texture characteristics.

Early investigators proposed that the improved baking qualities resulting from treatment with oxidizing agents were attrituable to inhibition of the proteolytic enzymes present in flour. However, a more recent belief is that dough improvers, at an appropriate time, oxidize sulfhydryl groups (-SH) in the gluten to yield an increased number of intermolecular disulfide bonds (-S-S-). This cross-linking would allow gluten proteins to form thin, tenacious networks of protein films that comprise the vesicles for leavening. The result is a tougher, drier, more extensible dough that gives rise to improved characteristics in the finished products. Excessive oxidation of the flour must be avoided since this leads to inferior products with gray crumb color, irregular grain, and reduced load volume.

The addition of a small amount of soybean flour to wheat flour intended for yeast-leavened doughs has become a common practice. The addition of soybean lipoxygenase (see Chaps. 5 and 7) is an excellent way to initiate the free radical oxidation of carotenoids [12]. Addition of soybean lipoxygenase also greatly improves the rheological properties of the dough by a mechanism not yet elucidated. While it has been suggested that lipid hydroperoxides become involved in the oxidation of gluten -SH groups, evidence indicates that other protein–lipid interactions are also involved in dough improvement by oxidants [12].

Inorganic salts incorporated into dough conditioners include ammonium chloride (NH_4Cl), ammonium sulfate [$(NH_4)_2SO_4$], calcium sulfate ($CaSO_4$), ammonium phosphate [$(NH_4)_3PO_4$], and calcium phosphate ($CaHPO_4$). They are added to dough to facilitate growth of yeast and to aid in control of pH. The principal contribution of ammonium salts is to provide a ready source of nitrogen for yeast growth. The phosphate salts apparently improve dough by buffering the pH at a slightly lower than normal value. This is especially important when water supplies are alkaline.

Other types of material are also used as dough improvers in the baking industry. Calcium stearoyl-2 lactylate [$(C_{17}H_{35}COOC(CH_3)HCOOC(CH_3)HCOO)_2Ca$] and similar emulsifying agents are used at low levels (up to 0.5%) to improve mixing qualities of dough and to promote increased loaf volume [53]. Hydrocolloid gums have been used in the baking industry to improve the water-holding capacity of doughs and to modify other properties of doughs and baked products [31]. Carrageenan, carboxymethylcellulose, locust bean gum, and methylcellulose are among the more useful hydrocolloids in baking applications. Methylcellulose and carboxymethylcellulose have been found to retard retrogradation and staling in bread, and they also retard migration of moisture to the product surface during subsequent storage. Carrageenan (0.1%) softens the crumb texture of sweet dough products. Several hydrocolloids (e.g., carboxymethylcellulose at 0.25%) may be incorporated into doughnut mixes to significantly decrease the amount of fat absorbed during frying. This benefit apparently arises because of improvement in the dough and because a more effective hydrated barrier is established on the surface of the doughnuts.

12.16 ANTICAKING AGENTS

Several conditioning agents are used to maintain free-flowing characteristics of granular and powdered forms of foods that are hygroscopic in nature. In general, these materials function by readily absorbing excess moisture, by coating particles to impart a degree of water repellency, and/or by providing an insoluble particulate diluent. Calcium silicate ($Ca\ SiO_3 \cdot XH_2O$) is used to prevent caking in baking powder (up to 5%), in table salts (up to 2%), and in other foods and food ingredients. Finely divided calcium silicate absorbs liquids in amounts up to 2½ times its weight and still remains free flowing. In addition to absorbing water, calcium silicate also effectively absorbs oils and other nonpolar organic compounds. This characteristic makes it useful in complex powdered mixes and in certain spices that contain free essential oils.

Food-grade calcium and magnesium salts of long-chain fatty acids, derived from tallow, are used as conditioning agents in dehydrated vegetable products, salt, onion and garlic salt, and in a variety of other food ingredients and mixes that exist in powder form. Calcium stearate is often added to powdered foods to prevent agglomeration, to promote free flow during processing, and to insure freedom from caking during the shelf life of the finished product. Calcium stearate is essentially insoluble in water but adheres well to particles and provides a partial water-repellent coating for the particles. Commercial stearate powders have a high bulk density (about 27 kg/m^3) and possess large surface areas that make their use as conditioners (0.5–2.5%) reasonably economical. Calcium stearate is also used as a release lubricant (1%) in the manufacture of pressed tablet-form candy.

Other anticaking agents employed in the food industry include sodium silicoaluminate, tricalcium phosphate, magnesium silicate, and magnesium carbonate. These materials are essentially insoluble in water and exhibit variable abilities to absorb moisture. Their use levels are similar to those for other anticaking agents (e.g., about 1% sodium silicoaluminate is used in powdered sugar). Microcrystalline cellulose powders are used to prevent grated or shredded cheese from clumping. Anticaking agents are either metabolized (starch, stearates) or exhibit no toxic actions at levels employed in food applications [16].

12.17 GASES AND PROPELLANTS

Gases, both reactive and inert, play important roles in the food industry. For example, hydrogen is used to hydrogenate unsaturated fats (Chap. 5), chlorine is used to bleach flour (see bleaching agents and dough improvers in this chapter) and sanitize equipment, sulfur dioxide is used to inhibit enzymic browning in dried fruits (see sulfites and sulfur dioxide in this chapter), ethylene gas is used to promote ripening of fruits (Chap. 16), ethylene oxide is used as a sterilant for spices (see epoxides in this chapter), and air is used to oxidize ripe olives for color development. However, the functions and properties of essentially inert gases used in food will be the topics of primary concern in the following sections.

12.17.1 Protection from Oxygen

Some processes for oxygen removal involve the use of inert gases such as nitrogen or carbon dioxide to flush a headspace, to strip or sparge a liquid, or to blanket a product during or after processing. Carbon dioxide is not totally without chemical influence because it is soluble in water and can lead to a tangy, carbonated taste in some foods. The ability of carbon dioxide to provide a dense, heavier-than-air, gaseous blanket over a product makes it attractive in many processing applications. Nitrogen blanketing requires thorough flushing followed by a slight positive pressure to prevent rapid diffusion of air into the system. A product that is thoroughly evacuated, flushed with nitrogen, and hermetically sealed will exhibit increased stability against oxidative deterioration [27].

12.17.2 Carbonation

The addition of carbon dioxide (carbonation) to liquid products, such as carbonated soft drinks, beer, some wines, and certain fruit juices causes them to become effervescent, tangy, slightly tart, and somewhat tactual. The quantity of carbon dioxide used and the method of introduction varies widely with the type of product [27]. For example, beer becomes partially carbonated during the fermentation process, but is further carbonated prior to bottling. Beer usually contains 3–4 volumes of carbon dioxide (1 volume of beer at 16°C and 1 atm pressure contains 3–4 volumes of carbon dioxide gas at the same temperature and pressure). Carbonation is often carried out at lowered temperatures (4°C) and elevated pressures to increase carbon dioxide solubility. Other carbonated beverages contain from 1.0 to 318 volumes of carbon dioxide, depending upon the effect desired. The retention of large amounts of carbon dioxide in solutions at atmospheric pressure has been ascribed to surface adsorption by colloids and to chemical binding. It is well established that carbamino compounds are formed in some products by rapid, reversible reactions between carbon dioxide and free amino groups of amino acids and proteins. In addition, formation of carbonic acid (H_2CO_3) and bicarbonate ions (HCO_3^-) also aids in stabilizing the carbon dioxide system. Spontaneous release of carbon dioxide from beer, that is gushing, has been associated with trace metallic impurities and with the presence of oxalate crystals, which provide nuclei for nucleation of gas bubbles.

12.17.3 Propellants

Some fluid food products are dispensed as liquids, foams, or sprays from pressurized aerosol containers. Since the propellant usually comes into intimate contact with the food, it becomes an incidental food component or ingredient. The principal propellants for pressure dispensing of foods are nitrous oxide, nitrogen, and carbon dioxide [27]. Foam and spray type products are usually dispensed by nitrous oxide and carbon dioxide because these propellants are quite soluble in water and their expansion during dispensing assists in the formation of the spray or foam. Carbon dioxide is also employed for products such as cheese spreads where tanginess and tartness are acceptable characteristics. Nitrogen, because of its low solubility in water and fats, is used to dispense liquid streams in which foaming should be avoided (catsup, edible oils, syrups). The use of all of these gases in foods is regulated, and the pressure must not exceed 100 psig at 21°C or 135 psig at 54°C. At these conditions, none of the gases liquify and a large portion of the container is occupied by the propellant. Thus as the product is dispensed, the pressure drops, and this can lead to difficulties with product uniformity and completeness of dispensing. The gaseous propellants are nontoxic, nonflammable, economical, and usually do not cause objectionable color or flavors. However, carbon dioxide, when used alone, imparts an undesirable taste to some foods.

Liquid propellants also have been developed and approved for food use, but environmental concerns regarding ozone depletion in the upper atmosphere has led to restrictions of these substances. Those approved for foods are octafluorocyclobutane or Freon C-318 (CF_2-CF_2-CF_2-CF_2) and chloropentafluoroethane or Freon 115 ($CClF_2$-CF_3). When used, these propellants exist in the container as a liquid layer situated on top of the food product, and an appropriate headspace containing vaporized propellant is also present. Use of a liquified propellant enables dispensing to occur at a constant pressure, but the contents first must be shaken to provide an emulsion that will foam or spray upon discharge from the container. Constant pressure dispensing is essential for good performance of spray-type aerosols. These propellants are nontoxic at levels encountered and they do not impart off flavors to foods. They give particularly good foams because they are highly soluble in any fat that may be present, and they can be effectively emulsified.

12.18 TRACERS

Substances in this category are added to food constituents that are difficult to detect in finished foods when routine monitoring of the constituent's concentration in a food is desired and when normal analytical procedures are inadequate. Although the concept seems attractive from a regulatory standpoint, this practice has become unnecessary because analytical procedures of adequate specificity and sensitivity for individual substances in finished foods have been satisfactorily developed in response to needs. Currently in the United States, titanium dioxide (0.1%) must be added to vegetable proteins that are used as meat extenders. However, removal of this requirement has been requested because gel electrophoresis techniques can quantitatively distinguish between soy and meat protein fractions in foods [15]. Adequate regulation of soy protein additions to meat should be possible by monitoring inventory and processing records and by analyzing meat products with electrophoretic techniques.

12.19 SUMMARY

Summarized in Table 5 are various kinds of food additives and their functions in food.

TABLE 5 Selected Food Additives

Class and general function	Chemical name	Additional or more specific function	Source of discussion (chapter)
Processing additives			
Aerating and foaming agents	Carbon dioxide	Carbonation, foaming	12
	Nitrogen	Foaming	12
	Sodium bicarbonate	Foaming	12
Antifoam agents	Aluminum stearate	Yeast processing	—
	Ammonium stearate	Beet sugar processing	—
	Butyl stearate	Beet sugar, yeast	—
	Decanoic acid	Beet sugar, yeast	—
	Dimethylpolysiloxane	General use	—
	Dimethylpolysilicone	General use	—
	Lauric acid	Beet sugar, yeast	—
	Mineral oil	Beet sugar, yeast	—
	Oleic acid	General use	—
	Oxystearin	Beet sugar, yeast	—
	Palmitic acid	Beet sugar, yeast	—

	Petroleum waxes	Beet sugar, yeast	—
	Silicone dioxide	General use	—
	Stearic acid	Beet sugar, yeast	—
Catalysts (including enzymes)	Nickel	Lipid hydrogenation	5
	Amylase	Starch conversion	4, 7
	Glucose oxidase	Oxygen scavenger	7
	Lipase	Dairy flavor developer	7
	Papain	Chill-proofing beer	7
	Pepsin	Meat tenderizer	7
	Rennin	Cheese production	7
Clarifying and flocculating agents	Bentonite	Absorbs proteins	12
	Gelatin	Complexes polyphenols	12
	Polyvinylpyrrolidone	Complexes polyphenols	12
	Tannic acid	Complexes proteins	12
Color control agents	Ferrous gluconates	Dark olives	—
	Magnesium chloride	Canned peas	10
	Nitrate, nitrite (potassium, sodium)	Cured meat	10, 12
	Sodium erythorbate	Cured meat color	10
Freezing and cooling agents	Carbon dioxide	—	12
	Liquid nitrogen	—	—
Malting and fermenting aids	Ammonium chloride	Yeast nutrients	12
	Ammonium phosphate	—	—
	Ammonium sulfate	—	—
	Calcium carbonate	—	—
	Calcium phosphate	—	—
	Calcium phosphate (dibasic)		—
	Calcium sulfate	—	—
	Potassium chloride	—	—
	Potassium phosphate	—	—
Material handling aids	Aluminum phosphate	Anticaking, free flow	12
	Calcium silicate	Anticaking, free flow	12
	Calcium stearate	Anticaking, free flow	12
	Dicalcium phosphate	Anticaking, free flow	12
	Dimagnesium phosphatre	Anticaking, free flow	12
	Kaolin	Anticaking, free flow	12
	Magnesium silicate	Anticaking, free flow	—
	Magnesium stearate	Anticaking, free flow	—
	Sodium carboxy-methylcellulose	Bodying, bulking	12
	Sodium silicoaluminate	Anticaking, free flow	12
	Starches	Anticaking, free flow	4
	Tricalcium phosphate	Anticaking, free flow	12
	Tricalcium silicate	Anticaking, free flow	12
	Xanthan (other gums)	Bodying, bulking	4, 12
Oxidizing agents	Acetone peroxides	Free radical initiator	12
	Benzoyl peroxide	Free radical initiator	12
	Calcium peroxide	Free radical initiator	12

TABLE 5 Continued

Class and general function	Chemical name	Additional or more specific function	Source of discussion (chapter)
	Hydrogen peroxide	Free radical initiator	—
	Sulfur dioxide	Dried fruit bleach	12
pH control and modification agents			
Acidulants (acids)	Acetic acid	Antimicrobial agent	12
	Citric acid	Chelating agent	5, 12
	Fumaric acid	Antimicrobial agent	12
	δ-Gluconolactone	Leavening agent	12
	Hydrochloric acid	—	12
	Lactic acid	—	12
	Malic acid	Chelating agent	12
	Phosphoric acid	—	12
	Potassium acid tartrate	Leavening agent	12
	Succinic acid	Chelating agent	12
	Tartaric acid	Chelating agent	12
Alkalies (bases)	Ammonium bicarbonate	Carbon dioxide source	12
	Ammonium hydroxide	—	—
	Calcium carbonate	—	—
	Magnesium carbonate	—	—
	Potassium carbonate	Carbon dioxide source	—
	Potassium hydroxide	—	—
	Sodium bicarbonate	Carbon dioxide source	12
	Sodium carbonate	—	—
	Sodium citrate	Emulsifier salt	12
	Trisodium citrate	Emulsifier salt	12
Buffering salts	Ammonium phosphate (mono-, dibasic)	—	12
	Calcium citrate	—	—
	Calcium gluconate	—	—
	Calcium phosphate (mono-, dibasic)	—	—
	Potassium acid tartrate	—	—
	Potassium citrate	—	—
	Potassium phosphate (mono-, dibasic)	—	—
	Sodium acetate	—	—
	Sodium acid pyrophospate	—	—
	Sodium citrate	—	—
	Sodium phosphate (mono-, di-, tribasic)	—	—
	Sodium potassium tartrate	—	—
Release and antistick agents	Acylated monoacylglycerols	—	5, 12

	Beeswax	—	5, 12
	Calcium stearate	—	5, 12
	Magnesium silicate	—	5, 12
	Mineral oil	—	—
	Mono- and diacylglycerols	Emulsifiers	5
	Starches	—	4
	Stearic acid	—	5, 12
	Talc	—	—
Sanitizing and fumigating agents	Chlorine	Oxidant	—
	Methyl bromide	Insect fumigant	—
	Sodium hypochlorite	Oxidant	—
Separation and filtration aids	Diatomaceous earth	—	—
	Ion-exchange resins	—	—
	Magnesium silicate	—	—
Solvents, carriers, and encapsulating agents	Acetone	Solvent	—
	Agar-agar	Encapsulation	4
	Arabinogalactan	Encapsulation	4
	Cellulose	Carrier	4
	Glycerine	Solvent	5, 12
	Guar gum	Encapsulation	4
	Methylene chloride	Solvent	—
	Propylene glycol	Solvent	12
	Triethyl citrate	Solvent	—
Washing and surface removal agents	Sodium dodecylbenzene sulfonate	Detergent	—
	Sodium hydroxide	Lye peeling	—
Final product additives			
Antimicrobial agents	Acetic acid (and salts)	Bacteria, yeast	12
	Benzoic acid (and salts)	Bacteria, yeast	12
	Ethylene oxide	General sterilant	12
	p-Hydroxybenzoate alkyl esters	Molds, yeast	12
	Nitrates, nitrites (K, Na)	*Clostridium botulinum*	10, 12
	Propionic acid (and salts)	Mold	12
	Propylene oxide	General sterilant	12
	Sorbic acid (and salts)	Mold, yeast, bacteria	12
	Sulfur dioxide and sulfites	General	12
Antioxidants	Ascorbic acid (and salts)	Reducing agent	5, 8
	Ascorbyl palmitate	Reducing agent	12
	BHA	Free radical terminator	5, 12
	BHT	Free radical terminator	5, 12
	Gum guaiac	Free radical terminator	5, 12
	Propylgallate	Free radical terminator	5, 12

TABLE 5 Continued

Class and general function	Chemical name	Additional or more specific function	Source of discussion (chapter)
	Sulfite and metabisulfite salts	Reducing agents	5, 12
	Thiodipropionic acid (and esters)	Peracid decomposer	12
Appearance control agents			
Colors and color modifiers	Annatto	Cheese, butter, baked goods	10
	Beet powder	Frosting, soft drinks	10
	Caramel	Confectionary	10
	Carotene	Margarine	10
	Cochineal extract	Beverages	10
	FD&C no. 3	Mint jelly, beverages	10
	FD&C no. 3 (erythrosine)	Canned fruit cocktail	10
	Titanium dioxide	White candy, Italian cheeses	10
	Turmeric	Pickles, sauces	10
Other appearance agents	Beeswax	Gloss, polish	5
	Glycerine	Gloss, polish	5
	Oleic acid	Gloss, polish	5
	Sucrose	Crystalline glaze	4
	Wax, carnauba	Gloss, polish	—
Flavors and flavor modifiers			
Flavoring agents[a]	Essential oils	General	11
	Herbs and spices	General	11
	Plant extractives	General	11
	Synthetic flavor compounds	General	—
Flavor potentiators	Disodium guanylate	Meats and vegetables	11
	Disodium inosinate	Meats and vegetables	11
	Maltol	Bakery goods, sweets	11
	Monosodium glutamate	Meats and vegetables	11
	Sodium chloride	General	—
Moisture control agents	Glycerine	Plasticizer, humectant	4, 11
	Gum acacia	—	—
	Invert sugar	—	4
	Propylene glycol	—	11
	Mannitol	—	4, 11
	Sorbitol	—	4, 11
Nutrient, dietary supplements			
Amino acids	Alanine	—	6
	Arginine	Essential	6
	Aspartic acid	—	6
	Cysteine	—	6

	Cystine	—	6
	Glutamic acid	—	6
	Histidine	—	6
	Isoleucine	Essential	6
	Leucine	Essential	6
	Lysine	Essential	6
	Methionine	Essential	6
	Phenylalanine	Essential	6
	Proline	—	6
	Threonine	Essential	6
	Valine	Essential	6
Minerals	Boric acid	Boron source	9
	Calcium carbonate	Breakfast cereals	9
	Calcium citrate	Cornmeal	9
	Calcium phosphates	Enriched flour	9
	Calcium pyrophosphate	Enriched flour	9
	Calcium sulfate	Bread	9
	Cobalt carbonate	Cobalt source	9
	Cobalt chloride	Cobalt source	9
	Cupric chloride	Copper source	9
	Cupric gluconate	Copper source	9
	Cupric oxide	Copper source	9
	Calcium fluoride	Water fluoridations	—
	Ferric phosphate	Iron source	9
	Ferric pyrophosphate	Iron source	9
	Ferrous gluconate	Iron source	9
	Ferrous sulfate	Iron source	9
	Iodine	Iodine source	9
	Iodide, cuprous	Table salt	9
	Iodate, potassium	Iodine source	9
	Magnesium chloride	Magnesium source	9
	Magnesium oxide	Magnesium source	9
	Magnesium phosphates	Magnesium source	9
	Magnesium sulfate	Magnesium source	9
	Manganese citrate	Manganese source	9
	Manganese oxide	Manganese source	9
	Molybdate, ammonium	Molybdenum source	9
	Nickel sulfate	Nickel source	9
	Phosphates, calcium	Phosphorus source	9
	Phosphates, sodium	Phosphorus source	9
	Potassium chloride	NaCl substitute	—
	Zinc chloride	Zinc source	9
	Zinc stearate	Zinc source	9
Vitamins	*p*-Aminobenzoic acid	B complex factor	8
	Biotin	—	8
	Carotene	Provitamin A	8
	Folic acid	—	8
	Niacin	—	8
	Niacinamine	Enriched flour	8
	Pantothenate, calcium	B complex vitamin	8
	Pyridoxine hydrochloride	B complex vitamin	8

TABLE 5 Continued

Class and general function	Chemical name	Additional or more specific function	Source of discussion (chapter)
	Riboflavin	B complex vitamin	8
	Thiamine hydrochloride	Vitamin B_1	8
	Tocopherol acetate	Vitamin E	8
	Vitamin A acetate	—	—
	Vitamin B_{12}	—	8
	Vitamin D	—	8
Miscellaneous nutrients	Betaine hydrochloride	Dietary supplement	8
	Choline chloride	Dietary supplement	8
	Inositol	Dietary supplement	8
	Linoleic acid	Essential fatty acid	5
	Rutin	Dietary supplement	8
Sequestrants (chelating agents)	Calcium citrate	—	12
	Calcium disodium EDTA	—	12
	Calcium gluconate	—	—
	Calcium phosphate (monobasic)	—	—
	Citric acid	—	12
	Disodium EDTA	—	12
	Phosphoric acid	—	12
	Potassium citrate	—	—
	Potassium phosphate (mono-, dibasic)	—	—
	Sodium acid pyrophosphate	—	9
	Sodium citrate	—	12
	Sodium gluconate	—	—
	Sodium hexametaphosphate	—	—
	Sodium phosphate (mono-, di-, tri-)	—	—
	Sodium potassium tartrate	—	—
	Sodium tartrate	—	—
	Sodium tripolyphosphate	—	9
	Tartaric acid	—	—
Specific gravity control agent	Brominated vegetable oil	Increase density of oil droplets	—
Surface tension control agents	Dioctyl sodium sulfosuccinate	—	—
	Ox bile extract	—	—
	Sodium phosphate (dibasic)	—	—
Sweeteners			
Nonnutritive	Acesulfame K	—	11, 12

	Ammonium saccharin	—	11, 12
	Calcium saccharin	—	11, 12
	Saccharin	—	11, 12
	Sodium saccharin	—	11, 12
Nutritive	Aspartame	—	11, 12
	Glucose	—	4
	Sorbitol	—	4, 12
Texture and consistency control agents			
Emulsifiers and emulsifier salts	Calcium stearoyl-2-lactylate	Dried egg white, bakery	5, 12
	Cholic acid	Dried egg white	5
	Desoxycholic	Dried egg white	5
	Dioctyl sodium sulfosuccinate	General	—
	Fatty acids (C_{10}–C_{18})	General	5
	Lactylic esters of fatty acids	Shortening	5
	Lecithin	General	5
	Mono- and diacylglycerides	General	5
	Ox bile extract	General	—
	Polyglycerol esters	General	—
	Polyoxyethylene sorbitan esters	General	5
	Propylene glycol, mono-, diesters	General	5
	Potassium phosphate, tribasic	Processed cheese	12
	Potassium polymetabisulfite	Processed cheese	12
	Potassium pyrophosphate	Processed cheese	12
	Sodium aluminum phospate	Processed cheese	12
	Sodium citrate	Processed cheese	12
	Sodium metaphosphate	Processed cheese	12
	Sodium phosphate, dibasic	Processed cheese	12
	Sodium phosphate, monobasic	Processed cheese	12
	Sodium phosphate, tribasic	Processed cheese	12
	Sodium pyrophosphate	Processed cheese	12
	Sorbitan monooleate	Dietary products	5
	Sorbitan monopalmitate	Flavor dispersion	5
	Sorbitan monostearate	General	5
	Sorbitan tristearate	Confection coatings	5
	Stearoyl-2-lactylate	Bakery shortening	5
	Stearoyl monoglyceridyl citrate	Shortenings	5
	Taurocholic acid	Egg whites	5

TABLE 5 Continued

Class and general function	Chemical name	Additional or more specific function	Source of discussion (chapter)
Firming agents	Aluminum sulfates	Pickles	12
	Calcium carbonate	General	12
	Calcium chloride	Canned tomatoes	12
	Calcium citrate	Canned tomatoes	12
	Calcium gluconate	Apple slices	12
	Calcium hydroxide	Fruit slices	12
	Calcium lactate	Apple slices	12
	Calcium phosphate, monobasic	Canned tomatoes	12
	Calcium sulfate	Canned potatoes, tomatoes	12
	Magnesium chloride	Canned peas	10
Leavening agents	Ammonium bicarbonate	Carbon dioxide source	12
	Ammonium phosphate (dibasic)	—	12
	Calcium phosphate	—	12
	δ-Gluconolactone	—	12
	Sodium acid pyrophosphate	—	12
	Sodium aluminum phosphate	—	12
	Sodium aluminum sulfate	—	12
	Sodium bicarbonate	Carbon dioxide source	12
Masticatory substances	Paraffin (synthetic)	Chewing gum base	12
	Pentaerythritol ester of rosin	Chewing gum base	12
Propellants	Carbon dioxide	—	12
	Nitrous oxide	—	12
Stabilizers and thickeners	Acacia gum	Foam stabilizer	4, 12
	Agar	Ice cream	4, 12
	Alginic acid	Ice cream	4, 12
	Carrageenan	Chocolate drinks	4, 12
	Guar gum	Cheese foods	4, 12
	Hydroxypropyl-methylcellulose	General	4, 12
	Locust bean gum	Salad dressing	4, 12
	Methyl cellulose	General	4, 12
	Pectin	Jellies	4, 12
	Sodium carboxy-methylcellulose	Ice cream	4, 12
	Tragacanth gum	Salad dressing	4, 12
Texturizers	Carrageenan	General	4, 12
	Mannitol	—	4, 12

	Pectin	—	4, 12
	Sodium caseinate	—	6
	Sodium citrate	—	12
Tracers	Titanium dioxide	Vegetable protein extenders	12

Note: For additional information see Refs. 6, 14, 18, 25, and 33; also see Bibliography.
[a]Individual members comprising flavoring agents are too numerous to mention. See Refs. 18 and 25 for comprehensive listings; also see Bibliography.

BIBLIOGRAPHY

Akoh, C. C., and B. G. Swanson (eds.) (1994). *Carbohydrate Polyester as Fat Substitutes*, Marcel Dekker, New York.

Branen, A. L., and P. M. Davidson (eds.) (1983). *Antimicrobials in Foods*, Marcel Dekker, New York.

Committe on Food Protection, Food and Nutrition Board (1965). *Chemicals Used in Food Processing*, Publ. 1274, National Acadmy of Sciences Press, Washington, D C.

Food Chemicals Codex (1981). Food and Nutrition Board, National Research Council, National Academy Press, Washington DC. 3rd ed.

Furia, T. E. (ed.) (1972). *Handbook of Food Additives*, CRC Press, Cleveland, OH.

Furia, T. E., and N. Bellecana (eds.) (1976). *Fenaroli's Handbook of Flavor Ingredients*, CRC Press, Cleveland, OH.

Furia, T. E. (ed.) (1980). *Regulatory Status of Direct Food Additives*, CRC Press, Boca Raton, FL.

Gould, G. W. (1989). *Mechanisms of Action of Food Preservation Procedures*, Elsevier Applied Science, Londson.

Heath, H. B. (1981). *Source Book of Flavors*, AVI, Westport, CT.

Lewis, R. J., Sr. (1989). *Food Additives Handbook*, Van Nostrand Reinhold, New York.

Lueck, E. (1980). *Antimicrobial Food Additives*, Springer-Verlag, Berlin.

Molins, R. A. (1991). *Phosphates in Food, CRC Press*, Boca Raton, FL.

Nabors, L. O., and R. C. Geraldi (eds.) (1991). *Alternative Sweeteners*, Marcel Dekker, New York, 2nd. ed.

Phillips, G. O., D. J. Wedlock, and P. A. Williams (eds.) (1990). *Gums and Stabilizers for the Food Industry*, IRL Press, Oxford University Press, Oxford.

Taylor, R. J. (1980). *Food Additives*, John Wiley and Sons, New York.

REFERENCES

1. Artz, W. E., and S. L. Hansen (1994). Other fat substitutes, in *Carbohydrate Polyesters as Fat Substitutes* (C. C. Akoh and B. G. Swanson, eds)., Marcel Dekker, New York, pp. 197–236.
2. Brusick, D., M. Cifone, R. Young, and S. Benson (1989). Assessment of the genotoxicity of calcium cyclamate and cyclohexylamine. *Environ. Mol. Mutagen. 14*:188–199.
3. Bryan, G. T., and E. Erturk (1970). Production of mouse urinary bladder carcinomas by sodium cyclamate. *Science 167*:996–998.
4. Caputi, A., and R. G. Peterson (1965). The browning problem in wines. *Am J. Enol. Viticult 16*(1):9–13.
5. Chichester, D. F, and F. W. Tanner (1972). Antimicrobial food additives, in *Handbook of Food Additives* (T. E. Furia, ed.), CRC Press, Cleveland, Ohio, pp. 115–184.
6. Committe on Food protection, Food and Nutrition Board (1965). *Chemicals Used in Food Processing, Publ. 1274*, National Academy of Sciences Press, Washington, DC.
7. Considine, D. M. (ed.) (1982). Masticatory substances, in *Foods and Food Production Encyclopedia*, Van Nostrand Reinhold, New York, pp. 1154–1155.
8. Crowell, E. A., and J. F. Guyman (1975). Wine constituents arising from sorbic acid addition, and

identification of 2-ethoxyhexa -3,5-diene as source of geranium-like off-odor. *Am. J. Enol. Viticult. 26*(2):97–102.
9. Dahlstrom, R. W., and M. R. Sfat (1972). The use of polyvinylpyrrolidone in brewing. *Brewer's Dig. 47*(5):75–80.
10. Ellinger, R. H. 91972). Phosphates in food procesisng, in *Handbook of Food Additives* (T. E. Furia, ed.), CRC Press, Cleveland, OH, pp. 617–780.
11. Etchells, J. L., T. A. Bell, and L. J. Turney (1972). Influence of alum on the firmness of fresh-pack dill pickles. *J. Food Sci. 37*:442–445.
12. Faubian, J. M., and R. C. Hoseny (1981). Lipoxygenase: Its biochemistry and role in breadmaking. *Cereal Chem. 58*:175–180.
13. *Food Chemicals Codex*. (1981). 3rd ed., Food and Nutrition Board, National Research Council, National Academy Press, Washington, DC.
14. *Food Chemical News*.(1983). USDA petitioned to remove titanium dioxide marker requirements for soy protein. March 21, pp.37–38.
15. Freese, E., and B. E. Levin (1978). Action mechanisms of preservatives and antiseptics, in *Developments in Industrial Microbiology* (L. A. Underkofler, ed.), Society of Industrial Microbiology, Washington, DC, pp. 207–227.
16. Furia, T. E. (1972). Regulatory status of direct food additives, in *Handbook of Food Additives* (T. E. Furia, ed.), CRC Press, Cleveland, OH, pp. 903–966.
17. Furia, T. E. (1972). Sequestrants in food, in *Handbook of Food Additives* (T. E. Furia, ed.), CRC Press, Cleveland, OH, pp. 271–294.
18. Furia, T. E., and N. Bellanca (eds.) (1976). *Fenaroli's Handbook of Flavor Ingredients*, 2 vols., Chemical Rubber Co., Cleveland, OH.
19. Gardner, W. H. (1972). Acidulants in food processing, in *Handbook of Food Additives* (T. E. Furia, ed.), CRC Press, Cleveland, OH, pp.225–270.
20. Goddard, S. J., and B. L. Wedzicha (1992). Kinetics of the reaction of sorbic acid with sulphite species. *Food Additives Contam.* 9:485–492.
21. Griffin, W. C., and M. J. Lynch (1972). Polyhydric alcohols, in *Handbook of Food Additives* (T. E. Furia, ed.), 2nd. ed., CRC Press, Cleveland, OH, pp. 431–455.
22. Griswold, R. M. (1962). Leavening agents, in *The Experiment Study of Foods* (R. M. Griswold, ed.), Houghton Mifflin, Boston, pp. 330–352.
23. Haddon, W. F., M. L. Mancini, M. McLaren, A. Effio, L. A. Harden, R. L. Degre, and J. L. Bradford (1994). Occurrence of ethyl carbamate (urethane) in U.S. and Canadian breads: Measurements by gas chromatography-mass spectrometry. *Cereal Chem. 71*:297–215.
24. Hamm, R. (1971). Interactions between phosphates and meat proteins, in *Symposium: Phosphates in Food Processing* (J. M. DeMan and P. Melnychyn, eds.), AVI, Westport, CT, pp. 65–84.
25. Heath, H. B. (1981). *Source Book of Flavors*, AVI, Westport, CT.
26. Ishii, H., and T. Koshimizu (1981). Toxicity of aspartame and its diketopiperazine for Wistar rats by dietary administration of 104 weeks. *Toxicoloy 21*:91–94.
27. Joslyn, M. A. (1964). Gassing and deaeration in food processing, in *Food Processing Operations: Their Management, Machines, Materials, and Methods* (M. A. Joslyn and J. L. Heid, eds.), vol. 3, AVI, Westport, CT, pp. 335–368.
28. Kabara, J. J., R. Varable, and M. S. Jie (1977). Antimicrobial lipids: Natural and synthetic fatty acids and monoglycerides. *Lipids 12*:753–759.
29. Karahadian, C., and R. C. Lindsay (1988). Evaluation of the mechanism of dilaurly thiodipropionate antioxidant activity. *J. Am. Oil Chem. Soc. 65*:1159–1165.
30. Khandelwal, G. D., Y. L. Rimmer, and B. L. Wedzicha (1992). Reaction of sorbic acid in wheat flour doughs. *Food Additives Contam.9*:493–497.
31. Klose, R. E., and M. Glicksman (1972). Gums, in *Handbook of Food Additives* (T. E. Furia, ed.), CRC Press, Cleveland, OH, pp. 295–359.
32. Labuza, T. P., N. D. Heidelbaugh, M. Silver, and M. Karel (1971). Oxidation at intermediate moisture contents. *J. Am. Oil Chem. Soc. 48*:86–90.
33. Lewis, R. J., Sr. (1989). *Food Additives Handbook*, Van Nostrand Reinhold, New York.

34. Lucca, P. A., and Tepper, B. J. (1994). Fat replacers and the functionality of fat in foods. *Trends Food Sci. Technol. 5*:12–19.
35. Mahon, J. H., K. Schlamb, and E. Brotsky (1971). General concepts applicable to the use of polyphosphates in red meat, poultry, seafood processing, in *Symposium: Phosphates in Food Processing* (J. M. DeMan and P. Malnychyn, eds.), AVI, Westport, CT, pp. 158–181.
36. Miller, M. S. (1994). Proteins as fat substitutes, in *Protein Functionality in Food Systems* (N. S. Hettiarachchy and G. R. Ziegler, eds.), Marcel Dekker, New York, pp. 435–465.
37. Miller, W. T. (1987). The legacy of cyclamate. *Food Technol. 41*(1):116.
38. Minifie, B. W. (1970). *Chocolate, Cocoa and Confectionary: Science and Technology*, AVI, Westport, CT, pp. 38–41.
39. Peng, S. K., C. Taylor, J. C. Hill, and R. J. Marin (1985). Cholesterol oxidation derivatives and arterial endothelial damage. *Arteriosclerosis 54*:121–136.
40. Pomeranz, Y., and J. A. Shallenberger (1971). *Bread Science and Technology*, AVI, Westport, CT.
41. Pratter, P. J. (1980). Neohesperidin dihydrochalcone: An updated review on a naturally derived sweetener and flavor potentiator. *Perfumer Flavorist 5*(6):12–18.
42. Price, J. M., C. G. Biana, B. L. Oser, E. E. Vogin, J. Steinfeld, and H. L. Ley (1970). Bladder tumors in rats fed cyclohexylamine or high doses of a mixture of cyclamate and saccharin. *Science 167*:1131–1132.
43. Ramarathnam, N., and L. J. Rubin (1994). The flavour of cured meat, in *Flavor of Meat and Meat Products* (F. Shahidi, ed.), Blackie, London, pp. 174–198.
44. Salant, A. (1972). Nonnutritive sweeteners, in *Handbook of Food Additives* (T. E. Furia, ed.), 2nd ed., CRC Press, Cleveland, OH, pp. 523–586.
45. Scharpf, L. G. (1971). The use of phosphates in cheese processing, in *Symposium Phosphates in Food Processing* (J. M. deMan and P. Melnychyn, eds.), AVI, Westport, CT, pp. 120–157.
46. Scott, U. N., and S. L. Taylor (1981). Effect of nisin on the outgrowth of *Clostridium botulinum* spores. *J. Food Sci. 46*:117–126.
47. Sebranek, J. G., and R. G. Cassens (1973). Nitrosamines: A review. *J. Milk Food Technol. 36*:76–88.
48. Singleton, V. L. (1967). Adsorption of natural phenols from beer and wine. *MBAA Tech. Q. 4*(4):245–253.
49. Smith, R. E., J. W. Finley, and G. A. Leveille (1994). Overview of Salatrim, a family of low-calorie fats. *J. Agric. Food Chem. 42*:432–434.
50. Stahl, J. E., and R. H. Ellinger (1971). The use of phosphates in the baking industry, in *Symposium: Phosphates in Food Processing* (J. M. Deman and P. Melnychyn, eds.), AVI, Westport, CT., pp. 194–212.
51. Stauffer, C. E. (1983). Dough conditioners. *Cereal Foods World 28*:729–730.
52. Stukey, B. N. (1972). Antioxidants as food stabilizers, in *Handbook of Food Additives* (T. E. Furia, ed.), CRC Press, Cleveland, OH, pp. 185–224.
53. Thompson, J. E., and B. D. Buddemeyer (1954). Improvement in flour mixing characteristics by a steryl lactylic acid salt. *Cereal Chem. 31*:296–302.
54. van Buren, J. P., J. C. Moyer, D. E. Wilson, W. B. Robinson, and D. B. Hand (1960). Influence of blanching conditions on sloughing, splitting, and firmness of canned snap beans. *Food Technol. 14*:233–236.
55. van der Wel, H. (1974). Miracle fruit, katemfe, and serendipity berry, in *Symposium: Sweeteners* (G. Inglett, ed.), AVI, Westport, CT., pp. 194–215.
56. Weast, R. C. (ed.) (1988). *Handbook of Chemistry and Physics*, CRC Press, Boca Raton, FL. pp. D161–D163.
57. Wedzicha, R. L. (1984). *Chemistry of Sulphur Dioxide in Foods*, Elsevier, London.
58. Wedzicha, R. L. (1992). Chemistry of sulphiting agents in foods. *Food Additives Contam.9*:449–459.
59. Wesley, F., F. Rourke, and O. Darbishire (1965). The formation of persistent toxic chlorohydrins in foodstuffs by fumigation with ethylene oxide and with propylene oxide. *J. Food Sci. 30*:1037–1042.

13

Toxic Substances

MICHAEL W. PARIZA
Food Research Institute, University of Wisconsin—Madison, Madison, Wisconsin

13.1 INTRODUCTION

We should begin with a few words about "food toxicology." It does not refer to foods per se but rather to the study of toxic substances—particularly naturally occurring toxic substances—that may be associated with food [9]. Certain nonconventional foods (the kinds you don't usually find in supermarkets, for example, puffer fish and bracken fern) contain levels of naturally occurring toxins that many would consider unacceptably high. Conventional foods (the kinds found typically in supermarkets) are on rare occasions contaminated with unacceptably high levels of naturally occurring toxins, but this is not the usual case. Hence, food toxicologists are not generally concerned with foods themselves but rather with toxic substances that may sometimes be found in foods. The two general types of toxicants—inherent toxicants, and contaminants—are described in Table 1.

That being said, we should now distinguish between "toxic substance" and "toxic effect"

TABLE 1 Types of Food Toxicants

Inherent toxicants
Metabolites produced via biosynthesis by food organisms under normal growth conditions
Metabolites produced via biosynthesis by food organisms that are stressed
Contaminants
Toxicants that directly contaminate foods
Toxicants that are absorbed from the environment by food-producing organisms
Toxic metabolites produced by food organisms from substances that are absorbed from the environment
Toxicants that are formed during food preparation

[9, 23]. It goes almost without saying that toxic substances induce toxic effects. What is not said as often is that a toxic substance will induce a toxic effect only if its concentration is sufficiently high. Hence the terms toxic substance (toxicant) and toxic effect are not synonymous.

This distinction takes the form of a principle that underlies the science of toxicology. It was first presented as follows: "Everything is poison. There is nothing without poison. Only the dose makes a thing not a poison. For example, every food and every drink, if taken beyond its dose, is poison" [10, 15].

This observation is remarkable enough on its merits, but perhaps most surprising is that the author, an alchemist named Paracelsus, penned it in 1564!

What we have learned of toxicology in the intervening 450 years has served only to reinforce Paracelsus' extraordinary insight. We recognize today that all chemicals are potentially toxic, and that the dose alone determines whether or not a toxic effect occurs. Indeed, toxicant (or toxin) was recently defined as a "substance that has been shown to present some significant degree of possible risk when consumed in sufficient quantity by humans or animals" [16]. We differ with Paracelsus only in reserving the term "poison" for the most potent substances (those that induce adverse effects at exposure levels of a few milligrams per kilogram body weight).

One expects drugs and household chemicals to exhibit at least some degree of toxicity; hence these materials are ordinarily treated with the deference due them. However, even the most sophisticated might find it difficult to fully accept the notion that potentially toxic substances are commonly present, much less possibly desirable, in conventional foods. And yet we know that some required nutrients, for example, vitamins A and D, induce toxic effects at intake levels not too much greater than those needed for optimal health. The safety margins for these and many other naturally occurring dietary substances are far smaller than the 100-fold safety factor customarily employed for synthetic food additives [9].

The purpose of this chapter, then, is to provide an overview of toxicology as it relates to food. It is hoped that the reader will carry from this an understanding that while all food chemicals (naturally occurring as well as synthetic) are potentially toxic, it is rare indeed to encounter a toxic effect induced as a result of consuming conventional foods.

13.2 FOOD TOXICOLOGY AND THE LAW

One cannot fully appreciate the relationship between food safety and the science of toxicology without first understanding why such research is done. It is done for two basic reasons: the pursuit of new knowledge, and the evaluation of safety. The former is a true scientific goal, while the latter is a hybrid of science, politics, and law [16].

By law, substances that are intentionally added to food must be shown to be safe under their intended conditions of use. The methods and procedures for evaluating safety are matters of science; however, additional considerations outside the purview of science also bear on the ultimate decision.

For example, substances may be classified GRAS (generally regarded as safe). The decision as to whether something is GRAS is left to the scientific community—specifically, those persons qualified as experts by training and experience. However, the Food and Drug Administration (FDA) reserves the right to challenge or affirm a GRAS determination [16].

There are special considerations in the U.S. Food, Drug and Cosmetics Act (FD&C Act) for color additives, pesticide residues, prior sanctioned ingredients (those in use prior to 1958, the year in which the FD&C Act was last amended), and substances for which FDA has issued formal food additive regulations. There is also the "constituents policy," which governs unavoidable contaminants.

Perhaps the best known provision of the FD&C Act is the "Delaney Clause," so-named for the Congressman who chaired the Congressional subcommittee that added the amendment to the FD&C Act. The Delaney Clause, which bars the use of food additives shown to cause cancer in humans or animals, actually appears in the FD&C Act three times. They are Sections 409 (Food Additives), 706 (Color Additives), and 612 (Animal Feeds and Drugs).

Pesticide residues on raw agricultural products are not subject to the Delaney Clause. However, concentrated residues that remain after a food is processed are subject to this legislation. In other words, it is legal to sell an apple that contains traces of a carcinogenic pesticide approved by the Environmental Protection Agency (EPA), but one could not sell applesauce made from that apple. It is not even necessary to show that the applesauce actually contains the pesticide residue. Under the law, a processed food may be banned if a concentrated pesticide residue might theoretically be present.

By contrast, the Delaney Clause does not apply to unavoidable contaminants that are present in a food or a food additive, even if the contaminants are carcinogens. Examples are traces of aflatoxin in peanuts, and certain minor impurities that may occur in some chemically synthesized colors. For such contaminants, the courts have held that the FDA may use risk assessment methodologies coupled with a legal concept called "*de minimis*." (*De minimis* means that the law does not deal with trifles [16, 21].)

The result of all this is that an additive, for example, saccharine, that is very weakly carcinogenic in certain animal experiments but for which there is no evidence of human carcinogenicity would be banned. (In fact, the only reason saccharine wasn't banned is because of a special congressional exemption.) At the same time, theoretically "riskier" naturally occurring or synthetic carcinogens, which unavoidably contaminate food, or otherwise innocuous food or color additives, are exempt from Delaney. For these, FDA sets limits on the amount of the contaminant that is to be tolerated.

Virtually all foods contain traces of carcinogenic substances, mostly from natural sources [1, 9, 13]. Examples are urethane in fermented beverages, and certain naturally occurring components of spices such as safrole (the natural root beer flavor, banned as an intentional food additive in the 1960s), which is present in sassafras, sweet basil, and cinnamon [20]. Hence, a practical consequence of the federal law is that foods may be legally sold in the United States even if some of the constituents they contain cannot be intentionally added to a food.

13.3 TERMS IN FOOD TOXICOLOGY

Like all scientific disciplines, toxicology has a vocabulary of its own. Terms used by food toxicologists include acute toxicity, chronic toxicity, subchronic feeding test, MTD, NOAEL, and ADI.

Acute toxicity refers to a toxic response, often immediate, induced by a single exposure. A lethal dose of hydrogen cyanide (50–60 mg) induces death within minutes; cicutoxin, the toxic principal of water hemlock, kills so rapidly that cattle that have eaten it often die before the implicated feedstuff has passed beyond the esophageal groove [24]. These are poignant examples of acute toxicity. The acute toxicity of a substance is defined by its LD_{50}, the dose of the substance that will kill 50% of a group of exposed animals.

Chronic toxicity refers to an effect that requires some time to develop, for example, cancer. Testing for chronic toxicity involves continuously feeding the test substance to rodents for 20–24 months. By analogy to LD_{50}, the amount of a carcinogen required to induce cancer in 50% of a group of exposed animals is referred to as the TD_{50} (tumor $dose_{50}$) [13].

Subchronic feeding test is a "ninety day toxicology study in an appropriate animal species" [16]. It is often used to define the MTD and, for noncarcinogens, the NOAEL.

MTD is the acronym for "maximum tolerated dose." It is the highest level of a test substance that can be fed to an animal without inducing obvious signs of toxicity [6]. In chronic toxicity tests, the test substance is typically fed at its MTD and perhaps one or two lower doses. The MTD concept has been criticized because it is based soley on gross measures of toxicity, for example, weight loss. Subtle biochemical indicators of cellular toxicity, which may occur at lower doses, are ignored [22].

NOAEL is the acronym for "no observable adverse effect level." For substances that induce a toxic response (other than cancer) in chronic feeding tests, the NOAEL is used to determine an acceptable daily intake (ADI) [21, 22].

ADI is the acronym for "acceptable daily intake." By convention, for noncarcinogens, it is set at 1/100th of the NOAEL.

Potential food additives found to induce cancer are of course subject to the Delaney Clause. The practical result is that such substances are assumed, legally, not to have a NOAEL. Hence, no amount can be added to food. (Whether this assumption is scientifically accurate is open to question.)

13.4 EVALUATING SAFETY: THE TRADITIONAL APPROACH

Subchronic and chronic rodent feeding tests are at the heart of traditional safety evaluations for new food ingredients. These tests were developed for the assessment of single chemicals, which could be fed at large multiples (100-fold or more) of proposed human consumption. High-dose feeding was needed, so the thinking went, to compensate for the relatively small number of test animals that practical considerations dictate must be used. (Small, that is, relative to the total number of potentially exposed humans. For example, the population of the United States is about 250 million persons, yet one would be hard pressed to conduct a study with more than a few hundred rats. Hence each rat subject is a surrogate for a very large number of potentially exposed humans.)

The trade-off of feeding relatively small numbers of animals subjects relatively large amounts of test substances begged another question: How high should the exposure to test materials be? Thus was born the MTD, the greatest amount of a test substance that can be fed to a rat or mouse before overt signs of toxicity appear. Overt signs of toxicity were defined by a set of gross measurements, for example, failure of test animals to maintain at least 90% of the weight of controls [6].

The traditional approach, then, is to determine the MTD for the test substance, and feed the test substance at its MTD to rodents for 2 years. If the test group develops cancer in excess of the controls, then the substance may be judged a carcinogen and banned as a food additive under the Delaney Clause.

If the substance is found not to induce cancer (or to induce it through secondary mechanisms that are not relevant to human experience), then its NOAEL is determined. The substance may then be permitted as a food additive at a concentration that is low enough so that humans will consume no more than 1/100th of the rodent NOAEL.

13.5 EVALUATING SAFETY: NEW APPROACHES

The procedures discussed to this point were designed to evaluate the safety of substances, typically single chemicals but sometimes simple mixtures, that are intended to be used in food at relatively low levels. The qualification "relatively low levels" is critical in that the test substance is fed to experimental animals at large multiples of projected human exposure, usually 100-fold or more. Hence, by definition, the projected human exposure cannot exceed 1% of the diet, since at that point animals receiving a 100-fold multiple of human exposure would be consuming only the pure compound and nothing else. (In practice, the projected human dietary exposure must be considerably less than 1%, since feeding a substance to animals at more than a few percent of their diet risks creating a nutritional imbalance.) Obviously, a new approach is required for the evaluation of novel foods and "macro ingredients" that are likely to represent 1% or more of the diets of at least some persons (for example, genetically engineered tomatoes and novel edible oils).

In evaluating such materials, plausible substitutes for the standard high-dose animal feeding tests are required. One approach has its origins in the safety evaluation of enzymes used in food processing [23]. Enzymes and their use in food processing are discussed in Chapter 7; here we are considering the issue of safety evaluation.

Food-grade enzymes are in fact complex mixtures, not single entities. Typically there is little purification, so virtually everything the source organism produces may be present in a commercial enzyme preparation. Given this, one may reasonably ask how safety evaluation should be conducted. Should one focus on the active enzyme, or on all the components that comprise the product as it is marketed? Is it necessary to consider all possible adverse effects, or can the task be simplified by concentrating on those possible adverse effects that might reasonably occur as a result of the use of the specific source organism?

Fortunately, there is a sizable data set for microbial toxins on which to base an evaluation system. It is possible, then, to evaluate a microbial enzyme preparation for its potential to produce adverse effects that might be anticipated given the production organism [23].

A safety evaluation system based on these considerations [23] begins with a thorough literature review to determine what (if any) adverse effects have been associated with the source organism and related organisms of the same species or genus. Particular attention is directed to finding reports of toxins active via the oral route (enterotoxins and certain neurotoxins). If such reports are found, it is then necessary to ensure that the enzyme preparation is free of the undesirable material.

The usual case, however, is that the production organism will have been selected from those species with a prior history of safe use in food or food ingredient manufacture. In this case it is most probable that no adverse effects will have been reported for the organism. Limited animal testing may still be required, however, to ensure that the proposed production organism does not produce unknown enterotoxins or other undesirable substances active via the oral route.

With slight modification, this decision tree approach can accommodate enzymes produced by genetically engineered organisms [16]. Moreover, the proposition of focusing on potentially adverse substances that might be produced by the source organism (in contrast to a broader and largely unworkable approach of searching for any conceivable adverse effect via high-dose animal feeding tests) also forms the basis of safety evaluation systems for genetically engineered crops endorsed by government and industry [16, 19].

13.6 COMPARATIVE TOXICOLOGY AND INHERENT TOXICANTS

It is important to keep in mind the historical framework of the science underlying the "traditional" and "new" approaches to safety evaluation, as described earlier. The traditional approach is based on concepts developed in the middle of this century, some of which are now difficult to justify [21, 22]. In particular, a basic assumption of the traditional approach, that dietary carcinogens are rare and therefore avoidable, is decidedly untenable [1, 13, 22].

In contrast, the nontraditional approaches are more recent attempts to wed safety evaluation to the emerging realization that a quest for absolute safety is not only scientifically naive but also counter to the need to rationally prioritize public health objectives in light of funding constraints [1, 13, 22].

The focus of the new approaches proposed by the International Food Biotechnology Council [16] and the FDA [19] is *comparative toxicology*. In this context, comparative toxicology refers to comparing the concentration of *inherent toxins* (e.g., those that are naturally and endogenously present) in the new food with the concentration of inherent toxins in the traditional counterpart of the new food. This is a fundamental conceptual shift from the traditional apprach, which stresses the absence of toxicity over a wide safety margin (set, arbitrarily at 100-fold). The new approach allows inherent toxicants in the new food if their levels do not exceed those contained in the traditional counterpart of the new food (the traditional counterpart is the food in current use that the new food will replace). Safety margins for the inherent toxicants would not ordinarily be of concern as long as the traditional counterpart is considered safe to eat.

Obviously, the credibility of this approach [16, 19] depends on the existence of reliable, comprehensive databases of toxicants that occur naturally in commonly consumed foods. Fortunately such databases exist, thanks to basic research that was conducted largely for the pursuit of knowledge without regard to the legal and regulatory aspects of food toxicology [16]. For example, we know that many kinds of inherent toxicants occur in plants (Table 2), and we also know that some of these substances are carcinogens in animal experiments (Table 3). However, we also know that only a few such inherent toxicants have actually caused harm to persons consuming normal human diets (Table 4).

Inherent toxicants are of course a subset of the naturally occurring chemicals, a broad category that includes all substances except those that are not of biosynthetic origin and that are intentionally or accidently introduced into food through human activity [16]. By this definition mycotoxins are naturally occurring, even if they developed through mold growth due to careless handling of stored grain, but residues of synthetic pesticides used to kill insects harboring the mold spores are not naturally occurring. Arsenic is naturally occurring when in seafood (it is a natural component of seawater), but arsenic is not naturally occurring when present in produce from a field to which the element was intentionally applied as a pesticide (a custom no longer practiced in the United States).

Most chemicals in the environment (and of course this includes food) are naturally occurring in that they are not intentional products of synthetic chemistry. Very few substances are solely synthetic in that they are not known to exist in nature. Hence the number, variety, and concentration of naturally occurring chemicals in food far outweighs, by many orders of

TABLE 2 Examples of Inherent Toxicants in Plants

Toxins	Chemical nature	Main food source	Major toxicity symptoms
Protease inhibitors	Proteins (mol. wt. 4000–24,000)	Beans (soy, mung, kidney, navy, lima); chick-pea; peas, potatoes (sweet, white); cereals	Impaired growth and food utilization; pancreatic hypertrophy
Hemagglutinins	Proteins (mol. wt. 10,000–124,000)	Beans (castor, soy, kidney, black, yellow, jack); lentils; peas	Impaired growth and food utilization; agglutination of erythrocytes in vitro; mitogenic activity to cell cultures in vitro
Saponins	Glycosides	Soybeans, sugarbeets, peanuts, spinach, asparagus	Hemolysis of erythrocytes in vitro
Glycosinolates	Thioglycosides	Cabbage and related species; turnips; rutabaga; radish; rapeseed; mustard	Hypothyroidism and thyroid enlargement
Cyanogens	Cyanogenic glucosides	Peas and beans; pulses; linseed; flax; fruit kernels; cassava	HCN poisoning
Gossypol pigments	Gossypol	Cottonseed	Liver damage; hemorrhage; edema
Lathyrogens	β-Aminopropionitrile and derivatives	Chick-pea; vetch	Neurolathyrism (CNS[a] damage)
Allergens	Proteins?	Practically all foods (particularly grains, legumes, and nuts)	Allergic responses in sensitive individuals
Cycasin	Methylazoxymethanol	Nuts of *Cycas* genus	Cancer of liver and other organs
Favism	Vicine and convicine (pyrimidine β-glucosides)	Fava beans	Acute hemolytic anemia
Phytoalexins	Simple furans (ipomeamarone)	Sweet potatoes	Pulmonary edema; liver and kidney damage
	Benzofurans (psoralins)	Celery; parsnips	Skin photosensitivity
	Acetylenic furans (wyerone)	Broad beans	
	Isoflavonoids (pisatin and phaseollin)	Peas, french beans	Cell lysis in vitro
Pyrrolizidine alkaloids	Dihydropyrroles	Families Compositae and Boraginaccae; herbal teas	Liver and lung damage carcinogens
Safrole	Allyl-substituted benzene	Sassafras; black pepper	Carcinogens
α-Amantin	Bicyclic octapeptides	*Amanita phalloides* mushrooms	Salivation; vomiting; convulsions; death
Atractyloside	Steroidal glycoside	Thistle *(Atractylis gummifera)*	Depletion of glycogen

[a]Central nervous system.
Source: Ref. 27.

TABLE 3 Some Naturally Occurring Carcinogens Inherent in Food

Rodent carcinogen	Plant food	Concentration (ppm)
5-/8-Methoxypsoralen	Parsley	14
	Parsnip, cooked	32
	Celery	0.8
	Celery, new cultivar	6.2
	Celery, stressed	25
p-Hydrazinobenzoate	Mushrooms	11
Glutamyl *p*-hydrazinobenzoate	Mushrooms	42
Sinigrin (allyl isothiocyanate)	Cabbage	35–590
	Collard greens	250–788
	Cauliflower	12–66
	Brussels sprouts	110–1560
	Mustard (brown)	16,000–72,000
	Horseradish	4500
Estragole	Basil	3800
	Fennel	3000
Safrole	Nutmeg	3000
	Mace	10,000
	Pepper, black	100
Ethyl acrylate	Pineapple	0.07
Sesamol	Sesame seeds (heated oil)	75
α-Methylbenzyl alcohol	Cocoa	1.3
Benzyl acetate	Basil	82
	Jasmine tea	230
	Honey	15
	Coffee (roasted beans)	100
Caffeic acid	Apple, carrot, celery, cherry, eggplant, endive, grapes, lettuce, pear, plum, and potato	50–200
	Absinthe, anise, basil, caraway, dill, marjoram, rosemary, sage, savory, tarragon, and thyme	>1000
	Coffee (roasted beans)	1800
	Apricot, cherry, peach, and plum	50–500
Chlorogenic acid (caffeic acid)	Coffee (roasted beans)	21,600
Neochlorogenic acid (caffeic acid)	Apple, apricot, broccoli, brussels sprouts, cabbage, cherry, kale, peach, pear, and plum	50–500
	Coffee (roasted beans)	11,600

Source: Ref. 1.

magnitude, synthetic substances. None of this will surprise anyone who has examined the other chapters in this textbook.

A vast number of naturally occurring chemicals are present in food. No one knows precisely how many naturally occurring chemicals a typical consumer ingests each day, but an estimate of one million compounds is probably conservative. Cooking no doubt amplifies the

TABLE 4 Plant Toxicants Documented as Causing Harm in Normal Human Diets

Substance (category/name/number of substances)	Plant source	Methods of risk reduction
Honey toxicants (7)	Rhododendron/andromeda/azalea family	Monitoring; prohibition of beekeeping
Acetylandromedel, andromedol, anhydroandromedol, and desacetylpireistoxin B		
Gelsamine	Yellow jasmine	
Tutin	Tutu tree	
Hyenanchin		
Forage and meat/milk toxicants (4)		Proper grazing and forage practices; avoidance
Cicutoxin	Water hemlock	
Coniine	Hemlock	
Methylconiine		
Conhydrine		
Toxicants from poor choice, handling, or processing of local diet (5+)		
Hypoglycin A	Ackee fruit (immature)	Avoidance
Linamarin and Lotaustralin	Lima beans and cassava root	Selection and breeding (and proper processing for cassava)
β-*N*-Oxalylamino-L-alanine	Chick-pea	Reduced usage
(-)-Sparteine and related alkaloids	Lupine	Proper processing
Plant genetic factors/poor handling (1)		Selection and breeding, monitoring, proper handling
Solanine	Potato	
Human genetic factors (2)	Fava bean	Reduced usage
Vicine		
Convicine		
Other (2)		
Cucurbitacin E	Squash, cucumber	Breeding isolation
Nitrates	Spinach and other green, leafy vegetables	Proper fertilizing practices and handling; monitoring
Total (21+)		

Source: Refs. 14, and 16.

total number so that the more one cooks, the more chemicals one encounters. Again, none of this will surprise anyone familiar with the subject.

Earlier we noted that all chemicals, whether synthetic or naturally occurring, exhibit toxicity at some level of exposure. Even pure water drunk to excess will kill through the induction of an electrolyte imbalance; there is, in fact, a condition of the elderly called "water toxicity" that is due in part to decreased renal capacity to excrete water [5]. Hence, the innumberable naturally occurring chemicals in food are all potentially capable of inducing toxic effects, but few are ordinarily present in sufficient concentration to actually do so [9, 16].

By one standard, a substance is of concern if it occurs in food within a narrow margin of safety, say 25 or less (in other words, if a toxic effect results from ingesting no more than 25 times the amount typically found in food) [16]. By this standard only a tiny fraction of food constituents—estimated to be less than 0.1%—are "toxicants."

Even so, the list of such toxicants is daunting. Table 2 is but a partial depiction; the reader is referred to a comprehensive list [16] of 209 plant-derived food constituents qualifying by these criteria as toxicants and/or carcinogens. Many of these are inherent toxicants that form the basis for the comparative toxicology safety evaluation proposals [16, 19].

Hall [14] discusses 21 plant-derived toxicants that are documented to have harmed humans who were eating "ordinary" diets (as opposed to the consumption of toxic herbs or poisonous mushrooms, which are not ordinary dietary items). These are shown in Table 4. Of these 21 toxicants, only five are inherent toxicants of foods that are common in Western countries: solanine in potatoes (arguably the most important), cucurbitacin E in squash and cucumber, nitrites in spinach and other green leafy vegetables, and linamarin and lotaustralin in lima beans.

13.7 CONTAMINANTS

Inherent toxicants are, of course, not the only undesirable food constituents. One must also be concerned with contaminants. Food contaminants may be synthetic or naturally occurring; FDA makes few distinctions between these in terms of regulation. However, there is general agreement among experts that the most important are naturally occurring contaminants, in particular those produced by microorganisms [22]. This is not to minimize the potential consequences of contamination of food by synthetic toxicants, but rather to emphasize that contamination by pathogenic microorganisms or their toxins occurs far more frequently.

Two general kinds of microbial toxins are especially important: mycotoxins (mold toxins), and bacterial food poisoning toxins.

Mycotoxins are relatiely low-molecular-weight organic molecules; almost all are smaller in mass than 500 D. The chemical structures of four important mycotoxins are shown in Figure 1. In some cases the common name (e.g., patulin) signifies a single substance, while in other cases the common name (e.g., trichothecenes) signifies a class of chemically related substances.

Mycotoxins induce a wide range of acute and chronic toxic effects in humans and livestock, depending on the compound or compound class, the level of exposure, and the duration of exposure. Among the effects induced by different mycotoxins are tremors and hemorrhaging, immunological suppression, kidney toxicity, fetal toxicity, and cancer [3, 7].

Toxigenic molds may enter the food and feed supply at numerous points, including production, processing, transport and storage [3]. Substrate, moisture, pH, temperature, and crop stress are the major environmental factors that affect mold growth and mycotoxin production [3]. Corn, peanuts, and cotton are the principal crops affected by mycotoxin contamination in the United States [3].

Numerous mold species, belonging to more than 50 genera, are reported to produce toxic metabolites [16]. However, most of these toxic metabolites have not been associated with human or animal disease [3]. Species belonging to three genera produce the most important mycotoxins known to cause illness in humans and animals. Those genera are *Aspergillus, Fusarium,* and *Penicillium* [3].

A great number of mycotoxins have been identified; however, they are not all equally

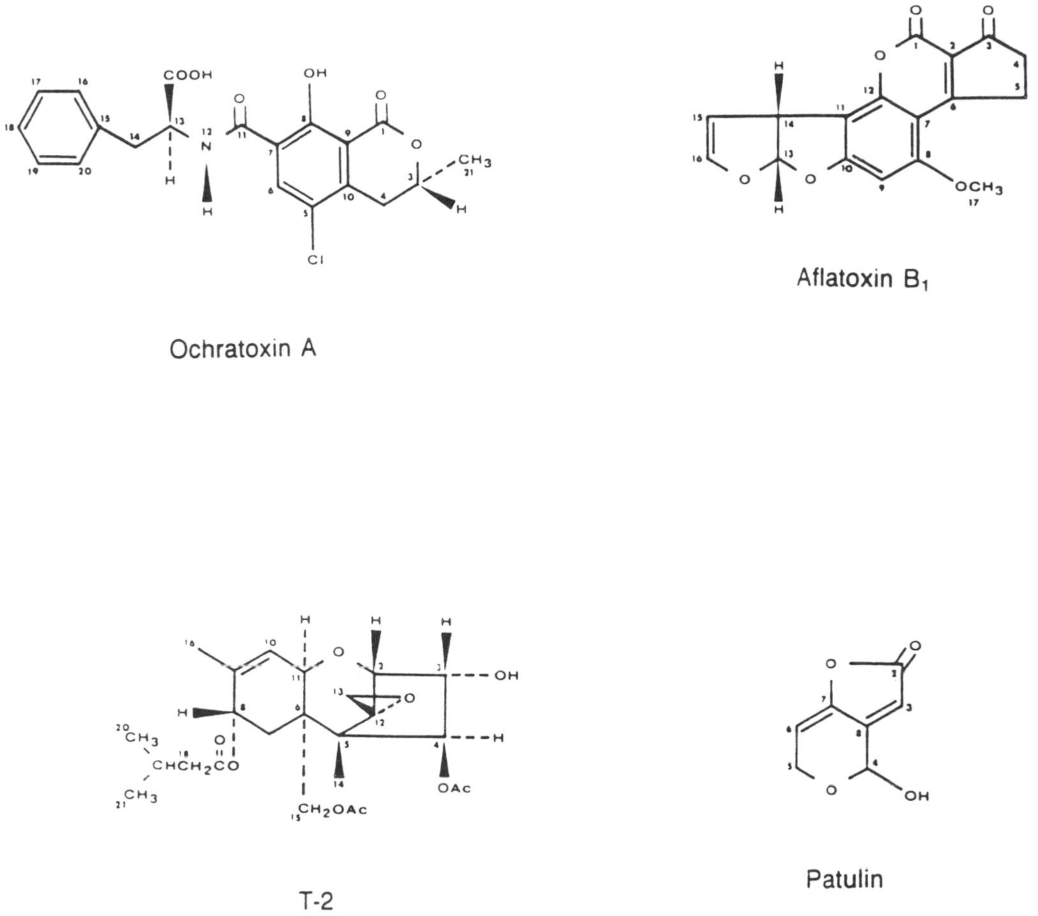

Ochratoxin A

Aflatoxin B_1

T-2

Patulin

FIGURE 1 Chemical structures of four important mycotoxins. (From Ref. 3.)

important in terms of degree of toxicity and likelihood of human or animal exposure. Those that pose the greatest potential risk to human and animal health [7] are listed in Table 5.

In the United States and other countries with advanced food production systems, incidences of human illness induced by mycotoxins are extemely rare and virtually never traced to commercially processed foods [3]. This is because food processors typically set strict specifications for the amounts mycotoxins that will be accepted on grains and other ingredients used in food manufacture. The specifications are based in part on tolerances established by regulatory agencies. Acceptable levels of mycotoxin contamination are determined using extremely conservative risk assessment analyses; hence public health is protected by wide safety margins.

By contrast, the chance contamination of animal feeds with harmful levels of mycotoxins is documented to occur in the United States and other developed countries [3]. Crop stress is an important contributing factor. For example, drought-induced stress enhances the probability of aflatoxin contamination of corn, whereas cold wet weather that slows harvest may lead to the contamination of grain with harmful levels of trichothecenes. Grain judged to be unacceptable for human consumption may end up being fed to livestock.

There is concern that mycotoxins in feeds might pass through to humans who consume animals products. For example, aflatoxin M is found in milk from cows that are fed aflatoxin-

Table 5 Natural Occurrence of Selected Common Mycotoxins

Mycotoxins[a]	Major producing fungi	Typical substrate in nature	Biological effect
Alternaria (AM) mycotoxins	Alternaria alternata	Cereal grains, tomato, animal feeds	M, Hr
Aflatoxin (AF) B_1 and other aflatoxins	*Aspergillus flavus, A. parasiticus*	Peanuts, corn, cottonseed, cereals, figs, most tree nuts, milk, sorghum, walnuts	H, C, M, T
Citrinin (CT)	*Penicillum citrinum*	Barley, corn, rice, walnuts	Nh, (C?), M
Cyclopiazonic acid (CPA)	*A. flavus, P. cyclopium*	Peanuts, corn, cheese	Nr, Cv
Deoxynivalenol (DON)	*Fusarium graminearum*	Wheat, corn	Nr
Cytlochlorotine (CC)	*P. islandicum*	Rice	H, C
Fumonisins (FM)	*F. moniliforme*	Corn, sorghum	H, Nr, C(?), R
Luteoskyrin (LT)	*P. islandicum, P. rugulosum*	Rice, sorghum	H, C, M
Moniliformin (MN)	*F. moniliforme*	Corn	Nr, Cv
Ochratoxin A (OTA)	*A. ochraceus, P. verrucosum*	Barley, beans, cereals, coffee, feeds, maize, oats, rice, rye, wheat	Nh, T Nr, C(?), D, T
Patulin (PT)	*P. patulum, P. urticae, A. clavatus*	Apple, apple juice, beans, wheat	Nr, C(?), M
Penicillic acid (PA)	*P. puberulum, A. ochraceus*	Barley, corn	Nr
Penitrem A (PNT)	*P. palitans*	Feedstuffs, corn	Nr
Roquefortine (RQF)	*P. roqueforti*	Cheese	H, T
Rubratoxin B (RB)	*P. rubrum, P. purpurogenum*	Corn, soybeans	H, C, M
Sterigmatocystin (ST)	*A. versicolor, A. nidulans*	Corn, grains, cheese	D, ATA, T
T-2 Toxin	*F. sporotrichioides*	Corn, feeds, hay	D, Nr
12-/13-Epoxytrichothecenes (TCTC) other than T-2 and DON	*F. nivale*	Corn, feeds, hay, peanuts, rice	
Zearalenone (ZE)	*F. graminearum*	Cereals, corn feeds, rice	G, M

Note: ATA, Alimentary toxic aleukia; C, carcinogenic; C(?), carcinogenic effect still questionable; Cv, cardiovascular lesion; D, dermatoxin; G, genitotoxin and estrogenic effects; H, hepatotoxic; Hr, hemorrhagic; M. mutagenic; Nh, nephrotoxin; Nr, neurotoxins; R, respiratory; T, tetratogenic.

[a]The optimal temperatures for the production of mycotoxin are generally between 24 and 28°C, except for T-2 toxin, which is generally produced maximally at 15°C.

Source: Ref. 7.

contaminated feed. The potential or actual extent of human harm from aflatoxin M or other mycotoxin metabolites in edible animal tissues is not known.

Bacterial food poisoning toxins are different from mycotoxins in three important ways. First, bacterial food poisoning toxins with rare exception are proteins, not low-molecular-weight organic molecules. Second, bacterial food poisoning toxins are acute toxins. The onset of symptoms occurs a few hours to a few days after exposure. There is no counterpart among bacterial food poisoning toxins to, for instance, the induction of liver cancer in rats by aflatoxin B_1. Finally, while bacterial food poisoning toxins are important causes of human illness, such toxins are rarely associated with animal feeds. (Feeds can, however, harbor pathogenic bacteria that may infect livestock and be passed on to humans.) No doubt the absence of bacterial toxins in feeds has to do with low moisture content. Bacteria require higher moisture levels for growth and toxin synthesis than do molds [8].

Bacterial toxins that cause food poisoning when ingested are listed in Table 6.

Staphylococcus aureus is the most common toxigenic food-borne bacterium. This organism is estimated to be responsible for over 1 million cases of food poisoning in the United States each year [4]. (*Salmonella* are estimated to produce almost 3 million cases of foodborne illness, but they do so via infection, not intoxication. *Staphylococcus aureus* is also an important infectious agent of humans and other mammals, but staphylococcal food poisoning is caused by the ingestion of food containing preformed enterotoxin.)

Staphylococcal enterotoxin is not a single entity but rather a group of more than seven single-chain proteins with molecular mass ranging from 26,000 to 29,000 D [2]. The enterotoxins are antigenic, and are detected using serological methods. Toxigenic strains of *S. aureus* may produce more than one enterotoxin. Symptoms of intoxication often include severe diarrhea and vomiting in addition to abdominal pain, headache, muscular cramping, and sweating. The mechanism(s) of action of the enterotoxins is not known.

The amount of a staphylococcal enterotoxin necessary to cause illness is not accurately known but is probably no more than a few hundred nanograms [2]. In general, it requires a million or more staphylococci to produce this amount of toxin. Foods most at risk for staphylococcal growth and enterotoxin production are those that are rich, moist, and subject to temperature abuse—in particular custard or cream-filled baked goods, cured meats such as ham, and salads containing egg, seafood, or meat.

The most feared toxigenic food-borne bacterium is *Clostridium botulinum,* the cause of botulism. Fortunately it is relatively uncommon, estimated at less than 300 cases per year in the United States [4].

Clostridium botulinum food poisoning is caused by a series of neurotoxins; the symptoms of botulism include paralysis and death. Seven different botulinal neurotoxins have been identified; each is a protein with a molecular weight of about 150,000 and with subunits linked

Table 6 Bacterial Toxins That Cause Food-Borne Illness

Toxin	Estimated cases of illness per year in United States
Staphylococcal enterotoxins	1,155,000
Bacillus cereus enterotoxins	84,000
Botulinal neurotoxins	270
All food-borne causes	12,581,270

Source: Ref. 4.

by disulfide bonds. Botulinal neurotoxins are the most potent acute poisons known: Toxic potency ranges from 10^7 to 10^8 mouse LD_{50} units per milligram of protein [25].

Clostridium botulinum is an anaerobic spore former, so it survives moderate heat treatments that will kill vegetative cells. By contrast, the neurotoxins are readily destroyed by heating. Early in this century the canning industry developed procedures to ensure the complete destructon of *C. botulinum* spores. This was a necessary prerequisite to producing safe canned foods. In the United States today botulism is almost always associated with mishandling that occurs in food service establishments or the home [4].

Bacillus cereus is a food borne pathogen that produces two distinct enterotoxins. One of the enterotoxins causes diarrhea, while the other induces vomiting. The mechanisms of action of these toxins are not known [17].

As its name implies, *B. cereus* is a common soil microorganism that frequently contaminates cereal products. It is not a major cause of food poisoning in the United States [4] but appears to be more common in Europe, particularly Great Britain.

Some less common forms of food poisoning are associated with bacterial metabolites that are not proteins. Examples include histamine poisoning, nitrite poisoning, and bongkrek food poisoning (caused by a toxic fatty acid produced by *Pseudomonas cocovenenans* and *Flavobacterium farinofermentans*) [26].

There are also several food borne bacterial pathogens that produce toxins in association with gastrointestinal infections. Examples include certain strains of *Escherichia coli* [11] and *Clostridium perfringens* [8]. In these cases toxin that is preformed in food does not appear to be associated with illness. Rather, illness is induced wholly or in part by toxin synthesized in situ following ingestion of the pathogenic microorganism.

13.8 FOOD TOXICOLOGY AND PUBLIC HEALTH: SETTING PRIORITIES

It is estimated that 12.5 million cases of food-associated illness occur each year in the United States, at a cost of about 8.5 billion dollars [4]. The great majority of these are due to infectious agents: pathogenic bacteria like *Salmonella*, viruses such as hepatitis A virus, and parasites such as *Toxoplasma gondii*. Food borne toxins are estimated to account for no more than 1.5 million of these cases. Of these cases, more than 1 million are due to the staphylococcal enterotoxins [4]. Hence the science of food toxicology indicates that while we should be vigilant in protecting foods against contamination by naturally occurring and synthetic toxins, by far the most common food borne toxicologic risks in the United States and other developed countries come from certain bacterial toxins. Moreover, a great deal of scientific data is available to assist in controlling such contamination.

That being said, there still remains the dilemma of public perception, often fueled by activist groups, that the toxicologic risk to our food supply is much greater than these scientifically derived estimates indicate. Concern over perceived risks typically focuses on synthetic contaminants, pesticide residues, and food additives [12].

There have been few attempts to address these concerns in a systematic, science-based manner. One such approach was developed by Ames and colleagues [13]. It involves ranking carcinogenic risks according to a ratio known by the acronym HERP (human exposure rodent potency). A HERP can be calculated for any carcinogen so long as its human exposure level and potency in a rodent bioassay are known. In calculating a HERP, the human exposure level (mg/kg body weight) is divided by the TD_{50} (mg/kg body weight), and the result is then multiplied by 100 [13].

TABLE 7 Prioritizing Commonly Encountered Carcinogen Exposures

Daily human exposure	HERP (%)
PCBs, daily dietary intake	0.0002
EDB[a], daily dietary intake	0.0004
Alar[a], 1 apple (230 g)	0.0002
Tap water, 1 L (chloroform)	0.001
Cooked bacon, 100 g (nitrosamines)	0.003
Worst well water in Silicon Valley (trichloroethylene)	0.004
Peanut butter (one sandwich) (aflatoxin)	0.03
One glass of wine (ethanol)	4.7

Note: EDB refers to ethylene dibromide, a pesticide.
[a]Banned because of health concerns; HERP value based on exposure prior to ban.
Source: Ref. 13 (data).

Some HERP values for common exposures to food substances are given in Table 7. As is evident, synthetic pesticide residues and contaminants represent a far lower risk than many naturally occurring carcinogens. This is largely a function of concentration. Ames and colleagues [1, 13] estimate that 99.99% of the total mass of carcinogens typically ingested is of natural origin.

It is important that these conclusions be interpreted correctly. One should not infer that the level of exposure to naturally occurring dietary carcinogens is necessarily excessive, especially in developed countries. Rather, the level of exposure to synthetic pesticide residues and contaminants is ordinarily so low as to be of no health significance whatever.

These considerations are of great importance in properly applying the results of food toxicology science to reducing public health risk.

BIBLIOGRAPHY

Committee on Food Protection (1973). *Toxicants Occurring Naturally in Foods,* 2nd ed., National Academy of Sciences, Washington, DC.

Council on Agricultural Science and Technology (1989). *Mycotoxins: Economic and Health Risks.* Task Force Report No. 116, Council on Agricultural Science and Technology, Ames, IA.

Council on Agricultural Science and Technology (1994). *Foodborne Pathogens: Risks and Consequences.* Task Force Report No. 122, Council on Agricultural Science and Technology, Ames, IA.

Gold, L. S., T. H. Slone, B. R. Stern, N. B. Manley, and B. N. Ames (1992). Rodent carcinogens: Setting priorities. *Science 258*:261–265.

Hall, R. L. (1992). Toxicological burdens and the shifting burden of toxicology. *Food Technol. 46*(3):109–112.

International Food Biotechnology Council (1990). Biotechnologies and food: Assuring the safety of foods produced by genetic modification. *Regul. Toxicol. Pharmacol. 12*(3), part 2 of 2:S1–S196.

Pariza, M. W. (1992). A new approach to evaluating carcinogenic risk. *Proc. Natl. Acad. Sci. USA 89*:860–861.

REFERENCES

1. Ames, B. N., M. Profet, and L. S. Gold (1990). Dietary pesticides (99.99% all natural). *Proc. Natl. Acad. Sci. USA 87*:7777–7781.
2. Bergdoll, M. S. (1990). Staphylococcal food poisoning, in *Foodborne Diseases* (D. O. Cliver, ed.), Academic Press, San Diego, pp. 85–106.

3. Council on Agricultural Science and Technology (1989). *Mycotoxins: Economic and Health Risks.* Task Force Report No. 116, Council on Agricultural Science and Technology, Ames, IA.
4. Council on Agricultural Science and Technology (1994). *Foodborne Pathogens: Risks and Consequences.* Task Force Report No. 122, Council on Agricultural Science and Technology, Ames, IA.
5. Chernoff, R. (1994). Thirst and fluid requirements. *Nutr. Rev. 52*(8, part 2):s3–s5
6. Chhabra, R. S., J. E. Huff, B. S. Schwetz, and J. Selkirk (1990). An overview of prechronic and chronic toxicology/carcinogenicity experimental study designs and criteria used by the National Toxicology Program. *Environ. Health Perspect. 86*:313–321.
7. Chu, F. S. (1995). Mycotoxin analysis, in *Analyzing Food for Nutrition Labeling and Hazardous Contaminants* (I. J. Jeon and W. G. Ikins, eds.), Marcel Dekker, New York, pp. 283–332.
8. Cliver, D. O. (ed.) (1990). *Foodborne Diseases,* Academic Press, San Diego.
9. Coon, J. M. (1973). Toxicology of natural food chemicals: A perspective, in *Toxicants Occurring Naturally in Foods,* 2nd ed., National Academy of Sciences, Washington DC, pp. 573–591.
10. Deichmann, W. B., D. Henschler, B. Holmstedt, and G. Keil (1986). What is there that is not poison? A study of the Third Defense by Paracelsus. *Arch. Toxicol. 58*:207–213.
11. Doyle, M. P., and D. O. Cliver (1990). *Escherichia coli,* in *Foodborne Diseases* (D. O. Cliver, ed.), Academic Press, San Diego, pp. 209–215.
12. Foster, E. M. (1990). Perennial issues in food safety, in *Foodborne Diseases* (D. O. Cliver, ed.), Academic Press, San Diego, pp. 127–135.
13. Gold, L. S., T. H. Slone, B. R. Stern, N. B. Manley, and B. N. Ames (1992). Rodent carcinogens: Setting priorities. *Science 258*:261–265.
14. Hall, R. L. (1992). Toxicological burdens and the shifting burden of toxicology. *Food Technol. 46*(3):109–112.
15. Holmstedt, B., and G. Liljestrand (1981). *Readings in Pharmacology.* Raven Press, New York, p. 29.
16. International Food Biotechnology Council (1990). Biotechnologies and food: Assuring the safety of foods produced by genetic modification. *Regul. Toxicol. Pharmacol. 12*(3), part 2 of 2:S1–S196.
17. Johnson, E. A. (1990). *Bacillus cereus* food poisoning, in *Foodborne Diseases* (D. O. Cliver, ed.), Academic Press, San Diego, pp. 127–135.
18. Johnson, E. A. (1990). *Clostridium perfringens* food poisoning, in *Foodborne Diseases* (D. O. Cliver, ed.), Academic Press, San Diego, pp. 1229–240.
19. Kessler, D. A., M. R. Taylor, J. H. Maryanski, E. L. Flamm, and L. S. Kahl (1992). The safety of foods developed by biotechnology. *Science 256*:1747–1749, 1932.
20. Miller, E. C., A. B. Swanson, D. H. Phillips, T. L. Fletcher, A. Liem, and J. A. Miller (1983). Structure-activity studies of the carcinogenicities in the mouse and rat of some naturally occurring and synthetic alkenylbenzene derivatives related to safrole and estragole. *Cancer Res. 43*:1124–1134.
21. Pariza, M. W. (1992). Risk assessment. *Crit. Rev. Food Sci. Technol. 31*:205–209.
22. Pariza, M. W. (1992). A new approach to evaluating carcinogenic risk. *Proc. Natl. Acad. Sci USA 89*:860–861.
23. Pariza, M. W., and E. M. Foster (1983). Determining the safety of enzymes used in food processing. *J. Food Prot. 46*:453–468.
24. Smith, R. A. (1994). Poisonous plants, in *Foodborne Disease Handbook* (Y. H. Hui, J. R. Gorham, K. D. Murrell, and D. O. Cliver, eds.), Marcel Dekker, New York, pp. 187–226.
25. Sugiyama, H. (1990). Botulism, in *Foodborne Diseases* (D. O. Cliver, ed.), Academic Press, San Diego, pp. 85–106.
26. Taylor, S. L. (1990). Other microbial intoxications, in *Foodborne Diseases* (D. O. Cliver, ed.), Academic Press, San Diego, pp. 159–170.
27. Wogan, G. N., and M. A. Marletta (1985). Undesirable or potentially undesirable constituents of foods, in *Food Chemistry* (O. R. Fennema, ed.), Marcel Dekker, New York, pp. 694–695.

14

Characteristics of Milk

HAROLD E. SWAISGOOD
North Carolina State University, Raleigh, North Carolina

14.1 INTRODUCTION

For the young of mammals, including humans, milk is the first and, for most, only food ingested for a considerable period of time. With the domestication of animals, it became possible to include milk in the diet of adult humans as well. For much of the world, particularly in the West, milk from cattle (*Bos taurus*) accounts for nearly all the milk processed for human consumption. In the United States the dairy industry is primarily based on cow's milk. Therefore, this discussion focuses on the properties of bovine milk.

The total production of milk and major processed dairy products for selected years between 1970 and 1992 is given in Table 1. Most of the milk produced is processed as fluid or beverage milk (36–46%). However, the consumption pattern of individual products has changed considerably during the past 20 years. In 1970 plain whole milk represented 78% and plain low-fat milk 12% of the beverage milk sold, but in 1992 the consumption pattern was 37% plain whole milk, 46% plain low-fat milk, and 12% plain skim milk. Furthermore, although the amount of beverage milk consumed has remained relatively constant, its percentage of the total milk consumed in all forms has fallen. Other noteworthy trends are the large increases in cheese and yogurt consumption.

During the past 20 years a new fluid milk product known as ultra-high-temperature (UHT) processed and aseptically packaged milk has appeared in the U.S. market. This product, which is commercially sterile and can be distributed and merchandized without refrigeration, has not yet gained acceptance in the United States; however, it represents a large share of the fluid milk market in many other countries. The biological origin and chemical nature of milk are considered next.

14.2 MILK BIOSYNTHESIS

In the early years of dairy chemistry, the product of the mammary gland was more thoroughly investigated and characterized than the gland itself because of the economic importance of milk products. More recently, however, with advances in molecular and cellular biology, the mammary gland has been the subject of intensive research. Several excellent reviews on milk biosynthesis have appeared and the reader is referred to these for more complete information [27, 30, 34, 42, 45]. Some people have argued that animal agriculture will disappear as the human population expands and competition for available grain intensifies. The fallacy of this argument, however, lies in the presence of a rumen in the cow's digestive system. The rumen allows the cow to synthesize nutrients from crude cellulosic and fibrous plant materials and from simple forms of nitrogen, such as urea. Consequently, these animals should be viewed as a unique means of producing high-quality foods from otherwise unusable feedstocks, rather than as inefficient and unnecessary.

The bovine mammary gland, resulting from years of genetic selection, is an amazingly productive organ for biosynthesis. In 1993 the average cow in the United States produced 7055 kg (15,554 lb) of milk in a 305-day lactation [2]. Some high producing animals yielded as much as 22,700 kg of milk (over 50,000 lb). The cellular machinery of this gland is shown at various levels of magnification in Figure 1. Milk originating in the secretory tissue collects in ducts that increase in size as the teat region is approached. The smallest complete milk factory, which includes a storage area, is the alveolus. It is a roughly spherical micro-organ consisting of a central storage volume (the lumen) surrounded by a single layer of secretory epithelial cells, which is connected to the duct system. These cells are directionally oriented such that the apical end with its unique membrane is positioned next to the lumen, and the basal end is separated from blood and lymph by a basement membrane. Consequently, a directional flow of metabolites

TABLE 1 U.S. Milk Production and Amounts of Dairy Products for Selected Years Between 1970 and 1992

	Value-added products							
Year	Milk production (kg in billions)	Beverage milk[a] (kg in billions)	Cheese[b] (kg in billions)	Yogurt (kg in billions)	Dried milks (kg in billions)	Canned milk (kg in billions)	Butter (kg in billions)	Frozen desserts (L in billions)
1970	53.1	24.0	1.5	0.08	0.69	0.58	0.52	4.0
1975	52.4	24.2	1.7	0.20	0.48	0.42	0.45	4.3
1980	58.4	24.1	2.3	0.27	0.57	0.33	0.52	4.3
1985	65.0	24.5	2.7	0.44	0.69	0.29	0.57	4.6
1990	67.4	24.9	3.1	0.47	0.48	0.27	0.59	4.5
1991	67.5	25.1	3.1	0.48	0.45	0.25	0.61	4.6
1992	68.9	25.1	3.3	0.49	0.47	0.27	0.62	4.6

[a]Includes whole, lowfat, skim, and flavored milks.
[b]Includes all types of cheeses.
Source: Adapted from Ref. 66.

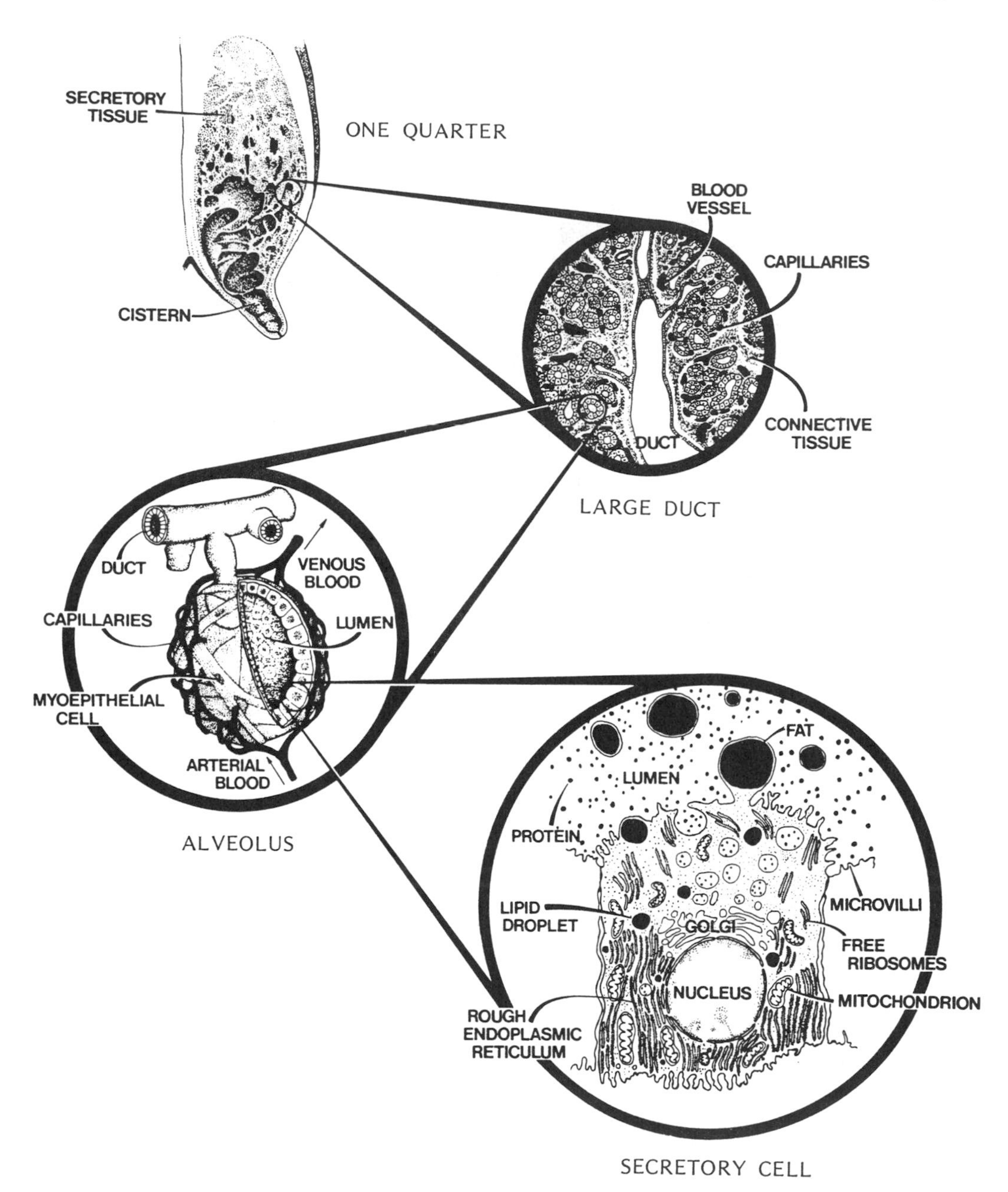

FIGURE 1 Bovine mammary gland at increasing levels of magnification.

occurs through the cell, with the building blocks of milk entering from the blood through the basolateral membrane. These basic components are synthesized into milk components on the production lines of the endoplasmic reticulum, which is supplied with energy by oxidative metabolism in the mitochondria. The components are then packaged in secretory vesicles by the Golgi apparatus or as lipid droplets in the cytoplasm. Finally, the vesicles and lipid droplets pass through the apical plasma membrane and are stored in the lumen. The layer of secretory epithelial cells surrounding the lumen of the alveolus is in turn surrounded by a layer of myoepithelial cells and blood capillaries. When oxytocin, a pituitary hormone circulating in the blood, binds to the myoepithelial cells, the alveolus contracts, expelling the milk stored in the lumen into the duct system.

It is currently thought that the intracellular membranes (endomembranes) also exhibit directional flow to the apical plasma membrane, with concurrent transformation of membranes from endoplasmic reticulum to Golgi, to secretory vesicles, to apical plasma membrane within the epithelial secretory cell. The endoplasmic reticulum, the endomembrane production line where active synthesis occurs, appears to be like a cellular plumbing system, with the inside of these tubes, known as the cisternae, emptying into the Golgi apparatus. The Golgi apparatus transforms into Golgi vesicles, the packages that carry the aqueous phase milk components to the apical plasma membrane.

Ribosomes, the machinery for protein synthesis, exist both free in the cytoplasm and bound to the endoplasmic reticulum. Near the basolateral membrane, the endoplasmic reticulum is covered with ribosomes, making this membrane appear "rough." However, moving in the direction of the apical end, amino acids are depleted and synthesis slows, allowing ribosomes to dissociate. Hence, this membrane becomes smooth as it is transformed into Golgi membrane. Synthesis is completed in the lumen of the Golgi; for example, proteins are glycosylated and phosphorylated and lactose is synthesized. It is here and in the Golgi vesicles that casein micelles first appear. Furthermore, this membrane is impermeable to lactose, so that the major secretory products of the cell are now segregated from all other cellular constituents. Secretion of the products of synthesis is completed when the Golgi vesicles merge with the apical plasma membrane, fuse to become part of that membrane, and empty the contents of the package into the alveolar lumen for storage. Note that in the membrane transformation, the inside or lumen side of the vesicle membrane becomes the outside of the cell plasma membrane. The caseins, β-lactoglobulin, and α-lactalbumin are synthesized in the mammary epithelial cells. Serum albumin and immunoglobulins are not synthesized in these cells, but specific receptors for these proteins appear on the basolateral membrane. Hence, these proteins are transported from blood to the alveolar lumen by internalization of the protein–receptor complex and passage through the cell in membrane vesicles.

Fat, which is also synthesized by the endoplasmic reticulum, is directed to the cytoplasmic side of the membrane where it collects as lipid droplets. These droplets move to the apical plasma membrane, where they are expelled into the alveolar lumen by pinocytosis, thus acquiring on their surface a coat of plasma membrane. The presence of this membrane has important consequences for the processing characteristics of milk products. Apical plasma membrane, lost in this process, is continually replenished, at least in part, by the fusion of vesicles as previously described. Electrolytes in milk appear to be transported from blood by pumps or channels in the basolateral membrane. The proportion of sodium and potassium in milk resembles the composition of the cytoplasm. However, because milk is isosmotic with blood plasma and because a large portion of that osmolality is contributed by lactose, the concentration of electrolytes is lower in milk than in the cytoplasm or in plasma.

It should be appreciated that many, if not all, of the characteristics of milk and its constituents are a consequence of the mechanism of synthesis and secretion.

14.3 CHEMICAL COMPOSITION

The composition of milk reflects the fact that it is the sole source of food for the very young mammal. Hence, it is composed of a complex mixture of lipids, proteins, carbohydrates, vitamins, and minerals. In addition, milk contains minor components derived from the cellular synthetic "machinery." The average composition of milk with respect to major classes of compounds and the range of average values for milks of Western cattle are given in Table 2. The greatest variability in composition is exhibited by the lipid fraction. In the past, breeders have selected cows for high fat production because of the economic value of this constituent. Currently, more importance is being placed on protein content, and breeders are beginning to select for higher protein/fat ratios and higher protein yield.

Because of the major contribution of lactose and milk salts to osmolality, and the required matching of milk's osmotic pressure with that of blood, very little variability is observed in the sum of these constituents. It should be noted that ash does not truly represent milk salts because organic salts are destroyed by ashing; for example, various salts of citrate are principal components of the milk salt system. The pH of freshly drawn milk is 6.6–6.8, which is slightly lower than that of blood.

Most of the constituents of milk are not present as individual molecules in solution. Instead, they exist in large complex associated structures (Table 3). This is especially true for the caseins, which are present as large spherical micelles, and the lipids, which form even larger spherical globules. Because of the spherical shape and the reduction in the effective number of molecules (kinetic units), both the viscosity and the osmotic pressure are much less than would be observed if these structures were not formed.

14.3.1. Milk Proteins

Milk contains 30–36 g/L of total protein and it rates very high in nutritive quality. There are six major gene products of the mammary gland: α_{s1}-caseins, α_{s2}-caseins, β-caseins, κ-caseins, β-lactoglobulins, and α-lactalbumins (Table 4). Each of these proteins exhibits genetic polymorphism because they are products of codominant, allelic, autosomal genes. Milk proteins are classified as either caseins or whey proteins. All the caseins exist with calcium phosphate in a unique, highly hydrated spherical complex known as the casein micelle. Such complexes vary in size from 30 to 300 nm in diameter, with a small percentage approaching 600 nm.

Milk proteins can be readily separated into casein and whey protein fractions. Casein comprises 80% of the bovine milk proteins; consequently, curd formed by agglomeration of casein micelles during cheese manufacture retains most of the total milk protein. The other proteins pass into the cheese whey, hence the designation whey protein. Historically, separation of caseins from other proteins by means of precipitation at their isoelectric point (around pH 4.6), yielding so-called "acid casein," or by enzyme induced agglomeration of casein micelles (renneting), as in cheesemaking, has formed the basis for many milk products such as cheese, whey protein products, and other protein ingredients. In the past whey was often discarded;

TABLE 2 Composition of Bovine Milk from Western Cattle

Component	Average percentage	Range for Western breeds[a] (average percentages)
Water	86.6	85.4–87.7
Fat	4.1	3.4–5.1
Protein	3.6	3.3–3.9
Lactose	5.0	4.9–5.0
Ash	0.7	0.68–0.74

[a]Western breeds include Guernsey, Jersey, Ayshire, Brown Swiss, Shorthorn, and Holstein.
Source: Adapted from Ref. 9.

TABLE 3 Numbers and Sizes of Major Milk Constituents

Constituent	Size (diameter, nm)	Number/ml
Lactose	0.5	10^{19}
Whey protein	4–6	10^{17}
Casein micelles	50–300	10^{14}
Milk fat globule	2000–6000	10^{10}

however, it is now economically feasible to concentrate or isolate the whey proteins, which exhibit excellent functionality and nutritional properties.

In addition to β-lactoglobulin and α-lactalbumin, which are gene products of the mammary gland, whey also contains serum albumin and immunoglobulins that are derived from blood. Milk also contains several "protein" components that are actually large polypeptides. These result from posttranslational proteolysis of milk proteins by the indigenous milk proteinase plasmin, which is derived from blood. Thus, the γ-caseins that fractionate with the caseins, and most of the proteose-peptones that are present in whey, are derived by limited proteolysis of β-casein.

The amino acid compositions and covalent molecular weights of the major milk proteins are very accurately known from amino acid analyses and from determination of primary structure by chemical sequencing and by inference from gene sequencing (Table 5). A key distinction of the caseins is their large content of phosphoseryl residues and their moderately large proline content. Although the caseins are rather low in half-cystine, the whey proteins contain considerably more of this residue.

For a more complete discussion of the characteristics of individual milk proteins, the reader is referred to several excellent reviews [7, 8, 21, 32, 33, 35–37, 53, 60–62, 64, 68, 69].

TABLE 4 Concentrations of the Major Proteins in Milk

Protein	Concentration (g/L)	Approximate percentage of total protein	
Caseins	24–28		80
α_s-Caseins	15–19	42	
α_{s1}	12–15	34	
α_{s2}	3–4	8	
β-Caseins	9–11	25	
κ-Caseins	3–4	9	
γ-Caseins	1–2	4	
Whey proteins	5–7		20
β-Lactoglobulins	2–4	9	
α-Lactalbumin	1–1.5	4	
Proteose–peptones	0.6–1.8	4	
Blood proteins			
Serum albumin	0.1–0.4	1	
Immunoglobulins	0.6–1.0	2	
Total		100	100

Table 5 Amino Acid Composition of the Major Proteins in the Milk of Western Cattle

Acid	Number of amino acids in								
	α_{s1}-Casein B-8P	α_{s2}-Casein A-11P	κ-Casein B-1P	β-Casein A^2-5P	β-Casein A^2-1P(f29-209) (formerly γ_1)	β-Casein A^2-1P(f106-209) (formerly γ_2)	β-Casein A(F108-209) (formerly γ_3)	β-Lacto-globulin A	α-Lact-albumin B
Nonpolar									
Pro	17	10	20	35	34	21	21	8	2
Ala	9	8	15	5	5	2	2	14	3
Val	11	14	11	19	17	10	10	10	6
Met	5	4	2	6	6	4	4	4	1
Ile	11	11	13	10	7	3	3	10	8
Leu	17	13	8	22	19	14	14	22	13
Phe	8	6	4	9	9	5	5	4	4
Trp	2	2	1	1	1	1	1	2	4
Polar, neutral									
Asn	8	14	8	5	3	1	1	5	12
Thr	5	15	14	9	8	4	4	8	7

Ser	8	6	12	11	10	7	7	7	7
Gln	14	16	14	20	20	10	10	9	5
Gly	9	2	2	5	4	2	2	3	6
Tyr	10	12	9	4	4	3	3	4	4
Cystine	0	1	1	0	0	0	0	2	4
Cysteine	0	0	0	0	0	0	0	1	0
Pyr Glu	0	0	1	0	0	0	0	0	0
Polar, acidic									
Asp	7	4	3	4	4	2	2	11	9
Ser P	8	11	1	5	1	0	0	0	0
Glu	25	24	12	9	2	5	5	16	8
Polar, basic									
Lys	14	24	9	11	10	4	3	15	12
His	5	3	3	5	5	4	3	2	3
Arg	6	6	5	4	2	2	2	3	1
Total residue	199	207	169	209	181	104	102	162	123
Molecular weight	23,623	25,238	19,006	23,988	20,523	11,824	11,559	18,363	14,174

Source: Adapted from Ref. 61 and 62.

TABLE 6 Lipid Composition of Bovine Milk

Lipid	Weight percent	g/L[a]
Triacylglycerols (triglycerides)	95.80	30.7
1,2-Diacylglycerols (diglycerides)	2.25	0.72
Monoacylglycerols (monoglycerides)	0.08	0.03
Free fatty acids	0.28	0.09
Phospholipids	1.11	0.36
Cholesterol	0.46	0.15
Cholesterol ester	0.02	0.006
Hydrocarbons	Trace	Trace

[a]Based on the usual butterfat percentage of commercial pasteurized whole milk, 3.2%.
Source: Adapted from Ref. 23.

14.3.2 Milk Lipids

Bovine milk contains the most complex lipids known. For detailed characteristics of the various lipids and a discussion of their biosynthesis, the reader should consult one of the reviews [3, 6, 22–24, 45]. Triacylglycerols (triglycerides) represent by far the greatest proportion of the lipids, comprising 96–98% of the total (Table 6). In milk, the triacylglycerols are present as globules 2–6 μm in diameter surrounded by membrane material derived from the apical cell membrane. The concentrations listed in Table 6 are representative of fresh milk. Some lipolysis occurs during storage, giving higher concentrations of free fatty acids and mono- and diacylglycerols.

Triacylglycerols containing three different fatty acids have a chiral carbon at the *sn*-2 position of the glycerol skeleton. Over 400 different fatty acids have been identified in bovine lipids. Consequently, if positional isomers are considered, 400^3 or 64 million triacylglycerol species are theoretically possible. However, only 13 fatty acids are present at concentrations exceeding 1% (w/w) (Table 7), corresponding to a theoretical maximum number of isomers of $13^3 = 2197$. If only compositional species, rather than positional isomers, are considered, then the possible number of triacylglycerols with different fatty acid compositions is $n!/3!(n-1)! + n(n-1) + n$, where n is the number of different fatty acids. With 13 different fatty acids, this results in 455 triacylglycerols with different fatty acid compositions. Recently [16], 223 triacylglycerols with different compositions accounting for 80% of the total triacylglycerols were isolated by reversed-phase liquid chromatography and identified. The total mole percent of each

TABLE 7 Major Fatty Acid Constituents of Bovine Milk Fat

Fatty acid	Weight percent	Fatty acid	Weight percent
4:0	3.8	15:0	1.1
6:0	2.4	16:0	43.7
8:0	1.4	18:0	11.3
10:0	3.5	14:1	1.6
12:0	4.6	16:1	2.6
14:0	12.8	18:1	11.3
		18:2	1.5

Source: Adapted from Refs. 23 and 65.

class of triacylglycerols and eight of the most abundant compositional species (concentrations exceeding 2 mol%) are listed in Table 8. Saturated and monounsaturated triacylglycerols comprise the bulk of the lipids (65 mol%), and the eight most prevalent triacylglycerols represent nearly one-fourth of the total.

A nonrandom distribution of fatty acids in the triacylglycerols is observed. The nonrandomness results from the specificities of the acyltransferase enzymes involved in synthesis. Thus, long-chain acylcoenzyme A is the preferred substrate for the acyltransferase specific for *sn* positions 1 and 2. Conversely, esterification by short-chain fatty acids is much faster with the enzyme specific for *sn*-3. As a result, triacylglycerols with one short-chain and two medium- or long-chain fatty acids are preferentially synthesized, whereas simple or mixed saturated long-chain triacylglycerols are present at low concentrations. These observations are clearly supported by the data in Table 9. For example, 85% of butyrate and 58% of caproate are at position *sn*-3, while only 13% of palmitate and 16% of stearate are located in this position.

In addition to exhibiting the greatest percent variability, milk lipids and fatty acids in particular are the most subject of all milk constituents to alteration by environmental factors, such as diet. Because synthesis of short-chain fatty acids and biohydrogenation occurs in the rumen, the ratio of saturated to unsaturated fatty acids does not vary greatly. However, if unsaturated fatty acids are fed in a protected form by encapsulation in denatured protein, they will pass through the rumen unaltered, resulting in greater incorporation of unsaturated acids in the triacylglycerols. Such alteration of triacylglycerol composition is currently attractive because of the reported health benefits of unsaturated fatty acids; however, the possible effects on the oxidative stability of the lipid and the resulting flavor instability have not been fully assessed.

The phospholipid and cholesterol fraction of milk lipids exist in the membrane material

TABLE 8 Composition of Major Triacylglycerols in Milk

Type	Carbon number	Molecular species[a]	Mole percent
Saturated triacylglycerols (total)[b]			32.4
	34	4:0 14:0 16:0	3.1
	36	4:0 16:0 16:0	3.2
	38	4:0 16:0 18:0	2.5
Monounsaturated triacylglycerols (total)[b]			32.6
	38	4:0 16:0 18:1	4.2
	40	6:0 16:0 18:1	2.0
	48	14:0 16:0 18:1	2.8
	50	16:0 16:0 18:1	2.3
	52	16:0 18:0 18:1	2.2
Diunsaturated triacylglycerols (class 011)(total)[c]			10.6
Diunsaturated triacylglycerols (class 002)(total)[c]			2.5
Polyunsaturated triacylglycerols (all classes)(total)[c]			5.2

[a]Carbon chain length: number of double bonds present. Position of acyl chains in the glycerol skeleton was not determined.

[b]This row gives the total mole percent of the specified triacylglycerol type. Sublistings under each type are specific triacylglycerols present in concentrations exceeding 2.0 mol%.

[c]Class 011 indicates two monounsaturated fatty acids in the triacylglycerol, while class 002 indicates one diunsaturated fatty acid present.

Source: Adapted from Ref. 16. Note that 80% of the triacylglycerols were identified in this study; for example, triacylglycerols containing odd-numbered fatty acids were not quantified.

TABLE 9 Positional Distribution of the Individual Fatty Acids in Bovine Milk Triacylglycerols

	sn Position (mol%)		
Fatty acid	1[a]	2[a]	3[a]
4:0	5.0 (10)	2.9 (5)	43.3 (85)
6:0	3.0 (16)	4.8 (26)	10.8 (58)
8:0	0.9 (17)	2.3 (42)	2.2 (41)
10:0	2.5 (21)	6.1 (50)	3.6 (29)
12:0	3.1 (25)	6.0 (47)	3.5 (28)
14:0	10.5 (27)	20.4 (54)	7.1 (19)
16:0	35.9 (45)	32.8 (42)	10.1 (13)
16:1	(48)	(36)	(16)
18:0	14.5 (58)	6.4 (26)	4.0 (16)
18:1	20.6 (42)	13.7 (28)	14.9 (30)
18:2	1.2 (28)	2.5 (60)	0.5 (12)

[a]Number in parentheses represents the percent of a particular fatty acid occupying the designated position.
Source: Refs. 20, 23, 45, and 47.

introduced into the milk during pinocytosis of lipid droplets through the plasma membrane into the lumen of the aveolus. Lipids of the milk fat globule membrane (Table 10), which are similar to those of cellular plasma membrane, consist of more than 20% phospholipids, the major types being phosphatidylethanolamine (22.3%), phosphatidylcholine (33.6%), and sphingomyelin (35.3%).

Due to a correlation between cholesterol and atherosclerosis, the content of this sterol in foods has received much attention. Milk contains relatively little cholesterol; for example, an 8-oz glass of milk (227 g) contains 27 mg cholesterol. To put this in perspective, one large egg contains 275 mg cholesterol, 10 small shrimp contain 125 mg, and 100 g of freshwater fish contain 70 mg. Because cholesterol occurs in the fat globule membrane, its concentration in a dairy food is related to fat content.

TABLE 10 Composition of the Lipid Portion of the Milk Fat Globule Membrane

Component	Percentage of membrane lipids
Carotenoids	0.45
Squalene	0.61
Cholesterol esters	0.79
Triacylglycerols[a]	53.41
Free fatty acids	6.30
Cholesterol	5.20
Diacylglycerols	8.14
Monoacylglycerols	4.70
Phospholipids	20.40

[a]Contains a large portion of high-melting glycerides (melting point, 52–53°C).
Source: Ref. 63.

14.3.3 Milk Salts and Sugar

The salts in milk consist principally of chlorides, phosphates, citrates, and bicarbonates of sodium, potassium, calcium and magnesium [19, 48]. Thus, both inorganic and organic salts are present in milk, and these should not be confused with quantities given for "ash", which represents oxides of the minerals resulting from combustion. Milk salt complexes have sizes ranging from those that are ultrafilterable, including free ions and ion complexes, to those that are of colloidal size. Some of the latter type participate in the structure of casein micelles (Table 11).

The ultrafilterable species, which are in equilibrium with colloidal forms, may be obtained in the permeate by dialysis or ultrafiltration. Analysis of the latter permeate provided the basis for formulation of a "simulated milk salt ultrafiltrate" [19], which is commonly used to simulate the milk environment for research on milk components. The multivalent ions Ca^{2+} and Mg^{2+} exist principally as complexes, including large amounts of Ca·citrate and Mg·citrate and lesser quantities of CaH_2PO_4 [19]. Thus, only 20–30% of the total Ca and Mg present as ultrafilterable species exists as free divalent cations. For example, milk contains only 2–3 m*M* Ca^{2+}. Likewise, most of the citrate is present as complex ions, whereas most of the phosphate exists as $H_2PO_4^-$ and HPO_4^{2-}. Univalent ions such as Na^+, K^+, and Cl^- are present almost entirely as free ions.

Colloidal species of the milk salts bind to milk proteins both as individual ions and as complex structures in casein micelles. These interactions affect the stability and functionality of milk proteins; hence, milk salts play an important role in the properties of dairy foods.

Because of the biosynthetic requirement of isosmolality with blood, one would expect a reciprocal relationship between milk salts and lactose. Such an inverse relationship has been documented between sodium and lactose contents and between sodium and potassium contents [19, 48]. Consequently, milk has an essentially constant freezing point (–0.53 to –0.57°C), and this colligative property is used to detect illegal dilution with water.

Lactose (4-*O*-β-D-galactopyranosyl-D-glucopyranose) is the predominant carbohydrate in bovine milk, accounting for 50% of the solids in skim milk. Its synthesis is associated with that of a major whey protein, α-lactalbumin, which acts as a modifier protein for UDP galactosyltranferase. Thus, α-lactalbumin changes the specificity of this enzyme such that the galactosyl group is transferred to glucose rather than to glycoproteins. Lactose occurs in both α and β forms,

TABLE 11 Concentration and Size of the Principal Salts and Lactose in Milk

Component	Total in milk (mg/100 ml)		Percentage ultrafilterable	Percentage colloidal
	Mean	Range of values		
Total calcium	121	114–124	33	67
Calcium ion	8	6–16	100	0
Magnesium	12.5	11.7–13.4	64	36
Citrate	181	171–198	94	6
Inorganic Phosphorus	65	53–72	55	45
Sodium	60	48–79	96	4
Potassium	144	116–176	94	6
Chloride	108	92–131	100	0
Lactose	4800	4600–4900	100	0

Source: Refs. 11, 19, and 67.

with an equilibrium ratio of $\beta/\alpha = 1.68$ at 20°C [43]. The β form is far more soluble than the α form, and the rate of mutarotation is rapid at room temperature but very slow at 0°C. The α-hydrate crystal form, which crystallizes under ordinary conditions, occurs in a number of shapes, but the most familiar is the "tomahawk" shape, which imparts a "sandy" mouth feel to dairy products, as in "sandy" ice cream. Lactose, with a sweetness about one-fifth that of sucrose, contributes to the characteristic flavor of milk.

14.3.4. Enzymes

Although present in small amounts, enzymes can have important influences on the stability of dairy foods. Their effects will become even more important as the industry moves toward high-temperature or ultra-high-temperature processing and long storage periods sometimes at room temperature. These conditions favor survival of enzyme activity or reactivation, and the longer period of storage allows more time for enzyme-catalyzed reactions to occur. The discussion here is limited to enzymes indigenous to milk, but it should be recognized that enzymes are also introduced into milk as a result of microbial growth. The more important enzymes that have been identified are listed in Table 12. For a more complete discussion the reader should consult several reviews [13, 25, 55].

Many of the enzymes are associated primarily with the membranes in milk, such as the fat globule membrane or skim milk membrane vesicles, or with casein micelles. For example, xanthine oxidase, sulfhydryl oxidase and γ-glutamyltransferase are primarily associated with the membrane fractions, plasmin and lipase are associated with casein micelles, and catalase and superoxide dismutase occur primarily in the milk serum. However, the distribution of enzymes between these constituents is affected by processing and storage conditions, and this can have substantial effects on their activities. Thus, cold-induced lipolysis of milk may be caused by transfer of lipase from micelles to fat globules, and cold storage may cause some dissociation of proteinase from casein micelles. These enzymes can have significant effects on flavor and protein stability in dairy foods. The oxidoreductases also can have effects on flavor stability, particularly in the lipid fraction, because of their influence on the oxidative state.

Milk also contains many vitamins, which are enzyme cofactors or precursors of cofactors, and these are discussed in a later section.

14.4 STRUCTURAL ORGANIZATION OF MILK COMPONENTS

14.4.1 Structure of Milk Proteins

The primary structures of all of the major milk proteins have been determined; thus our scientific knowledge of milk holds a unique position among foods [7, 60–62, 68, 69]. Moreover, the three-dimensional structures of the major whey proteins are also known. Although the relationship between protein structure and food functionality is not well understood, significant progress in this area is occurring. This relationship must be born in mind by the food scientist while examining the structures of food proteins.

14.4.1.1 The Caseins

Caseins exhibit unique interactions with calcium ions and calcium salts, which may be the most important feature of their physiological and nutritional role. Because of their unique primary and tertiary structures, these proteins undergo posttranslational phosphorylation. Such modification results in the formation of anionic clusters in the "calcium-sensitive" caseins, while a single

TABLE 12 Enzymes Indigenous to Bovine Milk, Partial List

Oxidoreductases[a]	
Xanthine oxidase (xanthine:O_2 oxidoreductase)	Catalase (H_2O_2:H_2O_2 oxidoreductase)
Sulfhydryl oxidase (protein, peptide-SH:O_2 oxidoreductase)	Diaphorase (NADH:lipoamide oxidoreductase)
Lactoperoxidase (donor:H_2O_2 oxidoreductase)	Cytochrome *c* reductase (NADH: cytochrome *c* oxidoreductase)
Superoxide dismutase (O_2^-:O_2^- oxidoreductase)	Lactate dehydrogenase (L-lactate:NAD oxidoreductase)
Glutathione peroxidase (GSH:H_2O_2 oxidoreductase)	
Transferases[a]	
UDP-Galactosyltransferase (UDP galactose:D-glucose-1-galactosyltransferase)	
Ribonuclease (polyribonucleotide 2-oligonucleotide transferase)	
γ-Glutamyltransferase	
Hydrolases[a]	
Proteinases (plasmin, thrombin, aminopeptidase) (peptidyl peptide hydrolase)	β-Esterase (carboxylic ester hydrolase)
Lipase (glycerol ester hydrolase)	Cholinesterase (acylcholine acylhydrolase)
Lysozyme (mucopeptide *N*-acetylneuraminyl hydrolase)	α-Amylase (α-1,4-glucan-4-glucanohydrolase)
Alkaline phosphatase (orthophosphoric monoester phospholydrolase)	β-Amylase (β-1,4-glucan maltohydrolase)
Phosphoprotein phosphatase or acid phosphatase (phosphoprotein phosphohydrolase)	5′-Nucleotidase (5′-ribonucleotide phosphohydrolase)
ATPase (ATP phosphohydrolase)	*N*-Acetyl-b-D-glucosaminidase
Lyases[a]	
Aldolase (fructose-1,6-diphosphate D-glyceraldehyde-3-phosphate lyase)	Carbonic anhydrase (carbonate hydrolyase)

[a]The systematic name is given in parentheses.
Source: Compiled in part from Refs. 13, 25, and 55.

residue is phosphorylated in κ-casein (calcium-insensitive). The calcium-sensitive caseins, α_{s1}-, α_{s2}-, and β-caseins, are so designated because of their extremely low solubilities in the presence of Ca^{2+}. Examples of the primary structure of these anionic clusters in calcium-sensitive caseins are given in Figure 2. Binding of Ca^{2+} in the clusters results in a discharging and dehydration of this region, altering the balance of hydrophobic interactions and electrostatic repulsive forces. It is believed that these sequences reside in short exons, allowing their duplication and rapid evolution among the calcium-sensitive caseins [62]. Most likely, the calcium-sensitive caseins evolved from a common ancestral gene, whereas κ-casein arose from a different gene.

Caseins have not been and apparently cannot be crystallized; therefore, their three-dimensional structures are not known. Nevertheless, their primary structures reveal another unique characteristic of these proteins, namely, the distribution of polar and hydrophobic residues. Clustering of polar residues and hydrophobic residues in separate regions of the primary structure suggests the formation of distinct polar and hydrophobic domains and a resulting amphipathic structure [60–62]. In solutions of the isolated individual protein, both β-caseins and κ-caseins self-associate to form large spherical complexes much like a detergent micelle.

The polar domains of the calcium-sensitive caseins are dominated by the phosphoseryl residues and thus carry a large net negative charge at the pH of milk (Fig. 2). For example, the 40-residue polar domain of α_{s1}-casein encompassing residues 41–80 exhibits a net charge of –20.6 at pH 6.6, which represents almost the total net charge of the entire molecule (Table 13). The N-terminal 40 residues and the C-terminal 100 residues are rather hydrophobic. Although the net charge for the entire molecule of α_{s2}-casein is less than that of α_{s1}-casein, it contains several regions of very high net charge density (Table 13 and Fig. 2). Thus, α_{s2}-caseins are more hydrophilic than other caseins and contain three regions of anionic clusters. Hydrophobic regions are confined to residues 90–120 and 160–207. Furthermore, although the C-terminal region is hydrophobic, it also exhibits a large net positive charge, +9.5, while the N-terminal 68-residue region has a net negative charge, –21. Hence, the properties of this protein are very sensitive to ionic strength. β-Caseins, on the other hand, are the most hydrophobic of the caseins. The protein has a small 21-residue N-terminal polar domain with a net charge of –11.5 and a single anionic cluster (Table 13 and Fig. 2). The remainder of the molecule is rather hydrophobic. Hence, this molecule is very amphipathic with a polar domain comprising one-tenth of the total residues in the chain but carrying one-third of the total charge at pH 6.6 (the net charge of this polar domain is nearly equal to that for the entire molecule). The large hydrophobic domain of this molecule makes the overall properties of β-caseins very sensitive to temperature.

The calcium-insensitive κ-caseins are also very amphipathic with distinct hydrophobic and polar domains; however, the polar domain of κ-caseins does not contain an anionic cluster. κ-Casein is essential to the formation of casein micelle structure. Hence, the physiological functional restraints placed on the structure of this protein are (a) the requirement for interaction with calcium-sensitive caseins to form micelles, thus stabilizing calcium-sensitive proteins in the presence of Ca^{2+}, (b) an amphipathic structure with an inert polar domain that is not precipitated by Ca^{2+}, and (c) a recognition sequence that is specific for limited proteolysis by chymosin,

PROTEIN	CHARGE
α_{S1} — CASEIN	
Glu • Ala • Glu • SerP • Ile • SerP • SerP • SerP • Glu • Glu	- 12
α_{s2} — CASEIN	
Glu • His • Val • SerP • SerP • SerP • Glu • Glu	- 9
SerP • SerP • SerP • Glu • Glu • SerP • Ala • Glu	- 11
β — CASEIN	
Glu • SerP • Leu • SerP • SerP • SerP • Glu • Glu	- 11

FIGURE 2 Anionic clusters of calcium-sensitive caseins.

Table 13 Charge Characteristics of the Milk Proteins

Protein	Charge at pH 6.6	Isoionic pH
α_{s1}-Casein B-8P	−21.9	4.94
α_{s2}-Casein A-11P	−13.8	5.37
β-Casein A^2-5P	−13.3	5.14
κ-Casein B-1P	−2.0	5.90
β-Lactoglobulin B	−10.0	5.34
α-Lactalbumin B	−2.6	4.8

Note: The nomenclature of the caseins indicates the genetic variant and the number of phosphoseryl residues. For example, α_{s1}-casein B-8P indicates the B genetic variant containing eight phosphoseryl residues.
Source: Refs. 61 and 62.

allowing release of the polar domain and coagulation of micelles. Instead of the anionic phosphoseryl clusters characteristic of calcium-sensitive caseins, κ-casein's polar domain contains seryl and threonyl residues that often are glycosylated. This posttranslational modification results in attachment of tri- or tetrasaccharide moieties that include anionic *N*-acetylneuraminic acid (AcNeu) residues (Fig. 3). There are no cationic residues in the C-terminal 53-residue polar domain, and the nonglycosylated domain has a net charge of −11 at pH 6.6; however, κ-casein A-1P(3AcNeu), representing the A variant with one phosphoseryl residue and three AcNeu residues, has a net charge of −14. Because κ-casein does not contain phosphoseryl anionic clusters, it does not bind Ca^{2+} like the calcium-sensitive caseins. The polar domain does, however, have many polar amino acids, a large negative charge, and eight evenly spaced prolyl residues creating a highly dehydrated, open and flexible structure. This polar domain is attached to a large, very hydrophobic domain with many sites that are potentially able to interact with other caseins. Moreover, the Phe-Met peptide bond at position 105–106, which is the N-terminus of the polar domain, is very susceptible to proteolytic cleavage by chymosin, probably because prolyl residues on either side cause these residues to protrude from the surface and increase chain flexibility.

Although the three-dimensional structures of the caseins have not been directly determined, global aspects of their structures can be inferred from their physico-chemical characteristics and their primary structures. Spectral properties, including circular dichroisim (CD) and Raman spectra, as well as predictions based on the primary structures, indicate that all of the caseins possess significant quantities of secondary structure [62]. In fact, supersecondary struc-

$$\text{AcNeu} \xrightarrow{\alpha\text{-}2,3} \text{Gal} \xrightarrow{\beta\text{-}1,3} \text{GalNAc} \xrightarrow{\beta\text{-}1} \text{Thr}$$

$$\text{AcNeu} \xrightarrow{\alpha\text{-}2,3} \text{Gal} \xrightarrow{\beta\text{-}1,3} \text{GalNAc} \xrightarrow{\beta\text{-}1} \text{Thr}$$

(GalNAc is also linked α-2,6 to a second AcNeu.)

Figure 3 Trisaccharide and tetrasaccharide structures typical of the carbohydrate moieties attached to κ-casein.

tures have also been suggested such as β α β motifs in the hydrophobic domain of κ-casein, a turn–β-strand–turn motif centered on the Phe-Met region of κ-casein, and a helix–loop–helix motif with the loop containing the anionic clusters of the calcium-sensitive caseins. Extended secondary structures may also contribute to casein interactions, leading to submicelle formation. However, these structures may be marginally stable and may actually fluctuate during the flexing of the casein chains in solution. Furthermore, physicochemical properties, as well as the proteolytic susceptibilities of isolated caseins, all suggest that their tertiary structures are rather open and flexible. The prevalence of prolyl residues, which interrupt secondary structure and promote turns, may in part be responsible for this openness and flexibility.

14.4.1.2 The Whey Proteins

The structures of β-lactoglobulin and α-lactalbumin are typical of those of other globular proteins [5, 17]. Like the caseins, they have a net negative charge at the pH of milk (Table 13); however, unlike the caseins, the sequence distribution of hydrophobic, polar, and charged residues is rather uniform. Consequently, these proteins fold intramolecularly, thereby burying most of their hydrophobic residues so that extensive self-association or interaction with other proteins does not occur. Their three-dimensional structures have been determined by x-ray crystallography and are shown in Figures 4 and 5.

The tertiary structure of β-lactoglobulin contains a β-barrel structural motif, similar to that of retinol-binding proteins, and a single short α-helix lying on its surface (Fig. 4). The center of the β-barrel forms a hydrophobic pocket, and a pocket may also exist on the surface between the α-helix and the β-barrel. As a result, β-lactoglobulin binds many small hydrophobic molecules with varying affinities.

The functionality of β-lactoglobulin is governed by these hydrophobic pockets and surface patches, and perhaps even more importantly, by the presence of disulfide bonds and a partially buried sulfhydryl group. Thus, under appropriate conditions this protein readily participates in sulfhydryl–disulfide interchange reactions with itself or with other proteins such as κ-casein.

The interactions of β-lactoglobulin are dependent upon pH. At the pH of milk it forms dimers with a geometry resembling two impinging spheres, below pH 3.5 the dimer dissociates to a slightly expanded monomer, between pH 3.5 and 5.2 the dimer tetramerizes to give an octamer, and above pH 7.5 the dimer dissociates with a concomitant conformational change to give an expanded monomer.

The structural stability of β-lactoglobulin is such that irreversible structural changes occur in the temperature range of many thermal processes for milk and dairy foods, and hence its functionality is very sensitive to precise control of these treatments.

The three-dimensional structure of α-lactalbumin (Fig. 5) is very similar to that of lysozyme [5]. It is a very compact, nearly spherical globular protein containing four α-helices, several 3_{10}-helices, and an antiparallel β-sheet. The recent discovery that α-lactalbumin is a calcium metalloprotein has identified the most unique aspect of its structure; thus, a single Ca^{2+} is bound with high affinity in a "calcium binding elbow" composed of 10 residues in a helix–turn–helix motif with coordinating oxygens provided by side-chain carboxyls of Asp 82, 87, and 88 and by peptide carbonyl groups of Lys 79 and Asp 84. In vitro folding studies have indicated that the reduced protein, which when oxidized contains four disulfide bonds, does not fold properly in the absence of Ca^{2+}. Also, the holoprotein has less surface hydrophobicity than the apoprotein. Hence, the Ca^{2+} binding is believed to assist protein folding in vivo and to aid release of the protein from the endoplasmic reticulum.

With the disulfide bonds intact, as the protein occurs in milk, the tertiary structure unfolds and refolds reversibly. Although α-lactalbumin denatures at a lower temperature than β-lactoglobulin, the transition is reversible except at very high temperatures. Thus, α-lactal-

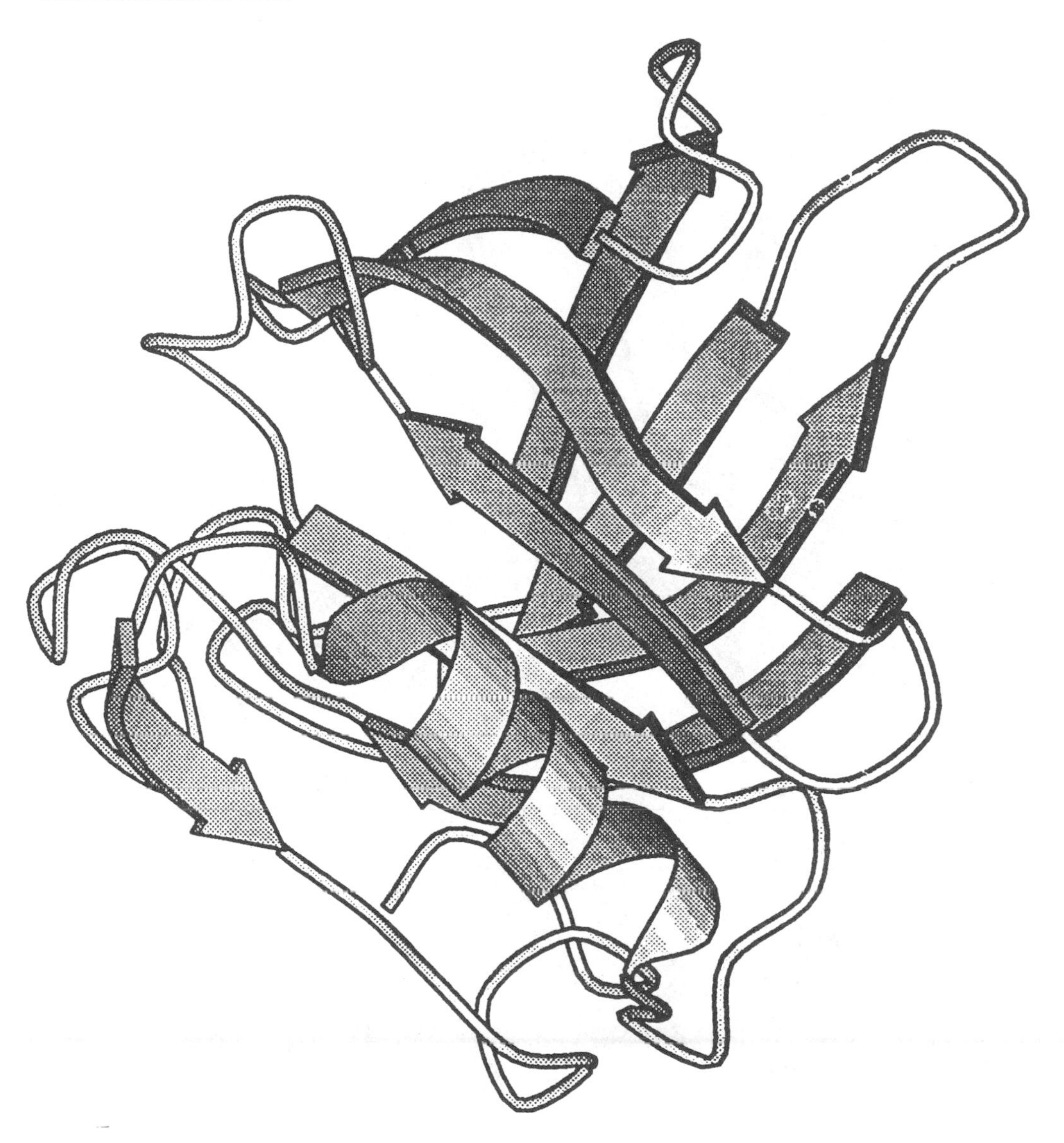

FIGURE 4 Three-dimensional structure of β-lactoglobulin. The entrance to the central binding calyx is from the top left, behind the upper β-sheet. This structure was kindly provided by Professor Lindsay Sawyer and was drawn using the MOLSCRIPT program developed by Kraulis [26].

bumin, unlike β-lactoglobulin, is not irreversibly thermally denatured under most milk processing conditions.

Physiologically, α-lactalbumin is known to function as a modifier protein of galactosyltransferase, converting it to lactose synthase [5]. Its role as a K_M modulator is performed by reversibly binding to the catalytic domain of the integral membrane enzyme galactosyltransferase, which protrudes into the lumen of the Golgi. In the absence of α-lactalbumin, this enzyme transfers galactose from UDP-galactose to form a β-linkage with the 4-hydroxyl group of β-linked *N*-acetylglucosamine in N-linked oligosaccharides of glycoproteins. When α-lactalbu-

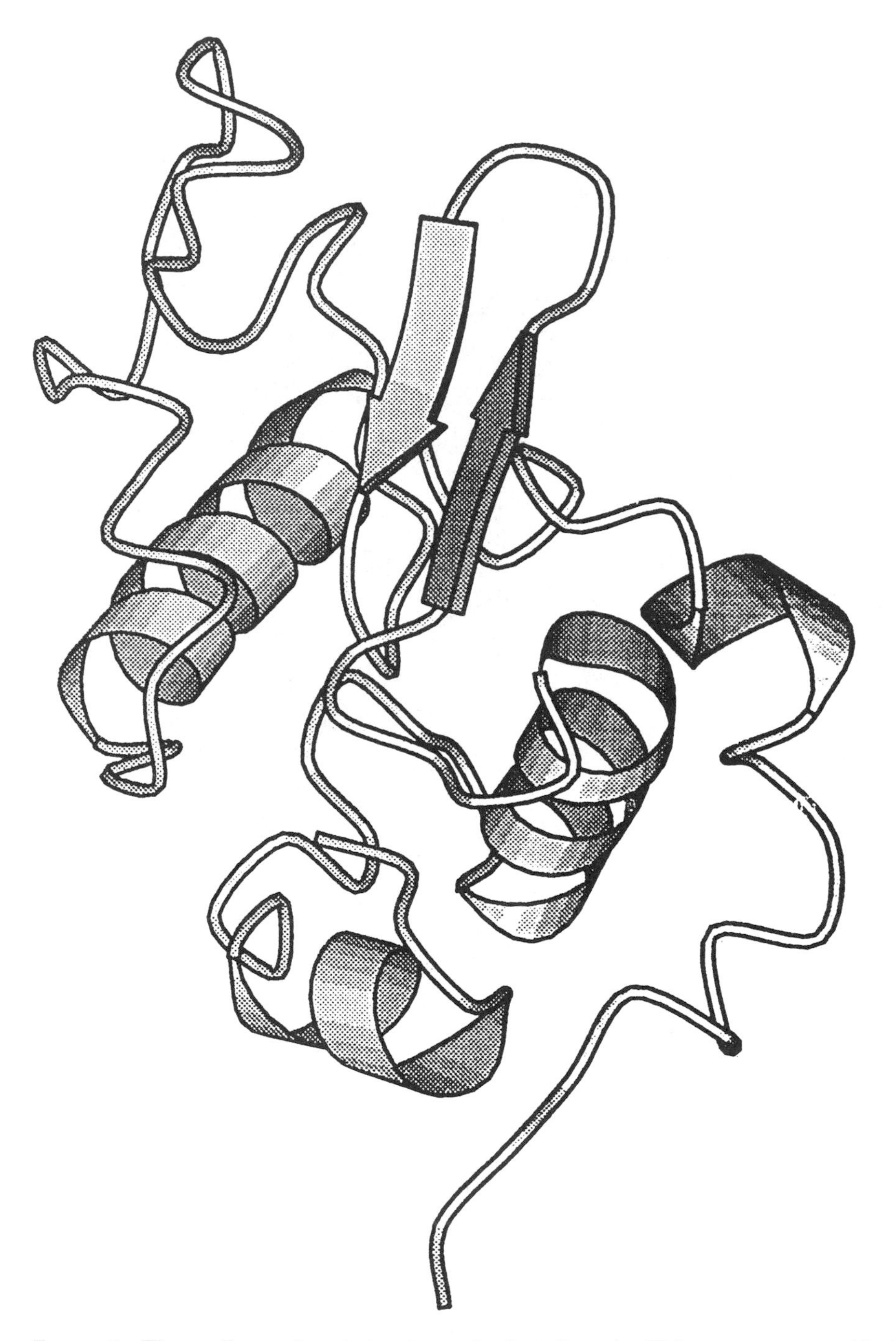

FIGURE 5 Three-dimensional structure of α-lactalbumin. This structure was kindly provided by Professor Lindsay Sawyer and was drawn using the MOLSCRIPT program developed by Kraulis [26]. The coordinates used were obtained from the Brookhaven Protein Data Bank [4] and represent the crystal structure of α-lactalbumin from baboon milk.

min is bound, however, the specificity for glucose is increased 1000-fold by reducing the K_M for this substrate. Consequently, lactose is synthesized at the physiological concentrations of glucose in the Golgi.

14.4.2 Casein Micelles and Milk Salts

As a result of their phosphorylation and amphiphilic structure, caseins interact with each other and calcium phosphate to form large spherical micelles of varying size (Table 3) with mean diameters of approximately 90–150 nm. Light scattering by these complexes is responsible for the white appearance of milk. Most evidence suggests that they are formed by association of nearly spherical submicelles having diameters of 10–20 nm. Hence, electron micrographs of casein micelles have a raspberry-like appearance (Fig. 6). Micelles contain 92% protein, composed of α_{s1}:α_{s2}:β:κ-caseins in an approximate mole ratio of 3:1:3:1, and 8% milk salts composed primarily of calcium phosphate but also significant amounts of Mg^{2+} and citrate. The characteristics of micelles determine the behavior of milk and milk products during industrial processes and storage; therefore, the properties of natural micelles and model micelle systems have received considerable study [10, 46, 51, 54, 57–59].

Micelles have a porous, "spongy" structure with a large voluminosity, approximately 4 ml/g of casein, and exceptional hydration of 3.7 g H_2O/g casein. This hydration is an order of magnitude larger than that of typical globular proteins. Hence, large molecules, even proteins, have access to and can equilibrate with all parts of the micelle structure. All components of the micelle apparently are in slow equilibria with milk serum. Thus, under appropriate conditions, various caseins and milk salts can be reversibly dissociated from the micelle. Surprisingly, such dissociation may occur to a limited extent without any apparent change in micelle size. Lowering the temperature to near 0°C causes some β-casein, κ-casein, and colloidal calcium phosphate to reversibly dissociate. However, at physiological temperature the amount of individual caseins or submicelles in the serum is extremely small.

Because the structure of the micelle has not been directly determined, the precise location of individual caseins is not known. Nevertheless, all evidence points to a predominately surface location for κ-casein, while α_s- and β-caseins are predominately in the interior. The distribution is not exclusive, however, because the calcium-sensitive caseins are also accessible on the surface. Thus, the amount of κ-casein in the micelle increases linearly with the surface/volume ratio, whereas the amount of β-casein decreases linearly (Note: The surface/volume ratio for a sphere increases with decreasing size.) Because serum components slowly equilibrate with the micelle, addition of κ-casein causes a decrease in micelle size, while addition of β-casein causes micelle size to increase. A variable composition of the submicelle is supported by these and other observations, such that submicelles rich in the calcium-sensitive caseins have a predominate interior location.

A fundamental characteristic of micelles is their resistance to irreversible association upon close approach; for example, pellets formed by sedimentation redisperse spontaneously upon standing. Hence, the micelle surface must be highly solvated and unreactive, a key to its stability under many harsh processing conditions.

Recently, a "hairy" layer has been identified on the surface of the micelle, and this layer is presumed to represent the polar domain of κ-casein. The presence of this flexible, highly hydrated polar polypeptide chain provides an inert surface and steric stabilization to the micelle. Cleavage of this domain from κ-casein by chymosin and release of this macropeptide to the milk serum completely changes the surface characteristics of the micelle, causing its destabilization. The modified surface is very active, leading to association of micelles and formation of gels, such as those in cheese.

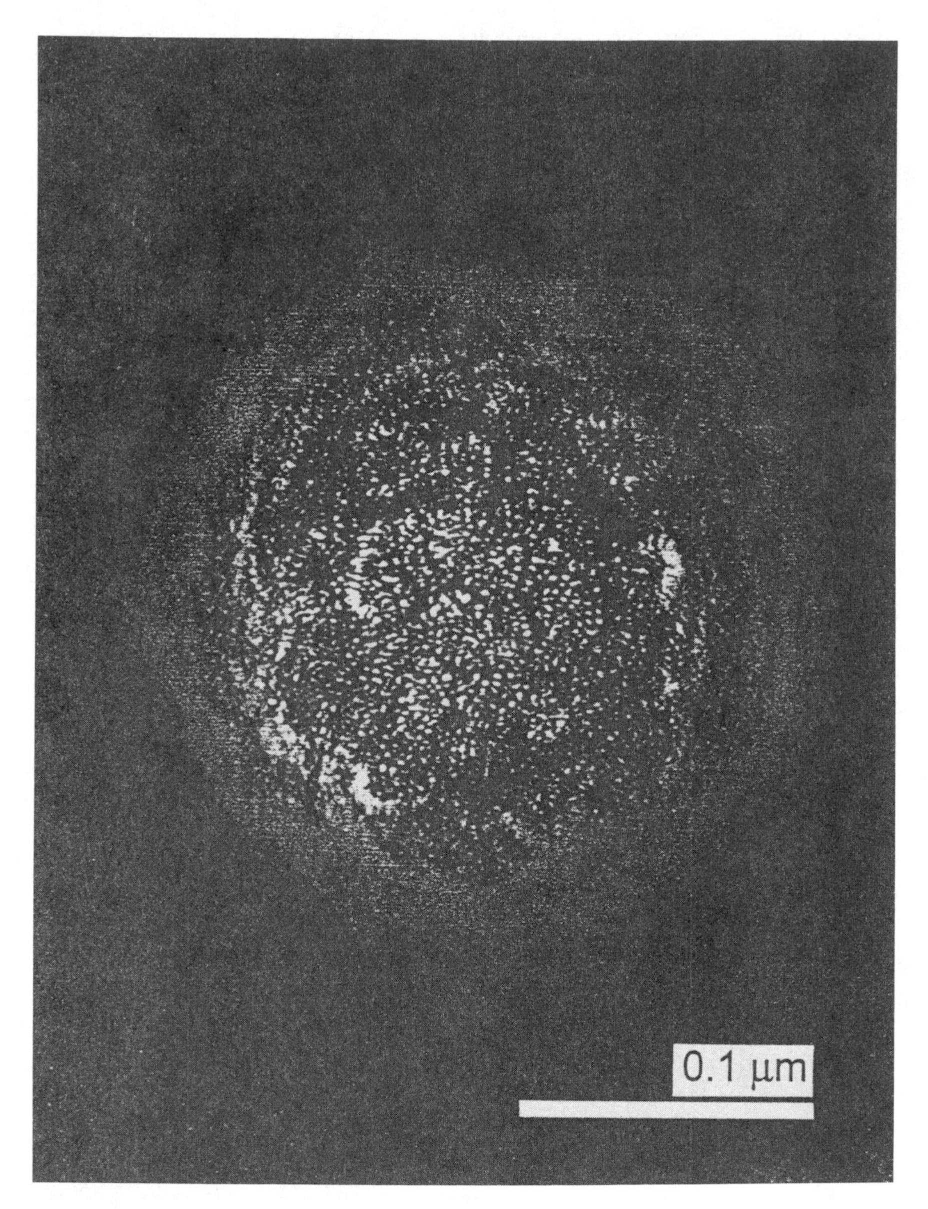

Figure 6 Electron migrograph of a casein micelle. The picture is a transmission electron micrograph of a micelle fixed with 2% glutaraldehyde. This photomicrograph was kindly provided by Dr. M. Kalab.

Calcium phosphate is the other key component required for formation of natural casein micelles [19, 51]. Although micelle-like complexes can be formed with just Ca^{2+}, inorganic phosphate is required to provide temperature stability. The structure of colloidal calcium phosphate, composed of Ca, inorganic phosphate, Mg, and citrate in fairly fixed proportions, has long been a subject of controversy. Recent x-ray absorption and infrared spectroscopy studies have shown that its structure in micelles closely resembles that of brushite [19]. Ester phosphates in the anionic clusters of submicelles may be an integral part of the calcium phosphate–Mg^{2+}–citrate matrix.

Various models of micelle structure have been proposed to account for experimental observations and stimulate further investigation. In general, these can be classified as core coat, internal structure, and subunit models [51]. The first subunit model was proposed by Morr. This model was modified by Slattery and coworkers, by Schmidt and coworkers, and finally by Walstra. Current thinking based on all available observations favors a subunit model in which the submicelles have a varying composition. Hydrophobic interactions are the predominant force binding the individual caseins together in the submicelle. This may occur through specific interactions between β-sheet structures in the hydrophobic domains of κ-casein and the calcium-sensitive caseins. Hydrophobic interactions cannot be the only force holding the submicelle together because they do not dissociate at low temperatures where such interactions are greatly weakened. A nonuniform distribution of κ-casein in the submicelles allows for interior submicelles rich in calcium-sensitive caseins and surface submicelles rich in κ-casein. When the polar domain of κ-casein, which cannot interact with calcium phosphate, occupies a sufficient amount of the micelle surface area, micelle growth stops; thus, a larger number of submicelles rich in κ-casein yields smaller micelles. In this model, the submicelles interact by incorporation of the anionic phosphoseryl clusters into the brushite-like structure of the colloidal calcium phosphate. However, the relative importance of the salt bridge and protein–protein interactions between submicelles is not clear. It appears that the salt bridge interaction between submicelles is the predominant one; however, the relative importance is probably dependent on temperature.

14.4.3 The Fat Globule

To lessen the surface free energy of lipids in an aqueous medium, the lipid molecules associate to form large spherical globules, thereby minimizing the interfacial area. Milk fat globules are the largest particles in milk (Table 3), ranging in diameter from 2 to 6 μm. Light scattering by these large globules is responsible for the "creamy" appearance of whole milk. Fat globules in milk have a mean diameter approximately 25-fold larger than the mean diameter of casein micelles. During secretion of fat globules through the plasma membrane, the globules acquire a coat of the plasma membrane, which serves to stabilize the oil-in-water emulsion. This membrane coat contains cell membrane proteins, including the enzymes, and nearly 70% of the phospholipid and 85% of the cholesterol in milk. Many of the enzymes having activities of importance to milk properties are associated with the fat globule membrane. These include alkaline phosphatase, xanthine oxidase, 5′-nucleotidase, sulfhydryl oxidase, and phosphodiesterase.

The milk fat globule therefore consists of a lipid core composed almost exclusively of triacylglycerols, a layer of some cytoplasmic proteins adsorbed prior to secretion, and an outer covering of plasma membrane. Apparently, however, the plasma membrane does not completely cover the globule because some of the cytoplasmic proteins also appear in the outer surface.

Because fat globules have a lower density than the aqueous phase of milk, they rise and cause "creaming" of unhomogenized milk. However, creaming occurs much more rapidly than would be predicted from the size of individual fat globules. This increased creaming rate is due to the clustering of globules caused by interaction of membrane proteins with immunoglobulin (IgM) and protein in the skim milk membrane vesicles. The large clusters, as theory predicts, rise much more rapidly than single globules.

14.5 USE OF MILK COMPONENTS AS FOOD INGREDIENTS

14.5.1 Effects of Processing on Milk Components

14.5.1.1 The Fat Phase

Milk fat is utilized in three major forms: (a) homogenized, as in whole milk, (b) concentrated, as in creams obtained by centrifugal separation, and (c) isolated, as in butter fat obtained by churning. Excellent reviews of the fat phase and products derived from it are available [6, 23]. Homogenization prevents the fat globules from forming a cream layer. The globules are reduced in size from 3–10 μm to less than 2 μm in diameter by forcing the liquified fat through restricted passages under high pressure (~2500 psi) at high velocity. As a result the surface area is increased 5- to 10-fold from 2 m^2/g to 10–20 m^2/g. The newly formed surface exhibits high interfacial free energy and rapidly adsorbs protein by hydrophobic interaction. Thus, casein micelles or submicelles and, to a lesser extent, whey proteins are adsorbed (approximately 10 mg/m^2), and these substances prevent coalescence of the newly formed small globules. As a result, these homogenized fat globules are somewhat like large casein micelles with respect to their surface properties. For example, any operation that will cause micelles to aggregate, such as treatment with chymosin (renneting in cheesemaking), acidification, or excessive heating, will also cause these fat globules to aggregate. Consequently, fermented dairy foods made with homogenized milk have different rheological properties than those of comparable foods made from unhomogenized milk, and the heat stability of homogenized cream is less than that of unhomogenized cream and is, therefore, more prone to feathering in coffee.

The major benefit of homogenization is the prevention of creaming. This results from a reduction in fat globule size and a decreased tendency for fat globule coalescence. Resistance to coalescence occurs because of the previously mentioned adsorption of casein micelles, casein submicelles, and whey proteins to the fat globule surface, and the denaturation of immunoglobulins and disruption of skim milk membrane vesicles. A secondary benefit of homogenization is increased whiteness that occurs because the increased number of fat globules scatters light more effectively.

Homogenization also has several disadvantages in addition to the previously mentioned reduced heat stability of homogenized cream as compared with unhomogenized cream. Homogenized milk has a blander flavor, fat globules are more susceptible to light-induced oxidation and to lipolysis when active lipase is present, and protein gels have a lower curd tension.

Cream can be obtained by centrifugal separation of cold (5–10°C) or warm unhomogenized milk. Cold separation is less disruptive of the fat globule membrane, and the resulting cream contains more immunoglobulins and is more viscous than cream obtained at a higher temperature. Churning of cream causes disruption of the fat globule membrane, resulting in clumping and coalescence of the exposed triacylglycerol globules. About 50% of the membrane is released into the buttermilk phase during churning. Clumping and coalescence of newly exposed fat globules does not occur properly when the fat is either all solid or all liquid; therefore, the temperature of churning must be carefully controlled to give an appropriate ratio of solid to liquid fat. The churning process causes a phase inversion from an oil-in-water emulsion to a water-in-oil emulsion (butter). The continuous fat phase contains some intact fat globules amounting to 2–46% of the total fat. Butter also contains 16% water as finely dispersed droplets.

14.5.1.2. The Protein Micelle and Milk Salt System

Conditions imposed by processing, such as changes in temperature, pH, and concentration, alter equilibria within the salt system, within the protein system, and between salts and proteins. Often these changes are not completely reversible even if the original conditions are restored, so the final characteristics of a dairy food depend upon the processing conditions.

The milk serum is saturated or supersaturated with respect to various calcium phosphates and calcium citrate; consequently, small changes in environmental conditions cause significant shifts in these equilibria [19]. Micellar casein is much less stable than sodium caseinate. In general, changes that decrease micelle voluminosity or increase micellar calcium phosphate will cause a decrease in stability. For example, micelles become less stable in the presence of alcohol or with an increase in Ca^{2+} activity. Pasteurization (71.7°C for 15 sec) or UHT processing (142–150°C for 3–6 sec) irreversibly increases the amount of colloidal calcium phosphate at the expense of both soluble and ionized calcium and soluble phosphate. Consequently, the pH also decreases, due to release of protons from primary and secondary phosphates. The calcium transformed to tertiary calcium phosphate does not come entirely from the serum because heating also causes dissociation of calcium bound to protein. Thus, pasteurization, and especially sterilization, affects the size distribution of micelles, leading to an increase in the abundance of both large and small micelles. Cooling to 0–4° has the opposite effect, causing an increase of soluble calcium and phosphate and an increase in pH. As previously noted, some β- and κ-casein also dissociates from the micelle upon cooling. However, these changes induced by cooling are largely reversible.

Individual milks have been classified according to their pH-dependent heat stability (duration of heating required for protein coagulation) [56]. Type A milk exhibits an optimum in the stability curve between pH 6.6 to 6.9, while Type B milk steadily increases in stability as the pH is raised above pH 6.6. This phenomenon appears to be related to the ratio of β-lactoglobulin to κ-casein and the heat-induced interaction between these two proteins. Addition of κ-casein will convert a Type A milk to Type B, and the reverse can be achieved by addition of β-lactoglobulin to Type B milk.

Pasteurization or UHT processing causes partial, irreversible unfolding of β-lactoglobulin, thereby exposing hydrophobic surface and the lone sulfhydryl group. The subsequent interaction with κ-casein, stabilized by a sulfhydryl–disulfide interchange, alters the surface properties of casein micelles. Preheating unconcentrated milk at 90°C for 10 min reduces its stability to subsequent treatments at higher temperatures. However, the same treatment, when applied to concentrated milk, nearly doubles the ability of the micelles to resist coagulation during subsequent processing at higher temperature. Concentrating milk lowers pH and increases both Ca^{2+} activity and the concentration of colloidal calcium phosphate, thus lowering micelle stability. In concentrated milk, the destabilizing effect of micelle interaction with denatured β-lactoglobulin is more than compensated for by heat-induced reduction of Ca^{2+} activity. Thus, concentrated milks cannot be sterilized without a preheat treatment.

Heat treatments of increasing severity are accompanied by increased production of dehydroalanyl residues, due to β-elimination of disulfide bonds and phosphoseryl residues, increased deamidation of aspartamyl and glutaminyl residues, and increased Maillard browning. Crosslinking of protein during heating can result from reaction of dehydroalanyl residues with ε-amino groups of lysyl residues to form lysinoalanine, or reaction with sulfhydryl groups of crysteinyl residues to form lanthionine. Continued heating (e.g., 20–40 min at 140°C) destabilizes micelles, leading to gel formation. Micelle destabilization results from a combination of factors, the most important being a decrease in pH. This decrease is caused by degradation of lactose, a shift in equilibria from primary and secondary phosphates to hydroxyapatite, and hydrolysis of phosphoserine [14]. The shift in calcium phosphate equilibria occurs rapidly, lowering the pH from about 6.7 to 5.5. Lactose degradation and phosphoserine hydrolysis continue more slowly, yielding an estimated pH of 4.9 at the point of milk coagulation. The colloidal calcium phosphate that forms upon heating is not the brushite-like calcium phosphate-magnesium citrate that occurs in natural micelles; consequently, this change in salt equilibria is mostly irreversible. It should be noted that at pH 4.9 the original colloidal calcium phosphate would be completely dissolved and the high temperature (140°C) and declining pH would cause progressive dissociation of the micelle. Thus, the coagulum that eventually forms most likely consists of a matrix of interacting individual protein chains or submicelles, rather than whole micelles as occurs in rennet curd. Sulfhydryl–disulfide interchange and covalent cross-linking may contribute to the stability of such protein–protein interactions.

Addition of salts, acids, or alkalis to milk affects Ca^{2+} activity and hence the calcium phosphate equilibria. This alters the binding of Ca^{2+} and colloidal calcium phosphate to the micelle, which changes micelle stability. For example, addition of secondary phosphate stabilizes the micelle system by (a) increasing pH, which increases micellar net charge, (b) reducing the soluble calcium concentration by shifting the equilibria toward tertiary calcium phosphate, and (c) competitive displacement of micellar calcium by sodium. Consequently, this salt may be used as a milk stabilizer. Likewise, polyvalent anions, such as citrate or polyphosphates, that bind and increase micellar net charge and lower soluble calcium are excellent milk stabilizers. Conversely, addition of calcium salts increases Ca^{2+} activity and hence colloidal calcium phosphate, which results in a lower pH and less stable micelles. As previously noted, concentration of milk has a similar effect because the salt equilibrium shifts toward colloidal calcium phosphate and the pH falls; for example, 11-fold concentration of milk increases the soluble calcium and phosphate concentration only 8-fold, and the Ca^{2+} concentration only 6-fold, while the pH is lowered from 6.7 to 5.9.

These principles must be taken into account in the processing of milk and in the incorporation of milk into formulated foods. Because of the instability of micelles in concentrated milk, stabilizers are usually added prior to concentration. Also, milk is preheated prior to concentration or drying to shift the calcium phosphate equilibria (thereby reducing Ca^{2+} activity) and to promote micelle interaction with denatured β-lactoglobulin. On the other hand, curd-setting properties of milk can be improved by adding calcium salts.

14.5.1.3 The Whey Proteins

Because denaturation of whey proteins occurs rapidly at temperatures above 70°C, normal commercial heat treatments denature at least a portion of these proteins. Major whey proteins exhibit thermostability to structural unfolding in the order α-lactalbumin < albumin < immunoglobulin < β-lactoglobulin. However, thermal unfolding of α-lactalbumin is reversible so that denaturation as measured by irreversible changes indicates an order of increasing thermostability of IgG < serum albumin < β-lactoglobulin < α-lactalbumin. Denatured whey proteins, particularly β-lactoglobulin, are considerably less soluble and more sensitive to precipitation by calcium ions than are their native counterparts. Hence, denatured whey proteins become partially incorporated into cheese curd; however, extensive binding of β-lactoglobulin to micellar κ-casein interferes with chymosin-catalyzed curd formation, so conditions of cheese making must be carefully chosen.

As noted previously, heat-induced association of whey protein (especially β-lactoglobulin) with casein micelles alters micelle properties and increases heat stability. The functionality of whey proteins is very sensitive to the extent of denaturation. For example, if whey proteins have not been heat denatured they are quite soluble at acid pH, a characteristic that facilitates their incorporation into carbonated beverages. The extent of denaturation is also extremely important and thus must be carefully controlled to optimize their performance as fat mimetics.

14.5.1.4 Lactose

Reaction of the aldehyde group of lactose with the ε-amino group of lysine (onset of Maillard reactions) occurs even under very mild heat treatment, and the reaction continues slowly during storage. The degree of Maillard reaction and attendant browning is very sensitive to the severity of heat treatment, and the effects are desirable in some products and undesirable in others. Very severe heat treatments, such as those used for an in-can sterilization of concentrated milks, causes lactose to partially degrade via Maillard reactions to yield organic acids, principally formic acid. This source of acidity is one of the major causes of protein destabilization in such products.

The very low solubility of the α-anomer of lactose can result in its crystallization in frozen products. Crystallization of α-lactose in frozen milk is typically accompanied by prompt destabilization and precipitation of casein, the reasons for which are not fully understood. Soluble lactose may have a direct stabilizing effect on casein, and if so, this is lost with lactose crystallization. Lactose crystallization may also have an indirect effect on casein stability, because its removal from the unfrozen phase decreases solute concentration, resulting in additional ice formation. The effect is similar to that observed during concentration of milk. Thus, calcium ion concentration in the unfrozen phase increases and more tertiary calcium phosphate precipitates, yielding a decline in pH and casein instability.

A common defect in ice cream, known as sandiness, is caused by the formation of large α-lactose crystals. This defect has been largely overcome by the use of stabilizers of plant origin that impede formation of lactose nuclei.

14.5.2 Use in Formulated Foods

14.5.2.1 Functionality of Proteins

Proteins from traditional sources are being increasingly utilized as ingredients in a growing number of formulated foods. The benefits of milk proteins as ingredients in other foods stem from their excellent nutritional properties and their ability to contribute unique and essential functional properties to the final foods [12, 38, 39, 41]. The functional properties of a food protein are obviously related to its structure; however, the relationship is currently not understood in sufficient detail to allow the design of protein structure to achieve a specific functionality. Nevertheless, some general features of the relationship between structure and functionality can be outlined.

Protein functional properties are related to several general molecular characteristics such as hydration, surface activity, and the type of protein–protein interactions favored by partially unfolded structures. The functional characteristics associated with these molecular properties are listed in Table 14. Hydration is a particularly important parameter because solubility is a requirement for many excellent functional properties and water binding is an essential function in many foods. Surface activity is a complex function of the protein's surface hydrophobicity

TABLE 14 Relationship Between Protein Molecular Properties and Functionality

Molecular property	Associated functional properties
Hydration	Solubility, dispersibility, swelling, viscosity, gelation, water absorption
Surface activity	Emulsification, fat adsorption, foaming, whipping
Protein–protein interactive potential	Aggregation, cohesion, texturization, gelation, elasticity, extrudability
Molecular structure or architecture yielding organoleptic properties	Color, flavor, odor

Source: Adapted from Ref. 39.

and its flexibility, which allows it to unfold and spread at an interface. The order of surface activity for various milk protein components is β-casein > monodispersed casein micelles > serum albumin > α-lactalbumin > α_s-casein–κ-casein > β-lactoglobulin [41]. The intermolecular interactions that occur between partially unfolded protein structures are extremely important to functionality, but they are a complex function of protein stability and structure and the type of surface or surface structure exposed in the partially unfolded protein. Caseins are unique because their flexible structures allow interaction with many partially unfolded structures of other proteins by hydrophobic interaction and/or by extension of secondary structure. Thermally unfolded proteins appear to be "molten globules" lacking tertiary structure but having secondary structure. Caseins are known to compete with chaperonins for binding of these molten globule states.

Advances in biotechnology have triggered interest in designing protein functionality for specific applications either by genetic manipulation or by enzymatic modification. Potentially, protein functionality can be altered, and perhaps designed, by specific limited proteolysis. Of course, this is exactly what occurs in the coagulation of milk by chymosin. It is extremely important to carefully control the extent of such reactions to optimize desirable properties and prevent off flavors. Control of the reaction can be most readily accomplished by using bioreactors containing immobilized enzymes.

Because milk proteins have evolved as the primary source of nutrition for very young mammals, excellent nutritional and flavor properties and freedom from antinutritional factors are ensured.

14.5.2.2 The Caseins

Caseins have excellent solubility and heat stability above pH 6. These proteins also have very good emulsifying properties because of their amphiphilic structure. Example compositions of several types of commercial casein products are given in Table 15, and the types of products that use casein as a protein ingredient are given in Table 16. The functionality of these commercial products depends upon the product type; hence careful consideration of the requirements for the

TABLE 15 Compositions of Some Casein and Whey Protein Products Commercially Available

	Composition (% dry weight)			
Protein products	Protein	Ash	Lactose	Fat
Casein products				
Sodium caseinate	94	4.0	0.2	1.5
Acid casein	95	2.2	0.2	1.5
Rennet casein	89	7.5	—	1.5
Coprecipitate	89–94	4.5	1.5	1.5
Whey protein concentrates				
Normal UF WPC	59.5	4.2	28.2	5.1
Neutral UF/DF WPC	80.1	2.6	5.9	7.1
Whey protein isolates				
Spherosil QMA WPI	77.1	1.8	4.6	0.9
Spherosil S WPI	96.4	1.8	0.1	0.9
Vistec WPI	92.1	3.6	0.4	1.3

Source: Adapted from Refs. 39 and 40.

TABLE 16 Some Applications of Milk Protein Products in Formulated Foods

Food category	Specific food in which casein/caseinates/coprecipitates are useful	Specific foods in which whey proteins are useful
Bakery products	Bread, biscuits/cookies, breakfast cereals, cake mixes, pastries, frozen cakes, pastry glaze	Breads, cakes, muffins, croissants
Dairy-type foods	Imitation cheeses, coffee creamers, cultured milk products, milk beverages	Yogurt, cheeses, cheese spreads
Beverages	Chocolate, fizzy drinks and fruit beverages, cream liqueurs, wine aperitifs, wine and beer	Soft drinks, fruit juices, powdered or frozen orange beverages, milk-based flavored beverages
Desserts	Ice cream, frozen desserts, mousses, instant pudding, whipped topping	Ice cream, frozen juice bars, frozen dessert coatings
Confectionery	Toffee, caramel, fudges, marshmallow and nougat	Aerated candy mixes, meringues, sponge cakes
Meat products	Comminuted products	Frankfurters, luncheon rolls, injection brine for fortification

Source: Adapted from Ref. 41.

ingredient must enter into its selection. Four types of casein products are available; acid caseins, rennet caseins, caseinates, and coprecipitates. Roughly 250,000 metric tons of casein products are produced annually worldwide.

Acid (hydrochloric, lactic, sulfuric) caseins are simply isoelectric precipitates (pH 4.6), and they are not very soluble. Likewise, rennet casein, which is produced by treatment with chymosin (rennet), is not very soluble, especially in the present of Ca^{2+}, because the polar domain of κ-casein has been removed. Rennet casein has a high mineral content because colloidal calcium phosphate–citrate is included in the clotted micelles, whereas, acid casein has a lower mineral content because the calcium phosphate is solubilized and passes into the whey. Coprecipitates of casein and denatured whey proteins are more soluble than acid or rennet casein but are not as soluble as the caseinates. Solubilization of coprecipitates can be achieved by adjustment to alkaline pH and addition of polyphosphates. Because of their low solubility, these products are best suited for products such as breakfast cereals, snack and pasta products and baked foods, to which they contribute texture and dough-forming characteristics.

The caseinates (sodium, potassium, and calcium) are prepared by neutralizing acid casein with the appropriate alkali prior to drying. These isolates, especially sodium and potassium caseinates, are very soluble and extremely heat stable over a wide range of conditions. Because of their amphiphilic structure, these proteins are useful for emulsification, water binding, thickening, whipping/foaming, and gel formation. Thus, these isolates have found wide acceptance as emulsifiers and water binders in formulated meat products, texturized vegetable protein products, margarine, toppings, cream substitutes, and coffee whiteners.

14.5.2.3 The Whey Proteins

Whey protein concentrates (WPC) or isolates (WPI) are very desirable as nutritional ingredients because of their high concentration of sulfur-containing amino acids. The annual worldwide production of whey protein products is roughly 600,000 metric tons. In the past, these products [lactalbumin (not to be confused with the protein α-lactalbumin), protein–metaphosphate complex, or protein–carboxymethylcellulose complex] had limited functionality because of extensive protein denaturation or the presence of precipitating reagents. Commercial adoption of membrane technology, including ultrafiltration (UF) and diafiltration (DF), and ion-exchange adsorption (IEA) has resulted in development of a variety of products with varied and excellent functionalities [40, 41]. Three currently used IEA processes are the Vistec process, based on a cellulosic ion exchanger, the Spherosil S process, based on a porous silica cation exchanger, and the Sperosil QMA process, which uses a porous silica anion exchange material. Compositions of these whey protein products are listed in Table 15. Example applications of whey protein ingredients in formulated foods are given in Table 16.

Key factors that determine the functionality of a product are the amount of protein denaturation, the lactose content, and the amounts of lipids and minerals present. Generally, protein denaturation lowers the solubility and adversely affects functions requiring high surface activity such as emulsification and foaming. Thus, UF WPC, UF/DF WPC, and Spherosil QMA WPI have excellent solubilities and emulsifying properties consistent with low values of protein denaturation. The presence of lipid interferes with foaming/whipping; thus, UF WPC must be defatted to function well, while Spherosil QMA WPI is a very good foaming/whipping product. More than 50% of the protein is denatured in Vistec and Spherosil S WPIs; however, Vistec WPI is a good emulsifier, perhaps because β-lactoglobulin when thermally denatured at alkaline pH remains soluble at pH 6. Differences between Vistec and Spherosil S WPI point to the importance of the structure and/or composition of the denatured state in determination of functionality. Because undenatured forms of whey proteins are soluble under acid condition, they should have applications in acid-type food formulations requiring solubility, such as carbonated fortified beverages.

In some applications, aggregation of the protein is important to texture and rheological properties, for example, in baking or textured food products. Again, however, the importance of the structure of the denatured, associated state must be emphasized. The recent development of fat mimetics using whey proteins illustrates the importance of control of the extent and type of protein denaturation to achieve a desired functionality in the final state.

14.5.2.4 Lactose

Uses of lactose depend on its low relative sweetness, protein stabilizing properties, crystallization habit, ability to accentuate flavor, nutritional attributes (inclusion of lactose in the diet improves utilization of calcium and other minerals), and ability to engage in Maillard browning [44]. The relative sweetness of lactose is commonly reported as being one-sixth that of sucrose; however, more recent studies indicate that its relative sweetness varies with concentration, ranging from one-half to one-fourth that of sucrose. Since sugars are often used to increase viscosity and/or mouth feel, or to improve texture, lactose can function in these capacities without being too sweet. Thus, lactose can replace a portion of the sucrose in toppings and icings to improve appearance and stability and to reduce sweetness. Unlike many other food sugars, addition of lactose does not reduce the solubility of sucrose. Furthermore, as the concentration of lactose is increased, the crystal habits of both sucrose and lactose change. Crystals of both become smaller, thereby yielding a softer, smoother product. Hence, the quality of certain candy and confectionery products can be substantially improved by the addition of lactose.

One of the principal functions of lactose in baked goods is to improve crust color and toasting qualities via the Maillard browning reactions. When added to biscuits, it also increases volume and tenderness and improves texture. Moreover, in baked goods containing yeast, lactose is not utilized by the yeast, thus remaining available for browning.

In common with other polyols, lactose stabilizes protein structure in solution. Thus, it has been used, for example, to reduce insolubilization of lipovitellin during freeze-drying and to preserve the activity of such enzymes as chymosin during spray-drying.

14.5.2.5 Functionality of Milk Lipids

Milk lipids contribute unique characteristics to the appearance, texture, flavor and satiability of dairy foods. Consequently, it is difficult to find suitable substitutes for milkfat in nonfat or low-fat foods. Currently, protein- or carbohydrate-based fat mimetics provide some but not all of the functions of the fat replaced. Fat mimetics that possess desirable flavor and satiability are especially difficult to find.

Structures of triacylglycerols in a composite fat are responsible for its melting point, crystallization behavior, and rheological properties. Consequently, alteration in the triacylglycerol composition will affect these properties. Fractionation of milk fat, for example, by supercritical fluid technology, can yield lipid fractions with unique functionalities desirable for specific applications in products such as spreads, pastries, or confectionery products. Fatty acid composition of the triacylglycerol can be altered by interesterification in the presence of specific added fatty acids. This reaction, catalyzed by immobilized lipase, and perhaps combined with lipid fractionation, offers an opportunity for modifying milk fat to achieve specific functionalities.

14.6 NUTRITIVE VALUE OF MILK

14.6.1 Nutrient Composition of Milk and Milk Products

Dairy foods make a significant contribution to the total nutrient diet of the U.S. population. For example, Americans obtain one-fourth or more of their protein, calcium, phosphorus, and riboflavin from dairy foods (Fig. 7). The nutrient composition of milk is shown in Table 17. The composition listed, which applies to fresh raw or pasteurized whole milk, does not include vitamin D because this nutrient is usually added to fluid milk at a level of 41 IU per 100 g, which is equivalent to about 35% of the RDA (recommended dietary allowance) per 250 ml.

Calculation of nutrient density, that is, the percent of RDA provided in a specified caloric portion or weight of a food product, provides a meaningful method of comparing the nutritive quality of various foods [18, 31]. Choosing 2000 kcal (8372 kJ) as the basis for comparison, which is roughly the average recommendation for adults in the U.S. population, the percentages of the RDA contained in 2000-kcal portions of various milk products can be calculated. Both 2000 and 2500 kcal (10,465 kJ) are used as a basis of nutrient density calculations on food labels (Fig. 8). Although several nutrients are no longer listed on the label, milk is obviously an excellent source of protein, riboflavin, vitamin B_{12}, calcium, and phosphorus, and a good source of vitamin A, thiamin, niacin equivalents, and magnesium. Removal of fat from milk removes most of the fat-soluble vitamins originally present (A, D, and E), but the nutrient density of other nutrients is significantly increased. In the preparation of cheese, the water-soluble vitamins are significantly lowered due to their partial elimination in whey. Calcium, however, is not reduced in renneted cheeses, as opposed to acid cheeses, because it remains in the clotted casein micelles.

The nutritional quality of a protein is usually a more important consideration than its quantity. Milk proteins correspond very well to human requirements and therefore are regarded as a high quality. The biological value of protein in raw milk is 0.9, based on a value of 1.0 for

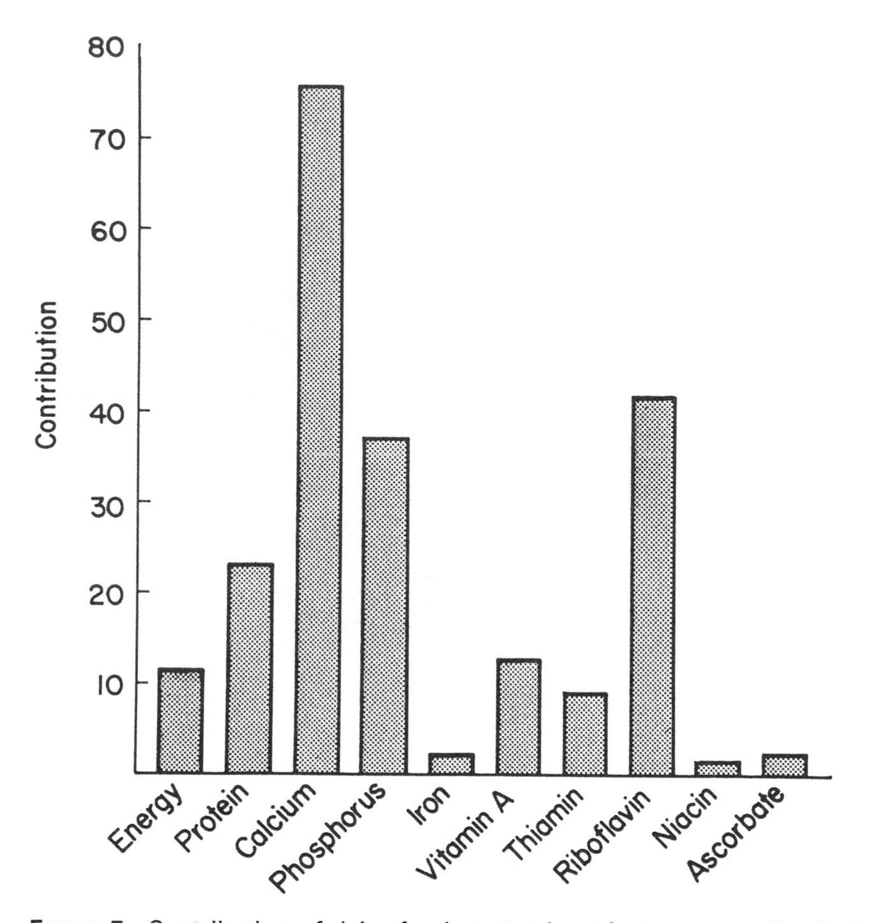

FIGURE 7 Contribution of dairy foods to total nutrient consumption in the United States. (From Ref. 31.)

whole egg protein [52]. Milk proteins are slightly deficient in the sulfur amino acids, methionine and cysteine, causing the biological value to be slightly less than ideal. Since sulfur amino acids are present in greater amounts in whey proteins than in caseins, the former have a higher biological value (1.0) than the latter (0.8). It is also noteworthy that caseins are readily digestible as a result of their structure.

14.6.2 Lactose Tolerance

A rather high percentage of African and Asian populations produce less intestinal β-galactosidase (lactase) than do Europeans or North Americans. Consequently, lactose maldigestion is encountered more frequently in African and Asian populations than in other populations. Symptoms of lactose maldigestion are diarrhea, bloating, and abdominal cramps [18]. Lactase deficiency, however, is usually one of degree rather than totality [15, 18, 31]. Since the common lactose intolerance test is conducted by administering a very large dose of 1–2 g lactose/kg body weight, the reported frequency of lactose intolerance is overestimated. Thus, intolerance to a 240-ml glass of milk is infrequently observed among individuals classified as lactose intolerant [15]. Furthermore, in certain cases milk intolerance is apparently not caused by lactose maldigestion. Nevertheless, a significant percentage of the population benefits from preconsumption hydrolysis of lactose in milk. The means for doing this industrially, using soluble or immobilized microbial

Table 17 Nutrient Composition of Whole Milk (3.3% Fat)

Nutrient[a]	Amount in 100 g	%RDA[b] in 250 ml
Protein	3.29 g	17.2
Vitamin A	31 RE[c]	8.9
Vitamin C	0.94 mg	4.2
Thiamine	0.038 mg	8.2
Riboflavin	0.162 mg	30.0
Niacin	0.85 NE[d]	13.9
Vitamin B_6	0.042 mg	6.5
Folacin	5 μg	6.4
Vitamin B_{12}	0.357 μg	46.1
Calcium	119 mg	32.0
Phosphorus	93 mg	25.0
Magnesium	13 mg	10.2
Iron	0.05 mg	0.9
Zinc	0.38 mg	6.5

Note: Values calculated from the recommended daily allowances (RDA) given by the National Academy of Sciences [49] and the composition given in USDA Handbook No. 8-1 [1].
[a]Includes all nutrients for which RDA are available, with the exception of vitamin D, vitamin E, and iodine, for which compositional data were not given in USDA Handbook No. 8-1.
[b]Average RDA for all males and females above age 11. A 250 mL volume is slightly more than 1 cup.
[c]Retinol equivalents: 1 μg retinol or 6 μg β-carotene.
[d]Niacin equivalents: 1 mg niacin or 60 mg dietary tryptophan. Only 10% of the NE in milk is in the form of niacin.

lactase, have been developed. Because relatively few persons are intolerant to lactose in low doses, and because those who are can be provided with either lactose-hydrolyzed milk or commercially available lactase in liquid or tablet form, milk products should be regarded as an important part of a varied and balanced diet and as an excellent food for remedying dietary deficiencies on a worldwide basis.

14.6.3 Effect of Processing on Nutritive Value

14.6.3.1 Effect on Proteins

Heating is involved in most types of processing, including pasteurization, sterilization, concentration, and dehydration. Effects of processing can be divided into two categories: (a) those that alter secondary, tertiary, and quaternary structure of proteins, and (b) those that alter the primary structure of proteins. The former effects, which lead to protein unfolding, may actually improve the biological value of a protein because peptide bonds become more accessible to digestive enzymes. However, modification of the primary structure may lower digestibility and produce residues that are not biologically available. Heat treatment of milk can cause β-elimination of cystinyl and phosphoseryl residues forming dehydroalanine. This substance readily reacts with lysyl residues to give lysinoalanine cross-links in the protein chain. Lysinoalanine is not biologically available, and the cross-linking lowers digestibility of the protein. Furthermore, because the nutritive value of milk proteins is limited by the low content of sulfur amino acids, such changes are particularly significant. Fortunately, pasteurization or UHT

WHOLE MILK

Nutrition Facts
Serving Size 1 Cup (240 mL)

Amount Per Serving	
Calories 150	Calories from Fat 70
	% Daily Value*
Total Fat 8g	13%
Saturated Fat 5g	26%
Cholesterol 35mg	11%
Sodium 120mg	5%
Total Carbohydrate 12g	4%
Dietary Fiber 0g	0%
Sugars 12g	
Protein 8g	
Vitamin A 6%	Vitamin C 4%
Calcium 30%	Iron 0%

* Percent Daily Values are based on a 2,000 calorie diet. Your daily values may be higher or lower depending on your calorie needs:

		Calories: 2,000	2,500
Total Fat	Less than	65g	80g
Sat Fat	Less than	20g	25g
Cholesterol	Less than	300mg	300mg
Sodium	Less than	2,400mg	2,400mg
Total Carbohydrate		300g	375g
Dietary Fiber		25g	30g

LOW FAT MILK (1%)

Nutrition Facts
Serving Size 1 Cup (240 mL)

Amount Per Serving	
Calories 110	Calories from Fat 20
	% Daily Value*
Total Fat 2.5g	4%
Saturated Fat 1.5g	7%
Cholesterol 10mg	3%
Sodium 130mg	5%
Total Carbohydrate 12g	4%
Dietary Fiber 0g	0%
Sugars 12g	
Protein 9g	
Vitamin A 10%	Vitamin C 4%
Calcium 30%	Iron 0%

* Percent Daily Values are based on a 2,000 calorie diet. Your daily values may be higher or lower depending on your calorie needs:

		Calories: 2,000	2,500
Total Fat	Less than	65g	80g
Sat Fat	Less than	20g	25g
Cholesterol	Less than	300mg	300mg
Sodium	Less than	2,400mg	2,400mg
Total Carbohydrate		300g	375g
Dietary Fiber		25g	30g

CHEESE (CHEDDAR)

Nutrition Facts
Serving Size 1 Oz (28g)
Servings Per Container 16

Amount Per Serving	
Calories 110	Calories from Fat 80
	% Daily Value*
Total Fat 9g	14%
Saturated Fat 6g	30%
Cholesterol 30mg	10%
Sodium 170mg	7%
Total Carbohydrate 0g	0%
Dietary Fiber 0g	0%
Sugars 0g	
Protein 7g	
Vitamin A 6%	Vitamin C 0%
Calcium 20%	Iron 0%

* Percent Daily Values are based on a 2,000 calorie diet. Your daily values may be higher or lower depending on your calorie needs:

		Calories: 2,000	2,500
Total Fat	Less than	65g	80g
Sat Fat	Less than	20g	25g
Cholesterol	Less than	300mg	300mg
Sodium	Less than	2,400mg	2,400mg
Total Carbohydrate		300g	375g
Dietary Fiber		25g	30g

YOGURT

Nutrition Facts
Serving Size 1 Cup (225g)

Amount Per Serving	
Calories 230	Calories from Fat 20
	% Daily Value*
Total Fat 2.5g	4%
Saturated Fat 1.5g	8%
Cholesterol 10mg	3%
Sodium 130mg	5%
Total Carbohydrate 43g	14%
Dietary Fiber 0g	0%
Sugars 34g	
Protein 10g	
Vitamin A 2%	Vitamin C 2%
Calcium 35%	Iron 0%

* Percent Daily Values are based on a 2,000 calorie diet. Your daily values may be higher or lower depending on your calorie needs:

		Calories: 2,000	2,500
Total Fat	Less than	65g	80g
Sat Fat	Less than	20g	25g
Cholesterol	Less than	300mg	300mg
Sodium	Less than	2,400mg	2,400mg
Total Carbohydrate		300g	375g
Dietary Fiber		25g	30g

FIGURE 8 Typical new food labels for several dairy foods.

processing does not result in significant formation of lysinoalanyl residues; however, in-can sterilization or boiling does.

Even mild heat treatment initiates Maillard reactions, forming lactulose-lysine and other compounds that reduce the amount of available lysine. Loss of available lysine is not significant during pasteurization (1–2%) or UHT sterilization (2–4%); however, more severe treatments, such as high-temperature evaporative concentration or in-can sterilization, can cause losses of more than 20% [50]. Storage of UHT products for long times at temperatures above 35°C can also significantly reduce available lysine. Because milk proteins contain an abundance of lysine, small losses are not nutritionally significant except in cases in which milk products are used to compensate for lysine-deficient diets.

14.6.3.2 Effect on Vitamins

The fat-soluble vitamins in milk, A (also carotene), D, and E, and the water-soluble vitamins, riboflavin, pantothenic acid, biotin, and nicotinic acid, are quite stable [50]. Accordingly, these vitamins do not sustain any detectable loss during pasteurization or UHT sterilization (Table 18). However, thiamine, B_6, B_{12}, folic acid, and ascorbate (vitamin C) are more susceptible to heat and/or oxidative degradation. Vitamins C, B_{12}, and folic acid are particularly susceptible to oxidative degradation during processing and subsequent storage. The first oxidation product of vitamin C (dehydroascorbate; has vitamin C activity) is very heat sensitive, whereas ascorbate is quite heat stable. Hence, methods that exclude or remove oxygen during processing, and packages that exclude oxygen during storage, serve to protect these vitamins.

In general, pasteurization and UHT sterilization of milk under proper conditions cause much less vitamin loss than those that occur during normal household food preparation. Furthermore, if the product is stored for long periods, the type of packaging and storage conditions are very important. In addition to exclusion of oxygen, exclusion of light is also

TABLE 18 Effects of Processing on Nutrients in Milk

Nutrient	Pasteurization[a]	Ultra-high-temperature sterilization[b]	Spray-drying[c]	Evaporated milk[d]
Vitamin A	0	0	0	0
Thiamin	10	10	10	40
Riboflavin	0	0	0	0
Nicotinic acid	0	0	0	5
Vitamin B_6	0	10	0	40
Vitamin B_{12}	10	10	30	80
Vitamin C	10–25	25	15	60
Folic acid	10	10	10	25
Pantothenic acid	0	0	0	0
Biotin	0	0	10	10

Note: The values listed in the table are the percentage losses resulting from processing.
[a]Heated at 71–73°C for 15 sec.
[b]Heated at 130–150° for 1–4 sec.
[c]Pretreated by heating at 80–90°C for 10–15 sec, homogenized, and evaporated at reduced pressure. Spray-dried by exposing a fine mist of milk concentrate to air at 90°C for 4–6 sec.
[d]Pretreated by heating at 95°C for 10 min, concentrated by evaporation at 50°C under reduced pressure, and sterilized in the can by autoclaving at 115°C for 15 min.
Source: Ref. 52.

important, not only to protect against development of off flavor, but also to prevent loss of riboflavin.

Severe heat treatments, such as in-can sterilization or dehydration, can cause significant losses of many of the vitamins. This is of special concern for vitamin B_{12}, because milk is an important dietary source of this vitamin. Knowledge of the causes of milk degradation and the use of methods to minimize degradation, such as the use of high-temperature-short-time processing and exclusion of oxygen and light during processing and storage, should, in the future, lead to milk products that are shelf stable, microbiologically safe, and almost unchanged with respect to original nutrients.

BIBLIOGRAPHY

Fox, P. F. (1982). *Developments in Dairy Chemistry. 1. Proteins*, Applied Science, London.

Fox, P. F. (1983). *Developments in Dairy Chemistry. 2. Lipids*, Applied Science, London.

Fox, P. F. (1985). *Developments in Dairy Chemistry. 3. Lactose and Minor Constituents*, Elsevier Applied Science, London.

Fox, P. F. (1989). *Developments in Dairy Chemistry. 4. Functional Milk Proteins*, Elsevier Applied Science, London.

Fox, P. F. (1992). *Advanced Dairy Chemistry. 1. Proteins*, Elsevier Applied Science, London.

Walstra, P., and R. Jenness (1984). *Dairy Chemistry and Physics*, John Wiley & Sons, New York.

REFERENCES

1. *Agriculture Handbook* No. 8-1 (1976). *Composition of Foods, Dairy and Egg Products*, U.S. Department of Agriculture, Agricultural Research Service.
2. Anonymous (1994). *Hoards Dairyman 139*(6):271.
3. Bauman, D. E., and C. L. Davis (1974). Biosynthesis of milk fat, in *Lactation—A Comprehensive Treatise*, vol. II (B. L. Larson and V. R. Smith, eds.), Academic Press, New York, pp. 31–75.
4. Bernstein, F. C., T. F. Koetzle, G. J. B. Williams, E. F. Meyer, Jr., M. D. Brice, J. R. Rodgers, O. Kennard, T. Shimanouchi, and M. Tasumi (1977). The protein data bank: A computer-based archival file for macromolecular structures. *J. Mol. Biol. 112*:535–542.
5. Brew, K., and J. A. Grobler (1992). α-Lactalbumin, in *Advanced Dairy Chemistry 1. Proteins* (P. F. Fox, ed.), Elsevier Applied Science, London, pp. 191–229.
6. Brunner, J. R. (1965). Physical equilibria in milk: The lipid phase, in *Fundamentals of Dairy Chemistry* (B. H. Webb and A. H. Johnson, eds.), AVI, Westport, CT, pp. 403–505.
7. Brunner, J. R. (1981). Cow milk proteins: Twenty-five years of progress. *J. Dairy Sci. 64*:1038–1054.
8. Brunner, J. R., C. A. Estrom, R. A. Hollis, B. L. Larson, R. McL. Whitney, and C. A. Zittle (1960). Nomenclature of the proteins of bovine milk—First revision. *J. Dairy Sci. 43*:901–911.
9. Corbin, E. A., and E. O. Whittier (1965). The composition of milk, in *Fundamentals of Dairy Chemistry* (B. H. Webb and A. H. Johnson, eds.), AVI, Westport, CT, pp. 1–36.
10. Creamer, L. K., G. P. Berry, and O. E. Mills (1977). A study of the dissociation of β-casein from the bovine casein micelle at low temperature. *N. Z. J. Dairy Sci. Technol. 12*:58–66.
11. Davies, D. T., and J. C. D. White (1960). The use of ultrafiltration and dialysis in isolating and in determining the partition of milk constituents between the aqueous and disperse phases. *J. Dairy Res. 27*:171–190.
12. De Wit, J. N. (1989). Functional properties of whey proteins, in *Development in Dairy Chemistry. 4. Functional Milk Proteins* (P. F. Fox, ed.), Elsevier Applied Science, London, pp. 285–321.
13. Farkye, N. Y. (1992). Other enzymes, in *Advanced Dairy Chemistry. 1. Proteins* (P. F. Fox, ed.), Elsevier Applied Science, London, pp. 339–367.
14. Fox, P. F. (1981). Heat-induced changes in milk preceding coagulation. *J. Dairy Sci. 64*:2127–2137.
15. Garza, C. (1979). Appropriatenss of milk use in international supplementary feeding programs. *J. Dairy Sci. 62*:1673–1684.

16. Gresti, J. M., M. Bugant, C. Maniongui, and J. Bezard (1993). Composition of molecular species of triacylglycerols in bovine milk fat. *J. Dairy Sci. 76*:1850–1869.
17. Hambling, S. G., A. S. McAlpine, and L. Sawyer (1992). β-Lactoglobulin, in *Advanced Dairy Chemistry. I. Proteins* (P. F. Fox, ed.), Elsevier Applied Science, London, pp. 141–190.
18. Hansen, R. G. (1974). Milk in human nutrition, in *Lactation—A Comprehensive Treatise*, vol. III (B. L. Larson and V. R. Smith, eds.), Academic Press, New York, pp. 281–308.
19. Holt, C. (1985). The milk salts: Their secretion, concentrations and physical chemistry, in *Developments in Dairy Chemistry. 3. Lactose and Minor Constituents* (P. F. Fox, ed.), Elsevier Applied Science, London, pp. 143–181.
20. Jenness, R. (1974). The composition of milk, in *Lactation—A Comprehensive Treatise*, vol. III (B. Larson and V. R. Smith, eds.), Academic Press, New York, pp. 3–107.
21. Jenness, R., B. L. Larson, T. L. McMeekin, A. M. Swanson, C. H. Whitnah, and R. McL. Whitney (1956). Nomenclature of the proteins of bovine milk. *J. Dairy Sci. 39*:536–541.
22. Jensen, R. G., J. G. Quinn, D. L. Carpenter, and J. Sampugna (1967). Gas-liquid chromatographic analysis of milk fatty acids: A review. *J. Dairy Sci. 50*:119–125.
23. Jensen, R. G., A. M. Ferris, and C. J. Lammi-Keefe (1991). The composition of milk fat. *J. Dairy Sci. 74*:3228–3243.
24. Kinsella, J. E., and J. P. Infante (1978). Phospholipid synthesis in the mammary gland, in *Lactation—A Comprehensive Treatise*, vol. IV (B. L. Larson and V. R. Smith, eds.), Academic Press, New York, pp. 475–502.
25. Kitchen, B. J. (1985). Indigenous milk enzymes, in *Developments in Dairy Chemistry. 3. Lactose and Minor Constituents* (P. F. Fox, ed.), Elsevier Applied Science, London, pp. 239–279.
26. Kraulis, P. (1991). MOLSCRIPT: A program to produce both detailed and schematic plots of protein structures. *J. Appl. Crystallogr. 24*:946–950.
27. Larson, B. L. (ed.) (1978). *Lactation—A Comprehensive Treatise*, vol. IV, Academic Press, New York.
28. Larson, B. L. and V. R. Smith, (eds.) (1974). *Lactation—A Comprehensive Treatise*, vol. 1, Academic Press, New York.
29. Larson, B. L., and V. R. Smith (eds.) (1974). *Lactation—A Comprehensive Treatise*, vol. II, Academic Press, New York.
30. Larson, B. L., and V. R. Smith, (eds.) (1974). *Lactation—A Comprehensive Treatise*, vol. III, Academic Press, New York.
31. Leveille, G. A. (1979). United States food and nutrition issues—Roles of dairy foods. *J. Dairy Sci. 62*:1665–1672.
32. Lindquist, B. (1963). Casein and the action of rennin, Part 1. *Dairy Sci. Abstr. 25*:257–265.
33. Lindquist, B. (1963). Casein and the action of rennin, Part 2. *Dairy Sci. Abstr. 25*:299–308.
34. Lizell, J. L., and M. Peaker (1971). Mechanism of milk secretion. *Physiol. Rev. 51*:564–597.
35. McKenzie, H. A. (1967). Milk proteins. *Adv. Protein Chem. 22*:55–234.
36. McKenzie, H. A. (ed.) (1970). *Milk Proteins*, vol. I, Academic Press, New York.
37. McKenzie, H. A. (ed.) (1970). *Milk Proteins*, vol. II, Academic Press, New York.
38. Morr, C. V. (1979). Utilization of milk proteins as starting materials for other foodstuffs. *J. Dairy Res. 46*:369–376.
39. Morr, C. V. (1982). Functional properties of milk proteins and their use as food ingredients, in *Developments in Dairy Chemistry. I. Proteins* (P. F. Fox, ed.), Applied Science, London, pp. 375–399.
40. Morr, C. V. (1989). Whey proteins: Manufacture, in *Developments in Dairy Chemistry. 4. Functional Milk Proteins* (P. F. Fox, ed.), Elsevier Applied Science, London, pp. 245–284.
41. Mulvihill, D. M. (1992). Production, functional properties and utilization of milk protein products, in *Advanced Dairy Chemistry. 1. Proteins* (P. F. Fox, ed.), Elsevier Applied Science, London, pp. 369–404.
42. Neville, M. C. (1990). The physiological basis of milk secretion. *Ann. NY Acad. Sci. 586:1–11*.
43. Nickerson, T. A. (1965). Lactose, in *Fundamentals of Dairy Chemistry* (B. H. Webb and A. H. Johnson, eds.), AVI, Westport, CT, pp. 224–260.
44. Nickerson, T. A. (1976). Use of milk derivative, lactose, in other foods. *J. Dairy Sci. 59*:581–587.
45. Patton, S., and R. G. Jensen (1976). *Biomedical Aspects of Lactation*, Pergamon Press, New York.

46. Payens, T. A. J. (1979). Casein micelles: The colloid-chemical approach. *J. Dairy Res. 46*:291–306.
47. Pitas, R. E., J. Sampugna, and R. G. Jensen (1967). Triglyceride structure of cow's milk fat. I. Preliminary observations on the fatty acid composition of positions 1, 2 and 3. *J. Dairy Sci. 50*:1332–1336.
48. Pyne, G. T. (1962). Some aspects of the physical chemistry of the salts of milk. *J. Dairy Res. 29*:101–130.
49. *Recommended Dietary Allowances* (1989). 10th ed., National Academy of Sciences, Washington, DC.
50. Renner, E. (1980). Nutritional and biochemical characteristics of UHT milk, in *Proceedings of the International Conference on UHT Processing and Aseptic Packaging of Milk and Milk Products*, Department of Food Science, North Carolina State University, Raleigh, pp. 21–52.
51. Rollema, H. S. (1992). Casein association and micelle formation, in *Advanced Dairy Chemistry. 1. Proteins* (P. F. Fox, ed.), Elsevier Applied Science, London, pp. 111–140.
52. Rolls, B. A. (1982). Effect of processing on nutritive value of food: Milk and milk products, in *Handbook of Nutritive Value of Processed Food,* vol. I (M. Rechcigl, ed.), CRC Press, Boca Raton, FL, pp. 383–399.
53. Rose, D., J. R. Brunner, E. B. Kalan, B. L. Larson, P. Melnychyn, H. E. Swaisgood, and D. F. Waugh (1970). Nomenclature of the protein's of cow milk: Third revision. *J. Dairy Sci. 53*:1–17.
54. Schmidt, D. G. (1980). Colloidal aspects of casein. *Neth. Milk Dairy J. 34*:42–46.
55. Shahani, K. M., W. J. Harper, R. G. Jensen, R. M. Parry, Jr., and C. A. Zittle (1973). Enzymes in bovine milk: A review. *J. Dairy Sci. 56*:531–543.
56. Singh, H., and L. K. Creamer (1992). Heat stability of milk, in *Advanced Dairy Chemistry. 1. Proteins* (P. F. Fox, ed.), Elsevier Applied Science, London, pp. 621–656.
57. Slattery, C. W. (1976). Review: Casein micelle structure; and examination of models. *J. Dairy Sci. 59*:1547–1556.
58. Slattery, C. W. (1979). A phosphate-induced sub-micelle equilibrium in reconstituted casein micelle systems. *J. Dairy Res. 46*:253–258.
59. Slattery, C. W., and R. Evard (1973). A model for the formation and structure of casein micelles from subunits of variable composition. *Biochim. Biophys. Acta 317*:529–538.
60. Swaisgood, H. E. (1973). The caseins. *CRC Crit. Rev. Food Technol. 3*:375–414.
61. Swaisgood, H. E. (1982). Chemistry of milk protein, in *Developments in Dairy Chemistry. I. Proteins* (P. F. Fox, ed.), Applied Science, London, pp. 1–59.
62. Swaisgood, H. E. (1992). Chemistry of caseins, in *Advanced Dairy Chemistry. I. Proteins* (P. F. Fox, ed.), Elsevier Applied Science, London, pp. 63–110.
63. Thompson, M. P., J. R. Brunner, C. M. Stine, and K. Lindquist (1961). Lipid components of the fat-globule membrane. *J. Dairy Sci. 44*:1589–1596.
64. Thompson, M. P., N. P. Tarassuk, R. Jenness, H. A. Lillevik, U. S. Ashworth, and D. Rose (1965). Nomenclature of the proteins of cow's milk-second revision. *J. Dairy Sci. 48*:159–169.
65. Timmen, H., and S. Patton (1989). Milk fat globules: fatty acid composition, size and in vivo regulation of fat liquidity. *Lipids 23*:685–689.
66. U.S. Department of Agriculture (1993). *Dairy S&O Yearbook*, DS-441, August, Washington, D.C., pp. 1–39.
67. White, J. C. D., and D. T. Davies (1958). The relation between the chemical composition of milk and the stability of the caseinate complex. *J. Dairy Res. 25*:236–255.
68. Whitney, R. (1977). Milk proteins, in *Food Colloids* (H. D. Graham, ed.), AVI, Westport, CT, pp. 66–151.
69. Whitney, R. M., J. R. Brunner, K. E. Ebner, H. M. Farrell, R. V. Josephson, C. V. Morr, and H. E. Swaisgood (1976). Nomenclature of the proteins of cow's milk: Fourth revision. *J. Dairy Sci. 59*:785–815.

15

Characteristics of Edible Muscle Tissues

E. ALLEN FOEGEDING AND TYRE C. LANIER
North Carolina State University, Raleigh, North Carolina

HERBERT O. HULTIN
University of Massachusetts, Amherst, Massachusetts

15.1 INTRODUCTION

The discovery in Florida of 12,200-year-old mastodon bones showing signs of butchering is evidence that meat has been part of the human diet since prehistoric times. With the progression of time, meat animals were domesticated and animal husbandry, rather than animal hunting, became the predominant way to put meat on the table. Today fish is the only meat source that is predominantly hunted rather than raised. However, rapidly growing aquaculture practices, and concerns about depleting various fish species, may change our source and variety of consumable fish. As with any biological tissue, the food quality of meat is highly connected with biological function and loss of homeostasis in converting muscle to meat. Muscles are designed to provide locomotion. The process of contraction consumes large amounts of energy (ATP) and is precisely regulated so that contraction/relaxation cycles produce motility.

Individual muscles are specialized for contraction speed and/or force, such that there are more similarities in muscles found in coinciding anatomical locations among species than there are among different muscles within one species. The specialized functions of various muscles are achieved by differences in the proteins repsonsible for contraction and structure, and by differences in the metabolic pathways and primary substrates (lipids or carbohydrate) used to generate energy. Biological function is the first major factor affecting meat quality. This becomes evident when comparing the white and dark muscles of chickens and turkeys. Meat animals that are poikilothermic, such as fish and crustaceans, also have to adjust biochemical factors to accomodate changes in body temperature. This is accomplished by periodically removing old muscle proteins and replacing them with new ones that may differ slightly in function. The process of protein turnover requires an array of proteinases that must be highly regulated to prevent autolysis in the live animal.

When viewed as a whole, the biological image of muscle is one of a dynamic tissue that is highly specialized to provide individual contraction requirements in a variety of environments. This results in a variety of sensory properties for muscles within and among species.

The majority of the biological controls are removed with death and bleeding (exsanguination) of the animal. Therefore, the second major factor affecting meat quality is the loss of biological regulation upon death. The chemical processes active in the conversion of muscle to meat determine the quality of fresh meat and the functional properties of meat used to manufacture processed meat products. Addressed in this chapter are the major chemical reactions associated with muscle contractions, converting muscle to meat, and manufacturing meat products. The term "meat" is used to indicate the contractive tissues from all species of animals used for food. While this omits animal products such as kidney and liver, it is inclusive for beef, pork, lamb, turkey, chicken, fish, and crustaceans. The ultimate goal is to understand the chemistry relevant to the texture, color, flavor, nutritional attributes, and processing properties of meat.

15.2 SIMILARITIES AND DIFFERENCES AMONG MUSCLES OF VARIOUS SPECIES

Throughout the animal kingdom muscle function is similar; that is, it provides locomotion. There are, however, important differences among species and, indeed, among muscles of the same animal, especially when muscle function is different. Nowhere is the species difference more obvious than for fish versus land animals and birds. The differences observed are due to three basic factors. First, the fish body is supported by water; thus, fish do not require extensive, strong connective tissues to maintain and support the muscles. Second, since most commercially important fish are poikilothermic animals and they live in a cold environment, the proteins of fish muscles have properties different from those of warm-blooded species. Finally, the structural arrangement of fish muscle is markedly different from that of land animals and birds. This arrangement is related to the peculiar movement of fish.

Generally speaking, food scientists have studied those properties of muscle tissues that have important influences on their value as food. For example, in land animals and birds, texture and appearance are the major properties that can be controlled during slaughtering and post-slaughtering procedures. Since textural properties of these muscles are due in part to the state of contraction of the myofibrillar proteins, considerable attention has been directed toward understanding contraction-related problems in these muscles postmortem. The muscle proteins of land animals and birds are also greatly influenced by the rate of change of pH postmortem, and this aspect has received considerable attention. On the other hand, textural properties of fresh fish are less of a problem than they are with warm-blooded animals; thus, comparatively little information is available on the postmortem biochemistry of fish as related to texture.

When considering frozen animal tissues, a quite different picture emerges. Important textural changes occur in fish during frozen storage, and these changes result from the intolerance of fish proteins to conditions induced by freezing, such as increased salt concentrations and decreased pH values. Consequently, textural changes in frozen fish have received considerable study. On the contrary, the proteins of land animals and birds are reasonably stable during frozen storage and research in this area is somewhat limited.

The importance of collagen also differs among species. In mammals and birds, both the amount and type of collagen have important influences on textural properties of the muscle. In fish, however, collagen is readily softened by normal cooking procedures and does not have an important influence on the textural properties of the final product.

Once meat is removed from the bone and is subjected to further processing the species identity becomes less apparent. This can create a regulatory compliance problem, especially when fish are processed on-ship to a point where species-specific morphological properties are lost. One way to differentiate among muscle tissue from various species is to look for species-specific proteins or probe the protein or DNA for slight differences in sequence. Methods based on differences in DNA sequences are especially attractive because DNA is quite stable to processing treatments and therefore likely to be present in a form that can be identified.

Flavor changes are very important, especially in fish, and extensive studies have been conducted on degradation of lipids and amino compounds postmortem since these degradation compounds contribute greatly to flavor. This area is less important for most land animals and birds, with the exception of "grass-feed" flavor in beef. This flavor problem arises when animals are slaughtered with minimal or no feeding of grains in the several week period just prior to slaughter. Flavor problems can arise in all species due to absorption of off-flavor compounds in the fat. In living monogastric animals this can arise from the feed, or in all slaughtered animals this can occur through absorption of undesirable compounds from the environment.

15.3 NUTRITIVE VALUE

In addition to its aesthetic appeal, meat is important because of its excellent nutritive properties. The composition of lean muscle tissue is given in Table 1, and it is similar for a wide variety of animals. Lipid is the most variable component of meat. The muscle tissues of fatty fish show large seasonal variations in lipid, based in large part on the reproductive cycle of the fish. For example, the lipid content of mackerel is known to vary between 5.1 and 22.6% during the course of a year [48]. Fluctuations in lipid content in fish are compensated for by changes in water content; that is, the percentage of lipid plus water remains essentially constant in fatty fish [55]. The composition of meat varies considerably depending on the amount of fat, bone, and skin included in the sample.

As shown in Table 1, about 18–23% of the lean portion of meat is protein. A significant exception to this occurs in starving fish or in fish in certain stages of the reproductive cycle. The latter is a form of starvation since sufficient dietary protein cannot be provided to compensate for protein synthesis in gonadal tissue. Fish utilize muscle tissue during starvation, and it is replaced by water. The muscle from a living but severely starved cod can contain as much as 95% water [55]. In meat animals starvation is a problem only in animals that are hunted.

The protein content of muscle tissue is very large and the quality of this protein is very high, containing kinds and ratios of amino acids that are similar to those required for maintenance and growth of human tissue. Of the total nitrogen content of muscle, approximately 95% is protein and 5% is smaller peptides, amino acids, and other compounds.

Lipid components of muscle tissue vary more widely than do the amino acids [3]. As seen in Table 2, there are differences among species, between muscles of one species, and among animals of different ages. In fish muscle, the differences have been "institutionalized" in the concept of lean versus fatty fish. In lean fish, storage fat is carried in the liver. Muscle of lean fish contains less than 1% lipid, mostly phospholipid located in membranes. In fatty fish, depot fat apparently occurs as extracellular droplets in the muscle tissue [42]. In white muscle, fat is apparently well dispersed outside the muscle cells; in red muscle, distinct fat droplets exist within the cells. In addition to species variations, the lipid components of muscle can be markedly influenced by diet.

The lipids of mammalian and avian muscle can be categorized as to location, that is, that in muscle tissue and that in adipose tissue. The composition of lipids in these two locations can be quite different. Lipids in muscle contain greater contents of phospholipids than do lipids in

TABLE 1 Composition of Lean Muscle Tissue

	Composition (%)			
Species	Water	Protein	Lipid	Ash
Beef	70–73	20–22	4–8	1
Pork	68–70	19–20	9–11	1.4
Chicken	73.7	20–23	4.7	1
Lamb	73	20	5–6	1.6
Cod	81.2	17.6	0.3	1.2
Salmon	64	20–22	13–15	1.3

Source: Ref. 95.

TABLE 2 Lipid Content of Various Meats

Species	Muscle or type	Lipid (%)	Content[a]	
			Neutral lipids (%)	Phospholipids (%)
Chicken	White	1.0	52	48
	Red	2.5	79	21
Turkey	White	1.0	29	71
	Red	3.5	74	26
Fish	White	1.5	76	2
(Sucker)	Red	6.2	93	7
Beef	dorsi	2.6	78	22
		7.7	92	8
		12.7	95	5
Pork	Lattissimus dorsi	4.6	79	21
	Psoas major	3.1	63	37
Lamb	Lattissimus dorsi	5.7	83	10
	Semitendinosus	3.8	79	17

[a]Percentage of gross muscle composition
Source: Ref. 3.

adipose tissue. Lean muscle contains about 0.5–1% phospholipids, and the fatty acids of phospholipids are more unsaturated than those of adipose triacylglycerols. Oxidation of the highly unsaturated fatty acid constituents found in the membranes of muscle can be very important contributors to the deterioration of meat.

Within a species, red muscle contains more lipid than white muscle (Table 2). The type of fatty acids present in muscle tissue is also species dependent. Fatty acids in cold-blooded fish are much less saturated than those in avian and mammalian muscles (Table 3). Presumably, the high concentration of polyunsaturated fatty acids in fish is necessary to keep lipids fluid at the low muscle temperatures sometimes encountered. The much greater percentage of polyenic fatty acids in lean fish (cod) compared with that of fatty fish (mackerel) reflects differences in

TABLE 3 Degree of Saturation of Fatty Acid Components of Lipid from Muscle Tissue of Various Species

Species	Fatty acids (%)		
	Saturated	Monenoic	Polyenoic
Beef	40–71	41–53	0–6
Pork	39–49	43–70	3–18
Mutton	46–64	36–47	3–5
Poultry	28–33	39–51	14–23
Cod (lean fish)	30	22	48
Mackerel (fatty fish)	30	44	26

phospholipid–triacylglycerol ratios, as mentioned earlier. Poultry fat is more unsaturated than pork fat, which in turn is more unsaturated than beef or mutton fat.

Muscle tissue is an excellent source of some of the B-complex vitamins, especially thiamin, riboflavin, niacin, B6, and B12. However, the B-vitamin content of muscle varies considerably depending on the species and on the type of muscle within a species. In addition, the levels of the B vitamins are also influenced by breed, age, sex, and general health of the animal. Less work has been performed on the fat-soluble vitamins in meat than on the water-soluble vitamins in meat, but the levels of vitamins D, E, and K in meats are generally rather low, although the amount of vitamin E (tocopherols) can be influenced by the diet of the animal. The content of vitamin A in muscle foods is somewhat greater than that of the other fat-soluble vitamins. Ascorbic acid is present in meat only at very low levels.

Meat is a good source of iron and phosphorus and a rather poor source of calcium, except in certain deboned or mechanically separated meats in which the calcium content may be greatly increased due to the presence of small bone fragments in the edible product (Table 4). Meat typically contains from 40 to 90 mg sodium per 100 g and about 250–420 mg potassium per 100 g lean tissue. Minerals and the water-soluble B-complex vitamins are found in the lean portion of the meat. The concentration of these substances in the whole muscle is therefore dependent on the amount of fat tissue and bone in a particular cut of meat, as well as on the cooking process.

15.4 MUSCLE STRUCTURE AND COMPOSITION

15.4.1 Structure of Skeletal Muscle

Skeletal muscle is composed of long, narrow, multinucleated cells (fibers) that range from a few to several centimeters in length and from 10 to 100 μm in diameter. Although there are differences in muscle fibers with regard to the amount of sarcoplasm and the amount and location of cellular membrane components, there is a close similarity, at the cellular level, of skeletal muscles from a wide variety of organisms. There are, however, different arrangements of muscle cells and connective tissue depending on the means of locomotion of the animal and its environment.

Figure 1 represents a typical arrangement of muscle components in mammals and birds. The fibers are arranged in parallel fashion to form bundles, and groups of bundles form a muscle. Surrounding the whole muscle is a heavy sheath of connective tissues, called the *epimysium*. From the inner surface of the epimysium, other connective tissues penetrate the interior of the muscle, separating the groups of fibers into bundles. This connective tissue layer surrounding groups of fibers is termed the *perimysium*, and extending from this are finer sheaths of connective tissue that surround each muscle fiber. These last sheaths are termed *endomysium*. The connective tissue sheaths merge with large masses of connective tissue tendons at the termini of the muscle, with these merger points serving to anchor the muscle to the skeleton. The long components of the circulatory system are located in the perimysium, whereas the smaller units (capillaries) are within the endomysium.

The arrangement of muscle fibers in fish is quite different from that in birds and mammals and is based on the need to flex their bodies for propulsion through water. The arrangement of the muscle tissue in a typical bony fish is shown in Figure 2. The W-shaped segments (Fig. 2) are called myotomes, and they have one forward and two backward flexures. These myotomes are not perpendicular to the vertical midplane of the fish but typically intercept this plane at sharp angles [45]. Thus, if a cut is made at right angles to the central skeleton, multiple myotomes are exposed. The myotomes are one cell deep, and the muscle cells are roughly perpendicular to the

TABLE 4 Mean Mineral Content of Selected Avian, Bovine, and Porcine Muscles and Three Typical Composite Meat Products

	Minerals (mg/g meat)						
Muscle and meat product	Calcium	Phosphorus	Magnesium	Potassium	Sodium	Zinc	Iron
Chicken							
Red (leg)							
6-week-old	3.90	181	20.2	252	72.7	1.44	1.06
1-year-old	3.30	170	20.5	237	77.2	1.51	1.32
White (breast)							
6-week-old	2.83	200	25.9	265	42.8	0.62	0.64
1-year-old	3.21	214	30.5	294	54.6	0.41	0.99
Beef							
Semimembranosus	4.62	202	26.9	380	70.4	2.99	2.78
Semitendinosus	4.28	216	27	417	55.2	2.92	2.00
Pork							
Semimembranosus							
Boars	5.47	167	21.3	283	82.8	2.09	1.54
Sows	5.53	211	26.7	398	68.2	1.87	1.50
Semitendinosus	5.87	190	21.5	341	88.7	5.47	3.00

Source: Ref. 99.

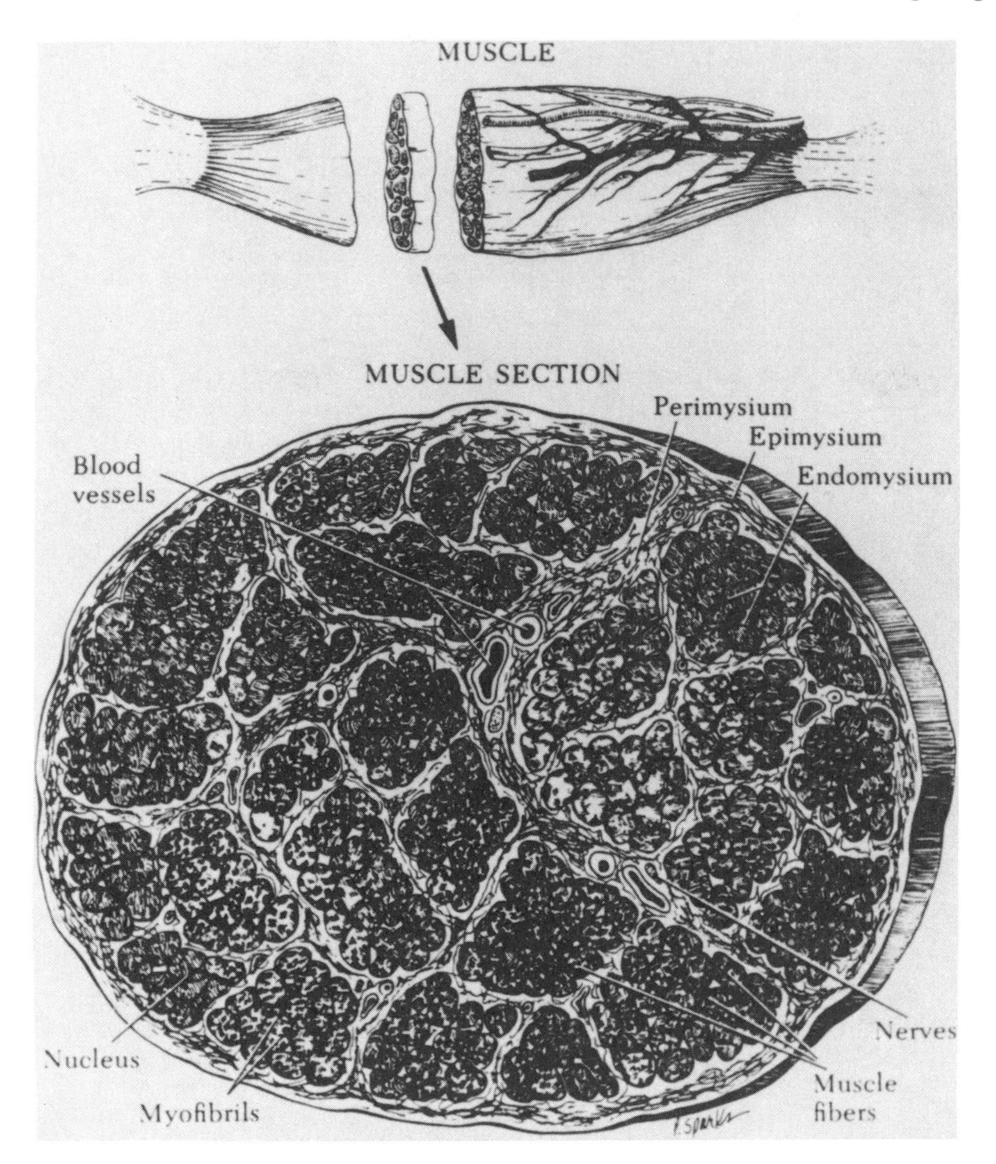

FIGURE 1 Muscle cross section, illustrating the arrangement of connective tissue into epimysium, perimysium, and endomysium and the relationship to muscle fibers and fiber bundles. The typical position of blood vessels is also shown.

surface of the myotome. The myotomes are connected one to another by thin layers of collagenous connective tissue called myosepta (myocommata).

In squid, a cephalopod, the mantle (external body wall) is made up of circular muscle fibers. At more or less regular intervals, these fibers are interrupted by sections of radially oriented fibers that function to "thin" the mantle and cause the mantle cavity to expand following the power stroke of the circular fibers [9]. In some species of squid, superficial fibers also run longitudinally along the mantle. The myofibrils of squid mantle muscle are helically wound and are referred to as obliquely striated [36].

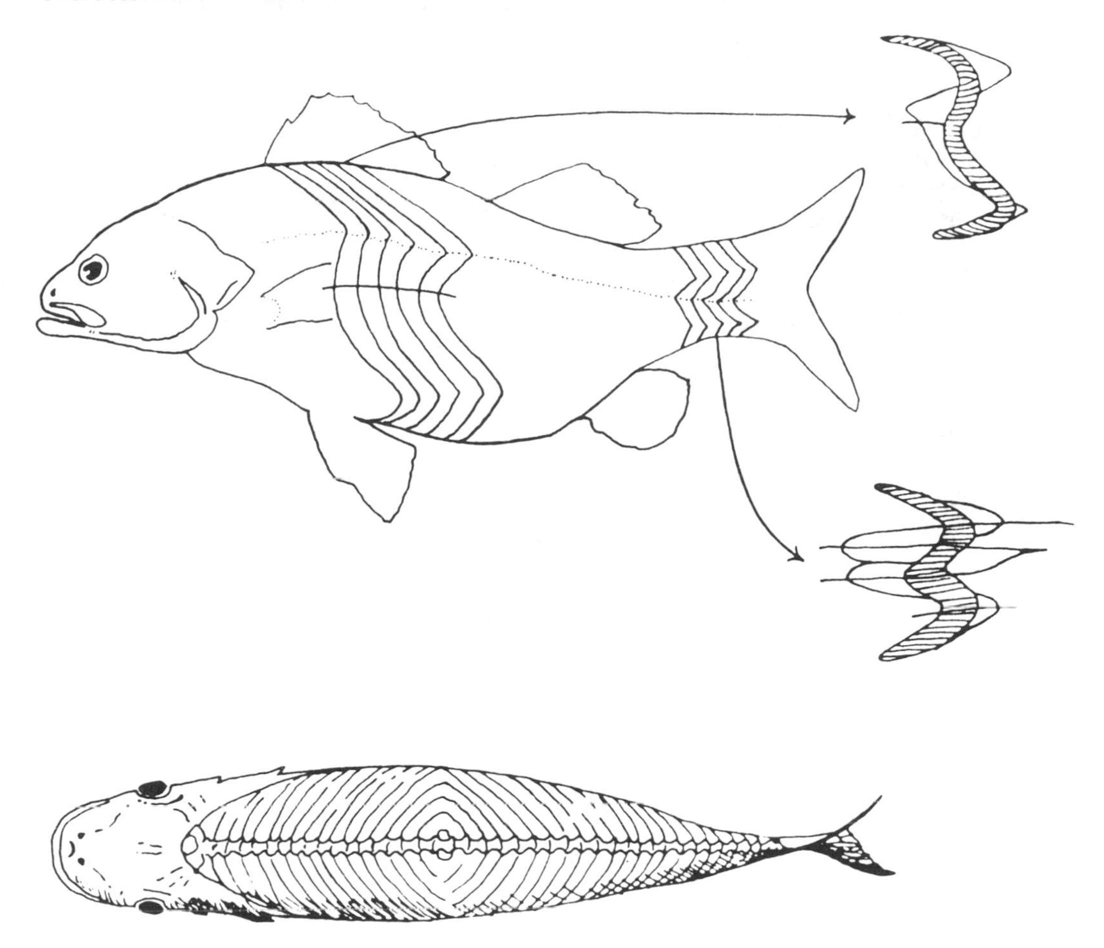

FIGURE 2 Myotome pattern of musculature of bony fish, with detailed lateral views of a single myotome.

Figure 3 is a diagrammatic view of a typical muscle fiber, with the major components being the sarcolemma, the contractile fibrils, the cell fluid (sarcoplasm), and the organelles. The surface of the muscle fiber consists of three layers: an outer network of collagen fibrils, a middle basement lamina composed of type IV collagen and proteoglycans, and an inner plasma membrane called the sarcolemma. Invaginations of the sarcolemma form the transverse system called T tubules. The ends of the T tubules meet in the interior of the cell close to two terminal sacs of the sarcoplasmic reticulum. This is shown in Figure 4. The sarcoplasmic reticulum is a membranous system located in the cell (fiber) and generally arranged parallel to the main axis of the cell. The meeting of the T system and the sarcoplasmic reticulum (triad) occurs at different intrafiber locations in different muscles. The triadic joint appears to exist most frequently at or near the Z disks or lines (Fig. 5) in fish and frog muscle and at the junction of the A and I bands in reptiles, birds, and mammals. The terminal cisternae of the longitudinal elements (sarcoplasmic reticulum) surround each sarcomere like a hollow collar. The collar is perforated with holes and is termed a *fenestrated collar*.

The T tubules extend the sarcolemma into the interior of the muscle cell, and it is this phenomenon that allows the muscle cell to respond as a unit (essentially no lag period in the interior of the cell). Depolarization of the sarcolemma and its intracellular extension (T tubules)

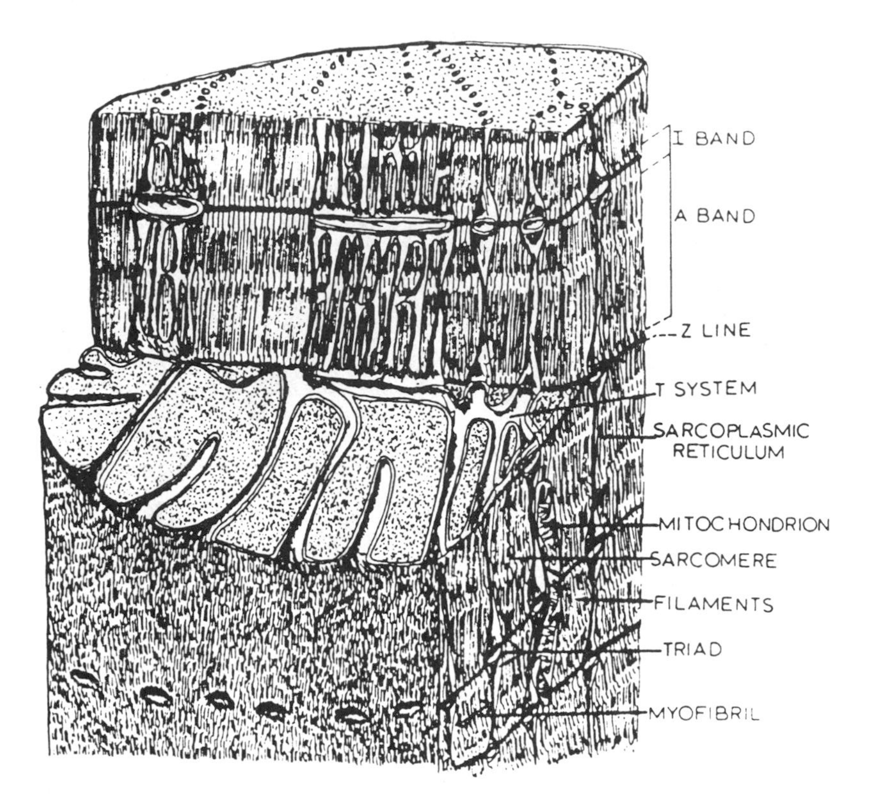

FIGURE 3 Cutaway of a muscle fiber showing the outer membrane and its invaginations (T system), which run horizontally and meet with two terminal sacs of the longitudinal sarcoplasmic reticulum in the triad. Repetitive cross sections are also indicated. (Adapted from Ref. 75).

triggers liberation of calcium from terminal sacs of the sarcoplasmic reticulum via a protein commonly called the calcium channel. This liberation of calcium activates ATPase of the contractile proteins and allows contraction to occur. Relaxation is achieved in part by a reversal of the process and by sequestering of calcium by the sarcoplasmic reticulum [37].

Mitochondria serve as prime energy transducers for the muscle cell, and these organelles are located throughout the cell. In some cases there is a concentration of mitochondria near the Z line or near the plasma membrane. Nuclei are distributed near the surface of the muscle cell and have an important role in protein synthesis.

Lysosomes are subcellular organelles that contain large quantities of hydrolytic enzymes and serve a digestive role in the cell. Lysosomes exist as part of the muscle cell per se and also originate from phagocytic cells present in the circulatory system. The proteolytic enzymes of these organelles are termed *cathepsins* and various cathepsins, with differing activities, have been isolated. Glycogen particles and lipid droplets also occur in some muscle cells, depending on the state of the muscle. All these cellular components are bathed in the sarcoplasm, a semifluid material that contains soluble components, such as myoglobin, some enzymes, and some metabolic intermediates of the cell.

The major inner components of the muscle fiber are the fibrils (myofibrils), which constitute the contractile apparatus. The myofibrils are surrounded by the sarcoplasm and some of the elements discussed earlier, such as mitochondria, the T tubules, and the sarcoplasmic reticulum.

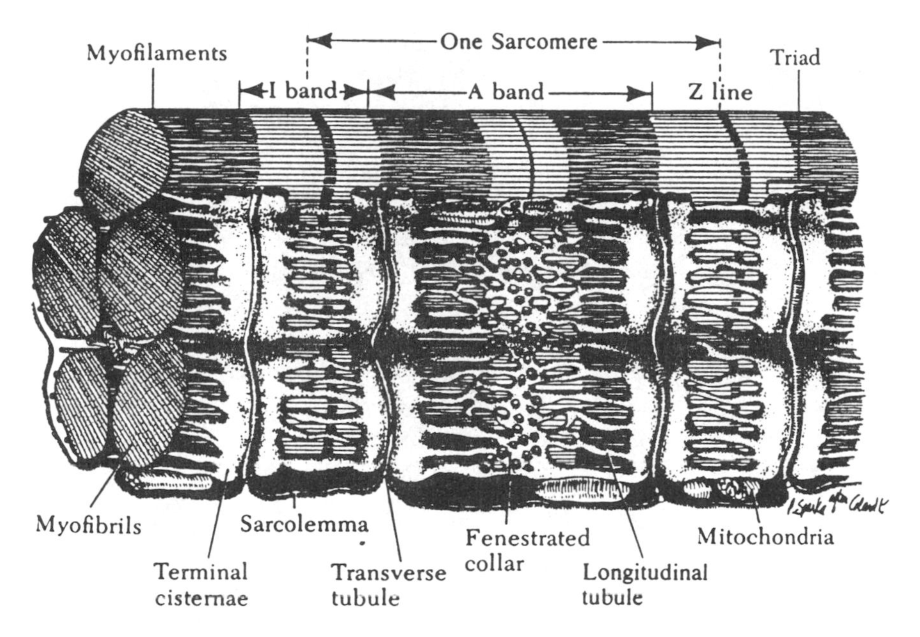

FIGURE 4 Diagram showing myofibrils and surrounding sarcoplasmic reticulum and transverse tubule membranous systems. The terminal cisternae, fenestrated collar, and longitudinal tubule are components of the sarcoplasmic reticulum. (From Ref. 25.)

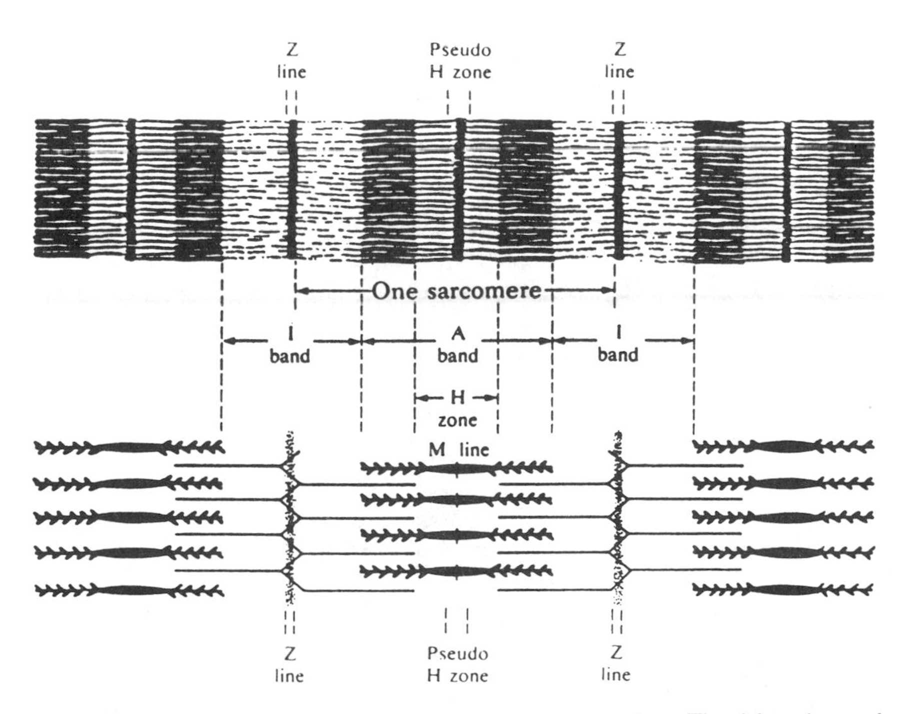

FIGURE 5 Striated muscle fibril in longitudinal section. The I-band consists only of thin filaments. The A-band is darker where it consists of overlapping thick and thin filaments and lighter in the H-zone, where it consists solely of thick filaments. The M-line is caused by a bulge in the center of each thick filament, and the pseudo-H-zone is a bare region on either side of the M-line. (From Ref. 25).

The characteristic striated appearance of skeletal muscle is due to a specific repetitive arrangement of proteins in the myofibrils. This arrangement is shown in Figure 5. The dark bands of the fibrils are anisotropic or birefringent when viewed in polarized light. Being anisotropic, they are termed "A bands." The bands that appear lighter are isotropic and are therefore termed "I bands." In the center of each of the I bands is a dark line, called the Z line or Z disk. The term Z is derived from the German *zwischen*, meaning between. In the center part of each A band is a light zone, and this is called the H zone. Frequently, at the center of the H zone there exists a darker M line. The contractile unit of the fibril is termed the *sarcomere,* defined as the material located between and including two adjacent Z disks.

A sarcomere is comprised of thick and thin longitudinal filaments. The A band is comprised of thick (mostly myosin) and thin filaments (mostly actin), whereas the I band is composed of thin filaments. The thin filaments extend outward from the Z disks in both directions, and the thin filaments overlap in the thick filaments in parts of the A band. The lighter zone of the A band, the H zone, is that area where thin filaments do not overlap thick filaments. The contractile state of the muscle has an important bearing on the size of these various bands and zones, since during contraction the thin and thick filaments slide past each other. During contraction, the length of the A band remains constant, but the I band and the H zone both shorten.

Thin filaments are imbedded in or connected to the Z disk, and because of this, the Z disk presumably serves as an anchor during the contractile process. Each thin filament in the I band is linked to the four closest thin filaments in the adjacent sarcomere. This linkage laterally offsets the thin filaments from one sarcomere to the next. The major protein found in the Z disk is alpha-actinin.

The M line is located in the area of the myosin thick filaments where projections of the myosin headpieces are not present (Fig. 5). The M line probably serves to keep the filaments in the correct geometric position.

Striated muscle from other vertebrates as well as invertebrates has essentially the same structure as that discussed already for mammalian muscle. However, there may be some differences in the arrangement of myofibrils, amount of sarcoplasm, relationship of nuclei and mitochondria to other components in the muscle cell, arrangement of the sarcoplasmic reticulum, and location of the triadic joint. In fish muscle cells, peripheral myofibrils are perpendicular to the sarcolemma (surface) but the inner myofibrils have a normal arrangement parallel to the long axis of the cell. Some molluskan muscles, such as scallop adductor and squid muscles, are obliquely striated instead of cross-striated [66].

15.4.2 Structure of Smooth or Involuntary Muscle

Smooth muscle fibers do not show the characteristic striations of voluntary or skeletal muscle. Certain organs containing smooth muscle, such as the gizzards of birds, intestinal tissue, and the meat of many mollusks (clams, oysters), are used for food.

15.4.3 Structure of Cardiac (Heart) or Striated Involuntary Muscle

Heart tissue is used as food directly or may be incorporated into sausage products. The myofibrillar structure of heart muscle is similar to that of striated skeletal muscles, but cardiac fibers contain larger numbers of mitochondria than do skeletal fibers. The fiber arrangement of cardiac muscle is also somewhat less regular than that observed in skeletal muscle.

15.4.4 Proteins of the Muscle Cell

The proteins/enzymes of muscle can be categorized based on biological function or chemical properties. Biological classes relate to energy metabolism, contraction, and structure. Chemical classes are water soluble, salt soluble, and insoluble. The water-soluble fraction can be extracted from muscle with water or dilute salt solutuion and is made up of enzymes, such as those involved in glycolysis, and the muscle pigment myoglobin. Many of the proteins are connected to the cytoskeleton [8] and are not soluble per se; however, they are easily extractable. The contractile proteins have classically been considered only soluble in salt solutions of high ionic strength, and not in water. However, some contractile proteins can be extracted by water, and the required extraction conditions are species dependent [85].

Extractability of salt soluble proteins depends strongly on salt concentrationa and pH. The insoluble fraction remains after treatment with salt solutions of high ionic strength, and it consists of connective tissue proteins, membrane proteins, and usually some unextracted contractile proteins.

15.4.4.1 Contractile Proteins

Proteins involved in the physical process of contraction are those contained in the sarcomere. These can be divided based on location, such as thick or thin filaments, or on function, such as force-generating or regulating proteins. The scientific study of muscle has progressed from an almost exclusive focus on skeletal muscle to a strong interest in motility processes. The latter includes not only cardiac and smooth muscle but also invertebrate contraction systems. This has led to the discovery of new muscle proteins; therefore, Table 5 provides only a partial listing of the proteins involved in motility.

Myosin

The major protein of the thick filaments is myosin, which comprises 45% of the myofibrillar proteins. It is an elongated protein molecule about 160 nm in length with a molecular mass of approximately 480,000 D. Myosin contains a total of six polypeptide chains, two heavy chains and four light chains. Myosin heavy chains have "head" and "tail" regions, reflecting the respective globular and rod portions of the molecules. The biological functions of myosin reside in the heavy chains.

The six polypeptide chains of myosin are assembled in a quaternary structure that resembles a stick (tail) with two pear-shaped heads (Fig. 6). The tail region consists of two alpha-helical heavy chains coiled together into a coiled-coil alpha-helical supersecondary structure. This structure terminates at the head region. The myosin head is 16.5 nm long, 6.5 nm wide, and approximately 4.0 nm thick [76]. The main secondary structure in the head is alpha-helix, accounting for approximately 48% of the amino acids. The top of the head contains the actin binding site; then following down the molecule are the ATPase site, alkali light chain site, and DTNB [(5,5′-dithiobis)-2-(nitrobenzoic acid)] light chain site. The light chains bind to the alpha-helical regions of the heavy chain. The tail portion of the heavy chain molecule is responsible for its association into thick filaments, and the head region contains actin binding and ATPase sites.

Associated with each myosin head section are two light chains, so that four light chains are associated with each myosin molecule. The light chains exist in two distinct chemical classes. One class consists of DTNB light chains, so called because treatment of myosin with the thiol reagent DTNB causes its removal. It is also called the regulatory light chain because it is responsible for calcium-mediated regulation of contraction in molluscan muscles, and phosphorylation-mediated regulation in smooth muscle. The second class of light chain consists of alkali

TABLE 5 Sarcomeric Proteins

	Muscle type	Number of amino acids	Molecular weight
	Thin filament		
Actin	Rabbit	374	42,000
β-Actinin			
BI	Chicken/fast twitch		35,000
BII			31,000
Tropomyosin			
α-tropomyosin	Chicken/fast twitch	284	33,000
β-tropomyosin	Chicken/fast twitch	284	36,000
Troponin			
C	Chicken/fast twitch		18,000
I	Rabbit	179	20,900
T	Chicken/fast twitch	274	30,500
	Thick filament		
Myosin			
Heavy chain	Chicken/fast twitch	1938	220,000
Alkali light chain			16,000–22,000
DTNB light chain			18,000
C-Protein	Human/fast twitch	1142	128,000
H-Protein	Rabbit		69,000
X-Protein	Rabbit		152,000
	Sarcomere structure		
α-Actinin	Rabbit/fast twitch		100,000
Desmin			55,000
Nebulin	Rabbit		700,000–900,000
Titin	Rabbit		1,000,000
Myomesin			170,000
I-Protein	Chicken/fast twitch		50,000

Source: Ref. 78 and several other sources.

light chains, so called because they are released under alkaline conditions. Alkali light chains were once considered essential for ATPase, and were therefore termed "essential" light chains. It was later discovered that the isolation process was inactivating ATPase associated with the heavy chains.

Each globular head of heavy-chain myosin contains one alkali light chain and one DTNB light chain. The two types of light chains are made from different genes. The DTNB light chain has a molecular mass of 18,000–19,000D. The alkali light chains are produced from a single gene that is differentially spliced to produce two polypeptides with different molecular weights. Alkali light chain-1 and light chain-3 from rabbit fast muscle have respective molecular masses of 20,950 and 16,660 D [61].

The large size and salt solubility of myosin made it difficult for early researchers to conduct molecular studies. Therefore, they developed ways to split myosin into smaller parts, which were more amenable to analytical techniques. Myosin can be cleaved in the middle region by proteolytic enzymes, such as trypsin, producing two fractions of the protein. One of these is called light meromyosin (Fig. 6) and the other, which contains the globular head structures of the myosin molecule, is called heavy meromyosin. Separated heavy meromyosin retains its

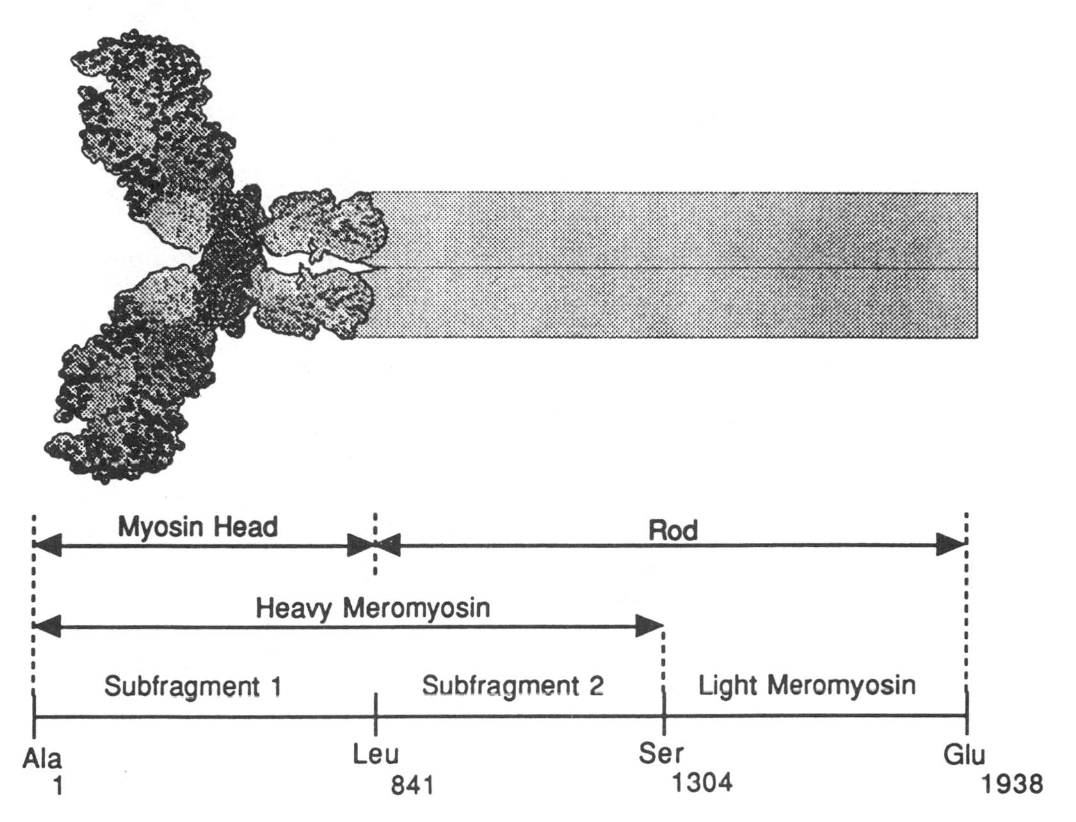

FIGURE 6 Myosin. Top shows three-dimensional structure of the myosin subfragment 1. The bars attached to each subfragments 1 represent the coiled-coil α-helical rod region. The bottom shows segments of the molecule created by limited proteolysis at specific amino acids from the chicken myosin heavy chain. (From Ref. 57.0)

ability to interact with actin and its ATPase activity. The head section of heavy meromyosin, designated subfragment 1, S1 or SF1, is attached to the rod section with rather flexible linkages, and can assume a wide variety of positions in relation to the rod section.

When separated from subfragment 1, the rod-like tail section of heavy meromyosin, called subfragment 2, S2 or SF2, shows no tendency to associate with itself or with light meromyosin. This indicates that light meromysin (which associates readily) is responsible for forming the backbone structure of the thick filament. The SF2 portion of myosin is free to swing out from the surface to allow the globular heads to make contact with the thin filaments.

Each thick filament contains some 400 molecules of myosin. These molecules polarize when they interact, joining in head-to-tail fashion in two directions, as illustrated in Figure 7. It is undoubtedly this polarity that allows contraction to occur.

Actin

The major protein of the thin filaments is actin, which comprises 20% of myofibrillar protein of muscle [78]. Actin is bound to the structure of the muscle much more firmly than is myosin. Extraction of actin can be accomplished by prolonged exposure of a muscle powder (obtained by acetone extraction) with an aqueous solution of ATP.

Actin's shape can be described as two peanut-shaped domains of equal size lying side-

FIGURE 7 Possible arrangement of myosin molecules in a filament. Globular regions are shown as zig-zags and tails as lines. The polarities of the myosin molecules are reversed on either side of the center, but all molecules on the same side have the same polarity.

by-side. Actin monomers, called globular actin or G-actin, are assembled in a double-helical structure called fibrous actin, or F-actin (see Fig. 8). This constitutes the main portion of the thin filament. G-actin is composed of 374–375 amino acids and has a molecular mass of 42,000–48,000 D. It is stable in water, where it can also exist as a dimer. Globular actin binds ATP very firmly and, in the presence of magnesium, spontaneously polymerizes to form F-actin with the concurrent hydrolysis of bound ATP to give bound ADP and inorganic phosphate. Globular actin also polymerizes in the presence of neutral salts at a concentration of approximately 0.15 *M*. Filaments of F-actin interact with the head portion of myosin. When negatively stained and examined in the electron microscope, myosin or its subfragments bound to F-actin appear as long strings of arrowheads pointing in the same direction. This indicates that the actin monomers in the F-actin filament have polarity. Actin filaments are assembled with opposite polarity, relative to myosin binding, on each side of the Z-disk, which allow the thick filaments in adjacent sarcomeres to move toward the Z-disk.

Actomyosin

When actin and myosin are mixed in vitro, a complex, called actomyosin, is formed. This complex can be dissociated by addition of ATP, as would be the case in living muscle. Actomyosin is the main state of actin and myosin in postmortem muscle because ATP is depleted by postmortem metabolism. However, unlike prerigor muscle, simple addition of ATP and other solubilizing compounds such as Mg^{2+} to meat does not dissociate all the myosin from actin, and extraction of myosin from postrigor meat is therefore difficult. Myosin and actomyosin are found in extracts of postmortem muscle.

Tropomyosin and Troponin

Tropomyosin, representing 5% of myofibrillar protein, is composed of two alpha-helical polypeptides wound together into a two-stranded, coiled-coil supersecondary structure. It resembles

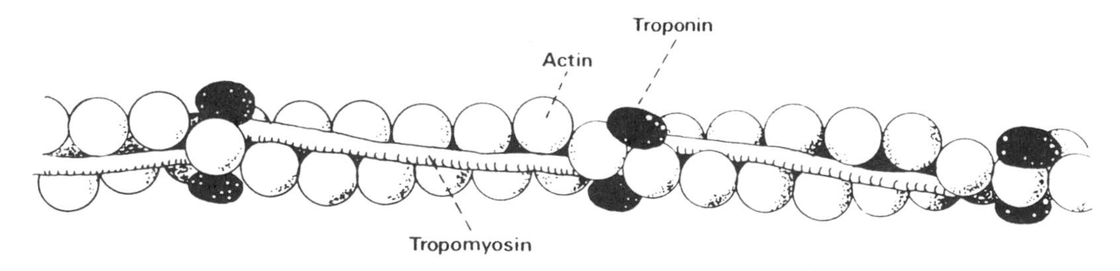

FIGURE 8 Thin filament complex with tropomyosin lying in grooves of the actin double helix. Troponin complex is spaced at 400 Å. From Ref. 20, p. 364.)

the tail or rod portion of the myosin molecule. In skeletal muscle two polypeptides, alpha- and beta-tropomyosin, can combine to form a tropomyosin dimer. The alpha- and beta-tropomyosin polypeptides have molecular masses of 37,000 and 33,000 D, respectively. They are found in muscle as the alpha–alpha or beta–beta homodimers and the alpha–beta heterodimer. Tropomyosin aggregates end-to-end and binds to actin filaments along each groove of the actin double helix such that each molecule interacts with seven G-actin monomers (Fig. 8). Tropomyosin and troponin are combined in a complex that regulates interactions of myosin with the thin filament.

Troponin, accounting for 5% of myofibrillar protein, consists of three subunits designated troponin C (for calcium binding), troponin I (for inhibitory), and troponin T (for binding with tropomyosin). Each subunit of troponin has distinct functions. Troponin C is a calcium binding protein and confers calcium regulation to the contractile process via the thin filament. Troponin I, when tested without the other subunits, strongly inhibits the ATPase activity of actomyosin. Troponin T functions to provide a strong associaiton site for binding of troponin to tropomyosin.

Thick Filament Proteins

Myosin is the major protein of the thick filament and has a primary function in muscle contraction. However, other proteins with less definitive biological functions are also found in thick filaments. C-, H-, and X-proteins are found in bands spaced periodically on the thick filament. They appear to be wrapped around the thick filament surface with their major axes in a plane normal to that of the thick filament. This has provoked the thought that these protein rings may protect the filament against destruction by tensile forces or changes in the ionic environment; however, this has not been established as a biological function. Phosphofructokinase (F-protein) is also found on thick filaments. This enzyme is responsible for the conversion of fructose 6-phosphate to fructose 1,6-diphosphate in the glycolytic pathway.

Sarcomeric Structural Proteins

Alpha-actinin is located exclusively in the Z disk. It has a molecular mass of 200,000 D and consists of two polypeptide subunits of similar mass. Alpha-actinin is a member of a family of F-actin-binding proteins. The apparent function of alpha-actinin is to provide a structure to the Z line.

Beta-actinin is a dimeric protein made up of polypeptides of 35,000 and 31,000 D. It binds to F-actin and inhibits the recombination of fragmented F-actin. It is located at the free end of the thin filaments and may be involved in regulating the length of thin filaments.

Desmin is a 55,000 D protein that forms filaments that are 10 nm (intermediate) in length. This protein is part of the cellular framework known as the cytoskeleton [8]. The desmine intermediate filaments are found at the periphery of each Z line and they form the connecting link between adjacent myofibrils.

The main protein in the M-line is myomesin. It is 165,000 D and accounts for 3% of the myofibrillar protein [78]. The proposed function of the M-line is holding thick filaments in proper register.

One of the proteins of the M line has turned out, suprisingly, to be creatine kinase, an enzyme involved in ATP regeneration. That such a catalytic agent may also be an important structural component of the myofibril has important implications with respect to operating mechanisms of the contractile protein system.

Titin (connectin) is a high-molecular-mass, ~1,000,000 D, elastic protein that is long enough (0.9–1 μm) to extend from the Z line to the M line, spanning half of the sarcomere. It is the third most abundant myofibrillar protein, representing 10% of the total. It interacts with myosin, M-line proteins, and alpha-actinin. Titin is thought to have a role in regulating the length of thick filaments and in maintaining mechanical continuity of myofibrils of striated muscle.

Titin-containing filaments appear to be identical to the "gap filaments" observed between thin and thick filaments when muscle is stretched beyond the point of filament overlap [51].

Nebulin is another large, 700,000–900,000 D, myofibrillar protein, which accounts for 5% of myofibrillar protein. This protein has multiple actin binding sites, and various lengths in accord with the length of thin filaments. It appears to function in regulating the length of the thin filament.

Paramyosin is found in muscles of mollusks and other invertebrates, accounting for as much as 50% of the total structural protein of these muscle fibers. It has a molecular mass of 220,000 D and is essentially 100% alpha-helical. It forms the central core of the myosin-containing thick filaments in these systems.

15.4.4.2 Contraction and Relaxation

Contraction

Although it is not the purpose of this chapter to delve into the physiology of living muscle, it is necessary to have some knowledge of the normal functioning and interactions of contractile proteins in the living cell to understand postmortem changes. Indeed, the chemical reactions that occur in postmortem muscle are the same as those in the live animal, except biological regulation is lacking.

Contraction occurs when muscle is stimulated by an electrical nervous impulse that depolarizes the muscle cell membrane (sarcolemma). The transverse tubular system of the muscle is an extension of the sarcolemma, providing a link between the outside and inside of the cell. The stimulus (depolarization) is transmitted to the interior of the muscle cell so that the whole cell can react as a unit. The transverse tubules are adjacent to, but not connected with, the calcium-containing sarcoplasmic reticulum which surrounds the myofibrils. The depolarization of the transverse tubular system is conveyed to the sarcoplasmic reticulum by the dihydropyridine receptor protein. This protein is a member of a group that is generally termed "calcium channel proteins". It contains voltage and calcium binding portions. Calcium is released from the sarcoplasmic reticulum by a calcium channel protein termed the ryanodine receptor [53]. This increases the calcium concentration in the sarcoplasm from 0.1 to 10 μM. The calcium binds to troponin C, causing a structural transition. This changes the troponin–tropomyosin–actin complex so that tropomyosin moves into the groove of the F-actin superhelix. By this proces the thin filament is "turned on", that is, it is able to interact with a thick filament and cause contraction. It was previously thought that the troponin–tropomyosin complex regulated contraction by blocking interaciton between actin and myosin. This has been disproved, and it is now believed that the actin filament has a direct role in contraction, implying that troponin–tropomyosin may function by altering F-actin flexibility.

Roles of Thick and Thin Filaments in Contraction

Our current understanding of muscle contraction originates with the "sliding filament hypothesis" proposed in 1954 [38, 40]. It was observed that the length of the A-band remains constant during contraction while the H-zone and I-bands decrease (Fig. 5). Furthermore, electron micrographs had shown that the sarcomere is composed of thick and thin filaments that, when viewed in cross section, show six thin filaments around each thick filament. From these and other observations, the sliding filament hypothesis was formulated: Muscle contraction involves chemical interactions between thick and thin filaments, causing them to slide past each other in a manner that shortens the sarcomere.

Since the origin of the sliding filament hypothesis, literally thousands of experiments have been done to better understand the molecular mechanism for muscle contraction. Each finding has caused the model for contraction to be revised. Major contributions that allowed the sliding filament hypothesis to be refined are the elucidation of the three-dimensional structures of that portion of the myosin molecule responsible for contraction (the head), and of the actin (thin) filament [35, 76]. The current model for contraction relies heavily on knowledge of these structures. Each myosin head interacts with two actin monomers of the thin filament at strong and weak binding sites. Starting with myosin strongly bound to actin in a rigor complex at the end of a contraction cycle, the triphosphate portion of an ATP molecule binds to the ATPase site on the myosin head. This causes an opening (widening) of two segments of the myosin head, which results in decoupling at the strong binding site while maintaining associations at the weak binding site. The myosin head continues to move and closes around the ATP molecule, resulting in dissociation of myosin from actin. ATP is hydrolyzed and the ADP and gamma-phosphate remain bound to the myosin head. Myosin then starts to recombine with actin, first with weak ionic interactions. As more of the myosin head associates with actin the binding progresses from weak to strong, and with release of the gamma-phosphate, the contraction (power) stroke is initiated. At the end of the power stroke ADP is released and the myosin head is back to the starting state. Thin filament regulation by troponin–tropomyosin most likely involves the transition from weak to strong binding of myosin to actin.

Relaxation

Relaxation in most skeletal muscle involves “turning off” the thin filament. This is done by decreasing sarcoplasmic calcium concentration from 10 to 0.1 μM, and requires active transport of calcium back into ther sarcoplasmic reticulum. The “calcium pump” is a 100,000 D protein that exists in the sarcoplasmic reticulum. One molecule of ATP and two calciums bind to the pump proteins on the sarcoplasmic side. Through structural transitions and ATP hydrolysis, calcium is transported to the interior of the sarcoplasmic reticulum where it is bound to calsequestrin, the protein responsible for calcium storage. Without this binding, more calcium could not be transported into the sarcoplasmic reticulum, and the sarcoplasmic concentration of calcium could not be lowered.

It is important to note that energy (ATP) is required for contraction, via myosin ATPase, and to stop contraction, by action of the calcium pump. When energy is depleted postmortem, the muscle goes into a state of rigor mortis.

15.4.4.3 Soluble Components of the Muscle Cell

Soluble proteins of the muscle cell are known by various terms, for example, myogen, which is a water extract of muscle, or simply sarcoplasmic extract. In any case, the soluble proteins of the muscle cell constitute a very significant portion of the proteins of the cell, usually from 25 to 30% of the total proteins.

Most of these soluble proteins are enzymes, principally glycolytic enzymes, but other enzymes are present as well, including those of the pentose shunt and such auxiliary enzymes as creatine kinase and AMP deaminase. The oxygen storage component of the muscle cell, myoglobin, is also a water-soluble protein. The high viscosity of the sarcoplasm is probably attributable to the presence of soluble proteins in concentrations as high as 20–30%. In some species, glyceraldehyde-3-phosphate dehydrogenase comprises approximately 10% of this soluble protein. As noted, some of these proteins are attached to the cytoskeleton, in which case they are not soluble in vivo but are easily solubilized.

Other soluble nonprotein constituents of the sarcoplasm include various nitrogen-containing compounds, such as amino acids and nucleotides; some soluble carbohydrates (intermediates in glycolysis); lactate, which is the major end product of glycolysis; a number of enzyme cofactors; and inorganic ions, including inorganic phosphate, potassium, sodium, magnesium, and calcium.

15.4.4.4 Insoluble Components of the Muscle Cell

Some components of the muscle cell are not soluble in water or in dilute or concentrated salt solutions that are encountered during muscle foods processing. This category includes unextractable contractile proteins, such as titin and desmin, portions of membrane systems, glycogen granules, fat droplets, and the largest fraction, connective tissue proteins. The principal neutral lipids in muscles are triacylglycerols and cholesterol. Proteins of the membrane fractions are usually very insoluble unless they are treated with detergents (surface-active agents). These fractions are therefore usually not extractable with most aqueous solvents.

Most of the lipid material in the cell is associated with membranes, and these lipids are typically rich in phospholipids. The phospholipid content varies among different membranes. Phospholipids comprise approximately 90% of the lipids in mitochondria and about 50% of the lipids in plasma membrane. The total lipid content of the muscle cell is small, only about 3–4% of the total weight of mammalian cells and less than 1% of the muscle cells of lean fish [1]. These lipids are, however, extremely important because they are involved in the structure and function of membranes, and as we shall see later, they are very important in some deteriorative reactions.

The major phospholipids of muscle are lecithin (phosphatidylcholine), phosphatidylethanolamine, phosphatidylserine, phosphatidylinositol, and some of the acidic glycerol phosphatides, such as cardiolipin (Table 6). The amounts of each of these substances vary depending on the particular subcellular fraction examined. This is true also to a somewhat lesser degree for the kinds of fatty acids that make up the phospholipids. Furthermore, compositional differences of the sort just mentioned also exist among muscles of a given species and among comparable muscles of different species.

15.4.5 Muscle Fiber Types

Individual muscles are historically the "smallest unit" used in assigning taxonomical classification relative to physiological activity. According to this approach all muscles are classified as either fast-twitch or slow-twitch. Fast-twitch (white or light in appearance) muscles have a more rapid progression to peak tension and decline to resting tension than do slow-twitch (red or dark appearance) muscles. Fast-twitch muscles are found in anatomical locations where forceful movement is needed, wherease slow-twitch muscles are associated with long sustained contraction, that is, for maintenance of posture. Their appearance, red versus white, is related to their energy metabolism. Fast-twitch muscles require a rapid source of energy, so glycolysis is the predominanat pathway used by these muscles. Slow-twitch muscles rely primarily on oxidative metabolism. This requires a large amount of myoglobin for oxygen storage, and these muscles are therefore redder (or darker) than fast-twitch (white) muscles. Although the differences between white (fast-twitch) breast muscles of poultry and red (slow-twitch) thigh muscles are easy to distinguish based on appearance, twitch types of muscles are more difficult to distinguish in beef. Indeed, a white beef muscle may appear as dark as a red poultry muscle.

Individual fibers can be classified histochemically with respect to enzyme activity (ATPase) related to twitch speed, and types of energy metabolism. Three classes are generally recognized: type I, slow twitch oxidative; type IIA, fast-twitch oxidative; and type IIB, fast-twitch glycolytic [74]. Alternative names for these classes are red or βR for type I, intermediate or βW for type

TABLE 6 Intramuscular Phospholipid Distribution

	Phospholipid distribution[a]						
	Turkey		Fish				
Phospholipid	White	Dark	White	Dark	Pig	Beef	Sheep
Phosphatidylethanolamine	20.8	26.0	22.5	18.0	28.1	26.6	21.3
Phosphatidylinositol+ Phosphatidylserine	10.3	9.0	8.0	19.0	9.5	9.7	6.4
Phosphatidylcholine	49.7	48.6	50.7	27.9	51.0	46.5	42.2
Sphingomyelin	12.2	11.2	4.8	11.7	5.8	4.5	6.4
Lysophophatidyl choline	7.0	5.2	2.9	5.8	—	0.7	—
Phosphatidic acid	—	—	2.2	4.9	—	0.3	—
Cardiolipin	—	—	3.8	9.11	5.0	8.9	7.0
Phosphatidylglycerol	—	—	—	—	—	0.3	—
Lysobisphosphatidic acid	—	—	—	—	—	0.1	—
Total	100	100	94.9	96.4	99.4	97.6	83.3

[a]Expressed as a percentage of total intramuscular phospholipid
Source: Ref. 3.

IIA, and white or αW for type IIB. Turkey breast meat contains 95% IIB fibers, whereas muscles from the leg and thigh have 50–70% type IIA and I fibers combined [96]. Many muscles used for meat contain a variety of fibers; therefore the distinct differences seen between poultry breast and thigh meat should be considered an exception rather than the rule.

As compared to white fibers, red fibers (type I) tend to be smaller; richer in mitochondria, myoblogin, and lipid; more generously supplied with blood; and more dependent on the tricarboxylic acid cycle and oxidative metabolism. Therefore, glycogen and many of the enzymes related to glycolysis tend to be less abundant in red fibers than in white. In addition, red fibers have a thicker sarcolemma, much less extensive and more poorly developed sarcoplasmic reticulum, less sarcoplasm, and a less granular appearance than do white fibers. The sarcoplasmic reticulum is more developed in white fibers to accomodate the rapid-twitch respone. The Z disks of red muscle fibers also tend to be thicker than those in white muscle fibers. The Z disks of white (pectoralis major) and red (semitendinosus) turkey sarcomeres are 39.0 and 53.0 nm thick, respectively [16].

Histochemical stains have commonly been used to detect biochemical differences for fiber classification, because measurement of twitch speed was impossible. It is now known that there are specific fast and slow isoforms for the majority of sarcomeric proteins. Isoforms are proteins with similar structure, names, and functions, but with small differences in the primary sequence of amino acids. For many myofibrillar proteins, there are several fast or slow isoforms. This allows an organsim to have a continuum of contraction speeds and precise regulation of movement, but makes classification of fibers more difficult. Adult rat muscle contains three fast myosin heavy-chain isoforms (types IIa, IIb, and IId) and one slow isoform (type I)[84]. These four isoforms are found in various combinations in individual fibers, producing 10 different fiber types. The number of fiber types are even greater if one considers differences in energy metabolism.

In fish muscle, fiber types are more distinctive than in mammalian or avian muscles. The arrangement of these muscles in codfish is shown in Figure 9. The dark muscle is concentrated superficially, particularly near the lateral line—a group of skin sensory organs extending in a single row, from head to tail, along the surface of each side of the body. In a few fish, such as tuna, which are very active and fast swimmers, deep-seated dark muscle also is present. A thin layer of intermediate (type IIA) muscle fibers separates the dark from the light fibers. In some species of fish, particularly salmonids, a few isolated dark fibers are scattered throughout the main bulk of the white muscle. Some deviation from the pattern in Figure 9 is not uncommon.

Generally, the proportion of dark muscle in fish increases toward the tail region. The fraction of red muscle fibers in the total muscle tissue varies greatly among species. Migratory (fatty) fish have abundant red tissue. At a point one-third the way forward from the tail, red fibers constitute 15–30% of the total. For bottom-feeding lean fish, the percentage of red fibers is low, ranging from 2 to 12% at this same location [29], and some fish species have no detectable red fibers at this location.

White muscle from fish is very uniform in composition no matter where it is located. Dark muscle, however, varies in composition as a function of its location, containing more lipid in the anterior part of the fish and more water and protein in the posterior part [55]. Also, when fish fatten, lipids accumulate mostly in the anterior dark muscle. A larger fraction of dark muscle tissue is present in fish that live in a low-tmeperature environment than in fish accustomed to higher temperatures [55]. Like other animals, the white muscle in fish appears to be used mostly for short, intense expenditures of energy, whereas the dark muscle is used for longer term exertion (swimming), which is why one finds more dark muscle tissue in migratory fish.

It has been suggested that red muscle tissue in fish serves many of the same functions as the liver in warm-blooded vertebrates. although this hypothesis is not widely accepted, the red muscle tissue of fish does have characteristics that give this hypothesis some validity. for example, unlike warm-blooded vertebrates, the content of glycogen is greater in red muscle

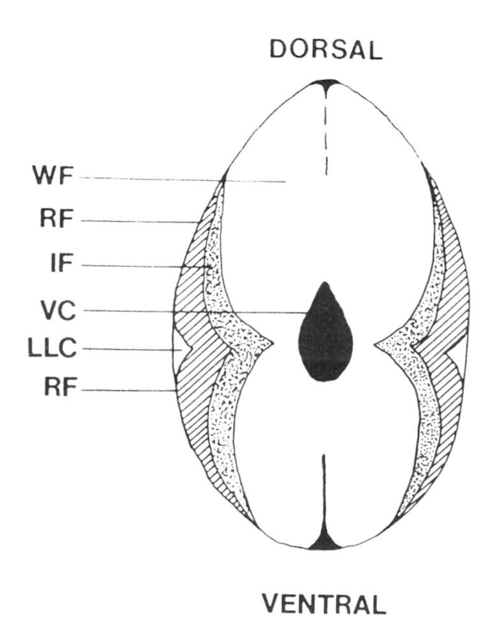

FIGURE 9 Transverse section of cod at point of maximal flexure: WF, white fibers; RF, red fibers; IF, intermediate fibers; VC, vertebral column, LLC, lateral line canal (From Ref. 29, p. 229.)

tissue than in white. Another characteristic worth noting is that the content of trimethylamine oxide in migratory fish is greater in dark muscle than in white, but the reverse is true in the muscle of nonmigratory lean fish [55].

Muscle fiber types can be changed by exercise. This is most evident in humans who spend extensive time lifting weights. Exercise can be used to change the fiber type in meat animals such as sheep [2]. Shown in Table 7 are contraction properties and myosin isoforms found in four rabbit muscles, and changes in these properties caused by chronic electrical stimulation. A comparision of the psoas and tibialis anterior fibers shows that a change in twitch speed of type IIB fibers is associated with an increased ratio of LC3 to LC1 in the myosin head. Chronic stimulation of tibialis anterior fibers changes the myosin heavy chain from type IIB to type IIA and changes the ratio of LC1 to LC3.

An understanding of the differences among muscle fibers and factors that control fiber types is important in food science since many of the characteristics of muscle postmortem are a function of fiber type. For example, muscle color and susceptibility to cold shortening (Sec. 15.6) are influenced greatly by fiber type; that is, muscles with a preponderance of white fibers generally are much less susceptible to cold shortening than are muscles that have a majority of red fibers.

The molecular switch that turns on different muscle genes is a family of proteins called helix–loop–helix (indicating their secondary structure) transcription factors. If these could be controlled by breeding or recombinant DNA approaches, one could modify animals with respect to meat quality.

15.4.6 Connective Tissue

Connective tissue consists of various fibers, several different cell types, and amporphous ground substances (also called basement lamina). Connective tissues hold and support the muscles by means of the component tendons, epimysium, perimysium, and endomysium. The relationship between connective tissue and muscle cells is so intimate that most likely all substances passing in or out of the muscle cell diffuse through some type of connective tissue.

Several types of cells are found in the connective tissue. These include fibroblasts, mesenchyme cells, macrophages, lymphoid cells, mast cells, and eosinophilic cells.

TABLE 7 Contraction Properties and Myosin Isoforms in Rabbit Muscle

	Muscle and fiber type				
Properties	Psoas (IIb)	Tibialis anterior (IIb)	Tibialis anterior (IIa) (chronic stimulation)[a]	Vastus intermedius (IIa)	Soleus (I)
Velocity[b]	2.07	1.63	0.86	0.98	0.41
Force (N/m^2)	12.45	11.97	10.00	10.88	10.68
Myosin heavy chain isoform	IIb	IIb	IIa	IIa	I
LC-1f content (%)	51.6	62.5	76.2	74.2	
LC-3f content (%)	48.4	37.5	23.8	25.8	

[a]10 Hz, 8 hr/day for 7 weeks.
[b]Fiber lengths/sec.
Source: Ref. 87.

15.4.6.1 Collagen

The major fraction of connective tissue is collagen. This component is important because it contributes significantly to toughness in mammalian muscle. Also, the partially denatured product of collagen, gelatin, is a useful ingredient in many food products because it serves as the functional ingredient in temperature-dependent gel-type desserts. Collagen is abundant in tendons, skin, bone, the vascular system of animals, and the connective tissue sheaths surrounding muscle. Collagen comprises one-third or more of the total protein of mammals. About 10% of mammalian muscle protein is collagen [6], but the amount in fish is generally much less [82]. Some of the collagen is soluble in neutral salt solution, some is soluble in acid, and some is insoluble.

The collagen monomer is a long cylindrical protein about 2800 Å long and 14–15 Å in diameter. It consists of three polypeptide chains wound around each other in a suprahelical fashion. Collagen occurs in several polymorphic forms. the most common collagen type is type I collagen and it contains two identical polypeptide chains, designated $\alpha 1(I)$, and a third chain, $\alpha 2$, which has a different amino acid sequence [83]. Each chain has a molecular mass of about 100,000 D, yielding a total molecular mass of about 300,000 D for collagen. The three chains are held together by hydrogen bonding. A second type of collagen, type III, is made up of three identical chains designated $\alpha 1(III)$. This type of collagen is unusual in that intramolecular disulfide bonds exist in the nonhelical, carboxy-terminal peptide.

A third type of collagen molecules, type IV, is about a third greater in length than types I and III and, like type III collagen, contains oxidizable cysteine residues. Type IV collagen is rich in hydroxyproline and hydroxylysine. It has a nonhelical, globular region at its C terminus. It is thought to form gossamer lattice sheets of a "chicken wire" structure by interaction of two C-termini and the overlapping of four adjacent N-termini through disulfide bonding [6].

A fourth type of collagen is type V collagen. Like type IV, it is rich in hydroxyproline and hydroxylysine content but contains no cysteine. It is about the same length as types I and II (the latter not found in meat) and consists of two or three different types of polypeptide chains.

The presence of type I collagen in the epimysium of muscle, types I and III in the perimysium, and types III, IV, and V in the endomysium has been reported [83]. Type III collagen appears to have an especially important role in imparting toughness to mammalian muscles. In some instances, the single strands of polypeptides that constitute collagen are cross-linked with covalent bonds. When two of the peptides are joined in this fashion it is called a β component; however, when all three are so joined, the product is known as a γ component. The solubility of collagen decreases as intermolecular cross-linking increases.

Polypeptides of collagen are mostly helical with the exception of the few residues at each end. However, the helices differ from the typical α-helix due to the abundence of hydroxyproline and proline, which interfere with α-helical structure. Collagen molecules link end to end and adjacently to form collagen fibers, as shown in Figure 10. There is a periodicity in the cross-striations of collagen at about 640–700 Å intervals. Collagen fibers are sometimes arranged in parallel fashion to give great strength, as in tendons, or they may be highly branched and disordered as in skin.

The amino acid composition of collagen is nutritionally unbalanced and is unusual in several other respects. Collagen is almost devoid of tryptophan; it is rich in glycine, hydroxyproline, and proline; and it is one of the few proteins that contains hydroxylysine. Glycine represents nearly one-third of the total residues, and it is distributed uniformly at every third position throughout most of the collagen molecule. The repetitive occurrence of glycine is absent in the first 14 or so amino acid residues from the N-terminus and the first 10 or so from the C-terminus, with these end portions being termed "telopeptides" [37]. Collagen is the only protein that is rich in hydroxproline (up to 10% in mammalian collagen); however, fish collagens contain less of this

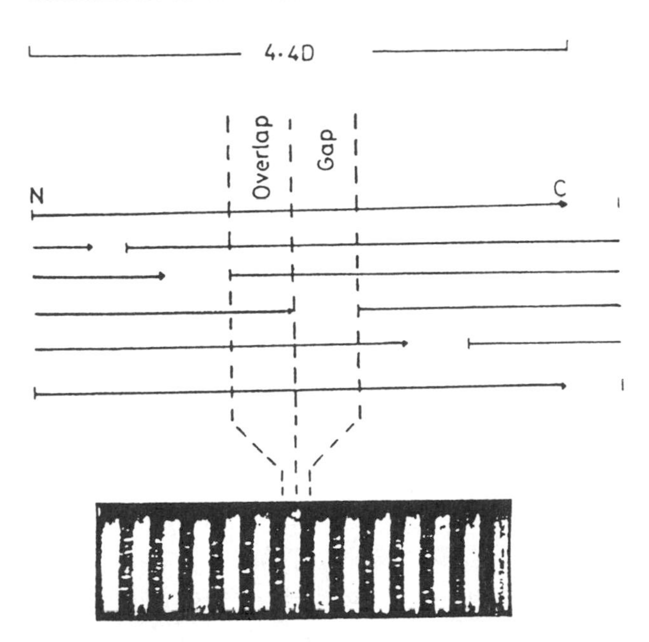

FIGURE 10 Illustration of the overlap structure of the collagen fiber responsible for the banding pattern of a negatively stained collagen fiber as seen by electron microscopy. (From Ref. 6.)

amino acid than do mammalian collagens. Because hydroxyproline is present in significant amounts in so few other proteins, it is often used as a measure of the amount of collagen in food samples.

The presence of proline stabilizes the helix structure by preventing rotation of the N-C bond. Hydroxyproline also stabilizes the collagen molecule, and collagens that contain small concentrations of both of these imino acids denature at lower temperatures than do those with large concentrations (Fig. 11). The imino acid content of fish collagens, and therefore their thermal stability, correlates with the water temperature of their normal habitat. Proteoglycan and glycoprotein contents can also affect collagen thermal stability.

Two important oxidations take place in collagens. These are conversion of proline to hydroxyproline and conversion of lysine to hydroxylysine. This latter compound and lysine can be oxidized, in some instances to 1-amino-1-carboxy-pentan-5-al. Oxidation of proline and lysine to their hydroxylated residues is catalyzed by proline hydroxylase and lysine hydroxylase, respectively. These enzyme systems utilize molecular oxygen, α-ketoglutarate, ferrous ion, and a reducing substance, such as ascorbate. Production of the 1-amino-1-carboxy-pentan-5-al is catalyzed by a copper-containing enzyme called lysyl oxidase; this compound is involved in collagen cross-linking.

The covalent cross-links involved in the β and γ components of collagen and the intermolecular cross-links between collagen molecules form spontaneously by the condensation of aldehyde groups. This may involve an aldol condensation-type reaction, or formation of a Schiff base when the aldehyde reacts with an amino group. When hydroxylysine reacts with hydroxylysine aldehyde, the reaciton product undergoes an Amadori-type rearrangement to form a "keto" structure, hydroxylysino-5-keto-norleucine. Examples of these reactions are shown in Figure 12.

As animals age, collagen cross-links are converted from a reducible form to a more stable nonreducible form. The nature of the "mature" nonreducible cross-link is not known, although

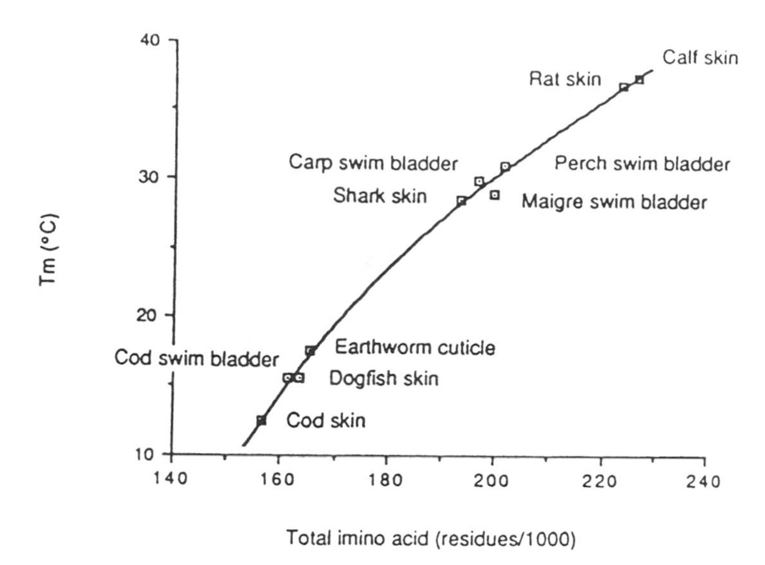

FIGURE 11 Relationship between the total imino acid content (proline + hydroxyproline) of various collagens from different species and the molecular melting temperatures (T_m) of these collagens. (From Ref. 6.)

several hypotheses have been proposed [6]. It is likely that nonreducible cross-links involve several functional groups and an extensive polymeric network. Intermolecular crosslinks are confined to the end-overlap region involving a lysine aldehyde in the telopeptide of one chain and a hydroxylysine in the helical region of an adjacent chain (28 nm from the N- or C-terminus). Noncollagenase proteinases, because they can attack the telopeptide region of collagen, can effectively break these cross-links.

The number of cross-links in collagen also increases with increasing age, and this may partially explain why meat from older mammalian animals is tougher than that from younger animals, even though muscles from younger animals generally contain more collagen. In fish, the situation is very different. Collagens of the myocommata of older fish are weaker and have fewer cross-links than do the collagens of younger fish. On the other hand, older fish contain more collagen (evident as thicker myocommata) than younger fish.

As cross-linking of collagen increases it becomes less soluble in a variety of solvents, such as salt and acid solutions. With advancing age the amount of insoluble collagen often increases manyfold in mammalian muscle, whereas the amount ot insoluble collagen of cod increases only slightly with age and the amount of soluble collagen actually increases [55]. Starving fish produce more collagen and collagen with a greater degree of cross-linking than do fish that are well fed.

Most proteolytic enzymes have little activity against native collagen, although they will readily degrade denatured collagen. Collagenase has been identified in several animal tissues. Its presence is difficult to detect because it exhibits low activity, apparently because of control mechanisms operative in the tissue.

For the most part, animal collagenases cleave a single bond in each of the three chains of the collagen molecule. Several microorganisms, particularly *Clostridium* species, produce collagenases. These enzymes differ from the animal collagenases in that they degrade collagen extensively. Noncollagenase proteinases can cleave the collagen molecule in the telopeptide region. They can contribute to destabilization of the collagen molecule by disrupting the region in which intermolecular cross-links are formed [10]. The importance of collagenase enzymes in

(a)

−NH−CH−CO−
|
$(CH_2)_3$
|
HC=O

\+

HC=O
|
$(CH_2)_3$
|
−NH−CH−CO−

→

−NH−CH−CO−
|
$(CH_2)_3$
|
HC
‖
C−C_H=O
|
$(CH_2)_2$
|
−NH−CH−CO−

2 α-amino-adipic-δ-semi-aldehydes → α, β-unsaturated aldol

(b)

−NH−CH−CO−
|
$(CH_2)_2$
|
CHOH
|
CH_2
|
NH_2
hydroxylysine

\+

HC=O
|
$(CH_2)_3$
|
−NH−CH−CO−

→

−NH−CH−CO−
|
$(CH_2)_3$
|
CHOH
|
CH_2
|
N
‖
HC
|
$(CH_2)_3$
|
−NH−CH−CO−

α-amino-adipic-δ-semi-aldehyde → Schiff base dehydro-hydroxylysinonor-leucine

(c)

−NH−CH−CO−
|
$(CH_2)_2$
|
CHOH
|
CH_2
|
NH_2
hydroxylysine

\+

C=O
|
CHOH
|
$(CH_2)_2$
|
−NH−CH−CO−
hydroxylysine aldehyde

Schiff base → Amadori →

−NH−CH−CO−
|
$(CH_2)_2$
|
CHOH
|
CH_2
|
NH
|
CH_2
|
C=O
|
$(CH_2)_2$
|
−NH−CH−CO−
hydroxylysino-5-keto-norleucine

FIGURE 12 Cross-link formation in collagen by side chain groups. (a) Aldol condensation followed by loss of water. (b) Schiff base formation. Lysine reacts in a manner analogous to hydroxylysine. (c) Schiff base formation followed by Amadori rearrangement.

the breakdown of collagen in situ is not known, but they seem to be of some importance in postmortem tenderization.

15.4.6.2 Conversion of Collagen to Gelatin

Collagen fibrils shrink to less than one-third their original length at a critical temperature, known as the shrinkage temperature, characteristic of the species from which the collagen is derived [37]. This shrinkage involves a disassembly of fibers and a collapse of the triple-helical arrangement of polypeptide subunits in the collagen molecule. Essentially the same type of molecular changes occur when collagen is heated in solution, but at a much lower temperature. The midpoint of the collagen-to-gelatin transition is defined as the melting temperature (see Fig.

11). During the collagen to gelatin transition, many noncovalent bonds are broken along with some covalent inter- and intramolecular bonds (Schiff base and aldo condensation bonds) and a few peptide bonds. This results in conversion of the helical collagen structure to a more amorphous form, known as gelatin [93]. These changes constitute denaturation of the collagen molecule but not to the point of a completely unstructured product. If the latter event happens, glue instead of gelatin is produced.

After gelatin is produced and the temperature is lowered to below the critical value, there is a partial renaturation of the collagen molecule, involving what is called the "collagen fold" [93]. Apparently, those parts of collagen that are rich in proline and hydroxyproline residues regain some of their structure, following which they can apparently interact. When many molecules are involved, a three-dimensional structure is produced that is responsible for the gel observed at low temperatures (see Chap. 3). This collagen fold is absolutely dependent on temperature. Above the melting temperature, the structure degrades and so does the gel. The strength of the gel formed is proportional to the square of the concentration of gelatin and directly proportional to molecular weight [93]. Ionic strength and pH normally have only small influences on gel structure over the range of values encountered in food products.

Processing of collagen into gelatin involves three major steps. There is first the removal of noncollagenous components from the stock (skin and bones), then the conversion of collagen to gelatin by heating in the presence of water, and finally recovery of gelatin in the final form. The molecular properties of gelatin depend on the number of bonds that are broken in the parent collagen molecules. Removal of noncollagenous material can be done with acid or with alkali, and the procedures used influence the properties of the gelatin. Acid hydrolysis is a milder treatment that will effectively solubilize collagens of animals slaughtered at a young age, such as pigs. Gelatins produced in this manner are termed type A gelatins. Type B gelatins are produced by alkali hydrolysis of beef materials and results in deamidation and a greater range of molecular weight species. Type A and B gelatins exhibit isoelectric points of pH 6–9 and about 5, respectively.

Conversion of collagen to gelatin occurs during normal cooking of meat, and this accounts for the gelatinous material that is sometimes evident in meat after heating and cooling. An aspic type of product, for example, results from the conversion of collagen to gelatin. Also, conversion of the collagen molecule to the much less structured gelatin can play an important role in tenderness of meat, especially in meat of poor quality. This kind of meat (e.g., pot roast or stew meat) has a high content of collagen and is usually cooked for a long time in liquid to achieve the desired tenderness, that is, the desired conversion of collagen to gelatin. The tenderness and flakiness of cooked fish are due to the relative ease with which their collagen is converted to gelatin.

15.4.7 Fat Cells

The fat cell deserves special attention. Fat cells appear to arise from undifferentiated mesenchyme cells, usually developing around the blood vessels in muscle. As a fat cell develops, it accumulates droplets of lipid. These droplets grow by coalescence, and eventually, in the mature fat cell, only a single large drop of fat exists. Cytoplasm surrounds the fat droplet, and subcellular organelles are located in this thin cytoplasmic layer. Fat cells are located outside the primary muscle bundles, that is, in the perimysial spaces or subcutaneously, but not in the endomysial area. Lipid in the latter area is usually associated with membranes or it exists as tiny fat droplets distributed in the muscle fiber itself.

The amount of fat that accumulates in an animal depends on age, the level of nutrition, exercise, and other physiological factors that function in the specific muscle in question. Initially, fat tends to accumulate abdominally or under the skin, and only later does it accumulate around the muscle. The latter accumulation leads to the so-called marbling of meat. Since this is the last fat to be deposited, a large quantity of feed is necessary to develop marbling. The amount of fat surrounding and within muscles is not constant but can increase or decrease depending on the food intake of the animal and other factors. The amount of fat surrounding the major muscles influences the appearance or "finish" of the meat.

The membranes surrounding fat cells contain small amounts of phospholipids and some cholesterol, but the most abundant lipids in adipose tissue are triacylglycerols and free fatty acids. This fat can be extracted from the carcass by rendering or heating to melt the fat and rupture fat cells. This fat can be used as such, as lard, for example, or it can be chemically modified for use in the manufacture of shortenings and margarines. The fat component of meat is very important since some is needed to produce a satisfactory flavor and mouth feel; however, excessive amounts decrease the usable lean portion without improving these quality attributes.

15.5 BIOCHEMICAL CHANGES IN MUSCLE POSTMORTEM

15.5.1 Biochemical Changes Related to Energy Metabolism

Muscle is a highly specialized tissue, and its functioning represents a classic example of conversion of chemical to mechanical energy in living systems. Muscle requires a large outlay of energy to rapidly operate the contractile apparatus. This energy is rapidly derived from the high-energy compound ATP. Resting muscle is normally about 5 m*M* in ATP, which is not enough for a long sustained contraction. The storage form of energy in muscle is creatine phosphate. Muscle contains 20 m*M* creatine phosphate, and this compound can rapidly transfer its high-energy phosphate to ADP to prevent excessive decreases in ATP levels during periods of vigorous muscle activity. Resynthesis of ATP from ADP and creatine phosphate is catalyzed by the enzyme creatine kinase. The enzyme adenylate kinase converts two molecules of ADP to one molecule of ATP and one molecule of AMP (see Fig. 14); this reaction also serves as a source of ATP in muscle cells.

For long-term activity, the living muscle must maintain a physiological level of ATP, and this is accomplished by oxidation of storage substrates, usually carbohydrate or lipids. Lipid metabolism seems to be an especially important source of utilizable energy in muscles accustomed to sustained activity, for example, those that support prolonged running of animals or prolonged flights of birds. Carbohydrates are also an important energy source for muscle. Glycogen is the most important source of carbohydrate energy. In red muscles and in muscles that are not working extremely hard, it is likely that most of the energy is supplied via the tricarboxylic acid cycle and the mitochondrial electron transport system (aerobic respiration). This system provides a large quantity of ATP molecules per molecule of substrate utilized and allows for the complete conversion of substrate to carbon dioxide and water.

The mitochondrial system, however, requires oxygen and in some instances, when the muscle is under heavy stress, oxygen is not available in sufficient amounts to maintain mitochondrial function. The anaerobic glycolytic system then becomes predominant. This is especially likely in white muscles, which are generally involved in sporadic bursts of activity requiring very large amounts of energy.

During anaerobic glycolysis, glycogen is converted through a series of phosphorylated six-carbon and three-carbon intermediates to pyruvate, which is then reduced to lactate. This system requires the cofactor NAD^+, and it continually regenerates the NAD^+ required. The terminal enzyme of the sequence, lactate dehydrogenase, is principally responsible for regeneration of NAD^+. In anaerobic glycolysis, ATP production is much less efficient than it is in aerobic respiration. For example, anaerobic glycolysis yields only 2 or 3 moles ATP per mole of glucose, whereas aerobic respiration yields 36 or 37 moles ATP. Also, anaerobic glycolysis results in incomplete oxidation of substrates and accumulation of lactate. Lactate can penetrate the cell membrane, and much of it is removed to the blood, where it goes to the liver and is used in the resynthesis of glucose. The glucose is then carried back to the muscle, where glycogen is eventually produced. Lactate transport out of fish muscle is very slow, and a very long period is required to deplete it [55].

Anaerobic glycolysis is favored only under conditions in which the mitochondrial system cannot function, that is, under conditions of limited oxygen. The amount of anaerobic glycolysis that occurs in a muscle is, therefore, dependent on many factors, including the vascular system, the type of muscle, the amount of myoglobin, and the number of mitochondria present.

These reactions should be of interest to food chemists because many of them proceed after the death of the animal and have an important bearing on food quality. Muscle tissue is often referred to as "dead," as opposed to postharvest fruits and vegetables, which are said to be "alive." This distinction arises mainly because of differencees in the physiology and morphology of muscle tissue as compared with plant tissue. Animal tissues are more highly organized than plant tissues, and life processes of animals rely heavily on a highly developed circulatory system. After death, all circulation ceases, and this rapidly brings about important changes in the muscle tissue. The principal changes are attributable to a lack of oxygen (anaerobic conditions) and the accumulation of certain waste products, especially lactate and hydrogen ions. Plant tissue is less dependent on a highly developed circulatory system than animal tissue, and although certain substances are no longer available to the fruit or vegetable after harvest (see Chap. 16), oxygen can diffuse in, carbon dioxide can diffuse out, and waste products can be removed from the cytoplasm by accumulaiton in the large vacuoles present in mature parenchymatous tissue. In contrast, enzymes of muscle postmortem are faced with an entirely different environment.

It is believed that cells in general try to maintain a high energy level. This is true for the muscle cell postmortem; however, it has restrictions imposed on it because the circulatory system has been disrupted. In a short time postmortem, the mitochondrial system ceases to function in all but surface cells, since internal oxygen is rapidly depleted. Oxidative metabolism of some substrates, such as lipids, ceases at this time. ATP is gradually depleted through the action of various ATPases, with some contributed by the contractile proteins but most coming from the membrane systems. Some ATP is temporarily regenerated by conversion of creatine phosphate to creatine and the transfer of its phosphate to ADP. The adenylate kinase reaction discussed earlier may also function here. After the creatine phosphate has been used up, which occurs fairly rapidly postmortem, anaerobic glycolysis continues to regenerate some ATP with the end product, lactate, accumulating. Glycolytic activity may cease because of exhaustion of substrate or, more likely, because of the decrease in pH caused by ATP hydrolysis. Glycogen content of chicken breast decreases from about 9.4 mg/g muscle at slaughter, to about 0.2 mg/g muscle 24 hr after slaughter [47]. As glycolytic activity slows down, ATP concentration decreases, with most of the ATP being depleted in 24 hr or less, depending on the species and the circumstances. Although the hydrogen ions generated in muscle come form hydrolysis of ATP and not from production of lactate, there is a close correlation between the amount of lactate produced and the pH decline. This relationship occurs because there is an almost linear relationship between ATP produced by the glycolytic system (and hence the ATP that may be hydrolyzed later) and the amount of lactate produced.

The decrease in pH that accompanies postmortem glycolysis has an important bearing on meat quality, which is discussed in Section 15.6.2. The rate and extent of pH decline are affected by metabolic rate and the buffering capacity of muscle. Generally, white muscles have relatively large contents of the amino acids carnosine and serine, and these probably function as buffering agents in the muscle cell. Under ordinary circumstances, the factor that limits the extent of postmortem glycolysis is pH. When the pH is low enough, certain critical enzymes, especially phosphofructokinase, are inhibited and glycolysis ceases. The final pH attained is called the "ultimate pH," and this value has an important influence on the textural quality of meat, its water-holding capacity, its resistance to growth of microorganisms, and its color.

Glycogen is depleted by several stress conditions, and in general, it is desirable to minimize these conditions as much as possible. Stress conditions include exercise, fasting, hot and cold temperatures, and fear [49]. Increased glycogenolysis and lipolysis generally accompany stress, but not all animals react in the same way or to the same degree to any one of these stresses. It is very difficult, for example, to deplete cattle of glycogen by exercising, whereas significant lowering of glycogen in pig muscles occurs with relatively mild exercise. Cattle must be subjected to both psychological and physical stress to achieve important modification of meat properties. Within a given species, the response to stress may vary among breeds or between sexes. It is possible to literally frighten some stress-susceptible animals sufficiently to cause death. Heat stress affects the quality of chicken breast muscle. Heat-stressed birds exhibit lower ultimate pH values and are tougher than unstressed birds [47], and they have lower water-holding properties [65].

The ultimate pH of fish is not related to stress or the exercise to which a fish is subjected prior to death. In living fish, lactate produced during exercise, such as struggling during harvest, is only very slowly removed from the muscle; thus, when a fish struggles extensively before death, much lactate can be produced and this lactate, for the most part, is present in the muscle postmortem. On the other hand, a fish that struggles only slightly prior to harvest will contain only a small amount of lactate at the point of death, but normal postmortem glycolysis will cause muscle lactate to increase to essentially the same level as that existing in fish that struggled vigorously prior to death. Since the amount of lactate formed is roughly proportional to the amount of ATP produced by anaerobic glycolysis, and since all the ATP in postrigor fish is broken down to produce hydrogen ions, the net effect is that struggling or stress prior to death does not affect its ultimate pH.

15.5.2 Consequences of ATP Depletion

The consequences of ATP depletion in a muscle cell are in part similar to those that presumably occur in any cell in which energy sources become depleted. Biosynthetic reactions come to a halt, and there is a loss of the cell's ability to maintain its integrity, especially with respect to membrane systems. There is an additional unique phenomenon that occurs with muscle cells. ATP is required to disrupt the actin–myosin complex and fuel muscle relaxation by pumping calcium from the sarcoplasm reticulum. The irreversible interaction of actin and myosin causes muscle to pass into a state known as rigor mortis. This means literally the "stiffness of death." The stiffness is due to the tension developed by antagonistic muscles. Individual muscles may or may not become stiff, but all become inextensible when ATP is depleted. When there is sufficient ATP and/or ADP, muscle can be stretched considerably before it breaks. In the state of rigor, however, muscle ATP has been depleted and it consequently cannot be stretched significantly without breaking.

The rate of loss of ATP postmortem is directly and causatively related to the decline in pH and is affected by factors such as species, animal-to-animal variation within a species, type of muscle, and temperature of the carcass. The ultimate pH of muscle varies with species and muscle type. Fast-twitch muscles from beef, chicken, and turkey have respective ultimate pH values of 5.5, 5.5, and 5.7, whereas slow-twitch muscle from these respective species have ultimate pH values of 6.3, 6.1, and 6.4. The ultimate pH of red meat fish, like tuna and mackeral, is around 5.5; that of lean white fish is 6.2–6.6. At the moment of slaughter or harvest, average ATP content of many muscles is in the range 3–5 mg/g of lean tissue, and the time at which most of the ATP is degraded can be used as rough indicator of the time of rigor onset. The time postmortem for rigor onset (incipient loss of extensibility) under normal processing conditions is < 0.5 hr for chicken, < 1 hr for turkey, 0.25–3 hr for pork, 6–12 hr for beef muscle, and up to 50 hr for whale muscle [37]. The time for rigor onset is strongly dependent on temperature, and, in general, extreme cold or heat will decrease the time from death to rigor.

The relationships between several chemical and physical properties of muscle postmortem are illustrated in Figure 13. The decrease in ATP, pH, and glucose proceed as the tissue tries to maintain a relaxed state. These reactions are clearly faster at 25°C than 5°C. Finally a point is reached where the muscle contracts as it progresses into rigor mortis.

15.5.3 ATP Breakdown to Hypoxanthine

Another important consequence of loss of ATP in muscle postmortem is its conversion to hypoxanthine, as shown in Figure 14. Certain 5′-mononucleotides, intermediates in the production of hypoxanthine, are important flavor enhancers in muscle foods. For these compounds to have flavor-enhancing properties, the ribose component must be phosphorylated at postion 5′ and the purine component must be hydroxylated at position 6. Compounds that fall into this category are IMP (phosphorylated inosinic acid) and GMP (phosphorylated guanylic acid). ATP is first converted to ADP and then to AMP by a disproportionation reaction. AMP is then deaminated to IMP (desirable), which can then degrade to inosine and eventually to hypoxanthine. Hypoxanthine has a bitter flavor, and it is suspected as a cause of off flavors in stored fish.

Hypoxanthine also may cause off flavors in unheated irradiated beef since some enzymes instrumental in its formation may remain active (enzymes are more difficult to destroy by ionizing radiations than are microorganisms). Irradiated beef therefore should be exposed to a heat treatment sufficient to inactivate enzymes so that hypoxanthine production and other detrimental enzyme-catalyzed changes do not occur.

The kinetics of ATP breakdown to IMP, inosine, and hypoxanthine in cod muscle at 0°C are shown in Figure 15. In this figure, it is also possible to compare the aforementioned changes with those of glycogen, lactate, and pH at any given time. The production of IMP from ATP (through AMP) is rapid, that of inosine is somewhat less so, and production of hypoxanthine from inosine is slow. Thus, measurement of the levels of different breakdown products at a given time can give an indication of the degree of spoilage. The breakdown products can be measured individually; however, in some cases it is advantageous to know the concentrations of more than one component. For example, the *K* value, which is derived from values of inosine, hypoxanthine, and the total amount of ATP-related compounds, is often used to determine the freshness of fish. The *K* value also may have significance in canned muscle tissue, such as herring, mackerel, or salmon. Since enzymic reactions are stopped during processing, the *K* value should provide an indication of the quality of the raw fish used. The use of levels of ATP degradation products to assess freshness has proven less successful for muscle tissues from warm-blooded animals than for those from cold-blooded animals.

15.5.4 Loss of Calcium-Sequestering Ability

One reason for the tendency for muscle to contract postmorten is a loss in the ability of the sarcoplasmic reticulum and mitochondria to sequester calcium. The sarcoplasmic reticulum and mitochondria release calcium during aging postmortem. The inability of these organelles to

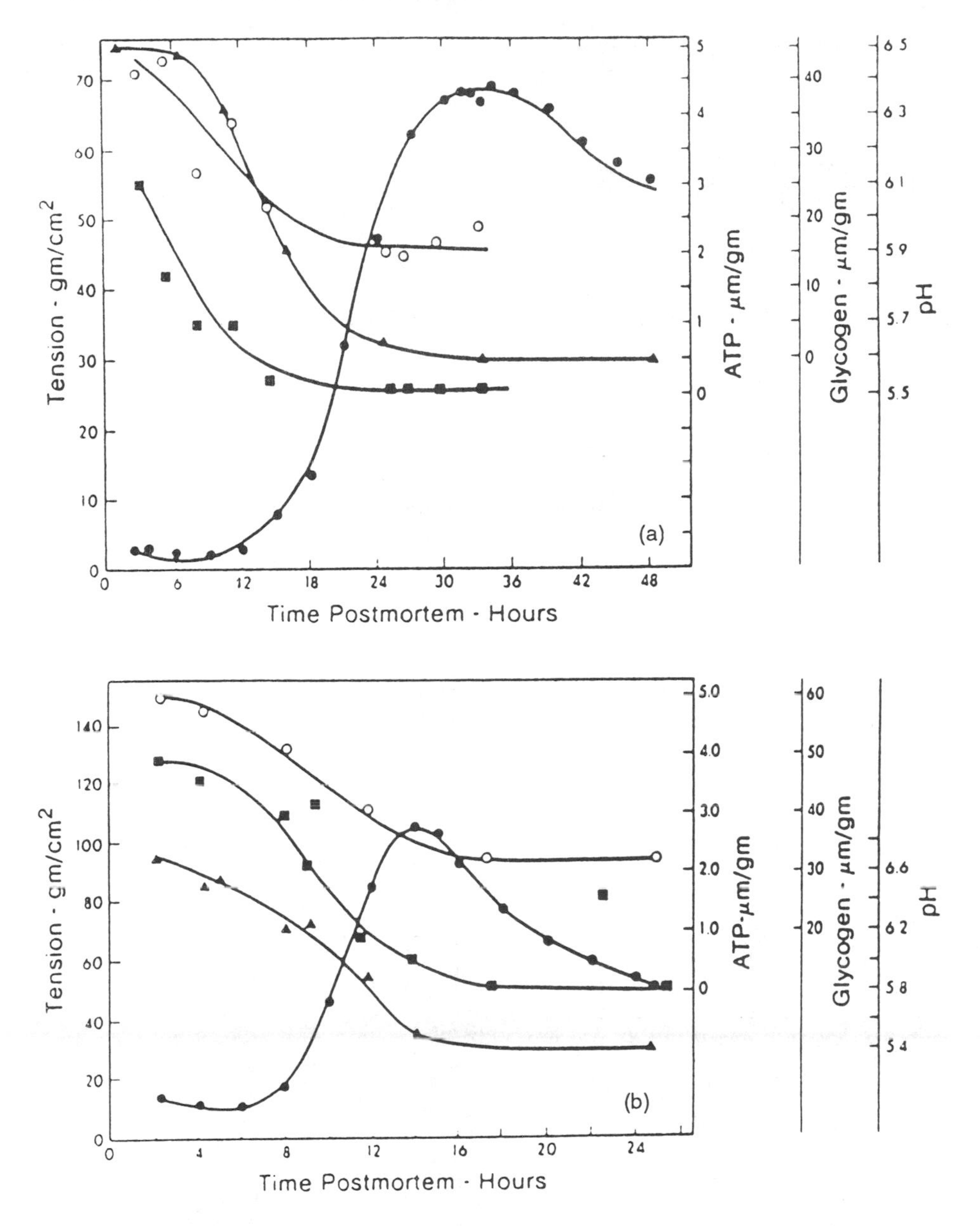

FIGURE 13 Changes in tension, pH, ATP and glycogen concentration in postmortem beef biceps femoris muscle. These changes are shown at (a) 5 and (b) 25°C. (From Ref. 66.)

control the calcium content of the sarcoplasm postmortem is probably both due to a deterioration in the energy-generating systems of the muscle and to an increase in permeability that the sarcoplasmic reticulum and mitochondria exhibit with postmortem age under anoxic conditions (this permits greater leakage of calcium from the membrane). Increased levels of calcium in the sarcoplasm cause contractile proteins to more fully display their ATPase potential and initiate activity of ATPases in membrane systems. This, in turn, causes increased activity of the glycolytic system and a more rapid decline in pH. The sarcoplasmic reticulum of postmortem chicken, however, does not lose its ability to sequester calcium [34].

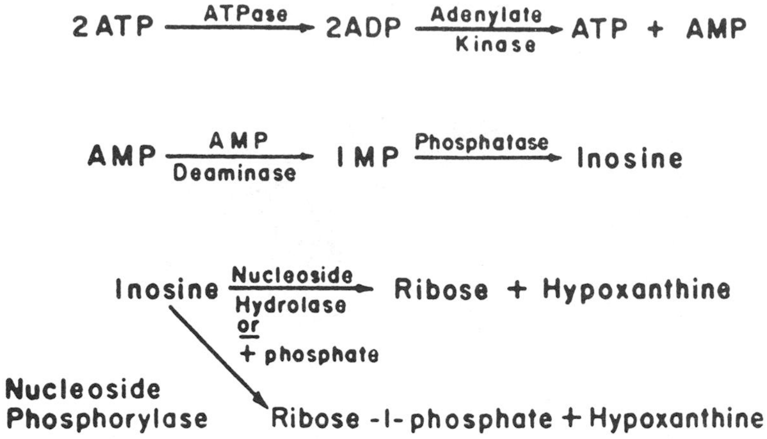

2ATP
ATPase
2ADP
Adenylate
Kinase
ATP + AMP
AMP
AMP
Deaminase
IMP
Phosphatase
Inosine
Inosine
Nucleoside
Hydrolase
or
+ phosphate
Ribose + Hypoxanthine
Nucleoside
Phosphorylase
Ribose -1- phosphate + Hypoxanthine

FIGURE 14 Conversion of ATP to hypoxanthine in postmortem muscle.

6.0
5.0
4.0
3.0
2.0
1.0
0
μ mole/g
ATP
IMP
Ino
Hx
ADP
(a)
0
4
6
8
10
12
ONSET
OF RIGOR
500
400
300
200
100
0
mg/100 g
LACTATE
GLYCOGEN
pH
7.5
7.0
6.5
6.0
pH
(b)
0
4
6
8
10
12
ONSET
OF RIGOR
DAYS AT O C

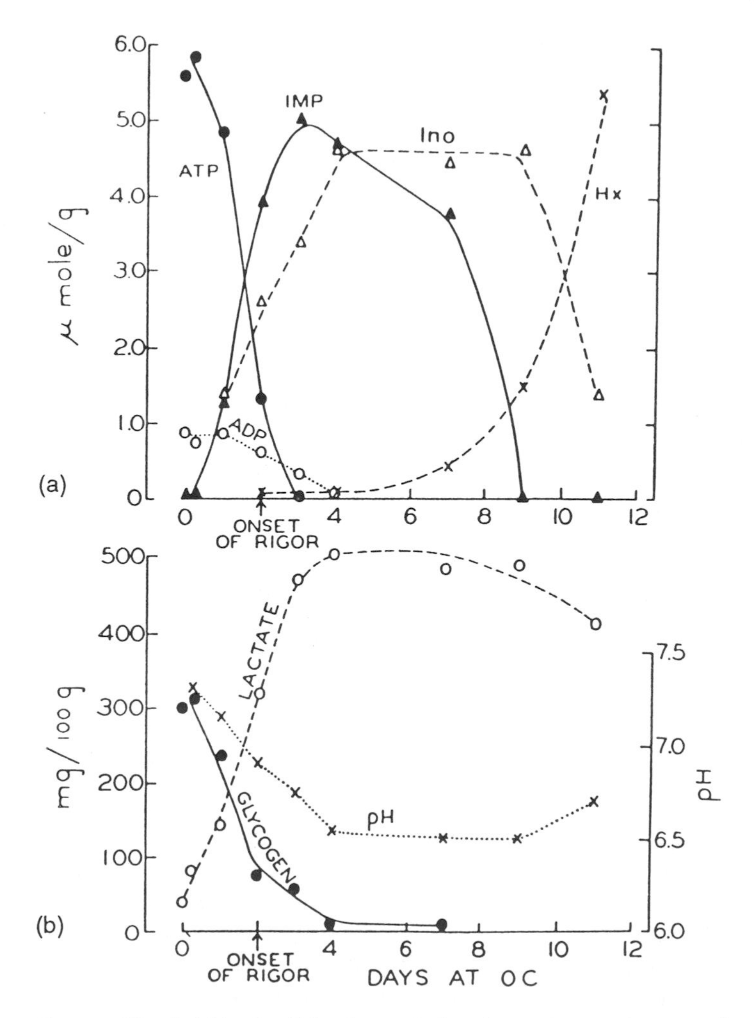

FIGURE 15 (a) Nucleotide degradation in cod muscle stored at 0°C: ATP, •; ADP, ○; IMP, ▲; inosine, △; x. (b) Glycolytic changes. (From Ref. 26, p. 1839)

15.5.5 Loss of Enzymic Regulation

15.5.5.1 Lipolysis and Oxidation of Lipids

Changes occur in the lipid components of muscle tissue postmortem, and these changes have been more extensively studied in fish than in warm-blooded animals. This is probably true for the following reasons: (a) fatty acids of fish lipids are much more unsaturated than those of mammals and birds and thus would be expected to undergo more rapid oxidation with associated development of off-odors and flavors; (b) free fatty acids in fish tissues are expected to have a greater effect on contractile proteins since these proteins are less stable in fish muscle than they are in warm-blooded animals; and (c) a larger proportion of fish tissue is marketed in the frozen state. Freezing can facilitate lipid oxidation, partly because the competing reactions of microbiological spoilage are avoided and partly because of concentration effects. Thus, lipid oxidation is relatively more important in frozen muscle tissue than in fresh tissue.

One important change that occurs in fish muscle lipids postmortem is hydrolysis of glycerol-fatty acid esters with release of free fatty acids. This is catalyzed by lipases and phospholipases. In general, lipase activity is greater in red muscle than in white muscle of the same fish species. In lean fish, the enzyme of principal concern is phospholipase, since most of the lipid in this type of tissue is phospholipid. It is also possible that lipases and phospholipases produce free fatty acids during the early stages of cooking. Hydrolysis of phospholipids appears to occur more rapidly than hydrolysis of triacylglycerols.

Oxidation of lipids also occurs during postmortem storge of muscle tissue, especially in fish muscle. The extent of oxidation depends on the concentration of pro-oxidants, such as endogenous ferrous iron, and the fatty acid composition of the meat [17]. Meats such as fish and poultry contain a high concentration of polyunsaturated fatty acids and are therefore more susceptable to oxidation. Furthermore, the concentration fo ferrous iron, and the ability of that iron to be active in the lipid oxidation reaction, will be a key factor causing differences among species and cuts of meat. In general, dark meats tend to have more reactive iron [17]. Other constituents of meat can accelerate oxidation. Enzymatic (microsomal) and non-enzymatic (ascorbate) reducing systems convert iron from the inactive ferric form to the active ferrous state, thereby promoting oxidation. Lipoxygenase can also accelerate oxidation [14].

Meat also contains natural antioxidants. Compounds such as tocopherol, histidine-containing dipeptides (e.g., carnosine), and enzymes such as glutathione peroxidase contribute to the endogenous antioxidant activity of meats [14]. As with any food, lipid oxidation in meat can be inhibited by addition of exogenous antioxidants. For example, cattle fed a diet rich in vitamin E (α-tocopherol acetate) will yield meat with an elevated concentration of this vitamin and greater resistance to lipid oxidation [22].

15.5.5.2 Proteolytic Enzyme Activity

It has long been known that muscle tissue from freshly slaughtered animals is more tender prior to onset of rigor mortis than it is immediately after. When muscle is allowed to stand postrigor at refrigerated temperatures it becomes more tender than muscle in rigor. This tenderization process, often called "resolution of rigor," is extremely important and has been studied extensively. The main factor in resolution of rigor appears to be loss of biological regulation of proteinases that function in muscle protein turnover in the live animal. These proteinases hydrolyze muscle proteins, apparently weakening the structure of the myofibril and promoting tenderness.

Many proteinases exist in muscle, and defining the precise role that each has, if any, in the tenderization process has been a challenging task. Proteinases in meat can be endogenous to muscle, or they may originate from other cells associated with muscles such as fibroblasts, or could come from bacteria or residual blood cells. The activity of these enzymes in postmortem muscle will depend on the amount of active (undenatured) enzyme; the presence of inhibitors, activators, and required cofactors; and solution pH. For example, a large amount of enzyme might be present but it could be inactive because of its association with a proteinase inhibitor. Activity, not quantity, is the key property that determines the importance of proteinases.

Cathepsins are lysosomal proteinases that show optimal activity at acid pH. Those known to occur in muscle and to act on muscle proteins are (type, pH optimum) cathepsin B1, pH 3.5–6.0; cathepsin H, pH 6.0; cathepsin L, pH 5.0; and cathepsin D, pH 3.0–5.0.

Another group of proteinases found in muscle are the calcium-activated neutral proteinases called calpains. These enzymes were discovered in rabbit skeletal muscle in 1972. It has since been shown that calpains are a ubiquitous class of proteinases expressed in almost every tissue. There are two types: μ-calpain (calpain I), which requires 50–70 μM of Ca^{2+} for activity, and m-calpain (calpain II), which requires 1–5 mM of Ca^{2+} for activity. Muscles from various meat animals are expected to show variation in amount and types of calpains.

Trypsin-like serine proteinases are found in fish [91]. These proteinases cause the undesirable breakdown of surimi-based fish products.

15.5.5.3 Reaction Products from Amino Acids

Microorganisms can utilize amino acids in muscle postmortem, leading to a large number of breakdown products, such as hydrocarbons, aldehydes, ketones, sulfides, mercaptans, and amines (see Chap. 6). Microbial conversion of amino acids to a variety of di- and polymines is a particular problem in certain fatty fish, because these products have unpleasant sensory characteristics and have been implicated in an allergenic type of food poisoning called scombroid poisoning, the symptoms of which often involve nausea and headaches. Although these symptoms are usually attributed to histamine, there is some evidence that other di- and polyamines also may be involved [77].

15.5.5.4 Breakdown of Trimethylamine Oxide

Marine fish, and indeed most animals from the marine environment, contain a significant concentration of trimethylamine oxide (TMAO). In certain species of fish the content of trimethylamine oxide in the aqueous phase of the muscle tissue can exceed 0.1 M. In postmortem fish muscle, microorganisms can reduce TMAO to trimethylamine, a compound responsible in part for the "fishy" odor that develops in fish on storage.

Trimethylamine oxide also decomposes in fish to dimethylamine and formaldehyde. This reaction is catalyzed by an enzyme, and the rate is particularly great in muscle tissue from gadoid species, such as cod, haddock, and hakes. The rate of this reaction is also increased by freezing or disrupting the muscle. It has been hypothesized that the formaldehyde produced in this reaction cross-links the muscle proteins, contributing to toughening that occurs in gadoid fish during frozen storage.

15.6 NATURAL AND INDUCED POSTMORTEM BIOCHEMICAL CHANGES AFFECTING MEAT QUALITY

15.6.1 Physicochemical Basis for Meat Texture and Water-Holding Capacity

Lean meat is about 75% water. Retention of this water is important for several reasons. First, loss of water is of economic importance because it is economically equivalent to loss of meat. Second, water loss during storage of fresh or cooked meat (known as "purge") is unattractive to consumers when it accumulates in the product package. Third, soluble nutrients are lost in the exuded fluid if this fluid is not collected and consumed with the final product. Amino acids and vitamins, especially those of the B complex, are most affected. Fourth, water retention is important to meat texture. Higher water content in muscle reduces its mechanical strength, other factors being equal.

Since myofibrils occupy ~70% of the volume of lean muscle, most of the tissue water is located in the myofibrils [31]. Thus, lateral expansion or shrinking of the filament lattice would greatly influence water holding ability of the muscle (Fig. 16). Migration of water out of the muscle cell, as can occur during freezing, facilitates loss of water from muscle. Cooking induces water loss due to shrinkage of both the filament lattice and the collagen of the endomysial sheath.

Meat toughness arises partly from water loss during cooking. During heating, at 40–50°C an increase in toughness occurs, evidenced by an increase in shear force required to rupture the muscle (Fig. 17). This is caused by aggregation of the denatured myofibrillar proteins, primarily actomyosin. Although this aggregation results in shrinkage of the muscle fibers, the water released is still retained within the endomysial sheath. At 60–70°C a second increase in shear value occurs due to shrinkage of the endomysium and (primarily) perimysium collagens. Heat-induced collagen contraction generates a tension, the magnitude depending on the nature and extent of collagen cross-linking. The older the animal (for species other than fishes), the higher the proportion of heat-stable cross-links and the greater the tension generated on shrinkage. This shrinkage of the perimysium and endomysium forces water out and increases toughness. Although lateral shrinkage of muscle fibers and connective tissue occur independently, their longitudinal shrinkage is functionally coupled because of linkages between them [68]. Extended cooking decreases toughness, and this is associated with conversion of collagen to gelatin and degradation of myofibrillar proteins.

At least 60% of the resistance of cooked mammalian muscle to shear is due to connective tissues and the remainder to myofibrils [50]. The contribution of the myofibrils to toughness is greater in raw muscle than in cooked. However, both collagen content and its degree of cross-linking markedly affect texture of raw and cooked meat. For example, a high content of collagen (near 2%) renders fish muscle too tough for consumption as sashimi (raw fish meat) [11].

However, in fish muscle, where collagen is more soluble, less thermally stable, and present at lower levels than in mammalian muscle, connective tissue differences do not relate to cooked texture. The relative firmness of cooked fish meats has been related primarily to muscle pH and water content, and to species differences in muscle fiber size and the amount of coagulated sarcoplasmic proteins in the interstitial region between individual muscle fibers. Smaller fibers, especially when they are immobilized in a coagulum of exuded sarcoplasmic proteins, impart a firm cooked texture [33].

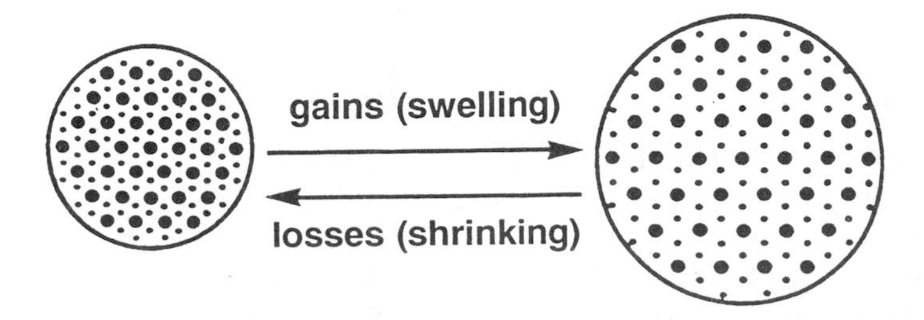

FIGURE 16 Proposed mechanism of water-holding in muscle. On the left is shown schematically a transverse section of a myofibril; on the right, the same myofibril is shown with an expanded filament lattice. (From Ref. 68.)

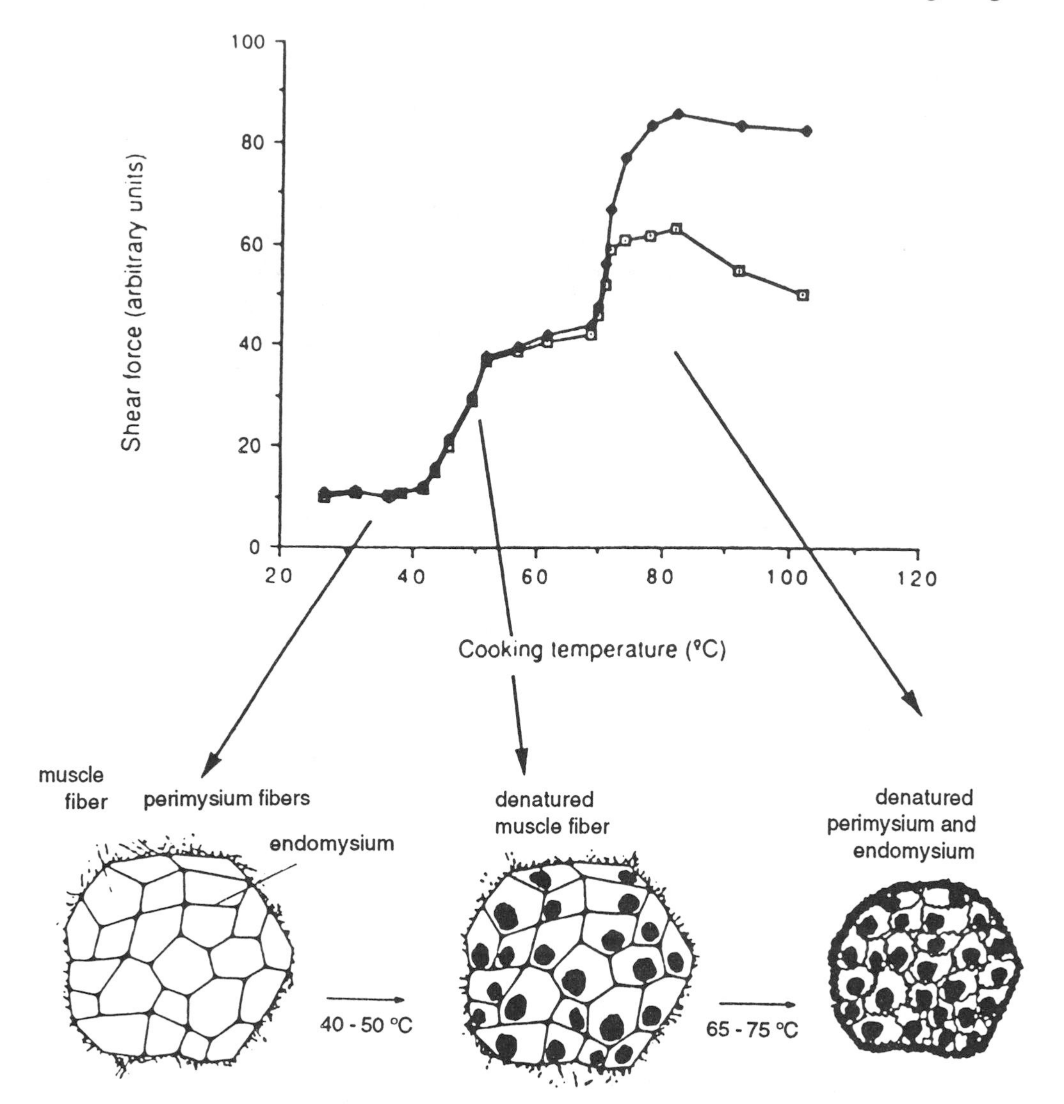

FIGURE 17 Diagrammatic representation of the changes in the muscle fibers and connective tissues during cooking in a cross section of muscle, as related to the change in maximum force during shearing. □, Calf muscle; ◆, ox muscle. (From Refs. 5 and 6.)

15.6.2 Effects of Rate and Extent of pH Change on Quality

One of the important biochemical events in muscle tissue postmortem is the reduction in pH brought about principally by ATP hydrolysis. Both the extent and the rate of pH change are important. The rate of change of pH is important because if the pH drops low enough while the temperature of the carcass is still high, for example pH 6.0 and 35°C [52], considerable denaturation of contractile and/or sarcoplasmic proteins may occur. If the sarcoplasmic proteins are denatured they may adsorb to contractile proteins, thus modifying the physical properties of the contractile proteins. This phenomenon, along with denaturation of the contractile proteins, decreases the ability of contractile proteins to bind water. An excessively rapid decrease in pH postmortem is associated with the pale-soft-exudative (PSE) condition that is especially troublesome in pork. A similar condition can occur in red-meat fish, like tuna. The Japanese term this "yakeniku." Characteristic attributes of this condition are a soft texture, poor water-holding capacity, and a pale color.

Also, lysosomal enzymes are released at high (ambient) temperatures, and if this is accompanied by low pH these lysosomal enzymes can act on contractile or connective tissue proteins producing a tenderizing effect. The contraction that may occur at elevated temperatures (see later) and the resulting increase in toughness will usually obscure the tenderizing by lysosomal enzymes in a low pH–high temperature environment.

The "ultimate" or final pH that a muscle tissue attains is also important. A low ultimate pH in beef provides resistance to the growth of microorganisms and a normal color. Sometimes a high ultimate pH occurs. This may be produced by some antemortem stress, usually prolonged in nature, that depletes glycogen reserves and limits postmortem glycolysis. This is fairly common in beef, causing the meat to be dark, firm, and dry (DFD). These "dark-cutting" muscles exhibit excellent water-holding properties but have poor resistance to the growth of microorganisms.

The ultimate pH of fish meat is generally higher than that of land animals and birds. It is affected by season, primarily because feeding patterns change with the season. Low ultimate pH values in muscle are occasionally observed, and this generally occurs in fish that recently resumed feeding following a period of starvation. This usually occurs in early summer and may lead to a defect known as "gaping."

Gaping (Fig. 18) occurs when the myotomes of fish muscle separate. Normally, the myotomes (see Fig. 2) are conected to the collagen fibers of the myocommata by collagenous microtubules. When these microtubules break, gaping occurs. Gaping is caused by one or more of several independent factors. Low pH and/or high temperature weaken the collagen at the myotome–myocommata junction. Onset of rigor at a high temperature leads to elevated tension in the muscle and the myocommata, and an increased tendency to gape. Gaping is particularly troublesome when fish are frozen, thawed, and then filleted, since freezing may produce ice crystals that are physically disruptive to the connective tissue [56, 54]. Depleted (starved) fish, which exhibit relatively higher moisture and lower protein contents, have no tendency to gape when fresh, nor do fillets that are removed from the skeleton of such fish prior to rigor mortis. Larger fish, which have proportionately thicker myocommata, gape less than smaller fish of the same species and physiological condition. Fish species that exhibit a stronger skin, like catfish, also have less tendency to gape, indicating that the collagens of these species are more highly cross-linked than that of other species [54].

The texture of cooked fish is closely related to the postmortem pH of the flesh; that is, the lower the ultimate pH, the tougher the texture. The pH apparently exerts its effect on the texture of fish muscle by influencing the contractile elements, since fish collagen is disrupted by normal cooking. In meats of all species, decreasing meat pH results in shrinkage of the filament lattice due to the equalization of charge. Attachment of the myosin heads to actin at rigor causes further shrinkage. Denaturation of myosin at lower pH (near its isoelectric point of about pH 5.5) can account for an additional 12% reduction in myofibrillar volume [68].

Color of muscle is affected principally by pH. The pH controls the physical state of the myofibrils and hence the scattering of light from muscle. The swollen fibers of high-pH DFD meat scatter less light than normal, while the shrunken fibers of PSE meat scatter more light than normal; hence the "dark" and "pale" aspects of DFD and PSE meats, respectively. The dark color of high-pH meat may also be due in part to the continuance of mitochondrial activity for a longer than normal period postmortem. Active mitochondria might compete with myoglobin for oxygen at the surface of the meat, thus reducing the amount of oxymyoglobin (bright red) formed. If the metmyoglobin-reducing systems in muscle are inactivated by low pH, more rapid oxidation of

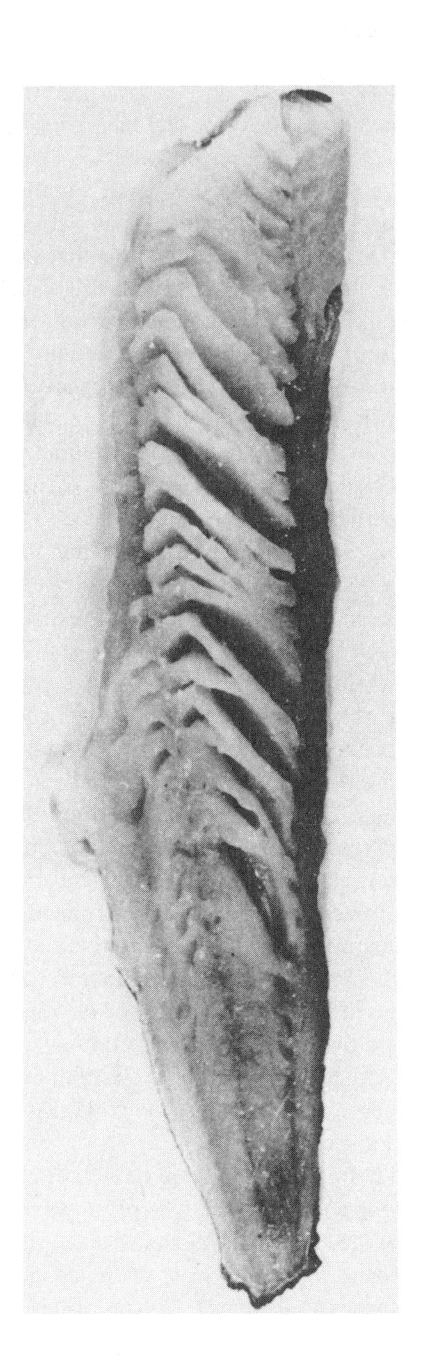

FIGURE 18 Cod fillets exhibiting gaping (right) and no gaping (left). (From Ref. 15.)

bright red oxymyoglobin to brown metmyoglobin can be expected. Low pH can also result in denaturation of myoglobin and hemoglobin, rendering the heme group susceptible to oxidation.

During most slaughtering procedures, some hemoglobin from the blood remains in the muscle. Estimates of the amount range from 5 to 40% of the total pigment. With respect to meat color, the principal difference between myoglobin and hemoglobin is that binding of O_2 to hemoglobin is pH sensitive, whereas with myoglobin it is not. The colors formed from each type of pigment and the reactions required to produce these colors are identical. Total heme pigment content can vary with muscle type and with animal activity; for example, the color of dark muscle intensifies when fish become more active [55]. Meat pigments are discussed more fully in Chapter 10.

15.6.3 Effects of Actin–Myosin Interaction and Contraction on Quality

A major phenomenon affecting the texture of meat is the interaction of actin and myosin after the depletion of ATP and ADP. If meat is cooked and eaten before it passes into rigor, that is, before actin and myosin interact, it is very tender. However, once actin and myosin interact, the meat exhibits increased toughness. Meat is sometimes aged to partially overcome this effect (see following section).

The development of toughness in meat is particularly influenced by the state of the sarcomere when actomyosin is formed, that is, the extent to which thin and thick filaments overlap. This is controlled by several factors in the early postmortem period. Since changes in sarcomere length require ATP to provide energy and to enable thick and thin filaments to slide along each other, it is appropriate to focus on the prerigor period when ATP is still abundant. Three temperature conditions of prerigor muscle can adversely affect the state of the sarcomere and toughness of the muscle: low nonfreezing temperatures (cold shortening), prerigor freezing (causes thaw rigor), and failure to cool the muscle immediately postmortem (high-temperature rigor).

15.6.3.1 Cold Shortening

If muscle is excised in a prerigor state, it is susceptible to rapid contraction. The extent of shortening depends on many factors, including the nature of the specific muscle, elapsed time between death and excision, and the physiological state of the muscle at the time of death. At high physiological temperatures, postmortem contraction is great. As the temperature is reduced, contraction of excised, prerigor muscle decreases, and for some muscles, minimal contraction occurs at about 10–20°C. If these muscles are exposed to still lower temperatures in the range 0–10°C, increased contraction again occurs (Fig. 19) [52]. This behavior, termed "cold shortening," is particularly noticeable and of commercial importance with muscles that consist primarily of red fibers, such as beef and lamb. Muscles that are highly susceptible to cold shortening (beef and lamb) should be maintained at temperatures above 10°C until they pass into rigor. Otherwise, the shortening results in undesirable increases in toughness.

The occurrence of a minimum in the contraction–temperature curve indicates that at least two processes are involved in the phenomenon of cold shortening. One of these is the normal decrease in contraction that is expected as the temperature is lowered, similar in nature to the behavior of most chemical reactions. At some point, however, it appears an additional factor comes into play that overcomes the normal effect of temperature. This phenomenon is believed to be the release of calcium into the sarcoplasm in sufficient amounts to induce contraction. It has been suggested that calcium ions are released from the mitochondria as a response to anoxia postmortem. In addition, low, nonfreezing temperatures may impair the ability of the sarcoplasmic reticulum to regulate calcium concentration in the sarcoplasm, thus allowing calcium levels to rise to activating levels.

It has been shown (Fig. 20) that toughness increases as shortening increases, up to a point at which the muscle is approximately 60% of its rest length (40% contraction) [58]. This length represents the point at which the myosin filaments just touch the Z disk, that is, the I-band just disappears. At this state of contraction, 94–100% of the myosin heads are bound to actin in rabbit skeletal muscle. With greater shortening, the muscle loses toughness rapidly. The reason for this

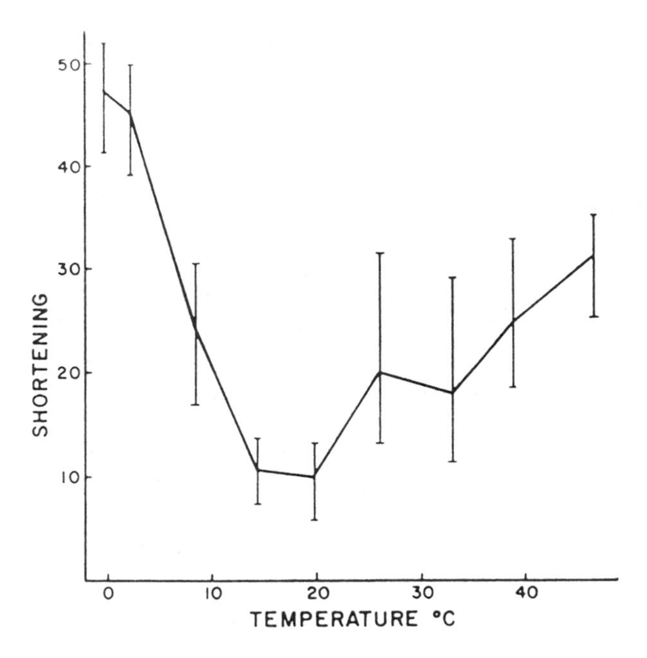

FIGURE 19 Relation between shortening of excised beef muscle and storage temperature postmortem. Vertical lines represent standard deviations (From Ref. 52, p. 788.)

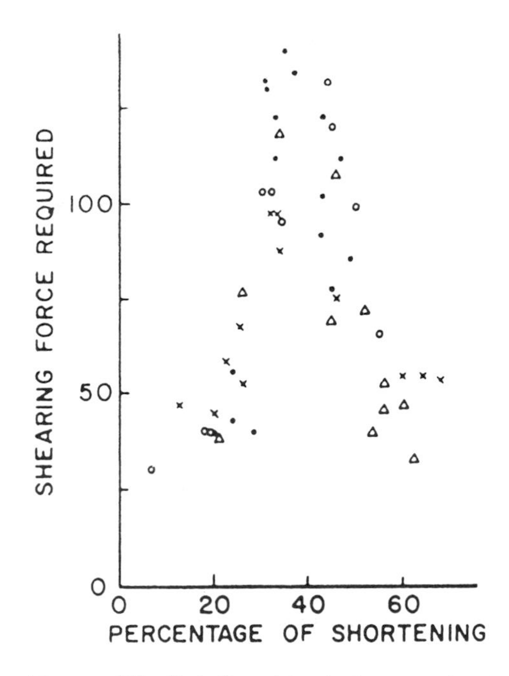

FIGURE 20 Relationship between shortening of prerigor beef muscle and tenderness following cooking. Shortening was produced by treatments of varying temperature and time and by freezing–thawing. (From Ref. 58, p. 635.)

tenderization is not clear, but it may result because thin and thick filaments slide past each other to an extent exceeding that which occurs in vivo, thus leading to severe, irreversible disruption of the sarcomere.

Muscles attached to the skeleton also can undergo some cold shortening. The extent to which this occurs depends on whether the muscle has been severed during slaughter (e.g., sternomandibularis or neck muscle), whether the temperature is conducive to cold shortening, whether a temperature gradient exists along the muscle during prerigor cooling causing some parts to contract and other parts to stretch, whether the muscle is naturally anchored at both ends or at one end only (e.g., longissimus dorsi), the species being handled, and how the carcass is positioned during the prerigor phase. With regard to the last point, improvements in tenderness can be accomplished by holding the carcass so that muscles are stretched during the postmortem prerigor period. The success of this approach is in accord with the known existence of a strong positive correlation between sarcomere length and degree of tenderness. Excision of a muscle prerigor, of course, removes the physical restraint to contraction, and the muscle is then much more susceptible to substantial shortening, sometimes leading to final lengths only 30–40% of the original rest length (60–70% contraction) [59]. This results in muscle that is highly disrupted (tender), misshapen, highly exudative (poor water-holding capacity), and of poor consumer acceptability.

Data relating muscle shortening and "drip" (inverse of water-holding capacity) are shown in Figure 21. Drip losses do not become excessive until shortening exceeds about 40% of the original length. This behavior is in accord with the previously stated belief that severe disruption of the muscle fiber occurs when contraction exceeds about 40%.

To prevent cold shortening, it is important that rigor occur before the carcass is cooled to temperatures that would stimulate contraction of sensitive, prerigor tissue. This may be achieved by subjecting the carcass to an electrical current before the muscle goes into rigor, a procedure

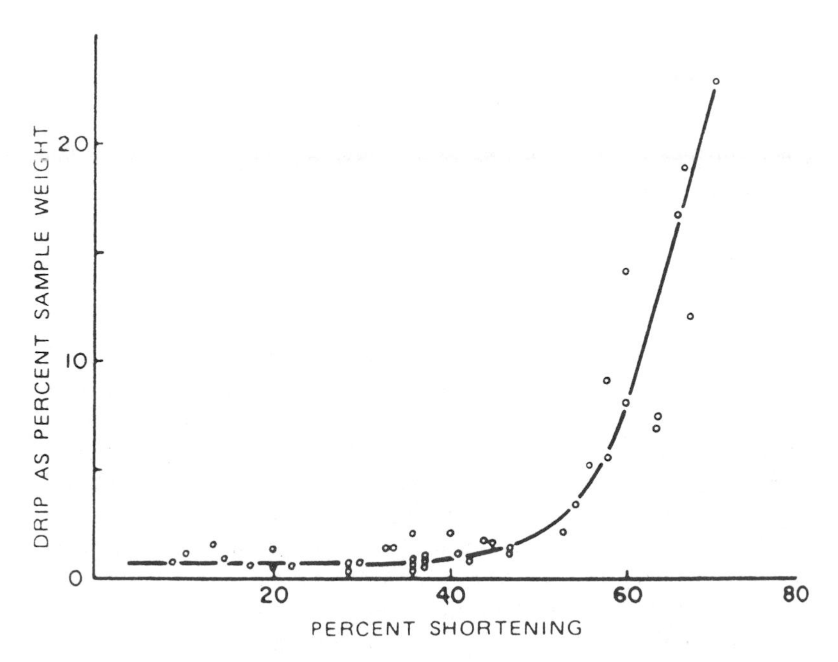

FIGURE 21 Relation between "drip" exudation (as percentage of frozen sample weight) and extent of thaw shortening (as percentage of initial excised length). (From Ref. 59, p. 456.)

known as electrical stimulation. Typical conditions are 500–600 V, 2–6 A, applied in 16–20 pulses of 1.5–2.0 sec duration within 1 hr after slaughter. The electrical current functions in the same way as an electrical nerve impulse in the living animal and causes strong contraction of the muscles. This process leads to faster development of rigor mortis, a more rapid postmortem decline in pH, and improved color and tenderness of some muscles. High-energy phosphate imtermediates, such as creatine phosphate, ATP, and ADP, decrease more rapidly during electrical stimulation [12]. Electrically stimulated muscles display contracture bands in certain areas and stretched sarcomeres in other areas.

Three major theories have been advanced to explain the increased tenderness of muscle tissue that has been electrically stimulated: (a) the stimulation occurs before the temperature of the carcass is lowered; thus, cold-shortening is markedly reduced; (b) the more rapid glycolysis causes the pH to fall rapidly, causing release of lysosomal hydrolases that can then weaken structural components of the muscle; and (c) there is a physical disruption of the myofibrils caused by supercontractions, and this disruption prevents the toughening that would ordinarily occur with contraction. These hypotheses are not mutually exclusive, and more than one phenomenon may occur simultaneously.

It has also been reported that electrical stimulation lowers the shrinkage temperature of collagen, which is an indication of a reduction in its thermal stability [41]. White fibers are more susceptible to electrical stimulation than red fibers, and intermediate fibers respond in an intermediate fashion [86].

Electrical stimulation is used principally for those species that undergo cold shortening, that is, beef and lamb. If carcasses are hot boned, electrical stimulation lessens toughening and the increased drip (before and after subsequent freezing) that could easily occur due to contraction of the muscle without skeletal restraint. Hot boning is the removal of warm meat from the skeleton. It can be accomplished after a short period of storage postmortem (minimum of about 3 hr), which allows for partial cooling of the carcass (to 14–20°C, at which shortening is minimal) and the occurrence of early postmortem biochemical events. This practice results in energy savings during cooling as compared to the cooling of intact carcasses. The holding time at 14–20°C has a beneficial effect on tenderness (possibly due to proteolysis) that tends to overcome the small degree of toughening that usually accompanies hot boning. This small amount of toughening could be due to "rigor shortening" that occurs just before depletion of ATP in muscle held at a temperature that minimizes cold-shortening. This rigor shortening is thought to be caused by the inability of the sarcoplasmic reticulum to maintain low calcium levels as the ATP concentration begins to rapidly decrease. Enough ATP remains, however, for some contraction to occur. Electrical stimulation can also be useful in preventing thaw rigor, discussed next.

15.6.3.2 Thaw Rigor

A phenomenon related to cold shortening is freeze–thaw contraction, or "thaw rigor." If a muscle is frozen prerigor it often undergoes considerable shortening during thawing. This can be accompanied by toughening of the muscle (if contraction is less than 40%) and poor water-holding capacity (if contraction exceeds 40%). For this reason meat that is to be frozen generally should be in rigor or postrigor if optimal quality is desired. However, thaw rigor can be avoided in fish frozen prerigor if the product is held at normal temperatures of frozen storage for about 2 months prior to thawing. Thaw rigor in frozen prerigor lamb, beef, or poultry can be avoided by electrical stimulation prior to freezing, or by holding the frozen muscle at –3°C for a time equal to that normally used during chilling. Thaw rigor presumably occurs because ice crystals, by direct or indirect means, disable the sarcoplasmic reticulum and/or mitochondria in prerigor muscle, thus destroying the muscle's ability to maintain low levels of calcium in the sarcoplasm and causing contraction to occur during thawing.

15.6.4 Changes Induced by Proteolysis

Postrigor beef muscle is sometimes held at refrigerated temperatures for 1–2 weeks to improve tenderness, a process known as aging or conditioning. Tenderization due to aging has been attributed to proteolytic degradation of the muscle structure. A relationship among proteolytic activity, changes in myofibrillar protein or collagen structure, and mechanical properties of meat is intuitively reasonable but has not been conclusively established. Disruption of the Z disk is one of the primary processes in the development of tenderness in poultry and in beef. Disintegration of the Z disk renders myofibrils more susceptible to fragmentation during homogenization [88], and results of this technique, expressed as the myofibril fragmentation index, are used as an indicator of meat tenderness. Destruction of the Z disk comes about by the action of calpain. Treatment of muscle or isolated myofibrils with calpain leads to the disappearance of the Z disk and release, but not degradation, of α-actinin.

Dissolution and disintegration of the Z disk, while related to increased tenderness of meat during postmortem aging, is probably not the only factor of importance, since the Z disk is intact in tender prerigor muscle. Furthermore, disintegration of the Z disk appears to occur well in advance of substantial tenderization during aging. Changes in the interaction of the thin and thick filaments or changes in the networks of desmin and connectin filaments are also likely involved in postmortem tenderization. Titin, nebulin, and desmin are degraded during postmortem holding of muscle from chicken, pork, and beef. Since these proteins are intimately involved with the structure of myofibrils, they may be important in the tenderizing process.

Calpains appear to be more important than cathepsins in typical postmortem tenderization reactions. The activity of calpain is greater in beef muscles, which show an age-dependent increase in tenderness [43], and loss of calpain activity upon prolonged storage of beef is correlated with a decrease in the tenderizing effect [19]. In contrast, the activity of cathepsins B, L, and H remains constant while tenderization occurs at different rates in different muscles during aging. Furthermore, infusion of calcium chloride into lamb muscles at slaughter causes a major increase in tenderness with no change in cathepsin B, H, and L activity [44]. The susceptibility of muscle to calpain degradation varies in direct relation to its thermal stability; thus myosin from fishes, especially those of cold water origin, is more easily degraded by calpain [62]. While calpain is the main enzyme responsible for tenderness in the normal procesing of beef and lambs, it may not be as important in other species, or under food processing conditions that favor cathepsin activity.

In contrast to the action of proteinases on myofibrillar proteins, proteases have more subtle effects on connective tissues of most species during aging. In unaged meat, the perimysium is the last structure to break, however this may not be so in aged muscle [68]. Very limited proteolytic cleavage of cross-links in the nonhelical regions of collagen would partially depolymerize the fiber, and this would probably result in less tension development during cooking and improve tenderness [6]. Electrophoretic studies have shown that proteolytic cleavage of collagen does occur during aging even when there is little detectable effect on its solubility. Effects of aging on solubility of collagen can vary markedly among muscles. For example, the increase in collagen solubility is very small in gastrocnemius muscle during postmortem aging whereas in psoas major the increase is usually substantial (Table 8). This may reflect differences in proteolytic activity, final muscle pH, susceptibility of collagen to proteases, or other factors [6].

Postmortem proteolysis of collagen by collagenases seems to contribute to softening of fillets from some fish species. Breakdown of type IV (basement membrane) collagen has been shown to occur in several species concomitant with tenderization. Type I collagen is not degraded during iced storage of rockfish [13], but at temperatures above 15–20°C (near the temperature of collagen melting in this species) degradation is quite rapid and complete, and this results in flesh softening. Increased collagen solubility during iced storage, rather than being

Table 8 Solubility of Perimysium Collagen in Fresh and Aged Bovine Muscles

	Solubilized collagen (%)	
	Fresh	Aged
Gastrocnemius	0.8	1.7
Psoas major	0.8	10.5
Pectoralis profundus	0.4	1.3
Gluteus medius	0.6	2.3
Supraspinatus	0.4	4.6

Source: Ref. 6.

attributable to collagenase activity, is more likely associated with degradation of polysaccharide chains of proteoglycans which comprise the extracellular matrix [30]. Increased collagen insolubility in hake flesh during frozen storage has been associated with increased toughness. This may result from protein cross-linking induced by formaldehyde, with the latter compound arising from breakdown of trimethylamine oxide [11].

Postrigor meat subjected to high pressure (150 MPa) at elevated temperatures (40–60°C) undergoes rapid tenderization (within 30–60 min). The effect has been attributed entirely to proteolytic degradation of myofibrillar proteins, and is sufficient to eliminate the toughening of meat that normally occurs upon subsequent heating at 60–80°C [32]. Presumably the elevated pressure denatures myofibrillar proteins to some extent, increasing their susceptibility to proteolysis.

Proteolytic enzymes are often incorporated into marinades or sauces applied to meats before and during cooking. The thermostable protease papain is often used as a meat tenderizing agent, exhibiting its greatest activity during the early stages of cooking the meat. A limitation of using currently available proteases for meat tenderization is their inability to selectively hydrolyze bonds that are critical for proper tenderization.

Proteolysis and nucleotide metabolism may produce compounds that contribute significantly to meat flavor or that react on cooking to produce flavor [73]. Five oligopeptide hydrolases, active at neutrality, are present in rabbit muscle. These enzymes, if present in other muscles, could contribute substantially to the development of flavor in cooked meat by releasing amino acids from oligopeptides. The major flavor components of haddock muscle have low molecular weights, similar to those of amino acids [89].

15.6.5 Changes in Lipids and Interaction of Lipids with Other Components

If fatty acids are produced during meat storage (e.g., by the action of lipases or phospholipases), they can cause off-flavors and they can also interact with contractile proteins and denature them. The latter occurrence may be an important cause of toughening of fish muscle during frozen storage. Oxidation products of fatty acids, including free radicals, can also interact with proteins, causing their insolubilization by intermolecular cross-linking, thereby decreasing water-holding ability and increasing muscle toughness.

The development of oxidized "cold-store" flavor in fish is attributable mainly to the compound *cis*-4-heptenal. This compound is produced by oxidation of the phospholipid fraction of fish muscle in both lean and fatty fish. When a lean fish, such as cod, is starved, some of the phospholipid is utilized, presumably for energy. This reduces the concentration of the C22:6 fatty acid, the source of the *cis*-4-heptenal. Therefore, off-flavor development is less severe in starved cod than in well-fed cod. Meat flavors are also discussed in Chapter 11.

15.7 CHEMICAL ASPECTS OF MEAT STABILITY

The purpose of many food-processing techniques is to slow down or prevent deleterious changes in food materials. These deleterious changes are often caused by contaminating microorganisms, by chemical reactions among natural components of the food tissues, or by simple physical occurrences (e.g., dehydration). The reactions may be chemically catalyzed by enzymes indigenous to the tissue.

The procedures used to prevent changes in muscle foods caused by microorganisms or chemical and physical reactions are (a) removal or immobilization of water, (b) removal of other active components, such as oxygen or glucose, (c) addition of chemical additives, (d) lowering of temperature, (e) input of energy, and (f) packaging. Very often when one desirable objective is achieved other undesirable consequences occur. This is especially true when the principal objective of the processing procedure is to inactivate the entire microbial population.

15.7.1 Heating

Heating of muscle tissue brings about extensive changes in its appearance and physical properties, and these changes are dependent on the time-temperature conditions imposed. α-Actinin is the most heat-labile muscle protein, becoming insoluble (in muscle of warm-blooded animals) at 50°C. The heavy and light chains of mammalian myosin become insoluble at about 55°C, and actin at 70–80°C. Tropomyosin and troponin are the most heat-resistant myofibrillar proteins of muscle, becoming insoluble at about 80°C.

Heating inactivates some enzymes, while others survive short times at temperatures of 60–70°C. Because proper cooking is important to the safety of meat, several methods for the determination of endpoint cooking temperature have been developed based on inactivation of enzymes. These include asays of residual activity of catalase, peroxidase, lactate dehydrogenase, pyruvate kinase, glutamic-oxaloacetic transaminase, and glutamic-pyruvic transaminase. Other methods of assessing endpoint temperature include use of electrophoresis, differential scanning calorimetry, near infrared spectroscopy, color analysis, and enzyme-linked immunosorbent assays (ELISAs).

The enzymic activities of myosin and actomyosin are destroyed during conventional cooking. Myosin ATPase activity is a sensitive indicator of myosin denaturation. Its heat stability varies with habitat temperature of fishes (Table 9), and it is most stable in muscle of warm-blooded animals. Acclimation of fishes to water temperatures different from that of their native habitat has been shown to alter the stability of myosin and actomyosin [92] but, interestingly, has no effect on the thermal stability of collagen [46]. This is odd given the relationship that exists between thermal stability of collagen and habitat temperature (Fig. 11). Actomyosin from Antarctic krill has very low thermal stability, about 100 times lower than that of Alaska pollack [64].

Myoglobin also undergoes denaturation during heating. The susceptibility of the heme pigment to oxidation in the denatured protein is much greater than in undenatured myoglobin. On heating, therefore, red meat generally turns brown due to the formation of the oxidized pigment, hemin (see Chap. 10).

With severe heating of meat, there are further changes in proteins and free amino acids

TABLE 9 Rate Constants (k_D) of Thermal Inactivation of Myofibrillar Ca^{2+}-ATPase of Fishes Caught in Fishing Grounds Differing in Bottom Temperature

Number of species	Number of Fish	Temperature (°C)	k_D[A] ($\times 10^{-5}$/sec)
32	79	28	20.3
16	26	25	18.6
18	54	17	48.3
4	11	9	63.5
9	33	2	283.00

Source: Ref. 92.

with production of some volatile breakdown products. Sulfur-containing compounds are produced, including hydrogen sulfide, mercaptans, sulfides, and disulfides, as well as aldehydes, ketones, alcohols, volatile amines, and others. Lipid components may also break down into volatile products, such as aldehydes, ketones, alcohols, acids, or hydrocarbons (see Chap. 5). Some of these volatile compounds, in both the fat and the lean portions of the meat, contribute to the flavor and odor of cooked meat.

Commercial heating generally has moderately detrimental effects on the vitamin content of meat. Thiamin is sensitive to heat, and it is partially destroyed during cooking or thermal processing. Certain amino acids may interact with glucose and/or ribose of meat (Maillard reaction), and the nutritional values of these amino acids can be impaired when this occurs (see Chap. 4 and 6). Lysine, arginine, histidine, and methionine seem to be particularly susceptible to degradation via this route. However, the conditions used during commercial sterilization generally have little detrimental effect on the nutritive value of proteins and lipids.

15.7.2 Refrigeration

Refrigeration is the most widely used technique to preserve meat. It preserves muscle tissue by retarding the growth of microorganisms and by slowing many chemical and enzymic reactions. However, the greater rate of glycolysis in prerigor beef muscle at 0°C than at 5°C and the occurrence of cold shortening in some muscle should be adequate warning that biological systems do not always obey simple rules.

15.7.2.1 Effect of Temperature on Enzymic Activity

It should not be assumed that lowering the temperature to a value close to 0°C necessarily reduces the activity of all enzymes in muscle tissue. Although a reduction in muscle temperature from about 40°C (close to that of living mammals or birds) to 0°C lowers the rate of most enzyme-catalyzed reactions, it should be realized that some fish live in a cold environment (body temperature equals that of the environment) and that their enzymes will be more active at low temperatures than will those of warm-blooded animals. This adaptation is not obvious, however, unless one measures the activity of the enzyme at low substrate concentrations, similar to those that exist in tissue. Thus, the Michaelis constant K_m, not the maximum initial velocity V_{max} is the appropriate value to monitor.

It has also been demonstrated that "adaptation" of enzymes to low temperature sometimes occurs in warm-blooded animals. For example, it has been demonstrated that the Michaelis constant of lactate dehydrogenase from chicken breast muscle is much smaller at refrigerated temperatures than it is at the body temperature of the bird, which is around 40°C [37]. Since a small Michaelis constant means that the enzyme retains a high proportion of its maximal activity at low substrate concentrations, lowering the temperature from 40 to 4°C combined with a decline in K_m may have little effect on the activity of the enzyme in situ. This behavior may have a bearing on the temperature dependence of cold shortening in beef muscle (shortening is minimal between 14° and 20°C [52]) and on the rate of ATP loss in chicken muscle (rate is minimal at 20°C [18]).

It should also be remembered that low temperature can trigger internal release of calcium, which not only stimulates contraction but also increases the activity of calpains. The major point is that decreases in rates of enzyme-catalyzed reactions in situ may not always correlate well with decreases in temperature.

15.7.2.2 Modified and Controlled Atmospheric Storage

Fresh meat (including fish) held at a refrigerated temperature has a limited shelf life, primarily because of microbial growth. It is possible to slow down this growth by storing muscle foods in modifed or controlled atmospheres. The term "modified atmosphere" refers to an initial adjustment in the composition of the atmosphere surrounding the product, with no further control other than the package. In contrast, the term "controlled atmosphere" indicates that the atmosphere is continuously controlled.

Some studies have been conducted on animal tissue stored under hypobaric conditions, and this approach is effective but expensive. Most studies have involved modification of atmospheric composition at normal pressures.

The preservative effect of CO_2 on stored muscle tissue has been known for a century. The effect of CO_2 is not due to a simple displacement of oxygen, since nitrogen atmospheres do not offer the same extension of storage life that CO_2 does. It is known that CO_2 lowers the internal pH of tissues, and it probably causes intracellular acidification of spoilage bacteria [72]. This could upset the major H^+/K^+ transport systems in the cell, which would have marked effects on gross metabolism. One of these effects could be a shift in equilibrium of the decarboxylating enzymes, such as isocitrate dehydrogenase, making them rate limiting. A decrease in surface pH has been observed with rock cod fillets stored in an atmosphere containing 80% CO_2 [72], but a decrease in pH was not observed in red hake and salmon flesh when an atmosphere containing 60% CO_2 was used [23].

At high concentrations of CO_2 surface browning of red meat is often observed. This presumably occurs because the myoglobin and hemoglobin pigments undergo oxidation to their ferric states. The most desirable concentration of CO_2 to use for modified atmosphere storage of meat is a compromise between that needed to inhibit microorganisms (the more CO_2 the greater the inhibition) and that which will not cause product discoloration.

Oxygen has an inhibitory effect on the breakdown of trimethylamine oxide to dimethylamine and formaldehyde in refrigerated red hake muscle. Thus, in species in which this is a potential problem, such as gadoids, conditions that exclude oxygen, whether hypobaric, modified atmosphere, or vacuum storage, could be potentially harmful to the quality of the product (see Sec. 15.5.5.4).

CO_2 and its hydrated bicarbonate ion can directly affect membranes, causing hydration or dehydration of the surface. This can result in changes in intermolecular distances of membrane components, making the membranes either more or less permeable [72]. This in turn could have major effects on spoilage microorganisms.

15.7.2.3 Ionizing Radiation in Conjunction with Refrigeration

Although ionizing radiation has not been approved for processing of most food products, it does constitute a potentially useful means of extending their refrigerated shelf life. Aside from its desirable ability to inactivate microorgansims, it also has the undesirable effect of altering meat pigments. If raw or cured meat is exposed to large doses of ionizing radiation, it turns brown. The color of cooked meat is not affected by ionizing radiation if air is present, but a pink color develops in the absence of oxygen. However, on exposure to air the pink color reverts to the original brown color. The pink color of irradiated cooked meat is attributable to a denatured globin-hemochrome pigment. In a model system, irradiation of NO-myoblobin causes denitrosylation and formation of metmyoglobin. If the radiation dose is sufficiently high, the metmyoglobin is, in turn, degraded to other products, including a ferriperoxide derivative and small amounts of choleglobin-type pigments.

Sterilizing doses of ionizing radiation also result in the breakdown of various lipids and proteins to compounds that have distinct and often undesirable odors. Tenderization of muscle may also occur during this treatment. The odors vary with the particular muscle tissue treated and range all the way from extremely unpleasant odors in beer to rather mild odors in pork and chicken. Furthermore, irradiated beef steak is sometimes bitter, and this may result from the conversion of ATP to hypoxanthine. Many of these reactions can be prevented by lowering the temperature during irradiation. Temperatures of –80°C or below greatly lessen undesirable side effects without seriously decreasing destruction of microorganisms. This occurs because microorganisms are primarily destroyed by direct interaction with ionizing radiation, whereas most of the undesirable changes in chemical components result from the indirect effects of free-radicals produced during irradiation.

Generally, enzymes are not completely inactivated by irradiation treatments sufficient to sterilize the product. Therefore, for long-term storage it is necessary to heat the meat to approximately 70°C, that is, cook it prior to irradiation and storage.

15.7.3 Freezing

Freezing is an excellent process for preserving the quality of meat and fish for long periods. Its effectiveness stems from internal dehydration (ice crystal formation) or immobilization of water and lowering of temperature (see Chap. 2). Although some microorganisms survive storage at very low temperatures, there generally is no opportunity for growth of microorganisms if recommended storage temperatures are maintained. Quality retention (or loss) during freeze preservation of animal tissues depends on how the freezing process is conducted and/or on innate characteristics of the tissue.

15.7.3.1 Effect of Freezing Rate

A major problem arising from improper freezing (slow) and storage procedures (too long or too high a temperature) is excessive thaw exudate. When muscle tissue is frozen rapidly, small ice crystals form both intra- and extracellularly. When muscle is frozen slowly ice crystals form first in the extracellular space, presumably because the freezing point is higher there (concentration of solutes is lower) than it is inside the fibers. Since the vapor pressure of ice crystals is lower than that of supercooled water, there is a tendency during further cooling for the water inside fibers to migrate to ice crystals that already exist in extracellular areas, resulting eventually in very large extracellular ice crystals and compacted muscle fibers. The increase in salt concentration and the concomitant change in pH may cause extensive denaturation of muscle proteins

[71]. If exposure to the high salt concentrations and unfavorable pH values occurs at a relatively high subfreezing temperature, the process of denaturation can be acelerated, with a resulting decrease in water-holding capacity of the tissue. This loss in water-holding capacity of the protein, along with mechanical damage to cells by ice crystals, is responsible in large part for the thaw exudate.

The rate at which fish muscle is frozen influences the degree of protein denaturation [60]. Although rapid freezing generally results in less denaturation than slow freezing, intermediate freezing rates can be more detrimental than slow freezing as judged by textural changes and solubility of actomyosin. Cod fillets frozen at intermediate rates develop intracellular ice crystals just large enough to damage cellular membranes.

Freezing rate also can influence the color of animal tissues in the frozen state. Rapid freezing causes tissues to become very pale. This is undesirable for red meats but desirable for poultry. This effect is apparently related to the number and size of ice crystals produced and the thickness and nature of the tissue matrix. Once thawed, the appearance of rapidly frozen and slowly frozen tissues cannot be distinguished.

15.7.3.2 Effects of Freezing on Enzymic Activity

Enzymes generally are not inactivated by the freezing process, and some may continue to function in frozen animal tissue. Enzymes with low activation energies retain considerable activity in the frozen state, often more than expected by extrapolation of Arrhenius plots from above-freezing temperatures. Freezing damage to the tissue (enzyme delocalization), changes in environmental conditions (pH, ionic strength, and substrate concentration), and changes in the Michaelis constant may account for the retention of moderate enzymic activity, which sometimes exceeds that in the unfrozen tissue.

Freeze delocalization of mitochondrial isoenzymes of glutamate-oxaloacetate transaminase has been used to differentiate frozen–thawed beef, pork, and chicken from the unfrozen products. However, this procedure may not be suitable for pork since procedures other than freezing and thawing can damage pig muscle mitochondria and cause redistribution of this enzyme [37]. Freezing and thawing also cause release of cytochrome oxidase from muscle mitochondria. The soluble activity of this enzyme increases more than twofold in frozen and thawed trout muscle and about fourfold in frozen and thawed beef muscle as compared to the activity in unfrozen samples. Measurement of this enzyme might be useful in developing an improved method for distinguishing unfrozen from frozen–thawed meat and fish [7].

A problem unique to marine species during frozen storage is the enzyme-catalyzed breakdown of trimethylamine oxide to yield dimethylamine and formaldehyde. As was mentioned, formaldehyde is believed to cause cross-linking of muscle proteins and an increase in muscle toughness. It also has been suggested that the formation of new disulfide bonds in fish muscle proteins may contribute to denaturation [60]. Cod actomyosin solutions, after sufficient frozen storage, form aggregates that are insoluble in a solution of sodium dodecyl sulfate and a disulfide reducing agent. This indicates that non-disulfide covalent bonds also may contribute to aggregation of the protein during frozen storage, possibly arising through the action of endogenous transglutaminase. An integrated scheme for some of the reactions that can cause denaturation of fish proteins and toughening of fish muscle is shown in Figure 22.

Enzymes that act on lipids can contribute to quality deterioration of animal tissues during long-term frozen storage. Lipases and phospholipases release free fatty acids that can react with proteins. Reactions of oxidized lipids with proteins may be a major cause of the undesirable toughening that codfish muscle undergoes during frozen storage (Fig. 22). There is evidence that the major source of free fatty acids in lean fish is phospholipids of the membrane system, not triacylglycerols [37].

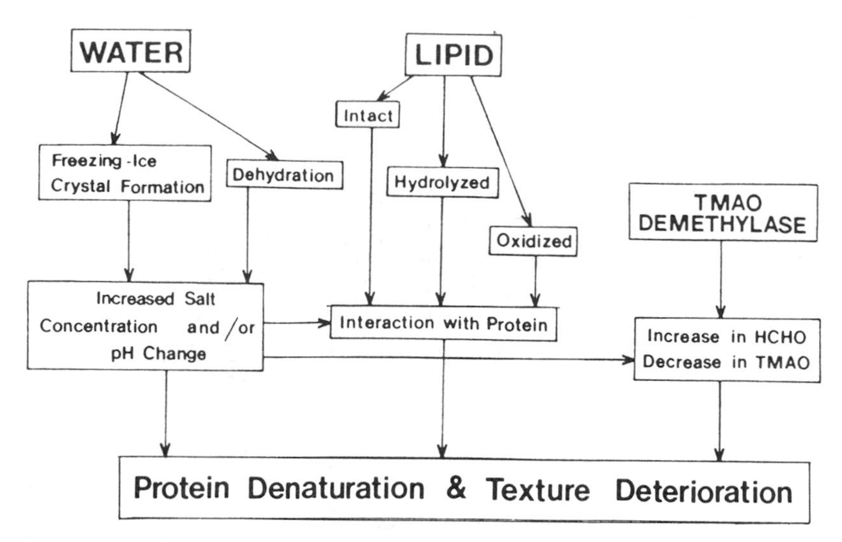

FIGURE 22 Some factors influencing the denaturation of fish proteins during frozen storage. (From Ref. 79.)

Generation of free fatty acids in either fish or beef inhibits the rate of oxidation [80]. The reason for this anomalous behavior is not clear, but it might be that enzyme-catalyzed oxidation of membrane lipids is slowed either by direct inhibition of the enzyme with the fatty acids or, indirectly, by disruption of the membrane structure [81].

15.7.3.3 Other Effects of Freezing on Meat Quality

Fish may undergo extensive textural changes during frozen storage because of the very high sensitivity of fish proteins to the denaturing factors active during frozen storage (Fig. 22). Meat of warm-blooded animals is also subject to freeze-induced denaturation; salt-extractable proteins have been reported to decrease as much as 50% during frozen storage simulating commercial conditions [71]. During frozen storage, tropomyosin is the most stable protein, actin is the next most stable, and myosin is the least stable. Note that this order of stability is the same as that toward heat [4].

Myofibrillar proteins can be stabilized during freezing and frozen storage by the infusion of cryoprotectants, these usually being sugars, sugar alcohols, or starch hydrolysis products. Vacuum infusion shows promise as a means of effectively distributing low-molecular-weight carbohydrates throughout intact muscle.

As discussed in Section 15.6.3.2, the meat of warm-blooded animals generally should not be frozen prior to the development of rigor, in order to avoid thaw rigor. However, when the meat is to be used for production of gelled meat products, such as frankfurters, comminution of the meat from the frozen state in the presence of sufficient salt will avoid rigor development. Binding of salt ions by proteins and the resulting increase in electrostatic repulsion between proteins apparently prevent the proteins from interacting after ATP has been depleted. If rigor occurs prior to salt addition, then the proteins are irreversibly bound and cannot be dissociated by treatment with salt. Salt addition prior to freezing is not advised, as this destabilizes the proteins during frozen storage [71].

Dehydration of the surface of animal tissue during frozen storage (freezer burn) can occur if improper packaging techniques are employed. Freezer burn occurs when the equilibrium water vapor pressure above the surface of the meat is greater than that in the air, causing ice crystals to sublime. Oxidation and darkening of the heme pigments usually accompany freezer burn.

15.7.4 Dehydration

Drying can be successfully employed for both raw and cooked meat; however, the quality of the final reconstituted product is better when the meat is cooked prior to dehydration. The reason for this is not clear, but it may be that protein denaturation that occurs during heating in the presence of the original moisture is less severe than that which occurs during the combined process of heating and moisture removal.

The method of dehydration also has a very significant effect on the quality of the final product. Conventional air drying, which is done at a relatively high temperature, is more detrimental to muscle tissue than is freeze dehydration, the latter process employing a low temperature and sublimation of ice. Maximal denaturation of myofibrillar proteins occurs when the temperature is high and the moisture content of the tissue is reduced to a value below about 20–30% [79]. In addition to the difference in temperature between two kinds of dehydration, there is also a difference in salt distribution. During drying an increase in salt concentration occurs. As moisture evaporates, more moisture is drawn to the surface, bringing salts with it. The resulting increase in ionic strength (especially at the surface) and change in pH adversely affect protein stability. During freeze dehydration the increase in solute concentration is more evenly distributed, and the temperature is low. The extent of denaturation is greater when muscle proteins are exposed to high salt concentrations or unfavorable pH. Another low-temperature drying method involves compressing meat strips between highly absorbent contact-dehydrant sheets.

Even freeze-dried meat, once rehydrated, shows a lower water-holding capacity than fresh tissue. If the flesh of freshly slaughtered animals is freeze-dried before the onset of rigor, the breakdown of ATP and glycogen can be inhibited. During rehydration of this tissue, rapid hydrolysis of ATP occurs along with rapid glycolysis. This can result in strong rigor contraction and a considerable loss of water-holding capacity. However, if the muscle tissue is ground and salted prerigor and then freeze-dried, rigor is prevented during rehydration and the high-water-holding capacity of the tissue is retained. Unfortunately, salt additon lowers the stability of the proteins, limiting storage life.

Texture is usually severely altered by dehydration. The change in texture is determined principally by the extent of denaturation of the muscle proteins, and this in turn is highly dependent on the pH of the meat at the start of dehydration. Even a very good freeze-drying procedure results in some lowering of the water-holding capacity of muscle if the process is carried out near the isoelectric points of the proteins (about pH 5.5). If one starts at a high ultimate pH (about 6.7), a relatively small decrease in the water-holding capacity of muscle occurs during dehydration. As with freezing, an even distribution of a lyoprotectant can stabilize myofibrillar proteins during drying and subsequent storage [4]. Measurements of rehydration, extractability of proteins, and ATPase activity all have been used to estimate the amount of denatured protein in dehydrated muscle.

Lipid oxidation is a major factor limiting the shelf life of dehydrated muscle tissue. The highly unsaturated nature of pork lipids and lipids from fatty fishes causes these products to be unstable in dehydrated form unless some kind of special protective treatment is applied. The addition of antioxidants and the use of appropriate packaging techniques are the most common approaches to retarding oxidation.

Lipid oxidation results principally in undesirable flavors that can make the animal tissue unacceptable. However, other effects may also occur: some destruction of oxidizable nutrients, such as essential fatty acids, some amino acids, and some vitamins; oxidation of heme pigments (discoloration); and protein cross-linking.

During high-temperature drying, tissue fat melts. Fat in excess of about 35–40% (fresh weight basis) usually cannot be retained in dehydrated tissue.

Another major type of deteriorative reaction in dehydrated muscle tissues is nonenzymic browning. Carbonyl sources for this reaction include glucose, phosphorylated sugar derivatives, and other aldehydes and ketones; amine groups for this reaction come mainly from either free amino acids or the ε-amino groups of lysine residues. The Maillard reaction (discussed in Chap. 4 and 6) results in dark pigments and decreased nutritive value of amino acids and proteins. Moreover, when sugars react with proteins, the physical properties of the proteins can change, resulting in a toughening or hardening of the texture. Nonenzymic browning can also lead to desirable changes in flavors, some of which are typical of cooked meats.

15.7.5 Pressure

High pressures, applied isostatically, can have effects similar to heating on the destruciton of microorganisms and denaturation of meat proteins. The mechanism whereby microbial destruction or protein denaturation is achieved by pressure treatment differs from that of heating, however. While the intramolecular hydrophobic interactions that stabilize protein native structure are strengthened with temperature increases up to 60°C, these are weakened by increasing pressure. Pressure-induced denaturation can be beneficial in creating new types of gelled meat products, but its occurrence is undesirable if the primary purpose of pressure treatment is to stabilize fresh meat for storage.

Pressure treatment may also aid in freezing and/or thawing of muscle foods. As pressure increases, the freezing point temperature drops and the glass transition temperature (see Chap. 2) increases. Attempts to successfully freeze human organs have shown that high pressure, in concert with infusion of a cryoprotectant that additionally increases the glass transition temperature, can effectively repress ice formation during freezing, frozen storage, and thawing [21]. Damage to meat structure and water-holding properties caused by ice crystal formation, as well as the associated freeze concentration of salts and hydrogen ions (pH decrease) that can contribute to protein denaturation (see Fig. 22), can be avoided by the use of high pressure. To prevent pressure-induced denaturation of proteins, sugars and polyols are added as baroprotectants. These compounds minimize the denaturing effects of heat, freezing, and pressure by enhancing intramolecular hydrophobic interactions that stabilize protein secondary and tertiary structure (see Sec. 15.8.4).

It was previously mentioned that pressures in the range of 150 MPa (1500 × atmospheric pressure) can induce rapid proteolytic tenderization of muscle at 40–60°C [32]. At the present time the biggest barriers to commercial use of high pressures for meat processing are the high cost, low throughput, and potential hazards of such equipment on a commercial scale.

15.8 CHEMISTRY OF PROCESSED MEATS

The term "processed" implies that meat has been subjected to some process that adds value, such as precooking, portioning, and/or packaging. More specifically, however, the term "processed meats" commonly denotes products that have been pickled or cured, and/or possibly restructured or blended with other ingredients to yield products much different from cooked, intact muscle. Pickled meats are stabilized in a salt and/or acid brine, whereby the salt and/or low pH and organic acids inhibit microbial growth. Such products are generally refrigerated to inhibit mold growth, stabilize flavor, and insure against growth and toxin production by pathogens.

15.8.1 Curing

Curing refers to the application of sodium chloride, usually in conjunction with a nitrite or nitrate salt. Curing is not used as much today for preservation of meats (except in very salty items like country ham and proscuitto). Rather it provides for the development of a characteristic flavor and color (see Chap. 10 and 11), and also provides protection against *Clostridium botulinum* outgrowth.

During formation of the cured color, endogenous low-molecular-weight components in the sarcoplasm facilitate the reaction of nitrite with myoglobin to yield nitric oxide myoglobin, which is stable and has a characteristic cured-meat color [69]. Reduced glutathione is active in this respect, and its effectiveness is enhanced by ATP, IMP, and ribose. Nitrite decomposition is also favored during this time. Curing appears to have little effect on either the quality of proteins or the stability of B vitamins.

Salt diffusion throughout the meat causes swelling of the myofibrillar matrix, though this is restricted in intact muscle by the sarcolemma. The presence of transition elements in salt used in cured meat enhances oxidation of lipid components, thus considerably reducing shelf life. Ascorbic acid in the presence of a metal chelating agent, such as polyphosphate, can effectively forestall this oxidative reaction. Antioxidative activity is also provided by nitric oxide myoglobin and *S*-nitrosocysteine, a compound formed during the curing process:

$$NO_2^- + HS-\underset{NH_2}{\overset{H}{C}}H_2-C-COOH \xrightarrow{H^+} ONS-\underset{NH_2}{\overset{H}{C}}H_2C \quad COOH$$

Many cured products are also smoked, or contain soluble components of wood smoke, mainly to add flavor. Many components of smoke are effective antioxidants.

Concerns have been expressed over the potential hazard of using nitrite in cured meats since this compound can react with amines, especially secondary amines, to form *N*-nitrosamines, which may be carcinogenic:

$$R_2NH + NO_2^- \xrightarrow{H^+} R_2NNO + H_2O$$

Although *N*-nitrosamines are not detectable in meat as marketed if the concentration of nitrite does not exceed U.S. federal regulations, commercial samples of frankfurters have occasionally been found that contain these compounds at low levels. In several instances, it is suspected that improper mixing may have resulted in high local concentrations of nitrite, thereby favoring formation of nitrosamines. High temperatures can also induce nitrosamine formation so a product such as bacon, which is subjected to very high temperatures, is especially vulnerable. Use of nitrites at concentrations just high enough to inhibit *C.botulinum* is the current practice. Even lower levels of nitrite might suffice for this purpose if compounds in addition to NaCl are present to help inhibit clostridia. Sorbate has been suggested as one such compound. Reducing agents, such as ascorbate, and antioxidants, such as tocopherol, are effective in reducing conversion of nitrite to nitrosamines and are often added to cured meat products.

15.8.2 Chemistry of Adhesion Between Meat Particles or Pieces

Pieces of meat, varying in size from chunks to flakes to particles, can be bonded together to simulate the appearance and texture of whole muscle products, or to create new products with unique textures. In all cases, adjacent meat surfaces are bonded by a gel network. The gel may consist of gelatin, added or derived from collagenous tissues during cooking of the meat, as in aspic products. Patented processes for forming "restructured" steaks involve use of either the calcium-alginate gelling reaction or clotting of plasma fibrinogen by a calcium-dependent transglutaminase in blood (factor XIII), activated by the protease thrombin. These are all processes for cold-forming of restructured products, although the latter two methods form heat-stable gels that will not melt upon subsequent cooking.

More commonly, the bond between meat pieces in restructured or reformed processed meats is effected by heat-induced gelation of soluble myofibrillar protein. Salt (sodium chloride) and possibly polyphosphates are added to meat to solubilize myofibrillar proteins prior to cooking. Products made from finely comminuted muscle, such as bologna-type sausages, frankfurters and imitation crabmeat, also rely on heat-induced gelation of myofibrillar proteins to build their structure and texture. These may contain added fat, starches/hydrocolloids, and non-meat proteins, and thus can be regarded as multicomponent gels, in which a gel network formed primarily by denatured myosin or actomyosin entraps other ingredients and plays the dominant role in determining their textural properties.

15.8.2.1 Emulsion and Gel Theory as Applied to Finely Comminuted Muscle Tissue

The first step in preparing fine-particle meat products typically involves comminution of the muscle tissue in the presence of salt. Meats comminuted with salt consist of a matrix of swollen fibers, myofibrils, and solubilized myofibrillar proteins. Other meat components, notably fat and connective tissue, plus any other ingredients added to the product formulation are embedded in this matrix. Traditionally, products such as bologna and hot dogs had high fat contents ($\geq 30\%$). Retention of this fat in the cooked product was often a problem, and a means of improving fat retention (stabilization) was sought. It was found that the ability of different meats to stabilize fat was strongly correlated with the fat emulsification properties of their salt-solubilized myofibrillar protein. This led to the theory that such products can be regarded as emulsions, and microscopy indicated the presence of fat globules surrounded by an adsorbed protein layer [28]. It is likely that comminution produces a protein film around fat globules, similar to a classical emulsifier layer; however, upon heating the fat becomes entrapped in part of a continuous gelled protein network. This entrapment of fat thus prevents fat migration.

15.8.2.2 Multicomponent Gels

If one accepts that comminuted fat-containing meats are correctly viewed as multicomponent gels rather than as emulsions, it is important to understand how nonmeat components affect the properties of the gelled myofibrillar matrix. Some studies on the subject have shown that the mechanical properties (measured both at small and fracture-causing deformations) of a composite gel are strongly dependent on the degree of interaction between the continuous (i.e., myofibrillar) protein phase and the dispersed phase (i.e., fat, nonmeat proteins and hydrocolloids), and on the distribution and phase volume of the dispersed phase within the gel matrix [27].

The composite filled gel is not the only possible model for the distribution of components in a muscle protein gel. Composite systems termed mixed gels [90] may also exist. In these, the filler (nonmyofibrillar) component may form a second continuous matrix within the muscle protein matrix, or may interact directly with the muscle protein to form a single mixed-composition matrix. Because of the difference in gelling temperature between muscle and most nonmyofibrillar gelling components (such as vegetable or milk proteins), cogelation of unlike proteins is less likely to occur than is the formation of an interpenetrating mixed gel.

The properties of mixed gels are influenced by the interaction energies of the component polymers [63]. When the net energy of interaction is favorable (exothermic) the polymers may associate into a single gel phase or precipitate together. However, the interaction between two polymer types is usually far less intensive than that between segments of the same type. Therefore, there is a tendency for unlike polymers to form separate domains, resulting in an increase in the effective concentration of each phase since neither is diluted to a large extent by the other. Water can be unequally partitioned between the two phases. Little is known about the phase compatibility of muscle proteins with other proteins or food ingredients or the consequences of these interrelationships on the quality of comminuted meat products.

15.8.3 Functionality of Muscle Proteins in Gel Formation

The ability of the myofibrillar constituents of meat to form a strong heat-induced gel requires good dispersion of these proteins and surface reactivity, that is, the availability of sites for favorable protein–protein interactions. Dispersion is dependent upon swelling and dissolution of the proteins by salt, and use of suitable comminution/blending equipment. Gels that exhibit an irregular, coagulative structure are not as strong and cohesive as those that display a more uniform matrix. A two-dimensional analogy of this is the appearance of a fish net with knots at the junctions compared with that of a tightly woven fabric.

A reactive protein surface seems to depend on maintaining the muscle in a near physiological state prior to the comminution step with salt. The ability of muscle tissue to bind water and function well in comminuted animal tissue is greater when actomyosin is in the dissociated state. This is the normal condition of prerigor muscle, and prerigor muscle has a very high gelling ability and water-holding capacity.

The specific properties of any gelled meat product depend on many factors including source of the muscle, biochemical state of the muscle, and processing variables. Myosin is the most important component of muscle tissue with respect to gel-forming ability. Actin assists in this process by forming F-actomyosin, which in turn interacts with free myosin [98]. The tail and head portions of the myosin molecule play different roles during heat-induced gelation of muscle minces. The head portion of the myosin molecule undergoes irreversible aggregation involving hydrophobic interactions and oxidation of -SH groups. This aggregation contributes to formation of the three-dimensional protein network. The tail portion of the myosin molecule undergoes a partially irreversible helix-to-coil transition during the heating process and then participates in the formation of a three-dimensional network.

Proteins of several fish species uniquely exhibit gelation of salted pastes or batters at temperatures below 40°C. When this gel is exposed to higher temperatures it exhibits greater strength than a gel cooked directly form the raw paste. The low-temperature gelling reaction is attributed to the action of an endogenous Ca^{+2}-dependent transglutaminase, which catalyzes an acyl transfer reacton in which the γ-carboxyamide groups of peptide-bound glutamine residues are the acyl donors. Primary amino groups in a variety of compounds, including the ε-amino groups of peptide-bound lysine residues, may function as acyl acceptors. When the latter reaction occurs, it yields an ε-(γ)-glutamyl) lysyl bond. The inability of mammalian and avian muscle pastes to gel in this way may not reflect absence of the enzyme, but rather lower reactivity (greater stability) of the myosin of homeotherms at temperatures below 40°C.

The formation of a meat gel involves protein denaturation under specific conditions of heating. Denaturation of the muscle proteins, especially myosin, under other conditions, such as in the original tissue, can interfere with gel formation. Thus, any condition that encourages incidental denaturation, such as low pH, freeze denaturation, or dehydration, will result in a weak heat-induced gel.

15.8.3.1 Stabilizaton of Muscle Proteins by Additives

Minced fish is often stabilized for frozen storage by removal of water-soluble components and the addition of stabilizing agents. The product obtained by washing and adding stabilizers is called surimi, and is used in the manufacture of imitation crabmeat and Japanese kamaboko products. When water-soluble proteins are not removed by washing they apparently associate with the contractile proteins as they denature, thereby interfering with the latter's ability to form a strong gel. Antioxidants and polyhydroxy compounds, such as sorbitol and sugar, are used as protein stabilizers. These compounds promote the preferential hydration of proteins which results in a positive free energy change. The positive free energy change increases with an increase in surface area of the protein. Therefore, the denatured form, which has a larger surface area, is thermodynamically less stable than the undenatured forms and so the latter is the preferred structure [4].

Polyhydroxy compounds are only one class of substances that impart stability to proteins under adverse conditions, such as the low water and high salt concentrations encountered during freezing. Others stabilizing compounds are some free amino acids and amino acid derivatives, such as taurine and β-alanine, and certain nitrogen compounds, such as methylamines, including trimethylamine oxide, betaine, and sarcosine [97]. Presumably these compounds exert stabilizing effects whether present naturally or added during processing. High-molecular-weight compounds, such as polydextrose or maltodextrin, have also been shown to protect fish myofibrillar proteins in frozen storage. These compounds may exert their cryoprotective effect by raising the glass transition temperature high enough to effectively immobilize reactants at conventional cold-store temperatures (see Chap. 2).

There has been much interest in drying muscle proteins so they may compete in the dry protein ingredients market. Disaccharides such as sucrose and trehalose exhibit the ability to protect proteins in both the frozen and dried states. These compounds apparently bind to proteins in the dry state and stabilize their tertiary structure [4].

15.8.3.2 Importance of Muscle Protein Integrity to Gelling Ability

Proteolytic degradation, occurring either prior to or during processing, reduces the strength and cohesiveness of meat gels, and also lowers fat and water binding. This can be a problem during manufacture of surimi-based foods, due to the presence in surimi pastes of heat-stable muscle proteases that become active during heating (optimally at 55–60°C). Problems with proteases vary with fish species, and many different proteolytic enzymes may be involved. Beef plasma has been found to contain effective inhibitors of proteases, and this is a common additive to surimi derived from problematic species. Other potentially useful food-grade inhibitors have been found in egg white, white potatoes, rice, bell pepper, and other natural sources [94].

15.8.4 Effects of pH and Ionic Ingredients on Gelling of Meat Proteins

During the comminution step of preparing a meat gel, the endomysium and sarcolemma are ruptured and the myofibrillar matrix is free to swell. The added salt partially solubilizes the myofibrillar proteins and yields a thick sol, or paste. It is theorized that the chloride ion selectively neutralizes positively charged sites on the protein molecules, effectively shifting the isoelectric pH to a lower value, thus enhancing protein solubility at the existing pH [31]. Sodium chloride also has a "salting-in" effect on myofibrillar proteins based on its effects on the system free energy, as designated by the lyotropic series (see Chap. 6).

If muscle tissue goes into rigor in either the intact or minced form, it loses much of its ability to bind water and exhibits poor gel-forming ability, and this loss cannot be completely restored by the addition of salt. However, if salt is added to minced muscle tissue before rigor occurs, the water-holding capacity of the proteins is maintained at a high level. Binding salt to muscle proteins results in electrostatic repulsion among molecules, thus loosening the protein network and allowing it to bind more water. The protein interactions that occur during onset of rigor, that is, after the loss of ATP, hinder this swelling effect of salt. The extent to which muscle sarcomeres shorten is irrelevant. Thus, very satisfactory sausage products can be made from prerigor muscle, but it must be ground and salted before the ATP level declines sufficiently to allow rigor mortis.

The gelling ability of postrigor meat can be improved by the addition of phosphates, particularly pyrophosphate. Apparently the primary effect of the phosphate is to partially decouple the actin–myosin complex formed at rigor. Postrigor meat stored frozen with an effective cryoprotectant additive performs equally well to fresh prerigor meat when polyphosphate is added during the comminution step [70].

A secondary effect of phosphate addition is to slightly increase the pH and move it away from the isoelectric point of the proteins (about pH 5.5). While the water-holding ability of meats improves as the pH is moved away from the isoelectric point toward neutrality, gelling ability (in terms of gel strength and cohesiveness) is optimum near pH 6.0–6.4 for muscle of homeotherms. Apparently good gelation depends on adjusting the degree of charge repulsion to achieve good protein dispersion and structure formation. The ability of fish pastes to undergo transglutaminase-linked gelation at low temperatures decreases as the pH is reduced below 7.0. Also, the innate ability of fish proteins to form heat-induced gels is optimal near neutrality.

15.8.5 Processing Factors

The successful production of comminuted meat products depends on the proper dispersion and gelation of the myofibrillar proteins. Therefore, manufacturing operations must maximize solubilization and swelling of the myofibrillar matrix early in the process with a minimum of protein denaturation. Added ingredients must also be well dispersed and of the optimum particle size to maximize their particular filler effect on the composite gel. In the case of higher fat-content products, fat particle size must be reduced, but continued comminution can lead to heat buildup unless cooling is provided. Excessive heating will lead to melting of the fat prior to protein gelation, with resultant poor fat stabilization.

Too high a heating rate during cooking can cause a similar problem. Gel formation is very sensitive to heating rate in that slower cooking generally allows a more rigid gel matrix to develop [24]. An exception to this is in the cooking of fish paste gels, where a high heating rate is desired to inactivate heat-stable proteases that are often present. Fish pastes should always be kept cold (< 10°C) during comminution with salt to avoid premature gelation of the meat via transglutaminase activity. Meat and poultry batters are generally chopped to relatively high temperatures to insure attainment of proper fat particle size and coating by proteins.

If dry components, such as hydrocolloids and nonmeat proteins, are added at levels that result in competition for water between the components, there is poor solubilization of myofibrillar proteins and/or poor hydration of the added dry ingredients. This causes a decrease in the quality of the cooked product. Hydrocolloids of exceptionally high water-holding ability, such as xanthan gum or pregelatinized starch, greatly interfere with development of a strong myofibrillar protein gel.

BIBLIOGRAPHY

Arakawa, N., Y.-J. Chu, M. Matsubara, and M. Takuno (1988). Effect of dehydration on meat.

Cross, H. R. (1979). Effects of electrical stimulation on meat tissue and muscle properties—A review. *J. Food Sci. 44*:509–514.

Gros, F., and M. Buckingham (1987). Polymorphism of contractile proteins. *Biopolymers 26*:177–192.

Haard, N. F. (1992). Technological aspects of extending prime quality of seafood: A review. *J. Aquatic Food Prod. Technol. 1*:9–27.

Hamm, R. (1981). Post-mortem changes in muscle affecting the quality of comminuted meat products, in *Developments in Meat Science* (R. Lawrie, ed.), Vol. 2., Applied Science, London, pp. 93–124.

Ledward, D. A. (1981). Intermediate moisture meats, in *Developments in Meat Science*, vol. 2 (R. Lawrie, ed.), Applied Science, London, pp. 159–194.

Sikorski, Z. E. (1994). The myofibrillar proteins in seafoods. In *Seafood Proteins* (Z. E. Sikorski, B. S. Pan, and F. Shahidi, (eds.), Chapman and Hall, New York, pp. 40–57.

Syrovy, I. (1987). Isoforms of contractile proteins. *Prog. Biophy. Molec. Biol. 49*::1–27.

REFERENCES

1. Ackman, R. G. (1980. Fish lipids, Part 1, in *Advances in Fish Science and Technology* (J. J. Connell, ed.), Fishing News Books, Surrey, England, pp. 86–103.
2. Aalhus, J. L., and M. A. Price (1991). Endurance-exercised growing sheep: I. Postmortem and histological changes in skeletal muscles. *Meat Sci. 29*:43–56.
3. Allen, C. E., and E. A. Foegeding (1981). Some lipid characteristics and interactions in muscle foods-a review. *Food Technol.* 253–257.
4. Arakawa, T., S. J. Prestrelski, W. C. Denney, and J. F. Carpenter (1993). Factors affecting short-term and long-term stabilities of proteins. *Adv. Drug Del. Rev. 10*:1–28.
5. Bailey, A. J. (1988). Connective tissue and meat quality. *Proc. 34th Intl. Cong. Meat Sci. Technol.*, Brisbane, Australia, pp. 152–160.
6. Bailey, A. J., and N. D. Light (1989). *Connective Tissue in Meat and Meat Products*, Elsevier Applied Science: New York.
7. Barbagli, C., and G. S. Crescenzi (1981). Influence of freezing and thawing on the release of cytochrome oxidase from chicken's liver and from beef and trout muscle. *J. Food Sci. 46*:491–493.
8. Bershadsky, A. D., and J. M. Vasiliev (1988). *Cytoskeleton*, Plenum Press, New York.
9. Bone, Q., A. Pulsford, and A. D. Chubb (1981). Squid mantle muscle. *J. Mar. Biol. Ass. U.K. 61*:327–342.
10. Bornstein, P., and W. Traus (1979). The chemistry and biology of collagen, in *The Proteins*, vol. IV, 3rd ed. (h Neurath and R.L. Hill, eds.), Academic Press, New York, pp. 411–632.
11. Bremner, H. A. (1992). Fish flesh structure and the role of collagen—Its post-mortem aspects and implications for fish processing. In *Quality Assurance in the Fish Industry* (H.H. Huss et al., eds.), Elsevier Science, New York, pp. 39–62.
12. Calkins, C. R., T. R. Dutson, G. C. Smith and Z. L. Carpenter (1982). Concentration of creatine phosphate, adenine nucleotides and their derivatives in electrically stimulated and nonstimulated beef muscle. *J. Food Sci. 47*:1350–1353.
13. Cepeda, R., E. Chou, G. Bracho, and N. Haard (1990). An immunological method for measuring collagen degradation in the muscle of fish. In *Advances in Fisheries Technology and Biotechnology for Increased Profitability* (M.N. Voigt and J. R. Botta, eds.), Technomic Pub.
14. Chan, K. M., and E. A. Decker (1994). Endogenous skeletal muscle antioxidants. *Crit. Rev. Food Sci. Nutri, 34*:403–426.
15. Connell, J. J. (1975). *Control of Fish Quality*, Fishing News Lts., Surrey, UK.
16. Dahlin, K. J., C. E. Allen, E. S. Benson, and P. V. J. Hegarty (1976). Ultrastructural differences induced by heat in red and white avian muscles in rigor mortis. *J. Ultrastructure Res. 56*:96–106.
17. Decker, E. A., and H. O. Hultin (1992). Lipid oxidation in muscle foods via redox iron, in *Lipid Oxidation in Food* (A. J. St. Angelo, ed.), American Chemical Society, Washington, DC, pp. 34–54.

18. DeFremery, D., and M. F. Pool (1963). Biochemistry of chicken muscle as related to rigor mortis and tenderization *Food Res. 25*:73–87.
19. Dransfield, E., D. J. Etherington, and M. Taylor (1992). Modelling post-mortem tenderization—II: Enzyme changes during storage electrically stimulated and nonstimulated beef. *Meat Sci. 31*:75–84.
20. Ebashi, S., M. Endo, and I. Ohtsuki (1969). Control of muscle contraction. *Q. Rev. Biophys.* 2:351–384.
21. Fahy, G. M., D. R. MacFarlane, C. A. Angelland, and H. T. Meryman (1984). Vitrification as an approach to cryopreservation. *Cryobiology 21*:407–426.
22. Faustman, C., R. G. Cassens, D. M. Schaffer, D. R. Buege, S. N. Williams, and K. K. Scheller (1989). Improvement of pigment and lipid stability in holstein steer beef by dietary supplementation with vitamin E. *J. Food Sci. 54*:858–862.
23. Fey, M. S., and J. M. Regenstein (1982). Extending shelf-life of fresh wet red hake and salmon using CO_2-O_2 modified atmosphere and potassium sorbate ice at 1°C. *J. Food Sci.47*:1048–1070.
24. Foegeding, E. A., C. E. Allen, and W. R. Dayton (1986). Effect of heating rate on thermally formed myosin, fibrinogen and albumin gels. *J. Food Sco. 52*:104–112.
25. Forrest, J. C., E. D. Aberle, H. B. Hedrick, M. D. Judge, and R. A. Merkel (1975). *Principles of Meat Science,* W. H. Freeman, San Francisco.
26. Fraser, D. I., J. R. Dingle, J. A. Hines, S. C. Nowlan, and W. J. Dyer (1967). Nucleotide degradation, monitored by thin-layer chromatography, and associated postmortem changes in relaxed cod muscle. *J. Fish Res. Bd. Can. 23*:1837–1841.
27. Goodwin, J. W., J. Hearn, C. C. Ho, and R. H. Ottewill (1974). Studies on the preparation and characterization fo monodisperse polystyrene lattices. *Colloid and Polymer Sco. 252*:464–471.
28. Gordon, A., and S. Barbut (1992). Mechanisms of meat batter stabilization: A review. *Crit. Rev. Food Sci. Nutr. 32*:299–332.
29. Greer-Walker, M. (1970). Growth and development of the skeletal muscle fibers of the cod (Gadus morhua L.). *J. Cons. Int. Explor. Mer. 33*:228–244.
30. Haard, N. F. (1992). A review of proteolytic enzymes form marine organsims and their application in the food industry. *J. Aquatic Food Prod. Technol. 1*(1):17–35.
31. Hamm, R. (1986). Functional properties of the myofibrillar system and their measurements, in *Muscle as Food* (P.J. Bechtel, ed.), Academic Press, New York, pp.130–200.
32. Harris, P., and W. R. Shorthouse (1988). Meat texture, in *Developments in Meat Science—4*, (R. Lawrie, ed.), Elsevier Applied Science, New York, pp. 245–296.
33. Hatae, K., F. Yoshimatsu, and J. J. Matsumoto (1990) Role of muscle fibers in contributing firmness of cooked fish. *J. Food Sci. 55*:693–696.
34. Hay, J. D., R. W. Currid, F. H. Wolfe, and E. J. Sanders (1973). Effect of postmortem aging on chicken breast muscle sarcoplasmic reticulum. *J. Food Sci. 38*:700–704.
35. Holmes, K., D. Popp, W. Gebhard, and W. Kabsch (1990). Atomic model of the actin filament *Nature 347*:44–49.
36. Howgate, P. (1979). Fish, in *Food Microscopy* (J. g. Vaughan, ed.), Academic Press London, pp. 343–392.
37. Hultin, H. O. (1976). Characteristics of muscle tissue, in *Principles of Food Science)*, Part I, *Food Chemistry* (O. R. Fennema, ed.), Marcel Dekker, New York, pp. 577–617.
38. Huxley, A. F., and R. Niedergerke (1954). Structural changes in muscle during contraction: interference microscopy of living muscle fibres. *Nature 173*:971–973.
39. Huxley, H. E. (1963). Electron microscope studies on the structure of natural and synthetic protein filaments from striated muscle. *J. Mol. Biol. 7*:281–308.
40. Huxley, H. E., and J. Hanson (1954). Changes in the cross striations of muscle during contraction and stretch and their structural interpretation. *Nature 173*:973–976.
42, Judge, M. D., and E. D. Aberle (1982). Effects of chronological age and postmortem aging on thermal temperature of bovine intramuscular collagen. *J. Anim. Sci. 54*:68–71.
42. Ke., P., R. G. Ackman, B. A. Linke, and D. M. Nash (1977). Differential lipid oxidation in various parts of frozen mackerel. *J. Food Technol. 12*:37–47.

43. Koohmaraie, M., S. C. Seideman, J. E. Schollmeyer, T. R. Dutson, and A. S. Babiker (1988). Factors associated with the tenderness of three bovine muscles. *J. Food Sci. 53*:407–410.
44. Koohmaraie, M., A. S. Babiker, A. L. Schroeder, R. A. Merkel, and T. R. Dutson (1988). Acceleration of postmortem tenderization in ovine carcasses through activiation of CA^{2+}-dependent proteases. *J. Food Sci. 53*:1638–1641.
45. Lagler, K. F., J. E. Bardack, R. R. Miller, and D. R. M. Pasino (1977). *Ichthyology,* 2nd ed., John Wiley and Sons, New York.
46. Lavety, J., O. A. Afolabi, and R. M. Love (1988). The connective tissues of fish. IX. Gapin in farmed species. *Int. J. Food Sci. Technol. 23*:23–30.
47. Lee, Y. B., G. L. Hargus, E. C. Hagberg, and R. H. Forsythe (1976). Effect of antemortem environmental temperatures on postmortem glycolysis and tenderness excised broiler breast muscle. *J. Food Sci. 41*:1466–1469.
48. Leu, S.-S., S. N. Jhaveri, P. A. Karakoltsidis, and S. M. Constantinides (1981). Atlantic mackerel (*Scomber scombrus*, L): Seasonal variation in proximate composition and distribution of chemical nutrients. *J. Food Sci. 46*:1635–1638.
49. Lister, D., N. G. Gregory, and D. D. Warriss (1981). Stress in meat animals, in *Developments in Meat Science*, vol. 2 (R. Lawrie, ed.), Applied Science, New York, pp. 61–92.
50. Locker, R. H., and W. A. Carse (1976). Extensibility, strength and tenderness of beef cooked to various degress. *J. Sci. Food. Agric. 27*:891–901.
51. Locker, R. H., G. J. Daines, W. A., Carse, and N. G. Leet (1977). Meat tenderness and the gap filaments. *Meat Sci. 1*:87–104.
52. Locker, R. H. and C. J. Hagyard (1962). A cold-shortening effect in beef muscle. *J. Sci. Food Agric. 14*:787–793.
53. Louis, C. F., W. E. Rempel, and J. R. Mickelson (1994). Porcine stress syndrome: Biochemical basis of this inherited syndrome of skeletal muscle. *Proc. 46th Annu. Reciprocal Meat Conf. National Live Stock and Meat Board*, Chicago, pp. 89–96.
54. Love, R. M. (1988). *The Food Fishes. Intrinsic Variation and Its Practical Implications*, Farrand Press, London.
55. Love, R. M. (1980. *The Chemical Biology of Fishes*, vol. 2, *Advances 1968–1977*, Academic Press, London.
56. Love, R. M., J. Lavety, and F. Vallas (1982). Unusual properties of the connective tissues of cod (*Gadus morhua*), in *Chemistry and Biochemistry of Marine Food Products* (R. E. Martin, G. J. Flick, C. E. Hebard, and D. R. Ward, eds.), AVI, Westport, CT pp. 67–73.
57. Maita, T., E. Yajima, S. Nagata, T. Miyanishi, S. Nakayama, and G. Matsuda (1991). The primary structure of skeletal muscle myosin heavy chain: IV. Sequence of the rod, and the complete 1,938-residue sequence of the heavy chain. *J. Biochem. 110*:75–87.
58. Marsh, B. B., and N. G. Leet (1966). Resistance to shearing of heat-denatured muscle in relation to shortening. *Nature 221*:635–636.
59. Marsh, B. B., and N. G. Leet (1966). Studies in meat tenderness. III. The effects of cold-shortening on tenderness. *J. Food Sci. 31*:450–459.
60. Matsumoto, J. J. (1979). Denaturation fo fish muscle proteins during frozen storage in *Proteins at low Temperatures* (O. Fennema, ed.), Advances in Chemistry Series 180, American Chemical Soc., Washington, DC, pp. 205–224.
61. Muller, B., D. Maeda, and A. Wittinghofer (1990). Sequence of the myosin light chain 1/3 isolated from a rabbit fast skeletal muscle lambda library. *Nucleic Acids Res. 18*:6688.
62. Muramoto, M., Y. Yamaoto, and N. Seki (1988). Comparison of calpain susceptibility of various fish myosins in relation to their thermal stabilities. *Nippon Suisan Gakkaishi 54*:917–923.
63. Morris, E. R. (1990). Mixed polymer gels. In *Food Gels*, (P. Harris, ed.), Elsevier, London, pp. 291–359.
64. Nishita, K., Y. Takeda, and K. Arai (1981). Biochemical characteristics of actomyosin from Antarctic krill. *Bull. Jpn. Soc. Sci. Fish. 47*:1237–1244.
65. Northcutt, J. K., E. A. Foegeding, and F. W. Edens (1994). Water-holding properties of thermally preconditioned chicken breast and leg meat. *Poultry Sci. 73*:308–316.

66. Nunzi, M. G., and C. Franzini-Armstrong (1981). The structure of smooth and striated portions of the adductor muscle of the valves in a scallop. *J. Ultrastruct. Res. 76*:134–148.
67. Nuss, J. I., and F. H. Wolfe (1981). Effect of post-mortem storage temperatures on isometric tension, pH, ATP, glycogen and glucose-6-phosphate for selected bovine muscles. *Meat Sci. 5*:201–213.
68. Offer, G., P. Purslow, R. Almond, T. Cousins, J. Elsey, G. Lewis, N. Parsons, and A. Sharp (1988). Myofibrils and meat quality. *Proc. 34th Int. Cong. Meat Sci. Technol.*, Brisbanc, Australia, pp. 161–168.
69. Okayama, T., N. Ando, and Y. Nagata (1982). Low-molecular weight components in sarcoplasma promoting the color formation of processed meat products. *J. Food Sci. 47*:2061–2063.
70. Park, J. W., T. C. Lanier, and D. H. Pilkington (1993). Cryostabilization of functional properties of pre-rigor and post-rigor beef by dextrose polymer and/or phosphates. *J. Food Sci. 58*:467–472.
71. Park, J. W., Lanier, T. C., Keeton, J. T., and Hamann, D. D. (1987). Use of cryoprotectants to stabilize functional properties of prerigor salted beef during frozen storage. *J. Food Sci. 52*:537–542.
72. Parkin, K. L. and W. D. Brown (1982). Preservation of seafood with modified atmospheres, in *Chemistry and Biochemistry of Marine Food Products* (R. E. Martin, G. J. Flick, C. E. Hebard, and D. R. Ward, eds.), AVI, Westport, CT pp. 453–465.
73. Pearson, A. M., A. M. Wolzak, and J. I. Gray (1982). Possible role of muscle proteins in flavor and tenderness. *J. Food Biochem. 7*:189–210.
74. Pette, D., and R. S. Staron (1990). Cellular and molecular diversities of mammalian skeletal muscles fibers. *Rev. Physiol. Biochem. Pharmacol. 116*:1–76.
75. Porte, K. R., and C. Franzini-Armstrong (1965). The sarcoplasmic reticulum. *Sci. Am. 212*:72–78,80.
76. Rayment, I., W. R. Rypniewski, K. Schmidt-Base, R. Smith, D. R. Tomchick, M. M. Benning, D. A. Windelmann, B. Wesenberg, and H. M. Holdon (1993). Three-dimensional structure of myosin subfragment-1: A molecular motor. *Science 261*:50–58.
77. Ritchie, A. H., and I. M. Mackie (1980). The formation of diamines and polymines during storage of mackerel (*Scomber scombrus*), in *Advances in Fish Science and Technology* (J J. Connell, ed.), Fishing News Books, Surrey, England, pp. 489–494.
78. Robson, R. M., and T. W. Huiatt (1983). Roles of the cytoskeletal proteins desmin, titin and nebulin in muscle. *Proc. 36th Ann. Reciprocal Meat Conf. National Live Stock and Meat Board*, Chicago, pp. 116–124.
79. Shenouda, S. Y. K. (1980). Theories of protein denaturation during frozen storage of fish flesh, in *Advances in Food Research*, vol. 26 (C.O. Chichester, ed.), Academic Press, New York, pp. 275–311.
80. Shewfelt, R. L. (1981). Fish muscle lipolysis—A review. *J. Food Biochem. 5*:79–100.
81. Shewfelt, R. L., and H. O. Hultin (1983). Inhibition of enzymic and non-enzymic lipid peroxidation of flounder muscle sarcoplasmic reticulum by pretreatment with phospholipase A_2. *Biochim. Biophiys. Acta 751*:432–438.
82. Sikorski, Z. E., and J. A. Borderias (1994). Collagen in the muscles and skin of marine animals, in *Seafood Proteins* (Z. E. Sikorski, B. S. Pan and F. Shahidi, ed.), Chapman and Hall, New York, pp. 58–70.
83. Sims, T. J., and A. J. Bailey (1981). Connective tissue, in *Developments in Meat Science*, vol. 2 (R. Lawrie, ed.), Applied Science, London, pp. 29–59.
84. Staron, R. S., and D. Pette (1993). The continum of pure and hybrid myosin heavy chain-based fibre types in rat skeletal muscle. *Histochemistry 100*:149–153.
85. Stefansson, G., and H. O. Hultin (1994). On the solubility of cod muscle proteins in water. *J. Agric. Food Chem. 42*:2656–2664.
86. Swatland, H. J. (1981). Cellular heterogeneity in the response of beef to electrical stimulation. *Meat Sci. 5*:451–455.
87. Sweeney, H. L., M. Kushmerick, K. Mabuchi, F. Sreter, and J. Gergely (1988). Myosin alkali light chain and heavy chain variations correlate with altered shortening velocity of isolated skeletal muscle fibers. *J. Biol. Chem. 263*:9034–9039.
88. Takahashi, K., T. Fukazawa, and T. Yasui (1967). Formation of myofibrillar fragments and reversible contraction of sarcomeres in chicken pectoral muscle. *J. Food Sci. 32*:409–413.
89. Thompson, A. B., A. S. McGill, J. Murray, R. Hardy, and P. F. Howgate (1980). The analysis of a

range of non-volatile constituents of cooked haddock (*Gadus aegle-finus*) and the influence of these on flavor, in *Advances in Fish Science and Technology* (J. J. Connell, ed.), Fishing News Books, Surrey, England, pp. 30–34.

90. Tolstogusov, V. B. (1986). Functional properties of protein-polysaccharide mixtures, in *Functional Properties of Food Macromolecules* (J. R. Mitchell and D. A. Ledward, eds.), Elsevier, London, pp. 385–415.
91. Toyohara, H., M. Kinoshita, Y. Shimizu, and M. Sakaguchi (1991). A group of novel latent serine proteinases degrading myosin heavy chain in fish muscle. *Biomed. Bloc. 50*:717–720.
92. Tsuchimoto, M., N. Tanaka, T. Misima, S. Yada, T. Senta, and M. Yasuda (1988). The influence of habitat water temperature on the relative thermostability of myofibrillar CA^{2+}-ATPase in fishes collected in the waters from tropical to frigid zones. *Nippon Suisan Gakkaishi 54*:787–793.
93. Veis, A. (1965). *The Macromolecular Chemistry of Gelatin*, Academic Press, New York.
94. Wasson, D. H., (1992). Fish muscle protease and heat-induced myofibrillar degradation: A reviw. *J. Aquatic Food Product Technol. 1*:23–41.
95. Watt, B. K., and A. L. Merrill (1963). *Composition of Foods*, Agriculture Handbook No. 8, U.S. Department of Agriculture, Washington, DC.
96. Wiskus, K. J., P. B. Addis, and R. T. Ma (1976). Distribution of βR, αR and αW fibers in turkey muscles. *Poultry Sci. 55*:562–57.
97. Yancey, P. H., M. E. Clark, S. C. Hand, R. D. Bowlus, and G. N. Somero (1982). Living with water stress: Evolution of osmolyte systems. *Science 217*:1214–1222.
98. Yasui, T., M. Ishioroshi, and K. Samejima (1982). Effect of actomyosin on heat induced gelation of myosin. *Agric. Biol. Chem. 46*:1049–1059.
99. Zarkadas, C. G., W. D. Marshall, A. K. Khalili, Q. Nguyen, G. C. Zarkadas, C. N. Karatzas, and S. Khanizadeh (1987). Mineral composition of selected bovine porcine and avian muscles and meat. *J. Food Sci. 52*:520–525.

16

Characteristics of Edible Plant Tissues

NORMAN F. HAARD
University of California, Davis, California

GRADY W. CHISM
The Ohio State University, Columbus, Ohio

16.1 INTRODUCTION

Plant tissues, through the agency of photosynthesis, are ultimately the supplier of all food for humans. An estimated 150 billion tons of carbon are fixed annually by more than 300,000 taxonomically distinct plant species. Worldwide, plant tissues directly provide about 70% of the protein consumed by humans. In affluent countries, like the United States, about 30% of dietary protein comes directly from plants. The annual food consumption of plant products in the United States remains somewhat less than that of animal products [21]. Per capita consumption of plant food products in the United States from 1980 to 1992 is shown in Figure 1. Consumption of plant products increased by more than 100 lb to 671 lb between 1980 and 1992. During this same period of time, overall intake of animal products (i.e., red meats, poultry, eggs, fish, shellfish, and dairy products) by American consumers remained constant at just under 900 lb per capita. Although thousands of crops fit into this dietary picture, some 100–200 species are of major importance in world trade. The major cultivated species produced in the world include sugar cane, wheat, maize, rice, potato, sugar beet, barley, cassava, sweet potato, and soybean. These crops directly or indirectly provide an estimated 80% of human dietary energy and protein.

The basic taxon of plants, the species, may be represented by hundreds or thousands of agricultural varieties or cultivars developed by plant breeders. An estimated 50,000 cultivars of

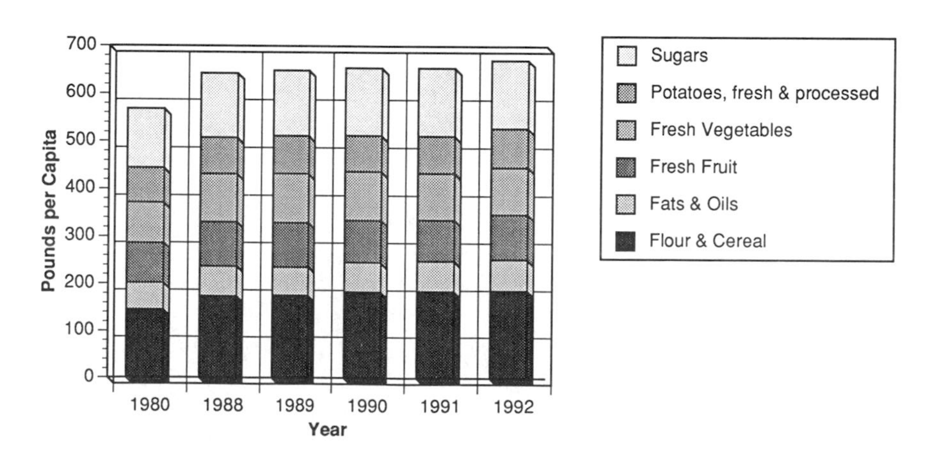

FIGURE 1 Annual food consumption of food groups derived from plant products. Data are on a retail weight basis. Fresh vegetable data are for lettuce, tomatoes, onion, carrots celery, corn, broccoli, asparagus, artichokes, cabbage, cucumbers, eggplant, garlic, green beans, green peppers, and cauliflower. (Data from Ref. 21).

wheat have been derived from a few species of genus *Triticum*, namely, *T. monococcum*, *T. dioccum*, *T. durum*, and *T. aestivum*. Different species of plants are also grouped on the basis of taxonomic families [40]. Taxonomic families often have common biochemical characteristics. For example, members of the Cruciferae or "mustard" family contain aromatic sulfur compounds derived from thioglucosides. Examples of some important families of plants that enter man's diet are listed in Table 1. Special mention should be made of the Gramineae or "grass" family, which includes the major cereal crops wheat, corn, rice, and barley. The sustenance provided by cereals is frequently referred to in the Bible, and they are by many other criteria the most important group of crops produced in the world. Cereal grain contains 10,000–15,000 kJ/kg, about 10–20 times more energy per unit weight than most succulent fruits and vegetables. In 1986 world production of the cereals was almost 2 billion metric tons. They are estimated to directly provide 50% of human calories in the world and 70–80% of all food calories consumed in China and India. Worldwide, cereals are the direct or indirect source of about 75% of our dietary protein.

In addition to supplying protein and energy, plant products are also an important source of other dietary nutrients, as shown by the relatively high "index of nutritional quality"(INQ) of several succulent crops [78]. The sum of INQ values for seven major nutrients is about 10 times greater for spinach, broccoli, or lettuce than it is for whole milk, egg, or meat (Table 2).

Many volumes have been written to describe the characteristics of edible plant tissues, and it is beyond the scope of this chapter to provide complete details on individual commodities, much less all these crops. The intent of this chapter is to outline the compositional and physiological diversity of crop plants and to relate these properties to postharvest quality, storage, and processing. Postharvest physiology remains one of the most important and challenging sectors of science and technology.

16.2 CHEMICAL COMPOSITION

Detached plant parts, like all living organisms, contain a wide range of different chemical compounds and show considerable variation in composition. The proximate composition for the major economic groups of edible plants is summarized in Table 3. Apart from the obvious interspecies differences, an individual plant organ, largely composed of living tissues that are metabolically active, is more or less constantly changing in composition. The rate and extent of chemical change can depend on the growing conditions prior to harvest, the physiological role of the plant part, the genetic pool of the cell, and the postharvest environment.

16.2.1 Water

Given an unlimited supply of water, the moisture content of viable plant tissue assumes a characteristic maximum value associated with a state of complete turgor of the component cells. At turgor, the internal pressure (up to 900 mpa or more) developed by osmotic forces is balanced by the inward pressure of the extended cell wall [55]. Accordingly, the maximum water content of a given tissue is influenced by its structural and chemical characteristics as well as extrinsic factors. The susceptibility of a commodity to water loss is dependent on the water vapor deficit of the surrounding atmosphere and the extent to which the external surface of the tissue is structurally and chemically modified to reduce transpiration (water loss). Water generally represents about 80–90% of the fresh weight of "wet" crops and less than 20% of the weight of "dry" crops such as cereals, nuts, and pulses. Apart from the influence of water loss on wilting and weight loss, one must recognize that moisture balance can indirectly evoke undesirable or desirable physiological changes in some crops. The interaction of water with other molecular constituents and its role in cellular metabolism are discussed in Chapter 2.

TABLE 1 Some Families in the Plant Kingdom That Are Important Sources of Human Food

Family	Examples	Characteristics
Gramineae	Grass family—cereals, sugar cane, bamboo	Directly and indirectly provide about 75% of food protein and energy; energy dense with 10,000–15,000 kJ/kg
Leguminosae	Legume family—peas, lentils, soybean, peanuts	Legume seeds average twice as much protein as cereals; soybeans are single largest cash crop in United States
Palmaceae	Palm family—coconut, date, palm	Source of tropical oils
Rosaceae	Rose family—many fruits, e.g., apple, pear, peach, apricot, plum, cherry, raspberry, strawberry, rose hips	Includes several fruit structures, e.g., pome fruits, stone fruits, and aggregate fruits; develop distinctive aromatic flavors and sweet taste during ripening
Curcurbitaceae	Melon family—squash, cantaloupe, watermelon, cucumber, pumpkin	History of cultivation is > 9000 years; fruit do not tend to sweeten as they ripen
Cruciferae	Mustard family—cabbage, cauliflower, radish, turnip, canola, horseradish, mustard	Characterized by pungent sulfur-containing compounds or "mustard oils" that are formed enzymatically from thioglucosides
Labiatae	Mint family—basil, marjoram, oregano, peppermint, rosemary, sage, savory, thyme	Herb plants, most of which are native to Mediterranean region; hair-like oil glands on leaves and stems contain "essential oils"
Umbelliferae	Parsnip family—celery, carrot, parsley, caraway, dill, fennel, coriander, anise, cumin	Herb and spice plants; seed-fruit are rich in "essential oils"
Solanaceae	Nightshade family—potato, tomato, eggplant, pepper	Characterized by toxic steroidal alkaloids in some nonedible species
Lilliacae	Lily family—alliums (onion, shallot, chive, garlic, leek), asparagus	Alliums characterized by lachrimatory, flavorful sulfur compounds that are formed enzymatically from *S*-alkylcysteine sulfoxides
Compositae	Daisy family—lettuce, endive, dandelion, chicory	Latex contains triterpenoid alcohols that act as soporific herbs
Rutaceae	Citrus family—lemon, orange, grapefruit, lime, citron, kumquat, mandarine	Flavanoid glycosides and terpenes contribute to characteristic flavor

Table 2 Index of Nutritional Quality (INQ) for Seven Major Nutrients in Selected Plant- and Animal-Derived Food Products

Food	INQ > 1, number of nutrients	Total INQ (7 nutrients)
Spinach	7	185
Broccoli	7	180
Lettuce	7	148
Orange	7	63
Apricot	7	45
Apple	2	8
Beans, navy	5	14
Potatoes	4	16
Whole wheat bread	5	8
Whole milk	4	17
Ground beef	4	13
Eggs	4	16
Tuna	3	17

Note: The INQ is the percent of nutrient need provided by the food divided by the percent of caloric need provided from food.
Source: Data from Ref. 78.

16.2.2 Carbohydrates

In general, approximately 75% of the solid matter of plants is carbohydrate. The total carbohydrate content can be as low as 2% of the fresh weight in some fruits or nuts, more than 30% in starchy vegetables, and over 60% in some pulses and cereals. Total carbohydrates are generally regarded as consisting primarily of simple sugars and polysaccharides, but they also properly include pectic substances and lignin. In plants, carbohydrates are localized in the cell wall and intracellularly in plastids, vacuoles, or the cytoplasm.

16.2.2.1 Cell Wall Constituents

The principal cell wall constituents are cellulose, hemicelluloses, pectins, and lignin. Cellulose is one of the most abundant substances in the biosphere, is largely insoluble, and is indigestible by human beings. The hemicelluloses are a heterogeneous group of polysaccharides that contain numerous kinds of hexose and pentose sugars and in some cases residues of uronic acids. These polymers are classified according to the types of sugar residues predominating and are individually referred to as xylans, arabinogalactans, glucommannans, etc. Pectin is generally regarded

Table 3 Ranges for Proximate Composition of Crops Represented in Major Economic Groups of Edible Plants

Group	Water, %	Carbohydrates, %	Lipids, %	Protein, %
Fruit	80–90	5–20	0.1–0.5	0.5–3
Fleshy vegetables	80–90	2–20	0.1–0.3	5–7
Pulses, seeds	10–50	6–60	1–18	5–25
Nuts	3–50	10–40	2–70	3–25
Cereals	12–14	65–75	2–6	7–12

as α-1,4-linked galacturonic acid residues esterified to varying degrees with methanol. Pectins found in the primary wall have a higher degree of methylation than do pectins in the middle lamella. Interaction of the carboxyl groups of pectin with calcium can form ionic cross-links between the polymer chains. Purified pectin invariably contains significant quantities of covalently linked nonuronide sugars, such as rhamnose, arabinose, and xylose, which result in the "hairy" region of the molecule. The albedo (white spongy layer) of citrus peel is an especially rich source of pectin, containing up to 50% of this constituent on a dry weight basis. Commercially, apple and citrus peel are major sources of pectin, with the latter being used primarily as a gelation agent. Lignin, always associated with cell wall carbohydrates, is a three-dimensional polymer comprised of phenyl propane units, such as syringaldehyde and vanillin, and these are linked through aliphatic three-carbon side chains. Lignification of cell walls, notably those of xylem and sclerenchymatous tissues, confers considerable rigidity and toughness to the wall. While generally ignored, cell walls also contain small amounts of structural proteins and enzymes. The best characterized structural protein is extensin, which occurs in the wall, but not the middle lamella [19]. Most of the enzymes that have been identified in the wall are peroxidases and hydrolases.

Cell wall constituents are the primary components of "dietary fiber." The biological availability of protein and other nutrients [83] can be reduced by these constituents, but there is considerable evidence for the beneficial role played by fiber in health and disease [3]. The structure and chemistry of these substances are discussed in more detail in Chapter 3.

In spite of rapid advances in analytical methodology and molecular biology, the precise molecular details of cell walls that are needed to explain softening during ripening and expansion during growth are still lacking. Pectin occupies a central location in the middle lamella (see Fig. 18) located between the primary cell walls of adjacent cells. Hydrolysis of pectin by enzymes (see Chap. 6) can reduce intercellular adhesion and thus soften the tissue. Nonripening tomato mutants [43] and transgenic tomatoes with drastically reduced polygalacturonase [18] demonstrate that polygalacturonase is necessary for significant softening to occur. Solubilization of pectin [85] and the release of low-molecular-weight uronides from cell walls occur concomitantly with softening in many fruits. Identification of a xyloglucan endotransglycosylase that increases during the ripening [72] of kiwifruit suggests that degradation of hemicellulose may also play an important role in softening. Calcium–pectin interactions also contribute substantially to the rigidity of the cell walls [27] (Section 16.5.1.4).

Cellulose is an important component of cell walls, but relatively little is known about its structure within the cell wall. The cellulose in apple cell walls is composed of triclinic Iα and monoclinic β crystalline forms [60], with little amorphous cellulose. This combination is consistent with the crisp texture of an apple, as amorphous cellulose gives rise to toughness. Cellulases have been implicated via gene expression experiments as important parts of ethylene-induced softening, but chemical measurements of cellulose have yet to confirm the role of cellulose degradation in softening of fruits [61].

The relative proportion and contents of cell wall constituents vary considerably among species, among cell types, with maturity at harvest, and with elapsed time and storage conditions after harvest. Hydrolysis of these components by carbohydrases produced during ripening along with removal of calcium leads to softening, while lignification and influx of calcium leads to "hardening."

16.2.2.2 Starch and Other Polysaccharides

The principal carbohydrate of plant tissues, which is not associated with cell walls, is starch, a linear (α-1,4) or branched (α-1,4;1,6) polymer of D-glucose (see Chap. 3). In the developing plant, amylose is synthesized from adenosine diphosphoglucose (ADPG) or uridine diphosphoglucose (UDPG) by the enzyme starch synthetase. This occurs in organelles called amyloplasts.

UDPG is formed from sucrose by the enzyme sucrose-UDP glucosyl transferase and ADPG, and ADPG is formed from glucose-1-P by the enzyme ADPG-phosphorylase [75]. A branching enzyme, called Q-enzyme, introduces α-1,6 glucosidic linkages into amylose chains to form amylopectin. Although most starches contain amylose and amylopectin in a ratio of 1:4, high-amylose and high-amylopectin ("waxy") cultivars of cereals have been developed. Waxy starches are valued as industrial products for their bright transparent appearance, which is attributed to the absence of retrogradation. Likewise, short-grain rice is moist, firm, and sticky when cooked due to the high proportion of amylopectin in the starch. In mature plants, starch is localized in intercellular plastids or granules that have species-specific shape, size, and optical properties. Industrially, starch is obtained from such crops as potato, sweet potato, or cereals that may contain up to 40% starch on a fresh weight basis [71]. Although starch contributes more calories to the normal human diet than any other single substance, it is absent or negligible in most ripe fruits and many vegetables.

The amount of nonstarch polysaccharides in edible plants is not accurately known because of limitations with analytical methods [23, 65]. Examples of other polysaccharides that occur as intracellular constituents of fruits and vegetables include α-D-1,4-linked glycopyranose of sweet corn, β-glucans of mango, and fructans. D-Fructose in the furanose form is the main unit of fructans and the units are linked β(2 → 1) or β(2 → 6). Inulin is the main carbohydrate reserve found in plants that do not accumulate starch (e.g., Jerusalem artichoke, sweet potato). An initial glucose on inulin indicates that a primer sucrose molecule is built up by successive transfer of fructose units (40–100 residues) from sucrose (Fig. 2).

Oats, barley, and rye are examples of cereals that contain a relatively high percent (5–25% of total carbohydrates) of nonstarch polysaccharides in the flour. The pentosan fraction of cereals is a complex mixture of branched polysaccharides with an arabinoxylan backbone containing small amounts of glucose and ferulic acid (Fig. 3). Some pentosans are conjugated with protein, and they can be separated into water-soluble and water-insoluble fractions. Rye flour contains a relatively high content (6–8%) of pentosans; 15–25% of rye pentosans are water soluble. Pentosans contribute to gluten development in wheat flour and are especially important to the

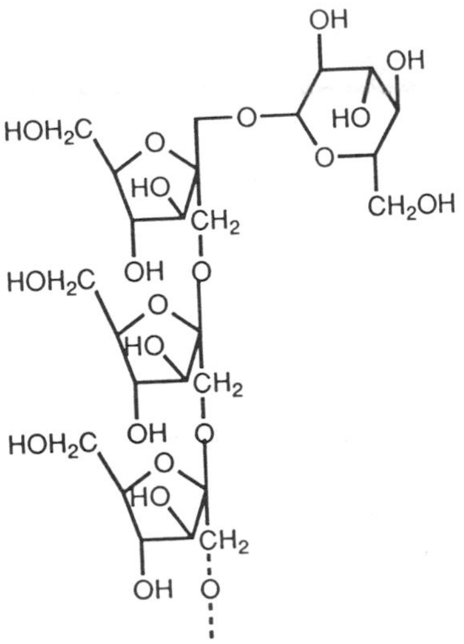

FIGURE 2 Inulin, a nonstarch carbohydrate reserve found in some plants. The β(2 → 1) fructan is the major polysaccharide found in Jerusalem artichoke. The β(2 → 6) fructans have also been identified in some plants.

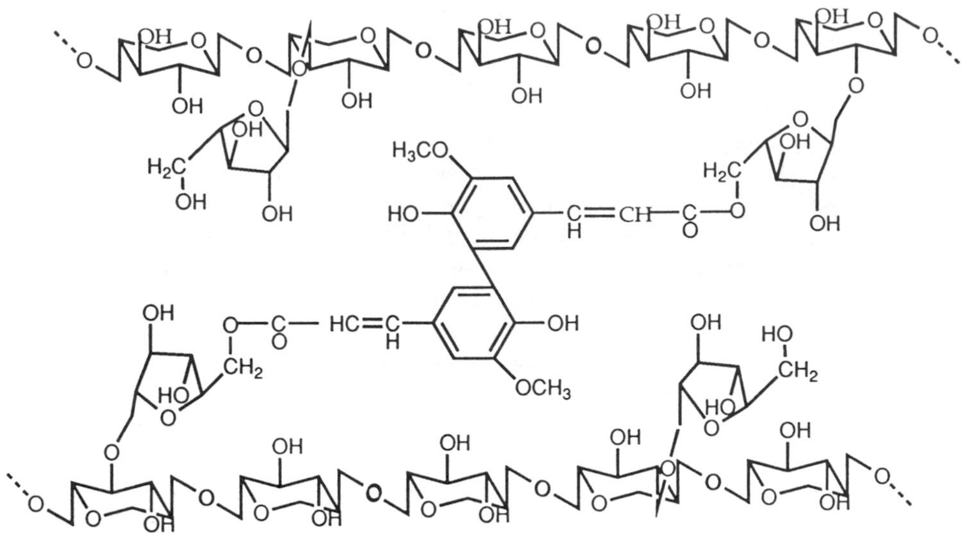

FIGURE 3 Portion of a pentosan with ferulic acid cross-link. The arabinoxylan consists of D-xylopyranose units with L-arabinofuranose glycosidic linkages in positions 2 and 3. Covalent cross-links of this type may occur between cell wall proteins and between polysaccharides and proteins.

bread-making characteristic of rye flour because of their ability to absorb large amounts of water and to form gels. Insoluble pentosans appear to be formed by peroxidase-catalyzed cross-linking of ferulic acid residues [48] and "oxidative gelation." Other nonstarch polysaccharides found in cereal flours are $\beta(1 \rightarrow 3)$ and $\beta(1 \rightarrow 4)$ linear polymers of D-glucopyranose called β-glucans or lichenins. These slimy substances predominate in oats and barley (6–8%), and their high viscosity can cause wort filtration problems in the brewing industry. Glucofructans are also present in significant amounts in durum wheat. These polysaccharides are branched chain, water soluble, and contain α-D-glucopyranose$(1 \rightarrow 2)$, β-D-fructofuranose, β-D-fructofuranose$(1 \rightarrow 2)$, and β-D-fructofuranose$(6 \rightarrow 2)$ linkages.

Gums and mucilages are also obtained from some plants as tree exudates, seed gums or as algae extracts (Table 4). These carbohydrates are chemically related to the hemicelluloses, and some, such as carrageenan, gum arabic, karaya gum, locust bean gum, guar gum, fuicoidans, laminarins, carob gum, and gum tragacanth, are isolated and used as food additives for stabilization, emulsification, thickening and gelation (see Chap. 3).

16.2.2.3 Simple Sugars and Related Compounds

The sugar content of fruits and vegetables varies from negligible (e.g., avocado) to over 20% (e.g., ripe banana) on a fresh weight basis. Sucrose, glucose, and fructose are the principal sugars of most commodities. In general, fruits and vegetables contain more reducing sugars than sucrose, although in many cases the reverse is true. Other sugars, such a xylose, mannose, arabinose, galactose, maltose, sorbose, octulose, and cellobiose, may also be present, and in some instances they may constitute a major portion of the total sugars. For example, avocado fruit contains significant amounts of heptuloses, heptose, octose, octulose, and nonulose sugars. Branched sugars, such as D-adiose (Fig. 4), are sometimes constituents of plant glycosides, such as in parsley. Fruit tissues also contain sugar alcohols (alditols), notably sorbitol (Fig. 4) and

TABLE 4 Examples of Nonstarch Polysaccharides Obtained from Plants for Use as Food Additives

Polysaccharide	Source	Composition
Gum arabic	Acacia tree exudate	Complex, branched, polymer of D-galactose, L-arabinose, L-rhamnose, and D-galacturonic acid
Carob	Carob tree seeds	Polymannose polymer, β(1 → 4) with single galactose branch every 4 or 5 mannose residues
Alginates	Brown algae	D-Mannuronic acid with β(1 → 4) bond or L-guluronic acid with α(1 → 4) bond or copolymer
Carrageenans	Red algae	Linear polymer galactose and sulfated galactose molecules, with β(1 → 4) and α(1 → 3) bonds

D-Adiose

Sorbitol

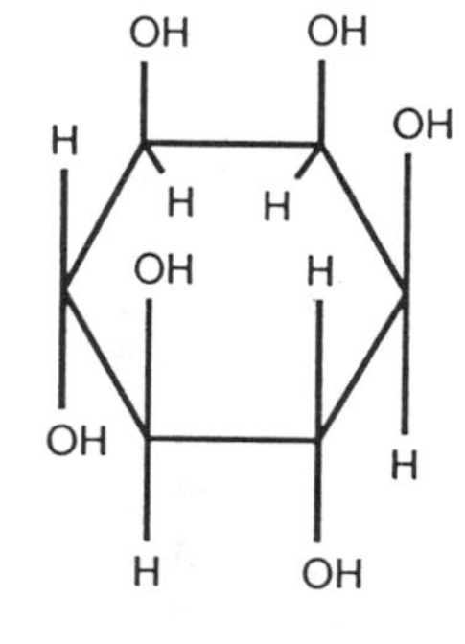

Myo-inositol

FIGURE 4 Examples of monosaccharide derivatives found in some plant tissues.

xylitol. For example, D-sorbitol is present in significant quantities in fruits of the Rosaceae family, and certain mushrooms contain more mannitol than monosaccharides.

Pulses frequently contain significant amounts (1–3%) of raffinose and stachyose, which contain $\alpha(1 \rightarrow 6)$ linkages between galactose and glucose. These oligosaccharides are a cause of flatulence when ingested because humans do not produce the α-galactosidase needed to hydrolyze the galactose–glucose bond. Penta- and hexahydroxylic cyclitols are also found in the plant kingdom. Myoinositol (Fig. 4) is combined with galactose in sugar beet and with phosphoric acid in various edible seeds. The hexaphosphoric ester of myoinositol is called phytic acid. Phytic acid forms complexes with divalent metal ions and reduces the bioavailability of minerals in many grains and legumes (see Chap. 8).

16.2.3 Proteins and Other N Compounds

The protein content varies greatly among various fruits and vegetables, although it generally represents only a small percentage of the fresh weight. Certain crops, such as cereals, pulses, tubers, and bulbs, may contain appreciable quantities of protein (Tables 3 and 5). When appreciable quantities of proteins accumulate, they are normally referred to as "storage proteins," although it is not clear that they are synthesized by the plant for the purpose of storage. These substances may function to lower the osmotic pressure of the amino acid pool, bind ammonia without changing in situ acidity, or serve as macromolecular shields to protect other compounds against enzymatic action [81]. Regardless of the role of storage proteins in the physiology or metabolism of the plant, they are very important to the food scientist because of their contribution to nutritive value and functional properties of various crops. Plant storage proteins are generally categorized based on their solubility, but this type of classification is somewhat arbitrary. It also should be recognized that the extraction and separation of plant proteins can be hindered by the presence of tannins and phenolic oxidizing enzymes. A general feature of storage proteins from seeds and tubers is the presence of amide linkages. For example, rapeseed protein has an isoelectric point of pH 11, and almost all the aspartic and glutamic acid residues are

TABLE 5 Storage Protein, Lipid, and Carbohydrate in Various Seed Crops and Tubers

Group	Example	Protein, %	Carbohydrate, %	Lipid, %
High carbohydrate	Wheat	12	80	2.0
	Corn	12	79	6.5
	Rice	9	82	2.0
	Potato	9	82	0.4
	Yam	9	82	1.0
	Cassava	4	85	1.0
High protein	Broad bean	26	57	1.5
	Haricot bean	21	58	1.5
	Lentil	26	56	1.0
	Pea	25	70	2.0
	Chick pea	21	71	5.0
High protein and lipid	Peanut	28	15	50
	Rape	30	20	40
	Soybean	35	25	21
	Sunflower	35	5	45
	Cotton	50	12	30

Note: Values are expressed as percent of dry matter and vary with cultivar.

amidated. Enzymatic deamidation of soybean protein with glutaminase can enhance solubility, emulsifying properties, and foaming [31]. Generally, plant storage proteins decrease in molecular weight during senescence. This decrease in molecular weight is partly due to dissociation of larger complexes. Finally, it should be noted that the storage proteins from different cultivars of a given species may exhibit different electrophoretic patterns. Accordingly, electrophoresis or isoelectric focusing can be used to "fingerprint" or identify cultivars of unknown origin [52].

16.2.3.1 Cereal Proteins

The protein content of wheat is about 12%, and extensive research has shown that the baking performance of wheat flour is related to the properties of certain storage proteins. Early workers divided the proteins of wheat into four solubility classes: albumins, which are soluble in water; globulins, which are soluble in salt solutions but insoluble in water; gliadins, which are soluble in 70–90% alcohol; and glutenins, which are insoluble in neutral aqueous solutions, saline solutions, or alcohol. In wheat, the principal storage proteins are the gliadin and glutenin fractions. These proteins represent about 80–85% of the endosperm protein.

The gliadins are a heterogeneous mixture containing as many as 40–60 components. The molecular masses of these proteins are approximately 36 kD, the amino acid sequences are homologous, and they appear to exist as single polypeptide chains rather than as associated units. Glutamine is the most prevalent amino acid in the gliadin fraction [92], and with glutamic acid they account for more than 40% of the amino acids on a mole basis. Proline constitutes about 15%.

Glutenin proteins consist of at least 15 distinct molecular mass fractions ranging from 12 to 133 kD for dissociated components. Associated forms of glutenin have average molecular masses of 150–3000 kD. Subunits associate by intra- and intermolecular disulfide bonds. The glutenin fraction has somewhat less glutamine (~34%) and proline (~11%) than the gliadin fraction, but still these account for about half of the amino acids present.

The albumin and globulin protein fractions have relatively low molecular masses of approximately 12 kD and remain unchanged in size following reduction of disulfide bonds.

Although considerable progress has been made toward understanding the molecular basis of functionality of wheat proteins in foods, we still do not have a detailed understanding of the contributions of individual components of wheat to the properties of wheat doughs and baked goods. Nonetheless, rapid advances in genetic technology have led to the identification of genes (and their chromosomal location) that are correlated with functional characteristics in foods, like cooking quality of pasta [5]. Advances in the understanding of the genetic map for wheat will not only provide for improved and unique food ingredients, but will also provide research tools that will advance our understanding of the functionality of this complex system at the molecular level.

The basics of dough formation are covered in Chapter 5. The major reactions in dough formation are sulfhydryl–disulfide interchange (intra- and intermolecular), hydration of the protein, and other protein–protein interactions. Oxidizing agents are usually added during commercial bread making to improve the quality of the dough. Addition of oxidizing agents like azodicarbonamide, bromate, and iodate promotes the formation of disulfide linkages, and addition of reducing agents like cysteine, bisulfite, and gluthathione promotes the breakage of disulfides. Ascorbic acid acts like an oxidizing agent because it is rapidly converted to dehydroascorbic acid by ascorbic acid oxidase. In this state it serves to dissociate disulfide bonds. Depending on the agent used and when it is added to the flour, varying affects of these agents can be seen on dough strength, mixing tolerance, mechanical properties of the dough, and final product quality. Dough viscosity and elasticity can also be modified by the addition of proteolytic enzymes. The use of such additives to control mixing times and product uniformity is standard practice in continuous bread-making operations.

16.2.3.2 Tuber Proteins

Approximately 70–80% of the extractable proteins in potato tuber (about 2–4% protein and amino acids on a fresh weight basis) are classified as storage protein. Up to 40 individual proteins have been separated from potatoes by electrophoresis. In mature potato tubers, the main group of proteins has subunit molecular weights of 17–20 kD. The heterogeneity of these storage proteins appears to be due to charge differences, and it is likely that this is a reflection of different amounts of amide in the polypeptide chain. Protein patterns obtained by isoelectric focusing on polyacrylamide gel can also be used to differentiate potato cultivars.

16.2.3.3 Pulse Storage Proteins

Pulses of major economic importance include peas, beans, lentils, peanuts, and soybeans. Legume seeds are characterized by a relatively high content (20–40%) of protein. The limiting essential amino acids in legumes are methionine and cysteine. The storage proteins in legumes or pulses, such as soybean, are also characterized by heterogeneity in molecular weight and charge. Separation of legume proteins by Osborne fractionation yields mostly globulins (60–90%). The albumins or water-soluble proteins have a molecular mass range of approximately 20–600 kD. Soybean proteins do not contain prolamines or glutelins, although peanuts, peas, and broad beans contain 10–15% glutelins. About 90% of soybean proteins are classified as globulins based on their solubility in salt. Soybean globulins can be separated into four fractions by sedimentation in the ultracentrifuge or by gel permeation chromatography. The small globulins (2S fraction) represent 6–8% of total protein and include two trypsin inhibitors, Kunitz's inhibitor (21.7 kD) and Bowman–Birk inhibitor (7.9 kD). Pulses also contain proteins and glycoproteins called hemagglutinins or lectins. Certain glycan sites on erythrocytes bind to lectins, causing agglutination or precipitation of the cells.

The 7S fraction (190 kD) of soybean, about one-third of total soybean protein, contains a glycoprotein called conglycinin. This fraction consist of three glycoprotein subunits, which are devoid of cysteine and contain about 5% carbohydrate as mannose and glucosamine. Other pulses have a 7S globulin similar to soybean conglycinin, that is, phaseollin in beans and vicillin in peas.

The 11S globulin fraction (350 kD) represents about 40% of soybean protein and contains a protein called glycinin. This storage protein contains six basic and six acidic subunits that show a complex pattern of association–dissociation involving disulfide bonds and hydrophobic bonding of the quaternary structure. Dissociation of subunits may be affected by appropriate conditions of ionic strength, pH, or solvent [91]. The solubility and food-related functional properties of soybean proteins can be improved by limited hydrolysis with proteolytic enzymes, such as trypsin. Excessive hydrolysis of soybean proteins is normally undesirable because it gives rise to bitter-tasting peptides (see Chaps. 5 and 6).

16.2.3.4 Nonprotein Nitrogen

Plant tissues may contain appreciable quantities (i.e., 20–75%) of nonprotein nitrogen. For example, more than two-thirds of the nitrogen in potato tuber or apple fruit may be in the form of free amino acids and other constituents. More than 60 different amino acids that are not found in proteins have been identified in plants [12]. Although amino acids are the main nonprotein nitrogen components of most fruits and vegetables, plants may contain appreciable quantities of amines, purines, pyrimidines, nucleosides, betaines, alkaloids, porphyrins, and nonproteinogenic amino acids (Fig. 5). The contribution of nonprotein nitrogen compounds to quality and postharvest changes is incompletely understood. Some of their roles in fruits and vegetables are known, however. For example, free amino acids probably contribute to the taste of fruits and vegetables, elevated levels of steroidal alkaloids (solanine and chaconine) in some cultivars of

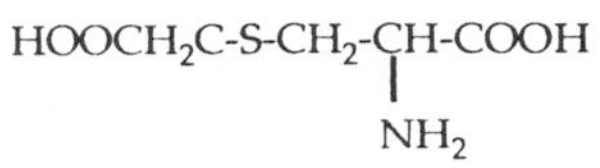

$HOOCH_2C\text{-}S\text{-}CH_2\text{-}CH(NH_2)\text{-}COOH$ — S-(Carboxymethyl)cysteine, Radish

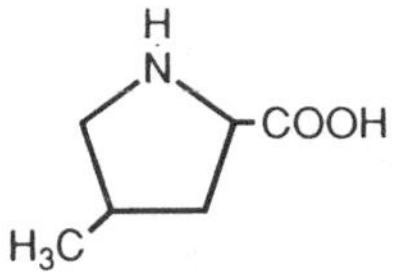

4-Methyl-L-proline — Apple

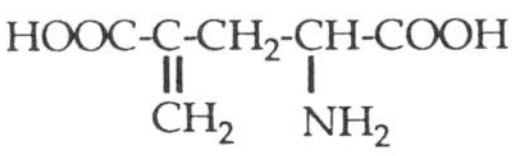

$HOOC\text{-}C(=CH_2)\text{-}CH_2\text{-}CH(NH_2)\text{-}COOH$ — 4-Methylene glutamic acid, Peanut

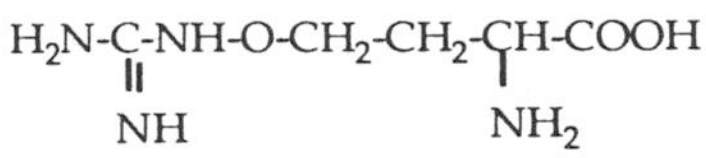

$H_2N\text{-}C(=NH)\text{-}NH\text{-}O\text{-}CH_2\text{-}CH_2\text{-}CH(NH_2)\text{-}COOH$ — Canavanine, Soybean

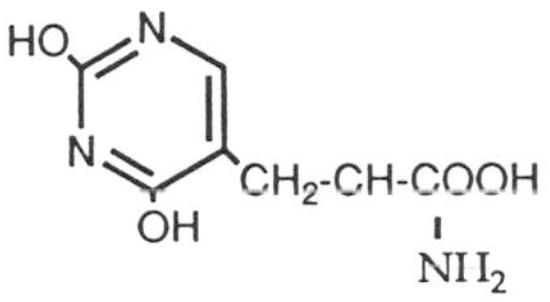

3-(2,6-Dihydroxypyrimidine-5-yl)-alanine — Garden pea

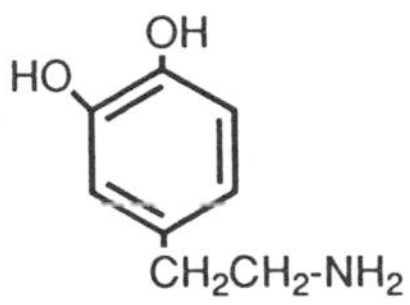

Dopamine — Banana

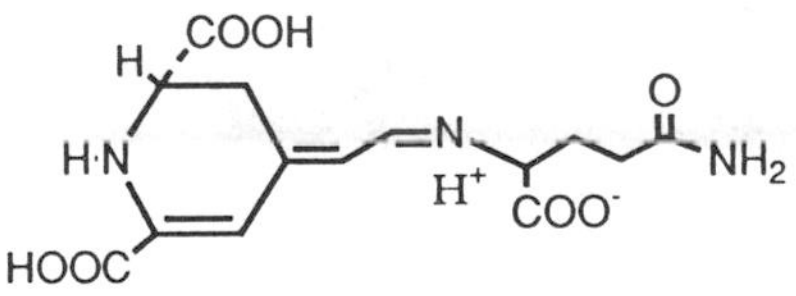

Isobetanidin aglycone — Red beet

FIGURE 5 Examples of nonprotein nitrogen compounds found in edible plants.

potato contribute to bitter taste and toxicity, betanins are red pigments in red beets, and several amino acids and amino acid derivatives are important precursors of aroma compounds in fruits and vegetables. An interesting recent development is that polyamines, such as spermidine and spermine, increase during low-temperature storage of zucchini squash and appear to increase resistance to chilling injury through their ability to stabilize membranes [87].

16.2.4 Lipids and Related Compounds

16.2.4.1 Polar Lipids

Lipids constitute less than 1% of the fresh weight of most fruits and vegetables (Table 3). These lipids are mostly polar, major examples being membrane phospholipids and glycolipids (Table

6 and Fig. 6). The increase in respiration rate of climacteric fruit is accompanied by a large increase in phosphatidyl choline and phosphatidic acid, and during the postclimacteric stage there is sudden breakdown (peroxidation) of total lipids [28].

On the other hand, cereals and pulses can accumulate significant amounts of polar lipids, especially in the endosperm. Soybeans contain up to 8% phospholipid. Total oat grain lipid includes 8–17% glycolipids and 10–20% phospholipid. The content of polar lipids is greater in oats than other cereals because they contain a relatively high content of lipid in the endosperm portion of the seed. The polar lipids of wheat flour include several glycolipids and phospholipids in both the starch and nonstarch fractions (Table 6). The glycolipids play an important role in gluten development during bread making [22]. Digalactosyl diacyglyceride appears to stabilize the starch-gluten network in wheat flour dough [68] (Fig. 7).

16.2.4.2 Triacyglycerols

Triacyglycerols of the neutral lipid fraction are the main reserve material in plant tissues such as nuts, oil seeds, pulses and cereals (Tables 3 and 5). Reserve triacylglycerols occur mainly as lipid droplets (< 10 μm) in the germ (embryo and associated tissue) and bran (aleurone surrounding endosperm) portions of the seed. The crude fat content of milling fractions from two cereals is shown in Table 7. Neutral fats account for 85–90% of rice bran oil and 60% of rice endosperm lipids. In wheat, neutral lipids predominate in the germ (80%) and aleurone (72%), while polar lipids predominate in the nonstarch endosperm (67%) and starch endosperm (95%)

Table 6 Lipid Composition of Wheat Flour

Fraction	Total flour (mg%)	Nonstarch (mg%)	Starch (mg%)
Triacylglycerides	709	674	35
Diacylglycerides	92	86	6
Free fatty acids	129	110	19
Sterol esters	90	90	18
Glycolipids	582	519	63
Esterified monogalactosyl DG and MG	73	66	7
Esterified steryl glycoside	77	71	6
Monogalactosyl DG	93	87	6
Monogalactosyl MG	30	23	7
Digalactosyl DG	226	214	12
Digalactosyl MG	83	58	25
Phospholipids	1026	242	728
N-Acylphosphatidyl ethanolamine	72	72	—
N-Acyllysophosphatidyl ethanolamine	34	34	—
Phosphatidyl ethanolamine and phosphatidyl glycerol	19	13	3
Lysophosphatidyl ethanolamine and lysophosphatidyl glycerol	69	10	6
Phosphatidyl choline	104	66	38
Lysophosphatidyl choline	693	36	657
Phosphatidyl serine and related compounds	35	11	24
Total lipids	2628	1703	925

Note: MG, monoacylglyceride; DG, diacylglyceride.
Source: Adapted from Ref. 22.

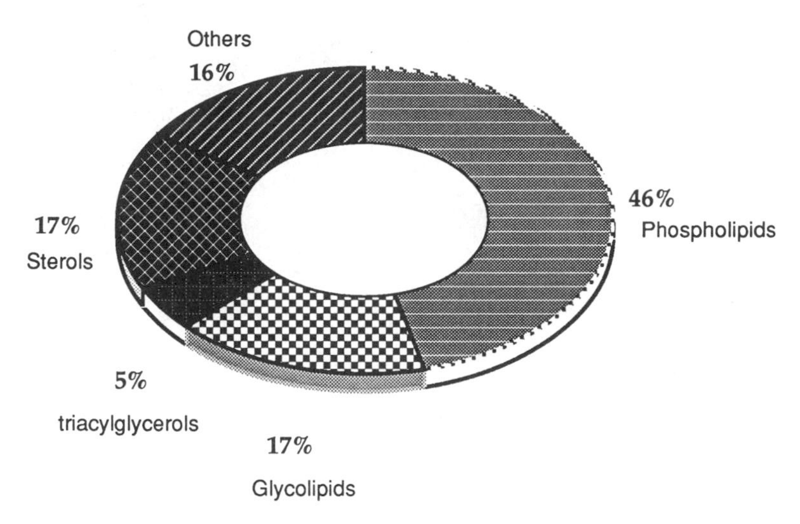

FIGURE 6 Lipid fractions of apple flesh (percent of total lipids).

fractions. Corn oil consists of about 95% triacylglycerols with 1.5% phospholipids, 1.7% free fatty acids, and 1.2% sterols. The major fatty acids in cereal grain and pulse lipids are linoleic ($C_{18:2}$), oleic ($C_{18:1}$), and palmitic (C_{16}). Palm and copra seed oils are unique in containing a high percentage of lauric (C_{12}), and myristic (C_{14}) along with palmitic and oleic acids. Cocoa fat contains a high percentage of stearic along with palmitic and oleic acids. The distribution of principal fatty acids, as percent of the total, in various vegetable oils is shown in Table 8.

16.2.4.3 Other Lipid Substances

Unsaponifiable lipids in seed oil include sterols, higher alcohols, and hydrocarbons like squalene. Several milligrams of carotenoids per 100 g of fresh weight are present in some fruits (e.g., pineapple, cantaloupe, orange) and vegetables (e.g., tomato, red pepper, carrot). Different kinds of carotenoids accumulate in plant crops (discussed in Chap. 9). During plant senescence, chloroplast carotenoids are degraded while chromoplast carotenoids are synthesized [38]. Bitter tasting triterpenoids occur in the seeds and flesh of Rutaceae fruits (limonin) and Curcurbitaceae fruits (Curcabitacins) (Fig. 8). Phytosterols include compounds that are structurally similar to cholesterol, such as sitosterol, campesterol, and stigmasterol (Fig. 8). The ratio of phytosterols to other sterols is characteristic of plant species and has been used to detect adulteration of plant oils.

Lipid substances are also prominent in the protective epidermal cells of certain plant organs. Some examples of these protective substances are illustrated in Figure 9. Waxes that protect the surfaces of leaves and seeds from dehydration include esters of long-chain alcohols and fatty acids. The waxy cuticle consists of a complex polymer of hydroxyl fatty acids called cutin (Fig. 9).

Other lipid or lipophilic substances, although present in trace quantities, may contribute to the characteristic flavor of edible crops. Odoriferous substances of fruits are mainly oxygenated compounds (esters, alcohols, acids, aldehydes, and ketones), many of which are derivatives of terpenoid hydrocarbons or lower aliphatic acids and alcohols. The range of such volatile constituents present in vegetables is generally more limited than that found in fruits. In some instances, these substances are dissolved in terpenoid hydrocarbons and localized in special oil sacs (e.g., essential oils of peppermint leaf). In other cases, the components are formed as a result of cellular decompartmentation during wounding, cooking, or chewing the tissue.

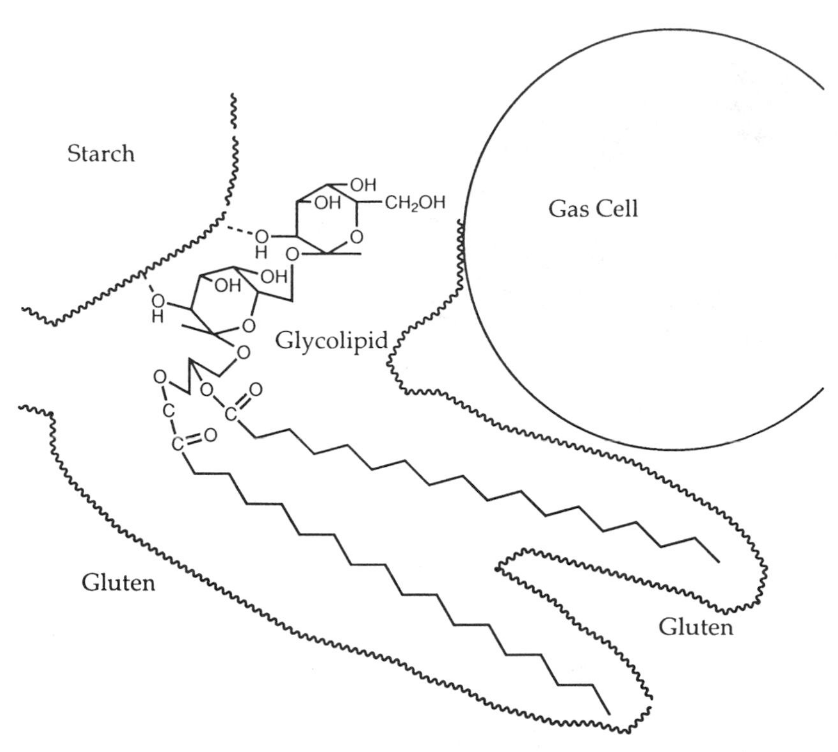

FIGURE 7 Model showing proposed role of digalactosyl diacyglyceride in stabilizing starch–protein network in bread dough.

TABLE 7 Crude Fat Contents of Rice and Wheat and Their Milling Fractions

Grain fraction	Percent crude fat (dry matter basis)
Rice	
Whole grain (brown rice)	2–4
Milled (endosperm)	<1
Bran	15–22
Embryo	15–24
Polish	9–15
Wheat	
Whole grain	2
Pericarp	1
Aleurone	9
Starch endosperm	1
Germ	10

Source: Adapted from Ref. 67.

TABLE 8 Principal Fatty Acids of Vegetable Oils and Fats

Source	$<C_{12:0}$ (%)	$C_{12:0}$ (%)	$C_{14:0}$ (%)	$C_{16:0}$ (%)	$C_{18:0}$ (%)	$C_{18:1}$ (%)	$C_{18:2}$ (%)	$C_{18:3}$ (%)
Cereals								
Corn	—	—	—	6	2	44	48	—
Wheat	—	—	3	18	7	31	57	4
Rye	—	—	6	11	4	18	35	7
Rice	—	—	1	28	2	35	39	3
Pulses								
Soybean	—	—	—	11	2	20	64	3
Peanut	—	—	—	9	6	51	26	—
Chick peas	—	—	1	9	2	35	51	2
Lentils	—	—	1	23	5	36	21	2
Tropical								
Palm	8	50	15	8	2	15	1	—
Copra	15	46	18	9	3	8	1	—
Cocoa	—	—	—	24	34	28	—	—
Coconut	—	45	20	5	3	6	—	—
Other								
Olive	—	—	—	15	—	75	10	—
Cottonseed	—	—	—	23	—	32	45	—
Canola	—	—	—	5	2	55	20	8
Sunflower	—	—	—	6	4	31	57	—

Note: Values are percent of total lipid.

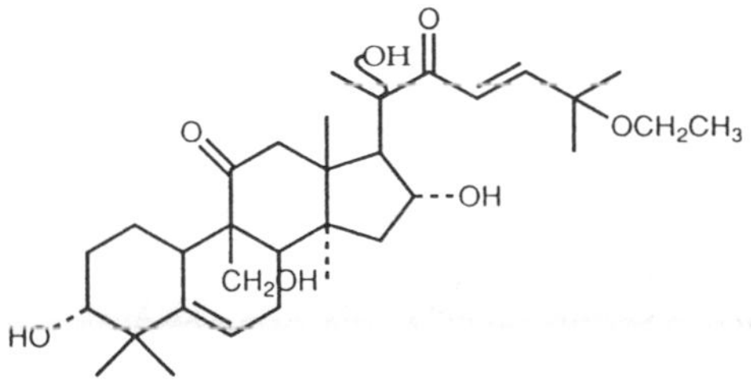

Curbitacin (cucumber)

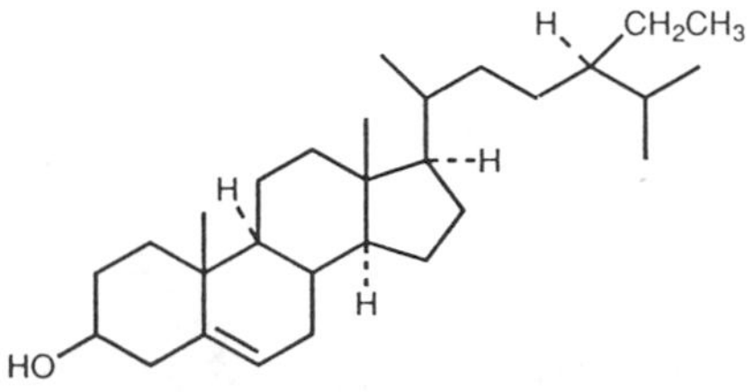

Sitosterol (24-α-ethyl cholesterol)
Common plant steroid

FIGURE 8 Examples of triterpenoids found in the total lipid fraction of plants.

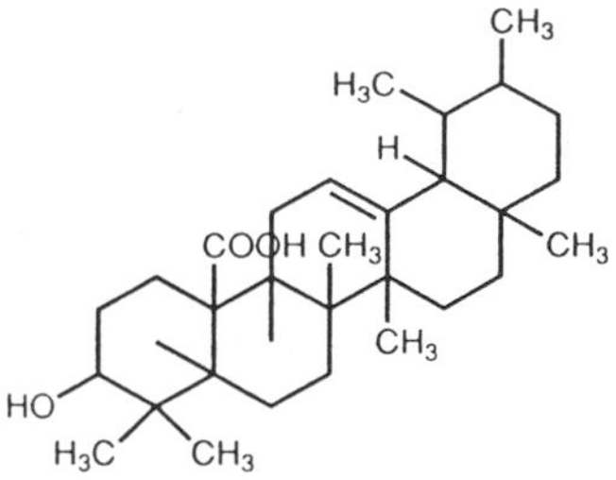

Ursolic acid (epidermal wax)

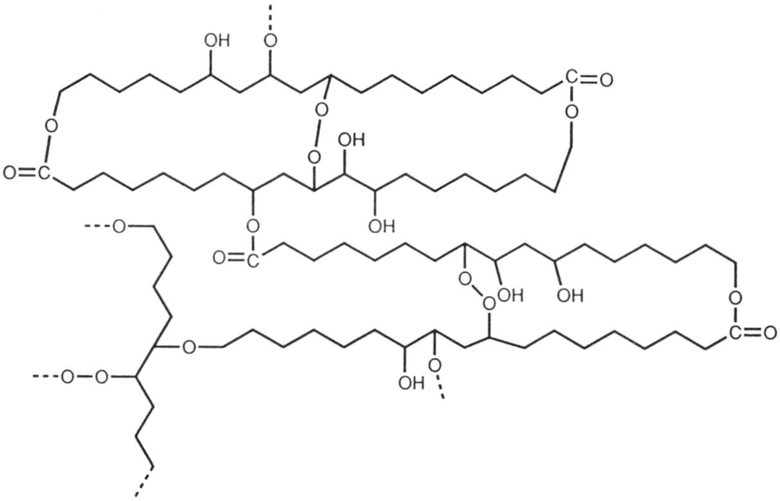

Segment of Cutin (Waxy cuticle)

FIGURE 9 Waxy substances associated with plant epidermal cells.

16.2.5 Organic Acids

Small quantities of organic acids occur in plants as metabolic intermediates (e.g., in the tricarboxylic acid cycle, glyoxylate cycle, or shikimic acid pathway) and they may also accumulate in vacuoles. The accumulation of organic acids gives rise to an acidic or sour taste. Acid levels range from very low (e.g., sweet corn) to high in such crops as currant, cranberry, or spinach. An acidic crop may contain over 50 mEq acid per 100 g of tissue, and a pH of less than 2.

16.2.5.1 Aliphatic Plant Acids

The most widely occurring and most abundant acids in plants are citric and malic, each of which can constitute up to 3% of the tissue on a fresh weight basis. Instances in which neither malic nor citric acids predominate include grapes and avocado (tartaric), spinach (oxalic), and blackberry (isocitric). The total acidity of many fruits declines during ripening, although specific acids may actually increase.

16.2.5.2 Carbocyclic Plant Acids

Various aromatic acids are also found in plant tissues (Fig. 10). The alicyclic acids quinic and shikimic are widely distributed in the plant kingdom as key intermediary metabolites. Some acids, such as chlorogenic, are important food constituents because of their contribution to both enzymatic and nonenzymatic browning reactions. Ferulic acid is involved with cross-linking of pentosans in cereal flours (Fig. 3). Others, such as benzoic, are important antifungal agents in crops such as cranberry.

16.2.5.3 Phenolic Acids

Phenolic compounds in plant foods include a wide range of compounds and a broad spectrum of functional activities. Traditionally, these compounds have been considered important in plant foods because of their flavor and color (particularly enzymatic browning; see Chap. 6), but there is substantial current interest in their potential health benefits [35], antioxidant activity [70] (see Chap. 4), and antimicrobial [24] effects. In the most simple chemical terms, phenolic compounds include a hydroxylated aromatic ring such as phenol, *p*-cresol, and 3-ethylphenol. Phenolic acids like caffeic, coumaric, and ferulic occur widely in the shikimic acid pathway of plant tissues,

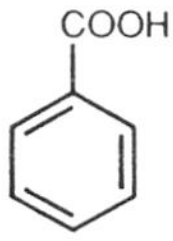

Benzoic (e.g., cranberry)

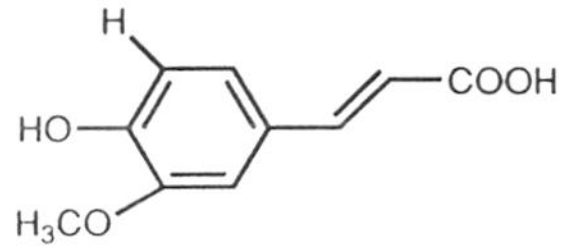

Ferulic (e.g., cereals)

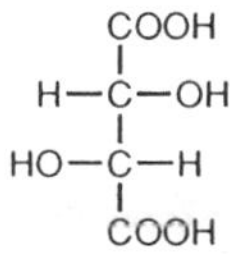

Tartaric (grapes)

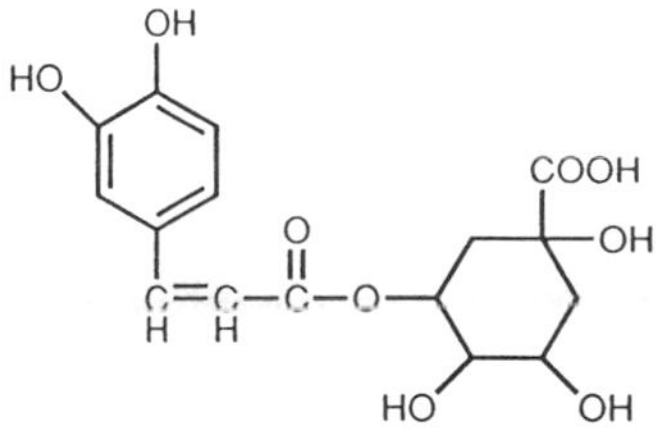

Chlorogenic (e.g., pears)

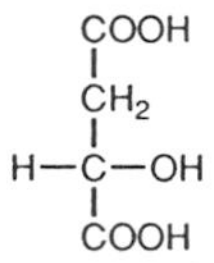

Malic (e.g., apple)

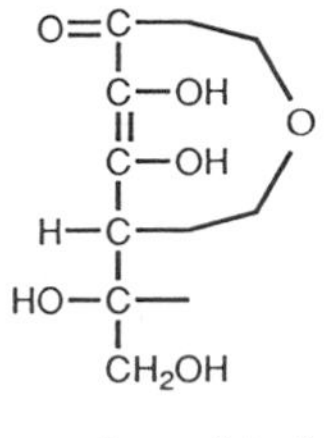

Ascorbic (e.g., potato)

FIGURE 10 Examples of organic acids in plant tissues.

which begins with condensation of phosphoenolpyruvate and erythrose 4-phosphate (Fig. 11). These compounds often occur as esters with sugars or other phenolics, or as a part of tannins, Chlorogenic acid is the ester of caffeic and quinic acid, occurs very widely, and is the major phenolic substance in apples and pears [80]. Phenylalanine and tyrosine are synthesized in the shikimic acid pathway and they can serve as important precursors for the formation of phenolic acids, the latter resulting in increased lignin biosynthesis and enzymatic browning.

Most of these compounds can serve as substrates for enzymatic browning, and they can also contribute to darkening by forming complexes with metal ions like copper and iron. Sequestrants like ethylenediamine tetraacetic acid (EDTA) or phosphates are frequently used to reduce the formation of these undesirable metal-phenol complexes. Differences in rates and extent of browning in different plant tissues are attributable to several factors, including availability of substrate (oxygen or the phenol), enzyme activity, compartmentation of enzymes and substrates, and available metal ions. Changes in phenolic substrates during ripening and storage can affect browning in a given tissue.

Fruits typically show a decline in phenolic compounds with ripening, and an increase in response to stresses like bruising and fungal infection. The extent and magnitude of these changes varies widely depending on the plant material and the conditions of storage. Shredded carrots accumulate *trans*-5′-caffeolyquinic acid rapidly when stored in air, but this is inhibited by high concentrations of carbon dioxide or low concentrations of oxygen [6]. For example, shredded lettuce accumulates phenolics more slowly when stored in the type of atmosphere just mentioned [54].

16.2.5.4 Tannins (Polyphenolics)

Polyphenolics have traditionally been separated into "condensed" and "hydrolyzable" tannins. These terms are somewhat confusing because both groups can be hydrolyzed. "Condensed" tannins are more correctly referred to as flavan-3,4-diol-derived tannins or proanthocyanidins, and the "hydrolyzable" tannins as gallotannins or ellagitannins (Fig. 12). Polyphenolics provide a number of different functionality's in foods, including color (see Chap. 9) and astringency. Tannins are a diverse group of molecules that range up to 3000 D and are formed from carbocyclic acids, phenolic acids, and sugars. The exact structures of the larger molecules are not known.

The name tannin comes from the use of plant extracts to treat hides to make leather. Tannin–protein interactions are also important in foods like beer where the tannins from the hops react with protein to form "haze." Complexation of tannins with proteins has been thought to reduce the nutritive value of tissues with high tannin content [76], but there is evidence to suggest that the antinutritional effects may be caused by a more direct physiological effect of the tannins [17]. Tannin–protein complexes involve both hydrogen bonds and hydrophobic interactions and may generate covalent links through amide and sulfhydryl groups when the phenolics are oxidized [33]. Proteins with high proline content have a greater affinity for tannins than do other proteins. The astringency of many fruits is thought to be caused by tannins interacting with the mucous membranes of the mouth. Tannins also form complexes with caffeine resulting in the formation of a precipitate known as "tea cream" [69]. Gelatin, because of its amino acid composition, can be used to complex tannins and reduce the astringency of wines.

Tea is the most widely consumed beverage in the world, and many of its characteristics are related to tannin content [7]. When converting freshly picked tea leaves into tea, the major concern is controlling oxidation of polyphenols. For green teas, the leaves are blanched at the beginning of the process to reduce oxidation of polyphenols. Production of black teas, on the other hand, requires oxidation of polyphenols, and thus the processing steps are designed to enhance and control the enzymatic oxidation of polyphenols. The processing sequence includes reducing the moisture content of the leaves to concentrate the phenols (withering), followed by

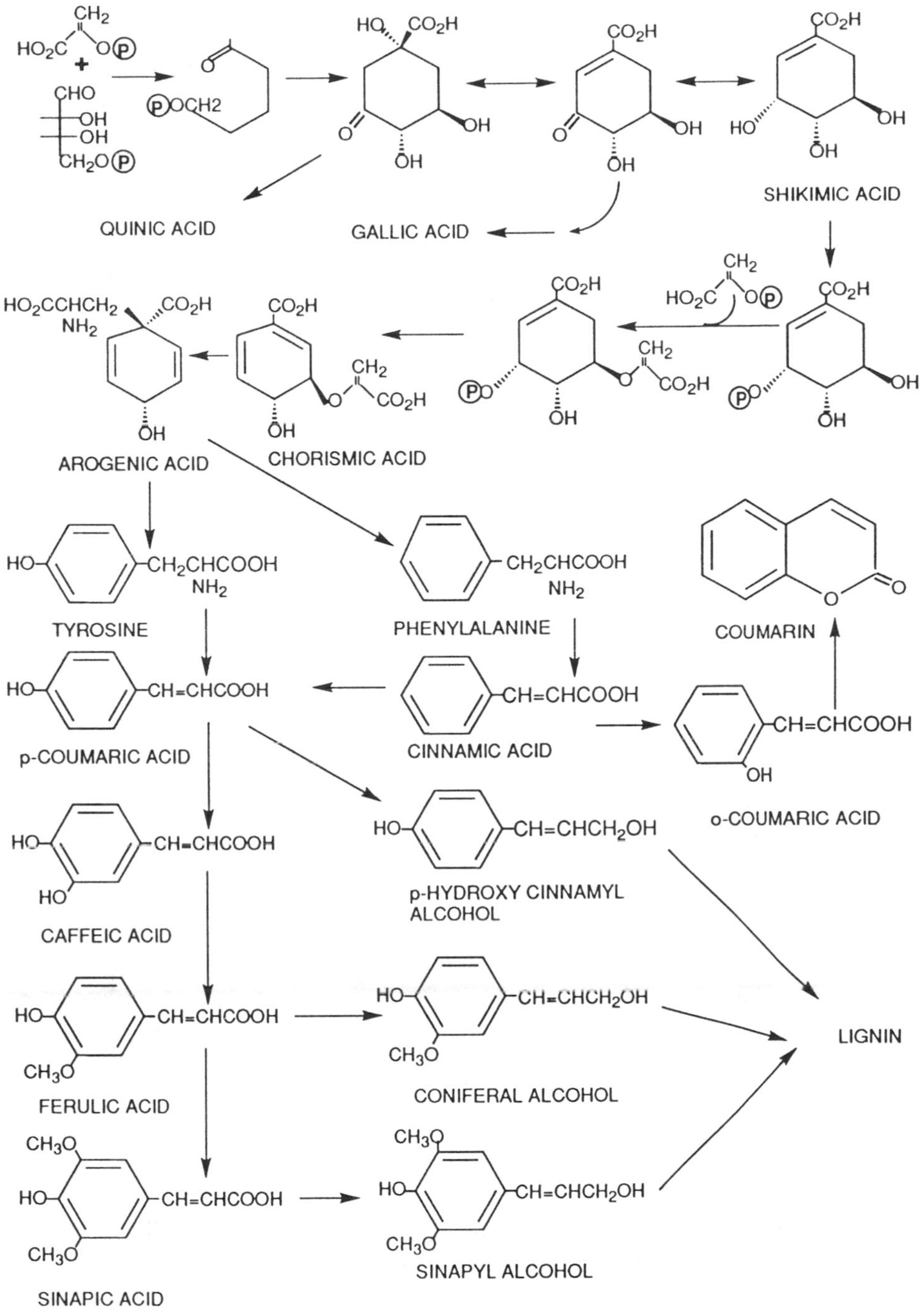

FIGURE 11 Biosynthesis of some phenolic acids (shikimic acid pathway).

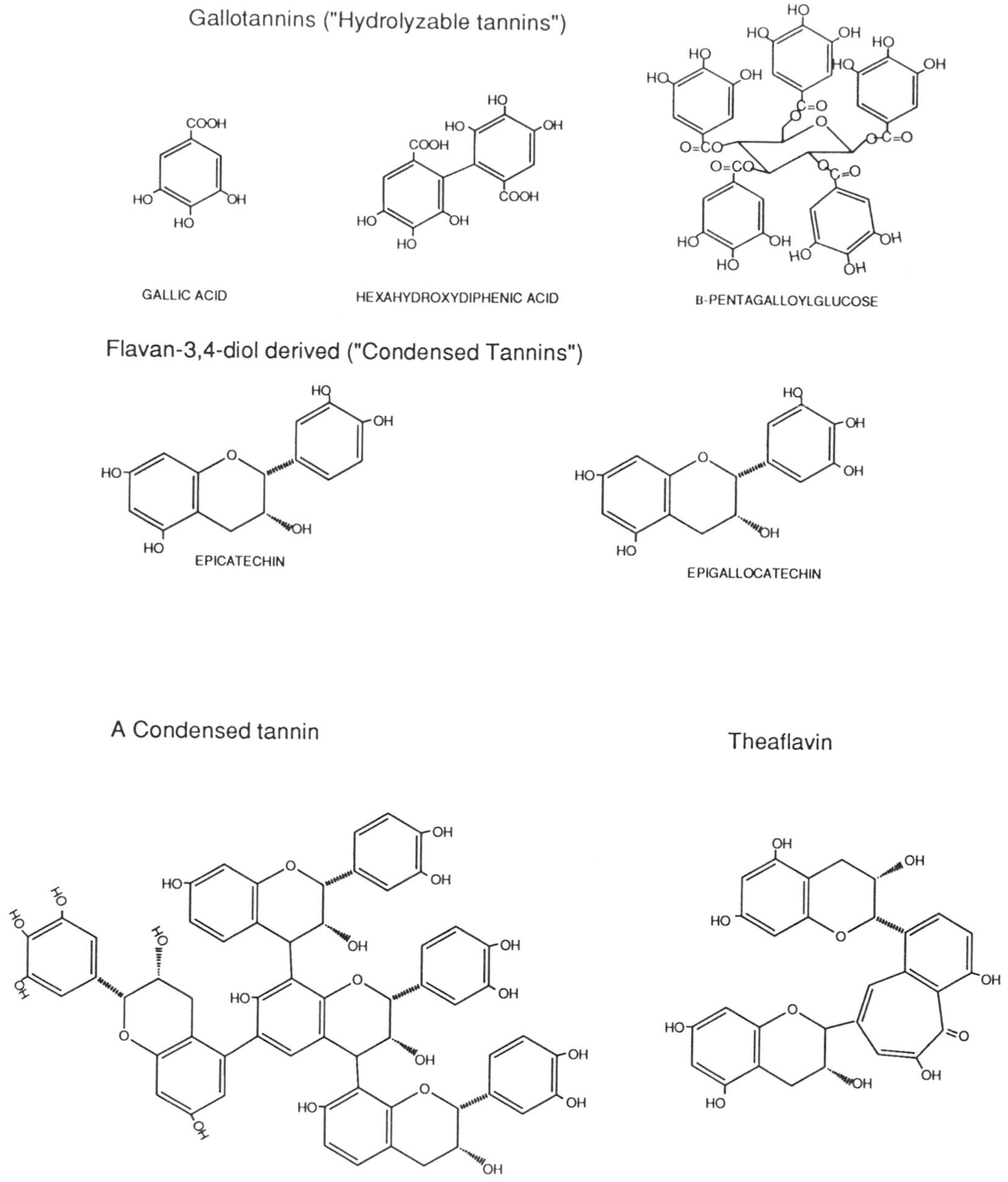

Gallotannins ("Hydrolyzable tannins")

GALLIC ACID

HEXAHYDROXYDIPHENIC ACID

B-PENTAGALLOYLGLUCOSE

Flavan-3,4-diol derived ("Condensed Tannins")

EPICATECHIN

EPIGALLOCATECHIN

A Condensed tannin

Theaflavin

Figure 12 Structure of tannins found in plant tissues and products. Tannins are normally classified as "condensed" and "hydrolyzable" (see text).

a crushing step that breaks down the cellular compartmentation and allows access of oxidizing enzymes to substrates. After the desired degree of oxidation, it is terminated by blanching. For oolong tea (mildly oxidized) the oxidation process is stopped soon after crushing.

Fresh green tea leaves contain about 40% (dry basis) polyphenols with epigallocatechin gallate, epigallocatechin, and gallocatechin being the major types. Oxidation of catechins and gallotannins with subsequent condensation leads to the formation of the flavins, which are responsible for the reddish color of black teas. Epitheaflavic acids arise similarly from epicatechins and gallic acid. The arubigens are a complex group of polymeric procyanidins that are responsible for the brown color of tea.

16.2.6 Pigments

The principal pigments of plant tissues—the chlorophylls, carotenoids, and flavonoids—are discussed in Chapter 9. The extent of pigment synthesis and degradation in detached fruits and vegetables can be influenced by storage conditions, such as light, temperature, and relative humidity, and by volatile substances such as ethylene. Ethylene is formed in plant tissue during ripening, following wounding, or by exposure of the plant to air pollutants. By virtue of its hormone action, ethylene initiates the degradation of chlorophyll in mature plant tissues like fruit and leafy vegetables. The exact mechanism by which ethylene action leads to chlorophyll degradation is not completely known. Chlorophyllase (EC 3.1.1.14) is a thylakoid membrane glycoprotein that catalyzes hydrolysis of the phytol side chain of chlorophyll (no change in color), although it appears that hydrolysis is followed by oxidative degradation of the tetrapyrrole involving a peroxidase (chlorophyll:H_2O_2 oxidoreductase). The participation of different lipoxygenase isoenzymes has also been implicated in chlorophyll and carotenoid degradation in growing plants and seeds [42]. For example, soybean lipoxygenase 1, which is specific for free fatty acids, has been implicated in chlorophyll degradation in wheat [45]. The transformation of chloroplast to chromoplast is a major metabolic event in ripening fruit. In citrus, this transformation appears to be influenced by the balance of nitrogen and sucrose in the peel tissue [37].

16.2.7 Mineral Elements

The total mineral content of plant tissues is sometimes expressed as ash content (residue remaining after incineration). The ash content of plant tissue varies from less than 0.1% to as much as 5% of the fresh weight. Mineral content is influenced by species characteristics as well as by agronomic practices. The mineral composition of some edible plant tissues is shown in Table 9. The distribution of particular minerals in a given tissue is known to be nonuniform. In peas, phosphorus is many times richer in cotyledons than in testa, whereas the difference is reversed for calcium. The most abundant mineral elements in plants are potassium, calcium, magnesium, iron, phosphorus, sulfur, and nitrogen. Potassium is the single most abundant element in most plants; for example, parsley contains over 1% on a fresh weight basis. Mineral

TABLE 9 Mineral Composition of Major Elements in Some Edible Plants

	Mineral (mg/100 g fresh weight)							
Crop	K	P	Mg	Ca	Na	Fe	Cl	Zn
Pulses								
Soybean	1400	490	210	210	6	7	6	1
Peanut	710	370	160	59	5	2	7	3
Vegetables								
Red beet	336	45	1	29	86	1	—	1
Tomato	297	26	20	14	6	1	60	—
Green beans	256	37	26	51	2	1	36	—
Cereals								
Wheat	454	433	183	45	—	—	—	—

elements occur mainly as salts of organic acids. During storage of fruits and vegetables, active exchange of minerals occurs between different physiologically active centers (e.g., stem, inflorescence, storage organs, leaves) [88].

The mineral content of a given species can influence physiological disorders that arise pre- and postharvest. For example, if calcium is increased by agronomic practices or by postharvest treatments, this can result in improved storage life and product quality.

16.2.8 Vitamins

Plant tissues are an important source of several vitamins. The contribution of edible plant tissues to the recommended daily allowance of some vitamins is summarized in Table 10. The vitamin content of a given species can vary considerably with variety, growing conditions, postharvest storage conditions, and processing. For example, the vitamin C content of spinach varies from 40 to 155 mg/100 g fresh weight. Oilseeds, notably soybean, are rich in vitamin E (> 10 mg/100 g fresh weight). Some fruits and vegetables are a particularly rich source of vitamin C (e.g., guava, black currant, strawberry, kale) and provitamin A (e.g., lettuce, spinach, carrots). Cereal grains are very good dietary sources of vitamins B_1, B_2, and B_6.

Vitamins are usually distributed nonuniformly in plant tissues. The old adage, "It's a sin to eat the potato and throw away the skin," makes reference to the greater concentrations of thiamin and ascorbic acid in the cortex of the tuber as compared with the interior. Separation of the starchy endosperm from the germ and aleurone during milling of cereals results in substantial loss of B vitamins and minerals. The influence of postharvest storage on the vitamin content of different crops is incompletely understood, although it is known that levels of ascorbic acid and β-carotene can fluctuate considerably.

16.3 BIOLOGICAL STRUCTURE

Knowledge of the structure of edible plant tissues is important to the food chemist for several reasons. It is important to recognize that an excised mass of tissue is not necessarily homogeneous with respect to cell type, cellular organization, and distribution of chemicals. In addition, the nature and extent of chemical changes that occur in the postharvest tissue are partly, if not wholly, dependent on cellular organization. Each tissue is structurally adapted to carry out a particular function. Much of the metabolic activity of the plant is carried out in relatively unspecialized tissue called parenchyma, which generally makes up the bulk of edible plant tissues. The outer layer of a plant is called the epidermis and is structurally adapted to provide protection against biological or physical stress. Specialized tissues, called collenchyma and

TABLE 10 Contribution of Edible Plants to Vitamin Consumption in the U.S. Diet

	Percent contribution to total intake				
Group	Vitamin A	Thiamin	Riboflavin	Niacin	Ascorbate
Fruits	7.3	4.3	2.0	2.5	35.0
Potatoes	5.7	6.7	1.9	7.6	20.9
Vegetables	36.4	8.0	5.6	6.8	38.3
Pulses	TR	5.5	1.8	7.0	TR
Cereals	0.4	33.6	14.2	22.7	0.0
Total plant	49.8	58.1	25.5	46.6	94.2

sclerenchyma, provide structural support. Water, minerals, and solute molecules are transported through the vascular tissues, xylem and phloem.

16.3.1 Organ Structure

Consideration is given to four categories of edible plant tissues, with categorization based on their appearance: roots, stems, leaves, and fruits. Foods are also classified on the basis of their economic use, for example, fruits, vegetables, nuts, grains, berries, bulbs, or tubers. However, the latter type of classification is based on custom rather than on botanical systematics, and accordingly varies in different cultures, Indeed, there are legal precedents that the tomato is a vegetable and not a fruit, much to the dismay of the botanist.

16.3.1.1 Roots

The economically important root crops include sugar beets, sweet potatoes, yams, and cassava. Some crops that appear to be root tissue are mostly modified hypocotyl (e.g., radish) or stem (e.g., potato) tissue. The basic anatomical structure of root tissue is illustrated in Figure 13. Fleshy roots are formed by secondary growth of the cambia (Fig. 13b, c). The secondary vascular tissues of such crops consist of small groups of conducting elements scattered throughout a matrix of parenchyma tissue. Multiple cambia are formed in certain tissues, such as in beet root (Fig. 13c), and sweet potato.

16.3.1.2 Stems

Examples of stem crops are potato, sugarcane, asparagus, bamboo shoot, rhubarb, and kohlrabi. Underground stems (rhizomes and bulbs) are modified structures and are exemplified by potatoes and onions. Topographically, a stem consists of four distinct regions: the cortex, located between the epidermis and the vascular system, and the pith, which forms the central core of cells (Fig. 14a). The secondary tissues exhibit different forms as illustrated in Figure 16.14b

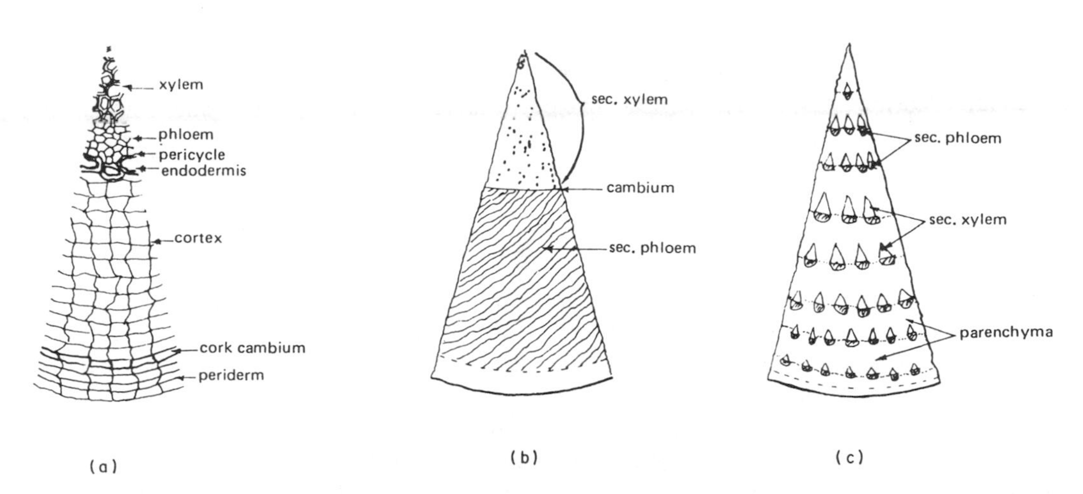

FIGURE 13 (a) Diagrammatic illustration of anatomical structure of root tissue in cross section. Fleshy roots are formed by secondary growth of meristematic tissue called cambia. (b) A typical example is carrot, in which the cambia forms phloem on the outside and xylem on the inside. The conducting elements are scattered throughout a mass of parenchyma tissue. (c) Beet root shows a different type of secondary growth, in which a series of concentric cambia are formed.

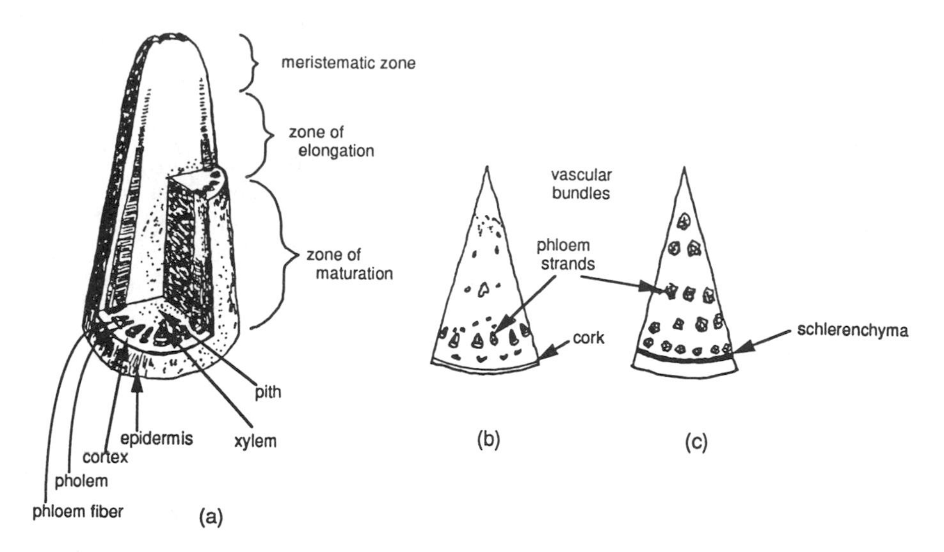

FIGURE 14 (a) Diagrammatic illustration of anatomical structure of stem tissue. (b) In dicotyledonous stem structures, such as potato, the vascular bundles are arranged in a single ring, as seen in tranverse section. (c) Monocotyledons, represented by asparagus, show a characteristic scattering of vascular bundles throughout the parenchymatous ground tissue.

and c. In bulbs, such as onion and garlic, the organ consists of a short stem that bears a series of fleshy leaves above and around the stem.

16.3.1.3 Leaf Crops

Lettuce, mustard greens, tea, chard, spinach, watercress, parsley, and cabbage are some examples of leaf crops. A leaf is typically a flat and expanded organ, primarily involved with photosynthesis, but deviations in morphology and function may occur. The leaf is composed of tissues that are fundamentally similar to those found in stem and root, but their organization is usually different. The epidermis contains a well-developed cuticle in most instances and stomata are characteristically present. Beneath the epidermal layer lies a series of elongated, closely packed palisade cells rich in chloroplasts. The vascular system consists of netlike veins in dicots and parallel veins in monocots.

16.3.1.4 Fruits

Botanically, a fruit is a ripened ovary or the ovary and adjoining parts; that is, it is the seed-bearing organ. The flesh of the fruit may be developed from the floral receptacle, from carpellary tissue, or from extrafloral structures, such as bracts. Whatever its origin, it is generally largely composed of parenchymatous tissue. The anatomical features of some fruit types are illustrated in Figure 15. Fruit vegetables include crops like squash, tomato, green bean, and cucumber. Culinary fruits are usually eaten raw as a dessert item, and they possess characteristic aromas due to the presence of various organic esters.

Cereals are a type of fruit called a caryopsis in which the thin fruit shell is strongly bound to the seed coat (Fig. 16). Sunflower seeds are actually a complete fruit called an achene. True nuts (e.g., acorns, hazelnuts, beechnuts, sweet chestnuts) are one-seeded fruit with a hardened fruit layer. Culinary nuts are processed and marketed as the seed portion only (e.g., Brazil, walnut, almond and pine nuts) or are not botanical nuts (e.g., peanut is a legume seed). Other

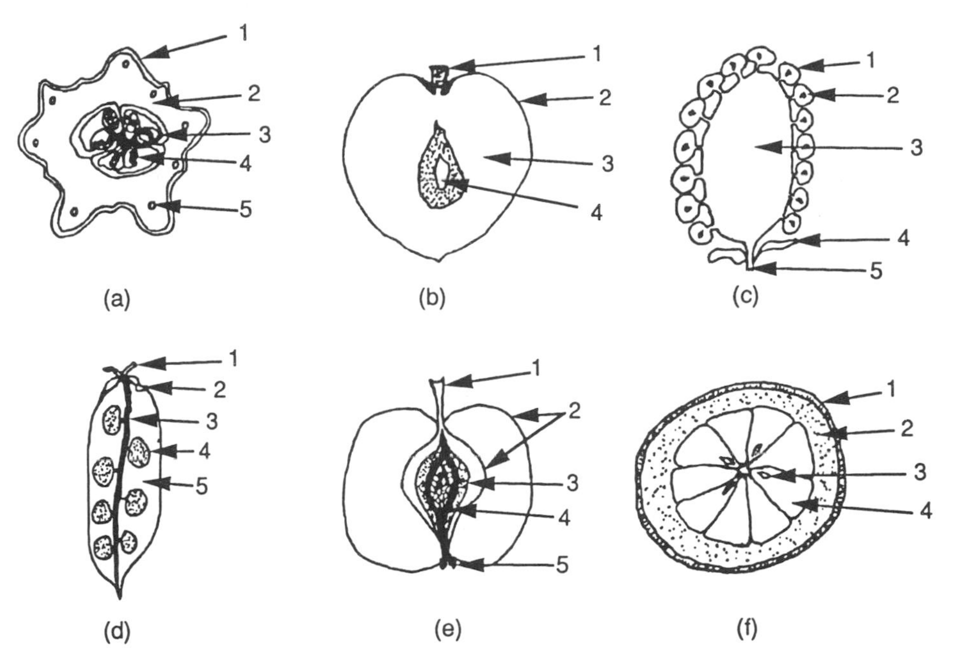

FIGURE 15 Diagrammatic illustrations of anatomical structures of different types of fruit: (a) Pepo (cucumber, squash, and pumpkin) in cross section: (1) rind (receptacular), (2) flesh (ovary wall), (3) placenta, (4) seed, and (5) vascular bundle. (b) Drupe (cherry, peach, and plum) in longitudinal section: (1) pedicel, (2) skin (ovary wall), (3) flesh (ovary wall), pit (stony ovary wall), and (5) seed. (c) Aggregate (raspberry, strawberry, and blackberry) in longitudinal section: (1) fleshy ovary wall, (2) seed (stony ovary wall plus seed), (3) fleshy receptacle, (4) sepal, and (5) pedicel. (d) Legume (pea, soybean, and lima bean) in longitudinal section: (1) pedicel, (2) sepal, (3) vascular bundles, (4) seed, and (5) pod (ovary wall). (e) Pome (apple and pear) in longitudinal section: (1) pedicel, (2) skin and flesh (receptacle), (3) leathery carpel (ovary wall), (4) seed, and (5) calyx (sepals and stamens). (f) Hespiridium (citrus) in cross section: (1) collenchymatous exocarp (the falvedo), (2) parenchymatous mesocarp (the albedo), (3) seed, and (4) endocarp of juice sacs formed by breakdown of groups of parenchyma-like cells.

legumes that are processed and marketed as the seed portion of the fruit include lentils, mung beans, soybeans, and chick peas. Soybean seeds have three major parts: the seed coat, embryo, and the cotyledons or seed leaves (Fig. 16). In developing legume seeds, the endosperm is absorbed by the embryo, which repackages the nutrients in the cotyledons. The hull is made up of an outer layer of palisade cells, a layer of hourglass cells, and smaller compressed parenchyma cells.

16.3.2 Tissue and Cell Structure

Some of the food-related properties of plant organs are intimately related to their substructure. Accordingly, some of the salient features of substructure are discussed briefly.

16.3.2.1 Parenchyma Tissue

The Greek word *parenchyma*, meaning "that which is poured in beside," is descriptive of the soft or succulent tissue that surrounds the hardened or woody tissues. This tissue is the most abundant type present in edible plants. The mature parenchyma cell possesses a thin cell wall,

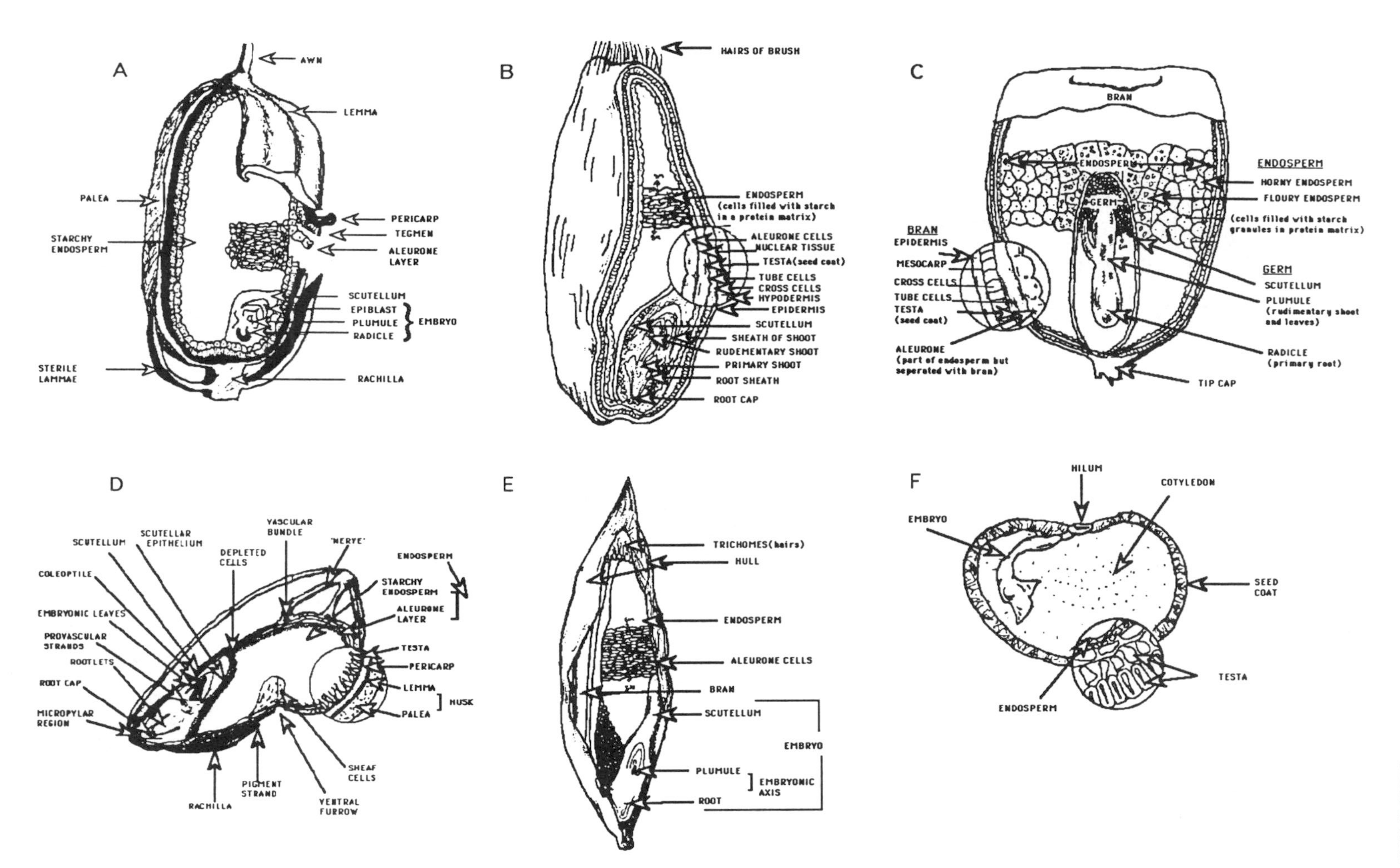

FIGURE 16 Diagrammatic illustrations of cereal grains called caryopsis fruit: (A) rice, (B) wheat, (C) corn, (D) barley, and (E) oat. An example of a true seed crop that does not contain a thin fruit shell is soybean. (F). See also Fig. 15 (d), (4).

is occupied by large vacuoles, and has a size and shape that vary with different species. The extent of cellular contact or lack of same also varies with different commodities and has an important influence on texture. The spaces between adjacent cells can contain air (e.g., 25 vol% in some apples) or pectic substances. Pectins serve as intracellular cement, and the degree of polymerization, methylation, and interaction with wall components can account for the smooth (e.g., peach) or coarse characteristics of different parenchyma tissues.

16.3.2.2 Protective Tissues

Protective tissues develop at the surface of plant organs, forming the "peel" or "skin" of the commodity. The outermost multilayer region of cells is called the epidermis. These cells usually fit together compactly with no air spaces, and their outer walls are usually rich in wax or cutin (Fig. 9). An extracellular layer, called the cuticle, is usually present on the outer surface of the epidermis (Fig. 17). More elaborate structural modification of epidermal tissue is found in seed, such as cereals (Fig. 16), and in pericarps of some fruits, such as citrus and banana. The practical importance of protective tissues of fruits and vegetables is evident when postharvest crop losses are measured as a function of surface abrasion. Accordingly, it is important to "wound heal" certain commodities (e.g., potatoes) under suitable conditions prior to storage.

16.3.2.3 Supporting Tissues

Collenchyma tissue is characterized by elongated cells that have unevenly thickened walls. This type of tissue is found particularly in petioles, stems, and leaves, occupying the ridges that are often situated on the surface of the organ (e.g., strands on the outer edge of the celery petiole). Thickened primary walls are especially rich in pectin and hemicelluloses, giving the walls unusual plasticity. Collenchyma tissue, although relatively soft, resists degradation during cooking and mastication.

Sclerenchyma cells are uniformly thick-walled cells that are normally lignified. Mature sclerenchyma tissue is nonliving and consists mainly of cell walls. The two main types of sclerenchyma cells are "fibers" and sclereids. These cells are not perturbed by cooking, and they give rise to stringy textures in asparagus and green beans or gritty textures in fruit, such as quince or pear. In certain commodities, rapid lignification is stimulated by the physical stresses associated with harvest, handling, or storage—hence, the folklore that "water in the pot should be boiling before going to the garden for certain vegetables."

16.3.2.4 Vascular Tissue

The vascular tissues, xylem and phloem, are complex tissues composed of simple and specialized cells. Although it is not clear whether vascular tissue has significant direct impact on food quality, it is possible that the vascular system has an influence on physiological events (e.g., storage disorders) in postharvest crops.

16.3.3 Subcellular Structure

The main components of typical plant cells are illustrated in Figure 18. The reader should refer to books on botany and plant physiology for detailed information on the structural characteristics of organelles. Organelles such as nuclei, mitochondria, microsomes, plastids, vacuoles, and lysosomes have been studied to varying degrees in the context of postharvest events, and important findings from these studies are discussed in the next section.

Some plants also contain organelles that store reserves of polysaccharide, protein, or lipid. Amyloplasts are the site of amylose and amylopectin in the developing plant, and they form the

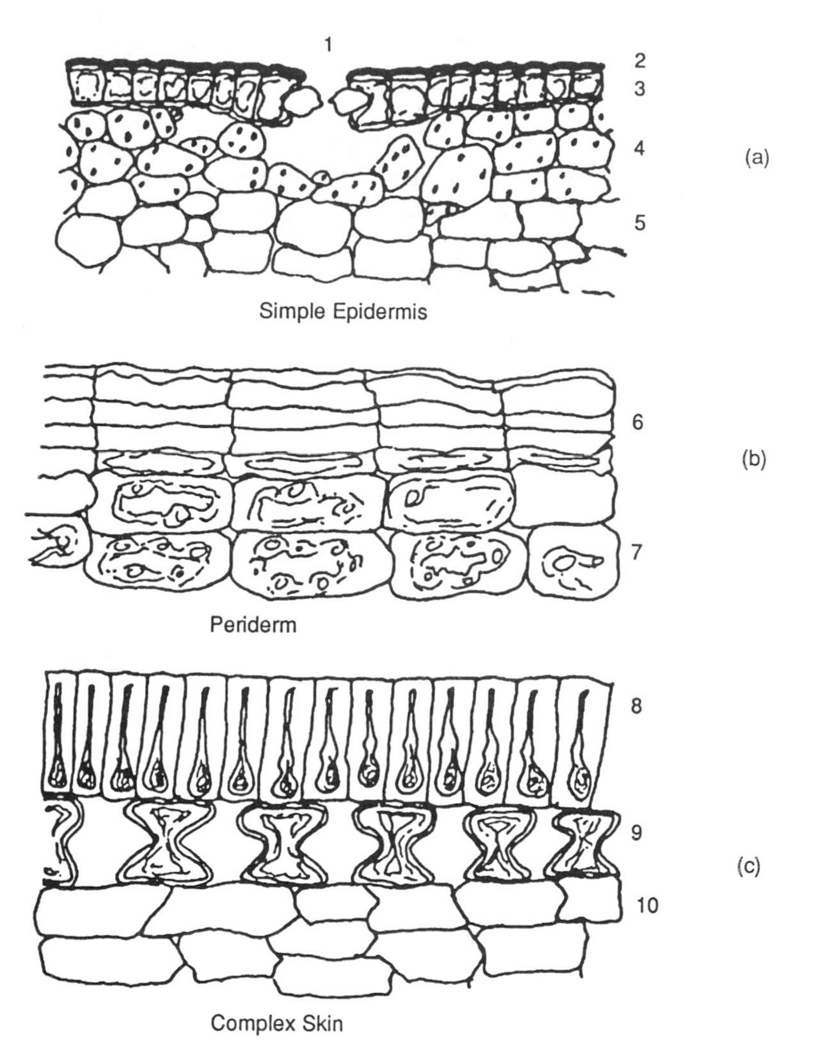

FIGURE 17 Diagrammatic illustration of protective tissues. (a) Simple epidermis, such as asparagus stem. (b) Periderm, such as potato tuber. (c) Complex "skin," such as testa of a pear: (1) stoma, (2) cuticle, (3) epidermis, (4) chlorpolast containing parenchyma, (5) unspecialized parenchyma, (6) corky tissue, (7) parenchyma cells containing developed starch granules, (8) compact layer of elongated epidermal cells with heavily thickened walls, (9) hypodermal layer of hourglass-shaped cells with large intercellular spaces, and (10) parenchymatous inner layer.

starch granules found in mature plants. The shape and size of the starch granule varies with plant species and these characteristics are sometimes used to detect adulteration. Storage protein in endosperm and aleurone layer of seed tissue is synthesized in protein bodies that form from the association of golgi derived vesicles. As the cereal grain matures, these protein bodies coalesce to form a protein matrix in the endosperm [63]. In legumes, protein bodies can vary from 2–20 μm in diameter. Spherosomes are small (0.2–0.5 μm) triacylglycerol-rich droplets found in the aleurone layer and endosperm.

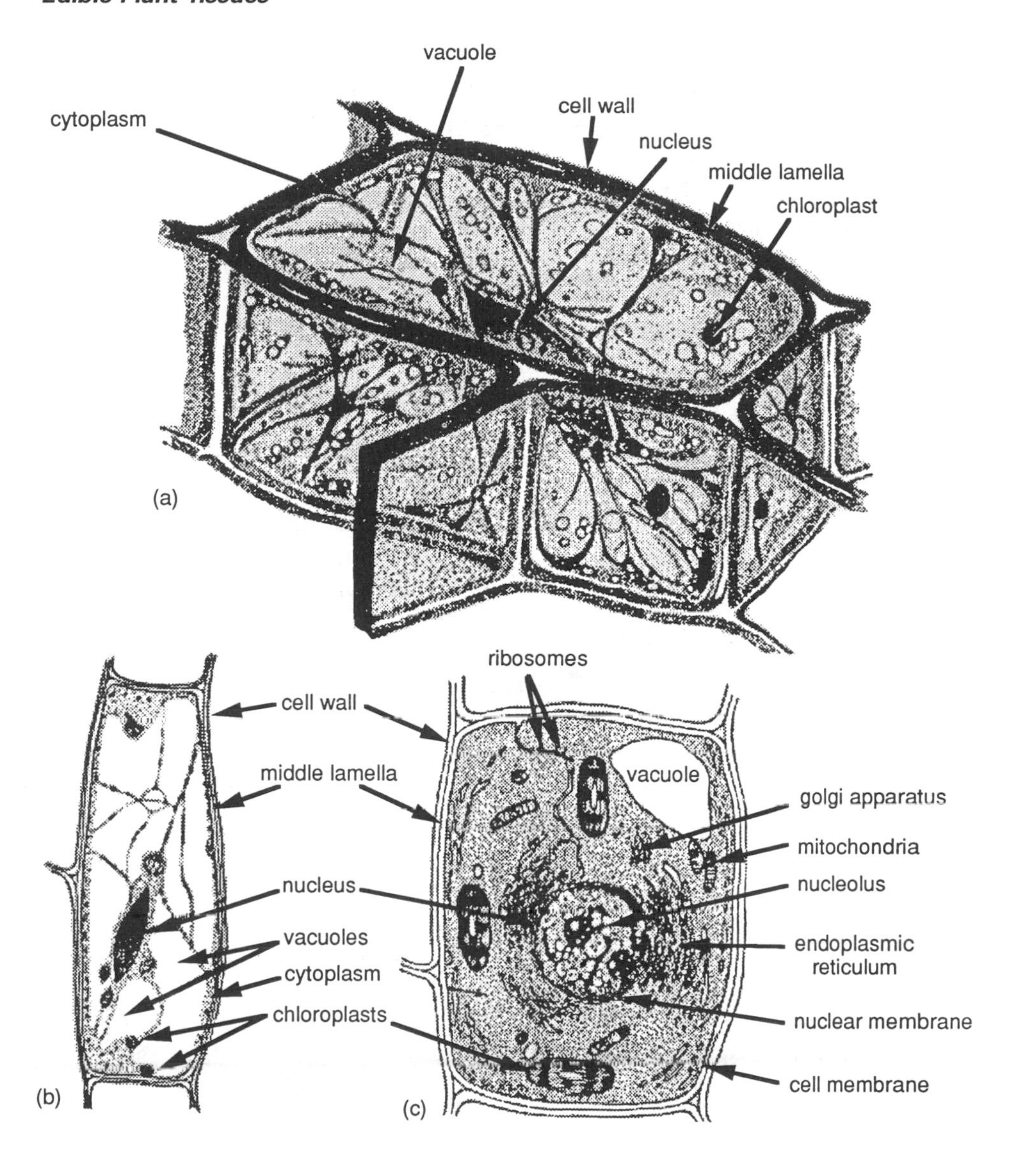

FIGURE 18 Diagrammatic representation of a mature plant cell. (a) Light microscope, (b) light microscope section, (c) electron microscope image. (From Ref. 40.)

16.4 PHYSIOLOGY AND METABOLISM

If we imagine that a circle encloses the set of all biochemical reactions occurring in plant tissues, we can, on the basis of current knowledge, define a subset consisting of reactions that are basically identical in all plant tissues. This common subset of reactions includes basal events, such as activation of amino acids preparatory to protein synthesis, mitochondria-linked respiration, and anabolism of certain cell wall polysaccharides, In addition to the reactions common to all detached plant parts, we can imagine additional subsets of reactions common to large groups of organs, such as photosynthesis in green plants, organic acid metabolism, shikimic pathway, and the isoprene (mevalonic acid) pathway. Finally, there may be aspects of metabolism that are highly specific to individual families (Table 1), genera, species, or cultivars. Two examples of the latter type of subset are biogenesis of aromatic sulfur compounds in the Lilliacae (Fig. 19) and biogenesis of mustard oils (thiocyanates and isothiocyanates) in the Cruciferae.

16.4.1 Respiration

Respiration is an aspect of basal metabolism that is of primary concern to postharvest physiologists. Plant respiration in detached plant parts can involve several pathways, and can be described by the overall reaction:

$$C_6H_{12}O_6 + 6O_2 + n\text{ADP} + n\text{P}_i \rightarrow 6CO_2 + 6H_2O + n\text{ATP} + \text{heat}$$

Theoretically, 60% of the bond energy (2870 kJ/mol glucose) is lost as heat with the formation of 37 mol adenosine triphosphate (ATP). However, calorimetry studies have shown that respiration in postharvest tissues often results in even more dissipation of energy as heat loss (90% or more) and less ATP synthesis. Heat of respiration is a primary consideration in designing appropriate ways to store fruits and vegetables [34]. The stoichiometry of oxygen consumption and carbon dioxide formation may also vary (see later discussion) depending on the oxidation and decarboxylation of other substrates (fatty acids, amino acids, organic acids).

16.4.1.1 Respiration and Storage Life

The importance of respiration in postharvest crops is best illustrated by comparing the relationship between respiratory rate and storage life of different crops (Fig. 20). Commodities that exhibit rapid rates of carbon dioxide evolution or oxygen consumption are generally quite perishable, whereas those with slow respiratory rates may be stored satisfactorily for relatively

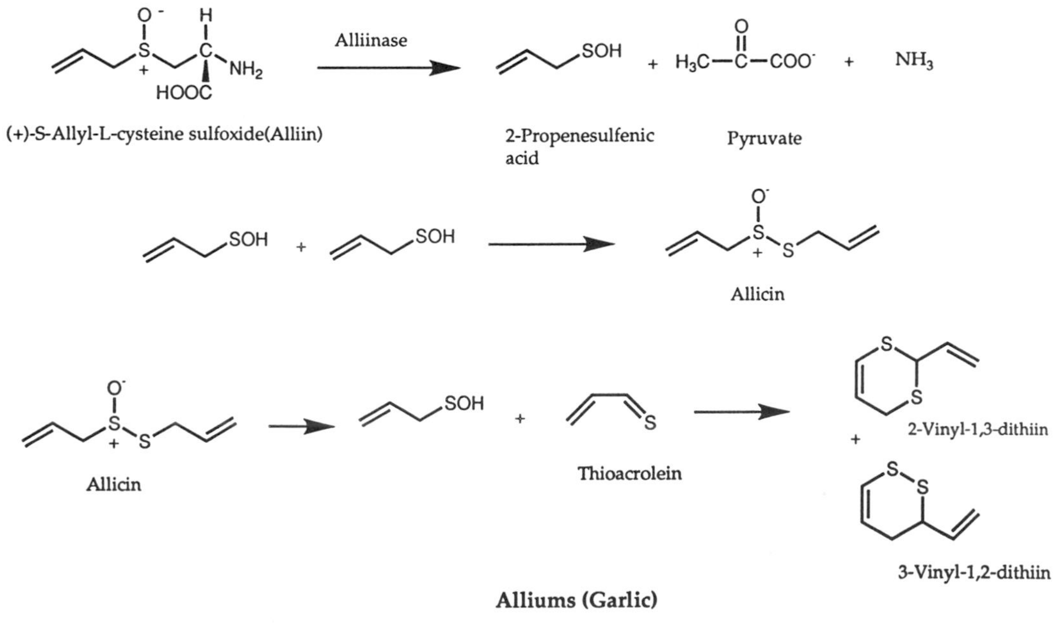

(+)-S-Allyl-L-cysteine sulfoxide(Alliin) —Alliinase→ 2-Propenesulfenic acid + Pyruvate + NH_3

2-Propenesulfenic acid + 2-Propenesulfenic acid → Allicin

Allicin → 2-Propenesulfenic acid + Thioacrolein → 2-Vinyl-1,3-dithiin + 3-Vinyl-1,2-dithiin

Alliums (Garlic)

FIGURE 19 Biogenesis of aroma compounds in garlic. In onion, the precursor is *trans*-(+*S*-(1-propenyl)-L-cysteine sulfoxide, a positional isomer of alliin, and the primary product is 1-propenesulfenic acid.

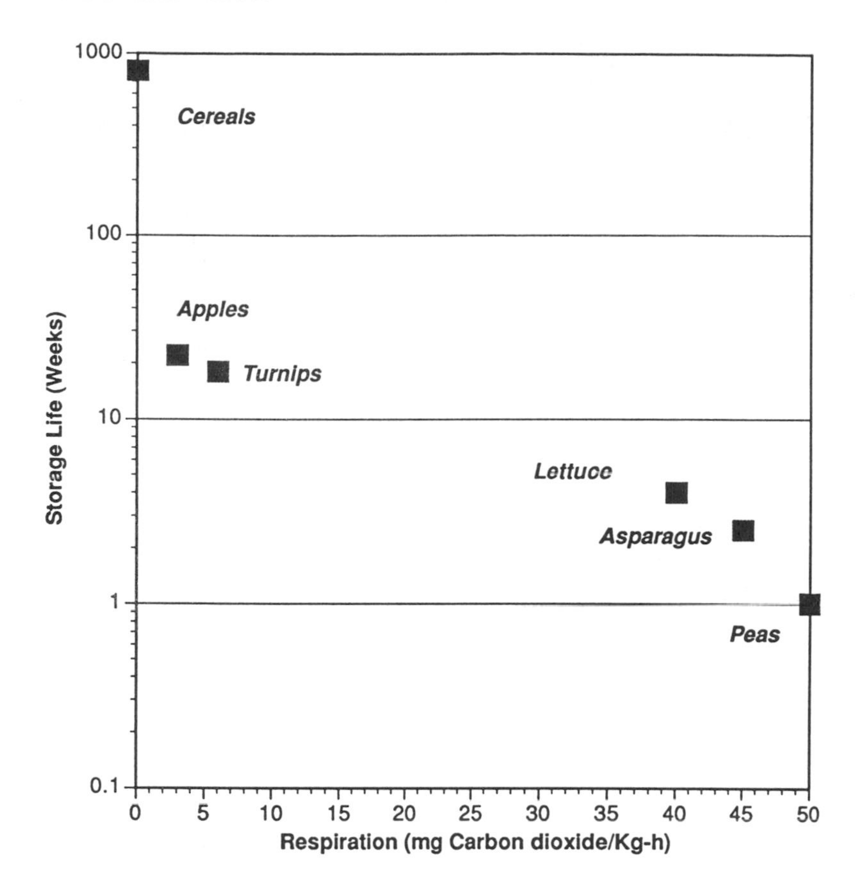

FIGURE 20 Relationship between initial respiration rate and storage life to unsalable condition at 5°C.

long periods of time. Moreover, the shelf life of a given commodity can be greatly extended by placing it in an environment that appropriately retards respiration (e.g., refrigeration, or controlled atmosphere).

16.4.1.2 Metabolic Pathways and Plant Respiration

The relative importance of different oxidation cycles (Krebs cycle, glyoxylate shunt, Embden Meyerhof pathway, and pentose phosphate cycle) appears to vary among species, among organs, and with the ontogeny of the plant. Metabolic pathways involving glycolysis and Krebs cycle oxidation are most common, although pentose phosphate (hexose monophosphate shunt) oxidation can account for approximately one-third of glucose metabolism in some instances (Fig. 21). The relative importance of the pentose cycle appears to increase in mature tissues, and it is known to be operative in such fruits as pepper, tomato, cucumber, lime, orange, and banana. It is possible that the reducing power (NADPH) generated by the shunt is required for various biosynthetic reaction (e.g., pigment and flavor biogenesis). Studies with ripening banana indicate that glycolysis and gluconeogenic carbon flow via the sucrose phosphate synthase reaction occur simultaneously [36]. Hexose phosphates arise from sucrose and ultimately exceed sucrose concentration. Thus, sucrose biosynthesis during ripening constitutes a significant sink for

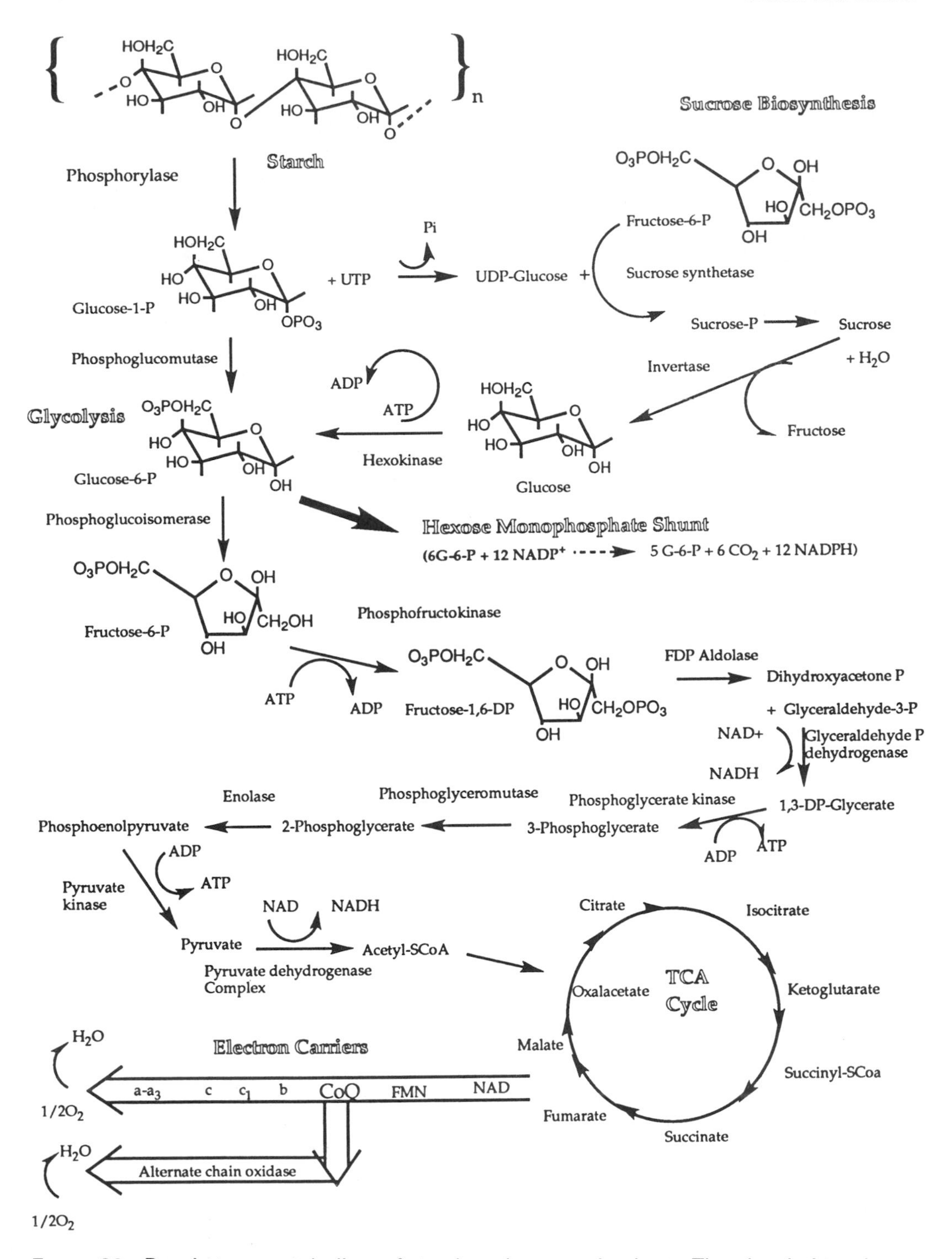

FIGURE 21 Respiratory metabolism of starch and sucrose in plants. The glycolytic pathway converts 1 mole of glucose or glucose 1-phosphate to 2 moles of pyruvate. The hexose monophosphate shunt reduces 12 moles of $NADP^+$ by oxidation of 1 mole of glucose 6-phosphate to carbon dioxide. The tricarboxylic acid cycle results in the conversion of pyruvate to carbon dioxide, with formation of ATP, NADH, and $FADH_2$. The electron

TABLE 11 Respiratory Quotients for Some Detached Plant Parts

System	Respiratory quotient[a]
Leaves rich in carbohydrate	1.00
Germinating starch seeds	1.00
Wheat seedlings in 5–20% O_2	0.95
Wheat seedlings in 3% O_2	3.34
Germinating linseeds	0.64
Germinating buckwheat seeds	0.48
Mature linseeds	1.22
Apples	
Preclimacteric apples	1.00
Climacteric apples	1.50
Postclimacteric apples	1.30
Apples treated with HCN	2.00
Apples treated with $CHCl_3$ vapor	0.25

[a]Ratio of respiratory CO_2 production to O_2 consumption.

respiratory ATP. Ripening fruit also exhibit cyanide insensitive respiration via an alternate oxidase [49, 50]. Since the alternate electron transfer chain is not coupled to oxidative phosphorylation (ATP synthesis), it is sometimes called "thermogenic" respiration.

The respiratory quotient (RQ) is a useful qualitative guide to changes in metabolic pathways or the emergence of nonrespiratory decarboxylase or oxygenase systems (Table 11). The RQ is defined as the volume ratio of carbon dioxide evolved to oxygen consumed during a given period of respiration. The RQ for fruit respiration increases at the climacteric stage of ripening. The rise in RQ during climateric respiration is indicative of increased decarboxylation or decreased carboxylation. Decarboxylation has been attributed to the increased activity of malic enzyme and pyruvate carboxylase in several ripening fruits. In ripening avocado, malate is oxidized by malic dehydrogenase via cytochrome oxidase, or by malic enzyme via the cyanide-insensitive alternate pathway [57].

$$\text{Malate} + \text{NADP}^+ \xrightarrow{\text{Malic enzyme}} \text{Pyruvate} + CO_2 + \text{NADPH} + H^+$$

Although respiratory rates of different commodities differ widely, it is not always clear why they do so. Cellular respiration may be coupled to various metabolic reactions related to wounding, harvesting, normal cellular maintenance, or the normal ontogeny of the organ. Catabolism of sugars is clearly important in commodities such as sweet corn and peas, which undergo a dramatic loss in sweet taste within hours after harvest. Similarly, anabolic reactions (such as synthesis of lignin and sucrose), which are dependent on energy made available by respiration, are equally important in postharvest systems.

---

transport system links the oxidation of NADPH and $FADH_2$ with the reduction of molecular oxygen to water and the synthesis of ATP from ADP and P_i at three phosphorylation sites. The alternate electron transport system branches from the electron transport system at ubiquinone (CoQ) and is not coupled to oxidative phosphorylation of ADP to ATP.

Although much of the activity in postharvest plants is related to the energy-conserving reactions of oxidative phosphorylation, some of this activity results in generation of heat (thermogenesis). For example, the respiratory burst in senescing flowers and in wounded storage organs, such as potato tuber, is not completely coupled to ATP synthesis. Thermogenesis appears to operate via an alternate electron transfer chain, which is insensitive to cyanide and other inhibitors of cytochrome oxidase (Fig. 21). The plant hormone ethylene and cyanide evoke cyanide-insensitive respiration in a variety of postharvest crops [50, 79]. In ripening fruit, increasing amounts of cyanide are produced concomitantly with ethylene synthesis. Evocation of the alternate electron transfer chain may function to decontrol respiration, causing (a) rapid catabolism of storage polysaccharides, (b) electron transfer without expenditure of reducing power, or (c) generation of heat. The latter event can differentially affect chemical reactions (e.g., accelerate autolytic reactions) or cause volatilization of biologically active molecules.

16.4.1.3 Respiratory Patterns

Most fleshy fruits show a characteristic rise in respiratory rate more or less coincident with the obvious changes in color, flavor, and texture that typify ripening. Fruit that exhibit this respiratory rise are classified as "climacteric fruit" (Table 12). This respiratory climacteric has been the subject of considerable study because it marks the end of optimum ripeness and the onset of senescence (deterioration or old age [47]). The relative magnitude of the climacteric burst varies considerably in different fruits, and many fruits (and all vegetables) fail to exhibit an autonomous rise after harvest [15] (Fig. 22). This latter group has traditionally been categorized as nonclimacteric, although it may be that some of these fruits are

TABLE 12 Classification of Edible Fruits According to Respiratory Patterns

Climacteric	Nonclimacteric
Apple	Blueberry
Apricot	Cacao
Avocado	Cherry
Banana	Cucumber
Breadfruit	Grape
Cherimoya	Grapefruit
Feijoa	Java plum
Fig	Lemon
Guava	Litchi
Mammee apple	Melon
Mango	Olive
Muskmelon cantaloupe	Orange
Papaya	Pineapple
Papaw	Rin-tomato
Passion fruit	Strawberry
Peach	Tamarillo
Pear	
Plum Sapote	
Tomato	
Watermelon	

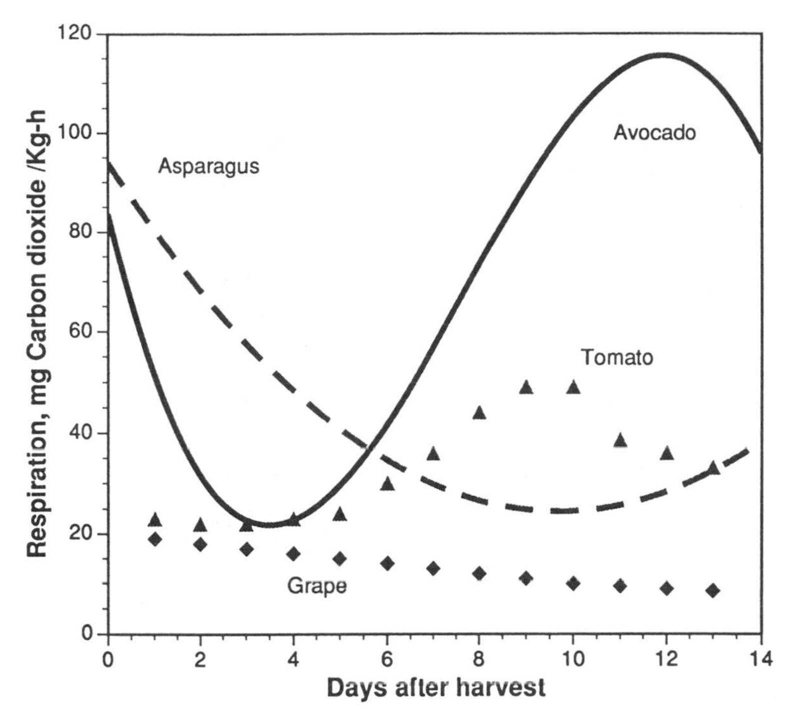

FIGURE 22 Respiratory patterns after harvest of a stem (asparagus), climacteric fruit (tomato and avocado), and nonclimacteric fruit (grape), The spontaneous respiratory burst in climacteric fruit is normally triggered by low concentrations of endogenous ethylene (Sec. 16.4.1.3 and 16.4.2). Exogenous ethylene and wounding can stimulate respiratory bursts (not true climacteric) in many detached plant organs.

postclimacteric when harvested. Confusion is avoided when the climacteric is defined as the period of enhanced metabolic activity during the transition from the growth phase to senescence. Mutant forms of tomato that do not exhibit a climacteric and do not fully ripen have been isolated [82].

Some metabolic differences have been demonstrated between climacteric and nonclimacteric fruits. Ripening in fruits categorized as nonclimacteric often proceeds more slowly. In climacteric fruit, a threshold concentration (~0.1–1 ppm) of ethylene initiates the climacteric rise, but ethylene does not influence respiratory rate once the climacteric is initiated. This is in contrast to nonclimacteric fruits whose respiration rates are dependent on the concentration of ethylene; that is, the rate increases as a function of ethylene concentration and declines when ethylene is removed. The relation between climacteric rise and ripening indices is not always perfect. Although the climacteric rise and ripening occur simultaneously in many fruits under a wide variety of conditions, evidence exists that these events can be separated.

Detached stem, root, and leaf tissues normally respire at a steady rate or show a gradual decline in rate with onset of senescence under constant environmental conditions. Mechanical injury (impact bruising) of mature green tomatoes causes a sustained doubling of the respiration rate. Although stress—imposed by mechanical injury, temperature extremes, chemicals, or biological agents—can result in a burst of respiratory activity, a climacteric coordinated with ripening is not always apparent.

16.4.1.4 Factors Influencing Respiration Rate

Several factors can influence the respiration rate of postharvest commodities, as illustrated for crisp head lettuce [41] (Fig. 23). The respiration process in plant tissues is markedly influenced by temperature. Within the "physiological" temperature range for a species, the rate of respiration usually increases as the temperature is raised, and the extent of the change in respiration rate can be expressed in terms of Q_{10} value. Q_{10} is defined as the rate of a reaction at T°C divided by the rate at $T - 10$°C. For most fruit and vegetable species Q_{10} values range from 7 to less than 1, although values between 1 and 2 are most common. Lowering the temperature of a climacteric fruit delays the time of climacteric onset, as well as its magnitude. In certain commodities, such as potato tuber, transfer of the tissue from a cold to a warm environment results in a temporary burst in respiratory activity followed by equilibration to an intermediate level of activity. The cold sweetening of some potato cultivars does not appear to be caused by the lower respiration rate [9].

In general, either a reduction in oxygen tension (below 21%) or an increase in carbon dioxide tension (above 0.03%) slows respiration and associated deteriorative reactions. Certain commodities can be successfully held at very low oxygen tension (e.g., less than 1%) or at a very high partial pressure of carbon dioxide (e.g., greater than 50%), but most are adversely affected by such extremes. Furthermore, it should be recognized that the concentration of a given gas inside the tissue may differ considerably from that of the surrounding environment because of solubility, resistance to diffusion, and metabolic characteristics of the tissue. Cytochrome oxidase, the final electron acceptor in "normal" respiratory metabolism, has a relatively high affinity for oxygen and is saturated at relatively low concentrations of oxygen. On the other hand, the affinity of the alternate chain oxidase for oxygen is less than that of cytochrome oxidase, and accordingly respiration via the alternative oxidase is influenced more profoundly by low con-

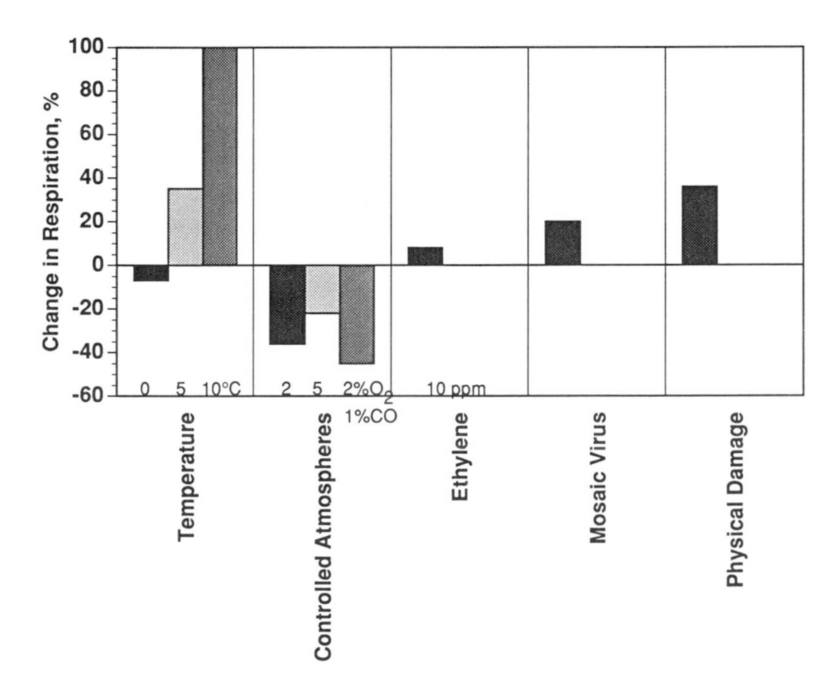

FIGURE 23 Relative effects of temperature, reduced oxygen, CO, C_2H_4, lettuce mosaic virus, and physical damage on respiration rate of crisphead lettuce in comparison to rate in air at 2.5°C. (Adapted from Ref. 41.)

centrations of oxygen in situ. Low concentrations of oxygen may cause accumulation of anaerobic end products that are phytotoxic. Pyruvate (Fig. 21) is converted to acetaldehyde and CO_2 by the enzyme pyruvate carboxylase and the cofactor thiamin pyrophosphate; acetaldehyde is converted to ethanol by alcohol dehydrogenase. The oxygen concentration at which a shift from aerobic to anaerobic respiration occurs is called the "extinction point" and varies with plant species.

Certain chemical treatments have also been successfully employed to restrict aerobic respiration of postharvest plant organs, and this lessens deteriorative reactions. Ventilation, scrubbing of storage atmospheres, or hypobaric storage to minimize atmospheric and tissue ethylene concentration can also be an effective means of maintaining low respiratory rates.

The respiration rate of different plant organs generally follows the following pattern: root, tuber, and bulb vegetables (e.g., potato, onion, sweet potato) < mature (unripe) fruit (e.g., tomato, apple) < immature fruit (e.g., green beans, egg plant) < growing stems and floral tissue (e.g., asparagus, broccoli). The nature of surface coatings (Sec. 16.3.2.2) and other morphological characteristics, such as the ratio of surface area to volume, can also influence respiration rate. Cereals and seed crops have very low respiration rates; however, the respiration rate increases dramatically when the moisture content rises above about 14%.

16.4.1.5 Biochemical Control of Respiration

When cells shift gears in metabolism, changes occur in the concentration of metabolites. These changes in the cell serve as signals to regulate the activity of key enzymes in metabolic pathways in order to keep supply in pace with demand. The signals controlling the flow of respiration are numerous. According to the "energy charge" hypothesis, a cell shifting into a state of high ATP consumption will be characterized by increased levels of ADP and AMP.

$$\text{energy charge} = \frac{[\text{ATP}] + \frac{1}{2}[\text{ADP}]}{[\text{ATP}] + [\text{ADP}] + [\text{AMP}]}$$

According to this hypothesis, cells are capable of maintaining an energy charge value within a narrow range of about 0.8–0.9. Key regulatory actions are activation of phosphorylase by ADP, or inhibition by ATP and inhibition of phosphofructokinase and pyruvate kinase by ADP and AMP (Fig. 21). Starch phosphorylase is also inhibited by glucose-6-P. Other agents of respiratory control include the concentrations of NADH and NAD^+.

The control of the climacteric rise has been the subject of considerable research. An early theory focused on the notion that release of phenolics and other substances from "leaky" membranes caused an "uncoupling" of oxidative phsophorylation in the mitochondrial electron transfer chain. Agents that uncouple oxidative phosphorylation facilitate a rapid respiratory rate. Since the late 1960s it has become well known that protein synthesis attends the ripening process [16] and that mitochondria from postclimacteric fruit retain the ability to synthesize ATP when protective agents are employed during isolation [30]. According to a second theory, the increase in respiration during the climacteric is caused by an increase in the turnover of ATP during protein synthesis. More recent studies indicate that sucrose biosynthesis is responsible for considerable ATP (or UTP) utilization in ripening fruit (Fig. 21) [36]. A third theory to explain the climacteric rise is the evocation of "thermogenic" or cyanide-resistant respiration (Fig. 21) [57, 79]. However, alternate chain respiration appears to make a small contribution to net oxygen consumption in ripening fruit [73].

Respiratory control in climacteric fruit has also been linked to the activation of malic enzyme through the alternate pathway [57] and to the increased concentration of fructose 1,6-bisphosphate. According to one study [39], leakage of P_i and citrate from vacuoles in ripening tomato causes dissociation of oligomeric to monomeric forms of phosphofructokinase.

Another study [13] showed that fructose 1,6-bisphosphate accumulation can result from an enzyme activated by fructose 2,6-bisphosphate, pyrophosphate:fructose-6-phosphate phosphotransferase, which is activated during the early stages of climacteric respiration.

16.4.2 Gene Expression and Protein Synthesis

Expression of genetic information is an event of general and fundamental importance in postharvest crops. Physiological processes such as sprouting of tubers and bulbs, germination of seeds, fruit ripening, and plant senescence are developmental cascades resulting from transcription of DNA and translation of mRNA to proteins (enzymes). Increased levels of specific enzymes have been noted to occur in numerous plant organs following harvest. For example, aldolase, carboxylase, chlorophyllase, phosphorylase, peroxidase, phenolase, transaminase, invertase, phosphatase, *o*-methyltransferase, catalase, and indoleacetic acid oxidase are among the enzymes known to emerge in ripening fruit. Many of these enzymes arise as a result of de novo synthesis and have a relatively short half-life in situ. It is generally observed that plant senescence is accompanied by increased RNA synthesis and an increased rate of transcription. Indeed, there appears to be an absolute requirement of continued protein synthesis for ripening fruit. For example, cycloheximide, an inhibitor of protein synthesis, also inhibits degreening, softening, ethylene biosynthesis, and ripening of ethylene-treated banana fruit [16]. Increased protein synthesis and nucleic acid synthesis are most pronounced at the early stages of fruit ripening. The propagation of slow-to-ripen or ripening-impaired mutant varieties of fruit is consistent with findings that fruit ripening is under genetic control [82]. Moreover, genes for key enzymes that promote plant senescence have been cloned and their expression blocked by anitsense technology [64] (Fig. 24). These findings show that postharvest deterioration of plant tissues can be controlled by manipulation of gene expression.

16.4.2.1 Recombinant DNA (rDNA) Technology

The discovery of DNA restriction (cut) and ligase (join) enzymes in the 1970s opened the way to extensive application of rDNA methods to alter the genome of plants. rDNA is DNA obtained

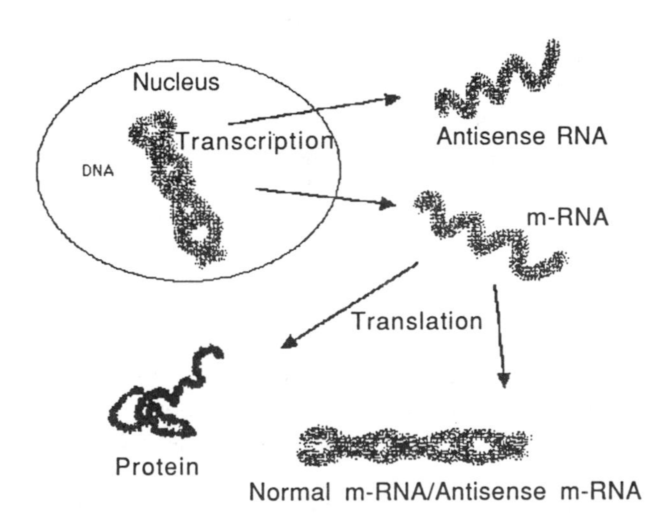

FIGURE 24 Protein synthesis and prevention by antisense technology. The first genetically engineered food approved for use in the United States was developed by applying this technique to the enzyme polygalacturonase in tomato fruit.

by cutting and recombining DNA molecules from different sources (Chap. 13). Isolated genes can be modified by site-directed mutagenesis to achieve a target protein with different properties (Chap. 7). Introduction of foreign DNA into an organism, such as a plant, is called transformation, and the new plant is called a transgenic. In recent years there has been intensive interest in genetically engineering plants for crop improvement [11, 26]. The basic steps in cloning a gene are (1) identify a target protein (e.g., enzyme or storage protein), (2) determine the sequence of the protein, (3) determine the corresponding DNA sequence from the triplet genetic code, (4) construct a DNA probe (single strand of DNA, with radiolabeled nucleotides, that encodes a complimentary section of the target gene), (5) screen a genome library (pool of DNA segments that is likely to contain the gene of interest), (6) insert the isolated gene in a vector (a self-replicating high-molecular-weight DNA sequence that can be absorbed by the host cell and its progeny), (7) incorporate the vector in a microbial host that is easily transformed, such as *Escherichia coli*, (8) propogate the microbial host and verify the DNA sequence of the cloned gene, and (9) transform a host plant to obtain a transgenic plant. The latter step is most often accomplished with a vector (disarmed plasmid) from the plant-tumor-inducing bacterium *Agrobacterium tumefaciens*. For plants that are difficult to transform with the *A. tumefaciens* plasmid, such as cereals and other monocots, a technique called "biolistics" has been used to shoot DNA-coated microprojectiles directly into plant cells.

An alternate approach to gene cloning that is often used with plant and other eukaryote genes is the isolation of complementary DNA. With this method, mRNA is isolated from the cytoplasm and is used as a template in conjunction with the enzyme reverse transcriptase to make a single strand of DNA. DNA polymerase is then used to make the complementary DNA strand. An advantage of the cDNA method is that the reverse transcriptase used does not copy introns or transcription signals (e.g., promotors) and therefore the isolated gene can be transcribed by bacteria (step 7). The first genetic transformation of a plant occurred in 1983 and involved introduction of the phaseollin gene from bean into sunflower plant.

16.4.2.2 Application of Genetic Engineering to Plant Crops

Various genetic engineering strategies have been used to improve plant crops. These include (1) antisense technology (a gene with reversed orientation with respect to its regulatory sequence so that it blocks normal gene expression) (Fig. 24), (2) gene amplification (introducing multiple gene copies with a combination of promotors to increase product yield), (3) metabolic pathway engineering (e.g., introducing genes that code for a regulatory enzyme that will increase or decrease a key metabolite, e.g., pigment biosynthesis), and (4) site-directed mutagenesis to modify the gene and gene product (e.g., alter the temperature coefficient or some other property of an enzyme).

In a May 1994 milestone decision, the U.S. Food and Drug Administration approved Calgene's Flavr Savr tomato, the first genetically engineered food. Tomatoes supplied to northern regions in the winter are normally harvested and transported in a mature green stage of development, prior to the initiation of ripening. On arrival at local markets the fruit are treated with ethylene gas to initiate ripening. Such fruit do not develop the intense flavor and pigmentation that are characteristic of vine-ripened fruit. This practice has been necessary since the ripening fruit have a limited market life and are readily damaged during shipment. Falvr Savr tomatoes are genetically engineered to contain the anitsense gene for polygalacturonase. Polygalacturonase is one of several enzymes synthesized during the ripening of tomato fruit. It catalyzes breakdown of pectin (Sec. 16.2.2.1) by hydrolysis of α-1,4-galacturonate linkages and thus causes softening of the tissue. During ripening of Flavr Savr the normal expression of the tomato polygalacturonase gene is blocked (>95%) and the softening process is delayed by several days, all the extra time that is needed to ship "breaker" (initial ripening stage) fruit to markets. Calgene did extensive testing of the tomato to address concerns about consumer and environmental safety of genetically engineered crops. Important points established during the testing and approval were (1) the genetic modification was stable, (2) the possibility of gene flow was remote, (3) the chemical composition (e.g., vitamins, glycoalkaloids) was within the normal range for tomatoes, (4) results of rat feeding studies yielded no adverse effects, (5) agronomic traits of the crop were satisfactory, and (6) control over quality assurance was satisfactory. Many other applications of genetic engineering to improve the sensory, nutritive, and processing characteristics of fruits and vegetables are under development or review for approval (Table 13).

16.4.3 Other Metabolic Processes That Influence Food Quality

16.4.3.1 Changes in Cell Wall Constituents

Because the cell wall is the primary structural element in plant tissues it is central to the textural quality of plant foods. In some cases, like apples, celery, and lettuce, crispness is desired, while in others, like peaches, strawberries, and tomatoes, a soft texture is preferred. Changes in the cell wall have a major influence on softening of plant tissues. Softening may also occur from a lack of turgor, but in most fruits and vegetables softening is the result of changes in cell wall polysaccharides cause by depolymerization, demethylation, or loss of calcium. Hardening may also occur as a result of liginin formation in the wall; examples include asparagus, celery, and pears. In asparagus and celery, walls of vascular tissue are the primary sites for lignin synthesis, whereas in pears it occurs mostly in the stone cells.

Softening is generally accompanied by solubilization of pectin and the release of polyuronides. Pectin methylesterase, which catalyzes removal of the methyl ester from pectin, occurs widely in plants and fungi. In the case of tomato fruit 90% of the pectin is methylated in "mature green" fruits, and as pectinesterase increases during ripening, this declines to about 30% in ripe fruit [44]. Pectin esterase in banana also increases with ripening. In apple, guava, mango, and strawberry, however, pectin esterase decreases during ripening. Endopolygalacturonases, which would be expected to have a greater effect on pectin structure than the exopolygalacturonases (see Chap. 6), are not found universally in fruits and are notably absent in apples [10]. Apples do contain exopolygalacturonases. In most fruits, polygalacturonase and cellulase increase during ripening. Polygalacturonase and pectinesterase increase during ripening of tomato fruits, but this does not totally account for the softening that occurs. Hemicellulases and cellulases are also present during ripening and are thought to play an important role in softening. While most of the information on cell wall changes during ripening is related to degradation, it must be noted that some synthesis also occurs [56].

The amounts and types of pectic enzymes in fruit have an important bearing on yield, clarity, and cloud stability of fruit juices and other beverages derived from them. In citrus and tomato juices, in which a stable colloidal suspension is desired, the presence of pectin esterases and polygalacturonases is undesirable. In citrus juices, obtaining a stable cloud is made more difficult because one form of the pectinesterase is quite stable to heat [89]. The classic "hot break" process for tomato juice is based on inactivating pectinesterase before cellular disruption because the latter would otherwise result in substantial deesterification of pectin. Polygalacturonase is added in some processes to solubilize pectin and increase the yield of soluble solids.

16.4.3.2 Starch–Sugar Transformations

Synthesis of starch and its degradation to simple sugars are important metabolic events in postharvest commodities. In potatoes, sugars are undesirable since they can cause poor texture after cooking, an undesirable sweet taste, and/or excessive browning during frying. Conversion

TABLE 13 Recombinant DNA Techniques to Improve the Quality Characteristics of Edible Plants

Crop	rDNA strategy	Improvement(s)
Tomato	Antisense DNA for polygalacturonase (PG) gene	Decreased PG activity; increased viscosity, firmer texture, extended market life
Climacteric fruit	ACC deaminase gene or antisense DNA for ACC synthase	Decreased ACC synthase activity and ethylene biosynthesis, extended market life
Potato and other crops	Thaumatin gene	Add sweet tasting protein, enhance flavor
Seed crops, e.g., rice and canola	Storage protein gene(s)	Improve protein content and amino acid score
Seed crops, e.g., corn	Modify existing gene for storage protein, e.g., α-zein	Improve lysine content of major storage protein
Oilseeds, e.g., canola	Antisense DNA for steroyl-acyl carrier protein thioesterase gene	Increase stearic acid content of oil, substitute for cocoa butter
Oilseeds, e.g., canola	Antisense DNA for lauryl-acyl carrier protein thioesterase gene	Increase lauric acid content of oil, use as tropical oil substitute, avoid need for hydrogenation process
Seeds, tubers, roots, e.g., potato	Bacterial gene for ADP glucose pyrophosphorylase	ADPG pool and starch content increase since bacterial enzyme not subject to feedback inhibition; potatoes absorb less oil during frying
Seeds, tubers, roots	Antisense DNA for starch synthetase	Amylose-free potatoes
Potato	Antisense DNA for invertase, etc.	Prevent reducing sugar accumulation ("cold sweetening"); less Maillard browning during frying
Fruits and vegetables	Bacterial "ice nucleating" gene	Increased number of ice crystals during freezing; less tissue damage during freezing

of starch to sucrose and reducing sugars occurs in most potato cultivars when they are stored at nonfreezing temperatures below 5°C. Current evidence indicates that biosynthesis of sucrose, catalyzed by sucrose synthase (Fig. 21), at low temperature serves to utilize hexose 6-phosphates (via fructose-6-P), which would otherwise cause feedback inhibition of glycolysis. Glucose 6-phosphate is an inhibitor of phosphorylase-catalyzed degradation of starch. At a storage temperature above 10°C, sucrose, or perhaps sucrose phosphate, functions to limit starch hydrolysis via feedback control on phosphorylase or by promotion of starch synthesis via ADP glucose-starch glucosyltranferase. In developing potato tubers preharvest, phosphorylase catalyzes synthesis of α-1,4-glycosidic linkages, whereas in mature tubers postharvest it functions primarily to catalyze starch breakdown. Amylases, which are important in starch catabolism in germinating tubers, do not appear to be very active in the dormant tuber. Potato tubers contain a protein inhibitor of the enzyme invertase (Fig. 21) that becomes inactive at low, nonfreezing temperatures (cold stress). Thus sucrose is partially converted to fructose and glucose. In Norchip tubers stored for 6 weeks at 4°C, sucrose increased from 0.70 to 3.27 mg/g fresh weight, while glucose increased from 0.05 to 1.31 mg/g and fructose increased from 0.18 to 1.48 mg/g [9]. Theories to explain the reason for cold sweetening in potato cultivars include the cold lability of phosphofructokinase and possibly pyruvate kinase [4], bypassing phosphofructokinase via the pentose phosphate cycle [86], and loss of starch granule integrity [8].

Inhibition of starch synthesis at low temperature may also play a role in cold sweetening. Blocking the expression of ADP-glucose pyrophosphorylase (ADPGP) with an antisense gene in transgenic potatoes results in the abolition of starch formation and massive accumulation of sucrose (30% of dry weight) [58]. ADPGP catalyzes formation of the nucleotide sugar (ADPG) that is incorporated into the polyglucan chain during amylose synthesis:

$$\text{Glucose-1-P} + \text{ATP} \xrightarrow{\text{ADPG phosphorylase}} \text{ADPG} + \text{PP}_i$$

$$\text{ADPG} + (\text{Glucose})_n \xrightarrow{\text{ADPG-starch synthetase}} (\text{Glucose})_{n+1} + \text{ADP}$$

The transgenic tubers also shows a major increase in the mRNA for the enzyme sucrose phosphate synthase.

Regardless of how sugars accumulate during cold storage of plant commodities, this does represent a problem during commercial storage of potatoes. In practical situations, this problem is overcome by "reconditioning" (holding for several days at warm temperatures to reduce the sugar content through catabolism and conversion to starch) or by leaching out unwanted sugar by exposure of cut potatoes to water.

In some commodities, notably seeds (peas, corn, and beans) and underground storage organs (sweet potato, potato, and carrot), synthesis of starch, rather than degradation, may predominate after harvest. Starch synthesis is generally optimal at temperatures above ambient. Diminution of sugars accompanies starch synthesis in such crops and is normally detrimental to quality. For example, sucrose may serve as a substrate for sugar nucleotides used in starch synthesis as follows:

$$\text{Sucrose} + \text{UDP} \xrightarrow{\text{Sucrose-UDPG glucosyltransferase}} \text{UDP-glucose} + \text{fructose-1-P}$$

$$\text{Fructose-1-P} \xrightarrow{\text{Phosphoglucosisomerase,}} \xrightarrow{\text{Phosphoglucomutase}} \text{Glucose-1-P}$$

$$\text{Glucose-1-P} + \text{ATP} \xrightarrow{\text{ADPG-phosphorylase}} \text{ADPG} + \text{PP}_i$$

The principal enzymes in starch synthesis appear to be those in photosynthetic tissues, namely, fructose transglycosidase, adenosine diphosphate glucose:starch glucosyltransferase, UDPG pyrophosphorylase, sucrose-UDPG glucosyl transferase, and ADPG pyrophosphorylase. ADPGP is inhibited by pyrophosphate and PP_i, and sprays containing these substances are sometimes applied to sweet corn to retard the conversion of sugar to starch [2]. Decreases in sugars may also result from their oxidation via mitochondria-linked reactions. Details of starch synthesis in postharvest commodities are not completely understood.

16.4.3.3 Metabolism of Organic Acids

Organic acids are in a constant state of flux in postharvest plant tissues and tend to diminish during senescence. Much of the loss is attributable to their oxidation in respiratory metabolism as suggested by the increase in respiratory quotient. The respiratory quotient is approximately 1.0 when sugars are substrates, increases to 1.3 when malate or citrate are substrates, and further increases to 1.6 when tartrate is the substrate. Some suggested pathways associating respiratory metabolism with organic acid metabolism are shown in Fig. 25. Certain enzymes (e.g., malic enzyme and pyruvate decarboxylase) and the respiratory quotient increase concomitantly with climacteric respiration of certain fruits.

The metabolism of aromatic organic acids via the shikimic acid pathway (Fig. 26) is important because of its relationship to protein metabolism (aromatic amino acids), accumulation of precursors of enzymic browning (e.g., chlorogenic acid), and lignin deposition (phenyl propane residues). It also should be mentioned that acetyl coenzyme A (acetyl-CoA) participates in the synthesis of phenolic compounds (Sec. 16.2.5.3), lipids, and volatile flavor substances. Apart from their general importance to flavor, texture, and color, certain organic acids appear to modify the action of plant hormones and are thereby implicated in control of ontogenic changes. For example, the finding that chlorogenic acid modifies the activity of indole-3-acetic acid oxidase may well relate to the increased concentration of this acid in the green blotchy areas of defective tomatoes. The biosynthesis of flavor substances in fruits and vegetables sometimes involves organic acids. For example, the formation of isoamyl alcohol and glutamic acid in ripening tomato involves the enzyme L-leucine:2-ketoglutarate amino transferase (Fig. 27).

16.4.3.4 Lipid Metabolism

Although lipid components of most fruits and vegetables are normally present at relatively low levels, their metabolism is important during postharvest storage of plant tissues, especially when handling and processing conditions are not ideal. Changes in membrane lipids are also important in ontogenic events, such as stress response, ripening, and senescence [47]. The aging process in plant tissues is associated with a decline in polyunsaturated fatty acids, and such changes apparently accompany autolysis of membranes and loss of cell integrity [53]. Phosphatidyl-linolely (-enyl) cascades in plants also lead to formation of a prostaglandin-like substance, jasmonic acid. Free radicals resulting from lipid oxidation act at sensitive subcellular sites and appear to play an essential role in normal plant development [90]. Free radicals are kept in check by scavengers such as superoxide dismutase, vitamins C and E, cytokinins, catalase, and glutathione peroxidase. Oxidative enzymes, such as lipoxygenase, hydroperoxide lyases, and hydrolytic enzymes, such as lipase and phospholipase, have also been shown to be important in postharvest metabolism. In fruits and vegetables, hydroperoxide lyases degrade specific hydroperoxides to form flavorful C_6 and C_9 volatile aldehydes and other products. Examples of flavor compounds formed by lipoxygenase cascades in injured and senescing plants include *t*,6-*cis*-nonadienal in cucumber, 1-octen-3-(*R*)-ol and 10-oxo-*t*-8-decenoic acid in mushroom, 9-hydroxy-*t*-10,*cis*-12-octadecadienoic acid in oats, and *cis*-3-hexenal and other compounds in tomatoes. Processes such as freezing and dehydration can activate lipoxygenase in unheated plant tissues

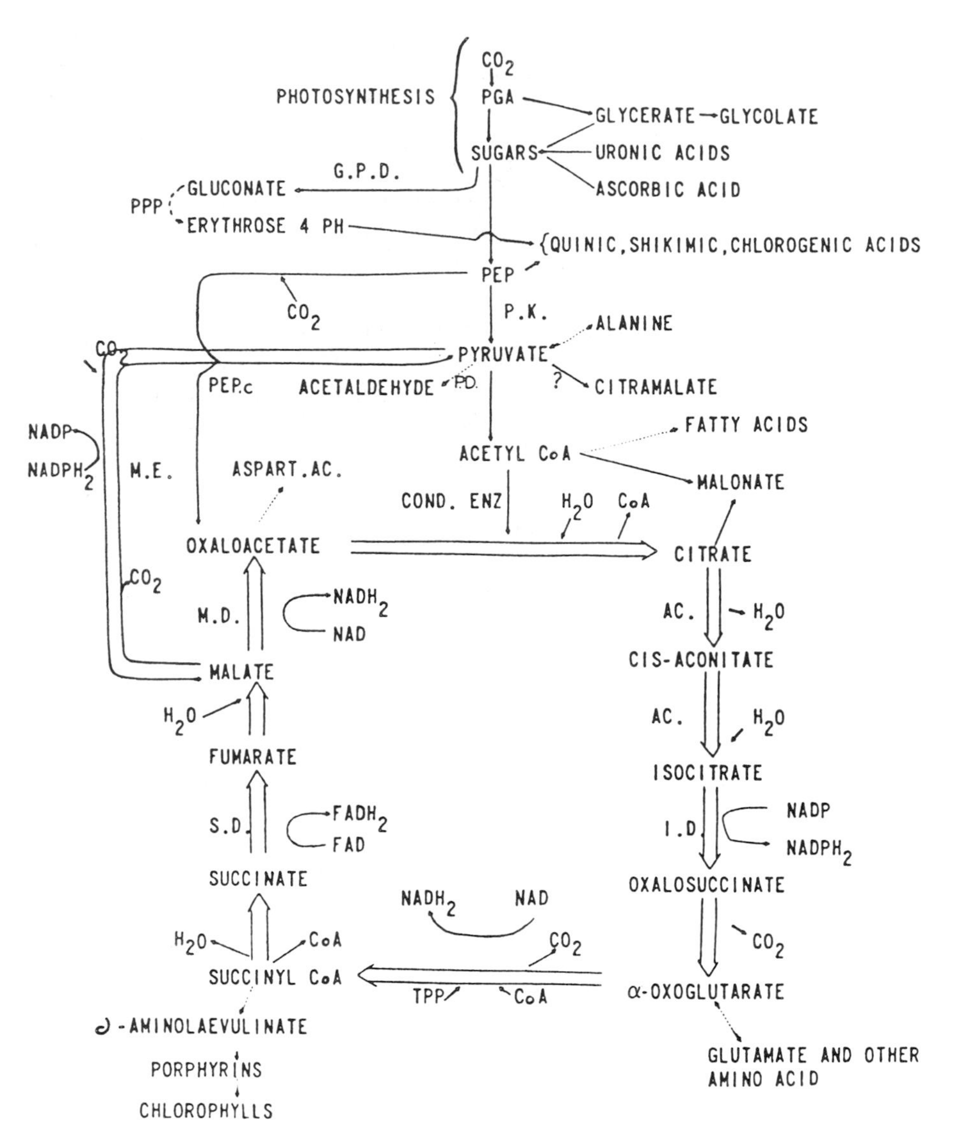

FIGURE 25 Krebs cycle (⇒) with some other modes of organic acid synthesis (→) or synthesis of other important constituents (→). Abbreviation: AC, aconitase; Cond. enz., condensing enzyme; GPD, glucose-6-phosphate dehydrogenase; ID, isocitric dehydrogenase; MD, malate dehydrogenase; ME, malic enzyme; PEPc, phosphoenolpyruvate carboxykinase; PGA, phosphoglyceric acid; PD, pyruvate decarboxylase; PK, pyrvuate kinase; PPP, pentose phosphate pathway; SD, succinic dehydrogenase; TPP, thiamine pyrophosphate. (From Ref. 15, p. 89.)

and thus lead to off-flavor development. Hydroperoxides formed by lipoxygenase can also react with other constituents, such as proteins, and cause bleaching of carotenoids and chlorophylls.

The reserve lipids (e.g., in palm, olive, and avocado) have compositions markedly different from those that are part of the functional cell framework. Recent studies using rDNA techniques have helped explain metabolic control of triacylglycerol accumulation in developing plant tissues [84]. Enzymic hydrolysis of triacylglycerols in postharvest oilseeds has an important bearing on

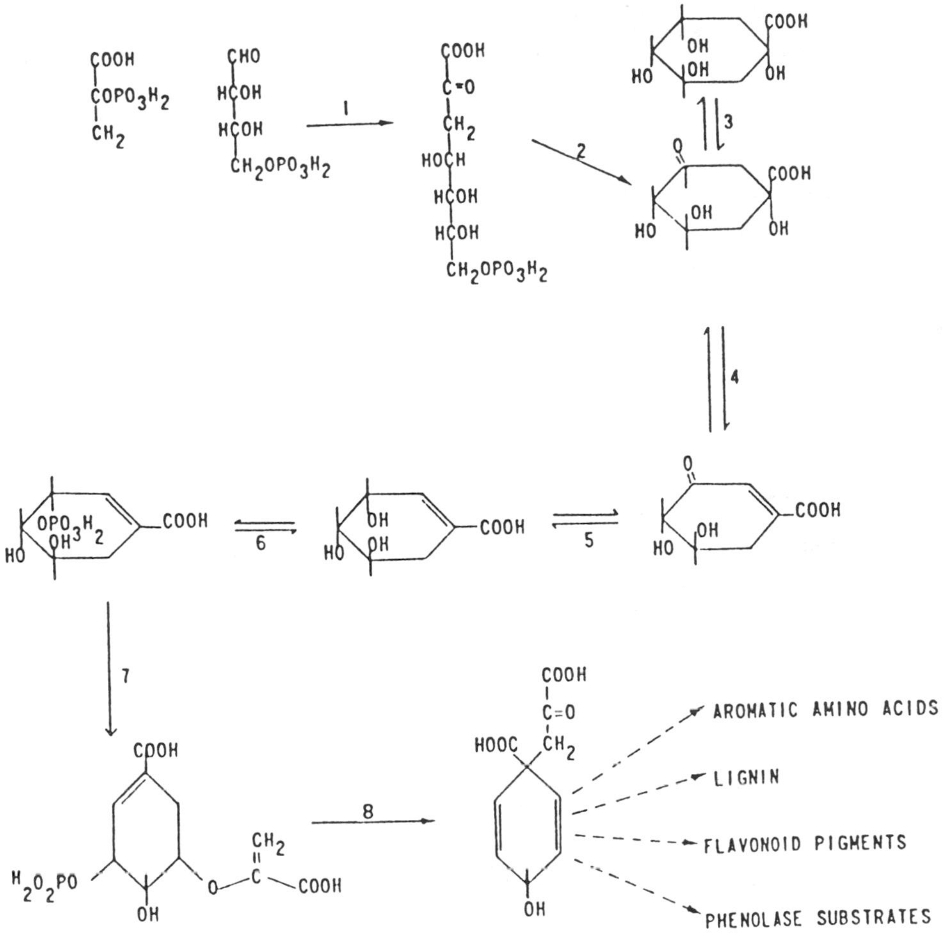

FIGURE 26 Biosynthesis of some important cell constituents via the shikimic acid pathway. (1) Phosphoenolpyruvate + D-erythrose 4-phosphate to 3-deoxy-D-arabinoheptulosonic acid 7-phosphate, (2) to 5-dehydroquinic acid, (3) to quinic acid, (4) to 5-dehydroshikimic acid, (5) to shikimic acid, (6) to 5-phosphoshikimic acid, (7) to 3-(enolpyruvate ether) of phosphoshikimic acid, (8) to prephenic acid. This role of this pathway in the formation of phenolic compounds is detailed in Fig. 11.

their quality. In comparison with other cereals, oats contain a high level of lipase, which together with lipoxygenase gives rise to bitter-tasting compounds (Fig. 28).

16.4.3.5 Pigment Metabolism

Extensive biosynthesis of carotenoids and related terpenoids occurs in many edible plant tissues. Biosynthesis of these compounds occurs through the universal system for isoprene compounds, and it involves the formation of mevalonic acid from acetyl-CoA. Most carotenoids are made up of eight isoprene units (C_{40}). Both biosynthesis and catabolism occur in detached plant parts, and these reactions may be influenced by storage conditions such as oxygen concentration, light, and temperatures. Carotenoid metabolism is affected by hormones, such as ethylene and abscisic acid. Some fruits and vegetables, such as mandarin orange, accumulate significant amounts of apocarotenoids (C_{30}). The most common catalytic agents for carotenoid destruction appear to be

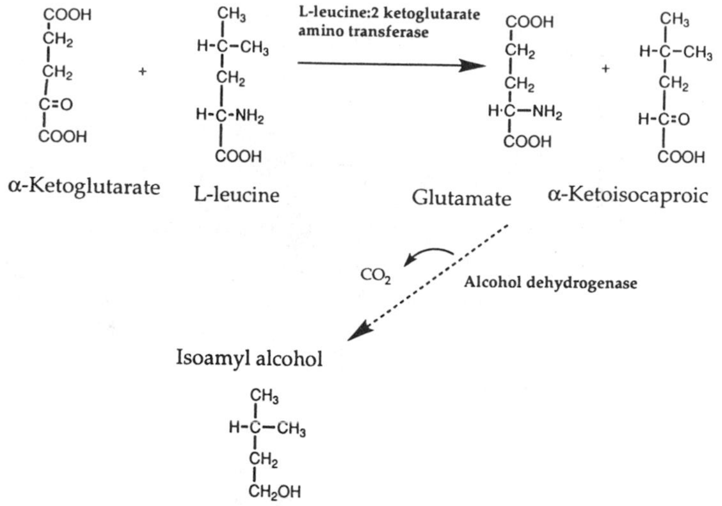

L-leucine:2 ketoglutarate amino transferase

α-Ketoglutarate + L-leucine → Glutamate + α-Ketoisocaproic

CO_2

Alcohol dehydrogenase

Isoamyl alcohol

FIGURE 27 Biosynthesis of a tomato flavor compound from an organic acid and an amino acid.

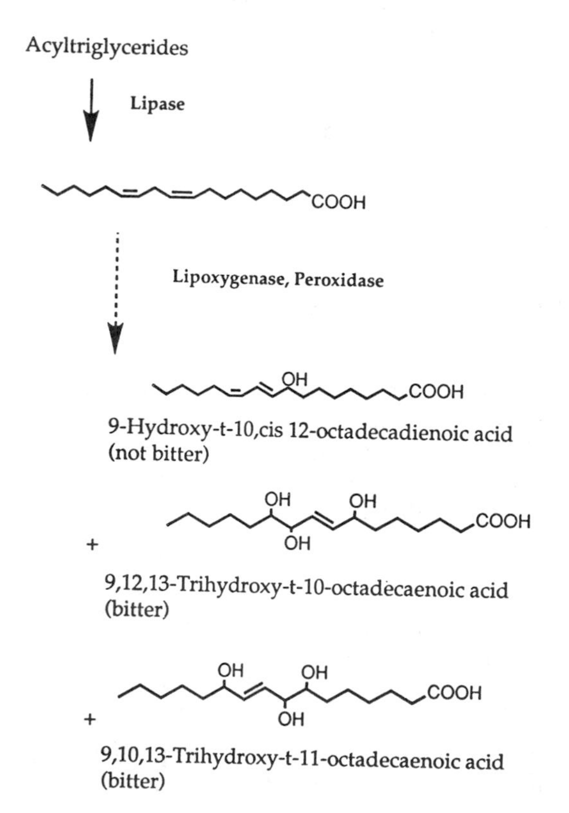

FIGURE 28 Formation of bitter trihydroxy fatty acids from acyltriglyceride in processed oats.

the lipoxygenases (indirectly via lipid oxidation) and the peroxidases (apparently promote β-carotene degradation directly).

The pathway for synthesis of the basic C_6:C_3 structure that comprises the A phenyl ring of the flavonoids is shown in Fig. 26. Relatively little is known about reactions leading to formation of specific flavonoid pigments, although there have been recent advances in this field. Synthesis of anthocyanins, colorants related to flavonoids (Chap. 9), occurs in postharvest plant organs and is stimulated by light and influenced by temperature. The purple anthocyanin pigments of red cabbage are synthesized and accumulate in cabbage stored below 10°C. Preharvest treatments, such as application of *N*-dimethylaminosuccinamic acid, can induce early formation of anthocyanins in certain crops.

Catabolism of anthocyanin pigments in situ is not very well understood. In processed foods—for example, during storage of frozen tissue that has not been blanched—anthocyanins can be co-oxidized by degradation products resulting from polyphenol oxidase or peroxidase-catalyzed oxidation of phenolic compounds. Decolorization of anthocyanins normally occurs following deglycosylation, and this is catalyzed by endogenous or fungal glycosidases called anthocyanases.

One of the more obvious changes that occurs in senescing plant tissues containing chlorophyll is the loss of a characteristic green color. Peel and sometimes pulp degreening is also associated with ripening of most fruit, and "yellowing" is a characteristic of senescing stem and leaf tissues consumed as vegetables. Chlorophyll degradation in situ is accompanied by synthesis of other pigments in several kinds of detached plant tissues. Chlorophyll catabolism is markedly influenced by environmental parameters, such as light, temperature, and humidity, and the effects of these factors are specific for the tissue. For example, light accelerates degradation of chlorophyll in ripening tomatoes and promotes formation of this pigment in potato tubers.

Chlorophyll degradation in living plant tissues often can be promoted by application of parts per million levels of the hormone ethylene. It is common commercial practice to utilize this principle for degreening citrus fruit and other commodities such as banana and celery. Chlorophyllase emergence in plant tissues is associated with the buildup of endogenous ethylene. However, the relationship between chlorophyllase action and degreening is not clear. Chlorophyllase is a hydrolytic enzyme that converts chlorophylls a and b to their respective chlorophyllides (see Chap. 9). Present evidence suggests that chlorophyllase does not catalyze the initial step in chlorophyll degradation in senescing plant tissues. The initial steps in this process appear to require molecular oxygen, and removal of the phytol side chain occurs in later steps. Lipoxygenase and peroxidase (chlorophyll: H_2O_2 oxidoreductase) are known to contribute to the loss of chlorophyll in senescing plants as well as in frozen vegetables. Chlorophyll conversion to pheophytin is acid catalyzed (Chap. 9). This reaction is of primary importance in heat-processed foods, but it also can occur in living vegetables when they are stored in a carbon dioxide-enriched atmosphere (Sec. 16.6.3).

16.4.3.6 Aroma Compounds—Biogenesis and Degradation

The characteristic flavor of fruits and vegetables is determined by a complex spectrum of organic compounds that form during maturation, senescence, and also sometimes during processing or wound injury (Sec. 16.5.1). Aroma compounds in fruits and vegetables are generally a complex mixture of esters, aldehydes, ketones, terpenes, and others (Chap. 10). For example, more than 300 volatile compounds contribute to strawberry flavor. In some fruits and vegetables a single compound or class of compounds has the characteristic aroma, such as, 2,6-nonadienal in cucumber, amyl acetate in banana, sulfides in onion, and phthalides in celery. Biosynthesis of fruit and vegetable flavors involves many different reactions that can include the metabolism or aromatic and S-containing amino acids (e.g., Figs. 18 and 27), carbohydrates and their derivatives (Fig. 29), and unsaturated fatty acids (Fig. 28; Sec. 16.4.4).

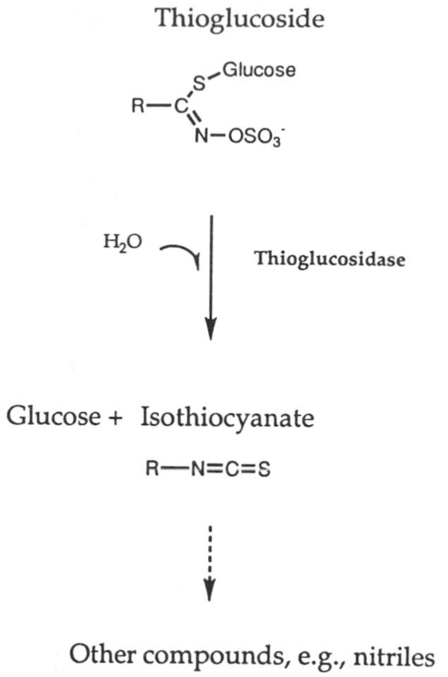

Thioglucoside

$$R-C(=N-OSO_3^-)-S-Glucose$$

H_2O → Thioglucosidase

Glucose + Isothiocyanate

$$R-N=C=S$$

↓

Other compounds, e.g., nitriles

FIGURE 29 Biosynthesis of pungent mustard oils, characteristic of the *Cruciferae* family (Table 1).

16.5 CONTROL MECHANISMS

Chemical changes in postharvest fruits and vegetables are subject to a wide array of biochemical control mechanisms. The physiological orchestration of chemical change at the cellular level involves several strategies. These include biological clocks and environmental factors that elicit gene expression, metabolic feedback and feed-forward sites in pathways, changes in cell compartmentation of reactants, and inhibitors and activators of enzyme-catalyzed reactions. In addition to cellular control mechanisms, factors like temperature, oxygen concentration, pH, and cellular integrity can have an important influence on chemical changes in postharvest and processed plant tissues. The mere presence of an enzyme does not alone indicate it functions in association with a physiological event. For example, invertase is present in potato tubers, but its activity is sometimes arrested by a proteinaceous invertase inhibitor. Flavor precursors are not necessarily converted to flavor compounds in situ, despite the presence of appropriate flavorases, since the reactants may be physically separated within the cell microstructure. Similarly, subtle differences in microconstituents, such as inorganic ions, can profoundly influence the direction of metabolism by modifying the activity of a particular enzyme. For example, calcium ions appear to influence association–dissociation of peroxidase from cell walls and thereby limit peroxidase-catalyzed reactions associated with lignification of cell walls. Control factors for enzyme-catalyzed reactions are discussed in more detail in Chapter 6.

16.5.1 Cell Disruption

The importance of cell disruption in the biochemistry of processed fruits and vegetables cannot be over emphasized [77]. In particular, cell disruption influences the formation of flavors, off flavors, and browning reactions.

16.5.1.1 Flavor Precursors

In certain tissues, such as *Allium* species and *Cruciferae,* flavor precursors are enzymically converted to flavor compounds when the cells are disrupted by chewing or other means of mechanical injury (Figs. 18 and 29). A patented process exists whereby flavor-potentiating enzymes are added to processed foods, just prior to consumption, to regenerate the fresh flavor that was lost during heating or dehydration. This method takes advantage of the fact that precursor molecules may be stable to processing, whereas flavor compounds and enzymes are often labile. Lipoxygenase cascades also give rise to "green," "grassy," "melon-like," and other aromas. The spectrum of compounds formed varies with different crops and depends on type of fatty acid(s) and on positional specificity of lipoxygenases, hydroperoxide lyases, and/or peroxidases (Sec. 16.4.4; Fig. 29). The presence of flavor precursors and their conversion to flavor compounds also has an important bearing on characteristic flavors that may develop as a consequence of different cooking and food preparation techniques (see also Chap. 10).

16.5.1.2 Tea Fermentation

Black tea is made by withering leaves to reduce the moisture content from 75% to ~60%, and rolling the withered leaves to gently crush the cells and attain release and uniform distribution of enzymes, notably polyphenol oxidase (Sec. 16.2.5.4). The resulting cell disruption leads to a complex enzymatic and nonenzymatic cascade that includes oxidation of phenolics (~30% of tea leaf solids) by polyphenol oxidase. The disruption also leads to enzymic-catalyzed protein hydrolysis, yielding amino acids, enzymic-catalyzed transamination of amino acids to form keto acids, degradation of chlorophyll (Sec. 16.4.5.3), and lipoxygenase cascades that result in co-oxidation of carotenoids to form various aromatic compounds. *o*-Quinone products of the polyphenolase reaction also participate in a variety of coupled reactions that lead to formation of characteristic tea flavors and colors.

16.5.1.3 Enzymatic Browning

Although enzymatic browning is a desired reaction in processes like tea fermentation, drying of dates, and cocoa fermentation, it is a cause of discoloration when other tissues are subject to cutting, bruising, disease, freezing, or other causes of cell disruption. Substrates for this reaction include simple phenols (e.g., catechol and gallic acid), cinnamic acid derivatives (e.g., chlorogenic acid and dopamine; Fig. 16.5), and flavonoids (e.g., catechin and epicatechin, Fig. 12). These reactions are discussed in detail elsewhere (Chap. 6). The propensity of a given tissue to brown varies considerably from one cultivar of a crop to another. These differences appear to relate to variations in enzyme content and sometimes to the kinds and amounts of phenolic substrates present. Methods used by the food processor to minimize this reaction include exclusion of oxygen, application of acidulants, heat inactivation (blanching), and use of inhibitors like sulfites.

16.5.1.4 Calcium Redistribution

Moderate heating, for example, at 60–70°C, of tissues causes disruption of membranes and decompartmentation of cell constituents. For some vegetables (e.g., potatoes, green beans) mild heating leads to firming of texture. Heat-induced firming is believed to start with damage to cell membranes that causes an increase in permeability. This leads to (1) liberation of Ca^{2+} (e.g., from starch phosphate in potato) and its diffusion to the cell wall/middle lamella, (2) activation of pectin methylesterase and deesterification of pectin, and (3) formation of Ca^{2+} (and/or Mg^{2+}) ionic cross-linkages between carboxyl groups of pectin. Ca^{2+} redistribution may also be involved in the "hard to cook" problem of legume seeds. Although not proven [14], extended storage of beans at high temperature and relative humidity may allow phytate to catalyze conversion of calcium phytate to myoinositol (Fig. 4) and the migration of free Ca^{2+} to the middle lamella, where it causes cross-linkages to form. Other studies have suggested that postprocess softening of canned apricots is caused by relocation of structural calcium from the cell wall to organic acids, notably citric acid [25].

16.5.2 Plant Hormones

Specific mention should be made of the plant hormones, which are important regulators of physiological events in detached plant organs. Of the five major categories of plant hormones (four are shown in Fig. 30), ethylene has received the most attention by postharvest physiologists. The ability of this agent to "trigger" ripening of fruit and generally promote plant senescence in plant tissues has led to the descriptive name "ripening hormone." Ethylene can also induce sprouting in potato tubers and is also described as a "wound hormone" because stresses like physical or chemical damage, fungal elicitation, and γ-irradiation stimulate ethylene production. Ethylene formation following wounding results from stimulation of existing mRNA for 1-aminocylopropane 1-carboxylic acid (ACC) synthase [46] (Fig. 31). Because ethylene biogenesis is autocatalytic, it is sometimes necessary, to avoid early fruit ripening, to remove even trace quantities of this gas from the storage environment. Climacteric fruit show a large increase in ethylene production during their ripening. It has long been recognized that ethylene acts to stimulate respiration, and this is of both practical and theoretical interest (Sec. 16.4.1; Fig. 23). Stimulation of respiration increases heat output of produce in storage and hence increases refrigeration requirements. The development of transgenic plants that do not synthesize ethylene [62, 64] has unequivocally shown that ethylene is a trigger rather than a by-product of ripening. Ethylene is widely used commercially to ripen bananas and tomatoes and to "degreen" oranges. Gibberellin, which increases amylase production in grains, is the only other plant hormone that has commercially important use with postharvest foods.

16.5.2.1 Ethylene Biosynthesis

Ethylene biogenesis in plant tissues involves conversion of *S*-adenosylmethionine (SAM) to 1-aminocyclopropane 1-carboxylic acid (ACC) by the enzyme ACC synthase, and the subsequent conversion of ACC to ethylene by ACC oxidase. Methionine is regenerated by the methionine cycle [93] (Fig. 31). ACC oxidase has been purified [66] and its gene has been cloned [32]. The cyanide that is produced in conjunction with ethylene is detoxified in a reaction catalyzed by cyanoalanine synthase [49].

$$\text{L-Cysteine} + \text{cyanide} \xrightarrow{\text{cyanoalanine synthase}} \beta\text{-cyanoalanine} + H_2O$$

16.5.2.2 Ethylene Action

The membranes of specific organelles, as well as the plasma membrane, may be sites of ethylene action. However, determining the specific mechanism or mechanisms by which ethylene gas exerts its physiological effects remains a challenging problem to postharvest physiologists. The structural requirements for biological activity of ethylene suggest that metal ion is present in the ethylene receptor site [1]. Cu(I) can apparently serve this role (Fig. 32). It has been postulated that bound ethylene is converted to ethylene glycol in the presence of oxygen and water.

16.5.2.3 Interaction of Ethylene with Other Hormones

Fruit ripening is not always accompanied by a clearly defined increase in ethylene to threshold levels. The efficacy of exogenous ethylene to stimulate such events as ripening can depend on the receptiveness of the tissue. For example, the concentration of ethylene required to initiate fruit

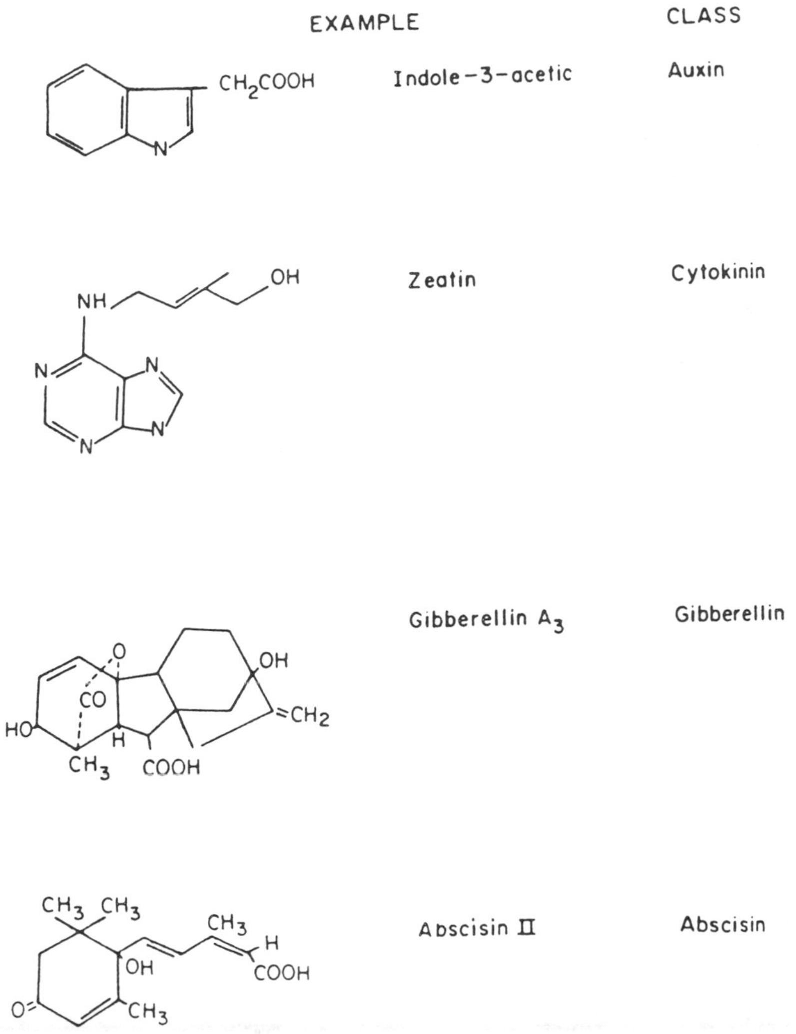

EXAMPLE	CLASS
Indole-3-acetic	Auxin
Zeatin	Cytokinin
Gibberellin A_3	Gibberellin
Abscisin II	Abscisin

FIGURE 30 Some plant hormones and their structures. Other hormone and hormone-like substances such as jasmonic acid, have recently been identified in plant tissues. (Sec. 16.3.4).

ripening declines as the fruit approaches full growth. Similarly, fruit attached to the plant (e.g., banana and avocado) display less sensitivity to ethylene than detached fruit. Also, certain cultivars of pear are sensitized to ethylene only after the harvested fruit are stored at low, nonfreezing temperatures. These observations indicate that "juvenility factors" desensitize fruit to ethylene and may otherwise control physiological events in immature fruits and vegetables. Application of exogenous sources of hormones and plant growth regulators, including auxins, cytokinins, or gibberellins, can simulate the action of juvenility factors in desensitizing plants to exogenous or endogenous ethylene.

There is also evidence that the abscisins, which play a role in leaf or organ abscision from the parent plant, also act to promote other senescence phenomena such as fruit ripening. As with other physiological events, such as seed germination, it appears that plant hormones act in concert to control physiological events in postharvest commodities. Although it is clear that hormones frequently exert an influence on gene expression, it has not yet been possible to establish a definite corollary between a control site of DNA metabolism or protein synthesis and hormone action.

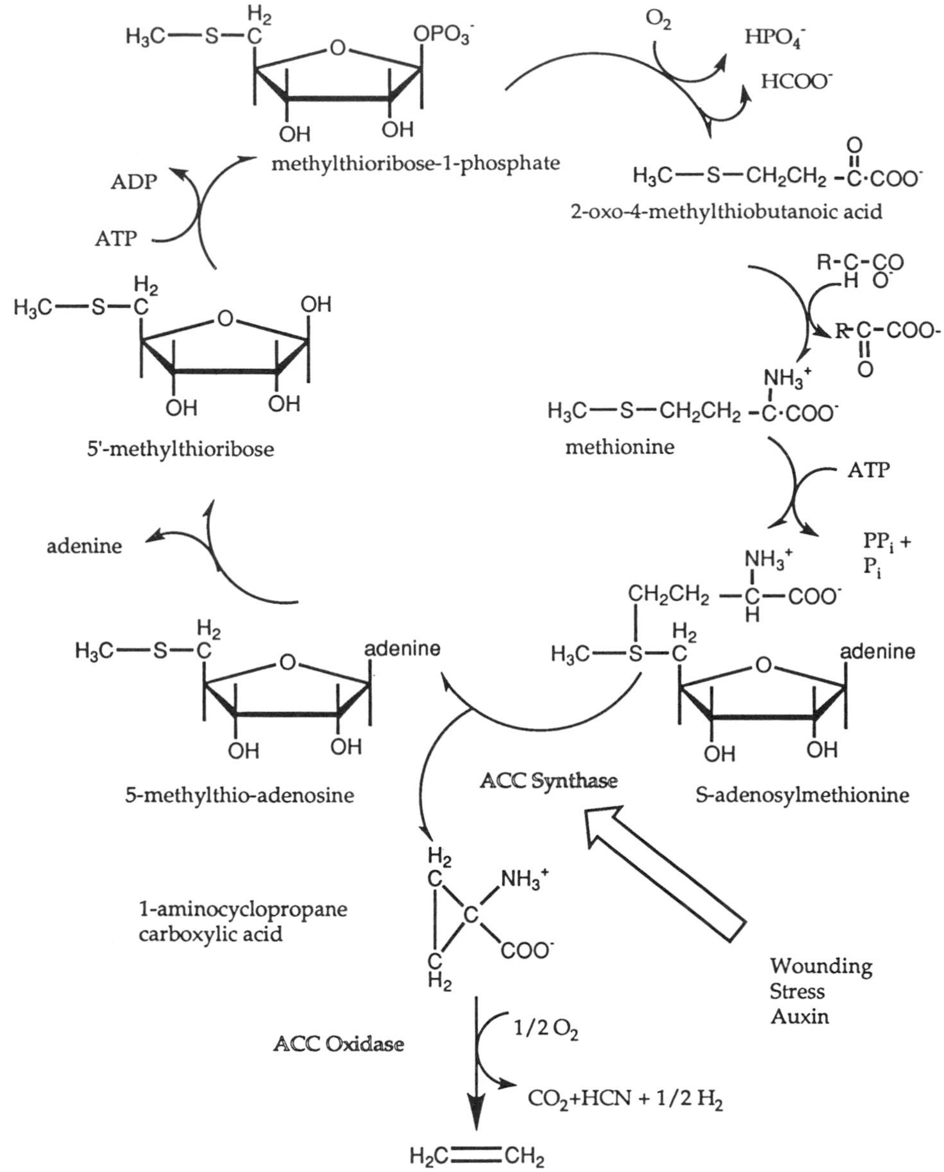

FIGURE 31 Biogenesis of ethylene. The amino acid methionine is recycled and is the carbon source for the key intermediate 1-aminocyclopropane carboxylic acid (ACC). Ethylene biosynthesis in fruits and vegetables can be decreased by incorporation of antisense rDNA for ACC synthase or ACC oxidase into the plant genome (Sec. 16.4.3.1).

16.5.2.4 Ethylene and Other Physiological Processes in Postharvest Crops

One of the important general effects of ethylene is to bleach the green color of leafy vegetables. A controlled atmosphere helps retain the green color of vegetables, presumably through competitive inhibition of ethylene action by the elevated concentration of carbon dioxide, and inhibition

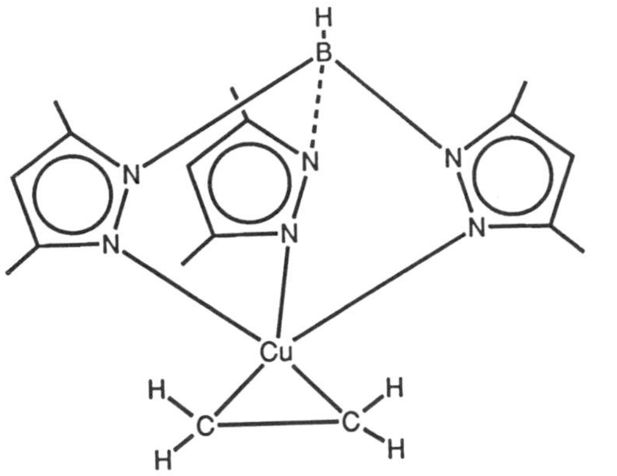

FIGURE 32 Interaction of Cu(I) with potassium hydrotris (3,5-dimethyl-1-pyrazolyl) borate, producing a neutral complex with ethylene.

of ethylene synthesis by the reduced concentration of oxygen. There is little evidence that ethylene directly affects pigments other than chlorophyll.

Ethylene also stimulates formation of isocoumarin and bitter taste in carrots, lignification of *Brassica* species and asparagus, production of phytoalexins in diseased crops, production of stress metabolites in sweet potato and white potato, onset of russet spotting in lettuce, and onset of hard core in sweet potato. It also reduces the incidence of chilling injury in muskmelons, and in cabbage it accelerates losses of weight and sugar and promotes changes in organic acids.

16.6 HANDLING AND STORAGE

16.6.1 Cereals, Pulses, and Oilseeds

Unlike fruits and vegetables, cereals, pulses, and oilseeds are relatively stable during storage, provided the moisture level is below a critical value—typically about 14%, but this varies somewhat with the crop. Above this moisture content, respiration, mold growth, and sprouting become problems. The higher the moisture content above the critical value, the greater are the problems. Changes in quality of these crops during storage under proper conditions are quite subtle and sometimes may not be obvious. The hard-to-cook phenomenon in dried beans, which results in beans that need substantially more cooking time than normal beans, is the result of membrane changes that occur during storage at higher than optimum humidities [74].

Milling, crushing, or extraction greatly reduces storage stability of cereals, pulses, and oilseeds. Destruction of cellular compartmentation results in mixing of enzymes and substrates and exposure of cellular contents to oxygen and microorganisms. For products with a moisture content yielding optimum overall stability, the rate of lipid oxidation (see Chap. 4) is the limiting factor for storage. The quality of wheat flour for baking increases after milling due to oxidative changes in the protein, but once optimum quality is achieved, further storage causes a slow reduction in quality [67].

16.6.2 Fruits and Vegetables

Throughout its life cycle, the intact plant must respond to various physical and chemical challenges from the environment. A plant's reaction to stress or environmental change are normally in the direction of protection, for example, prevention of water loss, adaptation to temperature extremes, or development of physical and chemical barriers to pathogens. When the challenge is such that the tissue cannot successfully adapt, the response may be one of hypersensitivity manifested as gene-directed death of particular groups of cells. It is important to recognize that species of plants may differ markedly in response to physical, chemical, and biological challenges. This is important because it means that no single condition or treatment is ideal for extending the storage lives of all fruits and vegetables. The following sections deal with the influence of physical and chemical factors on the physiological behavior and storage life of fresh fruits and vegetables.

16.6.2.1 Mechanical Injury

Wounding of plant organs generally results in a temporary, localized burst of respiration and cell division, formation of ethylene, a rapid turnover of certain cellular constituents, and sometimes accumulation of specific secondary metabolites or products that appear to have a protective function. In flesh organs, wounded cells commence synthesis of messenger RNA and proteins. Although the impact of wound physiology on plant food commodities is not fully understood, sufficient examples exist to illustrate the general and practical significance of these events.

16.6.2.2 Protective Substances

The response of plant organs to stress may be accompanied by "wound healing," that is, formation of a physical barrier of protective substances. Formation of a waxy or suberized barrier, or of a lignified layer, is important in certain commodities since these barriers can effectively prevent invasion by saprophytic microorganisms and subsequent spoilage. Accordingly, following harvest of certain crops, such as potatoes, it is desirable to cure or store the crop under environmental conditions conducive to rapid wound healing. These conditions usually differ from those yielding maximum storage life after wound healing is complete. Wound healing has become of increasing concern with the advent of mechanical harvesting techniques, since these procedures can result in extensive surface abrasion. In certain crops, such as asparagus, the cut injury associated with harvesting may be especially undesirable since the resulting stimulation of lignin deposition is directly associated with the losses in tenderness and succulence.

16.6.2.3 Stress Metabolites

Wound injury can also lead to synthesis and accumulation of a diverse family of substances called "stress metabolites" [29]. In certain cases, accumulation of a stress metabolite occurs only when mechanical injury is accompanied by other conditions such as the presence of ethylene (sec. 16.4.1.4) or exposure to chilling temperatures. Stress metabolites encompass a broad family of chemical compounds, including isoflavonoids, diterpenes, glycoalkaloids, and polyacetylenes (Fig. 33), and they appear to serve a protective function for the plant by virtue of their antibiotic properties. However, they may also detract from food quality because of their bitter taste (e.g., coumarins in carrot root, furanoterpenes in sweet potato root, or glycoalkaloids in white potato tuber) and because some have toxic properties. Certain stress metabolites are also precursors of enzymic browning reactions.

16.6.2.4 Temperature

Some of the reactions, such as respiration, that take place in postharvest crops are essential for maintenance of tissue integrity, and some are more or less incidental to this primary purpose. The latter category of reactions may, however, influence food quality in desirable (e.g., flavor biogenesis) or undesirable (e.g., loss of free sugars) ways. These reactions may have slightly different temperature dependencies, thus making the influence of temperature change on overall quality quite complex. Moreover, other tissue attributes that indirectly influence biochemical

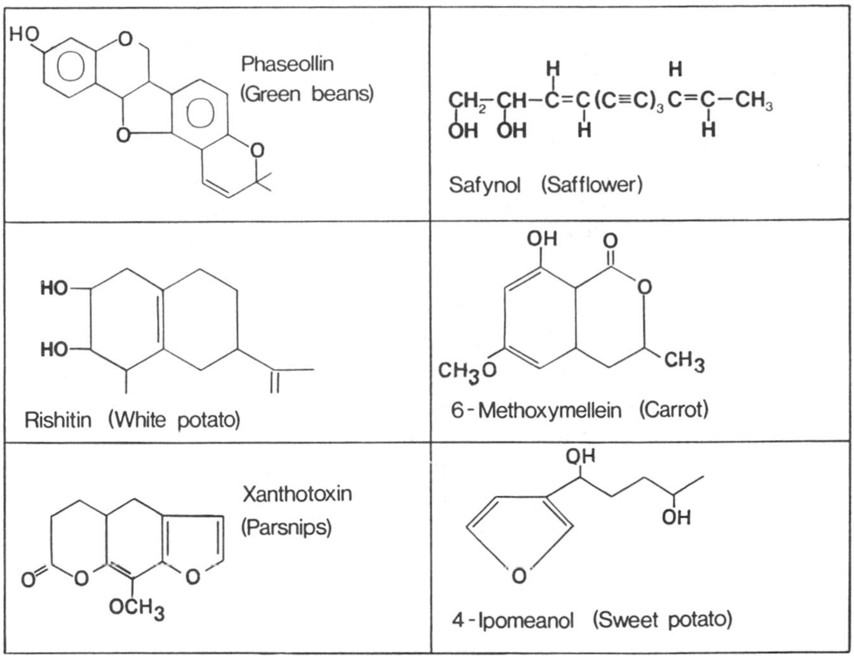

Phaseollin (Green beans)

Safynol (Safflower)

Rishitin (White potato)

6-Methoxymellein (Carrot)

Xanthotoxin (Parsnips)

4-Ipomeanol (Sweet potato)

FIGURE 33 Examples of stress metabolites identified in edible plants. The compounds shown are representative of many other compounds that accumulate in stressed plant tissues and are called "phytoalexins" (to ward off) by plant pathologists because of their antibiotic activity. In sufficient quantities some of these compounds are toxic to mammals.

transformation, such as membrane integrity and oxygen solubility and diffusivity, are influenced by temperature. Within certain limits of temperature change, vegetative tissues accept the resulting balances, and though the metabolism alters, there is no obvious adverse reaction to this stress. Within these temperature limits, changes associated with ripening, senescence, or breaking of dormancy proceed normally, albeit at different rates. Hence, lowering of temperature can retard some undesirable physiological reactions as well as spoilage caused by growth of microorganisms. Generally, commodities that have relatively fast respiratory rates respond best to lowered temperature, provided the nonfreeezing temperature remains above that which provokes low-temperature injury in susceptible commodities. In such crops, precooling prior to shipping or storage can greatly extend storage life.

16.6.2.5 Low-Temperature Injury

Exposure of a susceptible commodity to a temperature outside the tolerable range for more than short periods causes injury and a decrease in both quality and shelf life. The tolerable range varies significantly depending on the commodity; for example, for some the range may be 0–35°C and for others 15–35°C. The disorders known as "low-temperature injuries" or "chilling injuries" (CI) occur in some plant tissues at temperatures above the freezing point of the tissues but below the minimum recommended storage temperature [51]. CI can result in substantial postharvest losses. The manifestations of CI vary with the crop, are numerous and include surface lesions, water-soaking of tissues, woolly texture, internal discoloration, failure of fruits to ripen, accelerate senescence, increased susceptibility to decay, and compositional changes that effect flavor. Some plant tissues, such as pear and lettuce, do not normally undergo CI at any temperatures above freezing, but for most warm-season crops (sweet corn is an exception) and several temperate crops (e.g., asparagus, cranberries, apples, and potatoes) there is a critical, nonfreezing temperature below which chilling injury occurs. For a given species or cultivar, sensitivity to CI can be influenced by growing conditions and stage of physiological development. The extent of CI is dependent on both temperature and duration of exposure. Recovery is known to take place from short exposures to potentially harmful temperatures. In some cases, intermittent warming, controlled atmospheres, and hypobaric storage can help prevent symptoms of CI.

Whether or not there is a universal biochemical mechanism of CI is not known. CI is, nonetheless, generally attributed to an imbalance in metabolic reactions resulting in underproduction of certain essentials and overproduction of metabolites that become toxic to the tissue. For example, accumulation of ethanol, acetaldehyde, and oxaloacetic acid are associated with CI of certain fruits. Alteration of the integrity and permeability properties of biological membranes has been proposed as the primary evidence of injury. There appears to be a relationship between sensitivity to CI and fluidity of membranes as temperature is reduced. These changes in cellular membranes have been observed to occur coincidentally with the onset temperatures for CI in intact plant organs. In some instances these phase changes may cause an alteration in the flow of metabolites through the membrane. In other instances, there is a direct relationship between a membrane phase change and the rate of a specific enzymic reaction. Oxidation of succinic acid via membrane-associated mitrochondrial enzymes is coupled to the electron transfer chain and respiratory metabolism. It is therefore reasonable to suggest that the influence of temperature on the fluidity of biological membranes is one factor that can cause an imbalance in cellular metabolism. An increase in unsaturation of membrane phosphatidylglycerol fatty acids has been achieved by over expression of glycerol-3-phosphate acyltransferase, and this has been shown to result in reduced incidence of CI [59]. In view of the diverse ways in which CI can occur in postharvest crops, it is likely that the specific mechanisms underlying this disorder may vary in different tissues.

16.6.2.6 High-Temperature Injury

Exposure of plant tissues to abnormally high temperatures can also produce characteristic injuries in specific commodities. For example, storage of many fruits above 30°C results in failure to ripen normally, and exposure of apples to 55°C for a few minutes gives the fruit an appearance similar to that resulting from CI.

16.6.3 Controlled Atmospheres

Controlled atmosphere (CA) storage usually refers to storage in which the composition of the atmosphere has been altered with respect to the proportions of oxygen and/or carbon dioxide, and in which the proportions of these gases are carefully controlled, usually within ±1% of the desired value. Modified atmosphere storage does not differ in principle from CA, except that control of gas concentration is less accurate. For example, carbon dioxide may be allowed to simply accumulate in railroad cars, either by normal respiratory activities of plant tissue of by sublimation of dry ice (solid CO_2) used for cooling.

Hypobaric storage is another form of controlled atmosphere storage in the sense that the partial pressure of the gaseous environment is precisely controlled by reducing the pressure, and by continuously administering small amounts of gas (usually air). Typical pressure conditions prevailing in the chamber are 10–30 kPa air. In addition, hypobaric storage creates additional benefits by providing a saturated relative humidity, and by increasing diffusivity of gases within the internal spaces of the tissue.

The use of CA for maintenance of high-quality produce during storage and transportation grew out of research by Kidd and West [15], who discovered over 70 years ago that certain varieties of apples remained in better condition in atmospheres that contained less oxygen and more carbon dioxide than air. Kader [41] described three levels of barriers to gas exchange between the product and air: (1) the structure of the commodities dermal system and added surface coatings, (2) packages with semipermeable properties, and (3) the gas tightness of storage rooms or transit vehicle (determines composition of the atmosphere in these chambers) (Fig. 34).

16.6.3.1 Benefits

The efficacy of CA techniques for extending storage life has been demonstrated for many crops, but their commercial use has been somewhat limited, either for economic reasons or because rapid turnover of a commodity in the marketplace lessens the need for improved storage procedures. Conventional CA storage (large facilities) is used routinely for apples and pears, and use of modified atmospheres in retail packages has increased. The latter approach is achieved by product generated gases and the use of packaging materials with desired gas permeability.

Hypobaric storage has been employed successfully by commercial growers of cut flowers, but at the time of this writing it has not been employed commercially for the storage or transportation of fruits or vegetables.

Collectively, these techniques appear to retard respiration, reduce the autocatalytic production, accumulation, and activity of ethylene, and reduce decay caused by microorganisms. In effect, they generally retard ripening and senescence-related processes, such as degreening of vegetables.

16.6.3.2 Disorders

Closely related species, and even different cultivars of the same species, have different, very specific, and often unpredictable tolerances to low concentrations of oxygen and high concen-

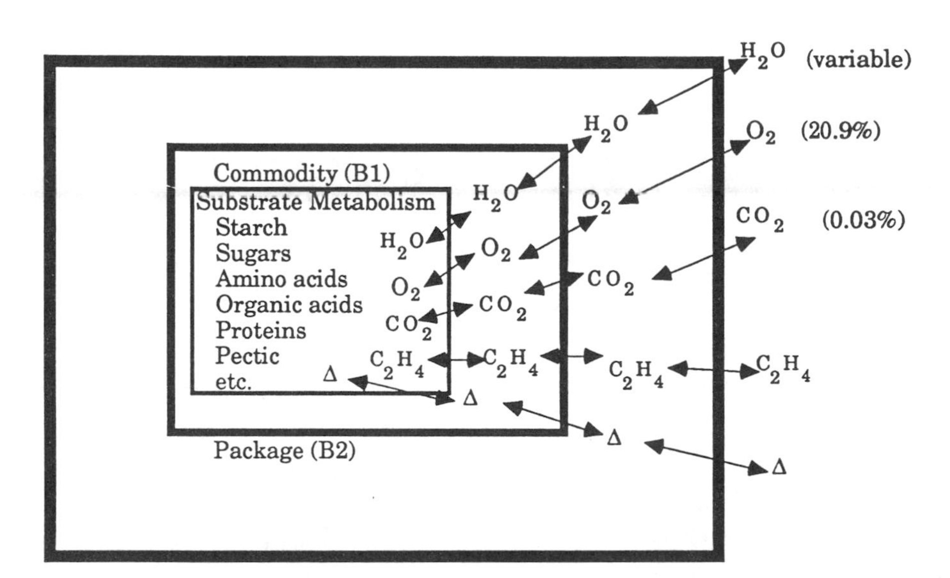

FIGURE 34 Model showing three levels of barriers to gas exchange in controlled atmosphere storage of a commodity. B1 = commodities dermal system and added barriers (e.g., waxing, film wrap, edible coating), B2 = packaging material or CA storage room, B3 = storage room or external environment. (Adapted from Ref. 41.)

trations of carbon dioxide. The maturity of the tissue at harvest and the temperature of the tissue during storage can also influence the response to CA conditions. With respect to maturity, McIntosh apples, when stored in an atmosphere containing 1% oxygen, exhibit less internal browning when immature than when mature.

If the composition of the storage atmosphere is not properly controlled it can have disastrous results, since the produce may develop physiological injuries. The degree of susceptibility to injury and the specific symptoms vary, not only among cultivars, but even among the same cultivars grown in different locations or different years in the same location. Some crops, such as sweet cherries, plums, and peaches, are not very susceptible to carbon dioxide injury and can be stored at carbon dioxide concentrations up to at least 20%, whereas other crops, such as lettuce, can show discoloration ("brown stain") when exposed to carbon dioxide concentrations of 2% or greater. Similarly, atmospheres containing large amounts of carbon dioxide and small amounts of oxygen interfere with wound healing of potato tubers. Thus, wounded potatoes, when exposed to CA immediately after harvest, decay very rapidly.

The mechanism by which physiological disorders are induced by extremes in atmospheric composition is not entirely clear. However, available evidence indicates that the same principles discussed in reference to CI also apply to carbon dioxide injury or damage caused by oxygen deprivation. Studies have shown that toxic levels of metabolites, such as succinic acid, ethanol, and acetaldehyde, accumulate in certain commodities prior to injury symptoms.

A further similarity to CI is that CO_2 injury in pears has been attributed to alterations in the structure and function of organelle membranes. It is, of course, very difficult to distinguish primary sites of injury from secondary consequences of primary action. However, it is reasonable to believe that alterations in respiratory metabolism and the inability of the tissue to cope with the resulting imbalances are a primary event in such injuries.

Although the final electron acceptor for respiratory metabolism (cytochrome oxidase) has a relatively high affinity for oxygen concentration, it is important to recognize that the gaseous concentrations in situ are not the same as those existing in the storage atmosphere. Thus, the diffusivity and solubility of a critical gas in the tissue may be determining factors. In view of recent findings that injury caused by a low oxygen concentration is minimal when carbon dioxide levels are retained at low levels, it appears that damage previously attributed to oxygen deprivation may, in fact, be caused by elevated levels of carbon dioxide. For example, apples can be successfully stored in an atmosphere consisting of only 1% oxygen and 99% nitrogen. Hybobaric storage also results in a large decrease in oxygen with an accompanying reduction in the partial pressure of carbon dioxide. Reports that hypobaric storage extends the storage life of many commodities beyond that possible with conventional CA storage (low oxygen and high carbon dioxide) may relate to the markedly different concentrations of carbon dioxide used in the two techniques. However, other differences also exist; namely, plant tissues stored under hypobaric conditions contain very low concentrations of ethylene, exhibit improved diffusion properties for gases, and are exposed to very high relative humidities.

16.6.4 Humidity

It is generally recommended that fruits and vegetables be stored at relative humidities sufficiently high to minimize water loss and to maintain cell turgor, but not so high as to cause condensation and accompanying microbial growth. The tendency of a commodity to lose moisture at a given relative humidity is related to the nature and amount of specialized cells at the tissue surface. Techniques that use an atmosphere saturated with water without condensation have been developed and employed successfully for vegetable storage in Canada and North European countries. Although such techniques are relatively expensive, they can be very effective in minimizing both physiological deterioration and microbial decay. Successful use of storage atmospheres that are saturated with water vapor requires a minimal temperature gradient throughout the storage chamber so that all atmospheric water remains in the vapor state.

Low relative humidity can influence physiological processes of postharvest crops. For example, when held in an atmosphere at 25% relative humidity, banana fruit fails to ripen and undergo climacteric respiration, whereas green tomatoes and pears exhibit a bimodal climacteric and intensification of ripening indices [40]. Atmospheric moisture levels can also influence the severity of chilling injury in certain fruits. Core breakdown and scald of apples is minimized by storage at low relative humidity, whereas CI of citrus fruits is lessened when fruits are wrapped in waxed paper to minimize moisture loss.

16.6.5 Ionizing Radiation

Low doses of ionizing radiation have been successfully applied experimentally and commercially to prolong the storage life of fresh fruits and vegetables. Inhibition of sprouting in potatoes, onions, and carrots, as well as destruction of insect pests in stored grains, can be accomplished with 0.1–0.25 kGy doses. Control of mold growth requires doses in excess of 1 kGy, and destruction of pathogenic bacteria necessitates doses of 5–10 kGy. Kilogram doses of ionizing radiation can temporarily or permanently prevent the onset of fruit ripening. Studies indicate that delay in ripening is associated with the uncoupling and eventual de novo biosynthesis by mitochondria. High doses of radiation, which irreversibly prevent normal fruit ripening, appear to interfere with the ability of the tissue to repair radiation damage. Low doses of irradiation cause a temporary burst in wound ethylene but they also inhibit the subsequent transcription-dependent stimulation of ethylene production and the activity of ACC oxidase [46].

Unfortunately, the quality of fruits and vegetables is usually adversely affected by doses of ionizing radiation in excess of 2 kGy. Moreover, low doses employed for sprout inhibition can interfere with wound healing, and this has created problems with commercial attempts to use this technique to prevent sprouting of potatoes.

The physiological changes induced by irradiation, especially at high doses, can bring about the death of cells. Irradiated commodities often show characteristic symptoms of damage. For example, apples exposed to 10 kGy exhibit a ring of necrotic tissues, not unlike core-flush, adjacent to the core. Citrus and banana exhibit peel lesions at doses as low as 0.25 kGy. Many commodities also exhibit extensive pulp softening, and this has been related to activation of pectic enzymes. Despite these difficulties, there appears to be renewed interest in the use of ionizing radiation for preservation of fresh fruits and vegetables. As with other techniques, the preceding limitations can probably be overcome when the influences of variables such as maturity, temperature, and gaseous environment are more fully controlled.

16.6.6 Other Factors

Postharvest treatments with fungicides, bactericides, and senescence inhibitors can be applied to fruits and vegetables to indirectly or directly minimize decay. Various experimental studies indicate that senescence inhibitors, such as hormone analogs (e.g., cytokinins and auxins) and respiratory inhibitors (e.g., benzimidazole and vitamin K analogs), are effective in prolonging the storage life of certain crops, although they are not approved for use commercially. Calcium treatment inhibits normal ripening of several fruits and prevents the development of various physiological disorders. Calcium in plants is involved in a number of fundamental physiological processes that influence the structure of cell walls, membranes, and chromosomes, and the activity of enzymes. In addition, some studies indicate that endogenous calcium may also serve an important regulatory function in fruit ripening. Little is known about how calcium exerts a beneficial effect when applied to detached plant tissues.

Exposure of fruits and vegetables to various other chemicals can influence physiological processes in an undesirable way. For example, postharvest commodities show different susceptibilities to air pollutants, and this can be of practical concern in urban areas. Automobile emissions, especially ethylene, can accelerate tissue necrosis at parts per million levels. The volatile components of photochemical smog (O_3, peroxyacyl nitrates, and NO_2) are known to create undesirable injury symptoms in a wide array of fruits and vegetables. Other stresses to plants are caused by SO_2 (e.g., from coal-fired plants for generating electricity) and by ammonia from leaks in refrigeration systems.

16.7 EFFECTS OF PROCESSING ON FRUITS AND VEGETABLES

The basic purpose of food processing is to curtail the activity of microorganisms and retard chemical changes that would otherwise adversely affect the edible quality of food, and to do so with minimal damage to quality attributes. Food processing techniques differ in principle from the preservation techniques described in the previous section in that viability of the living tissue cells is lost during processing. An important advantage of processing is that it enables food to be stored in edible condition for longer periods than would generally be possible by other means. However, because of the severity of these processes, the product usually undergoes rather extensive changes, and as a result the quality characteristics of a given commodity are often changed. Such alterations may be desirable (e.g., inactivation of antinutritional factors by heat, softening of hard or tough tissues, creation of flavors) or undesirable (e.g., loss of vitamins by heat, loss of color, texture changes, production of off flavors) and may be acceptable to consumers depending on local customs and eating preferences.

16.7.1 Chemical Changes

The rate of a chemical reaction in a foodstuff is a function of many factors, mainly reactant concentration, availability, and mobility; temperature; pH; oxidation–reduction potential; and inhibitors and catalysts. Therefore, the effects of processing on chemical changes in fruits and vegetables vary widely with the nature of the process and composition of plant tissue. Even within the same species, substantial differences may occur among different cultivars and for a single cultivar that has been grown or stored under different conditions. The reactions that occur in plant tissues are not unique to plants except for those that occur while the tissue is still "alive" and those that involve components that are specific to plants. Thus, chemical alterations in starch that occur during canning of whole potatoes are not fundamentally different from those occurring when potato starch is canned as a part of a soup. However, there are important physical differences that are related to whether or not the starch granule is a part of a complex cellular organization or simply dispersed in water and these physical differences may affect the rate and extent of chemical reactions. Some examples of reactions that occur during processing and storage of fruits and vegetables are shown in Table 14. Because of the complexity of fruit and vegetable tissues it is often difficult to predict the consequences of processing and storage conditions on chemical reactions that influence quality. The details of many of these chemical reactions can be found in the chapters dealing with the individual components.

16.7.2 Enzyme-Catalyzed Reactions

The effects of cellular disruption on chemical reactions in living plant tissues were discussed previously (Sec. 16.4.1). Even when foods are to be thermally sterilized, cellular disruption during peeling, size reduction, etc. before the enzymes are inactivated contributes significantly to the chemical changes that occur. Many processing sequences are designed to minimize these changes by reducing the time between disruption and enzyme inactivation. This is one of the functions of blanching prior to thermal processing or freezing. The "hot break" process for tomatoes, which inactivates polygalacturonase before significant losses of viscosity occur, is a good example of this type of strategy. However, in the processing of orange juice, some of the polygalacturonase are quite heat stable, and substantial changes in flavor occur when all of the polygalacturonase is inactivated by heat. Failure to inactivate polygalacturonase in orange juice results in an unstable "cloud" and a poor appearance.

16.7.3 Cell Disruption

Cellular disruption also contributes to nonenzymatic changes. Disruption of the central vacuole releases acid, which changes the pH of the tissue and alters the rates of numerous pH-dependent reactions. Color changes (see Chap. 10) are especially noticeable examples of this. Disruption may also increase the rate of nonenzymatic oxidation due to the increased concentration of oxygen. When freezing plant tissues, it is generally agreed that ice formation directly causes physical disruption of the relatively rigid plant cell structure and indirectly causes damage due to the concentration of solute molecules in the unfrozen phase (see Chap. 2). Tissues vary widely in their susceptibilities to these kinds of damage. Differences in freezing damage to a leaf or tomato as compared to that in a potato or wheat kernel illustrates this point. Due to the effects of temperature, rates of reaction following cellular disruption during freezing are quite different from those that occur following disruption at higher temperatures where enzymatic and chemical reactions are rapid.

16.7.4 Changes in Molecular Structures

Chemical reactions during processing are influenced in many ways other than just cell disruption. Thermal processing denatures proteins (see Chap. 6) and alters the interaction of proteins with lipids, water, etc. Inactivation of enzymes is essential for producing foods with long shelf lives. Also, membrane structures like chloroplasts are irreversibly altered by heat processing and the components (e.g., chlorophyll, lipids, proteins) become more available to participate in reactions. Destruction of the native environment surrounding pigments results in color changes.

Polysaccharides in plant cells are highly ordered in their native state, and the structural integrity of the cells is dependent on this order. Heating generally causes these large molecules to absorb water, to "swell," and to become more mobile and more available to participate in chemical reactions. Swelling may also disrupt the structure enough to cause changes in adjacent components. Starch granules are a familiar example of this, but cell walls undergo similar changes, although to a much less extent.

16.7.5 Other Factors Influencing Reaction Rates

In addition to structure-related changes discussed earlier, thermal processing has a wide range of effects on individual components, and these effects depend on the time and temperature of processing and the activation energies of the individual reactions. Examples of these reactions are discussed throughout this volume. Drying of plant tissues causes an increase in concentration

TABLE 14 Examples of Nonenzymic Reactions That Can Affect the Quality of Processed Fruits and Vegetables

Reactant(s)	Product(s)/result	Importance
Chlorophyll	Pheophytin	Conditions of low pH and high temperature favor this reaction, which results in olive-brown discoloration of canned green vegetables.
Glutamine	Pyrrolidone carboxylic acid	PCA formed during thermoprocessing of canned vegetables is believed to contribute to acid-catalyzed pheophytin formation and off flavor.
All-*trans*-carotenoids	*cis, trans*-Carotenoids	Heat, light, or dilute acid conditions cause isomerization of all-*trans*-carotenoids to various cis isomers. The reaction results in loss of vitamin A activity and changes in color.
Thiamin	Pyrimidine and thiazole	Heating at $pH < 6$ results in cleavage of the methylene bridge of vitamin B_1 to form indicated products.
Ascorbic acid	Dehydroascorbic acid, diketogulonic acid, etc.	Oxidation in the presence of molecular O_2 results in loss of vitamin C activity and may be coupled to other reducing reactions, such as disulfide bond formation during gluten development, or conversion of ethanol to acetaldehyde in wine aging.
Ascorbic acid	3-Deoxypentosulose, 2-furaldehyde	High-acid foods like lemon juice undergo nonoxidative degradation of ascorbic acid. The reaction appears to contribute to "ascorbate browning" of these products.
Glycosides	Hydrolysis products	Mild acid conditions and heat cause hydrolysis of glycosidic linkages. Phenolic glycosides are also hydrolyzed under mild alkaline conditions. Such reactions can influence the texture and flavor of fruits and vegetables.
Reducing sugar, amino acid	Maillard browning	Concentration of reactants, high temperature, and alkaline pH can favor this important reaction that can impact nutrition (adverse), color, flavor, and safety of processed fruits and vegetables.
Organic acid, Ca^{2+}	Ca^{2+} chelate	Phytic or citric acids can sequester Ca^{2+} from pectate and thereby cause softening of canned fruit.
Amylose	Crystallization	Alignment of linear starch chains can form insoluble precipitates that contribute to processes like bread staling.
Methyl-methionine	Dimethylsulfide	Thermal degradation of sulfur-containing compounds gives rise to many important flavor compounds.
Hydrogen sulfide	Discoloration	Iron sulfide appears as black spots both on the can and in potatoes.
Fe^{2+}	Fe^{3+}	Increased lipid oxidation and reduced iron absorption are two important consequences that can occur when iron is oxidized.
Organic acids	"Detinning"	Coating of cans with enamels greatly reduces solubilization of metal by acid.
Anthocyanin, SO_2	Decolorization	The addition of SO_2 to the 4-position of anthocyanins to form a bisulfite addition product causes decolorization.

of solutes, as well as migration of solutes, and, depending upon the type of the drying process utilized, may expose the tissue to high temperatures and oxygen. All of these changes favor increased rates of chemical reactions. Furthermore, migration and concentration of solutes can have a profound effect on macromolecules and dramatically alter the rehydration behavior of these products.

16.8 CONCLUSION

As the world population increases and utilizable agricultural lands diminish, the problems of food availability are certain to increase and botanical sources of food will tend to supplant animal sources. Furthermore, increased emphasis on consumption of plant tissues for general and specific health reasons is likely to substantially increase the demand for foods as derived from plants. There are many gaps in our knowledge regarding postharvest physiological attributes of edible plant tissues, and continued research is needed to improve our ability to better utilize postharvest crops and to deliver the highest quality plant foods to consumers. Totally new approaches to plant food production, quality improvement, and preservation will also evolve from basic research on plant physiology. Recombinant DNA technology offers the potential to dramatically improve the quality of may plant products, but these advances will require a better understand the biochemical determinants of quality.

BIBLIOGRAPHY

Eliasson, A.-C., and K. Larsson (1993). *Cereals in Breadmaking*, Marcel Dekker, New York.

Haard, N.F. (1984). Postharvest physiology and biochemistry of fruits and vegetables. *J. Chem. Educ. 61*:277–290.

Herregods, M. (ed.) (1989). *International Symposium on Postharvest Handling of Fruits and Vegetables*, Leuven, Belgium, August 29–September 2, 1988, International Society for Horticultural Science, Wageningen, The Netherlands.

Kays, S.J. (1991). *Postharvest Physiology and Crop Protection*, Plenum, New York.

Phan, C.T. (1984). *Symposium on Post-Harvest Handling of Vegetables* Montreal, Quebec, Canada, 3–8 June 1984, ISHS Working Group, Postharvest Handling of Vegetables.

Richardson, T., and J.W. Finley (eds.) (1985). *Chemical Changes in Food During Processing*, AVI, Westport, CT.

Robinson, T. (1991). *Organic Constituents of Higher Plants*, 6th ed., Cordus Press, North Amherst, MA.

Snowdon, A.L. (1989). *A Color Atlas of Post Harvest Diseases and Disorders of Fruit and Vegetables*, CRC Press, Boca Raton, FL.

Weichmann, J. (1987). *Postharvest Physiology of Vegetables*, Marcel Dekker, New York.

Wilson, C. L., and E. Chalutz (eds.) (1991). *Biological Control of Postharvest Diseases of Fruits and Vegetables*, Workshop proceedings, Sherpherdstown, WV, September 12–14, 1990, U.S. Department of Agriculture, Agricultural Research Service, Beltsville, MD.

REFERENCES

1. Ainscough, E. W., A. M. Brodie, and A. L. Wallace, (1992). Ethylene—An unusual plant hormone. *J. Chem. Education 69*: 315–318.
2. Amir, J. and Cherry, J. H. (1971). Chemical control of sucrose conversion to polysaccharides in sweet corn after harvest. *J. Agric. Food Chem.*:954.
3. Anderson, J. W., D. A. Deakins, T. L. Floore, B. M. Smith, and S. E. Whitis (1990). Dietary fiber and coronary heart disease. *CRC Crit. Rev. Food Sci. Nutr. 29*:95–147.
4. Ap-Rees, T.; Dixon, W. L.; Pollack, C. J. and Franks, F. (1981). Low temperature sweetening of higher

plants, in *Recent Advances in the Biochemistry of Fruits and Vegetables* (J. Friend and M. J. C. Rhodes, eds.), Academic Press, New York, pp. 41–60.

5. Autran, J. C. (1993). Recent perspectives on the genetics, biochemistry and functionality of wheat proteins. *Trends Food Sci. Technol. 4*:358–364.
6. Babic, I., M. J. Amiot, T. C. Nguyen, and S. Aubert (1993). Changes in phenolic content in fresh ready-to-use shredded carrots during storage. *J. Food Sci. 58*:351–356.
7. Ballentine, D. A. (1992). Manufacturing and chemistry of tea, in *Phenolic compounds in food and their effects on heatlh I Analysis, Occurrence & Chemistry* (C. T. Ho, C. Y. Lee, and M. T. Huang, and C. T. Ho, eds.), American Chemical Society, Wahsington, DC, pp. 102–117.
8. Barichello, V., R. Y. Yada, R. H. Coffin, and D. W. Stanley (1990). Low temperature sweetening in susceptible and resistant potatoes. Starch structure and composition. *J. Food Sci. 55*:1054.
9. Barichello, V., R. Y. Yada, R. H. Coffin, and D. W. Stanley (1990). Respiratory enzyme activity in low temperature sweetening of susceptible and resistant potatoes. *J. Food Sci. 55*:1060–1063.
10. Bartley, I. M., and M. Knee (1982). The chemistry of textural changes in fruit during storage. *Food Chem. 9*:47–58.
11. Beck, C. I., and T. Ultrich (1993). Biotechnology in the food industry. *Bio/Technology 11*:895–902.
12. Belitz, H.-D., and W. Grosch (1987). *Food Chemistry*, 1st English ed., Springer-Verlag, Berlin, p. 774.
13. Bennett, A. B., G. M. Smith, and B. G. Nichols (1987). Regulation of climacteric respiration in ripening avocado. *Plant Physiol. 83*:973–976.
14. Bernal-Lugo, I., G. Prado, C. Parra, E. Moreno, J. Ramirez, and O. Velazco (1989). Phytic acid hydrolysis and bean susceptibility to storage induced hardening. *J. Food Biochem. 14*:253–261.
15. Biale, J., and R. E. Young (1981). Respiration and ripening of fruits retrospect and prospect, in *Recent Advances in the Biochemistry of Fruits and Vegetables* (J. Friend and M. J. C. Rhodes, eds.), Academic Press, New York, p. 89.
16. Brady, C. J., P. B. H. O'Connel, J. K. Palmer, and R. M. Smille (1970). An increase in protein synthesis during ripening of the banana fruit. *Phytochemistry 9*:1037–1047.
17. Butler, L. G. (1988). Effect of condensed tannins on animal nutrition, in *Chemistry and Significance of Condensed Tannins* (R. W. Hemingway and J. J. Karchesy, eds.), Plenum Press, New York, pp. 343–396.
18. Carrington, C. M. S., L. C. Greve, and J. M. Labavitch (1993). Cell wall metabolism in ripening fruit. VI. Effect of antisense polygalacturonase gene on cell wall changes accompanying ripening in transgenic tomatoes. *Plant Physiol. 103*:429–434.
19. Cassab, G. I. and J. E. Varner (1988). Cell wall proteins. *Annu. Rev. Plant Physiol. Plant Mol. Biol. 39*:321–353.
20. Chung, O. K. (1991). Cereal lipids, in *Handbook of Cereal Science and Technology* (K. J. Lorenz and K. Kulp, eds.), Marcel Dekker, New York, Vol. 41, pp. 497–554.
21. Dunham, D. (1993). Food Cost Review, 1992, U. S. Department of Agriculture, Economic Research Service, Washington, DC.
22. Eliasson, A.-C., and K. Larsson, (1993). *Cereals in Breadmaking*, Marcel Dekker, New York, p. 376.
23. Englyst, H. N. V. Anderson,, and J. H. Cummings (1983). Starch and non-starch polysaccharides in some cereal foods. *J. Sci. Food Agric 34*:1434–1440.
24. Faith, N. G., A. E. Yousef, and J. B. Luchansky (1992). Inhibition of Listeria monocytogenes by liquid smoke and isoeugenol, a phenolic component found in smoke. *J. Food Safety 12*:263–314.
25. French, D. A., A. A. Kader, and J. M. Labavitch (1989). Softening of canned apricots: A chelation hypothesis. *J. Food Sci. 54*:86–89.
26. Gaser, C. S., and R. T. Fraley (1989). Genetically engineering plants for crop improvement. *Science 244*:1293–1299.
27. Glenn, G. W., and B. W. Poovaiah (1990). Calcium mediated postharvest changes in texture and cell wall structure and composition in golden delicious apples. *J. Am. Soc. Hort. Sci. 115*:962–968.
28. Guclu, R., A. Paulin, and P. Soudain (1989). Changes in polar lipids during ripening and senescence of cherry tomato (*Lycopersicon esculentum*): Relation to climacteric and ethylene increases. *Physiol. Plant. 77*:413–419.

29. Haard, N. F. (1983). Stress metabolites, in *Postharvest Physiology and Crop Protection* (M. Lieberman, ed.), Food and Nutrition Press. Westport, CT, pp. 433–448.
30. Haard, N. F. and H. O. Hultin (1968). An improved method for isolation of mitochondria from plant tissue. *Anal. Biochem. 24*:299–304.
31. Hamada, J. S. and W. E. Marshall (1989). Preparation and functional properties of enzymatically deamidated soy proteins. *J. Food Sci. 54*:595–601, 635.
32. Hamilton, A. J., M. Bouzayen, and D. Grierson (1991). Identification of a tomato gene for the ethylene-forming enzyme by expression in yeast. *Proceedings of the National Academy of Science 88*:7434–7437.
33. Haslan, E., T. H. Lilley, E. Warminski, H. Liao, Y. Cai, R. Martin, S. H. Gaffney, P. N. Goudling and G. Luck (1992). Polyphenol complexation: A study in molecular recognition, in *Phenolic compounds in Food and their Effects on Health I. Analysis, Occurrence & Chemistry* (C. T. Ho, C. Y. Lee, and M. T. Huang, eds.), American Chemical Society, Wahsington, DC, pp. 8–50.
34. Hayakawa, K. I. (1987). Temerature:energy flux in stores, in *Postharvest Physiology of Vegetables* (J. Weichmann, ed.), Marcel Dekker, New York, pp. 181–202.
35. Ho, C. T., C. Y. Lee, and M. T. Huang (1992). *Phenolic Compounds in Food and Their Effects on Health I. Analysis, Occurrence, & Chemistry*, American Chemical Society, Washington, DC.
36. Hubbard, N. L., D. M. Pharr, and S. C. Huber (1990). Role of sucrose phosphate synthase in sucrose biosynthesis in ripening bananas and its relationship to the respiratory climacteric. *Plant Physiol 94*:201–208.
37. Huff, A. (1984). Sugar regulation of plastid interconversion in epicarp of citrus fruit. *Plant Phsyiol 76*:307.
38. Ikemefura, J. and J. Adamson (1985). Chlorophyll and carotenoid changes in ripening palm fruit, *Elacis guineesis, Phytochemistry 23*:1413.
39. Isaac, J. E. and M. J. C. Rhodes (1987). Phosphofructokinase and ripening in *Lycopersicum esculentum* fruits. *Phytochemistry 25*:649.
40. Janick, J. R. W. Schery, F. W. Woods, and V. M. Ruttan (1969). Food and human needs, in *Plant Science* (J. Janick, R. W. Schery, F. W. Woods, and V. M. Ruttan, eds.), W. H. Freeman, San Francisco, pp. 37–52.
41. Kader, A. A. (1987). Respiration and gas exchange of vegetables, in *Postharvest Physiology of Vegetables* (J. Weichmann, Ed.,) Marcel Dekker, New York, pp. 25–43.
42. Klein, B. and S. Grossman (1985). Co-oxidation reaction of lipoxygenases in plant systems. *Adv. Free Radical Biol. Med. 1*:309.
43. Koch, J. L. and D. E. Nevins (1990). The tomato fruit cell wall. II. Polyuronide metabolism in a non-softening tomato mutant. *Plant Physiol. 92*:642–647.
44. Koch, J. L. and D. J. Nevins (1989). Tomato fruit cell wall I. Use of purified tomato polygalacturonase and pectinmethylesterase to identify developmental changes in pectins. *Plant Physiol. 91*:816–822.
45. Kockritz, A., T. Schewe, B. Hieke, and N. Haas (1985). The effect of soybean lipoxygenase-1 on chloroplasts from wheat. *Phytochemistry 24*:381.
46. Larrigaudiere, C., A. Latche, J. C. Pech and C. Triantaphylides (1989). Short term effects of γ-irradiation on 1-aminocyclopropane-1-carboxylic acid metabolism in early climacteric cherry tomatoes. *Plant Physiol. 92*:577–581.
47. Lesham, Y. Y. (1988). Plant senescence processes and free radicals. *Free Radical Biology and Medicine 5*:39–49.
48. Liyama, K., T. B.-T. Lam, and B. A. Stone (1994). Covalent cross-links in the cell wall. *Plant Physiol. 104*:315–320.
49. Lurie, S. and J. D. Klein (1989). Cyanide metabolism in relation to ethyelene production and cyanide insensitive respiration in climacteric and nonclimacteric fruits. *J. Plant Physiol. 135*:518–521.
50. Lurie, S., B. Shapiro, and S. Ben-Yehoshua (1989). Ethylene and cyanide insensitive respiration in ripening apples: The effect of calcium, in *Biochemical and Physiological Aspects of Ethylene Production in Lower and Higher Plant* (H. Clisters and R. Marcelle, eds.), Kluwer, Amsterdam, pp. 81–90.

51. Lyons, J. M. and R. W. Briedenbach (1987). Chilling injury, in *Postharvest Physiology of Vegetables* (J. Weichmann, ed.), Marcel Dekker, New York, pp. 305–326.
52. Macko, V. and H. Stegmann (1969). Mapping of potato proteins by combined electrofocusing and electrophoresis. *Hoppe-Seyler's Z. Physiol. Chem. 350*:917–919.
53. Maquire, Y. P. and Haard, N. F. (1975). Fluorescent product accumulation in ripening fruit. *Nature 258*:599–600.
54. Mateos, M., D. Ke, M. Cantwell, and A. A. Kader (1993). Phenolic metabolism and ethanolic fermentation of intact and cut lettuce exposed to CO_2-enriched atmospheres. *Postharvest Biol. Technol. 3*:225–233.
55. McGarry, A. (1993). Influence of water status on carrot (*Daucus carota* L.) fracture properties. *J. Hort Sci. 68*:431–437.
56. Mitcham, E., K. Gross, and T. Ng (1989). Tomato fruit cell wall synthesis during development and senescence. *Plant Physiol 89*:477–481.
57. Moreau, F., and R. Romani (1982). Malate oxidation and cyanide insensitive respiration in avocado mitochondria during the climacteric cycle. *Plant Physiol. 70*:1385.
58. Muller-Rober, B., U. Sonnewald, and L. Willmitzer (1992). Inhibition of the ADP-glucose pyrophosphorylase in transgenic potatoes leads to sugar-storing tubers and influences tuber formation and expression of tuber storage protein genes. *EMBO J. 11*:1229–1238.
59. Murata, M., O. Ishizaki-Nishizawa, S. Higashi, H. Hyashi, Y. Tasaka, and I. Nishida (1992). Genetically engineered alteration in the chilling sensitivity of palnts. *Nature 356*:710–713.
60. Newman, R. H., M. A. Ha, and L. D. Melton (1994). Solid state 13C NMR investigation of molecular ordering in the cellulose of apple cell walls. *J. Agric. Food Chem. 42*:1402–1406.
61. O'Donoghue, E. M., and D. J. Huber (1992) Modification of matrix polysaccharides during avocado fruit ripening: an assessment of the role of Cx-cellulase. *Physiol. Plant 86*:33–42.
62. Oeller, P. W., L. M. Wong, L. P. Taylor, D. A. Pike and A. Theologis (1991). Reversible inhibition of tomato fruit senescence by antisense RNA. *Nature 254*:437–439.
63. Pernollet, J. C. and J. Mosse (1983). Structure and location of legume and cereal seed storage proteins, in *Seed Proteins* (J. Daussant, J. Mosse, and J. Vaughan, eds.), Academic Press, New York, pp. 155–191.
64. Picton, S., S. L. Barton, M. Bouzayen, and A. J. Hamilton (1993). Altered fruit ripening and leaf senescence in tomatoes expressing an antisense ethylene-forming enzyme transgene. *Plant J. 3*:469–481.
65. Pilnik, W. and A. G. J. Voragen (1984). Polysaccharides and food. *Gordian 70*:144–175.
66. Pirrung, M. C., L. M. Kaiser, and J. Chen (1993). Purification and properties of the apple fruit ethylene-forming enzyme. *Biochemistry 32*:7445–7450.
67. Pomeranz, Y. (1992). Biochemical, functional, and nutritive changes during sotrage, in *Storage of Cereal Grains and Their Products* (D. B. Sauer, ed.), American Association of Cereal Chemists, St. Paul, MN, pp. 55–141.
68. Pomeranz, Y. and O. K. Chung (1978). Interaction of lipids with proteins and carbohydrates in breadmaking. *J. Am. Oil Chem. Soc. 55*:285–289.
69. Powell, C., M. N. Clifford, S. C. Opie, M. A. Ford, A. Robertson, and C. L. Gibson (1993). Tea cream formation: the contribution of black tea phenolic pigments. *J. Sci. Food Agric. 63*:77–86.
70. Pratt, D. E. (1992). Natural antioxidants from plant material, in *Phenolic compunds in Food and Their Effects on Health I Analysis, Occurrence, & Chemistry* (C. T. Ho, C. Y. Lee, and M. T. Huang, eds.), American Chemical Society, Washington, DC, pp. 54–71.
71. Radley, J. A. (1968). *Starch and Its Derivatives*, Chapman and Hall, London.
72. Redgewell, R. J. and S. C. Fry (1993). Xyloglucan endotransglycosylase activity increases during kiwi fruit ripening. *Plant Phsyiol. 103*:1399–1406.
73. Rich, P. R., A. Boveris, W. D. Bonner, and A. L. Moore (1976). Hydrogen peroxide generated by the alternate oxidase of higher plants. *Biochem. Biophys. Res. Commun. 3*:695.
74. Richardson, J. C. and D. W. Stanley 91991). Relationship of loss of membrane functionality and hard-to-cook defect in aged beans. *J. Food Sci. 56*:590–591.
75. Robyt, J.F. (1984). Enzymes in the hydrolysis and synthesis of starch, in *Starch: Chemistry and*

Technology, 2nd ed. (R. L. Whistler, J. N. BeMiller, and E. F. Paschall, eds.), Academic Press, New York, pp. 87–123.
76. Salunke, D. K., S. J. Jadhav, S. S. Kadam, and J. K. Chavan (1982). Chemical, biochemical and biological significance of polyphenols in cereals and legumes. *CRC Crit. Rev. Food Sci. Nutr. 17*:277–305.
77. Schwimmer, S. (1978). Enzyme action and modification of cellular integrity in fruits and vegetables: Consequences for food quality during ripening, senescence and processing, in *Postharvest Biology and Biotechnology* (H. O. Hultin and M. Milner, eds.), Food and Nutrition Press, Westport, CT, p. 37.
78. Snook, J. T. (1984). Nutrition, Prentice-Hall, Englewood Cliffs, NJ, p. 521.
79. Solomos, T. and G. Laties (1976). Induction by ethylene of cyanide-resistant respiration. *Biochem. Biophys. Res. Commun. 70*:663–671.
80. Spanos, G. A. and R. E. Wrolstad (1994). Phenolics of apple, pear and white grape juices and their changes with processing and storage—A review. *J. Agric. Food Chem. 40*:1478–1487.
81. Stegemann, H. (1975). Properties and physiological changes in storage proteins, in *The Chemistry and Biochemistry of Plant Proteins* (J. B. Harborne and C. F. van Sumere, Eds.), Academic Press, New York, Chap. 3.
82. Tigchelaar, E. C., W. B. McGlasson, and M. J. Franklin (1978). Genetic regulation of tomato fruit ripening. *HortScience 13*:508.
83. Torre, M., and A. R. Rodriguez (1991). Effects of dietary fiber and phytic acid on mineral availability. *CRC Crit. Rev. Food Sci. Nutr. 30*:1–22.
84. Voekler, T. A. (1992). Fatty acid biosynthesis redirected to medium chains in transgenic oilseeds. *Science 257*:72–74.
85. Wade, N. L., S. H. Satyan, and E. E. Kavanagh (1993). Increase in low molecular size uronic acid in ripening banana fruit. *J. Sci. Food Agric. 63*:257–259.
86. Wagner, A. M., T. J. A. Kneppers, B. M. Kroon, and L. H W. Plas (1987). Enzymes of the pentose phosphate pathway in callus forming potato discs grown at various temperatures. *Plant Sci. 57*:159.
87. Wang, C. Y. (1993). Relation of chilling stress to polyamines in zucchini squash. *Acta Hort. 343*:288–289.
88. Weichmann, J. (1987). Minerals, in *Postharvest Physiology of Vegetables* (J. Weichmann, ed.), Marcel Dekker, New York, pp. 481–488.
89. Wicker, L., M. R. Vassallo, E. J. Echeverria (1988). Solubilization of cell wall bound, thermostable pectinesterase from valencia orange. *J. Food Sci. 53*:1171–1174.
90. Williams, R. J. P. (1985). The necessary and desriable production of radicals in biology. *Phil. Trans. Royal Soc. Lond. B 311*:593–603.
91. Wolfe, W. J. (1972). Purification and properties of soybeans, in *Soybeans Chemistry and Technology* (A. K. Smith and S. J. Circle, eds.), AVI, Westport, CT, pp. 93–143.
92. Wrigley, C. W. and J. A. Bietz (1988). Proteins and amino acids, in *Wheat Chemistry and Technology* (Y. Pomeranz, eds.), American Association of Cereal Chemists, St. Paul, pp. 179–275.
93. Yang, S. F., and N. E. Hoffman (1984). Ethylene biosynthesis and its regulation in higher plants. *Annu. Rev. Plant. Physiol. 35*:155.

17

Summary: Integrative Concepts

Petros Taoukis
National Technical University of Athens, Athens, Greece
Theodore P. Labuza
University of Minnesota, St. Paul, Minnesota

17.1 INTRODUCTION

The need for fast assimilation and judicious synthesis of the state of the art scientific knowledge, from all food science related fields, is ever more apparent. An informed and assertive global food consumer, at the threshold of the 21st century, requires maximum "benefit," that is, organoleptic superiority, high nutritional value, healthy image, convenience, and extended shelf life, with minimum "cost," that is, less processing, minimum additives, and environmentally friendly packaging.

The objective of this final chapter is to illustrate the methodology through which the material presented in this book can be integrated in a systematic approach to product development and food quality optimization. This multidisciplinary approach requires input from fields such as food chemistry, food microbiology, analytical chemistry, physical chemistry, food engineering, polymer science, and food regulations.

17.2 CHEMICAL CHANGES DURING PROCESSING AND STORAGE

Food processing subjects food to controlled conditions with the objective of altering it to a state of safety, durability and convenience, and this is done by affecting targeted biological, physical and chemical parameters. Unavoidably, undesirable changes also occur [13].

Changes that can be classified as desirable include:

1. Development or preservation of organoleptic properties such as color, flavor, and texture. Complex chemical reactions such as lipid oxidation, the Maillard reaction, Stecker degradation, caramelization, and enzyme-catalyzed reactions are major contributors to altered organoleptic properties. Chemical alteration of polymers by hydrogen bonding, hydrophobic clustering, and cross-linking via multivalent ions, can also have a profound effect on texture.
2. Improvement of food ingredient functionality, such as gelatinization and chemical modification of starch, glucose isomerization to fructose, and alkali processing of soy proteins.
3. Control of enzymes, mainly by heat denaturation but also by pH control, chemical inhibition, reactant removal or modification, and cofactor sequestering.
4. Improvement of digestibility and nutritional value, and inactivation of antinutritional agents.

The undesirable changes often involve:

1. Degradation of color, flavor, and texture. Notable examples are cooked notes in ultra-high-temperature (UHT) milk, chlorophyll and texture loss in canned or dehydrated vegetables, and toughening of frozen fish and seafood muscle.
2. Deterioration of functional properties of ingredients, such as loss of water binding, emulsifying, or foaming capacity of heated proteins.
3. Loss of nutritional value and development of potentially toxic compounds. Certain vitamins are susceptible to heat (e.g., C, thiamin, folate, B_6), to oxidation (e.g., C,D,E,A), or to photodegradation (riboflavin) (Chap. 8). Proteins, carbohydrates, and lipids can undergo numerous reactions, discussed in detail in previous chapters. Depending on the severity of conditions and the coexisting constituents, these can lead to a reduction of nutritional value and undesirable byproducts, sometimes of toxicological significance. These reactions are schematically summarized in Figures 1, 2, and 3. A solid understanding of their implications is a prerequisite for successful optimization of processes.

Similar reactions continue to occur postprocessing, at a rate that is determined by the inherent properties of the food, the type of packaging, and conditions of storage and distribution. These factors determine the shelf life of the food. The fundamentals of assessing food quality and designing shelf-life tests make-up the rest of this chapter.

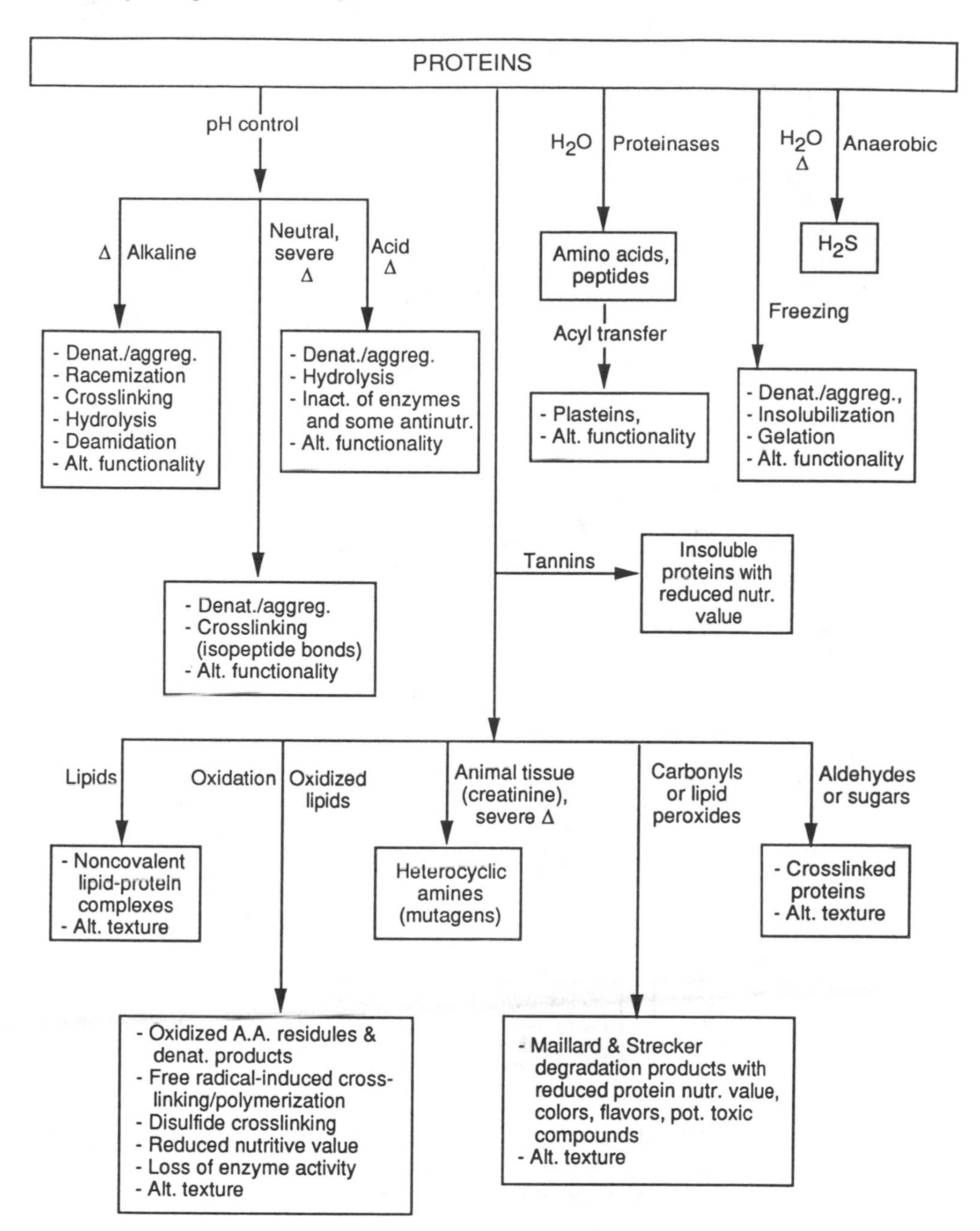

FIGURE 1 Major reactions that proteins can undergo during the processing and handling of foods. Aggreg., aggregation; alt., altered; antinutr., antinutrient; denat., denaturation; inact., inactivation; nutr., nutritive; pot., potentially. Prepared by O. Fennema.

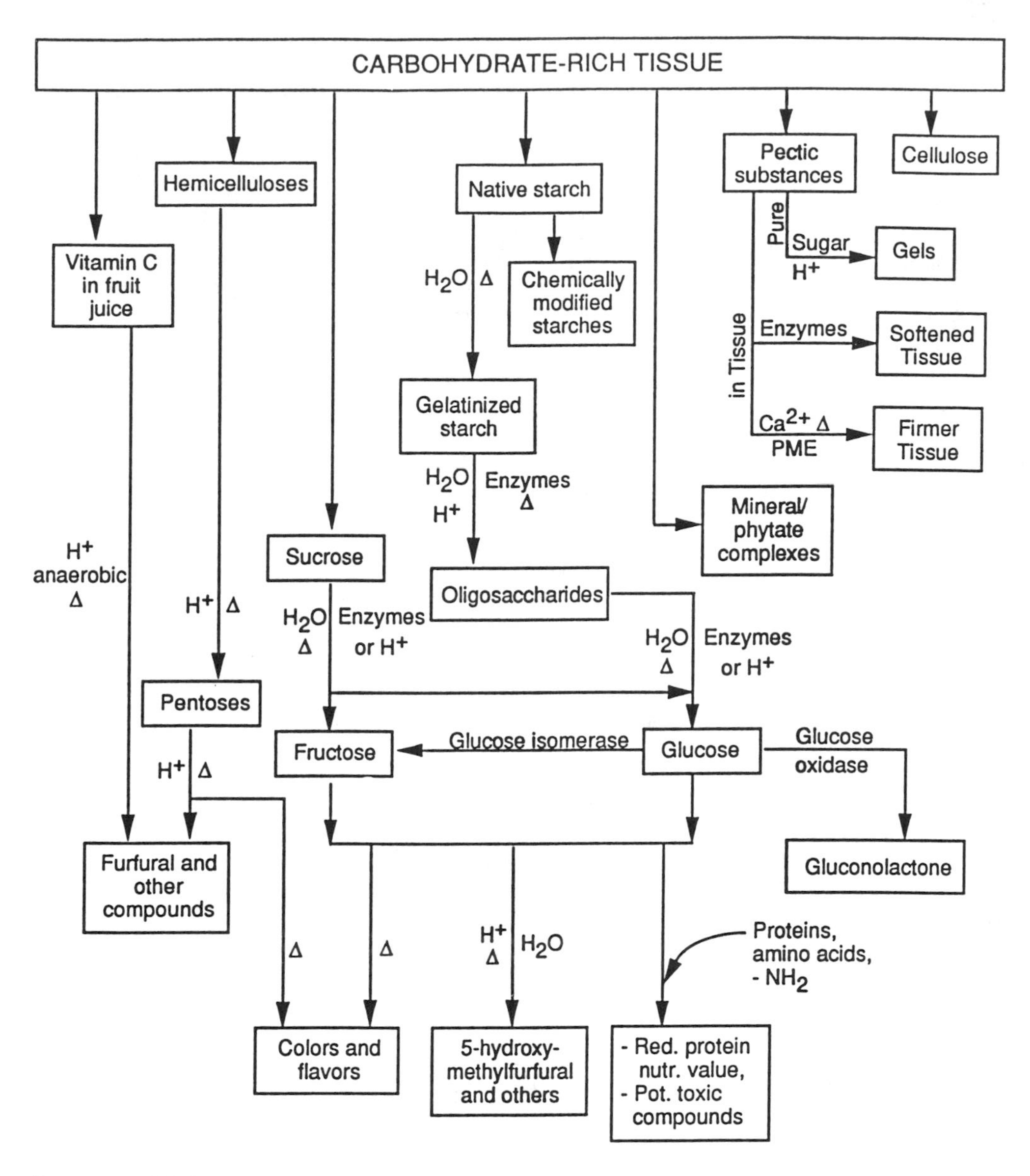

FIGURE 2 Major reactions that carbohydrates can undergo during the processing and handling of foods. Red., reduced; PME, pectin methylesterase; other terms are defined in the legend of Fig. 1. Prepared by O. Fennema.

17.3 APPLICATION OF KINETICS TO SHELF-LIFE MODELING

17.3.1 Principles

Despite the complexity of food systems, the systematic study of their deteriorative mechanisms can lead to sound methods of determining shelf life. What should be made clear is that prediction of shelf life can be approached in two general ways. The most common method involves selecting some single abuse condition, exposing the food product to this condition for several

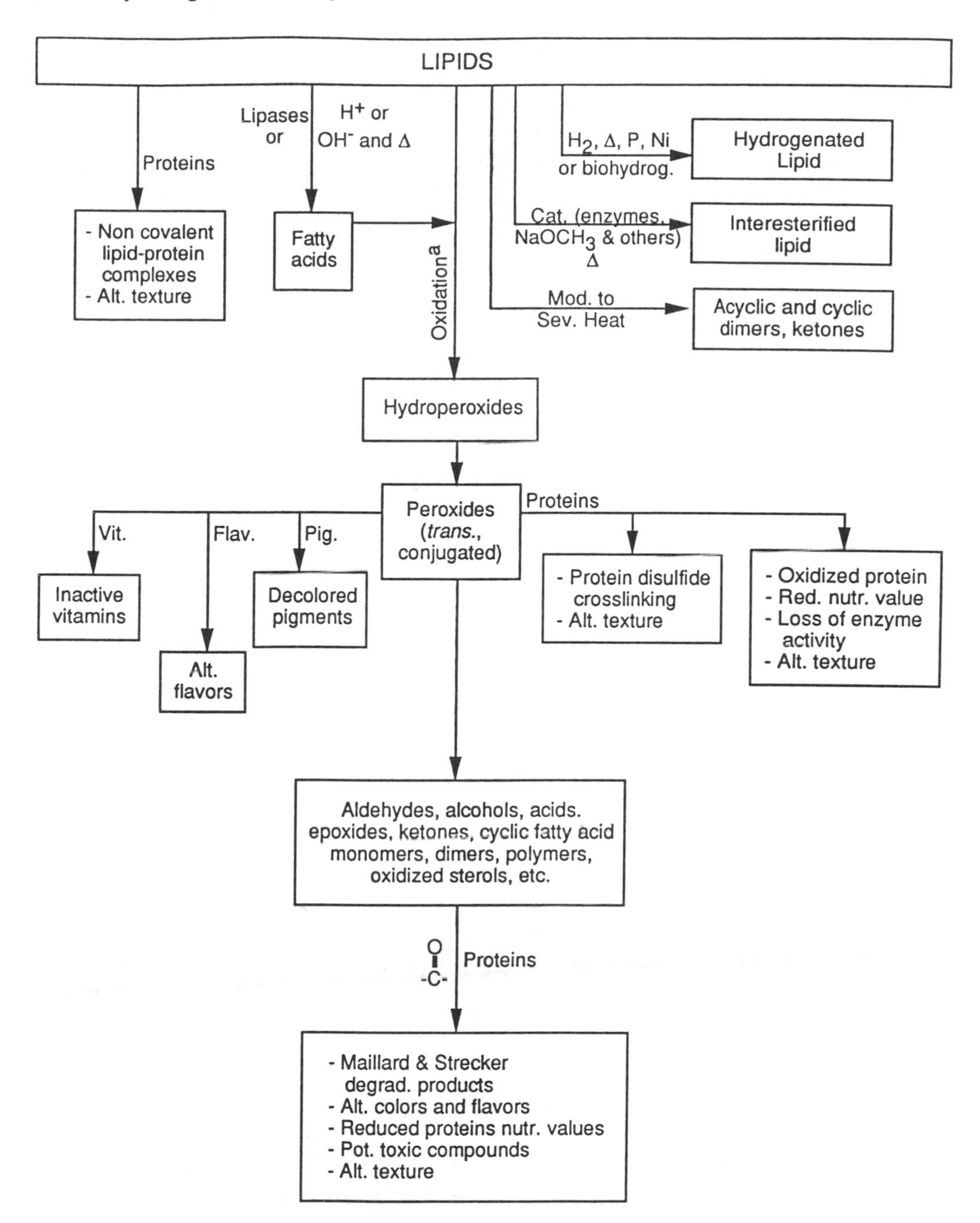

FIGURE 3 Major reactions that lipids can undergo during the processing and handling of foods. [a]Oxidation as catalyzed by enzymes, metals, myoglobin, chlorophyll, and/or irradiation. Degrad., degradation; flav., flavors; mod., moderate; P., pressure; pig., is pigments; sev., severe; vit., vitamins A, C, D, E, and folate; red., reduced; other terms are defined in legend of Fig. 1. Prepared by O. Fennema.

storage times, evaluating quality, generally by sensory methods, and then extrapolating the result (usually empirically) to normal storage conditions. The alternative approach is to utilize a more elaborate design based on principles of chemical kinetics and to determine the actual temperature dependency of various quality attributes. This is initially more costly but is likely to result in more accurate predictions, more effective product formulation, and an ability to optimize the process. A detailed review of the fundamentals of food kinetics is outside the scope of this chapter and is available elsewhere [4, 37, 44]. Only the key principles necessary for the effective use of the proposed methodology are given in this chapter.

17.3.2 Food Quality Function—Order of Reaction

Quality loss can be represented by the loss of a quantifiable desirable quality attribute A (e.g., a nutrient or characteristic flavor) or by the formation of an undesirable attribute B (e.g., an off flavor or discoloration). The rates of loss of A and of formation of B are expressed by the following equations:

$$\frac{-d[\mathrm{A}]}{dt} = k[\mathrm{A}]^n \tag{1}$$

$$\frac{d[\mathrm{B}]}{dt} = k'[\mathrm{B}]^{n'} \tag{2}$$

where k and k' are the reaction rate constants and n and n' are the apparent reaction orders. Both Equations 1 and 2 can be integrated to the expression:

$$F(A) = kt \tag{3}$$

That is, A (or B), with the appropriate transformation, can be expressed as a linear function of time t. The term $F(A)$ is called the quality function [67] of the food and its expression for different apparent reaction orders is given in Table 1.

Most shelf-life data for change in a quality attribute, if based on a characteristic chemical reaction or microbial growth, follow zero-order (e.g., frozen foods overall quality, Maillard browning) or first-order kinetics (e.g., vitamin loss, oxidative color loss, microbial growth and inactivation). For zero-order data, a linear plot is obtained using linear coordinates (Fig. 4), whereas for first-order data, a logarithmic ordinate is needed to produce a linear plot (Fig. 5). For second-order data, a plot of $1/A$(or $1/B$) versus time produces a linear relationship. Thus, on the basis of a few measurements the order can be defined by the plot that gives the best linear fit, and the components of Equation (3) can be determined. One can then extrapolate to the value of A_S (or B_S), the value the attribute reaches at the specified end of the shelf life, t_S. The time, t, it takes for A to reach any specific value can be calculated or vice versa. Caution is advised in deciding the appropriate apparent reaction order and quality function [37]. For example, when the reaction is not carried far enough (less than 50% conversion) both zero- and first-order kinetics might be indistinguishable from a goodness-of-fit point of view, as is illustrated in

TABLE 1 Form of Food Quality Function for Different Reaction Orders

Reaction order	0	1	n
Quality function $F(A)$	$A_o - A$	$\ln(A_o/A)$	$\frac{1}{n-1}(A^{1-n} - A_o^{1-n})$

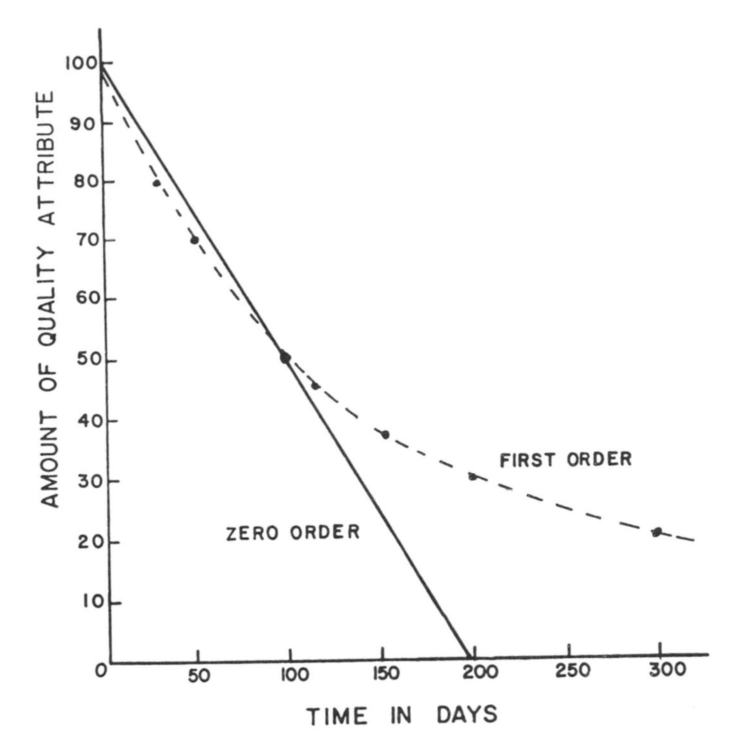

FIGURE 4 Change in quality versus time showing effect of order of reaction on rate of change.

Figure 4. On the other hand, if the end of shelf life occurs within less than 20% conversion, as is often the case [36], either model is adequate for practical purposes.

To determine the value of k (the slope) from the appropriate plot, various line-fitting statistical methods can be used. The better the precision of the method for measuring the attribute value, the smaller the extent of change needed in the attribute value to arrive at an accurate rate value of k. Generally, precision is defined as

$$\text{Precision} = \pm \sigma / X \ 100\% \tag{4}$$

where σ is the standard deviation and X is the mean value of replicate analyses. Based on the calculations of Benson [4] in Table 2, it is apparent that one must allow a sufficient change in the reactant species monitored or good precision to obtain a reliable estimate of the rate constant. Because of normal variabilities, many food reactions, such an nonenzymic browning, have a precision error on the order of ±10%. Erroneous estimates of shelf life are often obtained if these matters are not taken into account, especially if the data are used to extrapolate to longer times. Unfortunately, reactions involved in food quality loss are often not monitored for a time (extent of reaction) that is sufficient to allow accurate determination of the rate constant and the reaction order. Thus, much of the data that have been collected on food quality deterioration are of limited usefulness for accurately predicting shelf life.

17.3.3 Temperature Effects—Arrhenius Approach

Kinetic equations for shelf life are specific to the food studied and the environmental conditions used. Of the noncompositional factors that can affect the reactions, that is, temperature, relative humidity, partial pressure of packaging gases, light, and mechanical stresses, the one usually

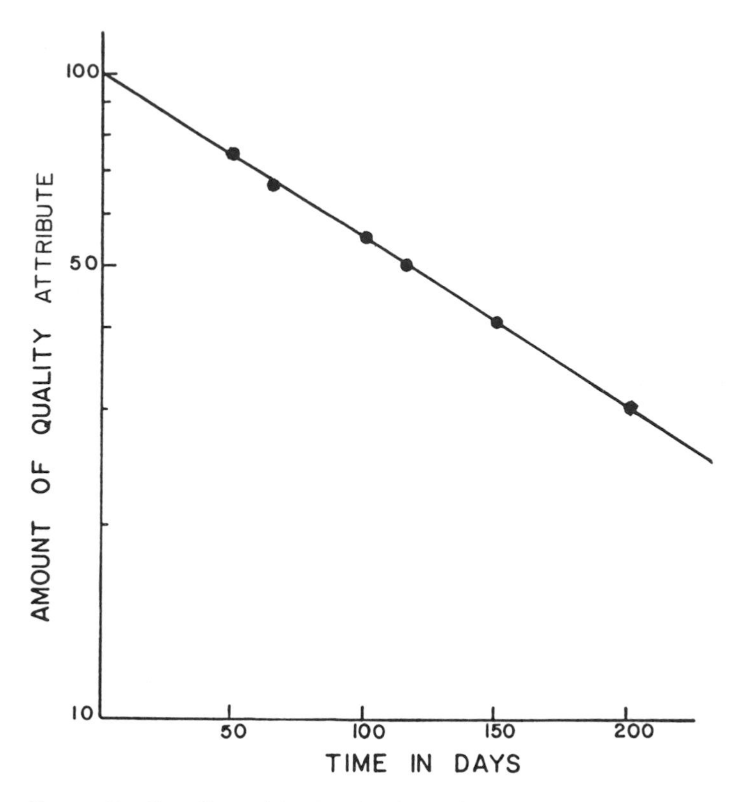

FIGURE 5 Semilogarithmic plot for a first-order reaction of quality loss versus time.

incorporated in the shelf-life model is temperature. It strongly affects reaction rates and is the only factor among those mentioned that is not influenced by the type of food packaging.

Most physical chemistry textbooks provide the fundamentals of the Arrhenius approach that interrelates temperature and the rate of chemical reactions. Labuza [34] has applied this approach to rates of deteriorative reactions in foods. The well-known Arrhenius relation is

TABLE 2 Effect of Analytical Precision of the Method on the Estimation of the Rate Constant for Chemical Reactions

Analytical precision (%)	Percent error in reaction rate constant *k* at given percent change in reaction species monitored						
	1%	5%	10%	20%	30%	40%	50%
±0.1	14	2.8	1.4	0.7	0.5	0.4	0.3
±0.5	70	14	7	3.5	2.5	2	1.5
±1.0	>100	28	14	7	5	4	3
±2.0	>100	56	28	14	10	8	6
±5.0	>100	>100	70	35	25	20	15
±10.0	>100	>100	>100	70	50	40	30

Source: From Ref. 4.

$$k = k_A \exp(-\frac{E_A}{RT}) \tag{5}$$

where k_A is the Arrhenius equation constant, E_A, in joules or calories per mole, is the activation energy (the excess energy barrier that attribute A needs to overcome to proceed to degradation products [or B to form]), T is absolute temperature (K), and R is the universal gas constant (1.9872 cal/mol · K or 8.3144 J/mol · K). In practical terms it means that if values of k are obtained at different temperatures and ln k is plotted against reciprocal absolute temperature, $1/T$, a straight line is obtained with a slope of $-E_A/R$. Under some instances, if two critical reactions occur in a food and they have different rates and activation energies, it is possible that one predominates above some critical temperature T_c and the other below this temperature (Fig. 6). The main value of the Arrhenius plot is that one can collect data at high temperatures (low $1/T$) and then extrapolate for the rate constant at some lower temperature, as shown in Figure 7.

When applying regression techniques, statistical analysis is used to determine the 95% confidence limits of the Arrhenius parameters. If only three k values are available, the confidence range is usually large. To obtain meaningfully narrow confidence limits in E_A and k_A, rates at more temperatures are required. It was proposed [48] that five or six experimental temperatures give the practical optimum ratio of accuracy versus amount of work. If one is limited to three experimental temperatures because of cost and/or the difficulty of obtaining incubators at six constant temperatures, one can use several statistical methods to increase accuracy [39], one of which is to consider each data point as an independent estimation of k. This approach has worked well in several studies [5, 45, 62]. Alternatively, a linear regression including the 95% confidence limits of the reaction rates as independent points will give narrower confidence limits for the

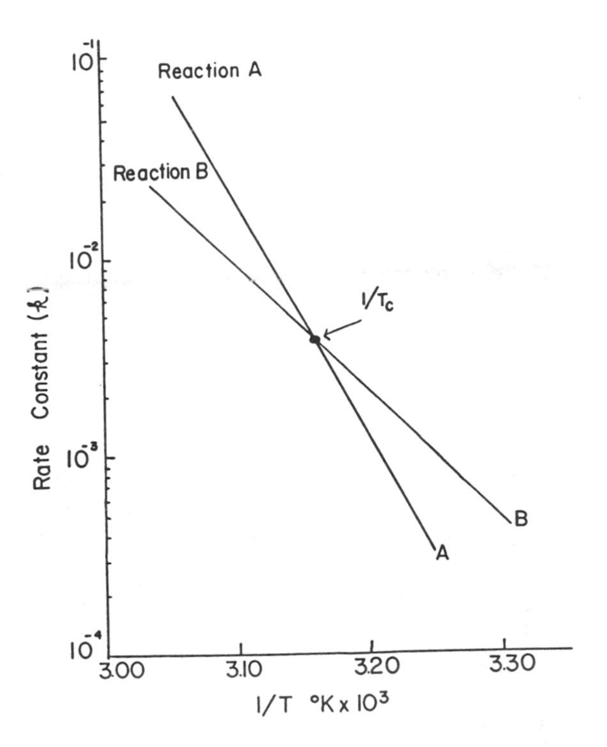

FIGURE 6 Arrhenius plot of reaction rate constant k versus inverse absolute temperature for two reactions exhibiting crossover at T_c.

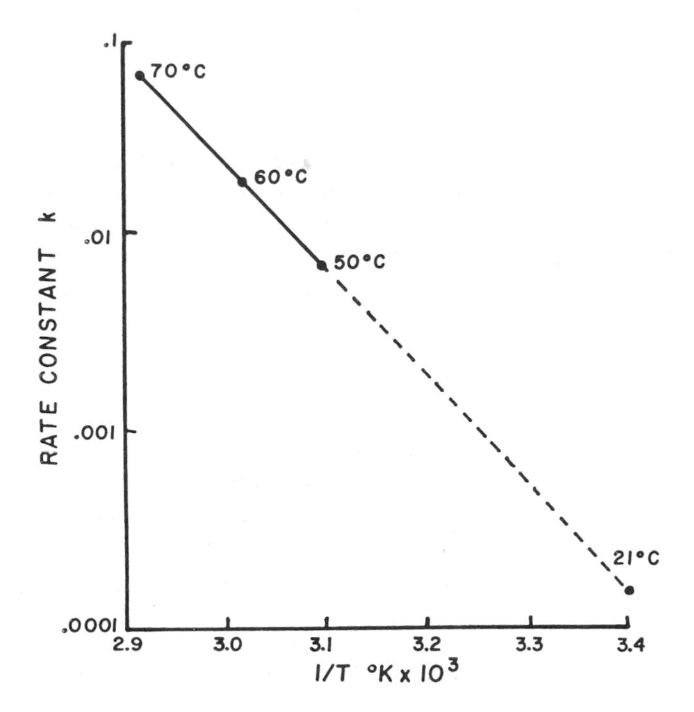

FIGURE 7 Prediction of shelf life by extrapolation from a higher to a lower temperature on an Arrhenius plot.

Arrhenius parameters [39]. Also, one can use a one step method that requires nonlinear regression of the equation that results from substitution of Equation (5) in Equation (3) [1, 10, 19].

It should be noted that it is often possible to derive an E_A value, with a satisfactory coefficient of determination ($r^2 > .95$), from three temperatures. However, the 95% confidence range obtained, especially if the aforementioned techniques are not used, is usually large (on the order of E_A itself) because of the large $t_{95\%}$ value. This does not mean the data are useless. In these cases the mean E_A is probably an adequate measure of the temperature dependence of the reaction. Whether using one's own generated data or data from the literature, it is important to consider the range and number of temperatures used to be able to estimate the validity of the data. One should also note that both k_A and E_A can depend on water activity [25, 35] thus, a_w must be held constant.

Additionally, attention should be given to factors that can cause deviation from an Arrhenius behavior. Labuza and Riboh [41] discussed potential causes for nonlinear Arrhenius plots. These include (a) a change in a_w or moisture, (b) a change of physical state such as ice formation or a glass transition, (c) a change in the critical reaction with a change in temperature, as noted in Figure 6, (d) a change in pH with a change in temperature, (e) a decrease in dissolved oxygen as the temperature is raised (slowing oxidation), (f) a partitioning of reactants between two phases, such as a concentration of reactants upon freezing, and (g) temperature history effects, that is, changes at high temperatures that irreversibly affect the rate of the reaction, when the food is subsequently cooled [40].

Further comments regarding glass transitions are appropriate (see Chap. 2). Above the glass transition temperature, T_g, in the rubbery state, the slope of the Arrhenius plot may gradually change. Williams, Landel, and Ferry [74] introduced the WLF equation to empirically model the temperature dependence of mechanical and dielectric relaxations within the rubbery state. It has been proposed [65] that the same equation may describe the temperature dependence of chemical reaction rates within amorphous food matrices, above T_g. In diffusion-controlled systems where diffusion is free volume dependent, reaction rate constants can be expressed as a function of temperature by the WLF equation [63]:

$$\log\left(\frac{k_{ref}}{k}\right) = \frac{C_1(T - T_{ref})}{C_2 + (T - T_{ref})} \tag{6}$$

where k_{ref} is the rate constant at the reference temperature $T_{ref}(T_{ref} > T_g)$ and C_1, C_2 are system-dependent coefficients. Williams et al [74], for $T_{ref} = T_g$ and using data then available for various polymers, calculated mean values of the coefficients: $C_1 = -17.44$ and $C_2 = 51.6$. In the literature it has become common practice to use these as mean values to establish the applicability of the WLF equation for different systems. This simplification is problematic [8, 14, 55] and one should try to use system-specific values whenever possible.

Recently the relative validity of the Arrhenius and WLF equations as they apply to the rubbery state, namely in the range 10–100°C above T_g, has been debated [27]. Viscosity-dependent changes in food quality (e.g., crystallization, textural changes) fit the WLF model. However, chemical reactions may be either kinetically and/or diffusion limited. In general the effective reaction rate constant can be expressed as $k/(1 + k/\alpha D)$ (where D is the diffusion coefficient and α is a constant independent of T). In most cases, k exhibits an Arrhenius-type temperature dependence and D has been shown in many studies either to follow the Arrhenius equation with a change in slope at T_g, or to follow the WLF equation in the rubbery state and especially in the range 10–100°C above T_g. The value of the ratio $k/\alpha D$ defines the relative influence of k and D. If $k/\alpha D < 0.1$, the deteriorative reaction can be successfully modeled by a single Arrhenius equation for the whole temperature range of interest. Otherwise, and depending on which of the aforementioned temperature dependencies D follows, a break in the slope of the Arrhenius plot of the effective reaction rate constant occurs at T_g with either constant slope above T_g or with a gradually changing slope. In the latter case the WLF equation is appropriate for the range 10–100°C above T_g. For complex food systems involving multiple reaction steps and phases, application of either model has to be viewed as a practical tool rather than as a descriptor of the actual phenomenon or mechanism.

Based on the Arrhenius model developed from several constant-temperature shelf-life experiments, quality attribute loss can be predicted for any variable-temperature exposure. Controlled-temperature functions like sine, square, and linear (spike) wave temperature fluctuations can be used to verify the shelf-life model [5, 25, 45, 60, 61, 68]. For example, for lipid oxidation in potato chips, the prediction is good to ±3% [5]. Accumulated shelf-life loss for known temperature conditions can be calculated using a series expansion approximation to calculate the value of the quality function at time t.

$$F(A)_t = k_A \sum_{i=0}^{m} \exp\left(\frac{-E_A}{RT_i}\right) \Delta t_i \tag{7}$$

17.3.4 The Simple Shelf-Life Plot Approach

For a given extent of deterioration, translated to a value of the quality function (Eq. 3), the rate constant is inversely proportional to the time to reach a specified degree of quality loss [29]. This holds also to t_s, the time for the quality to reach an unacceptable level, that is, the shelf life. Thus, a plot of log shelf life (log t_s) versus $1/T$ is a straight line, as shown in Figure 8A. Furthermore, if only a small temperature range is considered, then one finds that most food data

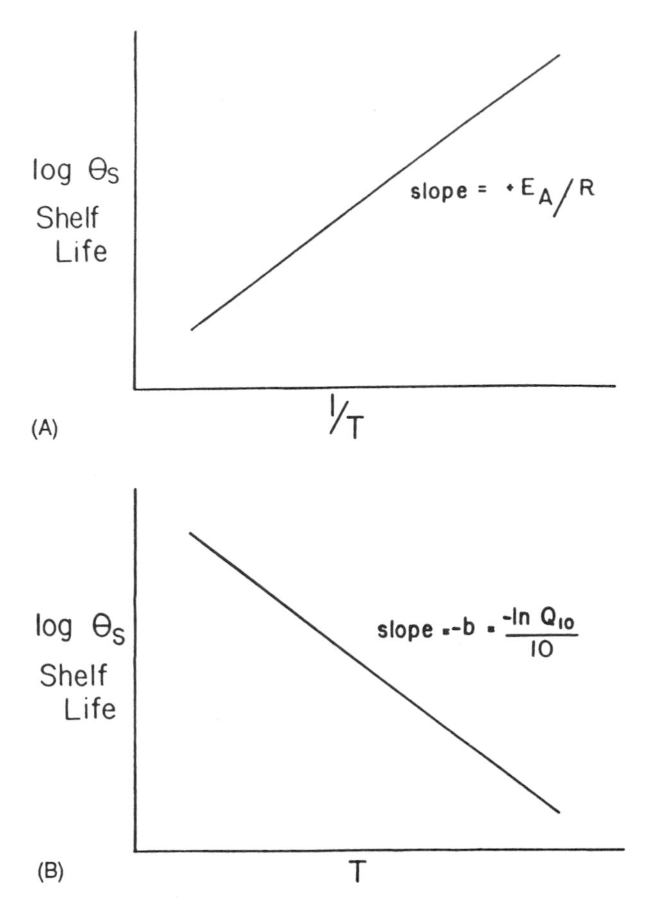

FIGURE 8 Plot of the logarithm of shelf life (A) versus inverse absolute temperature and (B) versus temperature (absolute or in degrees C).

yield a linear plot when log t_s is plotted versus temperature T (Fig. 8B). The equation of this shelf-life plot is

$$t_s = t_{s_0} e^{-bT} \tag{8}$$

where t_s is shelf life at absolute temperature T, the constant t_{s_0} is "shelf life" at the y-intercept, and b is the slope of the shelf-life plot. It should be noted that Equation 8 also holds using temperature in degrees Celcius (or Fahrenheit), which is often more convenient for conceptual and presentation purposes. In these cases care should be taken to use the appropriate value and unit conversions (for degrees C, slope b is the same as in Eq. 8 and t_{s_0} is value of t_s at 0°C; for degrees F, slope is equal to $b/1.8$ and t_{s_0} is value of t_s at 0°F).

Temperature dependence often is expressed as Q_{10}, the ratio of the reaction rate constants at temperatures differing by 10°C, or the change of shelf life t_s when the temperature of the food is raised 10°C. The majority of the early food literature contains endpoint data rather than the preferred complete kinetic modeling of quality loss. The Q_{10} approach is equivalent to the shelf-life plot approach of Equation 8. The relation between the aforementioned kinetic parameters is expressed as

$$Q_{10} = \frac{t_s \text{ at } T}{t_s \text{ at } (T + 10)} = exp(10b) = \exp\left[\frac{E_A}{R} \frac{10}{T(T + 10)}\right] \qquad (9)$$

The validity of the simple shelf-life plot and the Q_{10} approach is usually limited to narrow temperature ranges [33, 41]. The Q_{10} values based on a large temperature range will often be somewhat inaccurate. In addition, the factors that cause deviations in the Arrhenius plot also influence the shelf-life plot. The value of Q_{10} is temperature dependent, as seen in Equation 9. This dependence is more pronounced with greater temperature "sensitivity" of the reaction that is, with higher activation energy. Shown in Table 3 are the dependence of Q_{10} on temperature and E_A and important types of food reactions that correspond to various values of Q_{10} and E_A.

The practical use of the shelf-life plot is shown in Figure 9. Suppose that the targeted shelf life of a product under development is at least 18 months at 23°C(73°F) and one wants to determine, using an accelerated test at 40°C(104°F), whether this has been achieved. By placing a point on the plot at 18 months and 23°C and drawing a straight line to the 40°C vertical line, using slopes dictated by the Q_{10} values, a shelf life of 1 month is indicated if the Q_{10} is 5, and 5.4 months if the Q_{10} is 2. Caution should be exercised when estimating the true Q_{10}. According to Labuza [36], canned products have Q_{10} values ranging from 1.1 to 4, dehydrated foods from 1.5 to 10, and frozen foods from about 3 to as much as 40. Thus, use of a mean Q_{10} for any of these categories, as is often done in the food industry, can result in highly inaccurate predictions. The only way to achieve reliable results is to determine the Q_{10} by conducting shelf-life studies at two or more temperatures. One must also not forget to hold a_w constant for dehydrated foods, since a_w affects Q_{10} values.

Even when proper shelf-life testing is conducted, one should be aware of how much the accuracy of the shelf-life prediction can be influenced by the uncertainty of the Q_{10} value. As shown in Table 4, if a product has a Q_{10} of 2.5 and a measured shelf life of 2 weeks at 50°C, the predicted shelf life at 20°C is 31 weeks. If the actual Q_{10}, however, is 3, then the predicted shelf life at 20°C is 54 weeks—a difference of about 5 months. Thus, substantial deviations in the predicted shelf life at low temperatures can arise if slightly inaccurate Q_{10} values are used to make the extrapolation. A 0.5 difference in Q_{10} for the 40–50°C temperature range corresponds to a difference in E_A of about 20 kJ/mol (at the 80 kJ/mol range). In the many studies in which a good estimate of the statistical variation of E_A has been made, it has a 95% confidence limit, at best, of ±15–20 kJ/mol. Thus, although this simple approach to shelf-life prediction is sound, accuracy of the estimates is sometimes, poor. The inaccuracy is much smaller when extrapolating from lower temperatures. In practice one usually compromises among time, cost, and accuracy. Additionally accelerated tests at temperatures of >40°C are not recommended because of the aforementioned possibility that the reactions of critical importance may change between the high temperature and the normal storage temperature of the product.

TABLE 3 Q_{10} Dependence on E_A and Temperature

E_A, kJ/mol (kcal/mol)	Q_{10} at 5°C	Q_{10} at 20°C	Q_{10} at 40°C	Typical food reactions
41.8 (10)	1.87	1.76	1.64	Diffusion controlled, enzymic, hydrolytic
83.7 (20)	3.51	3.10	2.70	Lipid oxidation, nutrient loss
125.5 (30)	6.58	5.47	4.45	Nutrient loss, nonenzymic browning
209.2 (50)	23.1	20.0	12.0	Vegetative cell destruction

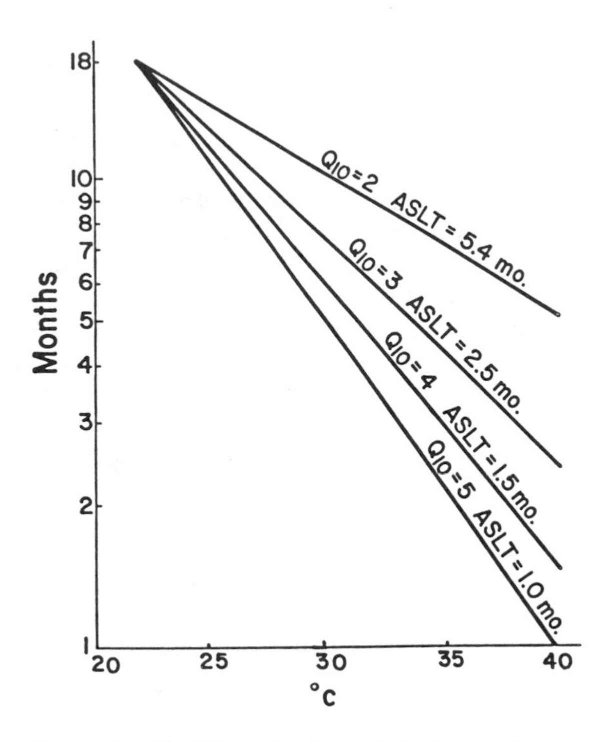

FIGURE 9 Shelf-life plot for a dehydrated food with desired shelf life of 18 months at 23°C and equivalent accelerated shelf-life test (ASLT) times at 40°C.

The thermal death time approach (TDT), which is widely used for expressing the temperature dependence of microbial inactivation during thermal processing and sometimes for food quality loss [20, 57], also deserves mention. The z-value, determined from a plot of TDT (log) versus temperature (C or F), is the temperature change that causes a 10-fold change in TDT. As in the case of Q_{10}, z depends on a reference temperature. It is related to b and E_A by the following equation:

$$z = \frac{\ln 10}{b} = \frac{(\ln 10)RT^2}{E_A} \tag{10}$$

TABLE 4 Effect of Q_{10} on Shelf Life

	Shelf life (weeks)			
Temperature (°C)	$Q_{10} = 2$	$Q_{10} = 2.5$	$Q_{10} = 3$	$Q_{10} = 5$
50	2[a]	2[a]	2[a]	2[a]
40	4	5	6	10
30	8	12.5	18	50
20	16	31.3	54	250

[a]Arbitrarily set at 2 weeks at 50°C. Shelf lives at lower temperatures are calculated on this arbitrary assumption.

Equation 10 allows z values for microbial or enzyme inactivation, and E_A values for quality loss, to be interconverted [53].

Other forms of the $k(T)$ function have been proposed [31], like linear, power, and hyperbolic equations, but over a wide range of temperatures, the Arrhenius equation gives the best correlation.

The same kinetic principles described in this section have been employed recently to study and establish a database on the temperature behavior of microbial growth, a field referred to as *predictive microbiology* [7, 17]. Several temperature-dependence equations have been used for modeling with the two most prevalent being the Arrhenius and the square root model [11, 46, 58, 59, 75],

$$\sqrt{k} = b(T - T_{min}) \tag{11}$$

where k is growth rate, b is slope of the regression line of $\sqrt{k}$ versus temperature, and T_{min} is the hypothetical growth temperature where the regression line cuts the T axis at $\sqrt{k} = 0$. The relation between Q_{10} and this expression is

$$Q_{10} = \left(\frac{T - T_{min} + 10}{T - T_{min}}\right)^2 \tag{12}$$

17.3.5 Effects of Other Factors

Besides temperature, moisture content and water activity (a_w) are the most important factors affecting the rate of deteriorative reactions at above-freezing temperatures (Chap. 2). Moisture content and water activity can influence the kinetic parameters (k_A, E_A), the concentrations of reactants, and in some cases even the apparent reaction reaction order, n. Most relevant studies have modeled either k_A as a function of a_w [35], or E_A as a function of a_w [49, 50]. The inverse relationship of E_A with a_w (increase in a_w decreases E_A and vice versa) can be theoretically explained by the proposed phenomenon of enthalpy–entropy compensation. The applicability of this theory and data that support it have been discussed by Labuza [34]. Additionally, a_w affects T_g (direct relationship), thus affecting chemical reactions as discussed in the previous section [52].

In many cases the effect of a_w is not taken into account. Food tested for shelf life is sometimes placed into environmental cabinets at different temperatures without holding relative humidity constant. This is poor practice if the effect of a_w is important. On the other hand, if the effect of a_w is known or the technical facilities and the scientific knowledge are available for such studies, shelf life at normal conditions can be accurately predicted from data collected at high temperature and high humidity conditions [52].

Mathematical modeling for predicting moisture gain with time in a packaged food originated with classic studies of the 1960s and 1970s [26, 38, 47, 49, 56, 64]. Criteria for evaluating the validity of these models using a dimensionless number, L, were recently introduced [22, 69]. If the rate of loss of quality is known as a function of temperature and a_w, and the barrier properties of the film are known, the shelf-life loss with time can be predicted for any temperature–humidity conditions [9, 51].

Other factors that affect reaction rates in foods are pH, gas composition and partial pressures, and total pressure. The effect of pH has been studied for several food reactions and food systems. Enzymatic and microbial activity are strongly affected by pH, each having a pH range for optimum activity and limits above and below which activity ceases. The functionality and solubility of proteins also depend on pH, with solubility usually having a minimum near the isoelectric point (Chap. 6), thus directly affecting their behavior in reactions. The effect of pH on numerous microbial, enzymatic, and protein reactions has been studied in model nutrient, biochemical, or food systems, but most studies neglect the possible interactions between factors, most importantly pH and temperature. Acid-base-catalyzed reactions important to foods, such as nonenzymatic browning and aspartame decomposition, are strongly pH dependent. Nonenzymatic browning of proteins shows a minimum near pH 3–4 and high rates in the strongly acidic and alkaline range [12]. Weissman et al. [73] demonstrated an approach to accelerated shelf-life testing for a Maillard browning system, whereby both pH and temperature were included as accelerating factors in the kinetic prediction model. Aspartame decomposition shows a minimum at pH 4.5 [21], but the relative rate also depends on the buffering capacity of the system and the specific ions present [72]. Studies on the effects of temperature, a_w and pH on aspartame degradation highlight the importance of these factors and their interactions and also the predictive errors that can occur in this system if these interactions are neglected [2, 3].

Gas composition is an additional factor that can play a significant role in some quality loss reactions. Oxygen availability is very important for oxidative reactions and can affect both the rate and apparent reaction order, depending upon whether it is limiting or in excess [32]. Vacuum packaging and nitrogen flushing are based on slowing down undesirable reactions by limiting the availability of O_2. Further, the presence and relative amount of other gases, especially carbon dioxide, strongly affect biological and microbial reactions in fresh meat, fish, fruits, and vegetables. The mode of action of CO_2 has not been completely elucidated but is partly connected to surface acidification [54]. Different commodities have different optimum O_2–CO_2–N_2 gas composition requirements for maximum shelf life. Excess CO_2 in many cases is detrimental. Other important gases are ethylene and CO. Successful controlled and modified atmosphere packaging applications are based on this knowledge. Kinetic modeling of CAP/MAP systems has been reviewed by Labuza et al. [46]. Total pressure is usually not an important factor except in the case of hypobaric storage, an alternative technology to CAP/MAP.

Recently, very high pressure technology (100–1000 MPa) has been used experimentally to achieve inactivation of microorganisms, modification of biopolymers (protein denaturation, enzyme inactivation or activation, degradation), increased product functionality and quality retention [23, 28]. Kinetic studies of changes occurring during high-pressure processing and their effects on shelf life of the foods are an area of current research.

17.3.6 Steps in Shelf-Life Testing—ASLT Principles

Based on the fundamentals of reaction kinetics as applied to foods, knowledge of food chemistry principles covered in this book, and knowledge from other fields of food science, one can design and conduct scientifically sound shelf-life testing. The goal is to obtain maximum information in minimum time and at least cost. This is achieved using the principles of accelerated shelf-life testing (ASLT) described by Schmidl and Labuza [42] and in guidelines recently published by the IFST (Institute of Food Science and Technology, UK) [24].
The following steps should be followed in designing a shelf-life test for quality loss in food.

1. Determine the microbiological safety and quality parameters for the proposed formulation and process.
2. Determine from an analysis of the ingredients and the process which chemical reactions are likely to be the major causes of quality loss. A literature search is mandatory. If major potential problems exist at this stage, the formula should be changed, if possible.
3. Select the package to be used for the shelf-life test. Frozen and canned products can be packaged in the final container. Dry products should be stored open in chambers at a predetermined percent relative humidity (%RH) or in sealed jars or impermeable pouches at the desired moisture and a_w.

4. Select the storage temperatures (at least two). Commonly, the following are good choices.

Product	Test temperatures (°C)	Control (°C)
Canned	25, 30, 35, 40	4
Dehydrated	25, 30, 35, 40, 45	−18
Frozen	−5, −10, −15	<−40

5. Using the shelf-life plot (Fig. 9) and knowing the desired shelf life at the mean distribution temperature, determine how long the product must be held at each test temperature. If no information is available on the probable Q_{10} value, then more than two temperatures are needed.
6. Decide which tests will be used and how often they will be conducted at each temperature. A good rule of thumb is that the time interval between tests, at any temperature below the highest temperature, should be no longer than

$$f_2 = f_1 Q_{10}{}^{\Delta T/10} \tag{13}$$

where f_1 is the time between tests (e.g., days, weeks) at the highest test temperature T_1; f_2 is the time between tests at any lower temperature T_2; and ΔT is the difference in degrees Celsius between T_1 and T_2. Thus, if a dry product is held at 45°C and tested once a month, then at 40°C (i.e., $\Delta T = 5$) and a Q_{10} of 3, the product should be tested at least every 1.73 months. Of course, more frequent testing is desirable, especially if the Q_{10} is not accurately known. Use of needlessly long intervals may result in an inaccurate determination of shelf life and thus a danger of rendering the experiment totally useless. At each storage condition, at least six data points are needed to minimize statistical errors; otherwise, the statistical confidence in t_s is significantly diminished.
7. Plot the data as they are collected to determine the reaction order and to decide whether test frequency should be increased or decreased. All too often the data are not analyzed until the experiment is over and then the scientist finds that nothing can be concluded.
8. From each test storage condition, estimate k or t_s, make the appropriate shelf-life plot, and estimate the potential shelf life at the desired (final) storage condition. Of course, one can also store product at the final condition and determine its shelf life to test the validity of the prediction. In academia this is commonly done; however, in industry this is uncommon because of time constraints. If a properly generated shelf-life plot indicates that the product meets stability expectations, then the product has a good chance of performing satisfactorily in the marketplace.

17.4 CASE STUDIES OF SHELF LIFE

In this section representative cases of shelf-life determination are presented. Attention is given to the factors influencing stability, and the design of the shelf-life test. The case of an artificially sweetened, dry dessert mix illustrates the straightforward application of the ASLT approach and its obvious effectiveness in situations where a dominant, quantifiable quality attribute exists and the corresponding quality function can be defined. The methodology is further put into perspective with two other actaul shelf-life problems of increasing complexity: the case of another dehydrated product, mashed potato flakes, and frozen pizza.

17.4.1 Dehydrated Products

17.4.1.1 Dry Dessert Mix

The chosen product is a gelatin-based dessert mix, colored with a natural colorant (e.g., riboflavin; Chap. 10), flavored with a fruit flavor (e.g., strawberry; Chap. 11), and sweetened with the intense sweetener aspartame. The shelf-life goal is 18 months at 22°C. With proper packaging that keeps a_w below 0.4, the basic modes of deterioration are loss of flavor intensity and loss of sweetness. The rate of the latter far exceeds the rate of flavor loss, so aspartame deterioration is the reaction that limits shelf life. Aspartame loss can be easily and accurately quantified by high-performance liquid chromatography (HPLC) [66]. Alternatively (or additionally), organoleptic techniques (focusing on acceptability from a sweetness point of view) can be used. A particularly efficient and suitable methodology for obtaining shelf-life data that can be used in an ASLT scheme is a graphical method called *Weibull hazard analysis* or *maximum likelihood* procedure [16, 43, 44].

The initial product is overcompensated in aspartame, so it retains acceptable sweetness when only 60% of the initial quantity remains undegraded. Literature data indicated that aspartame degradation has a Q_{10} of about 4. This would result in a shelf life of a little more than 40 days at 40°C (see Fig. 9). A testing frequency of 5 days at this temperature will give enough data points for accurate determination of reaction order and rate. According to Equation 13 the appropriate testing frequency at 35, 30 and 25°C will be 10, 20, and 40 days, respectively. Shown in Figure 10A are simulated aspartame degradation data for a dehydrated system at a_w of 0.33 and pH 5 [2]. In Figure10B the quality function $F(A) = \ln(A/A_0)$ is plotted against time, and good linearity occurs at all temperatures ($r^2 > .99$). This indicates apparent first-order reaction kinetics and allows determination of the reaction rate constants ($k = -0.0120$, -0.00620, -0.00297, and -0.00145 day^{-1} at 40, 35, 30, and 25°C, respectively). Constructing the Arrhenius plot (Fig. 11) of ln k versus $1/T$ gives a very satisfactory fit for the pertinent temperature range. The estimated activation energy is 110 kJ/mol (26.2 kcal/mol). Based on this temperature dependency the shelf

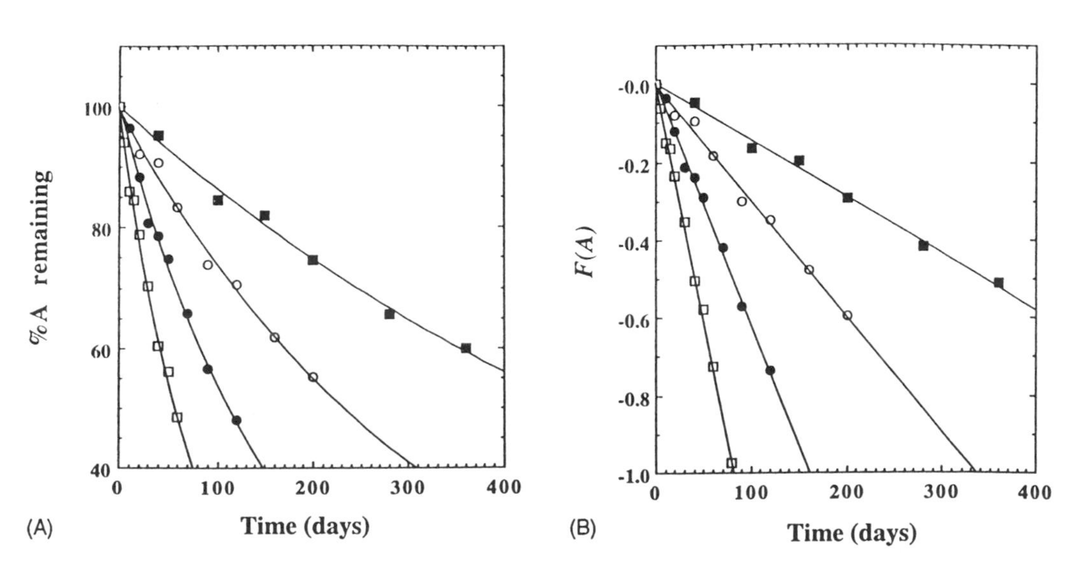

FIGURE 10 (A) Plot of percent aspartame remaining (%A) versus time at various temperatures. (B) Plot of quality function $F(A) = \ln(A/A_0)$ versus time: □ at 40°C, ● at 35°C, ○ at 30°C, and ■ at 25°C.

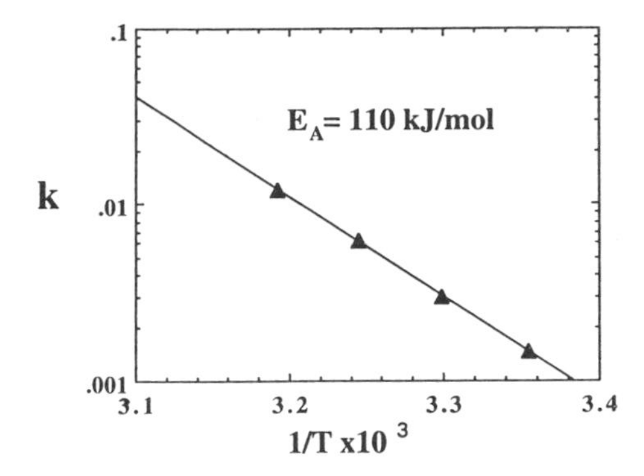

FIGURE 11 Arrhenius plot of aspartame degradation in a dry mix.

life at 22°C (i.e., time for 60% remaining aspartame) is predicted as 552 days, which satisfies the requirement of 18 months.

This case demonstrates the effectiveness of the ASLT methodology when an appropriate, dominant, and accuratcly quantifiable quality attribute can be identified. Note that, based on available information on aspartame degradation, showing a good fit to Arrhenius kinetics for a wide range of temperatures, one could reliably test using only two temperatures (40 and 35°C), in which case a testing period of less than 3 months would suffice for determination of shelf life.

It should also be noted that for many food systems indirect indices (not primary quality factors, but ones whose changes correlate well with a primary quality factor) can be used for modeling quality loss, for example, loss of vitamins can be used for frozen foods [36], and oxidation of *d*-limonene for natural lemon flavor [70].

17.4.1.2 Dehydrated Mashed Potatoes

Dehydrated potatoes are made by drum drying a slurry of previously cooked, mashed potatoes that may include added sulfite and antioxidants. Dehydrated potatoes are susceptible primarily to browning at temperatures above 30–32°C, and primarily to lipid oxidation below 30–32°C. In addition, the rate of vitamin C loss generally exceeds the rates of either of the other aforementioned reactions if the a_w is above the monolayer value (see Chap. 2).

Quality Aspects

Microbiological Changes Since this is a dehydrated product, a_w should be in the range of 0.2–0.4, and growth of microorganisms should not occur (see Chap. 2). Growth of typical spoilage organisms does not occur unless the product reaches an a_w of at least 0.7. Most likely if product a_w does increase to this level during storage, chemical reactions would render the product unsatisfactory before microbial growth became a factor of importance.

Chemical Changes Potatoes may contain up to about 4% free reducing sugars, and almost 50% of the protein nitrogen is in the form of free amino acids. Thus, nonenzymic browning occurs, unless the product is dried and kept near or below the monolayer value. Because free amino acids are abundant, both the Shiff's base–Amadori rearrangement sequence and the Strecker degradation pathway occur. The latter reaction produces CO_2 and other volatile compounds, such as 2-methylpropanal, 2- and 3-methylbutanal, and isovaleraldehyde, which are detectable in the parts per billion range as extremely charred off odors [6, 62]. The browning reaction can utilize vitamin C as one of the substrates and it can also reduce the subsequent hydratability of the potato.

Oxidation of unsaturated lipids occurs at or below about 32°C, especially if the product is dried below the monolayer value. Dried potatoes contain only 1% lipid, but only 2–3 ppm hexanal, an oxidative degradative product, renders them unacceptable [15, 62]. Since about 60–70% of the fatty acids are polyunsaturated, production of volatile off flavors occurs readily unless antioxidants and vacuum packaging are used. Gas chromatography–mass spectrometry (GC-MS) is an easy, effective method to detect and to correlate the changes in headspace compounds with sensory acceptability. Generally, only about 5–10% of the unsaturated fat present needs to be oxidized to produce an unacceptable product.

Kreisman [30] has published a shelf-life plot for dehydrated potatoes (Fig. 12) that correlates shelf-life data with lipid oxidation and nonenzymic browning. The a_w values for these studies varied between 0.2 and 0.4, and this accounts, in part, for the breadth of the confidence limits.

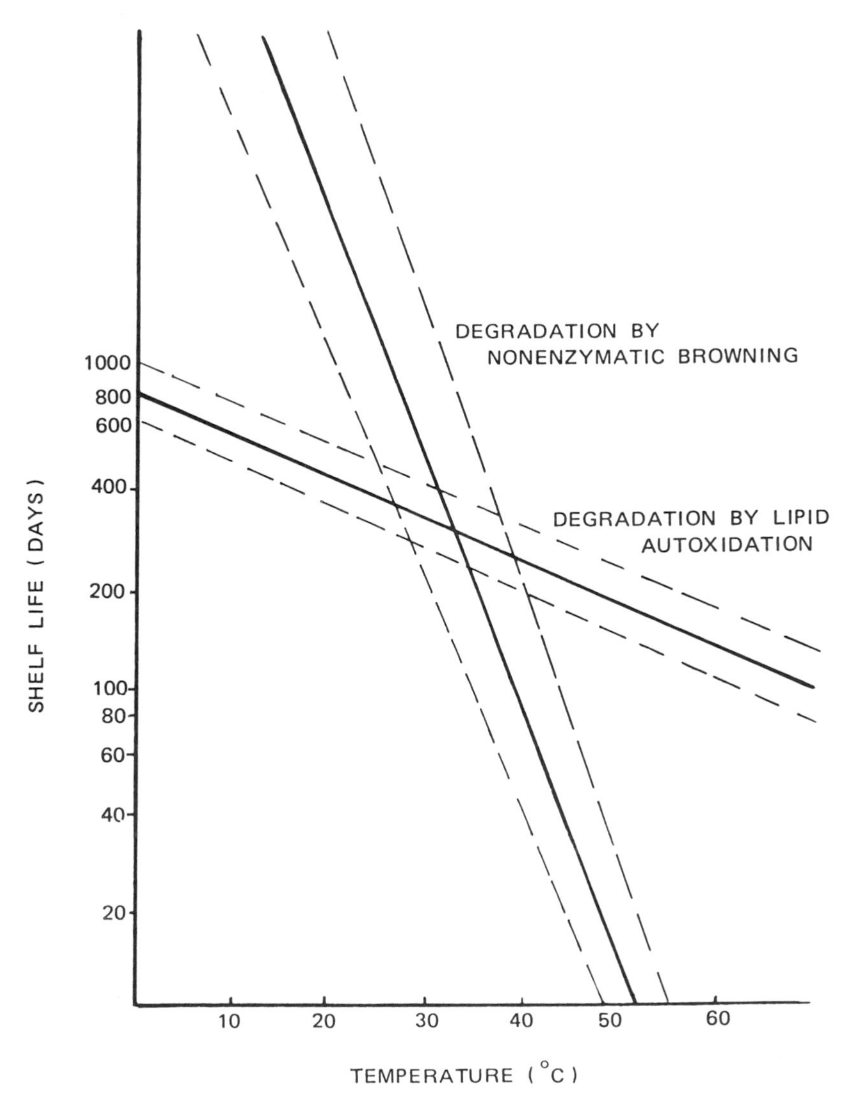

FIGURE 12 Shelf-life plot for dehydrated potatoes with quality loss resulting from nonenzymic browning and lipid oxidation.

Physical Changes In general, no physical changes should occur in stored dehydrated potatoes except loss of rehydratability. However, the extent of browning or oxidation needed to produce poor rehydration probably would make the dry product unacceptable from a flavor or color standpoint.

Shelf-Life Design

It is obvious from Figure 12 that the Arrhenius or shelf-life approach cannot be used over a wide temperature range because the chemical attribute that is shelf-life limiting changes at about 35°C. Thus, sensory acceptability must be used for determining shelf life. The following test conditions are reasonable when dehydrated potatoes are stored in moisture-impermeable pouches (*sensory test; chemical tests at all times).

Temperature (°C)	Test times (months)
25	6, 9, 12,* 18*
30	6, 8, 10,* 12,* 14
35	1, 2, 3,* 4*
40	1, 2,* 3*

The frequency factor equation (Eq. 13) cannot be used with dehydrated potatoes because of the change in degradative mechanisms with a change in temperature. Obviously, 25°C is too low to achieve significant acceleration of the reactions compared with normal storage conditions, but since the primary mode of degradation changes at about 35°C, tests at 25°C are also advisable. Control samples in this instance should be nitrogen flushed, sealed in cans, and held at 0°C. Other than sensory testing, GC-MS measurements of specific volatiles, and perhaps measurement of brown pigment development, are all that is needed.

17.4.2 Frozen Pizza

17.4.2.1 General Aspects of Quality Loss

A frozen pizza consists of a crust and a topping containing tomato sauce, cheese, spices, vegetables, and sausage. The product is then partially baked, cooled, shrink-wrapped, and blast-frozen. The major problems that occur during storage are development of in-package ice, development of off flavors, loss of the deep-red tomato color, loss of spice impact, and development of crust sogginess. Each of these changes influences shelf life and must be considered in any test design.

17.4.2.2 Specific Changes That Limit Shelf Life

Microbiological Changes

Both the cheese and sausage, if naturally fermented, have high total counts of bacteria. Thus, standard plate counts are not good predictors of whether off flavors will arise from microbial causes. Since the product is prebaked and then frozen, the numbers of vegetative microorganisms decline until thawing occurs. Unfortunately, pathogens such as *Staphyloccus aureus* are not totally inactivated by these treatments. If the product is then abused during distribution (repeated freeze and thaw), the pathogens may grow; however, the product would need to exceed 7°C near the surface for this to occur. Obviously, even moderately good control of the distribution system will avoid this problem.

In a shelf-life study, one can easily determine the potential magnitude of the problem with pathogens. This can be achieved by preinoculating the meat and cheese with organisms of public health significance and abusing the product by cycling it through a freeze and thaw sequence. A typical test sequence used by some food companies is as follows:

Day	Abuse temperature cycle	No. packages remaining
1	24 hr at –18°C	5
2	20 hr at –18°C, 4 hr at 38°C	4
3	20 hr at –18°C, 4 hr at 38°C	3
4	20 hr at –18°C, 4 hr at 38°C	2
5	20 hr at –18°C, 4 hr at 38°C	1

The study begins with five packages placed at –18°C for 1 day. At the end of the day a single package is removed and is tested for microbiological indicators. All the other packages are returned to –18°C for the next 20 hr and then abused at 38°C. Another package is then removed for testing, and the cycle is repeated for the other packages. If there is no significant increase in spoilage or pathogenic organisms after day 5, the food is deemed safe microbiologically. Of course, since it takes some time before the analysis can be done, these products should not be tested for taste. This abuse sequence is rather severe, and some product development people feel that the tests can be terminated after the third day.

Chemical Changes

When pizza is frozen, temperature lowering usually causes reaction rates to decrease dramatically. The Q_{10} values for frozen foods have been reported to vary from 3 to 40 [36]. However, as noted in Chapter 2, reactants may concentrate in the unfrozen aqueous phase, and this increase in concentratrion may cause some reactions to increase in rate or to decrease in rate less than expected as temperature is lowered. This type of behavior is most likely to occur just below the freeze–thaw point of the pizza (–5 to –3°C). This means that data collected under abuse conditions, for example at –5°C, may not abide by a straight-line relationship assumed for shelf life and Arrhenius plots, and that straight-line extrapolation of –5°C data to lower temperatures may result in substantial errors. In addition, if the product is freeze–thaw abused, an inordinately great time is spent at or near –3 to –5°C, where uncommonly large reaction rates are often observed. This propensity for the product to tarry at –3 to –5°C occurs because a large fraction of the ice–water transformation occurs in this range.

Another accelerating factor can arise during freezing. Because of freeze damage to cellular membranes, enzymes that are normally compartmentalized can be released into cellular fluids (see Chap. 7). These enzymes can then contact substrate and accelerate loss of quality.

In pizza, enzymes of the greatest potential concern are lipases and lipoxygenases. Lipases exist in natural cheese, and these enzymes might produce sufficient amounts of fatty acids to cause a hydrolytic rancid flavor, especially if butyric acid is released (from milk-based products). One way to avoid this problem is to use an imitation cheese made with soybean oil. Lipases are also present in vegetables (green peppers) and in the crust. Prebaking of the crust generally is not sufficient to destroy lipase, so the crust could develop a soapy flavor or an off odor from the free fatty acids produced. The vegetables and tomato sauce should not be a source of enzymes if they are blanched or cooked; however, spices contain many enzymes, including lipases. It should be noted that the standard catalase or peroxidase tests used to determine the adequacy of vegetable blanching are not adequate tests of lipase activity. One approach to monitoring lipase activity is to measure total free fatty acids during storage of pizza. Alternatively, volatile fatty acids can be isolated and characterized by gas–liquid chromatography (GLC) or high-performance liquid chromatography techniques. The big problem is sampling. The food scientist must decide whether to sample each ingredient, a slice, and/or an entire pizza.

Lipoxygenase, present in spices, sausage, wheat flour, and vegetables, catalyzes oxidation of unsaturated fats, producing peroxides as well as volatile breakdown products. As with lipase, it is quite heat stable and may survive baking or precooking. Several tests can be used to determine the consequences of lipoxygenase activity, including determination of peroxide value and measurement of volatile aldehydes or ketones by GLC (e.g., hexanal or pentanal). If peroxides are measured, this should be done at least twice weekly; otherwise, substantial increases in oxidation may be missed. This is true since peroxides increase to a maximum and then fall to near zero soon thereafter. Thus, a low peroxide value does not necessarily mean a nonrancid food. The minimum peroxide value associated with a rancid taste varies from food to food. Because sausage is cured, it may become rancid first. Thus, the addition of antioxidants may be desirable, and should be done if allowed. Addition of ethylenediamine tetraacetic acid (EDTA) may also be helpful.

In addition to causing off odors, oxidative rancidity can lead to a bleaching of the deep-red tomato color, giving rise to an orangish color. Thus, one should test for color changes either with a color meter or by extracting the pigments and measuring them in a spectrophotometer (see Chap. 10). Obviously, to retard color loss and oxidative rancidity, lipoxygenase should be destroyed during processing. Unfortunately, heat treatments needed to destroy lipoxygenase may also damage desirable flavors and textures, so a balance must be sought. Vacuum packaging or nitrogen flushing is also helpful [64]. However, because of the irregular surface of a pizza (large headspace), this is difficult to adequately accomplish at high speeds.

Another cause of deterioration in pizza quality is loss of spice impact. This can result from volatilization and/or chemical reactions. Since many flavor compounds are aldehydes and ketones, they can react with proteins or amino acids, thereby decreasing flavor. Others may be oxidized directly by oxygen or indirectly by free radicals produced during oxidative rancidity. Although sensory testing is the best approach to monitoring loss of spice impact, this approach is expensive. If available, GLC analysis of the headspace coupled with mass spectroscopy to identify key volatile spice flavor compounds is a rapid, useful, but not inexpensive alternative.

The last major chemical reaction leading to off flavors or color changes in pizza is nonenzymatic browning (NEB), and the cheese component is especially susceptible. NEB does occur at significant rates at subfreezing temperatures, but is not likely to be the reaction that limits shelf life. It should, however, be monitored to determine its importance. Browning can be measured easily by extracting the pigments and measuring optical density at 400–450 nm, or by measuring the amount of available lysine by the fluorodinitrobenzene procedure [61]. Browning also could be significant in imitation cheese if high-fructose corn syrup is used as the sugar source.

In summary, tests for the following constituents or properties should be considered for monitoring chemical changes in pizza during frozen storage:

Total free fatty acids
Specific volatile free fatty acids by GLC
Peroxides
Oxidative volatiles (e.g., hexanal) by GLC
Spice volatiles by GLC
Lysine
Color (decrease in red color or increase in brown)
Sensory properties: taste and flavor

Nutrient Loss

Chemical reactions can also cause a loss of vitamins. Of special concern, if nutritional labeling is done, are vitamins C and A, both of which are subject to oxidation. Since vitamin C is quite unstable at pH values above 5.0, its loss is sometimes used as an index of loss of overall quality. For example, when 15–20% of the vitamin C is lost from frozen vegetables, they become unacceptable from a sensory standpoint [36].

Physical Changes

One of the major complaints consumers have about frozen pizza is that the crust is not as crisp as that of fresh pizza. The reason for this is not entirely clear. If the crust has a high enough moisture content it certainly freezes when held at –18°C. Since all the other ingredients also contain ice, then at any temperature below about –18°C all pizza components have the same a_W (see Chap. 2) and there should be no moisture migration from the sauce to the crust. At higher temperatures (–7 to –1°C), however, the crust thaws and achieves a lower a_W than the other ingredients, and might absorb the water from the sauce by capillary suction, leading to a higher moisture and loss of crispness.

For dry foods, crispness is lost at a_W above about 0.4–0.5. Unfortunately, a pizza crust with that a_W would not have a desirable chewiness. One way to impart chewiness while retaining crispness is to fry the crust in fat. Research is needed to determine if the starch in flour can be chemically modified to yield a more desirable texture without losing other desirable functional properties. If successful, this would overcome the need to fry the crust in fat.

Another undesirable physical change in pizza is a loss of meltability and functionality of the cheese. As the water in the cheese freezes, concentration of the lactose and salts can cause irreversible chemical and structural changes in the protein. In addition, nonenzymic browning can lead to loss of protein functionality. These reactions reduce the ability of the protein to bind water after thawing. Furthermore, irreversible protein–protein interactions, occurring at the high solute concentrations in frozen cheese, tend to reduce the flow (meltability) of the cheese when heated. The concentrated solutes also affect the pectin and starches in the tomato sauce, resulting in loss of water binding and, thus, weeping (exudation of water) upon thawing.

Finally, in-package ice is a problem in stored frozen pizza. As it goes through inadvertent freeze–thaw cycles, the air space in the package warms up faster than the product, and the air then can hold more water. Water acquired by the air comes from the frozen ingredients. As the air temperature declines again, the water vapor crystallizes out as ice particles on the product surface. Repeated cycling causes the crystals to grow, become visible, and eventually convey an undesirable appearance to the product. A freeze–thaw cycling test, such as for the microbiological test, is generally adequate to determine if in-package ice will be a problem. The abuse temperature in this case should be at 20°C instead of 38°C.

To prevent in-package ice, the wrapper should be as tight as possible to minimize air spaces, something not easily done with pizza. The wrapper should also contain a release agent to prevent the topping from sticking to it.

17.4.2.3 Shelf-Life Test

Temperature Range

If a minimum shelf life of 12 months at –18°C is desired the product must, depending on the Q_{10} value, retain satisfactory quality for the following minimum times at –4°C.

Q_{10}	Time at –4°C that equates to 12 months at –18°C
2	4.5 months
5	1.2 months
10	14 days
20	6 days

Most published results suggest that Q_{10} values for vitamin C loss and quality loss in frozen vegetables range from 2 to 20 and that the shelf life of vegetables is only 6–8 months at –18°C. Considering these Q_{10} values, a product that does not retain good quality for 4.5 months at –4°C may not retain good quality for 12 months at –18°C. Given the preceding calculations, the following sampling frequency is proposed (*sensory test times; chemical tests at all times):

Temperature	Test times
–3.9°C (25°F)	1, 2*, 3, 4, 5, 8, 12, 14, 16*, 20* weeks
–6.7°C (20°F)	2, 4*, 10, 15*, 20* weeks
–9.4°C (15°F)	4*, 10, 15*, 20* weeks

The various temperatures are necessary to ascertain whether the Arrhenius relationship is conformed to. All simple tests should be conducted at each sampling time; sensory testing should be concentrated mainly toward the end of the test sequence, with a few near the beginning. In addition to this simple, constant-temperature study, a variable-temperature (cycling) study, as described before, should be done to determine freeze–thaw effects. Data should be analyzed as previously discussed.

Controls

Whenever sensory analysis is to be used or rates of change of some quality factors are to be determined, both initial values and values for stored control samples are needed. It is desirable, from a statistical standpoint, to perform numerous measurements (replicates) on the initial product so the precision of the method can be accurately established. Studies [5, 39, 45, 60, 61] suggest that 6–10 initial samples should be tested, but this is not reasonable for sensory testing because of the time and expense. For sensory tests involving a comparison triangle method, stored control samples are also needed. Whereas one can use suitably processed, refrigerated controls for shelf-life studies on dehydrated and canned foods, this is not appropriate for frozen foods. A fresh product is also not appropriate for this purpose since it has not been frozen and thawed. Thus, pizza held at –40 to –70°C (to minimize change during storage) is a suitable control for shelf-life studies of frozen pizza. Temperature fluctuations in the freezer used for control samples must also be kept to a minimum.

17.5 DISTRIBUTION EFFECTS: TIME–TEMPERATURE INDICATORS

Determination of the quality function of a food product and its kinetic parameters not only allows estimation of shelf life, but also permits the effect on quality of any actual or assumed variable-temperature exposure to be calculated. Thus, the remaining shelf life at any point of a monitored distribution can be estimated using Equation 7. Unfortunately, distribution conditions are not controlled by the manufacturer and are difficult to monitor. The effects of the presumed conditions that a packaged food can possibly face during the distribution chain were studied by Kreisman [30]. Assuming average storage times at each location of the assumed distribution routes of a dehydrated product (mashed potatoes), the shelf-life loss ranged from 50% to about 300% for best and worst temperature scenario.

What would be desirable is a cost-effective way to individually monitor the conditions of the products during distribution and a means to signal their remaining shelf life. This would lead to effective control of distribution, optimized stock rotation, reduction of waste, and meaningful information on product "freshness," which is demanded by the consumer. Time–temperature indicators (TTIs) are a potential solution.

A TTI can be defined as a simple, inexpensive device exhibiting an easily measurable time–temperature-dependent response that reflects the full or partial temperature history of the product to which it is attached. It is based on mechanical, chemical, enzymatic, or microbiological systems that change (usually in the form of a visual signal) irreversibly from the time of their activation. The rate of change has a temperature dependence similar to the described physicochemical reactions that determine quality in foods. A TTI can be used to monitor the temperature history of cases or pallets up to the time they are displayed at the supermarket and individual packages up to the time of product consumption. With quality loss being a function of temperature history and with a TTI giving a measure of that history, their response can be correlated to the quality level of the food. Thus, the TTI can serve as an inventory management and stock rotation tool at the retail level. Also, TTIs attached to individual packages can serve as dynamic or active shelf life devices. The TTI would indicate to the consumers whether the products were properly handled and what percent of the original shelf life remains. A thorough review of TTI systems and uses is given by Taoukis et al. [71].

A general scheme has been proposed that allows the correlation of the measurable response of a TTI, X, to the quality change of a food product (i.e. the value of the quality attribute A) at any particular time during distribution [67]. This approach, based on the kinetic principles described in this chapter, does not require previous simultaneous testing of the indicator and the food to preestablish a statistical correlation between X and A. In analogy to the food quality function ($F(A)$, Eq. 3), the response function of the TTI, $R(X)$, is defined as

$$R(X)_t = kt = k_I \exp(-E_A/RT)t \tag{14}$$

where t is the storage time, k is the response rate constant of the TTI, and the constant k_I and the activaton energy E_A are the Arrhenius parameters of the TTI. For a variable-temperature distribution, $T(t)$, an effective temperature, T_{eff}, is defined as the constant temperature that causes the same response or change as the variable temperature $T(t)$. For a TTI of established kinetic parameters going through the same temperature distribution, $T(t)$, as that of the monitored food, the measured response X allows calculation of the value of $R(X)$ and subsequently of T_{eff} (from Eq. 14, for $T = T_{eff}$). The T_{eff} and knowledge of the kinetic parameters of deterioration of the food allow the estimation of $F(A)$ and the variable quality loss of the product (Eqs. 5 and 3). The reliability of the TTI under variable temperature conditions has been asssessed and was judged satisfactory [68]. Application of the scheme was illustrated for refrigerated and dehydrated products [70], and these cases provide good examples of how the principles described in this chapter and throughout this book can be effectively and practically used.

BIBLIOGRAPHY

Benson, S. W. (1960). *Foundations of Chemical Kinetics,* McGraw-Hill, New York.

Charalambous, G. (1986). *Handbook of Food and Beverage Stability,* Elsevier, Amsterdam.

Labuza, T. P. (1980). Temperature/enthalpy/entropy compensation in food reactions. *Food Technol. 34*(2):67.

Labuza, T. P. (1982). *Shelf Life Dating of Foods,* Food and Nutrition Press, Westport, CT.

Labuza, T. P. and P. S. Taoukis (1990). The relationship between processing and shelf life, in *Foods for the 90's* (G. G. Birch, G. Campbell-Platt, and M. G. Lindley, eds.), Developments Series, Elsevier Applied Science, New York, pp. 73–105.

Okos, M. R. (1986). *Physical and Chemical Properties of Foods,* American Society of Agricultural Engineering, St. Joseph, MI.

Richardson, T. and J. W. Finley (1985). *Chemical Changes in Food During Processing,* AVI, Westport, CT.

Villota, R., and J. G. Hawkes (1992). Reaction kinetics in food systems, in *Handbook of Food Engineering* (D. R. Heldman and D. B. Lund, eds.), Marcel Dekker, New York, pp. 39–143.

REFERENCES

1. Arabshashi, A. and D. B. Lund (1985). Considerations in calculating kinetic parameters from experimental data. *J. Food Proc Eng. 7*:239–251.
2. Bell, L. N., and T. P. Labuza (1991). Aspartame degradation kinetics as affected by pH in intermediate and low moisture food systems. *J. Food Sci. 56*:17–20.
3. Bell L. N., and T. P. Labuza (1994). Influence of the low-moisture state on pH and its implications for reaction kinetics. *J. Food Eng. 22*:291–312.
4. Benson, S. W. (1960). *Foundations of Chemical Kinetics,* McGraw-Hill, New York.
5. Bergquist, S., and T. P. Labuza (1983). Kinetics of peroxide formation in potato chips undergoing a sine wave temperature fluctuation. *J. Food Sci. 43*:712.
6. Boggs, M. M., R. G. Buttery, D. W. Venstrom, and M. L. Belote (1964). Relation of hexanal in vapor above stored potato granules to subjective flavor estimates. *J. Food Sci. 29*:487.
7. Buchanan, R. L. (1993). Predictive food microbiology. *Trends Food Sci. Technol. 4*:6–11.
8. Buera, P., and M. Karel (1993). Applicaton of the WLF equation to describe the combined effects of moisture, temperature and physical changes on non-enzymatic browning rates in food systems. *J. Food Proc. Preserv. 17*:31–47.
9. Cardoso, G., and T. P. Labuza (1983). Effect of temperature and humidity on moisture transport for pasta packaging material. *Br. J. Food Technol. 18*:587.
10. Cohen, E., and I. Saguy (1985). Statistical evaluation of Arrhenius model and its applicability in prediction of food quality losses. *J. Food Proc. Pres. 9*:273–290.
11. Davey, K. R. (1989). A predictive model for combined temperature and water activity on microbial growth during the growth phase. *J. Appl. Bacteriol 67*:483–488.
12. Feeney, R. E., and J. R. Whitaker (1982). The Maillard reaction and its preservation. In *Food Protein Deterioration, Mechanisms and Functionality* (J. P. Cerry, ed.), ACS Symposium Series 206, ACS, Washington, DC, pp. 201–229.
13. Fennema, O. (1985). Chemical changes in food during processing—An overview, in *Chemical Changes in Food during Processing* (T. Richardson and J. W. Finley, eds.), AVI, Westport, CT., pp. 1–16.
14. Ferry, J. D. (1980). *Viscoelastic Properties of Polymers,* 3rd ed., Wiley, New York.
15. Fritsche, C. W., and J. A. Gale (1977). Hexanal as a measure of oxidative deterioration in low fat foods. *J. Am. Oil Chem. Sco. 54*:225.
16. Fu, B. and T. P. Labuza (1993). Shelf-life prediction: Theory and application. *Food Control* 4:125–133.
17. Fu, B., P. S. Taoukis, and T. P. Labuza (1991). Predictive microbiology for monitoring spoilage of dairy products with time temperature indicators. *J. Food Sci. 56*:1209–1215.
18. Guadagni, D. G. (1968). Cold storage life of frozen fruits and vegetables as a function of temperature

and time, in *Low Temperature Biology of Foodstuffs* (J Hawthorne and E. J. Rolfe, eds.), Pergamon Press, New York.

19. Haralampu, S. G., I. Saguy, and M. Karel. (1985). Estimation of Arrhenius model parameters using three least squares methods. *J. Food Proc. Pres. 9*:129–143.
20. Hayakawa, K. (1973). New procedure for calculating parametric values for evaluating cooling treatments applied to fresh foods. *Can. Inst. Food Sci. Technol 6(3)*:197.
21. Holmer, B. (1984). Properties and stability of aspartame. *Food Technol. 38*:50.
22. Hong Y. C., C. M. Koelsch, and T. P. Labuza (1991). Using the L number to predict the efficacy of moisture barrier properties of edible food coating materials. *J. Food Proc Pres. 15*:45–62.
23. Hoover, D. G. (1993). Pressure effects on biological systems. *Food Technol. 47(6)*:150–155.
24. Institute of Food Science and Technology (UK) (1993). *Shelf Life of Foods—Guidelines for Its Determination and Prediction,* IFST, London.
25. Kamman, J., and T. P. Labuza (1981). Kinetics of thiamine and riboflavin loss in pasta as a function of constant and variable storage conditions. *J. Food Sci. 46*:1457.
26. Karel, M. (1967). Use-tests. Only real way to determine effect of package on food quality. *Food Can. 27*:43.
27. Karel, M. (1993). Temperature-dependence of food deterioration processes [letter to the editor]. *J. Food Sci. 58*:(6):ii.
28. Knorr, D. (1993). Effects of high-hydrostatic-pressure processes on food safety and quality. *Food Technol. 47*(6):155–161.
29. Kramer, A. (1974). Storage retention of nutrients. *Food Technol. 28*:5.
30. Kreisman, L. (1980) Application of the principles of reaction kinetics to the prediction of shelf life of dehydrated foods. M. S. thesis, University of Minnesota, Minneapolis.
31. Kwolek, W. F., and G. N. Bookwalter (1971). Predicting storage stability from time-temperature data. *Food Technol. 25*(10):51.
32. Labuza, T. P. (1971). Kinetics of lipid oxidation of foods. *CRC Rev. Food Technol. 2*:355.
33. Labuza, T. P. (1979). A theoretical comparison of losses in foods under fluctuating temperature sequences. *J. Food Sci. 44*:1162.
34. Labuza, T. P. (1980). Temperature/enthalpy/entropy compensation in food reactions. *Food Technol. 34*(2):67.
35. Labuza, T. P. (1980). The effect of water activity on reaction kinetics of food deterioration. *Food Technol. 34*:36.
36. Labuza, T. P. (1982). *Shelf-life Dating of Foods,* Food & Nutrition Press, Westport, CT.
37. Labuza, T. P. (1984). Application of chemical kinetics to deterioration of foods. *J. Chem. Educ. 61*:348–358.
38. Labuza, T. P., and R. Contreras-Medellin (1981). Prediction of moisture protection requirements for foods. *Cereal Foods World 26*:335–343.
39. Labuza, T. P., and J. Kamman (1983). Reaction kinetics and accelerated tests simulation as a function of temperature, in *Applications of Computers in Food Research* (I. Saguy, ed.), Marcel Dekker, New York, Chap. 4.
40. Labuza, T. P., and J. O. Ragnarsson (1985). Kinetic history effect on lipid oxidation of methyl linoleate in model system. *J. Food Sci. 50*(1):145.
41. Labuza, T. P., and D. Riboh (1982). Theory and application of Arrhenius kinetics to the prediction of nutrient losses in food. *Food Technol. 36*:66–74.
42. Labuza, T. P., and M. K. Schmidl (1985). Accelerated shelf-life testing of foods. *Food Technol. 39*(9):57–62, 64.
43. Labuza, T. P. and M. K. Schmidl (1988). Use of sensory data in the shelf life testing of foods: Principles and graphical methods for evaluation. *Cereal Foods World 33*(2):193–206.
44. Labuza, T. P. and P. S. Taoukis (1990). The relationship between processing and shelf life, in *Foods for the 90's* (G. G. Birch, G. Campbell-Platt, and M. G. Lindley, eds.), Developments Series, Elsevier Applied Science, New York, pp. 73–105.

45. Labuza, T. P., K. Bohnsack, and M. N. Kim (1982). Kinetics of protein quality loss stored under constant and square wave temperature distributions. *Cereal Chem. 59*:142.
46. Labuza, T. P., B. Fu, and P. S. Taoukis (1992). Prediction for shelf life and safety of minimally processed CAP/MAP chilled foods. *J. Food Prot. 55*:741–750.
47. Labuza, T. P., S. Mizrahi, and M. Karel (1972). Mathematical models for optimization of flexible film packaging of foods for storage. *Trans. Am. Soc. Agric. Eng. 15*:150.
48. Lenz, M. K., and D. B. Lund (1980). Experimental procedures for determining destruction kinetics of food components. *Food Technol. 34*(2):51.
49. Mizrahi, S., T. P. Labuza, and M. Karel (1970). Computer aided predictions of food storage stability. *J. Food Sci. 35*:799–803.
50. Mizrahi, S., T. P. Labuza, and M. Karel (1970). Feasibility of accelerated tests for browning in dehydrated cabbage. *J. Food Sci. 35*:804–807.
51. Nakabayashi, K., T. Shimamoto, and H. Mina (1981). Stability of package solid dosage forms. *Chem. Pharm. 28*(4):1090; *29*(7):2027, 2051, 2057.
52. Nelson, K., and T. P. Labuza (1994). Water activity and food polymer science: Implications of state on Arrhenius and WLF models in predicting shelf life. *J. Food Eng. 22*:271–290.
53. Norwig, J. F., and D. R. Thompson (1986). Microbial population, enzyme and protein changes during processing, in *Physical and Chemical Properties of Foods* (M. R. Okos, ed.), American Society of Agricultural Engineering, St. Joseph, MI.
54. Parkin, K. L., and W. D. Brown, (1982). Preservation of seafood with modified atmospheres, in Chemistry and *Biochemistry of Marine Food Products* (R. E. Martin. et al., eds.), AVI, Westport, CT. pp. 453–465.
55. Peleg, M. (1990). On the use of WLF model in polymers and foods. *Crit. Rev. Food Sci. Nutr. 32*:59–66.
56. Quast, D. G., and M. Karel (1972). Computer simulation of storage life of foods undergoing spoilage by two interactive mechanisms. *J. Food Sci. 37*:679.
57. Ramaswamy, H. S., F. R. Van de Voort, and S. Ghazala (1989). An analysis of TDT and Arrhenius methods for handling process and kinetic data. *J. Food Sci. 54*:1322–1326.
58. Ratkowsky, D. A., J. Olley, T. A. McMeekin, and A. Ball (1982). Relationship between temperature and growth rate of bacterial cultures. *J. Bacteriol. 149*:1–5.
59. Ratkowsky, D. A., R. K. Lowry, T. A. McMeekin, A. N. Stokes, and R. E. Chandler (1983). Model for bacterial culture growth rate throughout the entire biokinetic temperature range. *J. Bacteriol. 154*:1222–1226.
60. Riboh, D. K. and T. P. Labuza (1982). Kinetics of thiamine loss in pasta stored in a sine wave temperature condition. *J. Food Proc. Pres. 6*(4):253.
61. Saltmarch, M., and T. P. Labuza (1982). Kinetics of browning and protein quality loss in sweet whey powders under steady and non-steady state storage conditions. *J. Food Sci. 47*:92.
62. Sapers, G. M. (1970). Flavor quality in explosion puffed dehydrated potato. *J. Food Sci. 35*:731.
63. Sapru, V., and T. P. Labuza (1992). Glassy state in bacterial spores predicted by glass transition theory. *J. Food Sci. 58*:445–448.
64. Simon, I., T. P. Labuza, and M. Karel (1971). Computer aided prediction of food storage stability: Oxidation of a shrimp food product. *J. Food Sci. 36*:280.
65. Slade, L., H. Levine, and J. Finey (1989). Protein–water interactions: Water as a plasticizer of gluten and other protein polymers, in *Protein Quality and the Effects of Processing* (R. D. Phillips and J. W. Finlay eds.), Marcel Dekker, New York, pp. 9–123.
66. Stamp, J. A., and T. P. Labuza (1989). An ion-pair high performance liquid chromatographic method for the determination of aspartame and its decomposition products. *J. Food Sci. 54*:1043.
67. Taoukis, P. S. and T. P. Labuza (1989). Applicability of time-temperature indicators as shelf-life monitors of food products, *J. Food Sci. 54*:783–788.
68. Taoukis, P. S. and T. P. Labuza (1989). Reliability of time-temperature indicators as food quality monitors under non isothermal conditions. *J. Food Sci. 54*:789–792.
69. Taoukis, P. S., A. El Meskine, and T. P. Labuza (1988). Moisture transfer and shelf life of packaged

foods, in *Food and Packaging Interactions* (J. H. Hotchkiss, ed.), ACS Symposium Series No. 365, pp. 243–262.

70. Taoukis, P. S., G. A. Reineccius, and T. P. Labuza (1990). Application of time-temperature indicators to monitor quality of flavors and flavored products, in *Flavors and Off-Flavors' 89* (G. Charalambus, ed.), Developments in Food Science Series, Elsevier Applied Science, New York, pp. 385–398.
71. Taoukis, P. S., B. Fu, and T. P. Labuza (1991). Time-temperature indicators. *Food Technol. 45(10)*:70–82.
72. Tsoumbeli, M. N. and T. P. Labuza (1991). Accelerated kinetic study of aspartame degradation in the neutral pH range. *J. Food Sci. 56*:1671–1675.
73. Weissman, I., O. Ramon, I. J. Kopelman, and S. Mizrahi (1993). A kinetic model for accelerated tests of Maillard browning in a liquid model system. *J. Food Proc. Pres. 17*:455–470.
74. Williams, M. L., R. F. Landel, and J. D. Ferry (1955). The temperature dependence of relaxation mechanisms in amorphous polymers and other glass-forming liquids. *J. Chem. Eng. 77*:3701–3707.
75. Zwietering, M. H., J. De Koos, B. E. Hasenack, J. C. DeWilt, and K. Van't Riet (1991). Modeling of bacterial growth as a function of temperature. *Appl. Environ. Microbiol. 57*:1094–1101.

Appendices

APPENDIX A: INTERNATIONAL SYSTEM OF UNITS (SI), THE MODERNIZED METRIC SYSTEM

SI is the form of the metric system that is preferred for all applications. All units associated with the various cgs systems (measurement systems based on centimeter, gram, and second) are to be avoided. The primary source of the following information is Ref. 2, but two English translations [1, 5] of the primary source and two ISO publications [3, 4] were also used to prepare this appendix.

Three Classes of SI Units

Base Units

These well-defined units are regarded as dimensionally independent, and only one base unit is defined for each basic quantity.

	SI base unit	
Quantity[a]	Name	Symbol[b]
Length	meter	m
Mass	kilogram	kg
Time	second	s
Electric current	ampere	A
Thermodynamic temperature[c]	kelvin	K
Amount of substance	mole	mol
Luminous intensity	candela	cd

[a]Measurable attributes of phenomena or matter.
[b]Unit symbols do not change in the plural.
[c]Use of degree Celcius (°C) is common and this is the SI unit used for the Celsius scale. °C = $T - 273.15$, where T is temperature in kelvins. The unit "Celsius" is equal to the unit "kelvin." An interval or difference in temperature can be expressed in either kelvins or degrees Celsius, the former being customary in all thermodynamic expressions.

Supplementary SI Units

Quantity	SI supplementary unit	
	Name	Symbol
Plane angle	radian	rad
Solid angle	steradian	sr

Derived Units

Derived units are formed by combining base units, supplementary units, and other derived units, and are expressed algebraically in accord with the relations existing among the corresponding quantities. The derived units are said to be coherent when they do not contain factors other than one. Because base units and coherently derived units do not always have a convenient size, subunits are sometimes constructed by using factors other than one (e.g., milligrams per cubic meter). These derived units are said to be noncoherent and the accepted prefixes and symbols are shown later in this section. Some coherently derived units along with their special names are listed next.

Quantity[a]	SI derived unit			
	Name	Symbol	Expression in SI base units	Expression in other units
Activity (of a radionuclide)	becquerel	Bq	s^{-1}	
Area	square meter	m^2		
Celsius temperature	degree Celsius	°C	See footnote c of base unit table	
Dose (absorbed)	gray	Gy	$m^2 \cdot s^{-2}$	J/kg
Dose rate (absorbed)	gray per second	Gy/s	$m^2 \cdot s^{-3}$	
Electric conductance	siemens	S	$A^2 \cdot s^3 \cdot kg^{-1} \cdot m^{-2}$	A/V
Electric field strength	volt per meter	V/m	$m \cdot kg \cdot s^{-3} \cdot A^{-1}$	
Electric potential, potential difference, electromotive force	volt	V	$kg \cdot m^2 \cdot s^{-3} \cdot A^{-1}$	W/A
Electric resistance	ohm	Ω	$kg \cdot m^2 \cdot A^{-2} \cdot s^{-3}$	V/A
Electricity (quantity), electric charge	coulomb	C	$A \cdot s$	
Energy, work, quantity of heat	joule	J	$kg \cdot m^2 \cdot s^{-2}$	$N \cdot m$
Entropy	joule per kelvin	J/K	$m^2 \cdot kg \cdot s^{-2} \cdot K^{-1}$	
Force	newton	N	$kg \cdot m \cdot s^{-2}$	

Frequency (of a periodic phenomenon)	hertz	Hz	s^{-1}	
Heat capacity	joule per kelvin	J/K	$m^2 \cdot kg \cdot s^{-2} \cdot K^{-1}$	
Heat flux density	watt per square meter	W/m^2	$kg \cdot s^{-3}$	
Mass density	kilogram per cubic meter	kg/m^3		
Molar energy	joule per mole	J/mol	$m^2 \cdot kg \cdot s^{-2} \cdot mol^{-1}$	
Molar entropy	joule per mole kelvin	J/(mol · K)	$m^2 \cdot kg \cdot s^{-2} \cdot K^{-1} \cdot mol^{-1}$	
Molar heat capacity	joule per mole kelvin	J/(mol · K)	$m^2 \cdot kg \cdot s^{-2} \cdot K^{-1} \cdot mol^{-1}$	
Moment of force	newton meter	N · m	$m^2 \cdot kg \cdot s^{-2}$	
Permittivity	farad per meter	F/m	$m^{-3} \cdot kg^{-1} \cdot s^4 \cdot A^2$	
Power, radiant flux	watt	W	$kg \cdot m^2 \cdot s^{-3}$	J/s
Pressure, stress	pascal	Pa	$kg \cdot m^{-1} \cdot s^{-2}$	N/m^2
Specific heat capacity	joule per kilogram kelvin	J/(kg · K)	$m^2 \cdot s^{-2} \cdot K^{-1}$	
Speed, velocity	meter per second	m/s		
Surface tension	newton per meter	N/m	$kg \cdot s^{-2}$	
Thermal conductivity	watt per meter kelvin	W/(m · K)	$m \cdot kg \cdot s^{-3} \cdot K^{-1}$	
Viscosity, dynamic	pascal second	Pa · s	$m^{-1} \cdot kg \cdot s^{-1}$	
Viscosity, kinematic	square meter per second	m^2/s		
Volume	cubic meter	m^3		

[a]Measurable attribute of phenomena or matter.

Prefix Names and Symbols

Prefix names and symbols representing decimal multiples and submultiples of SI units, except for kilogram, are listed below (partial listing). SI prefixes should be used to indicate orders of magnitude, thus eliminating nonsignificant digits and leading zeros in decimal fractions, and providing a convenient alternative to the powers-of-ten notation often used in computation. It is preferable to chose a prefix so that the numerical value lies between 0.1 and 1000. Care should be taken to minimize the variety of prefixes used even if this violates the previous statement.

Multiplication factor	Prefix	Symbol
10^{12}	tera	T
10^9	giga	G
10^6	mega	M
10^3	kilo	k
10^2	hecto[a]	h

Multiplication factor	Prefix	Symbol
10^1	deka[a]	da
10^{-1}	deci[a]	d
10^{-2}	centi[a]	c
10^{-3}	milli	m
10^{-6}	micro	μ
10^{-9}	nano	n
10^{-12}	pico	p
10^{-15}	femto	f
10^{-18}	atto	a

[a]Generally to be avoided (exceptions apply to expressions of area and volume).

Some Rules of Style and Usage for SI Unit Symbols

1. Unit symbols should be printed in upright type, generally lower case, unless the unit is derived from a proper name, when the first letter of the symbol is upper case. The exception is the symbol for liter, where either "l" or "L" is acceptable.
2. Unit symbols remain unaltered in the plural.
3. Unit symbols are not followed by a period except when used at the end of a sentence.
4. When a quantity is expressed as a numerical value and a unit symbol, a space should be left between them, for example, 35 mm, not 35mm. Exceptions: No space is left between the numerical value and symbols for degree, minute, and second of plane angle.
5. When a quantity is expressed as a number and a unit and is used in an adjectival manner, it is preferable to use a hyphen instead of a space between the number and the symbol, for example, a 35-mm film. Exception: a 60° angle.
6. Units formed by multiplication or division. For division, oblique strokes (/) or negative exponents can be used, but if oblique strokes are used, these strokes must not be repeated in the same group unless ambiquity is avoided by a parentheses (e.g., m/s/s is forbidden); m/s, $\frac{\mathrm{m}}{\mathrm{s}}$, $\mathrm{m} \cdot \mathrm{s}^{-1}$ are equally acceptable. For multiplication, two or more units may be expressed in either of the following ways: Pa · s or Pa s.
7. When names of units are used in quotients, use the word per and not an oblique stroke, for example, meter per second, not meter/second.
8. To avoid ambiguity in complicated expressions, symbols are preferred over words.
9. When writing numerals less than one, a zero should appear before the decimal marker.
10. For American usage, the dot is used for the decimal marker instead of the comma, and the spellings "meter," "liter," and "deka" are used instead of "metre," "litre," and "deca."

References

1. ASTM (1991). *Standard Practice for Use of the International System of Units (the Modernized Metric System)*, E380-91a, American Society for Testing Materials, Philadelphia, PA.
2. Bureau International des Poids et Mesures (BIPM) (1991). *Le Système International d'Unités* (SI), 6th ed., BIPM, Pavillon de Breteuil (Parc de Saint-Cloud), Paris.
3. ISO (1992). *Quantities and Units—Part 0: General Principles,* International Standard ISO 31-0, International Organization for Standardization, Geneva.

4. ISO (1992). *SI Units and Recommendations for the Use of Their Multiples and of Certain Other Units,* International Standard ISO 1000, International Organization for Standardization, Geneva.
5. Taylor, B. N. (ed.). (1991). *The International System of Units (SI),* National Institute of Standards and Technology Special Publication 330, U.S. Government Printing Office, Washington, DC (English translation of Ref. 2).

APPENDIX B: CONVERSION FACTORS (NON-SI UNITS TO SI UNITS)

To convert from	to (SI units)	Multiply by [a]
Area		
ft^2	square meter (m^2)	9.290304*E-02
hectare	square meter (m^2)	1.000000*E+04
in^2	square meter (m^2)	6.451600*E-04
yd^2	square meter (m^2)	8.361274 E-01
Energy		
British thermal unit (mean)	joule (J)	1.05587 E+03
calorie (kilogram, International Table)	joule (J)	4.186800*E+03
calorie (kilogram, thermochemical)	joule (J)	4.184000*E+03
electron volt	joule (J)	1.60219 E-19
erg	joule (J)	1.000000*E-07
kW · h	joule (J)	3.600000*E+06
Force		
dyne	newton (N)	1.000000*E-05
kilogram-force	newton (N)	9.806650*E+00
poundal	newton (N)	1.382550 E-01
Length		
angstrom	meter (m)	1.000000*E-10
foot (U.S. survey)	meter (m)	3.048006 E-01
inch	meter (m)	2.540000*E-02
mil	meter (m)	2.540000*E-05
Mass		
carat (metric)	kilogram (kg)	2.000000*E-04
grain	kilogram (kg)	6.479891*E-05
gram	kilogram (kg)	1.000000*E-03
ounce (avoirdupois)	kilogram (kg)	2.834952 E-02
ounce (troy or apothecary)	kilogram (kg)	3.110348 E-02
pound (lb avoirdupois)	kilogram (kg)	4.535924 E-01
pound (troy or apothecary)	kilogram (kg)	3.732417 E-01
Pressure or stress (force per unit area)		
atmosphere (standard)	pascal (Pa)	1.013250*E+05
bar	pascal (Pa)	1.000000*E+05
$dyne/cm^2$	pascal (Pa)	1.000000*E-01
inch of mercury (32°F)	pascal (Pa)	3.38638 E+03
inch of mercury (60°F)	pascal (Pa)	3.37685 E+03
inch of water (39.2°F)	pascal (Pa)	2.49082 E+02
inch of water (60°F)	pascal (Pa)	2.4884 E+02
psi	pascal (Pa)	6.894757 E+03
torr (mm Hg, 0°C)	pascal (Pa)	1.33322 E+02

To convert from	to (SI units)	Multiply by [a]
Radiation units		
curie	becquerel (Bq)	3.700000*E+10
rad	gray (Gy)	1.000000*E-02
rem	sievert (Sv)	1.000000*E-02
roentgen	coulomb per kilogram (C/kg)	2.580000*E-04
Temperature		
degree Celsius	kelvin (K)	$T_K = t_C + 273.15$
degree Fahrenheit	degree Celsius (°C)	$t_C = (t_F - 32)/1.8$
degree Fahrenheit	kelvin (K)	$T_K = (t_F + 459.67)/1.8$
degree Rankine	kelvin (K)	$T_K = T_R/1.8$
kelvin	degree Celsius (°C)	$t_C = T_K - 273.15$
Viscosity		
centipoise (dynamic viscosity)	pascal second (Pa · s)	1.000000*E-03
centistokes (kinematic viscosity)	square meter per second (m^2/s)	1.000000*E-06
poise	pascal second (Pa · s)	1.000000*E-01
stokes	square meter per second (m^2/s)	1.000000*E-04
Volume (includes capacity)		
ft^3	cubic meter (m^3)	2.831685 E-02
gallon (Canadian liquid)	cubic meter (m^3)	4.546090 E-03
gallon (U.K. liquid)	cubic meter (m^3)	4.546092 E-03
gallon (U.S. liquid)	cubic meter (m^3)	3.785412 E-03
in^3	cubic meter (m^3)	1.638706 E-05
liter	cubic meter (m^3)	1.000000*E-03
ounce (U.K. fluid)	cubic meter (m^3)	2.841306 E-05
ounce (U.S. fluid)	cubic meter (m^3)	2.957353 E-05
pint (U.S. liquid)	cubic meter (m^3)	4.731765 E-04
quart (U.S. liquid)	cubic meter (m^3)	9.463529 E-04
yd^3	cubic meter (m^3)	7.645549 E-01

[a]Factors with an asterisk are exact.
Source: American Society for Testing and Materials (1991), E380-91a, ASTM, Philadelphia, PA.

APPENDIX C: GREEK ALPHABET

Greek form			
Upper case	Lower case	Name	Roman or English equivalent
Α	α	alpha	a
Β	β	beta	b
Γ	γ	gamma	g
Δ	δ	delta	d
Ε	ε	epsilon	e (short)
Ζ	ζ	zeta	z
Η	η	eta	e (long)
Θ	θ	theta	th

Ι	ι	iota	i
Κ	κ	kappa	k, c
Λ	λ	lambda	l
Μ	μ	mu	m
Ν	ν	nu	n
Ξ	ξ	xi	x
Ο	ο	omicron	o (short)
Π	π	pi	p
Ρ	ρ	rho	r
Σ	σ, ς	sigma	s
Τ	τ	tau	t
Υ	υ	upsilon	u, y
Φ	φ	phi	ph
Χ	χ	chi	kh, ch
Ψ	ψ	psi	ps
Ω	ω	omega	o (long)

Index

About the Editor

Owen R. Fennema is a Professor of Food Chemistry in the Department of Food Science at the University of Wisconsin—Madison. He is coauthor of the books *Low Temperature Foods and Living Matter* (with William D. Powrie and Elmer H. Marth) and *Principles of Food Science, Part II: Physical Principles of Food Preservation* (with Marcus Karel and Daryl B. Lund) (both titles, Marcel Dekker, Inc.), and the author or coauthor of over 175 professional papers that reflect his research interests in food chemistry, low-temperature preservation of food and biological matter, the characteristics of water and ice, edible films and coatings, and lipid-fiber interactions. A consulting editor for the *Food Science and Technology* series (Marcel Dekker, Inc.), he is a Fellow of the Institute of Food Technologists and of the Agriculture and Food Chemistry Division of the American Chemical Society, and a member of the American Institute of Nutrition, among other organizations. Dr. Fennema received the B.S. degree (1950) in agriculture from Kansas State University, Manhattan, and the M.S. degree (1951) in dairy science and Ph.D. degree (1960) in food science and biochemistry from the University of Wisconsin—Madison.